An entirely new way to EXPLORE LIFE

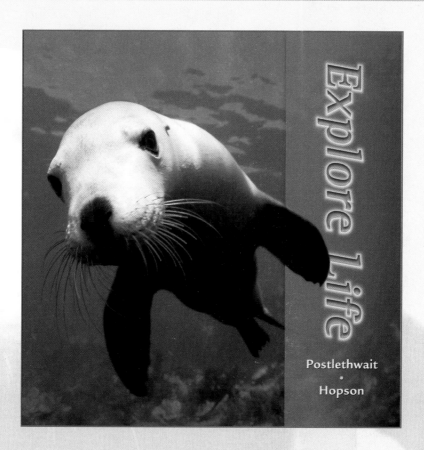

Explore Life

Postlethwait
·
Hopson

Welcome to an innovative and integrated learning experience . . .

Postlethwait
·
Hopson

THOMSON

BROOKS/COLE

Successful learning happens when biology becomes a personally relevant learning experience.

From the Authors . . .

Explore Life is a fresh approach to teaching non-science majors that is based on two sets of beliefs: First, that students majoring in English, business, art, and other nonscience disciplines absorb biology best when it is presented in the context of personal experience or human relevance. Second, that today's electronic media—video, CD-ROMs, and the Internet—offer unprecedented opportunities to bring people, imagery, biological processes, hands-on interactivity, and up-to-the-minute data into the study of biology.

We have created a topical, case-history approach to teaching life science fundamentals that fully integrates electronic media so that students can explore relevant biology in ways suited to their own learning styles. That integration is immediately obvious in our innovative book layout. We have used the outside column on each book page to present visual connectors—icons that relate to the CD exercises, small photos, tantalizing quotes from the videos, and integrated study questions—that link the text with the other media.

Our collaboration has spanned 20 years and has produced six textbooks aimed at four different segments of the audience for introductory biology. **Explore Life is different**. The supporting media are integrated into the course line by line, not just attractively packaged and offered as ancillaries. This makes the preparation of the text you are about to view and read far more complex, but, we hope, far more valuable to instructors and students alike.

John H. Postlethwait   **Janet L. Hopson**
University of Oregon

2

The authors

John H. Postlethwait

is professor of Biology at the University of Oregon where he has taught general biology since 1971, to both biology majors and non-majors, as well as upper division courses in embryology and genetics. His research on the genetic mechanisms of embryonic development and the origin of vertebrate genomes is supported by the National Institutes of Health and the National Science Foundation. Publication of this work began appearing in 1969 and encompasses more than 120 research publications in the journals *Science, Nature, Development, Developmental Biology, Genetics, Molecular and Developmental Evolution, Cell, Proceedings of the National Academy of Sciences,* and others. For three one-year periods, Dr. Postlethwait conducted research supported by Fullbright grants at the Institüt für Molekular Biologie in Salzburg, Austria; the Laboratoire de Génétique Moléculaire des Eucaryotes in Strasbourg, France; and the Imperial Cancer Research Fund in Oxford, England. A recipient of the Ersted Distinguished Teaching Award, Dr. Postlethwait encourages active participation of undergraduates in research and many of his publications include undergraduate students as coauthors. His love of teaching stems from his undergraduate days at Purdue University, where for three years he was a teaching assistant in General Biology.

Janet L. Hopson

has authored or coauthored seven books: a popular book on human pheromones, another on human brain development and enrichment, and five different college biology textbook titles for students of various levels. She has taught science writing on two campuses of the University of California and holds B.A. and M.S. degrees from Southern Illinois University and the University of Missouri. Ms. Hopson has won awards for magazine writing, and her articles have appeared in *Smithsonian, Psychology Today, Science Digest, Science News, Outside,* and others. She is married and enjoys pets, tennis, golf, skiing, swimming, and traveling.

A new way to EXPLORE the world

Explore Life takes an entirely new approach to introductory biology. Each chapter of the text springs from a **Case Study**— an engaging, personal narrative that introduces the questions and the biological problems raised by chapter concepts. The **Case Study** is then revisited throughout the chapter and serves as the framework for the **Explore Life** system of learning by tying together the text, videos, and e-media components. These components are all integrated line by line to create a biology program that gives students the maximum opportunity for success.

CHAPTER **5** PATTERNS OF INHERITANCE

A Devastatingly Common Illness

When Jonathan Zuckerman was a teenager growing up in Boston, he'd never heard of the disease he would someday study and treat in patients like Shana Reif: cystic fibrosis (CF). With one affected child in every 3,000 births, it is the most common lethal inherited illness among Caucasians. It causes a constellation of breathing, digestion, and other medical problems, but it has never been widely understood by the public. That may be, in part, because an affected child born in 1960—Dr. Zuckerman's year of birth—was unlikely to survive past his or her teens.

Today, thanks to improvements in conventional approaches like antibiotics, diet, and physical therapy, cystic fibrosis patients are living to their thirties, on average, and their care, as adults, has become a new medical specialty. Zuckerman was a physician and researcher at the University of Pennsylvania in Philadelphia until 1999 and has since worked at the Maine Medical Center in Portland. In both institutions, he has headed specialized programs for adults with cystic fibrosis. By applying modern genetic discoveries, he and others in his field are pioneering treatments to help keep CF patients alive and feeling as well as possible.

During and after medical school, Zuckerman focused on lung diseases and became a pulmonologist. Like most doctors in that specialty, he tended to see older patients who, through cigarette smoking, had developed lung cancer or emphysema. When he started working with cystic fibrosis pa-

tients, he found their youth, courage, and motivation "very inspirational." They are attempting, he explains, "to live a normal life and take care of their cystic fibrosis problem but still be out in the world" studying or working at their jobs. For the first time, there are accountants, lawyers, artists, and musicians with cystic fibrosis. "Three quarters of our patients," says Zuckerman, "are out there making a real contribution." And Shana, a registered nurse and cystic fibrosis patient in Philadelphia, was one of them.

Shana was born with cystic fibrosis and diagnosed on her second day of life. Her younger sister and brother are carriers of the disease but don't suffer the many symptoms she has faced: shortness of breath, recurrent lung infections, and CF-related diabetes. Every day is a struggle to stay well with breathing and chest-thumping treatments,

Test of Lung Capacity

▲ **Help for Cystic Fibrosis.**
Jonathan Zuckerman, Shana Reif, and Shana in the adult cystic fibrosis clinic.

Each chapter begins with a brief **Case Study** that poses an interesting problem that students can solve by studying the chapter. **Case Studies** draw readers into the material and make it meaningful to their own lives.

Case Studies cover such timely and intriguing topics as the likelihood of passing a genetic condition on to one's children, the latest methods of cancer therapies, and genetically modified foods.

The body of the text presents a lively investigation of the concepts and principles needed to understand the problem posed by the **Case Study**. As each chapter unfolds, students come to a greater understanding of the biological issues involved in the **Case Study**.

OVERVIEW

E xplorer 5.0

Patterns of Inheritance

aerosol inhalants, insulin injections, treadmill exercises, yoga, special diets, and sometimes intravenous antibiotics and hospital stays. Despite all of this, Shana has worked in hospitals herself, lives independently with her husband, and helps maintain their home. "I have goals and dreams for the future," she says, "but I have to focus on living one day at a time."

The powerful inspiration of Shana's life and others with cystic fibrosis is magnified when you consider the illness they face—essentially a disease of clogged ducts. A child who inherits one broken cystic fibrosis gene from each parent will produce a faulty version of a protein. This protein lies embedded in cell membranes, especially the membranes of **epithelial cells,** which tend to line ducts leading from the lungs, stomach, pancreas, sweat glands, and reproductive organs. The protein is involved in salt and water movement across cell membranes, and being defective, it prevents normal fluid transport. The walls of the ducts and the protective coatings they secrete tend to dry out, creating a thick, sticky mucus layer. This, in turn, clogs narrow passageways and ducts in the organs just mentioned. Because of this abnormal gene sequence, an individual who inherits cystic fibrosis usually has difficulty breathing. He or she repeatedly contracts dangerous bacterial infections in the lungs, suffers stomachaches and a diarrhea-like condition due to poor absorption of fats in the diet. Most also exude a salty secretion on the skin. And affected males are almost always sterile because of blocked ducts leading from the testes.

Patients like Shana usually take special pancreatic enzymes with their meals to help them digest fats, along with medicines to open the airways and antibiotics to prevent lung infections. They also require three or more "percussion sessions" per day during which a family member or friend claps them on the chest—front and back—for 20 minutes or so at a time to help dislodge gummy mucus and allow easier breathing. "My parents used to call them 'bangies,'" says Shana, "'Come on! It's time for your bangies!'" Recently, inflatable vests that emit air pulses have become available to do the thumping mechanically and Shana uses one regularly. But for her and others with cystic fibrosis, all the needed

therapies are time-consuming and must be juggled with studying, working, hobbies, and family demands.

Aside from the difficulties and dangers it poses, cystic fibrosis has become, in a sense, a "model" disease. Its molecular genetic basis, explains Zuckerman, was uncovered in the late 1980s just as physicians were starting to understand the disease's many effects on the body. With their working knowledge of the underlying gene and protein deficiencies, medical researchers can plan sophisticated, highly targeted treatments rather than simply treating symptoms separately. Experimental gene therapy going on at Johns Hopkins University, Cornell, the University of North Carolina, and elsewhere is one such targeted approach, aimed at inserting a healthy, functioning gene into patients' airways.

Identifying the cystic fibrosis gene has also enabled physicians to look for the inherited factor in potential parents—and they have found it in one of every 30 Caucasians. Finally, doctors and genetic counselors can apply the principles of genetics to predict what will happen if a person carrying one or two copies of the cystic fibrosis gene produces children with another person carrying zero, one, or two copies. For example, Shana married her high school sweetheart, Kurt, a few years ago and wanted to start a family. But she and Kurt wondered whether their children would be born with cystic fibrosis. By applying genetic principles, Dr. Zuckerman was able to advise them, as we'll see in the video, in the introductory Explorer for Chapter 5, and later in these pages.

Our goal in this chapter is to help you understand inheritance patterns like these—patterns that determine who will display lethal disease symptoms and who won't. Inheritance patterns underlie each of our thousands of traits—eye color, hair color, height, and so on—and those of all other living organisms. The science of **genetics** explores the nature of genes and how they are organized on chromosomes; how genes govern our appearance, physical functioning, and even behavior; and how medical researchers can manipulate genes to treat diseases like cystic fibrosis.

As you explore Chapter 5, you'll find the answers to these questions:

Explorer icons in the margins direct students to the **Explore CD-ROM**, where they'll discover dynamic and interactive new ways to understand concepts. **Explorer** icons include:

- **Overview**—Outline of major concepts for each chapter

- **Get the Picture**—Animated book art with brief quizzes

- **Read On**—Supplemental information, often with Web links

- **Solve It**—Investigative exercises that utilize the scientific way of problem solving

A new way to EXPLORE life in action

Many of the concepts found on the pages of the text are vividly brought to life through 26 original videos. Each video introduces the **Case Study**, poses the problem, and animates key concepts from the chapter. Developed specifically for **Explore Life**, these videos use interviews with cutting-edge scientists, physicians, patients, and others. They also include live-action footage and animations to bring people, places, and situations into the classroom in an unprecedented way. These videos are available in standard VHS format, as well as on a CD-ROM that allows instructors to utilize them for online courses.

The video for Chapter 5, *Patterns of Inheritance*, for example, focuses on a young Philadelphia woman who has the most common inherited illness among Caucasians—cystic fibrosis. A physician and researcher explains the genetics of this common disease, and how knowledge of its inheritance patterns can help current patients and may prevent some future cases. Building on this story, the chapter explains the history of Mendel and his experiments; how he discovered the rules governing inheritance in plants, people, and other animals; how organisms inherit individual and mutual traits; chromosomes and sex; and how molecular research is changing the way medicine predicts and treats diseases.

26 original videos

Each 10- to 12-minute video contains two places where instructors can stop the film. The first video stop follows the initial statement of the problem and allows students to investigate the material and devise a solution. At this point, the professor might want to lecture on the problem, and let students gather information about possible solutions from the textbook or other resources. Then the professor might want to play the rest of the video. The second stop poses a question of a social, political, or ethical nature that arises from the material. This format is intended to inspire class discussions and debates.

What is the role of a genetic counselor?

only from the mother. These maternal genes occur in the mitochondrion, the cell's energy "powerhouse." In a human, each mitochondrial genome has exactly 37 genes, while each nucleus has about 40,000 genes. Since sperm donate only a cell nucleus and usually no mitochondria to the offspring, an animal usually inherits all of its mitochondria from its mother through the egg. For this reason, regardless of your sex, you can trace your mitochondria to your mother, maternal grandmother and so on back through time. As you might expect, mutations in mitochondrial genes can decrease a cell's ability to make ATP, and so their effects are felt mainly in high-energy tissues, like the brain, heart, muscles, and eyes. A disease called Leber optic atrophy, for example, begins at about age 30 as a loss of vision in the central part of the visual field; this can spread over a couple of years and cause varying degrees of blindness. Leber optic atrophy comes about through mutations that disrupt proteins in the electron transport chain (review Fig. 3.15) and is passed only from a mother to her children. We'll explore more about mitochondrial genes, human evolution, and the relatedness of all eukaryotes in later chapters.

An Allele's Phenotype Can Depend on Other Genes and on the Environment

As Dr. Jonathan Zuckerman points out, until fairly recently people with cystic fibrosis rarely survived past infancy, but today, patients are living well into their 30s and beyond. This extended survival is due to advances in several environmental factors, including appropriate diet, special respiratory therapy, and aggressive antibiotic treatment against infection. The influence of environmental factors applies to many other genetic traits as well, with part of that environment being the interaction of seemingly unrelated genes. An example of this combined effect is breast cancer.

About 90 percent of breast cancers are sporadic, that is, they occur in people with no familial history of breast cancer. Geneticist Marie-Claire King, however, noticed that in some cases, breast cancer seemed to run in families. She collected pedigrees for a number of families in which several people (almost always women) had developed breast cancer and narrowed her study to those with early onset of the disease, before the person's mid-40s. King found pedigrees like the ones in Figure 5.20a. Look at this evidence and see if you can tell whether breast cancer appears to be caused by a dominant or recessive allele. Many affected people have at least one affected parent, so it looks like a dominant al-

lele. But for some affected individuals, neither parent had breast cancer. Most often, the unaffected parent is a male. So is this an *X*-linked trait? Apparently not, because there is father-to-son transmission in Family D (Fig. 5.20a). So this type of breast cancer is inherited as an autosomal dominant, but other genes affect its expression, especially the gene on the *Y* chromosome that causes male phenotype (*SRY*). Geneticists use the term **epistasis** (*epi* = over + *stasis* = standing) for cases in which one gene affects the expression of an entirely different gene. Hereditary breast cancer not only illustrates epistasis but another principle, as well: **sex-limited traits**—conditions shown by one sex or the other simply because of their maleness or femaleness. Breast cancer is sex limited in that it occurs almost exclusively in women. One male in Figure 5.20a did develop the disease, but less than 1 percent of breast cancer patients are male. Pattern baldness is another sex-limited trait and affects mainly males.

Despite the pattern Marie-Claire King saw in the pedigrees, they do show that some phenotypes vary for unknown reasons that may be related to people's environment. In Family D, for example (Fig. 5.20a), individual III6 doesn't have breast cancer but she apparently passed it to her daughter. The pedigrees were key to an even more important finding: with them, Dr. King was able to pinpoint a gene that predisposes people to breast cancer. This gene, called *BRCA1*, for "breast cancer 1," is located about halfway down the long arm of chromosome 17. Finding the exact gene is crucial for investigating diseases at the molecular level and developing treatments. This approach—mapping exact genes and then learning to counteract or correct their action—is changing the face of medicine. Let's see how in our final section.

5.6 How Is Genetics Changing Our World?

Shana, now in her mid-20s, is far from rare among people with cystic fibrosis. The majority are now surviving into midlife, although there has been a leveling off of life expectancy over the last few years. To move beyond the environmental remedies now available (diet, airway clearing therapies, and antibiotics), new therapies are needed to extend patients' lives further. To this end, Dr. Zuckerman and other researchers are mapping genes, learning what the genes do in healthy and diseased cells, and working on molecular solutions. The

Video captures in the text margin directly link the material being discussed in the book to the **Case Study** found on the corresponding video.

New ways to EXPLORE with innovative technology

The unique and text-specific **Explore CD-ROM** provides an in-depth and interactive exploration of text topics. Animations, video clips, audio, and exercises bring even the most complex concepts into sharp focus.

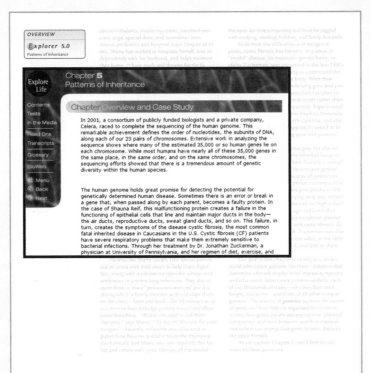

Every **Explore CD-ROM** chapter begins with an overview that outlines the major biological concepts and their relationship to the case study for that chapter, including a brief outtake from that chapter's video.

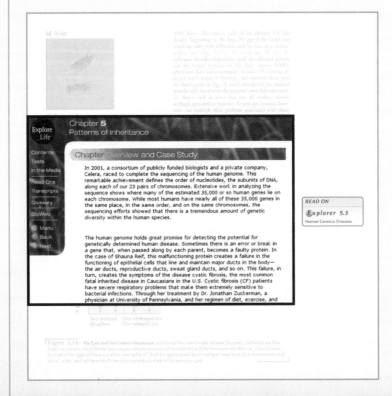

Read On Explorers provide expansions of material in the text with interactive tables, Web links, and investigative topics for student assignments and group activities. Each **Read On Explorer** offers clickable links that provide definitions of key terms or take students to relevant Web sites.

Get the Picture Explorers help students to understand complex concepts through animated presentation. Narration reinforces material presented in the text, and each **Explorer** is followed by several interactive quizzes for student self-assessment.

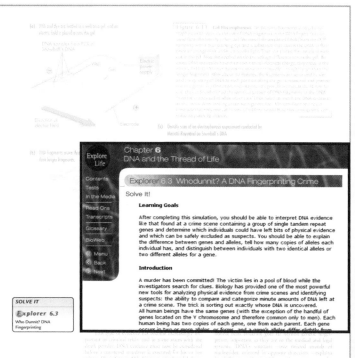

Solve It Explorers, often in the form of an interactive table or diagram, provide investigations, laboratory simulations, and experiments designed to encourage problem solving and critical thinking.

Turn the page to learn about **EXPLORE LIFE'S** integrated Web component . . .

New ways to EXPLORE on the Web

You'll find even more opportunities to discover and learn beyond the pages of the text with **Explore Life's** dedicated Web site. More than a mix of links, this site provides directed but open-ended opportunities for students to delve into the borders of current knowledge.

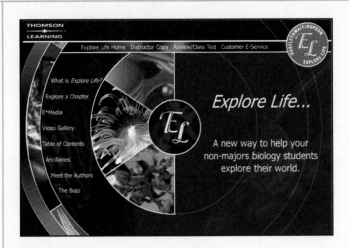

Also on the EXPLORE LIFE Web site:

■ An electronic **Glossary** that covers the key terms found in the book's integrated print, video, and electronic media components.

■ **Community Resources** that offer ideas for more effective learning and teaching and links that pertain to each chapter's material.

■ A comprehensive array of **download-able databanks** for instructors, including an image bank, test question bank, activities bank, and teaching tips.

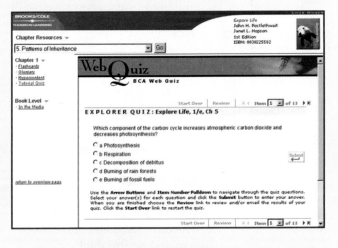

Chapter Quizzes assess student understanding of chapter materials at a variety of cognitive levels. Students get electronic scores on questions from the textbook, along with additional questions that address content from the **Explorers**. Scores and responses can be submitted to instructors via e-mail.

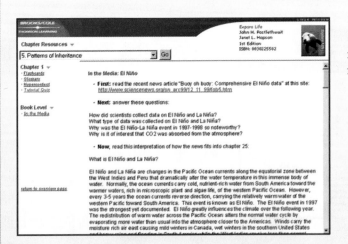

In the Media is a regularly updated news feature that relates to concepts and vocabulary from each chapter of the text.

New ways to EXPLORE with art

Explore Life's rich illustrations make the material accessible to visual learners. The book's colorful and detailed art program, including orienting icons, helps students understand macroscopic or unfamiliar biological structures, classic experiments, and biological processes.

becomes far saltier than normal. And the exit ducts for sperm become blocked, leading to sterility in males. Most human genetic diseases are pleiotropic like cystic fibrosis. Tracing the primary defect to a single gene and protein can often help provide therapies that relieve several symptoms at once.

5.4 What Are the Rules Governing How Organisms Inherit Multiple Traits?

So far, we have discussed single genes and their alleles, whether for flower color or cystic fibrosis. But organisms have thousands of genes. A single-celled yeast has about 6,000 genes, a soil-living roundworm has about 19,000, and you have about 40,000. What happens when a geneticist studying the cross between two individuals focuses on more than one gene at the same time in the same cross? For example, about 25 years ago, biologists wondered about the inheritance of ABO blood types in families with some members who had cystic fibrosis. How might these two traits be inherited? Would children with cystic fibrosis inherit only a particular blood type? Or, might there be no relationship at all between the inheritance of the two genes? Some of Gregor Mendel's experiments allow us to interpret and predict the result of such human crosses.

Formation of Gametes

Recall that Shana is homozygous for the cystic fibrosis trait and has two recessive alleles: CF^-CF^-. She also has blood type A, and is homozygous dominant for this

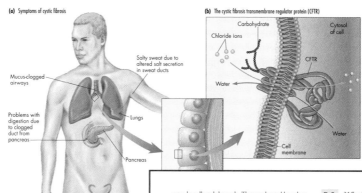

(a) Symptoms of cystic fibrosis

- Mucus-clogged airways
- Salty sweat due to altered salt secretion in sweat ducts
- Problems with digestion due to clogged duct from pancreas
- Lungs
- Pancreas
- Testis
- Infertility in males due to clogged sex ducts

(b) The cystic fibrosis transmembrane regulator protein (CFTR)

- Carbohydrate
- Cytosol of cell
- Chloride ions
- CFTR
- Water
- Water
- Cell membrane

peas, he collected the seeds. These seeds would produce the next generation, called the **first filial (F$_1$) generation,** meaning the first generation in the line of descent. Planted in the spring of the second year, the F$_1$ seeds of the long-stem/short-stem cross all grew into plants with stems just as long as the original long-stemmed parent (Fig. 5.3b). Mendel repeated this type of experiment for other traits—flower color (purple vs. white), seed shape (round vs. wrinkled), and so on—and he found that in each case, only one alternative of each trait appeared in the F$_1$ hybrid generation. It was as if one of the traits had totally disappeared. The trait that appears in the F$_1$ hybrid (such as long stems in peas) is said to be **dominant,** while the trait that does not show in the hybrid (such as short stems in peas) is referred to as **recessive.**

Now came the crucial part of the experiment. What happened to the recessive characteristic in the hybrid? Did it blend with the dominant characteristic? Did it disappear completely and forever? Or did it remain intact but hidden in the F$_1$ generation? To find out, Mendel allowed the long-stemmed F$_1$ hybrid plants to self-fertilize, and the next spring he planted the seeds of the **second filial (F$_2$) generation.** When the second generation of pea plants grew up, most of them had long stems, but significantly, some plants had short stems. Again, there were no stems of intermediate length (Fig. 5.3c). The reappearance of plants with stems just as short as the stems of the original short-stemmed parents, and the absence of any intermediates were the results predicted by the particulate model of heredity and dramatic disproof of the blending model of heredity.

5.2 What Rules Govern the Inheritance of a Single Trait?

Before we return to cystic fibrosis, Shana, and Dr. Zuckerman, we need to learn more about Mendel's simple experiments. Being a careful and inquisitive person, Mendel was not satisfied with just saying that "some" of the F$_2$ plants had short stems and therefore the blending hypothesis was wrong. He wanted to understand what he saw. So he counted the plants and by analyzing the numbers, was able to infer the mechanisms that hid the short-stemmed trait in the F$_1$ and its reappearance in the F$_2$. Figure 5.4 lets you collect data from an actual experimental cross to practice gathering, analyzing, and interpreting scientific data. This cross uses corn plants because the results are readily visible in a single ear of corn and each ear has enough individual offspring (the kernels) to allow meaningful inferences. Pea pods, by contrast, have only about six seeds (too few per pod for analysis) and must be opened to reveal the offspring (peas).

Collecting Genetic Data

Figure 5.4b shows a corn ear with kernels of different colors. The kernels on this corn ear are the F$_2$ progeny of the cross shown in Figure 5.4a. To make that cross, we first bought two types of seeds, one from a stock that yields only purple seeds and another from a stock that yields only yellow seeds. In spring, we planted the two types of seeds and allowed the corn plants to ma-

Figure 5.4 **Collecting Data from a Cross.** The matings in (a) gave rise to the F$_2$ corn ear shown in (b). Count the yellow and purple kernels. What ratio do you obtain? How do you interpret your results?

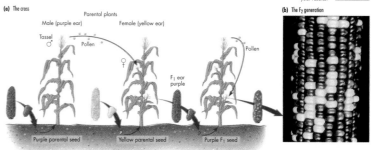

(a) The cross

- Parental plants
- Male (purple ear)
- Female (yellow ear)
- Tassel
- Pollen
- Pollen
- F$_1$ ear purple
- Purple parental seed
- Yellow parental seed
- Purple F$_1$ seed

(b) The F$_2$ generation

New ways to EXPLORE teaching

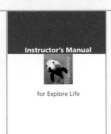

Instructor's Manual
ISBN: 0-03-022118-8

The ideal teaching companion! **Explore Life's Instructor's Manual** offers suggestions for lesson plans and enrichment activities. Also included is a *Video Viewing Guide* and extensive suggestions for ways of integrating the various aspects of the program (print, video, and multimedia) into the course.

Multimedia Presentation Manager for Biology:
A Microsoft® PowerPoint® Presentation Tool
ISBN: 0-534-40608-4

This CD-ROM contains everything instructors need for dynamic lectures. It includes simulations for lectures, key art pieces from the text, and other items suitable for classroom presentation. Line illustrations are available in four different formats for maximum flexibility and can be imported into Microsoft® *PowerPoint*® lecture outlines.

Test Bank
ISBN: 0-03-022117-X

The test bank provides an outstanding set of questions at various levels, including simple recall, but with all questions emphasizing working knowledge, application, evaluation, and critical thinking. Questions are labeled according to their level.

Overhead Transparencies
ISBN: 0-03-022124-2

This set includes approximately 250 full-color transparencies that cover key concepts from the text.

WebTutor™ Advantage on WebCT & Blackboard
WebCT ISBN: 0-534-40609-2 • Blackboard ISBN: 0-534-40610-6

A great study tool . . . a great course management tool . . . a great communication tool. **WebTutor Advantage** is filled with pre-loaded, text-specific content (including interactive simulations, *PowerPoint*® files, and much more) and is ready to use as soon as you and your students log on. At the same time, you can customize the content in any way you choose, from uploading images and other resources, to adding Web links, to creating your own practice materials.

For students, **WebTutor** offers real-time access to a full array of study tools, including flashcards (with audio), practice quizzes, online tutorials, and Web links. Use **WebTutor** to provide virtual office hours, post your syllabi, set up threaded discussions, track student progress with the quizzing material, and more. **WebTutor** provides rich communication tools, including a course calendar, asynchronous discussion, "real time" chat, a whiteboard, and an integrated e-mail system. Professors who have tried **WebTutor** love the way **WebTutor** allows students—even those in very large classes—to actively participate in class discussions online. This student-to-student interaction has enormous potential to enhance each student's experience with the course content.

RESOURCES

ExamView®
Computerized Testing
ISBN: 0-03-022119-6

Create, deliver, and customize tests and study guides (both print and online) in minutes with this easy-to-use assessment and tutorial system! ExamView's *Quick Test Wizard* guides you step-by-step through the process of creating and printing a test, while its "WYSIWYG" capability allows you to see the test you are creating on the screen exactly as it will print. You can build tests of up to 250 questions using up to 12 question types, including true/false, modified true/false, multiple choice, matching, yes/no, essay, case, problem, short answer, numeric response, and fill-in-the-blank.

Using **ExamView's** complete word processing capability, you can enter an unlimited number of new questions or edit existing questions. To create online *Tests* and *Study Guides,* simply choose *"Save as Internet Test"* after you have built a test. *Online Test Wizard* allows you to first format a test for Internet delivery over local area network (LAN) or standalone computer, then control who takes the test, when it may be completed, and what information (results, responses, answers, etc.) is provided to students. *Online Test Player* allows you to include graphics, audio, video, and animations in your tests. Call-in testing is also available.

ExamView is a registered trademark of FSCreations, Inc. Used herein under license.

CNN® Today: Biology Videos
Volume I ISBN: 0-314-22478-5
Volume II ISBN: 0-534-52317-X
Volume III ISBN: 0-534-52318-8
Volume IV ISBN: 0-534-38112-X

Good news is in hand! Now you can integrate the up-to-the-minute programming power of CNN and its affiliate networks right into your course when you adopt **Explore Life**. Updated yearly, **CNN Today Videos** are course-specific videos that can help you launch a lecture, spark a discussion, or demonstrate an application—using the top-notch business, science, consumer, and political reporting of the CNN networks. With accompanying critical-thinking question worksheets, the possibilities for teaching are endless. After all, these are **CNN Today Videos**. And any way you view them, it's good news for education.

New ways to EXPLORE learning

Study Guide

ISBN: 0-03-022572-8

Even more resources for mastering biology . . . at your fingertips! This **Study Guide** includes tutorial, practice quizzes, and much more.

Explore CD-ROM

Packaged FREE with every new copy of this text! This integral part of the **Explore Life** learning system features interactive explorations and quizzing, allowing students to take their understanding of biology to new levels. See pages 8 and 9 for further information on this valuable resource.

Explore Life Web Site

http://info.brookscole.com/postlethwait1

See page 10 for more information on this dynamic Web site.

InfoTrac® College Edition

. . . the online library!

Turn to **InfoTrac® College Edition** for the latest news and research articles online—updated daily and spanning four years! Packaged FREE with every new copy of **Explore Life**, this unparalleled resource gives you four months of free access to an easy-to-use online database of reliable, full-length articles (not abstracts) from thousands of top academic journals and popular sources. Journals available 24 hours a day, seven days a week include:

- *Annual Review of Genetics* ■ *Annual Review of Microbiology* ■ *Biological Bulletin*
- *BioScience* ■ *Human Biology* ■ *Life Science Today* ■ *Perspectives in Biology and Medicine*
- *Quarterly Review of Biology* . . . and thousands more!

The possibilities for incorporating **InfoTrac® College Edition** into your course are virtually limitless, from putting together reading assignments to using articles to launch lectures, ignite discussions, or open whole new worlds of information and research for students. If you want your students to have access to a custom, no-cost set of online **InfoTrac® College Edition** readings that feature the latest research and findings in biology, you can work with your Thomson•Brooks/Cole representative to create an up-to-the-minute resource.

InfoTrac® College Edition is available to college and university students only. Journals subject to change.

The Brooks/Cole Biology Resource Center

http://biology.brookscole.com

Access to this online resource is FREE to text adopters and their students. This outstanding site features descriptions of degrees and careers in biology, a student feedback site, biological games, and cool clip art. For instructors, there are ideas for teaching on the Web and an instructors' forum where you can share thoughts on teaching with colleagues.

RESOURCES

Also Available:

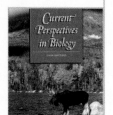

Current Perspectives in Biology
ISBN: 0-314-20638-8

by **Shelly Cummings**. This collection of 46 articles comes from a variety of publications, both scholarly and popular. Each article begins with a brief summary of the ideas presented in the selection and ends with a set of questions to help identify the key points of discussion. The answers are in the back of the book.

Classic and Modern Readings in Biology
ISBN: 0-03-097108-X

by **Randy Moore**. This versatile collection of modern and classic essays stresses reading and writing tools for learning about biology. Students are carefully guided to help them discover and communicate information.

Cooperative Learning:
Making Connections in General Biology
ISBN: 0-534-37605-3

by **Mimi Bres** and **Arnold Weisshaar**. This booklet contains hands-on, cooperative learning activities for biology students to do in the classroom or in the lab, in less than 10-15 minutes. The booklet includes activity worksheets correlated to the text for students to complete. An instructor's manual is also available that includes tips for incorporating cooperative learning into your course and answers to the cooperative learning activities.

Thinking Toward Solutions:
Problem-Based Learning Activities for General Biology
ISBN: 0-03-025033-1

by **Deborah Allen** and **Barbara Duch**. The book, a product of over 10 years of teaching experience, is filled with innovative student problem-solving activities. Designed to help provide the ideal science learning model as set forth by the Project Kaledioscope Committee, this book seeks to enmesh the learner in a community of learners, make the learning experience personal, and establish connections that place the content in context.

The many ways to EXPLORE LIFE ...

For an at-a-glance view of how the various components of the teaching and learning package work together, take a look at the visual summary below.

 Video CD-ROM Web Site

A Devastatingly Common Illness

When Jonathan Zuckerman was a teenager growing up in Boston, he'd never heard of the disease he would someday study and treat in patients like Shana Reif: cystic fibrosis (CF). With one affected child in every 3,000 births, it is the most common lethal inherited illness among Caucasians. It causes a constellation of breathing, digestion, and other medical problems, but it has never been widely understood by the public. That may be, in part, because an affected child born in 1960 — Dr. Zuckerman's year of birth — was unlikely to survive past his or her teens.

Today, thanks to improvements in conventional approaches like antibiotics, diet, and physical therapy, cystic fibrosis patients are living to their thirties, on average, and their care, as adults, has become a new medical specialty. Zuckerman was a physician and researcher at the University of Pennsylvania in Philadelphia until 1999 and has since worked at the Maine Medical Center in Portland. In both institutions, he has headed specialized programs for adults with cystic fibrosis. By applying modern genetic discoveries, he and others in his field are pioneering treatments to help keep CF patients alive and feeling as well as possible.

During and after medical school, Zuckerman focused on lung diseases and became a pulmonologist. Like most doctors in that specialty, he tended to see older patients who, through cigarette smoking, had developed lung cancer or emphysema. When he started working with cystic fibrosis patients, he found their youth, courage, and motivation "very inspirational." They are attempting, he explains, "to live a normal life and take care of their cystic fibrosis problem but still be out in the world" studying or working at their jobs. For the first time, there are accountants, lawyers, artists, and musicians with cystic fibrosis. "Three quarters of our patients," says Zuckerman, "are out there making a real contribution." And Shana, a registered nurse and cystic fibrosis patient in Philadelphia, was one of them.

Shana was born with cystic fibrosis and diagnosed on her second day of life. Her younger sister and brother are carriers of the disease but don't suffer the many symptoms she has faced: shortness of breath, recurrent lung infections, and CF-related diabetes. Every day is a struggle to stay well with breathing and chest-thumping treatments,

Help for Cystic Fibrosis.
Jonathan Zuckerman, Shana Reif, and Shana in a cystic fibrosis clinic.

 CASE STUDY
Organizes and integrates the text, video, CD-ROM, and Web activities

 EXPLORER LINKS
Visual connectors between the text and its multimedia

VIDEO CAPTURES
References to key concepts that are highlighted on the chapter video and/ or CD-ROM

What is the role of a genetic counselor?

  ART
Illuminating visualization of concepts; often animated on the CD-ROM and Web site

---

3500 boys. The muscle cells of an affected boy die slowly, beginning in the legs. By age 5 the child can stand up only with difficulty, and he rises in a characteristic way (Fig. 5.17a). At about age 20, the diaphragm muscles degenerate, and the affected person can no longer breathe. In the fight against DMD, physicians have taken immature muscle cells from an affected boy's father or brother, and injected them into the boy's arms or legs. A small number of the normal muscle cells fused with the patients' own defective muscle fibers, and in some but not all studies, muscle strength appeared to increase. As you can imagine, however, the multiple allele problem associated with tissue incompatibilities has resulted in limited success. Researchers hope that through drug therapy, muscle-cell injections, and other future discoveries, successful treatments will emerge for this debilitating recessive condition.

Observe the pedigree for Duchenne muscular dystrophy (Fig. 5.17b); it shows the typical inheritance pattern of *X*-linked genes. These traits are more frequent in boys, and they are passed to boys from their mothers, who are heterozygous carriers. *If a pedigree shows that a boy inherits a trait from his father, it is unlikely to be an X-linked condition.* Another *X*-linked recessive allele causes hemophilia and another causes color blindness. You can learn more about sex-linked traits, including an *X*-linked dominant disease called Charcot-Marie-Tooth Disease, in Figure 5.18 and Explorer 5.3.

Chromosomes and Sex Determination

The historic experiments with sex chromosomes in fruit flies showed that the expression of male and female characteristics depends on chromosomes. They did not show, however, how sex determination works. Consider this: In flies as well as in people, *XX* individuals are females and *XY* individuals are male. But in both cases, males differ from females in two factors: the presence or absence of a *Y*, and the number of *X* chromosomes. Which is more important? To answer, geneticists have studied individuals with unusual numbers of sex chro-

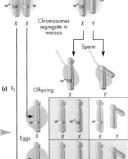

(a) Fly eyes

(b) The parents (P)
Heterozygous female   Normal male

Chromosomes segregate in meiosis

Sperm

(c) F₁   Offspring

Eggs

Two red-eyed daughters   One white-eyed son   One red-eyed son

READ ON
Explorer 5.3
Human Genetics Diseases

Figure 5.16 Fly Eyes and Sex-Linked Inheritance. (a) Normal flies have bright red eyes (bottom), and white eye flies (top) are mutant. (b) A female heterozygous for the recessive white allele (w) and the dominant red allele (w⁺) has red eyes. (c) Half of her eggs will have a w allele, and half a w⁺. Half the sperm made by her red eyed mate have an X chromosome with the w⁺ allele, and half have the Y chromosome with no allele of the eye color gene.

Explore Life

John H. Postlethwait
University of Oregon

Janet L. Hopson

BROOKS/COLE

™

THOMSON LEARNING

Australia   Canada   Mexico   Singapore   Spain
United Kingdom   United States

Biology Editor: Nedah Rose
Development Editor: Gabrielle Goodman
Editorial Assistant: Rebecca Eisenman
Assistant Editor: Christopher Delgado
Technology Project Manager: Keli Sato Amann
Marketing Manager: Ann Caven
Media Managers: Cathleen Petree/Carol Pritchard-Martinez
Advertising Project Manager: Stacey Purviance
Sr. Production Manager, Editorial Production: Charlene Catlett Squibb
Print/Media Buyer: Kristine Waller
Production Service: P. M. Gordon Associates

Text Designer: Ruth Hoover
Sr. Art Director: Caroline McGowan
Photo Researcher: Amy Dunleavy
Videographer: Leighton Images
Copy Editor: Joanne Fraser
Illustrator: Rolin Graphics, Inc.
Cover Designer: Caroline McGowan/Lawrence Didona
Cover Image: ©ZEFA Biotic/Photonica
Cover Printer: The Lehigh Press, Inc.
Compositor: Progressive Information Technologies
Printer: RR Donnelley & Sons

For more information about our products, contact us at:

Thomson Learning Academic Resource Center
1-800-423-0563

For permission to use material from this text, contact us by:

Phone: 1-800-730-2214
Fax: 1-800-730-2215
Web: http://www.thomsonrights.com

Library of Congress Control Number: 2002101493
ISBN: 0-03-022558-2

Brooks/Cole—Thomson Learning
511 Forest Lodge Road
Pacific Grove, CA 93950
USA

Asia
Thomson Learning
60 Albert Street, #15-01
Albert Complex
Singapore 189969

Australia
Nelson Thomson Learning
102 Dodds Street
South Melbourne, Victoria 3205
Australia

Canada
Nelson Thomson Learning
1120 Birchmount Road
Toronto, Ontario M1K 5G4
Canada

Europe/Middle East/Africa
Thomson Learning
Berkshire House
168-173 High Holborn
London WC1V 7AA
United Kingdom

To Nita.

To Jerry, my continuing inspiration.

Acknowledgments

Explore Life has been, for us, an odyssey to the frontiers of publishing. It started with some wishful thinking while we were hiking together one day: "Wouldn't it be great if we could get case studies about real people and real situations into a biology program? Then we could merge those case studies into a textbook that is completely integrated with a video series, a CD-ROM, and the Internet." We agreed that yes, it would be great. In time, it came not just from our own imaginings but from an Acquiring Editor, Edith Beard-Brady, and a top-notch educational book house, Saunders College Publishing, led by Emily Barrosse. As part of Harcourt College Publishers, they had the vision and the wherewithal to make it happen.

That was in 1997. From the beginning, we knew we would be bushwhacking through uncharted and tangled territory. Modern biology textbook programs, including four of our own previous projects, may have some videos available to adopters, CD-ROMs for the students, and some kind of Internet add-on. But these elements are virtually always created at different times by different teams. First the authors write the book, then an electronic media team comes onboard to fashion ancillaries to include in the advertised and marketed package.

When we were daydreaming about real integration, we meant working on all the materials simultaneously. Finding fascinating case history subjects. Writing about them. Filming them for the videos. Using still shots of them and their stories in the book and CD-ROM. Developing art for the book that also appears in the videos. Developing animations for the videos that appear in the CD-ROM. Creating Internet exercises and virtual field trips, and discussing them in the book. Generating a glossary for the book that would also pop up on command in the CD-ROM and Internet materials. The potential for student participation and understanding was thrilling. The reality was beyond daunting.

It meant two things. We would need a highly professional and creative team—a big one. And we would all have to move forward at the same time on each piece of this intertwined mass of traditional and state-of-the-art media. Edith Beard-Brady brought in Gabrielle Goodman as our well-qualified and highly energetic project manager and Developmental Editor. Together they added more players: Experienced independent filmmakers Geoffrey Leighton and Anita Clearfield of Leighton Images who criss-crossed North America for the interviews, along with their staff, Jennifer van der Werf, Jennifer Caritas, and later Stephanie Holmes, who designed great graphics for the videos. Cathleen Petree, a well-connected Consulting Editor who could help us find and persuade researchers, physicians, patients, and other case study subjects to appear in *Explore Life*. The very talented and imaginative science illustrator Elizabeth Morales as Art Development Editor for diagrams and figures. Chip Price, Amy McCorkle, David Shaw, and later Arati Nagaraj of VPG Integrated Media to develop the CD-ROM and Internet software. Biologist Steve Brewer of the University of Massachussets, Amherst, to create innovative electronic study materials. Travis Moses-Westphal, Dickson Musselwhite, and Natalie Peretti to help plan our electronic components and Erik Fahlgren followed by Kathleen McClellan as Marketing Managers.

Along the way, Edith Beard-Brady changed companies and Nedah Rose very ably took over the important role of managing the burgeoning team. Consulting multimedia expert Ian List came on board to create CD-ROM "Get the Picture" Explorers under the supervision of Cathleen Petree, who by this time had become our E-media Coordinator. Biologist Karin Jegelian came in to write selected storyboards. Amy Dunleavy, a consulting Photo Researcher, began finding hundreds of photos for the text and selecting stills from the growing video series for both the text and CD-ROM. Biologists Betsy Ott of Tyler Junior College and David Tapley of Salem State College joined the team as reviewers of the text and electronic media. Betsy also began contributing questions for various program components. Biologists Felicia Goodrum and Alex Flood of Princeton University started contributing CD-ROM Explorers as well, and writing "In the Media" articles for the *Explore Life* Web site.

Consultant Joanne Butler in Chicago began planning text ancillaries such as the all-important *Instructor's Manual with Video Guide* and *Study Guide.* Robin Bonner managed the long production process and when she moved on to a new job, Ellen Sklar picked up seamlessly. Caroline McGowan designed our handsome graphics program and cover art, and Rolin Graphics skillfully rendered final line illustrations. Freelance copyeditor Joni Fraser combed both book and e-media programs for grammatical gaffs and equally importantly, for inconsistencies in presentation between the various media.

Gabe Goodman took maternity leave and editor Lee Marcott stepped in, also seamlessly, to steer our now sizeable team as we inched forward with our multifaceted but tightly entwined efforts. Lee moved on to a new company and Gabe returned. Julie Naef edited glossary materials and Megan Thynge carried on after her. Emily DiTomo assisted Gabe and Lee, followed by Kim Paschen. Nedah and Gabe brought in extremely competent and careful reviewers for the text, the videos, and the CD-ROM and Internet materials. They found faculty members from across the country to participate in focus groups and workshops and to class test our materials as our team developed them. Vital market research was conducted both in-house by Rosemarie Console and her team, and by marketing consultant Marjorie Waldron.

Our team continued to grow and change faces as Harcourt College Publishers became part of Thomson Learning. An extremely able consultant for electronic media, Carol Pritchard-Martinez, coordinated the last stages of CD-ROM and Internet development. Our new Marketing Manager, Ann Caven, planned a campaign to introduce our unique program to the biology community. Under the direction of consulting editor Liz Covello, innovative educators Donald Cronkite of Hope College and Jewell Reuter of Louisiana Virtual School invented Internet materials that will unite our users into one interactive teaching community.

Every major player in this ambitious project knew that as we forged new ground in educational publishing, their job would be challenging, vexing, at times even maddening. *Explore Life* did not disappoint. Yet all of these focused professionals with their diverse talents maintained an admirable geniality while pushing forward through jumbled terrain like some giant literate plasmodium. Our gratitude is unending: Thank every one of you for adopting our vision as your own. For enduring the pressures and complexities of this project. For putting up with our foibles. And for helping us to make *Explore Life* a truly new and significant contribution to biology education.

Our thanks extend to several groups outside the immediate publishing team, as well.

Dozens of researchers, physicians, patients, educators, students, and others agreed to phone interviews and video tapings for our case studies. We understand the sacrifice of time and privacy this represented and truly appreciate your willingness and cooperation. The interviewees included:

Chapter 1: What is Life (and Why Study It)?
Dr. Margaret Race, NASA
Dr. Pamela Giles Conrad, Research Scientist Astrobiology Group
Chapter 2: Cells and the Chemistry of Life
Dr. Jay Levy, University of California-San Francisco
Dan Foley, Financial Analyst
Chapter 3: How Cells Take In and Use Energy
Dr. William Calder, University of Arizona
Hannah Griscom, Undergraduate Assistant Field Biologist
Peter Groblewski, President, Bates College Cycling
Chapter 4: The Cell Cycle
Dr. Ervin Epstein, San Francisco General Hospital
Patricia Hughes, Senior Technical Support Specialist
Chapter 5: Patterns of Inheritance
Dr. Jonathan Zuckerman, Maine Medical Center
Larry Culp, Editor Cystic Fibrosis Newsletter
Shana Reif, Registered Nurse
Kurt Reif, Husband, Patient
Chapter 6: DNA and Thread of Life
Dr. Marilyn Menotti-Raymond, National Cancer Institute
Victor David, Geneticist National Institute of Health
Chapter 7: Gene Function and Manipulation
Dr. Harry Meade, Genzyme Transgenics
Chapter 8: Reproduction and Development
Dr. Francis Batzer, Women's Institute for Fertility, Endocrinology, and Menopause
Katherine Go, Ph.D., Embryologist
Stephen, Patient
Marcy, Patient
Chapter 9: Mechanisms of Evolution
Dr. Fred Tenover, Centers for Disease Control and Prevention
Dr Joshua Fierer, Head Division of Infectious Diseases, University of California-San Diego
Wayne Chedwick (pseudonym), Patient
Chapter 10: Life's Origins and Biodiversity
Dr. Robert Robichaux, University of Arizona
Chapter 11: Single-Celled Life
Dr. Daniel Goldberg, Washington University School of Medicine
Mark Stover, Medical Student
Chapter 12: Fungi and Plants: Decomposers and Producers
Dr.Paul Bosland, New Mexico State University
Chapter 13: Animals: the Great Consumers
Dr. Bruce Gill, Canadian Food Inspection Agency
Chapter 14: Body Function, Survival and Steady State
Dr. Bernard Harris, NASA
Chapter 15: Circulation and Respiration
Dr. Robert Winslow, Sangart, Inc.
Chapter 16: Immune System
Dr. Jack Gwaltney, University of Virginia Medical Center
Adrienne Wilson, Patient
Chapter 17: Nutrition and Digestion
Dr. Robyn Barbiers, Lincoln Park Zoo, Chicago
Sue Crissy, Brookfield Zoo, Chicago
Michael Brown Palsgrove, Zookeeper
Chapter 18: Hormones: messengers of Change
Dr. Paul Plotsky, Emory University
Wade Yandell, Student
Chapter 19: Nervous System
Dr. Helen Neville, University of Oregon
Sherry Greer, Deaf Subject
Johanna Larson-Muhr, American Sign Language Interpreter
Chapter 20: Muscles and Skeleton: The Body in Motion
Dr. Marjorie Woollacott, University of Oregon
Billy Harper, Long Distance Runner
Mary Walrod, Dr. Woollacott's Subject

Nicole Commissione, Sprinter
Chapter 21: Plant Life: Form and Function
Geof Kime, President Hempline
Gordon Scheifele, Research Coordinator for Field Crops, University of Guelph, Ontario Canada
Peter Dragla, Research Agronomist, Hemp Breeder
Chapter 22: How Plants Grow
Dr. Raymond Rodriguez, University of California-Davis
Chapter 23: Dynamic Plant
Dr. Joseph DiTomaso, University of California-Davis
Casey Stone, Rancher
Chapter 24: Ecology of Populations and Communities
Dr. Michael Sissenwine, Northeast Fisheries Science Center, Wood Hole
Lendall Alexander Jr., Fisherman
Chapter 25: Ecology of Ecosystems and Biosphere
Dr. Camille Parmesan, University of Texas-Austin
Chapter 26: Animal Behavior
Dr. Frans de Waal, Yerkes Primate Center, Emory University

A long list of thoughtful and helpful biology educators agreed to read and critique the textbook manuscript, early versions of the videos, and our e-media materials including CD-ROM and Internet components, and we are indebted to them for helping us improve our accuracy and presentation:

Douglas Allchin, University of Texas-El Paso
Jane Aloi-Horlings, Saddleback College (CA)
David Arnold, University of Texas-Austin
C. Warren Arnold, Allan Hancock College (CA)
Sarah Barlow, Middle Tennessee State University
Michael C. Bell, Collin County Community College (TX)
Rudi Berkelhamer, University of California-Irvine
Charles Biggers, University of Memphis (TN)
Lorena Blinn, Michigan State University
Susan K. Blizzard, Community College of Southern Nevada
Sara Brenizer, Shelton State Community College (AL)
Kimberly Brown, Mississippi Gulf Coast Community College-Jackson County
Arthur Buikema, Jr., Virginia Polytechnic Institute and State University
Ruth Chesnut, Eastern Illinois University
William Coleman, University of Hartford (CT)
Linda Crow, North Harris Montgomery Community College (TX)
Stan Dalton, Jones County Junior College (MS)
Juville Dario-Becker, Central Virginia Community College
Garry Davies, University of Alaska-Anchorage
Paul G. Deceiles, Johnson County Community College (KS)
Brent de Mars, Lakeland Community College (OH)
Jean de Saix, University of North Carolina-Chapel Hill
Miriam del Campo, Miami-Dade Community College (FL)
Kathleen Dillon, Brookdale Community College (NJ)
Linda Dixon, American River College (CA)
Gary Donnermeyer, Kirkwood Community College (IA)
James F. Duke, Calhoun Community College (AL)
Jamin Eisenbach, Eastern Michigan University
Dr. Thomas Emmel, University of Florida
Susan Evarts, University of St. Thomas (MN)
Gerald Farr, Southwest Texas State University
Alex Flood, Princeton University
Kevin Fox, State University of New York-Fredonia
Kathy Gallucci, Elon College (NC)
Felicia Goodrum, Princeton University
Jack Grubaugh, University of Memphis (TN)
Jennifer Gruber, Harrisburg Area Community College (PA)
Cheryl Hack, Anne Arundel Community College (MD)
Debby Hanmer, University of Wisconsin-La Crosse
Robert Harms, Saint Louis Community College (MO)
David Hedgepeth, Valdosta State University (GA)
Bob Herrington, Georgia Southwestern State University
Beatrice Holton, University of Wisconsin-Oshkosh
Dr. James Horowitz, Palm Beach Community College (FL)
Michael S. Hudecki, State University of New York-Buffalo
Georgia Ineichen, Hinds Community College (MS)

Charles W. Jacobs, Henry Ford Community College (MI)
Isidore Julien, Purdue University
George Karleskint, Saint Louis Community College (MO)
Arnold Karpoff, University of Louisville (KY)
Marlene Kayne, College of New Jersey
Bobbi Kervin, Presentation College (SD)
John Killian, Virginia Western Community College
Stacey Kiser, Lane Community College (OR)
Robert Kitchin, University of Wyoming
Brenda Knotts, Eastern Illinois University
Phyllis Laine, Xavier University
Tom Lancraft, St. Petersburg Junior College (FL)
Alicia Lesnikowska, Georgia Southwestern State University
Barbara Liedl, Central College (IA)
David Loring, Johnson County Community College (KS)
Ann Lumsden, Florida State University
Bonnie Lustigman, Montclair State University (NJ)
Charles H. Mallery, University of Miami (FL)
Kenneth Mason, University of Kansas
Ric Matthews, San Diego Miramar College (CA)
Susan McMahon, Pellissipi State Technical Community College (TN)
Stephen Mech, University of Memphis (TN)
Pam Miljak, McHenry County College (IL)
Neil A. Miller, University of Memphis (TN)
Kenneth D. Nadler, Michigan State University
Mark Newton, San Jose City College (CA)
John Osterman, University of Nebraska-Lincoln
Betsy Ott, Tyler Junior College (TX)
Greg Paulson, Shippensburg University of Pennsylvania
Frank Pearce, West Valley College (CA)
William Pegg, Frostburg State University (MD)
Marjorie Plummer, Faulkner University (AL)
Jeffrey Pommerville, Glendale Community College (AZ)
Leonard Pysh, Roanoke College (VA)
June Ramsey, Pensacola Junior College (FL)
Laurel Roberts, University of Pittsburg (PA)
Joseph Russin, Lane Community College (OR)
Robert Schoch, Boston University
Patricia Shields, George Mason University (VA)
Beth Shields, California State University
Mark Shoop, Tennessee Weslyan College
Marilyn Shopper, Johnson County Community College (KS)
Bill Simco, University of Memphis (TN)
Mark Smith, Fullerton College (CA)
Allan Smits, Quinnipiac University (CT)
Sally Sommers-Smith, Boston University
Judy Stewart, Community College of Southern Nevada
Brett Strong, Palm Beach Community College (FL)
Gerald Summers, University of Missouri
Pam Tabery, Northampton County Area Community College (PA)
David Tapley, Salem State College (MA)
Salvatore Tavormina, Austin Community College (TX)
Stephen Timme, Pittsburg State University (KS)
Lance Urven, University of Wisconsin-Whitewater
Mark Venable, Appalachian State University (NC)
Jennifer Warner, University of North Carolina-Charlotte
Jacqueline Webb, Villanova University (PA)
Dan Wivagg, Baylor University (TX)
Calvin Young, Fullerton College (CA)
Henry Ziller, Southeastern Louisiana University

Our focus group and workshop participants were invaluable in supplying ideas and perspectives for developing the *Explore Life* program:

Mike Bell, Collin County Community College (TX)
Paul Billeter, Charles County Community College (MD)
Mimi Bres, Prince George's Community College (MD)
Sandra Bobick, Community College of Allegheny County (PA)
Jennifer Chase, Northwest Nazarene University (ID)
Bill Coleman, University of Hartford (CT)

Brent de Mars, Lakeland Community College (OH)
Jean de Saix, University of North Carolina-Chapel Hill
Kathy Dillon, Brookdale Community College (NJ)
Christopher Dobson, Idaho State University
Cathy Donald-Whitney, Collin County Community College (TX)
Gary Donnermeyer, Kirkwood Community College (IA)
Steve Fifield, University of Delaware
Kristin Florista, University of Florida
T. Ford, Indiana University of Pennsylvania
Cheryl Hack, Anne Arundel Community College (MD)
Frank Hanson, University of Maryland Baltimore County
Ellen Holtman, Virginia Western Community College
James Horwitz, Palm Beach Community College (FL)
Isidore Julien, Purdue University
Kris Hueftle, Southwestern Oregon Community College
Elijah Kihanya, Essex Community College (MD)
Tom Lancraft, St. Petersburg Junior College (FL)
Roger Luckenbach, Fresno City College (CA)
Anne Lumsden, Florida State University
Ric Matthews, San Diego Miramar College (CA)
Cynthia Moore, Illinois State University
Alison Morrison-Shetlar, Georgia Southern University
Betsy Ott, Tyler Junior College (TX)
William Pegg, Frostburg State University (MD)
Charles Pumpuni, Northern Virginia Community College-Alexandria
Lyndell Robinson, Lincoln Land Community College (IL)
Laurel Roberts, University of Pittsburg (PA)
John Rushin, Missouri Western State College
Phillip Shelp, Brookhaven College (TX)
Bruce Sundrud, Harrisburg Area Community College (PA)
Pam Tabery, Northampton County Area Community College (PA)
Doug Ure, Chemeketa Community College (OR)
Jerry Waldvogel, Clemson University (SC)
Patricia Walsh, University of Delaware
Jaqueline Webb, Villanova University (PA)
Fredella Wortham, Brookhaven College (TX)

And we are indebted to the professors and their students who class-tested *Explore Life* and provided excellent feedback. You were pioneers at the frontiers of biology publishing along with us:

Maala Allen, Fullerton College (CA)
Charlotte Bacon, University of Hartford (CT)
Paul Billeter, Charles County Community College (MD)
Dr. H. Delano, Black University of Memphis (TN)
Susan Bower, Pasadena City College (CA)
Arthur Buikema, Jr., Virginia Polytechnic Institute and State University
Mary Colavito, Santa Monica College (CA)
Brent de Mars, Lakeland Community College (OH)
Jean de Saix, University of North Carolina-Chapel Hill
Miriam del Campo, Miami-Dade Community College (FL)
Linda Dixon, American River College (CA)
Gary Donnermeyer, Kirkwood Community College (IA)
Jim Ellinger, Bellevue Community College (WA)
Robert Fields, Southwestern Oregon Community College
Stacy Kiser, Lane Community College (OR)
Richard Lance, University of Memphis (TN)
Ann Lumsden, Florida State University
Rick Millis, Charles County Community College (MD)
Greg Paulson, Shippensburg University of Pennsylvania
Joseph Shannon, Northern Arizona University
Don Thomas, University of Memphis (TN)
Greg Thorn, University of Wyoming
Carol Wake, South Dakota State University

Janet L. Hopson
John H. Postlethwait
March, 2002

Table of Contents

Full Circle from Earth to Mars

A huge headline in *Time* magazine screamed unambiguously: "LIFE ON MARS." In August 1996, NASA scientists created an international media frenzy by claiming that a 4-pound, potato-shaped rock contained the remains of Martian life.

The blackened lump in question is a meteorite from Mars that entered Earth's atmosphere as a brilliant shooting star about 13,000 years ago during our planet's Ice Age. Prehistoric hunters and farmers probably watched that arrival blaze across the night sky. But the stone fell anonymously to the frozen surface of Antarctica and lay undisturbed for 13 millennia until scientists discovered it in 1984. Twelve years later, a team from Johnson Space Center in Houston published evidence that the charred chunk contains microscopic wormlike structures formed nearly 4 billion years ago. This evidence, they claimed, was compatible with life on Mars. Within months, many scientists had rebutted the claim: Nothing but mineral structures, they said.

Debate still simmers, even years later. But an answer may be on its way. NASA began a series of Mars missions in 1998 that will retrieve a canister of Martian soil and rocks by 2008. And recent photographic data suggest that Mars may have liquid water below its surface capable of sustaining life. Once NASA has retrieved the sample of Martian soil, biologists will immediately begin searching it for signs that organisms lived—or still live—on our neighboring red planet. Proof of life on Mars could be one of the most exciting discoveries in the history of life science. It could also be the scariest. What if Martian germs started to multiply and spread to Earth's plants and animals? What if the new organisms caused a deadly, worldwide epidemic? What if they acted like the monsters from the movie *Alien,* which the *Washington Post* parodied recently: "First you start coughing and then a slime-flecked weasel from Hell bursts out of your chest cavity"? The search for life on Mars may be exciting, but we need a protector— a planetary protector—to make sure this mission won't backfire on us.

Enter Margaret Race. As a young girl, Margaret Race was a self-proclaimed "astronaut groupie." Her dad helped design engines for the types of helicopters that plucked the earliest astronauts out of the ocean along with their Apollo and Mercury space capsules. Starting in the eighth grade, Race sent letters and greeting cards to her favorite astronaut, Scott

▲
The Search for Life on Mars.
Margaret Race; Mars yard at Jet Propulsion Laboratory; JPL worker in a high-containment suit studies meteorites.

How could we recognize extraterrestrial life?

Carpenter (who flew in 1962), and they became fast friends. "He and I joke," she says, "that after all this time, I'm his longest surviving fan!"

A successful academic career, marriage, and a busy family life swept away her secret dreams of space exploration. Race earned a doctoral degree in biology, studying the introduction of **alien** (nonnative) animals and plants into San Francisco Bay. Her ability to understand the scientific, policy, and public issues surrounding the release of such alien or **"exotic"** species then led her to a pivotal role at the University of California during one of the first deliberate releases of a gene-spliced organism. In the late 1980s, researchers were poised to plant the first genetically engineered strawberries into test fields. They turned to Margaret Race to help explain the deliberate release of these "aliens" to a worried public, which she did.

In 1991, a phone call from NASA brought Margaret Race full circle to her fascination with space. At that time, NASA officials were just beginning to plan the missions that will bring back rocks and soil from Mars and to ponder the risks: How would we recognize martian organisms? What dangers could these exotics pose for us and our planet? How could we protect ourselves from them—and them from us? Martian species would be the ultimate "exotics." With her background, Dr. Race was already primed for these problems and for the many issues underlying the search for life elsewhere in the universe: How safe is this quest? How does the public feel about bringing samples back from other planets? How should this whole enterprise be planned, managed, monitored, and explained? Dr. Race was soon working full time on the problems, and for the past decade, she has been funded by NASA as a consultant working for the Office of Planetary Protection for the ongoing Mars missions.

Race has worked with astronomers and aerospace engineers on a difficult challenge: Where's the best site for finding evidence of life on Mars? In a dry lake bed? At the frigid poles? Near a volcanic vent? And how can NASA drop a roving lander precisely *there* while "both planets are hurtling through space, going around their orbits and rotating." These NASA workers, laughs Race, "really *are* rocket scientists! And I can barely find a planet in the night sky!"

The biological questions she and other life scientists face in planning the Mars missions are actually no less complex. The scientific controversy over the martian meteorite stimulated new debates: How small can a living thing be on Earth? Could it be smaller on Mars? Should we test the returning martian soil and rock samples in the same way as the martian meteorite? Or do we need new tests? If we get confirmation of a living cell while the soil sample is still on Mars, should we even bring it back? If we do, should we try to grow more of the martian cells and keep them in an alien "zoo"? What about the possibility of releasing a Doomsday bug that would destroy life here? And what about false positives? As Margaret Race explains this last question, we need ways to determine whether evidence of life was "accidentally sent up from [Cape Canaveral] Florida, or whether it actually originated on Mars."

If there was ever a time in our history to understand the central enigma beneath the search for extraterrestrial organisms, it is now and it is this: *What is life?* This chapter begins our multimedia exploration of biology's central puzzle. As you read along, you'll learn why it's so important for each of us to understand life, and how biologists study every aspect of living organisms, their environments, and their interrelationships. You'll discover a lot more about Margaret Race and the search for life on Mars. And you'll find the answer to these questions:

1. What characteristics do all living things share?

2. Which characteristics relate to gathering and using energy?

3. Which characteristics relate to reproduction?

4. Which characteristics relate to evolving and adapting?

5. Which characteristics relate to the physical environment?

6. How do biologists study life and how will we explore it in this course?

7. How can the study of life help us solve societal and environmental problems?

1.1 Characteristics of Living Organisms

The search for life on Mars and other planets requires that we recognize life when we see it. This recognition is central to Margaret Race's job of protecting Earth from any Martian organisms that might return in the soil and rock samples. Race and her colleagues have needed to identify the general properties of life on our planet (Fig. 1.1) and then speculate about the similarities and differences they might encounter while searching on Mars and elsewhere in the solar system and universe (Fig. 1.2). After centuries of study and thought, biologists have come to a consensus about when an entity is alive, formerly alive, or nonliving, and these ideas support NASA's planetary protection program.

The problem of recognizing life is clearly fundamental to the search for alien organisms. But it has other important applications, too. Today's biologists and physicians have unprecedented abilities. These include sustaining the human body and individual organs on life-support machines, freezing human and animal embryos for later use, and changing and merging hereditary traits of microbes, plants, and animals. Perhaps one day this list will extend to generating life in a test tube and to creating hybrids between computers and living things. To manipulate life's most fundamental properties, biological engineers need to know exactly what the boundaries are, how far they can be stretched, and what changes would be desirable, practical, and worth pursuing.

At the same time, the public needs to be fully aware of the benefits and risks of manipulating life so they can be informed watchdogs and consumers of these biotechnologies. In the broadest sense, every course in biological science and every experiment, no matter how simple or complex, probes the question "What is life?" It takes this entire course of study to provide an answer. The citizen who learns about biology—including the college student majoring in some other field like business, physical education, psychology, or English—also discovers a realm of intricacy and beauty that helps them understand their environment, their health, their day-to-day functioning, their children's growth and development, and the issues they see in the news, including cancer treatments, impotence drugs, habitat destruction, species loss, and global warming.

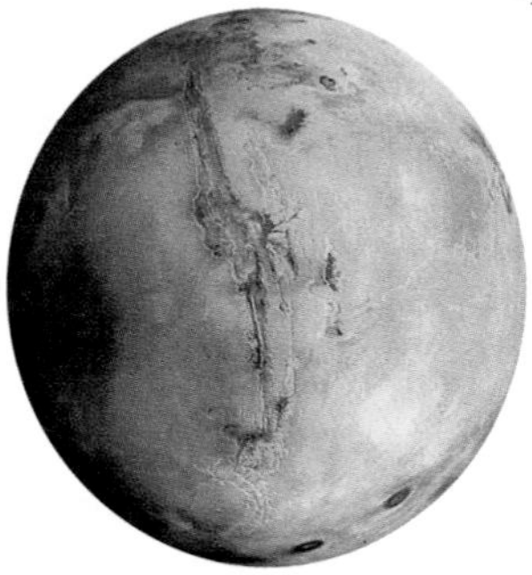

Figure 1.2   Mars: A Haven for Past Life? An aerial view of Mars looks bleak, dry, cold, and lifeless. But at one time water flowed on its surface, its temperatures were warm, and it may have harbored life.

What, then, is life? You may be surprised to learn that no one, neither basic biologist, bioengineer, nor planetary protector, has a thumbnail definition that lays out the essence of the living state. Instead, they focus on the characteristics of life as a collective, descriptive definition.

Life's Characteristics

Think for a minute about puppies, roses, dinosaur bones, and motorcycles. Why those things? Puppies are clearly alive, frolicking, rolling around, and begging for dog biscuits, as directed by their highly organized brains. A rose is obviously alive, too, although it hardly moves and never devours dog treats. Still, a rose plant soaks up sunshine and soil nutrients, and it makes beautiful flowers with seeds that can produce a new generation of roses. A dinosaur bone is certainly not alive, but its close appearance to the bones of living animals confirms that it was once alive. And the motorcycle? It's highly organized and requires energy from the environment. It moves under its own power and responds by going faster when the throttle is turned up. It must be alive, too, right? Of course not. The last time we checked, motorcycles couldn't reproduce, at least not without the help of an assembly line and some good mechanics.

Comparisons like these underlie the list of characteristics that living things share, whether on Earth or, conceivably, on Mars (Table 1.1). Living systems have internal order, or a high degree of organization. To maintain that order, living things carry out metabolism: they use energy to transform and organize materials. Living things also use energy to move under their own power, a trait biologists call motility. They also use energy to react to outside stimuli, a trait called responsiveness.

While living things do all of the above, they also do more. Living things have the ability for self-replication or reproduction. They show growth and

Figure 1.1   Earth: A Haven for Life Today. Some of Earth's special characteristics are visible from space: swirling white clouds of water vapor, brownish land masses, and vast blue oceans.

Table 1.1   Characteristics of Life

Life Characteristic	Property
1. Order	Each structure or activity lies in a specific relationship to all other structures and activities.
2. Metabolism	Organized chemical steps break down and build up molecules, making energy available or building needed parts.
3. Motility	Using their own power, organisms move themselves or their body parts.
4. Responsiveness	Organisms perceive the environment and react to it.
5. Reproduction	Organisms give rise to others of the same type.
6. Development	Ordered sequences of progressive changes result in an individual acquiring increased complexity.
7. Heredity	Organisms have units of inheritance called genes that are passed from parent to offspring and control physical, chemical, and behavioral traits.
8. Evolution	Populations of organisms change over time, acquiring new ways to survive, to obtain and use energy, and to reproduce.
9. Adaptations	Specific structures, behaviors, and abilities suit life-forms to their environment.

development or the expansion of young organisms in size and complexity. Living things are related by heredity; that is, organisms give rise to like organisms (not dinosaurs from roses or roses from puppies). Finally, living things evolve or change over many generations and they adapt or change to better fit shifting environments. Let's see a motorcycle do that!

We can look around the nonliving world and see many of life's characteristics in action: Waves move, flames use energy, crystals grow. Only living organisms, however, display *all* the characteristics we just discussed at some point during their individual life cycle or species history. For example, rose petals move as the rosebud unfolds; the rose plant captures and uses energy from the sun; the plant originally emerged from a seed, then grew, and developed flowers. Rose bushes can evolve and adapt to changing climates. All in all, they're alive. And Harleys aren't.

Themes That Recur As We Explore Life

Some of the life characteristics we just listed occur again and again as themes that guide our exploration of biology. Living things use energy at several levels of biological organization: in the smallest cells, in individual organisms, and in large groups of organisms called biological communities. The members of a species reproduce and adapt to the environment over generations by means of evolution, and these themes help explain why organisms act as they do. A fun part of biology is learning how people have explored and discovered exactly what organisms do. These discoveries almost always use a special system of investigation called the scientific method. We'll see the five themes of energy, reproduction, environment, evolution, and the scientific method interwoven throughout all of our discussions, and these themes will help to organize the sprawling subject at hand, biology. Recognizing the characteristics of life on Earth—and, in turn, how to protect our planet from extraterrestrial life—is one good way to introduce these themes.

1.2   Life Characteristics Relating to Energy

Once NASA scientists suspected that the potato-shaped meteorite found in Antarctica was actually from Mars, they started studying slices of it, looking for hints of life on Mars and perhaps new answers to the question "What is life?" They first sought to confirm that the meteorite indeed originated on Mars. They did this by showing that tiny bubbles trapped inside the rock contain air with the same chemical composition as the atmosphere of Mars—a mix that was measured directly by the Viking mission in 1976 and is distinctly different from Earth's. The mixture of chemicals in the rock itself suggested that the meteorite probably formed 4.5 billion years ago, shortly after Mars solidified as a planet. So how did a rock that old from Mars get here, only to be discovered on the Antarctic ice cap in 1984? NASA geologists and astronomers surmise that a huge asteroid slammed into the Martian surface about 16 million years ago, blasting dirt and rocks high enough into the atmosphere that some escaped. This material orbited the sun independently for millions of years, and some of it eventually got tugged firmly enough by Earth's gravity to streak into our atmosphere as fiery meteorites that fell to the surface. The NASA researchers were excited by the prospect of looking for signs of life in the Martian rock. But what approach would they take to detecting the unmistakable signatures of living things in this traveling chunk of Mars?

Order

First, they could look for **order** or structural and behavioral complexity and regularity, because living things possess a degree of order far greater than that of the

(a)

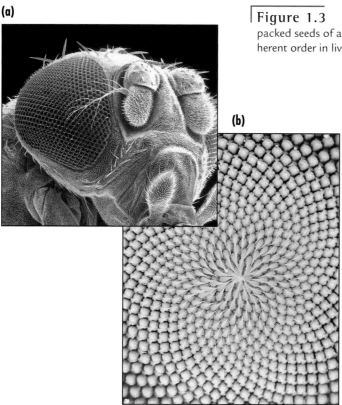

(b)

Figure 1.3 **Order in Life.** (a) A fruit fly's eye and (b) the tightly packed seeds of a sunflower form geometrical arrays that reveal the inherent order in living things.

finest Swiss clock, the fastest racing car, or anything else in the nonliving world. The eye of a fly, for example, and the spiral-packed seeds of a sunflower head both consist of highly organized units repeated and arranged in precise geometric arrays (Fig. 1.3). We know that order is important because disorder quickly leads to death in a living thing: most weapons of murder, in fact—

clubs, knives, guns, and poisons—will disorganize you beyond repair.

Knowing that order is a hallmark of life, NASA investigators searched their slices of the Mars rock for organized structures and found tiny, regular tubes (Fig. 1.4). The researchers became convinced that these forms are "microfossils," the small, preserved bodies of ancient organisms, and published a scientific paper claiming so. Many other biologists, however, disagree that the so-called microfossils are evidence of past or present Martian life because they are smaller than the smallest known organisms on Earth. At a workshop in October 1998, scientists concluded that life as we know it could not survive in a package any smaller than a sphere 200 nm (200 billionths of a meter) in diameter. The tubules in the Mars meteorite, however, were half that long and one-tenth that wide. For this reason, many think they must be simply mineral formations. The tubules indeed looked highly ordered, but that alone doesn't confirm that they were once alive.

A Hierarchy of Order

Martian organisms, if they ever existed, are a near-total mystery, including the degree of order they might possess. But organisms on Earth have an order that is readily apparent at several levels. Biologists define an **organism** as an independent individual possessing the

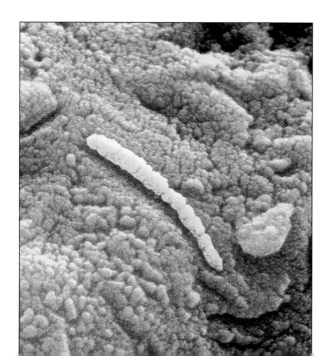

Figure 1.4 **Could This Be a Fossilized Martian?** This tubelike form, seemingly divided off into cells, was found in a Martian rock. Biologists have wondered if it could represent a fossilized Martian bacterium, even though it is about 100 times smaller than very small bacteria on Earth.

Organ System: A group of body parts that carries out a particular function in an organism

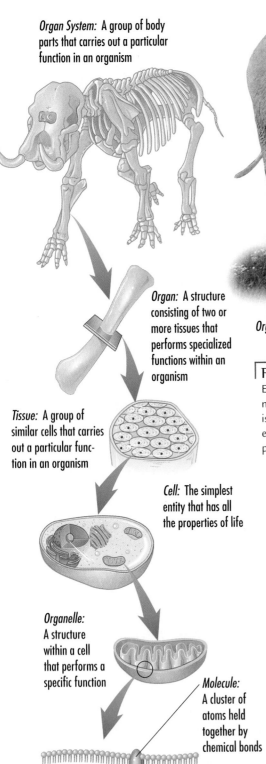

Organ: A structure consisting of two or more tissues that performs specialized functions within an organism

Tissue: A group of similar cells that carries out a particular function in an organism

Cell: The simplest entity that has all the properties of life

Organelle: A structure within a cell that performs a specific function

Molecule: A cluster of atoms held together by chemical bonds

Organism: An individual, independent living entity

Population: A group of individuals of a particular organism that inhabits a given region and interbreeds

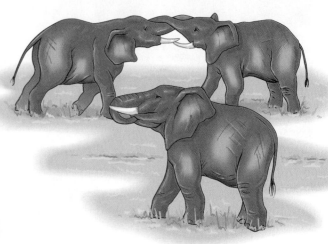

Figure 1.5 **Order Reigns at Every Level in the Living World.** An African elephant on the savanna in Eastern Kenya symbolizes levels of biological organization. The elephant's body is made up of highly ordered molecules, cell parts, cells, tissues, organs, and organ systems that function together smoothly. This individual is part of a population within a diverse community, including the egrets searching for insects stirred up by the elephant's heavy footsteps and the grasses the pachyderm grazes on and tramples. The community, in turn, is part of the savanna ecosystem, with its expansive plain, mountain range, dry-adapted trees, and arid climate.

characteristics of life. The elephant in Figure 1.5 and **E**xplorer 1.1 is an individual organism. Each organism, in turn, is made up of **organ systems,** groups of body parts arranged so that together they carry out a particular function within the organism. The skeletal system, for example, supports an elephant's body.

Organ systems are made up of **organs,** sets of two or more tissues that together perform specialized functions for the organ system. An example is a single bone that supports part of an elephant's leg. Each organ is made up of **tissues,** groups of similar cells that carry out the function of the organ. For example, bone tissue—made up of several kinds of cells functioning collectively—provides physical support to the elephant's leg. Tissues are made up of **cells,** the simplest entities that have all the properties of life. Cells contain within them small structures known as **organelles,** which perform the functions necessary for the life of the cell. Finally, organelles consist of **biological molecules,** the building blocks of all biological structure and activity. The tubules in the Martian meteorite yielded no visual evidence of organelles. Researchers, however, did find some subunits of biological molecules that on Earth can be associated with organelles. No one is sure, of course, whether the biological molecules were contaminants from Earth organisms or arrived in the meteorite itself.

Community: Various living organisms that interact in a given region

Ecosystem: A living community and its physical environment

Biosphere: The portion of the earth that contains living organisms

Metabolism

Scientists looking at Mars rocks could—and did—look for evidence of energy use. Living things maintain order in their organelles, cells, and organs through **metabolism:** they take energy from the environment and use it, along with materials, in a series of consecutive chemical steps, for repair, maintenance, and growth. By taking energy and materials from the environment and using them for repair, growth, and other survival processes, metabolism helps to combat the disorganization that occurs with time. If you scrape your knee in a fall, for example, metabolism in your cells helps to repair the damage and to generate new, healthy skin, nerves, and blood vessels.

In 1976, NASA sent a lander called Viking to Mars to photograph the planet's surface and test a scoop of Martian soil. The highlights of that mission were several experiments conducted remotely and designed to detect something in the soil—something alive, perhaps—that could use energy to metabolize (to transform and organize materials) as living things do on Earth. The Viking mission created quite a stir because the "right" byproducts (including carbon dioxide, CO_2) were released during the experiments—as if life were present and actively metabolizing. However, the *ways* the products were generated cast doubt on a biological origin. Most biologists are now convinced that the results from the 1976 Mars experiments were due solely to nonliving soil chemistry and not to metabolism by living cells.

The chemical reactions of metabolism require water, for reasons we'll see in Chapter 3. The Viking mission found no surface water on Mars. But other missions, including high-resolution photography by the Mars Global Surveyor in 1999, did document geological features (such as deep channels, flood plains, wave patterns in sand, and erosion in sedimentary rocks) consistent with standing and flowing water in the past. Recent findings even suggest that water may have flowed within the past few thousand years. These discoveries have encouraged biologists to believe that life may once have flourished on Mars and that remnant populations may still survive today in areas that harbor small amounts of liquid water.

Motility

There was a NASA joke in 1976 that while the Viking lander was focusing in on chemical evidence for microbes in the Martian soil, it might miss bigger evidence like footprints or little green aliens walking by. Self-propelled movement, or **motility,** would certainly have been as good an indicator of life on Mars as it is here. Even organisms as simple as bacteria can move on their own. Plants, which cannot move from place to place, do

GET THE PICTURE

*E*xplorer 1.1
Hierarchy of Life

Figure 1.6 **Living Organisms Display Movement.** Based on cell growth in stems, the floral heads of these sunflower plants face the sun throughout the day.

show various subtle movements based on growth. For example, the little organelles that capture sunlight in plant cells are in constant motion. The flowers of some plants open in the morning, trace the sun's arc through the sky, then close at night (Fig. 1.6). Animals, of course, have elevated movement to an art form in their pursuit of food, displays of dominance, and escape from enemies.

Responsiveness

If you poke a sea slug, it withdraws. If you turn a houseplant around, its leaves move imperceptibly until, in a day or two, they're once again oriented toward a window. Organisms are **responsive:** they respond to changes in their environment involving temperature, food, water, enemies, mates, or other elements. The reaction to the change can be instantaneous: A moth hears the high-pitched whine of a swooping bat and zigzags away on a midnight breeze, and a Venus flytrap snaps shut on a tiny, unsuspecting frog. The response can be gradual, as well. A trumpeter swan detects the shortening days of autumn and responds by feeding more heavily and then migrating south. Or,

a daffodil reacts to the lengthening days of spring by forming flowers.

By metabolizing, moving, and responding, organisms obtain energy and materials from the environment and use them to maintain order in their bodies. Nevertheless, aging inevitably sets in—whether in hours or days for a microbe, a century for a tortoise, or a millennium for a bristlecone pine. An aged organism can no longer stem the resulting disorganization and death. Likewise, an old motorcycle will rust and its fenders fall off. But life continues to exist because organisms reproduce.

1.3 Life Characteristics Relating to Reproduction

Like begets like—that's a central feature of life. And this concerns an expert like Margaret Race working with the Office of Planetary Protection. If Martian organisms do exist, and if NASA brings them back to Earth, could the aliens reproduce here and multiply

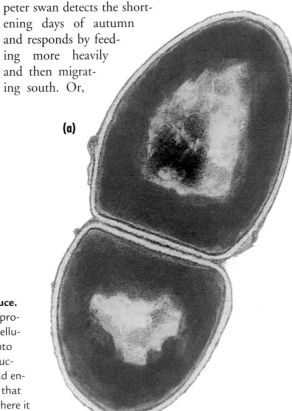

(a)

(b)

Figure 1.7 **Organisms Reproduce.** (a) Asexual reproduction. Bacteria reproduce by making new copies of each cellular part and then distributing them into two new offspring. (b) Sexual reproduction. This bucket orchid from Trinidad entices bees with drops of sweet nectar that can knock a bee into the "bucket," where it picks up pollen. The bee can carry the pollen to other flowers and inadvertently cross-pollinate them.

out of control? If so, how might we recognize and stop the process? Let's look at the life characteristic of reproduction, and two related ones, development and heredity.

Reproduction

Organisms give rise to others of the same kind—roses to roses, robins to robins—by means of a defining life process, **reproduction,** or the means by which individuals give rise to other individuals of the same type. In **asexual reproduction,** a single parent produces offspring identical to it and each other. One-celled microbes, for example, reproduce asexually by splitting into two identical daughter cells (Fig. 1.7a).

Most complex organisms reproduce by **sexual reproduction,** with genetic information coming from two parents and combining in offspring that are very similar but not identical to the parent or each other. Organisms sometimes go to great lengths for sexual reproduction to occur. The bucket orchid (Fig. 1.7b), for example, forms a large drop of sweet nectar in a faucet-shaped structure; when the drop falls, it can knock a visiting bee into the bucket. The bee must then crawl through a small tunnel in the flower to escape and in doing so collects pollen (sex cells) from the flower. If the bee visits another flower of the same type, it can then deliver the pollen to the other flower's sex organ, facilitating the combining of genes and the flower's reproduction.

Margaret Race and others in NASA's planetary protection program must plan for the possibility—extremely remote as it is—that Martian organisms brought back to Earth could escape and start to reproduce here through some complex series of processes we do not understand. Recognizing this life characteristic of reproduction in an alien organism would be a first step; blocking it would be a second, based entirely on the means of reproduction.

Development

Young organisms like the zebrafish in Figure 1.8 usually start out smaller and simpler in form than their parents. The offspring then grow in size and increase in complexity, a process known as **development.** Eventually, the organism may reach sexual maturity and become a parent itself.

Heredity

One of the most intriguing questions in all of biology is how a fertilized egg develops into the millions of cells of various types that function as a viable organism. The answer lies in the remarkable process of **heredity,** the transmission of genetic characters from parents to offspring.

Do you know any sets of identical twins? Twins are proof that some type of hereditary information directs each individual's development with such amazing precision—so much that two separate organisms can go through all the steps of growth and increasing complexity over years of time and still wind up looking virtually

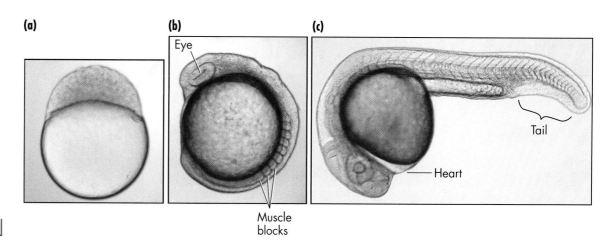

Figure 1.8 **Development: Sequence of Growth, Consequence of Reproduction.** (a) A zebrafish embryo with several hundred cells. Development continues, and (b) by 14 hours, the eye and muscle blocks (see leaders) are clearly visible. (c) By 24 hours, the rudiments of a little fish have formed, including heart and tail.

alike (Fig. 1.9). Contrast identical twins with their different-looking brothers and sisters, however, and you can see that hereditary information must also contain variations so that the offspring in one family can have similar noses but different heights, or similar eyes but different hair color. Biologists have identified the units of inheritance that control an organism's traits and call them **genes.**

Genes, made of a remarkable molecule called DNA (deoxyribonucleic acid), determine whether a person's hair is red, black, brown, blond, or gray. As we will see, genes also direct the day-to-day metabolic activities within cells. If Martian organisms exist, would they have DNA? No one knows. But DNA is so crucial to Earthly life that some of the upcoming NASA tests of the Mars sample will look for evidence of DNA and similar molecules.

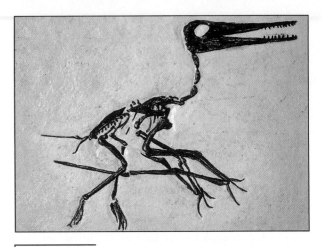

Figure 1.10 **Fossils: Evidence of Changing Life.** This long-extinct pterodactyl with its heavily toothed beak lived over 140 million years ago in what is now Germany.

1.4 Life Characteristics Relating to Evolution

Four and a half billion years ago, Earth and Mars were newly formed planets and these neighbors—the third and fourth "rocks" from the sun—were probably very similar. The radius of Mars is only about half that of Earth, but the two planets have similar compositions and both probably had stable water near the planet surfaces for much of their history. Biologists think that if life did arise on Mars, it was probably similar to early life here. Fossil evidence tells us that life has changed or evolved over the millennia on our planet (Fig. 1.10). So would the same be true for Martian life? Once again, no one knows. But NASA will be looking for any minute evidence of fossils or changes in cell-like structures in the Martian soil and rock samples.

Life Changes Over Time

Over time, life forms change. Biologists call this descent with modification **evolution,** and it is based on changes in the frequencies of genes within populations over time. We can tell that life evolves partly from the fossilized imprints of early organisms. The older a fossil, the less similar it is likely to be to present-day forms. This dissimilarity is good evidence not only of change, but of *continued* change in living species. Using fossils, DNA analysis, and other evidence of changes in gene frequencies, biologists can trace an organism's family tree. For example, modern day house cats and tigers (young branches on the feline tree) are closely related. Reaching farther back in time, cats and dogs would share a common ancestor, and going back to life's early history, cats would share a common ancestor with snakes, fish, beetles, mushrooms, trees, and bacteria. Tracing the evolutionary tree back to life's first beginnings on Earth, all organisms would eventually share a common ancestor in the distant past.

Figure 1.9 **Genes: Hereditary Factors Control an Organism's Form and Function.** Identical twins and their very different-looking siblings show that genes can control the development of traits with amazing precision yet can contain variations, even within the same family.

Classification of Living Things

Evolution is the mechanism that produced the huge variety of Earth's life forms, including hundreds of tropical fish species with their bright colors and fantastical shapes. Put on a mask and snorkel and plunge into the warm, blue waters off Hawaii, and you can count dozens of fish species in just a few minutes—including, perhaps, butterfly fish, bird wrasses, striped Moorish idols with streaming back fins, or intensely turquoise parrot fish. This same dazzling variety is evident in the colorful songbirds of eastern forests, the butterflies of Midwestern weed fields, and the wildflowers of the Rocky Mountains.

The process of evolution and the tracing of family lineages back in time can help explain the immense diversity of life, which some biologists estimate at upwards of 50 million species. To help make sense of this vast diversity, biologists have created a system for categorizing organisms into groups according to their similarities. Brightly colored tropical frogs, for example, are more similar to bullfrogs than to elephants, but frogs and elephants are more similar to each other than to mushrooms or grass. If they discover Martian or other extraterrestrial organisms, biologists can begin to catalog them as well, perhaps based on our own existing system for Earth organisms.

Species

Species are groups of individuals with similar structures that descended from the same initial group and that have the potential to breed successfully with one another in nature. House cats are one species, and ocelots, small jungle cats—whose habitats range from steamy Amazon rain forests to the dry chaparral of Texas—are a different, but related species (Fig. 1.11).

Genus

A **genus** (plural, *genera*) contains several related and similar species. Biologists refer to each species by a two-part name beginning with a term denoting the genus, followed by a separate term denoting the species. This two-word system for naming genus and species is called "binomial nomenclature." For example, the house cat *Felis catus* is related to but clearly distinct from the ocelot in the same genus *Felis* but the different species *pardalis*. (Together, then, the official name is *Felis pardalis*.) A lion (Fig. 1.11), which is obviously still a cat but quite different in size, coloration, and habits from house cats and ocelots, is in a different genus, *Panthera,* and its species name is *Panthera leo*. After once mentioning the complete two-part name, biologists often abbreviate the genus, referring to *F. catus* or *P. leo*. You've probably seen the term *E. coli* in newspaper articles about outbreaks of food poisoning. In this case, the *"E."* stands for the unwieldy bacterial genus name *Escherischia.*

Just as biologists group related species into genera, they also group similar genera into **families,** similar families into **orders,** similar orders into **classes,** similar classes into **phyla** (or, in plants, **divisions**), similar phyla into **kingdoms,** and similar kingdoms into **domains.** You can investigate this hierarchy of classification in Explorer 1.2 and see the levels to which people, puppies, and rose bushes belong. Biologists recognize just three domains, each containing millions of life-forms (Fig. 1.12). Two of the domains, Bacteria and Archaea, consist of microscopic, mostly single-celled organisms that differ in fundamental ways (see Chapter 11 for details). The domain Bacteria includes the species that cause strep throat, for example, and that recycle decaying matter in soil and at the bottom of ponds. Members of the domain Archaea often live in harsh environments that are very hot, cold, acidic, or salty, such as thermal springs, salty lakes, or Antarctic ice. After Earth formed, environments like these would have been quite common, and the Archaea alive today probably share similarities with some of Earth's earliest organisms. (The name "Archaea" reflects the supposed "archaic" nature of these cells.)

GET THE PICTURE

Explorer 1.2

The Diversity of Life

Figure 1.11 **A Hierarchy of Cat Species.** The house cat, *Felis catus,* is a separate species from the ocelot, *Felis pardalis.* They share the same genus, however. The lion, *Panthera leo,* is a different species in a different genus. The figure shows a family tree for these cats.

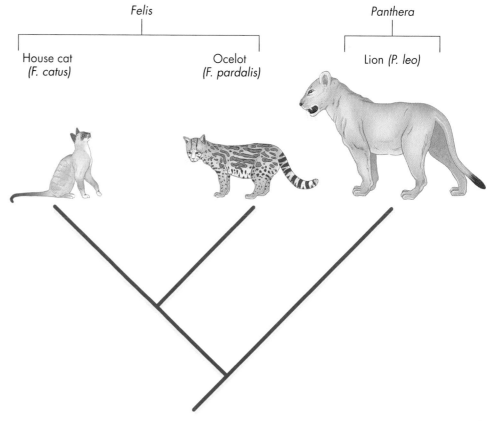

Felis

House cat (*F. catus*)   Ocelot (*F. pardalis*)

Panthera

Lion (*P. leo*)

How can life be both diverse and unified?

Mars and Earth had similar beginnings, but Mars is very different now, with its frigid surface temperatures of −15 °F during the day and −125 °F at night. (The Martian day, by the way, is 24 hours and 37 minutes long.) Some biologists think that if Martian organisms once existed, or exist now, they would probably resemble members of the domain Archaea. Those terrestrial organisms therefore are a model in many ways for the search for extraterrestrial life. In addition, recent research on soil samples from many meters below Earth's surface turned up living archaea that had been dormant for over a million years and then started dividing and growing again when provided with appropriate nutrients. This, too, has exciting implications for the Martian soil samples NASA is planning to retrieve.

Earth's third domain of living organisms, Eukarya (*eu* = true + *karya* = nucleus), consists of larger, more complex cells containing a nucleus, a special compartment that contains the cell's DNA. The domain Eukarya contains four kingdoms: Plantae, Animalia, Fungi, and Protista (Fig. 1.12). You're familiar with the plant and animal kingdoms, of course. Fungi include mushrooms, molds, and yeast. And protists are less familiar because they are often microscopic, and include amebas and other organisms with a single, but complex, cell.

The Unity and Diversity of Life

With 50 million living species or more, life on Earth is obviously diverse, but the fact that all this multiplicity arose from a single group of ancestral cells present at the dawn of life gives it *unity* as well—unity of origin, of cell structure, of genetic material, and of basic day-to-day functioning. How can life be both diverse and unified? What mechanisms can foster not only vast diversity of form but also unity at the level of genes, cells, and basic function? The answer is evolution, the unifying theme for all life science.

Adaptation

Different species have different ways of extracting energy and materials from their surroundings. Think, for a minute, about organisms living in the bitter cold environment of Antarctica in the icy continent's McMurdo Dry Valleys (Fig. 1.13a). This area receives very little moisture, and temperatures are below freezing nearly all year round. The rocks that line the valleys, however, are somewhat porous, and they trap water. As sunlight hits the rocks, they can warm up above freezing, and the water can become liquid. Amazingly, some bacterial species have become adapted to live *within* the rocks, extracting energy from the sunshine, minerals from the rocks, and molecules necessary for life from the air (Fig. 1.13b). Specializations that help an organism adapt to its own special way of life are called **adaptations.** Some biologists hypothesize that if life did evolve on Mars in the distant past, Martian organisms may have become adapted to ways of living very similar to Earth's antarctic life-forms. Some think it's even possible

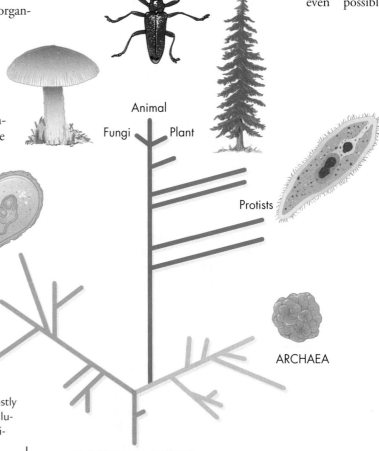

EUKARYA

Animal

Fungi   Plant

Protists

BACTERIA

ARCHAEA

UNIVERSAL ANCESTOR

Figure 1.12   **The Domains of Life.** Two domains, Bacteria and Archaea, are mostly single-celled microbes. The domain Eukarya encompasses all single-celled and multicellular organisms with a true nucleus in each cell, including the organisms in kingdoms Animalia, Plantae, Fungi, and Protista. Biologists are still sorting out the names and characteristics of the many kingdoms in the other domains.

(a)

Figure 1.13
Life in the Rocks.
(a) While the Mc-
Murdo Dry Valleys are perpetually cold and dry and appear lifeless, bacteria survive in the ice that covers the lakes. (b) Remarkably, small living cells also exist within the rocks. This section through one of these Antarctic rocks reveals a layer turned green by organisms able to use sunlight to make sugars from water and molecules in the air.

(b)

that organisms may still live in Martian rocks or deep below the soil. Where would we look to find them?

Many adaptations relate to taking in energy and materials, but there are others, too, that improve an organism's ability to grow, to reproduce, to move, to live in a group, or to attract a mate more successfully.

Natural Selection

Adaptations usually arise by **natural selection,** a "weeding out" process that selects for hereditary differences in individuals' abilities to reproduce and to obtain energy. To see how this works, let's go back to the example of cells living in McMurdo's Dry Valleys. Some of these bacteria may have hereditary factors that allow them to manufacture sugars at a slightly lower temperature than other nearby cells. In such a bitterly cold climate, this ability could help these particular bacterial cells survive and reproduce under conditions in which other cells become dormant or die. As time passes, this advantage may mean that most or all of the cells in this better-surviving population and their descendants have the favorable trait while the nearby cells, lacking the trait, are out-competed. In 1859, Charles Darwin published the landmark book *On the Origin of Species,* and coined the term natural selection, for the mechanism underlying change in living species over time. He called it "natural selection" because nature was "select-

ing" the individuals with the most suitable variations to survive and become the parents of the next generation. He likened this to the way a farmer selects cows that give the most milk as mothers for the next generation, gradually increasing the milk yield of the entire herd. A contemporary of Darwin's, Alfred Russell Wallace, also published papers naming natural selection as the prime mechanism of evolution.

During the century and a half since Darwin's and Wallace's work, biologists have delved deeply into both the principles of, and the evidence for, evolution by natural selection. Some of the most important advances pinpointed the sources of variation from which nature selects individuals: Variations usually arise through mutations or alterations in gene structure. Largely because of modern molecular genetics, today's biologists can explain both the diversity of life and the unity of its origins and shared characteristics. They can account for life's remarkable diversity because different environments require unique adaptations. Looking at foxes, for example, some of the characteristics that help an arctic fox survive in the frozen north are very different from those that help a kit fox survive in the desert. The arctic fox has small ears and a short muzzle that help conserve body heat and a white coat that serves as camouflage against snow and ice (Fig. 1.14a). In contrast, the kit fox has large ears and a long muzzle that radiate extra body heat and a tan coat that disappears against the surrounding sand and rocks (Fig. 1.14b). An arctic fox in the desert would overheat, stand out against the earth-toned background, and likely exhaust itself before catching rabbits or other prey. A kit fox in the arctic would have frost-bitten ears, would stand out against the snowy background, and would probably die of exposure or hunger before catching enough prey. The diversifying action of the environment, working through natural selection over millions of years, could have produced the

Figure 1.14 Fox Adaptations to Arctic and Desert.
(a) With its thick white coat, stubby ears, and short muzzle, an arctic fox is well-adapted to a snowy northern terrain. (b) The very different kit fox is equally well adapted to life in the desert.

(a)

(b)

remarkable variety of life-forms that have existed on our planet.

Likewise, evolution—descent with changes over time—can account for life's unity, including the common characteristics we're surveying in this chapter. Biologists can trace the ancestry of both arctic and kit foxes back to a single fox species that lived in North America millions of years ago. They can trace that fox back to a common ancestor with all dogs. They can trace this group, the canids, back to a common ancestor with all mammals, and so on back to the origin of life itself. As we will see in more detail in later chapters, the metabolic machinery within all living cells is very similar despite wide differences in cell shape, size, and function. Biologists think all organisms alive today inherited this underlying machinery from cells that appeared at the dawn of life, and this helps explains the characteristics they share today.

Evolution by natural selection is so grand an organizing principle for all biology that it will resurface repeatedly in this book, in the many magazine and newspaper articles you may read on life science, and in any future biology courses you may take. You can see the principles animated in *E*xplorer 1.3.

GET THE PICTURE

*E*xplorer 1.3

The Scientific Method: Barn Swallows

1.5 Life Characteristics Relating to Environment

Was life widespread on Mars early in its history? People used to believe the planet was covered in canals built by little green aliens until modern space probes showed that at least some of the "canals" were the remains of giant river beds. We know from the geological evidence of flowing water and the current absence of liquid water on the planet's surface that Mars was once wet and must have changed drastically to its current extreme aridness. If life was present and if it survived at all, then it would have had to change too. **Ecology** is the branch of biology that studies the relationships between living organisms and their environment, and it is an interesting and pervasive part of exploring life.

The Hierarchy of Life

Organisms interact with their living and nonliving environments at several different levels. These levels extend the continuum of order we discussed earlier: molecules to cells to organisms. Take, for example, life in the African savanna environment—a splendid collection of plants, animals, fungi, microbes, and habitats that includes elephants, acacia trees, tussock grasses, and arid plains. The hierarchy of life on the savanna proceeds from small to large in the following sequence (Fig. 1.5): Organisms, as we saw earlier, are individual, independent living things; an elephant is an organism and so is an acacia tree. Groups of a particular type of organism that live in the same area and actively interbreed with one another are called **populations,** for example, an elephant herd or a field of grass. All the populations that live in a particular area, including the plants, animals, and other organisms that share the savanna, for example, make up a **community.** The living community together with its nonliving physical surroundings is called an **ecosystem.** The savanna ecosystem includes elephants, the egrets that pick insects off their skin, and the coarse grass they chew and trample, as well as the water in clouds, the sandy soil underfoot, and the hot African sunshine. All the ecosystems of the Earth make up the **biosphere,** that portion of the Earth on which life exists, including every body of water; the atmosphere to a height of about 10 km (6 mi); the Earth's crust to a depth of many meters; and all living things within this collective zone. The biosphere encompasses unimaginably remote places that nevertheless still teem with life, such as the deepest parts of the ocean floor, deep-sea vents spewing superheated water (Fig. 1.15), the frozen ice of Antarctic lakes, and porous hot rocks a mile and a half (2.7 km) down toward the center of the Earth. A huge variety of organisms exists in these and the more familiar and hospitable forests, meadows, lakes, marshes, and grasslands we know in our own surroundings. The surprising abundance of life-forms in Earth's extreme environments, as well, gives scientists like Margaret Race good reason to think

Figure 1.15 **Deep-Sea Vents: An Extreme Habitat on Earth.** These giant worms live alongside crabs and mussels in the sunless, superheated waters in deep-sea vents—cracks in the deep ocean floor that spew superheated water. Some cells in the domains Bacteria and Archaea obtain energy from chemicals in the vent solution and survive at temperatures above 100 °C (212 °F). The giant worms may get their nourishment secondarily from such bacteria. Do similar communities exist today on Mars, based on similar environments and similar principles?

that Mars could also harbor life. The search for life throughout the Solar System and beyond is fueled by our understanding of life's diversity and tenacity.

We've seen that organisms take their energy from the environment, and that over time, evolution fits organisms to their environments. Because this is just as true for Earth's extreme environments as for its temperate ones, there is good reason to think that at least some environments on ancient Mars were inhabited by living organisms.

1.6 How Biologists Study Life

So far, we've examined the basic life characteristics on our planet. But we haven't talked much about how biologists coax, prod, and pry the secrets of nature from living organisms. It's important to keep in mind that behind every fact and concept in this course, there were people like Margaret Race in laboratories or field stations engaged in the often joyful and exciting, but sometimes tedious and frustrating, pursuit of knowledge about living things.

Natural Causes and Uniformity of Nature

A lightning bolt flashes in a cloud-darkened sky. A man eating in a restaurant becomes enraged and abusive when the waitress tells him they are out of apple pie. Modern scientists assume that events like these are due to natural causes. The ancient Greeks, on the other hand, believed that thunderbolts arose when the god Zeus hurled them at the Earth, and that mental illness was due to evil spirits. Today's scientists do not yet fully understand what causes Alzheimer's disease, for instance, or the El Niño climate fluctuation. But they firmly believe these are based upon natural causes they will someday discover by applying the scientific process.

Scientists consider the fundamental laws of nature to be uniform and to operate the same way at all places and at all times. For example, biologists assume that the fixed speed of light, the laws of gravity, and the properties of chemical elements work the same way today in Columbus, Ohio, as they did in East Africa 1 million years ago or on Mars 3 billion years ago. The events that led to life's origin and diversity on Earth, and perhaps on Mars and other planets, occurred long before humans lived to observe them. Yet biologists are confident that today's natural laws functioned the same way at the dawn of time, as life began, and all during its evolution. If they didn't have this conviction, it would be hard to justify the multimillion-dollar NASA search for fossils and living cells on Mars. With its current dry,

frigid climate, the chances are slim that we will find organisms alive today. But their fossil remains—laid down millions of years ago by processes similar to those on Earth—would still be sensational finds that confirm our theories about life's origins and evolution.

The Power of Scientific Reasoning

The search for life on Mars reveals the two kinds of scientific reasoning biologists use. In one type, the biologist collects specific cases and then generalizes from them to arrive at broad principles. For example, after observing that cells can live inside rocks in Antarctica's frigid deserts, biologists proposed a generalization: that Mars rocks could perhaps harbor life, as well. The instant when the scientist's mind leaps from previously isolated facts to a broad, unifying generalization is a creative, intuitive, exciting moment every bit as original as writing a sonnet or sculpting a form from clay.

The second type of reasoning starts with general principles and then goes in the opposite direction to analysis of specific cases. For example, biologists knew that living organisms had turned up in places previously thought to be lifeless. Based on that knowledge, they reasoned that living cells might be found within the permanent ice that covers Antarctic lakes. They took ice samples like the one in Fig. 1.13a, tested them, and found that, indeed, bacteria do survive around dark specks of dust incorporated in the ice. These particles absorb sunlight, warm up, and melt a small halo of water around themselves. This moisture and the pale Antarctic summer sunlight are enough to provide a brief growing period each year for the bacteria.

These two types of reasoning—from the specific to the general, and from the general to the specific—help shape how scientists think, but they aren't unique to science. What *is* unique to the scientific process is a particular approach to testing generalizations. The steps may sound regimented, but they're really just organized common sense.

How do biologists approach scientific puzzles?

Testing Generalizations: The Scientific Method

You may not know it, but you already use scientific reasoning. Say you come home late one night and flip on the light switch in the hall, but the overhead fixture stays dark. You think to yourself, "I guess the bulb's burned out." On the basis of that hypothesis (guess), you predict that a bulb you know to be working (because it lights up a nearby floor lamp) will fix the hall light. You test your hypothesis by screwing the working bulb into the hall socket and flip the switch again. Still no light. You have just disproven your initial hypothesis ("burned-out bulb") and need a new hypothesis ("broken switch" or "broken socket"). Biologists use this approach of hypothesis and testing—the **scientific method**—in much the same way.

First, they *ask a question* or identify a problem to be solved based on observations of the natural world. Your observation was a dark hall, and your question was, "I wonder why the light won't turn on?"

Second, they *propose* a **hypothesis,** a possible answer to the question or a potential solution to the problem. A hypothesis is a guess. Yours was "burned-out bulb."

Then they *make* a **prediction,** a statement of what they will observe in a specific situation if the hypothesis is correct. You predicted that a working bulb would fix the problem.

They *test the prediction* by performing an experiment or making further observations. You tested a bulb in a floor lamp to make sure it glowed, then screwed that working bulb into the hall light socket and flipped the switch again. The floor lamp provided a **control,** a standard for comparison based on keeping all factors the same except for the one being tested. The hall light socket provided the **experimental** situation—the carefully planned and measured test of the hypothesis.

Finally, they *draw a conclusion.* If the hypothesis predicts incorrectly, then they must discard it as wrong. In your case, you said, "Nope. Not the bulb." If the hypothesis predicts correctly—let's say the light did go on—then they devise more tests to see whether the hypothesis might still be incorrect in some way. If they can never design a situation that shows the hypothesis to be wrong, then they begin to accept it. (Here's where we differ from scientists in our daily lives. If the bulb goes on, we think, "Solved!" and go about our business. We don't dream up more tests for *why* it worked!)

The Scientific Method at Work

How do biologists approach scientific puzzles? Let's look at two examples of how biologists apply the scientific method to questions relating to Martian life: Does life exist on Mars? And what happens when an introduced "alien" species drives out a native species, a danger not unlike that posed by the space flights between Earth and Mars?

Searching the Mars Rock for Signs of Life

NASA researchers investigating the Mars rock found in Antarctica (see Fig. 1.4) began by posing a question: Are there signs of past life inside this Martian meteorite? Next they stated an assumption they were making: that

How could experimenters set up controls for their tests of Martian rocks?

Martian life is or was similar to Earthly life. Without knowing anything about possible life on Mars, they had to make *some* assumptions about what it would be like, and what they'd be looking for, and so logically they chose the characteristics of life on Earth. (Explicitly stating underlying assumptions is an important, but sometimes overlooked, part of the scientific method.) Then they created a hypothesis that life did exist at one time in the Martian rock and left fossilized remains. Based on that hypothesis and their assumption, they carried out several tests: They looked for shapes similar to fossilized Earthly bacteria; they searched for traces of chemicals similar to those formed by life on Earth; and they studied grains of a magnetic substance in the Mars rock similar to grains found inside certain bacteria that can orient the cells to Earth's magnetic field. Their hypothesis predicts each of these factors, and indeed, the researchers found these items in the rock.

We saw that to interpret any experiment, scientists need a control or known standard for comparison. In tests of the Martian rock, NASA researchers found little tubular structures that look like fossilized bacteria (Fig. 1.4). For the control, they chose fossilized Earthly bacteria of a size and shape similar to the tubular structures and in rocks of a similar age and composition to the Martian meteorite. Despite their best choices thus far, however, all of the fossilized bacteria from Earth have been much larger than the Martian tubular "fossils." Because of such comparisons with Earthly controls, most biologists doubt that the tubules in the Martian rock are fossilized cells, since there would have been so little space inside them for genes and the machinery of life as we know it.

In their second test, the NASA scientists did find molecules in the Martian meteorite similar to those produced by Earthly life. Their hypothesis predicted this finding. Remember, though, that *while an incorrect prediction disproves a hypothesis, a correct prediction does not automatically prove it to be correct.* For example, the chemicals they found in the Mars rock can be produced by living organisms but they can also be produced by purely chemical means, as well. In our previous light bulb example (page 16), screwing in a working bulb didn't fix the overhead fixture. But if it had, it would have been *consistent* with the hypothesis (burned-out bulb) yet still not *proven* it. Other hypotheses could have been true, as well—for example, maybe the original bulb still worked, but was not screwed in tightly enough. How would you test *that* hypothesis?

In their third test, the NASA researchers found tiny crystals of an iron compound called magnetite that some bacteria on Earth use as an internal compass. Crystals of this exact shape and size are not known to be formed outside of living cells. Other shapes are, however, and many researchers think that the magnetite crystals in the Mars rock may have a nonliving origin. Again, the NASA team's hypothesis predicted correctly, but they were not able to rule out other competing hypotheses.

The search for life in the Mars rock illustrates several points about the scientific method, but differs in a couple of key ways from how biologists often learn about the natural world. First, the scientists could not produce controls or do direct experiments, since they were observing a unique specimen. Second, scientists usually carry out the same experiment or observation many times before drawing a firm conclusion. To do this would require searching for life on many different Earthlike planets and/or getting many samples from Mars. The return of Martian soil and rocks in 2008 may help provide controls and duplicate the experiments. In the meantime, let's look briefly at another example here on our home planet.

Alien Invaders in San Francisco Bay

Margaret Race's current job is helping to protect the planet from possible Martian invaders returning with soil from an exploratory space mission. But earlier in her career, she conducted experiments that illustrate the application of the scientific method as well as showing the effect of alien organisms on natural environments. In a beautiful set of simple experiments, she demonstrated for the first time how an introduced species can drive a native species out of its home. Even though she was only studying mudsnails at the time, this experience helped build the expertise on alien invasions that NASA would call upon for the U.S. space program.

As a graduate student, Race learned that for thousands of years, small brownish snails inhabited the tidal mudflats, marshes, and creeks around San Francisco Bay. Around 1905, however, some oyster farmers had accidentally introduced an "alien" species, the Atlantic mudsnail, to the area. Today, these "exotics" (as ecologists call an introduced non-native species) occupy the rich feeding grounds in the low-lying mudflats and creeks, while their native Pacific counterparts live only in the higher, poorer feeding grounds of the marshlands. As insignificant and unseen as they may seem, Atlantic mudsnails can have a major ecological impact when introduced into non-native areas. Perhaps you have heard about more visible "exotics" like the leafy kudzu vines taking over parts of the American southeast and choking out native plants or the European starlings that have spread across the United States and reduced populations of many native songbirds.

After observing both kinds of snails in San Francisco Bay, Race hypothesized that the Atlantic snails directly outcompeted the Pacific snails, took over the richer mudflats and creeks, and relegated the native snails to the poorer marsh portion of their natural range. She tested her hypothesis (Fig. 1.16) by building cages that would let water pass through but would keep snails trapped inside. She placed the cages so that one end would lie submerged in the creek and the other end would sit up on the bank of the creek. Then she put several dozen Pacific mudsnails into each cage; she

(a)

(b)

Figure 1.16 **Solving a Problem Using the Scientific Method: The Pacific Mudsnail.** Margaret Race observed that "alien" Atlantic mudsnails invade and capture territory from native Pacific mudsnails in San Francisco Bay, and her experiments proved this. (a) In a cage containing only Pacific mudsnails, the animals preferred to live in the wetter areas (equivalent to the richer feeding grounds of coastal mudflats and creeks) rather than in the drier areas (equivalent to the poorer feeding grounds of coastal marshes). (b) In a cage containing both Atlantic and Pacific snails, however, "aliens" crawled on top of the natives, and only those natives retreating to the drier areas survived.

added several dozen Atlantic snails into most of the cages, as well. As controls, she left a few cages with only native Pacific mudsnails (Fig. 1.16a). After two weeks, she found that native snails were twice as likely to be living in the high "marshy" equivalents of the cages when the Atlantic exotics were present in the cages (Fig. 1.16b) as they were when only natives were present. Why did the Pacific mudsnails tend to live in the poorer feeding areas? Race had watched the two species interact in the wild and had observed the aliens crawling over the tops of the natives (especially in the wetter parts of the habitat) and the natives then retracting into their shells. Race thinks that, hiding that way, the Pacific mudsnails weren't able to forage and probably starved to death in all but the higher, drier, less desirable areas where the aliens were less likely to settle in. (By working through **E** 1.3, you can apply the scientific method to her experiment and to others, as well.)

Margaret Race's well-designed experiments established her as an expert on the interactions of alien and native species and allowed her, years later, to predict how alien organisms from Mars might affect animal, plant, and microbial populations on Earth.

A Word About Theories

Eventually, a theory can emerge from a broad general hypothesis that is tested repeatedly but never disproved. But what is a theory? A **theory** is a general principle about the natural world, like the theory of gravity, the cell theory, or the theory of evolution. People often say "It's just a theory," meaning something that's an untested idea. But scientists don't use "theory" in that way; to them, a theory is a highly tested and never disproven principle that explains a large number of observations and experimental data.

The scientific method is a powerful tool for understanding the natural world but it does not apply to matters of religion, politics, culture, ethics, or art. These valuable systems for approaching the world rely on different lines of inquiry and experience. There will always be a place for scientific reasoning, though, because so many of the world's complex problems have underlying biological bases, and we can't solve them without biological facts and principles.

1.7 Biology Can Help Solve World Problems

The search for life elsewhere in the universe is an obvious application of biology and of the question, "What is life?" But biology can do more than just prepare us to look for life on other worlds. It can contribute solutions to a long and growing list of problems here on Earth. The human population is expanding rapidly. We are causing drastic changes all over our planet, including depleting its forests, destroying many of its species, and perhaps heating up its climate. There is famine, war, crime, drug addiction, AIDS, cancer, heart disease, pollution, ozone depletion, and acid rain. These problems tend to have multiple roots. But many are rooted in biology, and biological solutions may help us deal with them effectively. People in all fields—writers, lawyers, auto mechanics, managers, dancers—must understand the biological bases of the world's problems and the broad outlines of the solutions science can provide.

Many of our most vexing problems stem from our species' enormous and burgeoning population (Fig. 1.17). As people celebrated the new millennium in January 2000, few stopped to think that our species' popu-

Figure 1.17 The Enormous Human Population. Shoppers and double-decker buses jam Oxford Street in London. Urban congestion is common in most countries and is emblematic of our species' overpopulation.

lation had just passed the 6 billion mark. Like all other species, ours is highly successful at obtaining energy and materials, and at reproducing. Our human abilities to reason, communicate, and manipulate the physical world are such successful adaptations that our single species is busily exhausting the limited resources that support all life on our small planet. Despite the threat of nuclear war hanging over our societies for the last half-century, many see the burden of humanity on our natural resources as a far greater threat to future security and quality of life. A planetary protector like Margaret Race can address the issues surrounding extraterrestrial life, and what would happen if it arrived on Earth and reproduced out of control. But it is up to each of us to become biologically educated citizens so that we may find ways to protect Earth's other life-forms from our own species' frightening "success."

In the chapters that follow, you'll explore how biology is helping to solve world problems. We are, in fact, in the midst of a revolution in the biological sciences, with exciting new information surfacing weekly in the fights against cancer, heart disease, AIDS, infertility, and obesity. Re-searchers are making rapid advances in gene manipulation to create new drugs, crops, and farm animals; in exercise physiology to improve human performance; in the diagnosis of genetic diseases; and in the transplantation of organs, including brain tissue. The discoveries are so frequent and fast-moving, in fact, that many of them will appear only on your *Explore Life* Web site and not in this text. Across all frontiers of biological science, at all levels of life's organization—from molecules to the biosphere—scientists are learning the most profound secrets of how living things survive day to day and reproduce new generations. You're about to embark on an adventure of exploration and discovery that will not only excite your imagination and enrich your appreciation of the natural world, but will also allow you to contribute intelligently to the difficult choices all human societies must make in the future. Some of the most intriguing choices will come in just a few years, when NASA collects soil and rocks from the Martian sites most likely to yield life. Should we bring back the canister? Should we open it? If we find organisms, should we try to grow them? What have we got to gain . . . or lose?

Chapter Summary and Selected Key Terms

Introduction   In planning missions to search for life on Mars and other planets, NASA has to consider planetary protection: safeguarding Mars against accidental release of Earth organisms, and safeguarding Earth against the possible retrieval and escape of **alien** (non-native) organisms from Mars. An important part of planetary protection is recognizing the characteristics of life on Earth and predicting what life might be like elsewhere in the universe.

Characteristics of Life   Biologists ask "What is life?" and although they can't answer with a simple definition, they can describe the common characteristics all living things share at some point in their life histories: order, metabolism, motility, responsiveness, reproduction, development, heredity, evolution, and adaptation.

Characteristics Relating to Energy Use
Living things possess a degree of **order** (p. 4), or structural and behavioral complexity and regularity. Biologists categorize the levels of that order as independent individuals, or **organisms** (p. 5); groups of functioning body parts, or **organ systems** (p. 6); groups of two or more tissues that together perform a certain specialized function, or **organs** (p. 6); groups of similar cells, or tissues (p. 6); the basic units of life, **cells** (p. 6); the specialized parts within cells, or **organelles** (p. 6); and the biochemical compounds that make up organelles and other cell components, or **biological molecules** (p. 6). Living things carry out **metabolism** (p. 7), taking energy from the environment and using it to organize materials. Living things have self-propelled movement, or **motility** (p. 7). Living things are **responsive** (p. 8)—they respond to changes in their environment.

Characteristics Relating to Reproduction   Living organisms carry out **reproduction** (p. 9), either **asexual** or **sexual** (p. 9). Living organisms usually **develop** (undergo **development**) (p. 9); that is, they usually grow in size and increase in complexity. **Heredity** (p. 9) is the transmission of genetic characters from parent to offspring. The hereditary information contained in **genes** (p. 10) is passed on to subsequent generations.

Characteristics Relating to Evolution
Species (p. 11) descend from the same initial group and can breed successfully with one another in nature. A **genus** (p. 11) is a group of related species. Higher orders include **families** (p. 11) of similar genera; **orders** (p. 11) of similar families; **classes** (p. 11) of similar orders; **phyla** (p. 11) or **divisions** (p. 11) of similar classes; **kingdoms** (p. 11) of similar phyla; and **domains** (p. 11) of similar kingdoms. Living organisms change over time in a process called **evolution** (p. 10). Living organisms display **adaptations** (p. 12), or specializations that help them adapt to their own special way of life. Adaptations usually arise through **mutations** (p. 13) and nature "selects" the fittest individuals in populations via **natural selection** (p. 13).

Characteristics Relating to Environment   **Ecology** (p. 14) is the branch of biology that studies the relationships of living organisms and their living and nonliving environments. Organisms interact with their environments at several different levels. Actively interbreeding groups in a particular area are called **populations** (p. 14). The populations in a particular area make up **communities** (p. 14). Communities and their nonliving physical surroundings are **ecosystems** (p. 14). All of Earth's ecosystems make up the **biosphere** (p. 14).

How Biologists Study Life   Biologists sort out events based on natural causes and consider the fundamental laws of nature to be uniform. Biologists use two kinds of reasoning, forming generalizations from many specific cases, and applying general principles to specific cases. Biologists test their generalizations through the **scientific method** (p. 16): They ask a question, propose a **hypothesis** (p. 16), make a **prediction** (p. 16), design a **control** (p. 16), and an **experimental procedure** (p. 16), test the prediction, and then draw a conclusion. Our examples of the scientific method included the search for signs of life in Martian rocks, and Dr. Margaret Race's study of how an exotic species drove out a native species in San Francisco Bay. Based on numerous applications of the scientific method, scientists sometimes develop a **theory** (p. 18), a general principle about the natural world.

Biologists Can Help Solve Problems

Our species is drastically changing Earth's landscapes, species, even its temperatures and climates. The burgeoning human population is a major contributor to global problems, including famine, diseases, ozone depletion, and the destruction of species. Biological advances can help society grapple with and solve many of these problems.

> All of the following question sets also appear in the Explore Life **E** electronic component, where you will find a variety of additional questions as well.

Test Yourself on Vocabulary and Concepts

In each question set below, match the description with the appropriate term. A term may be used once, more than once, or not at all.

SET I

(a) order (b) metabolism (c) motility (d) responsiveness (e) reproduction (f) development (g) genes (h) evolution (i) adaptation (j) all of these

1. The defining characteristics of life
2. The chemical reactions characteristic of living things
3. The hereditary units
4. Self-propelled movement
5. The acquisition or modification of structures or of functions that work well in a particular environment

SET II

(a) adaptation (b) evolution (c) mutation (d) natural selection (e) kingdom (f) none of these

6. A category of classification of living things
7. A change in the structure of a gene
8. A structure or behavior in an organism that relates to its special way of life
9. The gradual accumulation of mutations that leads to changes in the kinds of organisms living on Earth
10. The main mechanism behind adaptations

SET III

(a) protists (b) fungi (c) plants (d) animals (e) all of these (f) none of these

11. A kingdom that includes mushrooms and yeast
12. Composed of complex cells
13. Motile and multicellular
14. A kingdom that includes mostly single-celled, microscopic organisms such as amoebae

Integrate and Apply What You've Learned

15. A student wonders if his or her philodendron plants really need a weekly dose of nitrogen fertilizer, as the fertilizer package recommends. Explain how this student might find out, using the scientific method.
16. Poisonous species of butterflies often are brilliantly colored. Many nonpoisonous species resemble poisonous ones very closely. Explain how natural selection might account for this phenomenon of mimicry.
17. What is the most important mechanism within organisms that leads to evolutionary change, and what role, if any, does the environment play?
18. Suppose that a particular organism is unable to extract either energy or material substances from its environment. What would happen to its growth, reproduction, waste disposal, and survival, and why?
19. Suppose you have formed a hypothesis and performed an experiment that supports it. Which of the following procedures would be most likely to produce further support, and why?

    (a) Repeat your previous experiment.
    (b) Have someone else repeat your previous experiment.
    (c) Repeat the experiment, but change the controls.
    (d) Repeat the experiment under different conditions, such as place, temperature, light, and humidity.
    (e) Make a new prediction based on your hypothesis and test it by a different experimental procedure.

Analyze and Evaluate the Concepts

20. What characteristics would you test for in an entity brought back to Earth from Mars to determine whether or not it was once alive, or is alive now?
21. An automobile uses energy, is highly organized, is motile, and responds when the accelerator is pushed. Why wouldn't you consider it to be alive?
22. By applying the rules of taxonomy, determine which two of the three following organisms are more closely related, even if their names are unfamiliar to you: *Felis domesticus, Musca domesticus, Felis concolor.* Defend your choice.
23. Darwin presented three main ideas: that species could change over time, that evolution occurs by descent with modification, and that natural selection underlies adaptation. How do those three ideas relate to each other?

PART 1 Cells, Genes, and Life's Perpetuation

CHAPTER 2
Cells and the Chemistry of Life

"I've been a science nerd about this and I believe that it probably saved my life."

CHAPTER 3
How Cells Take in and Use Energy

"Here is an animal that weighs less than a dry tea bag, yet can migrate across the Gulf of Mexico."

CHAPTER 4
The Cell Cycle

"I can easily have 10 new carcinomas every visit."

CHAPTER 5
Patterns of Inheritance

"I have goals and dreams for the future, but I have to focus on living one day at a time."

CHAPTER 6
DNA: The Thread of Life

"That had to be real surprise to the family when the Mounties showed up with a warrant for their cat."

CHAPTER 7
Genes, Proteins, and Genetic Engineering

"It's like having a bioreactor that lives on hay."

CHAPTER 8
Reproduction and Development

"We never gave up hope."

A Threat to Cells and Lives

Battling a Nonliving Enemy: Jay Levy, Dan Foley, A Close-up of HIV.

A tiny cut on his hand nearly cost Dan Foley his life. For this young financial analyst, a deadly battle began unnoticed on a quiet evening spent at home with his partner. The two were well aware of and observed the precautions needed to prevent sexually transmitted diseases. For an instant, though, the partner's body fluids contacted a barely noticeable paper cut on Dan's finger. Unbeknownst to either, particles of the **human immune deficiency virus (HIV)** passed through the opening and into Dan's bloodstream. The next spring, early in 1992, Dan felt ill and went for a blood test. The doctor discovered that his T cells—white blood cells that provide immune protection—had dropped below a fifth of their former population count. He was on his way to a full blown case of **acquired immune deficiency syndrome, or AIDS.**

The doctor prescribed the drug AZT, and Dan began taking it immediately. But he had apparently contracted an especially dangerous strain of the HIV virus with resistance to AZT and a tendency to cause infected cells to clump together. His disease grew steadily worse. A year later, bluish-purple lesions called Kaposi's sarcoma (KS)—a type of cancer induced by HIV—had appeared all over his skin and even internally, inside his stomach. Dan began chemotherapy to fight the cancer and was forced to stop working for several months. His T-cell count dropped below one-fiftieth its normal level, and he grew so utterly exhausted he could barely climb the stairs to his bathroom and bedroom. "I was skin and bones," he remembers. "Any food in my stomach was excruciatingly painful because of the KS lesions. I was anemic," he says, "and emotionally it was horrible." An AIDS patient is really "fighting on two fronts," Dan explains, "a biological battle but also an emotional one." He experienced intense anxiety, feelings of isolation and humiliation, and a terrible fear of the "long, lingering, unpleasant death" he might face. At that point, he recalls, "the outlook was quite grim."

By the end of 1995, however, biologists had learned enough about AIDS to provide more effective chemotherapy agents and anti-viral drugs, and Dan began treatment with a combination of them. From the beginning of his ordeal, Dan had kept himself meticulously informed about research developments in the field through the Internet and other avenues. "I've been a science nerd about this," he jokes now, "and I believe that it probably saved my life." He volunteered for clinical trials on several new drugs, and his T-cell count started to rebound. The Kaposi's sarcoma went into remission, and the purplish lesions disappeared, as well. "I think 1997 was the first time I'd been able to wear a short-sleeved shirt for over three years," he says.

Dan has also contributed blood, lymph nodes, skin tumors, and other tissue samples to Dr. Jay Levy for his basic AIDS research. Levy was one of the first researchers to isolate and study the HIV virus back in 1981. By now, Levy has spent more than half of his long scientific career trying to understand the mechanisms by which HIV, a nonliving enemy, enters and takes over human cells.

In his years at the University of California at San Francisco, Levy has watched AIDS grow from a seemingly isolated threat into a global epidemic infecting more than 33 million people. "I can't believe where the time has gone," he says, "and that today, we still don't have a really long-lasting treatment." That, he says, is because the virus keeps evolving drug-resistant forms. By now, some of Levy's research subjects, like Dan, have been able to stay relatively healthy for many years. For his laboratory team, Levy explains, "that's very uplifting." But in other patients initially helped by various combinations of antiviral medicines, "the virus is coming back and there are no more drugs to use. I think these are going to be sad cases," he says, "and we're seeing up to 40 percent of the people at San Francisco General Hospital with this."

Patients like Dan Foley, researchers like Jay Levy, and the ongoing fight against HIV and AIDS make an ideal case study for this chapter. Our subjects here are the chemistry of life and the structure and function of cells, the fundamental units of life. By

studying the HIV virus, you will see how virus particles contain different kinds of building blocks called biological molecules. We will compare viruses, which are nonliving, to bacterial, plant, and animal cells to reveal their basic differences and unique characteristics. By following the entry of an HIV particle into a human cell, as Levy and others have investigated in great detail, you will get an intimate tour of a functioning cell and its many internal organelles. From this, you will see for yourself why AIDS is such a deadly disease and come to understand the efforts now underway to control this global threat through drugs and vaccines.

As we go through the chapter, you'll find out why you can pick up a cold virus—but not HIV—from a doorknob. You'll see why health experts fear that AIDS could annihilate 40 percent of the population of Africa before a solution is found. You will also find the answers to these questions:

1. What's the structure of the HIV virus? (Chapter 11 presents our broader discussion of viruses.)

2. How are atoms structured and how do they function?

3. How do atoms bond to each other and form molecules?

4. What are the special properties of water?

5. What are the main kinds of biological molecules and their roles?

6. How is a living cell different from an HIV virus?

7. How do the various cell parts function and how does HIV sabotage them?

8. How did HIV infect Dan's cells?

2.1   What Is HIV?

Biologists consider HIV, the human immune deficiency virus that attacked Dan, to be *nonliving* because it has some but not all of the life characteristics we discussed in Chapter 1. A virus has internal order, based on the same groupings of atoms or **biological molecules** that

make up living things. Viruses have genetic material that can change over time, and thus they can evolve. Finally, parts of the virus particle can move, so we can consider them to display the living trait of motility. However, a virus has no metabolism, it is unresponsive in the biological sense, and it lacks the ability to reproduce (without help from living cells). While virus parti-

(a) HIV viewed by electron microscopy

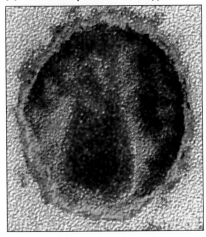

(b) The structure of HIV

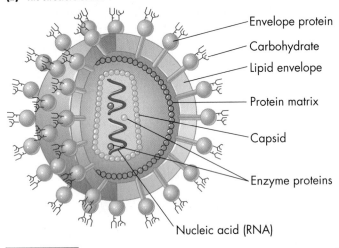

Envelope protein

Carbohydrate

Lipid envelope

Protein matrix

Capsid

Enzyme proteins

Nucleic acid (RNA)

Figure 2.1 **HIV: A Simple But Deadly Group of Molecules.** (a) HIV particles as viewed in an electron microscope, with colors added to improve clarity. (b) HIV contains the four main types of large biological molecules: carbohydrates, lipids, proteins, and nucleic acids.

cles do assemble themselves, this construction requires the machinery inside a host cell and so cannot be considered true growth or development. Virus particles are clearly not alive by our definitions in Chapter 1. Yet their structure and behavior inside the human body are still governed by the same set of chemical and physical laws that determine how all biological molecules form and act and how all living cells function. Let's look more closely, then, at our lethal yet nonliving enemy HIV, and at the atoms and molecules that make up all viruses, cells, and larger organisms.

The Structure of HIV

Viewed in a high-powered electron microscope (Fig. 2.1a), HIV particles look rather simple and they are, compared to an oak tree, an elephant, or even a bacterial cell. Ten thousand of these viruses lined up side by side would only stretch across a single letter in this book. Yet HIV is made of the same four types of biological molecules—proteins, carbohydrates, lipids, and nucleic acids—that compose our own cells. You've no doubt heard these terms before or read some of them on food labels. For instance, the muscle fibers in meat are mostly protein. The starch in flour is carbohydrate. And the fat in butter is a type of lipid. Nucleic acids carry genetic information—the so-called code of life. Using HIV, let's look at where these four types of chemicals occur in a virus. We'll return to the biological molecules in more detail later.

From the outside, HIV looks like a basketball studded with beads (Fig. 2.1b). The beads are a specific protein that allows the HIV particle to enter a cell and infect it. Extending away from each "bead" are branched thread-like chemicals that also help HIV infect a cell. These threads are a type of carbohydrate, the second major type of biological molecule. The stalk end of the bead is embedded into the smooth surface of the basketball. This surface, called an envelope, is made of

the third class of biological molecules, the lipids. This lipid envelope binds the rest of the HIV components together into a single sphere. The other components include the **matrix,** an inner sphere of protein that further strengthens the envelope, and inside the matrix, a cone-shaped object, the **capsid,** also made of protein. The cone-shaped capsid surrounds two kinds of active proteins called enzymes. We'll see later how these enzymes help the HIV duplicate, and how antiviral drugs attack those particular proteins. Finally, the capsid surrounds the virus particle's nucleic acid, the fourth type of biological molecule. HIV's nucleic acid is RNA, and it contains the genetic information to build new virus particles. Our cells and all other living cells also contain RNA, but have their genetic blueprints in the related nucleic acid DNA.

It is remarkable that a simple, nonliving particle made up of just a few chemicals can wreak such havoc in the lives of infected people like Dan Foley. Our goal in this chapter is to explore the chemical nature of HIV in enough detail to understand how current AIDS drugs and treatments work and how more effective therapies may be available in the near future. HIV's chemical nature stems from its atoms and molecules, and the same, in fact, is true for all matter in the universe.

2.2 What Are Atoms?

All matter, including HIV particles and the cells they infect, are based on atoms of distinguishable types. That's where the elements come in and they take us back to some of history's earliest students of life and matter.

Elements

Ancient Greek philosophers realized that some materials, such as rocks, wood, and soil, are composed of more than one substance, while other materials, such as chunks of iron, gold, and sulfur appear to be pure materials. Chemists call pure substances like these that can't be broken down further into different constituents **elements.** Chemists also assign each known element a chemical symbol; for example, the symbols for the main elements found in an HIV particle are C (carbon), H (hydrogen), and O (oxygen).

Chemists have discovered 118 elements. Of these, 89 occur in nature, while scientists have created the rest in the laboratory. The properties of different elements vary widely. For example, carbon is a black solid, sulfur is a yellow solid, and helium is a colorless, odorless gas. Although the Earth contains dozens of elements, only seven elements, headed by oxygen, silicon, and aluminum, make up about 98 percent of Earth's surface layer. Researchers have found more than three dozen elements in living things, but most occur only in traces. Just three elements make up 98 percent of the body of a human or a fern—hydrogen, oxygen, and carbon. In our later discussion of water and carbon, we'll see why living tissue is a unique and special form of matter. Science fiction writers delight in inventing life forms based on the common crustal elements silicon, aluminum, and oxygen. It's unlikely, though, that life as we know it, with its huge structural diversity, based on the millions of molecules carbon can form, could be based on anything else.

Atoms and Molecules

Never satisfied by superficial discoveries, early scientists wondered what makes each element distinct: How is gold, for example, fundamentally different from oxygen? (Knowing this might, among other things, have helped them turn iron or carbon into gold—or so they hoped!) In the 1800s, the English chemist John Dalton concluded that each element is composed of identical particles called **atoms** (Greek, *atomos* = indivisible). Atoms are the smallest particles of an element that still display that element's chemical properties. A **molecule** is the chemical combination of two or more atoms. In a molecule of water, for example, two hydrogen atoms are combined with one oxygen atom.

Atoms are extremely tiny. About a million carbon atoms could sit side by side on the period ending this sentence. A small gold nugget consists of billions of gold atoms. Tiny as they are, though, atoms themselves have an internal structure.

Structure of Atoms

Whether found in a lifeless rock or in a biological entity, such as a person with AIDS, all atoms are composed of protons, neutrons, and electrons (Fig. 2.2). A **proton** is a subatomic particle with a positive electrical charge, and a **neutron** is a particle with no electrical charge. **Electrons** are much smaller (have less mass) and they have a negative electrical charge.

A Model of the Atom

Think of an atom as resembling a miniature solar system. The **atomic nucleus** at the center contains protons and neutrons and accounts for most of the atom's mass. A specific number of electrons, equal to the number of protons, orbit the nucleus at a relatively great distance. Figure 2.2a shows the simplest atom, hydrogen, with its single proton, single electron, and no neutron. If a somewhat larger atom like carbon (Fig. 2.2b) were the size of the Houston Astrodome, the nucleus would be a small marble on the 50-yard line.

What Gives Atoms Their Properties?

Why is a chunk of the element carbon black and solid, while the element oxygen is a clear, colorless gas? The answer is that each type of atom contains a unique number of protons in its nucleus. All carbon atoms have six protons, for example, and all oxygen atoms have eight protons (Fig. 2.2b and c). The number of protons affects the atom's mass and its attraction for electrons, and these two features, in turn, determine the atom's physical and chemical properties.

What gives substances—including the compounds in AIDS drugs—their essential properties?

Electrons and Energy Levels

The attraction between the positively charged nucleus and the orbiting electrons, with their

(a) Hydrogen atom

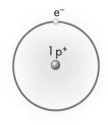

1 p⁺

(b) Carbon atom

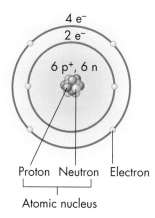

4 e⁻
2 e⁻
6 p⁺, 6 n

Proton Neutron Electron

Atomic nucleus

(c) Oxygen atom

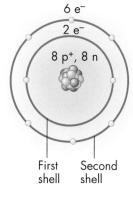

6 e⁻
2 e⁻
8 p⁺, 8 n

First shell Second shell

Figure 2.2 **Models of an Atom.** (a) Hydrogen, (b) carbon, and (c) oxygen atom. In each atom, negatively charged electrons (represented by small gold dots) orbit the positively charged nucleus. The nucleus contains protons (red spheres) and neutrons (tan spheres). There is one proton for each electron, and the number of neutrons is usually close to the number of protons. Electrons move in specific orbits around the nucleus. The system of static rings shown here represents energy levels and not the actual positions of electrons in space.

negative charge, sets up conflicting forces: The opposite charges pull the electrons toward the nucleus, but their rapid circling tends to throw them outward, away from the nucleus, the way a rock tied to a twirling string pulls outward.

Electrons are too small to be seen with the eye or most instruments, of course, but scientists picture them whizzing about in **energy shells** at specific distances from the nucleus, with higher energy levels the farther they orbit from the nucleus. Each shell can contain a certain maximum number of electrons, and the bonding of atoms into molecules depends on the order of shell-filling. The ring or electron shell nearest the nucleus can hold either one electron, as in hydrogen, or two, as in helium. The second shell can accommodate up to eight electrons, and will be filled before any electrons appear in the higher-energy third shell, which also holds up to eight electrons. Subsequent shells also become filled with set numbers of electrons, and tend to fill in order.

Variations in Atomic Structure

Slight exceptions to the standard structure of atoms—either in the number of neutrons or the number of electrons—help explain phenomena as diverse as atomic bombs, acid rain, and the actions of your nerve cells.

An atom of a given kind contains a set number of protons, but the number of neutrons can vary. Atoms with the same number of protons but different numbers of neutrons are different **isotopes** of the same element. The most common carbon isotope is ^{12}C with six neutrons and six protons. Other carbon isotopes are ^{13}C and ^{14}C, with seven and eight neutrons, respectively. Both ^{12}C and ^{13}C are stable, nonradioactive forms, but ^{14}C is radioactive—it tends to break down and emit radiation. In 1991, hikers high on a ridge in the Swiss Alps found an unfortunate hiker's head and shoulders sticking out of a chunk of melting ice. To everyone's surprise, when they retrieved the man from the ice, they found fur and leather clothing instead of Polartec®, and a flint dagger instead of a pocket knife. Researchers used radioactive carbon 14 (^{14}C) to determine the age of the ice man. When an organism dies, it stops incorporating ^{14}C from the environment, and the isotope begins to decay into the isotope nitrogen 14 (^{14}N). (^{14}C has six protons and eight neutrons, while ^{14}N has seven protons and seven neutrons.) As a result of this decay, the concentration of ^{14}C relative to ^{12}C decreases. One-half of the original amount of ^{14}C decays in 5730 years; this is known as the **half-life** of ^{14}C. Through ^{14}C dating, scientists concluded that the ice man lived about 5300 years ago, and that he was the oldest well-preserved human body yet found.

While in the case of isotopes, neutron numbers vary, in **ions,** electron numbers vary. This means the entire atom has a specific positive or negative electrical charge; the number of electrons in an ion does *not* equal the number of protons. For example, the most common form of the hydrogen atom has one proton and one electron: Because the electrical charges cancel each other out, the atom has no net charge (Fig. 2.3a). A hydrogen ion, on the other hand, is missing its electron; as a result, it has only one proton, and is positively charged (Fig. 2.3b). One can think of a hydrogen ion as nature's simplest chemical—nothing but a proton, one of the three major subatomic particles. An acidic solution, such as a cola drink or

⌐**Figure 2.3** **Ions.** (a) A hydrogen atom has one negative electron and one positive proton, and hence the atom has no net electrical charge. (b) A hydrogen ion has lost its electron, and so consists of a single proton, with a net charge of plus one. (c) The number of electrons equals the number of protons in a sodium atom. (d) A sodium ion has lost the single electron in its outer energy shell, leaving it with one fewer electron than protons. The result is a positively charged ion possessing an outer shell filled with eight electrons.

(a) Hydrogen atom (H)

(b) Hydrogen ion (H⁺)

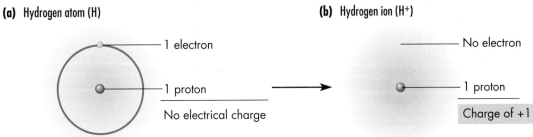

(c) Sodium atom (Na)

(d) Sodium ion (Na⁺)

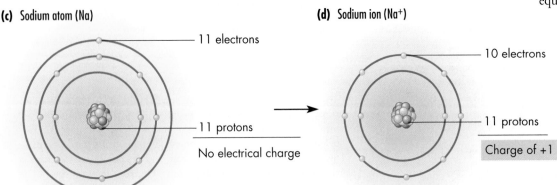

acid rain, has a high concentration of hydrogen ions. Hydrogen ions are also crucial to the ways cells harvest and use energy, as we'll see in Chapter 3. A chlorine ion (Cl^-), on the other hand, has a negative charge. You may have heard it said that positive ions can build up inside a closed car on a hot day and cause headaches, while negative ions (including those of oxygen) build up around waterfalls, fountains, and showers, and tend to invigorate bystanders and bathers.

It's clear that the properties of the elements emerge from both the structure of the atomic parts and the way those parts are arranged. As we'll see in the next sections, this idea of emergent properties also holds for the way atoms make up molecules and the way molecules make up living things.

2.3 How Do Atoms Form Molecules?

The atoms of life—carbon, hydrogen, oxygen, and the others—are joined in tens of thousands of combinations to form the molecules in your food, in other animals and plants, in your bodies, and in HIV particles. How atoms combine to form different kinds of molecules helps determine their properties in living things.

In molecules, two or more atoms are linked by an attractive force called a **chemical bond.** The bonds that link atoms are not actual physical connections, like the couplings between railroad cars. Instead, they are links of energy acting like "energy glue," often based on shared or donated electrons. Bonds act like invisible springs; once a bond forms between two atoms, it requires energy to pull the atoms apart or to push them closer together. We'll see three kinds of bond in our exploration of biology: covalent bonds, hydrogen bonds, and ionic bonds.

Covalent Bonds

When two atoms share a pair of electrons, the most common type of chemical bond forms, a **covalent bond** (Fig. 2.4a). In a water molecule (chemical formula, H_2O), two hydrogen atoms each share a pair of electrons with one oxygen atom. As each hydrogen atom approaches the oxygen atom, its positively charged nucleus begins to attract electrons orbiting the other nucleus. Eventually, the electron orbits overlap

and fuse, and the two atoms—the hydrogen atom and the oxygen atom—share a pair of electrons.

In some molecules, the electrons spend as much time orbiting one nucleus as the other, and the electrical charge is evenly distributed about both ends, or *poles,* of the molecule. A molecule with this equal sharing of charge is said to be **nonpolar.** In a molecule like water, however, the electrons spend more time orbiting the oxygen than the hydrogen. This leaves the oxygen pole of the molecule with a slightly negative charge, and the hydrogen pole of the molecule with a slightly positive charge, making H_2O a **polar** molecule.

Hydrogen Bonds

With their charged ends, some polar molecules can form another kind of chemical bond—a hydrogen bond. In liquid water, for example, a hydrogen from one water molecule can electrically attract an oxygen from an adjacent water molecule (Fig. 2.4b). The attraction of a hydrogen atom to an atom (usually oxygen or nitrogen) in another molecule is called a **hydrogen bond.**

Hydrogen bonds are much more easily broken and reformed than covalent bonds. Some of water's unusual properties (such as the tendency for ice to float) are based on hydrogen bonds, and some important biological molecules are held together by hydrogen bonds. Hydrogen bonds, for example, are involved in the interaction between the bead-like proteins on the surface of HIV particles and the surface of a human cell about to become infected. Hydrogen bonds also hold together the nucleic acids that give DNA molecules their "meaning" as the code of life. (We'll encounter this code in Chapter 6.)

Ionic Bonds

In the third type of chemical bond, electrons from one atom are completely transferred to another atom rather than shared. Salt (NaCl) is a good example: a sodium ion is positively charged (Na^+), and a chlorine ion is negatively charged (Cl^-) (Fig. 2.4c). These oppositely charged ions can attract each other, rather like magnets, and an **ionic bond** forms between the Na^+ and Cl^- and holds the atoms together. Ionic bonds are much stronger than hydrogen bonds, but still not as strong as covalent bonds. That explains why ionically bonded compounds dissociate (break down) into their component ions when dissolved in water, as when table salt dissolves in a pot of soup.

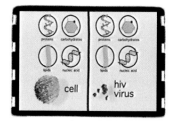

How do the molecules in cells and viruses depend on chemical bonds?

Figure 2.4   Atomic Bonding. (a) Hydrogen and oxygen form covalent bonds in a water molecule. The oxygen nucleus with its six protons attracts electrons from the two hydrogen molecules, the outer shells overlap, and the atoms share electron pairs in two covalent bonds. This sharing fills the shells of all three atoms. The water molecule is polar: Electrons tend to "hang out" around the oxygen atom, giving the oxygen "end" a slightly negative charge and leaving the hydrogen end with a slightly positive charge. (b) A hydrogen bond or shared hydrogen atom (indicated here by a dashed red line) is weak and easily broken. Nevertheless, it helps explain surface tension, floating ice, stable temperatures, and other properties of water that are crucial for living organisms (see Appendix A for more details). (c) The regular three-dimensional latticework of sodium and chloride ions in a salt crystal (like those pictured here, greatly magnified) results from ionic bonds. A chloride ion has gained one electron, leaving a net charge of minus one, while the sodium ion has lost an electron, leaving a net charge of plus one. The positive and negative charges of the two ions attract each other, forming an ionic bond.

(a) Covalent bond

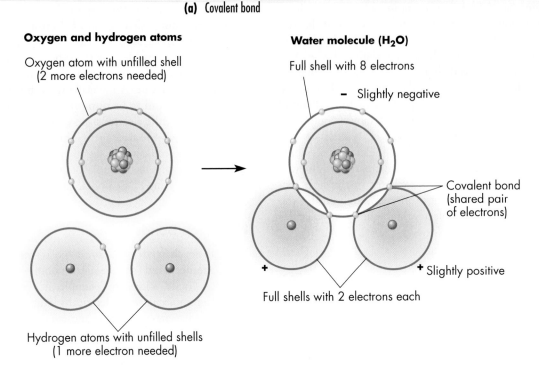

Oxygen and hydrogen atoms

Oxygen atom with unfilled shell
(2 more electrons needed)

Hydrogen atoms with unfilled shells
(1 more electron needed)

Water molecule (H_2O)

Full shell with 8 electrons

− Slightly negative

Covalent bond (shared pair of electrons)

Full shells with 2 electrons each

+ Slightly positive

(b) Hydrogen bonds

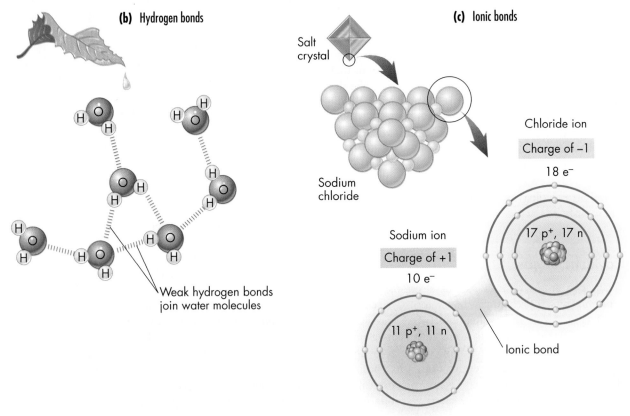

Weak hydrogen bonds join water molecules

(c) Ionic bonds

Salt crystal

Sodium chloride

Chloride ion

Charge of −1

18 e−

17 p+, 17 n

Sodium ion

Charge of +1

10 e−

11 p+, 11 n

Ionic bond

2.4 What Makes Water So Special for Life?

Recall from our case history that Dan Foley started to improve when he began to take a "cocktail" of different drugs to combat his HIV infection. One of those drugs targeted a particular HIV protein (called **protease**). The virus needs protease in order to mature, to exit one infected cell, and then to infect another. The function of protease depends on a specific water molecule held in a certain place in the protein by hydrogen bonds. The highly publicized drugs called protease inhibitors disturb this special water molecule and therefore the structure and action of the protease. This, in turn, blocks HIV's infectiousness. Let's look at the many properties of water molecules that make them so special for life.

Physical and Chemical Properties of Water

The tendency of water molecules to form hydrogen bonds gives water several of its important physical characteristics—all of which are important for living organisms. For example, hydrogen bonds make water molecules stick to each other and to soil, glass, and other substrates. This explains the capillary action and transport processes that draw water up into plants—even towering trees. Water's stickiness also creates surface tension or a "skin" on liquid water that some insects can glide across. Hydrogen bonds in frozen water make ice float. And because of hydrogen bonds, it takes a large amount of heat to increase the temperature of water, and this helps living organisms sustain steady internal temperatures. Appendix A explains the physical properties of water in more detail.

Water also has chemical properties that explain why things dissolve and why some substances are acidic and some are basic (alkaline). In one sense, living things are forms of water moving about the planet: Our bodies, for instance, contain more than 60 percent H_2O, and so the movement and bonding of water molecules is crucial to blood flow, food digestion, and so on. Life itself would be very different if water's chemical behavior was other than it is. Let's see why.

Why Water Dissolves Things

Washing dishes is no fun. But it illustrates some of water's important chemical properties. Dishwater—in fact, all water—is a **solvent,** a substance capable of dissolving other molecules. Dissolved substances are called **solutes.** Water can dissolve polar compounds, such as table sugar, and most kinds of ionic compounds, such as table salt. When a polar solute such as sugar—say, syrup on dirty plates, or glucose molecules inside cells—becomes surrounded by water molecules, hydrogen bonds form. With an ionic solute like salt, the component ions dissociate, and each becomes surrounded by an oriented cloud of water molecules (Fig. 2.5a).

Compounds such as sugar and salt that dissolve readily in water are called **hydrophilic,** or "water-loving," compounds. In contrast, nonpolar compounds, such as cooking oils and animal fats on dirty dinner dishes, do not dissolve readily, and are called **hydrophobic,** or "water-fearing," compounds. Instead of dissolving in water, hydrophobic compounds form a boundary (or interface) with the water (Fig. 2.5b). As we will see shortly, the membranes that surround all living cells are just such boundaries.

Acids and Bases

Water has another chemical property with significant implications for living things: Its molecules have a slight tendency to break down into a positively charged hydrogen ion (H^+) and a negatively charged hydroxide ion (OH^-):

$$H_2O \longrightarrow H^+ + OH^-$$

This breakdown, however, is a relatively rare event in pure water. By definition, an **acid** is any substance that gives off hydrogen ions when dissolved in water, thereby increasing the H^+ concentration of the solution. Lemon juice and vinegar are acidic solutions, and contain many hydrogen ions. A **base** is any substance that accepts hydrogen ions in water. This property allows a base to reduce the H^+ concentration of a solution. An ammonia-containing window cleaner like Windex is a basic solution, with lots of hydroxide ions but few hydrogen ions.

The concentration of hydrogen ions is important to living cells because many of the chemical reactions that drive life's processes—the digestion of foods, for example—depend on specific concentrations of these ions. Biologists measure hydrogen ion concentration on the **pH scale,** which ranges from 1 to 14. On the pH scale, water has the neutral value of 7, in the middle of the scale. Acidic solutions like stomach acid or coffee have pH values between 0 and 7. Basic solutions like drain cleaners or baking soda in water have pH values between 7 and 14. The pH inside most cells stays fairly neutral, between about 6.5 and 7.5, and it is only within this narrow range that many vital cellular reactions take place at optimum speed. Figure 2.6

Why are acids and bases so crucial to living cells?

(a) Salt dissolves in water

(b) Oil and water don't mix

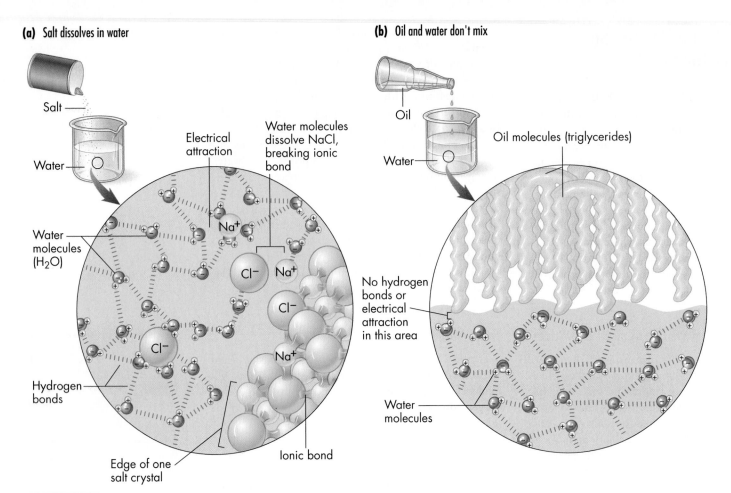

Figure 2.5 **Hydrophilic Substances Dissolve Well in Water, Hydrophobic Ones Don't.**
(a) Salt is hydrophilic, or "water loving," because its charged ions form bonds with water molecules. (b) Oil is hydrophobic, or "water fearing," because its chemical structure is such that hydrogen bonds do not usually form, and hence an oil-water boundary develops.

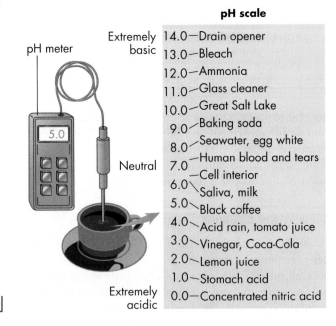

Figure 2.6 **The pH Scale.** This scale represents the hydrogen ion concentration in some common substances. The more hydrogen ions, the more acidic the substance; the fewer hydrogen ions, the more basic. Adjacent units on the scale differ in the concentration of hydrogen ions by a factor of ten. (See Appendix A for details.)

pH scale

Extremely basic	14.0—Drain opener
	13.0—Bleach
	12.0—Ammonia
	11.0—Glass cleaner
	10.0—Great Salt Lake
	9.0—Baking soda
	8.0—Seawater, egg white
	—Human blood and tears
Neutral	7.0
	—Cell interior
	6.0
	—Saliva, milk
	5.0—Black coffee
	4.0—Acid rain, tomato juice
	3.0—Vinegar, Coca-Cola
	2.0—Lemon juice
	1.0—Stomach acid
Extremely acidic	0.0—Concentrated nitric acid

shows the pH of some common solutions. A subject related to acids and bases concerns **buffers,** agents that soak up or dole out hydrogen ions and help control pH level. Antacids buffer your stomach acid, for example, and cells contain natural buffering agents.

Water is clearly a key to life, but so is the element carbon and the biological compounds it forms.

2.5 What Are Biological Molecules?

An HIV particle, the infected cells of an AIDS patient, and all the living things on our planet contain four main types of biological molecules (carbohydrates, lipids, proteins, and nucleic acids) as well as millions of smaller molecules, all based on the special properties of the carbon atom (*E*xplorer 2.1). Why are carbon atoms so crucial for living things? And how do its bonding properties allow the raw materials of life to form?

Carbon Compounds

Our bodies may be mostly water, but a full 18 percent of your weight comes from carbon atoms. For a large tree, the figure can approach 50 percent. Carbon and its chemical bonds are so interesting that chemists divide all molecules into two broad types—those that contain carbon (so-called **organic** molecules, often made by organisms), and those not primarily based on carbon, or **inorganic** (lifeless) molecules. (An interesting exception of great importance to biology is the inorganic molecule carbon dioxide [CO_2].)

Interestingly, there are far more organic than inorganic compounds. Why? Because carbon, with its unique structure, can form millions of different combinations with other atoms. This versatile bonding is carbon's key characteristic, and it explains why carbon is the "stuff of life," not silicon, aluminum, or any other science fiction fantasy.

Carbon Backbones

Carbon is "hungry" for electrons and can form covalent bonds with up to four atoms at a time (review Fig. 2.2b). A simple example of a "satisfied" carbon atom is methane gas (CH_4), sometimes called *swamp gas* (Fig. 2.7). Many biological compounds are much larger than methane and have a backbone of several carbon atoms bonded to each other in long, straight chains, branched chains, or rings—a sculptural property, again, based on forming up to four bonds per carbon atom.

Functional Groups

The chemical sculpturing of an organic molecule's shape contributes to its role in a cell, whether as water-tight seal, storage compound, messenger, protector, or reference library. But a molecule's specific activities often come from small clusters of atoms called **functional groups** that hang from the carbon backbone. Functional groups usually contain atoms other than carbon and hydrogen, and they give special properties to the molecules they are part of. The presence of a methyl group (CH_3), for example, prevents a molecule from quickly dissolving in water, and the hydroxyl group (OH) gives wood alcohol some of its properties like low boiling point and solvent activity. *E* 2.1 includes some biologically important functional groups and explains their properties.

The functional groups help explain why the four classes of large biological molecules—carbohydrates, lipids, proteins, and nucleic acids—form and act as they do. Together with a few other materials, these four types of compounds account for the diverse shapes, colors, and textures of organisms. They explain, for example, why we can easily break down and digest starch molecules but not the rigid carbohydrates in wood.

Carbohydrates

We said earlier that an HIV particle looks like a basketball studded with beads tipped with little threadlike branches (Fig. 2.8a). These branches are made of carbohydrates, and together they form a fuzzy coat that helps an HIV particle to recognize a target cell to infect. In addition to their role in recognition, carbohydrates are also crucial to cell structure and to energy storage. Carbohydrates include sugars and starches, and the term

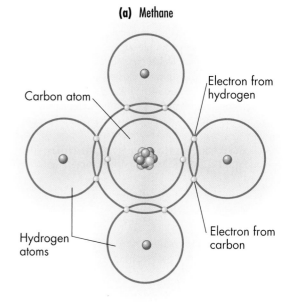

(a) Methane

Carbon atom

Electron from hydrogen

Electron from carbon

Hydrogen atoms

(b) Methane (ball-and-stick model)   **(c) Methane (letter model)**

$$H - C - H$$

Figure 2.7 **The Versatility of Carbon.** (a) A carbon atom has four electrons in its outer shell, and thus an additional four electrons will complete that shell with eight electrons. Each of the four hydrogen atoms in this methane molecule shares a pair of electrons with the carbon atom, thus filling the shell. Two other common ways of representing molecules are (b) a ball-and-stick model and (c) a model with just letters representing the atoms.

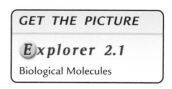

GET THE PICTURE

*E*xplorer 2.1

Biological Molecules

Figure 2.8 Carbohydrates. (a) Beadlike envelope proteins on the surface of HIV are coated with carbohydrate molecules that help the virus recognize and infect cells. HIV's carbohydrate coat consists of simple sugars such as glucose joined together into branched chains. (b) Each glucose molecule consists of six carbons in a ring formation to which are bonded hydrogen and oxygen atoms in clusters (called hydroxyl groups: —OH). (c) A simpler representation of glucose leaves out the hydrogen and oxygen atoms and assumes there is one carbon atom at each corner and one at the end of each protruding stick. (d) Fructose, or fruit sugar, like glucose has six carbons, but they are arranged differently in a ring formation. (e) A disaccharide forms by a condensation reaction; in this case, glucose and fructose join to yield sucrose. (f) Polysaccharides such as starch or the surface coating on HIV particles form by the addition of many simple sugars together in long, sometimes branched, chains.

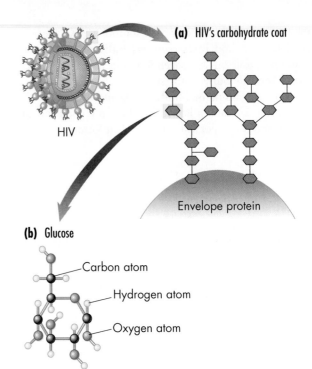

(a) HIV's carbohydrate coat

Envelope protein

(b) Glucose

Carbon atom

Hydrogen atom

Oxygen atom

(c) A simple representation of glucose   **(d)** Fructose, a monosaccharide

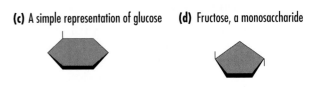

(e) Formation of a disaccharide

Glucose   Fructose

OH   HO

H_2O (water)

Sucrose

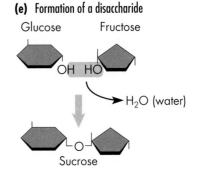

(f) A portion of a polysaccharide

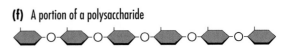

itself literally means "carbon-water." Indeed, carbohydrates usually contain carbon, hydrogen, and oxygen in a ratio of 1:2:1 (CH_2O). The formula for the sugar glucose, for example, is a multiple of that CH_2O subunit: $C_6H_{12}O_6$. The same generally holds true for both simple and more complex carbohydrates.

Simple Carbohydrates: Mono- and Disaccharides

The simple sugars, including glucose and fructose, contain three to six carbon atoms. They share the same molecular formula, $C_6H_{12}O_6$, but have slightly different properties because their —OH functional groups are attached in different places. As a group they are called **monosaccharides** ("single sugars"), and their simple structures form the subunits of more complex carbohydrates. Glucose (Fig. 2.8b, c) is the universal fuel that cells break down for the energy stored in its chemical bonds (Chapter 3). The fuzzy carbohydrate coat of an HIV particle contains about a dozen simple sugars joined together. Fructose ("fruit sugar") (Fig. 2.8d) gives many fruits their sweet taste.

A molecule with two simple sugars joined together is a **disaccharide** (*di* = two). For example, glucose bonded to fructose yields sucrose, or table sugar (Fig. 2.8e). A disaccharide forms by a **condensation reaction,** in which an enzyme removes a hydrogen atom from one sugar and an —OH from the other sugar and joins them together to make water, while linking the two sugars together. In this case, the joining of glucose and fructose makes the disaccharide sucrose. Sucrose is abundant in the saps of sugarcane, maple trees, and sugar beets—our major sources of sugar for refining.

Complex Carbohydrates: Polysaccharides

Many carbohydrates consist of thousands of simple sugar subunits joined into long chains or **polysaccharides** (Fig. 2.8f). A polysaccharide molecule is like a string of beads, where each bead is a simple sugar, often glucose. In general, a long molecular chain of similar subunits is called a **polymer,** and each individual subunit is called a **monomer.**

Starch is the polysaccharide stored as an energy reserve in plants. It is a polymer made up of glucose monomers. **Glycogen** is the polysaccharide storage molecule in animals. **Cellulose** and **chitin** (KYE-tin) are structural polysaccharides that give form and rigidity to plants and insects, respectively. *E* 2.1 explains these polysaccharides in more detail, including why the molecular structure of starch makes it soft and digestible to us, but the very similar structures of cellulose and chitin are rigid and, to us, indigestible.

Proteins

The complexities of our bodies and the rich diversity of life in general—the millions of organisms of different textures, colors, and life styles—depend on different types of proteins in different types of cells. Proteins come in such a wide variety of forms (at least 10 to 100 million different kinds in the spectrum of the earth's organisms) that they can easily explain the myriad shapes and functions of specific cells and whole living things.

Proteins have many functions: They can form structural parts of cells (e.g., the contractile machinery in a muscle cell). They can control cell processes (e.g., the thousands of individual steps involved in metabolism). They can act as messengers that move through fluids (e.g., hormones). They can carry other substances (e.g., the hemoglobin in red blood cells). They can protect animals from disease (e.g., antibodies). They can speed life processes (e.g., enzymes). And they can act as receptors on cell surfaces (e.g., the receptors for HIV). *E* 2.1 describes all these protein roles in more detail, as well.

Protein Structure: An Overview

Inside the "basketball" portion of an HIV particle sits an inner sphere of protein, the matrix (Fig. 2.9), that serves as a good introduction to protein structure and to a basic principle: The overall shape of a protein determines its function.

Hundreds of matrix proteins are packed together inside the spherical matrix, each a winding "ribbon" that makes several turns (Fig. 2.9). The protein ribbon has regions with two distinct forms. A protein region shaped like a spiral staircase is called an **alpha-helix** and provides rigidity. A **pleated sheet** region (often represented by a broad flat arrow) gives proteins flat, box-like sides. The two types of regions within the protein ribbons are connected by **disordered loops,** which are usually gently curved. The helix, sheet, and loop regions occur in specific positions along the length of the protein and help provide each protein's unique overall shape.

Take the protein keratin, for instance, that makes up your hair. Keratin has lengthy regions of alpha-helix,

about 300 amino acids long. Adjacent molecules wrap around and are covalently bonded to each other with sulfur atoms between the chains. Permanent wave lotion breaks these covalent bonds and causes new ones to reform in the coiled shape of the hair wound around a curler. When you simply dry your wet hair around a curler, you merely break and reform hydrogen bonds, and since these are weaker than covalent bonds, the wave or curl relaxes much more quickly.

Amino Acids: Building Blocks of Proteins

There is another level of organization below protein shape that also contributes to how a protein functions, whether that protein is part of an HIV particle or an antibody that fights the virus. The long ribbon of a protein not only has coiled, looped, and pleated regions, but it can also be represented in finer detail as beads on a string, where each bead is an **amino acid** with the general structure shown in Figure 2.10a and b and in *E* 2.1. The matrix protein of HIV, for example, contains 245 amino acids, and a keratin molecule in your hair contains 414 amino acids.

Regardless of how many hundreds of amino acids a protein molecule may contain, these subunits fall into just 20 different types, and each is a variation on a theme. Each amino acid has a portion that is the same in all 20 amino acids and a functional group, also known as a side chain, that is different (Fig. 2.10b). Different side chains have different properties; for example, some attract water, some repel water, some are acidic, and some are basic (Fig. 2.10c).

The portion that is the same in each type of amino acid is responsible for linking each amino acid "bead" to the next one on the string by means of covalent bonds called **peptide bonds.** The

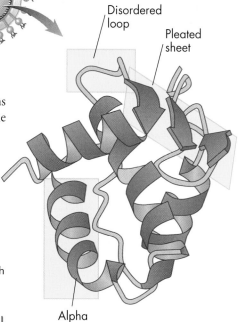

Figure 2.9   The Overall Shape of a Protein. The matrix protein of HIV is shaped like little balls attached to each other inside the viral envelope. In some parts of the protein, the amino acid chain forms an alpha-helix region, which looks like a spiral staircase. In other parts of the protein, the amino acid chain interact with their neighbors and form a pleated sheet, indicated by broad flat arrows. Regions of less specific but constant shape called disordered loops link other parts of the protein.

Matrix protein

Disordered loop

Pleated sheet

HIV

Alpha helix

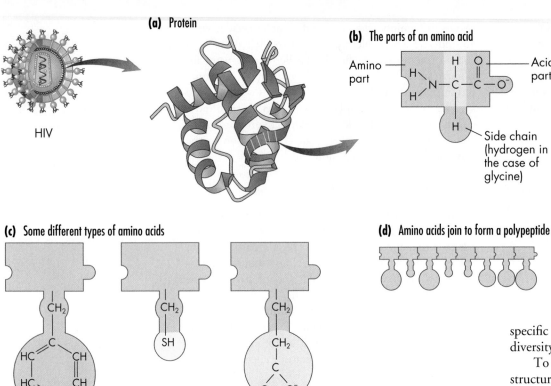

(a) Protein

HIV

(b) The parts of an amino acid

Amino part — Acid part

Side chain (hydrogen in the case of glycine)

(c) Some different types of amino acids

CH_2

CH_2 / SH

CH_2 / CH_2 / C / O O^-

Phenylalanine

Cysteine

Glutamic acid

(d) Amino acids join to form a polypeptide

particular way these helices, sheets, loops, and intervening parts pack together into a three-dimensional ball like the HIV matrix protein or the long, fibrous shape of the keratin protein. Some proteins consist of more than one amino acid chain—the oxygen-carrying protein in your blood, hemoglobin, is an example—and the protein's quaternary structure is the way these chains pack together. Within the human body, there are over 50,000 different proteins, each with a unique shape. In the living world collectively, there are tens of millions of unique sequences of amino acids, giving rise to an enormous array of specific proteins that make possible life's amazing diversity.

To some, the chemistry of biological molecules and structures may seem detached from the living world they inhabit, but in fact, it's actually the basis for life.

1. The variable side chains (functional groups) on each of the 20 different amino acids found in proteins determine how those amino acid "beads" behave within their protein chains.
2. The order of the amino acid beads determines the shape of the overall protein molecule.
3. The overall shape of the protein molecule determines its function in the organism.
4. Collectively, the functions of an organism's proteins determine what the organism looks like and how it lives.

For an HIV particle, amino acid properties and sequences determine the shape of the matrix protein, which in turn helps strengthen the virus particle. Other proteins with their own sequences and shapes protect the virus's genetic instructions and help make new copies that will infect additional cells and eventually destroy a patient's immune system. The virus contains not just protein but another category of biological molecules—lipids. So let's look at that type next.

Lipids

The smooth surface of the HIV "basketball"—beneath the protein "beads" with their fuzzy carbohydrate extensions—is made of lipid molecules (Fig. 2.11). Lipids are a class of biological molecules that tend not to dis-

Figure 2.10 Amino Acids. (a) Proteins like the HIV matrix protein consist of hundreds of amino acids strung together. (b) A single amino acid consists of an amino part, an acid part, and a side chain. The amino acid shown here with a single hydrogen atom as a side chain is called glycine. (c) The side chains of the 20 amino acids commonly found in proteins differ in chemical composition. Some side chains, like the one in the amino acid phenylalanine, are repelled by water. The amino acid side chain of cysteine contains sulfur; it is responsible for the bonds between keratin molecules that give hair its curly or straight texture. Glutamic acid has an acidic side chain that is also hydrophilic. (d) In cells, amino acids are joined together by linking the acid part of one amino acid to the amino part of another. Polypeptides and long proteins are made up of many linked amino acids.

order of the beads is distinctive in each protein—HIV matrix protein, keratin, and so on—and is determined by the organism's genes.

Just as simple sugar units are joined into a polymer called a polysaccharide, amino acid subunits are joined into a polymer called a **polypeptide** (Fig. 2.10d).

The 20 types of amino acids with their different side chains function as subunits in a biological alphabet, forming complex proteins much as the 26 letters of our alphabet can form a nearly infinite array of words. And just as the order of letters in a word determines the word's meaning, the identity of a protein—its shape, properties, and functions—depends on the exact order of its amino acid "letters." Biologists refer to the order of these letters as the protein's primary structure. They call the specific helical portions, sheets, and loops we discussed earlier the protein's secondary structure. A protein's tertiary structure is the

solve in water. Thus they can act as organic raincoats that keep water from rushing into cells and diluting their contents.

Lipids can serve as energy-storage molecules in plants and animals, just as carbohydrates do. Solid storage molecules are **fats;** these include bacon fat, lard, and butter. Liquid storage molecules are **oils,** such as corn oil and olive oil. **Waxes,** such as beeswax and the natural coatings on many leaves and fruits, are semisolid types of lipids. Another important class of lipids, the **steroids,** includes certain vitamins, some hormones, and cholesterol. *E* 2.1 discusses these various lipids in more detail. Here we focus on the general structure of lipids and the specific ones in the HIV envelope and in the membranes of living cells.

Fatty Acids

Just as simple sugars make up the subunits of carbohydrate polymers, and amino acids make up the polypeptide chains in proteins, the lipids in cell membranes are made up of fatty acids. **Fatty acids** have two main parts: a long "tail" of carbon and hydrogen atoms, and a functional group forming a "head" (Fig. 2.11a). The tail of fatty acid molecules is generally 12 to 24 carbons long, and the tail's length affects the molecule's melting point. (Butter, for example, has a lower melting point than bacon fat, because the fatty acids in butter are shorter than those in bacon fat. Hydrogens attached directly to carbons do not generally make hydrogen bonds with water; the fatty acid chains therefore give lipids that "waterproof" quality.

Some fatty acids, called **saturated fatty acids,** have only single covalent bonds between carbon atoms while other molecules, called **unsaturated fatty acids,** have adjacent carbon atoms linked by two covalent bonds, forming a double bond. **Polyunsaturated fats** have several double bonds in the molecule, while **monounsaturated fats** have just one. Saturated fatty acids pack together tightly and are solid at room temperature, as in the dense, white fat in a slab of bacon. In contrast, unsaturated fats bend and kink because of the double bond, so they cannot pack closely together; the result is a slippery liquid at room temperature. Nutritional research suggests that it is much healthier to cook with small amounts of vegetable oils such as canola, safflower, or olive oil (which are high in unsaturated fatty acids) than to consume saturated fats such as bacon fat or lard. Three fatty

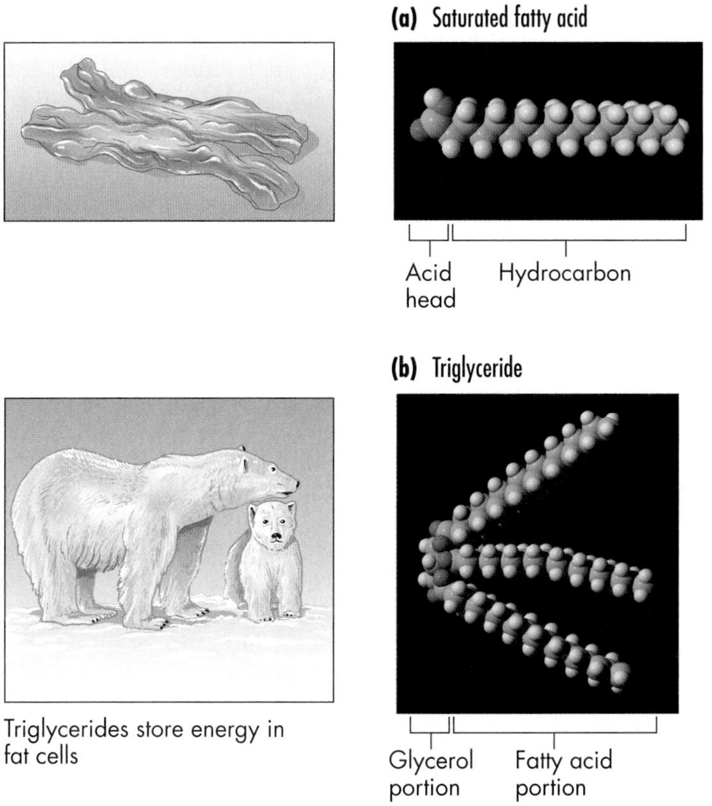

(a) Saturated fatty acid

Acid head   Hydrocarbon

(b) Triglyceride

Triglycerides store energy in fat cells

Glycerol portion   Fatty acid portion

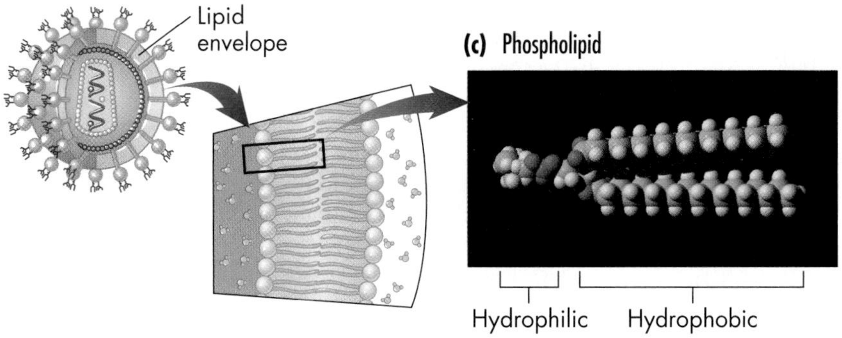

Lipid envelope

(c) Phospholipid

Hydrophilic   Hydrophobic

Figure 2.11   **Lipids.** (a) Fatty acids like those in lard or bacon fat consist of "tail-like" chains of carbon and hydrogen with an acidic functional group as a "head." (b) Triglycerides have three fatty acid subunits attached to a glycerol molecule. This is the storage form of fat in animal cells like those of plump polar bears. (c) Phospholipids have two hydrophobic fatty acids attached to glycerol, but the glycerol head has in it a hydrophilic phosphorus-containing group. In HIV's outer lipid envelope and in plasma membranes, phospholipids join together in sheets with their hydrophilic heads facing out and their hydrophobic fatty acid chains facing inward.

acids can join to a three-carbon molecule called glycerol, forming compounds called **triglycerides** (Fig. 2.11b). If you can pinch a "spare tire" around your midsection, you are pinching triglycerides, the major storage form of lipids in plant and animal cells.

Phospholipids

The class of lipids in HIV particles and cell membranes is the **phospholipids,** which contain just two fatty acid chains, not three. As with other lipids, these chains are hydrophobic, but the two chains are attached to one hydrophilic "head" that contains a glycerol molecule with a phosphorus-containing group (Fig. 2.11c). Overall, a phospholipid molecule looks a bit like a rounded head (the soluble, phosphorus-containing group) on a pair of tails (the insoluble fatty-acid chains). This two-faced nature of phospholipids is critical for forming the lipid envelope of an HIV particle, as well as the membranes of cell surfaces and internal cell parts. We'll see why shortly.

HIV

(a) RNA

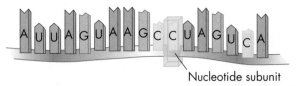

Nucleotide subunit

(b) ATP (adenosine triphosphate)

(c) DNA

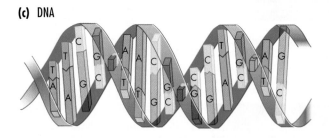

(d) A simple representation of RNA and DNA

RNA

DNA

Nucleic Acids and Nucleotides

An HIV particle is somewhat like a Russian doll with its smaller and smaller dolls inside. The HIV's lipid membrane surrounds its protein matrix, which surrounds its protein capsid, which encloses two identical copies of a molecule called RNA, a member of the fourth major class of biological molecules (Fig. 2.12a). **Nucleic acids,** including DNA and RNA, carry the chemical "code of life" and transmit genetic information from one generation to the next. The information necessary for building new HIV particles, with their lethal threat to patients like Dan Foley, lies in the viral RNA molecule.

Nucleic Acids: Molecules of Information Storage and Processing

Ribonucleic acid, or **RNA,** stores genetic information in HIV and certain other viruses. The nucleic acid **deoxyribonucleic acid,** or **DNA,** stores hereditary information in all living cells and most types of viruses. RNA is also present in living cells and helps in the processing and use of the information stored in DNA. (Chapter 6 covers RNA and DNA in detail; here we'll just explore enough to understand the basic story of HIV, AIDS, and the chemistry of living cells.)

Nucleotides: Building Blocks and Energy Transfer Compounds

Nature is good at exploiting successful approaches. Like carbohydrates, proteins, and lipids, nucleic acids are polymers made up of many subunits—in this case, **nucleotides.** The two RNA molecules in HIV are each about 9700 nucleotides long and their nucleotides are of four types abbreviated A,U,G, and C (Fig. 2.12a). Like RNA, DNA has four types of nucleotides abbreviated A,T,G, and C (Fig. 2.12a and Chapter 6). In comparison, a typical human cell contains 46 DNA molecules, each about 130 million nucleotides long. As in proteins, the order of these nucleotides in RNA and

Figure 2.12 **Nucleic Acids and Nucleotides.** (a) The core of the HIV particle contains two molecules of the nucleic acid RNA. An RNA molecule is a single-stranded chain of nucleotides of four different types with a unique order in each kind of RNA. (b) ATP is a nucleotide that functions as an energy carrier. It has three phosphate groups, each containing a phosphorus atom attached to four oxygens. (c) DNA is a double-stranded nucleic acid that stores genetic information in all cells. (d) A simpler way of showing RNA and DNA.

DNA carries information for building and running cells and organisms.

A major difference between RNA and DNA is that RNA has a single nucleotide chain (it's "single stranded"), while DNA is usually "double stranded"—it has two nucleotide chains bonded together like a ladder and twisted to form a double helix (see Fig. 2.12c).

Nucleotides have another biological job: They play a crucial role in energy transfer within cells. Adenosine triphosphate, or **ATP,** is an especially important nucleotide for energy exchange (Fig. 2.12b). It features a tail of three phosphate groups connected to each other by chemical bonds. When broken one by one, the bonds can release energy that cells use to fuel life processes (more about this in Chapter 3). The enormous American enterprise of growing, selling, cooking, and serving food is, in simple terms, nothing more than a tasty way to renew our cells' supply of ATP.

Nucleic Acids and HIV Infection

When Dan Foley was exposed to the HIV virus, viral RNA molecules entered his blood cells, and were copied into molecules of DNA (Fig. 2.13a, see also Chapter 6). This so-called reverse transcription of RNA to DNA was facilitated by the enzyme **reverse transcriptase.** (The process is called "reverse" because in living cells, genetic information is copied from DNA to RNA, not RNA to DNA.) The first, and still one of the best, AIDS therapies available, **AZT,** blocks this copying of RNA to DNA. Dan began taking doses of AZT as soon as he was diagnosed with AIDS. AZT is a nucleotide that biochemists have modified. The copying enzyme can add the AZT drug molecule onto a growing DNA chain, but this addition prevents the chain from becoming any longer (Fig. 2.13b). This shuts down the copying process like paper jamming a photocopy machine. For AZT to work, it must interact with reverse transcriptase, the enzyme that adds nucleotides to the growing DNA chain in HIV infection. Unfortunately, Dan contracted a nasty form of HIV that could recognize AZT's chemical trick, work around it, and continue copying its RNA into DNA molecules. These viral DNA molecules then could integrate into Dan's own DNA, and encode many more viruses. Because of his infection with resistant HIV, Dan's condition continued to deteriorate until research advances by Dr. Levy and others resulted in new and better therapies.

The RNA molecule in an HIV particle carries three major genes, laid out one after the other. Each gene provides all the instructions for making a single type of protein. The three proteins initially made are then chopped into the smaller proteins that make up the virus. Newer drugs that could prevent this chopping were the basis of Dan's recovery, as we'll see later. Chapter 6 explains how nucleic acids store the instructions for making proteins. For now, just remember that your DNA, by determining the shape of your proteins, has played a central role in shaping the way you look and the way your body functions, and the same is true for all plants, animals, and other living organisms.

How Big Are Molecules?

We've been talking about molecules as small as water, and as large as DNA, but the chemistry of life is an invisible realm and all those individual molecules are far below our ability to see except with the most sophisticated microscopes. (Appendix B introduces several kinds of modern microscopes and how biologists use them.) Figure 2.14 shows you the size relationships among several molecules we've been discussing. An amino acid has more than ten times the volume of a wa-

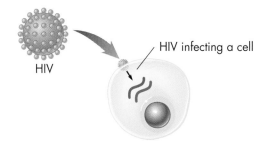

(a) Copying RNA of HIV into DNA

HIV RNA

DNA copy

Reverse transcriptase enzyme

The enzyme adds new subunits to the growing single-stranded DNA chain

(b) AZT blocks chain growth

AZT nucleotide

DNA chain cannot continue to grow

Figure 2.13 **Blocking RNA Replication: A Key Therapy for AIDS.** (a) Immediately after infection, viral RNA enters a cell along with the enzyme reverse transcriptase. The enzyme copies the RNA into a DNA replicate by joining together DNA nucleotides in the order specified by the nucleotides in RNA. (b) AZT is a chemically altered nucleotide that can join the growing DNA chain, but can't accept a nucleotide at its free end. AZT thus prevents the virus's DNA chain from elongating, and so inhibits the infection of the cell.

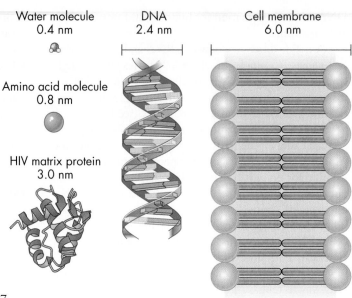

Figure 2.14 **How Big Are Different Molecules?** It's sometimes hard to keep in perspective the relative sizes of things so small that they are invisible to the naked eye. Here a water molecule, an amino acid, an HIV matrix protein, a segment of DNA, and a cross-section of a membrane are drawn to the same scale. In nature, all are far below the resolving power of the human eye.

ter molecule. Even a smallish protein like the HIV matrix protein is much larger than an amino acid—in fact this protein is 245 amino acids long stuffed into a ball-like shape. A typical protein molecule has a greater circumference than a DNA molecule, but at the magnification in Figure 2.14, the average DNA molecule in one of your cells would be 60 km (37 mi) long—a truly colossal molecule! Individual lipid molecules in a membrane are of moderate size, but when thousands line up together, they can enclose an entire cell and yet collectively still be way below the threshold of human sight. We couldn't put an average animal cell on the same figure with these molecules because at this magnification, it would be about 35 m (114 ft) in diameter, and an HIV particle about 1 m (39 inches) in diameter!

You've seen how atoms join to form biological molecules and how an HIV particle is built of all four kinds. But why do we keep insisting that HIV is nonliving, while cells are alive? What makes a cell a cell and a virus . . . different?

2.6 What Is a Living Cell, and What Makes HIV a Nonliving Enemy?

The HIV particles that passed through the paper cut into Dan's bloodstream were swept along and eventually infected some of his cells. His life has been dramatically altered ever since—and was nearly lost at one point. What makes cells alive and susceptible to being killed by HIV particles, which themselves are *not* alive and are so very difficult to fight? Answering this requires an explanation of cells—including those found in animals, plants, and single-celled organisms like bacteria and protozoa—as well as a discussion of viruses. How big are both entities? What are their parts? How do they function? Size is a good place to start, and that leads us to the story's 17th-century beginnings.

The Discovery of Cells

More than 300 years ago, English scientist Robert Hooke focused one of the very earliest microscopes on some everyday objects from his home: the point of a pin, the surface of a nettle leaf, and the body of a flea. Hooke was astonished by the fine detail he could make out in this new, previously unseen world. When Hooke looked at a thin slice of cork through his microscope, he saw what he called "cells," which reminded him of the small rooms inhabited by monks. We now know that Hooke was actually seeing only the remains of living cells—the tough cell walls that surround each plant cell.

Hooke was apparently the first person to publicize seeing cells, but he could not fully define what he was observing. Modern biologists know that a **cell** is the smallest entity completely surrounded by a membrane and capable of reproducing itself independent of other cells. It is also the smallest unit displaying all the properties of life listed in Chapter 1, including the orderly chemical activities of metabolism, the capacity of self-propelled motion, the ability to reproduce and develop, and the potential to evolve over many generations.

Cells Versus Viruses

Biologists have found that cells have three fundamental parts:

1. A surface envelope of lipid and protein (a so-called plasma membrane) that controls the passage of materials into and out of the cell.
2. A central genetic region that controls all the cell's functions and stores DNA, the repository of the cell's hereditary information.
3. A gel-like substance (called **cytoplasm**) that fills the cell between the surface envelope and the genetic storage region, and surrounds small uniquely structured compartments called **organelles** (see Chapter 1) that carry out specialized functions.

Given these principles, how does a virus like HIV differ from a cell and why is it nonliving rather than living? HIV has a membrane made up of lipid and protein, and it has a central genetic region that contains RNA rather than DNA. Both of these features embody the life characteristic of order. (In many viruses, the genes are DNA.) HIV can also evolve or change and adapt over time; recall that Dan's HIV had evolved resistance to the drug AZT.

However, as we saw earlier, viruses lack metabolism, responsiveness, and independent reproduction. HIV, for example, is not filled with gel-like cytoplasm, and this material is integrally involved in a living cell's ability to carry out metabolic functions such as harvesting energy from glucose, generating self-powered movement, and building proteins. Virus particles are tiny compared to the living cell. There's simply not enough space inside a virus for the machinery of the cytoplasm that gives cells their capacity for independent life, including metabolism, motion, and reproduction. Instead, viruses have evolved to use a cell's machinery. Viruses in fact may once have been complete cells that lost elements and became parasites on true cells.

Cell Size: An Import-Export Problem

Even though a cell is thousands of times bigger than an HIV particle, cells are still minuscule. The average cell in your body is just one-fifth the thickness of the paper in this book. Why are cells so small? The answer is that cells have an import-export problem, based on the physical relationship between surface area and volume.

A cell's active cytoplasm needs to take in materials to fuel activities and build cell parts, and it needs to get rid of wastes it produces as by-products—in general,

the more cytoplasm, the more materials and wastes. A cell imports materials and exports wastes across its surface envelope or **plasma membrane**—a boundary, gated wall, and raincoat all in one. The greater the surface area of this plasma membrane, the more rapidly the cell exchanges substances with its environment. When a cell increases in size, its volume increases more rapidly than its surface area (Fig. 2.15), and its import-export needs outstrip its ability to exchange these items with the surroundings. If a cell got much larger than a certain typical size (for a bacterium, under 10 micrometers; for an animal cell, 5 to 30 micrometers; for a plant cell, 35 to 80 micrometers), it couldn't meet its material and waste needs quickly enough to survive. (Note that the abbreviation μm is frequently used for a micrometer, one millionth of a meter.) This is why an elephant's liver is hundreds of times bigger than a mouse's liver, but its *cells* are the same size. There are just millions more of them. For an analogy to the surface-area-to-volume problem, think of a pile of wet laundry. If left in a heap, this soggy pile takes a long time to dry because its exposed surface area is small compared with its volume. But if you hang the items on a line to dry, the surface area is large, while the volume is unchanged, and the laundry can dry much faster.

Viruses have one solution to this surface-to-volume problem and cells have others. Since viruses don't metabolize, particles like HIV don't have to "worry" about their surface-to-volume ratio; living cells import, export, and stockpile all their needed materials, and the viruses simply have to enter and use it. This strategy clearly wouldn't work for most living cells, and so one primary means of solving their surface-to-volume problem is through altered cell shape or contents. A long, thin cell, such as a nerve cell that reaches from a giraffe's spine down to its hoof, can have the same volume as a round or cube-shaped cell, but a greatly expanded surface area. An egg yolk survives even though it is large and roundish because the active cytoplasm is flattened into a thin sheet just below the outer membrane, while the metabolically sluggish yolk inside consists mainly of storage lipids and protein. Oranges and grapefruits solve the problem in a similar way: each miniature sack inside a citrus fruit section is an elongated, spindle-shaped single cell with a very thin layer of cytoplasm surrounding a droplet of sugary juice.

The Cell Theory

Because cells are so small, large organisms such as people and apple trees consist of trillions of cells. About

What are the major differences between cells and viruses?

(a) Doubling the linear dimensions of a cell increases its surface area 4 times but increases its volume 8 times.

(b) A large cube made up of small cubes maintains the original surface-to-volume ratio.

Figure 2.15 **Why Are Cells So Small?** The life of a cell depends on the exchange of materials across its surface. The greater the cell volume, the more surface area required. Most cells are microscopic, and their surface-to-volume ratios are favorable. Doubling the linear dimensions of a cube increases its volume eight times, while increasing its surface area only four times. A large cube made up of many small cubes still has the same surface-to-volume ratio as each individual cube. That's why a large organism can survive—because it has more cells than a small organism, but the cells are roughly the same size.

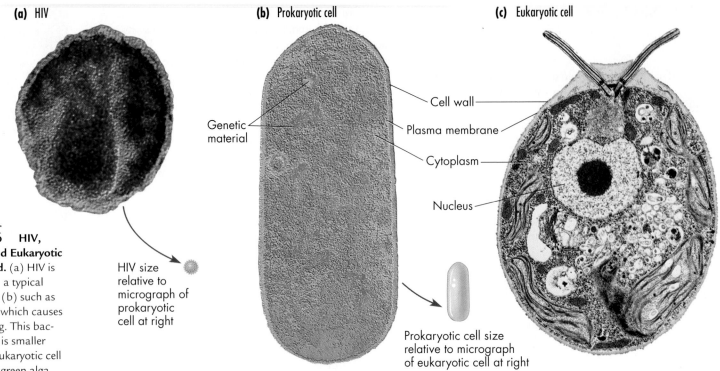

(a) HIV

(b) Prokaryotic cell

Genetic material

(c) Eukaryotic cell

Cell wall

Plasma membrane

Cytoplasm

Nucleus

HIV size relative to micrograph of prokaryotic cell at right

Prokaryotic cell size relative to micrograph of eukaryotic cell at right

Figure 2.16 HIV, Prokaryotic, and Eukaryotic Cells Compared. (a) HIV is far smaller than a typical prokaryotic cell (b) such as this bacterium, which causes blood poisoning. This bacterium, in turn, is smaller than a typical eukaryotic cell (c) such as this green alga. The only organelle HIV shares with living cells is the lipid envelope, which is very similar to the plasma membrane. In eukaryotic cells, DNA is contained in a membrane-bound nucleus; in prokaryotic cells, DNA is loose in the cell; in HIV, the genes are made of RNA rather than DNA.

160 years ago, biologists were trying to understand how cells could be so small and yet how billions of them could function in a coordinated way inside a plant or animal organ. Their efforts resulted in the **cell theory,** which illuminated the cell's significance to life for all of us who came later. According to the cell theory:

1. All living things are made up of one or more cells.
2. Cells are the basic living units within organisms, and the chemical reactions of life take place within cells.
3. All cells arise from pre-existing cells.

What makes these simple statements so important? First, there are no living organisms made up of anything other than cells. Organisms such as bacteria consist of just one cell, while people and trees contain trillions, but the living subunits are always still cells. The viruses that cause diseases like AIDS and influenza are considered particles not cells, because they don't carry on their own energy metabolism and can't reproduce independently. Most biologists consider them non-living and probably derived from living cells millions of years ago.

Second, cells are the basic units of life because the individual components that make up cells lack the complete properties of life. For example, if you take the nucleus out of a cell, it can't carry out life functions or replicate on its own anymore.

Third, new cells arise today only from pre-existing cells that divided into daughter cells. Each cell in your body can be traced back to a single fertilized egg cell generated when your mother's egg cell fused with your father's sperm cell. These sperm and egg, in turn, were produced by other cells in your parents' bodies; each of your parents arose from a single fertilized egg cell produced by your grandparents, and so on back in time. Virus particles like HIV do not arise from the division of a pre-existing HIV particle into two new particles. The end of this chapter explores not only how new HIV particles are assembled from parts like the manufacture of cars on an assembly line, but also, why a living cell is required for the assembly.

Cell Types in Life's Kingdoms and Domains

It's been three centuries since Hooke first saw cells. During the interim—along with industrialization, high technology, and all the other astonishing changes that have taken place—biologists have examined cells from thousands of plants, animals, mushrooms, algae, and other kinds of organisms. Despite the huge diversity of living things, though, biologists have never found a cell they can't assign to just one of two basic types: prokaryotic or eukaryotic.

A cell's most obvious distinguishing feature is the presence or absence of a cell nucleus. **Eukaryotic** cells (*eu* = true + *karyo* = nucleus) contain a prominent, roughly spherical, membrane-enclosed body called the nucleus, which houses DNA, the cell's hereditary material. In contrast, in **prokaryotic** cells (*pro* = before + *karyo* = nucleus) the DNA is loose in the cell's interior and not separated from the rest of the cell's contents by a membrane (Fig. 2.16). An organism made up of a prokaryotic cell is called a **prokaryote;** an organism made up of one or more eukaryotic cells, a **eukaryote.**

Our own cells are eukaryotic and so we humans are eukaryotes, belonging to the domain Eucarya (see Fig. 1.15). So are all the members of the animal and plant kingdoms, the mushrooms and other fungi, and the algae and other protists. The two remaining domains of life, the Eubacteria (also called simply Bacteria) and Archaea, contain thousands of species in which each individual is made up of a single prokaryotic cell lacking a cell nucleus. Since HIV and other viruses are not cells, they're neither prokaryotic nor eukaryotic.

Prokaryotic Cells

Prokaryotes are the smallest cells, and some bacteria are no more than 0.2 micrometer (μm, one millionth of a meter) in length, just twice as big as an HIV particle. For comparison, this page is about 100 μm, or 200 times thicker. A larger, more typical bacterium such as *Escherichia coli* (an inhabitant of the human intestinal tract) has a volume about 10 times greater than the smallest prokaryote and is about 2 μm or 10 times as long. Prokaryotes exploit environments not open to eukaryotes, with their more complex cells. Some members of the domain Archaea, for example, thrive in the boiling waters of Yellowstone National Park's hot springs. At the cooler edges of the spring, bright orange Eubacteria are able to survive and grow in a slimy mat.

Eukaryotic Cells

Because they house nuclei and other compartments or organelles that perform specific tasks, eukaryotic cells are generally much larger than prokaryotic cells. An average-sized animal cell (Fig. 2.17a) is about 20 μm in length, and as we saw earlier, about five animal cells could be lined up across the thickness of a sheet of paper. A typical plant cell (Fig. 2.17b) (also a eukaryote) is a bit larger at about 35 μm across. (See *E*xplorer 2.2 for an in-depth comparison of animal and plant cells.)

Multicellular organisms depend on a division of labor in which groups of eukaryotic cells carry out spe-cialized tasks, generally in tissues and organs. (Many cells of a similar type constitute a tissue, and several tissues usually join to make an organ. Review Fig. 1.5.) For example, the white blood cells that HIV attack are specialized for defense of the body. We have a good framework, now, for understanding where virus particles and human cells fit into the big picture of cell types. Now we can focus on how HIV commandeers the organelles inside a human (eukaryotic) cell and uses them to make more viruses.

2.7 What Are the Parts of the Cell, and How Does HIV Sabotage Them?

If the HIV that infected Dan Foley is not a cell, and is not alive, how can it destroy white blood cells, lead to bluish-purple cancers, and cause the other devastating consequences of AIDS? More specifically, how can HIV take over the various organelles inside the cell and replicate more deadly particles of itself? This section gives the answers by following the way HIV infects a cell, step by step, cell part by cell part. This will serve as an introductory tour of the cell and how it functions, while at the same time explaining the dangers of HIV and AIDS.

HIV Infection and the Cell Surface

HIV enters a cell by injecting its contents through the cell surface and into the cell's interior. An animal cell such as a human cell contains many organelles, including those responsible for its metabolism and reproduction (review Fig. 2.17a). But we'll start our tour of the eukaryotic cell at the cell's outer envelope, then see how HIV overcomes a component of that protective barrier to gain entry.

As we saw earlier, a cell's contents are separated from the surrounding environment by a flexible sheet of fatty material called the plasma membrane (Fig. 2.18 and *E* 2.2). This surface envelope of lipid and protein keeps most unnecessary or harmful materials out of the cell, while keeping most useful substances within. The plasma membrane is not, however, an impermeable seal around the cell. On the contrary, it regulates a constant flow of materials into and out of the cell, allowing water, ions, and certain organic molecules to pass through and enter the cell, while allowing toxic or useless waste products to exit.

READ ON

*E*xplorer 2.2

Components of the Cell

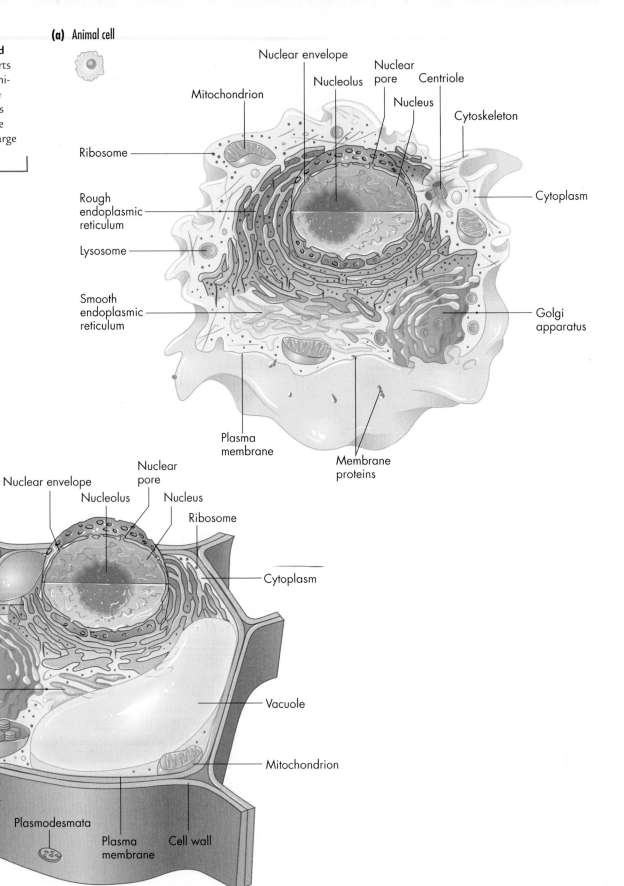

(a) Animal cell

Figure 2.17 **Generalized Animal and Plant Cells.** This figure shows the major parts of the cytoplasm and nucleus in a typical animal cell (a) and a typical plant cell (b). The most obvious differences are that plant cells are encased with a rigid cell wall outside the plasma membrane and have one or more large internal vacuoles, while animal cells have neither.

Nuclear envelope

Nucleolus

Nuclear pore

Centriole

Nucleus

Mitochondrion

Cytoskeleton

Ribosome

Cytoplasm

Rough endoplasmic reticulum

Lysosome

Smooth endoplasmic reticulum

Golgi apparatus

Plasma membrane

Membrane proteins

(b) Plant cell

Nuclear envelope

Nucleolus

Nuclear pore

Nucleus

Ribosome

Cytoplasm

Rough endoplasmic reticulum

Golgi apparatus

Smooth endoplasmic reticulum

Lysosome

Vacuole

Mitochondrion

Chloroplast

Plasmodesmata

Plasma membrane

Cell wall

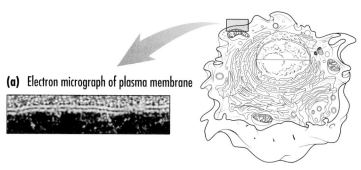

(a) Electron micrograph of plasma membrane

Figure 2.18 **The Plasma Membrane.** (a) An electron micrograph shows the side-by-side plasma membranes of two adjacent cells. Each plasma membrane is a lipid bilayer, two tiers of phospholipid molecules with their hydrophilic heads on the membrane's surface, and hydrophobic tails oriented toward the middle of the membrane. (b) A single plasma membrane is pictured as a fluid plane with floating protein "icebergs" that can extend through the membrane and project from either or both sides. Because of its shape, the protein on the far right (called CD4) takes part in the entry of an HIV particle into a cell. You'll see this protein again in Figure 2.21.

(b) Plasma membrane

Carbohydrate

Protein

Outside cell

Inside cell   Channel protein   Cholesterol

A Model of Membrane Structure

The plasma membrane is only 0.1 μm thick, about one-thousandth the thickness of a sheet of paper. Viewed with the most powerful electron microscope, a plasma membrane is a double layer (Fig. 2.18a), and each layer is a sheet consisting of phospholipid molecules. Biologists often refer to the membrane as a lipid bilayer (*bi* = two) because of this structure. The two phospholipid sheets are oriented "tail to tail," with the hydrophilic heads of the phospholipid molecules facing the watery cell contents on one side and the fluid outside the cell on the other. Phospholipids make the membrane both flexible and dynamic, and this flexibility is affected by other lipid components, such as cholesterol, often embedded in the membrane.

A similar two-layer type of boundary also encircles the organelles or specialized compartments inside cells. These internal compartments or organelles (literally "little organs") carry out specific functions such as making cell components, harvesting energy from organic molecules, and transporting materials inside the cell. The organelle membranes may have a different set of proteins and lipids than the plasma membrane, but their lipid bilayer structure is the same, and so is the way they act as organic "raincoats," regulating the flow of materials into and out of the cell.

Membrane Proteins

Embedded in the plasma membrane's fatty layers are irregularly shaped proteins (Fig. 2.18b). Some proteins protrude into the cell's interior and attach to cell parts, while other proteins pass all the way through and form channels like the one shown in the figure. Other membrane proteins attach cells to things in their environments—in fact, one such protein attaches cells to HIV and is necessary for infection to occur. Because the flexible, dynamic membrane has the consistency of motor oil rather than the solidity of cold lard, cell biologists envision the protein and lipid molecules as freely moving about within the plane of the membrane, with the proteins suspended like toy boats floating on the surface of a pond.

Surface Carbohydrates

Membrane proteins and lipids are sometimes festooned with carbohydrate molecules that stick out from the cell surface. Your red blood cells have a blood type because of such cell surface carbohydrates. People with blood type A, for example, have a different cell surface carbohydrate than people with blood types B or O. Surface carbohydrates on many cells act as receptors that bind to substances outside the cell, and this binding, in turn, transmits information from outside the cell to inside it.

Crossing Plasma Membranes

As a flexible fatty boundary studded with proteins and carbohydrates, the plasma membrane tends to keep the watery cell contents in and moisture, chemicals, and other elements of the external environment out. Recall, though, that nutrients must pass into cells and waste products must pass out. What's more, to be infective, HIV and other viruses must gain entry to the cell.

Clearly, the plasma membrane is *permeable* (penetrable) to certain substances, but not all. Biologists say that cell membranes are **selectively permeable.** So what accounts for the selectivity that allows nutrients, wastes, and viruses to pass through plasma membranes while most other substances are barred?

Fat-Soluble and Water-Soluble Molecules

Passage through plasma membranes depends first on size. Some very small inorganic molecules such as oxygen, carbon dioxide, and water pass through plasma membranes by simple **diffusion** (Fig. 2.19a). They move from a region of high concentration to one of low concentration. The diffusion of water from an area of higher concentration (of water molecules) across a semipermeable membrane to an area of lower concentration is **osmosis.** (Appendix 2.3 discusses this important biological concept in more detail.) The waste product carbon dioxide (CO_2) tends to build up inside the cell, for example, then diffuse out across the plasma membrane to the surrounding fluid, with its lower concentration of CO_2 molecules. Oxygen works in reverse, moving from outside the cell, where it's more plentiful, across the cell's boundary to the interior, where oxygen levels are lower.

Organic molecules tend to be larger, so instead of diffusing, they pass into cells in two other ways, depending on whether they dissolve in fat or water. Some

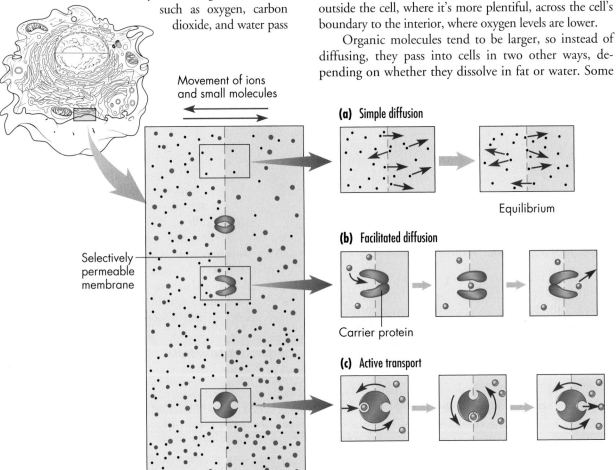

Movement of ions and small molecules

Selectively permeable membrane

(a) Simple diffusion

Equilibrium

(b) Facilitated diffusion

Carrier protein

(c) Active transport

Figure 2.19 **Movement of Molecules Into and Out of Cells.** (a) Being selectively permeable, cell membranes allow nonpolar molecules such as lipids and small polar molecules such as water to pass via simple diffusion. This is the movement of a substance from a region of higher concentration to one of lower concentration. At equilibrium, the concentration of a diffusing molecule will be the same inside and outside the cell. (b) Somewhat larger molecules, such as sugars, can't diffuse through the membrane without the action of special carrier proteins that span the membrane. The proteins change shape, allowing the molecule to cross the membrane, then spontaneously change back, ready for the next molecule. This process, called facilitated diffusion, allows some substances to pass across the membrane that otherwise would not. Because the substance always moves down a concentration gradient, the process is true diffusion, with no energy expenditure by the cell. (c) Active transport moves substances against their concentration gradient and requires energy expenditure by the cell. Special protein "pumps" move substances the cell needs. Examples are the movement of potassium ions into the cell and the movement of sodium ions out of the cell.

(a) A white blood cell engulfs a yeast cell

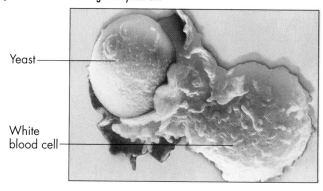

Yeast

White
blood cell

Figure 2.20 **Cell Engulfing Large Materials.** (a) A large white blood cell approaches and within about two minutes engulfs a yeast cell in a process called phagocytosis. (b) Endocytosis, the engulfment and taking in of fluids and/or materials of fluids (blue dots). Endocytosis can occur when receptor proteins in the cell's plasma membrane bind to the substance. The membrane then indents and forms a vesicle and the contents are released into the cell's cytoplasm. Cells can expel some substances by the reverse procedure, exocytosis.

(b) A cell engulfs and expels proteins

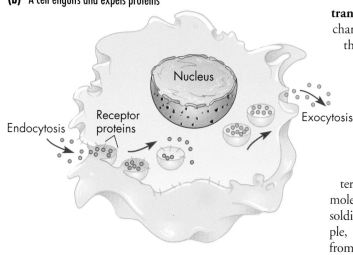

Nucleus

Receptor
proteins

Endocytosis

Exocytosis

organic substances are fat-soluble: they can dissolve in fats, and thus they can pass directly through the fatty plasma membrane. Examples are vitamin E and lecithin (a softening agent often listed among the ingredients on candy bar labels).

Most organic molecules, however, are water-soluble, including nutrients such as sugars and amino acids, and cellular wastes such as urea. A plasma membrane made of pure lipid molecules (fat or oil) would act like a perfect raincoat, blocking out all water-soluble materials and eventually starving or poisoning the cell. Instead, small water-soluble organic compounds enter the cell by passing through proteins floating in the plasma membrane. Some membrane proteins allow only particular sugars, ions, or amino acids to pass, often through a channel in the protein. If the protein helps the substance pass down its concentration gradient without the expenditure of energy, we call the process **facilitated diffusion** (Fig. 2.19b). If a substance is too large to simply diffuse through the membrane or slip through a channel, the cell may have to expend energy to pump materials in or out by means of a process called **active**

transport (Fig. 2.19c). This pumping process often changes the shape of the membrane protein and pushes the substance against its concentration gradient, that is, transporting it out of the cell even if there is already more outside than inside. Like rolling a rock up a hill, this requires energy expenditure.

Large Materials

A cell's outer membrane obviously has lots of "gates" and lots of traffic in and out. Really large materials, though, such as virus particles and large protein molecules usually can't get through those gates. Often, soldiers of the body's immune defense system—for example, large, mobile blood cells—engulf debris left over from a dying cell or surround whole parasites such as bacteria or viruses with in-pockets of the cell membrane. The engulfment of solid material is called **phagocytosis** (FAJ-oh-sigh-toesis) (literally "cell eating"; see Fig. 2.20a). Jay Levy was the first scientist to show that these large cells or **phagocytes** can themselves be infected by HIV.

Cells can move substances in across the plasma membrane by the import process of **endocytosis** (of which phagocytosis is one type) or out across the plasma membrane via the export process of **exocytosis.** Vesicles or little bubbles of membrane surrounding materials to be exported can fuse with the plasma membrane like two soap bubbles joining together and the contents of the vesicle can then be expelled from the cell (Fig. 2.20b and *E*xplorer 2.3). Some cells discharge wastes this way, or secrete proteins, such as hormones or digestive enzymes, into the bloodstream or a food-digesting organ like the stomach or small intestine.

How HIV Gets into a Cell

Recall that HIV looks like a basketball studded with beads and that the skin of the "basketball" is made of lipids while the beads are a type of envelope pro-

GET THE PICTURE

*E*xplorer 2.3

Crossing the Plasma Membrane

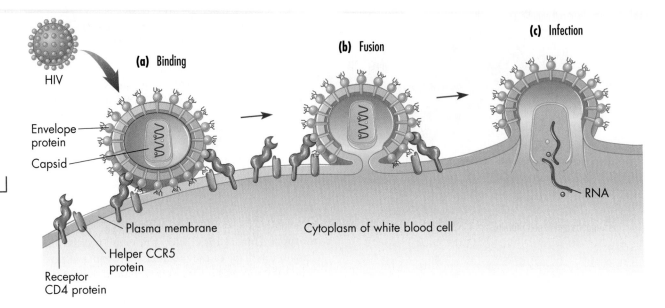

Figure 2.21   HIV Fuses with White Blood Cell and Infects It. (a) The HIV envelope protein binds to the CD4 receptor on a susceptible white blood cell. This binding process is aided by the helper protein CCR5. (b) The lipid membranes of HIV and the cell fuse together. (c) HIV spills its contents into the cell, and the cell is infected.

tein. One end of the envelope protein sticks through the lipid membrane toward the viral interior and the other end, the "bead," projects outward from the virus and first comes into contact with human white blood cells (Fig. 2.21a). The bead binds to a receptor protein called CD4 on the blood cell's surface. Normally, this cellular CD4 protein allows various kinds of white blood cells to communicate with each other. But HIV overcomes this protein and its helper protein (called CCR5) to gain entry into the cell. When HIV contacts the receptor, the lipid membrane surrounding the virus particle is suddenly able to fuse with the lipid membrane of the blood cell like two soap bubbles converging (Fig. 2.21b). The contents of the HIV particle can then spill into the cell and infect it (Fig. 2.21c). This process of binding, membrane fusion, and infection probably happened within minutes after Dan's blood came in contact with the virus through the cut in his finger.

AIDS researchers like Jay Levy have noticed that human cells lacking the CD4 receptor on their surfaces, such as muscle cells or red blood cells, do not become infected with the virus. They have also studied rare cases of people exposed to the virus who never develop AIDS from it. Some of them turn out to have a mutation that prevents their white blood cells from making the helper protein (CCR5). Without it, HIV can't enter their cells as easily and thus they have a measure of protection against AIDS. If you were trying to design a new AIDS drug, how could you use what you now know about the mechanisms of HIV entry into cells to help you save lives and perhaps win a Nobel prize?

Once the HIV membrane has fused with the cell membrane, the capsid inside the virus breaks open and releases its genetic material, RNA. This gets copied into the cell's own genetic language, DNA, and this new viral DNA enters the cell's central repository of genetic information, the nucleus. We'll take a closer look at this process as we examine the nucleus.

The Nucleus

When you look through a microscope at a typical animal or plant cell, often the most conspicuous organelle you will see is the **nucleus** (Fig. 2.22a). This roughly spherical structure contains genetic information that controls most of the cell's activities. Just as an enemy tries to take over an opposing army's central command post, HIV must get into the nucleus if it's going to take over the cell's activities and commandeer them for its own purpose: making new HIV virus particles.

Nuclear Envelope

To get into the nucleus, the viral DNA must pass through the **nuclear envelope,** a boundary consisting of *two* lipid bilayer membranes separated by a thin space. The nuclear envelope is perforated at dozens of points by **nuclear pores,** giving it the appearance of a dimpled golf ball (see Fig. 2.17). Each pore is a cluster of proteins that form a channel that regulates the passage of genetic molecules from the nucleus to the watery cytoplasm that fills the cell. Researchers still don't fully understand how HIV's genetic information moves toward the nucleus and gets through the nuclear envelope. Solving this puzzle is part of their ongoing investigation.

Chromosomes

Once inside the nucleus (Fig. 2.22a), the viral genes (in the form of DNA at this point, not RNA) can integrate into the DNA of the host cell with the help of an active viral protein called *integrase* (Fig. 2.22b). Whenever a cell is not actively dividing, its own DNA takes the form of long fibrils or threads. The loose, extended nature of these threads makes the nucleus appear grainy except when cells are dividing. When cells do divide, the DNA threads compact into **chromosomes** (literally, "colored bodies," because they easily take up certain dyes) (Fig. 2.22c). Chromosomes are microscopic structures that carry hereditary information.

Information Flow in the Nucleus

In the nucleus of all eukaryotic cells, including our own, the information in DNA is copied into RNA, the RNA then moves out of the nucleus into the cytoplasm, and its information is used to make proteins, which carry out the work of the cell. This flow of information can be diagrammed:

$$DNA \longrightarrow RNA \longrightarrow Protein$$

One type of RNA called messenger RNA carries the pattern for making proteins. Another type, ribosomal RNA, forms a "workbench" upon which proteins can be forged. Workbench ribosomal RNA is made in a dense area within the nucleus called the **nucleolus** (pl., **nucleoli**) (see Fig. 2.22a). A third type of RNA, transfer RNA, assists in making proteins. The messenger RNA that leaves the nucleus and enters the cytoplasm controls the rate of cell growth, cell division, and energy use. The information molecules made in the nucleus direct most of a cell's moment-to-moment and day-to-day activities. (Chapters 6 and 7 describe the structure and activity of DNA and RNA in detail.)

We've said that when HIV enters the cell, its RNA is copied into DNA and this viral DNA enters the cell nucleus. Once inside, however, and once integrated into the cell's chromosomes, it follows the DNA → RNA information flow characteristic of healthy, uninfected cells. Thus new viral RNA is once again made and moves back out of the nucleus into the watery cytoplasm to begin diverting the cell to act as an HIV-manufacturing factory. This takeover involves a number of cell organelles.

Cytoplasm and Organelles

We saw earlier that Dan's cells, your cells, and in fact, all living cells are filled with cytoplasm, a semifluid, highly organized pool of raw materials and fluid in which the cell's internal organelles are suspended. (See **E** 2.2 for an overview of cell organelles.) Cytoplasm is about 70 percent water, 20 percent protein molecules, and 10 percent carbohydrate, lipid, and other types of molecules. An average cell contains 10 billion or so protein molecules of about 10,000 different kinds!

(a) Nucleus

Decondensed chromosomes

Nucleolus

Nuclear envelope

(b) DNA from HIV into host cell DNA

HIV DNA

Integrase

Host cell DNA

Integration

(c) Chromosomes in dividing cell

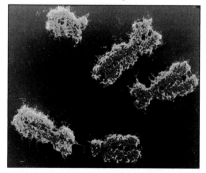

Figure 2.22 **The Nucleus: The Cell's Control Center.** (a) A cross-section of a nucleus reveals a large area of unraveled chromosomes, and the dark-staining nucleolus, which makes the RNA workbench on which proteins are constructed. The nuclear envelope is a double layer perforated with pores. (b) After an HIV particle's RNA enters a cell, it is copied into DNA. This DNA then moves to and enters the nucleus, where it integrates into the host cell's DNA. This integration is facilitated by integrase enzymes from the virus. The photo in (c) shows chromosomes in a dividing cell.

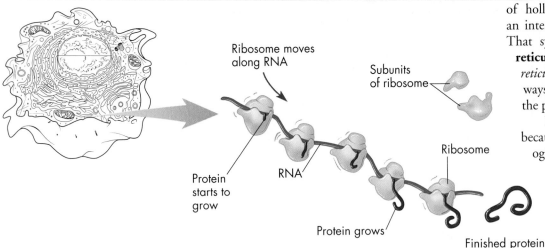

Ribosome moves along RNA

Subunits of ribosome

Protein starts to grow

RNA

Protein grows

Ribosome

Finished protein

Figure 2.23   **Ribosomes: Sites of Protein Manufacture.** Ribosomes bind to an RNA molecule and move down it like a bead moving along a string. As they go, the ribosomes synthesize a protein, adding one amino acid after another, according to instructions encoded in the RNA. When the ribosomes reach the end of the RNA, the newly made protein is complete, the ribosome falls off the end of the RNA, and it splits into its two subunits.

Many of these proteins are highly active **enzymes,** substances that speed biochemical reactions, while others are structural proteins that assemble into various cell parts. Suspended in the cytoplasm are numerous kinds of organelles—many of which HIV uses for its own ends.

Ribosomes

In order for new virus particles to form inside a cell, HIV needs new proteins. So it usurps the cell's protein-generating machinery, the ribosomes. Thousands of beadlike clusters called **ribosomes** are embedded throughout the cell's cytoplasm. The ribosomes serve as the workbench upon which proteins are synthesized (Fig. 2.23). There are no functional ribosomes inside the nucleus and so protein building occurs only in the cytoplasm. We'll see more about how HIV takes over ribosomes shortly.

Rough Endoplasmic Reticulum

Most of a cell's ribosomes float freely in the cytoplasm, but some are attached to flattened membranous sacs within the cell; these make certain proteins for the cell and HIV takes them over, as well. The flattened sacs are part of the cell's internal membrane system, a network

of hollow tubules, sheets, and reservoirs that form an interconnected set of channels throughout the cell. That system of channels is called the **endoplasmic reticulum,** or **ER** (*endo* = inside; *plasmic* = cell; *reticulum* = network), and its convoluted passageways extend all the way from the nuclear envelope to the plasma membrane (Fig. 2.24).

Part of the endoplasmic reticulum looks rough because it is so heavily studded with ribosomes; biologists call this the **rough ER** (Fig. 2.24a). This rough region makes many proteins that wind up being exported from the cell. For example, the rough ER helps produce the enzymes that digest most of the foods you eat. The HIV virus takes over the rough ER, and tracing how the membranous network makes and transports viral protein is a good way to see how the ER functions.

In a white blood cell infected with HIV, viral RNA made inside the nucleus attaches to floating ribosomes, and then the complex docks with the rough ER (Fig. 2.24b, Step 1). Next, the ribosome workbenches make some of HIV's bead-shaped proteins. As the "beads" are assembled, they pass through a pore in the rough ER membrane and enter the interior cavity of the rough ER (Step 2). Eventually, the newly synthesized proteins enter transport vesicles—little membranous bubbles that pinch off from the rough ER (Step 3). Most of these transport vesicles then fuse with membranes of another cell organelle, the Golgi apparatus (Step 4).

Golgi Apparatus

Named for the Italian cell biologist who first described it, the **Golgi apparatus** (GOAL-gee) is a series of flattened membranous sacs resembling a pile of nearly empty hot-water bottles (Fig. 2.24c). Some cells have just one Golgi apparatus, but others may have hundreds. The Golgi apparatus can further modify proteins altered in the rough ER. For example, it can subtly change the structure of the carbohydrates added to the HIV envelope proteins or "beads" (Fig. 2.24b, Step 5). Once modified, these carbohydrates can act like the colored tags added to luggage so that baggage handlers send it to the right airport.

The Golgi apparatus sends vesicles containing newly made proteins to various parts of the cell. For example, it sends vesicles carrying the new HIV proteins to the cell surface, where they fuse with the cell's plasma membrane (Step 6). The HIV proteins then embed in the cell's membrane, so that the infected cell's surface now has the bead-studded appearance that usually characterizes an HIV particle (Step 7). The Golgi apparatus acts like a traffic cop, directing different proteins to different parts of the cell, where they perform their functions. Curiously, the enzymes that digest your food follow basically the same pathway through the cells of your pancreas that HIV membrane proteins follow through an infected cell: rough ER → vesicles → cell surface. The difference is that the HIV proteins have their "tails" stuck into the cell membrane, so when they move to the surface, they remain embedded (Fig. 2.24 Step 7). Digestive proteins made in the pancreas are free inside the Golgi apparatus and vesicles, so when they reach the cell surface, they just diffuse away, enter the bloodstream, and participate in digestion in your intestines.

Figure 2.24   The Endoplasmic Reticulum and the Golgi Apparatus. The cell's internal membrane system makes some of a cell's proteins and lipids. We can understand how they work by following how HIV uses them to make copies of itself. (a) In an electron micrograph, the rough endoplasmic reticulum (RER) appears as a system of flattened sacs with ribosomes on the surface. The rough ER synthesizes membrane proteins and proteins for export. (b) RNA for these proteins, including the HIV envelope protein, binds to ribosomes and joins the rough ER (Step 1), where the protein is made (Step 2). Transport vesicles bud off the rough ER (Step 3), move through the cytoplasm, and join with the Golgi apparatus (Step 4). There, enzymes add carbohydrates to the protein (Step 5). Transport vesicles bud from the Golgi and move to the cell surface, where they fuse with the cell's plasma membrane (Step 6). This places the protein on the surface of the cell, with part sticking into the cell's internal environment (Step 7). (c) In an electron micrograph, the Golgi apparatus appears as short, flat stacks of membranous sacs.

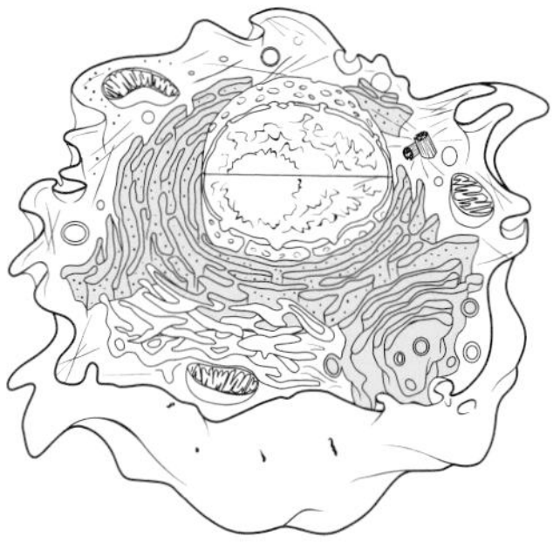

(a) Rough endoplasmic reticulum

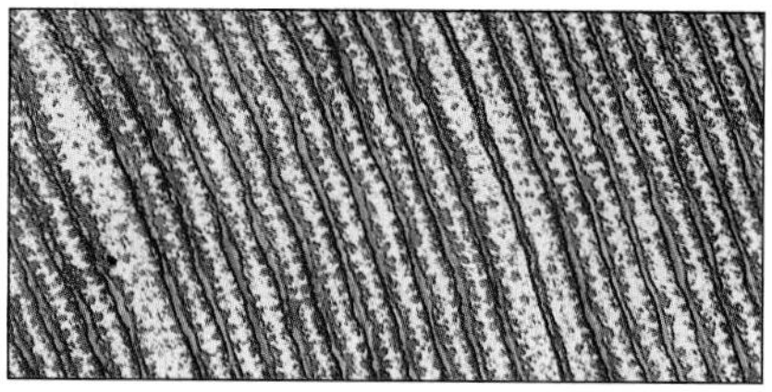

(b) Function of the RER and Golgi

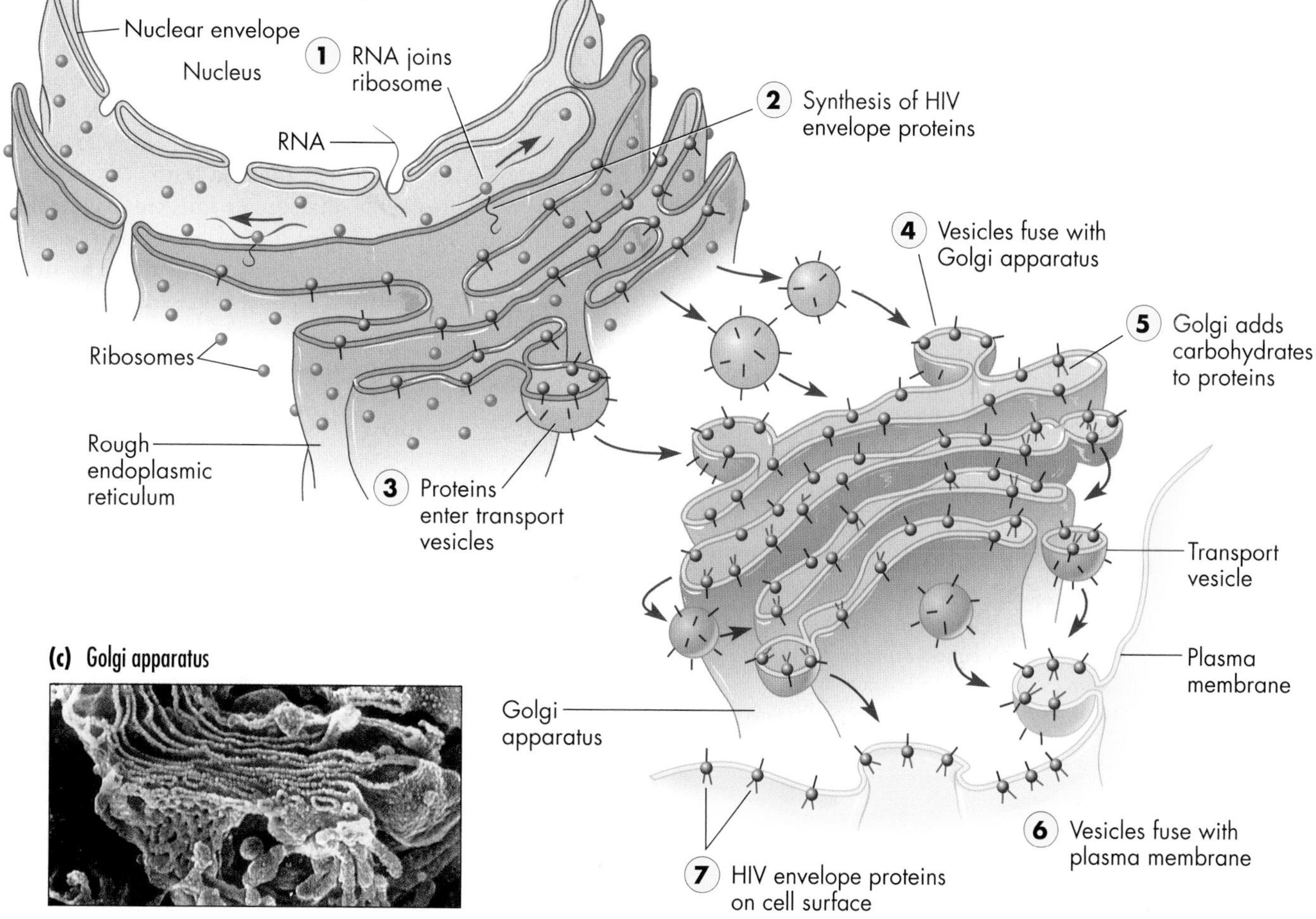

Nuclear envelope

Nucleus

1 RNA joins ribosome

RNA

2 Synthesis of HIV envelope proteins

4 Vesicles fuse with Golgi apparatus

5 Golgi adds carbohydrates to proteins

Ribosomes

Rough endoplasmic reticulum

3 Proteins enter transport vesicles

Transport vesicle

Plasma membrane

(c) Golgi apparatus

Golgi apparatus

6 Vesicles fuse with plasma membrane

7 HIV envelope proteins on cell surface

Smooth ER

Another part of the endoplasmic reticulum, the **smooth ER,** is folded into smooth tubes and small sacs (Fig. 2.25a). The smooth ER makes, and also detoxifies, substances that can dissolve in lipids.

Smooth ER makes the lipids that line up side-by-side in cell membranes. When commandeered, they also make the HIV particle's lipid "basketball" or envelope. Smooth ER makes other familiar lipids, as well, including cholesterol, and the steroid sex hormones estrogen and testosterone. In a person's skin, enzymes in smooth ER use the energy of sunlight to convert modified cholesterol into a precursor of vitamin D, a compound necessary for maintaining strong, healthy bones. Interestingly, some people get very little exposure to sunlight, including North African women of Bedouin tribes, who wear dark, full-length garments. As a result, their cells' smooth ER can't convert enough cholesterol into vitamin D, and the women's bones can grow soft and weak (Fig. 2.25b).

Smooth ER in liver cells is especially active at detoxifying harmful lipids, such as the sedative drug phenobarbital. Enzymes in the smooth ER render the drug molecules more water soluble so that they can eventually be excreted in the urine.

Lysosomes: The Cell's Recyclers

Cellular debris—dead cells burst by HIV particles, for example, or worn-out cell parts—needs to be cleaned up and recycled. Detritus like this is often devoured by large white blood cells acting like custodians (review Fig. 2.20). The cell takes in the refuse and surrounds it with a vesicle that fuses with a little digestive organelle called a **lysosome** ("loosening body"). **Lysosomes** are tiny spherical bags of powerful digestive enzymes that can digest invading bacteria or debris the cell has engulfed, or cell parts that have worn out internally. Digestive enzymes in the lysosome generally break down the refuse into smaller molecules, which then reenter the cytoplasm for reuse by the cell.

Lysosomes also sometimes act as seemingly self-destructive "suicide bags" that break open and spill their contents, literally digesting entire damaged or aged cells from the inside out. This process is sometimes programmed during development and is especially dramatic during the transformation of a tadpole with a tail into a tail-less frog. One by one, the tail cells die and are reabsorbed, making way for the adult, with its new form.

Cytoskeleton

The **cytoskeleton** is a three-dimensional structure of thin protein fibers forming a lattice throughout the cytoplasm, suspending the organelles and allowing cell parts to move. Protein fibers in the cytoskeleton also act as internal girders and cables that help maintain a cell's shape and transport cell organelles, including vesicles, much as skiers are pulled along on moving rope tows. Figure 2.26c shows some of the fibers that make up the cytoskeleton.

The cytoskeleton figures prominently in Dan's form of AIDS. Dan was infected with a particularly nasty variant of HIV that causes infected cells to fuse with other cells until they form a huge cell a hundred times bigger than normal, with many nuclei instead of one (Fig. 2.26a). Remarkably, these colossal cells can crawl through the body using a large, clear "false foot" (pseudopodium) that extends forward from the cell while the nuclei

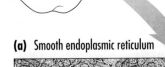

Figure 2.25 Some Bedoin Women Have a Smooth ER Problem. (a) In an electron micrograph of a testes cell, the smooth endocytoplasmic reticulum (SER looks like a series of smooth tubes made of lipid membrane. (b) Because this woman's clothing leaves little or no skin exposed to sunlight, her smooth ER may not be able to make enough of the vitamin D necessary to maintain strong, healthy bones.

(a) Smooth endoplasmic reticulum

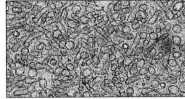

(b)

(a) Huge cell from Dan's variant of HIV

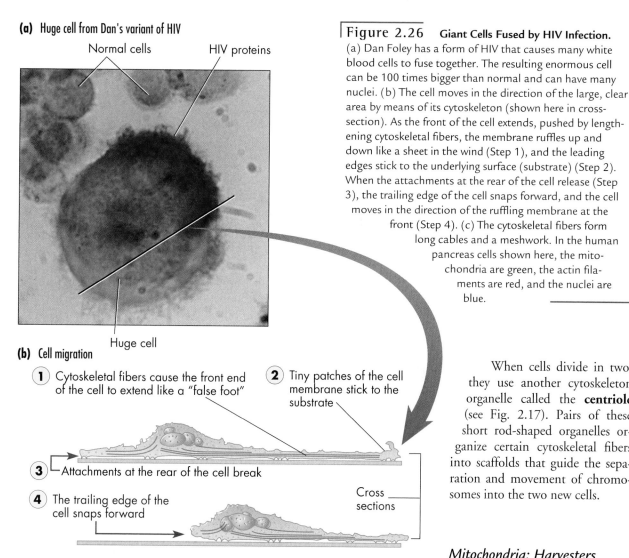

Normal cells   HIV proteins

Huge cell

(b) Cell migration

1. Cytoskeletal fibers cause the front end of the cell to extend like a "false foot"

2. Tiny patches of the cell membrane stick to the substrate

3. Attachments at the rear of the cell break

4. The trailing edge of the cell snaps forward

Cross sections

(c) Cytoskeletal fibers

accumulate in the rear (Fig. 2.26b). The fibers of the cytoskeleton allow the movement of these giant cells, as well as the movement and shape of normal cells.

Figure 2.26   **Giant Cells Fused by HIV Infection.** (a) Dan Foley has a form of HIV that causes many white blood cells to fuse together. The resulting enormous cell can be 100 times bigger than normal and can have many nuclei. (b) The cell moves in the direction of the large, clear area by means of its cytoskeleton (shown here in cross-section). As the front of the cell extends, pushed by lengthening cytoskeletal fibers, the membrane ruffles up and down like a sheet in the wind (Step 1), and the leading edges stick to the underlying surface (substrate) (Step 2). When the attachments at the rear of the cell release (Step 3), the trailing edge of the cell snaps forward, and the cell moves in the direction of the ruffling membrane at the front (Step 4). (c) The cytoskeletal fibers form long cables and a meshwork. In the human pancreas cells shown here, the mitochondria are green, the actin filaments are red, and the nuclei are blue.

When cells divide in two, they use another cytoskeleton organelle called the **centriole** (see Fig. 2.17). Pairs of these short rod-shaped organelles organize certain cytoskeletal fibers into scaffolds that guide the separation and movement of chromosomes into the two new cells.

Mitochondria: Harvesters of Energy

AIDS patients like Dan who take AZT sometimes feel weak and fatigued because of the medicine. These unfortunate side effects arise from the unique design of the **mitochondria** (sing., mitochondrion), cellular organelles that harvest energy from food molecules. We'll return to AZT's effects shortly.

Mitochondria provide chemical fuel for cellular activities such as building proteins, copying DNA, and moving cells and cell parts. They break down carbon-containing molecules and release energy packets, or ATP (see Fig. 2.12b), that quickly diffuses throughout the cell and fuels life processes. These chemical conversions require oxygen—the oxygen you breathe—as Chapter 3 explains.

As the root words suggest, mitochondria (Greek, *mito* = thread + *chondrion* = small grain) can indeed look like grains of rice or, at lower magnification, like

Figure 2.27 **Mitochondria: Power for the Cell.** (a) Mitochondria help eukaryotic cells harvest energy from foods. The mitochondria are stained red in this fluorescent image of a cell that lines the artery walls. (The nucleus of the cell is stained purple; the cytoskeleton, green.) (b) Higher magnification shows that mitochondria have internal membranes; upon these sheets the cell's energy currency, ATP, is generated. Mitochondria contain their own genetic material, DNA (not visible here).

(a) Cells from artery lining

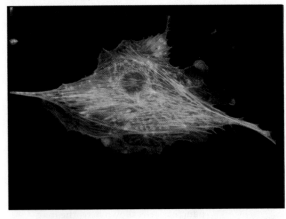

(b) Mitochondrion

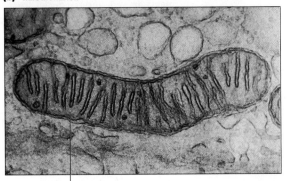

Mitochondrion

tiny threads (Fig. 2.27). Mitochondria have a semiautonomous existence in a cell: They have their own DNA and make some of their component proteins on their own ribosomes. The cell's DNA encodes the rest. They can also divide in half independently of the cell's normal division cycle. Mitochondria resemble bacteria in many ways, and this evidence convinces most biologists that the energy organelles originated first formed more than 2 billion years ago when larger cells engulfed ancient bacteria.

Each of us got all or virtually all of our mitochondria from our mother's egg, not our father's sperm, and they from their mothers, and so on back in time. This has led to the controversial notion of "mitochondrial Eve," the African woman from whom all people now alive supposedly descended.

Let's return to AZT and Dan's health. Recall that AZT blocks reverse transcriptase, the copying enzyme that translates HIV's RNA genes into DNA. Experiments show that AZT can also inhibit the enzyme that copies mitochondrial DNA before the organelle divides. Just as this blockage slows HIV replication, it also slows

mitochondrial DNA. This, in turn, may interfere with the way the energy organelles function and lead to the weakness and fatigue of AIDS patients taking AZT.

Specialized Organelles

So far, all the cell structures we've talked about can be found in human cells and in most other eukaryotic cells, and are commandeered by HIV and contribute to AIDS. No exploration of cell biology would be complete, however, without examining a few specialized organelles for movement and storage found only in certain cell types and that play no known role in HIV or AIDS. Let's discuss these before returning to HIV and the cell one last time.

Organelles of Cell Movement.   Some cells, such as the glowing green *Euglena* that live in pond water, pursue sunlight; others, like certain white blood cells, stalk and engulf invading microorganisms. For these and many other mobile cells throughout the kingdoms of life, movement is made possible by propelling extensions of the cytoskeleton. *Euglena* have a fine, whiplike organelle called a **flagellum** (Latin, small whip; pl., *flagella*) that extends from the cell surface and undulates, pulling or pushing the cell through its liquid medium (Fig. 2.28a–c). Many kinds of sperm rely on propulsion by flagella; human sperm can't fertilize an egg without a functional flagellum. Some Maori tribesmen from New Zealand, for example, are sterile because of a genetic defect that keeps their sperms' flagella from lashing properly.

Certain single-celled protists have thousands of projections called **cilia** (Latin, eyelashes; sing., *cilium*) that look and act much like flagella. Cilia beat in concert like the oars of a medieval galley ship, allowing the cell to swim quickly. In cells that line the human breathing passages, cilia sweep dust particles out toward the mouth and nose, where they are eventually expelled in mucus or swallowed. Cigarette smoke paralyzes these cilia, contributing to "smoker's cough" and greatly increasing a smoker's chances of developing lung cancer because damaged, inactivated cilia can't sweep mucus and harmful particles out of the lungs.

Links Between Cells.   Most animal cells are attached to neighboring cells by links called **intercellular junctions,** which help weld cells together into functional tissues and organs (Fig. 2.29a). These junctions also allow free communication between cells and the coordination of cells in tissues and organs. Some linkages allow materials to flow between cells; others prevent leaks between cells and help organs like the urinary bladder hold fluid.

Figure 2.28 **Flagella: Cellular Oars.** (a) A vivid green Euglena cell is propelled through pond water by waves moving down its flagellum. The flagellum contains a ring of microtubules surrounding an inner pair of microtubules (b), and when the microtubules move with respect to each other by means of the cross links visible in (c), the flagellum bends.

(a) A swimming *Euglena*

(b) A portion of a flagellum

(c) EM of a cross section of a flagellum

Flagellum

Microtubules

Plasma membrane

People with genetically defective junctions often lose large patches of skin from the slightest scrape.

Junctions at the base of many cells attach an external meshwork of fibrous proteins or **extracellular matrix** to polysaccharides that surround and support the cell and glue it to adjacent cells (Fig. 2.29b). The most common of these fibrous proteins is collagen, which has stiff, ropelike polypeptide chains wound around each other into fibrils. Collagen molecules constitute 25 percent of all the protein in a typical mammal, including most of the material in tendons, the cables that enable muscles to move bones. You can feel the tendons that run behind your knee by bending your leg and pulling your heel backward. Those cords are almost entirely made up of proteins in the extracellular matrix.

Specialized Organelles in Plant Cells. Plant cells have some special organelles that account for the successful stationary, light-harvesting life style of our neighboring green organisms. (*E* 2.2 compares plant and animal cells).

Plastids are oval organelles surrounded by a double-layered membrane. They harvest solar energy,

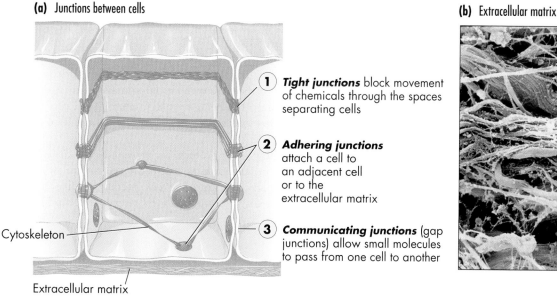

(a) Junctions between cells

1. **Tight junctions** block movement of chemicals through the spaces separating cells

2. **Adhering junctions** attach a cell to an adjacent cell or to the extracellular matrix

3. **Communicating junctions** (gap junctions) allow small molecules to pass from one cell to another

Cytoskeleton

Extracellular matrix

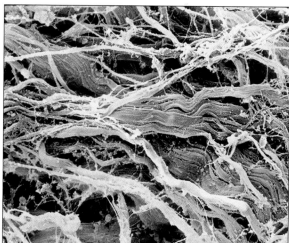

(b) Extracellular matrix

Figure 2.29 **Links Between Cells.** (a) Three types of junctions allow linkage and communication between cells. (b) The extracellular matrix is a meshwork of protein and carbohydrate fibers that cushions the cell and strengthens the tissue of which the cell is a part. This false-colored scanning electron micrograph reveals collagen and elastin fibers surrounding an animal cell.

manufacture nutrient molecules, and store materials. The plastids called **chloroplasts** ("green plastids") are membranous organelles that trap the energy of sunlight in a chemical form—generally, sugar molecules—in the process of photosynthesis (see details in Chapter 3). All animals and fungi, as well as most of Earth's other organisms, depend, ultimately, on photosynthesis—directly or indirectly—to supply the nutrients, materials, and energy compounds they need for survival.

Chloroplasts share several features with mitochondria, an organelle that plant cells also have. Both have an outer membrane and a convoluted inner membrane (see Fig. 2.17b). Both organelles house their own DNA molecules and make some of their own proteins. And biologists suggest that chloroplasts, like mitochondria, evolved from primordial bacteria that came to inhabit larger eukaryotic cells. A major difference, though, is that the convoluted innermost set of chloroplast membranes contains the green, light-absorbing pigment **chlorophyll.**

Other types of plant cell plastids contain yellow, orange, and red energy-trapping pigments; these give color to many fruits, vegetables, and flower petals, and are among the cancer-fighting plant compounds now recommended for our daily diets. Yet other plastids are colorless and store starch granules and sometimes proteins or lipids. Potatoes, for example, contain billions of starch-storing plastids.

Many plant cells have large fluid-filled sacs in the center called **vacuoles.** Plant vacuoles contain water and nutrients; they have a single surrounding membrane and can fill 5 to 95 percent of the total cell volume. That's a lot of space, and in fact, taking up space is the vacuole's main role. Essentially a bag of water that fills up a plant cell, the vacuole presses a small amount of cytoplasm and all the cell's organelles into a thin layer (see Figure 2.17b) with a favorable surface-to-volume ratio. The visible cells in an orange, we saw earlier, consist mostly of a vacuole filled with sweet juice. Full vacuoles keep plant cells plump and give firm shape to the leaves, stems, and other structures. The next time your houseplant wilts, you'll know that its vacuoles need refilling.

Cell walls surround most plant cells just outside the plant cell's plasma membrane. Cell walls are composed largely of cellulose. They remain stretchable and flexible until the cell has stopped growing and has begun to mature, then they harden. The porous wall allows water, gases, and some solid materials to pass through to the plasma membrane, and specialized junctions between plant cells called **plasmodesmata** (Fig. 2.17b) facilitate cell-to-cell communication. If you are

reading this book at a wooden desk, your work is being supported by thickened, dead cell walls—all that remains of a once-living tree. If you are dressed in a cotton garment today, you are wearing the spun and woven cell walls of little hairs that grew from a cotton plant's seeds.

Some mushrooms, protozoa, and algae have cellulose in their cell walls along with a wide variety of other molecules. Bacteria also secrete cell walls, but these aren't cellulose and are instead made up of complex materials that include sugars, lipids, and amino acids. The cell walls of archaebacteria contain unusual polysaccharides, which is one reason why they are classified in a distinct domain of life.

2.8 How Does HIV Complete Its Infective Cycle?

Our complete tour of cell parts puts us in position, now, to return to HIV and understand its whole cycle of infection; how it involves the biological molecules and cell parts we discussed in this chapter; how it harms a patient like Dan; and how the drugs called protease inhibitors are helping many people with AIDS.

We've already seen:

- How HIV works its way into the cell by linking with a receptor and a helper on the cell surface (Fig. 2.30 Step 1).
- How the viral RNA is copied onto DNA (Step 2).
- How that DNA enters the nucleus and is integrated into the cell's own chromosomes, which leads to new viral RNA being formed (Step 3).
- How the cell makes viral proteins on its own ribosomes (Step 4).
- How these proteins are moved along, tagged for export, and transported to the cell surface, where they bulge out from the cell membrane like the same "beads" that stud the HIV virus itself (Step 5).

We haven't seen, however, how new viruses are assembled. This last step is crucial because it provides the target for the protease inhibitor drugs that helped Dan return to relative health. Here, then, is the final part of the HIV infection story.

Assembly of new HIV particles involves the proteins tagged for export and now bulging out of the cell's surface, but it also requires additional large viral proteins. These are produced on cellular ribosomes floating

HIV

① Entry

Viral RNA

Reverse transcriptase

② Viral RNA copied to viral DNA

Integrase

③ Viral DNA integrates into cell chromosomes and makes more viral RNA

Viral RNA

④ Synthesis of HIV proteins

⑥ Protease cleaves large proteins into smaller ones

⑤ HIV envelope proteins come to cell surface

⑦ HIV assembles and buds from cell

free in the cytoplasm (not attached to ER membranes). After they're made, the active viral protein called protease cuts the big proteins into smaller proteins (Step 6), including those that form the inner containers (the matrix and capsid; see Fig. 2.1) of new viral particles. These proteins eventually surround the viral RNA, move toward the surface, and help assemble hundreds, even thousands of complete viral particles (Step 7). Eventually the new particles bud off from the cell membrane, sometimes killing the cell, moving through the surrounding blood or tissue fluid, and landing on new

cell surfaces to begin new rounds of infection. The death of cells helps explain why Dan's white blood cell count fell so low, leaving him vulnerable to infections and to the purplish Kaposi's sarcoma tumors. And the role of protease in chopping large proteins into smaller ones explains why blocking that process with protease inhibitor drugs helped Dan and so many other AIDS patients so dramatically. In *E*xplorer 2.4, you can see the whole HIV life cycle and help design the next generation of anti-HIV drugs, based on what you now know about cells, viruses, and AIDS.

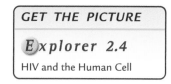

GET THE PICTURE

*E*xplorer 2.4

HIV and the Human Cell

Chapter Summary and Selected Key Terms

Introduction   Dan Foley had momentary contact with the **human immune deficiency virus (HIV)** (p. 22) and he contracted **acquired immune deficiency syndrome,** or **AIDS** (p. 22). Dan's story and HIV serve as a good introduction to the molecules of life, to the structure of the cell, and to the differences between living cells and nonliving virus particles.

What Is HIV?   HIV is a nonliving particle that infects human cells. It is made up of the same groups of **biological molecules** (p. 23) in all living things: carbohydrates, lipids, proteins, and nucleic acids, but it lacks metabolism and responsiveness, and it cannot reproduce independently. HIV has an inner sphere of protein, the **matrix** (p. 24), and within the matrix, a **capsid** (p. 24) containing enzymes.

What Are Atoms?   Pure substances that can't be broken down further into different chemical constituents are **elements** (p. 24), 89 of which occur in nature. The ele-

ments hydrogen, oxygen, and carbon make up 98 percent of most animal and plant bodies. Elements are composed of identical particles or **atoms** (p. 25), whereas **molecules** (p. 25) are combinations of two or more atoms. Atoms are composed of **protons** (p. 25), positively charged particles; **neutrons** (p. 25), particles with no charge; and **electrons** (p. 25), negatively charged particles. At the atom's center is the **atomic nucleus** (p. 25), which contains protons and neutrons. Orbiting the nucleus in **energy shells** (p. 26) are a number of electrons equal to the number of protons. This atomic structure and the unique number of particles in the atoms of each element give atoms their physical and chemical properties and in turn give elements their characteristics. **Isotopes** (p. 26) of the same element are atoms with the same number of protons but different numbers of neutrons. Carbon-dating is based on isotopes of carbon. In **ions** (p. 26), electron numbers vary, so the whole atom has a positive or negative charge. The strengths of acids are based on the levels of hydrogen ions.

How Do Atoms Form Molecules?

The atoms in molecules are linked by **chemical bonds** (p. 27). Atoms share electron pairs in **covalent bonds** (p. 27). When a molecule has an equal sharing of charge at both ends, it's called **nonpolar** (p. 27); when the sharing is unequal, as in the water molecule, it's called **polar** (p. 27). Polar molecules can form **hydrogen bonds** (p. 27), such as occur in the ice lattice. Hydrogen bonds also occur between HIV particles and human cells. Salt (NaCl) has **ionic bonds** (p. 27), or the attraction between oppositely charged ions. This follows the transfer of electrons between atoms rather than the sharing of electrons.

What Makes Water Special for Life?

Dan Foley takes drugs that target the viral protein called **protease** (p. 29), which depends on a water molecule held in place by hydrogen bonds in a strategic place in the proteins. Water's tendency to form hydrogen bonds allows the molecules to stick to each other, to soil, to glass, and to other substances and explains how water can "climb" inside plants (p. 29). Water can form a "skin," it can dissolve other substances (p. 29), and it holds heat more stably than many other compounds. It also tends to "break down" into H^+ and OH^- ions, and this contributes to forming an **acid** (p. 29) or a **base** (p. 29), which are measured on the **pH scale** (p. 29).

What Are Biological Molecules?

The human body is 18 percent carbon by weight; a tree, 50 percent carbon. Carbon can form covalent bonds with up to four atoms, and can form chains and rings (p. 31). Carbon-containing or organic molecules often contain **functional groups** (p. 31) that lend special properties to the molecules and help explain why the large biological molecules form and act as they do. Carbohydrates contain multiples of CH_2O and include the **monosaccarides** (p. 32), the **disaccharides** (p. 32), the **polysaccharides** (p. 32) such as **starch** (p. 33), **glycogen** (p. 33), **cellulose,** and **chitin** (p. 33). Long chains of

similar subunits are **polymers** (p. 33); each subunit is a **monomer** (p. 33). Proteins contain **amino acids** (p. 33) and have complex three-dimensional shapes including regions of **alpha-helices, pleated sheets,** and **disordered loops** (p. 33). There are 20 types of amino acids, which join via **peptide bonds** (p. 34) to form **polypeptides** (p. 34). The order of the amino acids and the overall shape of the molecule determine its properties. Lipids can be **fats** (p. 35), **oils** (p. 35), **waxes** (p. 35), **steroids** (p. 35), or **phospholipids** (p. 36), and contain saturated and/or unsaturated **fatty acids** (p. 35). **Triglycerides** (p. 36) are the main storage form in plant and animal cells. **Nucleic acids** (p. 36) include **DNA** and **RNA** (p. 36), which store genetic information. Nucleic acids are made up of **nucleotides** (p. 36) in an order that encodes information for protein structure. **ATP** (p. 37) is an important nucleotide that stores and tranfers energy within the cell. **AZT** (p. 37) is a chemically modified nucleotide that can prevent HIV from replicating.

Living Cells Versus Nonliving HIV

Robert Hooke first saw a **cell** (p. 38) in 1663. Cells contain **organelles** (p. 39) that carry out specialized functions. The small size of cells insures a beneficial surface-to-volume ratio and adequate speed of import and export. The 18th-century **cell theory** (p. 40) set out the basic roles and properties of cells. **Eukaryotic cells** (p. 40) contain DNA in a distinct nucleus, whereas **prokaryotic cells** (p. 40) have DNA loose in one region of the cell.

Cell Parts and HIV Sabotage

The flexible outer **plasma membrane** (p. 39) acts as a boundary separating the cell's contents from the surrounding environment. This membrane is often studded with proteins and carbohydrates (p. 43) and HIV uses these to enter the cell. Cells are **selectively permeable** (p. 43), allowing nutrients, wastes, and viruses to pass through by **diffusion** (p. 44) and/or **osmosis** (p. 44), but barring most other substances. Sometimes energetic pumping called **active transport** (p. 44) is involved. Sometimes, membrane proteins (carriers) assist transport via **facilitated diffusion** (p. 44). Sometimes the cell engulfs materials via **phagocytosis** (p. 44) (a type of **endocytosis** [p. 44]) or expels materials via **exocytosis** (p. 45).

The **nucleus** (p. 47) is roughly spherical and contains genetic information. It is bounded by a **nuclear envelope** (p. 47) perforated by **nuclear pores** (p. 47),

through which molecules can pass. DNA is packed into microscopic structures called **chromosomes** (p. 48). Information flows from DNA to RNA to protein. **Nucleoli** (p. 48) in the nucleus form one type of RNA. The **cytoplasm** (p. 39) is a semifluid ground substance acting as a pool of raw materials and containing a number of organelles. Within the cytoplasm, among other things, are thousands of **enzymes** (p. 48), highly active proteins that speed biochemical reactions. **Ribosomes** (p. 48) in the cytoplasm act as workbenches upon which new protein molecules are forged. The cells have an internal membrane system called the **endoplasmic reticulum** (p. 49) or **ER**. The **rough ER** (p. 49) is studded with ribosomes and makes proteins for export. The **Golgi apparatus** (p. 50) can modify proteins for export. The **smooth ER** (p. 50) makes lipids and detoxifies substances that can dissolve in lipids. **Lysosomes** (p. 51) are tiny bags of digestive enzymes. The **cytoskeleton** (p. 51) is a three-dimensional lattice structure throughout the cytoplasm involved in suspending other organelles and in cell movement. The **centriole** (p. 52) organizes cytoskeletal fibers and helps move chromosomes during cell division.

Mitochondria (p. 52) harvest cellular energy from food molecules. The **flagellum** (p. 52) extends from the cell surface and undulates, pushing or pulling the cell through its liquid medium. **Cilia** (p. 53) have a similar structure but occur in the hundreds or thousands and beat in concert. **Intercellular junctions** (p. 53) help hold cells together and allow free communication between cells and coordination of cells in tissues. The **extracellular matrix** (p. 53) is an external meshwork that surrounds and supports the cell and can act like glue. Plant cells have **plastids** (p. 53) that harvest solar energy, manufacture nutrient molecules, and store nutrients. **Chloroplasts** (p. 53) contain the pigment **chlorophyll** (p. 53) and trap the energy of sunlight in photosynthesis. A **vacuole** (p. 54) is a central bag of water and nutrients that fills up much of a plant cell. **Cell walls** (p. 54) made up mostly of cellulose surround most plant cells, just outside the plasma membrane.

The HIV Infective Cycle

HIV works its way into the cell through a receptor and helper; its viral RNA is copied into DNA; this DNA enters the nucleus and integrates into the cell's own chromosomes; the cell starts making viral proteins; these are transported to the cell surface as new viral particles; they bulge out and bud off from the cell membrane; and the particles finally move toward new cells to infect.

Test Yourself on Vocabulary and Concepts

In each question set below, match the description with the appropriate term. A term may be used once, more than once, or not at all.

SET I

(a) chemical (b) covalent (c) ionic (d) hydrogen (e) polar (f) nonpolar

1. The basic kind of bond that joins elements together to form compounds.
2. A molecule in which the constituent electrons tend to be more concentrated at one part of the molecule than at another.
3. A weak chemical bond that makes possible life as we know it.
4. The kind of bond that is responsible for most of the special properties of water.
5. The kind of bond that is formed by the sharing of electrons between two atoms.
6. A structural character of the water molecule due to the distribution of its orbital electrons.
7. The kind of bond that is formed between two atoms when one of them transfers one or more of its electrons to the other.

SET II

(a) carbohydrate (b) lipid (c) protein (d) nucleic acid

8. DNA
9. $C_6H_{12}O_6$
10. A polymer of amino acids
11. Triglycerides
12. Chitin
13. Enzyme
14. Can become modified to form ATP.

SET III

(a) ribosomes (b) cytoskeleton (c) smooth endoplasmic reticulum (d) rough endoplasmic reticulum (e) Golgi apparatus

15. This structure synthesizes some lipids and vitamin D.
16. This protein component of cells moves several other organelles around in the cell and also helps cells to maintain their shape.
17. This structure comprises a system of intracellular channels through which proteins travel and often are modified as they move.
18. These beadlike structures are composed of nucleic acids and proteins and are the site of protein synthesis.
19. This structure modifies, "packages," and releases for distribution molecules of protein, lipid, and/or carbohydrate.

SET IV

(a) nucleolus (b) lysosome (c) mitochondrion (d) cilium (e) collagen

20. The most abundant fibrous protein in animals; produced intracellularly but found within the extracellular matrix.
21. Contains a special type of RNA that is required for cellular synthesis of proteins.
22. Permanent cytoplasmic vesicles that contain digestive enzymes used to destroy intracellular debris.
23. Cellular projections that either help to move a cell through its environment or to move the environment over the cell.
24. Makes some of its own protein, is passed down from mother to offspring, and is the major site of aerobic metabolism in cells.

Integrate and Apply What You've Learned

25. Why would life not be possible if cells were bounded by a layer of completely water-soluble molecules instead of by phospholipids?
26. List the functions of proteins, and explain why this category of biomolecules is so versatile.
27. How do carbohydrates, lipids, proteins, and nucleic acids differ with respect to constituent atoms and functional groups?
28. Cellulose, starch, and glycogen are all polymers of glucose. How can you account for their different properties?
29. Could a single-celled, ameba-like creature ever be as big as a truck? Explain.
30. What are the *minimum* requirements for something to qualify as a cell?

Analyze and Evaluate the Concepts

31. How is the AIDS virus different from the human cell it infects? What makes the virus nonliving but the cell living?
32. Golds of different carat weight have different physical properties. For example, a 24-carat gold ring is shinier and much softer than an otherwise identical ring made from 14- or 10-carat gold. How can you explain this?
33. If you constructed a linear protein from 500 amino acid molecules, how many molecules of water would be produced?
34. If you wanted to observe as much detail as possible of the surface of a mitochondrion, which type of microscope would you choose and why?
35. Let's say that you are a family physician, and you have noticed that during the last few days many more patients have complained of stomach cramps and intestinal "flu" than you would expect for this time of year. You take samples from these patients and isolate a single-celled organism that has a nucleus, but few other membrane-bound bodies within its cytoplasm. To treat these patients, will you prescribe drugs effective against prokaryotic or eukaryotic organisms? Defend your decision to one of your patients.
36. The head of your local blood bank decides that blood supplies would be much more versatile if researchers could change blood types A and B into blood type O, the universal donor. You decide to take on the challenge and develop a way to change blood cell types. What part of the cell and what type of molecule would you alter?

The Highly Improbable Hummingbird

▲
Lessons in Metabolism: Banding a hummingbird; ready to fly.

A hummingbird hovering close by can make your heart race. That's partly because its loudly vibrating wings simulate the sudden dive-bombing of a giant honeybee, and partly because, in direct sunlight, the bird's metallic throat feathers can flash like 10-carat amethysts, rubies, or sapphires. For an instant, you don't know whether to swat this buzzing jewel or stare at it in wonder. But the more you learn about hummingbirds, the clearer the choice becomes. These tiny birds are equal parts improbability and superlative, and their vital statistics are truly dazzling:

- Here is an animal that weighs less than a dry tea bag or a plastic credit card yet can migrate 800 km nonstop across the Gulf of Mexico.
- Here is the only bird that can fly backward as well as suspend itself in mid-air, 1000 times a day or more, to sip nectar from flowers. Breaching whales are certainly impressive, and so are rampaging elephants and salmon leaping upstream. But the simple act of hummingbird hovering is the hardest, most energy-costly work any vertebrate (animal with a backbone) can perform.

- A hummingbird's miniature physique supports this mid-air suspension: Its short wings beat 40 to 80 times *per second*, depending on species, and its flight muscles make up one fifth to one third of its total weight. The flight muscle cells are jammed with so many power organelles (mitochondria) that any more would prevent normal muscle contraction.

- Proportionate to its body size, a hummingbird's heart is twice as big as any other bird's, and beats far faster—up to 1260 beats per minute. A breathing rate of 250 breaths per minute supplies this organic pump with oxygen, which is distributed to the legions of mitochondria through the circulating blood.

- With all its cellular power plants and all its physical activity, a hummingbird's cells burn sugar and fat 12 times faster than ostrich cells and 25 times faster than chicken cells. If you or I had such a blaze going in our cellular "furnaces," we'd have to eat twice our own body weight every 24 hours and our internal temperatures would soar to 400 °C (750 °F)!

It's not surprising that an animal this small, speedy, and powerful could fascinate a zoologist.

And Bill Calder has responded to the call with 30 years of his professional career. A professor at the University of Arizona, Calder spends every summer at the Rocky Mountain Biological Laboratory in Gothic, Colorado. Once the site of an Old West silver mining town, the area is now a reserve of valleys, peaks, and alpine meadows. Every year, the Laboratory acreage explodes with glacier lilies, blue bells, larkspur, Indian paintbrush, and honeysuckle, and the fragrant blossoms lure hundreds of individuals from four hummingbird species: Broad-tailed, Rufous, Calliope, and Magnificent. By gently trapping the birds in fine mist nets, then banding, weighing, and releasing them, Calder and colleagues have learned new and surprising details about these diminutive aerial acrobats. For instance:

- Hummingbirds are masters of weight control, rarely varying a fraction of a gram. The only exception comes before a long migration, when they can pack on an extra 50 to 60 percent of body weight, then expend it within a matter of hours.
- To meet their high energy needs primarily through watery flower nectar, hummingbirds must take in 1.6 to 3.3 times their own body mass in fluid each day. Processing all this water into urine consumes a big chunk of the energy they harvest from food. It also requires countless "perch breaks" during the day for elimination.
- A hummingbird will occasionally take in too little food to get all the way through the night with metabolic furnaces blazing. The bird doesn't feed at night because it requires daylight to see and visit flowers, and night-blooming flowers aren't usually rich enough in sugars for the animal's needs. Nevertheless, the bird's body can compensate by somehow calculating the shortfall ahead of time and triggering the onset of a state something like suspended animation. During this state, body temperature and energy consumption plummet, allowing the bird to survive until sun-up, when it once again refuels at nectar-rich flowers.

- Despite its high-velocity life style, a hummingbird can live a dozen years or more—twice as long as expected for an animal its size.

Calder's discoveries are related to, and require an understanding of, **energetics,** the study of energy intake, processing, and expenditure. So do each of the bird's other amazing adaptations. Energetics is part of our exploration of how animals, plants, bacteria, and other living things acquire energy and use it to fuel the ongoing enterprise of maintaining order and staying alive. Energy metabolism is a complex subject. But it helps us appreciate life's intricate beauty and its important biochemical symmetry and balance. It also allows us to explore other subjects later in the book, including how our cells build the proteins they need, how we digest the food we eat, how we burn that fuel during exercise, and—on a much bigger scale—what we must do to keep our global ecosystems healthy. In the course of this chapter, you'll find the answers to these questions:

1. What universal laws govern how the cells of hummingbirds, flowers, and human researchers gather and use energy?
2. What are the energy routes and carriers in living things?
3. How do oxygen-dependent organisms like hummingbirds harvest energy?
4. How do yeasts, bacteria, and other cells that can survive without oxygen harvest energy through alternative pathways?
5. How do we get the energy for exercise, and how do imbalances in eating and exercising lead to weight problems?
6. How do plants trap solar energy and form sugars and other biological molecules?
7. How does our planet, along with all its living inhabitants, cycle carbon, and how do our human activities affect that cycling?

3.1 What Universal Laws Govern How Cells Use Energy?

Hummingbirds have a problem when it comes to energy use: Their bodies have to be very small in order to hover in front of delicate flowers and thus extract an energy resource not readily available to most other species. Their small size, however, means that they lose heat faster than bigger animals because hummingbirds have a larger surface area relative to their body mass. To maintain a high body temperature and still generate enough energy to hover, their rate of food use and calorie expenditure—for their size, anyway—far exceeds that of any other animal with a backbone (Fig. 3.1). How do they do it? What mechanisms do they share with all other organisms for extracting energy from food? What special tricks do they have up their feathery sleeves that allow them to harvest energy fast enough so they can hover like nature's helicopters? The answers to these questions unfold as we move through the chapter.

All cells need energy to live. That's true whether they are muscle cells driving a hummingbird's wing beat, nectar-producing cells in the red columbine flower the hummingbird sips from, or bacterial cells decomposing leaves that have fallen off the plant. Regardless of type, however, no cell can "make" energy; instead, it must get energy from some outside source in the environment. Let's follow the flow of energy through a mountain ecosystem that includes hummingbirds and the flowers on which they depend (Fig. 3.2.)

Sunlight arriving from space strikes the leaves of the red columbine plant, and they carry out **photosynthesis,** the trapping of energy in a series of metabolic steps that eventually stores energy in the chemical bonds of sugars. The sugar molecules are transported inside the plant, which uses some of the energy to build the delicate, tubular blossoms that attract the hummingbird. Other sugar molecules become concentrated in the flower's nectar. After a hummingbird sips and swallows the nectar, the sugars in the sweet droplets leave the bird's stomach and intestines, flow through the bloodstream, and enter muscle, brain, lung, and other tissue cells. The millions of cells then break down the sugar molecules in the presence of oxygen and release some of the energy in the process called **cellular respiration.** That energy is now available to power the hummingbird's breathing, heartbeat, hovering, and other activities. This represents an energy transformation from solar energy to chemical energy. Such transformations are never 100 percent efficient, however, and some energy is inevitably lost as heat. As a result, the hummingbird warms up as it hovers over one columbine flower then another. The bird's excess body heat dissipates in the cool mountain air around it. Eventually, when the bird dies, bacteria and fungi break down the little body and the rest of the energy still trapped in its tissues fuels the decomposer's activity or is lost as heat. We will explore the processes of photosynthesis and cellular respiration in more detail later in this chapter. Our main message for now is that energy flows from the nonliving physical world to the living world and back, and it is essential for the activities and survival of all living cells. There are a few basic physical laws that underlie energy transactions in nature. Let's look at them now.

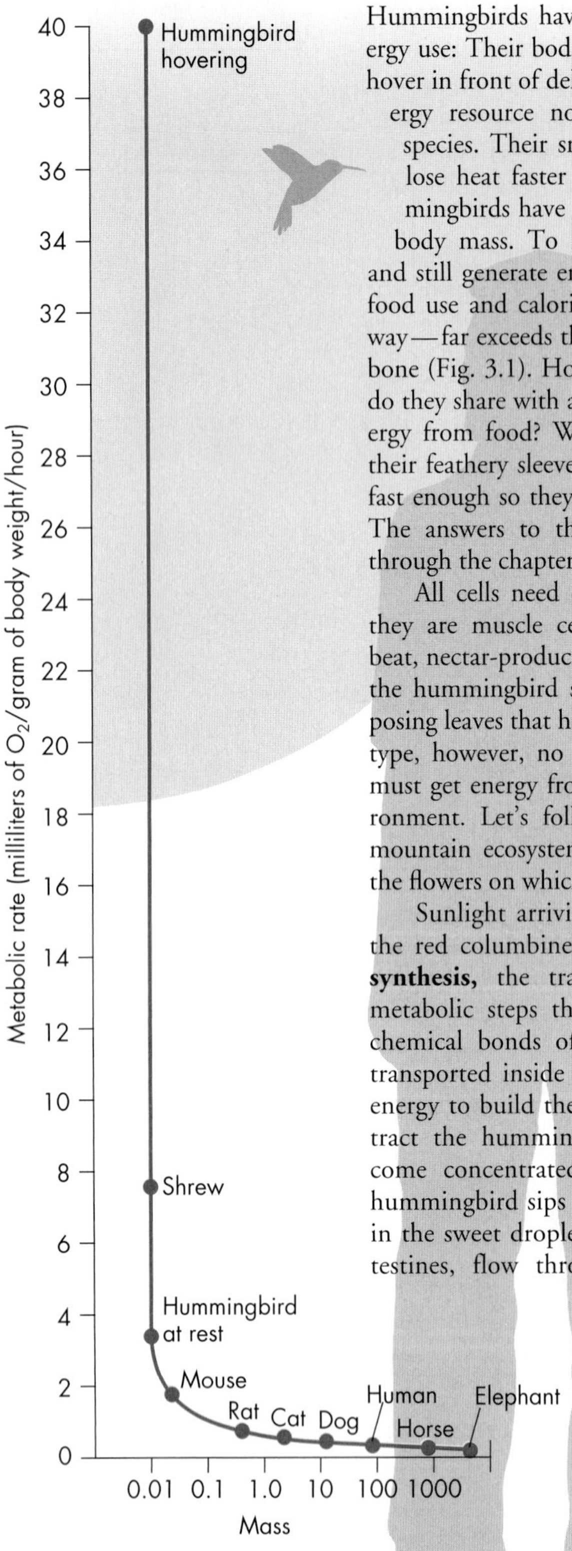

The Laws of Energy Conversions

Ready for an elementary energy demonstration? Lift this book and hold it parallel to the floor at head level. Now drop it. What seemed like nothing but a slam on the floor is really a series of energy conversions, **energy** being officially defined as the ability to perform work or to produce change.

States of Energy

The lifted book contained stored energy in the form of **potential energy**—energy that is available to do work. Likewise, water saved up behind a dam contains stored energy. So do the chemical bonds in firewood and sugar molecules. And so does a hummingbird soaring high in the air. This potential energy can be released and accomplish work, such as turning the turbines of a hydroelectric generator, heating pancakes on a griddle, or

Figure 3.1 **Metabolic Rates Demonstrate Life in the Fast Lane**. Smaller animals use more oxygen per unit of body mass than large animals do. The metabolic rate for a hovering hummingbird is greater than for any other animal with a backbone.

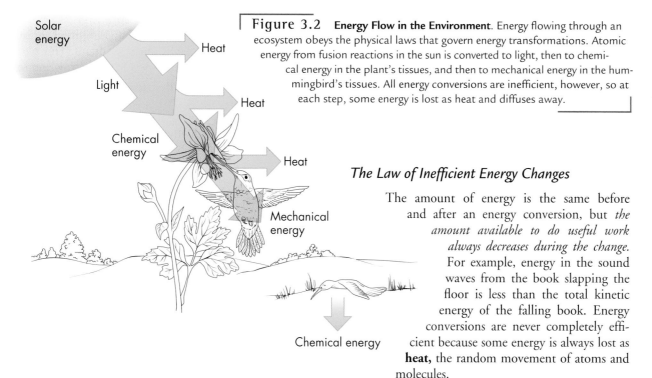

Figure 3.2 **Energy Flow in the Environment**. Energy flowing through an ecosystem obeys the physical laws that govern energy transformations. Atomic energy from fusion reactions in the sun is converted to light, then to chemical energy in the plant's tissues, and then to mechanical energy in the hummingbird's tissues. All energy conversions are inefficient, however, so at each step, some energy is lost as heat and diffuses away.

The Law of Inefficient Energy Changes

The amount of energy is the same before and after an energy conversion, but *the amount available to do useful work always decreases during the change.* For example, energy in the sound waves from the book slapping the floor is less than the total kinetic energy of the falling book. Energy conversions are never completely efficient because some energy is always lost as **heat,** the random movement of atoms and molecules.

Random movement is the opposite of order, so the inefficient conversion of energy from one form into another increases disorder in the system. Scientists use the term **entropy** (EN-tro-pee) to describe the disorder or randomness in a system. The more disorder, the greater the entropy. Anyone who has ever tried to keep a college dorm room tidy is already quite familiar with the concept of entropy: Without a constant input of energy

making a loud vibrating trill in the wing feathers as the male hummingbird plunges toward Earth in his power mating display. When you dropped the book, it began falling and potential energy was transformed into **kinetic energy,** the energy of motion. These two forms of energy, potential and kinetic, are the two major states of energy in the universe, and each can be converted into the other.

The Law of Energy Conservation

As the book fell and hit the floor, energy changed in form but not in amount. This is because *energy is conserved in energy transformations;* it is neither created anew nor destroyed. When the falling textbook loudly met the floor and abruptly stopped moving, the amount of energy released as sound waves in the air, plus the amount of energy that warmed the floor, book, and surroundings, exactly equaled the kinetic energy in the moving book. Likewise, the chemical energy used up as a hummingbird flaps its wings is equal to the energy of motion of the wings, the heat energy dissipated into the environment, and the sound energy of the whirring wings (Fig. 3.3). Because new energy can't be made, and existing energy can't be destroyed, the total amount of energy in the universe remains constant and simply changes forms. That's why scientists have to account for heat when they write equations for energy conversions from one form to another.

Chemical energy from sugar is transformed into...

Kinetic energy (in fluttering wings)

Plus heat energy (dissipated into environment)

Plus sound energy (of whirring hummingbird wings)

Figure 3.3 **Universal Laws of Energy Transformation**. Energy is neither created nor destroyed but can change from one form to another. Hummingbird muscles transform the chemical energy stored in sugar molecules into an equivalent amount of kinetic energy (of the fluttering wings) plus heat energy (escaping from the bird's body and warming the cool mountain air) and sound energy (which we hear as the whir of hummingbird wings).

to sweep up dust and put away clothes and books, entropy (disorder) tends to increase, and things get disorderly fast.

Energy Laws and Living Organisms

The energy laws have profound consequences for organisms and ecosystems. Let's go back to hummingbirds in a mountain meadow (Fig. 3.2). When sunlight enters the earth's atmosphere and strikes the leaves of a red columbine plant, most of the energy reflected back as heat simply radiates away into the atmosphere in a low-quality form unusable to living things. Leaves absorb a small amount of light energy and convert it into chemical energy stored as starch, nectar, and other molecules, but during this process, still more heat is lost. Finally, a hummingbird sips nectar from the columbine flower, and the bird breaks down the plant's complex molecules into simpler ones, releasing energy that is used to fuel life processes such as hovering, flying, perching, and sleeping. As always, the conversion is inefficient, and energy wastefully radiates into the air as body heat from the warm-blooded bird.

We can interpret the energy laws in a very immediate biological way. It takes many pounds of sugar in meadow wildflowers to make a few ounces of hummingbird tissue. That's because most of the energy consumed by the bird warms its body and is eventually lost as heat to the surrounding air. The same holds true when a hawk swoops down and eats the hummingbird. That's why a forest ecosystem—the living and nonliving components of the forest—can support many more pounds of plants than of hummingbirds, and many more pounds of hummingbirds than of hawks. What would the energy laws imply for the human diet? The answer is captured in a long-standing catch-phrase of the ecology movement, "Eat low on the food chain." In other words, according to the energy laws, our global environment will be stabler if we eat more plant products (fruits, vegetables, grains) and fewer plant-eaters (chicken, beef, and so on; see Chapter 25).

Cells and Entropy

The law of inefficient energy transformations most directly affects an organism's cells. Cells remain healthy only if they obtain enough energy to fuel all their synthesizing and repair activities, over and above the loss of energy to inefficiency and heat. A living cell is just a temporary island of order supported by the cost of a constant flow of energy. If energy flow is impeded, order quickly fades, disorder reigns, and the cell dies. Clearly, a cell's continued life, including its nonstop or-

What do people mean when they say "Eat low on the food chain"?

ganizational activity, bears a steep energy price tag. Hummingbirds have evolved a special mechanism, called **torpor,** that responds to energy drains. This state of torpor allows the animal's metabolism to slow and its body temperature to drop during the night, allowing the remaining energy stores in the tiny bird's body to last until morning, when the animal forages once again. With such a narrow margin between entropy and eternity, the hummingbird is truly living on the edge.

3.2 How Does Energy Flow in Living Things?

If a cell is a temporary island of order, then hummingbirds are temporary archipelagoes with millions of islands, all maintained and orderly—at least so long as the animal is alive. What maintains that order? The answer is chemical energy, trapped and released through chemical reactions. Recall from Chapter 2 that chemical bonds are a type of "energy glue" that joins atoms together. In a **chemical reaction,** energy in chemical bonds shifts, and atoms rearrange, forming new kinds of molecules. As a kid, did you ever play with a toy boat or bottle rocket powered by mixing vinegar with baking soda? The resulting chemical reaction propels the boat around the tub or the rocket over the back fence through the vigorous release of energy. The starting substances, or **reactants,** interact to form new substances, the **products.** In the toy boat, the reactants are hydrogen ions (H^+) from the acetic acid in the vinegar and bicarbonate ions (HCO_3^-) from the baking soda (Fig. 3.4). Chemical bonds break, allowing the OH^- group to leave the bicarbonate ion, which becomes CO_2, or carbon dioxide gas, bubbling in the water. The OH^- combines with a hydrogen ion (H^+), producing H_2O, water. This reaction takes place spontaneously, and releases energy in the form of heat because the energy conversion is inefficient. The entropy (disorder) of the system increases as the carbon dioxide molecules bubble off into the air randomly.

The reaction of vinegar and baking soda satisfies the law of inefficient energy exchanges: The products contain less energy and are more disordered than the reactants. Some reactions release energy like the one we just discussed, but need some energy input before they will proceed, just as a rock poised at the top of a hill needs an accidental shove before it will start rolling downhill and release potential energy. Even with this needed initial energy to get the reaction started, however, the overall result is a *release* of energy. Reactions of

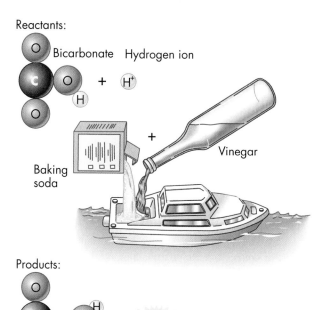

Reactants:

Bicarbonate   Hydrogen ion

O
C   O   +   H⁺
H
O

Baking
soda   +   Vinegar

Products:

O
C   +   O   +   Energy
H   H
O   Water

Carbon
dioxide

Figure 3.4   **Energy-Releasing Reactions: Home Experiments**. Vinegar and baking soda react in an energy-yielding reaction that jets a toy boat across a tub of water or sends a bottle rocket into the air. This reaction transforms reactants into products. A hydrogen ion (H^+) from the acetic acid in vinegar joins with an OH^- group from baking soda's bicarbonate ion, forming water (H_2O). Removal of the OH^- from bicarbonate leaves carbon dioxide (CO_2); the gas quickly bubbles away and generates the force that pushes the boat forward (or the rocket upward). The reaction of vinegar and baking soda releases energy as heat, and the gaseous products are more disorganized than the reactants.

reaction, and so on down a chain of linked reactions called a **metabolic pathway** (more on this shortly).

ATP: The Cell's Main Energy Carrier

Hummingbirds are the superathletes of the bird world, and their cells are vibrant with energy moving from one chemical reaction to another. But what form does that energy take? Light? Heat? Electricity? You'll never see a fast flyer with a long extension cord. That's because it turns out that most of the energy given off during an energy-releasing reaction is quickly trapped in the chemical bonds of a compound that can carry energy from one molecule to another. The most common energy-carrying molecule within a living cell is **ATP (adenosine triphosphate).** As we saw in Chapter 2, ATP is a nucleotide, a two-part molecule with a head composed of three molecular rings and a tail made up of three phosphate groups (PO_4^-, a phosphorus atom bonded to four oxygen atoms) (Fig. 3.6a). Living cells make the energy carrier ATP from **ADP (adeno-**

these two types that proceed spontaneously or need a starting "push" but eventually release energy are called **energy-releasing reactions.**

An entirely different set of reactions called **energy-absorbing reactions** will not proceed spontaneously and do not give off heat. Think about an egg cooking in a pan. The added energy (heat) causes the egg white proteins to form new chemical bonds with each other and to change into a white rubbery solid. Another example is the way red columbine or purple lupine leaves trap energy from sunlight and use it to power the building of carbohydrate molecules in flower nectar, cell walls, pollen, or other plant parts. Energy-absorbing reactions underlie most of the transformations that maintain order in the cell, such as the building of proteins and the replacement of worn-out cell parts.

So where, then, does the energy come from to fuel a cell's huge number of energy absorbing reactions? Whether that cell is in a hovering hummingbird, in an alpine flower, or in a curious scientist, the answer involves a beautiful symmetry: *The power for energy-absorbing reactions comes from the cell's energy-releasing reactions* (Fig. 3.5). The two kinds of reactions are energetically *coupled*, so that the leftover energy of a releasing reaction provides the energy needs of an absorbing

(a) Energy-releasing reactions

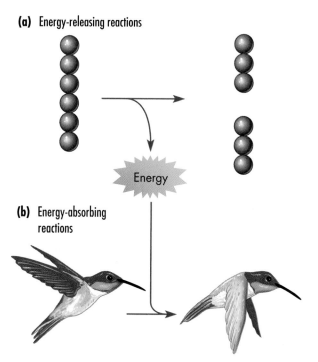

Energy

(b) Energy-absorbing reactions

Figure 3.5   **Energy-Coupled Reactions and Whirring Hummingbird Wings**. Living cells couple energy-releasing reactions to energy-absorbing reactions. The splitting of glucose (a) releases energy. This is quickly coupled to the energy-absorbing reactions that cause proteins in the hummingbird's wing muscles to shorten (b). This causes muscles to contract and the wings to flap.

sine diphosphate), a related molecule with only two phosphate groups (Fig. 3.6b). Cells can use the energy freed from an energy-releasing reaction to add a phosphate group to ADP and thus make ATP, with a higher level of stored energy in its chemical bonds.

Cells can then "spend" their ATP currency, because the release of a phosphate from ATP delivers energy in a controlled way. The cleavage of ATP releases ADP, a phosphate, and a small amount of energy (Fig. 3.6c). A cell can then use that energy to do work. A hummingbird flaps its wings 40 to 80 times per second and the muscle cells that contract to move those wings carry out millions of energy-absorbing reactions every second to keep up with that hard, fast work. An energetic phosphate released from ATP will often transfer to another molecule, energizing that molecule (Fig. 3.6d). When phosphate is split off an ATP, ADP forms; that ADP can itself lose another phosphate group and release more energy plus the molecule **AMP (adenosine monophosphate)** (Fig. 3.6e). Cells only use this second release and formation of AMP when energy demands are unusually high. The significance of ATP

and its less energetic relatives ADP and AMP is that they function as links between energy exchanges in cells.

Let's look at another good analogy for ATP: money. When farmers sell crops and then buy new tools, the intermediate for both exchanges is money. A cell "earns" ATP "coins" during energy-releasing reactions and "spends" them during energy-absorbing reactions. Cells continually break down food molecules, collect the energy in ATPs, and then spend them to offset the staggering energy costs of resisting entropy and staying alive. The human body, with its trillions of cells, cycles an estimated 40 kg (about 88 lb) of ATP molecules into ADP and back again each day, representing quadrillions of chemical reactions and energy exchanges daily. A hummingbird is far smaller, of course, but the exchange of ATP is disproportionately huge because it goes on far faster than in a person's cells to fuel the bird's rapid-fire breathing and heartbeat, its lightning-fast wing beats, and its constant heat generation and loss.

How Enzymes Speed Up Chemical Reactions

Chemical reactions rearrange chemical bonds and release or absorb energy. But how fast does this take place? For example, if you collected some nectar from red columbine blossoms and allowed it to dry in a cool, arid place, the high-energy food (sucrose) would sit there intact for decades, like sugar in a jar, rearranging its chemical bonds very slowly and releasing en-

Figure 3.6 **ATP: The Cell's Main Energy Carrier.** (a) ATP (adenosine triphosphate) consists of an adenine portion and three phosphate groups. (b) Cells make ATP by adding a phosphate group and energy to ADP (adenosine diphosphate). (c) ATP releases energy when its last phosphate is cleaved. This released energy can fuel muscle contraction and allow a hummingbird to fly. (d) If the terminal phosphate transfers to a molecule like glucose, the latter can become "supercharged." (e) Cleaving a phosphate from ADP produces the lower energy molecule AMP (adenosine monophosphate).

(a) ATP

Adenosine

Triphosphate

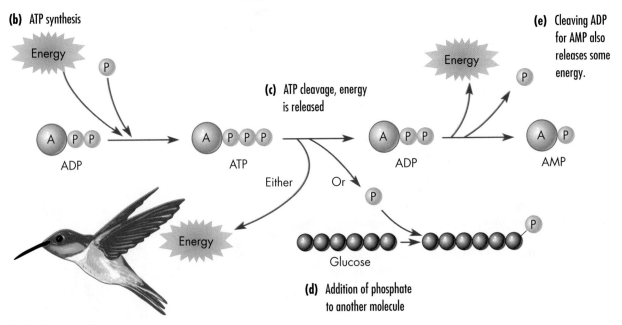

(b) ATP synthesis

Energy

ADP

ATP

(c) ATP cleavage, energy is released

ADP

AMP

(e) Cleaving ADP for AMP also releases some energy.

Either   Or

Energy

Glucose

(d) Addition of phosphate to another molecule

ergy at a nearly undetectable rate. When a hummingbird takes that same nectar into its digestive tract, however, special digestive proteins can break down the chemical bonds in the sucrose within seconds into simpler sugars. Clearly, these proteins can speed up chemical transformations. Let's look at how they do it.

Within living cells, most chemical transformations are facilitated by **enzymes.** Digestive enzymes are just one type. Enzymes are proteins that function as biological **catalysts,** agents that speed up specific chemical reactions without themselves being permanently changed by the reaction. The importance of enzymes becomes clear when we look at the many human diseases caused by defective enzymes. Some children, for example, are born without the ability to synthesize an enzyme needed to change a precursor molecule into vitamin D, a compound needed to build strong, healthy bones. Without active vitamin D, the children develop the disease called *rickets,* characterized by severely curved legs. Fortunately, if these children get enough of the active form of vitamin D in their diets, for example in fortified milk, they can bypass the need to synthesize the enzyme and can still develop and grow normally. Many other human genetic diseases are caused by the lack of individual enzymes.

Why do human cells require an enzyme to convert vitamin D to an active form? Why can't that conversion just occur spontaneously? Actually, vitamin D *does* change spontaneously from an inactive to an active form, but the conversion happens very slowly—so slowly that too little of the active form would be available to ensure strong bones and normal growth. When a specific enzyme is present, however, the reaction occurs far faster than without the enzyme. Without dozens of kinds of enzymes to speed the breakdown of sugars, hummingbirds would not be able to maintain their high metabolic rates.

Enzymes Lower Activation Energy

In chemical reactions, there is an energy barrier separating the reactants and products. For a pair of reactants such as bicarbonate and hydrogen ions to be converted into products such as carbon dioxide and water, the reactants must collide with each other hard enough to break chemical bonds in the reactants and form new bonds in the products. For a fleeting instant during the conversion, chemical bonds in the reactants are distorted like springs being stretched, and this fleeting intermediate state cannot be reached without a very energetic collision between the molecules. We call the momentary, intermediate springlike condition a **transition state.** Because energy input is required to achieve it, the transition state is often characterized as an energy hill or barrier that separates reactants and products (Fig. 3.7).

Most molecules jostling about and colliding randomly lack the energy of motion they need

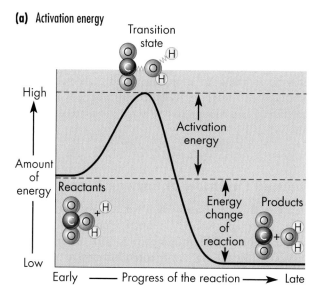

(a) Activation energy

(b) Analogy for enzyme action

Figure 3.7 **Activation Energy and Enzymes.** (a) Activation energy. Reacting molecules must collide with a certain minimum amount of energy before they can reach the springlike transition state. This "activation energy" resembles a hill on an energy graph. From the transition state, the reaction can continue forward, forming the products that have lower energy and/or higher entropy. (b) Enzyme action. Without itself being changed or used up, an enzyme lowers the activation energy as if boring a tunnel through the energy hill. This speeds biological reactions like those in hummingbirds, flowers, or people.

How fast could a hummingbird fly if its cells lacked enzymes?

to overcome the energy barrier; they simply bounce off each other. Some individual molecules, however, do jostle about with enough kinetic energy so that they achieve a productive collision, an impact that generates the springlike transition state, from which reactants change into products. An energy of impact great enough to cause molecules to cross the energy barrier is called the **activation energy** (Fig. 3.7).

In a living cell, most molecules need help in overcoming the energy barrier that prevents the start of many reactions. Heat, of course, speeds up colliding molecules, but this increase of speed requires an increase of temperature. High temperatures would speed *all* the reactions in a cell, not just those special ones that help a cell function. This overall temperature increase would disorganize the cell and kill it. Luckily, enzymes can lower the level of activation energy enough so that biochemical reactions can take place without an increase of temperature. The action of an enzyme, in effect, bores a

tunnel through the energy hill, allowing the reactants to become products without needing to roll like a heavy boulder up and over the high barrier represented by activation energy.

Enzymes not only allow reactions to proceed at the relatively low temperatures compatible with life processes, but they act only on specific reactions. For example, one specific enzyme speeds up the conversion of vitamin D into its active form. Another enzyme speeds up a reaction similar to the one shown in Figure 3.4 which generates carbon dioxide (CO_2). This reaction helps red blood cells transport CO_2 from tissues that produce it rapidly, such as active muscles, to the lungs, from which the waste gas can be exhaled. Each single molecule of this enzyme facilitates this change at the amazing speed of 600,000 times per second, nearly a million times faster than the reaction would occur without a catalyst. Biologists estimate that every second, a hummingbird's blood makes an entire circuit of its heart, lungs, and body and the animal breathes in and out four to eight times. Without this red blood cell enzyme working at near top speed, carbon dioxide produced in working cells could not possibly get transferred fast enough to the lungs to ensure the bird's survival.

Induced Fit Model of Enzyme Action

Enzymes are like virtually all other biological molecules in an important way: Their functioning relates directly to their form. A hummingbird's beak extends and curves in just the right way to collect nectar from the flowers it visits. In the same way, a protein's physical form relates closely to its vital activity.

Like other proteins, each enzyme has its own unique three-dimensional shape, but most share an important feature: a deep groove, or pocket, on the surface called the **active site.** You can see this groove in a computer graphic of the enzyme lysozyme (*lyso* = splitting, *zyme* = enzyme), which is found in the whites of eggs from hummingbirds and other birds (Fig. 3.8). A hummingbird's egg is huge for the size of the bird; the female usually lays a clutch of

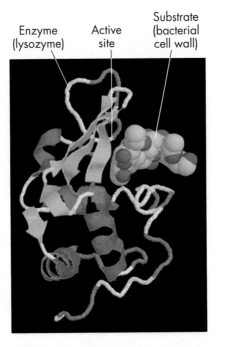

Enzyme (lysozyme)　　Active site　　Substrate (bacterial cell wall)

Figure 3.8　Lysozyme: The Egg White Enzyme. Lysozyme in shelled hummingbird eggs digests and destroys the cell walls of invading bacteria, thereby protecting the enclosed embryos from infection. The enzyme binds to a substrate—the bacterial cell wall (in grey and red)—in its central V-shaped cleft (visible on the right side of this model).

only two eggs and together, they equal about a third of her weight (equivalent to a woman having twins each weighing 20 pounds at birth!). Each egg white contains the enzyme lysozyme, which destroys bacteria that might otherwise infect and kill the incubating chick. The shape of the enzyme's active site fits a specific reactant. The reactant, or **substrate,** fits into the groove rather like a key in a lock forming an **enzyme-substrate complex** (Fig. 3.9a,b). In the case of lysozyme, the substrate would be a cell wall from an invading bacterium. The active site of the enzyme can change shape as the substrate binds to it, improving the "fit" (Fig. 3.9c). (For this reason, biologists sometimes call it the "induced fit" model of enzyme action.) This in turn orients the substrate and subtly changes its shape, straining its bonds and helping it to reach the transition state, where reaction is possible. This bond stretching lowers the activation energy, making it easier for the substrate to react and form products (Fig. 3.9d). In the hummingbird egg, the products of lysozyme action would be dismantled molecules from bacterial cell walls.

Enzymes accelerate a huge number of processes in living cells. Among these processes are the ones that allow organisms to transform energy in food molecules, light, or other nonliving sources into forms of energy that can do work for the cell. These enzyme-catalyzed reactions, along with all the other reactions of the body that sustain life, constitute **metabolism.** Metabolic reactions can break down or build up molecules, and can give off or use energy. As we saw earlier, metabolic reactions are interlinked within cells in metabolic pathways. These pathways allow living cells to subdivide a big chemical change into a number of smaller steps. One such change might, for example, release a tremendous amount of heat. This, in turn, would liberate energy in packets small enough for the cell to use efficiently. In the rest of the chapter, we'll follow the metabolic pathways that cells of hummingbirds, red columbines, people, and other organisms use to extract useful energy from sugars. We'll also examine the pathways that plants like columbines and lupines use to trap the energy of sunlight in chemical form.

3.3 How Do Living Things Harvest Energy?

Some of the busiest molecules in the living world are the enzymes within hummingbird muscles that keep those flight "motors" whirring. They have to be amaz-

ingly active to keep up with the animals' constant energy needs for hovering, flying, defending territory, and staying warm. The energy comes primarily from sugars present in flower nectar. Likewise, one particular enzyme called rubisco that helps capture light energy and store it in sugar molecules is more abundant than any other protein on the planet. In this section, we'll examine the mechanisms that cells use to obtain energy from food molecules. In the next section, we'll go on to see how plants, algae, and some bacteria trap solar energy in energy-rich molecules. These are challenging subjects, but they're central to all that goes on in cells and organisms and crucial to understanding biology. The subject is as personal as how your body is currently utilizing what you ate for your previous meal. The energy pathways within the cell also reflect the remarkable unity of life. Most of the metabolic reactions that go on in your cells also take place in other animal, plant, fungal, protist, and bacterial cells. This is a powerful reminder of life's descent from common ancestors. On a larger scale, the subject of cellular energy harvest is also a key to understanding ecology, including how energy flows in the environment as one organism feeds on another. We'll return to it in Chapter 25.

An Overview of Energy Harvest

A hummingbird lighter than a tea bag can migrate all the way across the Gulf of Mexico without stopping. Think about this animal's exertion as it migrates nonstop, then contrast it with the rapid burst of effort the bird would make as it accelerated to escape a swooping predatory hawk. These exertion levels are similar to you strolling for miles through the fall leaves versus running to catch the bus. They reflect two pathways of energy harvest taking place in animal muscle cells as well as in the cells of fall leaves and all other types of living cells: Energy harvest with plenty of oxygen, the **aerobic pathway,** and energy harvest with minimal oxygen or

(a) Enzyme binds substrate

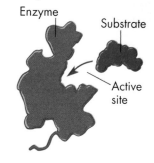
Enzyme
Substrate
Active site

(b) Enzyme-substrate complex

(c) Active site's new shape induces a better "fit"

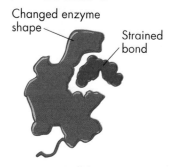
Changed enzyme shape
Strained bond

(d) Enzyme unchanged

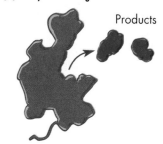
Products

Figure 3.9 **How an Enzyme Works**. (a) Enzymes have an active site that interacts with the substrates of the reaction. (b) The substrate binds to the enzyme. (c) This often induces a slight change in the enzyme's shape that in turn stresses some of the substrate's chemical bonds. (d) In this way, the enzyme facilitates the breaking or making of chemical bonds in the substrate and releases the products. The enzyme emerges unchanged from the reaction.

Figure 3.10 **An Overview of Energy Harvest**. The anaerobic and aerobic pathways of energy harvest involve only four processes: glycolysis, fermentation, the Krebs cycle, and the electron transport chain. They are shown here in the context of a muscle cell from a man's leg as he hikes on flat ground. The anaerobic pathway (orange arrows) requires no oxygen, takes place in the fluid portion of the cytoplasm, starts with glycolysis, and ends with fermentation. The aerobic pathway (yellow arrows) consumes oxygen, it also starts with glycolysis in the cytoplasm, and it is followed by the Krebs cycle and the electron transport chain in the mitochondrion. The aerobic pathway dismantles glucose, produces water, and generates a large number of ATP molecules ready to power cellular activity. The gray arrows show movement of carbon-containing molecules. This and subsequent figures show glucose and its breakdown products only by their carbon skeletons, not their more complex chemical shapes and formulas.

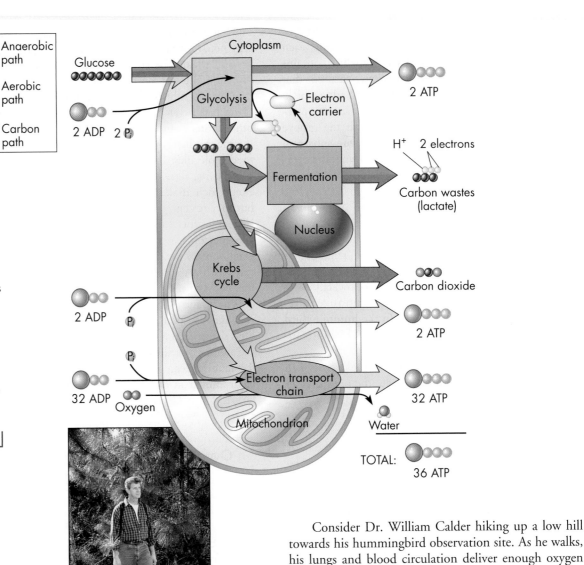

none at all, the **anaerobic pathway** (**E**xplorer 3.1). Many living cells are committed to one type or the other. Our brain cells, for example, and the cells of land plants are limited to the aerobic pathway, while tetanus-causing bacteria are restricted to the anaerobic one. Yeast cells, animal muscle cells, and certain bacterial cells are somewhat special in their ability to obtain energy from either pathway depending on oxygen availability at any given moment.

Consider Dr. William Calder hiking up a low hill towards his hummingbird observation site. As he walks, his lungs and blood circulation deliver enough oxygen to his muscles so that the cells can use the aerobic pathway (the yellow arrows in Fig. 3.10). During a more strenuous climb into the peaks surrounding the alpine meadows at Rocky Mountain Biological Laboratory, his heart, lungs, and blood vessels cannot provide oxygen as fast as muscle cells require it. For this reason, muscle cells switch to the alternative anaerobic pathway (the orange arrows in Fig. 3.10).

The Aerobic Pathway

The aerobic energy harvesting pathway, also called **cellular respiration,** consists of three main parts (follow the yellow arrows in Fig. 3.10). The first part, **glycolysis,** begins the breakdown of glucose and produces just a few ATP molecules. The second part of the aerobic pathway, called the **Krebs cycle,** completely dismantles glucose to individual carbon atoms while capturing high-energy electrons. The third part of the aerobic pathway, the **electron transport chain,** is where the oxygen comes in: The high-energy electrons captured in the Krebs cycle join with oxygen and hydrogen. This produces water molecules and a large number of ATP molecules—a big energy payoff for the cell. Most cells in animals, plants, algae, and fungi, and most kinds of bacteria carry out a strictly aerobic energy harvest.

The Anaerobic Pathway

The alternative anaerobic pathway consists of two main parts (follow the orange arrows in Fig. 3.10). The first, glycolysis, is the same as in the aerobic pathway. The second, **fermentation,** doesn't release any useful energy, but it does recycle materials necessary for glycolysis to continue. Without fermentation, a cell could not even use glycolysis to produce its meager amount of ATP. Animal muscle cells, yeasts in low-oxygen environments, and many kinds of bacteria harvest energy through this anaerobic pathway some or all of the time. Let's survey the broad outlines of the aerobic and anaerobic energy-releasing pathways; you can find more details in Ⓔ 3.1.

How Do Cells Harvest Energy When Oxygen Is Present?

How can the wing muscles of "superathletic" hummingbirds keep burning fuel and contracting quickly enough to allow the animals to hang suspended like miniature helicopters? And how can a tiny bird store enough energy to make a migration of 800 km or more? In other words, what accounts for the hummingbird's tremendous feats of strength and endurance—or for that matter, your own? The answers can be summarized with a simple equation for cellular respiration, the process of energy harvest utilizing oxygen:

Glucose + Oxygen + ADP + Phosphate $\longrightarrow$
Carbon dioxide + Water + ATP

This equation basically says that in the presence of oxygen, the energy of sunlight trapped in glucose molecules is transferred to ADP along with a phosphate ion, thereby producing the more readily usable energy carrier ATP. Water and carbon dioxide are byproducts of this ATP formation. The arrow in the equation above represents the cellular mechanisms that bring about the phases of this energy harvest: glycolysis, the

Krebs cycle, and the electron transport chain (yellow arrows in Fig. 3.10). (As we will see, hummingbirds also have a few special tricks to help the energy-harvesting process along.)

Glycolysis

Energy harvest begins with the breakdown of the simple sugar glucose in a sequence of reaction steps called glycolysis (*glyco* = sugar + *lysis* = splitting). Cells can convert other kinds of sugar, such as the sucrose (see Fig. 2.10e) in flower nectar, into glucose and use it as a basic fuel. Energy stored in the chemical bonds of glucose molecules originally came from the sun, and it was trapped in molecular form by photosynthesis taking place in green plants (such as the red columbine of mountain meadows) as well as in algae and certain kinds of bacteria.

The reactions of glycolysis split the six-carbon sugar glucose into two molecules of the three-carbon compound, **pyruvate** (ionized pyruvic acid; Fig. 3.11 and Ⓔ 3.1). Pyruvate contains some stored chemical energy and acts as an **intermediate,** a compound that serves as a product for one reaction and a reactant for the next in a metabolic pathway. The splitting of glucose makes available energetic electrons (charged atomic particles) and hydrogen ions (H^+, also called protons). These electrons and hydrogen ions are transferred to a special **electron carrier** molecule—a type of biochemical "delivery van" we'll discuss in detail later. The steps of glycolysis take place in the cell's liquid cytoplasm, rather than in an organelle, and they are facilitated by enzymes dissolved in the watery cytoplasmic solution. For each glucose molecule split during glycolysis, there is a net gain of two ATPs and two pyruvate molecules. The ATPs can move through the cytoplasm to places in the cell where energy is immediately needed. The pyruvate molecules leave the cytoplasm and enter the cell organelle called the mitochondrion.

The Mitochondrion

Highly active cells—your own heart muscle cells, for example—contain large numbers of sausage-shaped organelles called **mitochondria,** which we encountered in Chapter 2 and which act as the cell's powerhouses, generating ATP. A muscle cell or a large plant cell can contain more than 1000 mitochondria, and they can occupy nearly 20 percent of the cell's volume. In the flight muscles of hummingbirds, that figure can reach 35 percent, so many that if any more mitochondria

What accounts for human feats of strength and endurance?

Figure 3.11 The Aerobic Pathway Part 1: Glycolysis.

Within the watery components of the cytoplasm, glycolysis splits glucose. The enzymes of glycolysis "supercharge" glucose by spending two molecules of ATP and adding their phosphates to the six-carbon sugar molecule (Step 1). Enzymes chop the six-carbon molecule in half, and then strip two electrons and a hydrogen from each three-carbon compound and add them to an electron carrier (Step 2). This step also adds two phosphates to each three-carbon compound. In the final step, enzymes remove both phosphates from each three-carbon compound, and use them to make 4 ATPs (Step 3). This yields a net gain of 2 ATPs (remember, two were spent in Step 1). Yellow arrows show the energy path; gray arrows the carbon path; green arrows the path of phosphate groups.

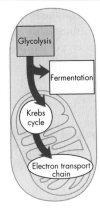

① The enzymes of glycolysis "supercharge" glucose by spending two molecules of ATP and adding their phosphates to the six-carbon sugar molecule.

② Enzymes chop the six-carbon molecule in half, and then strip two electrons and a hydrogen from each three-carbon compound and add them to an electron carrier while adding another phosphate to each three-carbon compound.

③ In the final step, enzymes remove both phosphates from each three-carbon compound and use them to make 4 ATPs, giving a net payoff of 2 ATPs.

Glucose

2 ATP expended → 2 ADP

2 empty electron carriers

2 phosphates

2 loaded electron carriers

4 ADP → 4 ATP

Pyruvate

---

were packed in, they would block proper cell contraction (Fig. 3.12)!

Mitochondria consist of two membranes, like a large, uninflated balloon folded up inside a smaller, fully inflated balloon (see the micrograph in Fig. 3.15 and ⓔ 3.1). A mitochondrion's *outer membrane* is directly bathed by the cell's cytoplasm. Large protein-bound pores perforate the membrane, and molecules up to the size of small proteins can pass through the openings. A mitochondrion's inner membrane is thrown into folds called **cristae** which are studded with en-

zymes and pigment molecules (see Fig. 3.15). Some of these enzymes take part in an energy bucket brigade described below, and one of the enzymes synthesizes ATP.

In all eukaryotic organisms—even lethargic slugs and slow-growing lichens—the mitochondrion's inner membrane is much less permeable than the outer membrane; hence, the area enclosed by the inner membrane, the **matrix,** is a compartment well isolated from the rest

---

Figure 3.12 How Hummingbirds Stay in the Fast Lane.

These micrographs compare cells from a rat's leg muscle (a) with cells from a hummingbird's flight muscle (b). How do the muscle cells differ with respect to oxygen supply and mitochondria? How might these factors contribute to the hummingbird's rapid ATP production and high metabolic rate?

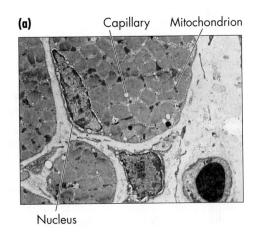

(a) Capillary   Mitochondrion

Nucleus

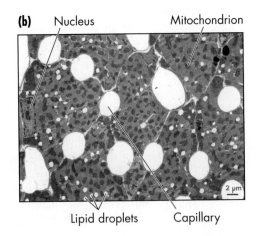

(b) Nucleus   Mitochondrion

Lipid droplets   Capillary

2 µm

of the cell. The matrix contains many enzyme molecules, including those that carry out the reactions of the Krebs cycle, which is so important in aerobic respiration. The matrix also contains several copies of the circular mitochondrial DNA molecule, and hundreds of mitochondrial ribosomes.

Inside the mitochondrion, the pyruvate that formed after glycolysis gets broken down by means of cellular respiration, the breakdown of nutrients and the production of ATP energy using oxygen. Aerobic respiration in mitochondria consumes oxygen and yields carbon dioxide and water plus a large harvest of ATP molecules. That energy harvest, involving the Krebs cycle and the electron transport chain, accounts for the mitochondrion's reputation as a cellular powerhouse, regardless of a cell's metabolic speed.

The Krebs Cycle

The Krebs cycle is part of a series of chemical reactions taking place inside mitochondria that break down pyruvate completely into carbon dioxide and water (Fig. 3.13 and **E** 3.1). In several steps, the carbons in pyruvate are cleaved off, one at a time, and released as carbon dioxide (CO_2). Your exhaled breath is the body's way of getting rid of carbon dioxide produced in the Krebs cycle. The Krebs cycle is important for metabolism in three ways:

1. It produces two ATPs for each glucose molecule that originally entered the aerobic pathway, adding to the two already produced in glycolysis.
2. The Krebs cycle strips off from pyruvate numerous high-energy electrons accompanied by hydrogen ions (H^+). These electrons and hydrogen ions are then delivered to electron carriers. Two important such carriers are called NADH and $FADH_2$. (These electron carriers undergo additional reactions, as we will see shortly.)

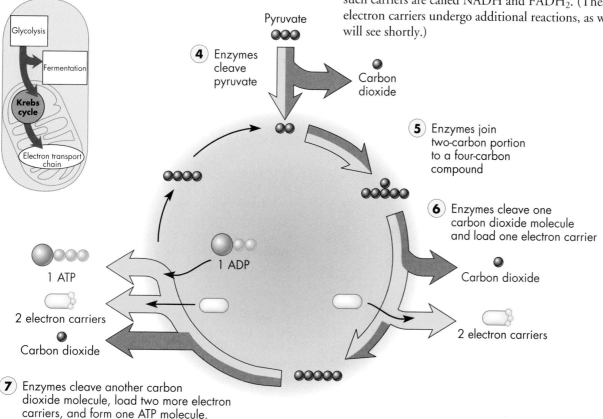

Figure 3.13 **The Aerobic Pathway Part 2: The Krebs Cycle**. The Krebs cycle is the source of the carbon dioxide we exhale. Pyruvate from glycolysis passes into the matrix of the mitochondrion where enzymes cleave pyruvate to one molecule of carbon dioxide plus a leftover two-carbon portion (Step 4; see Steps 1–3 in Fig. 3.11). This two-carbon portion enters the Krebs cycle as enzymes join it to a four-carbon compound to make a six-carbon molecule (Step 5). Other enzymes sequentially cleave two carbon dioxide molecules from this six-carbon molecule, and produce four energized energy carriers (two in Step 6 and two in Step 7), plus one ATP. These reactions simultaneously regenerate the original four-carbon compound, completing the cycle (Steps 6 and 7). Because each glucose gives two pyruvates, this cycle turns twice for each molecule of glucose.

3. The Krebs cycle produces the carbon skeletons for many important small biological molecules and acts as a clearinghouse or swap-meet for metabolism, a place where one metabolic intermediate can be converted into another or diverted toward a different pathway. Let's look at the action of the Krebs cycle in a bit more detail.

The Krebs cycle starts with the end product of glycolysis, the three-carbon substance pyruvate (review Fig. 3.11), which passes from the cell's watery cytoplasm into the mitochondrion. In the matrix of the mitochondrion, enzymes cleave pyruvate into a molecule of carbon dioxide plus a two-carbon portion (Step 4 of Fig. 3.13). Enzymes in the Krebs cycle join this two-carbon portion to a four-carbon compound to make a six-carbon molecule (Step 5). Other enzymes sequentially cleave two carbon dioxide molecules from this six-carbon molecule (a person exhales this in his or her breath), and produce four energized electron carriers. Notice the two electron carriers in Steps 6 and 7). In addition, enzymes in Step 7 regenerate the original four-carbon compound, completing the cycle. An additional step, not shown, results in one ATP.

The end result is that the Krebs cycle converts all the carbons from the original glucose into carbon dioxide and stores the energy in electron carriers.

Let's summarize the Krebs cycle. Because each glucose forms two pyruvate molecules, the net result from the Krebs cycle is six CO_2, eight energized electron carriers, and two ATP molecules for each molecule of glucose originally broken down. We eventually exhale the carbon dioxide, use the ATPs for energy-requiring reactions, and the electron carriers move to the electron transport chain, which we describe in a moment. First, though, let's briefly investigate the role of the Krebs cycle in building biological molecules.

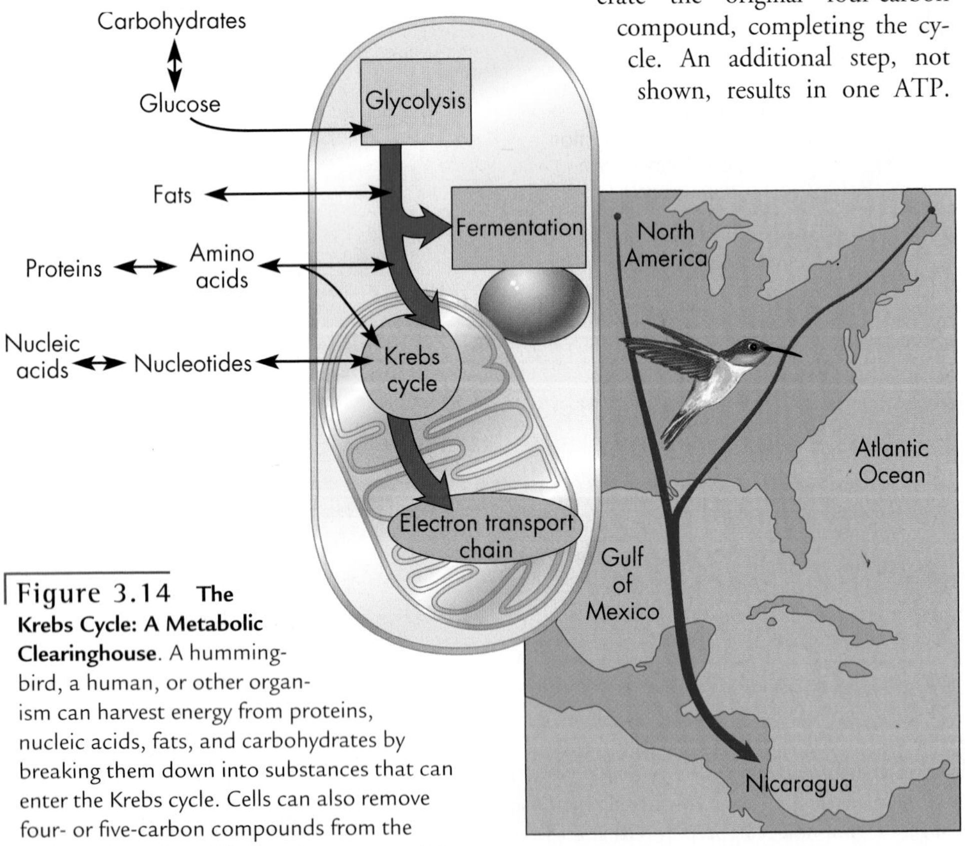

The Krebs Cycle as a Metabolic Clearinghouse

Let's return to hummingbirds for a minute and a topic we've already touched on: As winter approaches, a 3.5-gram hummingbird can store enough body fat to fuel its nonstop migratory flight across the Gulf of Mexico (over 800 km). Hummingbirds can do this because in the days before migration, they can put on 2 grams of fat (the equivalent of you gaining nearly half your body weight). Then, as they migrate, they burn this fat instead of sugar. You can see the fat cells (lipid droplets) in a cross section of a hummingbird's flight muscle (Fig. 3.12b) Their cells must have a mechanism for selecting the right fuel at the right time, and that mechanism—for hummingbirds and all other organisms—is the Krebs cycle.

The various compounds taking part in the Krebs cycle are intermediates, reactants in one metabolic reaction and products in the next. Cells can use these intermediates in several ways. The Krebs cycle can help a cell use lipids or proteins for energy in addition to carbohydrate. When an animal or plant cell's supply of sugar falls (or is insufficient for a task like nonstop migrating), the cell begins to break down lipids or proteins. Some of the subunits from these reactions can then be converted into pyruvate or into Krebs cycle intermediates. These compounds can then enter the Krebs cycle pathway at the appropriate stage and be dismantled for energy harvest (Fig. 3.14). If an organism is starving, this process can provide enough energy to keep the body alive, but the body is literally digesting itself to provide that energy, with the Krebs cycle making it possible. When food once more become available, intermediates from the Krebs cycle can become the raw materials for growth: They can serve as skeletons for the synthesis of new fats, proteins, and carbohydrates that restore those broken down during lean times, or that wear out and need replacing.

Figure 3.14 **The Krebs Cycle: A Metabolic Clearinghouse**. A hummingbird, a human, or other organism can harvest energy from proteins, nucleic acids, fats, and carbohydrates by breaking them down into substances that can enter the Krebs cycle. Cells can also remove four- or five-carbon compounds from the Krebs cycle and modify them into new useful materials, as indicated by the double-headed arrows. Hummingbirds can migrate long distances because they can convert fats and proteins to intermediates that can enter the Krebs cycle and produce ATPs by cellular respiration.

An important exception to the clearinghouse principle is the human brain cell. For complex reasons, it can use only glucose as fuel, and this fact has major implications: First, it explains why sugary foods are temporary mood elevators. Soon after you eat a candy bar or drink a sugary soda, glucose molecules are cleaved from sucrose (table sugar) and enter the bloodstream and brain. Infused with their favorite fuel, the brain cells function at peak efficiency, leading one to feel happier, smarter and livelier—at least for a time. Second, this property of the brain cell explains why dieters are warned to consume at least 500 calories in carbohydrates per day and not to restrict themselves to liquid or powdered protein diets: The brain alone needs at least 500 calories of glucose for normal functioning, and without it, a person can grow faint and even lapse into unconsciousness. (The brain and some other organs can make some glucose from three- and four-carbon precursors by a process called gluconeogenesis [*gluco* = glucose + *neo* = new + *genesis* = origination], essentially a reversal of glycolysis.)

Let's go back, now, to the electron carriers generated by the Krebs cycle and trace their final destination to the electron transport chain.

The Electron Transport Chain

Still inside the mitochondrion, eight electron carrier molecules per initial glucose molecule are loaded with electrons from the Krebs cycle. These carriers move to the **electron transport chain,** a group of enzymes and pigment molecules embedded in mitochondrial membranes (Fig. 3.15 and *E* 3.1). Each member of the chain passes electrons to the next member, like a bucket brigade. During each electron transfer, the electron loses a bit of energy (like a splash of water spilling from the bucket). Particular mitochondrial enzymes use this released energy to synthesize ATP from ADP. For each initial molecule of glucose that entered glycolysis, the yield from the electron transport chain is a whopping 32 ATPs.

The electron transport chain is the stage of the aerobic pathway that actually uses the oxygen because the last electron acceptor in the chain is oxygen. As electrons are added to oxygen atoms, and hydrogen ions follow along, the hydrogen and oxygen combine to form water, H_2O. Thus, the oxygen you are breathing in as you read these sentences will be converted to water in your mitochondria as oxygen accepts electrons and aerobic respiration takes place. The legend to Figure 3.15 describes this step by step.

If oxygen is not present to accept the electrons and hydrogen ions once they have passed down the electron

transport chain, the entire process quickly stops. The organism or cell, if strictly oxygen-requiring, then dies. That's how cyanide kills living things. The $C \equiv N$ chemical group (carbon triple-bonded to nitrogen) in sodium cyanide binds tightly to a certain molecule in the electron transport chain. This bonding prevents it from accepting an electron transfer. The result is that ATP formation halts, and the cell quickly starves for energy and dies.

Interestingly, biologists think that the speed and efficiency of the electron transport chain determine the outside limit to biological miniaturization. Another way to say this is that the smallest hummingbirds are close to as small and fast as warm-blooded animals can ever get because the electron transport chain in their cells produces ATP virtually as quickly as that system ever could. A warm-blooded animal smaller than the smallest hummingbird would need even more energy to keep warm, since its surface area, and hence heat loss, would be so large compared to its small internal volume.

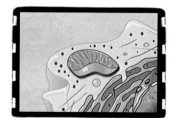

Which aspect of energy harvest has features in common with a bucket brigade?

An ATP Tally

Let's pause now and tally up the ATP yield from the aerobic pathway (see Fig. 3.10). From each initial glucose molecule metabolized in the cell, the net yield from glycolysis is 2 ATPs, the Krebs cycle results in 2 more, and the electron transport system produces 32. This gives a total net yield of 36 ATP molecules per glucose molecule. When you compare this yield to that of energy harvest when oxygen is absent, you'll discover why you can exercise aerobically for long periods of time, but can carry on very strenuous anaerobic exercise for only a short time before running out of energy and feeling exhausted.

Let's also return for a moment to the special tools hummingbirds make use of. Hummingbirds feed almost continuously, then their bodies convert nectar quickly into glucose, and billions of mitochondria break down that fuel via cellular aerobic respiration into ATPs at a high rate. To conserve energy, hummingbirds spend as much time as possible between flower forays resting on their favorite perches. And at night, hummingbirds often go into a state of torpor, using even less energy than sleeping requires. Finally, the bird's cells have another metabolic card to play, as the next section explains.

How Do Cells Harvest Energy Without Oxygen?

A hummingbird has a greater blood supply to its wing muscle cells than a mammal does to its wings or legs (see Fig. 3.12). Sometimes, though, the bird must exert

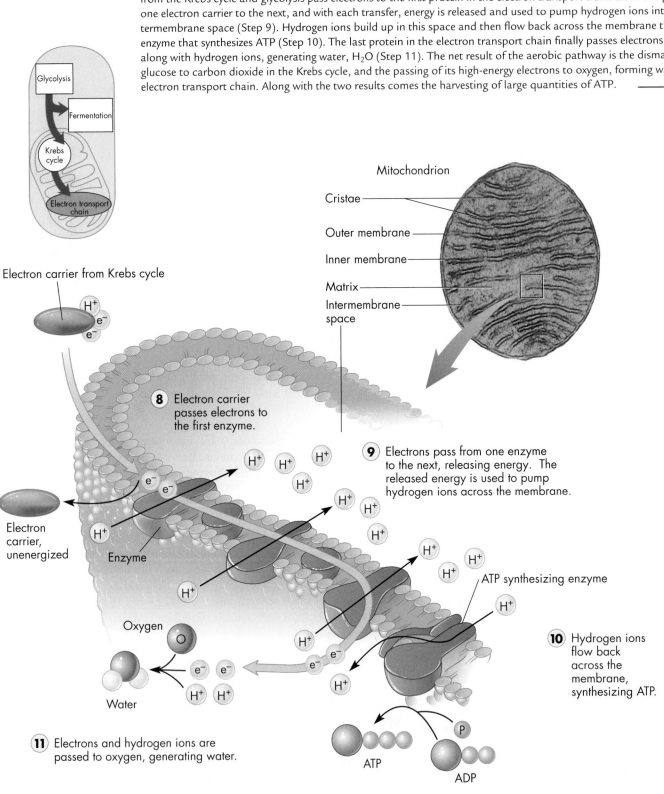

Figure 3.15 **The Aerobic Pathway Part 3: The Electron Transport Chain.** The electron transport chain in the mitochondrial membrane harvests large quantities of ATP. In Step 8 of the aerobic pathway, high-energy electron carriers from the Krebs cycle and glycolysis pass electrons to the first protein in the electron transport train. Electrons pass from one electron carrier to the next, and with each transfer, energy is released and used to pump hydrogen ions into the intermembrane space (Step 9). Hydrogen ions build up in this space and then flow back across the membrane through an enzyme that synthesizes ATP (Step 10). The last protein in the electron transport chain finally passes electrons to oxygen along with hydrogen ions, generating water, H_2O (Step 11). The net result of the aerobic pathway is the dismantling of glucose to carbon dioxide in the Krebs cycle, and the passing of its high-energy electrons to oxygen, forming water in the electron transport chain. Along with the two results comes the harvesting of large quantities of ATP.

Glycolysis

Fermentation

Krebs cycle

Electron transport chain

Mitochondrion

Cristae

Outer membrane

Inner membrane

Matrix

Intermembrane space

Electron carrier from Krebs cycle

H^+

e^-

e^-

8 Electron carrier passes electrons to the first enzyme.

e^- e^-

H^+ H^+ H^+

H^+

H^+ H^+

9 Electrons pass from one enzyme to the next, releasing energy. The released energy is used to pump hydrogen ions across the membrane.

H^+

Electron carrier, unenergized

H^+

Enzyme

H^+

H^+ H^+

H^+

ATP synthesizing enzyme

H^+

Oxygen

O

H^+

10 Hydrogen ions flow back across the membrane, synthesizing ATP.

e^- e^-

e^- e^-

H^+ H^+

H^+

Water

11 Electrons and hydrogen ions are passed to oxygen, generating water.

P

ATP

ADP

itself so strenuously that its furiously beating heart still cannot deliver oxygen to the muscles as fast as they are using the gas. As a result, the hummingbird's muscle cells have to rely for short periods of time on anaerobic energy harvest with its two phases, glycolysis and fermentation (the orange arrows in Fig. 3.10).

Glycolysis

Anaerobic energy harvest or the **anaerobic pathway** begins with glycolysis, the same process that starts the aerobic pathway (Fig. 3.16). Because glycolysis doesn't vary from one pathway to the other, all organisms— from the bacteria that decompose organic sludge at the bottom of an oxygen-depleted swamp to the green leaves of the columbine plant that supplies hummingbirds with nectar—possess the enzymes of glycolysis. This universal prelude to energy metabolism leads biologists to think that glycolysis probably evolved in earth's earliest cells over 3 billion years ago, and was passed down to all surviving organisms. This common biochemical thread underscores the evolutionary relatedness of all life forms.

Fermentation

Recall that energized electron carriers (NADH) and the 3-carbon compound pyruvate are end products of glycolysis. During the second phase of the anaerobic pathway, called **fermentation,** enzymes modify pyruvate in the absence of oxygen. The enzymes that speed the reactions of fermentation lie in the cell's cytoplasm, just like those that facilitate glycolysis. Depending on the organism, however, fermentation converts the pyruvate into various end products, such as ethanol and carbon dioxide, or lactic acid (Fig. 3.16). The intoxicating effect of wine and beer and the fragrant, homey aroma of baking bread come from ethanol, the alcohol produced during fermentation in yeast cells. Likewise, fermentation carried on by certain bacteria growing in milk releases the lactic acid that gives some cheeses their sharp taste. Your muscle cells also produce lactic acid (or the electrically charged form called *lactate*) when you exercise anaerobically. The "burn" that you feel when you work a muscle very hard comes from the liberation of lactate. Hummingbirds probably experience a "burn" sometimes, too.

Ironically, fermentation itself yields no ATP. The wastes—the ethanol or lactic acid—squander the energy that would be captured in ATP if the cell were using the aerobic pathway. You might wonder, then, why cells would carry out fermentation and produce toxic

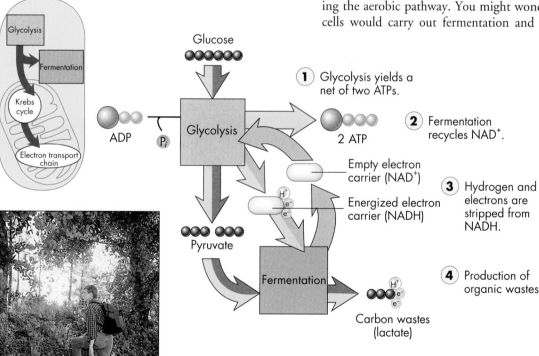

1. Glycolysis yields a net of two ATPs.

2. Fermentation recycles NAD⁺.

3. Hydrogen and electrons are stripped from NADH.

4. Production of organic wastes

Glucose

ADP   P_f

Glycolysis

2 ATP

Empty electron carrier (NAD⁺)

Energized electron carrier (NADH)

Pyruvate

Fermentation

Carbon wastes (lactate)

Figure 3.16 **The Anaerobic Pathway: Glycolysis and Fermentation**. Glycolysis takes place in the cytoplasm in the absence of oxygen (as in the man's hard-working leg muscles as he climbs a steep hill), and provides a net gain of two ATP molecules (Step 1). It also traps hydrogen and energetic electrons in the electron carrier NADH (Step 2). Fermentation recycles the less-energized electron carrier NAD⁺ by removing electrons and hydrogen from NADH (Step 3) and transferring them to pyruvate's three-carbon skeleton. The result is the waste product lactate (lactic acid) from our muscle cells and in certain kinds of bacteria (Step 4) or ethanol and carbon dioxide from yeast cells. Without the recycling of the electron carrier NAD⁺, glycolysis would grind to a halt.

waste products such as ethanol and lactate? The answer is that fermentation reactions recycle the electron carrier molecule needed for glycolysis, stripping away the electrons and hydrogens from the carrier and making the carrier available for a new round of glycolysis (blue arrows, Fig. 3.16). In fermentation, it is as if the electron carrier "delivery van" dumps its load unused just so it can return and pick up a new load. Without the recycling of the energy carrier, the cell would run out of this necessary "delivery" molecule, and glycolysis would cease.

The energy yield of the anaerobic pathway may seem meager, just 2 ATPs per glucose compared to 36 for the aerobic pathway. Anaerobic metabolism, however, is crucial to the global recycling of carbon and the stability of the environment. Organic matter from dead leaves, dead microorganisms, and other sources often sinks into an environment devoid of oxygen, such as the soft layers at the bottom of lakes or oceans. If it weren't for anaerobic decomposers—organisms capable of breaking down organic matter via anaerobic metabolism—most of the world's carbon would eventually be locked up in undecomposed organic material in these oxygen-poor environments. As a result, there would be too little carbon dioxide available as a raw material for photosynthesis, plants would be unable to generate new glucose molecules, and neither plants nor animals would survive.

Control of Metabolism

Hummingbirds can burn carbohydrates by either the aerobic or the anaerobic pathway if their activity demands it temporarily. As we saw, they can burn fats when they are migrating nonstop, and, thanks to the Krebs cycle clearinghouse, they can also use proteins for energy. These intricacies of metabolism bring up new questions: What makes a cell start to burn its stores of lipids or proteins when glucose runs out? Then, when the supplies of glucose return, what stops the cell from burning all its lipids or proteins, literally eating itself up from the inside and destroying its cellular structures? Finally, what triggers a cell to build molecules only when they are needed? Clearly, there is a great deal of internal coordination and control over the cell's harvesting of energy and building of needed materials. But what form does it take?

Here's a simple analogy for one way cells control metabolism. Imagine a shoe factory in which shoes pile up so high that they topple over, clog up a machine, and stop the assembly line; it is only after the shoes are removed and shipped out that the line is free to start up again and make more shoes. Likewise, levels of ATP can

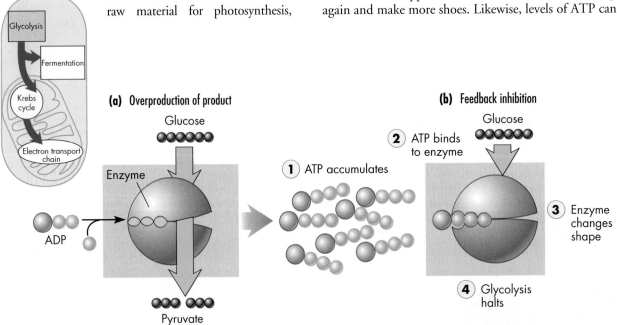

Figure 3.17 **Cells Regulate Energy Metabolism by Feedback Mechanisms**. (a) When glycolysis produces ATP more rapidly than it is used, ATP accumulates in the cell (Step 1). (b) The accumulated ATP can bind to an enzyme early in the pathway of glycolysis (Step 2) and cause the enzyme to change shape (Step 3) in a way that inhibits the enzyme's activity. (This is why it's called feedback inhibition.) Because the enzyme can't change substrate to product, the whole pathway stops (Step 4). Without more pyruvate, there is no fuel for the Krebs cycle; aerobic respiration, with its rapid production of ATP, therefore slows or stops. When the excess ATP is used up, it stops inhibiting the enzyme and the cell can resume energy harvest.

build up in a cell so high that the cell requires no more of the molecular fuel (Fig. 3.17a). When this occurs, the ATP binds to a special regulatory site on a specific enzyme that facilitates an early step in glycolysis. This binding shuts down enzyme activity (Fig. 3.17b). This, in turn, switches off the entire glycolytic pathway. If an early reaction does not function, no substrate is produced for the reactions that follow. The regulatory advantages of this system are obvious: When the cell already has high levels of ATP, the presence of the ATP molecule itself serves as a control to turn off its own production. Then, when ATP levels drop, the bound ATP leaves the enzyme "machinery," and the glycolysis "assembly line" resumes once more. This kind of metabolic regulation is called **feedback inhibition.** We'll see other examples of it in later chapters. This term is appropriate, since the accumulation of the product, in effect, feeds back and inhibits its own production.

Through feedback inhibition and other forms of control, the activity of a cell's metabolic enzymes is turned on or off so that the cell burns glucose when that sugar is available; it burns lipids or proteins when glucose is lacking; and it builds the appropriate biological raw materials just when they are needed for growth or maintenance activities. These control mechanisms ensure that order rather than metabolic chaos reigns within living organisms.

3.4 How Do We Get Energy for Exercise?

Compared to hummingbirds, people are lumbering giants. Nevertheless, we, too, rely on the energy-harvesting pathways discussed in this chapter. This is true whether the energy is needed for the explosive movements of a tennis serve, for a swimmer's backstroke, or for a hiker's long steady climb up a mountain trail. Energy for different forms of exercise, however, comes from different parts of the metabolic pathway.

Exercise physiologists have determined that three energy systems—the immediate system, the glycolytic system, and the oxidative system—supply energy to a person's muscles during exercise. The duration of physical activity and availability of oxygen dictate which system the body uses. The **immediate energy system** is instantly available for a brief explosive action, such as one bench press, one tennis serve, or one ballet leap, and the system has two components. One component is the small amount of ATP stored in muscle cells, immediately useful like the few coins you carry around in

your purse or pocket. This stored ATP, however, runs out after only half a second—barely enough time to heave a shot or return a tennis serve, let alone to trudge up a long hill. The second component of the immediate energy system is a high-energy compound called **creatine phosphate,** an amino acid–like molecule that has an energetic phosphate, like ATP. Muscle cells store creatine phosphate in larger amounts than ATP. Creatine phosphate is more like a handful of dollar bills than a few coins. When the ATP in a muscle cell is depleted, creatine phosphate transfers its phosphate to ADP; this regenerates ATP, and ATP can then fuel the muscle cell to contract and move the body. Many health food stores and bodybuilder magazines advertise and sell creatine under colorful brand names like "Kick Some Mass," and many weight lifters and football players take it. Without any evidence for the long-term safety of this compound, however, most football coaches have now stopped providing creatine to their players. Even the cell's store of creatine phosphate, however, becomes depleted after only about a minute of strenuous work; thus, muscles must rely on more robust systems than the immediate energy system to power longer term activities.

The **glycolytic energy system,** which depends on splitting glucose by glycolysis in the muscles, fuels activities lasting from about 1 to 3 minutes, such as an 800-m run or a 200-m swim. This storage form is more like an account at the bank than like dollars in your wallet, and it "purchases" more activity. Glycolysis going on in the muscle cell cytoplasm can cleave glucose in the absence of oxygen and generate a few ATPs. Fermentation takes the product of glycolysis a step further and makes the waste product lactic acid. Recall that there is a net yield of only two ATP molecules for each molecule of glucose from glycolysis. For this reason, the glycolytic energy system can sustain heavy exercise for only about 3 minutes, and all of it would be considered anerobic exercise, powered by the cells' anaerobic pathway. It's also why lactic acid begins to build up in muscles in just that short amount of time and can lead to a muscle cramp or "stitch" in your side.

Activities lasting longer than about 3 minutes—a jog around the neighborhood, an aerobic dance session, or a long uphill hike—require oxygen and employ the **oxidative energy system.** This system can supply energy for activity of moderate intensity and long duration (aerobic exercise), and is like money invested in stocks and bonds that give long-term steady income. The oxidative energy system is based on cellular respiration and includes the Krebs cycle and the electron transport chain. It uses oxygen as the final electron acceptor and generates many ATP molecules per molecule of glucose burned. Clearly, anyone interested in melting away body fat should engage in aerobic (oxygen-utilizing) activities like jogging, swimming, bicycling, or strenuous hiking which rely primarily on the oxidative energy supply system and its ability to use fats as fuel.

The glycolytic energy system can make ATP only from glucose or glycogen (cleaved to release glucose), but the oxidative energy system can produce energy by breaking down carbohydrates, fatty acids, and amino acids mobilized from other parts of the body and transported to the muscle cells by way of the bloodstream (review Fig. 3.14). The oxidative energy supply system can provide many more ATP molecules than creatine phosphate or glycolysis, but its supply rate is slower.

Mitochondria and Obesity: Why Are So Many Americans Overweight?

Recent surveys have shown that Americans are getting poorer at balancing food intake with energy expenditure. Data from the National Health and Nutrition Examination Surveys show that in the last three decades, the percent of obese Americans has nearly doubled to one in four adults (Fig. 3.18). To see where you stand on the thin-to-fat scale, calculate your body mass index (BMI, weight in kilograms divided by height in meters squared). Obesity (defined as a BMI of 30 or more) poses a serious threat to our health because it increases the likelihood of developing conditions such as cardiovascular disease or diabetes.

At the simplest level, the body accumulates fat when energy intake is greater than energy expenditure. But understanding the mechanisms that allow some people to accumulate weight more easily in response to this energy imbalance is complicated. The current American environment presents a ubiquitous and almost limitless supply of relatively inexpensive, good tasting, energy-dense foods for intake and a life style that requires little physical activity and its associated energy expenditure. Much of the problem has to do with our superb physiological mechanisms that defend against loss of body energy stores, but not against overaccumulation in times of plenty. We will talk more about these physiological mechanisms in a later chapter. Here, let's look at some recent research related to the mitochondria.

In certain situations, animals (humans included) can generate extra heat by "short-circuiting" their mitochondria, boosting the amount of energy they burn and effectively increasing their metabolic rate. This often occurs in special brown fat cells. Hibernating bears stay warm all winter and newborn babies heat themselves up after they emerge from the womb by means of stoked-up metabolism in such brown fat cells. A protein in these cells called an "uncoupling protein" dissociates (uncouples) the breakdown of food molecules from the production of ATP. Recall from Step 10 of Figure 3.15 that hydrogen ions normally flow across the mitochondrion's inner membrane through a protein that couples the passage of hydrogen ions to ATP production. In brown fat cells, however, the uncoupling proteins serve as holes in the inner mitochondrial membrane through which hydrogen ions pass *without* making ATP. The wasted energy is transformed into heat and this warms the bear, baby, or you. Observes one scientist who studies these proteins, "Its like jogging without jogging."

Recent work shows that adults have forms of the uncoupling proteins that may be activated by cold to turn up the metabolic rate. Researchers are testing the hypothesis that obese people have different forms of the uncoupling proteins than do people with higher metabolic rates who are naturally resistant to obesity. A biotech company in Massachusetts has patented the human gene for one of these uncoupling proteins and is developing a test to determine a person's obesity risk. As a thought exercise, try designing a potential diet pill based on uncoupling proteins. Some obesity drugs interfere with the nervous system's control of eating. Would there be a potential

How can the science of energy metabolism help us understand weight control?

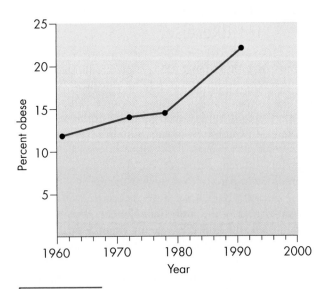

Figure 3.18 **The Fattening of America**. In the last 40 years, the frequency of obese Americans has nearly doubled. From a body mass chart you can determine whether your body mass index is in the recommended range below 25.

advantage to attacking obesity through uncoupling proteins instead of this nervous system alteration?

The science of energy metabolism can help us understand not just weight control but two of life's most profound phenomena: breathing in and breathing out. Breathing in supplies the oxygen necessary as the last acceptor of electrons in the electron transport chain. This in turn allows cells to harvest a large amount of ATP from the energy stored in glucose. Breathing out rids the body of carbon dioxide, which arises when Krebs cycle enzymes in the mitochondria chop carbon atoms from glucose breakdown products. The plants around you, such as houseplants, can then take up the carbon dioxide you exhale (plus the carbon dioxide the plant produces in its own mitochondria) and, with energy from sunlight, incorporate it into new carbohydrate molecules. The cycle of inhaling and exhaling keeps you alive, and the cycle of aerobic respiration and photosynthesis maintains a big part of the energy balance in the world as we know it. We've looked at cellular respiration. Now let's consider photosynthesis in more detail.

3.5 How Do Plants Trap Energy in Biological Molecules?

A hummingbird rockets from plant to plant, hovering in front of individual blossoms as its forked, grooved tongue juts forward to lap up nectar. It's amazing that a 3-gram bird can visit 1000 flowers a day, drink its body weight in sugary syrup, and burn up all that fuel in a blur of bejeweled feathers. No wonder hummingbirds never need to diet! We've been talking so far only about how the cells of birds, people, or red columbine flowers extract energy from biological molecules. But how did the energy get trapped in those biological molecules in the first place? It's less obvious, perhaps, but no less amazing that red columbine plants use photosynthesis to convert sunlight into chemical energy, store it in the bonds of organic molecules, and use those same molecules to build leaves, stems, flowers, and sweet nectar as well as to fuel the plants' own energy needs and the steps of aerobic respiration taking place in all the plant's cells.

Compare, for a moment, the way you gather energy with the way a columbine does it. The plant obtains its energy from the nonliving environment in the form of sunlight and inorganic raw materials, while you must get yours by eating plants or by eating other animals that ate plants. Biologists classify organisms that take in preformed nutrient molecules from the environment as **heterotrophs** (*hetero* = other + *troph* = feeder). Heterotrophs include many prokaryotes; most protists (such as protozoa); mushrooms, yeasts, and other fungi; and all animals, including humans and hummingbirds. In contrast to heterotrophs, **autotrophs** (*auto* = self) are organisms that take energy directly from the nonliving environment and use it to synthesize their own nutrient molecules. Autotrophs include photosynthetic organisms—green plants and certain protists and prokaryotes that obtain energy from sunlight. Autotrophs also include a small group of **chemosynthetic** organisms—a few kinds of prokaryotes that extract energy from inorganic chemicals such as hydrogen gas and hydrogen sulfide with its rotten-egg smell. We can see that, ultimately, autotrophs are the source of all energy that flows through living systems, since heterotrophs obtain nutrients either from autotrophs or from heterotrophs that once consumed autotrophs. By far the most important source of energy flowing through organisms begins with the sun's light. So let's take a look at the nature of light and of solar energy.

Physical Characteristics of Light

If you sit in a mountain meadow next to a waterfall, with mists of crystal stream water rising off the rocks, you'll probably see rainbows in the spray. That's because fine water droplets in the spray act like tiny prisms and separate sunlight into a spectrum of colors. Visible light is just a small part of the **electromagnetic spectrum,** which is the full range of electromagnetic radiation in the universe, from highly energetic gamma rays to very low–energy radio waves (Fig. 3.19). Such radiation travels through space behaving both as particles called **photons** and as waves. The amount of energy in a photon determines its wavelength, the distance it travels during one complete vibration.

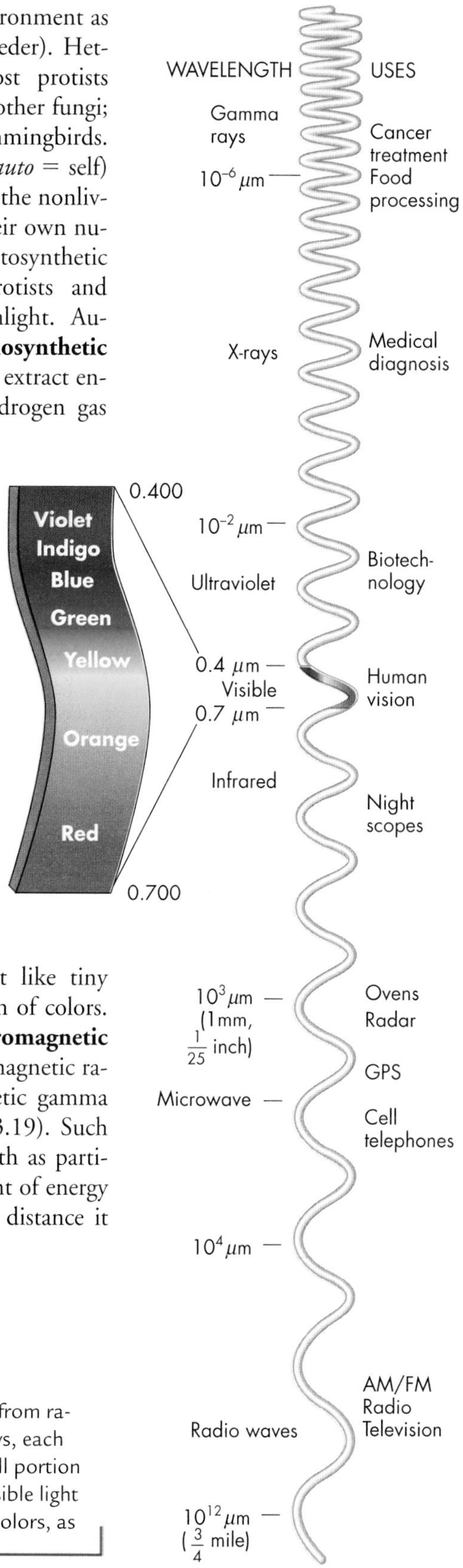

WAVELENGTH — USES

Gamma rays — 10^{-6} μm — Cancer treatment / Food processing

X-rays — Medical diagnosis

10^{-2} μm — Ultraviolet — Biotechnology

0.400

Violet
Indigo
Blue
Green
Yellow
Orange
Red

0.700

0.4 μm — Visible — Human vision
0.7 μm

Infrared — Night scopes

10^3 μm (1mm, $\frac{1}{25}$ inch) — Ovens / Radar

Microwave — GPS / Cell telephones

10^4 μm

Radio waves — AM/FM Radio Television

10^{12} μm ($\frac{3}{4}$ mile)

Figure 3.19 **Visible Light and the Energy Spectrum.** We are constantly bathed by energy, from radio waves to infrared rays (heat) and gamma rays, each with its own wavelength and energy. Only a small portion of the entire spectrum is visible to us as light. Visible light can be broken by a prism into light of different colors, as happens in a rainbow.

Photons of visible light have wavelengths in a narrow range: If the entire electromagnetic spectrum shown in Figure 3.19 wrapped once around Earth, the visible part would be the length of your little finger. The violet light we see bouncing off a hummingbird's iridescent throat feathers has a shorter wavelength than the red light bouncing off a columbine blossom. It is no coincidence that living things can absorb and use light within the restricted wavelength range of visible light.

Gamma rays are so short and energetic that they disrupt and destroy biological molecules they strike, while radio waves are so long and low in energy that they do not excite biological molecules. We perceive colors in sunlight because the pigments in the retinas of our eyes are excited by wavelengths within the visible range, and nerve impulses fired off to the brain are interpreted as red, blue, violet, and other colors. Light travels so quickly and the photosynthetic process takes place so fast that you can practically eat sunlight in a fresh picked leaf. If you pluck and immediately chew a growing lettuce leaf, you are consuming energy that left the sun just eight minutes earlier and was converted to the chemical energy in carbohydrate molecules almost instantly upon striking the plant.

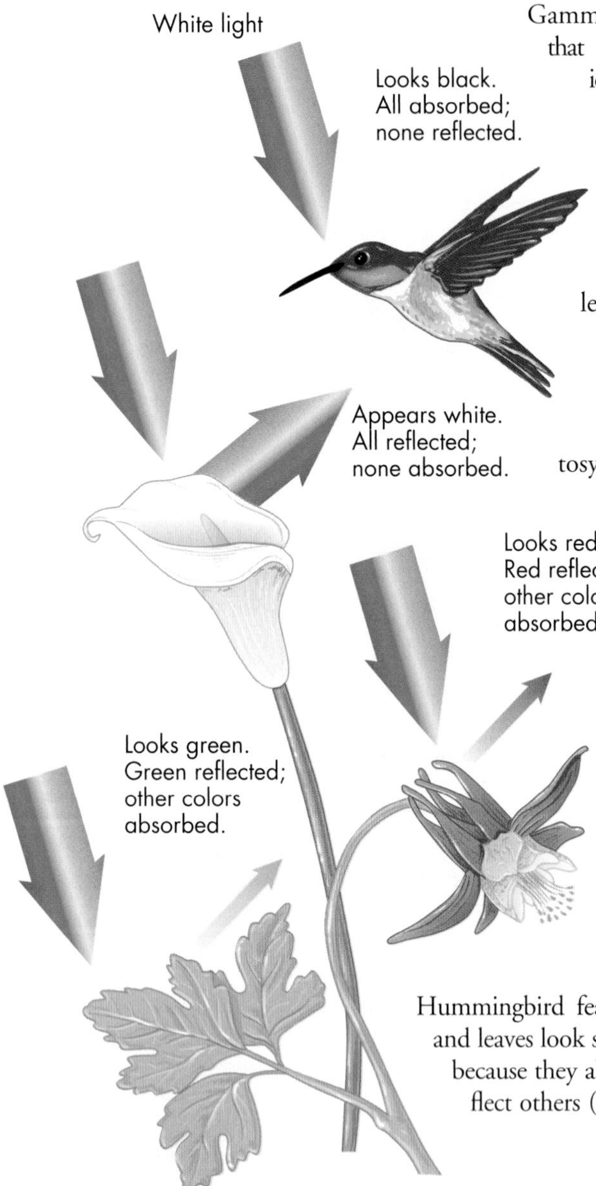

Chlorophyll and Other Pigments Absorb Light

Hummingbird feathers, bees's wings, flower petals, and leaves look sapphire, red, green, and other hues because they absorb certain colors of light and reflect others (Fig. 3.20). A bird's feathers are red because pigment molecules in the feather absorb light from various parts of the visible spectrum and reflect only red light. A columbine leaf looks green because its green pigments absorb violet, blue, yellow, and red light and reflect only green. This range of absorbed light is called the **absorption spectrum,** and is unique for each pigment. The green pigment in leaves is called **chlorophyll,** and it takes part in photosynthesis as well as giving a green leaf its color.

Chlorophyll is often accompanied by colorful **carotenoid pigments,** which absorb green, blue, and violet wavelengths and reflect red, yellow, and orange light. Carotenoids are generally masked by chlorophyll and thus tend to be unnoticed in green leaves. However, they give bright and obvious color to many non-photosynthetic plant structures, such as roots (carrots), flowers (daffodils), fruits (tomatoes), and seeds (corn kernels). Carotenoids are behind the glorious colors of autumn. As summer ends and the nights grow cool, chlorophyll begins to break down, allowing the gold and red carotenoids to show through and emblazon maple, oak, sumac, and other trees and vines. Chlorophyll absorbs all but green wavelengths of light, and carotenoids all but the red, orange, and yellow wavelengths. Functioning together in pigment complexes, chlorophylls and carotenoids can absorb most of the available energy in visible light. The importance of these pigments, of course, is not just that they absorb light, but what becomes of that captured energy.

The Chloroplast: Solar Cell and Sugar Factory

What gives a columbine or lupine plant growing in an alpine meadow the ability to gather sunlight in its leaves and to make the nutrients in flower nectar? The photosynthetic pigments in green leaves are concentrated in layers of green cells that carry out photosynthesis (Fig. 3.21a–c). These cells contain **chloroplasts,** green organelles in which both the energy-trapping and carbon-fixing reactions of photosynthesis take place (Fig. 3.21c,d). Each leaf cell may contain about 50 chloroplasts, and each square millimeter of leaf surface may contain more than half a million of the green organelles. Just as the mitochondrion's architecture is closely tied to its role as powerhouse for the cell, the chloroplast's structure underlies its important jobs. Chloroplasts are similar to mitochondria in several ways: Both are elongated organelles with an inner and outer membrane and interior flattened sacs (see Fig. 3.15 and Fig. 3.21d); both carry out energy-related tasks in the cell; and both

White light

Looks black.
All absorbed;
none reflected.

Appears white.
All reflected;
none absorbed.

Looks red.
Red reflected;
other colors
absorbed.

Looks green.
Green reflected;
other colors
absorbed.

Figure 3.20 **Pigments and Absorption Spectra**. The colors we see around us depend on which wavelengths of light are absorbed by the molecules in objects and which are reflected back to our eyes. Calla lily blossoms look white because all wavelengths are reflected. The dark feathers of a hummingbird's chin absorb light of all wavelengths and reflect little, so they look black to us. Chlorophyll in a columbine leaf absorbs many wavelengths of light, but green light is most strongly reflected, and thus the leaf looks green.

have their own DNA in circular chromosomes. However, while mitochondria are "powerhouses" that generate ATP, chloroplasts are both solar cells and sugar factories that capture sunlight and generate sugarphosphates and other carbohydrates.

Chloroplast Membranes

Chloroplasts have an outer membrane and inner membrane that lie side by side and that collectively enclose a space filled with a watery solution, the **stroma** (see Fig. 3.21d). A third membrane system lies within the inner membrane and forms the **thylakoids,** a complicated network of stacked, disklike sacs, interconnected by flattened channels. Each individual thylakoid disk has an internal compartment, the thylakoid space (Fig. 3.21e). Chlorophyll and other colored pigments are embedded in the thylakoid membrane and make this membrane the only part of an entire plant that is truly green. The fact that we see most leaves as green shows how in-credibly abundant thylakoids are in nature. In addition to pigments, the thylakoid membrane contains members of an electron transport chain and, in some areas, many copies of an ATP-synthesizing enzyme (Fig. 3.21f).

An Overview of Photosynthesis

There is a beautiful symmetry to the metabolic processes of respiration and photosynthesis that is revealed by their nearly opposite overall equations. Earlier, we saw that the aerobic cellular respiration taking place in cells of a hummingbird or columbine plant could be summarized like this:

Glucose + Oxygen + ADP + Phosphate $\longrightarrow$
$\qquad$ Carbon dioxide + Water + ATP

The process of photosynthesis taking place in a red columbine, lupine, or other plant is a nearly opposite equation:

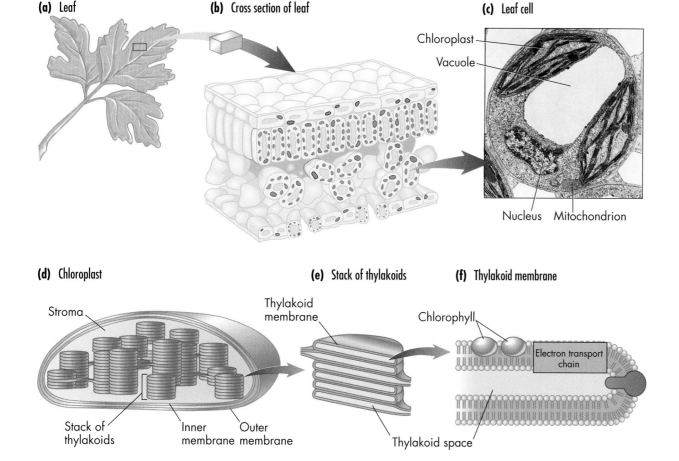

(a) Leaf

(b) Cross section of leaf

(c) Leaf cell

Chloroplast

Vacuole

Nucleus Mitochondrion

(d) Chloroplast

Stroma

Stack of thylakoids

Inner membrane Outer membrane

(e) Stack of thylakoids

Thylakoid membrane

(f) Thylakoid membrane

Chlorophyll

Electron transport chain

Thylakoid space

Figure 3.21 **Leaves and Photosynthesis**. (a,b,c) Cells in the interior of a leaf contain chloroplasts distributed throughout their cytoplasm. (d) Chloroplasts are surrounded by an outer and an inner membrane that enclose a space (the stroma) and a third set of membranes folded into disklike sacs called thylakoids. (e) Each thylakoid has its own membrane and internal space, and oftentimes, several sacs are stacked together. (f) The thylakoid membrane contains the green pigment chlorophyll and the proteins that conduct the light-dependent reactions of photosynthesis.

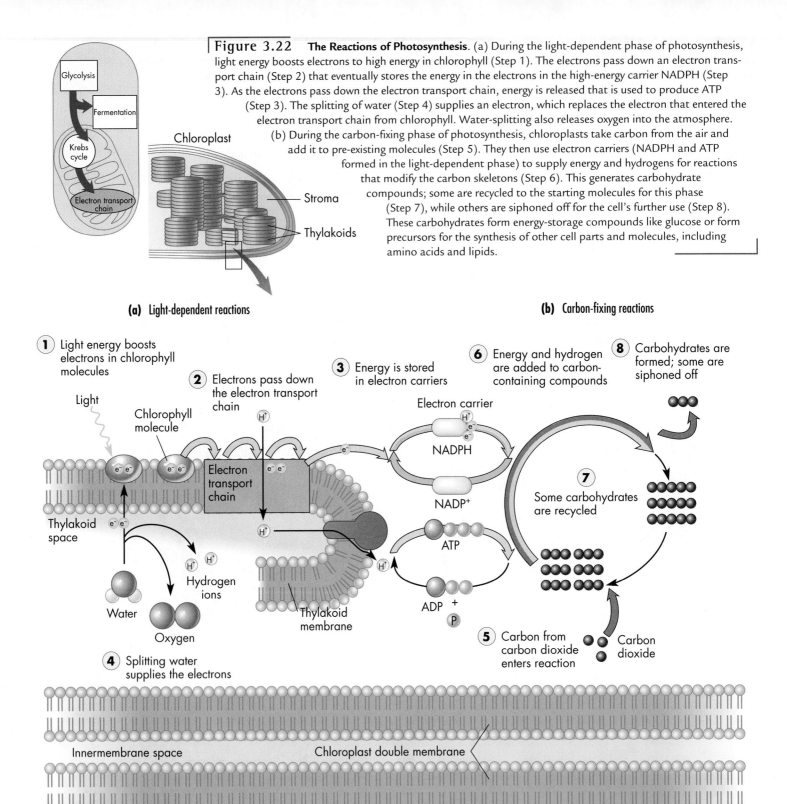

Figure 3.22 **The Reactions of Photosynthesis.** (a) During the light-dependent phase of photosynthesis, light energy boosts electrons to high energy in chlorophyll (Step 1). The electrons pass down an electron transport chain (Step 2) that eventually stores the energy in the electrons in the high-energy carrier NADPH (Step 3). As the electrons pass down the electron transport chain, energy is released that is used to produce ATP (Step 3). The splitting of water (Step 4) supplies an electron, which replaces the electron that entered the electron transport chain from chlorophyll. Water-splitting also releases oxygen into the atmosphere. (b) During the carbon-fixing phase of photosynthesis, chloroplasts take carbon from the air and add it to pre-existing molecules (Step 5). They then use electron carriers (NADPH and ATP formed in the light-dependent phase) to supply energy and hydrogens for reactions that modify the carbon skeletons (Step 6). This generates carbohydrate compounds; some are recycled to the starting molecules for this phase (Step 7), while others are siphoned off for the cell's further use (Step 8). These carbohydrates form energy-storage compounds like glucose or form precursors for the synthesis of other cell parts and molecules, including amino acids and lipids.

Glycolysis

Fermentation

Krebs cycle

Electron transport chain

Chloroplast

Stroma

Thylakoids

(a) **Light-dependent reactions**

(b) **Carbon-fixing reactions**

① Light energy boosts electrons in chlorophyll molecules

② Electrons pass down the electron transport chain

③ Energy is stored in electron carriers

⑥ Energy and hydrogen are added to carbon-containing compounds

⑧ Carbohydrates are formed; some are siphoned off

Light

Chlorophyll molecule

Electron carrier

NADPH

NADP$^+$

Electron transport chain

Thylakoid space

⑦ Some carbohydrates are recycled

ATP

ADP + P

Hydrogen ions

Water

Thylakoid membrane

Oxygen

④ Splitting water supplies the electrons

⑤ Carbon from carbon dioxide enters reaction

Carbon dioxide

Innermembrane space

Chloroplast double membrane

Carbon dioxide + Water + Light energy $\longrightarrow$ Glucose + Oxygen

Recall that when oxygen is present, aerobic respiration in mitochondria breaks down glucose into carbon dioxide and water and releases chemical energy that becomes stored in the expendable currency, ATP. Nearly the reverse takes place in photosynthesis: The chloroplast traps light energy, transforms it into chemical energy, and then uses that chemical energy to convert carbon dioxide and water into sugars, releasing *oxygen* as a waste product (**E** xplorer 3.2).

Clearly, living things must have both a source of energy and a means of releasing it, and for green plants and most other autotrophs, the direct energy source is sunlight. To understand photosynthesis, we'll follow the path of light striking a columbine leaf, and track the electrons whose energy level is boosted by sunlight. That pathway has two phases: a light-trapping phase and a carbon-fixing phase.

The Light-Trapping Phase of Photosynthesis

Our story of photosynthesis starts when sunlight falls upon a red columbine plant growing in a mountain meadow. Some of the solar energy strikes chlorophyll or other colored pigment molecules in the chloroplasts, and becomes trapped as it boosts electrons in the pigments to higher energy levels (Fig. 3.22a, Step 1, and **E** 3.2). Then, before the electrons drop back to their original energy levels, they leave the chlorophyll and pass down an electron transport chain much like the one in the mitochondrial membrane (Step 2). As the electrons travel down the electron transport chain, they release their energy bit by bit, as happened in the mitochondrion, and this energy is then stored in the chemical bonds of ATP and the supercharged electron carrier NADPH (Step 3). (NADPH is similar to the electron carrier NADH used by mitochondria, but it has an extra phosphate group, hence the "P" in NADPH.) The "hole" in the chlorophyll left by the energized electrons is filled by electrons stripped from a water molecule (H_2O) (Step 4). The hydrogens from the water molecule stay in the chloroplast, but the oxygen is released to the atmosphere, where humans, hummingbirds, and other organisms, including red columbines

and other plants, can use it in aerobic respiration. These events make up the first phase of photosynthesis, the **light-dependent** or **energy-trapping reactions.** The reactions are driven by light energy and can take place only when light is available, and they produce oxygen, ATP, and energized electron carriers.

The Carbon-Fixing Phase of Photosynthesis

Now we move to the second part of the story. The ATP and electron carriers produced by the energy-trapping reactions supply the energy needed for the second phase of photosynthesis, a biochemical cycle called the **carbon-fixing reactions** (Fig. 3.22b and **E** 3.2. (These are also called the **Calvin-Benson cycle** or sometimes the **light-independent reactions.**) "Carbon fixing" refers to a cell taking inorganic carbon from the air and joining it ("fixing it") to a biological molecule. The carbon-fixing reactions can go on day or night because they do not directly require light energy; they only require the energy carriers ATP and electron carriers produced by the light-dependent reactions.

During the carbon-fixing reactions, an enzyme in the stroma of the chloroplast first adds carbon dioxide from the air to a previously formed five-carbon compound, making a six-carbon compound that immediately breaks into two three-carbon compounds (Fig. 3.22b, Step 5). Then the chloroplasts transfer the energy stored in the bonds of ATP and electron carriers, and the hydrogens from the electron carriers, to the newly made three-carbon compounds (Step 6). Some of the newly formed three-carbon molecules are joined together and rearranged to regenerate the original starting molecules of the cycle (Step 7), while others can be siphoned off in energy-storing carbohydrate molecules (Step 8). Chloroplasts have to run this cycle with three carbon dioxide molecules to get out one three-carbon carbohydrate molecule. Cells can use the newly available carbohydrate to make the phosphorylated version of fructose (fruit sugar) in apples and the sucrose (table sugar) in some flower nectars. These compounds can then diffuse from the chloroplast into the fluid of the plant cell's cytoplasm, where they can be used in one of two ways. They can fuel the plant cell's own survival activities (and those of nonphotosynthetic plant parts such as roots) via the energy-harvesting steps of glycoly-

GET THE PICTURE

E xplorer 3.2

Photosynthesis

sis and aerobic respiration in the plant's mitochondria. Or the plant cell can make sugar as in flower nectar; cellulose, a structural material in cells walls, stems, leaves, and other plant parts; or starch, a form of long-term energy storage. Thanks to the formation of cellulose, we have the paper and wood we use daily. And the plant starch stored in rice, wheat, oats, potatoes, corn, and other crops are staples of the human diet.

3.6 What Is the Global Carbon Cycle and How Are We Affecting It?

It's hard to imagine how the metabolic pathways inside microscopic cells could possibly help perpetuate a cycle of planetary proportions. But a global carbon cycle moves vast amounts of carbon-containing compounds through the atmosphere, soil, water, and living organisms based on the carbon-fixing activities of autotrophs, the release of carbon dioxide by heterotrophs, and on geological phenomena such as erosion (as Chapter 25 will describe).

It is also hard to imagine that people could be altering that vast carbon cycle through their activities. But we are releasing millions of tons of extra carbon dioxide into the atmosphere each year by burning rain forests to clear new agricultural land and burning carbon-containing fossil fuels like coal and oil in factories and cars. This extra carbon dioxide causes extra heat to be trapped in the atmosphere. Scientists call this process the **greenhouse effect.** The greenhouse effect appears to be causing a slow but steady increase in average air and water temperatures called **global warming** (also discussed in Chapter 25). In addition, the destruction of hummingbird wintering grounds in Mexico not only contributes greenhouse gases, but also appears to be contributing to decreases in hummingbird populations.

As Figure 3.23a shows, we have experienced a steady rise in atmospheric carbon dioxide levels, largely due to our human agriculture, industry, and transportation. Why does this graph have a zig-zag line? It reflects yearly fluctuations of carbon dioxide levels, which are high in the winter and low in the summer. Plants photosynthesize more in the spring and summer; they take more carbon dioxide from the air in those seasons, and so the atmospheric levels of the gas

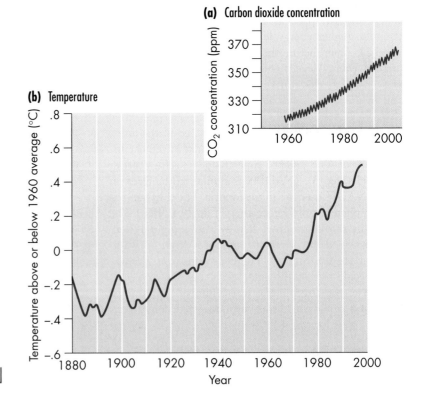

(a) Carbon dioxide concentration

(b) Temperature

Figure 3.23 **Atmospheric Carbon Dioxide and an Ominous Warming Trend.** (a) Measurements of carbon dioxide levels in the atmosphere have shown steady increases for many decades. Superimposed on that rise is a seasonal up-and-down fluctuation, high in the winter and low in the summer. What could cause such a seasonal pattern? (b) The world is much warmer today than it was a century ago. Many climatologists believe that the rise in carbon dioxide levels contributes to the warming of the globe due to the greenhouse effect.

go down a bit. Conversely, photosynthesis occurs at a slower rate in the winter, so carbon dioxide levels in the atmosphere go up. A graph like this shows quite convincingly how great an effect living organisms have on Earth and its atmosphere. But there is more evidence of that, too: The steady climb of the zig-zagging line shows the relentless increase of atmospheric carbon dioxide, which may be based mainly on our species' activities.

Aside from the changes in global air and water temperatures, wouldn't an increase in carbon dioxide *benefit* plants, since they take in the gas as a carbon source? Plant researchers have shown that they can add extra carbon dioxide to the plants growing in a controlled environment and measure an increase in the carbon-fixing reactions of photosynthesis and plants can grow faster. In an equivalent way, supplying more gasoline will speed up a car's engine. Whether plants actually grow faster or not in nature depends on leaf structure (Fig. 3.21). Gases are exchanged between the air and the inside of the leaf only through cell-lined pores in the leaf's lower surface (so-called **stomata**). If these pores open widely, more carbon dioxide diffuses in, but more water diffuses out, and this can lead to drying out. When carbon dioxide concentrations are higher in the air, the stomata can be partially closed and still admit the same amount of carbon dioxide. Partial closure has the advantage of retarding water loss. In regions where water is scarce, high carbon dioxide levels could make the difference between luxuriant and scraggly growth. This would seem like a boon to ecosystems, right? The problem is, some kinds of plants might respond better than others to the increased carbon dioxide, leading to imbalances as one plant species becomes overpopulated at the expense of others. If the benefit went mostly to desirable crop plants and not to the weeds that tend to choke them out, this might be a welcome outcome. Unfortunately, our limited understanding of natural ecosystems does not allow us to predict which plants would respond to heightened carbon dioxide levels and take over certain ecosystems.

Associated with the rapid increase in carbon dioxide in the atmosphere, we are experiencing an increase in global temperature as well (Fig. 3.23). In fact, all of the century's warmest years have occurred in the past decade. This warming trend appears to be affecting all kinds of plant and animal populations. As Chapter 25 explains in detail, the distributions of animals is changing, along with certain disease-causing organisms. What this next century of warming will mean to populations of hummingbirds in the Rocky Mountains—and to our own burgeoning numbers—we can only guess.

Chapter Summary and Selected Key Terms

Introduction
For their size and weight, hummingbirds have the highest rate of energy use among animals with backbones. Studying them and the flowers they feed on helps us understand **energetics** (p. 59): universal energy principles; how energy is transferred in the cell; and pathways for energy harvest and storage in living organisms.

Universal Energy Laws and Cellular Energy
Energy flowing from the sun is trapped via **photosynthesis** (p. 60) in the chemical bonds of sugar molecules. A hummingbird sips sugar concentrated in flower nectar and the bird's cells break down the sugar molecules in the presence of oxygen, releasing stored chemical energy; this process is called **cellular respiration** (p. 60). Energy transformations in cells are inefficient and energy is lost as heat with every step; thus it takes many pounds of sugary nectar to make and maintain a few ounces of hummingbird tissue. **Energy** (p. 60) is the ability to perform work or produce change. **Potential energy** (p. 60) is the energy available to do work. **Kinetic energy** (p. 61) is the energy of motion. Throughout the universe, energy can be transformed from one state to another, but the conversion is always inefficient and **heat** (p. 61) is produced. Heat is the random movement of atoms and molecules. **Entropy** (p. 61) is a measure of increasing disorder or randomness in a system.

Energy Flow in Living Things
In **chemical reactions** (p. 62), chemical bonds break and reform and atoms become rearranged into new molecules. The starting substances are **reactants** (p. 62); the new substances are **products** (p. 62). Some reactions release energy and are spontaneous; some won't proceed unless they absorb additional energy. The power for energy-absorbing reactions in the cell comes from energy-releasing reactions, and these two kinds are linked in chains of reactions called **metabolic pathways** (p. 63). **ATP** (p. 63) is the most common energy-carrying molecule in the cell. **ADP** (p. 63, 64) and **AMP** (p. 64) are forms that carry less energy. **Enzymes** (p. 65) are proteins that function as biological **catalysts** (p. 65); they facilitate and speed chemical reactions in the cell (p. 65). Enzymes bring molecules together with enough energy to reach the **transition state** (p. 65) and cross the energy barrier called the **activation energy** (p. 66). Enzymes have a deep groove or pocket, the **active site** (p. 66) where chemical reactants or **substrates** (p. 67) fit, forming an **enzyme-substrate complex** (p. 67). Together, all the enzyme catalyzed reactions and other reactions in the cell constitute **metabolism** (p. 67).

Energy Harvest
Metabolic pathways in the cell break down nutrient molecules and make energy

available for sustaining life. When oxygen is present, energy is harvested through an **aerobic pathway** (p. 67), also called **cellular respiration** (p. 68), that begins with the breakdown of glucose or **glycolysis** (p. 68), and continues with the **Krebs cycle** (p. 68). During this reaction series, six-carbon glucose molecules are dismantled, CO_2 is released, some energy is stored in ATP, and some in the **electron carriers** (p. 69, 71) NADH and $FADH_2$. During the final phase, the **electron transport chain** (p. 69, 73), electrons captured during the Krebs cycle in NADH and $FADH_2$ are joined with oxygen and water by moving down a chain of enzymes and pigment molecules located in the mitochondrion's inner membrane or **cristae** (p. 70). A large number of ATP molecules (36) form and move into a compartment of the mitochondrial **matrix**, enclosed by the mitochondrion's inner membrane (p. 70).

When oxygen is absent, energy is harvested through an **anaerobic pathway** (p. 68, 75) that begins with glycolysis and ends with **fermentation** (p. 69, 75). During the latter phase, the waste products lactic acid or ethanol and CO_2 form and the energy carrier $NADH^+$ is recycled. Energy metabolism is controlled by processes such as **feedback inhibition** (p. 77). The accumulation of ATP, for example, can build up and this feeds back and inhibits further production.

Energy for Exercise
During exercise, the muscles get quick energy through the **immediate energy system** (p. 77); longer term energy through the **glycolytic energy system** (p. 77); and energy for sustained activity through the **oxidative energy system** (p. 77). An imbalance in the amount of energy harvested and the amount used during exercise can contribute to obesity.

Plants Trap and Convert Light Energy to Sugars
Heterotrophs (p. 79) must take in preformed nutrient molecules, while **autotrophs** (p. 79) can make their own nutrient molecules through photosynthesis or, in a few cases, other chemical processes. Photosynthesis begins after **photons** (p. 79) of light leave the sun and reach earth's surface. The green pigments called **chlorophyll** (p. 80) and yellow/red/orange pigments called **carotenoids** (p. 80) absorb different parts of the color spectrum of light. Plants have green organelles called **chloroplasts** (p. 80) which contain chlorophyll in the membranes of **thylakoids** (p. 81) surrounded by the **stroma** (p. 81), a space filled with a watery solution. The **light-dependent** (p. 83) or energy-trapping reactions as well as the **carbon-fixing** (p. 83) reactions (also called the **Calvin-Benson cycle** or **light-independent reactions**) (p. 83) of photosynthesis both take place in chloroplasts.

Global Carbon Cycle
A planet-wide cycle circulates carbon-containing compounds through Earth's atmosphere, soil, water, and living organisms. Human activities like burning trees and fossil fuels are releasing extra carbon dioxide into the atmosphere, where it traps heat at the planet's surface in a **greenhouse effect** (p. 84) that is leading to **global warming** (p. 84).

All of the following question sets also appear in the Explore Life *E* electronic component, where you will find a variety of additional questions as well.

Test Yourself on Vocabulary and Concepts

In each question set below, match the description with the appropriate term. A term may be used once, more than once, or not at all.

SET I
(a) potential energy (b) kinetic energy (c) enzyme (d) entropy (e) activation energy

1. Heat-caused random atomic or molecular motion that is not available to do useful work
2. A stored form of energy, usually as a result of position or tension, as in a chemical bond or a boulder on the top of a hill
3. Protein molecules of specific shape that form a temporary union with one or more specific reactants
4. The energy of motion, released by the movement of atoms, molecules, or other objects of mass
5. Energy required to make the collision between reactants sufficiently forceful to affect the transition state

SET II
(a) transition state (b) active site (c) substrate (d) ATP

6. Reactants that participate in an enzymatically regulated reaction
7. The participant in an energy-releasing reaction that is widely used as a coupling agent with energy-absorbing reactions
8. The specific site within an enzyme to which the correspondingly specific site of a particular substrate molecule can attach by chemical bonding
9. The general term that includes enzyme-substrate complexes and other energetically unstable combinations of atoms or molecules

SET III
(a) ADP (b) ATP (c) $NADH^+$ (d) pyruvate

10. A three-carbon compound formed as an end product of glycolysis
11. A high-energy nucleotide that is a byproduct of glycolysis
12. An electron carrier in the process of fermentation
13. A nucleotide having two phosphate groups

SET IV
(a) aerobic pathway (b) anaerobic pathway (c) feedback inhibition (d) creatine phosphate (e) fermentation

14. The switching off of the glycolytic pathway that occurs when ATP levels become excessive is an example of this
15. A high-energy compound that can resupply ATP for about one minute of strenuous exercise
16. The set of reactions that produces ethanol or lactate and recycles an electron carrier
17. The set of reactions that culminates in the production of carbon dioxide and water

SET V
(a) chlorophyll (b) thylakoid (c) thylakoid membrane (d) thylakoid space (e) stroma

18. The only structural layer in a green plant that is truly green
19. A stack of sacs
20. Light-absorbing pigment molecules within a molecular complex
21. The region inside the chloroplast in which the carbon-fixing reactions occur
22. The region where the light-dependent reactions take place

Integrate and Apply What You've Learned

23. Our planet can support a larger number of vegetarians than meat eaters. Why?
24. Describe the connection between the electron transport chain and the actual synthesis of ATP.
25. When bread dough rises, it produces a characteris-

tic aroma. What is the relationship, if any, between the rising and the aroma?

26. If one wishes to lose weight, why is it better to concentrate on endurance-style exercise than sprint-style exercise?

27. What role does water play in the aerobic breakdown of glucose?

28. In what ways are the process of photosynthesis and the harvesting of energy dependent upon each other?

29. What happens to water during the light-dependent reactions of photosynthesis?

Analyze and Evaluate the Concepts

30. What adaptations allow the hummingbird to function as the vertebrate animal with the fastest metabolism?

31. Bread dough is usually made to rise with yeast, while cakes are leavened with the release of carbon dioxide from baking soda (sodium bicarbonate). If you made a cake and a loaf of bread using identical amounts of sugar and starch in both, which would end up tasting sweeter, and why?

32. We have used the analogy of currency for ATP. Using this same analogy, what substance(s) in living cells would a bank savings account represent?

33. If all the animals on Earth died, what would be the effect on the trees, flowers, and other plants? If all the plants in the world died, what would be the effect on people and the other animals?

34. A student friend of yours claims that plants cannot be poisoned by cyanide because plants do not employ mitochondria and the electron transport chain to harvest ATP, but instead obtain ATP by photosynthesis. Analyze this argument.

An Unavoidable Loss of Control

The puzzle of BCNS: Ervin Epstein, Patricia Hughes, and Hughes and Epstein in the clinic.

For Patricia Hughes, cancer is a fact of everyday life and has been so for the past 27 years. Before that, as a teenager in the midwest, she spent lazy summer days cooling off at the local swimming pool, splashing and sunbathing with friends. Her 17th summer was a turning point, though. That year began a long struggle with renegade skin cells that has claimed all of her adulthood so far and could consume another decade before a truly effective treatment comes along.

During the summer of 1972, Patricia's dermatologist found a small, pearly pink bump on her big toe and diagnosed it as a **basal cell carcinoma.** This patchy overgrowth of cells is the most common form of skin cancer—in fact, it is the most frequently diagnosed human cancer, with more than 600,000 new cases per year in the United States and rising. The typical skin cancer patient, however, is middle aged or older. So finding a lesion in a teenager was somewhat unusual—fair-

skinned though Patricia was and susceptible to sunburns. The discovery of a second basal cell carcinoma on her temple a few weeks later was more ominous. It inspired a detailed search of her entire skin surface—and the doctor found yet more tiny, pearly cancers.

Their multiple occurrence in a girl of her age was diagnostic of a disease called **Basal Cell Nevus Syndrome (BCNS)**—a rare inherited condition that starts in adolescence and is characterized by numerous skin cancers throughout life. Patricia's days in the sun were not the cause of her disease—but they contributed. She bought broad-brimmed hats, spread on sunscreen, and retreated inside long-sleeved clothing. With difficulty, she also learned to tolerate an unending series of surgeries and liquid-nitrogen freezing treatments on her endless string of new skin cancers.

Today, in her 40's, Patricia still sees a dermatologist every 4 to 6 weeks and, she says, she "can easily have 10 new carcinomas every visit." Her face, arms, chest, and back bear a disheartening record of un-pigmented zones, suture lines, and skin grafts from hundreds of excisions. "When you look at yourself in the mirror," she remarks, "and see this line and that line, you think, 'Gee, I've been through a lot. I feel kind of beat up on.'" But she's accepted her situation with remarkable grace. She actually feels lucky, since BCNS is not fatal and some of her close friends have had to face potentially lethal cancers. Patricia goes through her daily routine as a software product consultant and remains hopeful that "still in my lifetime, some treatment will come up" to alleviate the disease she's been fighting for so long.

That treatment could come through the study of BCNS patients at the University of California, San Francisco—a study Patricia has been part of for years. The study's principal investigator, Ervin Epstein, is also hopeful for Patricia and his other experimental subjects. And he is trying to find answers for the millions of people who will develop one or more basal cell carcinomas during their later years. If caught early enough, these cancers are "essentially 100 percent curable," Epstein explains. But his goal is to thoroughly understand how cells grow and divide at the level of genes, proteins, and other biological molecules. With that detailed knowledge of growth and division, he and his colleagues could possibly devise much more effective treatments and even preventative measures for skin cancers. Then there would be both hope and answers for people with more common types—cancers that spring up from too much sun over too many years—as well as those who suffer inherited diseases like BCNS with heartbreaking recurrences month after month. An important offshoot from any future progress toward skin cancer causes and cures could be ways to fight cancers in other parts of the body as well.

Patricia Hughes's long battle with skin cancer and Ervin Epstein's study of basal cell carcinoma are fitting subjects for this chapter. Here we explore the **cell cycle**—the stages of growth, duplication, and division in living cells. There are two basic types of cell division in humans and other eukaryotic organisms: mitosis and meiosis. **Mitosis** is the process of nuclear division and the formation of two daughter cells genetically identical to each other and the original parent cell. **Meiosis** takes place in the reproductive organs and is the type of cell division that

occurs during gamete (egg or sperm) formation; meiosis produces four cells with half the normal chromosome number. Of the two types, cancers involve the cell divisions of mitosis. As you explore the cell cycle, you'll see that a cancer starts when the normal controls over mitosis go haywire within a single cell. Instead of starting and stopping at appropriate times, cancerous cell division goes on and on until a mass of cells called a **tumor** results and displaces or invades other tissues. The key to understanding both cancer and the normal growth and reproduction of cells and organisms lies in the intricacies of the cell cycle. That makes it one of the more important and fascinating topics in modern biology.

As we move through this chapter on the cell cycle, you'll learn how radiation kills cells and organisms and how Down syndrome arises in a human embryo. You'll also encounter protein with the strange name of sonic hedgehog that is involved in cell to cell signalling and in turning on cancerous cell divisions. Finally, you'll find the answers to these questions:

1. What are the patterns of normal cell growth and cell division?

2. What is mitosis and what happens during cell division?

3. What controls the timing and location of cell division?

4. What is meiosis, the special cell divisions that precede sexual reproduction, and what happens during this process?

4.1 Patterns of Cell Growth and Cell Division

Because Patricia inherited genes that led to BCNS, her skin has produced hundreds of tumors. To understand how her skin cancers arise, we first need to understand normal cell growth and cell division in skin. Then we can compare these typical patterns to the rare case of genetically determined BCNS in which cell growth and division patterns are abnormal and the cell cycle goes on uncontrolled.

Where and When Do Skin Cells Divide?

Your skin is your body's largest organ, and it forms a wonderful protective barrier around the muscles and other tissues. The outer skin surface layer is continually abraded away by the rubbing of clothes or the scrubbing of a washcloth in the shower. We rarely notice the millions of tiny, scaly bits of skin we lose each day except, perhaps, when we see household dust on tables and shelves and realize that some of that dust is actually sloughed-off skin cells. Only when a flaked-off piece of scaly skin is a little bigger than normal and gets caught

(a) Normal skin

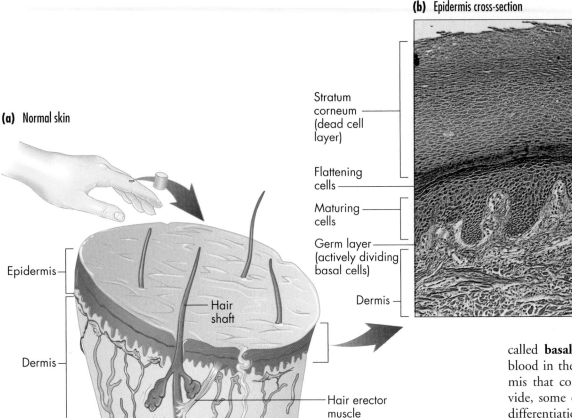

(b) Epidermis cross-section

Stratum corneum (dead cell layer)

Flattening cells

Maturing cells

Germ layer (actively dividing basal cells)

Dermis

Epidermis

Dermis

Subcutaneous tissue

Hair shaft

Hair erector muscle

Sensory receptor

Sweat gland

Blood vessels

Fat cell

Nerve fiber

Figure 4.1 Where Do Cells Divide in Skin? (a) A section of skin with its three zones: epidermis, dermis, and subcutaneous tissue that binds skin to underlying organs. Cells of the epidermis (b) arise from divisions in the basal cell layer (also called the germ layer). As they mature, they move towards the surface, accumulate keratin proteins, and then die, forming the tough, dry protective outer layer of skin (or stratum corneum).

in your hair do you recognize it as dandruff. How does the skin normally replace all this lost exterior material? The answer is a controlled replacement process of the outermost layers by deeper, dividing layers.

Skin consists of two main zones, a thin outer zone, or **epidermis,** and a thicker zone underneath, or **dermis** (Fig. 4.1a). Below the dermis is a **subcutaneous** layer, mostly fat (adipose) cells that bind the skin to underlying organs. The dermis contains tiny blood vessels, sweat glands, hair roots, and nerve endings sensitive to heat and touch. None of these structures plays a direct role in the tumors of BCNS.

The epidermis is the skin's protective zone and it consists of several layers (Fig. 4.1b). The layer nearest the dermis is a **germ layer** or dividing layer (also called the **basal layer**). Because this dividing region lies at the base of the epidermis, the cells there are sometimes

called **basal cells.** This dividing layer is close to the blood in the dermis and is the only part of the epidermis that contains reproducing cells. As these cells divide, some of them get pushed up into a layer of cell differentiation where they take on their mature characteristics. As they mature, the cells become tightly joined by cell junctions (review Fig. 2.29), they accumulate large quantities of the cytoskeletal protein keratin, and they become very flat, like miniature cookie sheets or pizza pans. As they move up further towards the body surface, their supply of nutrients becomes limited, and the cells die. The outermost portion of the epidermis is a dry layer consisting of sheet after sheet of flat, dead cells that are easily rubbed away.

Which skin layer, then, is altered—and in what way—to give rise to Patricia's skin tumors? To determine which cells give rise to a skin tumor, a surgeon can remove a basal cell carcinoma—say, on the palm of her hand—with a sharp instrument and then slice the cluster of cells very thinly to observe sections of it (Fig. 4.2a,b). Comparing Figure 4.2b to the normal skin section in Figure 4.1b, what do you see? You should notice a clump of cells in the dermis with the characteristics of cells that belong in the epidermis. What has happened is that basal cells from the dividing germ layer of the epidermis have divided too many times and made a mass that extends downward into the underlying dermis, an "overgrowth" process biologists call **proliferation** (an increase in cell numbers). To understand Patricia's skin tumors, we need to look more closely at these dividing cells. In the process, we'll discover more about normal cell division, as well.

(a) Tumor on skin

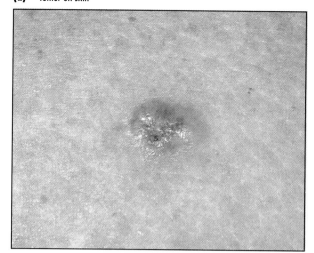

(b) Section of a BCNS tumor

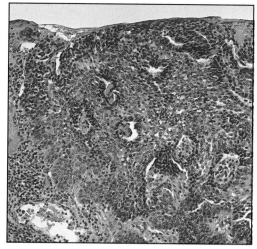

Figure 4.2 **Where Do Basal Cell Skin Cancers Arise?** (a) A basal carcinoma typical of the Basal Cell Nevus Syndrome (BCNS). (b) A cross section through a basal cell carcinoma (compare to Figure 4.1b). Note that the epidermal cells have proliferated, raising a bump at the skin surface.

The Cell Cycle: Action in the Dividing Layer

The cells in the skin's dividing layer share a general strategy of cell reproduction with other living cells: They take in nutrients, increase in size, and then divide into two daughter cells. These two daughter cells may then, themselves, go through a period of growth followed by division. This alternation of growth and division is called the **cell cycle** (Fig. 4.3). During the **division phase** of the cell cycle, microscopy reveals a partitioning of the cells' internal organelles, and then a dividing of the cell in two. The growth period between two division phases is called **interphase.**

As we saw in Chapter 2 (Fig. 2.17), the material inside of a cell is not a homogeneous mass like a loaf of bread that you can just slice to yield halves containing very similar materials. Cells contain various organelles in differing amounts and places. Some of these organelles, like the ribosomes, mitochondria, Golgi apparatus, and so on, are present by the thousands or millions in each cell. If the cell is just cleaved in two, each half will have sufficient numbers of these organelles to ensure smooth functioning. Some organelles, however, are present in a single copy and may be offset into one part of the cell. Special mechanisms have evolved that distribute these "singleton" organelles correctly to daughter cells. These organelles include the chromosomes inside the nucleus with its genetic contents and the centrioles in the cytoplasm.

Chromosomes

Patricia inherited her BCNS; this means her cells include the instructions for the disease in their genetic in-

formation. As we saw in Chapter 2, **chromosomes** contain a cell's hereditary information—the instructions the cell uses to construct itself—and thus the organism of which it is a part. We'll look at chromosomes in more detail in Chapter 6. The important thing, here, is that cells need an orderly distribution mechanism so that as they divide, each of the two daughter cells receives the same hereditary information and both have the same information as did the parent cell.

All organisms contain chromosomes, although the size, shape, and number of these hereditary structures differ from species to species. Eukaryotic cells have 2 or more chromosomes: a cell in a roundworm that might infect your puppy's intestines contains 4 chromosomes; the nucleus of a cell from a giant sequoia leaf contains 22; a goldfish nucleus contains 104; and the nucleus in a human skin cell has 46.

Chromosomes in the Cell Cycle

When a cell divides, each new offspring cell receives its own set of chromosomes containing an identical copy of the hereditary material. For this reason, the structure, duplication, and distribution of the chromosomes are central to the cell cycle. Chromosomes consist of protein and DNA, the long molecule that bears hereditary information. You can see chromosomes as distinct bodies by looking through a microscope, but they are visible only during a certain portion of the cell cycle, the division phase (review Fig. 4.3). At this time, the DNA is wound up into tight bundles we can see—bundles that resemble a kite string wound around a spool. The rest of the time, in interphase, the DNA is unwound and spread out, like a loose pile of kite string on the ground. It's impossible to visualize individual chromosomes in this state.

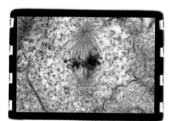

What's the importance of the DNA "packaging" in chromosomes?

Figure 4.3 **Overview of the Cell Cycle.** Cells cycle through a period of growth, called interphase, followed by a division phase in which the parental cell divides into two daughter cells.

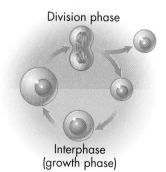

Division phase

Interphase (growth phase)

(a) The chromosome cycle

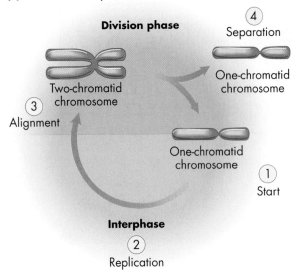

Division phase

④ Separation

Two-chromatid chromosome

One-chromatid chromosome

③ Alignment

One-chromatid chromosome

① Start

Interphase

② Replication

(b) Electron micrograph of a chromosome

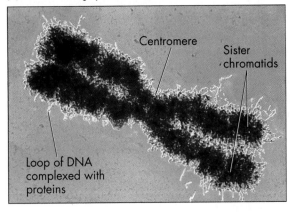

Centromere

Sister chromatids

Loop of DNA complexed with proteins

(c) Sketch of a chromosome

Each chromatid has a single DNA molecule

Sister chromatids

Kinetochore

Centromere

Spindle fibers

Figure 4.4 **The Chromosome Cycle and Chromosome Structure.** (a) During the cell cycle, the chromosomes also cycle. After the division phase, each chromosome is unreplicated (Step 1) and consists of a single chromatid. During interphase (Step 2), each replicates to become a two-chromatid chromosome (Step 3). Finally, during the next division phase, each chromosome aligns in the middle of the cell and the two chromatids separate (Step 4). (b) An electron micrograph reveals that a replicated chromosome consists of two very long DNA molecules complexed with proteins and organized into two sister chromatids joined at the centromere. (c) Each chromatid has a single DNA molecule and the centromere contains the kinetochore, to which spindle fibers attach and move the chromosome.

At the end of the division phase, each chromosome consists of a single long rod called a **chromatid** (Fig. 4.4a, Step 1). During interphase, the DNA in the chromosome undergoes **replication,** the copying of one DNA molecule into two identical DNA molecules (Step 2). When chromosomes become visible once again at the beginning of the division phase (Step 3), each chromosome consists of two rods (two chromatids), with each rod containing just one double-stranded DNA molecule. (Note that replication by itself does not change the number of chromosomes in a cell. Each chromosome simply doubles from one chromatid to two.) At this stage, chromosomes undergo **alignment,** aligning themselves in the middle of the cell as we will investigate shortly. Finally, in Step 4, **separation** takes place; the replicated chromosome with its two chromatids splits into two chromosomes, each with one chromatid made up of a double-stranded DNA molecule.

To summarize, the chromosome cycle involves one rod replicating to two rods during interphase, the replicated chromosome aligning in the cell, and then the two rods separating to different cells during the division phase.

Chromosome Structure

Now, let's look a bit closer at the anatomy of a chromosome. Figure 4.4b shows an electron micrograph of a two-chromatid chromosome in the division phase of the cell cycle. You can see that each individual rod is formed by great loops of DNA complexed with protein that extend out from the axis of the condensed chromosome. In these loops, the DNA winds around proteins like string wrapped around thousands of tiny spools. Because each individual rod, or chromatid, in a chromosome contains a single extraordinarily long double-stranded DNA molecule (Fig. 4.4c), this packaging decreases the tangling of DNA during chromosome alignment and separation in the division phase.

The chromosomes we have just seen consist of two identical rods, called **sister chromatids,** held together at a single point, the **centromere** (Fig. 4.4b,c). Located at the centromere is a group of proteins called the **kinetochore,** and this structure attaches to long fibers in the cell called **spindle fibers** (Fig. 4.4c). These spindle fibers move chromosomes around, for reasons we'll see shortly. Each chromatid has just one double-stranded DNA molecule (Fig. 4.4c), and the DNA molecules in two sister chromatids are identical. As cell division progresses, the two sister chromatids separate from each other (Fig. 4.4a, Step 4). One of the sister chromatids ends up in one of the daughter cells, and the other, genetically identical chromatid ends up in the other daughter cell. Thus, immediately after division, each chromosome consists of a single chromatid with a single DNA molecule.

The Cell Cycle

Cells proliferate unchecked in basal cell carcinomas like the ones Patricia Hughes develops every month. Because the alternating periods of growth and division are so central to understanding this proliferation, we must continue exploring the stages of the cell cycle. Recall that the cell cycle has a division phase and a period of growth, the interphase. Biologists divide interphase into three portions called G_1, **S,** and G_2. G_1 and G_2 stand for *Gap 1* and *Gap 2,* because they are gaps in the cell cycle that come between the time of chromosome division, and the time of chromosome replication. The S stands for synthesis of DNA. And biologists usually call the division phase **M,** for mitosis (Fig. 4.5).

Different cells require different amounts of time to pass through the entire cell cycle. Certain cells, like mature nerve cells of the brain, arise in the embryo and never divide again. In contrast, cells in your bone marrow replace worn-out blood cells by dividing every 18 hours or so. Cells that line your stomach divide in a cycle about 24 hours long. Most skin cells, even in the dividing basal layer (Fig. 4.1), replicate only every week or so. A basal cell carcinoma, however, can undergo new rounds of the cell cycle every 67 hours. Clearly, the cell cycle can accelerate, and we'll discuss the reasons later.

G_1: Active Growth

G_1 and G_2 are important growth phases. During the G_1 **phase,** cells manufacture new proteins, ribosomes, mitochondria, and other cell components in preparation for DNA synthesis and cell division. The length of the G_1 phase determines the length of the entire cell cycle. G_1 can be quite short or very long, depending on the type of cell, its role in the organism, and conditions in its environment. For example, skin cells normally have a long cell cycle and a long G_1 (a few days). That can change, however. If a wound removes some of those cells, the G_1 phase may shorten in the skin cells at the edge of the wound, speeding up growth and division, and enabling the wound to heal rapidly. (Skin cancer cells have an accelerated cycle a bit like a wound that never heals.) Plants can have an analogous process. In an aspen tree, if a deer or beaver nibbles away the bark, the bark-forming cells below can enter a shortened G_1 phase, rapidly producing new bark, which protects the damaged area. Unlike in cancer cells, however, in both normal skin and tree bark the faster cell cycling stops when the wound is healed. Following G_1, eukaryotic cells enter the synthesis phase, S.

S: Synthesis of DNA

After the G_1 phase, cells enter the S phase or **interphase,** during which enzymes replicate the double-stranded DNA molecule in each chromosome (see Chapter 6 for details). Cells also synthesize certain proteins necessary for maintaining chromosome structure as the S phase proceeds. When the S phase ends, each chromosome consists of two identical and parallel double-stranded DNA molecules packaged into two chromatids. Every stretch of DNA in a chromosome is copied once and only once during each S phase. After copying is complete, the cell enters G_2. As you might imagine, processes occurring during S phase provide physicians with important targets for cancer therapies. Nondividing cells don't go through S, so agents that block the events of the S phase affect only dividing cells, such as Patricia's skin tumors. Such therapeutic agents include substances that mimic DNA subunits and compounds that block DNA synthesis.

G_2: Preparation for Division

During the G_2 **phase,** the cell continues to synthesize many proteins. If a researcher artificially blocks this synthesis, the cell fails to divide, suggesting that some pro-

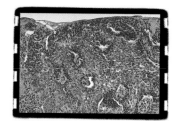

What happens if cells of the skin's basal layer keep dividing continuously?

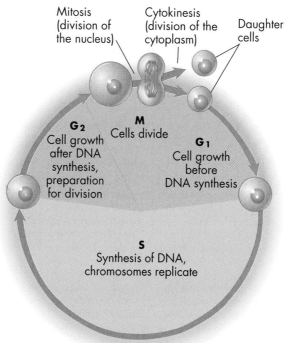

Figure 4.5 **Phases of the Cell Cycle.**

teins synthesized during G_2 promote mitosis, division of the nucleus. When all the necessary proteins have been synthesized, the cell leaves the final growth phase and begins to divide.

M: The Cell Divides

The **M phase** generally consists of two main events: mitosis, the division of the nuclear material, and **cytokinesis,** the division of the cytoplasm (Fig. 4.5). Both mitosis and cytokinesis are vitally important for the equitable distribution of genetic material and other cell components, so we'll talk about each one separately. But keep in mind what's at work: mechanisms to insure

that a cell gives rise to identical daughter cells with the same genetic information and role in the organism. Without these carefully controlled and timed divisions, nothing would keep a toe cell from becoming a liver cell, or a skin cell from becoming a tumor.

4.2 Mitosis and the Mechanisms of Cell Division

In the previous section, we've seen that cells in the lower, basal layer of the epidermis pass through the cell cycle,

Figure 4.6   Chromosome Choreography: The Stages of Mitosis in an Animal Cell. The photos are micrographs of mitosis in whitefish cells magnified 450 times.

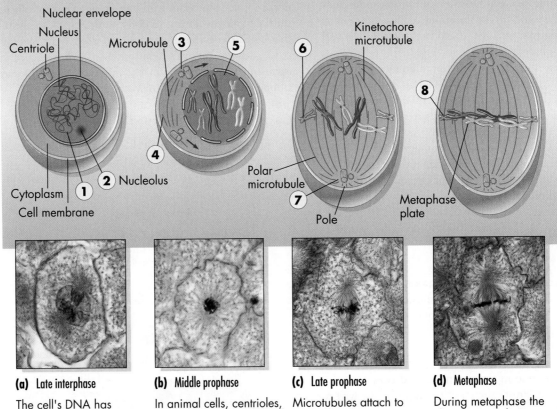

(a) Late interphase

The cell's DNA has already replicated during S in the previous interphase. As the cell enters the first part of mitosis, called prophase, the DNA changes from its diffuse and tangled state in interphase (1) to become more tightly packaged. Also, the nucleolus, a dense organelle within the nucleus (2), disperses.

(b) Middle prophase

In animal cells, centrioles, cylindrical groups of short microtubules, duplicate and organize spindle microtubules (3). Plant cells lack centrioles. Centrioles separate and move toward opposite ends, or poles, of the cell, spinning out the mitotic spindle (4). As the nuclear envelope disperses (5), the spindle invades the nuclear region.

(c) Late prophase

Microtubules attach to chromosomes by kinetochores (6). The chromosomes then jostle back and forth as the polar microtubules, which suspend and move the chromosomes, interact with the kinetochore microtubules, and the centrioles complete their migration to the poles (7).

(d) Metaphase

During metaphase the chromosomes become aligned on the metaphase plate (8), a plane lying halfway between each pole.

thereby generating new skin. If their growth is uncontrolled, the cells instead grow into basal cell carcinomas. The division phase, or M phase, is therefore a crucial part of the cell cycle. So let's see how a normal cell divides.

The Phases of Mitosis

Whether in the skin or other tissue, the mitotic dance of the chromosomes goes on continuously in a dividing cell. Biologists have named several prominent phases of mitosis to simplify its description. The events of these phases are summarized here and illustrated and explained in more detail in Figure 4.6. Recall from Figure 4.4 that chromosome replication has already occurred during the S phase, before mitosis begins.

Prophase

In **prophase** (*pro* = before), the chromosomes condense and become visible, the nucleolus (review Fig. 2.25) disappears, and a mitotic spindle forms (Fig. 4.6 a–c). The mitotic spindle is a bundle of certain filaments of the cytoskeleton (microtubules) that suspends and moves the chromosomes. In late prophase, a stage cell biologists call prometaphase (*meta* = middle), the nuclear envelope disappears, the spindle enters the nu-

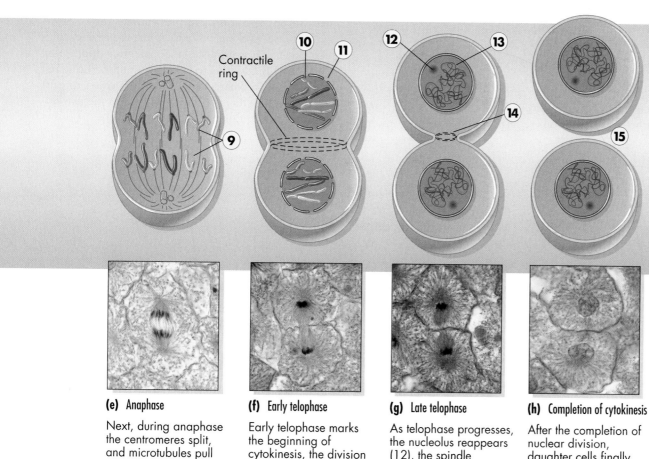

(e) Anaphase

Next, during anaphase the centromeres split, and microtubules pull sister chromatids apart, toward opposite poles (9).

(f) Early telophase

Early telophase marks the beginning of cytokinesis, the division of the cytoplasm. The daughter chromatids (now independent chromosomes) arrive at each pole (10), and the nuclear membrane re-forms around the chromosomes (11).

(g) Late telophase

As telophase progresses, the nucleolus reappears (12), the spindle dissolves, and the chromosomes reel out again into a tangled mass of DNA and protein (13). In addition, in late telophase in animal cells, a contractile ring tightens around the cell's midline where the metaphase plate had been, creating a furrow (14).

(h) Completion of cytokinesis

After the completion of nuclear division, daughter cells finally separate in cytokinesis (15).

clear region, and the chromosomes attach to the spindle at the centromere (Fig. 4.6c). Individual chromosomes jostle back and forth, as if they were involved in a tug-of-war between the two poles.

Metaphase

In **metaphase,** the spindle microtubules align the chromosomes in the middle of the spindle, each chromosome lined up independently of the others in a single plane, called the metaphase plate, in the middle of the spindle (Fig. 4.6d). The metaphase plate is a bit like the flat surface of a grapefruit that has been sliced in half.

Anaphase

In **anaphase** (*ana* = apart, opposed), the centromeres split and the spindle microtubules separate the chromatids (now called chromosomes) and pull them toward opposite poles (Fig. 4.6e). In a time-lapse movie (*E*xplorer 4.1), the chromosomes appear to be dragged through a viscous fluid by spindle fibers attached to the centromere. Figure 4.7 describes in more detail the role of the spindle in separating the chromatids during anaphase.

Telophase

In **telophase** (*telo* = goal), the chromosomes arrive at opposite poles of the cell, and the preparatory events are reversed: the nuclear envelope reappears, the spindle dissolves, and so on (Fig. 4.6f,g). Once telophase is over, the division of the cell nucleus (mitosis) is complete. Now the cell has two nuclei carrying identical sets of chromosomes. The M phase continues, however, with cytokinesis, the division of the cytoplasm (Fig. 4.6h).

There is a key difference between the replication of a cell involving mitosis and the reproduction of a person or a rose involving meiosis. When these large, complex organisms reproduce, the parent and offspring both generally continue to exist. In contrast, during cell reproduction, the parent cell (Fig. 4.6a) ceases to exist as an entity and its parts are distributed to the two offspring cells (Fig. 4.6h).

GET THE PICTURE

*E*x*plorer 4.1*

Chromosomes and Mitosis

(a) Spindle forms

(b) Sister chromatids separate and spindle elongates

(c) Fluorescence microscopy of spindle

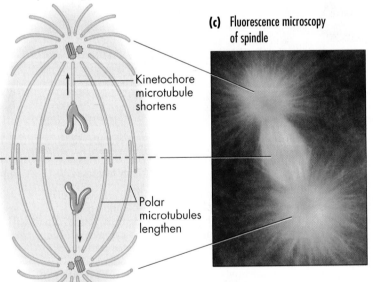

Figure 4.7 **What Moves the Chromosomes?** (a) As a spindle forms in animal cells, microtubules radiate from the centrioles, the cell's microtubule-organizing centers, like a star or aster. Longer polar microtubules extend from each pole toward the other and over-lap at the metaphase plate or cell's equator. Another set of microtubules, the kinetochore microtubules, lead from the centrioles at the poles to each chromatid, attaching at the kinetochore, a special microtubule attachment site at the centromere of each chromosome. (b) During anaphase, sister chromatids move towards the poles, and the cell lengthens. The chromosome's centromere splits, and the sister chromatids separate, moving toward opposite poles. Overlapping microtubules at the cell's center cause the two poles to move apart in preparation for division into two cells. (c) In this anaphase cell, the spindle fibers are labeled with a green fluorescent dye, and the chromosomes with a red fluorescent dye.

Labels in figure: Centriole; Polar microtubule; Kinetochore microtubule; Chromosome; Cell's equator; Kinetochore; Mitotic spindle; Aster; Kinetochore microtubule shortens; Polar microtubules lengthen

The Mitotic Spindle

Figure 4.6 shows where the chromosomes move during mitosis. It doesn't, however, reveal the mechanism that *causes* the chromosomes to move. This mechanism involves the **mitotic spindle,** a microscopic scaffolding of fibers made of microtubules that suspends and moves the chromosomes (Fig. 4.7). **Spindle fibers** extend from the **centrioles,** organelles consisting of two short cylinders that organize the cell's network of microtubules. Some spindle fibers attach directly to special structures (kinetochores) at the centromeres of chromosomes, and pull the two chromatids toward the poles (Fig. 4.7). Other spindle fibers overlap at the cell's center and cause the two poles to move apart from each other. Knowing how the spindle works is also important in designing anticancer drugs. The physician wants to block mitosis in tumor cells. But how?

Cytokinesis: The Cytoplasm Divides

In order for two cells to appear where there was once one, the original cell must cleave in two. Toward the end of mitosis, the cytoplasm of most plant and animal cells begins to divide by means of **cytokinesis** (literally, "cell movement") (Fig. 4.6f–h). In both animal and plant cells, new plasma membranes form at or near the place once occupied by the chromosomes during metaphase and separate the two nuclei into the two new cells. The details of cytokinesis vary because animal cells have a pliable outer surface, while plant cells have a rigid cell wall.

Cytokinesis in Animal Cells Animal cells divide from the outside in, as a circle of microfilaments called a **contractile ring** pinches each cell in two. Late in mitosis, a ring of filaments containing the contractile protein, **actin**, creates a furrow in the cell surface in much the same way that a purse string tightens around the neck of a purse (Fig. 4.6g–h). The furrow deepens, and eventually squeezes the cell in two. You can watch this take place in *E* 4.1.

Cytokinesis in Plant Cells Plant cells, with their rigid cell walls, retain their shape throughout the cell cycle, dividing from the inside out (Fig. 4.8). At the end of mitosis, vesicles filled with cell wall precursors collect in the center of the cell. The separate vesicles gradually fuse, forming a central partition, or **cell plate,** made of cell-wall material sandwiched between plasma membranes. This fusion completes the central partition and divides the plant cell into two identical daughter cells, which remain connected. Each cell now has its own nucleus and is ready to begin interphase.

Applying Our Knowledge of Mitosis to Cancer Therapy

Patricia's multiple skin cancers result from an overgrowth of cells. In her cancer and in all cancers, a cell starts growing and dividing by means of the phases we've just explored, but then continues unchecked, forming a tumor that can displace or invade other cells and tissues. The real key to cancerous cell division is understanding how the normal control of the cell cycle is lost, and we'll return to that shortly. In the meantime, some aspects of mitosis apply to the treatment of cancer once it arises.

The best way to treat most tumors, including BCNS lesions, is to surgically remove them. This makes sense for basal cell carcinomas because, being at the surface, they are readily accessible. Furthermore, while the masses they form crowd other cells, they do not grow especially aggressively, nor do they tend to invade and take over adjacent tissues. Once the skin tumor is removed or zapped through freezing with liquid nitrogen, it seldom returns. Certain other tumors, however, such as melanomas (pigmented skin tumors), and cancers of the breast, ovaries, and pancreas are more aggressive and invasive and often do require other methods in addition to surgery, such as chemotherapy and radiation.

Chemotherapy and the Spindle

We saw earlier (Fig. 4.7) that the spindle is crucial for moving chromosomes around in mitosis. Does this

What's the significance of the patched and sonic hedgehog genes?

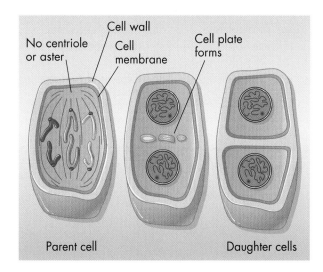

Figure 4.8 **Cytokinesis in Plant Cells.** A cell plate partitions a dividing plant cell from the inside out.

Labels in figure: Cell wall; No centriole or aster; Cell membrane; Cell plate forms; Parent cell; Daughter cells

open an avenue for cancer therapy? Researchers thought so. They reasoned that if they could block the action of the spindle, they could block cell division and hence slow cancer growth. In fact, several **chemotherapy** drugs now used to treat cancer attack the spindle. Taxol and related compounds, extracted from the bark of the Pacific yew tree, have proven useful in treating carcinomas of the cervix and ovary. Taxol binds to microtubules and blocks their breakdown into protein subunits. Thus, a cancerous cell gets stuck part way through mitosis and, with the spindle still in place, can't divide into new malignant daughter cells. Other drugs, like vinblastine from the periwinkle plant, stop the division of cancer cells by blocking the formation of the spindle in the first place. These chemotherapeutic drugs have powerful side effects, however, and so pharmaceutical researchers are always trying to develop new and better treatments.

Radiation Therapy

Have you known someone with cancer who received radiation therapy? During this treatment, technicians aim a radiation beam precisely at the patient's tumor and shield other parts of the patient's body from the beam. How does radiation therapy work? And why is it necessary to shield the patient's body?

Radiation can break chromosomes (Fig. 4.9a). If such breakage occurs in a cell that does not divide, the cell will still be able to carry on in a normal fashion because most body cells contain two copies of each chromosome. If chromosome breakage takes place in a cell that later divides—a skin cell, for example, or a tumor

cell—the consequences are often quite different, and your knowledge of mitosis can help explain why. Let's say radiation breaks a chromosome and leaves a fragment that is no longer attached to a centromere. What would happen to the chromosome fragment during cell division? It might not be distributed normally to the daughter cells. This is because the centromere is the only part of the chromosome that is directly attached to spindle microtubules, which pull each chromatid to one of the cell's poles during mitosis (review Fig. 4.7b). In a case of chromosome breakage like this, one cell might end up with an extra chromosome part, while the sister cell ends up missing this chromosome part (Fig. 4.9b). The resulting genetic imbalance can alter the cell's information and lead to its death.

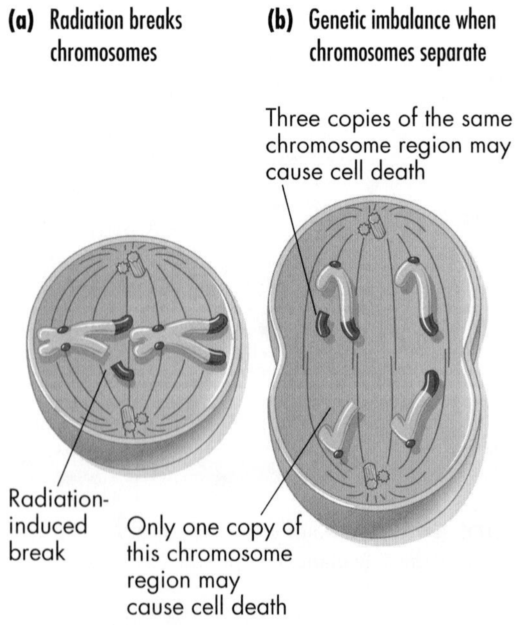

The most rapid cell division taking place in a cancer patient often involves the cancer cells themselves. Therefore, well-aimed radiation therapy damages and kills those cells more than it hurts the patient's surrounding cells. Like all humans, however, cancer patients have other rapidly dividing cells in addition to the cancerous cells. These include the precursors of red and white blood cells, skin cells, and cells of the intestinal lining. Not surprisingly, these cell types can be damaged during radiation therapy, and their disruption explains some of the short-term side effects of radiation therapy: anemia (too few red blood cells), susceptibility to infection (too few white blood cells), hair loss (damaged skin cells), and nausea (damaged intestinal cells). Ironically, long-term side effects include an increased risk of cancer, since, as we say, radiation can cause genetic mutations and these can lead to the uncontrolled divisions in a tumor. The disruption of rapidly dividing cell types also explains the destructive, often lethal effects of exposure to high levels of radiation from a hydrogen bomb or an accident at a nuclear power plant. So-called radiation sickness is like a severe set of symptoms from radiation therapy. These facts about mitosis underscore the devastating consequences nuclear war or nuclear accidents could have for people and most other life forms on Earth. They also highlight the restorative powers of mitosis, which keep most of us healthy most of the time.

(a) Radiation breaks chromosomes

(b) Genetic imbalance when chromosomes separate

Three copies of the same chromosome region may cause cell death

Radiation-induced break

Only one copy of this chromosome region may cause cell death

Figure 4.9 **How Radiation Kills Cells.** (a) Radiation, including the kind applied during cancer therapy, can break chromosomes. (b) A chromosome fragment unattached to a centromere will not move toward a pole when the centromeres separate. Thus, a daughter cell can have too few or too many copies of some chromosome parts. The resulting genetic imbalance can kill the cell.

4.3   What Controls Cell Division?

Let's return, now, to Patricia's Basal Cell Nevus Syndrome (BCNS) and the frequent formation of basal cell carcinomas. The growth of these tumors is evidence that some mechanism is missing that would normally stop her skin cells from dividing so rapidly and continuously. Some mechanism may be broken that normally acts like a car's brake to reduce the speed of the cell cycle. Or perhaps her cells have an "accelerator" that is working overtime. Normal cells in the skin and other tissues have some means of regulating the cell cycle. So what is the nature of that control? And what goes wrong with it in BCNS?

Stem Cells and Growth Control

Think, for a minute, about cell division in the skin. Normally after a basal cell in the dividing layer of the skin undergoes mitosis, then one of the daughter cells stops cycling and matures into a cell of the outermost layer of the skin (Fig. 4.10a). Although the process is entirely healthy, this cell is, in effect, on a suicide mission—it's "born to die," since it will never divide again and will eventually expire and flake off. The other daughter, however, can continue to divide. Cells with the ability to continue dividing are called **stem cells** and their progeny can behave in two ways, with one daughter maturing into a differentiated cell type and the other retaining the ability to divide.

How do stem cells in the dividing cell layer "know" when they should divide? Here's a simple thought experiment: If you were to scrape off the upper layer of the epidermis in a small patch of skin, the dividing layer below it would somehow "recognize" the thinness. More cells in that dividing basal layer (or germ layer) would then leave the G_1 phase of the cell cycle, enter S, and eventually complete a division. When the epidermis becomes thick enough once more, the basal layer "senses" this, too, and slows its rate of cell division.

Now, in a basal cell carcinoma, the signal to divide is somehow turned on but the signal to slow or cease dividing never comes, or if it does, it goes undetected. One possibility is that cells produce too much of the signal to divide; as a result, at any given time, more of the stem cells would undergo division than is needed to replace sloughed off skin. Here's another possibility: The daughter cell that should mature into an epidermal cell doesn't and instead, it too continues to divide (Fig. 4.10b). This is part of the overall question: What tells a cell to divide or stop dividing? Let's look at some possibilities for what is most likely happening.

The Molecular Basis of Basal Cell Carcinoma

People like Patricia have inherited genetic factors that predispose skin cells to form basal cell carcinomas. If she were to have children, about half of them would be likely to be affected with the condition (and Chapter 5 explains why). By investigating the inheritance of the disease gene in many families, Dr. Erwin Epstein and colleagues at the University of California at San Francisco were able, in 1996, to isolate the specific gene that is disrupted in BCNS. They were astonished at what they found. The gene turned out to be closely related to a gene already isolated from a fruit fly, *Drosophila melanogaster*. The gene is called *patched,* because fly embryos with the defective gene have abnormal little patches of hair-like structures in their skin. Epstein's group found that human families with BCNS had mutations in the human *patched* gene in all of their cells. They also found that people with individual, spontaneous basal cell carcinomas due to sun exposure and not inheritance often had mutations in the *patched* gene, but only in the cells of their skin tumor, not in *all* of their skin cells. It was clear that the *patched* gene normally functions in some way to prevent the growth of basal cell carcinomas and that mutations of the gene allow skin tumors to form.

By studying flies, workers had shown that the *patched* gene causes a certain protein to appear on the surface of cells. That protein, in turn, is necessary for the cell to interpret signals coming from the outside of the cell that stimulate cell division. The basic idea is that a substance in the cell's environment, often a protein called a **growth factor** made by a nearby cell, can act as a signaling protein telling other cells to divide. In flies, researchers showed that *patched* is the receptor for the signaling protein. In humans, the specific signaling protein is called sonic hedgehog because it is made by the *sonic hedgehog* gene. (This odd name comes from a mutant fly that, in this case, has prickly skin, making it look like a hedgehog.)

How Growth Factors Act

Still looking, then, at what instructs a cell to divide or stop dividing, let's see how growth factors might work. At the site of a cut or other wound, growth factors such as the human sonic hedgehog protein, may be released

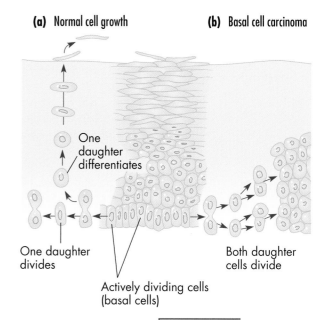

(a) Normal cell growth   **(b)** Basal cell carcinoma

One daughter differentiates

One daughter divides

Actively dividing cells (basal cells)

Both daughter cells divide

Figure 4.10 **Stem Cells and Skin Cancer.** (a) In normal growth, a stem cell in the basal (dividing) layer of the skin divides into two daughters, one of which matures into a skin cell, and the other of which remains as a dividing stem cell. (b) In basal cell carcinoma, perhaps both daughter cells continue to divide. Alternatively, more stem cells might be recruited to divide at the same time.

from dying cells (Fig. 4.11a, Step 1). The factors could diffuse locally, and bind to a **receptor,** a protein that chemically recognizes the specific growth factor. Receptors may be embedded in the membranes of nearby cells (Step 2). The binding of a growth factor from the wounded cell to the receptor of a nearby cell then stimulates an interior cascade of signals (Step 3). Some of those internal signals include **cyclins,** proteins whose levels rise and fall at different parts of the cell cycle and cause the cell to move from one phase to the next. The net result of this signaling (Step 4) is that the cells lining the wound are stimulated to divide. After enough cell division in the neighboring cell, the tear in the skin eventually fills in with new cells, growth factor levels fall, and the rate of cell division slows down.

Dr. Epstein's work showed that BCNS cells have a defective *patched* gene (Fig. 4.11b). This mutated gene makes an altered receptor that acts as if it is bound to the sonic hedgehog even when that signaling molecule is absent. Even without the sonic hedgehog "signal," the mutant patched protein stimulates the cascade of internal signals telling the cell to divide. This causes the cells in BCNS sites like those on Patricia's face and arms to enter S phase and divide inappropriately. Further investigations of this pathway in BCNS cells:

Sonic hedgehog protein signal ⟶
Patched receptor ⟶ Turned-on
cell division

may eventually allow Dr. Epstein to understand exactly why and how basal cells start dividing in a skin cancer but never stop. This, in turn, could lead to treatments for BCNS as well as for the more common basal cell carcinomas that some people get later in life from too much exposure to sunlight. **E**xplorer 4.2 summarizes and animates faulty growth control during cancer.

This research is proceeding. In the meantime, researchers have used growth factors to treat patients with wounds that wouldn't heal. In Oklahoma, for example, a man who was repairing an industrial refrigeration system accidentally splashed ammonia into his eye. After three weeks of conventional therapy, his cornea (the transparent protective covering of the eye) had not healed, and he could see only blurs and shadows. Then biologists at the University of Oklahoma treated his eye with drops containing a growth factor called epidermal growth factor. This turned on cell division in stem cells and in just four days, the cornea had healed. A few weeks later his vision had returned to nearly normal. Intensive research continues in this fascinating and important field of growth factors and their control of cell division.

Cancer as a Disease of Altered DNA

Recent evidence indicates that most cancers are related to changes in a cell's DNA. Numerous studies have shown that many substances

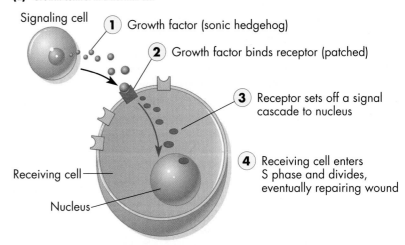

Figure 4.11 **Growth Factor Control, Normal Cell Division, and Cancer.** (a) In a normal cell, growth factors (Step 1) can bind to receptors (Step 2), thereby setting off a signal cascade (Step 3) that eventually tells the nucleus to divide (Step 4). (b) In a cancerous cell, the cell may turn itself on by producing its own growth factor signal; or a mutant receptor can turn division on without a signal, as in Patricia's multiple skin cancers; or the internal switch to divide can be thrown without input from a receptor. Regardless, the cancer cell enters the S phase and divides inappropriately.

in the environment, including ultraviolet light, industrial chemicals, radiation, the tar in cigarettes, and certain viruses, cause such changes in DNA structure. Some DNA changes can affect either growth factors or receptors for growth factors on the cell's plasma membrane, as Dr. Epstein found for BCNS. Surprisingly, some cancer cells may produce *both* the growth factor to turn on cell division *as well as* its specific receptor; the result is cells that are constantly stimulating themselves to divide. There is still much uncertainty about the causes of cancer, but it is clear that the answers will be found in the regulation of the cell cycle, and the stages of growth, division, and rest. In the meantime, it's wise to avoid the most common environmental hazards associated with cancers such as ultraviolet light and cigarette smoke.

4.4 Meiosis: The Special Cell Divisions That Precede Sexual Reproduction

If a BCNS patient like Patricia has children, there is a 50–50 chance that each child will inherit the disease of multiple skin cancers. Yet consider a teenager playing in the sun who experiences an ultraviolet light-induced change in the DNA of the *patched* gene in a cell on, say, the nose. This change or mutation could then lead to basal cell carcinoma arising on that person's nose. But this individual would *not* pass on to his or her children an increased likelihood of having either a single basal cell carcinoma or the multiple reoccurrences of BCNS. How can this be? What's the difference in their response? The answer revolves around the special cell population from which sex cells arise, and the special type of cell division that produces them.

Sexual Reproduction: Offspring from Fused Gametes

In **sexual reproduction,** parents (usually two, but sometimes one) generate specialized sex cells called **gametes** (Fig. 4.12). When gametes from two individuals (usually one male and one female) fuse during **fertilization,** they set into motion the life of a new individual.

Most of us are familiar with sexual reproduction and the human life cycle. As with most plants and animals, the human female's gametes are large cells that are incapable of spontaneous movement called **egg cells,** while the male gametes are small motile cells called **sperm,** which can move or be carried from the male to the egg (Fig. 4.12a). Eggs and sperm are usually pro-

duced in specialized organs. In flowering plants, they are produced by structures in the flowers (carpels and anthers; see Chapter 12); in animals, gametes are made by special organs called **gonads.** The female gonad is

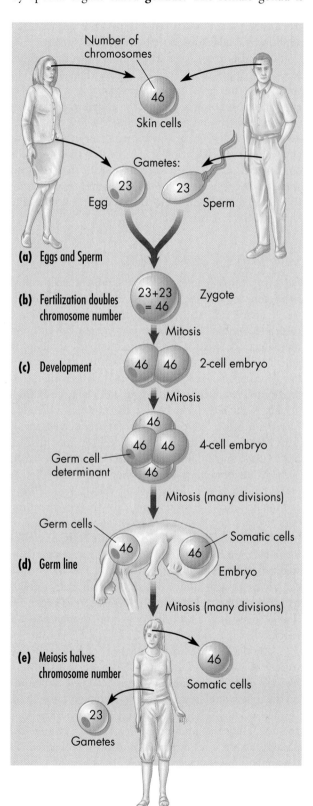

(a) Eggs and Sperm

(b) Fertilization doubles chromosome number

(c) Development

(d) Germ line

(e) Meiosis halves chromosome number

Figure 4.12 A Human Life Cycle. (a) A human egg and sperm each have 23 chromosomes. (b) When they join in fertilization, they create a zygote with 46 chromosomes. (c) The zygote undergoes mitotic divisions and develops into an embryo, with 46 chromosomes in all. (d) Within the embryo, a special set of cells is set aside, called germ cells, that will form egg or sperm. (e) When the child becomes sexually mature in adolescence, germ cells can complete meiosis, which decreases the number of chromosomes from 46 to 23, and gametes form. Thus each egg and sperm have 23 chromosomes. In many animal types, special proteins (shown here in brown) establish the germ line.

the **ovary,** which produces eggs, and the male gonad is the **testis,** which produces sperm.

In many complex organisms, including human beings and ginkgo trees, each individual produces just one kind of gamete, either egg or sperm. However, in pear trees, earthworms, and a number of other species, each adult individual can produce both types of gametes.

The fusion of egg and sperm, or fertilization, results in a single cell called the **zygote** (Fig. 4.12b). In the zygote, the hereditary information from both parents unites, creating a genetically unique combination of genes and chromosomes. The single-celled zygote undergoes *development,* usually a period of rapid mitosis and cellular specialization during which the new cells emerge and take on their specific roles in the organism (Fig. 4.12c). As a result of development, an immature form emerges, continues to grow, and eventually changes into a mature adult.

Germ Cells and Somatic Cells

Early in animal development, a group of cells called **germ cells** is set aside. Germ cells are like the stem cells we discussed earlier; they retain the potential to divide to produce more germ cells, or to differentiate into gametes (eggs or sperm), and to migrate to the developing gonad during development (Fig. 4.12d). Animal eggs have a particular group of proteins that become separated into certain cells as the egg divides by mitosis (Fig. 4.12c). Whichever cells receive these proteins develop into germ cells. The cells that don't receive these proteins become the rest of the body's cells in the muscles, brain, nose, and so on. These special proteins are called germ cell determinants, because they determine whether embryonic cells will become germ cells. Germ cell determinants form a continuous cell lineage extending back to your mother, her mother and so on back in time. The normal body cells, or **somatic cells,** of a multicellular animal can't form egg and sperm, and these somatic cells eventually die. If, however, an organism's egg or sperm cells unite with those of another individual of the same species, then in a sense, that germ cell lineage lives on in the offspring even though the somatic cells die.

Now let's return to the question of how a BCNS patient might pass on the condition to offspring, while a teenager that just happens to get a basal cell carcinoma on the nose would not pass that carcinoma on to a child. In the case of BCNS, the disease gene is in every body cell, including the germ cells (Fig. 4.12). It is therefore passed into the eggs or sperm and can be inherited. In the sunbathing teenager, however, the diseased copy of the *patched* gene brought about by solar radiation occurs only in one or more somatic cells (in this case, skin cells on the nose) and not in the germ cells. Therefore, the mutation could be passed on to the progeny of that one skin cell, but not to the person's future offspring.

An organism's life cycle comes full circle when its germ cells undergo the special type of cell division called meiosis. This division decreases the number of chromosomes present in germ cells by half, from 46 to 23 in the case of humans (Fig. 4.12e). The resulting cells can then become egg or sperm and lead to a new generation.

Meiosis: Halving the Chromosome Number

Why would an organism need a special type of cell division before producing gametes? Each of your somatic cells has 46 chromosomes. Let's say you produced gametes that had 46 chromosomes. How many chromosomes would your children's somatic cells have? Without a special mechanism, your child would get 46 chromosomes from you and another 46 from your mate, giving a total of 92. The next generation would have 184 and so on. Clearly this doesn't happen because each human baby has 46 chromosomes just like each parent. So what prevents the doubling of chromosomes in each generation?

A special type of cell division called meiosis prohibits this runaway increase in chromosome number. Meiosis ensures that gametes contain half as many chromosomes as normal body cells (Fig. 4.13). In other words, each sperm or egg cell you make has 23 chromosomes, not 46. And the fusion of gametes at fertilization restores the original parental chromosome number, 46.

Meiosis (literally, to make smaller) is a special type of cell division that produces gametes or, in some species, such as mushrooms and ferns, other specialized reproductive cells called **spores,** whose chromosome number is half that of other body cells (see Fig. 4.14i).

In terms of chromosome number, fertilization and meiosis play opposite roles in a life cycle involving sexual reproduction. Fertilization *doubles* the chromosome number, while meiosis divides it in *half.* As we will see later in the chapter, meiosis also increases genetic variation, which is a precondition for evolution.

Chromosome Sets

Before we can investigate how meiosis reduces the number of chromosomes, we must understand how many sets of chromosomes each cell contains at different

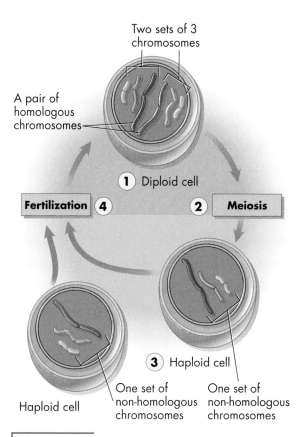

Two sets of 3 chromosomes

A pair of homologous chromosomes

1 Diploid cell

Fertilization 4 **2** **Meiosis**

3 Haploid cell

Haploid cell

One set of non-homologous chromosomes

One set of non-homologous chromosomes

Figure 4.13 **Haploid and Diploid Cells.** (1) The cell in Step 1 has three chromosome pairs, one long pair, one medium pair, and one short pair. Diploid cells have two sets of chromosomes. Each chromosome in a homologous pair is generally the same size and shape and has the same genes in the same order. When a diploid cell with two sets of chromosomes undergoes meiosis (Step 2), it produces haploid cells containing a single set of chromosomes. This haploid cell (Step 3) has three chromosomes, each of which is a different size and shape and therefore nonhomologous. (4) When two haploid cells, (each with one set of chromosomes) unite in fertilization (Step 4), the result is a diploid cell with two sets of chromosomes.

phases of the life cycle. As we have seen, the body cells of each species have a characteristic number of chromosomes: human beings have 46, chimpanzees 48, houseflies 12, onion plants 16, roundworms 4. What do all these chromosome numbers have in common? They are all even numbers. This is because in the body cells of most eukaryotes, *chromosomes are present in pairs.* A pair of chromosomes that have similar size, shape, and usually gene order are called **homologous chromosomes.**

A human skin cell contains 23 pairs of homologous chromosomes, for a total of 46 chromosomes.

One member of each pair came from the individual's mother, the other member of each pair came from the individual's father. Thus, a skin cell has two *sets* of chromosomes: a maternal set of 23 and a paternal set of 23. A cell such as this that has two sets of chromosomes is called a **diploid** cell (*di* = two). A cell that has just one set of chromosomes, one copy of each homologous pair, is called a **haploid** cell. Human gametes are examples of haploid cells.

Follow through the steps of Figure 4.13 to see these principles at work in an organism with just three pairs of chromosomes in each diploid cell.

The Cell Divisions of Meiosis

Because you've already explored the divisions of mitosis, you are in a good position to understand the variations that give rise to meiosis. Keep in mind that only somatic cells such as skin cells undergo mitosis and that the process leads to cell replacement and to growth and repair. Only germ cells like those in the gonads undergo meiosis and lead to the production of gametes and the possibility of future generations. Also remember that the result of a meiotic division is to reduce the number of chromosomes by half, changing a diploid stem cell into haploid cells that can become gametes. We'll also see how meiosis provides an enormous amount of genetic variation, and why that's important for the process of evolution.

The first point to notice about meiosis is that the change from diploid to haploid chromosome number involves *two* sequential cell divisions called meiosis I and meiosis II. In **meiosis I,** a diploid parent cell divides, forming two haploid daughter cells. In **meiosis II,** those two daughter cells divide again, resulting in four haploid cells. As in mitosis, chromosome *replication, alignment,* and *separation* are central concepts. Figure 4.14 charts chromosome movements during meiosis.

Replication: The Interphase Before Meiosis I

Let's return to our organism with three chromosomes in a haploid set, one long, one medium sized, and one short. As Figure 4.13 showed, the germ cell from which this haploid cell arose is initially diploid and so it has two long, two medium, and two short chromosomes. Each of the two chromosomes make a homologous pair (Fig. 4.14a).

We start with the cell just after its parent cell has divided (for example, the lower cell in Fig. 4.6f, with two sets of one-chromatid chromosomes). At this stage of meiosis, our cell is in the G_1 phase of the cell cycle (Fig. 4.14a). This germ cell leaves G_1 and enters S phase, and

Figure 4.14   The Major Chromosomal Events in Meiosis.

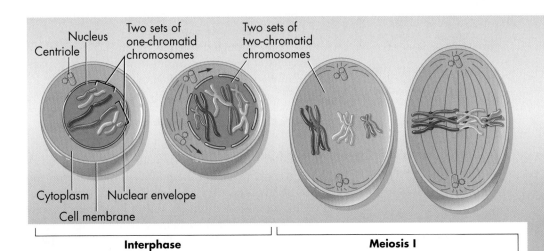

Interphase

Meiosis I

(a) End of previous cell cycle

This diploid cell has two sets of one-chromatid chromosomes.

(b) Replication

After replication the cell has two sets of two-chromatid chromosomes.

(c) Pairing

Homologous chromosomes pair up.

(d) Alignment

Homologous chromosome pairs align along the metaphase plate.

replicates each chromatid. Now, after replication, each chromosome has two chromatids (Fig. 4.14b).

Alignment: Homologous Chromosomes Pair in Meiosis I

As meiosis I begins, homologous chromosomes pair, lining up very close together: the two long chromosomes next to each other, the two short ones next to each other and so on (Fig. 4.14c). In fact, homologous chromosomes pair so closely together that they exchange genetic material, as we'll see later.

As meiosis I progresses, the paired homologous chromosomes align in the center of the cell, moved by the spindle fibers (Fig. 4.14d). The order of nonhomologous chromosomes on the spindle, however, is random.

Separation: Homologous Chromosomes Separate from Each Other in Meiosis I

As meiosis I continues, homologous chromosomes separate from each other (Fig. 4.14e). When the nucleus divides and cytokinesis is completed after meiosis I, the two resulting cells now have the haploid content of

chromosomes (Fig. 4.14f). Look at the figure to confirm that. Note that each cell in Figure 4.14f has three chromosomes: one long, one medium, and one short; these two cells, therefore, are haploid: Meiosis I has reduced the chromosome number from diploid to haploid. Notice, however, that each chromosome is a two-chromatid chromosome, not a one-chromatid chromosome as at the end of a mitotic division. Meiosis II takes care of that problem.

Replication: Chromosomes Do Not Replicate Between Meiosis I and II

After meiosis I, an interphase follows that is special in that it involves no DNA synthesis or chromosome replication. At the beginning of meiosis II, then, each cell still has a haploid set of two-chromatid chromosomes (Fig. 4.14f).

Alignment: Chromosomes Align Independently in Meiosis II

As meiosis II progresses, the chromosomes line up on the spindle independently of each other (Fig. 4.14g).

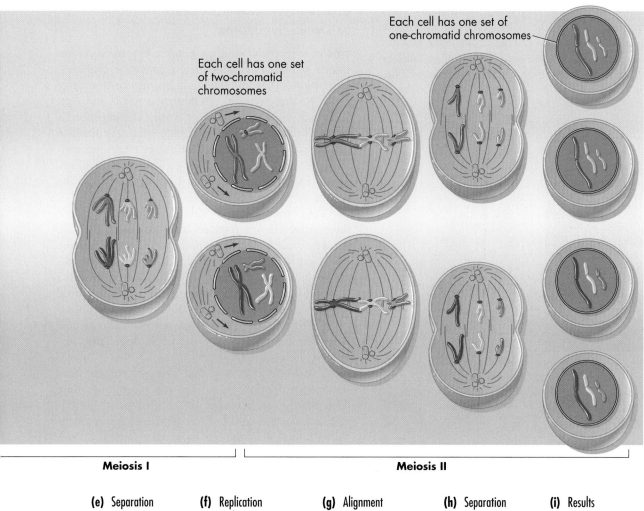

Each cell has one set of two-chromatid chromosomes

Each cell has one set of one-chromatid chromosomes

Meiosis I **Meiosis II**

(e) Separation

Homologous chromosomes separate.

(f) Replication

Chromosomes do not replicate.

(g) Alignment

Chromosomes align independently.

(h) Separation

Sister chromatids separate.

(i) Results

Four genetically non-identical haploid germ cells.

Separation: Sister Chromatids Separate in Meiosis II

Meiosis II then follows like a normal mitotic division, and the sister chromatids separate to opposite poles of the cell (Fig. 4.16h). The result, after cytokinesis (Fig. 4.14i), is now four cells, each of which is haploid, and each of which has one set of one-chromatid chromosomes.

Compare, now, the parental cell and the daughter cells of one complete meiotic division. While the parental cell had two sets of one-chromatid chromosomes (Fig. 4.14a), the daughter cells—four of them from the two successive divisions of meiosis—each have one set of one-chromatid chromosomes (Fig. 4.14i). Depending on the adult's sex, those cells can develop into either sperm or eggs.

The Fate of the Haploid Products of Meiosis

The four haploid products of meiosis can have different fates in different species or in males and females of the same species. For example in the human female, only one of the four products becomes the egg; that cell keeps nearly all of the cytoplasm. The other three cells are called **polar bodies** and eventually die. Geneticists sometimes use polar bodies during genetic tests because from them, they can determine the genes in the egg cell. In the human male, each of the four haploid products will become a sperm. In some species, the haploid products of meiosis become spores that can divide by mitosis. In the case of moss, ferns, and certain other plants, the spore can proliferate through mitotic divisions into a haploid plant. (Note that in this case, haploid body cells divide by mitosis into new haploid body cells, forming, for instance, the green leafy portion of the moss plant.) This, in turn, can eventually make gametes which unite, form a zygote, and grow into the familiar plant that carpets some shady forests and provides greenery for terrariums.

(a)

(b) Normal meiosis

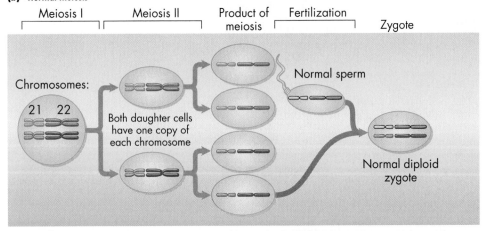

(c) Meiosis leading to Down syndrome

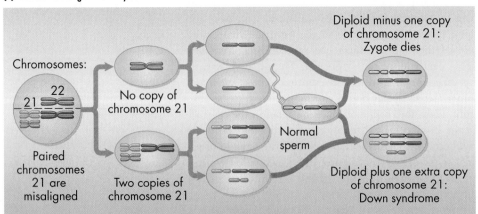

Errors in Meiosis Can Lead to Down Syndrome

One in every 1000 children born in United States has **Down syndrome** (Fig. 4.15a), a condition characterized by lower than average IQ, heart malformation, eyelid folds, short stature, foreshortened hands and feet, and palm prints with a unique crease. The likelihood a woman will have a child with Down syndrome increases dramatically with her age. One out of every 365 children born to women who have reached 35 years of age display Down syndrome, as do 1 out of every 12 babies born to women age 51.

Cells from children with Down syndrome have three rather than two copies of chromosome 21. The abnormal number of chromosomes in Down syndrome made geneticists suspect that a problem in meiosis might underlie the condition. At the time of birth, the germ cells in baby girl's ovaries are arrested in an early stage of meiosis. The cells remain arrested for many years until a hormone signal causes the egg to resume meiosis in preparation for fertilization. Somehow, during that long stage of arrested meiosis, an egg can become damaged in a way that prevents meiosis from resuming properly. The older a woman gets, the more likely this damage becomes.

One possible result of the damage is failure of the paired homologues to separate as they should. If both homologues move to the same pole in meiosis, the egg would then contain two copies of chromosome 21 (Fig. 4.15c). When fertilized by a normal sperm with its own copy of chromosome 21, the zygote would then have three copies of chromosome 21. Such an embryo has a gene imbalance in each cell, and this alters normal development and leads to the characteristic birth defects seen in Down syndrome.

Unraveling the cause of Down syndrome was an important scientific achievement. Once researchers re-

Figure 4.15   Down Syndrome and Meiotic Errors.
(a) A child with Down syndrome. (b) Normal meiosis leads to egg and sperm with one haploid copy each of chromosomes 21 and 22. (Only these two chromosomes are drawn here, but chromosomes 1 through 20 would behave like 22, as well.) (c) Abnormal meiotic divisions lead to Down syndrome. The paired homologous copies of chromosome 21 may move to the same pole in meiosis I. This leads to meiotic products that may either lack a copy of chromosome 21 or have an extra copy of chromosome 21. If the former becomes an egg and is fertilized, the zygote will die. If the latter becomes an egg and is fertilized, the zygote will have trisomy 21 (three copies of chromosome 21) or Down syndrome.

vealed the age connection, women could decrease their likelihood of producing a Down syndrome child by completing their families before age 40 or by seeking specific tests during pregnancy that reveal whether or not a fetus has chromosomal defects.

Meiosis Contributes to the Origin of Genetic Variation

Ever stop to think about just how different the children of one set of parents can be? They can differ in height, skin coloration, hair characteristics, willingness to take daredevil risks, and even in traits such as Basal Cell Nevus Syndrome, where one child can be affected while her brothers and sisters are not. Although some of these differences may have environmental causes, others are due to genetic differences among the siblings. Meiosis plays a large role in the genetic differences that exist in a group of brothers and sisters because during meiosis, there is a reshuffling of the maternal and paternal chromosomes. This reshuffling is called **genetic recombination,** and it occurs in two ways: crossing over and independent assortment.

Crossing Over: Homologous Chromosomes Exchange Parts

Crossing over occurs during meiosis I, when homologous chromosomes pair (Fig. 4.16a). While the chromosomes are paired, enzymes can break the DNA molecule in each homologue, switch corresponding regions of each chromosome (the actual **crossing over**), and then attach them to the new chromosome (Fig. 4.16b). After meiosis II, some of the individual haploid cells will contain **recombinants,** that is, they will contain chromosomes of mixed ancestry, as the dark and light blue colors in Figure 4.16 indicate. Note that in Fig. 4.16c, just one crossover event makes each haploid cell genetically unique.

Independent Assortment: Chromosome Pairs Align Randomly

A second mechanism also contributes to genetic variety after meiosis and requires no chromosome breaks. It is called **independent assortment,** a property based on the fact that nonhomologous chromosomes align independently during meiosis I (review Fig. 4.14d). The chromosomes in Figure 4.17a are arranged with all of the paternal copies of the chromosomes oriented toward the same pole of the cell. Because of this alignment, each of the haploid products of meiosis will contain ei-

(a) Crossover   **(b)** Regions exchanged   **(c)** Products of meiosis

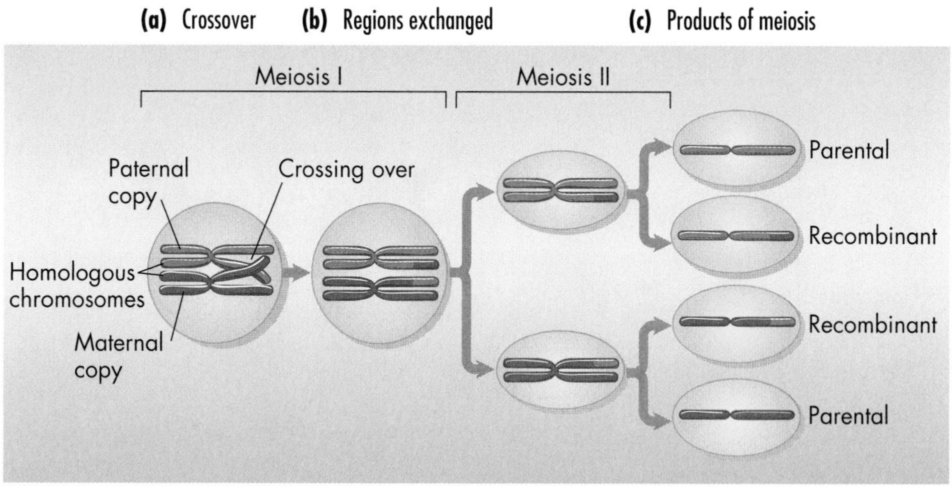

Figure 4.16 **Genetic Recombination Through Crossing Over.** Due to crossing over during meiosis I, the four haploid cells generated during meiosis II can be of two types: Two products have the chromosome arrangements of the original parent chromosomes (parentals), and two products have the new arrangement (recombinants).

ther all paternal copies or all maternal copies. Another possible arrangement is shown in Figure 4.17b, with one of the paternal chromosome copies oriented toward one pole, and the other toward the opposite pole. With this arrangement, each meiotic product will be recombinant, with one paternal and one maternal chromosome.

Take a moment to look at the set of eight gametes in Figure 4.17a and b. Count how many different types of gametes have formed. Each gamete pictured has a haploid set of chromosomes (one short and one long). But there are four combinations present based on the parental origin: (1) both paternal, (2) both maternal, (3) long maternal and short paternal, and (4) long paternal and short maternal. These four combinations differ not only from each other, but also from the diploid parent cells.

The independent assortment of chromosomes during meiosis can be compared to choosing from the menu of a restaurant where there are two choices for each course. Each course represents a chromosome, and the two choices—chicken or fish, rice or potatoes, apple pie or chocolate cake, and so on—represent the two homologous copies of each chromosome. Just as the potential number of different meals depends on the number of courses, the potential number of unique genetic combinations in the gametes depends on the number of chromosomes. From a menu with 3 courses and 2 choices per course, you could make up eight (= $2 \times 2 \times 2$) different meals. For each chromosome added, you'd have to multiply again by 2. So for a human being, that number would be 2^{23}, or a potential of 8 million different chromosome combinations of eggs or sperm in the nuclei. Because crossing over (review Figure 4.16) adds even more new combinations, the number of actual possible types of gametes made by a single

(a) One possible chromosome arrangement

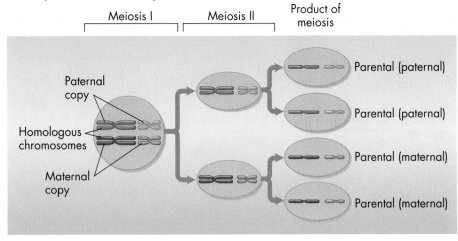

(b) Another possible chromosome arrangement

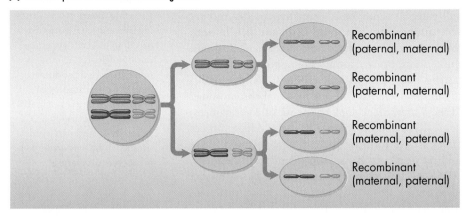

Figure 4.17 Genetic Recombination Through Independent Assortment. (a) Chromosomes may align in meiosis I with the paternal homologues oriented toward the same pole. (b) Alternatively, chromosomes may align in meiosis I with the paternal copy of one chromosome towards one pole and the paternal copy of the other chromosome towards the other pole. This can lead to recombinant type gametes.

READ ON

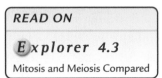

*E*xplorer 4.3

Mitosis and Meiosis Compared

person is astronomical. Finally, the random combination of maternal and paternal chromosomes in the zygote at fertilization further increases the number of genetic combinations.

The incredible genetic diversity that results from genetic recombination helps explain why an organism is highly unlikely to produce two genetically identical gametes, and why, in turn, cucumbers, prize bulls, puppies, and all other organisms resulting from sexual reproduction may resemble their parents but are never exactly like either parent.

The Evolutionary Significance of Genetic Recombination

Together, crossing over and independent assortment ensure that sexually produced offspring are genetically different from their parents. A litter of puppies, all with different fur color, fur pattern, and body size, is living evidence that genetic recombination took place during meiotic divisions in the gamete-producing reproductive organs of the parent dogs. The puppies resemble their parents in many ways, but differ in at least a few, and some of these differences—better hearing, let's say, or a stronger heart—may give some of the puppies a better chance of survival. And therein lies the evolutionary significance of genetic recombination during meiosis. (We'll return to this subject in Chapters 5 and 10.)

Mitosis and Meiosis Compared

By now you should have a pretty clear understanding of why out-of-control cell division can lead to a cancer. You should also know why inherited cancer, like the lesions of BCNS, can be passed on to offspring while an acquired cancer—based on too much sunbathing, for example—cannot. But because the details of mitosis and meiosis are relatively easy to confuse, it's worth one last comparison of the two (Fig. 4.18). A mitotic division can occur in either a haploid body cell, like the leafy part of a moss plant, or a diploid body cell, like one of your skin cells. Either type of body cell can then produce two cells genetically identical to each other and to the parent cell. Meiosis, however, can occur only in special diploid *reproductive* cells found in, for example, the ovaries or testes in human beings, the flowers of flowering plants, and special reproductive organs in moss plants. The products of meiosis are unique. They are haploid and can undergo fertilization and produce a new diploid individual.

Mitosis occurs only in somatic cells and allows for the growth of the organism and repair of its parts. Meiosis occurs only in germ cells and can result in the production of gametes that can take part in sexual reproduction. Mitosis accomplishes two things: (1) The reproduction of cells, and (2) the equal distribution of DNA to each new daughter cell. Meiosis accomplishes three things: (1) By reducing the chromosome number from diploid to haploid, meiosis prevents an increase in chromosome number that otherwise would occur at fertilization. (2) Crossing over during meiosis permits new combinations of maternal and paternal hereditary traits. (3) Independent assortment allows for the further random combination of maternal and paternal chromosomes.

You can explore further the differences in chromosome movements generated by differences in replication, alignment, and separation during mitosis and meiosis in *E*xplorer 4.3.

(a) Mitosis

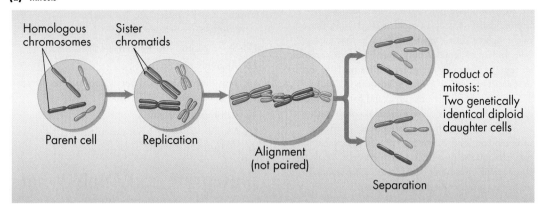

(b) Meiosis

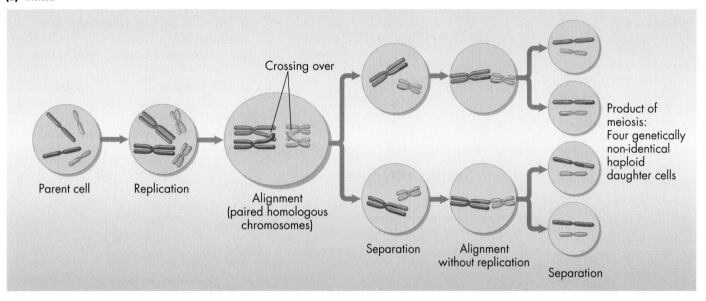

Figure 4.18 Mitosis and Meiosis Compared.

Chapter Summary and Selected Key Terms

Introduction   Basal Cell Nevus Syndrome is a disease of multiple **basal cell carcinomas** (p. 88) or cancers of the lower skin layer. This syndrome highlights the problem of renegade cells that have escaped the normal controls over growth and division. An understanding of cancer and effective new treatments will require a good understanding of the growth and division cycles or **cell cycles** (p. 89) that cells undergo, including **mitosis** (p. 89) and **meiosis** (p. 89), how each is governed, and how aberrant controls can lead to **tumor** (p. 89) formation.

Cell Growth and Cell Division   Skin cells are a good example of normal cell growth and division. Basal cells at the base of the **epidermis** (p. 90) or thin

outer zone divide regularly. The **dermis** (p. 90) is the thicker zone underneath. The layer nearest the dermis is the **germ layer** (p. 90) or **basal layer** (p. 90), with its **basal cells** (p. 90). One of the progeny remains behind, grows larger, but stays in an undifferentiated state. Eventually, it will divide again. The other progeny cell also grows, pushes upward, and takes on the mature flattened shape of an epidermal cell. Eventually it dies and flakes off. A growth of cells is called **proliferation** (p. 90). The alteration of normal growth and division is called the **cell cycle** (p. 91). During the **division phase** (p. 91), cell parts are duplicated and/or partitioned in two. The growth period between two division phases is called **interphase** (p. 91). **Chromosomes** (p. 91) contain a cell's hereditary information and each new daughter cell receives copies during cell division. Each replicated chromosome consists of two identical rods or **chromatids** (p. 92), each made up of a single long DNA molecule. Pairs of these rods called **sister chromatids** (p. 92) are held together at the **centromere** (p. 92). The

sister chromatids undergo **alignment** (p. 92) along the cell equator, then undergo **separation** (p. 92) as they separate during cell division with one set going to each daughter cell. Each chromatid **replicates** (p. 92) or becomes copied to two chromatids during interphase, and then the two chromatids separate to different daughter cells during the division phase. Interphase has three portions: the G_1 **phase** (p. 93) or active growth phase during which the cell manufactures new cell components in preparation for cell division; the **S phase** (p. 93) during which DNA is replicated and each chromosome now consists of two identical chromatids; the G_2 **phase** (p. 93, 94), a time of protein manufacture before cell division; and the **M phase** or **mitosis** (p. 93, 94), during which the nuclear material is divided, followed by the division of cytoplasm in **cytokinesis** (p. 94).

Mitosis and Cell Division   In **prophase** (p. 95, 96) the chromosomes condense and become visible. During **interphase** (p. 93), the DNA doubles and when chromosomes appear during mitosis, they consist of two sister chromatids held together by the centromere. During **metaphase** (p. 96) the replicated chromosomes align in the middle of the cell, then in **anaphase** (p. 96) sister chromatids separate and move to opposite ends (or poles) of the cell. The chromosomes move apart on a gossamer scaffolding, the **mitotic spindle** (p. 96, 97), the **spindle fibers** (p. 97) of which extend from two organelles called **centrioles** (p. 97). In **telophase** (p. 96), the chromosomes arrive at opposite poles. After the nucleus divides in mitosis, the cytoplasm divides by means of **cytokinesis** (p. 97). In animal cells, a **contractile ring** (p. 97) pinches the cell in two; in plant cells, a **cell plate** (p. 97) forms and partitions the cell in half. Some kinds of cancer **chemotherapy** (p. 98) attack the spindle of dividing cancer cells or prevent it from forming. Radiation can mutate or break the chromosomes of cancer cells.

Control of Cell Division
Cells like the basal cells of skin's epidermal layer can continue to divide, thus they are **stem cells** (p. 99). Studies of basal cell carcinomas have revealed that a mutated gene called *patched* (p. 99) allows the cells in basal cell tumors to go on dividing and dividing. **Growth factors** (p. 99) act as chemical signals telling cells when to start dividing, as when a wound is created and cells must divide to fill in the gap. These factors bind to **receptors** (p. 100) embedded in cell membranes and this stimulates a cascade of internal signals, which include **cyclins** (p. 100). The *patched* gene encodes a receptor for a signaling protein called *sonic hedgehog*. When the *patched* gene is mutant (defective), it can turn on cell division in basal cells, but it can't turn them off again, and the result is a basal cell carcinoma, a type of skin cancer.

Meiosis: Cell Division Preceding Sexual Reproduction
In **sexual reproduction** (p. 101), parents generate sex cells or **gametes** (p. 101). The fusion of gametes or **fertilization** (p. 101) results in a single cell, the **zygote** (p. 102) and initiates the life of a new individual. Female gametes are usually **eggs** (p. 101), male gametes are usually **sperm** (p. 101). Some species produce spores. In animals, gametes are formed in **gonads** (p. 101) such as the ovary and testes. **Germ cells** (p. 102) in the gonads produce gametes through the divisions of meiosis which halve the chromosome numbers in those sex cells. **Somatic cells** (p. 102) or body cells cannot form eggs and sperm and divide by mitosis. In somatic cells, chromosomes are present in pairs of similar size and shape and usually gene order called **homologous chromosomes** (p. 103). A cell with two sets of homologous chromosomes is **diploid** (p. 103); one member of each pair comes from the mother, the other from the father. A cell with just one set of chromosomes, one copy of each homologous pair, is **haploid** (p. 103). Body cells are generally diploid; gametes are haploid. In **meiosis I** (p. 103), a diploid parent cell divides, forming two haploid daughter cells. In **meiosis II** (p. 103), those two daughters divide again, resulting in four haploid cells. In human females, only one of the four becomes an egg; the other cells are **polar bodies** (p. 105). Errors in meiosis can lead to **Down syndrome** (p. 106). There is a reshuffling of maternal and paternal chromosomes during meiosis called **genetic recombination** (p. 107) involving either **crossing over** (p. 107) or **independent assortment** (p. 107). Genetic recombination explains the incredible genetic diversity resulting from sexual reproduction and leads to survival differences that fuel natural selection and evolution.

All of the following question sets also appear in the Explore Life *E* electronic component, where you will find a variety of additional questions as well.

Test Yourself on Vocabulary and Concepts

In each question set below, match the description with the appropriate term. A term may be used once, more than once, or not at all.

SET I
(a) chromatid (b) centromere (c) chromosome
(d) microtubules (e) cytokinesis

1. The mitotic spindle
2. During G_1, each chromosome is composed of one of these structures.
3. During prometaphase this structure attaches directly to a spindle fiber.
4. To determine the number of these structures, just count the number of centromeres.
5. Final stage of the M phase

SET II
(a) S (b) G_1 (c) G_2 (d) M (e) more than one of the preceding

6. Included in interphase
7. Primary stage of cell growth and metabolic function
8. Replication of DNA occurs during this stage.
9. The chromosomes are composed of two chromatids during the first part of this stage.
10. The stage during which sister chromatids separate

SET III
(a) prophase (b) metaphase (c) anaphase
(d) telophase (e) prometaphase

11. Spindle forms, chromosomes coil and shorten.
12. Alignment phase
13. Nuclear membrane disperses; centromeres attach to spindle fibers.
14. Chromatids separate.
15. Stage in which a cell has two nuclei

Integrate and Apply What You've Learned

16. What is the difference between mitosis and cytokinesis?
17. How does cytokinesis differ in plant and animal cells?
18. What critical events in meiosis promote an increase in genetic variation and halve the number of chromosomes in a cell?
19. How many DNA molecules are contained in the nucleus of one of your skin cells just after it has completed cell division? Justify your answer.

Analyze and Evaluate the Concepts

20. In what way do cancers like the basal cell carcinomas of BCNS demonstrate the normal cell cycle gone awry?
21. An oncologist (a physician specializing in the treatment of cancer) compares two tumors by looking at the fraction of cells found in the M phase of the cell cycle. In one tumor, 5 percent of the cells are in M phase; in the other, 1 percent are in M phase. Which tumor is growing more rapidly? Defend your choice.
22. Biologists have isolated a compound from Pacific yew trees that physicians have been able to use as an effective treatment for ovarian and breast tumors. Biologists have also isolated a compound from the Madagascar periwinkle that is effective against childhood leukemias, which are forms of cancer. In both cases, the molecules act by binding to the proteins of the spindle and interfering with the spindle's separation of chromosomes in mitosis. Why would these compounds be more harmful for cancer cells than normal cells?
23. Where in your body are cells dividing right now by mitosis? Where in your body are cells dividing by meiosis?

A Devastatingly Common Illness

▲
Help for Cystic Fibrosis. Jonathan Zuckerman, Shana Reif, and Shana in the adult cystic fibrosis clinic.

When Jonathan Zuckerman was a teenager growing up in Boston, he'd never heard of the disease he would someday study and treat in patients like Shana Reif: cystic fibrosis (CF). With one affected child in every 3,000 births, it is the most common lethal inherited illness among Caucasians. It causes a constellation of breathing, digestion, and other medical problems, but it has never been widely understood by the public. That may be, in part, because an affected child born in 1960 — Dr. Zuckerman's year of birth — was unlikely to survive past his or her teens.

Today, thanks to improvements in conventional approaches like antibiotics, diet, and physical therapy, cystic fibrosis patients are living to their thirties, on average, and their care, as adults, has become a new medical specialty. Zuckerman was a physician and researcher at the University of Pennsylvania in Philadelphia until 1999 and has since worked at the Maine Medical Center in Portland. In both institutions, he has headed specialized programs for adults with cystic fibrosis. By applying modern genetic discoveries, he and others in his field are pioneering treatments to help keep CF patients alive and feeling as well as possible.

During and after medical school, Zuckerman focused on lung diseases and became a pulmonologist. Like most doctors in that specialty, he tended to see older patients who, through cigarette smoking, had developed lung cancer or emphysema. When he started working with cystic fibrosis pa-

tients, he found their youth, courage, and motivation "very inspirational." They are attempting, he explains, "to live a normal life and take care of their cystic fibrosis problem but still be out in the world" studying or working at their jobs. For the first time, there are accountants, lawyers, artists, and musicians with cystic fibrosis. "Three quarters of our patients," says Zuckerman, "are out there making a real contribution." And Shana, a registered nurse and cystic fibrosis patient in Philadelphia, was one of them.

Shana was born with cystic fibrosis and diagnosed on her second day of life. Her younger sister and brother are carriers of the disease but don't suffer the many symptoms she has faced: shortness of breath, recurrent lung infections, and CF-related diabetes. Every day is a struggle to stay well with breathing and chest-thumping treatments, aerosol inhalants, insulin injections, treadmill exercises, yoga, special diets, and sometimes intravenous antibiotics and hospital stays. Despite all of this, Shana has worked in hospitals herself, lives independently with her husband, and helps maintain their home. "I have goals and dreams for the future," she says, "but I have to focus on living one day at a time."

The powerful inspiration of Shana's life and others with cystic fibrosis is magnified when you consider the illness they face—essentially a disease of clogged ducts. A child who inherits one broken cystic fibrosis gene from each parent will produce a faulty version of a protein. This protein lies embedded in cell membranes, especially the membranes of **epithelial cells,** which tend to line ducts leading from the lungs, stomach, pancreas, sweat glands, and reproductive organs. The protein is involved in salt and water movement across cell membranes, and being defective, it prevents normal fluid transport. The walls of the ducts and the protective coatings they secrete tend to dry out, creating a thick, sticky mucus layer. This, in turn, clogs narrow passageways and ducts in the organs just mentioned. Because of this abnormal gene sequence, an individual who inherits cystic fibrosis usually has difficulty breathing. He or she repeatedly contracts dangerous bacterial infections in the lungs, suffers stomachaches and a diarrhea-like condition due to poor absorption of fats in the diet. Most also exude a salty secretion on the skin. And affected males are almost always sterile because of blocked ducts leading from the testes.

Patients like Shana usually take special pancreatic enzymes with their meals to help them digest fats, along with medicines to open the airways and antibiotics to prevent lung infections. They also require three or more "percussion sessions" per day during which a family member or friend claps them on the chest—front and back—for 20 minutes or so at a time to help dislodge gummy mucus and allow easier breathing. "My parents used to call them 'bangies,'" says Shana, "'Come on! It's time for your bangies!'" Recently, inflatable vests that emit air pulses have become available to do the thumping mechanically and Shana uses one regularly. But for her and others with cystic fibrosis, all the needed therapies are time-consuming and must be juggled with studying, working, hobbies, and family demands.

Aside from the difficulties and dangers it poses, cystic fibrosis has become, in a sense, a "model" disease. Its molecular genetic basis, explains Zuckerman, was uncovered in the late 1980s just as physicians were starting to understand the disease's many effects on the body. With their working knowledge of the underlying gene and protein deficiencies, medical researchers can plan sophisticated, highly targeted treatments rather than simply treating symptoms separately. Experimental gene therapy going on at Johns Hopkins University, Cornell, the University of North Carolina, and elsewhere is one such targeted approach, aimed at inserting a healthy, functioning gene into patients' airways.

Identifying the cystic fibrosis gene has also enabled physicians to look for the inherited factor in potential parents—and they have found it in one of every 30 Caucasians. Finally, doctors and genetic counselors can apply the principles of genetics to predict what will happen if a person carrying one or two copies of the cystic fibrosis gene produces children with another person carrying zero, one, or two copies. For example, Shana married her high school sweetheart, Kurt, a few years ago and wanted to start a family. But she and Kurt wondered whether their children would be born with cystic fibrosis. By applying genetic principles, Dr. Zuckerman was able to advise them, as we'll see in the video, in the introductory Explorer for Chapter 5, and later in these pages.

Our goal in this chapter is to help you understand inheritance patterns like these—patterns that determine who will display lethal disease symptoms and who won't. Inheritance patterns underlie each of our thousands of traits—eye color, hair color, height, and so on—and those of all other living organisms. The science of **genetics** explores the nature of genes and how they are organized on chromosomes; how genes govern our appearance, physical functioning, and even behavior; and how medical researchers can manipulate genes to treat diseases like cystic fibrosis.

As you explore Chapter 5, you'll find the answers to these questions:

1. How did a 19th-century monk, working on his own, discover the universal principles of heredity?

2. What rules govern the way organisms inherit individual traits?

3. How do geneticists analyze inheritance patterns in people?

4. What rules govern the way organisms inherit several traits at the same time?

5. How does our sex influence how we inherit traits?

6. How is the study of genetics changing the way we predict and treat diseases?

5.1 How Did Scientists Discover the Universal Principles of Heredity?

When Dr. Zuckerman's patient, Shana, decided to get married, she and her husband Kurt naturally wondered how likely it would be that any or all of their children would be born with cystic fibrosis. To help them calculate the risk, Zuckerman applied the universal laws of heredity, the principles that govern how traits are passed from parents to offspring. These rules have a long and interesting history. They were discovered by a European monk named Gregor Mendel, who first made his discoveries public in 1865 (Fig. 5.1). Even today, in the 21st century, it is easiest to understand these rules of heredity if we learn how Mendel himself discovered them.

In Mendel's day, 140 years ago, most observers thought each individual's traits resulted from a blending of their parents' traits. Looking at organisms in nature, it is not hard to see why people believed that offspring were intermediates between their parents. Consider, for example, two monkey flower plants whose flowers have petals of vastly different sizes (Fig. 5.2). Let's say we mated plants with these different flower shapes—one with long petals to one with short petals. The result would be a **hybrid:** the offspring of two individuals with differing forms of a given trait. In this case, the hybrid's flowers had petals intermediate in length between those of the two parents. Such observations made people think that the hereditary "stuff" of a mother and father was

liquid and would *blend* to produce the characteristics found in the offspring, just as cream mixes with dark-brown coffee to produce the beige-colored café au lait. This idea became known as the **blending model of heredity.**

Mendel wondered whether the blending model could really explain what happened in heredity. Mendel was the son of a farmer and a housewife and grew up in a Moravian village in what is now the Czech Republic. Through the financial sacrifices of his parents and sisters, he was able to attend school at a very special monastery in the old city of Brünn, now called Brno. There Mendel was so successful in his studies that the school administrators sent him to the University of Vienna to take a qualifying exam to become a certified teacher of natural history and physics. Mendel flunked the examination. Twice! But while he was at the University, he learned from his professors that all matter is made up of discrete atoms and molecules. He wondered if heredity could also be governed by "particles" that retain their identity from generation to generation. He put his new **particu-**

Figure 5.1 **Gregor Mendel.** The scholarly monk (1822–1884) with his pea plants.

Figure 5.2 Casual Observations Can Lead to Wrong Conclusions: The Blending Hypothesis. Some monkey flowers have very small petals and are self-fertilizing (left). Others have large, showy petals that attract insect or bird pollinators (right). When flowers from the two populations are cross-pollinated, the offspring are hybrids with intermediate petal length (central). Observations such as this led early biologists to think that parental traits blend in hybrids.

late model of heredity to the test in a long-term study involving pea plants, controlled matings between them, and careful tabulations of the kinds of offspring each cross produced.

Genetics in the Abbey

The blending hypothesis predicted that, like café au lait, each hereditary factor would be permanently diluted in the hybrid. Mendel's particulate model, however, predicted that each hereditary factor would remain unchanged in a hybrid, like dark-brown and cream-colored marbles mixed in a bag. Mendel's key insight was that he could disprove one of these two models not by looking at the hybrid itself—the first generation of the mating—but by checking the *offspring* of hybrids—the second generation. If the original parental forms reappeared in the second generation, this would show that the hereditary factors had passed through the hybrids unchanged and remained as some kind of intact particles. If, however, the original forms *failed* to reappear in the hybrid's offspring, then the factors would appear to have been blended.

Mendel chose common garden pea plants as his test subject because peas have several advantages. From seed stores he could purchase strains of pea plants that showed clear alternative forms for single traits, such as stem length or flower color. For example, long-stem plants versus short-stem, or purple flowers versus white. By selecting strains that differed in only one trait such as height or flower color, he could study inheritance of one feature unconfused by all other variations. In addition, Mendel could also easily control which pea plant mated with which other pea plant. A pea flower normally **self-fertilizes** or mates with itself. But Mendel **cross-fertilized** plants. From a purple flower, for example, he could simply clip off the organ that produces pollen, the sources of the sperm, and dust the egg-containing organ of that flower with pollen from another plant (for example, a white flower). From the seeds of this cross fertilization or "cross," Mendel could grow a new generation of pea plants and watch to see which traits were expressed.

With his clearly stated hypotheses and well-chosen experimental system, Mendel was now ready to perform

the scientific tests that would lead to the rules of heredity, rules that medical doctors like Jonathan Zuckerman can still apply a century later to patients like Shana Reif.

Mendel Disproves the Blending Model

In one of Mendel's first crosses, he planted seeds from long-stem and short-stem plants early one spring and let them grow into the **parental (P) generation** (Fig. 5.3a). Later that spring, when the parental plants had flowered, Mendel cross-fertilized long-stemmed plants with pollen from the short-stemmed plants. In the summer, when the pods became swollen with plump

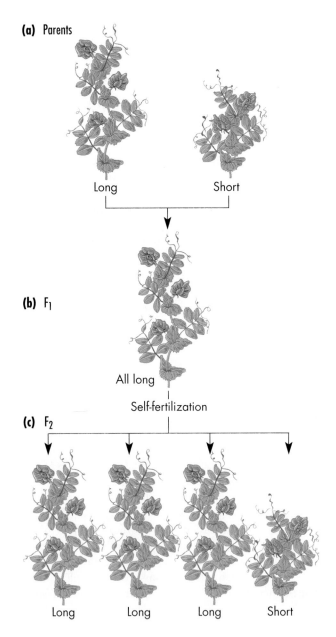

(a) Parents

Long   Short

(b) F₁

All long

Self-fertilization

(c) F₂

Long   Long   Long   Short

Figure 5.3 Mendel's Evidence for a Nonblending (Particulate) Model of Heredity. When Mendel crossed long- and short-stem pea plants (a), he got all long-stemmed progeny in the F₁ generation (b) but in the F₂ generation (c), short stems reappeared in some of the progeny although most were long-stemmed. Although the short-stemmed trait was hidden in the hybrid, it was unmixed in the progeny of the hybrid. This ruled out the blending hypothesis of heredity for this trait.

peas, he collected the seeds. These seeds would produce the next generation, called the **first filial (F₁) generation,** meaning the first generation in the line of descent. Planted in the spring of the second year, the F_1 seeds of the long-stem/short-stem cross all grew into plants with stems just as long as the original long-stemmed parent (Fig. 5.3b). Mendel repeated this type of experiment for other traits—flower color (purple vs. white), seed shape (round vs. wrinkled), and so on—and he found that in each case, only one alternative of each trait appeared in the F_1 hybrid generation. It was as if one of the traits had totally disappeared. The trait that appears in the F_1 hybrid (such as long stems in peas) is said to be **dominant,** while the trait that does not show in the hybrid (such as short stems in peas) is referred to as **recessive.**

Now came the crucial part of the experiment. What happened to the recessive characteristic in the hybrid? Did it blend with the dominant characteristic? Did it disappear completely and forever? Or did it remain intact but hidden in the F_1 generation? To find out, Mendel allowed the long-stemmed F_1 hybrid plants to self-fertilize, and the next spring he planted the seeds of the **second filial (F₂) generation.** When the second generation of pea plants grew up, most of them had long stems, but significantly, some plants had short stems. Again, there were no stems of intermediate length (Fig. 5.3c). The reappearance of plants with stems just as short as the stems of the original short-stemmed parents, and the absence of any intermediates were the results predicted by the particulate model of heredity and dramatic disproof of the blending model of heredity.

5.2 What Rules Govern the Inheritance of a Single Trait?

Before we return to cystic fibrosis, Shana, and Dr. Zuckerman, we need to learn more about Mendel's simple experiments. Being a careful and inquisitive person, Mendel was not satisfied with just saying that "some" of the F_2 plants had short stems and therefore the blending hypothesis was wrong. He wanted to understand what he saw. So he counted the plants and by analyzing the numbers, was able to infer the mechanisms that hid the short-stemmed trait in the F_1 and its reappearance in the F_2. Figure 5.4 lets you collect data from an actual experimental cross to practice gathering, analyzing, and interpreting scientific data. This cross uses corn plants because the results are readily visible in a single ear of corn and each ear has enough individual offspring (the kernels) to allow meaningful inferences. Pea pods, by contrast, have only about six seeds (too few per pod for analysis) and must be opened to reveal the offspring (peas).

Collecting Genetic Data

Figure 5.4b shows a corn ear with kernels of different colors. The kernels on this corn ear are the F_2 progeny of the cross shown in Figure 5.4a. To make that cross, we first bought two types of seeds, one from a stock that yields only purple seeds and another from a stock that yields only yellow seeds. In spring, we planted the two types of seeds and allowed the corn plants to ma-

Figure 5.4 **Collecting Data from a Cross.** The matings in (a) gave rise to the F_2 corn ear shown in (b). Count the yellow and purple kernels. What ratio do you obtain? How do you interpret your results?

(a) The cross

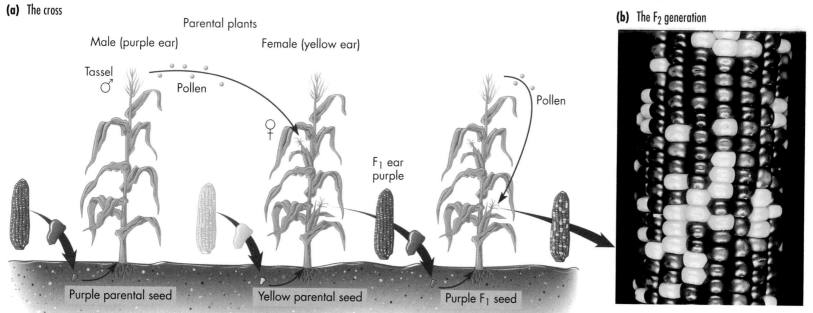

Parental plants

Male (purple ear)   Female (yellow ear)

Tassel
♂
Pollen

♀

Pollen

F_1 ear purple

Purple parental seed   Yellow parental seed   Purple F_1 seed

(b) The F_2 generation

ture. In July, when the corn was tall and bright green, we took pollen from the tassel of the plant that grew from the purple seed and dusted it on the female parts of the plant that grew from the yellow seed to create an F_1 generation. When the F_1 kernels matured on the ear, we found that all of them were purple. The next spring, we planted these hybrid purple F_1 kernels, raised the corn plants, then allowed them to self-fertilize to produce the F_2. Figure 5.4b shows the resulting F_2 ears.

Your task, now, is to look at the mottled ears and to count and record the number of purple and yellow F_2 kernels along with any kernels intermediate in color that occurred. What is the ratio of purple to yellow kernels on this cob? (Although we can show just a single cob, a geneticist would probably count the kernels on several to obtain a larger number for statistical comparisons.)

Now, let's compare your results with corn kernels to Mendel's results with a trait in pea plants—a trait like stem length. The good monk found that 787 of the F_2 plants he counted had long stems and 277 had short stems. These numbers showed a 787 : 277 ratio or approximately 3 : 1 ratio of long-stemmed to short-stemmed plants in the F_2 generation. (A perfect 3 : 1 ratio for 1064 plants would be 798 : 266, not much different from the 787 : 277 he actually observed.) Now, how do Mendel's results compare with our results from the corn experiment? What was the total number of kernels you counted? What is one quarter of this number? What would be a perfect 3 : 1 ratio for the number of corn kernels you counted? How does that compare to your actual results?

It turns out that the results for corn kernel color in the F_2 generation and Mendel's observations for pea stem length apply to many traits in eukaryotic organisms. This includes Shana's cystic fibrosis trait, and it helps a physician like Dr. Zuckerman predict the likelihood that a couple's offspring would have cystic fibrosis. The general finding is that with two clear alternative traits such as long versus short stems, purple versus yellow seeds, or presence of cystic fibrosis versus absence of the disease, the hybrid (the F_1 generation) shows only one trait, the dominant one. The mating of two hybrids (the F_2 generation) produces offspring in which three quarters show the trait that appears in the hybrid (the dominant trait), while one quarter show the trait that is hidden in the hybrid (the recessive trait). How can we understand the mechanism that causes such a result to occur?

Genes and Alleles

Mendel reasoned that because short stems reappeared in the F_2 plants, the hereditary factor that causes short

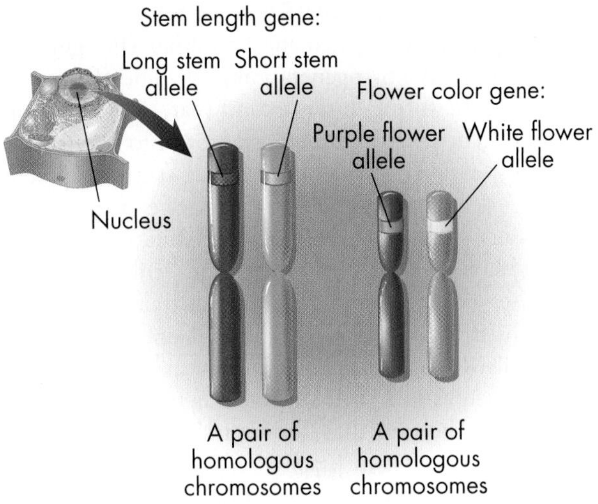

Figure 5.5 **Genes, Alleles, and Chromosomes.** Genes reside on chromosomes, and each gene is located at a specific place, or locus, on a specific chromosome. Different chromosomes have different sets of genes. Eukaryotic cells contain pairs of homologous chromosomes with two copies of each chromosome.

stems had to be an individual unit, like a particle, and not like a liquid that could be mixed with another liquid of a different color. Modern geneticists call this particulate factor a **gene.** While Mendel did not use that term, we will use it in the following discussion for clarity.

A gene influences a specific trait in an organism, such as the length of a pea stem, the color of a corn kernel, or the presence or absence of a hereditary disease like cystic fibrosis. The gene is not the trait itself. Instead it is a factor that causes the organism to develop a specific trait.

Mendel's insight was remarkable. Even though he had no knowledge of DNA or genes, he reasoned that hereditary "particles" must come in different forms. Nearly a century later, molecular researchers would show that genes do, in fact, have different forms, which are now called **alleles.** An allele (AL-eel) is an alternative form of a gene. In pea plants, the gene for stem length has two alleles, one causing long stems and one causing short stems. Likewise, modern geneticists know that one allele of the cystic fibrosis gene causes the disease, while another allele (alternative form) of the same gene is necessary for the normal functioning of the airways and other ducts.

Geneticists have also known for a half-century that a gene is a portion of a DNA molecule in a chromosome (Fig. 5.5 and Table 5.1). Although an individual

Table 5.1 Principles of Heredity

1. A hereditary trait is governed by a gene.

2. Genes reside on chromosomes, and are specific sequences of DNA in all cells, but are RNA in some viruses.

3. A gene for each trait can exist in two or more alternative forms called alleles. An individual's alleles, interacting with the environment, determine its external appearance, biochemical functioning, and behavior.

4. Most higher organisms have two copies of each gene in body cells (they are diploid). Gametes (eggs or sperm), however, have only one copy of each gene (they are haploid).

5. Homologous chromosomes are two chromosomes that are similar in size, shape, and genetic content.

6. A homozygote has two identical alleles of a gene; a heterozygote has two different alleles of a gene.

7. An individual's physical makeup (the way it looks and functions) is its phenotype; an organism's genetic makeup is its genotype.

8. In a heterozygote, generally only one of the two alleles shows in the phenotype, while the other allele is hidden. The allele that shows is the dominant allele, and the hidden allele is the recessive allele.

9. Pairs of alleles separate, or segregate, before egg and sperm formation, so each gamete has a single copy of each gene. At fertilization, sperm and egg combine randomly with respect to the alleles they contain, and the resulting zygote in general has two copies of all genes.

10. Genes on different chromosomes assort independently of each other into gametes.

11. Linked genes lie on the same chromosome and tend to be packaged into gametes together.

---

chromosome may contain thousands of genes controlling hundreds of different traits, each chromosome will have just one allele for any individual gene. Because eukaryotic cells contain pairs of homologous chromosomes (review Fig. 4.13), each individual pea plant or person generally has two alleles for each gene, which may be the same or different.

Dominant and Recessive Alleles

Mendel realized that the reappearance of short-stemmed plants in the F_2 generation meant that the short-stem allele was present but invisible in the F_1 hybrids. If the short-stem allele had not been present in the hybrid, it could not have been passed on to the F_2 offspring. Because the hybrids showed the long-stem trait, Mendel knew that the long-stem allele was also present in the hybrid. So Mendel concluded that a hybrid plant contains two copies, or alleles, of each gene, one visible and one invisible.

The allele whose trait shows in a hybrid is said to be *dominant.* The allele that is overshadowed each time it is paired with a dominant allele is said to be *recessive.* The long-stem allele of the stem length gene in peas was dominant to the recessive short-stem allele. Which allele was dominant and which recessive for the corn kernels? Can you guess which allele is dominant and which recessive for cystic fibrosis?

In his work with pea plants, Mendel reasoned that because each hybrid plant has two alleles of each gene, each pure-breeding parent plant must also have

two copies of each gene. (A pure-breeding organism always produces offspring with traits identical to its own.) In the case of the hybrid plant, the two alleles are different, one dominant and one recessive. But in the case of the pure-breeding parents, both alleles are identical, either both dominant or both recessive.

Genotype and Phenotype

Although the long-stemmed F_1 hybrid plants Mendel studied looked just like the long-stemmed pure-breeding plants of the parental generation, they were genetically different. Today, we refer to an organism's physical characteristics—stem length, kernel color, or airway functioning, for example—as its **phenotype.** We call the organism's specific alleles or genetic makeup its **genotype.** In the case of a long-stemmed phenotype, there are two possible genotypes. Some long-stemmed plants could have two identical dominant long-stem alleles, but other long-stemmed plants could have two different alleles, the visible dominant long-stem allele and the hidden recessive short-stem allele (Fig. 5.6a,b). Geneticists often indicate dominant alleles with uppercase (capital) letters and recessive alleles with lowercase letters. We can represent with L the dominant long-stem allele for stem length, and with l the recessive short-stem allele. In the case of long-stemmed plants, then, the genotype would be either LL or Ll. Organisms with two different types of alleles for a given trait are said to be **heterozygous** for that trait (Fig. 5.6b). Pure-breeding organisms, with a pair of identical alleles for a

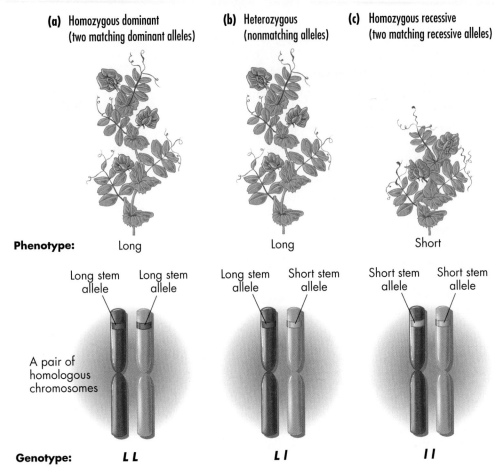

(a) Homozygous dominant
(two matching dominant alleles)

(b) Heterozygous
(nonmatching alleles)

(c) Homozygous recessive
(two matching recessive alleles)

Phenotype: Long   Long   Short

Long stem   Long stem   Long stem   Short stem   Short stem   Short stem
allele   allele   allele   allele   allele   allele

A pair of
homologous
chromosomes

Genotype: *L L*   *L l*   *l l*

Figure 5.6 **Phenotype Versus Genotype.** Underlying an organism's physical appearance (its phenotype) is pair of alleles for each gene (the genotype). (a) A homozygous dominant individual has two identical copies of a dominant allele, one on each homologous chromosome. (b) The dominant allele is defined by the phenotype of the heterozygote, an individual with two different alleles. The allele that shows in the heterozygote is the dominant allele. (c) A homozygous recessive has a recessive allele on each homologous chromosome.

given trait, are **homozygous** for that trait (Fig. 5.6a,c). A heterozygous individual is called a **heterozygote,** and a homozygous individual is called a **homozygote.** Again, geneticists following in Mendel's footsteps learned that pure-breeding long-stemmed and short-stemmed parents are homozygotes, while their hybrid offspring are heterozygotes.

In studying Shana and Kurt's chances of producing children with CF, Dr. Zuckerman knew that Shana was, like all cystic fibrosis patients, homozygous for a disease-causing allele of the cystic fibrosis gene. In other words, she has two alleles for cystic fibrosis. Her husband's genotype was unknown, however, so Dr. Zuckerman tested that. Genetic analysis revealed that Kurt has one disease-causing allele and one normal allele, thus he is heterozygous for the cystic fibrosis trait. We'll return to this young couple's family decision and its genetic significance later in the chapter.

Mendel's Segregation Principle

Mendel concluded that each individual has two copies of each factor (each gene)—two copies of the stem

length gene, two copies of the flower color gene, or two copies of the cystic fibrosis gene. Where did these two copies come from? Mendel suggested that each individual receives one allele from its mother and the other from its father for each of its many traits. Thus, the two alleles possessed by a parent must separate, or **segregate,** from each other so that only one allele of each gene goes into each egg and only one allele of each gene goes into each sperm. Recall from Chapter 4 that a cell with just one copy (allele) of each gene is said to be haploid and a cell with two copies (alleles) of each gene is diploid. Eggs and sperm are haploid, but all other cells in a human being and nearly all other cells in a pea plant are diploid.

If we generalize from Mendel's pea experiments, we can define his **law of segregation** this way: Sexually reproducing diploid organisms have two copies of each gene, which segregate from each other during meiosis without blending or being altered. When gametes form, they each contain only one copy of each gene.

Genetic Symbols and Punnett Squares

The segregation principle is probably easiest to understand using the upper and lowercase symbols L and l for the alleles of the stem length gene (Fig. 5.7). The heterozygous F_1 generation is then designated as Ll.

During meiosis in the Ll heterozygote, the alleles separate. As a result, half the gametes end up with the capital L allele and the other half with the lower case l allele (Fig. 5.7b). Mendel pointed out that to get the $3:1$ phenotypic ratio, eggs and sperm can come together totally at random with respect to the allele they carry (Fig. 5.7c). In other words, an egg cell with an l allele is just as likely to be fertilized by a sperm cell with an l allele as it is to be fertilized by a sperm carrying an L allele.

A good way to visualize the consequences of random fertilization is to draw an organized diagram called a **Punnett square,** as shown in Figure 5.7c. To construct a Punnett square for the mating of two heterozygous pea plants, draw a large square made up of four smaller squares. Along the top of the large square, write the two possible genotypes of the pollen (L and l) and along the left side of the square, write the two possible genotypes of the eggs (also in this case L and l). Then in the four empty boxes, fill in the genotypes of the offspring that result from the fertilization of each egg type with each pollen type. *E*xplorer 5.1 is a good place to see this animated and to practice with it.

Fig. 5.7c shows four F_2 genotypes: LL, Ll, lL, and ll. Because the order of alleles is not important, Ll and lL are equivalent; thus, there are really only 3 genotypes, found in the ratio 1 LL to 2 Ll to 1 ll. If we look at the

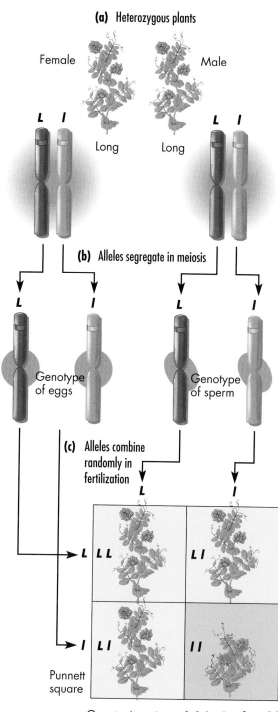

(a) Heterozygous plants

Female   Male

L l   **L l**

Long   Long

(b) Alleles segregate in meiosis

L   **l**   **L**   **l**

Genotype of eggs   Genotype of sperm

(c) Alleles combine randomly in fertilization

    L   **l**

	L	**l**
L	**LL**	**Ll**
l	**Ll**	**ll**

Punnett square

Genotypic ratio — 1 **LL** : 2 **Ll** : 1 **ll**
Phenotypic ratio — 3 long : 1 short

physical characteristics of the plants themselves, however, we find that the 1:2:1 genotypic ratio produces a 3:1 phenotypic ratio (3 long stem to 1 short stem). The reason is that the single *LL* genotype and both *Ll* genotypes have the same long-stem phenotype, because *L* is dominant to *l*. Recall that this 3:1 ratio is very close to

Figure 5.7   **Meiosis and Mendel's Principle of Segregation.** (a) Heterozygotes for the long (*L*) and short (*l*) alleles of the stem length gene are like the F₁ of Figure 5.3. (b) After chromosome replication, cells pass through meiosis and the two alleles separate, so that the resulting haploid cells each have a single allele of each gene. (c) During fertilization, alleles recombine at random. You can sketch this easily in a square as shown in the figure. The resulting offspring give a phenotypic ratio of 3 dominant to 1 recessive phenotype.

what Mendel observed in his experiment (787:277), and close to what you got when you counted the corn kernels.

Genetics and Probabilities

Mendel's segregation principle predicts a 3:1 phenotypic ratio in the offspring from a mating of two heterozygotes. But he actually found 2.84:1 for the mating we discussed, and 3.15:1 and 2.96:1 for other similar matings involving different traits. Why don't the figures come out to exactly 3:1? The answer is that the principles of genetics rely on the laws of chance and probability.

You can demonstrate the probability of obtaining the 3:1 relationship by tossing two different coins simultaneously. Let a penny represent sperm from a pollen grain, and let a nickel represent an egg. The head of each coin represents the dominant allele, and the tail represents the recessive allele. Note that each coin has an equal number of dominant and recessive alleles, just like the population of gametes from a heterozygote.

To model fertilization, flip both coins at the same time and record whether they land heads up or tails up. If both are heads, the genotype is homozygous dominant; if both are tails, the genotype is homozygous recessive; and if one coin is heads and the other is tails, the "offspring" will be heterozygous. Flip the pair of coins 20 times. How many times would you expect each of the three possible outcomes? Did you obtain exactly what you would expect? If not, how can you explain the discrepancy?

What would be the probable result if you tossed the coins many more times than 20, say, 1064 times, as Mendel did when he was experimenting with pea stem length? Like the toss of a coin, the combination of alleles in fertilization is governed by the laws of chance. In a low number of trials, as you conducted, the results may differ substantially from those predicted for random tossing, but as the number of trials increases, the results will come closer to the mathematically predicted values. This principle is especially important when doing human genetics because of small family size, as we will see in the next section.

> **GET THE PICTURE**
>
> *E*xplorer 5.1
>
> Mendel's First Law of Inheritance

5.3 How Do Geneticists Analyze Human Inheritance Patterns?

Like so many of history's geniuses, Gregor Mendel was ahead of his time. He published his results in 1865 in a journal that was circulated to about 100 scientific libraries. But very few scientists cited Mendel's findings and those that did clearly missed the importance of his ideas. Discouraged, Mendel eventually gave up breeding experiments and became abbot of his monastery.

Finally, 35 years after publication, Mendel's ideas were "rediscovered" in 1900 by 3 European botanists who performed experiments similar to those of Mendel. Only two years after that, at the turn of the 20th century, other biologists showed that Mendel's ideas were applicable to human hereditary diseases. With a bit more background, we'll be ready to see if we can apply Mendel's principles of heredity to the problem of how likely Shana and her husband Kurt would be to have children with cystic fibrosis.

Homo sapiens: **An Inconvenient Experimental Animal**

Even with Mendel's principles in mind, humans are uniquely difficult subjects for a geneticist to study. First of all, geneticists aren't matchmakers and can't convince people to choose mates and produce offspring just to satisfy their curiosity. Investigators must search for existing subjects and matings that happen to express traits of interest. In addition, there is never a true F_2 generation available for study because brothers and sisters rarely mate. Beyond that, individual human families are too small for statistical analysis; couples rarely produce more than ten children, and usually produce fewer than three. Finally, the human life cycle is too long. It could take an entire career to follow the traits in two human generations. So scientists and physicians such as Jonathan Zuckerman rely heavily on collecting and analyzing family histories.

Pedigrees: Family Genetic Histories

A major method in human genetics is to follow the inheritance of a trait through all the members of a family. Geneticists search out families with particular genetic traits, and then interview family members, check their medical records, and collect samples of blood or other tissues from as many family members as possible. For example, data for a set of 61 large, three-generation families has been collected by the Foundation Jean Dausset in France. From such records, the investigator draws up **pedigrees,** orderly diagrams that show family relationships, birth order, gender, phenotype, and, when possible, the genotype of each family member.

To see how a family pedigree works, let's consider the family tree of a Swedish family with the cystic fibrosis trait (Fig. 5.8). In a pedigree, each generation occupies a separate horizontal row, with the ancestors at the top and more recent generations below. Males are indicated by squares and females by circles. Symbols for people affected (not *in*fected!) with the trait are filled in. Geneticists designate each generation with a roman numeral and each individual with an arabic number from left to right. For example, the matriarch of the family, a woman born in 1859, is I1, and her two daughters are II2 and II3. The boy and girl VI1 and VI2 are the only family members with cystic fibrosis. As in many pedigrees, this one arranges a group of brothers and sisters in order from oldest (left) to youngest (right).

Another convention is that a horizontal line joins two parents, and the offspring are attached to the line

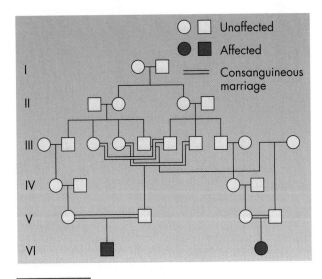

Figure 5.8 **Pedigree for a Family with Cystic Fibrosis.** This pedigree of a Swedish family shows the inheritance pattern of a condition controlled by a recessive allele. Notice the hallmarks of that pattern: Both affected and unaffected sons and daughters derive from parents who are themselves unaffected. The female I1 was born in 1859 and her cystic fibrosis was transmitted silently through at least six generations. Modern geneticists traced the trait through a tangled web of consanguineous (intrafamily) marriages (indicated here by double lines). As a practice exercise, write the genotypes for as many individuals as you can. (Source: Schaedel, C., Schwartz, M., Kornfalt, R., and Holmberg, L. Cystic fibrosis caused by homozygosity for the CFTR gene mutation *Scand. T. Acta Paediatr.* 84, 1199–1200, 1995.)

below. Parents II1 and II2, for example, produced two daughters and two sons, individuals III2 to III5. Geneticists sometimes omit from a pedigree parents who are unrelated and unaffected. They also tend to show consanguineous marriages (unions between blood relatives) with double horizontal lines like the ones in Figure 5.8.

Is Cystic Fibrosis Inherited as a Recessive or Dominant Trait?

A pedigree can look rather formidable, with its marching rows of grandparents, aunts, brothers, and sisters. Nevertheless, the rules for analyzing a pedigree follow Mendel's principles. Let's analyze our hypothetical family's pedigree more closely. By doing this, we can determine whether cystic fibrosis shows a pattern of dominant or recessive inheritance.

The pedigree (Fig. 5.8) shows that several of the siblings in the family lacked the cystic fibrosis trait; neither parent showed the trait, either. Let's represent the cystic fibrosis gene by CF, with CF^- representing the disease allele and CF^+ the healthy allele. Because our affected female VI2 has cystic fibrosis, she must have at least one CF^- allele and must have inherited it from one of her parents. That shows that at least one of the parents must carry at least one CF^- allele. Since neither parent shows the trait, we can conclude that each parent has at least one CF^+ allele, and that at least one parent is a heterozygote with one CF^+ allele and one CF^- allele. Because both parents are healthy, we must conclude that a person with the heterozygous genotype has a healthy phenotype. We saw earlier that in a heterozygote, the dominant allele shows. Therefore, the CF^+ (healthy) allele must be dominant to the CF^- (disease) allele. In other words, the allele that causes cystic fibrosis must be recessive and the genotype of VI2, Shana Reif, and other affected people is CF^-CF^-. To generalize this argument: If an offspring inherits a condition but neither parent shows the condition, the trait is usually recessive.

Because the disease allele of the cystic fibrosis gene is recessive, affected people, including Shana, must have two copies of the disease allele. One of those copies must have come from Shana's dad, and the other from her mom. Since both of her parents passed on the disease allele, but do *not* show its effects, they must be heterozygotes, or **carriers**. In carriers, the dominant normal allele masks the recessive allele, which is **mutant** (the result of a mutation).

Recall that tests showed Shana's husband, Kurt, is a heterozygote for the CF gene and he doesn't, himself, have cystic fibrosis but is a carrier for the disease, like Shana's parents. Based on the couple's genotypes for cystic fibrosis, Dr. Zuckerman was able to predict that half of their children would have cystic fibrosis, a 1:1 ratio. This was disappointing to the young couple, especially since no one in Kurt's family had ever developed cystic fibrosis. They were undecided about whether or not to take the risk of parenting, a risk both to future children and to Shana's own delicate health.

It turns out that many human genetic diseases are inherited as recessive traits like cystic fibrosis. Some, however, are inherited as dominants. Let's look at some of each.

A Pedigree for a Dominant Trait: Familial Hypercholesterolemia

We've seen the pedigree for cystic fibrosis, which is inherited as a recessive trait. So what would a pedigree look like for a condition inherited as a dominant trait? Consider the condition called familial hypercholesterolemia (FH) or hereditary high blood cholesterol level. People with FH—late night television's David Letterman is an example—have excess amounts of low density lipoprotein (LDL, so-called "bad" cholesterol). Fifty percent of affected males and 15 percent of affected females die before age 60, usually of a heart attack. The pedigree in Figure 5.9a shows the hallmarks of a dominant inheritance pattern. For dominant conditions, *affected individuals have at least one affected parent.* Furthermore, both males and females transmit the trait, and both sexes show the trait in about equal frequencies. Characteristically, when an affected person who is a heterozygote mates with an unaffected individual who is a homozygote, about half of the offspring are affected and half unaffected. (In small human families, however, ratios like these are not informative.) Figure 5.9b shows why. In contrast matings between two heterozygotes produce offspring in the phenotypic ratio of 3 affected to 1 unaffected. At the level of genotype, the affected individuals are either heterozygous or homozygous for the disease allele. Individuals with two copies of the disease allele are much more severely affected (Fig. 5.9c), and unfortunately, usually die of heart attacks as children.

How Alleles Interact

Experimenting in his quiet abbey garden, Mendel showed that each gene—for plant height, flower color, and so on—has two alleles, which are either dominant or recessive. Life, however, is not always so simple, as later geneticists found with more sophisticated experi-

What are Shana and Kurt's genotypes for the CF gene?

(a) Pedigree

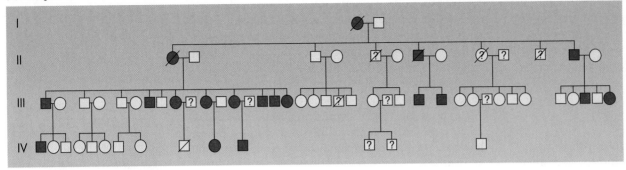

(b) Punnett square

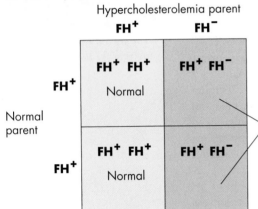

Hypercholesterolemia parent

FH⁺   FH⁻

	FH⁺	FH⁻
FH⁺	FH⁺ FH⁺ Normal	FH⁺ FH⁻ Hypercholesterolemia
FH⁺	FH⁺ FH⁺ Normal	FH⁺ FH⁻ Hypercholesterolemia

Normal parent

(c) Homozygote (FH⁻ FH⁻)

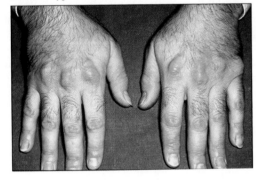

Figure 5.9 **Inheritance of a Dominant Trait: Familial Hypercholesterolemia.** (a) This pedigree shows a large family from the Aleutian Islands with familial hypercholesterolemia. The mother had 18 children, 7 of whom are shown here. The affected individuals (indicated by filled symbols) had very high cholesterol levels in their blood and tended to suffer heart attacks in their 50s and 60s. (b) A Punnett square for the mating of a heterozygote with a homozygous recessive person shows that half the offspring are affected. (c) Homozygotes for this trait show lumpy deposits of cholesterol in the skin and tendons, and generally die of a heart attack as children.

ments. The alleles of some genes fail to fall clearly into either the dominant or recessive category, and some genes have many more than two alleles.

Incomplete Dominance

In 1905, a young African American experiencing pains in his joints and abdomen, chronic fatigue, and shortness of breath consulted a Chicago physician. A blood test showed that the man had too few red blood cells (a condition called anemia) and that many of his blood cells were shaped like crescents, or sickles, instead of the normal disks (Fig. 5.10). The sickle-shaped blood cells tended to clump together, lodging in the smallest blood vessels and blocking blood flow. This blockage was especially pronounced in the joints and the spleen, a blood storage organ in the upper left side of the abdomen, causing the pain. Studies revealed sickle-shaped blood cells to be fragile and easily destroyed. This condition is called **sickle cell anemia,** and it's the most commonly inherited lethal disease among African Americans.

Sickle cell anemia affects about 60,000 people in the United States, about 1 in 400 African Americans. It is present in 1 in 50 West Africans, and is also frequent in parts of Saudi Arabia, India, and Asia. It is rare in many other populations, however. (Chapter 9 discusses why sickle cell anemia is distributed this way.)

A condition related to sickle cell anemia is called sickle-cell trait. In people with sickle cell trait, red blood cells form a sickle shape when deprived of oxygen in a test tube—a condition that fails to induce sickling in normal red blood cells. People with sickle cell trait are normal except when exposed to extreme conditions, such as high altitude or severe physical exertion. For example, several men with sickle cell trait living in low-altitude cities suffered severe spleen pain within two days of arrival in a part of Colorado with high altitudes. Sickle cell trait is thus intermediate in severity between full-blown sickle-cell anemia and normal health. What is its genetic basis?

Figure 5.10b shows a pedigree for sickle cell anemia in a family from Kingston, Jamaica. Notice that each person with sickle cell anemia has two parents that both display sickle cell trait. By examining a large number of families with sickle cell anemia, geneticists have found that the mating of two people with sickle cell trait produces offspring in a 1:2:1 ratio with 1/4 showing full blown sickle cell anemia, 1/2 showing sickle-cell trait, and 1/4 showing neither condition. The Punnett square in Figure 5.10c reveals the origin of this 1:2:1 phenotypic ratio. Sickle cell anemia displays **incomplete dominance:** the phenotype of heterozygotes—in this case individuals with sickle cell trait—is intermediate between the homozygous dominant and the homozygous recessive conditions. In incomplete dominance, the phenotypic and genotypic ratios are the same. The principle of incomplete dominance can help explain why early observers devised the incorrect blending hypothesis of inheritance (review Fig. 5.2). Several genes act together to control petal length in monkey flowers, for example, and these are inherited in typical Mendelian fashion — except that their alleles display incomplete dominance.

Codominance

We just discussed incomplete dominance, in which the phenotype of the heterozygote is intermediate between the two homozygotes. Another variation is called **codominance,** in which the phenotype of the heterozygote simultaneously shows *both* phenotypes. A familiar example of codominance is the blood-type gene called *ABO* (Fig. 5.11). For this gene, allele *A* causes a certain carbohydrate molecule called type A to appear on the surface of red blood cells, and a person with this allele may have blood type A. Shana, for example, has blood type A. Allele *B* causes a different carbohydrate molecule called type B, to appear on the surface of red blood cells, and produces blood type B. Someone who has two *A* alleles has only the A molecule, and a person with two *B* alleles has only the B molecule. But a heterozygote with one *A* allele and one *B* allele has both A and B molecules on the surface of red blood cells. That person has blood type AB, a codominant phenotype.

Figure 5.10 Sickle-Cell Anemia: A Pattern of Incomplete Dominance. (a) This scanning electron micrograph shows a sickled cell (left), then a normal disc-shaped cell, and then three more abnormally shaped red blood cells (right). (b) The pedigree of a family from Kingston, Jamaica, shows that people with the intermediate, sickle-cell trait phenotype are heterozygotes, and people with sickle-cell anemia are homozygotes for the sickle-cell allele. Individuals IV16 and IV17 are twins. (*Source:* Wainscoat, J., Thein, S., Higgs, D., Bell, J., Weatherall, D., Al-Awamy, B., and Serjeant, G. A genetic marker for elevated levels of haemoglobin F in homozygous sickle cell disease? *Brit. J. Haematology* 60, 261–268, 1985.) (c) A Punnett square for the offspring of two parents both with sickle-cell trait shows the origin of the 1:2:1 phenotypic ratio found in cases of incomplete dominance.

(a) Sickle cells

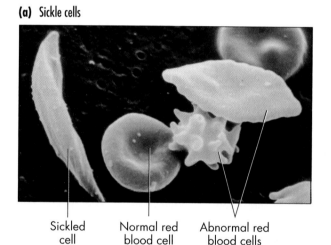

Sickled cell | Normal red blood cell | Abnormal red blood cells

(c) Punnett square

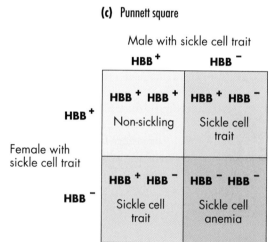

(b) Pedigree

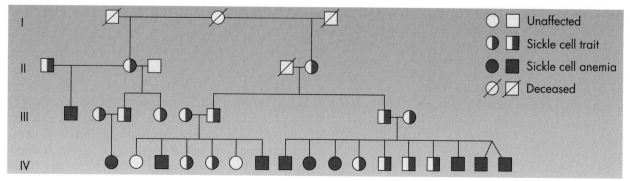

Students are sometimes confused by the difference between codominance and incomplete dominance. In codominance, *both* alleles are *fully* expressed in the heterozygote, while in incomplete dominance, the phenotype is intermediate. AB blood type is not intermediate between A and B: it is fully A *and* B. But people with sickle cell trait are nearly normal in phenotype, becoming ill only under extreme circumstances, and thus incomplete dominance is at work.

Multiple Alleles

If, like Kurt, your blood type is O, not A, B, or AB, then you are a good example of another genetic concept: Some genes have more than two alleles, and the human *ABO* blood group gene is an example. In addition to the two codominant alleles *A* and *B*, the *ABO* gene has a third allele that is fully recessive to both *A* and *B*. This recessive allele is called *o*. A person with two doses of *o* has neither the A nor B molecular marker and has blood type O. Because *o* is recessive, an *Ao* heterozygote has blood type A, and a *Bo* heterozygote has blood type B. Although there are three alleles of the *ABO* gene found in the human population, no one person can have all three at once, because each child gets only *one* allele of each gene from each parent, for a total of two copies of each gene. Figure 5.11 shows some pedigrees for the *ABO* gene.

The *ABO* gene has three alleles, but some genes have even more than that, and this is important to some of Dr. Zuckerman's patients with severe cystic fibrosis who have received transplanted organs. Because the effects of cystic fibrosis on the lungs and airways can be life-threatening, hundreds of cystic fibrosis patients have received transplanted lungs as a treatment of last resort. Unfortunately, as you have probably read in newspapers or magazines, tissue transplants from unrelated people are likely to be rejected, and the reason hinges on multiple alleles.

The **major histocompatibility complex (MHC)** is a group of genes that encode certain proteins on cell surfaces. These substances serve as identification markers that help the body distinguish its own cells from foreign substances like bacteria, viruses, or parasites that might otherwise successfully invade the body and cause disease. Because there are several genes in the MHC, and because each gene has many alleles, it is highly unlikely that two unrelated persons will have precisely the same combination of alleles. That is why a person with kidney or liver disease or lung damage from cystic fibrosis must often wait a long time before being matched with a suitable donor. If there are too many allelic differences between the tissues of donor and recipient, the immune system cells of the recipient will kill the cells of the donated organ. Even with immunosuppressant drugs that help stop tissue rejection, the multiple alleles for tissue types constitute a major obstacle to many life-saving transplants.

Pleiotropy: Multiple Effects of a Single Gene

Cystic fibrosis is another important exception to Mendel's rules and another example of gene interactions. Recall that the genes Mendel studied affected just a single aspect of the phenotype, such as plant height or flower color. In contrast, the cystic fibrosis gene appears to affect many phenotypes, including the airways, pancreas, sweat glands, and the male's reproductive ducts. Cases such as this, in which a single mutated gene affects several different phenotypes, are known as **pleiotropy** (*pleio* = more) (Fig. 5.12a). A geneticist who studies human genes must often try to find which single biochemical feature underlies all the observed effects on the phenotype. Researchers like Jonathan Zuckerman know now that in a patient with cystic fibrosis, a genetic mutation alters the structure of a single protein. This protein, CFTR, is involved in regulating the movement of chloride ions across the surfaces of cells that line organs and the ducts leading from them (Fig. 5.12b). Without a proper movement of ions and the water that follows by diffusion and osmosis, mucus in the lungs and its ducts, for example, becomes too sticky to be expelled readily. What's more, digestive enzymes made in the pancreas fail to move down clogged ducts to the intestine, interrupting food digestion. Altered chloride transport changes the composition of the sweat, which

(a) Blood types

Phenotype		Genotype
Blood type	Cell surface molecule	
A	Red blood cell	**AA** or **Ao**
B		**BB** or **Bo**
AB		**AB**
O	(Neither A nor B)	**oo**

(b) Family studies

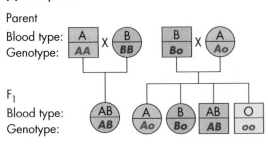

Parent
Blood type: A × B   B × A
Genotype: AA × BB   Bo × Ao

F₁
Blood type: AB   A   B   AB   O
Genotype: AB   Ao   Bo   AB   oo

Figure 5.11 **Blood Groups, Incomplete Dominance, and Multiple Alleles.** Blood transfusions must be matched because different people have different molecules on the surfaces of their red blood cells. (a) One class of these molecules is controlled by the *ABO* gene, which has three alleles (*A, B,* and *o*), four genotypes, and four blood types. (b) Two of the many possible matings among people with different blood types.

becomes far saltier than normal. And the exit ducts for sperm become blocked, leading to sterility in males. Most human genetic diseases are pleiotropic like cystic fibrosis. Tracing the primary defect to a single gene and protein can often help provide therapies that relieve several symptoms at once.

5.4 What Are the Rules Governing How Organisms Inherit Multiple Traits?

So far, we have discussed single genes and their alleles, whether for flower color or cystic fibrosis. But organisms have thousands of genes. A single-celled yeast has about 6,000 genes, a soil-living roundworm has about 19,000, and you have about 40,000. What happens when a geneticist studying the cross between two individuals focuses on more than one gene at the same time in the same cross? For example, about 25 years ago, biologists wondered about the inheritance of ABO blood types in families with some members who had cystic fibrosis. How might these two traits be inherited? Would children with cystic fibrosis inherit only a particular blood type? Or, might there be no relationship at all between the inheritance of the two genes? Some of Gregor Mendel's experiments allow us to interpret and predict the result of such human crosses.

Formation of Gametes

Recall that Shana is homozygous for the cystic fibrosis trait and has two recessive alleles: $CF^- CF^-$. She also has blood type A, and is homozygous dominant for this

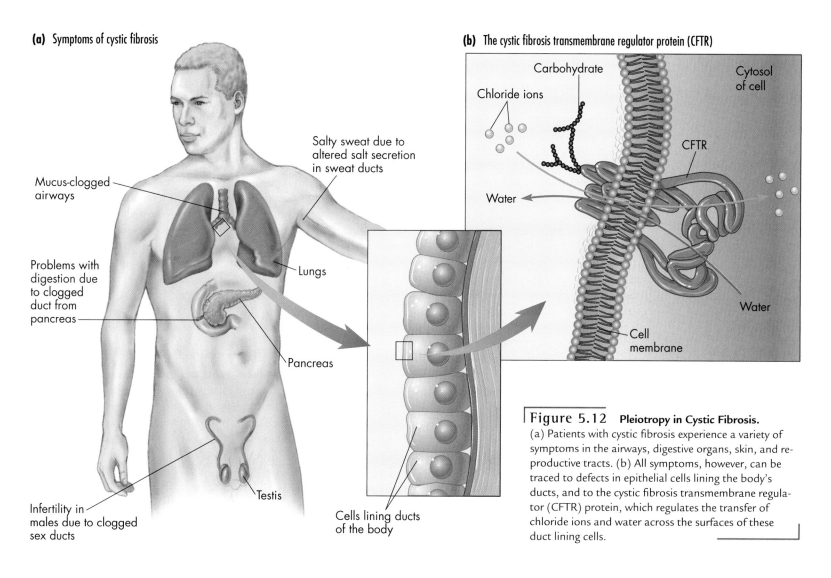

(a) Symptoms of cystic fibrosis

Mucus-clogged airways

Salty sweat due to altered salt secretion in sweat ducts

Problems with digestion due to clogged duct from pancreas

Lungs

Pancreas

Testis

Infertility in males due to clogged sex ducts

Cells lining ducts of the body

(b) The cystic fibrosis transmembrane regulator protein (CFTR)

Carbohydrate

Chloride ions

Cytosol of cell

CFTR

Water

Water

Cell membrane

Figure 5.12 **Pleiotropy in Cystic Fibrosis.** (a) Patients with cystic fibrosis experience a variety of symptoms in the airways, digestive organs, skin, and reproductive tracts. (b) All symptoms, however, can be traced to defects in epithelial cells lining the body's ducts, and to the cystic fibrosis transmembrane regulator (CFTR) protein, which regulates the transfer of chloride ions and water across the surfaces of these duct lining cells.

blood type, with the genotype *AA*. Kurt is heterozygous $CF^+ CF^-$, and he is homozygous recessive for blood type O, with genotype *oo*. Suppose that they decided to have a child, and produced a son who is simultaneously heterozygous for cystic fibrosis $CF^+ CF^-$, and has blood type A with the genotype *Ao* (one dominant *A* allele, one recessive *o* allele of the *ABO* blood group gene). This son's genotype for the two traits would then be $CF^+ CF^-$ *Ao*.

When their son grows up and begins to generate gametes, what types would he make? Recall that Mendel's principle of segregation says that each gamete gets one copy of each gene, and so each gamete must have one copy of the cystic fibrosis gene and one copy of the blood type gene. Mendel's principle further suggests that half of the sperm cells would get a CF^- allele and the other half a CF^+ allele. Likewise, half of the sperm would get an *A* blood type allele and the other half an *o* allele. So the son's sperm could be of four types:

$$\boxed{CF^- A} \quad \boxed{CF^- o} \quad \boxed{CF^+ A} \quad \boxed{CF^+ o}$$

But in what proportions will these four types of gametes actually form? Would the sperm possess only the parent's original genotypes, $CF^- A$ and $CF^+ o$? Geneticists call these **parental** types because they are like the original parents. Or would some of the sperm also possess the new combinations $CF^- o$ and $CF^+ A$? Geneticists call these types the **recombinant** types (see Chapter 4), because they are "recombined" and not present in the original parents.

In working with peas, Mendel found that alleles of different genes move independently into the gametes, a process he called **independent assortment.** What it means is that the segregation of a particular allele pair into separate gametes is independent of other allele pairs. As a result, all four types of gametes (such as those we just showed for cystic fibrosis and blood type) are equally likely, each occurring one quarter of the time.

One way of visualizing this is to use the following diagram:

Cystic fibrosis alleles	ABO alleles	gametes
1/2 CF⁻	1/2 A	1/4 CF⁻ A
	1/2 o	1/4 CF⁻ o
1/2 CF⁺	1/2 A	1/4 CF⁺ A
	1/2 o	1/4 CF⁺ o

This diagram shows the essential features of Mendel's principle of independent assortment. For genes that are inherited independently of each other, half of the gametes from an individual that is doubly heterozygous are of the parental type ($CF^- A$ and $CF^+ o$), and half are of the recombinant type ($CF^- o$ and $CF^+ A$). **E**xplorer 5.2 illustrates this.

Recall that Mendel always took his experiments through an F_2 generation. So let's consider what would happen if the theoretical son we've been discussing married a woman with exactly his same genotype for cystic fibrosis and blood type ($CF^+ CF^-$ *Ao*). Furthermore, since this is hypothetical, let's say that this couple had hundreds of children so we could obtain statistically meaningful results. What genotypes would the children have and in what proportions? Mendel carried out similar crosses, which he called **dihybrid crosses** (that is, crosses following two traits), with pea plants. He followed both long and short stems and purple and white flowers (represented by these alleles *LL, Ll, ll* and *PP, Pp, pp*). Figure 5.13 shows the results in the offspring, and we can see that the phenotypic ratio is 9:3:3:1. We can apply Mendel's results to the human situation as well, and get the same 9:3:3:1 ratio. What we would see is 9 double dominant showing both the healthy (noncystic fibrosis) phenotype and the A blood type; 3 with the recessive cystic fibrosis phenotype but the dominant A blood type; 3 with the dominant noncystic fibrosis phenotype but the recessive O blood type; and 1 double recessive with both the cystic fibrosis phenotype and the O blood type. (Note that 9 + 3 + 3 + 1 adds up to 16, which is the number of squares in the 4 × 4 Punnett Square in Figure 5.13.)

In summary, Mendel's second principle shows that different hereditary factors segregate into gametes independently of each other. As a consequence of independent assortment, we see the **9:3:3:1 ratio** in the F_2 generation.

Genes Are Located on Chromosomes

While Mendel's report was gathering dust in European libraries, other biologists were studying cells and advancing our understanding of their structure and activity. Particularly important were the observations some early biologists made with simple microscopes that showed how chromosomes move during mitosis and meiosis (see **E** 4.1, 4.2). Shortly after researchers rediscovered Mendel's principles around 1900, biologists realized that there are several parallels between the inheritance of genes and the distribution of chromosomes during meiosis:

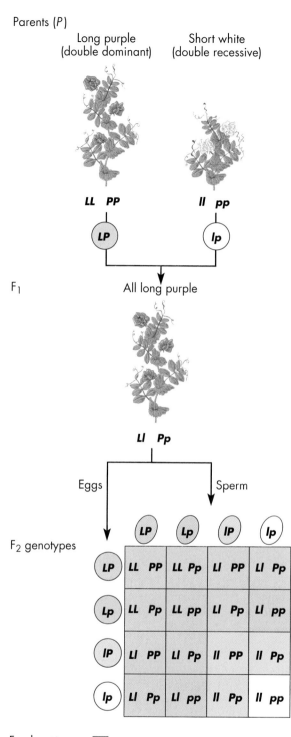

Parents (P)

Long purple
(double dominant)

Short white
(double recessive)

LL PP

ll pp

LP

lp

F₁

All long purple

Ll Pp

Eggs

Sperm

F₂ genotypes

	LP	Lp	lP	lp
LP	LL PP	LL Pp	Ll PP	Ll Pp
Lp	LL Pp	LL pp	Ll Pp	Ll pp
lP	Ll PP	Ll Pp	ll PP	ll Pp
lp	Ll Pp	Ll pp	ll Pp	ll pp

F₂ phenotypes

9	Long purple (dominant – dominant)
3	Long white (dominant – recessive)
3	Short purple (recessive – dominant)
1	Short white (recessive – recessive)

16 TOTAL

Figure 5.13 **Independent Assortment in Pea Plants and People.** Mendel's experiment with a dihybrid cross in pea plants, following the traits of stem length and flower color. A Punnett square reveals why this cross produces the four phenotypes in a 9:3:3:1 ratio.

1. Two copies of each gene and two copies of each chromosome exist in each body cell.
2. Pairs of alleles and pairs of homologous chromosomes both segregate during gamete formation.
3. Genes for different traits and nonhomologous chromosomes both assort independently when egg and sperm are formed.

These facts suggested that genes are physically linked to chromosomes. To test that possibility, investigators would have to locate individual chromosomes and show that when an organism inherits that chromosome, a specific trait is always transmitted with it. That became possible by investigating sex chromosomes.

5.5 How Does Sex Influence the Inheritance of Traits?

The pedigrees we looked at earlier for cystic fibrosis and familial hypercholesterolemia show about the same number of affected males as affected females. Many other genetic conditions, however, such as **color blindness** (the inability to see specific colors) and **hemophilia** (inability to form a blood clot) are much more prevalent in males than in females. By investigating traits influenced by sex, early 20th-century geneticists were able to show that genes are indeed located on chromosomes.

Sex Chromosomes

Have you ever wondered why there are roughly as many boy babies as girl babies (the actual ratio is about 106 boys to 100 girls)? A **karyotype** or arrayed set of chromosome photographs reveals why (Fig. 5.14). For 22 of our 23 chromosome pairs, both members are identical in size and shape. For the 23rd chromosome pair, however, males and females differ. Chromosome pairs in which both chromosomes look the same in both sexes are **autosomes,** while chromosome pairs with dissimilar members in males and females are **sex chromosomes.** Humans have 22 pairs of autosomes and one pair of sex chromosomes. In fruit flies and people, females have two identical sex chromosomes, called **X chromosomes,** and males have one X chromosome and another, often smaller chromo-

Figure 5.14 **A Set of Human Chromosomes.** A karyotype, an organized presentation of a set of chromosomes, is made from cells caught in metaphase of mitosis. Technicians treated these chromosomes, and then photographed and rearranged them from largest to smallest. This is from a male with one X and one Y chromosome.

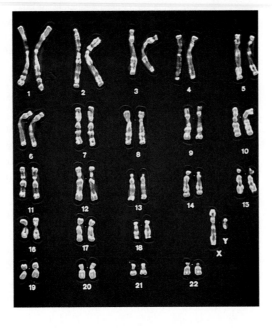

(XX). Note that a male's single X chromosome must be inherited from his mother. Becauses males and females have different chromosomes, we know that at least one trait—sex—is regulated by chromosomes. But are there any others?

Sex-Linked Traits

In 1910, Thomas Hunt Morgan and his associates at Columbia University began a series of experiments that would change genetics forever. Morgan wanted to find out about genes and chromosomes but did not have Mendel's monastic patience, so he chose the fast-breeding fruit fly, *Drosophila melanogaster.* No bigger than an "l" in this sentence, fruit flies are easy to raise and breed, and in just 12 days, an egg becomes a reproductive adult ready to produce hundreds of offspring.

One day, as Morgan observed fruit flies under the microscope, he noticed a fly with white eyes instead of the usual red (Fig. 5.16a). A **mutation**—a permanent change in the genetic material—had altered a gene for eye color from the normal red-eye allele (symbolized by fly geneticists as w^+) to the mutant white-eye allele *(w).*

From a series of crosses like the one shown in Figure 5.16b, Morgan realized that the gene for eye color is carried on the X chromosome. That gene is therefore a sex-linked gene, or more specifically, an **X-linked gene.**

Other genes were found on the fly X chromosome, including *yellow* body and *singed* bristles. After studying such genes, Morgan and his coworkers drew an important conclusion: The Y chromosome, being considerably smaller, carries no allele of the gene for eye color, or for most of the other X-linked genes. From this series of experiments, Morgan drew three important conclusions: (1) genes are located on chromosomes, (2) each chromosome carries many different genes, and (3) genes on the X chromosome have a distinct pattern of inheritance.

Duchenne Muscular Dystrophy: An X-Linked Recessive

The human genes we've been discussing, including cystic fibrosis, ABO blood groups, and hypercholesterolemia, all lie on autosomes (nonsex chromosomes). We usually call these genes "autosomal recessives," or "autosomal dominants" to reflect that fact. In contrast, other conditions are X-linked.

The most common X-linked recessive genetic disease is **Duchenne muscular dystrophy (DMD),** a degenerative muscle condition that strikes 1 out of every

some called a **Y chromosome.** Although sex chromosomes are common in animals, they are rarely found in plants, fungi, or protists.

The way sex chromosomes become distributed during meiosis explains the appearance of about equal numbers of males and females. An XY male is like the heterozygous parent, and an XX female is like the homozygous parent (Fig. 5.15). In the male, the X and Y segregate during meiosis, and as a result, one half of the sperm contain a Y chromosome and one half an X chromosome. In the female, the two X chromosomes segregate during meiosis; as a result, each egg contains one X chromosome. If the X and Y sperm randomly fertilize a group of eggs, then half of the zygotes formed will be male (XY) and half female

Parents

Female — Male

X X — X Y

Chromosomes segregate in meiosis

Sperm

Offspring

X — Y

Eggs

X | X X | X Y
X | X X | X Y

Two daughters — Two sons

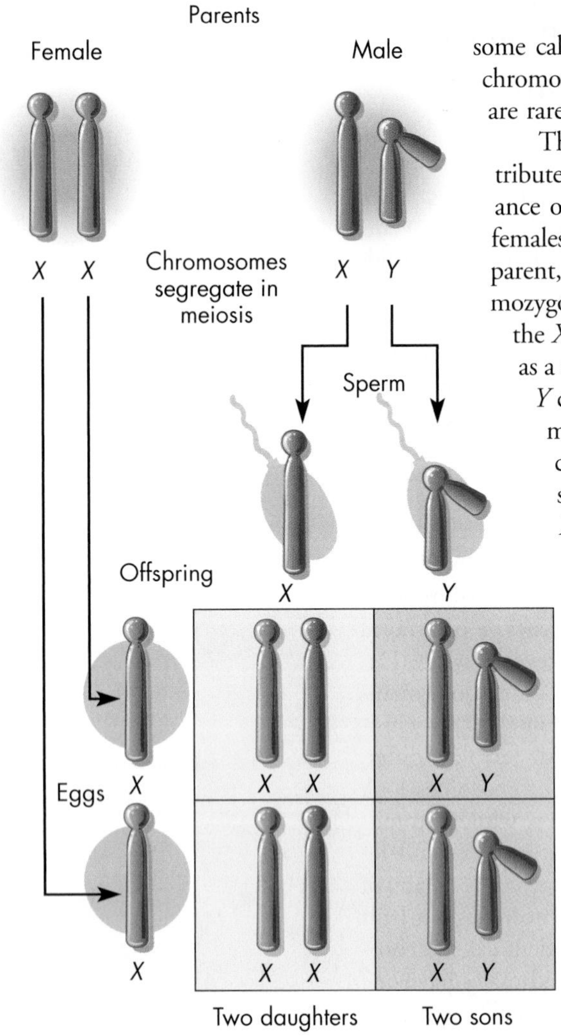

Figure 5.15 **Inheritance of the X Chromosome.** Women have two X chromosomes, but men have one X and one Y chromosome. Sex chromosomes separate when meiosis produces eggs or sperm. Daughters inherit an X from each parent, while sons inherit their X from their mother and their Y from their father.

(a) Fly eyes

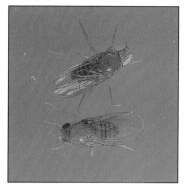

(b) The parents (*P*)

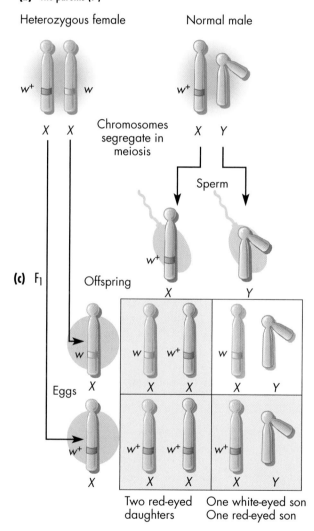

Heterozygous female   Normal male

w^+ w w^+

X   X   X   Y

Chromosomes segregate in meiosis

Sperm

w^+

X   Y

(c) F₁

Offspring

X   Y

Eggs

w — X

w^+ — X

	X	Y
w / X	w w^+ / X X	w / X Y
w^+ / X	w^+ w^+ / X X	w^+ / X Y

Two red-eyed daughters   One white-eyed son
One red-eyed son

3500 boys. The muscle cells of an affected boy die slowly, beginning in the legs. By age 5 the child can stand up only with difficulty, and he rises in a characteristic way (Fig. 5.17a). At about age 20, the diaphragm muscles degenerate, and the affected person can no longer breathe. In the fight against DMD, physicians have taken immature muscle cells from an affected boy's father or brother, and injected them into the boy's arms or legs. A small number of the normal muscle cells fused with the patients' own defective muscle fibers, and in some but not all studies, muscle strength appeared to increase. As you can imagine, however, the multiple allele problem associated with tissue incompatibilities has resulted in limited success. Researchers hope that through drug therapy, muscle-cell injections, and other future discoveries, successful treatments will emerge for this debilitating recessive condition.

Observe the pedigree for Duchenne muscular dystrophy (Fig. 5.17b); it shows the typical inheritance pattern of *X*-linked genes. These traits are more frequent in boys, and they are passed to boys from their mothers, who are heterozygous carriers. *If a pedigree shows that a boy inherits a trait from his father, it is unlikely to be an X-linked condition.* Another *X*-linked recessive allele causes hemophilia and another causes color blindness. You can learn more about sex-linked traits, including an *X*-linked dominant disease called Charcot-Marie-Tooth Disease, in Figure 5.18 and **E**xplorer 5.3.

READ ON

Explorer 5.3

Human Genetic Diseases

Chromosomes and Sex Determination

The historic experiments with sex chromosomes in fruit flies showed that the expression of male and female characteristics depends on chromosomes. They did not show, however, how sex determination works. Consider this: In flies as well as in people, *XX* individuals are females and *XY* individuals are male. But in both cases, males differ from females in two factors: the presence or absence of a *Y*, and the number of *X* chromosomes. Which is more important? To answer, geneticists have studied individuals with unusual numbers of sex chro-

Figure 5.16 **Fly Eyes and Sex-Linked Inheritance.** (a) Normal flies have bright red eyes (bottom), and white eye flies (top) are mutant. (b) A female heterozygous for the recessive white allele (*w*) and the dominant red allele (w^+) has red eyes. (c) Half of her eggs will have a *w* allele, and half a w^+. Half the sperm made by her red eyed mate have an *X* chromosome with the w^+ allele, and half have the *Y* chromosome with no allele of the eye color gene.

Figure 5.17
Duchenne Muscular Dystrophy, an *X*-linked Recessive Disease. (a) To rise from the floor, boys with this condition "climb up themselves" because their lower limbs grow weak before their upper bodies do. (b) A pedigree of a family with Duchenne muscular dystrophy shows the pattern of *X*-linked recessives: Most affected people are males, and a son never inherits the condition from his father. Spouses are omitted here.

(a) How a boy with DMD rises

(b) Pedigree of DMD, an *X*-linked gene

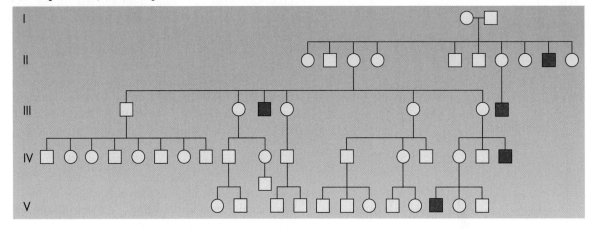

mosomes. Before reading on, look at the data in Table 5.2 to see if *Y* or *X* chromosomes are more important for flies. Then do the same for humans.

By analyzing the data in the table, you can see that any fly with at least two *X* chromosomes is a female, regardless of how many *Y* chromosomes she has. What about humans? Did your analysis of the table show that people with a *Y* chromosome develop as males, regardless of the number of *X* chromosomes? Good! This fact

indicates that there must be a genetic factor on the human *Y* chromosome that is essential for producing the male phenotype. Geneticists have isolated that gene, and named it **SRY,** for "sex-determining region, *Y* chromosome." They still don't know yet exactly how it works, but in some way, *SRY* turns on some or all of the genes that stimulate the male phenotype and suppresses those leading to the female phenotype.

More evidence of human sex determination comes

Figure 5.18 **Charcot-Marie-Tooth Disease, an *X*-Linked Dominant.** This condition causes the muscles in the hands and lower legs to atrophy. The pedigree shows that affected males always have an affected mother and males pass it to all of their daughters. (*Source*: Hanh, AF, Brown, W., Koopman, W., Feasby, T.) *X*-linked dominant hereditary motor and sensory neuropathy. *Brain* 113, 1511–1525, 1990.)

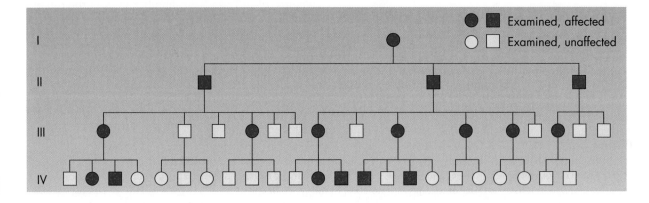

Table 5.2 Chromosomes and Sex Differentiation

Chromosome Constitution	Fruit Fly Sex	Human Sex
X	Male (sterile)	Female (Turner syndrome
Y	Lethal	Lethal
XY	Male	Male
XX	Female	Female
XXY	Female	Male (Klinefelter syndrome)
XYY	Male	Male
XXX	Female	Female
XXXY	Female	Male

from the Turner and Klinefelter syndromes. A person with one X and no Y chromosome (XO) is a sterile female with Turner syndrome (Fig. 5.19). About 1 in 2200 newborns show this condition, characterized by folds of skin along the neck, a low hairline at the nape of the neck, a shield-shaped chest, and later in life, failure to develop adult sexual characteristics at puberty. About 1 newborn male in 1000 has two X chromosomes and one Y chromosome (XXY), a condition called Klinefelter syndrome. Affected people develop as sterile males with small testes, long legs and arms, and somewhat diminished verbal skills, although their IQ scores are near normal. Most men with Klinefelter syndrome manage well in society, and many are unaware of their chromosomal abnormality until they marry and are unable to father a child. If identified before adolescence, a Klinefelter male can take male hormones to enhance his eventual sexual performance, prevent osteoporosis, and improve his mood and general sense of well-being.

Y-Linked Genes

We said earlier that the human Y chromosome was small, and you can see that on Figure 5.14. The Y actually contains only about 20 genes, in contrast to about 1000 on the X chromosome. A few Y chromosome genes have copies on the X chromosome, but most do not. Most of those 20 Y-chromosome genes are expressed only in the testes, where they are probably responsible for male fertility. Many sterile men who come to fertility clinics have mutations in a Y-linked gene. These mutations probably arose in a single sperm cell in the sterile man's dad, because a man whose Y chromo-some carries a sterile mutation will have no offspring. If a Y-linked gene had a phenotype other than sterility, then an affected man would pass the trait to all of his sons and none of his daughters.

X Chromosome Inactivation

An old song made these lyrics famous: "Thank heaven for little girls!" But a geneticist might sing, "Thank heaven for the little X!" Because as every XY male and every XO female proves, a person can get along quite well with only one X chromosome even though the absence of two copies of any other chromosome causes death before or shortly after birth. What's so special about the X? The answer is, no matter how many X chromosomes are present, *both sexes have only a single functional copy of it*. At the stage when a female human embryo (or other mammal) consists of only about a thousand cells, one of the X chromosomes in each of her cells becomes genetically inactive—it no longer reads out any genetic information. (Geneticists call this X-chromosome inactivation.) Hence, the genes on that inactivated chromosome can have no effect on the phenotype. After one of the X chromosomes in a female embryonic cell becomes inactive, all of the millions of daughter cells derived from it will have the same inactive X. Thus, a female mammal is a mosaic of cells containing active X chromosomes of maternal or paternal origin. You can see this mosaicism in the patches of black and orange fur in a calico cat (Fig. 5.19), as well as in women with a certain skin condition.

In a female with two Xs (the typical number), a geneticist can see the inactive X chromosome in cells scraped from the inside of the mouth as a small, dark spot on the edge of a nucleus. XY males, who have no inactive X, lack this dark spot in the nucleus. For several Olympic games before 1992, officials relied on this procedure as a test to certify the "femininity" of female athletes, regardless of the other sex chromosomes.

Maternal Inheritance

By now, you know that if you have a Y chromosome, you must have inherited it from your father. For some genetic conditions, however, the responsible genes come

Figure 5.19 Calico Cats and Mosaicism. A female mammal is a mosaic of cells with either the mother's or the father's X chromosome inactivated. A cat heterozygous for black and yellow alleles of an X-linked coat color gene shows patches where one or the other X is being used. The white fur is determined by an unrelated gene.

only from the mother. These maternal genes occur in the mitochondrion, the cell's energy "powerhouse." In a human, each mitochondrial genome has exactly 37 genes, while each nucleus has about 40,000 genes. Since sperm donate only a cell nucleus and usually no mitochondria to the offspring, an animal usually inherits all of its mitochondria from its mother through the egg. For this reason, regardless of your sex, you can trace your mitochondria to your mother, maternal grandmother and so on back through time. As you might expect, mutations in mitochondrial genes can decrease a cell's ability to make ATP, and so their effects are felt mainly in high-energy tissues, like the brain, heart, muscles, and eyes. A disease called Leber optic atrophy, for example, begins at about age 30 as a loss of vision in the central part of the visual field; this can spread over a couple of years and cause varying degrees of blindness. Leber optic atrophy comes about through mutations that disrupt proteins in the electron transport chain (review Fig. 3.15) and is passed only from a mother to her children. We'll explore more about mitochondrial genes, human evolution, and the relatedness of all eukaryotes in later chapters.

What is the role of a genetic counselor?

An Allele's Phenotype Can Depend on Other Genes and on the Environment

As Dr. Jonathan Zuckerman points out, until fairly recently people with cystic fibrosis rarely survived past infancy, but today, patients are living well into their 30s and beyond. This extended survival is due to advances in several environmental factors, including appropriate diet, special respiratory therapy, and aggressive antibiotic treatment against infection. The influence of environmental factors applies to many other genetic traits as well, with part of that environment being the interaction of seemingly unrelated genes. An example of this combined effect is breast cancer.

About 90 percent of breast cancers are sporadic, that is, they occur in people with no familial history of breast cancer. Geneticist Marie-Claire King, however, noticed that in some cases, breast cancer seemed to run in families. She collected pedigrees for a number of families in which several people (almost always women) had developed breast cancer and narrowed her study to those with early onset of the disease, before the person's mid-40s. King found pedigrees like the ones in Figure 5.20a. Look at this evidence and see if you can tell whether breast cancer appears to be caused by a dominant or recessive allele. Many affected people have at least one affected parent, so it looks like a dominant allele. But for some affected individuals, neither parent had breast cancer. Most often, the unaffected parent is a male. So is this an *X*-linked trait? Apparently not, because there is father-to-son transmission in Family D (Fig. 5.20a). So this type of breast cancer is inherited as an autosomal dominant, but other genes affect its expression, especially the gene on the *Y* chromosome that causes male phenotype (*SRY*). Geneticists use the term **epistasis** (*epi* = over + *stasis* = standing) for cases in which one gene affects the expression of an entirely different gene. Hereditary breast cancer not only illustrates epistasis but another principle, as well: **sex-limited traits**—conditions shown by one sex or the other simply because of their maleness or femaleness. Breast cancer is sex limited in that it occurs almost exclusively in women. One male in Figure 5.20a did develop the disease, but less than 1 percent of breast cancer patients are male. Pattern baldness is another sex-limited trait and affects mainly males.

Despite the pattern Marie-Claire King saw in the pedigrees, they do show that some phenotypes vary for unknown reasons that may be related to people's environment. In Family D, for example (Fig. 5.20a), individual III6 doesn't have breast cancer but she apparently passed it to her daughter. The pedigrees were key to an even more important finding: with them, Dr. King was able to pinpoint a gene that predisposes people to breast cancer. This gene, called *BRCA1*, for "breast cancer 1," is located about halfway down the long arm of chromosome 17. Finding the exact gene is crucial for investigating diseases at the molecular level and developing treatments. This approach—mapping exact genes and then learning to counteract or correct their action—is changing the face of medicine. Let's see how in our final section.

5.6 How Is Genetics Changing Our World?

Shana, now in her mid-20s, is far from rare among people with cystic fibrosis. The majority are now surviving into midlife, although there has been a leveling off of life expectancy over the last few years. To move beyond the environmental remedies now available (diet, airway clearing therapies, and antibiotics), new therapies are needed to extend patients' lives further. To this end, Dr. Zuckerman and other researchers are mapping genes, learning what the genes do in healthy and diseased cells, and working on molecular solutions. The

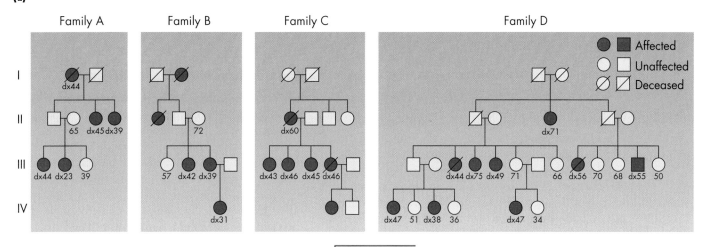

(a)

Family A   Family B   Family C   Family D

Affected
Unaffected
Deceased

I   II   III   IV

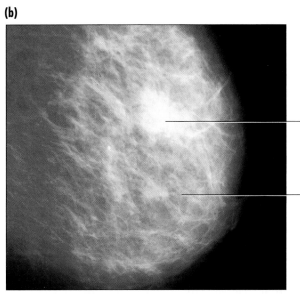

(b)

Growth in breast

Normal breast tissue

Figure 5.20 **Breast Cancer.** (a) Filled symbols represent people with breast cancer. Age of diagnosis is represented by the letters dx followed by an age. For unaffected people, their age is noted at the last interview. (b) A mammogram can reveal abnormal growths in the breast. About 7 percent of such growths turn out to be cancers.

history of gene mapping helps explain the pivotal importance of this technique.

Gene Mapping

Gene mapping, the assignment of genes to specific locations or **loci** (sing. *locus*) along a chromosome, was developed in fruit flies, like so many other genetic principles. Recall that Thomas Hunt Morgan used flies to show that a single chromosome can carry many genes—that is, to show genetic **linkage.** An enterprising undergraduate in Morgan's genetics laboratory at Columbia University, Alfred Sturtevant, was washing fly-raising bottles and eavesdropping on the lab talk. Soon Sturtevant began his own experiments aimed at understand-

ing how numerous genes are arranged on a single chromosome. He realized that, just as a long hose tangles and knots more easily than a short one, the greater the distance between two genes on a chromosome, the greater would be the likelihood of a crossover event between the two genes. (We discussed crossing-over in Figure 4.16). In 1913, Sturtevant showed that genes lie in a straight line along a chromosome, and that simple mating experiments can map the genes—that is, reveal their order and relative distance from each other.

Gene Mapping and the Human Genome Project

Human gene mapping was painstakingly slow for decades, and it wasn't until 1967 that geneticists were able to assign the first human gene to an autosome. In 1985, Dr. Francis Collins of the University of Michigan, used gene mapping information to identify all the base pairs in the cystic fibrosis gene *CFTR*, and how it works. Today Dr. Collins heads the **Human Genome Project** at the National Institutes of Health. His agency has funded the complete sequencing of the human genome in a spirited race with Dr. Craig Venter and other researchers at a private company called Celera.

READ ON

Explorer 5.4

The Human Genome Project

Their goal was to determine the sequence of the human genome, the precise order of As, Ts, Gs and Cs for all 23 human chromosomes. This would be the ultimate gene map, revealing the position of every human gene (**E** xplorer 5.4).

In a dramatic press conference in the year 2000, the rival teams together announced the sequencing of the entire human genome, all 3 billion nucleotides. As the completely aligned sequences of the two smallest human chromosomes (21 and 22) became available, biologists were amazed to learn that humans have far fewer genes than they had previously thought. For instance, chromosome 22 has just 545 genes and chromosome 21 has just 225 genes. This suggests that we humans have just 40,000 genes, rather than the 100,000 to 140,000 genes they once predicted. This is truly amazing when you consider that the nematode worm, *Caenorhabditis elegans,* has 19,000 genes and the fruit fly has 14,000 genes. How can humans have only two to three times as many genes as worms and flies? The answer may lie with the escalating ways genes can interact as their numbers increase, but a full understanding awaits further research.

Sequencing the human genome is important because researchers will have a far easier time now identifying the structure and activity of disease-related genes. As a result, they'll be better able to design new drugs to treat debilitating conditions. Knowing the sequence may also help geneticists determine why different people respond differently to therapies for serious diseases. Let's look, now, at some of the ways geneticists are applying their knowledge of the human genome for our benefit.

Isolating Disease Genes

Knowing a gene's nucleotide sequence can help reveal its function. By isolating and then analyzing the cystic fibrosis gene, for example, Francis Collins and other researchers showed that it causes cell membranes to make a particular protein. This protein, the cystic fibrosis transmembrane regulator, or CFTR protein, regulates the transport of sodium and chloride across the surfaces of the epithelial cells that line the body's narrow passageways and ducts (Fig. 5.12). As we saw earlier, an error or mutation in the cystic fibrosis gene, and hence the CFTR protein, causes aberrant ion transport in the ducts of the lungs, pancreas, sweat glands, and other organs leading to sticky mucus build up. The discovery in the early 1990s that people with cystic fibrosis have a problem with the CFTR protein gave them and their

doctors hope that we would someday have a cure for this, the most common fatal genetic disease in America today.

Should Human Genes Be Patented?

Gene mapping and isolating disease genes has become a big business. Biotechnology companies are patenting the sequences of specific human genetic fragments even before they know the exact genetic function—so that they "own" the rights to them and someday may be able to use the specific gene to develop treatments for human diseases. Some hail this as a medical boon, while others find it ethically repugnant that any single group can claim ownership of a human gene. The latter point out that if a company owns a particular disease gene, then independent scientists would have little or no incentive to study it. The patenters, however, counter that bringing a drug to market can cost millions of dollars and requires patent protection to insure their investment in expensive research and development. Without patents, they say, and the profit incentive they provide, few if any new drugs would appear. What would have happened if the first anatomist to discover the pancreas had been able to patent it and prevent others from investigating its function? For now, courts have decided that patenting pieces of human DNA is legal and appropriate. What do you think about this issue?

Detecting Genetic Disease

For some genetic diseases like cystic fibrosis, detection can start with something as simple as a kiss. Many parents of children with CF first suspect something is amiss with their tiny baby when a kiss reveals that he or she tastes unusually salty. Early disease detection is important, whether based on simple observations and senses or complicated medical tests because physicians must characterize the problem before they can treat it. Several genetic conditions appear only in adulthood, such as Leber optic atrophy, breast cancer, and Charcot-Marie-Tooth disease. The majority of genetic traits, however, including cystic fibrosis and Duchenne muscular dystrophy, become noticeable in early childhood.

Physicians can easily detect some genetic traits as soon as a baby emerges from the womb. Lack of pigment in the skin, hair, and eyes (albinism) is one example, and extra fingers and toes (polydactyly) is another. Obstetricians can also detect less obvious conditions right in the hospital nursery. An example is phenylketonuria (PKU), a condition that blocks mental development at about age two without treatment. (If you were born after 1962, you were probably tested for PKU soon after birth.) Body fluid from a baby with PKU, taken from the infant's heel or even a wet diaper, has 20 times the normal amount of the amino acid phenylalanine. Early detection is crucial in a condition like PKU, because a special diet very low in phenylalanine can prevent brain damage. Once a PKU patient passes age six, he or she can begin to tolerate small amounts of phenylalanine but must restrict the dietary amino acid throughout life.

Detecting Changes Within a Gene

The mapping and detection of diseases such as cystic fibrosis has another important application: revealing unaffected carriers (heterozygotes) of the disease. Mutations change DNA structure, and this fact can be used to help detect carrier status. If two people who are carriers know their status as heterozygotes for a disease gene, they can learn (often with the help of a genetic counselor) what the chance would be of passing the disease to their own child. In Shana's case, she learned she was homozygous for the cystic fibrosis gene and her husband-to-be was heterozygous, giving each of their potential children a 50 percent probability of receiving two alleles of the CFTR gene and thus showing the disease. What should they or any other couple do with such information? If heterozygotes could somehow select and use only eggs and sperm carrying nondisease alleles, then their babies would be homozygous for the normal allele and would be born disease-free. This kind of gamete selection might eventually be available, at least in females based on the patterns of cell division during meiosis (review Fig. 4.14). It is unlikely, however, to become widely used any time soon.

Here's a more common alternative. After conception, physicians test early-stage fetuses for genetic diseases, including the most common forms of cystic fibrosis. Using genetic tests such as **chorionic villus sampling** and **amniocentesis** (see Fig. 8.20), they can analyze the genetic makeup of fetal cells and detect many chromosome abnormalities, including Down syndrome (see Fig. 4.15). Women who know they are carrying a fetus with a genetic disease face the hard choice of abortion versus giving birth to a child with special needs. Couples who are both heterozygotes for a genetic disease must weigh the risks of getting pregnant and passing on the disease trait to a child. Genetic counselors try to assist people with these difficult dilemmas, including helping prospective parents in making a decision either to terminate the pregnancy or to prepare emotionally and financially to support a child with special needs.

Given that cystic fibrosis is the most common lethal disease in the United States, and that early detection can improve the care of a child with cystic fibrosis, shouldn't everybody be tested at birth for the condition? There are drawbacks to such an approach. First, while cystic fibrosis occurs in all ethnic groups, it is particularly high only in Caucasians (1 in 2500 births). In the Saguenay-Lac St. Jean region of Canada, the frequency at birth is 1 in 926, and 1 in 15 people are carriers (heterozygotes)! In African Americans, the rate is only 1 in 18,000, and in Asian Americans, 1 in 90,000. Should all groups be forced to pay for the test even if the likelihood of finding a homozygote is low? Should testing occur only in a family with a history of the disease? Who should have access to the results of the test? Will insurance companies and potential employers be able to exclude people based on genetic tests for this and other traits? How accurate are the tests? How often does someone test positive but not get the disease (a so-called false positive) or test negative and then develop the condition (a so-called false negative)? For CFTR and most other disease genes, there are many ways to break the gene, and each way requires different test reagents. Which mutations in each disease gene will be tested for? What if there are unknown mutations? Finally, how much would testing cost? What fraction of the population would have to participate to make it financially worthwhile?

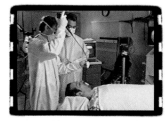

How could researchers and physicians use gene therapy to help a patient with cystic fibrosis?

Treating Genetic Disease

Dr. Zuckerman and his colleagues are looking for ways to exploit our knowledge of the CFTR gene to help improve the quality of life for patients like Shana. There are a number of possible avenues. The therapies for some diseases involve replacing the defective protein. That works well in certain cases, for example the injections of the protein insulin into diabetics, and the injection of blood clotting factors into people with hemophilia. But for cystic fibrosis, the missing CFTR protein is embedded in cell membranes and is not normally carried through the blood like insulin or clotting factors. For this reason, Dr. Zuckerman's colleagues at various institutions are testing a different alternative, **gene therapy,** the introduction of normal CFTR genes into cells with a disease genotype.

In 1990, biologists tried introducing the normal allele of the CFTR gene into a cystic fibrosis patient's cells growing in a test tube. This introduction corrected the defective transport of sodium and chloride and engendered hope that the same could be done in living patients. But how can you get DNA into airway cells? In research at the University of Pennsylvania, Dr. Zuckerman and colleagues reasoned that after millions of years of evolution, viruses have become masters of introducing foreign DNA into the cells they infect. Thus, they chose a modified cold virus, inserted a normal copy of the *CFTR* gene into it, and then introduced it into a few patients' lungs. Their first job during that study, and the job of cystic fibrosis researchers elsewhere, was to prove that the virus itself wouldn't harm the patient. Then they needed to determine whether the normal *CFTR* gene carried by the virus actually enters and remains in human lung cells. So far, Zuckerman and others have found that at high doses of the modified cold virus, the lungs grow inflamed but do fight back the infection and that a low level of the gene does indeed move from the virus into the patient's lung cells. One additional finding is potentially exciting: "All you need to do," says Zuckerman, is insert the correct gene in "about 10 percent of the duct-lining cells" to normalize the salt and water movement across the whole surface.

To date, researchers are cautiously searching for the right viruses or other gene-carrying vehicles, and gene therapy has helped only a few cystic fibrosis patients. The therapy currently works for only a short time because, unfortunately, the genetically corrected cells eventually slough off and are lost. There is certainly no cure for CF available based on gene therapy, and the same is true for other diseases. There have been groundbreaking attempts at curing disease, however. Two young girls received gene therapy in the late 1990s for a disease that blocked the development of their immune systems (Fig. 5.21). At the National Institutes of Health, Dr. W. French Anderson extracted immune system cells (white blood cells) from the girls and then exposed the cells to mouse leukemia viruses they had engineered to contain the protein (called adenosine deaminase, or ADA), which the girls lacked. They then dripped the retrofitted immune cells back into the girls' veins. Anderson repeated this procedure a dozen times on each patient until finally their bodies contained enough engineered cells to protect them from disease. As a precaution, however, both girls continued to receive the very expensive weekly injections of the ADA protein they were missing genetically. Both girls are still doing well today. On the negative side, a boy suffering from liver disease died during a gene therapy trial at the University of Pennsylvania in 2000, and all attempts at gene therapy were suspended there during a lengthy investigation.

Figure 5.21 **Gene Therapy for Immune Deficiency.**
Ashanti di Silva received gene therapy to insert the healthy gene for ADA (adenosine deaminase protein), which she lacked and which prevented her body from generating a normal immune response. To date, she is doing well.

Gene therapy using the ADA protein, the CTFR protein, and many others now underway are designed to modify existing body cells. In the case of cystic fibrosis, the target is the epithelial lining cells of the airways. This approach, called **somatic gene therapy,** is just an extension of current therapeutic practices—an attempt by a physician to modify the body of one patient at a time. Somatic gene therapy contrasts with **germ-line gene therapy,** which would substitute a gene into the sperm or egg of a person bearing the defective gene. This type of genetic alteration is not currently allowed in humans for ethical and moral reasons. Scientists are also cautious about unknown deleterious changes that could occur with the introduction of foreign genes into the human gene pool.

To Dr. Zuckerman and patients like Shana Reif, developing adequate treatments is an urgent problem and living with a genetic disease and making reproductive and other decisions are very real, day-to-day issues. The future of genetic research, based on Gregor Mendel's principles of heredity, is sure to be fascinating, powerful, and with any luck, life-extending.

Chapter Summary and Selected Key Terms

Introduction
Cystic fibrosis is an inherited disease, based on a mutated gene that brings about sticky mucus and clogged ducts in vital organs such as the lungs and pancreas. This chapter explains inheritance patterns like those for cystic fibrosis as well as many other aspects of **genetics** (p. 112).

Universal Principles of Heredity
Gregor Mendel, working in a quiet abbey garden in the mid-1800s, worked out the basic laws of heredity. He created **hybrids** (p. 113) by crossing pea plants with various pure-breeding traits and devised a **particulate model of heredity** (p. 113, 114) to replace the widespread belief at that time in the **blending model of heredity** (p. 113). In these hybrid experiments, the pure-breeding plants (for example, long-stemmed pea plants versus short-stemmed pea plants) formed the **parental (P_1) generation** (p. 114). By cross-fertilizing the two plant types, Mendel produced a **first filial (F_1) generation** (p. 115). He called the trait that showed up in the F_1 generation **dominant** (p. 115) and the one that did not **recessive** (p. 115). He allowed F_1 plants to self-fertilize to produce the **second filial (F_2) generation** (p. 115), then looked for the pattern of inheritance of the two traits.

Rules Governing Inheritance of Single Traits
The phenotypic ratio of dominants to recessives in the F_2 generation was 3:1. Mendel reasoned that particulate factors influence the traits and are not blended from one generation to the next. Modern geneticists call the factors **genes** (p. 116). Mendel also reasoned that the particulate factors must come in different forms, now called **alleles** (p. 116). These are either dominant or recessive and each trait is governed by two alleles. An organism's visible or measurable physical traits are its **phenotype** (p. 117) while its genetic make-up, based on its combination of alleles, is its **genotype** (p. 117). **Heterozygous** (p. 117) individuals or **heterozygotes** (p. 118) have two different alleles for a given trait (one dominant and one recessive) while **homozygous** (p. 118) individuals or **homozygotes** (p. 118) have identical alleles for a given trait (either two dominant or two recessive). Mendel concluded that an individual's two allele copies for each trait come from the mother and father and that these become segregated during the production of sex cells. This is Mendel's **law of segregation** (p. 118). Punnett squares are a useful tool for diagramming crosses and following allele segregations and combinations. Genetic principles rely on the laws of chance and probability.

Analyzing Human Inheritance Patterns
Our species is an inconvenient genetic subject because geneticists can't choose with whom we mate and produce offspring. Also, we produce too few offspring for good statistical analysis of probabilities, and our generation times are too long. Geneticists can draw up orderly diagrams of family genetic histories called **pedigrees** (p. 120). Using such pedigrees, geneticists discovered that cystic fibrosis is inherited as a recessive trait. Heterozygous individuals who pass on a disease allele but do not show its effects are called **carriers** (p. 121). A trait like sickle-cell anemia shows **incomplete dominance** (p. 122); carriers show a mild form of the disease. Human blood types are an example of **codominance** (p. 123), in which the heterozygous individual shows both phenotypes—for example, AB blood types. Some genes have more than two alleles, including human blood types. This explains why blood types can be A, B, O, or AB. Sometimes a single mutated gene causes several different effects (several different phenotypes). The single cystic fibrosis gene causes all the major underlying features of the disease; this is called **pleiotropy** (p. 124).

The Inheritance of Multiple Traits
Geneticists have learned to diagram the inheritance of more than one trait at a time. Offspring showing the same traits as parents are called **parental** (p. 126) types, while offspring with new combinations are called **recombinant** types (p. 126). Diagramming crosses involving two traits or **dihybrid crosses** (p. 126) on an enlarged **Punnett square** (p. 118) produces a **9:3:3:1** (p. 128) ratio of phenotypes. This basic analysis works because hereditary factors segregate into gametes independently of each other; this is Mendel's principle of **independent assortment** (p. 126). In the years since Mendel's experiments, biologists have confirmed that genes are located on chromosomes and undergo independent assortment during meiosis.

Sex and Inheritance
Males and females have several pairs of similar chromosomes or **autosomes** (p. 127), and one pair that can have dissimilar members, the **sex chromosomes** (p. 127), the **X-chromosome** (p. 127) and **Y-chromosome** (p. 128). Some genes are **X-linked** (p. 128), that is, they are carried on the X chromosome. A **mutation** (p. 128) is a permanent change in the genetic material. Duchenne muscular dystrophy is an X-linked recessive trait; Charcot-Marie-Tooth disease is an X-linked dominant trait. In people, XX individuals are females and XY individuals are males. $X0$ individuals are sterile females; individuals with XXY, $XXXY$, and so on, are males. The few Y-linked traits are probably responsible for male fertility. In both sexes, there is only one functional copy of the X chromosome; in females this is called X-chromosome inactivation. We inherit our mitochondrial chromosomes from our mothers. Environmental factors can influence gene expression.

When one gene affects the expression of a different gene, it is called **epistasis** (p. 132). Some traits, such as breast cancer, are **sex-limited** (p. 132); 99 percent of the cases affect women.

Genetics and Societal Change
Geneticists have learned to map individual genes to their specific locations, or **loci** (p. 133) on chromosomes and discovered that chromosomes show genetic **linkage** (p. 133): that is, they carry many genes. **The Human Genome Project** (p. 133) is a modern effort to determine the nucleotide sequence of all 23 human chromosomes. Gene mapping can help researchers to isolate the genes for specific diseases such as cystic fibrosis. Prenatal testing such as **amniocentesis** (p. 135) and **chorionic villus sampling** (p. 135) allows physicians to detect disease genes before a baby is born. One solution to genetic diseases will be **gene therapy** (p. 135), introducing normal genes into cells with a disease genotype. Introducing a normal cystic fibrosis gene into a patient's lungs is an example of **somatic gene therapy** (p. 136). **Germ-line gene therapy** (p. 136) would substitute a healthy gene into an egg or sperm cell, but is not currently allowed.

All of the following question sets also appear in the Explore Life *E* electronic component, where you will find a variety of additional questions as well.

Test Yourself on Vocabulary and Concepts

In each question set below, match the description with the appropriate term. A term may be used once, more than once, or not at all.

SET I

(a) phenotype (b) genotype (c) hybrid (d) pure-breeding (e) F_1

1. The physical manifestation of a genetic trait, usually expressed in words

2. A synonym for heterozygous, usually referring to the progeny of individuals of different homozygous strains

3. The symbolic representation of the allelic composition of an individual

4. In a formal genetic cross, the heterozygous offspring of two contrasting, homozygous parents are members of the _____.

5. A synonym for homozygous

SET II

(a) locus (b) multiple alleles (c) polygenic trait

6. More than two forms of a given gene exist

7. The position of a gene on a chromosome

8. A phenotype that is determined by many different genes

SET III

(a) carrier (b) karyotype (c) amniocentesis

9. An individual who is heterozygous for a recessive allele that may be passed on to another generation

10. A representation of a single cell's set of chromosomes, usually arranged in homologous pairs

Integrate and Apply What You've Learned

11. Show all the possible genotypes for a pea plant with long stems (dominant) and purple flowers (dominant). Use the letters *L* or *l* to symbolize the stem-length alleles and *P* or *p* for the flower-color alleles.

12. Show all the genotypes of the gametes produced by the hybrid pea plant *LlPp*.

13. In guinea pigs, black fur *(B)* is dominant to white fur *(b)*. How could an animal breeder test whether a given black guinea pig is homozygous or heterozygous for this color gene?

14. Can you always tell an organism's genotype by looking at its phenotype? Why or why not?

15. How are the sex chromosomes of sperm and eggs of mammals alike? How are they different?

16. Explain how it happens that the father, rather than the mother, determines the gender of a human baby.

17. Distinguish between pedigree analysis, amniocentesis, and karyotyping. Which tests reveal definitive information, and which indicate only the probability of either having or of being a carrier for some specific mutation or heritable trait?

18. If individuals who carry an autosomal dominant lethal gene die as a consequence, how do you account for the fact that such lethal genes, like that for Huntington's disease, are perpetuated in a population?

Analyze and Evaluate the Concepts

19. Explain why if Shana and Kurt had children each would have a 50–50 chance of inheriting cystic fibrosis.

20. In the Smith family, all the children have small, tightly connected earlobes, commonly called attached lobes. Their parents both have large, pendulous earlobes. What is the most likely explanation of the children's lobes?

21. The Lee children and one of their parents are all blood group B. The other parent could be what blood type?

22. Outline the major benefits and risks associated with the Human Genome Project. What is your stand on the issues you present?

23. Which of the following abnormalities could not be detected by karyotyping? Give the reason for your answer.

    (a) Down syndrome   (b) Turner syndrome
    (c) Klinefelter syndrome (d) Phenylketonuria

A Cat, A Crime, An Identity

Clues in DNA: Dr. Marilyn Menotti-Raymond in the lab; Snowball the cat; Canadian Mountie Roger Savoie (far right).

Prince Edward Island is a collection of sleepy farms, fishing villages, and seaside resorts just off the shore of Nova Scotia, Canada. With its damp green forests and red coastal cliffs, the island has been notable for geographic beauty as well as for something else: Anne Gilbert, the red-haired heroine of Lucy Montgomery's beloved novel, *Anne of Green Gables*. For decades, bus loads of tourists have jammed the island every summer to see the flowering orchards and shady lanes where the loquacious Miss Gilbert might have wandered—had she been real. But in 1995, the island drew international attention for a totally factual and far less wholesome story involving a disappearance, a murder, and a white-haired cat.

On October 3, 1994, a 32-year-old mother of five named Shirley Duguay went out and failed to return to her modest home in Richmond on the island's west side. Within days, the Royal Canadian Mounted Police found her car abandoned in the woods and splattered with blood, soon identified as Shirley's. Toward the end of October, soldiers on a military maneuver found a man's leather jacket hidden in underbrush and also spotted with Shirley Duguay's blood. Her family feared the worst but kindled hope until May 6, 1995, when a fisherman spotted an earthen mound in the woods and dug-up Shirley's decomposing body.

The detective in charge of what was now a murder investigation, Canadian mountie Roger Savoie, began amassing evidence. The chief suspect was Douglas Beamish, Shirley's estranged husband, who had been released from prison not long before

Shirley disappeared. Did the discarded jacket belong to Beamish? The crime lab found no blood, human hairs, saliva, or other traces that could have linked the clothing to the suspect. They did, however, find 27 white cat hairs in the jacket's lining, and since his release from prison, Beamish had been living with his parents and their white cat Snowball! Savoie arrested the suspect and contacted, of all places, the U.S. National Institutes of Health in Frederick, Maryland. Through the Internet, Savoie had learned that a laboratory group at NIH, headed by Dr. Stephen O'Brien, had for years been studying the DNA of felines—cheetahs, tigers, pumas, ocelots, and domestic cats. If anyone could tie the cat hairs to Beamish it was this team of scientists. But could *they*?

The analysis fell to Dr. Marilyn Menotti-Raymond and her technician Victor David. Among the hairs they found four with roots, each one made

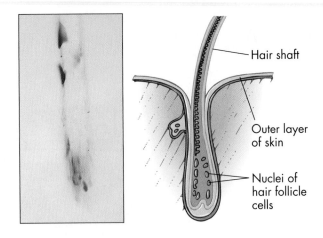

Hair shaft

Outer layer of skin

Nuclei of hair follicle cells

up of several skin cells and each cell containing DNA in its cell nucleus (Fig. 6.1). "We knew," says Menotti-Raymond, that you can get "only about one-tenth as much DNA from an animal hair root as from a human hair." So the task was to carefully extract the minute quantities of DNA, amplify it with a special copying procedure, and then identify its unique "fingerprint." They carried out the arduous extraction procedures again and again but were able to get DNA from just a single hair root. Luckily, they had a set of techniques called **PCR (polymerase chain reaction)** — a laboratory procedure which employs DNA's own natural copying mechanisms inside the cell to increase the amount of DNA in a sample. With PCR, they were able to make millions of copies of ten specific regions along the cat's DNA molecules. These ten places were known to have the two nucleotide bases thymine and guanine (G and T; see Chapter 2) repeated over and over: TGTGTGTGTGTGTGTG. "These bases are repeated anywhere from 12 to 30 times at each locus," says Menotti-Raymond. As we saw in Chapter 5, a **locus** (pl., *loci*) is a site on a chromosome. "But the number of repeats is different from individual to individual." By looking at the number of repeats at ten such sites, "you can come up with a unique pattern, sort of like a bar code in the grocery store." This pattern, Menotti-Raymond says, amounts to a DNA fingerprint that is "unique to a particular individual."

After months of testing and retesting, the team confirmed that the DNA from the hair root in the jacket's bloody lining matched Snowball's DNA exactly and would be unlikely to turn up by chance

even in millions of domestic cats of the genotypes on the island. Dr. Menotti-Raymond testified before the Supreme Court of Prince Edward Island in 1996 along with her boss Stephen O'Brien and technician Victor David. "This was really quite remarkable for all of us," she recalls. "We work at a lab bench and go to seminars. We don't have that kind of drama in our lives." The heavy responsibility of the situation came home to her, she says, when she saw the accused sitting in court and his family "sitting in their car afterwards and sort of glaring at us as we came out of the court room." But the lab work was careful and irrefutable. And the jury, convinced that the cat hairs were Snowball's and the jacket had been worn by Douglas Beamish, convicted the defendant of murder. *Anne of Green Gables* is still Prince Edward Island's most notable character but today, Snowball may be a close second.

The DNA that resides in a cat's cells and in our own cells carries a unique fingerprint of individual identity. But it does much more. DNA is a remarkable molecule with an elegant helical structure that encodes our genetic traits and instructs our day-to-day, cell-by-cell functioning. It is truly the thread of life: The structure and function of DNA help explain how cells survive and replicate. They help us understand how organisms change over time and how specific mutations and other mechanisms bring about this evolution by natural selection. Finally, changes in DNA help us understand how inherited diseases, cancers, and many other ailments arise.

In this chapter, you'll find a story within a story. Inside the case history of cat hairs and a murder trial, you'll see the race to discover the famous double-helix shape of DNA. You'll also see how that shape explained the molecule's information coding and copying. Along the way, you'll find answers to these specific questions:

1. How did scientists identify the nature of hereditary material?

2. How is DNA like a twisted ladder?

3. What is the exact chemical structure of DNA and how does it carry information?

4. How does that structure allow DNA to copy itself?

6.1 Identifying the Hereditary Material

The DNA that Marilyn Menotti-Raymond extracted from one of Snowball's hairs contained the animal's hereditary instructions. Along with environmental factors, it determined Snowball's unique combination of body size; hair color, texture, and length; eye color; disposition; and hundreds of other traits both obvious and subtle, morphological and biochemical. As in all living organisms, the cat's DNA contained its **genes,** the very same hereditary "factors" Mendel discovered over 150 years ago (see Chapter 5). At that time, however, Mendel had no idea what these hereditary factors might be made of or how they might work. The first step toward our present knowledge of genes was the insight (review Fig. 5.5) that genes occur in chromosomes. Biologists were able to determine in the 1930s that chromosomes in eukaryotes contain mainly protein, RNA, and DNA. They presumed that one of those substances had to make up the genes carrying genetic information. But which one?

In the 1940s, some geneticists argued that only proteins were versatile enough molecules to carry the complex information in genes. These biologists pointed out that proteins are constructed from 20 different subunits (amino acids) (see Fig. 2.10), while DNA has only four: the nucleotide bases adenine (A), thymine (T), guanine (G), and cytosine (C) (see Fig. 2.12). Clearly, 20 amino acids can form many more combinations—and hence carry much more information—than 4 bases, just as you can form more words from an alphabet of 20 letters than from an alphabet of 4 letters. In the mid-20th century, geneticists devised tests to determine whether genes are made of protein or nucleic acid, and these experiments led to a biological revolution.

Evidence for DNA: Bacterial Transformation

In 1928, British researcher Frederick Griffith found that with a chemical of unknown identity extracted from bacteria, he could transfer an inherited characteristic—the ability to cause pneumonia—from one strain of bacteria to another. Griffith had two strains of pneumococcus bacteria, one that grew into smooth colonies in the lab and could cause a lethal infection if injected into a mouse, and another that grew into rough colonies and did not cause mice to die when injected. Griffith injected mice simultaneously with live cells from the nonlethal bacterial strain *and* with cells from

of the disease-causing strain that had been killed with heat. These dually injected mice died. The formerly nonlethal bacteria had somehow acquired the ability to cause lethal pneumonia from an unknown chemical in the heat-killed cells. As a control, Griffith injected mice with only the chemicals from the heat-killed lethal strain and found that the mice survived. (Why is that control procedure important?) Significantly, the "transformed" cells—formerly nonlethal bacteria exposed to chemicals from the lethal strain—could pass the lethality trait to their daughter cells during cell division. Clearly, whatever the unknown chemical was, it could carry hereditary information.

In 1944, Oswald Avery and coworkers in New York sought to determine which chemical component was carrying hereditary information from the lethal bacteria to the nonlethal strain. To do so, they extracted the carbohydrates, proteins, and DNA from the lethal strain (Fig. 6.2a,c), incubated the nonlethal bacteria (Fig. 6.2b,d) with the various chemicals from the lethal strain, and checked to see which substance would transform the nonlethal cells (Fig. 6.2e). They observed that the nonlethal bacteria treated with either carbohydrate or protein remained nonlethal (Fig. 6.2c–f). In contrast, the nonlethal bacteria treated with DNA from the lethal cells picked up the ability to cause pneumonia in mice and passed that trait to their progeny (Fig. 6.2c–f). They concluded that DNA contains the genes for physical and biochemical traits, in this case, the genetic information for the ability to cause fatal pneumonia in mice. Researchers have a word for the transfer of an inherited trait by the uptake of DNA: **transformation.** This process is responsible for the transfer of genes for resistance to antibiotics from one bacterium to another, and is a serious problem in modern medicine (see Chapter 9).

Confirmation That Genes Are Made of DNA

The transformation experiments showed that DNA can carry the information for one trait in one species. But what about other traits in other organisms? In 1952, Alfred D. Hershey and Martha Chase got an answer by experimenting with viruses that infect and reproduce in bacteria (Fig. 6.3a). We'll refer to them as bacterial viruses, but their official biological name is **bacteriophages** (literally "bacteria eaters"; also called simply **phages**).

They chose to use bacterial viruses for several reasons. First, these viruses are easy and inexpensive to

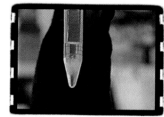

Why do biologists call DNA "the thread of life"?

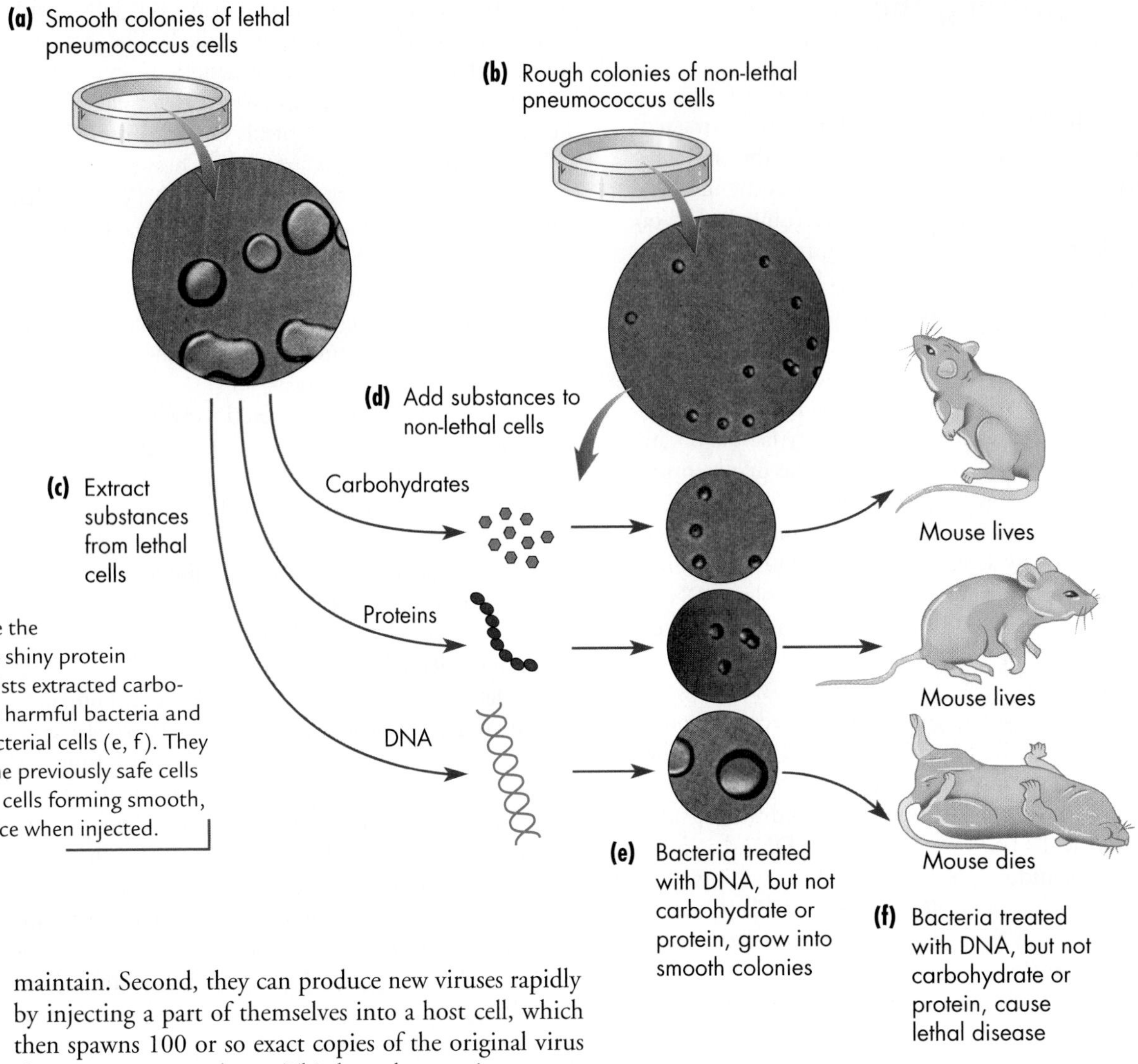

Figure 6.2 **The Chemical Composition of Genes.** (a) Colonies of lethal pneumococcal bacteria grow quickly and look shiny in a laboratory culture dish due to the polysaccharide coats the individual bacterial cells generate. These coats can protect the cells from an animal's immune defense system, and as a result, the cells can cause pneumonia. (b) In contrast, a closely related nonlethal strain of the same species grows small rough colonies, and cannot cause pneumonia, because the cells fail to generate protective coats. Each colony contains thousands of bacterial cells. (c) To determine which chemical in the lethal strain could cause the nonlethal strain to acquire the lethality trait (the ability to form the shiny protein coat and cause pneumonia), biologists extracted carbohydrate, protein, and DNA from the harmful bacteria and used them to treat the nonlethal bacterial cells (e, f). They found that only the DNA changes the previously safe cells that form rough colonies into lethal cells forming smooth, shiny colonies—cells that will kill mice when injected.

(a) Smooth colonies of lethal pneumococcus cells

(b) Rough colonies of non-lethal pneumococcus cells

(c) Extract substances from lethal cells

(d) Add substances to non-lethal cells

Carbohydrates

Proteins

DNA

Mouse lives

Mouse lives

Mouse dies

(e) Bacteria treated with DNA, but not carbohydrate or protein, grow into smooth colonies

(f) Bacteria treated with DNA, but not carbohydrate or protein, cause lethal disease

maintain. Second, they can produce new viruses rapidly by injecting a part of themselves into a host cell, which then spawns 100 or so exact copies of the original virus about 25 minutes later. Third, and most important, bacterial viruses consist simply of a core of DNA surrounded by a protein coat (Fig. 6.3a,b). This gave Hershey and Chase a clear shot at showing which component contains genes and is responsible for heredity. If only protein from the virus entered the host cell, then protein must be the hereditary material. But if only DNA from the virus entered the bacterial cell, DNA must be the molecule containing the genes. (What would you conclude if both substances entered the host cell? What if neither entered?)

Hershey and Chase knew that proteins contain sulfur but do not contain phosphorus, while DNA has the opposite constitution: it contains phosphorus but no sulfur. (Review protein and nucleic acid structure in Chapter 2.) They labeled the proteins and DNA with radioactive isotopes of sulfur and phosphorus, respectively, which give off detectable signals. By tracing these chemicals, Hersey and Chase were able to determine whether protein (bearing radioactive sulfur) or DNA (bearing radioactive phosphorus) entered the host cell and altered that recipient's genetic activity. The results showed that after infection, "hot" phosphorus was on the inside of the infected cell, and "hot" sulfur was on the outside (Fig. 6.3h). This demonstrated that DNA had entered the cell while protein did not, and that the DNA encoded new phage particles (Fig. 6.3f) that eventually killed the cell (Fig. 6.3g). The researchers concluded that DNA, not protein, was responsible for directing the genetic activity of the virus.

With these experiments, the researchers showed that DNA is definitely the hereditary material in a virus and a bacterium. But what about in more complex

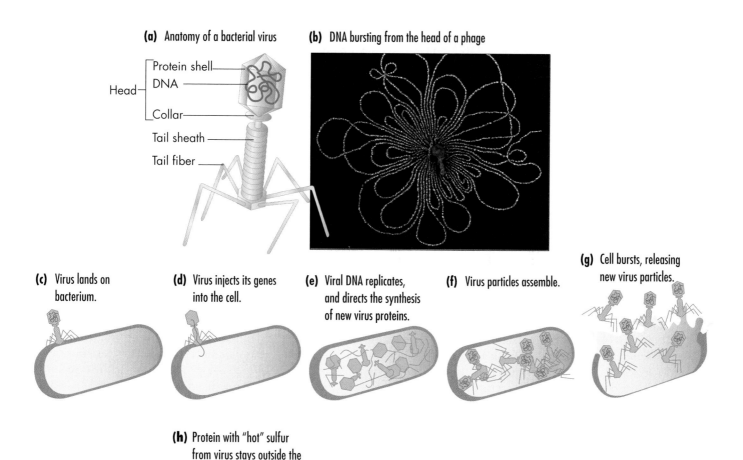

(a) Anatomy of a bacterial virus

Head
Protein shell
DNA
Collar
Tail sheath
Tail fiber

(b) DNA bursting from the head of a phage

(c) Virus lands on bacterium.

(d) Virus injects its genes into the cell.

(e) Viral DNA replicates, and directs the synthesis of new virus proteins.

(f) Virus particles assemble.

(g) Cell bursts, releasing new virus particles.

(h) Protein with "hot" sulfur from virus stays outside the bacterial cell while DNA with "hot" phosphorus — and genes — enters the cell.

Protein shell labeled with radioactive sulfur

DNA labeled with radioactive phosphorus

Figure 6.3 **Life Cycle of a Bacterial Virus.** (a) The bacterial virus called T_4 phage has a head and a tail. The head consists of a protein shell surrounding a core of DNA. The tail is made of proteins with tentacle-like protein tail fibers that attach the virus to the bacterium. (b) An electron microscopist treated the virus with a procedure that caused the viral head to burst and spew out its DNA. If this DNA could be stretched out, it would wind around the perimeter of the photo six times. (c) In the normal virus life cycle, the phage lands on a bacterial cell and injects its genes (d). In the normal viral life cycle, once inside the cell (e) the viral genes direct the cell to make new virus particles (f), which burst out and infect additional cells (g). When Hershey and Chase labeled protein with radioactive sulfur and DNA with radioactive phosphorus (h), they found that only the "hot" phosphorus entered the cell. They concluded that genes are made of DNA.

species? Research by others confirmed that even in complicated organisms, the answer is still DNA. (In a few viruses, however, including the virus that causes AIDS [see Chapter 2], genes can be made of RNA.)

These historic experiments changed our concept of the gene from Mendel's abstract hereditary "particle" to a tangible chemical that biologists could see and manipulate. Nowadays, molecular biologists such as Marilyn Menotti-Raymond break open cells, separate the DNA from the proteins, add ice-cold ethanol to the DNA, and put it in the freezer. A few hours later, a stringy white precipitate a bit like cotton fibers sits at the bottom of a test tube—pure DNA, pure genes.

Learning that genes are made of DNA was a huge step. But to be able to use DNA for sophisticated analyses such as identifying an individual cat from a couple dozen hairs, a researcher needs to know the precise structure of the hereditary molecule.

6.2 DNA: The Twisted Ladder

Both a cat and the mice it preys upon have genes encoded in molecules of DNA. DNA can replicate so perfectly that the offspring of two Siamese cats will

How could genetic evidence serve as legal evidence?

have the same creamy fur and dark colored "points" on the nose, ears, feet, and tail. DNA can still contain enough variation, however, so that a Siamese cat looks and acts different from Snowball, a tiger, or a mouse. The variability inherent in DNA is the basis of natural selection—the key to how organisms evolve—as well as the explanation for life's wonderful variety. This leaves us with an interesting puzzle, however. How does the structure of DNA account for both the *unity* of life—the shared traits and common descent of living things—and also its stunning *diversity*?

In the early 1950s, James D. Watson, an American postdoctoral fellow, and Francis H.C. Crick, a British researcher, met in Cambridge, England, and began their quest to understand the structure of DNA (Fig. 6.4a). They knew that American biochemist Linus Pauling, who had already done his Nobel Prize–winning work on the nature of the chemical bond, was also researching DNA structure, and they wanted to beat this outstanding scientist to the finish line. Watson and Crick also knew from published experiments that DNA is a linear molecule; that is, it is a very long thread. To visualize how long DNA is, look at the length compared to the width of the thread in a simple bacterial virus (see Fig. 6.3b). The host bacterium's own DNA molecule is 100 times longer than the virus's DNA molecule. And the longest DNA molecule in your cells can be 5000 times as long but the same width!

Watson and Crick knew two additional facts: that DNA is a chain of nucleotides (see Fig. 6.4b), and what those nucleotides are. Recall that each nucleotide consists of three parts—a sugar, a phosphate, and a portion called a base. Because of the shape of the sugar portion of the molecule, the nucleotide has a front and a rear end, like a car. The carbon atoms in the sugar are labeled 1′ (pronounced "one prime"), 2′, 3′, 4′, and 5′. At the rear end of the "car" is a "trailer hitch," the phosphate group, attached to the 5′ carbon of the sugar (labeled in Fig. 6.4b). The sugar portion acts like the car's chassis, and the base like the car's interior compartment. At the front is the "grill," representing the 3′ carbon. Watson and Crick also knew that DNA has nucleotides of four types (see Fig. 6.4b), identical except for the bases they contain. These are the **adenine (A), cytosine (C), guanine (G),** or **thymine (T)** we encountered in Chapters 2 and 5.

Geneticists often describe a section of DNA by the *sequence* of bases in the chain. For example, a 22-nucleotide portion of a single-stranded DNA chain might have bases in the sequence AGGAAAAT-GAAGTCAAGAAAATGG. As we will see later, this specific 22-base sequence played a role in helping to convict Douglas Beamish of killing Shirley Duguay.

But first, back to Watson and Crick. From the work of Erwin Chargaff, the team knew another significant detail about DNA: In any molecule of a cell's DNA, the bases A and T appear in equal amounts, and the bases C and G also exist in equal amounts. Watson and Crick were anxious to build a model of the DNA molecule that accounted for this peculiar regularity. But they couldn't do it until a last piece of information came from the British biophysicist Rosalind Franklin, who was working in the laboratory of Maurice H.F. Wilkins at King's College in London (Fig. 6.4c). Franklin took some particularly good images of DNA in an attempt to see how the four subunits—A, T, G, and C—were arranged (Fig. 6.4d). She made the images using **x-ray diffraction,** a process in which a beam of x-rays passes through a crystalline fiber made of many parallel strands of pure DNA. Within the DNA fiber, similar structures repeat, like the pattern of tiles on a floor. When x-rays pass through these repeated structures, the x-rays bend, just as light rays bend when they pass from air into water, causing a twig sticking out of a pond surface to appear bent. This bending results in spots cast on photographic film (Fig. 6.4d). By analyzing patterns of spots, biophysicists can map the relative positions of atoms in a molecule.

When Watson and Crick saw the x-ray photos, they noticed a certain symmetry, suggesting that the molecule might consist of *two* connected strands of DNA. The pictures further suggested that DNA was most likely a **helix,** a structure similar to a spiral staircase. Each loop of the helix consisted of ten nucleotides. Now the young researchers had to find out how the bases and sugar-phosphate backbones were arranged in three-dimensional space. They hoped that knowing the structure would also reveal how the molecule replicates itself and stores genetic information.

Excited by Rosalind Franklin's stunning data, Watson and Crick chose a model-building approach to testing their hypotheses about the helical structure. They arranged and rearranged pieces of Tinker Toy–like sticks and balls into various combinations to see which arrangement of basic components might best reproduce the evidence from Franklin's x-ray data. After building one incorrect structure after another, Watson and Crick were finally able to visualize the possible molecular structure. The structure was a **double helix,** which resembles a twisted ladder (Fig. 6.5). By combining their intuition, tinkering, and knowledge from pre-existing

Figure 6.4   Elucidating the Structure of DNA. (a) James Watson and Francis Crick worked out the structure of DNA by trying to arrange scale models of the molecule in ways that were consistent with all the data available at the time. (b) DNA is a chain of nucleotides. Each nucleotide consists of a phosphate attached to a sugar attached to a base. The carbons on the sugar are numbered 1′ (pronounced "one prime") to 5′. Each nucleotide has a front and rear end. The phosphate is analogous to a trailer hitch at a car's rear; the sugar is like the frame; the base is like the car's interior compartment; and the 3′ carbon is like the front grill. There are four types of bases in DNA: A, T, C, and G. The "train" of nucleotide "cars" is oriented from the top of the page to the bottom (follow the pale blue arrow). The short chain here has only four nucleotides. In the nucleus of each of your body cells, there are 46 DNA molecules, each containing on average 130 million nucleotides. (c) Rosalind Franklin, working with Maurice Wilkins, performed x-ray diffraction experiments that provided the key data required to determine the structure completely. (d) One of Franklin's x-ray diffraction images of DNA.

(a) Watson and Crick and the DNA model

(c) Rosalind Franklin

(d) A x-ray diffraction image of DNA

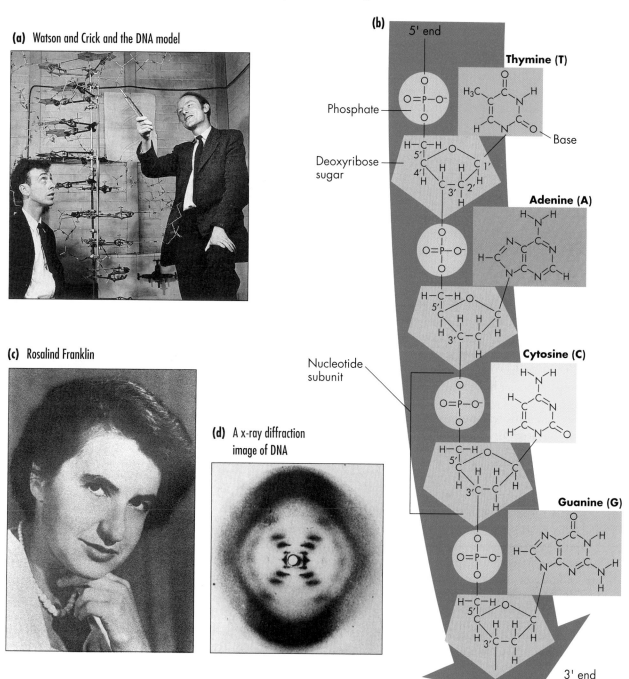

(b)

Figure 6.5   The Architecture of a Single DNA Strand. A computer rendering of Watson and Crick's model of DNA.

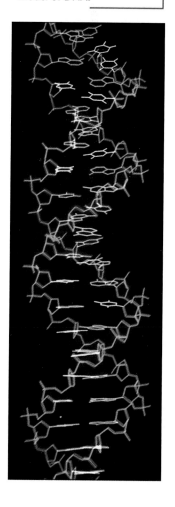

research, Watson and Crick arrived at the structure of DNA before any other researchers. In 1953, Watson and Crick published their findings on the structure of DNA in a two-page report in the international journal *Nature.* In a classic understatement, they wrote that it had "not escaped our notice" that the model immediately suggested ways in which DNA could fulfill its two major biological functions: replicating itself and storing information. The model's simplicity, plus its enormous power to explain observations of nature, led to its rapid acceptance by the scientific world.

In 1962, Watson, Crick, and Wilkins were awarded the Nobel prize for their assessment of the structure of DNA. Franklin had died four years earlier of breast cancer at the age of 37, and most observers agree that she had played a crucial, but underappreciated, role in the elucidation of the structure of DNA.

6.3   The Structure of DNA

What are the detailed features of the structure of DNA, first discovered by Watson and Crick, that explain its function? (These same features, three decades later, helped Marilyn Menotti-Raymond convict a murderer.)

1. A DNA molecule is composed of two nucleotide chains (Fig. 6.6a).
2. These two chains are oriented in opposite directions, like the northbound and southbound lanes of a highway (Fig. 6.6b). All the nucleotide "cars" in each opposing lane of the highway face in the same direction, with the 3′ sugar leading forward and the 5′ phosphate trailer hitch behind (review Fig. 6.4b). To emphasize the opposite orientation of the two strands, geneticists refer to them as "antiparallel."
3. The two sugar-phosphate chains form the outside of the molecule and are the uprights of the twisted ladder (Fig. 6.6c). In contrast, the bases attached to the backbones face inward, connecting in the middle like the rungs of a ladder (Fig. 6.6d).
4. The bases A and T pair with each other, and the C and G bases also pair; that is, A is complementary to T, and C is complementary to G—the two fit together like adjacent puzzle pieces (Fig. 6.6e). Bases are held together by hydrogen bonds (review Fig. 6.6), hydrogen atoms shared between an A and a T base or a C and a G base.

This **complementary base pairing** by hydrogen bonds is significant in three ways. First, it provides the

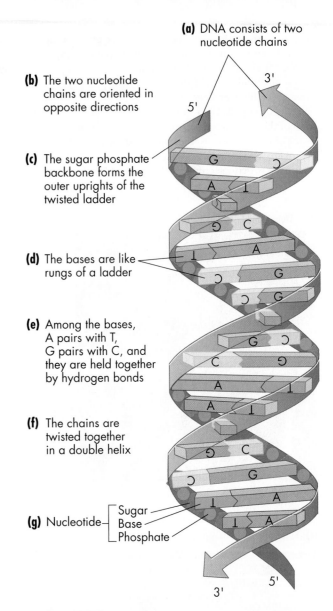

(a) DNA consists of two nucleotide chains

(b) The two nucleotide chains are oriented in opposite directions

(c) The sugar phosphate backbone forms the outer uprights of the twisted ladder

(d) The bases are like rungs of a ladder

(e) Among the bases, A pairs with T, G pairs with C, and they are held together by hydrogen bonds

(f) The chains are twisted together in a double helix

(g) Nucleotide— Sugar / Base / Phosphate

Figure 6.6   The Structure of Double-Stranded DNA. DNA consists of two oppositely oriented sugar-phosphate ribbons. Bases protrude from these ribbons toward the middle, holding the two strands together into a single molecule the ways rungs hold the two sides of a ladder together.

force that holds two single strands of DNA together into a double-stranded molecule. Because hydrogen bonds are relatively weak, however, they are easily broken by temperatures of 60 °C or so. This is important in the DNA-copying method called PCR that Menotti-

Raymond used in her studies of Snowball's DNA. Second, complementary base pairing by hydrogen bonds explains Erwin Chargaff's finding (see p. 144) that cellular DNA always has equal amounts of A and T and equal amounts of C and G: Whenever one strand has a T, the other has an A, and so on. Knowing the rules of base pairing, you should be able to finish writing out the sequence complementary to the 22 bases shown here:

5′ A G G A A A A T G A A G T C A A G A A A A T G G 3′
3′ T C C T _ 5′

Third, complementary base pairing is important for DNA's major biological activities: replication and information storage. We'll see why shortly. But we've already seen an example of base pairing's importance in replication. Dr. Menotti-Raymond used complementary base pairing by the 22 base-pair sequence just mentioned when she applied the PCR process to the tiny amount of DNA in the white cat's hair root. She did this to make millions of copies of a specific portion of Snowball's genetic "fingerprint," and this, in turn, allowed her to study the fingerprint closely.

5. A final feature of DNA structure is that the two strands of the DNA molecule twist together to form the double helix (Fig. 6.6f). This twist is important to at least two of DNA's features. It's crucial to the regulation of DNA expression, since specific proteins bind in the grooves of the double helix and control which genes a cell will use. And it's also crucial to DNA replication since a cell's DNA must unwind like the fibers of a rope and allow matching up of complementary base pairs. Figure 6.6g points out the parts of a nucleotide: a sugar, a base, and a phosphate. *E*xplorer 6.1 allows you to investigate the structure of DNA more closely.

Packaging DNA in Chromosomes

As you can see from Figure 6.1, there are very few cells in the follicle of a cat hair. As a result, Dr. Menotti-Raymond had to use micromethods to purify the DNA from those few cells and separate it from lipids, carbohydrates, and proteins in the hair root. Prying DNA away from proteins is especially difficult because the two kinds of biological molecules are intimately associated in chromosomes.

In the nucleus of a cat's cells and those of most other eukaryotes, each chromosome consists of a single, long, tightly wound DNA molecule. In the fruit fly, for example, the actual length of the DNA molecule from the largest of the fly's 4 chromosomes is more than an inch long—about 12 times as long as the 1-mm long fly itself! Despite the molecule's length, each of the fly's millions of cells contains two copies of this chromosome, as well as pairs of the other three chromosomes. Likewise, a cat has 19 chromosomes, each even longer than a fruit fly's. How do cells package such a huge genetic molecule into a structure as small as a chromosome? The enormous length of DNA in a eukaryotic cell can't just be wadded up haphazardly. If it were, the separation of DNA molecules during cell division would be as difficult as unraveling two tangled kite strings. What happens instead is that DNA, like a proper kite string, is wound in an orderly way (Fig. 6.7). Specifically, DNA is wound around spools of proteins called **histones** (see Fig. 6.7e). A single spool consisting of several histone molecules wrapped with two loops of DNA (140 base pairs long) is called a **nucleosome** (Fig. 6.7e). Adjacent nucleosomes pack closely together to form a larger coil, somewhat like a coiled telephone cord. This cord, in turn, is looped and packaged with scaffolding proteins into **chromatin,** the combined proteins and genetic material that constitute the substance of chromosomes.

Compare this understanding about DNA packing in chromosomes with what you already know about chromosome activity in mitosis. Recall that during the interphase portion of mitosis, individual chromosomes are invisible (review Figs. 4.5 and 4.9). This is because in interphase, the spools of DNA and protein are not packed closely together, allowing the DNA to spread diffusely in the nucleus. In contrast, during prophase (review Fig. 4.9), the spools compact together as shown in Fig. 6.7e, leading ultimately to a visible chromosome (Fig. 6.7a,b). The orderly packaging of DNA around proteins prevents massive DNA tangles during cell division. A molecular biologist like Marilyn Menotti-Raymond studying DNA separates the genetic molecule from these protein spools by using special solvents that destroy the structure of proteins, but leave DNA intact. After this, Menotti-Raymond's next step was to make millions of copies of certain regions of Snowball's chromosomes so she could investigate their genetic properties, identify the cat hairs, and eventually help solve the mystery of who killed Shirley Duguay.

How could cat DNA help solve a murder?

GET THE PICTURE

*E*xplorer 6.1

The Structure of DNA

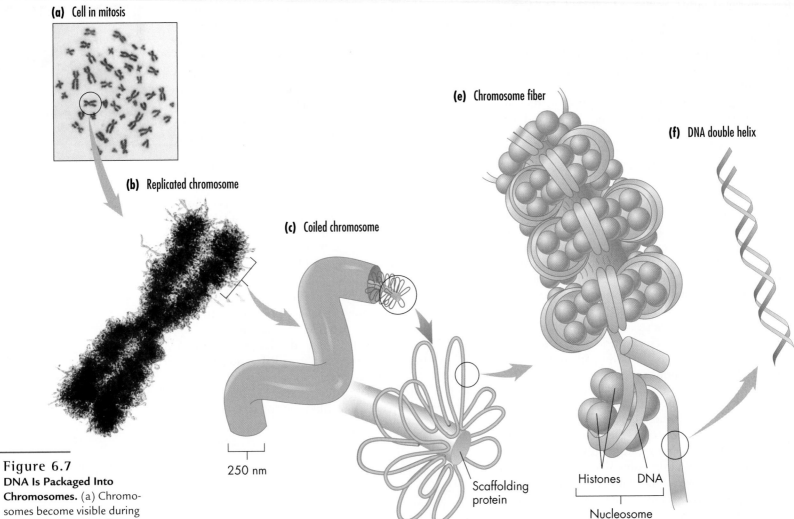

(a) Cell in mitosis

(b) Replicated chromosome

(c) Coiled chromosome

250 nm

(d) Chromosome scaffolding, or chromatin

Scaffolding protein

(e) Chromosome fiber

(f) DNA double helix

Histones   DNA

Nucleosome

Figure 6.7

DNA Is Packaged Into Chromosomes. (a) Chromosomes become visible during prophase of mitosis. (b) A replicated chromosome in late prophase is coiled and looks a bit like curly pasta noodles. The thickness of that coil (c) consists of scaffolding proteins (chromatin) in the middle with protruding loops of DNA and protein (d). These loops are made of protein spools called nucleosomes joined together (e). Each nucleosome consists of histone proteins around which a DNA double helix wraps about two times. This arrangement packages the enormous length of DNA (f) into a metaphase chromosome without tangling.

6.4   DNA Replication

As we've seen, the amount of DNA in a single hair follicle is very small. Because there is a single DNA molecule in each chromosome, any individual cat, for example, will have only two copies of any gene. (Recall from Chapter 5 that a gene is a unit of inheritance and is specifically a portion of a DNA molecule encoding an RNA and usually a polypeptide chain.) One copy of the gene on the DNA is inherited from the mother and the other copy on the DNA is inherited from the father. Recall that only four cat hairs in Douglas Beamish's coat had root cells attached. To get enough DNA to work with, therefore, Menotti-Raymond had to make many copies of specific regions of DNA by using a test tube version of the normal method of DNA replication. Let's first examine the ways in which cells copy DNA and why that's important. Then we can see how Menotti-Raymond harnessed the process of DNA replication to help solve the murder case.

Let's start by placing the biochemical process of DNA replication into the familiar context of the cell cycle. We saw in Chapter 4 that cells cycle through a period of growth (interphase) and a period of division (M phase) (see Fig. 4.5). Interphase, remember, has three parts: In G_1, cells have one double-stranded copy of each nuclear DNA molecule; in the S phase, DNA replicates; as a result, in G_2, there are two double-stranded copies of each DNA molecule. Finally, in mitosis, those two DNA copies separate into the two nuclei of the daughter cells. Let's now focus on the events of the S phase, the copying of DNA.

Steps in Replication

The replication of DNA occurs in the cell nucleus (see Chapter 4) and follows directly from the principle of complementary base pairing. It can be divided into three steps: (1) strand separation, (2) complementary base pairing, and (3) joining. Keep in mind that before the S (synthesis) phase of the cell cycle, DNA is present in the double helix form (Fig. 6.8).

1. *Separation.* For replication to begin, the two strands of the double helix must first unwind and then strands must separate from each other (Fig. 6.8, Step 1). In cells, the unwinding and separation of the strands are catalyzed by enzymes that help break the "rungs" of the ladder. Those rungs, remember, are the bases on each strand that are bound together by weak hydrogen bonds. In the test tube, Marilyn Menotti-Raymond and her colleagues used heat to "melt" the hydrogen bonds. After that, for a short time, the separated base pairs are unpaired. (An antibiotic called Cipro has been in the news because it is useful in treating anthrax-causing bacterium. Cipro works by inhibiting the enzyme in this bacterium that catalyzes the unwinding reaction. Because DNA can't unwind in the bacterial cells before DNA replication, the cells die and the infected person lives. The U.S. government bought and distributed millions of doses of Cipro to postal workers and others affected by terrorism in 2001.)

2. *Complementary base pairing.* The unpaired bases form new hydrogen bonds with free nucleotides (As, Ts, Gs, Cs) that happen to diffuse into the area (Fig. 6.8, Step 2). An A base on one DNA strand pairs only with a free T base (complete with its sugar-phosphate backbone), and an attached C pairs only with a free G. Likewise, attached Ts bond only to free As, and attached Gs bond only to free Cs. Thus, the sequence of bases in the original strand specifies the same

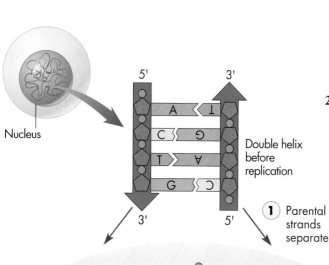

Nucleus

Double helix before replication

1 Parental strands separate

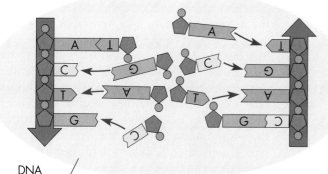

DNA polymerase

2 Free nucleotides diffuse in and pair up with bases on the separated strands

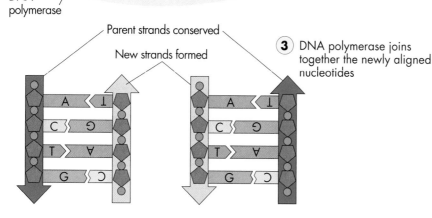

Parent strands conserved

New strands formed

3 DNA polymerase joins together the newly aligned nucleotides

Figure 6.8 **An Overview of DNA Replication.** Before the S (synthesis) phase of the cell cycle, DNA in the cell nucleus occurs as a double helix. As a first step in DNA replication, the strands separate (Step 1). Previously unattached nucleotides that have accumulated in the nucleus diffuse in and pair up with appropriate unpaired bases (Step 2). The enzyme DNA polymerase links the new bases together, forming new strands (Step 3). The two new double-stranded molecules are identical to the parent DNA molecule, and in each daughter molecule, one strand is inherited intact from the parent and one is newly formed.

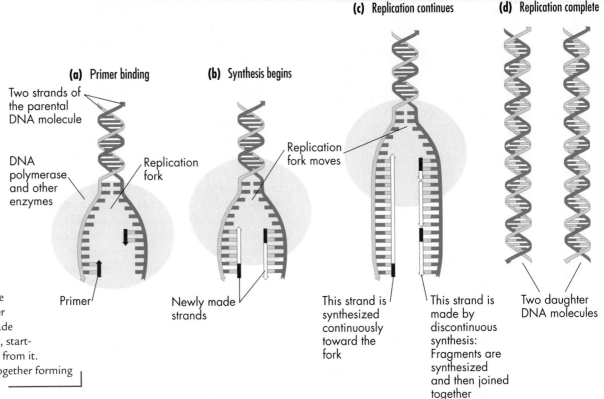

Figure 6.9 Action at the Fork: Details of DNA Replication. (a) DNA replication begins as two strands separate, making a replication fork. Then primers, short lengths of single-stranded DNA (or RNA), pair with a complementary region in the DNA to be replicated. (b) Replication begins as the enzyme DNA polymerase adds nucleotides to the end of the primer. DNA polymerase can add new nucleotides only in one direction, like adding new cars to a train only at the front. (c) When the replication fork has moved forward, the strand made in the direction the fork is moving (the left strand in the figure) can be synthesized continuously. The other newly made strand, however, is made discontinuously in short fragments, starting near the fork and moving away from it. (d) Enzymes then join fragments together forming two new daughter strands.

(c) Replication continues

(d) Replication complete

(a) Primer binding

Two strands of the parental DNA molecule

DNA polymerase and other enzymes

Replication fork

Primer

(b) Synthesis begins

Replication fork moves

Newly made strands

This strand is synthesized continuously toward the fork

This strand is made by discontinuous synthesis: Fragments are synthesized and then joined together

Two daughter DNA molecules

GET THE PICTURE

Explorer 6.2

DNA Replication

sequence in the new strand according to the rules of complementary pairing.

3. *Joining.* The joining together of the newly paired bases creates a new strand that is complementary to the parent strand, forming two new double helices that are identical to the original double helix (Fig. 6.8, Step 3). The joining, or **polymerization,** of the new double helices is catalyzed by an enzyme called **DNA polymerase.** This enzyme joins the phosphate group (the trailer hitch in our car analogy) of one nucleotide to the "grill" of the previous nucleotide. (In the PCR method of DNA copying, Dr. Menotti-Raymond used a DNA polymerase derived from a hot-springs bacterium to replicate cat DNA. This enzyme from a heat-loving bacterium is not destroyed by the heat she needs to apply in order to break the hydrogen bonds and separate DNA strands.)

Semiconservative Replication

The three steps of DNA replication—strand separation, base pairing, and joining—occur over and over again along the length of the DNA molecule and pro-duce two double-stranded DNA molecules identical to the parental molecule. Each new DNA has a base sequence identical to the base sequence of the original. You can see in Figures 6.8 and 6.9 and in **E**xplorer 6.2 that each of the two daughter DNA molecules has one strand intact from the original parent, while the other strand is completely new. Because only one of the two strands in the daughter molecule is inherited intact—or conserved—from the parent molecule, this type of replication is called **semiconservative replication.** Apparently, all living creatures share this mode of DNA replication. Semiconservative copying of DNA is very different from, say, copying a piece of paper on a photocopy machine. The machine produces a totally new copy but conserves the original fully intact. In contrast, during DNA replication, the original molecule ceases to exist, but one half of it becomes part of one offspring molecule and the other half becomes part of the other offspring molecule.

Some Finer Points about DNA Replication

Two further points about DNA replication will interest curious students. First, a molecule called a **primer** defines the point at which DNA synthesis begins. A primer is usually a short single-stranded RNA or DNA that binds to a specific region (Fig. 6.9a). The 22 base pair sequence mentioned earlier is an example. Dr. Menotti-Raymond used the primer in the PCR process to copy a specific portion of Snowball's DNA comprising part of Snowball's DNA fingerprint. Second, the two newly made DNA strands have certain differences because the polymerase enzyme can synthesize DNA in only one direction (that is, it can add more "cars" only at the front of the "train," not at the rear). One strand is synthesized continuously in one direction (Fig. 6.9b), but the other is made discontinuously, as described in more detail in the legend to Figure 6.9c.

While the details of DNA synthesis may seem a bit specialized to you, without this knowledge, biologists would not have been able to develop drugs that help AIDS patients live longer and healthier.

Accuracy of DNA Replication

It is a wonder that DNA molecules are ever copied correctly, considering the immense length of DNA molecules and the complexities of the unwinding, separation, base pairing, and polymerization required for semiconservative replication. Nevertheless, DNA synthesis is incredibly accurate: An error is made only about once in every 10^9 bases. To approach that level of accuracy you would have to type this entire book a thousand times with only one typing error! For many organisms, survival requires extreme accuracy of DNA replication. For example, the human genome (the total of all the genes in a single haploid egg or sperm cell) contains about 3×10^9 base pairs. On average, then, each egg or sperm will have about three new errors. If the rate were much higher, the genetic information would be so altered that the new organism resulting from fertilization could not function. Cells have enzymes that search out and correct genetic errors. Some errors will go undetected and can be harmful, even lethal. But some are not harmful and ultimately provide the genetic variability that fuels evolution.

The Principles of DNA Replication Applied to a Murder Case

Dr. Marilyn Menotti-Raymond used the principles of DNA structure and replication to positively identify the cat whose hairs clung to the lining of the discarded leather jacket alongside droplets of Shirley Duguay's

blood. The genetic "fingerprinting" process depends on two principles we've discussed in this chapter, the principle of base order along the DNA molecule and the process of DNA replication.

Base Order and Individual Identity

At thousands of places along the DNA of many species, the two nucleotides C and A are repeated over and over: CACACACACA. The opposite DNA strand, of course, has the sequence TGTGTGTGTG, which is the complementary sequence in the opposite orientation. For convenience, however, we'll talk about the sequence of just one strand, because given that ordering of base pairs, one can figure out the sequence of the other strand by applying the rules of base pairing.

Cats have 19 pairs of chromosomes. In comparison, we humans have 23 pairs, fruit flies have 4 pairs, and corn plants have 10 pairs. Studying Snowball the cat's DNA, Menotti-Raymond found an important spot along the DNA from one of the two copies of chromosome B3. She called this locus, this region of 115 base pairs, "FCA88" and it had the following sequence of bases: AGGAAAATGAAGTCAAGAAAATGGCTTAA TCCAAAGTCACACAGTACTTAATGTGTGTGTG TGTGTGTGTGTGTGTGTGTGTGTGTGTATGT GTGTAACGGGAAAAAGAAAAA.

Note that the sequence "TG" appears over and over again in this portion of the chromosome. Also notice that we've printed the sequence of just one of the two strands of the DNA double helix. What would the repeated portions "read" on the other strand? If you said, "CA" repeated over and over, you're right!

Dr. Menotti-Raymond carefully studied the repeated sequences in this FCA88 position on the chromosome. Areas of repeated base sequences are also called **genetic markers** because different numbers of repeats tend to occur in different individuals and this can serve as a specific recognizable section of the DNA that adds to the unique identity. Next, she found 28 copies of the TG repeat on one copy of chromosome number B3 and 25 copies of the TG repeat on the other copy of chromosome B3 from the hair root cell DNA left on the bloody jacket. (The differences are understandable if you consider that each individual has two copies of each chromosome, one from the mother and one from the father.) Once she saw this "fingerprint" in the DNA from the cat hairs on the bloody jacket, Menotti-Raymond could determine that it matched the number of TG repeats at this locus for the cat Snowball. She also had to test other cats on Prince Edward Island to see if this fingerprint was common or truly unique.

How do geneticists use genetic markers to construct a DNA fingerprint?

(a)

(1) Dr. Menotti-Raymond preparing samples for PCR.

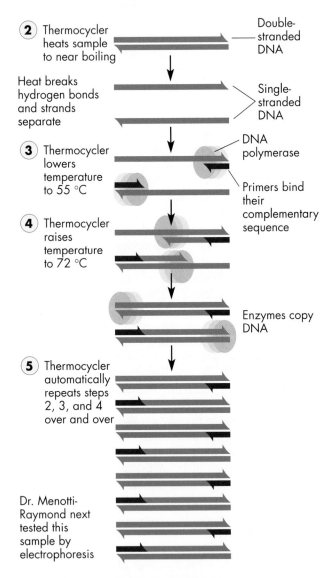

(2) Thermocycler heats sample to near boiling — Double-stranded DNA

Heat breaks hydrogen bonds and strands separate — Single-stranded DNA

(3) Thermocycler lowers temperature to 55 °C

DNA polymerase

Primers bind their complementary sequence

(4) Thermocycler raises temperature to 72 °C

Enzymes copy DNA

(5) Thermocycler automatically repeats steps 2, 3, and 4 over and over

Dr. Menotti-Raymond next tested this sample by electrophoresis

Menotti-Raymond examined DNA from 19 cats who lived on Prince Edward Island but were unrelated to Snowball, as well as another 9 cats from around the United States. As we just saw, Snowball's DNA had precisely 28 copies of the TG repeat at the FCA88 locus on copy 1 of chromosome B3 and 25 copies of the TG repeat at the FCA88 locus on copy 2 of chromosome B3. The researcher found that an unrelated cat—say, a Siamese cat living on Prince Edward Island—might have as few as 12 or as many as 29 TGs in a row at position FCA88 on chromosome B3. Only a few cats on the island had the same combination of 28 and 25 TG repeats at locus FCA88 as found in the DNA from the cat hairs taken into evidence. By studying repeated base sequences at nine more loci along the DNA retrieved from the cats as well as from the leather jacket, Menotti-Raymond came up with a set of sequences absolutely individual to one single cat. With it, she could rule out 219,999,999 of every 220 million cats living on Prince Edward Island. (This is a theoretical number, of course, since only a few thousand cats actually live there). She could also rule out 69,999,999 out of every 70 million cats on the U.S./Canadian mainland. This virtually guaranteed that the hairs in the discarded jacket came from the one cat, Snowball, who lived with Douglas Beamish and his parents.

Figure 6.10 **The Polymerase Chain Reaction Catches a Murderer.** (a) Dr. Menotti-Raymond isolated DNA from the roots of cat hairs found on the bloody jacket, then loaded the samples into a thermocycler (Step 1). DNA in the sample has two strands of the double helix oriented in opposite directions. The sample's temperature is raised to nearly boiling in the thermocyler machine (Step 2). This breaks the hydrogen bonds that hold the two DNA strands together and allows the two strands to separate from each other. The thermocycler lowers the temperature to 55 °C—cool enough so that the primers can bind to the chromosomal DNA by hydrogen bonds (Step 3). (Note that the two primers flanking the GT/CA repeat are oriented in opposite directions.) A temperature increase to 72 °C now creates an optimum temperature for the DNA polymerase (taken from a strain of hot springs bacteria)(Step 4). The enzyme replicates DNA, starting from the primers. The thermocycler repeats Steps 2 to 4 thirty times or so, doubling the number of copies of the sequence between the primers with each repetition (Step 5). After all five steps, Dr. Menotti-Raymond had millions of copies of a single short stretch of DNA (115 nucleotides long in this case). (b) The number of times TG/CA repeats in the fragment determines the final size of the replicated fragment. This is the actual sequence of primers and cat DNA for one of the markers Dr. Menotti-Raymond investigated.

(b)

TGAGGAAAATGAAGTCAAGAAAATGGCTTAATCCAAAGTCACACAGCTACTTAATGTGTGTGTGTGTGTGTGTGTGTGTGTGTGTGTGTGTGTATGTGTGTATTACGGGAAAAAGAAAAACACAT

CACATATGGTGCCCTTTTTTTTTCTTTTT

AGGAAAATGAAGTCAAGAAAATGG

ATGTGTGTACACACGATACACACACACACACGACTCTGTTCTTTTTTTCGACCTTTTAGGTTCCTTGAAGTTAAGGCTTCGACGGAAATTAGCCATTTTACTTCTTGACTTCTTTTTCCTCA

The above reasoning shows how forensic scientists use the order of nucleotides in DNA to produce genetic fingerprints and identify individuals, whether cats, dogs, or people. But base pairs are submicroscopic. How can geneticists actually count the number of repeats (whether TGs, GCs, or some other combination) in a portion of DNA?

Using DNA Replication to Detect Variations in DNA

Menotti-Raymond needed a method to detect the differences between, say, 25 copies of TG at each locus and 28 copies. To do this she used the process we mentioned earlier, PCR, or **polymerase chain reaction.** PCR allows a biologist to copy specific portions of a DNA molecule by applying the principles of DNA replication. Menotti-Raymond isolated DNA from the root of the cat hair found on the bloody jacket, and used the DNA polymerase enzyme (see Fig. 6.8) to make millions of copies of a specific portion of DNA. (Fig. 6.10 illustrates the actual steps.)

First she extracted DNA from chromosomes in the few cells at the bases of the cat hairs (Step 1). That DNA has the typical double helix form, with two strands oriented in opposite directions. She placed the DNA samples in a thermocycle instrument and raised the sample's temperature to near boiling; this broke the hydrogen bonds that hold the two strands together, allowing the two strands to separate from each other (Step 2). The thermocycler now lowered the temperature to 55 °C, which is cool enough so that primers could bind to the chromosomal DNA by hydrogen bonds (Step 3). The close-up in Figure 6.10b shows the actual sequence of primers and cat DNA for one of the markers Menotti-Raymond used. The temperature in the instrument then shot up to 72 °C (optimal for the DNA polymerase enzyme isolated from a hot springs bacterium, which is able to link nucleotides at high temperature) (Fig. 6.10a, Step 4). (Your own DNA polymerase, by the way, would congeal like a boiled egg at 72 °C.) At this point, the enzyme replicates DNA, starting from the primers. In Step 5, Dr. Menotti-Raymond programmed the thermocycler to repeat Steps 2, 3, and 4 30 times. Each repetition doubled the number of copies of the sequence between the primers. After 30 cycles, Dr. Menotti-Raymond had millions of copies of a single short stretch of DNA (115 nucleotides long in this case).

With the millions of copies of DNA based on the original evidence, Menotti-Raymond could now begin the process of counting the number of TG repeats at the FCA88 locus. To do this, she used **gel electrophoresis,** a procedure that separates molecules on the basis of size (Fig. 6.11). The researcher places a sample of the DNA (or other material, such as an unknown protein) into a soft, rubbery matrix similar to gelatin (hence the name "gel"). An electric field applied to the gel (with a positive pole on one end and a negative pole on the other) then causes molecules with their own natural electric charge to move through it. (Fig. 6.11 shows this technique.) Small DNA fragments move more quickly through the electrified gel than larger DNA molecules, just as 185-pound running backs move faster down the football field than 310-pound center guards, or as short worms crawl faster through a dense tangle of branches than long worms. After applying the technique for 30 minutes or so, Dr. Menotti-Raymond was able to separate all of the DNA fragments by size. Gel electrophoresis is sensitive enough to detect the size difference of a single nucleotide pair, for example, the difference between a molecule 115 nucleotides long with TG repeated 22 times, and one 117 nucleotides long with TG repeated 23 times. Figure 6.11c shows the wide variety in lengths of a specific TG repeat locus in house cats.

Menotti-Raymond's final step was to determine whether the genetic "fingerprint" she saw in the DNA from the bloody jacket was rare or common among cats. Recall that she saw 25 and 28 repeated TGs at locus FCA88 on the cat's two copies of chromosome B3 (in other words, it was heterozygous [see Chapter 5] at that locus). She also saw 12 and 15 repeated TGs at locus FCA43 on chromosome C2, and looked at several more loci on various chromosomes for a total of 10 loci. In her genetic comparisons, Menotti-Raymond might have found, for example, that half of the cats on Prince Edward Island have the 28/25 repeat genotype for locus FCA88, but only one in ten had the 12/15 genotype at locus FCA43. This would mean only 5 percent (50 percent $\times$ 10 percent) would be expected to have that same genotype at both loci. By adding more and more sites from different chromosomes, and by knowing the frequency of each type, Menotti-Raymond could finally reach the estimate we saw earlier: that only one in 220 million cats with the genotypes found on Prince Edward Island and one of 70 million cats with the genotypes found on the mainland could have the specific combination she identified in the hair follicle from the jacket. This evidence was sufficient for the jury to accept that the owner of the discarded leather jacket lived in the house with Snowball the cat. Because Douglas Beamish lived with Snowball, and the jacket had both the murder victim's blood and Snowball's hair, the jury

What did Snowball's DNA reveal?

(a) DNA and dye are loaded in a well on a gel, and an electric field is placed across the gel.

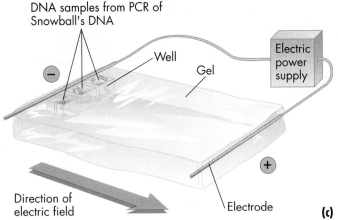

DNA samples from PCR of Snowball's DNA

Well

Gel

Electric power supply

Direction of electric field

Electrode

(b) DNA fragments move through the gel, shorter fragments faster than longer fragments.

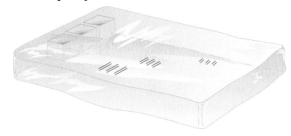

Figure 6.11 **Gel Electrophoresis.** Dr. Menotti-Raymond used gel electrophoresis to measure the size of DNA fragments in the DNA fingerprints isolated from the bloody jacket. (a) She mixed the amplified DNA (from the PCR samples) with a blue tracking dye and a substance that causes the DNA to fluoresce an orange-pink under ultraviolet light. Then she placed the sample in wells cut in the gel. Next she applied an electric voltage difference across the gel. Because DNA molecules have their own natural electrical charge, they move in the electric field. (b) Shorter fragments move more rapidly through the gel than the longer fragments. After about 30 minutes, the fragments are separated by size and the quantity of DNA at each position along the gel is measured and printed out in a graph. (c) One of Menotti-Raymond's gels shows that at the FCA88 locus, the cat Snowball had the identical pattern of DNA fragments as the DNA from the bloody jacket while other cats' DNA failed to match the DNA evidence at this locus. After looking at ten such genetic loci, Menotti-Raymond could conclude that only one cat in tens of millions would have this same genetic constitution purely by chance.

(c) Density scan of an electrophoresis experiment conducted by Menotti-Raymond on Snowball's DNA

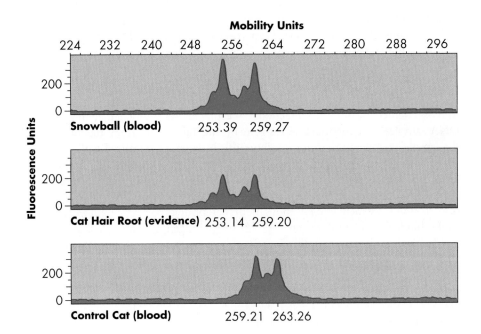

Mobility Units

Snowball (blood) 253.39 259.27

Cat Hair Root (evidence) 253.14 259.20

Control Cat (blood) 259.21 263.26

Fluorescence Units

SOLVE IT

E xplorer 6.3

Who Dunnit? DNA Fingerprinting

was inclined to return a guilty verdict. Combining this evidence with other details from the crime scenes, they did just that, and the defendant went to prison. E xplorer 6.3 gives you an opportunity to review DNA evidence in another case and arrive at your own decision regarding guilt or innocence.

This type of evidence has become increasingly important in criminal trials, and in some states with the death penalty, DNA evidence must now be considered before a convicted murderer is executed for his or her crime. The infamous O.J. Simpson murder trial, how-

ever, shows that juries sometimes find DNA fingerprint evidence less compelling than do biologists!

Our cat hair case study has demonstrated DNA's capacity for identifying individuals with unique "fingerprints" of repeated sequences at different loci along the chromosomes. But the significance of DNA's elegant double helix structure goes far beyond genetic fingerprints, important as they are to the medical and legal systems. DNA's structure—two twisted strands of nucleotides, oriented in opposite directions—explains its ability to replicate with amazing accuracy as the strands separate and enzymes synthesize two daughter molecules, each a complementary copy of the parent strand. DNA structure also explains the way the "thread of life" controls an organism's physical make up and abilities. Living things, of course, are made up of cells, and it is the collective form and function of those cells that determines the entire organism's appearance, its physical and chemical properties, and its day-to-day behaviors. In our next chapter, we'll see how the blueprints for form and function inscribed in DNA are actually decoded into proteins and cellular activities, and how molecular geneticists have learned to manipulate these blueprints for medicine, agriculture, and industry.

Chapter Summary and Selected Key Terms

Introduction
Dr. Marilyn Menotti-Raymond and colleagues at the National Institutes of Health helped solve a murder case by carrying out lab procedures to positively identify the DNA in cat hair roots. The hairs were left in the lining of a jacket also bearing the murder victim's blood. The unique structure of DNA as well as repeated sequences of nucleotides that vary at different **loci** (p. 140) along the chromosomes from individual to individual allow researchers to create genetic "fingerprints."

Identifying the Hereditary Material
Mendel discovered in the mid-1800s that hereditary "factors" determine an organism's physical, chemical, and functional characteristics. But what are these factors that today we call **genes** (p. 141)? Biologists in the 1940s thought only protein was versatile enough to carry complex information, since there is an "alphabet" of 20 amino acids forming the millions of kinds of proteins. Researchers devised a way to test whether protein or DNA could carry information. They extracted carbohydrates, proteins, and DNA from disease-causing bacteria (p. 141) and added these separate extracts to nonlethal bacteria. Bacterial cells treated with carbohydrates and proteins remained nonlethal but those treated with DNA became lethal—able to cause pneumonia in mice and other mammals. Researchers called this transfer of an inherited trait (in this case, the ability to cause disease) by the uptake of pure DNA **transformation** (p. 141). Alfred Hershey and Martha Chase also tested bacterial viruses or **bacteriophages** (p. 141) (sometimes called simply **phage,** p. 141) for this protein versus DNA question. They radioactively labeled the sulfur in protein and the phosphorous in DNA and watched to see which labeled element entered the bacterial host cell infected by a phage. Phosphorous entered but sulfur remained outside, once again confirming that DNA is the hereditary material.

DNA: The Twisted Ladder
The variability of DNA is the key to evolution by natural selection yet the stability of DNA determines an organism's resemblance to its parents. How does one molecule do both things? James Watson and Francis Crick knew several facts about DNA from other researchers' efforts. DNA is a very long, linear thread made up of a chain of nucleotides, **A,T,G,** and **C (adenine, thymine, guanine, and cytosine)** (p. 144). They knew A and T occur in equal proportions in DNA, and so do G and C. Finally, they had an **x-ray diffraction** (p. 144) photo from Rosalind Franklin that suggested two strands, probably in a **helix** (p. 144). Watson and Crick built models that would fit all the available evidence and came up with the actual structure, a **double helix** (p. 144).

DNA Structure
After decades of work, modern geneticists know that DNA has two nucleotide chains, oriented in opposite directions like the northbound and southbound lane of highway: they are antiparallel. Sugar-phosphate backbones form the uprights of the twisted ladder and A–T G–C pairs, joined by **complementary base pairing** (p. 146), form the ladder rungs. Each chromosome is made up of one long, tightly wound DNA molecule coiled around spools of protein called **histones** (p. 147). A **nucleosome** (p. 147) is one spool wrapped with two loops of DNA. Adjacent nucleosomes form coiled cords and the cords are packaged into **chromatin** (p. 147), the substance of a chromosome. During the interphase of mitosis, the spools are unwound and diffuse. During prophase, the chromatin is compacted and the chromosomes are visible.

DNA Structure and Replication
The steps of DNA replication are (1) separation, during which the double helix unwinds and enzymes break the hydrogen bonds between complementary base pairs; (2) complementary base pairing during which free-floating bases pair up and bond to the newly unattached bases; and (3) joining, during which two new double helices form. The joining or **polymerization** (p. 150) requires the enzyme **DNA polymerase** (p. 150).

The three-step process is also known as **semiconservative replication** (p. 150) because only one of the two strands in the DNA molecule is completely new. This contributes to the amazing accuracy of DNA replication.

Marilyn Menotti-Raymond was able to use the principles of DNA structure and replication to help identify the cat hairs on the jacket and thus to help solve the murder. Using **primers** (p. 150), she found a spot along the DNA called FCA88 with 28 repeats of the CA nucleotide sequence on one chromosome B3 (which came from one cat parent) and 25 repeats of CA on the other chromosome B3 (which came from the other cat parent). Elsewhere on the cat's DNA, she found nine additional loci with different numbers of nucleotide repeats which she used as **genetic markers** (p. 151). Together, the pattern of repeats is like a unique bar code or fingerprint and could only occur exactly the same way in one of every 220 million cats of the genotypes living on Prince Edward Island and 1 of every 70 million cats living on the mainland. She made use of the **polymerase chain reaction** or **PCR** technique (p. 140, 153) to make millions of copies of the DNA from the cat hairs on the jacket, and **gel electrophoresis** (p. 153) to identify specific portions of DNA, such as sections with varying numbers of base-pair repeats (say, 28 TGs versus 25 TGs). DNA structure explains not only unique identity but day-to-day functioning, as Chapter 7 explains.

> All of the following question sets also appear in the Explore Life *E* electronic component, where you will find a variety of additional questions as well.

Test Yourself on Vocabulary and Concepts

In each question set below, match the description with the appropriate term. A term may be used once, more than once, or not at all.

SET I

(a) double helix (b) cytosine (c) adenine (d) hydrogen

1. Thymine in one strand of DNA pairs with _____ in the other.

2. The pairing of complementary bases to form "rungs" of the DNA ladder occurs by means of _____ bonding.

3. Guanine in one strand of DNA pairs with _____ in the other.

4. DNA normally exists as a double-stranded molecule called a _____ .

SET II

(a) chromosome (b) circular DNA (c) chromatin

5. A "colored body" involved in inheritance.

6. The hereditary material of bacteria.

7. An organized collection of DNA, histone, and other proteins.

SET III

(a) semiconservative replication (b) bacterial transformation (c) DNA is the transforming principle (d) DNA is a double helix

8. Experiments that isolated all of the molecular constituents of the lethal strain of pneumococci showed this: _____.

9. Experimental evidence showed that one strand in a DNA duplex is inherited intact while the other strand is newly synthesized one nucleotide at a time. This is called _____.

10. Regularities in x-ray diffraction photographs showed this: _____.

11. The deliberate or accidental uptake of pure DNA by a cell can change that cell's phenotype. This is called _____.

Integrate and Apply What You've Learned

12. How can studying a specific DNA molecule help determine an organism's unique identity?

13. Contrast the meaning of the terms *complementary bases* and *semiconservative*.

14. Explain how experiments with viruses demonstrated that DNA, rather than protein or carbohydrate, is the carrier of hereditary information.

15. Why was the structure of DNA considered to be so important that several research groups engaged in a race to find a model?

16. Why were geneticists reluctant at first to accept the idea that genes are composed of DNA rather than protein? Can you find any flaw in their reasoning?

Analyze and Evaluate the Concepts

17. Why did Dr. Menotti-Raymond apply the techniques of PCR and gel electrophoresis to help create the DNA fingerprint that solved the Shirley Duguay murder case?

18. HIV, the virus that causes AIDS, consists of a lipid membrane surrounding a layer of protein that encloses two molecules of RNA (a nucleic acid related to DNA). When HIV infects a human cell, it replicates and kills the cell, and in the process, more virus particles are released. Design an experiment to determine which substance in HIV carries the hereditary material of the virus: lipid, protein, or RNA.

19. Let's say you decide to test whether or not the usual rules for base pairing apply to sea urchins. You remove an urchin from a tide pool and determine that 33 percent of the bases in its DNA are T. What percentage of its bases would you predict to be A? What percentage should be left over for the other two bases? What percentage of C should you expect?

20. A botanist discovers a new plant in the Amazonian rain forest. This plant's DNA bases are 18 percent adenine. What is its percentage of cytosine nucleotides?

Milking Mice and a Midnight Brainstorm

"You never know," laughs Harry Meade, "when you might be learning something you could use in your next career." Dr. Meade is the head of research at a Massachusetts biotechnology company called Genzyme Transgenics that is marketing the world's first commercial product from a genetically modified animal. Both Meade's job and the new blood-clotting drug his company makes are the result of intertwined threads as disparate as dairy farming, breast tumors, heart disease, and goats.

Harry Meade was raised on a dairy farm in western Pennsylvania where his parents and brothers still tend a small herd of cows. Meade loves dairy cows and still goes home sometimes to milk them. That love no doubt drew him, during his undergraduate college years, to work with a professor who was studying breast tumors and cancer viruses in mice. Improbable as it sounds, Meade says, "He was *milking mice*. So we were always talking about purifying proteins out of the milk."

Meade was in graduate school during the 1980s as the genetic engineering revolution surged. Like so many biologists, he grew interested in the possibility of transferring human genes into bacteria and producing quantities of desirable human proteins. A favorite example at the time was making human insulin in *Escherichia coli* bacteria; after years of scientific effort and millions of dollars of research and development funds, that eventually became commercially feasible on a large scale. At the time, Meade was working on a human protein called tissue plasminogen activator (tPA), that can dissolve blood clots during a heart attack or stroke. He found, however, that it was next to impossible to produce human tPA in bacteria. So instead, he started searching for other ways to engineer it.

"Late one night," he recalls, "I was talking with [fellow student] Neils Lonberg about making transgenic animals." This involves transferring foreign genes into other animal species and collecting proteins from them. "He said, 'We'll make things in the animal's blood,' and I said, 'No, no, I know how to do this! We'll make protein in milk!' The next thing you know, it's 15 years later and you're doing it for real!"

Meade and Lonberg won a patent for their two-part idea: First, finding the "promoter" for the most common protein in milk, casein. A **promoter** is a sequence of nucleotide base pairs that dictates when and where a gene will be turned on—in this case, the gene for casein. Their second idea was attaching

▲
Pharming AT3: Harry Meade; Purifying AT3; Transgenic Goat.

157

that promoter to a human gene of interest and inserting the combination of human gene, casein gene, and casein promoter into an animal host. Meade was partial to cows, of course. But their goal was to produce large quantities of human proteins in animal milk, and for that, goats were a much more practical choice. For the new business of "drug farming," goats mature to milking age much earlier than cows.

Meade and colleagues went after the specific promoter for the casein protein in goats. Then they attached it to a human gene for a protein that prevents blood clots—a protein called AT3 (or antithrombin III). "When people go onto a bypass machine during coronary artery bypass surgery," he says, "their normal AT3 levels drop" and this increases their risk of forming blood clots that can lead to a stroke or heart attack. If Meade's group could make a large supply of AT3 available for use in a commercial drug, physicians could transfuse the protein into patients during heart surgery to prevent them from forming the dangerous clots. The drug might also combat a complication called sepsis, or blood poisoning. During sepsis, Meade explains, the blood can quickly coagulate (clot) throughout the body.

In the 1990s, the Genzyme Transgenics company hired Meade, licensed the goat casein patent, and started a goat farm in central Massachusetts. Today they have over 1800 goats; dozens are already producing large amounts of AT3 in their milk, and dozens more are churning out various types of antibodies and other desirable human proteins. Competing biotech drug companies are still mak-

ing certain human proteins in bacteria or in tissue culture. However, "drug farming," or making pharmaceuticals in barnyard animals (also called "pharming") is a particularly fast and economical way to make large amounts of useful proteins. Harry Meade literally spawned a successful branch of the biotechnology industry from the intertwining of his improbable life threads.

This chapter explores two of the most important topics in modern biology: First, the elegant way in which living cells use the information inscribed in genes to direct the production of proteins for cell survival, growth, and division. Second, how biologists have learned to use these basic processes to splice foreign genes into so-called recombinant organisms to make useful proteins in new ways. You'll learn in this chapter how cells make proteins, how the cell controls this vital process, and about the tools and techniques of genetic engineering, including "pharming."

Along the way, you'll find answers to these questions:

1. How do genes run most of the cell's day-to-day activities?
2. How is DNA copied into RNA?
3. How is RNA translated into proteins?
4. How are genes controlled?
5. What is recombinant DNA technology?
6. What are the promises and problems posed by genetic manipulation?

7.1 How Do Genes Program Cell Activities?

Before Dr. Meade could engineer goats to produce human AT3 in their milk, he had to know and apply the principles of gene action: How do cells use the information stored in DNA to control day-to-day activities? How does DNA encode protein structure? And how are some genes turned on in some cells but not in others? For example, your liver cells normally make

AT3, but your salivary gland cells do not. How does each cell "know" which proteins to make? Dr. Meade's knowledge of gene function is based on many important discoveries from the second half of the 20th century. For this reason, we'll take a historical approach to these questions and their answers. We'll investigate classic experiments that taught geneticists like Dr. Meade the basics of gene action and indirectly, the way to engineer new kinds of organisms and new types of drugs that promise to save peoples' lives.

Using Mutations to Learn How Genes Work

Much of our modern understanding of gene activity and protein formation comes from studying mutations—changes in DNA structure. In one of the first and most important such investigations, geneticists George Beadle and Edward Tatum devised a clever series of experiments back in the 1940s to determine what genes do. The mutations they looked at were in the bread mold *Neurospora crassa,* which you might find growing as a fuzzy pink patch on a forgotten heel of bread in your kitchen cabinet.

Although *N. crassa* is simple in structure (Fig. 7.1a), it's actually a sophisticated "biochemist," able to construct for itself every compound it needs for growth and reproduction. It can make all of its needed amino acids, nucleotides, vitamins, and so on from a simple medium consisting of sugar, a source of nitrogen, some salts, and the vitamin biotin. This spartan diet is called a *minimal medium.* Beadle and Tatum's clever twist was to produce, isolate, and study mutant *N. crassa* molds that could no longer synthesize all of their necessary amino acids and vitamins, and then figure out why not.

Figure 7.1 shows how the researchers found a mutant form of the bread mold that could no longer make vitamin B_1 (thiamine). First, they irradiated a tuft of healthy mold with ultraviolet light to induce mutations (Fig. 7.1a). From the irradiated individual molds, Beadle and Tatum picked out spores (specialized reproductive cells) and placed them one by one into many test tubes partly filled with *complete medium,* a substance that contains all the amino acids and vitamins that *Neurospora* needs to grow (Fig. 7.1b.) Both normal and mutant cells can grow on complete medium because it supplies all needed substances, even to cells that can't generate them themselves. Next, the researchers transferred cells from the complete medium to a minimal medium, which lacked vitamin B_1, the amino acid histidine, and other substances (Fig. 7.1c). The mutant cells failed to grow on this medium because they couldn't generate the missing compounds themselves. The normal, nonmutant strain could still grow well on the same minimal medium, however, because normal *Neurospora* cells can synthesize their own vitamins and amino acids.

Next, they wanted to see exactly what substance the mutant cells needed in their diet to survive. So the researchers mixed up a batch of minimal medium, then added either vitamin B_1 (Fig. 7.1d), the amino acid histidine (Fig. 7.1e), or other individual nutrients. Finally, they inoculated the tubes with separate kinds of mutant molds. They found that some mutants, like Mutant 1

Figure 7.1 Mutant Mold and the One Gene – One Enzyme Hypothesis.
(a) To induce mutations in the pink bread mold *Neurospora crassa,* Beadle and Tatum irradiated mold spores with ultraviolet light. (b) Beadle and Tatum placed individual irradiated spores in vials containing complete medium providing all the complex molecules both normal and mutant *Neurospora* needed for growth. (c) They then transfered normal and mutant spores to minimal medium, which lacked many of the vitamins, amino acids, and other substances *Neurospora* cells normally make for their own growth. Mutant cells that could not synthesize the missing complex nutrients failed to grow on minimal medium. (d) Adding vitamin B1 allowed Mutant 1 to grow but did not help Mutant 2, suggesting that Mutant 1 must lack an enzyme involved in making vitamin B1. (e) Mutant 1 failed to grow on minimal medium plus histidine, and so its mutation could not involve histidine synthesis. Mutant 2, however, grew when supplemented with histidine, and so it must lack an enzyme needed for making histidine. Based on such experiments, Beadle and Tatum coined the one gene–one enzyme principle.

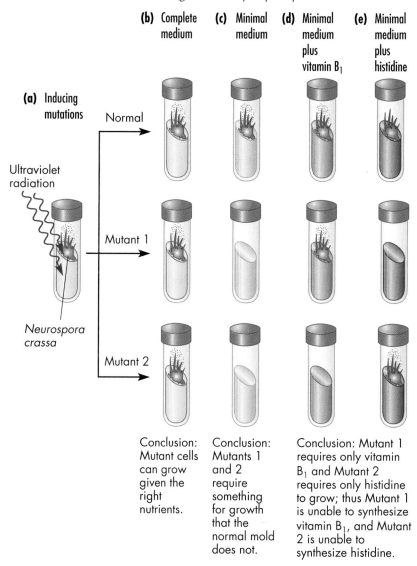

(b) Complete medium **(c) Minimal medium** **(d) Minimal medium plus vitamin B_1** **(e) Minimal medium plus histidine**

(a) Inducing mutations

Normal

Ultraviolet radiation

Neurospora crassa

Mutant 1

Mutant 2

Conclusion: Mutant cells can grow given the right nutrients.

Conclusion: Mutants 1 and 2 require something for growth that the normal mold does not.

Conclusion: Mutant 1 requires only vitamin B_1 and Mutant 2 requires only histidine to grow; thus Mutant 1 is unable to synthesize vitamin B_1, and Mutant 2 is unable to synthesize histidine.

in Figure 7.1, grow well on minimal medium plus vitamin B_1, but die on the minimal medium plus histidine. Other mutants, like Mutant 2, could grow on minimal medium plus histidine, but not on minimal medium supplemented with vitamin B_1. Beadle and Tatum concluded that some of the mutants failed to grow on minimal medium because they couldn't make specific needed growth substances for themselves. For Mutant 1, vitamin B_1 was the only substance the cells couldn't make on their own. For Mutant 2, the missing ability was making histidine. For still other mutants, it was making vitamin B_{12}, or the amino acid tryptophan, or other compounds. From these experiments, Beadle and Tatum concluded that ultraviolet light could destroy specific genes that the mold cells need for making their own vitamin B_1, histidine, or other substance.

Beadle and Tatum collected several different mutant molds that needed to be "fed" extra vitamin B_1 in their growth medium, and found that each was missing a different enzyme involved in making vitamin B_1. Each mutation, in other words, had destroyed at least one of the enzymes in the pathway for vitamin B_1 synthesis.

The One Gene–One Enzyme Hypothesis

Beadle and Tatum took stock of their results this way: Ultraviolet light caused gene mutations. If a mutant alternative of a gene (a mutant allele) causes the *absence* of enzyme function, then the normal allele is responsible for the *presence* of enzyme function. In other words, they reasoned, a gene must function by allowing a specific enzyme reaction to take place; more specifically, by causing a specific enzyme to be formed. A sound bite for their work is the phrase **one gene–one enzyme** hypothesis; this implies that each gene regulates the production of only one enzyme. Beadle and Tatum even suggested that because enzymes are proteins, each individual gene actually specifies a different specific protein. For their insightful work with *Neurospora crassa,* Beadle and Tatum won the Nobel Prize in 1958.

How could a contemporary researcher like Harry Meade apply this understanding of the gene in his search for a new blood-clotting drug? Meade knew from the work of various medical researchers that blood clots are caused when an enzyme called **thrombin** acts on a protein called **fibrinogen** to make **fibrin,** a fibrous protein (Fig. 7.2). Fibrin strands join together and make a thick, Brillo pad–like plug at the site of a wound. The protein antithrombin III (AT3) blocks the action of thrombin, prevents fibrin strands from forming, and so prevents blood clot formation. Applying the one gene–one enzyme hypothesis, Meade would predict that a single gene, the *AT3* gene, would code for the synthesis of the AT3 protein. So he would need to find and understand the human version of that gene for his research. (We'll get back to his search shortly.)

(a) How antithrombin 3 works

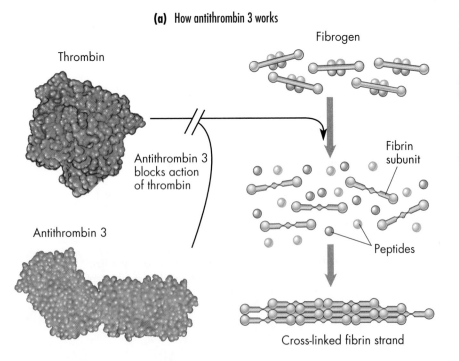

Thrombin

Fibrogen

Antithrombin 3 blocks action of thrombin

Antithrombin 3

Fibrin subunit

Peptides

Cross-linked fibrin strand

(b) Fibrin in blood clot

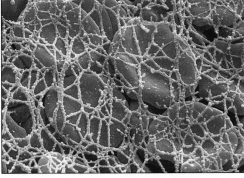

Figure 7.2 **Antithrombin III Protein and Blood Clotting.** (a) The enzyme thrombin cleaves small peptides from fibrinogen, allowing fibrinogen subunits to align in long filaments of the fibrous protein fibrin. (b) Fibrin fibers create a thick meshwork that captures blood cells, platelets, and other debris in a clot. Antithrombin III blocks the action of thrombin, and thus inhibits blood clotting.

One Gene–One Polypeptide

Beadle and Tatum showed that genes somehow code for the production of enzymes, but exactly *how?* The answer came from another classic set of genetic studies, this time of young African-American patients with an inherited blood disease.

In 1905, a young Chicagoan went to the doctor with pains in his joints and abdomen, chronic fatigue, and shortness of breath. A blood test showed that the man's blood had too few red blood cells and that many of the cells were shaped like crescents, holly leaves, or sickles, instead of the normal, round, lozenge-like disks (review Fig. 5.10). Medical researchers were learning then that sickle-shaped blood cells like his sometimes get stuck in logjams in the smallest blood vessels (capillaries), especially in the joints and abdomen. These logjams, in turn, block blood flow and cause pain. Sickle-shaped blood cells are also fragile and easily destroyed, leading to a shortage of red blood cells or **anemia.** In turn, because red blood cells carry oxygen from the lungs to the rest of the body, a shortage of these cells cuts down a person's oxygen supply and causes a feeling of breathlessness and fatigue.

You've probably already recognized that the patient had **sickle cell anemia,** a genetic condition that currently affects about 60,000 people in the United States (and about 1 out of 50 West Africans, 1 out of 400 African Americans, and higher-than-average frequency in parts of Southern Italy and Greece). We saw in Chapter 5 that sickle cell anemia is inherited as a recessive mutation: To suffer from sickle cell anemia, a child must inherit a sickle cell allele from each parent. In Chapter 9, you'll see why the disease is distributed the way it is in different places and populations.

What's important here is that sickle cell anemia led to a crucial piece of the puzzle of gene function. Scientists knew that red blood cells contain the protein hemoglobin, which carries oxygen from the lungs to the rest of the body (Fig. 7.3). They also knew that a single molecule of the hemoglobin protein consists of two pairs of amino acid chains or polypeptides: two alpha chains and two beta chains (Fig. 7.3c). If the one gene–one enzyme idea was correct, then a single gene would specify both types of chains in the hemoglobin protein. When researchers compared the amino acid sequence of normal hemoglobin with that of hemoglobin from sickled red blood cells, they found a subtle difference that caused a huge change in molecular structure. The sickled cells' beta chains differ from most regular beta chains by just 1 amino acid out of the 146 present in each chain. (The alpha chains are identical in sickle

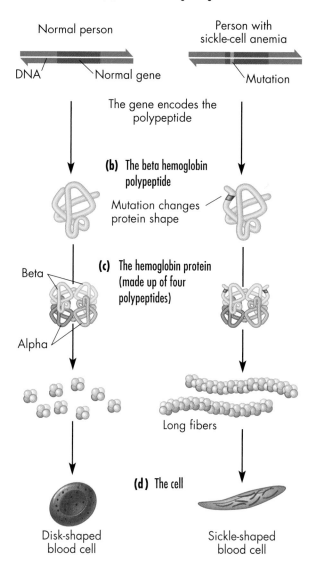

(a) The beta hemoglobin gene

Normal person

Person with sickle-cell anemia

DNA — Normal gene

Mutation

The gene encodes the polypeptide

(b) The beta hemoglobin polypeptide

Mutation changes protein shape

(c) The hemoglobin protein (made up of four polypeptides)

Beta

Alpha

Long fibers

(d) The cell

Disk-shaped blood cell

Sickle-shaped blood cell

Figure 7.3 **Studies of Sickle Cell Anemia Lead to the One Gene–One Polypeptide Hypothesis.** A change in the beta hemoglobin gene leads to an altered hemoglobin polypeptide, changed blood cell shape, and widespread effects on the body. This information revised the idea of one gene–one enzyme to one gene–one polypeptide, since a different gene encodes each polypeptide chain in a protein like hemoglobin containing several such chains.

cell hemoglobin and normal hemoglobin.) The substitution of one amino acid (glutamic acid for valine) is enough to distort the shape of the hemoglobin molecule, and the distortion causes groups of hemoglobin molecules to join together in long fibers instead of remaining separate. The fibers, in turn, change the shape of the red blood cells from normal to sickled and cause disease symptoms.

Recognizing this single amino acid substitution allowed physicians to diagnose sickle cell anemia earlier and to work toward better treatments. But in a larger genetic context, this work showed that for proteins made up of several polypeptide chains, there is generally a different gene for each *polypeptide.* This changed the one gene–one enzyme hypothesis to **one gene–one polypeptide.** Because the mutation changed the polypeptide's amino acid sequence, the work suggested something else, too: that genes specify a protein's amino acid sequence. Together, these studies revealed how a simple mutation can alter the shape of a specific polypeptide, interfere with its proper functioning, and cause the whole organism to suffer as a result.

Let's go back to Harry Meade now. Recall that during heart by-pass surgery, a patient's normal levels of the AT3 protein drop. Meade's goal was to provide doctors with a supply of that protein to help prevent patients from forming potentially fatal blood clots, one of the biggest causes of death during heart surgery. Studies have revealed that the *AT3* gene codes for the amino acid sequence in the AT3 protein's single polypeptide chain. If Meade could isolate this human *AT3* gene, he might be able to splice it into goat chromosomes and get large quantities of AT3 protein from the goat's milk. His manipulation of the gene, however, depended on a much deeper understanding of exactly *how* the *AT3* gene determines the amino acid sequence of the AT3 protein. This, in turn, rested on more basic information about how genes work as blueprints for proteins.

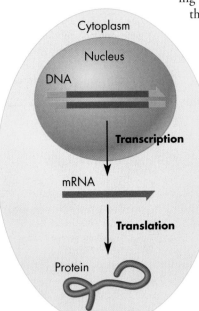

Figure 7.4 Information Flows from DNA to RNA to Protein. In transcription, enzymes copy DNA into an RNA molecule. In translation, ribosomes translate the genetic message in the type of RNA called messenger RNA (mRNA) into the amino acid chain of a protein.

7.2 How Does a Gene Specify the Amino Acid Sequence of a Protein?

Genes specify protein structure. This concept is so essential to understanding biology that in the second half of the 20th century, geneticists dubbed it the "central dogma." So let's take a look at that well-accepted truth—that the information stored in genes is played out in protein structure.

An Overview of Information Flow in Cells

Here's a shorthand way of understanding biology's central dogma: In cells, *genetic information generally flows from DNA to RNA and then from RNA to protein.* In the first step, called transcription, DNA codes for RNA. In the second step, called translation, RNA codes for the structure of a specific protein (Fig. 7.4).

(In a few instances, information can flow backwards, from RNA to DNA. This happens, for example, when the AIDS-causing virus HIV infects a human cell; see Fig. 2.30.)

DNA to RNA: Transcription

In the first step of information flow, **transcription,** enzymes copy a portion of a DNA molecule into an RNA molecule. RNA is an information-storing molecule with a structure somewhat similar to DNA, as we'll see. The word transcription implies that information in one "dialect" (the base sequence of DNA) is copied into a different "dialect" (the base sequence of RNA). Keep in mind, though, that both sets of information are "dialects" of the same language, the language of nucleic acids.

RNA to Protein: Translation

In the second step of information flow, **translation,** the informational content of one type of RNA molecule (called messenger RNA) is translated by ribosomes into specific sequences of amino acids that make up polypeptide chains. This step is called translation because genetic information is translated from the language of nucleic acids into the language of proteins.

Because transcription copies DNA to RNA, the copying takes place where DNA resides in the cell; in a eukaryote, the nucleus (Fig. 7.4). Likewise, because translation occurs on ribosomes, translation takes place where the ribosomes are located—the cell's cytoplasm. After an RNA is transcribed in the nucleus, it moves through the pores in the nuclear envelope to the cytoplasm. There the RNA binds to ribosomes, which translate the RNA into protein molecules. Cells then transport proteins to their place of action in the cell via the routes we've discussed (Fig. 2.24). Let's take just one example of this: human liver cells making the AT3 protein. After this protein is synthesized on ribosomes, tiny bubble-like vesicles carry the newly made molecules to the cell surface and release them. The AT3 molecules then travel in the blood to various body sites where AT3 regulates blood clotting.

Now that we've seen an overview of the DNA → RNA → protein process, let's take a closer look at each step. We'll start with transcription, which allows the hereditary information stored in DNA to leave the nucleus and reach the cytoplasm, where protein synthesis occurs.

7.3 Transcription: Copying DNA into RNA

A genetic engineer like Harry Meade deals with genes every day—for example, taking the *AT3* gene from a human liver cell and splicing it into a goat. So exactly what is a gene? What does he "take" and "splice"? Many biologists today would define a gene as a portion of a DNA molecule that is copied (or transcribed) into an RNA molecule plus the sequences that regulate when and where cells use the gene (Fig. 7.5). As we just saw, they, in turn, define transcription as the synthesis of a single-stranded RNA as directed by DNA. The structures of RNAs and DNA are clearly important to understanding what genes *do*. So let's compare their architectures.

Comparing DNA and RNA

Like DNA, RNA consists of a long string of nucleotides linked by sugar-phosphate backbones (Fig. 7.5). Unlike DNA, RNA can move from the nucleus to the cytoplasm for its role in protein synthesis. RNA also differs from DNA in four other ways.

1. RNA nucleotides contain the sugar ribose instead of deoxyribose, the sugar in DNA. Deoxyribose has one less oxygen atom than ribose (Fig. 7.5b). This is the only structural difference between the two sugars, but it leads to important differences in function.

2. RNA contains the base uracil (U) instead of the base thymine (T), which is found in DNA (Fig. 7.5b). Just as thymine can pair with adenine (A–T) in DNA, uracil can pair with adenine (A–U) in RNA.

3. RNA usually consists of a single strand of nucleotides (Fig. 7.5 a,b), whereas DNA usually consists of two strands. In some kinds of RNA, however, a single RNA molecule folds back on itself and forms short, double-stranded regions connected by complementary base pairs.

4. RNA molecules are much shorter than the DNA molecules that make up chromosomes. Each DNA molecule carries hundreds or thousands of genes, but an RNA molecule usually contains information from only one gene.

The Transcription Process

The transcription of a portion of DNA into RNA involves three basic steps: DNA strand separation, complementary base pairing, and nucleotide joining. Let's take the *AT3* gene as the portion of DNA and go through those three steps to see how it's transcribed into RNA in a person's liver cell. (*E*xplorer 7.1 animates these steps.)

1. **Strand separation.** As transcription of the *AT3* gene begins in the nucleus of a liver cell, enzymes unwind and separate portions of the DNA double helix near one end of the gene (Fig. 7.6, Step 1). For any given gene, only one of the two separated DNA strands serves as a template (or model) for making a complementary strand of RNA. The other DNA strand simply stays out of the way. As we saw in the chapter introduction, a certain sequence of nucleotides called the promoter defines where the RNA copying enzyme (RNA polymerase) binds to DNA and hence where a new RNA copy begins.

2. **Complementary base pairing.** Nucleotides containing the sugar ribose diffuse around the separated DNA and pair up with complementary nucleotides

GET THE PICTURE

*E*xplorer 7.1

Transcription

Figure 7.5 **The Structure of RNA.** (a) An RNA molecule consists of a ribbon of many nucleotides that can loop back and form complementary base pairs with itself. An A base pairs with U, and C pairs with G. (b) A nucleotide in RNA consists of a base (A, C, G, or U), the sugar ribose, and a phosphate group. The base shown here is U (uracil), which is found in RNA but not DNA. Compare the structure of U with that of T (see Fig. 6.4).

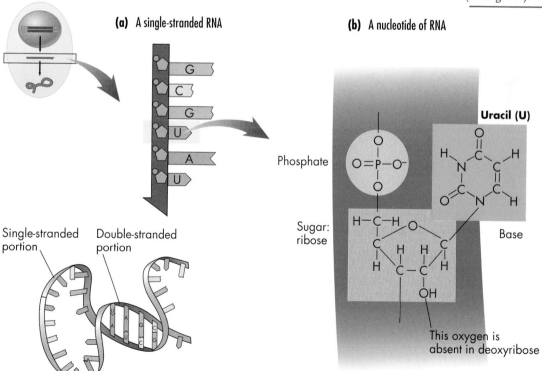

(a) A single-stranded RNA

Single-stranded portion   Double-stranded portion

(b) A nucleotide of RNA

Uracil (U)

Phosphate

Sugar: ribose

Base

This oxygen is absent in deoxyribose

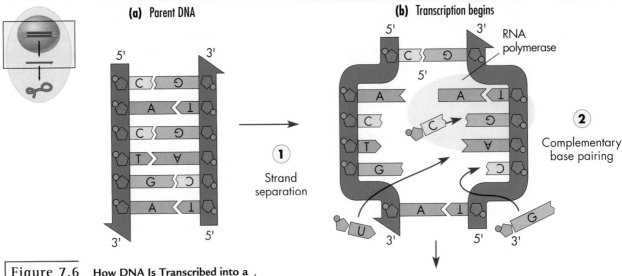

(a) Parent DNA

(b) Transcription begins

1 Strand separation

RNA polymerase

2 Complementary base pairing

Figure 7.6 How DNA Is Transcribed into a Messenger RNA.

(c) Transcription continues

Non-coding strand

Coding strand

3 Nucleotide joining

(d) Products of transcription

New RNA strand (actually several hundred base pairs long)

Parent DNA totally conserved

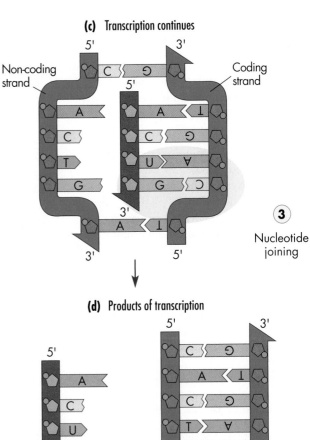

on that strand (Fig. 7.6, Step 2). (This is much like the replication of DNA we saw in Chapter 6.) In this case, however, the base A on the DNA pairs with a U nucleotide for RNA. As in DNA replication, T on the DNA pairs with an A nucleotide for RNA and C pairs with G.

3. **Nucleotide joining.** The enzyme RNA polymerase joins two adjacent RNA nucleotides together (Fig. 7.6, Step 3). Other RNA nucleotides diffuse in and pair with their complementary bases in the DNA strand, and the enzyme joins them together one by one, forming a single strand of RNA. Although the figure shows only a few nucleotides in the RNA molecule, a completed RNA molecule for the *AT3* gene actually will have a total of 1389 nucleotides. The enzyme machinery for joining new nucleotides to an RNA molecule adds about 30 nucleotides per second. So it would take a liver cell about 46 seconds to transcribe the *AT3* gene onto a single RNA molecule. The RNA polymerase finally stops making an RNA copy of the gene when it hits a "stop signal," a specific nucleotide sequence in DNA. As transcription is completed, the parental DNA rewinds. The DNA is now available to produce more copies of the same RNA. Some chemical modification may be made to the new RNA molecule (lower part of Fig. 7.6), but once completed, it passes out of the nucleus into the cytoplasm.

Comparison of Transcription and DNA Replication

If you compare Figure 6.8 to Figure 7.6, you can see that the transcription of DNA into RNA is somewhat similar to the DNA replication that takes place before a cell divides. In both cases, the two DNA strands unwind and separate, nucleotides diffuse to the site and line up by base pairing, and a polymerase enzyme joins the new nucleotides into a nucleic acid strand. Furthermore, the polymerase enzymes in both cases copy in the same direction (5′ to 3′) along a single DNA strand.

Transcription does differ from DNA replication, though, in several ways. Here are the differences in brief: (1) The entire DNA molecule—hundreds of genes—is copied during DNA replication. During transcription, however, only a portion of the DNA—often a single gene—is transcribed at any one time. (2) During DNA replication, both strands of a DNA molecule are copied. During transcription into RNA, however, only one of the two separated DNA strands is transcribed. (3) During DNA replication only one copy of each gene is made. In transcription, however, a single

gene may be copied thousands of times. The cell needs to make large quantities of certain proteins such as AT3, for example, so that they can move out of the cell to be carried throughout the body in the bloodstream. (4) Finally, as we discussed in Chapter 4, DNA replication occurs only in the S phase of the eukaryotic cell cycle. Transcription, however, occurs throughout interphase, in G_1, S, and G_2.

Introns and Exons

Harry Meade wasn't the first geneticist to isolate and study the human *AT3* gene, although he later worked with it extensively. Early workers found that the gene is located at a specific position on human chromosome 1 (Fig. 7.7a). The gene is initially transcribed into an RNA molecule about 19,000 base pairs long. This long RNA is called the primary transcript (Fig. 7.7b). Next, an amazing thing happens to this initially transcribed RNA—enzymes cut the primary transcript and discard parts of it while splicing other portions together, forming a spliced RNA (Fig. 7.7c). Biologists chose the term **introns** for the portions of the gene

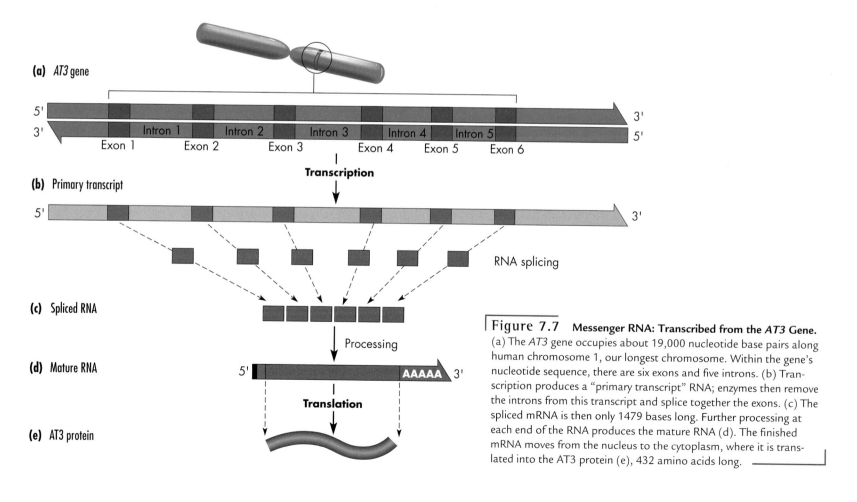

Figure 7.7 **Messenger RNA: Transcribed from the *AT3* Gene.** (a) The *AT3* gene occupies about 19,000 nucleotide base pairs along human chromosome 1, our longest chromosome. Within the gene's nucleotide sequence, there are six exons and five introns. (b) Transcription produces a "primary transcript" RNA; enzymes then remove the introns from this transcript and splice together the exons. (c) The spliced mRNA is then only 1479 bases long. Further processing at each end of the RNA produces the mature RNA (d). The finished mRNA moves from the nucleus to the cytoplasm, where it is translated into the AT3 protein (e), 432 amino acids long.

that are initially transcribed, but then are spliced out of the primary transcript. *AT3* has five introns. Biologists chose the separate term **exons** for the portions of the gene that are transcribed and then spliced together and appear in the final spliced RNA. *AT3* has six exons. Introns *intrude* into the gene but don't appear in the final RNA and so they aren't translated into protein. In contrast, exons are *expressed*—they are usually translated into protein. Because introns are removed, the spliced RNA from the *AT3* gene is only 1479 base pairs long—just 11 percent of the primary transcript's original length. Most genes in eukaryotic cells have introns. In contrast, prokaryotic cells generally don't have these extra DNA segments in their genes. (Surprisingly, some introns act as ribozymes, RNAs with the ability to catalyze their own splicing, like enzymes made of RNA instead of protein. We'll run into ribozymes again in Chapter 10 in our discussion of the origins of life.) Introns are important because, by facilitating the swapping of exons, they may help proteins evolve new functions. In the *AT3* gene, the spliced RNA undergoes further processing on each end and becomes the mature RNA. RNA then leaves the nucleus and heads for the cytoplasm, where it is translated into the AT3 protein.

Harry Meade had to keep all these details about introns, exons, and RNA splicing in mind while designing a new method to induce goats to make human AT3 protein. We'll see how he did all of this a bit later. First, though, let's explore the details of the translation process itself, starting with the types of RNA.

How does "pharming" employ genetic translation?

7.4 Translation: Constructing a Polypeptide

Transcription actually makes several types of RNA molecules and all three different types are involved in translation, the synthesis of polypeptide chains that become active proteins or parts of proteins. Here are the three types and what they do.

Types of RNA

Protein synthesis—say, the building of a blood-clot blocking protein like AT3—requires three kinds of RNA: messenger RNA, transfer RNA, and ribosomal RNA. All three are made in the nucleus by gene transcription and then move to the cytoplasm.

Messenger RNA

Messenger RNA, or mRNA, carries genetic information from DNA in the nucleus to the ribosomes in the cytoplasm, where it directs protein synthesis. We just saw, for example, how the *AT3* gene is transcribed into RNA. This RNA becomes mRNA when the introns are edited out (see Fig. 7.7). Mature mRNA then leaves the nucleus for synthesis of the AT3 protein in the cell cytoplasm. Because mRNA has a sequence of nucleotide bases that is complementary to a gene on the DNA molecule, it carries the same information as the gene but in complementary form. Messenger RNA molecules are generally from about 1000 to 10,000 nucleotides long, and this type of RNA makes up only 5 percent of the total RNA in the cell.

Transfer RNA

Transfer RNA, or tRNA, first picks up amino acids in the cytoplasm and then aligns them on a ribosome in the exact order specified by the information in the mRNA. These functions (picking up amino acids then ordering them) occur at opposite ends of the tRNA molecule. At one end of the tRNA, a specific enzyme catalyzes a reaction that attaches the tRNA to one of the 20 amino acids (Fig. 7.8a). At the other end of the tRNA, a group of nucleotides joins the tRNA to a specific position in the RNA (Fig. 7.8b).

Each tRNA molecule carries only one specific kind of amino acid. Because there are 20 different amino acids, there must be at least 20 different tRNAs (some amino acids are transported by more than one kind of tRNA). Molecules of tRNA are only about 75 nucleotides long, much shorter than mRNAs.

Codons and Anticodons

During protein synthesis, sequences of three adjacent bases on mRNA specify the insertion of a particular amino acid into the growing polypeptide chain (Fig. 7.8b). Biologists call a **codon** a set of three nucleotides in mRNA that specifies the position of an amino acid in a protein. For example, the first codon in the mRNA for antithrombin III protein is AUG, the second is UAU, and the third is UCC (Fig. 7.8b).

Since tRNA recognizes a codon in mRNA, it has to have a sequence complementary to the codon. This portion is called an **anticodon**, and is three adjacent nucleotides in tRNA that bind to a codon in mRNA (Fig. 7.8b). For example, a tRNA that carries the amino acid serine has, at the opposite end, the anticodon AGG. The AGG anticodon pairs up with the UCC codon on an mRNA. The pairing of codon with anticodon is the key to how amino acids line up in the right sequence during protein synthesis. (We'll get back to this shortly.) The order of the codons in mRNA specifies the order in which tRNAs will bring over and line up specific amino acids. The lined-up amino acids are then joined to each other by the ribosome, which contains the third type of RNA.

Ribosomal RNA and Ribosomes

Ribosomal RNA, or **rRNA,** is a key component of ribosomes, the microscopic "workbenches" on which proteins are forged. Ribosomal RNA works together with messenger RNA and transfer RNAs to translate genetic information into proteins. Eukaryotic ribosomes are about one-half protein and one-half rRNA. Ribosomal RNA is the most abundant type of RNA (80 percent) in the eukaryotic cell.

As the cell's workbenches, the ribosomes support mRNAs and tRNAs all the while genetic information is being translated. They also help to link amino acids to each other in a growing polypeptide chain. Approximately 75 proteins are wrapped around the rRNAs, forming the beadlike structures of ribosomes (Fig. 7.9).

Each ribosome consists of one large and one small subunit separated by a groove (Fig. 7.9a). An mRNA strand "threads" through the groove, and the ribosome moves along the mRNA like a pulley on a rope. As it moves, the ribosome joins amino acids to the growing polypeptide chain one by one in the order specified by codons in the mRNA. Recent evidence suggests that rRNA itself helps to catalyze the addition of each amino acid to the next. Usually, several ribosomes at a time move down a single mRNA molecule like cars in a caravan, each ribosome producing a single polypeptide chain (Fig. 7.9b).

Cells contain tens of thousands of ribosomes, located wherever proteins are being synthesized; this can be in the cytoplasm of eukaryotic cells, or in the DNA-containing region of prokaryotic cells. (In prokaryotes, translation of an mRNA begins at one end even before the other end of the mRNA has been completely transcribed.)

Translation: Protein Synthesis

Once James Watson and Francis Crick discovered the elegant double helix structure of DNA (see Chapter 6), the secret of the molecule's information storage capacity could be unlocked: it lies in the order of the nucleotide

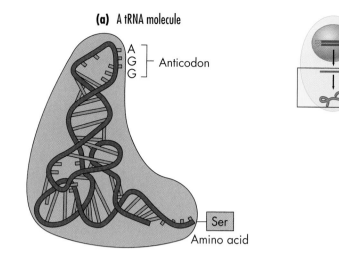

(a) A tRNA molecule

Anticodon: A G G

Ser — Amino acid

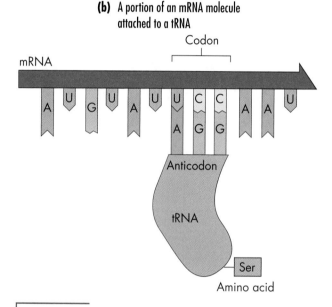

(b) A portion of an mRNA molecule attached to a tRNA

Codon

mRNA — A U G U A U U C C A A U

Anticodon — A G G

tRNA

Ser — Amino acid

Figure 7.8   Transfer RNA: Amino Acid Carriers.
(a) Each transfer RNA is a ribbon about 75 bases long that loops back on itself and folds in space into a boot-like form held together by complementary base pairing. At one end of each tRNA is its specific anticodon, in this case, AGG; at the other end is attached the amino acid corresponding to this anticodon, here, Ser (serine).
(b) The tRNA's anticodon forms base pairs with a codon in mRNA (in this case, UCC).

bases. Later work then showed that the sequence of nucleotides transcribed from DNA into mRNA determines the sequence of amino acids in proteins. Protein synthesis is one of the most important topics in biology, so let's explore it with the AT3 protein as our example.

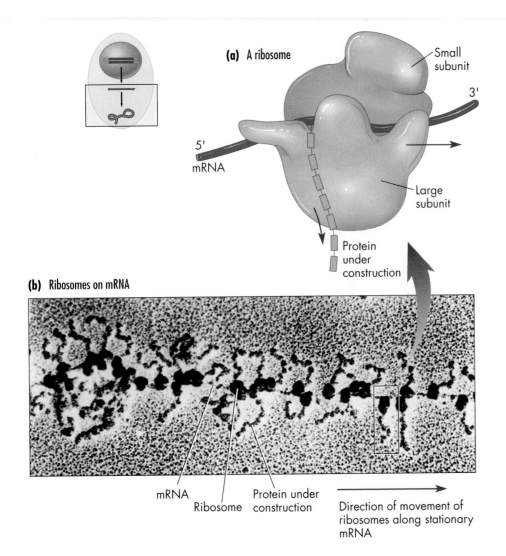

(a) A ribosome

Small subunit

3'

5'

mRNA

Large subunit

Protein under construction

(b) Ribosomes on mRNA

mRNA

Ribosome

Protein under construction

Direction of movement of ribosomes along stationary mRNA

Figure 7.9 **Ribosomes: Sites of Protein Synthesis.** (a) The large and small ribosomal subunits consist of RNAs and associated proteins. The ribosomal subunits join to an mRNA, then slide along the mRNA like a pulley on a rope, constituting a moveable "workbench" for adding amino acids into a polypeptide sequence. (b) The electron micrograph shows ribosomes sliding along an mRNA. An ever-elongating polypeptide chain dangles from each ribosome. This example of translation is from the salivary glands of *Chironomus,* a type of gnat (magnification 17,500×).

the first amino acid of most polypeptides. It binds to the **start codon,** AUG, on the mRNA.

During **elongation,** the next tRNA diffuses in, lining up its amino acid with the previous one (Fig. 7.10b), and the ribosome joins the amino acids together by peptide bonds (Fig. 7.10c). Simultaneously, the first tRNA is ejected from the complex as the ribosome moves one codon along the mRNA (Fig. 7.10c). Elongation continues, as the ribosome joins more and more amino acids together. Figure 7.10 shows just 3 amino acids being joined, but for the AT3 protein your liver produces, the elongation steps are repeated over and over until the polypeptide chain is 464 amino acids long. Amazingly, it takes only about 20 seconds for a protein of this size to be forged on the ribosomes in your liver cells.

Termination occurs when the ribosome reaches a stop codon in mRNA (Fig. 7.10d). In the case of AT3, the stop codon is the three nucleotides UAA. At that point, a termination factor protein binds to the ribosome–mRNA complex rather than a tRNA, and this brings the growth of the polypeptide chain to a halt.

In the final step, **disassembly,** the newly formed polypeptide falls away from the ribosome, and the workbench components disassemble (Fig. 7.10e).

You can see the central dogma at work in Figure 7.10: The sequences of nucleotide bases in mRNA specify the order of tRNAs, each one toting a particular kind of amino acid. Just as DNA carries a genetic code, so too does each mRNA, with bases complementary to those in DNA.

As you read this, cells in your liver are probably cranking out messenger RNAs bearing the genetic information from the *AT3* gene in your DNA that are ready to join with ribosomes and begin making the anti-blood-clotting protein. Here are the stages.

Stages of Protein Synthesis

The building of a polypeptide chain for a protein involves four stages: initiation, elongation, termination, and disassembly. Figure 7.10 pictures each of these stages and **E**xplorer 7.2 animates them.

During **initiation,** the machinery needed for protein synthesis assembles itself to form the workbench, the ribosome (Fig. 7.10a). Specifically, the two ribosomal subunits come together on the mRNA, and the first tRNA floats over and binds to the mRNA. This first tRNA totes along the amino acid methionine, which is

The Genetic Code

We've seen how a human liver cell can assemble an AT3 protein. But if Harry Meade splices the human gene for AT3 into a goat, will the goat cells translate the human mRNA into an identical protein? Do different

(a) Initiation

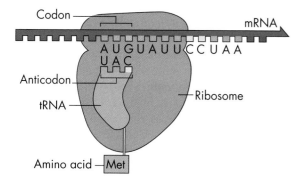

Codon

mRNA

A U G U A U U C C U A A
U A C

Anticodon

tRNA

Ribosome

Amino acid — Met

(b) Elongation

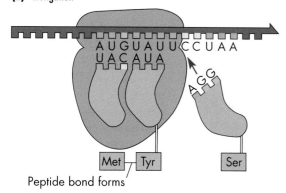

A U G U A U U C C U A A
U A C A U A

A G G

Met — Tyr   Ser

Peptide bond forms

(c) Elongation continues

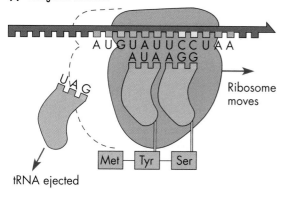

A U G U A U U C C U A A
A U A A G G

U A G

Ribosome moves

Met — Tyr — Ser

tRNA ejected

(d) Termination

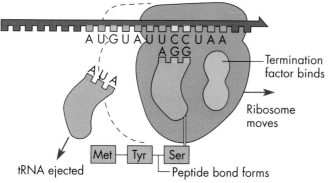

A U G U A U U C C U A A
A G G

A U A

Termination factor binds

Ribosome moves

Met — Tyr — Ser

tRNA ejected

Peptide bond forms

(e) Disassembly

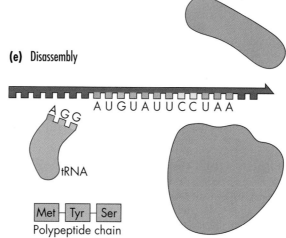

A G G

A U G U A U U C C U A A

tRNA

Met — Tyr — Ser

Polypeptide chain

Figure 7.10 **The Main Events of Protein Synthesis.** (a) Initiation. Several components assemble before translation begins, including an initiator tRNA carrying the amino acid methionine, the small and large ribosomal subunits, and the mRNA. The tRNA's anticodon (UAC) binds to the mRNA's start codon (AUG). (b) Elongation. A second tRNA carrying (in this case) tyrosine lines up and binds to the mRNA; now the first two amino acids methionine and tyrosine are side by side and each is still attached to its tRNA. (c) Elongation continues. The bond between the first tRNA and methionine breaks, and the ribosome forges a new bond between the methionine and serine. The first tRNA, now minus its amino acid, is ejected from the mRNA and recycled; the ribosome moves down the mRNA one codon. These elongation steps now repeat with more tRNAs ferrying in new amino acids for the growing chain until the complete polypeptide is formed. (d) Termination. At the end of the message, the ribosome reaches a stop codon along the mRNA such as UAA. Since there is no tRNA for the stop codon, a termination factor (a certain protein) binds in place of tRNA and this halts translation. (e) Disassembly. With protein synthesis completed, the "workbench" disassembles. The two ribosomal subunits separate from each other and release the mRNA, and the polypeptide is cleaved from the last tRNA molecule. The polypeptide chain can then float away to do its job in the cell, and the machinery can reassemble and start a new round of protein synthesis from another mRNA molecule.

species, in other words, speak the same genetic language? The answer is yes, and it's based on the nature of the genetic code.

The **genetic code** is like a dictionary cells use to define which amino acids will be translated from each sequence of three adjacent bases in mRNA—in other words, from each codon. We've seen that RNAs have an alphabet of only 4 letters (the bases A, C, G, and U), while proteins have 20 amino acids. How can these 4 nucleotides code for all 20 different types of amino acids found in proteins?

In trying to understand the genetic code, biologists reasoned that if each individual nucleotide coded for only one amino acid, then there would only be four possible amino acids in proteins. We know, though, that proteins have 20 different amino acids. If a series of two bases encoded each amino acid (for example, AU for one, CA for another, and so on), the system would encompass 16 amino acids (4 × 4), still not enough. Biologists realized that RNA "words" made up of combinations of 3 nucleotides would allow for 64 possible combinations (4 × 4 × 4), and this would be more than enough to code for the 20 amino acids observed in proteins. In other words, even though the RNA alphabet contains only 4 letters, it could spell out 64 different three-letter words (codons) just as the 26 letters of our alphabet can be combined to make hundreds of thousands of words.

Since there are 20 amino acids in nature, this assumption seemed correct. But was it? And how could biologists figure out which codon translates into which amino acid? Experimenters in the laboratory created artificial genetic messages with RNA and mixed them in test tubes with ribosomes and other translational machinery that they harvested from bacterial cells. They made the artificial message UUUUUUUUUUUU, for example, and found that the "machinery" translated it into the polypeptide Phe–Phe–Phe–Phe. From this they learned that the codon UUU specifies the amino acid phenylalanine. By working through the entire set of 64 codons, they were able to work out the specific amino acid that each codon encodes. This is the genetic code shown in Figure 7.11.

Look closely at that figure: Is each amino acid coded for by a single codon? Or are there amino acids that are encoded by more than one codon? Are there any codons that encode more than one amino acid? Do all codons encode amino acids? Through this short exercise, you've probably already discovered that serine (Ser), taking one example, is encoded by six different codons UCU, UCC, UCA, UCG, AGU, and AGC. Obviously, the genetic code is redundant, and this has an impact on certain genetic mutations. The mutation of a UCU codon to UCC, for example, would not change the amino acid encoded because both UCU and UCC encode serine. This mutation, then, would be unlikely to affect the organism.

Perhaps your inspection of the table also uncovered this fact: Although an amino acid can have more than one codon, there are no codons that encode more than one amino acid. Each codon "spells out" just one single amino acid. In addition, three codons—UAA, UAG, and UGA—spell "stop" instead of specifying amino acids; they are the stop codons we saw earlier, and they terminate a growing polypeptide chain (review Fig. 7.10d). As we also saw earlier, AUG is the start codon and it usually encodes methionine wherever it occurs.

Codon			Amino acid	Codon			Amino acid	Codon			Amino acid	Codon			Amino acid
U	U	U	Phe	A	U	U	Ile	C	U	U	Leu	G	U	U	Val
U	U	C	Phe	A	U	C	Ile	C	U	C	Leu	G	U	C	Val
U	U	A	Leu	A	U	A	Ile	C	U	A	Leu	G	U	A	Val
U	U	G	Leu	A	U	G	Met (START)	C	U	G	Leu	G	U	G	Val
U	C	U	Ser	A	C	U	Thr	C	C	U	Pro	G	C	U	Ala
U	C	C	Ser	A	C	C	Thr	C	C	C	Pro	G	C	C	Ala
U	C	A	Ser	A	C	A	Thr	C	C	A	Pro	G	C	A	Ala
U	C	G	Ser	A	C	G	Thr	C	C	G	Pro	G	C	G	Ala
U	A	U	Tyr	A	A	U	Asn	C	A	U	His	G	A	U	Asp
U	A	C	Tyr	A	A	C	Asn	C	A	C	His	G	A	C	Asp
U	A	A	STOP	A	A	A	Lys	C	A	A	Gln	G	A	A	Glu
U	A	G	STOP	A	A	G	Lys	C	A	G	Gln	G	A	G	Glu
U	G	U	Cys	A	G	U	Ser	C	G	U	Arg	G	G	U	Gly
U	G	C	Cys	A	G	C	Ser	C	G	C	Arg	G	G	C	Gly
U	G	A	STOP	A	G	A	Arg	C	G	A	Arg	G	G	A	Gly
U	G	G	Trp	A	G	G	Arg	C	G	G	Arg	G	G	G	Gly

Figure 7.11 **The Genetic Code.** This dictionary shows how just four ribonucleotide bases (U, C, A, and G) can combine to make 64 codons, each 3 nucleotides long. It also gives the genetic "meaning" for each one. For example, UCU equals Ser, or serine; AUG equals Met or methionine, the start codon, and starts the reading frame. Three codons are stop signals (UAA, UAG, and UGA). (Abbreviations of amino acids: Phe, phenylalanine; Leu, leucine; Ser, serine; Tyr, tyrosine; Cys, cysteine; Trp, tryptophan; Ile, isoleucine; Met, methionine; Thr, threonine; Asn, asparagine; Lys, lysine; Arg, arginine; His, histidine; Gln, glutamine; Val, valine; Ala, alanine; Asp, aspartic acid; Glu, glutamic acid; Gly, glycine.)

Reading the Genetic Message

Geneticists studying the genetic code got a surprise when they tested a genetic message with a repeating triplet motif, such as AGCAGCAGCAGCAGC. They expected this message to translate into one polypeptide, but in fact, it encoded three different ones, each made up of just one kind of amino acid: either all serine (codon AGC), all alanine (codon GCA), or all glutamine (codon CAG). This test tube experiment makes sense only if translation can start at any place along the message:

…AGC AGC AGC AGC AGC…

…A GCA GCA GCA GCA GC…

…AG CAG CAG CAG CAG C…

Experiments like this proved that once translation starts at one point, it does indeed read off bases in groups of three and without overlapping or skipping bases. The experiments also showed that the starting place for translation determines the meaning of the message. The genetic message AGCAGCAGCAGCAGC can be divided into codons in the three ways you just saw and each different way is called a **reading frame.** How does a reading frame determine the meaning of a message? Just consider for a moment howc hangesi nreadingf ramec ana lterthism essages pelledo utinEnglis hletters.

A cell needs to "know" where to start reading, and cells do have a special codon that signifies where translation should start; thus it establishes the reading frame. This start codon, AUG, also codes for the amino acid methionine (see Fig. 7.10a). In fact, methionine is the first amino acid in nearly every polypeptide chain.

The Genetic Code is Almost Universal

Returning to Dr. Meade's research problem, do human cells and goat cells have the same genetic code? Yes. Meade observed that the human *AT3* gene, transcribed and translated in goat cells, produces the exact sequence of human AT3 protein. Humans and goats clearly must use the same genetic code. Furthermore, earlier experiments in bacteria, plants, and in many other life forms revealed that nearly all organisms use that same genetic code. (A few codons do differ from the norm in certain mitochondria and some protists.) It is an amazing testament to the unity of all living things that our human cells "speak" exactly the same genetic language as bacterial cells. Each codon is translated into the very same amino acid, whether the codon resides in a human liver cell, a goat mammary gland cell, or even an *E. coli* bacterial cell. The nearly universal nature of the genetic code is expected if all living things hark back to the same ancestral cells that originated very early in earth's history. The genetic code was apparently developed at the dawn of life and has been inherited intact for billions of years.

Gene Mutation

Dr. Meade had been working for years on an anticlotting drug based on large quantities of the protein AT3. One use for the drug would be heart bypass patients who tend to form blood clots during surgery. But there are other potential uses, too, including the treatment of people with a mutation of the *AT3* gene. With our background of discussions to this point, we can now define a **mutation** as a change in the base sequence of an organism's DNA. People with a mutation in the *AT3* gene can't form the AT3 protein, and thus tend to develop blood clots at inappropriate locations. These clots can block the blood supply to various tissues and cause severe, life-threatening diseases.

Geneticists recognize two general categories of mutations. **Chromosomal mutations** affect large regions of chromosomes, or even entire chromosomes, and so alter the locations of many genes. Chromosomal mutations include changes in chromosome structure or number (as in Down syndrome; see Fig. 4.15). **Single-gene mutations,** or changes in the base sequence of a single gene, alter individual genes, sometimes disrupting gene function, but sometimes having no effect at all.

Kinds of Single-Gene Mutations

We covered several types of chromosome mutations in Chapters 4 and 5, so this section focuses on single-gene mutations.

Base Substitution Mutation   A common sort of mutation called a **base substitution mutation** occurs when one base pair replaces another. For example, in constructing the AT3 protein, the 129th codon is normally CGA for the amino acid arginine (Fig. 7.12a). In certain human families, however, this codon is changed to CAA, which encodes glutamine (Fig. 7.12b). This amino acid alteration changes the shape of the AT3 protein so that it does not inhibit clotting anymore and the family members often inappropriately form clots in veins. In certain other families, a base pair substitution in the same DNA codon changes the original CGA to TGA. The complementary codon in the mRNA then changes to UGA,

(a) Original DNA

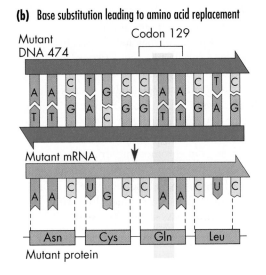

(b) Base substitution leading to amino acid replacement

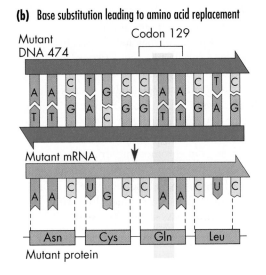

(c) Base substitution leading to premature termination

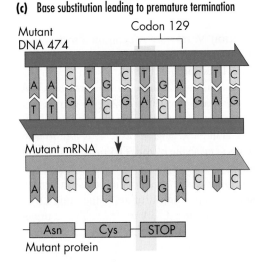

(d) Original DNA

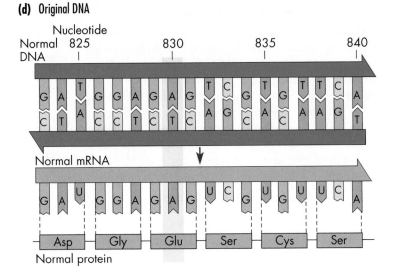

(e) Base deletion leading to a frame shift

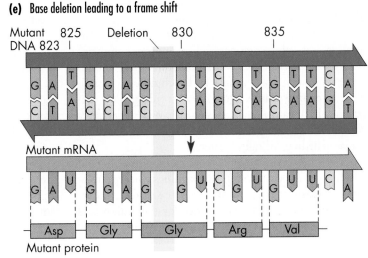

Figure 7.12 **Mutations in Genes.** A mutation can change the genetic message in a normal gene (a). In a base substitution (b), one base pair is replaced by another. For example, in most people's *AT3* gene, codon 129 is CGA encoding the amino acid arginine (Arg). In some, however, the CGA changes to CAA specifying Gln (glutamine) and disrupting the structure of the AT3 protein. No longer able to inhibit the thrombin, the altered protein can allow life-threatening clots to form. (c) Another type of base substitution in the same codon changes the original CGA into TGA. This becomes UGA in the mRNA—the stop codon—and it halts translation when it is reached. The result is an abnormally short AT3 protein and abnormal blood clotting. (d) Elsewhere in the *AT3* gene, the nucleotide at position 830 normally contains an A. (e) In some people, this A at position 830 is deleted, changing the reading frame and affecting the sequence of the rest of the protein as it forms. The resulting AT3 protein is very abnormal and it, too, can lead to abnormal clotting.

which means STOP (Fig. 7.12c). As a result, the AT3 protein they make is too short to do its job.

Base Deletions and Insertions We just saw how base substitutions change one base pair into another. In contrast, **base deletions** and **base insertions** remove or add base pairs to the gene. For example, some people have a single base pair deleted from their *AT3* gene, making the gene one base pair shorter than normal (Fig. 7.12d,e). Changes like this can alter the reading frame of an mRNA, garbling the amino acid sequence from the point of the mutation all the way to the end of the protein. Shifts in reading frame can be caused by the deletion or insertion of one or two base pairs, but not *three*. Can you see why?

The Origin of Mutations

Mutations can obviously be destructive. So where do they come from? Mutations arise either through spontaneous errors that occur as DNA is replicating, or later, through damage to the DNA by physical or chemical agents. As DNA is replicating, the wrong base can be inserted into the growing helix. Usually, a repair by the DNA polymerase enzyme edits out this kind of error. Sometimes, though, physical and chemical agents called **mutagens** change DNA structure. Mutagens include ultraviolet rays from the sun, chemicals in cigarette smoke, and even many natural substances from plants or fungi. For example, aflatoxin, a natural compound found in moldy peanuts, is an extremely potent mutagen.

Despite the mutagens in our environment, certain enzymes usually repair the type of DNA damage they cause. When these enzymes themselves are defective, however, the consequences can be dire. Some people inherit the skin disease xeroderma pigmentosum and have a mutation that blocks the body's ability to form a critical DNA repair enzyme. As a result, their cells quickly accumulate mutations. One young Navajo girl, for example, suffered from this condition and had dozens of skin cancers where UV light from the sun has mutated the DNA in her skin cells, and the cells weren't able to repair these mutations.

Cancer-causing substances are known as **carcinogens** and they often act by generating mutations as well. A mutagen has the potential to cause cancer if it affects a gene that controls cell growth. Because people and bacteria have such similar DNA metabolism, nearly anything that can cause mutations in bacteria can also cause mutations in people. Because of this link, researchers have invented a test for the carcinogenicity (cancer-causing ability) of chemical and physical agents that exposes certain bacteria to suspected carcinogens and then measures the bacterial mutation rates. Researchers have used this test, called the **Ames test,** to catalog thousands of substances that are mutagenic and also potentially carcinogenic. The list includes asbestos, ultraviolet light, nitrates in foods that become carcinogenic during cooking, aflatoxin from moldy peanuts, and other food contaminants. People who expose themselves to known mutagens, such as too much sunlight or the toxins in cigarette smoke, place a heavy burden on their DNA repair enzymes and magnify their risk of developing cancer.

If Dr. Meade's company succeeds in producing enough human AT3 protein for a new drug, it may benefit people who form too many blood clots due to mutations in their *AT3* genes. Meade must also be sure that in manipulating the human *AT3* gene, he and his colleagues don't cause new mutations that could lead to a defective protein. Mutation is an important element in both gene action and in genetic manipulation. And as we saw in Chapter 1, and will see again and again in the book, mutations are the source of genetic variation and of evolutionary change. But mutation is not the only difficult issue in genetic engineering. There is another one involving the basic control of gene action: How can you get a protein made in the human liver (AT3) to be made in a goat's mammary gland and then excreted in its milk?

How can kids make human AT3 protein?

7.5   How Are Genes Controlled?

Normally, the liver is the only tissue that makes large amounts of the anticlotting protein AT3. Before Meade could splice the gene into a goat and get the animal to make AT3, he needed a very thorough understanding of how genes are regulated in cells. Luckily, researchers have extensive knowledge of **gene regulation,** the process that controls how and when each gene is turned on and off in each living cell.

Gene Regulation in Prokaryotes

Biologists first studied gene regulation in bacteria and, as a result, understand gene control in those organisms in the greatest detail. They've learned that bacteria have streamlined and highly sophisticated mechanisms of controlling gene expression and this allows them to change the kinds of proteins they make quickly and frequently. Let's consider, for example, the *E. coli* bacteria growing in your intestine, and let's say you just finished eating a bowl of yogurt or ice cream. The bacteria in your intestine are now floating in a sea of lactose, the main sugar in milk. In order to use that sugar, the bacteria quickly make enzymes that help break down lactose once it is taken into the bacterial cell.

A bacterial cell that uses resources efficiently can grow and divide quickly. But an *E. coli* cell that makes unnecessary proteins—say, the enzymes for breaking down lactose when that sugar is not present—expends energy and materials that could go toward cell reproduction; thus, it grows and divides more slowly. As simple as bacterial cells are in structure and function, they have **operons,** elegant gene regulation systems consisting of regulated clusters of genes acting in a coordinated function. Operons allow the cells to make only the types and amounts of proteins they need when and where they need them.

The best-studied operon regulates the way cells break down lactose. Around 1950, biologists knew that an enzyme called beta-galactosidase in *E. coli* cells breaks down lactose into the two simple sugars glucose and galactose. *E. coli* cells growing in the absence of lactose (for instance, your intestines *before* you eat yogurt) contain none of this enzyme. However, in the presence of lactose (right after the yogurt), each bacterial cell produces thousands of beta-galactosidase molecules within about 20 minutes. Somehow, lactose *induces* the formation of the sugar-digesting enzyme as well as two other functionally related proteins.

In 1960, through elegant experiments, French geneticists Francois Jacob and Jacques Monod determined how *E. coli* bacteria regulate the synthesis of beta-galactosidase. They found that a certain protein called the **repressor** in the *E. coli* cell can sense lactose in its surroundings. If lactose is absent, the repressor protein prevents the transcription of mRNA for the beta-galactosidase enzyme (Fig. 7.13a). If lactose is present, how-

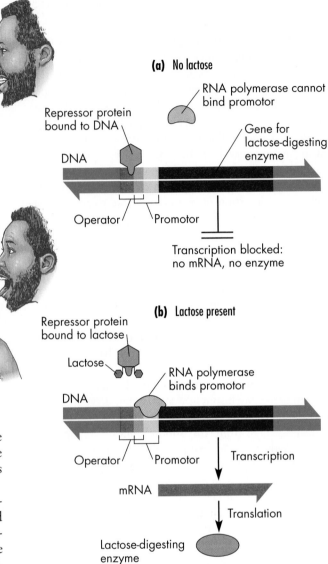

(a) No lactose

RNA polymerase cannot bind promotor

Repressor protein bound to DNA

Gene for lactose-digesting enzyme

DNA

Operator   Promotor

Transcription blocked: no mRNA, no enzyme

(b) Lactose present

Repressor protein bound to lactose

Lactose

RNA polymerase binds promotor

DNA

Operator   Promotor   Transcription

mRNA

Translation

Lactose-digesting enzyme

(c) The lactose-repressor protein bound to the operator

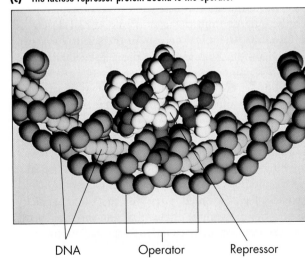

DNA   Operator   Repressor

Figure 7.13   **Gene Regulation in a Bacterial Cell.** A normal bacterial cell inside your digestive tract makes enzymes for digesting lactose only when that sugar is present in your body. (a) When there is no lactose, a repressor protein binds to the DNA inside the bacterial cell at the operator, a region of DNA near the gene for the digestive enzyme beta-galactosidase. The repressor physically blocks RNA polymerase from binding to the promoter, and transcribing the gene. Without transcription, no mRNA forms, and no protein (in this case, beta-galactosidase) can be made from this gene. (b) When lactose is present, however, it binds to the repressor protein, changing its shape and preventing the repressor from binding to the operator. After the repressor is out of the way, RNA polymerase can bind to the promoter, and transcribe the gene into mRNA. When this mRNA is translated, the cell makes beta-galactosidase, which can digest lactose. (c) A computer image shows how the repressor wraps an arm around one of the grooves in DNA and hence blocks transcription. Here DNA is shown in rust and gold, and the repressor protein in white, blue, red, and green.

ever, the repressor protein allows the DNA to start churning out copies of mRNA for the enzyme (Fig. 7.13b). Careful experiments showed that the repressor acts by binding to DNA. Specifically, the repressor, in the absence of lactose, binds to the **operator,** a series of base pairs in DNA near the gene that encodes the lactose-digesting enzyme (Fig. 7.13a,c). When the repressor binds to the operator DNA, it covers a portion of the DNA called the promoter, the base pairs to which RNA polymerase binds. (Recall that RNA polymerase is the enzyme that copies DNA into RNA.) Because RNA polymerase can't bind the promoter, there is no transcription. Without transcription, there can be no mRNA. And without mRNA, there can't be an extra lactose-digesting enzyme. But when the bacteria are growing on lactose, that sugar binds to the repressor (Fig. 7.13b), and the binding alters the shape of the repressor so that it can no longer bind to the operator. This then unblocks the promoter. Now RNA polymerase can bind to the promoter and transcribe the gene, and this transcription, in turn, is translated into the lactose-digesting enzyme. The bacterium can now use the energy from lactose for growth. Jacob and Monod gave the name "operon" to the protein signaler, the operator DNA sequence to which it binds, the promoter to which RNA polymerase binds, and the group of genes transcribed together in a single RNA in that part of the genome. This work was so fundamental to understanding how genes are controlled that Jacob and Monod won the Nobel Prize in 1965.

Gene Regulation in Eukaryotes

Back now to the question of how multicellular organisms turn on certain genes in certain tissues so that a person, let's say, makes AT3 in her liver but not in her mammary glands, and milk protein (casein) in her mammary glands but not her liver. All your body cells, in other words, have the same genes in the cell nucleus but only certain genes are available for transcription in each cell. Why? It turns out that eukaryotic cells like your own have regulatory mechanisms similar to the ones we just saw in bacteria: Regulatory proteins bind special regions of DNA called promoters, regulatory elements, and enhancers (Fig. 7.14a). This binding then alters the action and the rate of gene transcription in a way that is specific to each cell type. So by applying these principles of eukaryotic gene regulation, we can see that Harry Meade had to isolate the regulatory portion of the casein gene and somehow hook it up to the coding portion of the *AT3* gene (Fig. 7.14b).

We've finally arrived, now, at the place where Dr. Meade was able to use all his background understand-

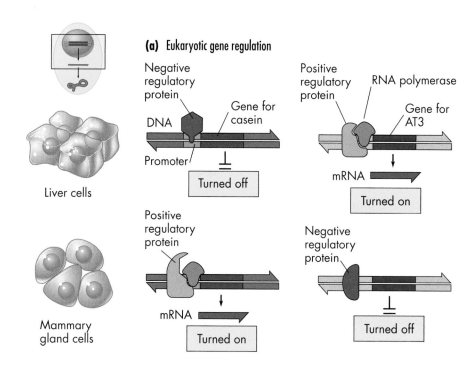

(a) Eukaryotic gene regulation

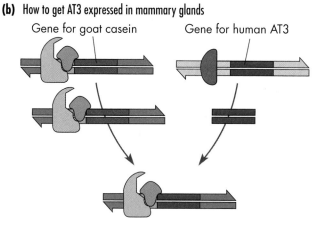

(b) How to get AT3 expressed in mammary glands

Figure 7.14 **Regulating Eukaryotic Genes: *AT3* and Goat's Milk.** Each cell in a multicellular organism has a complete set of genetic information but different cells use different sets of genes and within the same cell, different genes are switched on at different times during development. (a) Every human cell has both the genes for AT3 and for the milk protein casein, but casein is turned on only in a nursing mother's mammary gland cells and *AT3* is turned on only in liver cells. (b) Harry Meade combined the regulatory region of the goat casein gene with the protein encoding section of the human *AT3* gene.

ing on gene function, gene regulation, and protein synthesis along with the modern genetic tools of **recombinant DNA technology,** methods for combining DNA segments (usually from different species) in the laboratory. Let's see how he did it.

7.6 Recombinant DNA Technology

DNA molecules recombine with each other in nature (Chapter 4). Bottom line, that's what sex is about. But viable sexual reproduction exchanges corresponding DNA fragments *within a single species*. In the early 1970s, biologists realized that they could deliberately recombine DNA molecules from different species in the lab. Here's how.

Constructing Recombinant DNAs

The first step in any attempt to recombine DNA from different species is to produce millions of copies of a single desirable piece of DNA, that is, to **clone** the DNA. Because Harry Meade wanted to combine a human *AT3* gene with the goat casein promoter he started with molecular cloning.

Molecular cloning is accomplished by splicing the desired DNA fragment (a gene, a promoter region, or what have you) into a **vector,** which is usually a virus or plasmid that occurs in nature. A **plasmid** is a simple circle of DNA, about 3500 base pairs long, that replicates independently inside a bacterial cell (Fig. 7.15). That independent replication makes them handy for genetic engineers and so do several other traits: Plasmids contain genes that make a bacterial cell resistant to certain antibiotics, such as tetracycline and ampicillin. These can serve as a marker, as we'll see. Plasmids can also move readily from one bacterial cell to another. And they can accept pieces of foreign DNA, carry them along, and replicate them wherever they go. So a geneticist could insert a gene like *AT3* into a plasmid for cloning.

It's important to keep in mind that the same properties that make plasmids so useful in the genetics lab make them a disaster in hospitals, because they can carry antibiotic resistance genes from one bacterial species to another and one patient to another. As resistance spreads, antibiotics can very quickly become ineffective at preventing disease. Chapter 9 explores plasmids and antibiotic resistance in detail.

Plasmids and cloning are just two tools of recombinant engineers. They also have special enzymes to cut DNA molecules in specific places and paste the fragments together, as well as ways to insert new DNA into foreign cells.

Cutting the DNA

To cut DNA molecules, researchers use special molecular "scissors," proteins called **restriction enzymes** that they isolate from bacteria (Fig. 7.16b). Bacteria naturally produce restriction enzymes, and use them to cut certain viral DNAs into pieces; this restricts the types of viruses that can infect them. A restriction enzyme recognizes a few specific base pairs in a row wherever they occur in a DNA molecule. The enzyme then cleaves the DNA at a consistent place in or near the recognized sequence. Some restriction enzymes cut the DNA in a staggered fashion, so that one strand of the double helix sticks out beyond the other (Fig. 7.16). When the weak hydrogen bonds between the strands break, the two DNA fragments can separate, and single-stranded ends protrude (Fig. 7.16c).

Joining the DNA

Biologists call the little tails protruding from the two ends of a staggered cut "sticky ends." These ends act "sticky" in that they can easily re-form hydrogen bonds with complementary base pair sequences that stick out on the ends of other DNA molecules. In fact, any two DNA molecules with complementary sticky ends can join together, no matter how unrelated are the rest of the two DNA molecules. By making these sticky ends on pieces of DNA from organisms as different as humans, goats, and bacteria, biologists can recombine the DNA molecules from the different species. Figure 7.16c shows in dark blue a small portion of the human *AT3* gene cut by a restriction enzyme called *Eco*RI. The figure shows in pale blue a DNA fragment from a bacterial plasmid similarly cut by the same restriction enzyme.

Hydrogen bonds can quickly form between the AATT base sequence of the human DNA and the complementary TTAA sequence of the plasmid DNA, and can hold the two fragments together (Fig. 7.16d). The biologist then adds the enzyme **DNA ligase,** which acts like a glue dispenser to cause strong covalent bonds to form between the opposing ends of the two molecules. (In nature, DNA ligase repairs damaged DNA.)

Transgenic Goats that Make Human AT3 Protein in their Milk

Harry Meade and his collaborators had cloning, plasmids, molecular "scissors" and a "glue dispenser" to use

Figure 7.15   A Plasmid: The Gene Engineer's Beast of Burden. A plasmid is a circle of DNA that can proliferate in a bacterial cell. Plasmids float free in the bacterial cell, as you can see in the cut-away view. Harry Meade splices human or goat DNAs into plasmids. Once inside a bacterium, the foreign DNA will replicate along with the original plasmid DNA, thus making millions of copies of the mammalian gene.

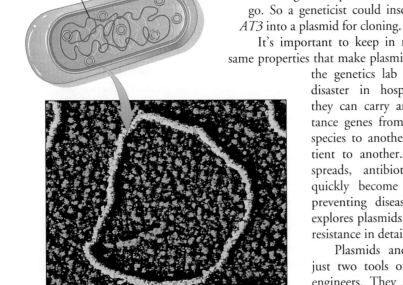

Plasmid DNA

Bacterial DNA

Figure 7.16 **How Geneticists Create Recombinant DNA Molecules.** Biologists make recombinant DNA molecules by cutting two different parental DNA molecules with molecular "scissors"—a restriction enzyme (a, b). Next, they mix DNA fragments, each with sticky ends (c), to allow pairing of complementary bases at the ends of the molecules. Finally, biologists use a molecular glue dispenser—the enzyme DNA ligase—to join the fragments together and create a recombinant DNA (d).

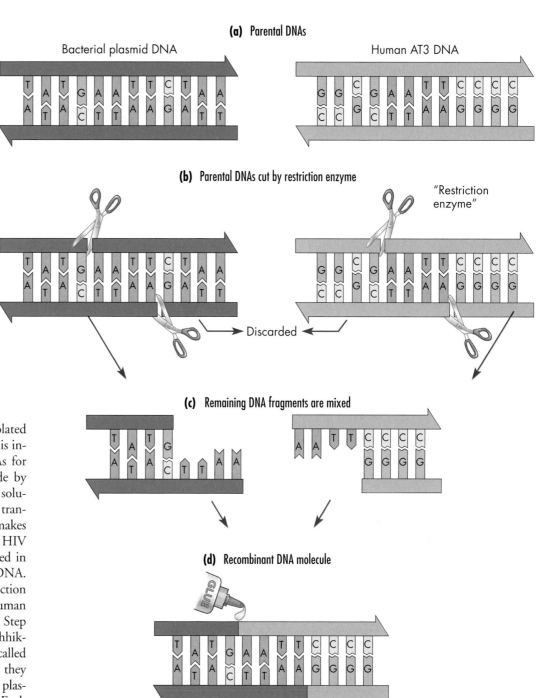

(a) Parental DNAs

Bacterial plasmid DNA

Human AT3 DNA

(b) Parental DNAs cut by restriction enzyme

"Restriction enzyme"

Discarded

(c) Remaining DNA fragments are mixed

(d) Recombinant DNA molecule

DNA ligase joins fragments

in constructing a recombinant DNA molecule, and their work required three phases. In the first phase, they cloned the *AT3* gene. In the second phase, they spliced the human *AT3* to goat casein promoter. In the third phase, they introduced the recombinant gene into goat cells and produced the desired human protein.

Phase 1: Cloning AT3

To clone *AT3*, Harry Meade and colleagues first isolated RNA from human liver cells (Fig. 7.17, Step 1). This included mRNA for the *AT3* gene, but also mRNAs for albumin and for thousands of other proteins made by liver cells. In Step 2, they mixed this mRNA with a solution of free base pairs and a special enzyme, reverse transcriptase. (Recall from Chapter 2 that this enzyme makes a DNA copy from an RNA template, as with the HIV virus in an infected cell; see Fig. 2.30.) This resulted in double-stranded complementary DNA copies, or cDNA. Next, they cut plasmids and the cDNA with a restriction enzyme to open the circle, and then inserted the human DNA fragments into plasmid molecules (Step 3). In Step 4, they transferred the plasmids, along with the hitchhiking human DNA, into bacterial cells, a process called **transformation** (review Fig. 6.2). Finally, (Step 5) they cultured the bacterial cells, which now contain the plasmid combined with human DNA, on a petri dish. Each cell divided repeatedly and produced a colony, a clone of cells with millions of identical copies, all descended from the same founder cell with the same DNA. Because there are thousands of different genes expressed in liver cells, each bacterial colony can have a cDNA from a different human gene. We said earlier that plasmids can carry antibiotic resistance. In this case, the researchers added an antibiotic to the bacterial growth medium so that any bacterial cell without the plasmid inside would die and not form a colony.

Eventually, thousands of bacterial colonies grew on the plate, and Meade and colleagues had to find the one (or ones) containing the *AT3* gene. In Step 6, they made a **probe** for the *AT3* gene, a radioactive nucleic acid that is complementary to the gene. They

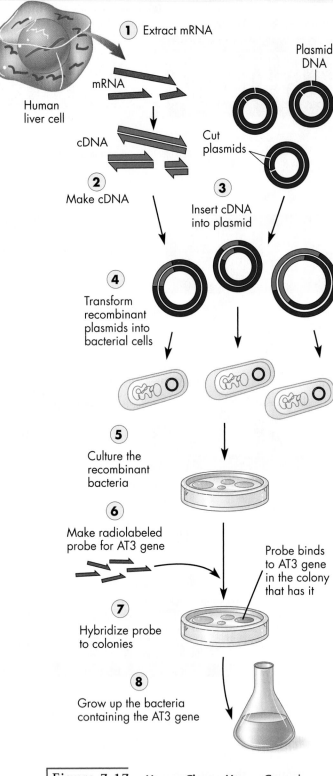

① Extract mRNA

Human liver cell

mRNA

cDNA

② Make cDNA

Plasmid DNA

Cut plasmids

③ Insert cDNA into plasmid

④ Transform recombinant plasmids into bacterial cells

⑤ Culture the recombinant bacteria

⑥ Make radiolabeled probe for AT3 gene

Probe binds to AT3 gene in the colony that has it

⑦ Hybridize probe to colonies

⑧ Grow up the bacteria containing the AT3 gene

Figure 7.17   How to Clone a Human Gene.

didn't know the gene's precise DNA sequence but because they had the AT3 protein, they knew its amino acid sequence. By using the genetic code in reverse, they could write down DNA sequences that could encode a specific portion of the AT3 amino acid sequence. They could then artificially synthesize a radioactive DNA with this sequence. In Step 7, they could let this labeled DNA bind (or hybridize) to its complementary DNA in the colonies, specifically that colony (and only that one) producing the *AT3* gene. Once they identified and isolated this colony (Step 8), they had millions of copies of the human *AT3* gene in the plasmids of millions of bacterial cells. They had cloned the gene.

Phase 2: Getting Proper Expression of the Cloned Gene

Using similar methods, Meade's group isolated and cloned the goat casein gene, including its promoter sequence. With molecular scissors and glue, they cut the genes to leave sticky ends on both, and then joined them together (see Fig. 7.16b–d). If they were lucky, the casein regulatory sequences would drive expression of the *AT3* gene during protein synthesis, so that when the goat made milk (including the most abundant protein in milk, casein), it would make human anticlotting protein as well.

Phase 3: Introducing Recombinant DNA into Goats

Now it was time to introduce the recombinant molecule into living goats. The researchers first isolated plasmids containing the casein–*AT3* recombinant gene from the bacteria, sucked up the recombinant DNA into a very small needle, and injected squirts of it into a number of goat embryos (each one a single cell and not yet divided into the first two, four, or eight cells of the embryonic animal). In preparation for the injection, they had flushed the embryos from the reproductive tracts of donor nanny goats (Fig. 7.18, Step 1). Next they transferred the DNA-injected embryos into the uteruses of surrogate nannies (recipient females) (Step 2), and allowed the embryos to develop into full-term baby goats. In Step 3, they tested the newborn kids for the presence of the injected DNA. Because injected DNA is rarely and randomly incorporated into goat chromosomes, most kids with the recombinant DNA will have it in one place in a single chromosome and hence be heterozygous for the desired DNA. About 5 to 10 percent of the offspring contained the recombinant

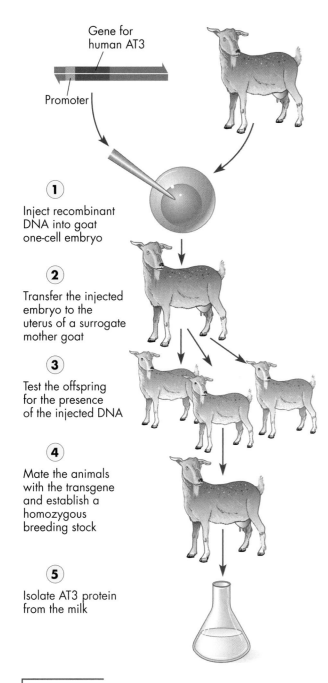

1 Inject recombinant DNA into goat one-cell embryo

2 Transfer the injected embryo to the uterus of a surrogate mother goat

3 Test the offspring for the presence of the injected DNA

4 Mate the animals with the transgene and establish a homozygous breeding stock

5 Isolate AT3 protein from the milk

Gene for human AT3

Promoter

| Figure 7.18   **Making Transgenic Goats.** |

DNA and could be called **transgenic** animals (animals bearing foreign genes inserted by recombinant DNA methodologies). They raised these transgenic goats and then mated pairs of them. Billy goats, obviously, can't produce milk like their sisters, but they're useful for breeding.

When the transgenic females became pregnant and gave birth, they started producing milk and many of them produced the human AT3 protein in their milk. Some of their newborn kids also had the recombinant gene in double doses—they were homozygous and were also transgenic—a second generation (Step 4). Finally, the research team started collecting milk from the lactating nannies and purifying the protein from the milk (Step 5). With enough of the AT3 protein, it's now possible to produce a prescription drug that can block inappropriate clotting during heart surgery or in other patients. So far, Genzyme Transgenics has over 1800 transgenic goats, each able to produce about 800 liters of goat's milk per year containing over 1 kilogram of AT3 protein. That's enough to create hundreds of doses of anticlotting medicine per goat per year. By using methods similar to those that produced Dolly the Cloned Sheep (details in Chapter 8), Meade hopes to generate an even bigger herd of AT3-producing goats.

7.7   Promises and Problems of Recombinant DNA

Harry Meade's work and that of other genetic engineers shows the tremendous promise of recombinant technology for humankind. Imagine a drug, available in great supply for only a few dollars, that can block the viruses that cause the common cold. Or an enzyme that can arrest a heart attack already in progress. Or bacteria that can make automobile fuel from discarded corn stalks. Or other bacteria that can protect crops from late-spring freezes. Imagine corn that contains proteins with the nutritional value of beef proteins. Or gene replacements to cure people and animals of crippling genetic diseases. Or engineered mice that can be used to screen for and identify cancer-causing agents.

These sound useful—even life-saving. But now consider the potential downsides to genetic engineering. Bacteria could be engineered to break down oil, for example, and then used to disperse oil slicks, but what if these hypothetical bacteria moved beyond oil slicks and began to devour the world's dwindling oil supplies? If we can engineer goats with desirable traits, will people try to engineer their children with traits they or society deem desirable? Will foods containing foreign genes be safe over many years of consumption? Many people worry about the social and ecological impacts of recombinant DNA research. So let's explore the promise, the performance, and the nagging questions surrounding this new field.

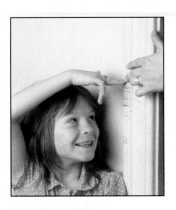

Figure 7.19 A Growth Spurt Thanks to Genetic Engineering. As the marks on the door frame show, this child grew several inches during treatment with growth hormone made through genetic engineering techniques.

How Genetic Engineering May Reshape Life

Genetic engineers can take three basic approaches. They can alter bacteria so that the prokaryotic cells produce a specific protein, such as human insulin, to treat disease. They can alter an organism to make it more useful, for example, generate goats that produce human AT3 protein in their milk, or engineer corn or cattle to grow faster and resist diseases better. Third, geneticists can alter the human genome itself to cure disease.

Mass-Producing Proteins

Many children, like the 10-year-old girl in Figure 7.19, are born unable to produce growth hormone, and until recently, they were usually destined to live out their lives as very short individuals. This girl, however, received human growth hormone and in just one year had grown about 12.6 cm (5 in). The growth hormone drug she received was produced by bacteria genetically engineered to make human growth hormone. Physicians hope someday to use the precious hormone to treat burns, slow-healing fractures, and bone-loss diseases, as well.

Biotechnologists have also engineered bacterial cells to make bovine (cow) growth hormone. When injected into dairy cows, this hormone increases milk production by 10 to 20 percent. The U.S. Department of Agriculture has certified that milk from treated cows is safe to drink, but some observers note that it requires more antibiotics to keep treated cows healthy and that some of the antibiotics, as well as some of the hormone itself, may end up in the milk. This advance might seem like a boon to the dairy industry, but using a product of recombinant DNA may have a significant economic downside. Agricultural economists have pointed out that farmers already produce a surplus of milk products. Given this surplus, competition from large milk-producing operations, which can more easily afford to use the expensive growth hormone drug, may drive more and more small family-owned dairies out of business.

Genetically altered cells are also producing a number of other medically useful proteins. These include: (1) alpha-1-antitrypsin, a protein that can help patients with the lung-destroying disease emphysema; (2) insulin, a protein that can help diabetics; (3) granulocyte colony stimulating factor, which can stimulate the body to form new white blood cells that help fight infection; and (4) interferon gamma, which can protect laboratory-grown human cells from infection by hepatitis and herpes viruses, and perhaps stimulate the growth of tumor-killing cells. This partial list of early successes only hints at the many useful products that genetically engineered microbes and transgenic animals may ultimately provide for us. Others that could be available soon include targeted antibodies (immune system "torpedoes") for fighting Crohn's disease, rheumatoid arthritis, nerve disorders, organ transplant rejection, and AIDS; and particular proteins for making artificial blood, fighting malaria, and treating diabetes in novel ways.

Improving Plant and Animal Stocks

For centuries, farmers have been genetically manipulating agricultural animals, but recombinant DNA does a similar thing in accelerated fashion. Farmers in ancient human societies domesticated cattle, pigs, goats, sheep, ducks, chickens, and other animals, selecting individuals with desirable traits to be parents of the next generation. These desirable traits arose by mutation and by naturally occurring DNA recombination. With modern genetic engineering technology, researchers can insert into livestock specifically selected genes for faster growth, leaner meat, stronger bones, higher milk production, and resistance to diseases, as well as other traits such as encoding particular therapeutic proteins like human AT3. These could increase our food supply substantially. So can the genetic engineering of more nutritious rice, corn, tomatoes, and other plants (which we cover in detail in Chapter 22).

Human Gene Therapy

Altering a person's genes to combat disease, a process called **gene therapy,** is a field still in its infancy. In theory, researchers could use gene therapy in either of two ways: (1) They could insert genes into the somatic (body) cells, or (2) they could insert genes into the germ cells (the cells that give rise to the sperm and eggs). The first procedure, somatic gene therapy (see Chapter 5) is the straightforward treatment of an individual's disease, a simple extension of current medical practice. In Chapter 5, we saw Dr. Zuckerman apply this approach to treating cystic fibrosis by having patients inhale gene solutions into their airways; from there, the genes could move into affected lung cells.

Physicians carried out the very first instance of somatic gene therapy years earlier in 1990 when they used the technique to treat a woman with malignant melanoma, a fast-growing cancer that often begins with a dark, irregular shaped mole that grows rapidly. The woman's cancer had spread through her bloodstream, and she had tumors in her chest wall, abdomen, and limbs that had failed to respond to normal cancer treatments. Physicians at the National Cancer Institute removed from her tumor a type of white blood cell (tumor-infiltrating lymphocyte, or TIL) that naturally homes in on a tumor mass. They then inserted a tumor killing gene (tumor necrosis factor, or TNF) into the white blood cells and infused them back into the patient's body. Within nine months of the first treatment, most of the tumors had either regressed or disappeared (Fig. 7.20). Results like these provide hope that **somatic cell gene therapy** like this and like the lung treatments for cystic fibrosis will help many kinds of patients.

Gene therapy has serious risks too, however. In 1999, 18-year-old Jesse Gelsinger died 4 days after receiving gene therapy as part of an experiment. Jesse had OTC deficiency, an inherited disease with a long name that blocks the normal processing of nitrogen during protein breakdown. Medical researchers thought they could help him by inserting the gene for OTC protein into his body cells. They used a virus as a vector to carry the gene for the OTC protein into Jesse's liver cells. But he apparently had an overwhelming immune response to the engineered virus. The unfortunate death of this courageous teenager led to a high profile investigation and has caused researchers to proceed with much more caution.

Most people have few ethical objections to this sort of gene therapy, because the altered lung, white blood cells, and liver cells are somatic cells. But in the second approach, **germ-line gene therapy,** the modified DNA goes into the sex cells and thus would also affect the genetic makeup of the patient's future children. Because germ-line gene therapy constitutes a totally new approach to medicine, it raises complex issues of safety and ethics. In germ-line gene therapy, medical researchers would inject recombinant DNA into human sex cells using the tools Harry Meade employed to make transgenic goats (see Fig. 7.18). This would potentially affect not only treated individuals but also their descendants. While researchers like Dr. Meade have succeeded with germ-line gene therapy in mice and goats, they must overcome numerous technical and ethical problems before trying it with human sex cells. So far, the success rate is low—only 6 successes out of 300 injected eggs in a typical mouse experiment—far too low to try with human embryos. In addition, the inserted DNA sometimes causes new mutations, possibly creating new defects—again, an unacceptable outcome in humans. Finally, the inserted normal gene (human *AT3*, for example) doesn't usually replace native defective genes, but is simply added to the genome. Defective genes, therefore, like Jesse's OTC deficiency, would still be inherited by the treated individual's children.

Right now, germ-line gene therapy is impossible for purely technical reasons. Someday, however, it will be achievable and will pose significant long-term ethical questions. Because its application would alter the genetic makeup of children not yet born, it could change the course of human evolution. Would a treatment that causes a heritable change infringe on the rights of future generations? Are the expected benefits of germ-line gene therapy worth the unknown risks? Could a problem be solved in more traditional ways, whose risks are understood and whose benefits are clearer? These and other questions have led scientists and lay observers to debate—at times heatedly—the potential misuse, accidental or deliberate, of recombinant DNA techniques.

Recombinant DNA: Environmental Risks

Some critics have suggested that introducing recombinant organisms into the environment poses significant risks and professional ecologists (scientists who study the interactions of organisms with each other and with their physical surroundings) agree that some dangers do exist. A dangerous new weed could conceivably be created if, say, a new variety of rice engineered with genes for salt tolerance escaped from cultivated fields and invaded the brackish water in the mouths of rivers. A weed might become a worse pest if it interbred with a genetically engineered domestic crop—say, sorghum and the related weed johnsongrass, or squash and wild gourds—and gained a gene for herbicide resistance. (Chapter 23 discusses this concern in detail.)

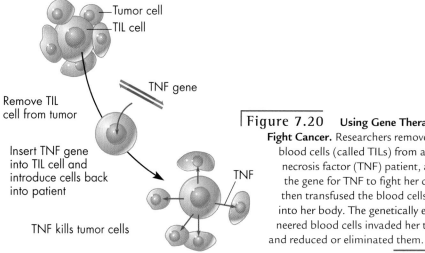

The process of gene therapy

Tumor cell
TIL cell

Remove TIL cell from tumor

TNF gene

Insert TNF gene into TIL cell and introduce cells back into patient

TNF

TNF kills tumor cells

Figure 7.20 **Using Gene Therapy to Fight Cancer.** Researchers removed white blood cells (called TILs) from a tumor necrosis factor (TNF) patient, added the gene for TNF to fight her cancer, then transfused the blood cells back into her body. The genetically engineered blood cells invaded her tumors and reduced or eliminated them.

Their caution stems from the many ecological disruptions that have occurred with natural, unengineered organisms. A prime example involves kudzu, a fast-growing, large-leafed Japanese vine that farmers introduced into the southeastern states to control soil erosion. Kudzu did this to some degree but it also ended up choking tens of thousands of acres of native and commercial forests (Fig. 7.21). Professional ecologists have urged that plants and animals containing foreign genes be carefully evaluated by government agencies before being released into the environment. A strong international protest movement has already reacted and tried to prevent the sale of genetically altered foods, although their use is already very common. By some estimates, over half of all products on supermarket shelves already contain genetically modified ingredients, mostly corn and soy engineered to resist insects and herbicides. In the fall of 2000, grocers pulled most taco shells from store shelves because they were mistakenly made with corn engineered to resist insects, but certified only for consumption by cattle. (Again, we'll talk about genetically modified foods in detail in Chapters 22 and 23.)

Gene Transfers: Ethical Dilemmas

Questions about both the safety and morality of recombinant DNA research have cast a shadow over its promise of a better future for the human race. Researchers have found that inserting a human growth hormone gene into mice causes them to grow larger. How long will it be before prospective parents ask physicians to insert additional growth hormone genes into their newly conceived embryo so they can produce an Olympic athlete or a commandingly tall politician?

Other parents will want children with greater intelligence or more winning personalities. We haven't a clue, at this point, how to use recombinant DNA to change complex traits like these, which are controlled by untold numbers of genes. Geneticists may *never* be able to find and clone all the genes that influence intelligence or personality. However, society should debate

Figure 7.21 **Novel Organisms, Unexpected Environmental Consequences.** Farmers brought the kudzu vine (which is not genetically engineered) to the American South and planted it widely to control erosion. The hardy plant competed with native vegetation and is now blanketing some forests and killing the trees.

these issues long before biotechnologists look for them. And if history repeats itself, the potential for human gene therapy will arrive far more quickly than we expect.

If we want higher yield crops, better livestock, and healthier humans through recombinant DNA technology with a minimum of risk, we're going to need free and open discussion for years to come and carefully arrived-at guidelines. You can participate in this debate yourself in *E*xplorer 7.3.

READ ON

*E*xplorer 7.3

Genetic Engineering Issues: Take a Stand

Chapter Summary and Selected Key Terms

Introduction Harry Meade used the tools of genetic engineering to help people with life-threatening blood-clotting disorders. He combined a human gene for the antithrombin III protein (AT3) with the control

sequence or **promoter** (p. 157) from a goat's milk protein gene. He and coworkers introduced the recombined DNA into the goat and the DNA expressed itself when the goat produced milk.

How Do Genes Control Cellular Activities?
Our current knowledge of gene activity was based on several historic experiments. Beadle and Tatum studied the bread mold *Neurospora crassa* to

learn what genes do. They induced gene mutations in the mold and then tested to see what capabilities got knocked out in the process. They generated nutritional mutants that couldn't make their own vitamin B_1 or histidine, and in this way, devised the **one gene–one enzyme hypothesis** (p. 160), the idea that each gene regulates the production of only one enzyme. This was further refined by work on sickle cell anemia, a disease caused by a mutation that affects one type of polypeptide chain in the hemoglobin protein in blood. Because one altered gene leads to one changed polypeptide, and not necessarily the entire enzyme or other protein, biologists adopted the **one gene–one polypeptide hypothesis** (p. 162) and it remains today. (Sometimes a protein is made up of just one polypeptide chain, but often it contains several.)

How Does a Gene Specify the Order of Amino Acids in a Protein?

Many important experiments led to biology's central dogma, a generalized picture of gene function in the cell: Genetic information flows from DNA to RNA to protein.

Transcription: DNA to RNA

Transcription (p. 162) is the synthesis of a single-stranded RNA under the direction of DNA. The DNA molecule is different from RNA in several ways. RNA is single stranded; has ribose instead of deoxyribose; contains uracil (U) instead of thymine (T); and RNA molecules are much shorter than DNA. RNAs usually contain genetic information from only one gene, while DNA carries information from hundreds or thousands of genes.

During transcription, the DNA strand separates and complementary nucleotide bases diffuse in and pair up (with U substituting for T). The nucleotides are joined and an RNA transcript forms. Usually, this transcript is far longer than the gene itself and contains **introns** (p. 165), or stretches of DNA that intervene between the **exons** (p. 166), or expressed gene sequences. The introns are cut out of the RNA transcript, leaving only the exon regions in the finished **messenger RNA (mRNA)** (p. 166). This molecule carries genetic information from the DNA in the nucleus to the cytoplasm, where it directs protein synthesis.

Translation or Protein Synthesis: RNA to Protein

After mRNA leaves the nucleus, it interacts with other types of RNA. **Transfer RNA (tRNA)** (p. 166) participates in **translation** (p. 162): It picks up amino acids and aligns them on a ribosome in the exact order specified by the mRNA. A sequence of three adjacent bases on an mRNA is called a **codon** (p. 166). The complementary sequence in a tRNA is an **anticodon** (p. 166). **Ribosomal RNA (rRNA)** (p. 167) is a key component of ribosomes, the workbenches on which proteins are forged. One large and one small subunit come together in the ribosome, and the total unit moves along the mRNA molecule, "reading" its information. The stages of translation are **initiation, elongation, termination,** and **disassembly** (p. 168).

During **protein synthesis** (p. 167–169) the two ribosomal subunits, an mRNA, and a tRNA carrying methionine bind together. During elongation, the amino acids join together by peptide bonds and this lengthens the polypeptide chain one link at a time. During termination, the ribosome reaches a stop signal in the mRNA. A termination factor binds to the ribo-some and this halts protein synthesis, after which the new polypeptide chain falls away.

Different species speak the same language, based on the **genetic code** (p. 169), a set of codons that translate only into particular amino acids. Codons must contain three bases, not one or two, in order to account for all 20 amino acids in nature. The code is redundant in that some amino acids are specified by more than one codon. Stop codons (UAA, UAG, UGA) spell stop and when they are reached in the mRNA, protein synthesis terminates. A **start codon** (p. 168) establishes the correct **reading frame** (p. 171) within the string of nucleotide bases.

A **mutation** (p. 171) is a change in the base sequence of an organism's DNA. **Chromosomal mutations** (p. 171) affect large regions of chromosomes. **Single-gene mutations** (p. 171) alter individual genes. These can involve **base substitutions** (p. 171), **base deletions** (p. 173), or **base insertions** (p. 173). Mutations can arise naturally as mistakes during DNA replication, or they can be caused by **mutagens** (p. 173) such as UV light or the chemicals in cigarette smoke. **Carcinogens** (p. 173) cause cancer in animals and are often mutagens, as well. This fact allowed researchers to create a mutagen test in bacteria, the **Ames test** (p. 173), to screen for potential mutagens and also cancer-causing substances.

How Are Genes Controlled?

A process called **gene regulation** (p. 173) controls each cell's gene activity. In prokaryotes, it involves **operons** (p. 174), or gene control systems that allow the cell to make proteins only when and where they are needed. Eukaryotes have promoters, repressors, operators, and other elements involved in turning gene transcription on and off at appropriate times (p. 175). These alter the act and rate of gene transcriptions so that genes are turned on and off at appropriate times in different kinds of cells.

Recombinant DNA Technology

Harry Meade used a special set of methods called **recombinant DNA technology** (p. 175) for combining DNA segments from different species in the laboratory. The first step is to **clone** (p. 176), or make millions of identical copies of, the desired DNA segment. Biologists often employ **plasmids** (p. 176) or circles of DNA from bacterial cells as a vehicle to carry cloned DNA. The genetic engineer must cut open the plasmid and must also cut out the desired gene sequence of DNA from some other source. To do this, he or she uses a **restriction enzyme** (p. 176), which cuts DNA and leaves sticky ends or unpaired sequences. The biologist then uses **DNA ligase** (p. 176), an enzyme that acts like a molecular glue dispenser, to rejoin cut DNA sequences.

To create a recombinant goat that can produce human AT3 protein, Harry Meade started by cloning the *AT3* gene, inserting it into plasmids, then transferring the plasmids into bacterial cells, a process called **transformation** (p. 177). Next he joined the goat casein promoter sequence to the human *AT3* gene, then injected the recombinant DNA into goat embryos to create **transgenic** animals (p. 179), animals bearing foreign genes. He and his colleagues created a large herd of transgenic goats and can now collect their milk and purify out the human AT3 protein.

Problems and Promises of Recombinant DNA

Recombinant DNA research has the capacity to mass produce useful proteins such as human growth hormone, AT3, insulin, interferon, and others. Biologists can use the technology to make agricultural animals and plants larger, faster-growing, more nutritious, and more resistant to disease. They can also use recombinant technology for **gene therapy** (p. 180), to insert genes into the body cells or into the germ cells. Medical researchers have already tried several instances of **somatic gene therapy** (p. 181) for patients with cystic fibrosis, malignant melanomas, OTC deficiency, and other diseases. Because of the risk to patients, however, researchers are proceeding with caution. Harry Meade used **germ-line gene therapy** (p. 181) to produce the transgenic goats, but this technique has not been applied to humans. Many observers object to its use in humans because it will change not only the patient's genes but his or her unborn children's genes, as well. Ecologists worry about releasing recombinant organisms into the environment for fear of creating super weeds, super predators, and other ecological imbalances. A widespread debate and careful regulation of the field are needed to prevent problems but allow the benefits of the technology.

All of the following question sets also appear in the Explore Life *E* electronic component, where you will find a variety of additional questions as well.

Test Yourself on Vocabulary and Concepts

In each question set below, match the description with the appropriate term. A term may be used once, more than once, or not at all.

SET I

(a) mRNA (b) tRNA (c) rRNA (d) codon

1. A sequence of three adjacent nucleotides in RNA that specifies an amino acid in a protein.

2. A nucleic acid that is found associated with several protein molecules and is essential to the construction of all kinds of cellular proteins.

3. A nucleic acid that is sometimes bonded to an amino acid.

4. A linear nucleic acid that is sometimes found paired with DNA and at other times paired with numerous smaller molecules of RNA.

SET II

(a) base substitution (b) reading frame (c) redundant (d) stop (e) deletion

5. A protein that is changed by the inclusion of one wrong amino acid is the result of a _____ mutation.

6. The genetic code is said to be _____ because there may be several tRNA molecules that have different anticodons but transport the same amino acid to the ribosome.

7. If a mutation results in the production of a protein that is significantly shorter than normal, a _____ mutation may have caused the change.

8. Deletion or insertion of one or two bases in DNA will most likely cause a change in the _____.

9. A sequence of three adjacent bases in mRNA for which no complementary tRNA pairing can occur is called a _____ codon.

SET III

(a) operon (b) repressor (c) exon (d) intron

10. A group of genes in bacteria that are adjacent and controlled in a coordinated fashion.

11. A length of DNA within a gene that codes for a section of mRNA that is removed after transcription and before translation.

12. A protein that has the ability to bind to a specific DNA sequence, thereby blocking the transcription of the adjacent gene.

13. A length of DNA within a gene that is transcribed into mRNA and remains intact and functional during translation.

SET IV

(a) restriction enzyme (b) DNA ligase (c) plasmid (d) clone (e) transgenic organism

14. A group of cells, all of which trace their ancestry to a single progenitor cell.

15. A protein that acts as scissors for cutting DNA.

16. A molecular glue gun that mends breaks in DNA.

17. A ring of genetic material that can infect a cell and become integrated in the cell's chromosome, and can be used as a vehicle for moving desired DNA fragments into or out of cells.

Integrate and Apply What You've Learned

18. Name two important differences between DNA replication and RNA transcription.

19. How does RNA differ from DNA, and how do the three different kinds of RNA differ from each other?

20. What are the four stages of protein synthesis?

21. Why did Harry Meade have to understand both gene transcription and gene translation in order to create a new anticlotting drug?

22. Why might cancer-causing agents (carcinogens) also be mutation-causing agents (mutagens)?

Analyze and Evaluate the Concepts

23. RNA molecules are rather unstable and tend to be broken down rapidly in cells. Let's say you want to compare the stability of RNAs in two different types of cells. You treat the cells with a transcription inhibitor from mushrooms. One cell type dies in 1 hour, and the other cell type dies in 12 hours. Which cell type is more likely to have more stable RNA molecules? Support your answer.

24. Use the genetic code (Fig. 7.12) to determine the amino acid sequence encoded by the following short stretch of bases in an mRNA. Use the start codon to set the reading frame.

CGAUGGUAUAGUCCCUGGUAGAGA

25. Suppose the ninth base in the sequence above mutates from a U into an A. What effect will this have on protein structure?

26. Outline the basic steps Harry Meade had to follow to clone the *AT3* gene, then transfer it into a goat.

27. Describe an important ethical issue in recombinant DNA research and list arguments on both sides of the issue.

28. An antibiotic-resistance gene is usually included in a plasmid that is to be used for gene transfer. Why?

29. Why did Harry Meade want his transgenic goats to express the human *AT3* gene in their mammary gland cells instead of in other tissues? Why isn't that human gene expressed in the human mammary gland?

30. What role does complementary base pairing play in transcription? In translation? In recombinant DNA methodology?

Medical Help with Conception

Marcy and Stephen were a couple with a plan. They met in college, dated for five years, and then became engaged one night in a horse-drawn carriage parked near the Liberty Bell in Philadelphia. Two years later, they exchanged vows at an early-December wedding that they recall as "spectacular," with its ice sculptures, autumn flowers, and elaborate buffets. They planned to travel, buy a house, and build their careers. Later, when Marcy turned 30, they would start their family. Everything went according to their careful plan except that six years into the marriage, they were still childless. Eventually, after month upon month of stinging disappointment from failed attempts to conceive, they sought out Dr. Frances Batzer, an infertility specialist at the Women's Institute for Fertility, Endocrinology, and Menopause in Philadelphia.

Nearly one in five American couples has difficulty conceiving a baby—that's more than 5.3 million couples in all—and the Women's Institute is only one of 340 clinics in the United States serving these would-be parents. In 40 percent of infertile couples, the woman has reproductive problems; in another 40 percent of cases, the man has them; in the rest, the infertility lies with both partners or is of unknown cause. By performing medical tests, Dr. Batzer found that Stephen had abundant, healthy sperm but that Marcy had a serious case of endometriosis—a disorder in which tissue that normally lines the inside of the uterus begins to grow elsewhere in the pelvis, blocking reproductive organs and preventing their normal function. Dr. Batzer re-

moved some of the endometriosis surgically and treated Marcy with drugs, but her infertility persisted. Despite this, the couple was still set on having their own baby rather than adopting one. Reproductive medicine offered one last chance to help them conceive—*in vitro* **fertilization:** literally, **fertilization,** or the fusion of egg and sperm, but in a glass dish. Before they found Dr. Batzer, they recalled having "no answers, no progress." Now, they had a new plan, "a path, and most of all, hope."

The principles of *in vitro* fertilization are fairly simple: Retrieve eggs from the would-be mother or a donor. Collect sperm from the prospective father or a donor, and then unite sperm and eggs in a laboratory dish to bring about fertilization. Allow some cell divisions to take place in the fertilized egg (zygote; see Chapter 4). Finally, transfer the very early embryo into the mother's womb or that of another woman (a

▲
Help with Conception. Marcy and Stephen discuss *in vitro* fertilization procedures with Dr. Batzer.

surrogate). When the embryo implants itself in the wall of the womb, pregnancy begins. The medical procedures are delicate and not always successful. But *in vitro* fertilization can sometimes circumvent inaccessible ovaries, blocked tubes, elderly eggs, scanty or slow-swimming sperm, and other common infertility problems. The result is the marvelous transformation from individual egg and sperm to zygote, embryo, fetus, and finally, after birth, a human baby. This transformation depends on **development,** the process by which offspring become bigger and more complex—from the single-celled stage to adulthood and sexual maturity. Development is not only a fascinating subject in itself, but it also forms an important bridge between the genes we have been discussing and the whole organs and organisms we'll talk about in the rest of the book. Through the unfolding process of development, genes direct the building of organs as complex and important as the human brain, hand, and eye.

As we move through this chapter, we'll see how Dr. Batzer helped Marcy and Stephen move past their profound disappointment. We'll focus mainly on human reproduction and development because—let's face it—whether in our day-to-day lives, or in books, movies, the media, even billboards, we humans are keenly interested in sexual attraction,

How can the study of development help infertile couples?

sexual activity, and reproduction. In fact, human sexuality shapes our bodies, our behavior, and our life-long experiences. Since sexuality reflects our evolution within the animal kingdom, we'll set the stage by discussing animal reproduction. Next we'll see where human anatomy and behavior fit in. Then we'll follow the normal sequence of development, from eggs and sperm all the way to a new baby, child, and adult. Along the way, you'll discover the answers to these questions:

1. How do animal mating behaviors maximize the chances that egg and sperm will unite?

2. What are the main parts and processes of the male and female reproductive systems and which hormones control them?

3. How do sperm and egg unite during human fertilization, and how can science assist it?

4. How do fertilized eggs become embryos?

5. How do an embryo's organs take shape in the proper places within the organism?

6. What are the events of human pregnancy and birth?

7. How do our growth, maturation, and aging complete the human life cycle?

8.1 Sexual Reproduction: Mating and Fertilization

In animals as different as zebras and zebrafish, the reproduction game is essentially the same: A tiny, mobile male gamete or **sperm** cell fuses with a relatively huge **ovum** (pl. *ova*) or egg cell and triggers the development of a new individual. The meeting of egg and sperm requires that individual organisms mate at an appropriate time and in a way that is typical to and successful for their species. Individual egg and sperm cells must make contact and fuse through a series of precise events based on the way the reproductive cells (gametes; see Chapter 4) are built and behave. So how does animal behavior ensure this mating and fusion?

Some animals, such as hydras, flat worms, and sponges sidestep the problem most of the time by reproducing asexually, with new individuals arising from old

through regeneration or budding. In others like earthworms and sea slugs, the animals are **hermaphrodites:** Each individual makes both eggs and sperm. As self-sufficient as this sounds, however, pairs of hermaphrodites usually get together and reciprocally exchange sperm and fertilize each other's eggs.

Cooperation and timing are obviously important for earthworms and sea slugs, and the same is doubly true for most other kinds of animals that make either egg or sperm, but not both. First, potential partners must recognize each other through sight, smell, or sound. The male peacock's tail, the lion's mane, the elephant seal's proboscis, and the stickleback fish's red stripe are all examples of visual cues that evolved and act as attractants to the opposite sex of their own species. Cricket chirps and bird songs are familiar auditory cues. And mammals and insects have evolved hundreds of odorous chemical compounds called **pheromones** (FEAR-o-moanz) that communicate sexual receptivity and attractiveness. (We'll talk more about pheromones, including the human variety, in Chapter 18.)

In many water-dwelling species, such as salmon and sea urchins, fertilization is external; that is, the animals deposit eggs and sperm directly into the surrounding environment (usually water), where some gametes meet by chance and fuse. In frogs and salamanders, the male clasps the female firmly for prolonged periods, and this touching stimulates the couple to release their gametes simultaneously (Fig. 8.1). This synchronized release makes it much more likely that healthy eggs and sperm will meet because unfertilized eggs survive only a short time. In land-dwellers, such as most mammals, birds, reptiles, insects, and snails, fertilization is internal: The male deposits sperm directly into the female's genital opening, and the gametes meet in a tube or a chamber. Internal fertilization helps ensure that sperm will be concentrated and protected within the female's body until viable eggs are available for fertilization.

Throughout the animal kingdom, sexual attractiveness, mating, and fertilization depend on very specific body structures. Our human structures, of course, are most familiar to us and help show how the physical characteristics of mature, adult organisms allow for the genesis of a new generation.

8.2 Male and Female Sexual Characteristics

Part of what drew Marcy and Stephen to each other was their physical differences and "Vive la difference!" as the motto of hopeless romantics goes (French and otherwise)—"Long live the difference!" Males and females have different sets of physical characteristics involved in sexuality and reproduction. Our primary sexual characteristics are our reproductive organs, which are capable of passing along part of an individual's set of genes to the next generation. In contrast, our secondary sexual characteristics are our external features, such as enlarged muscles in males and milk-producing breasts in females. These secondary characteristics are not directly involved in sexual intercourse, but they can play significant roles in attraction, nursing, and other reproductive behaviors. As a society, we already devote huge amounts of attention to these outward (secondary) traits and how to enhance them. So here we're going to concentrate on the reproductive organs themselves.

Each sex has a pair of **gonads**—reproductive organs that produce sex cells and sex hormones. Through mei-

Figure 8.1 External Fertilization. A male golden toad (*Bufo periglenes*) clasps the larger female. They then release eggs and sperm into the water. Unfortunately, this beautiful species may now be extinct in its native Costa Rica.

otic divisions (see Chapter 4), the male gonads—the **testis** (pl. *testes*)—produces sperm, and the female gonad—the **ovary**—produces eggs. Both sexes also have various ducts and glands that transport and store sex cells and, in the female, that nurture the developing embryo. All of these structures are involved in human fertilization, whether by the intimate old-fashioned method, or by the high-tech *in vitro* approach.

Male Reproductive System

As a typical male, Stephen's reproductive organs had a straightforward job: produce and transfer sperm into the female reproductive tract. (*E*xplorer 8.1 animates and describes these roles.) The gonads that produce the sperm cells, the testes, also secrete the primary male hormone, **testosterone.** Testosterone not only aids in sperm production but also helps bring about and control male secondary sexual characteristics, from facial and body hair growth to muscle enlargement, voice deepening, and (in some males) aggressive behavior.

Testes: Sperm-Producing Organs

Most males have a pair of testes that develop inside the body cavity. Shortly before birth, they descend into the **scrotum,** an external sac between the thighs (Fig. 8.2). The lower temperature in the scrotum, outside the body, is necessary for active sperm production, and an elevated temperature of this sac, and the testes inside, can cause sperm development to stop temporarily. A

GET THE PICTURE

*E*xplorer 8.1

How Hormones Control Sperm Production

Why is the ovum so huge relative to the sperm?

fever, for example, can kill hundreds of thousands of sperm cells. To help correct a low sperm count, Dr. Batzer tells her male patients to avoid long time periods in tight shorts or hot tubs, or on bicycle seats. (She also advises them, however, that these are not effective methods of preventing pregnancy.)

Each testis is an oval structure about 4 cm (1.5 in.) long. Packed inside are about 400 highly coiled, hollow tubes called **seminiferous tubules,** each around 70 cm (28 in.) long. The walls of the seminiferous tubules contain sperm-generating cells, or **spermatogenic cells.** These meiotic cells undergo the special cell divisions of meiosis that reduce the number of chromosomes from two sets (diploid) to one set (haploid) (see Chapter 4). The resulting haploid cells develop into male gametes (sperm). While sperm cells are developing within the tubule walls, other larger cells embedded nearby sustain, surround, and nourish the sperm. These support cells are called **Sertoli cells** (Fig. 8.2c). The tissue that encases the seminiferous tubules also houses **interstitial cells,** which produce the male steroid hormone testosterone, a chemical relative of cholesterol that stimulates male sexual characteristics and behavior.

Accessory Ducts and Glands: Sperm Delivery Route

Sperm travel from the seminiferous tubules in each testis into a coiled tube, the **epididymis,** attached directly to the top of the testis (Fig. 8.3). Here the sperm

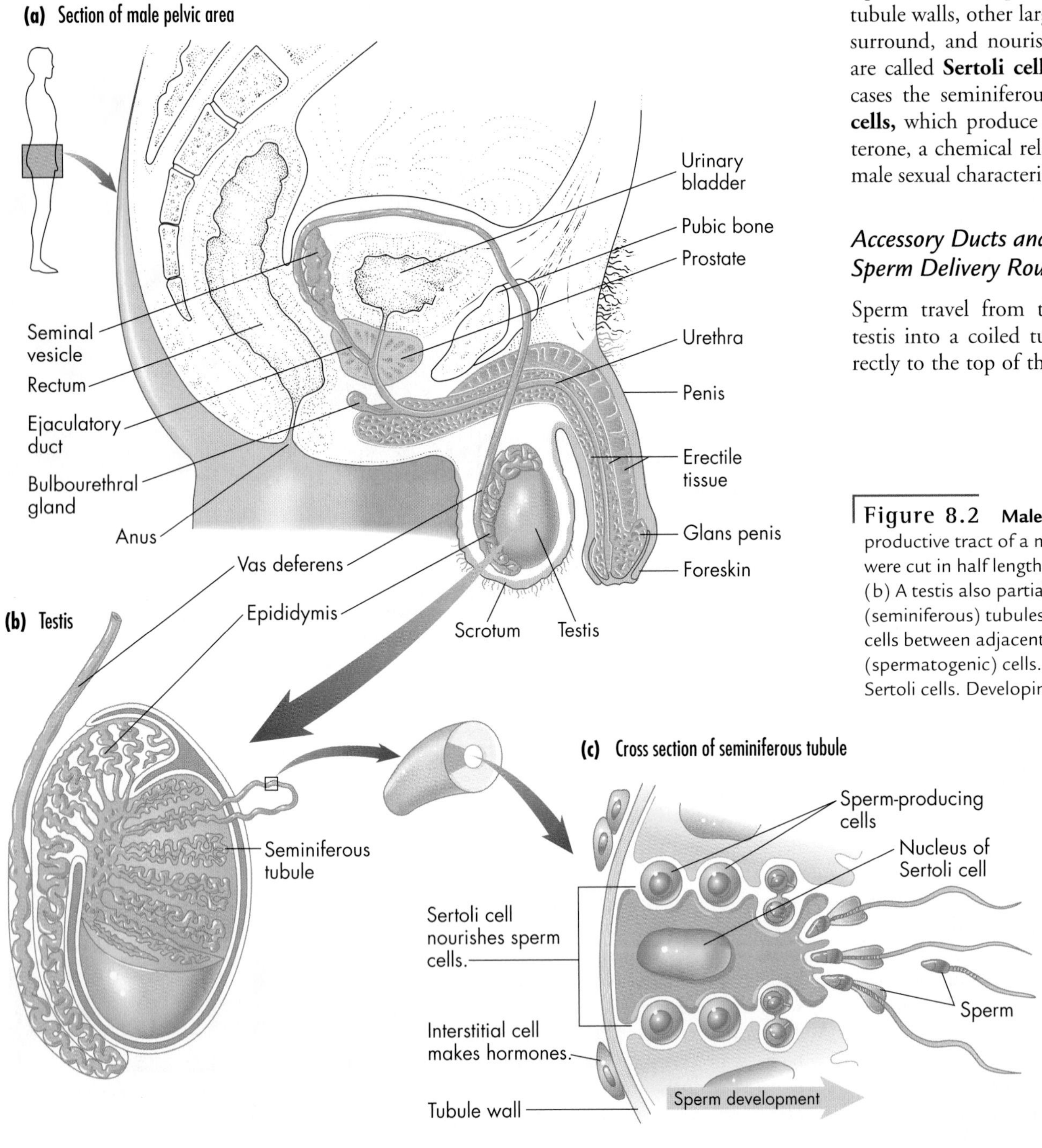

(a) Section of male pelvic area

- Urinary bladder
- Pubic bone
- Prostate
- Urethra
- Penis
- Erectile tissue
- Glans penis
- Foreskin
- Seminal vesicle
- Rectum
- Ejaculatory duct
- Bulbourethral gland
- Anus
- Vas deferens
- Scrotum
- Testis

(b) Testis

- Epididymis
- Seminiferous tubule

(c) Cross section of seminiferous tubule

- Sperm-producing cells
- Nucleus of Sertoli cell
- Sertoli cell nourishes sperm cells.
- Sperm
- Interstitial cell makes hormones.
- Tubule wall
- Sperm development

Figure 8.2 **Male Reproductive System.** (a) The reproductive tract of a male as it would look if the body were cut in half lengthwise and viewed from the side. (b) A testis also partially cut in half. (c) Sperm-bearing (seminiferous) tubules, when enlarged, reveal interstitial cells between adjacent tubules and sperm-producing (spermatogenic) cells. The latter are nourished by huge Sertoli cells. Developing sperm also are visible. *E* 8.1

mature and develop the ability to swim (motility) (Fig. 8.2a and b). When a male is sexually stimulated, sperm are washed rapidly from the epididymis down a system of ducts—like logs in a flume—and are forcefully spewed from the body. Contractions in the walls of the epididymis push sperm into a connecting tube, the **vas deferens** (45 cm, or 18 in., long), a sperm duct that also contracts and continues propelling the sperm (Fig. 8.2a and b). A physician (usually a urologist) severs this sperm duct when performing a **vasectomy;** this procedure permanently prevents sperm from exiting the body and so acts as a form of sterilization, even though it doesn't prevent the production of sperm or male hormones. (The hormones continue to circulate normally, and the sperm, blocked from exit, are broken down and reabsorbed by the reproductive tissues.) The vas deferens from each testis merges into the ejaculatory duct. Glands secrete buffering fluids that combine with sperm to become the semen that is ejaculated. Secretions from one set of glands, the seminal vesicles, regulate the pH of semen and stimulate muscular contractions in the female reproductive tract. Secretions from the chestnut-shaped **prostate gland** help neutralize the acidity of the female reproductive tract.

Brain

① Brain releases hormones.

⑤ Testes hormones inhibit release of brain hormones.

LH

FSH

Pelvic bones

② Brain hormones stimulate testes.

Seminiferous tubule

Testis

Sertoli cell

Sperm-forming cell

Testosterone

④ Increased testosterone secretion

Interstitial cells

③ Increased sperm production

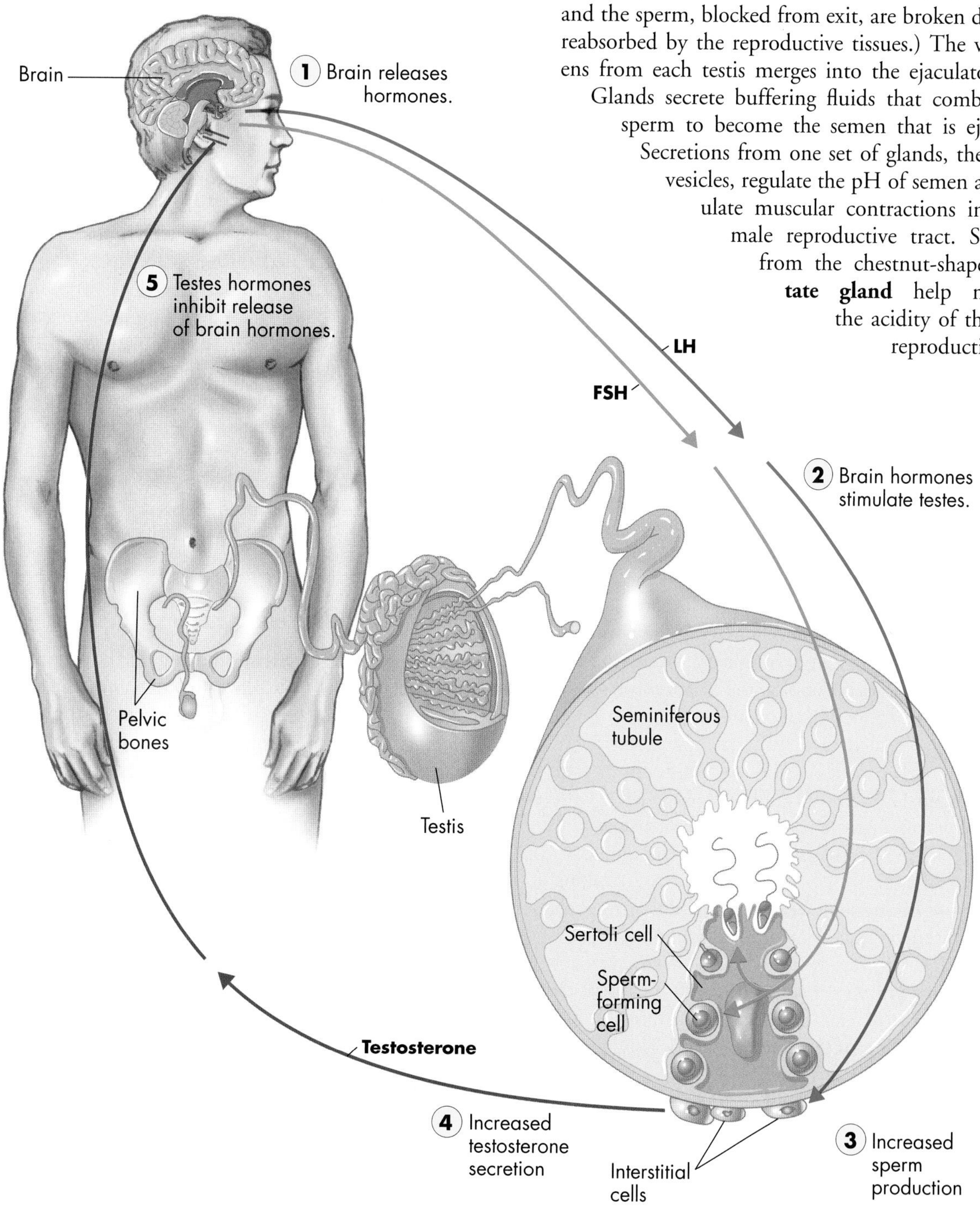

Figure 8.3 **How Hormones Control Sperm Production.** Hormones from the brain and testes control the timing of sperm production through interlocking feedback loops. **E** 8.1

This natural acidity protects a woman's delicate tissues from microorganisms but also tends to inhibit sperm swimming.

The prostate gland is a frequent cancer site: One in five of the men reading this book will develop cancer of the prostate during his lifetime. In the United States, this represents over 300,000 new cases per year. Because of this relatively high risk, many physicians recommend that men over the age of 40 be checked regularly for signs of prostate cancer.

A final set of glands, the bulbourethral glands, adds yet another alkaline fluid to the sperm, and when a man becomes sufficiently aroused, semen finally exits the **urethra,** a tube that runs through the penis.

The Penis

The **penis** has a dual role: transporting urine and semen to the outside and becoming erect for semen delivery. During erotic stimulation, cylindrical columns of erectile tissue in the penis fill with blood, and the penis becomes stiff like a balloon filled with water (Fig. 8.2a and b). This happens because cells in the penis release the gas nitric oxide (NO), which causes a second messenger, a nucleotide (see Chapter 2), to prompt muscles in the spongy erectile tissues to relax. This relaxation allows blood to "pool up" in the penis, enlarging and stiffening it. To sustain an erection, a man must produce the nucleotide faster than a naturally occurring enzyme can break it down. In many impotent men, however, the enzyme wins out. That's where the much-publicized drug Viagra comes in: Viagra blocks the enzyme's action for a while, tipping the balance in favor of the nucleotide and allowing erection.

An average ejaculation from the penis produces about 3 or 4 mL of semen (about a teaspoonful) and usually contains 120 to 400 million sperm. Men with fertility problems often have a sperm count below 100 million sperm per ejaculation. Some people are convinced that chemicals in the environment, including hormones used in agriculture, are contributing to such declines. Through delicate manipulation, an embryologist such as Dr. Go, who works with Dr. Batzer, can get around a man's low sperm count problem by capturing individual sperm cells and injecting them directly into eggs. We'll see details of that later.

How the Body Controls Sperm Production

How did Stephen's body "know" when to produce additional sperm? How does any male's body? The answer is that hormones from the brain and testes work together to regulate the timing of sperm production. If sperm are ejaculated or if the testes fall behind in their production of the wriggling gametes, the brain secretes **luteinizing hormone (LH)** and **follicle-stimulating hormone (FSH)** into the bloodstream (Fig. 8.3, Step 1). Blood vessels carry these brain hormones to the testes (Step 2). There, the brain hormones cause testicular tissue to step up production of both sperm and testosterone (Steps 3 and 4). When the testosterone concentration (also carried in the bloodstream) rises above a certain level, it helps block further release of the brain hormones (Step 5). The lower concentrations of LH and FSH then trigger a decline in the testes' production of sperm and testosterone. When ejaculation releases sperm, the hormone levels fall below the set point once again, and the cycle begins anew. This type of regulatory cycle is called a negative feedback loop because it maintains a fairly constant level of a substance (here, testosterone) through an opposing mechanism—when levels rise, it brings them back down; when they fall, it brings them back up (more on this in Chapter 14). (**E** 8.1 animates the process of sperm production.) Although the details of male reproductive anatomy differ between, say, an elephant, a crocodile, and a man, the principles of hormonal control are the same, as is the function—sperm delivery.

Female Reproductive System

Marcy's reproductive system, like any female's, is more complicated than a male's because it has many more jobs to do: Produce eggs and release them at appropriate intervals. Receive sperm. Nourish, carry, and protect the embryo as it develops. Eventually deliver a new organism into the world. The keys to all this feminine reproductive work are anatomy and hormones.

Production and Pathway of Eggs

The day Marcy was born, her body already contained about 2 million egg cell precursors, just like every other typical baby girl. The eggs reside in two solid almond-shaped organs called **ovaries,** which lie inside the body cavity just below the waistline (Fig. 8.4a and b). Each ovary contains immature egg cells called **oocytes,** which develop into mature eggs. Helper cells called **follicular cells** surround and supply materials to the oocytes (Fig. 8.4c). An oocyte surrounded by these helper follicular cells constitutes a unit called a **follicle** (Fig. 8.4c

(a) Section of female pelvic area

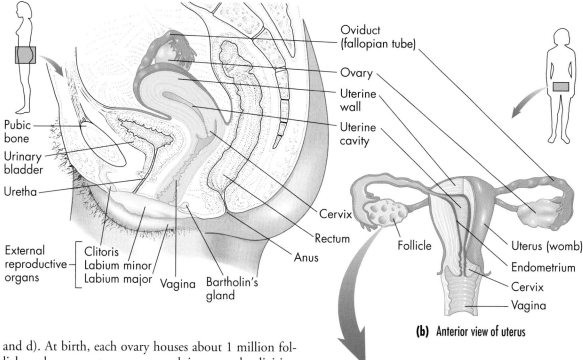

Oviduct (fallopian tube)

Ovary

Uterine wall

Uterine cavity

Pubic bone

Urinary bladder

Uretha

Cervix

Rectum

Anus

External reproductive organs
- Clitoris
- Labium minor
- Labium major

Vagina

Bartholin's gland

Figure 8.4 **Female Reproductive System.** The female reproductive system as it would look if it were cut in half lengthwise and viewed from the side (a) or viewed from the front (b). (c) Cross section through an ovarian follicle.

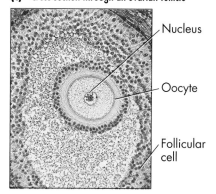

Follicle

Uterus (womb)

Endometrium

Cervix

Vagina

(b) Anterior view of uterus

(c) Cross section through an ovarian follicle

Nucleus

Oocyte

Follicular cell

and d). At birth, each ovary houses about 1 million follicles whose oocytes are arrested in an early division stage of meiosis.

When Marcy reached puberty at about the age of 12 or 13, a process called **ovulation** began to take place in her body every 28 days or so and will continue until she is about 50. During a woman's monthly cycle, one or more follicles in one of a woman's ovaries enlarge, the oocyte(s) matures, and at the time of ovulation (usually around day 14) one follicle ruptures, and a mature egg—now called an ovum—bursts into the body cavity (Fig. 8.5). (An ovum is the largest rounded human cell; it is about the size of the dot over this i.) (Ovulation is usually accompanied by an increase in **basal body temperature** or the body temperature at rest (such as first thing in the morning). Women seeking to avoid or assist conception often take their temperatures at mid-cycle to look for this indicator of ovulation.

The follicle cells left behind in the ovary after ovulation enlarge and form a new gland, the **corpus luteum** (Latin for "yellow body"). Most birth control pills work by preventing ovulation. (*E* xplorer 8.2 covers birth control methods in detail.) Because a baby girl has all of the egg cell precursors she'll ever have already in place, those cells age throughout her life. Genetic mistakes and malfunctioning organelles can accumulate as the years go by, however, and this explains why "old eggs" are one of the most important factors in infertility.

Every month, after a woman's ovum is released from the ovary, it floats toward the fringed opening of one of two tubes called **oviducts** (also known as **fallopian tubes** in humans) (Fig. 8.4b). The tubes have a small range of motion and actually assist conception by sweeping across the ovary surface and picking up an ovulated egg in a frilly-looking tube opening. In Marcy's case, endometriosis prevented this normal movement. Oviducts are lined with millions of cilia that assist conception in another way. They act like paddles, sweeping the fertilized ovum toward the **uterus,** the

READ ON

E xplorer 8.2

Birth Control Methods

thick-walled chamber where the embryo develops. Fertilization usually takes place in the oviduct as a phalanx of swimming sperm meets, and one penetrates a slowly moving ovum. Most barrier-type contraceptives, such as condoms and diaphragms, prevent this encounter between egg and sperm (see *E* 8.2 on birth control). Whether or not an ovum has been fertilized, cilia sweep it into the muscular, pear-shaped uterus. If the egg has been fertilized, it begins dividing and the very early embryo implants or burrows into the uterine wall, where it develops into an **embryo** and **fetus.** In many women who have difficulty conceiving a baby, the tubes are blocked by scar tissue from a prior infection. Dr. Batzer checked Marcy for this problem but found the endometriosis instead. Blocked fallopian tubes can some-

times be surgically cleared. *E*xplorer 8.3 describes sexually transmitted diseases and how they can lead to tubal scarring and other reproductive problems.

During nearly all of a woman's monthly cycles, the ovum remains unfertilized: Instead of implanting into the

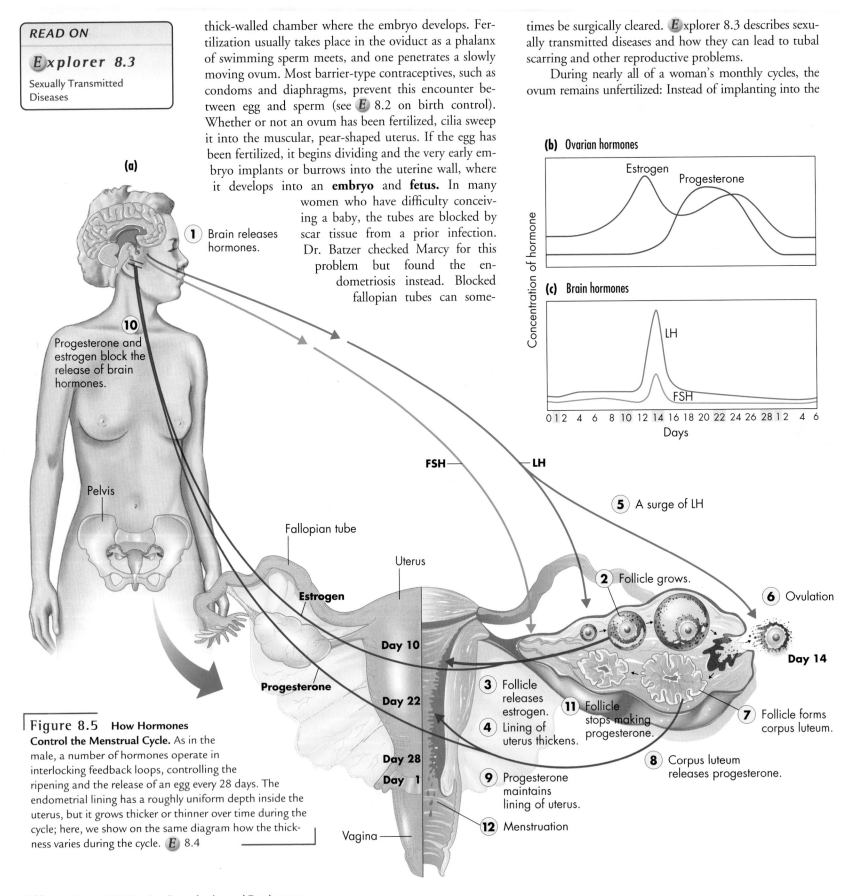

(a)

(b) Ovarian hormones

(c) Brain hormones

1 Brain releases hormones.

10 Progesterone and estrogen block the release of brain hormones.

Pelvis

FSH — LH

5 A surge of LH

Fallopian tube

Uterus

Estrogen

2 Follicle grows.

6 Ovulation

Day 14

Day 10

Progesterone

Day 22

3 Follicle releases estrogen.

11 Follicle stops making progesterone.

4 Lining of uterus thickens.

7 Follicle forms corpus luteum.

8 Corpus luteum releases progesterone.

Day 28

Day 1

9 Progesterone maintains lining of uterus.

12 Menstruation

Vagina

Figure 8.5 **How Hormones Control the Menstrual Cycle.** As in the male, a number of hormones operate in interlocking feedback loops, controlling the ripening and the release of an egg every 28 days. The endometrial lining has a roughly uniform depth inside the uterus, but it grows thicker or thinner over time during the cycle; here, we show on the same diagram how the thickness varies during the cycle. *E* 8.4

uterine wall, it degenerates. The remains of the egg may be discharged from the uterus through the cervix, the neck of the uterus. Just beneath the **cervix** lies the **vagina**—a hollow, muscular tube that receives the penis during intercourse, conveys uterine secretions (including monthly menstrual flow), and serves as the stretchable birth canal through which the fetus passes during childbirth.

The vaginal opening is surrounded by external reproductive structures that include protective tissues called the labia minor and labia major, as well as tissues sensitive to sexual stimulation, such as the clitoris. Also included are the lubricating Bartholin's glands (see Fig. 8.4a).

The Menstrual Cycle: Hormonal Control of the Ovaries and Uterus

Just as the moon and tides have regular, natural cycles, women's bodies have a regular monthly cycle called the menstrual cycle (Fig. 8.5). During each menstrual cycle, the inner lining of the uterus, the **endometrium,** thickens in preparation for pregnancy just before the ovary releases an egg (ovum). If the ovulated ovum is not fertilized, the uterine lining sloughs off, and menstrual bleeding occurs, beginning a new 28-day cycle.

As part of the menstrual cycle, the ovaries produce rising and falling levels of the female steroid hormones **estrogen** and **progesterone.** The female brain makes LH and FSH chemically identical to those produced in the male brain. Together, these hormones drive a feedback loop that produces the fairly regular monthly cycle.

Whole books have been written on the intricacies of the female hormonal cycle, but here we'll just give a quick overview. Refer to the numbered steps in Figure 8.5 as you read along: Hormones from the brain (LH and FSH; Step 1) cause a follicle to grow in the ovary (Step 2) and to release estrogen (Step 3); the estrogen, in turn, travels through the bloodstream and causes the lining of the uterus to thicken (Step 4). A surge of LH (Step 5) then induces ovulation (Step 6). The ovary's follicular cells mature into the corpus luteum (Step 7) and then make and release progesterone (Step 8), which maintains the uterine lining (Step 9). Progesterone and estrogen also block the release of the brain hormones in a negative feedback loop (Step 10). Soon, if there is no pregnancy, the corpus luteum stops making progesterone (Step 11), the uterine lining sloughs off, and menstruation begins (Step 12). When the progesterone level drops, the brain is once again free to release its hormones, and the cycle starts over again. You can learn more about this hormonal cycle in *E*xplorer 8.4 and

in *E*xplorer 8.5 conduct an investigation of how hormones help control fertility.

In some women who reduce their level of body fat to a very low amount by exercising strenuously or dieting, the menstrual cycle can become irregular or stop altogether. Apparently, a certain amount of body fat is necessary to produce appropriate levels of brain hormones. Once these women begin to exercise less or eat more and sufficient body fat accumulates, they start producing enough hormone to allow their menstrual periods to begin again.

Despite "la différence" between males and females, both share certain hormonal similarities: They both produce the same brain hormones that control reproduction. They both produce reproductive hormones that help make eggs and sperm available and bring about sexual development in teenagers. And in both sexes, hormones functioning in a negative feedback loop control the production of gametes.

8.3 Human Fertilization: The Odyssey of Eggs and Sperm

A human egg is the largest rounded cell in a female's body. An ostrich egg housed in its thick shell is not only the largest ostrich cell but the largest living cell so far discovered. Whether in woman or mother ostrich, these large cells are complicated and delicate. In women of reproductive age, an ovulated egg is receptive to sperm for only a day or so. If sexual intercourse takes place around this time and fertilization occurs, the fertilized egg is able to implant in the uterine wall, and pregnancy can begin. Given a year of unprotected intercourse, 90 percent of women between ages 15 and 40 will become pregnant—some immediately, others after a few months. The intricacies of fertilization are important to obstetricians, gynecologists, and infertility specialists like Dr. Batzer. But we all need to understand the basics of reproduction and development to be wise medical consumers and to make appropriate decisions about birth, birth control (see *E* 8.2), and related issues such as preventing sexually transmitted diseases (see *E* 8.3).

An egg cell's large size helps it perform its many jobs (Fig. 8.6):

- The egg donates a haploid nucleus containing one set of chromosomes to the new embryo. In some animals, the egg protects the developing embryo inside jellylike protein coatings, strong fertilization membranes, sacs of fluid, or hard or leathery shells.

GET THE PICTURE

*E*xplorer 8.4

How Hormones Control the Menstrual Cycle

SOLVE IT

*E*xplorer 8.5

Predicting Conception

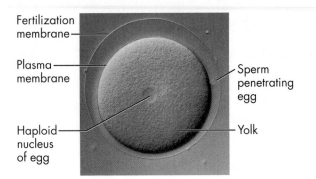

Fertilization membrane

Plasma membrane

Haploid nucleus of egg

Sperm penetrating egg

Yolk

Figure 8.6 **The Egg.** The egg cell of a sea urchin protects the new individual, instructs its development, and contributes half its chromosomes. Caught at the moment of fertilization, a sperm penetrates the egg from the right. Because this egg warehouses nourishment for the embryo, it is 1000 times the volume of the sperm, which donates only a haploid set of chromosomes and the centriole.

- **Yolk,** which contains rich stores of lipids, carbohydrates, and special proteins, nourishes the embryo and usually provides all the embryo's mitochondria.
- The egg supplies the cytoplasm for the fertilized egg (or zygote; see Chapter 4) and in it the machinery for making new proteins—a machinery that allows rapid cell division.

In many animals, the egg's cytoplasm contains special substances that control development. An ovum is like a computer loaded with a software program, ready to play out the actions of development as soon as the sperm's entry starts the program running.

Compared to the egg, a sperm is one of the smallest cells in a male's body, stripped down to just the essential elements needed to perform superb penetration (Fig. 8.7). These include: (1) a compact haploid nucleus; (2) several mitochondria, which provide energy; (3) a long, lashing flagellum (tail) that propels the sperm; and (4) a sac of enzymes, called the **acrosome,** that digests a path through the egg's protective outer coatings. Some men are nearly sterile because their sperm swim slowly and weakly; others are completely sterile because their sperm have abnormal tails or oversized heads or can't penetrate the egg's protective coat (Fig. 8.7b). A sperm's streamlined size and shape reflect its narrow function: to reach the egg, penetrate its coating, and deliver a haploid nucleus (containing one set of chromosomes) into the egg's cytoplasm. This penetration and delivery of the nucleus, together with the activation of the resting egg, make up the events of fertilization.

Fertilization Without Medical Help

For years, Marcy and Stephen tried for a successful fertilization in order to start their family and learned quite a bit about this microscopic penetration that initiates development in an embryo. You can see fertilization illustrated in Figure 8.8 and animated in Explorer 8.1. The major events are much the same across many animal species, ourselves included.

When the season is right and mating behavior brings animals and their sperm and eggs, into close proximity—either inside the female's body or in a watery environment—the sperm head for the eggs, probably following a chemical trail. A sperm's lashing flagellum propels the microscopic gamete headlong into the jellylike outer coating of the egg, which looms hundreds of times larger than the individual fishlike sperm. At the moment of contact, the acrosome at the tip of the sperm head releases enzymes that help the sperm to penetrate the coatings on the egg's surface. The plasma membranes of the egg and sperm fuse. Then the sperm head—including the nucleus and the centriole (see Chapter 4; the organelle that helps organize the cell's inner scaffolding)—plunges into the egg's cytoplasm. The sperm's tail, including its energy-producing mitochondria, remains outside the egg. (This is why the egg

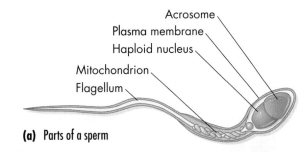

Acrosome
Plasma membrane
Haploid nucleus
Mitochondrion
Flagellum

(a) Parts of a sperm

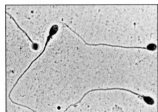 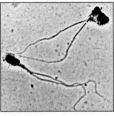

(b) Sperm from a normal man   **(c)** Sperm from a sterile man

Figure 8.7 **The Sperm.** This mobile cell contributes half of the chromosomes during fertilization and triggers development in the zygote. (a) The streamlined sperm cell contains its payload: the haploid nucleus; the acrosome, which allows it to recognize and penetrate the egg coverings; and mitochondria that power the lashing tail, or flagellum, allowing the sperm to swim. (b) Normal sperm. (c) Some men are infertile because their sperm cells have abnormal tails. Here, one of the sperm has an extra tail, and one has an abnormally large head.

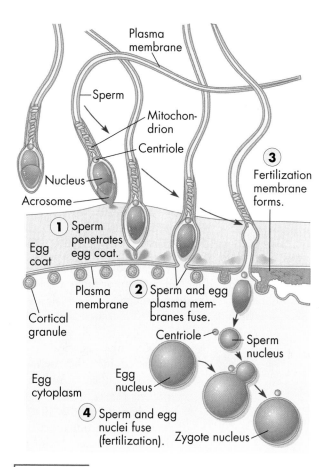

Figure 8.8 **Fertilization: Egg and Sperm Fuse.** During fertilization, the sperm penetrates the egg's outer coats, fuses with the plasma membrane, delivers its nucleus to the egg's cytoplasm, and triggers the raising of the fertilization membrane.

donates all of an embryo's mitochondria.) Beginning precisely at this site of sperm penetration, a wave of molecular signals sweeps over the egg and triggers the rapid construction of a barrier, the **fertilization membrane** (see Figs. 8.6 and 8.8). This essentially marks the victory of one sperm because the barrier prevents additional sperm from entering the egg. The sperm and egg nuclei fuse, establishing a novel genetic combination that is diploid (contains two sets of chromosomes)—and a new zygote.

In humans, fertilization normally occurs as the egg moves along the oviduct and swimming sperm encounter the huge target. After fertilization, the newly formed zygote continues its journey through the oviduct until it reaches the uterus four or five days later.

The process of egg release (ovulation) has already caused the ripe follicle (the corpus luteum) to make the hormone progesterone and in turn, through this chemical messenger, to maintain the monthly buildup of the uterine walls (see Fig. 8.5, Steps 6 to 9). As a result, the fertilized egg can implant in the well-prepared uterus and begin the remarkable transformation from a single fertilized egg cell into a complex animal—first an embryo and then a fetus—through the mechanisms of development.

In Vitro Fertilization: Medical Help with Conception

Marcy and Stephen were surprised and dismayed at their difficulty in conceiving a baby and they had reason to be surprised: In most women of childbearing age, nature strongly favors procreation. As we saw, 90 percent of reproductive-age women become pregnant within one year if they have intercourse without some form of birth control (**E** 8.2). But the 10 percent that don't conceive after a year of trying contribute to the 5.3 million infertile couples (nearly one couple in five). This "epidemic of infertility," as *Newsweek* magazine called it, has started to mount in the past decade. Experts say this upsurge is partly due to venereal diseases and scarred reproductive organs (**E** 8.3) and partly due to deferring parenthood to the 30s or 40s (for financial reasons, remarriages, and other factors), when fertility naturally declines. Finally, many researchers also suggest that environmental exposures to agricultural hormones, pesticides, and other human-made chemicals may be altering human eggs and sperm.*

Marcy and Stephen were in their 30s, but the endometriosis prevented conception for years before Dr. Batzer employed high-tech methods of *in vitro* fertilization to help them conceive a baby. After treating Marcy's endometriosis, Dr. Batzer gave her different hormones to stimulate the growth of several ovarian follicles simultaneously. Each morning during a month-long in vitro "cycle," Marcy had blood drawn and Dr. Batzer measured its estrogen content. If high, then several estrogen-making follicles had ripened. At an appropriate estrogen level, Dr. Batzer injected Marcy with a

*Halwell, Brian. Plummeting Sperm Counts Cause Concern. *The Futurist,* 1999, Vol. 33, No. 9, pp. 14–15.

Glausiusz, Josie. Positively Against Pollutants. *Discover,* May 2000. Vol. 21, No. 5, p. 16.

What is in vitro **fertilization?**

What are GIFT, ZIFT, and ICSI?

natural hormone called human chorionic gonadotropin (hCG), which mimics the action of LH (review Fig. 8.5). Precisely 34 hours later, the egg cells were mature and ready for harvest. Dr. Batzer gently suctioned up several ripe eggs through a needle that penetrated the vaginal wall and extended into the ovary. She placed the eggs in a laboratory dish (Fig. 8.9a), and the embryologist, Dr. Go, then added Stephen's sperm, which fertilized several of the eggs in the dish, much as they could have done in an oviduct. Over several days, each zygote cleaved into 2 cells, then 4, then 8. At about this stage, Dr. Batzer transferred embryos into Marcy's uterus, where they burrowed into the thick endometrial lining and began to grow into twin embryos—the couple's future sons.

Depending on a couple's infertility problems, a doctor can employ variations on this procedure. For example, physicians sometimes place harvested eggs and sperm directly into the fallopian tube, where fertilization usually occurs; this procedure is called gamete intrafallopian transfer, or GIFT. In other cases, fertilization occurs in a laboratory dish, but the developing zygote is placed into the fallopian tube; this is known as zygote intrafallopian transfer, or ZIFT. Finally, when the sperm are unable to swim (review Fig. 8.7), they may be sucked up into a very fine needle and delivered directly into the egg cytoplasm via intracytoplasmic sperm injection, or ICSI (Fig. 8.9b). The entire sperm is injected, including tail and mitochondria, but by some poorly understood mechanism, the male mitochondria are eliminated and the maternal ones (from the egg cytoplasm) take over.

After ICSI, the egg cleaves and is transferred into the uterus as described earlier. Additional variations can occur if someone other than the mother donates the egg, if someone other than the father donates the sperm, or if the mother donates her egg but a surrogate mother incubates the developing embryo and fetus.

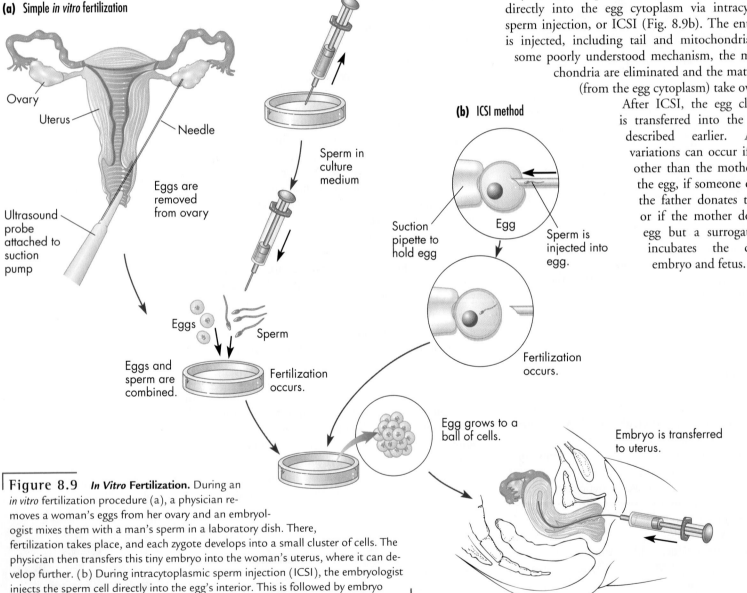

(a) Simple *in vitro* fertilization

Ovary

Uterus

Needle

Ultrasound probe attached to suction pump

Eggs are removed from ovary

Sperm in culture medium

Eggs Sperm

Eggs and sperm are combined.

Fertilization occurs.

(b) ICSI method

Suction pipette to hold egg

Egg

Sperm is injected into egg.

Fertilization occurs.

Egg grows to a ball of cells.

Embryo is transferred to uterus.

Figure 8.9 *In Vitro* **Fertilization.** During an *in vitro* fertilization procedure (a), a physician removes a woman's eggs from her ovary and an embryologist mixes them with a man's sperm in a laboratory dish. There, fertilization takes place, and each zygote develops into a small cluster of cells. The physician then transfers this tiny embryo into the woman's uterus, where it can develop further. (b) During intracytoplasmic sperm injection (ICSI), the embryologist injects the sperm cell directly into the egg's interior. This is followed by embryo growth in a dish and subsequent transfer into the uterus.

8.4 Embryonic Development

We've seen how sperm and egg form, and we've explored the marvelous instant of fertilization. But fertilization is over in seconds—and then what? What ingenious biological mechanisms start to unfold to turn a single fertilized egg into a baby with its intricate little ears, wrinkled hands, tiny nose, and sophisticated heart and brain? If you were watching the process unfold under a microscope, you'd see embryonic **development**—the process by which an offspring increases in size and complexity from a fertilized egg to a complex organism—and its four stages: cleavage, gastrulation, organogenesis, and growth (Table 8.1). So let's look at each stage now, then at the underlying mechanisms that bring them about and transform a fertilized egg into an embryo, fetus, and baby like Marcy and Stephen's twin boys.

Figure 8.10 depicts the main events of development in a frog. Each frog embryo develops into a tiny, wrig-

(a) Formation of egg and sperm

Sperm

Egg

Eggs are produced in ovaries, and sperm are produced in testes of mature females and males.

(b) Fertilization

Sperm and egg fuse.

(c) Cleavage

Mitotic cell divisions in the fertilized egg create a hollow ball of cells, the blastula.

Cross section of blastula

Figure 8.10 **An Overview of Developmental Stages and Processes in a Frog.** After egg and sperm form (a) and fuse (b) during fertilization, cleavage produces the blastula, a hollow ball of cells (c). Embryologists have mapped the location of cells that will eventually form the three primary tissue layers—the outer ectoderm (dark gray), the middle mesoderm (light gray), and the inner endoderm (yellow) (d). As gastrulation begins (d), the mesodermal cells leave the surface of the blastula and enter the cavity of the hollow through the site of the future anus. The endoderm follows. During midgastrulation, the future gut cavity enlarges and the blastula cavity becomes smaller. Late in gastrulation a three-layered embryo has developed with back and front, belly and rear axes clearly evident. Organogenesis generates the organs in their proper places (e). Differentiation brings about the specialization of cells for their roles in the organism, and growth enlarges the embryo (f) and young. *E* 8.6

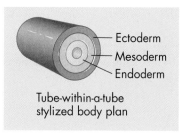

Ectoderm
Mesoderm
Endoderm

Tube-within-a-tube stylized body plan

(d) Gastrulation

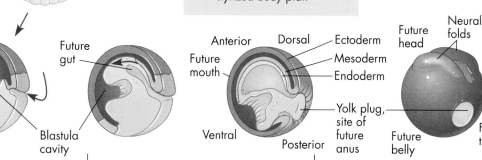

Future gut

Blastula cavity

Anterior

Future mouth

Ventral

Dorsal

Ectoderm
Mesoderm
Endoderm

Posterior

Yolk plug, site of future anus

Future head

Neural folds

Future belly

Future tail

The ectoderm folds into a tube and forms the spinal cord.

Cell rearrangements in the blastula cause three tissue layers to form by cells moving from outside to inside of embryo.

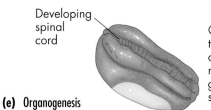

Developing spinal cord

Cells continue to divide, migrate, change shape, and rearrange themselves, generating organ shape.

(e) Organogenesis

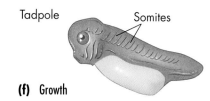

Tadpole

Somites

Different cells accumulate different sets of proteins and assume their specific functions. The embryo grows.

(f) Growth

197

gling tadpole inside a transparent jelly coat, and the stages are easily seen. But the processes are the same, in principle, for the embryos of worms, flies, frogs, humans, and most other animals. In fact, much of what biologists know about human development they learned first from animals in the lab. (Plant development is different in several important ways, as we'll see in Chapter 22.)

Four Stages of Development

Cleavage: Cell Division Begins

If you were watching the marvel of development unfold step by step, the most obvious thing you'd see is a change in cell number: One fertilized egg must become millions of cells in an embryo, billions in a fetus, and trillions in a child and adult. This enormous increase takes place through a series of cell divisions. In the first few division cycles, the fertilized egg cleaves or divides in two; these so-called daughter cells (see Chapter 4) cleave again to form four cells; those form eight, and so on. Based on this cleaving, the first few division cycles are called **cleavage** (Fig. 8.10c). In many animals, cleavage produces a ball, called a **blastula,** which is no bigger than the original fertilized egg but is made up of many cells that are smaller than the original egg (Fig. 8.10c). As the original egg cell contents become separated into different cells of the blastula, molecular information that will direct the shaping of the embryo sometimes ends up in different regions of the "ball." Because of this regional information and its subsequent control of embryonic shape, cleavage helps organize development at a very fundamental level.

Gastrulation: Three Layers Form and Migrate

If you were watching the steps of development, you'd see another obvious change, as well: cell position. Embryos and babies are not just random blobs of cells but are, instead, precisely shaped, and their cells are exquisitely organized in space. The most basic feature of this organization is that an early embryo has three main body layers that give rise to all the organs and tissues of the developing, enlarging individual. The process that converts a hollow ball, the blastula, into a three-layered embryo is **gastrulation,** which means, literally, "the formation of the stomach." Indeed, the three layers become arranged as a tube within a tube within a tube. The stomach and digestive tube lie on the inside; a tube of blood vessels, muscles, bones, kidneys, and other organs come to surround the digestive tube; and an outer

tube, the skin, covers both of the others. Cell movements in the hollow blastula give rise to these three nested tubes (Fig. 8.10d).

These cell movements of gastrulation are a living sculptural process. Cells at a particular spot on the blastula begin to dive inside the ball, pulling their neighboring cells in with them (Fig. 8.10d). The "diving-in" movements can be best understood by watching a time-lapse film 𝐸xplorer 8.6). The result is three primary tissue layers (Fig. 8.10d): The inner layer or **endoderm** ("inner skin") will produce the digestive tract and parts of the liver and lungs that are derived from it. The middle layer or **mesoderm** ("middle skin") will form the blood vessels, kidneys, and reproductive organs, as well as the body's muscles and most of the bones. The outer layer, or **ectoderm** ("outer skin") will become the outer parts of the skin and the nervous system. At the embryo's front end, the cells eventually punch through to form the mouth. At the opposite end, where the cells first in-pocketed, the anus forms. Now the embryo has an orientation—a head-tail axis—as well as its basic layers. Hard as it is to imagine, each of us went through this same sculpturing process in the second week of life.

Organogenesis: Organs Unfold

As gastrulation ends, the three embryonic layers are poised to generate the next phase of development, **organogenesis,** or the formation of the body's organs and tissues with proper shapes, positions, and functions (Fig. 8.10e). This stage can last several weeks and bring into existence all the embryo's systems for moving, digesting, exchanging air, expelling wastes, protecting itself from disease, and so on. Refinements and maturation of these organs and systems will continue throughout development, well into youth and adolescence. A cell-to-cell communication process (induction) helps determine where organs like the spinal cord and brain will form. A set of cell-sculpting processes (morphogenesis) brings about the proper shape of the organs once positioned. This, for example, allows the human brain to take on its hallmark contours. It also causes the **somites,** segmentally repeating blocks of mesoderm, to become your vertebrae (Fig. 8.10f). Pattern formation helps shape a whole region of the embryo, so several parts—such as the features of the face and regions of the brain—are in proper relationship to each other. And the cells of the organs take on their specialized functions through the differentiation process so that, for example, cells in the eye detect light and cells in the stomach wall secrete acid and not vice versa.

GET THE PICTURE

𝐸xplorer 8.6

Early Animal Development

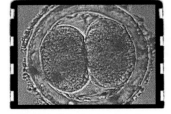

What are two outcomes of cleavage?

(a) The sheet begins to fold.

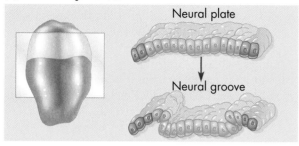

(b) The neural folds move together.

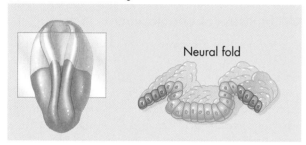

(c) The folds touch.

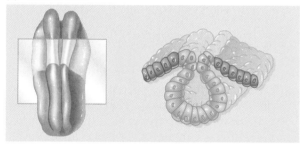

(d) The tube forms, and crest cells disperse.

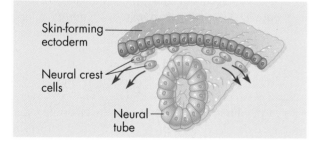

Figure 8.11 **Neural Tube Formation in a Frog Embryo.** The events of neural tube formation are envisioned by considering only the upper surface of this frog embryo cut in half (a, b, c). First, a region of tall ectoderm cells (a, b) takes shape on the embryo's dorsal (back) surface. Folds form as cells in the neural plate change shape from rectangles to wedges. This causes a wrinkle in the sheet (b, c) When the tops of the folds come together, the neural tube pinches off; the surface, which is the future skin, becomes a continuous sheet; and a special population of cells, the neural crest, begins to disperse (d).

We'll explore these mechanisms in the next section, but first let's look at the very earliest organ formation in the animal embryo, the origination of the nervous system, with its spinal cord, senses, nerves, and brain. That set of organs will one day control all our actions, reactions, thoughts, and feelings, but it begins just after gastrulation as nothing more than a sheet of nondescript ectoderm cells on the surface of the embryo (Fig. 8.11a). Suddenly, cells in this sheet begin to thicken and then fold up into a minute, hollow cylinder, the **neural tube.** First, the layer of cells folds, as if it were forced in from the sides like two wrinkles in a bed sheet (Fig. 8.11b, c). The two folds fuse at the top, and then bud off from the ectoderm to form the neural tube, an embryonic structure that will, in turn, form the spinal cord and brain (Fig. 8.11d). Cells forming the neural tube assume a wedgelike shape, helping the sheet roll up into a tube.

In the human embryo, the neural tube first closes in the mid trunk and then zips up toward the front and down toward the rear (Fig. 8.12a). This process is not complete until about 29 days after fertilization in a human embryo. (*E* 8.6 shows a time-lapse movie of a neural tube forming and closing in a frog embryo.) If the human neural tube does not completely close at the back end (posterior) of the embryo, spinal bones (vertebrae) do not grow to encircle the unclosed portion of the tube, and the spinal cord can squeeze out of the gap. The result is spina bifida, or open spine, the most common severe major birth defect among live-born

Table 8.1 Events of Development

Event	What Happens
Fertilization	Egg and sperm fuse, creating a new genetic combination.
Cleavage	Fertilized egg subdivides into a blastula (hollow ball) with many cells; developmental determinants end up in different cells.
Gastrulation	Cell rearrangements sculpt blastula into an embryo, with three cell layers and head/tail, back/belly axes.
Neurulation	Embryonic induction causes sheet of ectoderm to roll up into neural tube, the future brain and spinal cord.
Organogenesis	Body organs form as cells interact, differentiate, change shape, divide, die, and migrate.
Growth	Organs and organisms increase in size and cell number, attaining adult size and maturity.
Gametogenesis	Functioning eggs and sperm develop in gonads, making reproduction possible.

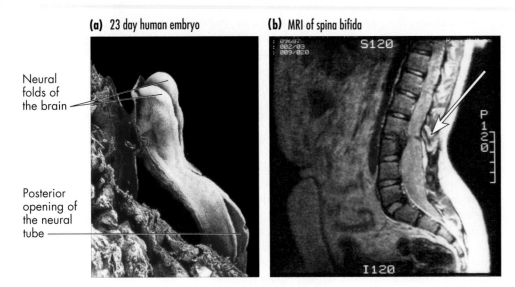

(a) 23 day human embryo   **(b)** MRI of spina bifida

Neural folds of the brain

Posterior opening of the neural tube

Figure 8.12   **Neural Tube Development and Spina Bifida.** (a) A human embryo at 23 days shows the open neural tube in the head and near the tail. The neural tube of all embryos remains open at both ends until about day 29. If the neural tube does not close in the back, then the result is spina bifida. (b) This MRI shows where the neural tube has failed to close completely, a condition that can leave the spinal cord open to the outside and can cause paralysis. Appropriate doses of some vitamins, for example folic acid, taken by the mother just before and during early pregnancy help reduce the risk of spina bifida.

infants, affecting one in every 2000 live births (Fig. 8.12b). Researchers have discovered that rapid DNA synthesis is needed to support the sculpturing of the neural tube, and that this fast production requires the B vitamin folic acid found in leafy green vegetables and certain other foods. The U.S. Public Health Service recommends that women consume about 400 micrograms of folic acid each day at least a month before conception and then during the first month of pregnancy as the neural tube–forming process (neurulation) takes place. Taking this amount will cut the risk of spina bifida by 50 to 70 percent.

Growth: The Organism Enlarges to Adult Size

Along with the emergence of embryonic organs comes **growth** or expansion in size. The cleavage that divides a fertilized egg into a ball of cells is only the beginning of cell division in the embryo. A major size increase is needed, and cell division—usually rapid—continues until hatching or birth and then throughout the development of the young organism. A blue whale, for example, increases in weight 200 millionfold from fertilized egg to adult. In three months' time as it goes from zy-

gote to adult, a rat embryo increases from one cell to 67 billion. And the human brain begins as part of the flat sheet of ectoderm we saw in Fig. 8.11a and becomes, by late adolescence, a mass of 100 billion nerve cells supported by one trillion helper cells. In the whale, rat, and human, as well as virtually all other organisms, growth is due to an increase in cell number, not the size of individual cells (see Chapter 2). Much of the growth goes on after birth or hatching when the limits to food and expansion room are removed.

Mechanisms of Development

Stephen called the *in vitro* fertilization and development of his twin boys "a miracle." Many people share that sense of awe at the transformation of one fertilized cell into a complex organism with tissues and organs in the appropriate relationships and with innate capacities and behaviors already in place. Let's look briefly at the mechanisms that underlie development and bring about the stages of cleavage, layer formation, organ formation, and growth we just saw.

Induction: Cells Talking to Cells

Development is like an elaborate ballet with cells dividing, moving, and changing shape to form emerging organs in the correct locations. How do cells "know" where to go, when to go there, and what to do in the developing embryo? In one of the most famous experiments in biology, almost 80 years ago, German embryologists Hans Spemann and Hilde Mangold tried to find out. They hypothesized that the embryo might have a specific group of cells responsible for organizing the rest of the embryo. To test this idea, they took embryos from a dusky-colored species of newt (the donor) in the early gastrula stage and removed groups of cells from various places in the embryo. Then they transplanted the dusky donor cells into embryos of the same early gastrula stage but from a newt species with pale coloring (the host). They placed the donor's pigmented graft near a region in the host that would normally form on the newt's belly (Fig. 8.13b, c). Cells from most positions in the donor embryo simply formed part of the host's belly skin and did not alter development. Amazingly, though, a small implant from one specific position—the region where cells first start to invaginate during gastrulation (Fig. 8.13b)—caused the second embryo's tissues to produce not a patch of belly skin but an entirely new embryo complete with a brain and spinal cord! This produced a bizarre set of conjoined twin newts connected belly to belly with most of the cells in the secondary embryo originating with the host (Fig. 8.13d). What do you think this strange finding meant?

Spemann and Mangold concluded that cells in the transplanted region had "talked" to cells in the host by means of some developmental signal that "told" the receiving newt embryo to organize the cells surrounding the patch into a second nervous system complete with several structures they normally wouldn't form. These tests showed Spemann and Mangold that one particular region of the embryo had the ability not only to **induce** (to cause) host cells to make a new nervous system but to make an entirely new duplicate of itself. Embryologists today are still trying to understand **induction,** the process by which cells cause their neighbors to embark upon a specific developmental pathway, and moreover, what the induction signals might be. They do know that groups of cells communicate with other cell groups in the embryo via signaling proteins and that these proteins "inform" receiving cells of what to do, when to do it, and where, such as making a nervous system in a given place. Some recent experiments point to at least one particular signaling protein at work in the embryonic nervous system. When the protein is altered by mutation or its normal action is blocked, an animal can be born with cyclopia, a bizarre set of nervous system defects including a single central eye and a missing portion of the brain (Fig. 8.14).

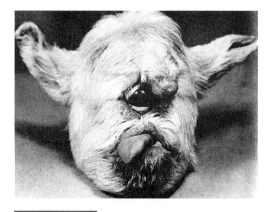

Figure 8.14   **Disrupted Signals and Developmental Anomalies.** Signals from the mesoderm induce the formation of the front of the face and front lower part of the brain. A blockage of the signal has caused this lamb to develop with a single central eye, a condition called cyclopia.

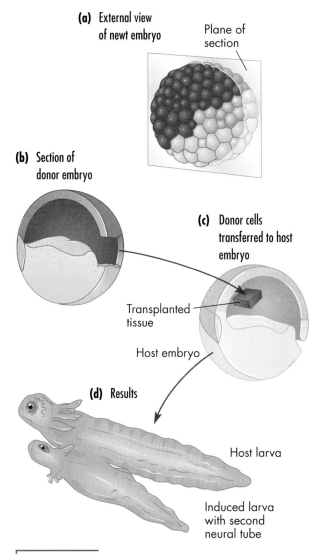

(a) External view of newt embryo

Plane of section

(b) Section of donor embryo

(c) Donor cells transferred to host embryo

Transplanted tissue

Host embryo

(d) Results

Host larva

Induced larva with second neural tube

Figure 8.13   **Induction of the Nervous System.** In their Nobel Prize–winning experiments, Spemann and Mangold transplanted cells from a special region of a dark donor newt embryo (a, b) to a position near the future belly of a light-colored host newt embryo (c) and watched what happened. In the experimental animal, two little embryos, each with its own nervous system, emerged joined belly to belly (d). This experiment shows that the special transplanted region signals the nearby tissue to reorganize and form a second nervous system.

Morphogenesis: Shaping Cells and Cell Populations

Once an organ or system is induced to form in a given place, another developmental process called **morphogenesis** ("shape origination") helps sculpt the forming organs. Morphogenesis includes changes in cell shape, cell migration, and even deliberate cell death.

Think back, for a moment, to the "wrinkles in the bed sheet" that squeeze toward each other, forming the embryo's neural tube (Fig. 8.11). This wrinkling takes place because rounded cells in the ectoderm layer begin to elongate and widen at one end, pushing the wrinkles up into a crest like the top of an ocean wave, the neural crest (Fig. 8.11b, c). Next, cells in the neural crest begin to migrate around the embryo, and this gives rise to cells and organs in distant locations (Figs. 8.11d and 8.15). Some neural crest cells, for example, migrate into the skin and form pigment cells. If you can see brown moles, freckles, or an overall dark skin pigment on your arm, you're looking at the results of embryonic pigment cells that migrated

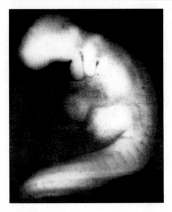

(a) About 25 days

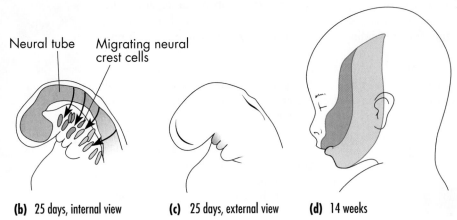

Neural tube   Migrating neural crest cells

(b) 25 days, internal view   **(c)** 25 days, external view   **(d)** 14 weeks

Figure 8.15 **Migrating Neural Crest Cells Help Form the Face.** In a human embryo at around 25 days (a), neural crest cells migrate from the top of the neural tube, around its sides (b), and cluster into tissue masses (c) that will later form the jaws and the lower part of the face (d).

from the neural crest near the top (back) of your spinal cord across your back and down your arm. Other neural crest cells migrate from the embryonic head, around both sides of the developing mouth, and form part of the teeth, as well as the muscles, bones, and cartilage of the lower face (Fig. 8.15d). The embryo may look a bit alien at this point, but a face is clearly emerging. As the migrating cells approach the front of the face, they grow together at the "trough" between the nose and upper lip. But there can be mistakes of morphogenesis: If the mass of cells fails to migrate and grow together, it leaves a space between the nose and mouth called a cleft lip, or it leaves a gap in the roof of the mouth, a cleft palate. Oral surgeons can usually correct these birth defects.

Finally, biologists have discovered that certain embryonic cells are programmed to die, and their elimination helps sculpt developing organs. The human hand, for example, starts as a tiny bud that flattens into a paddle (Fig. 8.16). Programmed cell death removes cells between the forming digits and helps sculpt the emerging fingers.

Pattern Formation: Establishing Spatial Relationship

The development of the human hand illustrates another mechanism of development: pattern formation. The hand not only has independent digits but these form a sequence from thumb to little finger with the thumb on one side of the arm, not the reverse. Biologists have learned that patterns must be established in the developing embryo so that the organs in entire body regions emerge in the right positions relative to each other. They've also learned that these patterns are established in the embryo by zones of cells that give off certain chemical signals. As the signals diffuse away from their original zones of production, they establish a so-called **gradient** or continuous series of concentrations from high to low.

In the human hand, for example, the chemical signal is retinoic acid, related to vitamin A and the orange pigment in carrots. The proximity of developing limb bud cells to the production zone—and thus to high concentrations of retinoic acid—determines whether it becomes a thumb, index finger, middle finger, ring finger, or little finger (Fig. 8.16). Human hands and feet are so marvelously dexterous and so emblematic of our human capacities that when Dr. Batzer was finally able to inform Marcy of her positive pregnancy test after three years of *in vitro* treatments, she used this metaphor: "It's not ten fingers and ten toes yet, but it's a start." "We'll take that start," Marcy replied; then months later, holding her twin baby boys, she mused, "It turned out to be 20 fingers and 20 toes!"

Pattern formation, based at least in part on retinoic acid, also helps determine the position of the various facial features, and alcohol consumed by a pregnant woman can interfere with the embryo's normal retinoic acid gradients and change the patterning of the neural crest, face, and brain. This explains the **fetal alcohol syndrome** brought on by heavy maternal drinking and why it can cause several devastating birth defects: mental retardation due to impaired brain development; low birth weight; slow growth; and abnormalities of the face and skull, including small head, small eyes, and under-development of the upper lip and jaw. Full-blown fetal alcohol syndrome occurs in one of about every thousand births. But recent studies suggest that low-level, intermittent drinking during pregnancy can affect the child's learning, emotions, and attention in subtle ways and that this is far more common.

Another drug, called Accutane, can also interfere with retinoic acid's normal patterning effects on the face and head. Released in 1982, Accutane helps people with severe cyst-forming acne on their faces, chests, and/or backs. Ironically, Accutane has a structure similar to retinoic acid. If a woman becomes pregnant while taking Accutane, her baby may be born with a variety of defects, similar to those of fetal alcohol syndrome. The drug carries strong warning labels, yet hundreds of babies have been born with Accutane-induced birth defects.

Differentiation: Development of Organ Function

We've seen embryonic cells dividing, migrating, infolding, and following patterns in a sculpting process that

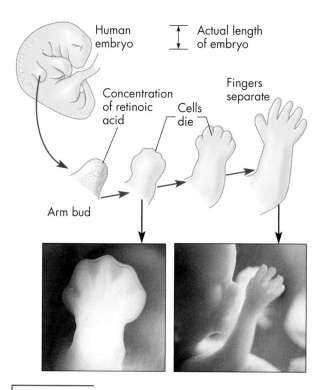

Human embryo

Actual length of embryo

Concentration of retinoic acid

Cells die

Fingers separate

Arm bud

Figure 8.16 **Development of the Hand.** (a) Your hand starts as a limb bud, then becomes a paddle. Fingers emerge when four rows of cells between the digits die away. In the arm bud stage, a gradient of retinoic acid, high near the little finger and low near the thumb, patterns the developing limb.

forms organs in the right places with the right shapes. But what about the function of those organs? Each type of cell in an animal's body performs a specific role: Red blood cells carry oxygen. Muscle cells move limbs and digits. Eye cells detect light, and so on. How does all of this come about? Research has shown that cells become specialized or differentiated for their particular roles because they activate specific genes and accumulate specific proteins; the **differentiation** process brings about this cell specialization. Red blood cells produce a protein that binds oxygen, muscle cells make proteins that act like little motors, and eye cells generate proteins that change shape when struck by light. Conversely, red blood cells don't make the light-sensitive protein or use the gene that encodes it, and eye cells don't make the oxygen-binding protein or use its gene.

One of the biggest mysteries in development has been this: What turns some genes on and others off in differentiated cells? Whatever the control signal, some biologists were convinced that the on-off control was permanent; in other words, that eye genes were permanently inactivated in muscle cells and muscle genes were eternally turned off in eye cells. But were they right? You may remember hearing about the lamb named Dolly a few years back; this little animal provided a test of this idea of permanent gene shut-down during development.

In 1997, Scottish researchers took the nucleus from a differentiated cell in an adult sheep's mammary gland (which is specialized to produce milk) and transplanted it into a sheep's egg, from which the egg nucleus had been removed. Eventually, that mammary cell nucleus programmed the cytoplasm of the sheep's egg and, with lots of technological support, a perfectly normal lamb, Dolly, developed and was born (Fig. 8.17). Clearly, the genes for cell types other than the mammary gland had not been permanently turned off in the nucleus they transplanted, even though it had come from a specialized (differentiated) cell in the milk-producing mammary gland. Since then, researchers have successfully cloned mice, pigs, and monkeys, as well.

Based on Dolly, other clones, and other kinds of evidence, biologists now conclude that virtually every differentiated cell retains all the genes present in the fertilized egg. Specialized cells just differ from other cell types in which sets of genes are turned on (expressed). In fact, proteins encoded by certain genes *select* all the other genes a cell will use. Biologists know this because these *selector* genes can become mutated, and this genetic alteration changes the function of an entire organ and transforms one body part into a totally different one. If you compare the two photos in Figure 8.18, you'll see the startling results of just such a mutation: An antenna on a fruit fly's head changes into a leg! Remarkably, you and I have a gene that is very similar in structure to this fly gene, and in its normal, nonmutated state it causes parts of the human brain to form in the right locations.

We've seen that the single fertilized egg can give rise to an entire human being with thousands of cell types. Yet a person's skin, liver, or kidney cell, when grown in a dish, will make only more skin or liver or kidney cells. Zygotes and adult cell types represent two extremes of potential. Between these lie **pleuripotent stem cells,** cells that can continue proliferating, and when given the appropriate signal, can differentiate into many of the cell types necessary to grow a complete individual, but not all. Pleuripotent stem cells reside in the inner cell mass (review Fig. 8.19 Step 5). As development proceeds, pleuripotent stem cells become more and more specialized into blood stem cells, or skin stem cells, or so on.

Figure 8.17 **Dolly Solves a Mystery.** Dolly, the famous cloned sheep developed from a single mammary gland cell taken from an adult sheep. Her successful cloning showed that differentiated cells can be reprogrammed to perform all of the functions needed during development.

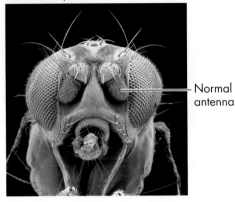

(a) Normal fly head

Normal antenna

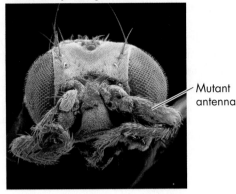

(b) Mutant fly with legs on head

Mutant antenna

Figure 8.18 **Selector Genes Control Development.** (a) A normal fruit fly head. (b) A mutant fruit fly head with legs replacing antennae. Here, the mutated gene changes which body parts form in which locations.

In the late 1990s, biologists began to obtain embryos produced in in vitro fertilization clinics—embryos that couples were not going to use and decided to donate to research. The experimenters isolated the inner cell mass cells, grew them, and found that some could differentiate into nerve cells, liver cells, kidney cells, and other cell types. They foresee testing new medications on these stem cell lines growing in culture instead of on human beings or other animals to streamline drug development and decrease the number of animals used in research. They also see using stem cells to provide replacement cells for patients with Parkinson's disease, Alzheimer's disease, spinal cord injuries, burns, stroke, heart disease, diabetes, or arthritis.

Much research is needed before these could become commonplace. Researchers must (1) come to understand what causes cells to differentiate into mature nerve, bone, or other cell types; (2) overcome the immunological rejection of transplanted cells derived from stem cells; (3) further investigate stem cells from adults to see if some could be coaxed to produce new cell types as do pleuripotent embryonic stem cells.

Much debate surrounds research on embryonic stem cells because of some people's religious beliefs about human embryos and in mid 2001, the U.S. government issued restrictions on their use in research, banning the use of new embryos and limiting research to stem lines already established from past embryos.

8.5 Human Development

To Marcy and Stephen, the arrival of their plump, dark-eyed baby boys Jack and Craig was a modern miracle made possible by *in vitro* fertilization. Human develop-ment is awesome, whether started by normal conception or helped along through medical intervention. But in many ways it is also utterly ordinary, logical, and explainable, and it is based on the same principles uncovered in animal research. Let's go through the major steps of development in babies such as theirs.

After natural conception, the fertilized human egg divides while it moves along the oviduct toward the uterus (Fig. 8.19 and *E* 8.4). In an assisted conception, the early divisions take place in a lab dish. But in both cases, by day 4 the embryo consists of about 30 cells, and a cavity has formed in its middle (Steps 4 and 5). The embryo, now called a blastocyst (Step 5), has two groups of cells: (1) the inner cell mass, which develops as the embryo proper, such as the frog embryo in Figure 8.10; and (2) the surrounding trophoblast, which forms nutritive tissue.

Six days after fertilization, the blastocyst consists of about 100 cells. By now, it has moved into the uterus (or a physician like Dr. Batzer has inserted it there directly); soon it attaches to the uterine wall, secretes enzymes that break down a small portion of the wall's outer layer, and burrows into it. This sequence establishes the first physical connection between mother and young and is called **implantation** (Fig. 8.19, Step 6). It marks the beginning of **pregnancy:** the development of an embryo in the uterus.

For infertile couples like Marcy and Stephen, conception and implantation are a physical and emotional struggle. But for most couples, these steps leading to pregnancy are easy—often, too easy. In fact, every minute of every day, 230 babies are born throughout the world, but only 90 people die. This leaves a net increase of 140 people per minute or 1.4 million additional people every week! For personal reasons as well as environmental ones, many people around the world seek to prevent unwanted pregnancies through **contraception,** or birth control. (*E* 8.2 investigates the many methods of birth control, how they prevent fertilization and implantation, and their relative effectiveness.)

Figure 8.19 **The Marvel of Human Development** ▶

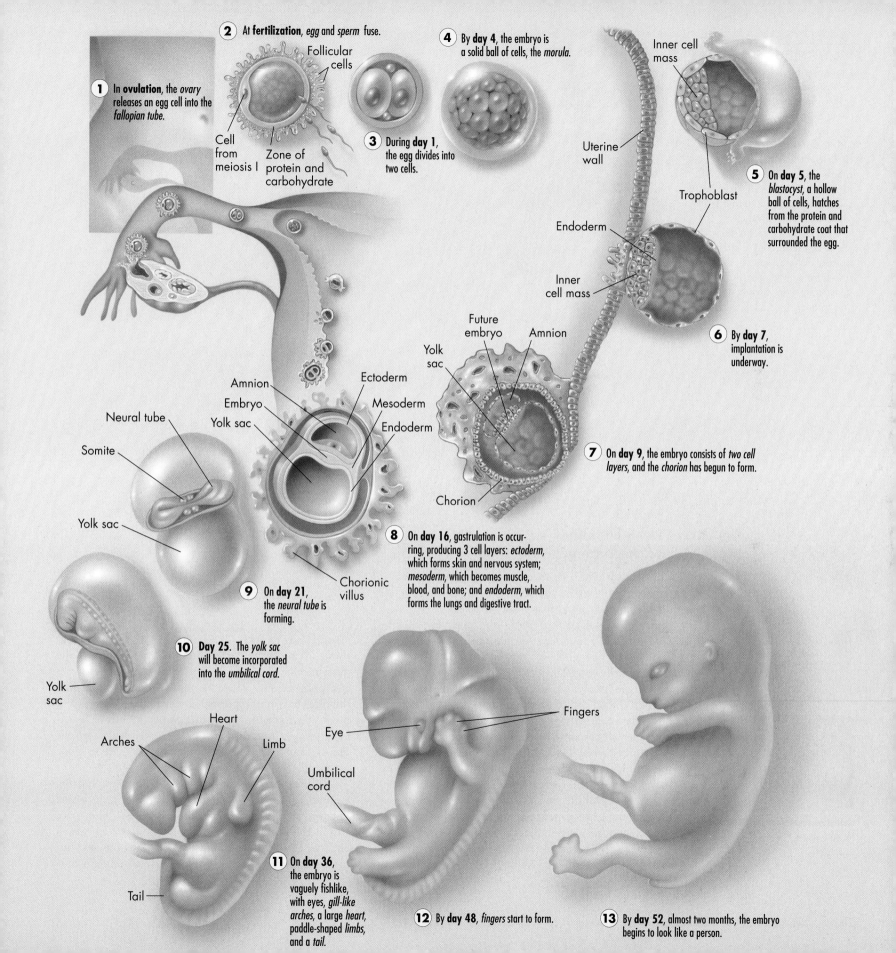

1 In **ovulation**, the *ovary* releases an egg cell into the *fallopian tube.*

2 At **fertilization**, *egg* and *sperm* fuse.

Follicular cells

Cell from meiosis I

Zone of protein and carbohydrate

3 During **day 1**, the egg divides into two cells.

4 By **day 4**, the embryo is a solid ball of cells, the *morula.*

Inner cell mass

Uterine wall

Trophoblast

Endoderm

Inner cell mass

5 On **day 5**, the *blastocyst*, a hollow ball of cells, hatches from the protein and carbohydrate coat that surrounded the egg.

6 By **day 7**, implantation is underway.

Future embryo

Amnion

Yolk sac

Chorion

7 On **day 9**, the embryo consists of *two cell layers*, and the *chorion* has begun to form.

Amnion

Embryo

Yolk sac

Ectoderm

Mesoderm

Endoderm

Neural tube

Somite

Yolk sac

Chorionic villus

8 On **day 16**, gastrulation is occurring, producing 3 cell layers: *ectoderm*, which forms skin and nervous system; *mesoderm*, which becomes muscle, blood, and bone; and *endoderm*, which forms the lungs and digestive tract.

9 On **day 21**, the *neural tube* is forming.

10 **Day 25**. The *yolk sac* will become incorporated into the *umbilical cord.*

Yolk sac

Arches

Heart

Limb

Tail

Eye

Umbilical cord

Fingers

11 On **day 36**, the embryo is vaguely fishlike, with eyes, *gill-like arches*, a large *heart*, paddle-shaped *limbs*, and a *tail.*

12 By **day 48**, *fingers* start to form.

13 By **day 52**, almost two months, the embryo begins to look like a person.

Figure 8.20 **Prenatal Diagnosis of Defective Genes and Chromosomes.** (a) In amniocentesis, fluid taken from the amniotic cavity around 15 to 16 weeks into the pregnancy contains cells from the fetus. In chorionic villus sampling (b), the physician removes cells from the chorion around 10 to 12 weeks into the pregnancy. With both sets of techniques, the cells are cultured in a laboratory dish and can be tested for a large variety of genetic and chromosomal conditions.

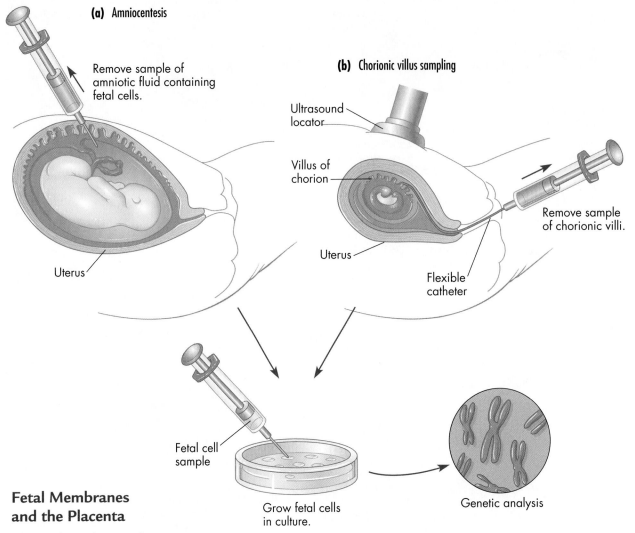

(a) Amniocentesis

Remove sample of amniotic fluid containing fetal cells.

Uterus

(b) Chorionic villus sampling

Ultrasound locator

Villus of chorion

Uterus

Remove sample of chorionic villi.

Flexible catheter

Fetal cell sample

Grow fetal cells in culture.

Genetic analysis

Fetal Membranes and the Placenta

The embryo has a three-way ticket to survival in the form of a double-layered, fluid-filled sac that surrounds and nurtures it with an outer layer or **chorion** and an inner layer or **amnion** (Fig. 8.19, Steps 7 and 8). The embryo's surrounding layer of trophoblast cells contributes to the chorion, and this layer does three things: (1) It absorbs nutrients from the mother's blood and passes them on to the rapidly dividing embryo; (2) it contributes to the dark, spongy placenta, which sustains the embryo throughout the nine months of pregnancy; and (3) it produces the hormone **human chorionic gonadotropin (hCG).** This hormone prevents the onset of a new menstrual cycle, which would flush the implanted embryo from the uterus. Home pregnancy tests use a simple but ultrasensitive system for detecting hCG, and 99 percent of the time these tests accurately reveal a pregnancy that is just two or three weeks old. As hCG enters the mother's

bloodstream (starting at about day 10 after fertilization), it causes the corpus luteum, the remains of the ovarian follicular cells, to continue to produce estrogen and progesterone. These hormones, in turn, maintain the uterine lining, prevent menstruation, and allow pregnancy to continue for the first two months. "Morning after pills" such as RU486 and newer drugs stop the uterus from receiving progesterone and thereby block pregnancy.

The chorion is the first layer to form around the embryo, but it is soon joined by a second layer inside the first, the amnion (Fig. 8.19, Step 8). The space within the amnion becomes the amniotic cavity, which encloses a watery, salty fluid that suspends the developing embryo in a relatively injury-free environment dur-

How can hCG reveal a pregnancy?

ing pregnancy (Fig. 8.19, Step 9). The embryo sloughs off cells into this amniotic fluid. Physicians can collect these cells (usually at weeks 15 to 16 of pregnancy) and analyze them for possible genetic abnormalities in a process called **amniocentesis** (Fig. 8.20a). As the chorion grows, it produces little fingers of tissue called **villi** (singular, villus) that become enmeshed with maternal tissue. By using an alternate diagnostic procedure called **chorionic villus sampling** earlier during pregnancy (at weeks 9 to 12), a physician can collect chorion cells from these fingers and look for genetic or chromosomal problems in the fetus (Fig. 8.20b). The enmeshment of chorion and maternal tissue forms the dark-red, spongy **placenta,** an exchange site where a thick tangle of embryonic blood vessels encounters blood-filled spaces in the uterine lining. The placenta, in turn, forms a vital link allowing the diffusion of CO_2, O_2, nutrients, and wastes between mother and embryo (Fig. 8.21). The umbilical cord is a lifeline connecting the embryo to the placenta.

In the placenta, the embryo's blood does not mingle with the mother's blood, but nutrients and oxygen pass from her blood across embryonic vessel walls into the embryo's blood supply. In the reverse direction, carbon dioxide and other wastes pass from the embryo into the placenta. After a few weeks, the placenta begins to produce estrogen and progesterone. These hormones maintain the uterine lining and block the production of LH and FSH. This blockage, in turn, prevents menstruation and the end of pregnancy. The corpus luteum then slowly degenerates and stops releasing estrogen and progesterone. At the 8-week stage, the major organ systems and external features have formed, and the embryo is now called the **fetus.** Fetal development continues until about 9 months. The three 3-month periods of pregnancy are called **trimesters.**

Mother's Contribution to the Fetal Environment

A woman must be especially careful throughout all three trimesters of pregnancy because the placenta forms such an intimate physical connection between her and her offspring. Most of the foods, drugs, and chemicals that the mother takes into her body pass through the placenta to the embryo—and later, the fetus—and can have a profound effect on it.

Figure 8.21 **Fetal Membranes and the Placenta.** By 60 days after fertilization (a), the placenta (b) is well established and produces the hormones that prevent menstruation for the remainder of the embryo's 9-month gestation. Narrow fingers of tissue, or chorionic villi, project from the chorion, and each eventually houses a tiny blood vessel. The maternal blood fills the spaces around the villi, and exchange of gases and materials takes place across the delicate layer separating the maternal and fetal blood supplies.

(a) A pregnant uterus at 2 months

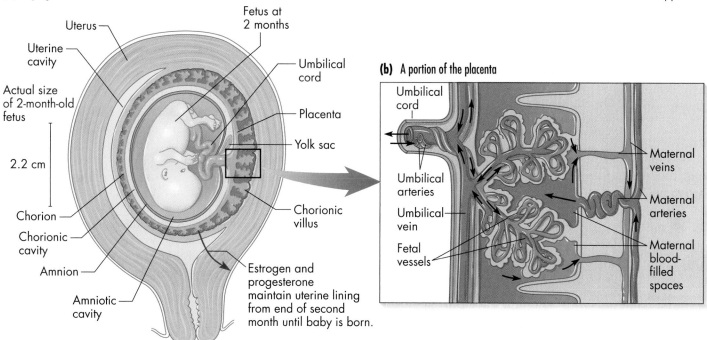

For proper growth of brain, muscles, and bones, the fetus requires a constant supply of protein and calcium, as well as fatty acids, carbohydrates, vitamins, and minerals. Because the mother's diet must provide these nutrients, obstetricians usually advise pregnant women to drink lots of milk, eat plenty of protein, and take vitamin and mineral supplements. As we saw earlier, folic acid is essential for normal neural development and can prevent the spina bifida we saw in Fig. 8.12. A maternal weight gain of 11 kg (25 lb) or so is now believed appropriate for most women to help prevent the birth of premature or underweight infants, who have a greater susceptibility to infections and breathing problems than do full-term infants.

A mother's smoking, drinking, or drug use can have severe repercussions for her fetus, including low birth weight, prematurity, and specific developmental problems. (We talked about the consequences of alcohol earlier.) Many obstetricians recommend that their patients avoid alcohol and cigarette smoking during pregnancy. In mothers who smoke, miscarriage (premature expulsion of the fetus) is much more likely, and surviving babies tend to be underweight and susceptible to respiratory disease and sudden infant death syndrome (suffocation during sleep). Mothers who take amphetamines or cocaine risk having infants with neurological defects, and mothers who take heroin or other narcotics often give birth to premature, addicted infants.

Some of the saddest chapters in modern medicine concern the fetal damage that can be caused by certain prescription drugs. The antinausea drug thalidomide calms adults' nerves, but it may also alter the embryonic nerves that help direct proper limb development and growth. Some pregnant women who took thalidomide in the early 1960s gave birth to mentally and emotionally normal babies who had shortened arms and deformed hands. Clearly, the drug had a major effect on the human embryonic limb bud, but surprisingly, had no such effect on mice or rats, the only animals tested initially. As a result, its devastating consequences remained undiscovered during premarket laboratory tests.

Premarket drug testing has become much more stringent in recent years, and more animal species have been included in them (rabbits and monkeys are susceptible to thalidomide). However, pregnant women must still be wary of exposure to drugs and other chemicals, as well as to certain viral diseases such as German measles (rubella). Pregnancy is clearly a time when a woman must take special care of her general health and nutrition—and, indirectly, her baby's.

Why is the first trimester so crucial and delicate?

Developmental Stages in a Human Embryo

Marcy and Stephen underwent several *in vitro* cycles at great emotional and financial expense. By the third week after Marcy's last *in vitro* procedure, a nurse from Dr. Batzer's office had called to tell the excited parents that a pregnancy was underway. The twin embryos were still smaller than the length of this "l," but by then gastrulation had already established their three-layered body plan (Fig. 8.19, Steps 8 and 9). During the next weeks and months, the nervous system and other organs would form, and the tiny masses would be transformed into organisms with a characteristic human shape.

The First Three Months

The formation of body organs begins as the neural tube rolls up midway through the third week of pregnancy. Primitive blood cells and blood vessels have already formed. During the fourth week (see Fig. 8.19, Step 10), the heart begins to pump blood. By the fifth week, the primitive brain looks like a miniature, lumpy inchworm, and its cells are dividing at such an inconceivably fast rate that 50,000 to 100,000 new neurons are generated each second. The limb buds are now visible, and the intestine is a simple tube. The liver, gallbladder, pancreas, lungs, eyes, nose, and brain begin to form. Although the embryo is only about 5 mm (less than 0.25 in.) in length from crown to rump, it is already about 10,000 times larger than the fertilized egg.

When the embryo (now officially called the fetus) reaches its eighth week (see Fig. 8.19, Step 13), most of its organs are present, and during the rest of gestation they simply enlarge and mature. The fingers and toes are well formed by the mechanisms of limb development we saw earlier, and the head begins to lift away from the chest. The first indications of skeletal bone formation can be seen during the ninth week, and by then the fetus can also bend its body, hiccup, and respond to loud sounds with increased movements. At ten weeks, it can move its arms, open its jaws, and stretch. At this point, the fetus is about 23 mm (1 in.) in length from crown to rump. The first trimester ends at 12 weeks; by this time, the fetal pulse is detectable, the fetus is about 40 mm (1.5 in.) long, and it can yawn, suck, swallow, and respond to touch with movements and quickened heartbeat.* The primary sex organs develop during the first

*For more details, see Hopson, Janet. Fetal Psychology. *Psychology Today*, September/October 1998.

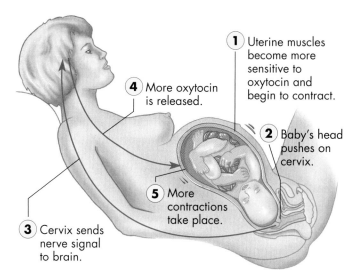

(1) Uterine muscles become more sensitive to oxytocin and begin to contract.

(2) Baby's head pushes on cervix.

(3) Cervix sends nerve signal to brain.

(4) More oxytocin is released.

(5) More contractions take place.

Figure 8.22 **Birth: Hormones Trigger Contractions and Delivery.** As birth approaches, uterine muscles become more sensitive to oxytocin (1) and they contract, pushing the baby's head into the mother's cervix and causing it to widen (2). The stretched cervix sends a signal via nerves to the brain (3), which causes the release of more oxytocin (4). This increase in oxytocin causes more uterine contractions, and the cycle escalates (5). The cycle continues until the baby is expelled through the birth canal.

trimester, but you can't tell a boy from a girl at that stage because the gonads, sex ducts, and external genitalia look the same in both sexes. The chromosomal tests we mentioned earlier (amniocentesis and chorionic villus sampling), however, can determine gender at this stage.

The Second Three Months

The second and third trimesters of pregnancy are devoted to the increase in size and maturation of the organs developed during the first trimester. During the 12th week, the mother feels her uterus enlarging. By the end of the 15th week, the fetus measures about 56 mm (2.25 in.) from crown to rump. By the 16th week, its face looks "human." From the 20th to the 24th weeks, the body becomes covered with downy hair, a stethoscope detects fetal heart sounds, and the fetus responds (through increased motion and heart rate) to its mother's voice. The lungs have formed but do not yet function, and the gripping reflex begins, along with tactile explorations of the umbilical cord, the amniotic sac, and the fetus's own face, feet, and toes.

Getting Ready for Birth

At the beginning of the third trimester, the fetus's eyelids open, its eyebrows and eyelashes form, and it can detect light. At this stage, with almost two months to go, a baby born prematurely would have at least a 10-percent chance of survival. The fetal brain grows rapidly throughout pregnancy, and eventually the outermost layer—the cerebral cortex—fills 80 percent of the skull, with two thirds of its surface area tucked into elaborate grooves and creases. During the final couple of weeks, fetal activity decreases because of its size and the lack of space in the mother's uterus. The fetus is said to have come to term, about 280 days after the mother's last menstruation before conception. When the mother is carrying twins or other multiples, the term of pregnancy is often shorter, and the babies are smaller.

Birth and Lactation

For Marcy, conception was extraordinary and highly technological, but the rest of her pregnancy and her delivery were typical. As labor and delivery approached, her two fetuses prepared themselves and her body for the coming events. Like all normal fetuses, they stored special brown fat around the neck and down the back to help generate heat after expulsion from the warm uterus. Special reserves of carbohydrates accumulated in the heart and liver as a source of nourishment until the babies could suckle milk. The placenta secreted a hormone that prepared Marcy's mammary glands to produce milk. The muscles of her uterus became more excitable, contracting periodically in "false labor," building strength as the time of birth approached.

During the last trimester, her uterus developed a 100-fold increase in sensitivity to the hormone oxytocin (Fig. 8.22). Oxytocin is produced in the hypothalamus in the brain and is transported to a pregnant woman's pituitary gland, which releases it. When sensitivity to oxytocin reaches a threshold, labor—the physical effort and uterine contractions of childbirth—begins. Nerves in the cervix signal the mother's brain to release more oxytocin, which causes the uterine muscles to contract even more. Oxytocin also stimulates the uterine wall to release prostaglandins—hormones that stimulate uterine contractions still further. This causes the release of even more oxytocin in a positive feedback loop that causes more and more oxytocin to be released until the

How does oxytocin help bring about birth?

final event, birth, ends the loop. (Chapter 18 gives more details on this loop.)

As contractions grow longer, stronger, and more regular, the cervix widens. Each contraction starts at the upper end of the uterus and moves downward toward the cervix, which the baby's head pushes farther and farther open as it moves toward the vagina, or birth canal. This period can span 12 hours or longer, or it can speed by in minutes (usually on second and subsequent births)—explaining why some babies are born in taxis! Uterine squeezing eventually causes the amniotic sac to burst (the "water breaks"), and when the cervical opening reaches a width of about 10 cm (4 in.), delivery is usually only minutes away. With considerable pushing of her abdominal muscles, the mother is able to force the baby's head past the pelvic opening, allowing the baby's shoulders and hips to emerge. Marcy's twins were delivered by caesarean section; that is, her babies were removed through an abdominal incision. (The choice of this procedure was unrelated to the *in vitro* conception.)

Marcy was soon nursing the twins—a process most mothers enjoy and one that links us humans to all other mammals. During pregnancy, estrogen and progesterone cause a woman's milk-secreting glands to develop, and *prolactin,* a brain hormone, causes milk production after she has given birth. Breasts contain an extensive drainage system that includes many lymph nodes in addition to the milk-producing mammary glands. These lymph nodes are susceptible to breast cancer (the most common form of cancer in females), and so women must regularly examine their breasts and have a physician check for lumps or growths yearly.

8.6 Growth, Maturation, and Aging

Marcy and Stephen have been thrilled with their young Craig and Jack, and have been fascinated by all the stages of growth (size enlargement) and development that follow conception, pregnancy, and birth. Says Stephen, "It's all been simply remarkable—a miracle!"

If development stopped at birth, we'd all be perpetual babies. Instead, development and growth continue throughout childhood and adolescence, ushering in significant change and maturation, especially in the brain and reproductive organs. At birth, the brain is the most developed organ and is 25 percent of its final adult weight. By age 3 it has reached 75 percent of its adult size and weight, and by age 10, 90 percent. Body weight lags far behind brain weight, advancing from 5 percent of adult weight at birth to only 50 percent at age 10. In girls between the ages of 11 and 13, and in boys between 13 and 15, the dramatic changes of sexual maturation or **puberty** take place; these include maturation of the reproductive system and development of the secondary sexual characteristics we talked about earlier.

With modern nutrition, sanitation, and health care, the physical maturity that starts at the end of adolescence begins an adult life that continues for 50 to 75 years or longer. The peak performance of all the body's organ systems usually occurs between the early twenties and the early thirties. Sometime after age 30, the process of **aging**—a progressive decline in the maximum functional level of individual cells and entire organs—begins to accelerate. At about age 50, females experience menopause, a cessation of the menstrual cycle, while males may experience a decrease in their ability to maintain an erection. Sometime after age 60 (or as late as 70 or 75 in active adults with healthy lifestyles), people reach **senescence,** or old age. The decline in cell and organ function is then less gradual and more profound.

What Causes Aging?

We know from watching our parents and grandparents—and even our cherished pets—that aging is natural and inevitable. But what makes it happen? Biologists have two main sets of ideas about aging: genetic clock hypotheses, and wear-and-tear hypotheses. Many experts support the idea of a genetic clock, arguing that there is a genetically specified timetable for aging and death. Just as genes regulate the timing of organ formation, they also influence the rate at which organ function slows. There is tantalizing evidence for the various genetic clock hypotheses. First, cells seem to have preset limits to the number of times they can divide. For example, skin cells taken from a human infant and grown in the laboratory divide about 50 times and then stop, while similar cells taken from a 90-year-old divide only a few times. Second, organs seem to age in a preprogrammed way. Each year, starting at about age 30, people (especially sedentary ones; see Fig. 8.23b) experience a 1-percent decline in the maximum function of various organs. This includes lung capacity, the amount of blood that the heart can pump, muscle strength and coordination, the filtering power of the kidneys, and the ability to hear high-pitched sounds.

Third, certain genetic conditions can bring about symptoms of aging. Children with progeria (early aging) begin to lose their hair, show wrinkles, and experience arthritis and heart attacks by age 5 or 6. These children, however, do not suffer the whole spectrum of age-related illnesses, which includes cataracts, diabetes, and cancer. While the complete genetic basis for diseases like progeria remains unknown, such conditions do seem to support the genetic clock concept.

Fourth, researchers have discovered genetically distinct strains of fruit flies and nematode worms that live twice as long as normal—the equivalent, in fact, to a person's living 150 years or longer. This suggests that aging is under genetic control. Exactly how the "Methuselah" genes extend the lives of flies or worms, however, will require further study. To other biologists, the evidence suggests that the most likely cause of aging is wear and tear: the accumulation of random errors in the replication and use of DNA and in the synthesis of proteins, or the accumulation of metabolic byproducts that disable enzymes, other proteins, and lipids. Informational errors may be due to environmental insults such as exposure to sunlight, radiation, or chemicals. One class of culprits may be toxic, chemically active, oxygen-containing molecules called free radicals that arise naturally during energy metabolism. The relentless piling up of metabolic garbage probably interferes with normal cell function, including the flow of information from DNA to RNA to protein, and perhaps contributes to decline in organ function with age. Evidence on the long-lived flies and worms we just discussed reveals that these animals have increased levels of an enzyme called superoxide dismutase, which breaks down free radicals of oxygen. This finding is extremely provocative in that it may lead to a theory unifying the genetic and the wear-and-tear hypotheses.

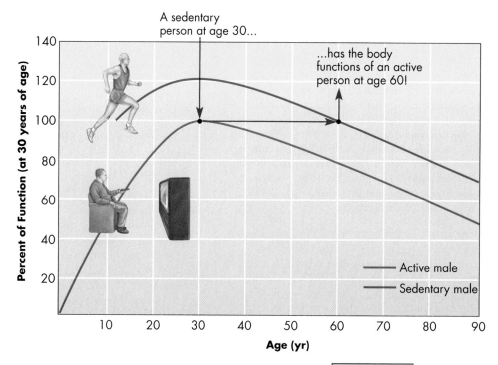

Figure 8.23 **Aging Gracefully.** Good habits, good health, and good times can help a person live a long and happy life. This graph shows how a sedentary lifestyle affects health and physiological function. A typical 30-year-old "couch potato" has the heart rate, lung capacity, and other physiological parameters of an active 60-year-old! The conclusion: Active people generally live significantly longer, healthier lives.

Unfortunately, researchers are still a long way from understanding the basic mechanisms of aging.

Regardless of what makes us age, studies show that certain health practices can significantly increase both our quality of life and our life expectancy. A man who eats regular meals (including breakfast), exercises regularly (Fig. 8.23), sleeps an adequate amount, maintains an ideal weight, does not smoke, and limits his alcohol consumption can live on average 11 years longer than one who follows only three or fewer of these practices. A woman who observes all six healthy practices can add seven years to her already longer life expectancy.

Chapter Summary and Selected Key Terms

Introduction   Some couples conceive the old-fashioned way, while others, like Stephen and Marcy, need the medical help of *in vitro* fertilization (p. 1) to unite egg and sperm in **fertilization** but outside the body in a glass lab dish. Regardless of the method, the zygote can become a human embryo, fetus, and baby through the mechanisms of **development** (pp. 2, 13).

Reproduction and Mating   Sexual reproduction in animals involves behaviors that promote the joining of a **sperm** (p. 2) and **ovum** (p. 2) in fertilization and the development of the fertilized egg into a new individual organism.

Sexual Characteristics   Each sex has a pair of **gonads** (p. 3) or organs that produce sex cells and sex hormones. In human males, the gonads are called **testes** (p. 3) and make sperm. These gametes are transferred through ducts via the **penis** (p. 6) into the female's reproductive tract. The large central tube through the penis is the **urethra** (p. 6). A negative feedback loop involving the hormone **testosterone** (p. 3) from the testes and two brain hormones, **luteinizing hormone (LH;** p. 6) and **follicle stimulating hormone (FSH;** p. 6), maintains a continual sperm supply.

Females make egg cells within gonads called **ovaries** (p. 3, 6). **Follicles** within the ovaries produce and release mature ova (eggs) via the process called **ovulation** (p. 7). The mature ovum bursts out and leaves behind a hormone-producing gland, the **corpus luteum** (p. 7). The ovum will move along one of the **oviducts** (p. 7) or **fallopian tubes** (p. 7) toward the womb or **uterus** (p. 7) where the **embryo** and later, the **fetus** (p. 8, 23) will develop before being delivered through the birth canal or **vagina** (p. 9). Hormonal feedback loops bring about the **menstrual cycle** (p. 9), the cyclic production and release of eggs and the preparation of the uterus for an embryo. Follicles within the ovaries produce the hormones **estrogen** (p. 9) and **progesterone** (p. 9). After ovulation, the corpus luteum's production and release of progesterone help build up the uterine lining.

Fertilization

Sexual intercourse can bring egg and sperm together in the **oviducts** (p. 7) and initiate the process of fertilization. If the egg goes unfertilized, the corpus luteum degenerates, and the uterine lining sloughs off in the menstrual flow.

Contraception (p. 20) is achieved by preventing fertilization or **implantation** (p. 20) or by terminating pregnancy. To overcome infertility, people like Marcy and Stephen may need hormone-like drugs, corrective surgery, *in vitro* fertilization, insemination with donor gametes, or the help of a surrogate.

Embryonic Development

Once conceived, human embryos pass through the same developmental stages that unfold in other vertebrate animals. Development involves a series of mitotic cell divisions (**cleavage** [p. 14]) that results in many cells' arising from the single fertilized egg. It also involves **gastrulation** (p. 14), which organizes three main cell layers: ectoderm, mesoderm, and endoderm. Development continues with the formation of body organs (**organogenesis** [p. 14]), such as the unfolding of the neural tube, and throughout development, an increase in body size (**growth** [p. 16]). Developmental biologists have learned a great deal about the mechanisms of development. These include induction as certain groups of cells induce others to form emerging organs in the correct locations (p. 16); **morphogenesis** or the shaping of cells, cell migrations, and programmed cell death (p. 17); the formation of developmental patterns by substances like retinoic acid that help establish organs in the proper spatial relationship to each other (p. 18); and **differentiation,** cells taking on specialized roles (p. 19). They've also learned about developmental mistakes that can lead to birth defects like spina bifida and fetal alcohol syndrome. The famous cloned sheep Dolly helped show that differentiated cells retain all their genes in a form that can be turned off and on again.

Human Development

A human **pregnancy** (p. 20) like Marcy's is a partnership between embryo and mother. After **implantation** (p. 20), the embryo develops a double-layered sac with an outer layer, the **chorion** (p. 22), which produces the hormone **human chorionic gonadotropin (hCG)** (p. 22); this prevents endometrial sloughing. The chorion grows and enmeshes with maternal tissue, forming the **placenta** (p. 23). A second, inner layer of the sac, the **amnion** (p. 22), surrounds the embryo and contains the cushioning amniotic fluid. The mother's smoking, drinking, and other lifestyle habits can affect fetal development. The fetus secretes hormones that prepare the mother's breasts for milk production and cause her body to release the signals that initiate labor and delivery.

Growth, Maturation, and Aging

Development continues after birth and changes accumulate throughout life. Infancy and childhood are times of phenomenal growth and mental development. **Puberty** (p. 26) involves additional growth and sexual maturation. During adulthood, physical performance peaks and then declines until **senescence** (p. 26) and **aging** (p. 26) set in. Biologists do not yet understand the causes of aging but have several theories.

All of the following question sets also appear in the Explore Life **E** electronic component, where you will find a variety of additional questions as well.

Test Yourself on Vocabulary and Concepts

In each question set below, match the description with the appropriate term. A term may be used once, more than once, or not at all.

SET I

(a) testis (b) follicle (c) brain (d) corpus luteum
(e) chorion (f) interstitial cell

1. Site of interstitial cells
2. FSH is made here
3. Secretes estrogen and progesterone
4. Secretes testosterone
5. A temporary structure that develops after ovulation
6. Structure that surrounds a maturing oocyte
7. Structure(s) that secrete hormones that block gonad-stimulating brain hormones
8. Secretes substance that maintains uterine lining

SET II

(a) trophoblast (b) neural crest (c) inner cell mass
(d) cleavage stage embryo (e) fetus

9. The cells of a very young mammalian embryo that will be the source of all the cells of all the body parts as development proceeds
10. Cells of the early mammalian embryo that will contribute to the formation of the placenta, but not to the embryo proper
11. Cells that wander far from their site of origin and give rise to certain structures in the face and nervous system and to the pigmented cells of the skin

SET III

(a) differentiation (b) organogenesis (c) implantation
(d) cleavage (e) neurulation

12. The process during which ectoderm rolls up to form a tube along the embryo's dorsal side
13. The acquisition of a specialized function due to particular cellular proteins
14. The first physical connection between mother and embryo; accomplishes pregnancy
15. Rapid cycles of cell division that transform a zygote into a ball of many cells

Integrate and Apply What You've Learned

16. Suppose a simple at-home test could indicate peaks in the levels of LH, estrogen, and progesterone. How could such a test be used in family planning and birth control?
17. Describe the events that occur immediately following fertilization. In what ways are they different from events following *in vitro* fertilization?
18. Draw a sketch to outline the feedback loop that maintains the sperm supply in males.

Analyze and Evaluate the Concepts

19. Which hormones do you think Dr. Batzer used to inject Marcy to prepare for *in vitro* fertilization? Justify your answer.
20. Which hormones do you think Dr. Batzer monitored when she took blood from Marcy each day?
21. Which do you think is the bigger problem in the world, infertility or overpopulation? Support your answer. Why do you think there are more U.S. businesses involved in infertility research than in contraception research?
22. Suppose you are the chief executive officer of a biotechnology firm. Suppose that one of your chemists had produced a substance that blocks the action of the hormone LH. How would you prepare a marketing campaign to explain to people the benefits of this drug and in what situations they might consider using it? How would you prepare your legal department for lawsuits?

PART 2 Evolution and Diversity

CHAPTER 9
The Mechanisms of Evolution

"They dropped to one medication and within 72 hours my foot was inflamed. It was a horrible mess!"

CHAPTER 10
Life's Origins and Biodiversity

"Where did life begin . . . in a shallow pool on a newly formed land mass?"

CHAPTER 11
Single-Celled Life

". . . Nigeria, West Africa. That's where I got malaria . . . it felt like someone was jack-hammering inside my skull."

CHAPTER 12
Fungi and Plants: Decomposers and Producers

"There seems to be a mystique to chiles . . . you don't see rutabagas on people's T-shirts."

CHAPTER 13
The Evolution and Diversity of Animals

"I saw this 10-lined June beetle racing across the streets of Vancouver and I thought, 'Wow, that is cool!'"

Resistance on the Rise

▲ **A Modern Instance of Evolution.** Wayne Chedwick with Dr. Fierer and with foot wounds visible; Dr. Fred Tenover; colonies of Staphylococcus aureus.

To Wayne Chedwick (a pseudonym), *Staphylococcus aureus* is an unseen enemy that robbed him of his livelihood, his ability to walk, and parts of his feet. This common bacterium grows in smooth, tidy clusters of spherical cells (see Fig. 11.4). To the naked eye, it forms shiny, yellowish, almost pleasant-looking "staph" colonies on culture plates. In any college lecture hall, movie theater, or ballpark, up to 10 percent of the crowd will have staph populations living harmlessly in their noses and throats. But *S. aureus* can quickly turn against people—regular carriers and others—and cause boils, ulcerated wounds, bone infections, even blood poisoning. Mr. Chedwick has had them all.

A 58-year-old former building contractor from Encinitas, California, north of San Diego, Wayne has suffered from adult-onset diabetes for over a decade. This blood-sugar disorder often damages blood vessels and nerves of the feet and legs. Wayne had simple blisters on his toes but couldn't feel them until staph infections set in, attacking and destroying soft tissue and then creeping into several toe bones. Eventually, the bacteria caused so much destruction that doctors had to amputate one toe on his left foot and three more on his right foot. In time, *S. aureus* cells infected ankle bones on that right foot, as well, and then found their way into his blood stream, causing a life-threatening case of blood poisoning, or toxemia. Doctors at the Veteran's Administration hospital in San Diego got the toxemia under control but the bone infection clung so tenaciously and painfully that Wayne became wheelchair-bound, unable to bear weight on his right foot. Once an active building contractor, he is, at this writing, staying home crafting hardwood espresso coffee carts. He prays that he won't lose his entire foot and that he can someday walk again.

Where, you might be wondering, did antibiotics come into this picture, and why didn't they help Wayne? His doctors prescribed various antibacterial drugs, of course. But the *Staphylococcus* strain colonizing his foot bones can't be killed by most of the usual antibiotics, including methicillin, erythromycin, and clindomycin. The reason? The strain he contracted has, over time, acquired genes for drug resistance: it has evolved new and dangerous genetic traits. Dr. Joshua Fierer, an infectious disease expert at the University of California, San Diego, has treated Wayne with a fairly new and unusual sulfa–antibiotic combination and as of this writing, is cautiously optimistic he can save Wayne's foot. Things could actually have been even worse. Wayne's staph infection could also have been resistant to vancomycin, a rarely used drug. Doctors have long considered vancomycin the failsafe weapon against *S. aureus* when all other antibiotics fail. But in recent years, physicians have recorded a handful of cases of resistance to vancomycin, as well.

In March, 1998, for instance, a 79-year-old retired detective from New York died when vancomycin-resistant *Staphylococcus* multiplied in his blood and couldn't be controlled. And two years earlier, a 4-month-old Japanese baby survived open heart surgery only to develop abscesses around his small heart from *Staphylococcus aureus* resistant to both methicillin and vancomycin. Only by administering several new antimicrobial drugs and by manually draining the abscesses was the medical team able to save the baby's life.

At the Federal Centers for Disease Control and Prevention in Atlanta, Dr. Fred Tenover has been carefully tracking these new cases of vancomycin resistance and considers them "worrisome" and potentially "a major threat." Each year, says Tenover, 2 to 3 million people pick up an infection while in the hospital. Over 70 percent of these "are resistant to at least one antibiotic," he explains, and the infections claim 90,000 lives annually. "There are some organisms, although very few," he goes on, "that are now resistant to *all* antimicrobial agents." Among these are a strain of typhoid fever bacteria in Southeast Asia where "we've run out of antibiotics," and tuberculosis strains "that are resistant to seven, eight, nine drugs."

These cases of near and total antibiotic resistance have profound medical implications, conjuring a world where simple infections take lives as they often did before penicillin became widely available in the 1940s. But the cases have obvious biological significance, as well. They are modern examples of evolution by natural selection, our subject in this chapter. People often think of evolution in terms of dinosaurs, fossil bones, and Charles Darwin's the-

ory. But chance mutations, and society's overuse of antibiotics in medicine and agriculture have brought about clear-cut examples of evolution in our own lifetimes. The results are these nearly invincible super germs that can cost people like Wayne Chedwick their appendages, if not their lives.

By exploring this chapter, you'll see how natural selection brings about antibiotic resistance, but also how it engenders Earth's stunning diversity of plant and animal life. You'll encounter Charles Darwin and Alfred Russell Wallace once again, as we did in Chapter 1. You'll see the patterns of evolution that established the dinosaurs and Ice Age mammals. You'll encounter peoples from various world regions. And you'll see how natural selection brings about new species, eliminates poorly adapted ones, and helps outfit organisms with successful survival tools such as genes for resisting antibiotics. Along the way, you'll find the answers to these questions:

OVERVIEW

Explorer 9.0

Antibiotic Resistance

1. How did naturalists develop the concept of evolution?

2. What is the evidence for evolution?

3. What are the various patterns of evolutionary change?

4. How does genetic variation arise?

5. What principles govern the inheritance of genetic variation in populations?

6. What are the mechanisms of evolution and how do they work?

7. How does natural selection fit populations to their environments?

8. What are species and how do they originate?

9. What principles can explain evolution above the level of the species?

9.1 Emergence of Evolution as a Concept

Dr. Tenover's explanation for how antibiotic resistance evolves in bacteria is clear, easy to grasp, and helps us understand **evolution,** or changes in gene frequencies in a population over time. Antibiotics kill virtually all of the cells in a population of, say, *Staphylococcus aureus*. In a few rare cells, however, mutated DNA may give the organisms resistance to one or more drugs. These rare mutants survive drug treatment, divide, and produce a new population with different gene frequencies than the original population. But does evolution operate the same way in other organisms? And how does evolution account for the tremendous diversity of life? There are millions of kinds of insects, hundreds of thousands of kinds of plants, and so on. Did evolution produce them all and if so, how? And how did biologists like Fred Tenover come to understand the mechanisms of evolution?

Until the end of the 18th and beginning of the 19th centuries, most naturalists believed that each species had been created separately and had remained unchanged from their creation to the current day (Fig. 9.1a). They thought, for example, that striped bass and sparrows were created at the beginning of the world and have remained exactly the same ever since.

About this same time, however, scientific exploration of the natural world was already uncovering facts that contradicted the notion of a single creation event and unchanging species. European explorers had witnessed many strange plants and animals in distant lands, such as the platypus in Australia and the dodo in Madagascar. If all types of organisms were created at a single place and at a single point in time, then *why were there different groups of organisms in Earth's different regions?* Besides that, geologists had discovered fossils that were quite different from organisms living today. If all organisms were created at one time, then *why would ancient extinct organisms be different from modern living species?* Furthermore, anatomists had found that the limbs of various mammals, such as whales, bats, and people, contain bones that are clearly modifications of a single basic plan, even though they are specialized to perform very different functions such as swimming, flying, and manipulating objects. If each species had been created individually and never changed, then *how could the same basic bone plan be the best design for swimming and flying and fine manipulation?* Faced with such puzzles, a few naturalists began to suggest that populations of organisms might have changed, or evolved, over time.

Lamarck: An Early Proponent of Evolutionary Change

French naturalist Jean Baptiste Lamarck (1744–1829) agreed with other scientists of his day that each species was created separately. According to Charles Darwin, whom we first encountered in Chapter 1, however, Lamarck, in 1801, was the first writer to attract much attention to the idea that species (including humans) are descended from other species (Fig. 9.1b). Lamarck also attributed this change to natural laws, not to miracles, and proposed a hypothetical mechanism for how this change could come about.

Lamarck was intrigued, for instance, by the graceful, towering giraffe with its extremely long neck and legs. He knew from the work of earlier naturalists that the giraffe's long neck allowed it to eat leaves from branches higher than wildebeests, zebras, elephants, and other large African grazing mammals could reach. Lamarck suggested two main points. First, he proposed that the physical needs of an animal such as the giraffe determine how its body will develop. Early giraffes, he wrote, must have stretched their necks trying to graze on the leaves of high branches. Second,

Figure 9.1 **How Did Species Arise? Three Hypotheses**. (a) Most 19th-century scientists believed that species were created individually and remained the same throughout time. (b) Lamarck proposed that created species could change over time. (c) Darwin, Wallace, and most later biologists theorize that all organisms descended with modifications from common ancestors.

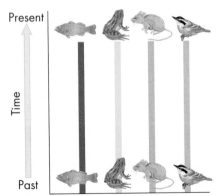

(a) Creation of unchanging species

(b) Creation of species that evolve

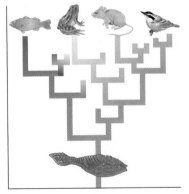

(c) Descent with modification

Lamarck proposed that changes in organ size caused by use or disuse would be inherited. For example, the longer neck a giraffe acquired through stretching would be passed on to its offspring. Later biologists called Lamarck's hypothesis the **inheritance of acquired characteristics.** With no understanding of inheritance at the cellular level, however, this hypothesis was impossible to test. Even Darwin agreed that use and disuse could influence evolution in, for example, the vestigial eyes of moles and cave-dwelling animals.

Only after the rediscovery of Mendel's laws in the early 1900s (Chapter 5) and detailed explanations for the mechanisms of development (Chapter 8) could biologists successfully challenge Lamarck's notion of the inheritance of acquired characteristics. Recall that in animals, the precursors of eggs and sperm (the germ cells) have a continuous lineage from one generation to the next, and are separate from the cells that make up the body (the somatic cells). Geneticists know today that genes in the germ-line cells determine the characteristics of each successive generation, but know of no biological mechanism that could transfer hereditary information from somatic cells (about the stretching of one's neck, for example) into the DNA of the germ line that instructs the building of features in the next generation (neck length, for instance).

Biologists in the mid-1800s did not yet understand such cellular mechanisms of heredity. But two English naturalists, Charles Darwin and Alfred Russell Wallace, were able to make detailed observations in various parts of the globe and devise a remarkable and elegant explanation for how living things evolve.

Voyages of Discovery

Darwin and Wallace lived in Victorian England, an era of courtly manners and rigid stratification by social class. In contrast to the stodgy social constraints of the day, however, many areas of science were advancing quickly. One area of excitement was the growing conviction that the natural world is in a constant state of change. Fossils discovered in the late 1700s and early 1800s convinced French zoologist Georges Cuvier that extinction is a fact and persuaded biologists Georges Louis Leclerc de Buffon, Jean Baptiste Lamarck, Erasmus Darwin (Charles's grandfather), and others that the forms of plants and animals had changed, or evolved, over time. One active fossil hunter from the era, Mary Anning, lived near the fossil-rich limestone cliffs of southern England (Fig. 9.2). She discovered the first plesiosaur, an extinct marine reptile, as well as many

other fossils that contributed to the debates over evolution. Geologists of that same period such as James Hutton and Charles Lyell also found evidence of physical evolution that contradicted Biblical timetables established for creation. They documented mountain building, erosion, and volcanic eruptions to prove that our planet must be millions, not thousands, of years old and must be undergoing slow but continuous change.

Darwin had also read the writings of economist Thomas Malthus (1766–1834), who explained that populations tend to grow geometrically (2, 4, 8, 16, 32, 64, and so on) and proposed that only wars, famine, and disease keep population growth in check. It was in this diverse social and scientific context that Darwin and Wallace travelled half a world away from England by ship. Independently, they observed unfamiliar organisms, and—unaware of each other's ideas—drew nearly identical conclusions about how the mechanisms of evolution might occur.

Darwin was a wealthy upper-class gentleman who had studied for the clergy at Cambridge University but was passionately interested in natural history (Fig. 9.3a).

Figure 9.2 Victorian Fossil Hunter Mary Anning.

Figure 9.3 **Darwin and Wallace.** (a) Charles Darwin circa 1840. (b) Alfred Russell Wallace circa 1865.

(a) Charles Darwin, circa 1840

(b) Alfred Russell Wallace, circa 1865

After graduation, he became the ship's naturalist for Captain Fitzroy of the HMS *Beagle* on a five-year voyage (1831–1836) to map the coastline of South America. At stops during the voyage, Darwin explored South American jungles, plains, and mountains. A seminal stop on the trip was the Galápagos, a cluster of black volcanic islands 1000 kilometers (600 miles) west of Ecuador inhabited by a small but unique set of plants and animals. Darwin was impressed by the variable shapes and colors the members of a single species could show, and he wondered how such variety could have arisen. He returned to England with notebooks full of sketches and observations, convinced that the living plants and animals he observed could illuminate the origin of species. But he continued to study variation in cows, pigeons, and other domesticated species back at home for nearly 20 years before going public with his ideas.

Darwin had written a short summary of his theory in 1842, but 16 years later, he had still not published it. That changed abruptly, however, when Alfred Russell Wallace's very similar thinking became public by 1858. Younger than Darwin, Wallace (Fig. 9.3b) had sailed to South America himself a decade earlier to collect plants, insects, and other natural specimens. Wallace visited Singapore and Borneo on another collecting trip. Like Darwin, he had read the works of Malthus and Lyell, and had noticed on his travels the striking variations among populations of individual species. Also like Darwin, Wallace had devised a theory of evolution by natural selection to explain what he saw. (He and Darwin even independently coined the same term for their major thesis, natural selection.) Unlike Darwin, however, Wallace was a working-class man, and perhaps because he had less to lose and fewer

fears of public disapproval, he was more eager to promote his ideas.

By 1858, Wallace had published several papers containing portions of a theory of evolution that included natural selection, the notion of changes over time from a common ancestor, and the idea of survival of the fittest. That same year, Wallace sent Darwin a paper for comments and criticisms and Darwin—realizing that his life's work was being scooped—rushed to write an abstract of his own work. Friends read this abstract before the Linnean Society (a British natural history organization) later that year along with Wallace's published work. Then the following year, Darwin released his monumental and well-documented book, *On the Origin of Species by Means of Natural Selection.*

In today's world of "publish or perish," gene patenting, and million-dollar research grants, two lab groups that simultaneously published nearly identical theories might end up in a lawsuit. But Darwin and Wallace were products of the gracious Victorian era and approached their situation with courtesy and generosity, each fully acknowledging the other's contribution at every opportunity. Darwin and Wallace, in fact, became personal friends; Darwin arranged for the impoverished Wallace to receive a government pension, and Wallace served as a pallbearer at Darwin's funeral.

Darwin and Wallace left the world a priceless legacy with their two key ideas: (1) **descent with modification**—the notion that all organisms are descended with changes from common ancestors; and (2) **natural selection,** the increased survival and reproduction of individuals better adapted to the environment. The latter was a biologically feasible mechanism for descent with modification in which nature selects well-adapted individuals to be parents of the next generation.

Descent with Modification

Darwin thought of evolution as a branching tree with the more recently evolved organisms at the tips of the outer branches and the more ancient or extinct species at the base and on lower branches (Fig. 9.1c). This contrasts sharply with the form predicted by Lamarck or by advocates of creation (Fig. 9.1a,b). It also contradicts the still widespread but wrong idea that evolution is a ladder from "lower forms," like worms and fish, to "higher forms," like birds, dogs, and humans. Species that exist today are *not* ancestors to other currently living species; the ancestral species are long extinct. The biggest conceptual advance Darwin and Wallace put forward was the *mechanism* for how descent with modification could occur.

Evolution by Natural Selection

Darwin's and Wallace's experiences in then-remote regions such as South America and the Galápagos Islands convinced them that evolution had occurred. But the earlier of the two travellers, Darwin, returned to England still puzzling over some mechanism that could cause evolution. He eventually put together two indisputable facts based on observations, and drew a far-reaching conclusion:

Fact 1. Individuals in a population (see Chapter 1) vary in many ways, and some of these variations can be inherited.

Fact 2. Populations can produce many more offspring than the environment's food, space, and other assets can possibly support. This causes individuals within the same population to compete with each other for limited resources.

Darwin's Conclusion. Individuals whose hereditary traits allow them to cope more efficiently with the local environment are more likely to survive and produce offspring than individuals without those traits. As a result, certain inherited traits become more common in a population over many generations.

Darwin (and later Wallace) used the term "natural selection" to describe the pressure exerted by environmental factors. Their term also applied to the greater reproductive success displayed by better adapted individuals. Both chose this term because nature "selects" the favorable traits that are passed on to the next generation. The favorable traits, or **adaptations,** include body parts, behaviors, and physiological processes that enable an organism to survive in its current environment. An example of an adaptive physiological process is the ability of a *Staphylococcus* bacterium to make an enzyme that destroys powerful antibiotics like methicillin or vancomycin. Favorable traits are modified or maintained as a result of natural selection because they improve an organism's chance of surviving and reproducing successfully. Biologists now accept the principle of natural selection as a main mechanism behind evolution in nature. It can explain how antibiotic resistance becomes prevalent in bacteria and it can just as easily explain the example that confounded Lamarck: the long necks of giraffes. We'll look at giraffes first, then return to bacteria and drug resistance.

Picture a population of giraffes browsing on trees in the African savanna hundreds of thousands of years ago. Some of those giraffes would have had longer necks than others, just as some people have longer necks than others. Researchers have shown that in various mammals, neck length is strongly determined by genes. (Interestingly, giraffes and humans share the same number of neck bones. Each bone is simply much longer in the giraffe.) If there were too few low-hanging leaves to feed all the giraffes in the population, then the longer-necked giraffes could reach more food, harvest more energy and materials, and survive and produce more offspring. A **population,** which we first encountered in Chapter 1, is by way of formal definition a local group of interbreeding members of a species. Many of these offspring, like their parents, would have genes for long necks and with more such genes around, the average giraffe neck would grow longer and longer over many generations until reaching its present-day length.

Giraffe's necks may seem like a nonserious topic. But evolution involves life or death issues, whether an animal's neck is too short to compete for leaves (and it starves) or whether a diabetic patient catches a painful bone infection that resists drugs.

Evolving Antibiotic Resistance

In modern-day hospitals, antibiotics are the "natural" agents that "select" which bacteria will contribute to the next generation. An ancient giraffe ancestor with long-neck genes (and the neck to go with them!) could harvest more high-growing leaves. In the same way, *Staphylococcus* bacteria with the genes to resist antibiotics like methicillin or vancomycin can survive—even in sick patients receiving the drugs—and therefore can leave more offspring that also bear those resistance genes. *E*xplorer 9.1 gives an overview of the antibiotic resistance problem.

Dr. Fred Tenover from the Centers for Disease Control and Prevention explains antibiotic resistance and evolution in a very concise way: The DNA in bacteria, like all DNA, can change or mutate (Fig. 9.4a). Most of these mutations alter proteins that have absolutely nothing to do with antibiotic resistance, and most mutations probably lessen the cell's ability to survive in a normal environment (Fig. 9.4b). However, those chance mutations that change proteins in ways that *do* improve the cell's resistance will give an organism a **selective advantage,** an increased likelihood of surviving the immediate environmental challenges. For example, antibiotic-susceptible organisms in a sick patient will

What mechanisms allowed antibiotic resistance to arise in Mr. Chedwick's infected foot?

GET THE PICTURE

*E*xplorer 9.1

Mechanisms of Antibiotic Resistance

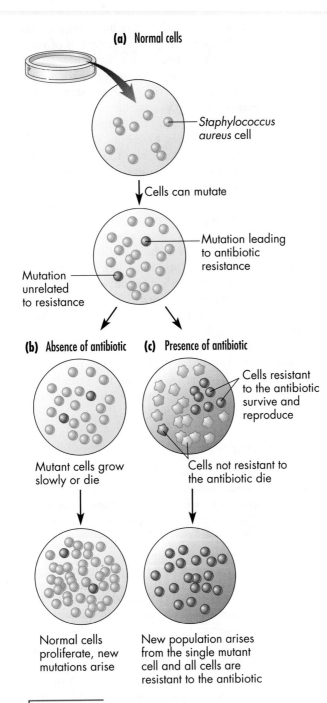

(a) Normal cells

Staphylococcus aureus cell

Cells can mutate

Mutation leading to antibiotic resistance

Mutation unrelated to resistance

(b) Absence of antibiotic

Mutant cells grow slowly or die

Normal cells proliferate, new mutations arise

(c) Presence of antibiotic

Cells resistant to the antibiotic survive and reproduce

Cells not resistant to the antibiotic die

New population arises from the single mutant cell and all cells are resistant to the antibiotic

Figure 9.4 **Natural Selection and Antibiotic Resistance.** (a) Mutations in DNA can alter proteins and sometimes the new proteins confer greater drug resistance. (b) In the absence of antibiotic, mutant cells resistant to the antibiotic will usually grow more slowly than unmutated cells. (c) In the presence of antibiotic, unmutated cells will die, but the cells with the mutations may survive the antibiotic's presence and with this selective advantage, give rise to new cell populations with new genotypes.

die when exposed to the drug. Those few cells with the new genetic information for resistance, however, will be *selected,* that is, they will survive a particular drug and live to divide and redivide, leaving a new population of cells that can flourish in spite of methicillin, say, or vancomycin (Fig. 9.4c).

Ironically, a hospital environment itself can select for resistance. About half of all hospital patients are receiving antibiotics at any given time. And the massive amounts of intricate medical equipment, the thousands of sink drains and mop heads, plus the often crowded conditions and overworked medical staff all combine to bring germs into constant contact with other germs and with antibiotics and so to select strongly for organisms with resistance. In Wayne Chedwick from Encinitas, in the retired detective from New York, and in the baby from Japan, the *Staphylococcus* populations were invincible, or nearly so. Let's see why in more detail.

Darwin's Fact 1 was that within populations, individuals vary in color, shape, size, and so on. We know today that much of this variation arises by mutations that crop up during DNA copying or when chemicals or radiation from the environment damage DNA. *Staphylococcus aureus* cells have several thousand genes encoding hundreds of traits. Looking at just one of them involved in cell wall construction can help us understand natural selection. Staph cell walls are built of tightly crosslinked strands, as if each fiber in a steel wool pad were linked to the neighboring fiber. A certain gene encodes an enzyme that puts little caps on the ends of these strands (Fig. 9.5a). The cell then uses the caps to make crosslinks between strands. A cell with the normal gene can grow well under most conditions, put the cap on the strands, and build a strong, tough wall that contains and protects the cell's contents. Mutations can occur in the gene, however, which can result in a mutant enzyme that makes a different kind of cap (Fig. 9.5b). This mutant cap leads to a weaker cell wall than normal, and so populations of these mutant cells grow more slowly than normal cells.

The environment now becomes key. In a regular cellular environment with no antibiotics present, staph cells with the normal gene grow quickly, cap their strand ends, crosslink them into strong cell walls, and outcompete staph cells with the mutant gene. If the drug vancomycin is present in the environment, however, the tables are turned on normal staph cells. In cells with a normal gene, vancomycin binds to the strand ends, prevents further cell wall formation, and leaves the cell with uncrosslinked weak spots in its wall (Fig. 9.5c). As water enters by osmosis, the cell

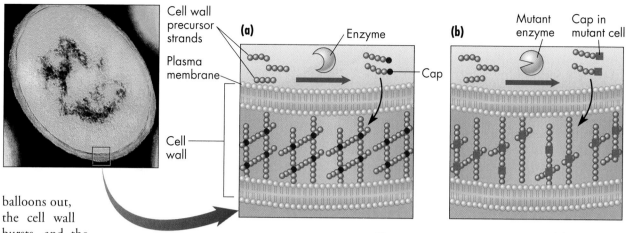

Cell wall precursor strands

Plasma membrane

Cell wall

(a) Enzyme — Cap

(b) Mutant enzyme — Cap in mutant cell

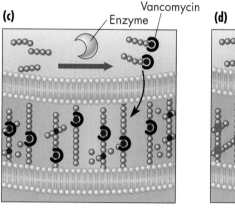

(c) Vancomycin — Enzyme

(d) Mutant enzyme

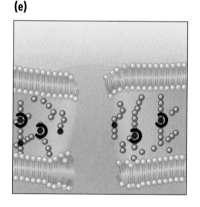

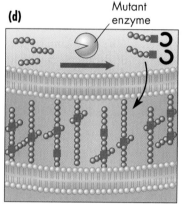

(e)

balloons out, the cell wall bursts, and the cell dies (Fig. 9.5d). The few surviving cells with the mutant gene, however, go right on slowly making cell walls with mutant strand ends unrecognized and unaffected by vancomycin (Fig. 9.5e). The cells divide and redivide and eventually constitute the entire local staph cell population. This is a clear case of natural selection with the antibiotic acting as a selection agent for the genetic mutation leading to vancomycin resistance. Unfortunately, the *Staphylococcus* cells' genetic success can be a tragedy if the evolutionary stage upon which the action takes place is Wayne Chedwick's infected foot bones, the bloodstream of a gravely ill retired detective, or the heart of a Japanese baby.

People sometimes get confused, thinking that natural selection *causes* the variations within a population. However, short necks, long necks, antibiotic resistance, and other traits already exist in populations as a result of mutations or, sometimes in bacteria, of picking up foreign genes (transformation). Natural selection simply results in the best-competing individuals—those with the traits best-adapted to current environmental conditions—surviving and producing more offspring in the next generation.

Later in the chapter, we'll return to a more detailed explanation of how natural selection favors certain genes. First, though, let's examine some of the evidence for evolution by natural selection amassed by Darwin, Wallace, and biologists after them.

Figure 9.5 **Staphylococcus, Vancomycin, and Evolution.** (a) In *S. aureus*, a gene encodes an enzyme that allows cells to cap the ends of wall-building strands, enabling them to crosslink strands in the cell wall. (b) Cells with mutant copies of the gene build walls with altered caps, and the cells grow slowly. (c) In cells with normal copies of the gene, vancomycin can bind to the caps on the cell-wall strands, blocking crosslinking and leading to weak cell walls that burst (d), causing the cell to die. (e) In contrast, mutant cells, with their abnormal strand ends, do not bind vancomycin. These cells then survive and the cell population slowly grows.

9.2 Evidence for Evolution

Evolution has such tremendous power to explain so many different aspects of biology that most biologists accept it as a fact. Researchers have found supporting evidence for evolution in the fossil record, in the anatomy of plants and animals, in molecular genetics, and in the geographical distribution of organisms. *E*xplorer 9.2 looks further into each type of evidence.

READ ON

*E*xplorer 9.2

Evidence for Evolution

Figure 9.6
Albertosaurus. Fossilized remains of animals such as Albertosaurus reveal that fierce meat-eating dinosaurs roamed the Earth 75 million years ago.

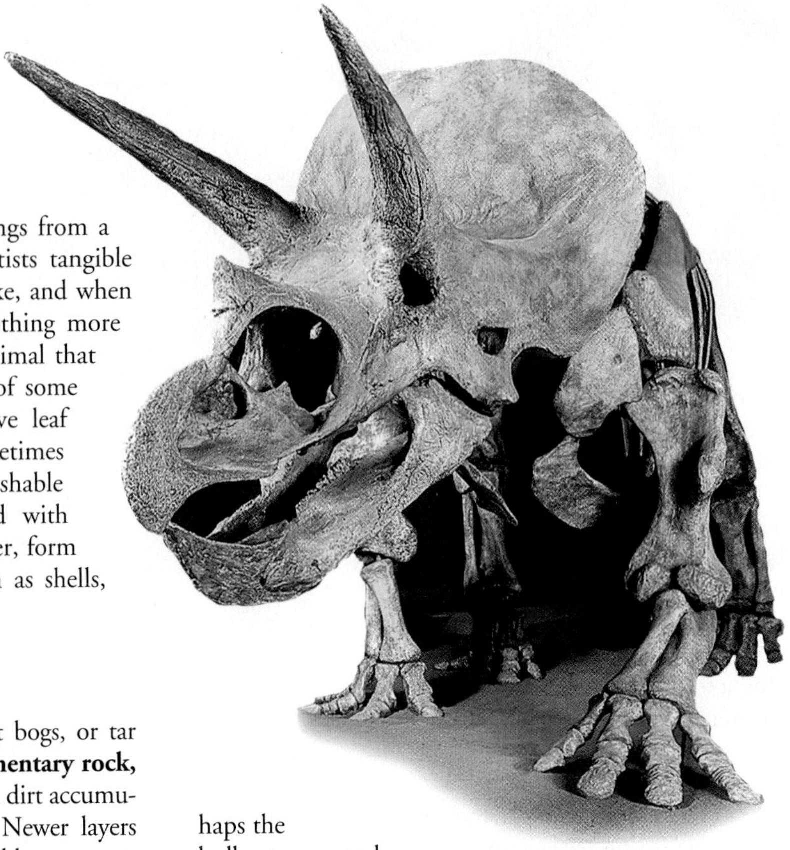

The Fossil Record

Fossils are traces or remains of living things from a previous geologic time. Fossils give scientists tangible evidence of what past organisms looked like, and when and where they lived. A fossil can be nothing more than the trail, preserved in rock, of an animal that once slithered across the muddy bottom of some ancient lake or sea. Other fossils preserve leaf prints, footprints, casts, outlines, or sometimes even soft body parts in rock after the perishable organic matter is removed and replaced with minerals. The most familiar fossils, however, form from hard, decay-resistant structures such as shells, bones, and teeth (Fig. 9.6).

Fossils and Sedimentary Rock

Although some fossils turn up in ice, peat bogs, or tar pits, fossil hunters find most fossils in **sedimentary rock,** which forms as layer upon layer of sand and dirt accumulate over thousands or millions of years. Newer layers press down on older ones, until pressure and heat cement the dirt, clay, and/or sand particles together, gradually changing them into rock.

A large Canadian deposit called the Burgess Shale, for example, contains sedimentary rock in which a strange and fascinating group of fossils lies embedded. One such fossil organism is the nightmarish *Hallucigenia* (Fig. 9.7), with its eight pairs of legs, sharp spines, and its ambiguous head: perhaps the bulb at one end but maybe the tube at the other. The Burgess Shale developed about 520 million years ago below many meters of seawater at the foot of the continental shelf off Canada's west coast. Seafloor dwellers living at the edge of the continental shelf were occasionally hurled over the edge and buried by sudden mudslides that created an anaerobic (oxygen-free) grave, safe from scavenging animals and rapid decay. The Burgess Shale demonstrates dramatically that organisms that lived long ago can be very different from today's living forms. And we could expect this difference, based on descent with modification.

Paleontologists discovered one of the richest fossil beds in North America in the Badlands region near the border of North and South Dakota, and learned from it priceless lessons about the evolutionary process (Fig. 9.8). Today, the Badlands region contains grasslands, high plateaus, and rock outcroppings. About 70 million

Fossil of Hallucigenia

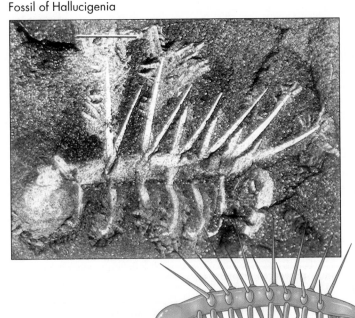

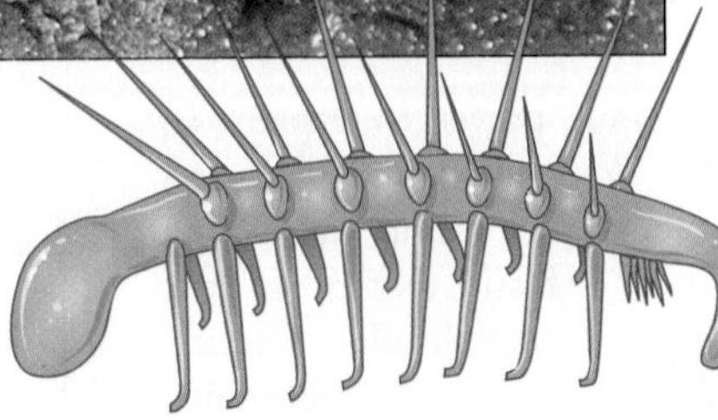

Figure 9.7
Hallucigenia. Hallucigenia was a curious organism just 1 inch (2.5 cm) long with a wormlike body and a head that lacked eyes, antennae, or other obvious sense organs. Jutting from its trunk were pairs of spines, flexible protrusions, and short tentacles.

years ago, however, the area lay beneath a shallow sea. The warm, expansive waters were densely populated by ammonites, extinct shelled relatives of squids and octopuses. By about 30 million years ago, the seas had retreated, and huge rhinoceros-like animals, the titanotheres, roamed what is now the Black Hills. A few million years after that, a three-toed, knee-high ancestor of the horse named *Mesohippus* grazed where the larger titanotheres had roamed (see **E** 9.2). Finally, fossils reveal that sheeplike oreodonts became abundant by about 25 million years ago. Paleontologists found each type of fossil at a particular level, suggesting that each of the different kinds of animals lived during a well-defined period.

If organisms evolved over immense time spans, then paleontologists ought to be able to find fossils intermediate in form between major groups of today's organisms (Fig. 9.9). And indeed they have. A famous example involves the 150-million-year-old crow-sized *Archaeopteryx*. This animal looked like a small dinosaur in nearly all respects, but it had feathers like a bird (Fig. 9.9c). A recently discovered series of fossils from China shows animals very similar to the velociraptors from the movie *Jurassic Park*. The tip of the animal's tail, however, was shaped like the feather-supporting tail stump in today's birds. Based on the similarities between birds and dinosaurs, some paleontologists now classify birds as a type of living dinosaur!

Fossil evidence supports evolution in several ways:

1. Different organisms lived at different times.
2. Past organisms were different from today's living organisms.
3. Fossils in adjacent rock layers are more similar to each other than to fossils in distant layers.
4. Many intermediate forms like *Archaeopteryx* have surfaced.
5. In general, older rocks contain simpler forms, and younger rocks contain more complex ones. As we saw before, though, there is no evolutionary rule that says that living things proceed from the simple to the complex. Instead, species have proceeded from being well-suited to their environments to being better suited—or else disappearing.

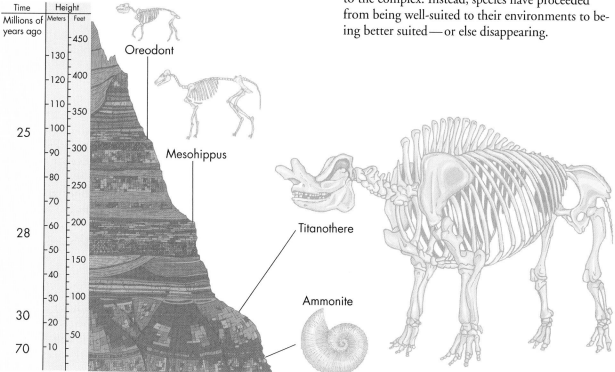

Figure 9.8 **Layers in Time.** Dozens of layers of sedimentary rocks lie exposed by erosion in Badlands National Park, South Dakota. Fossils within the layers show a procession of organisms that arose, adapted, and died out as the environment slowly changed from barren rock plateau to ocean to swamp to ice fields to dry, exposed hills. Shelled, squidlike ammonites lie entombed in the oldest (deepest) strata, while later layers contain huge titanotheres; small, early ancestors of the horse (Mesophippus); and sheeplike oreodonts.

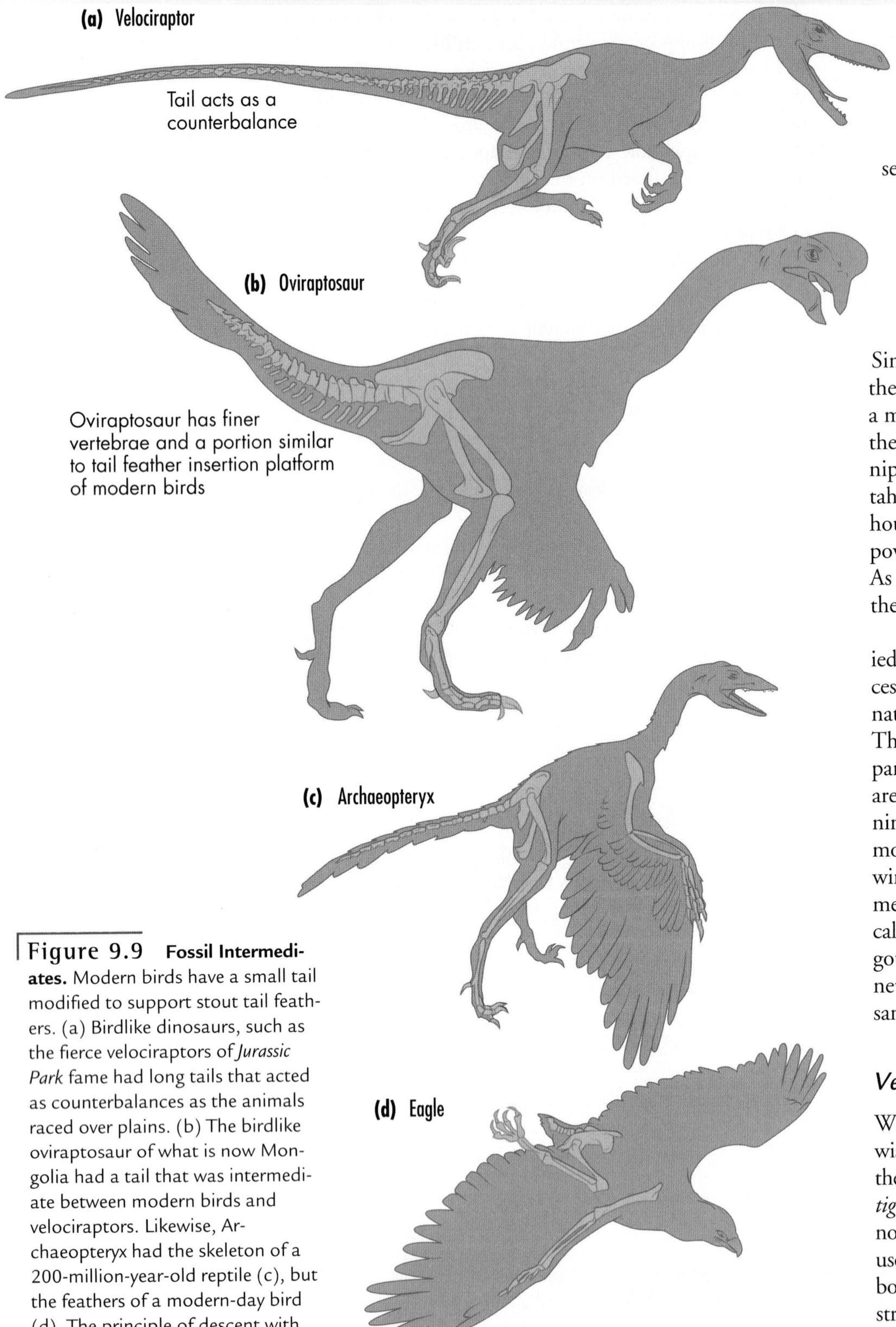

(a) Velociraptor

Tail acts as a counterbalance

(b) Oviraptosaur

Oviraptosaur has finer vertebrae and a portion similar to tail feather insertion platform of modern birds

(c) Archaeopteryx

(d) Eagle

Figure 9.9 **Fossil Intermediates.** Modern birds have a small tail modified to support stout tail feathers. (a) Birdlike dinosaurs, such as the fierce velociraptors of *Jurassic Park* fame had long tails that acted as counterbalances as the animals raced over plains. (b) The birdlike oviraptosaur of what is now Mongolia had a tail that was intermediate between modern birds and velociraptors. Likewise, Archaeopteryx had the skeleton of a 200-million-year-old reptile (c), but the feathers of a modern-day bird (d). The principle of descent with modification predicts fossil intermediates like these.

Evidence from Comparative Anatomy

If evolution occurs by descent with modification, that is, inheritance with changes, then one would predict the anatomy of living species to resemble that of extinct relatives, but with changes. And that is just what biologists see by studying two types of anatomical evidence: homologous structures and vestigial organs.

Homology: Organs with Similar Origins

Similar structures on different organisms help support the argument for descent with modification. Think, for a moment, about the forelimbs of various mammals and the functions they allow: Human hands can deftly manipulate keyboards, surgical tools, paintbrushes; a cheetah's front legs help it run 116 kilometers (72 miles) per hour for short sprints; a whale's flippers allow it to dive powerfully and gracefully; a bat's wings enable it to fly. As different as they are, each type of limb is made up of the same skeletal elements (Fig. 9.10).

Fossil evidence supports the hypothesis that the varied forelimbs of mammals arose from the forelegs of ancestral five-fingered amphibians and became modified by natural selection in ways that facilitated different tasks. The idea that this same set of bones—one in the upper part of the limb, two in the lower part, and five digits—are the very best arrangement for manipulation *and* running *and* swimming *and* flying seems unreasonable to most anatomists. Elements (such as legs, flippers, and wings) in different species that derive from a single element in the last common ancestor of those species are called **homologous** (*homo* = same) elements. Homologous organs can have different functions but similar genetic blueprints, and can be constructed in much the same way in the embryos of different species.

Vestigial Organs

What does your appendix have in common with your wisdom teeth and with a snake's hipbones? The answer is they're all **vestigial organs** (veh-STIHJ-uhl; Latin *vestigium* = footprint, trace): or rudimentary structures with no apparent use in the organism, but strongly resembling useful structures in probable ancestors. You may have lost both your appendix and wisdom teeth because these structures can be worse than useless if they become infected or impacted. We humans also have the vertebrae (spinal bones) for a tail and for muscles that can wiggle our ears, despite their uselessness today for our survival. In

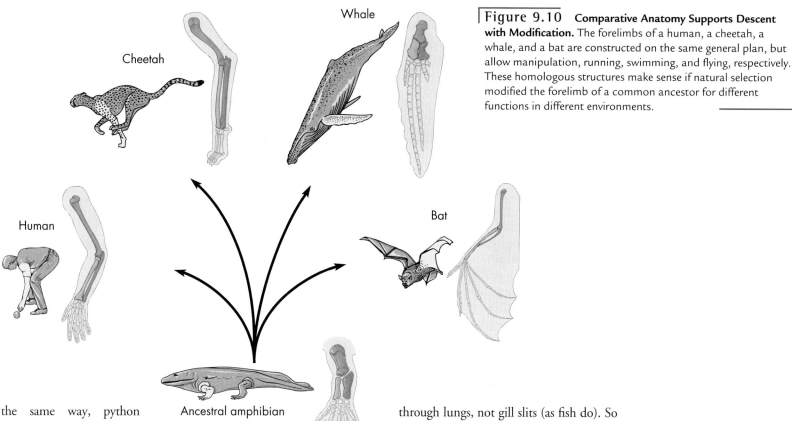

Cheetah

Human

Whale

Bat

Ancestral amphibian

Figure 9.10 **Comparative Anatomy Supports Descent with Modification.** The forelimbs of a human, a cheetah, a whale, and a bat are constructed on the same general plan, but allow manipulation, running, swimming, and flying, respectively. These homologous structures make sense if natural selection modified the forelimb of a common ancestor for different functions in different environments.

the same way, python snakes have vestigial hipbones and thighbones, even though the bones no longer function (Fig. 9.11). Recent experiments show that pythons fail to form forelimbs because homeotic genes (see Chapter 8) are expressed differently in them than in animals with front legs. Also, in python embryos developing inside their leathery eggs, a crucial limb-triggering region fails to form (see Fig. 8.16). Curiously, a snake embryo implanted with a chick limb bud starts making signals to build legs. This shows that as different as pythons are from chickens, they still share developmental pathways for homologous organs, in this case, wings (chickens) and useless leg bones (snakes). This type of sharing is a powerful argument for descent with modification. Chickens still have their wings. In snakes and their ancestors, however, natural selection has removed alleles of genes necessary for forming the limb bud, with its signaling center. Lacking that limb bud, snakes make only vestiges of limbs, or no limbs at all.

Comparative Embryology

Strange as it seems, as embryos, we humans pass through stages similar to those of fish embryos, during which our neck region forms grooves and pouches similar to gill slits (Fig. 9.12a,b). Like most vertebrates, we breathe

through lungs, not gill slits (as fish do). So why do human embryos form these pouches? People's embryonic neck pouches don't become mature gill slits but they (along with adjacent tissues) do form useful organs: one pouch forms the eustachian tube leading from the mouth to the ear, and tissue between the pouches forms the thymus gland, and the tonsils. In some people, a pouch fails to change or disappear during development, and these people are born with a tube leading from inside the mouth to the outside of the neck like a fish. This abnormality can be fixed surgically and does, at least, serve as evidence of evolution.

Evolutionary theory can easily explain why all vertebrate embryos look similar to fish embryos for a while, right down to the pouches. The last common ancestor of humans and ray-finned fish (such as bluegill or bass) lived 450 million years ago, and the embryos of this ancestor developed gill pouches like today's hu-

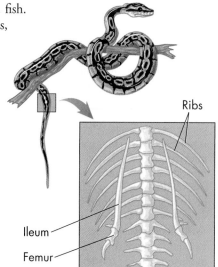

Ribs

Ileum

Femur

Figure 9.11 **Vestigial Organs.** A python has a rudimentary ileum (pelvis) and femur (thigh) bones. These were inherited from ancestors, modified, and sustained, even though they don't contribute to locomotion.

Figure 9.12 Comparative Embryology, Gills Slits, and Descent.

(a) Fish embryos develop pouches from the pharynx, the inside and back of the mouth, that break through to the neck and form the gill slits. The tissue between the slits forms the gills. (b) Human embryos also form these pouches, but instead of a gill, the first pouch becomes the eustachian tube, a passageway from the ear to the inside of the mouth, while the tissue around the second pouch becomes the tonsil and the slit disappears.

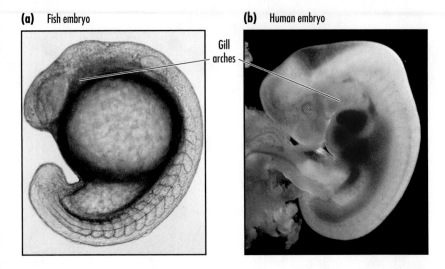

Gill arches

man and fish embryos. In the lineage giving rise to humans, changes in gene frequencies driven by natural selection caused the pouches and nearby tissues to develop into the eustachian tube, thymus, and tonsils. In fish, natural selection modified the same embryological structures into efficient oxygen-gathering gills. Embryology therefore supplies evidence that organisms often inherit features that originated in ancient ancestors but became genetically modified over time.

Evidence From Molecules

If all life evolved from a common ancestor that underwent gradual genetic changes, then all living organisms today should share basic features at the molecular level.

In fact, all organisms have the same four bases in DNA, the same 20 amino acids in proteins, virtually the same genetic code, and so on. By comparing the same protein or the same gene in dozens of different species, molecular geneticists can construct molecular family trees (Fig. 9.13). On such a molecular tree, they would place two very similar proteins or genes on branches diverging from each other more recently, and two less similar proteins or genes on branches that diverged longer ago. Biologists then assume that the organisms from which the

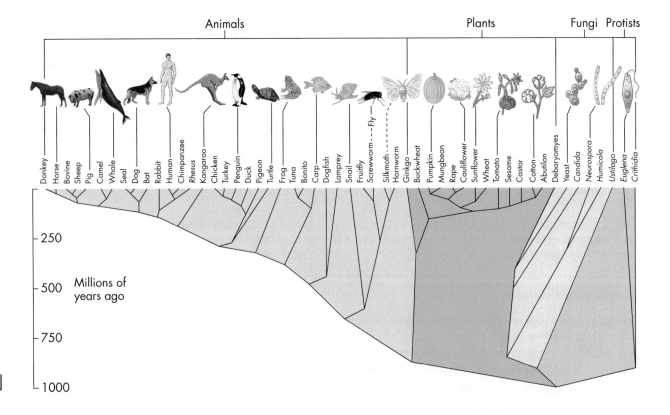

Figure 9.13 Molecular Evolution: Eukaryotic Family Tree.
Researchers compared a single gene from 53 different species of eukaryotic organisms, the gene for the mitochondrial protein cytochrome C. They then arranged the gene sequences by similarities. As you can see, a donkey (far left) is more closely related to a human than to a turtle, and is most distantly related to a protist like Euglena (far right). The branching patterns approximate those derived from the fossil record, even though the two data sets are completely independent and provide strong support for evolution.

genes or proteins came share the same evolutionary relationships as the molecules on the tree.

Molecular family trees powerfully support evolution because totally independent molecular and fossil data show common patterns of relationships between living and extinct species. Biologists have even been able to isolate DNA from some fossils. For example, researchers have extracted DNA from 30-million-year-old termites trapped in amber, the fossilized sap of ancient conifer trees. The DNA from the fossil termites was very similar to DNA from living termites that share the same physical features, and very different from DNA in living termites with radically different physical features. These results are exactly what evolutionary biologists expected to find—that ancient species are more similar to some of today's species and less similar to others.

Evidence From Biogeography

When Darwin and Wallace sailed separately to foreign lands, their first clue about evolution was **biogeography,** or the geographic distribution of organisms. In the Galápagos Islands, for example, Darwin was astonished by the unusual plants and animals he saw: finches using cactus spines to pry insect larvae out of dead wood; gigantic tortoises that varied in form from island to island; iguanas that swam in the sea and fed on algae instead of living exclusively on dry land. He saw no frogs or salamanders, however, and the only native mammals he observed were a few bats and one type of mouse.

Darwin compared these Galápagos organisms to the plants and animals he had seen while visiting the Cape Verde Islands off the coast of Africa. Even though the two island groups had similar climates, sizes, and the same dark volcanic soil, their organisms were of totally different species, genera, and taxonomic families. Rather than being related to *each other's* fauna and flora, each island's organisms were more closely related to those inhabiting the nearest *mainland*. Ironically, though, those mainland areas were very different from the neighboring islands in climate, soil, and other ecological details. So why were the island animals more closely related to those of the mainland than those on other similar islands?

Most 19th-century naturalists accepted the idea of special creation, and would explain the distribution of organisms by suggesting that each individual island had its own individual creation event related somehow to the creation event on the nearest mainland. But why would that creation event have discriminated against frogs, salamanders, and most mammals in a place like the Galápagos? Darwin thought it unlikely that creation overlooked amphibians. Instead, it seemed much more likely to him that the Galápagos harbored only organisms that had flown, swam, or rafted over from the nearest continental shore on floating vegetation to the newly forming volcanic islands. Encountering no competition on the young islands, these "colonists" could then diverge into new species as they adapted to different island environments through natural selection. Like Wallace, Darwin had begun his travels believing that every species was created in its special place, but saw evidence in biogeography that argued against it and the notion of unchanging species.

Throughout his travels in the Galápagos and Cape Verde Islands, Darwin also noted the powerful influence of the environment. For example, in both places, species of plants that grew as low, green herbs on the mainland had close relatives on the islands that were woody and treelike. He saw examples of unrelated organisms, as well, displaying similar characteristics in response to common environmental conditions. Australia, South America, and Africa, for instance, all have large, flightless birds with heavy bodies and long necks: the emu, the rhea, and the ostrich, respectively (Fig. 9.14). Why wouldn't the ostrich ever be created in South America, Darwin wondered? Why would an open grassland environment in Africa produce an ostrich but in South American a rhea and in Australia an emu? Why

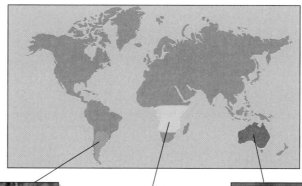

Figure 9.14 **Big Birds and Biogeography.** Darwin wondered why each southern continent has its own species of large flightless birds. Answering his own question, he wrote: "The bond is simply inheritance, that cause which alone, as far as we positively know, produces organisms quite like each other, or, as we see in the case of varieties, nearly alike." In other words, the South American rhea, the African ostrich, and the Australian emu all descended with variation from a common ancestor and adjusted to their specific environments.

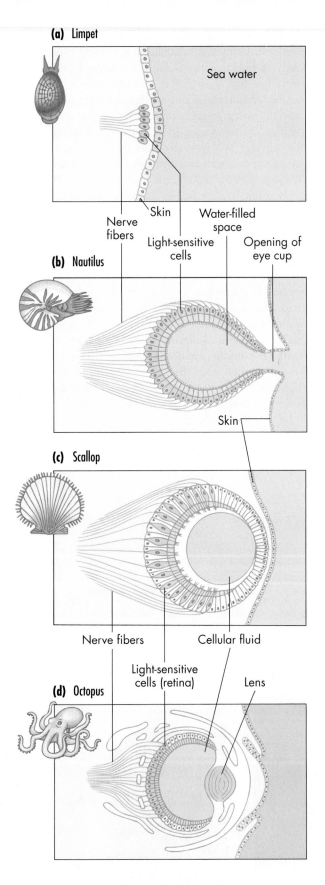

Figure 9.15 **Natural Selection and the Complex Eye.** (a) Animals that contain small groups of light-sensitive cells shaded by other cells would be able to detect directional light, and have an eye like those in limpets. (b) Additional mutations could cause light-sensitive cells to sink into cuplike depressions and thus improve on detecting light sources (as well as food and predators). A nautilus has this type of eye. (c) Further random mutations could cause the edges of the cup to join and fluid to accumulate in the cavity. If the fluid focused light, it would be easier for the organism to detect specific images. Scallops have an eye like this. (d) Finally, if mutations arose that caused tissue to grow over the eye and accumulate transparent proteins and form an effective lens, then a complex "camera" eye capable of forming tightly focused images would emerge. The squid and octopus have eyes with lenses similar to our own.

would three types of flightless birds have been created when surely one would have sufficed for the different continental grasslands? Darwin suggested that descent with modification led to the different species on the different continents, with similar physical and behavioral adaptations resulting from similar selection pressures.

Gradual Evolution of Complex Organs

Have you ever wondered how an eye, a complex organ beautifully adapted for detecting light, could have arisen "by chance"? This type of question confounds some nonbiologists, who point to it as evidence against evolution. The key to this conundrum, however, is to realize that *mutation and genetic recombination* occur by chance, whereas the development of an entire, complex functioning organ does not.

Mutations can occur randomly virtually anywhere in an organism's DNA. Natural selection, however, is distinctly *nonrandom*. If a certain genetic combination allows an organism to survive and reproduce a little bit more effectively than individuals without this genotype, then those genes will be favored and will increase in frequency. In the case of the eye, if a random mutation changes a gene, a protein, and ultimately a photoreceptor apparatus in a way that helps the individual survive and reproduce, then that mutation for an altered photoreceptor will tend to increase in the population by natural selection.

Small genetic changes that lead to altered organization in eye-forming cells can occur over and over. Such mutations producing favorable adaptations gradually led to complex animal eyes, such as those of today's mollusks (Fig. 9.15). The eyes shown in Figure 9.15 do not form an evolutionary series because all the organisms pictured are alive today and none gave rise to another. What this group of eyes does show, however, is that a light-gathering organ can be simple and still work well for an organism in its specific environment. The series also shows that modern mollusks' eyes could have resulted from a stepwise accumulation of mutations that alter eye shape and sharpness of vision. The evolution of eyes involves no "forethought," no structural "goal." Instead, individuals who happen to inherit genes that help them adapt better to their environments—even by a little—will, on average, reproduce more successfully than individuals lacking this accidental gene combination and the traits that go with it. Other sorts of mutations can arise, of course, but if they resulted in a less effective eye, natural selection would weed them out. (An exception would be an animal living in a dark environment—a cave dwelling fish, let's say—in which a better eye would be a *disadvantage*. For them, inheriting mutations to produce a more complex eye would use up precious energy and make them less fit for their lightless world.)

Sickle Cell Anemia and Natural Selection

The inherited human disease sickle cell anemia is an example of natural selection in action. In people that are homozygous for the recessive sickle cell allele, red blood cells change shape (review Fig. 7.3), clog capillaries, and are removed rapidly from the blood stream by the spleen, resulting in anemia (too few red blood cells). Before the days of modern medicine, the clogged capillaries, enlarged spleen, and other symptoms usually killed victims of sickle cell anemia before they reached reproductive age.

In some areas of Africa, 40 percent of the alleles of this hemoglobin gene in the local human population are the sickle cell allele. Among African Americans, the figure is only 5 percent, and among European Americans, it is only 0.1 percent. What could produce such a high frequency of a harmful allele in specific populations? Biologists began to understand this puzzle when they discovered that the sickle cell allele was most frequent in areas with a high incidence of malaria, an often fatal disease. They later determined that heterozygotes with one normal and one sickle cell allele are less likely to die of malaria than people with two copies of the normal allele. Individuals with one copy of each allele benefit from **heterozygote advantage,** wherein the heterozygote has greater fitness than either homozygote. Homozygotes for sickle cell allele die of sickle cell anemia, and homozygotes for the normal allele often die of malaria. When malarial parasites cause a heterozygote's red blood cells to sickle, the spleen destroys the resident parasites along with the sickled red blood cells, and this limits the infection.

We can see that natural selection caused the frequency of the sickle cell allele to increase in these populations. On the other hand, the allele remained rare in malaria-free regions, such as Europe, because it conferred little or no advantage there.

9.3 Pathways of Descent

We've seen all sorts of evidence for evolution from fields as different as paleontology, anatomy, embryology, genetics, and biogeography. Now we need to focus on *how* it supports the principles of evolution. This section focuses on how evolutionary change takes place over time. What are the patterns of evolutionary change? How fast do organisms evolve? And how do extinctions

affect evolution? The answers will help us understand the many pathways of descent over time.

Patterns of Evolution

The *Staphylococcus* bacteria infecting Wayne Chedwick's feet could have become resistant to antibiotics in several small steps as mutations accumulated or in one jump if the cells acquired genes for antibiotic resistance from other bacteria. In multicellular organisms, evolution can also take place at different rates and can follow a number of converging, diverging, or radiating patterns.

Gradual Evolutionary Change

One species can gradually change into a new species as genetic differences accumulate in a population slowly over many generations (Fig. 9.16a). A certain genus (*Globoratalia*) of marine protists called foraminifera is a good example. From about 10 million to 5.6 million years ago, the chalky shells secreted by these saltwater species changed very little in shape. Gradually, over the next 0.6 million years, they underwent rather rapid change in size and shape into what experts consider a new species, which remains alive and nearly unchanged in today's Indian Ocean.

In a bacterium like *Staphylococcus*, gradual change can occur as a species is exposed to a low level of antibiotic, and a random mutation results in an altered gene. This mutation could change a gene that, for example, encodes an enzyme that helps digest a compound related in structure to penicillin. Now altered slightly, the enzyme can slowly digest the antibiotic, too, allowing the cell to resist low (but not high) concentrations of the antibiotic. This mutated cell will divide and leave more daughter cells. In time, another mutation in the same gene might occur that enables the enzyme to digest the antibiotic even more rapidly, and thus allow the cell to resist higher and higher doses of the antibiotic. Eventually, the accumulating population of daughter cells could be classified as a different strain. If similar change happens in many different proteins, the cells might be classified as a new species.

Divergent Evolution

Sometimes, a single population splits into two or more populations and the isolated populations start to accumulate genetic differences gradually. This is **divergent evolution.** Millions of years ago, before the Colorado River cut a mile-deep gash through what is now the high plateau of northern Arizona, a single species of squirrel

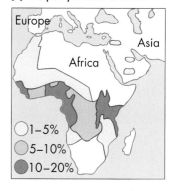

(a) Frequency of sickle-cell allele

○ 1–5%
◐ 5–10%
● 10–20%

(b) Distribution of malaria

Sickle-cell Anemia and Malaria: Natural Selection in Action. (a) The sickle-cell allele that encodes the beta chain of hemoglobin is found at surprisingly high frequencies in certain parts of Africa, the Mediterranean, and the Middle East. (b) The map highlights regions where the malarial parasite lives. The similar distribution of the sickle hemoglobin allele and malaria provided the first clue that malaria might be an agent of natural selection.

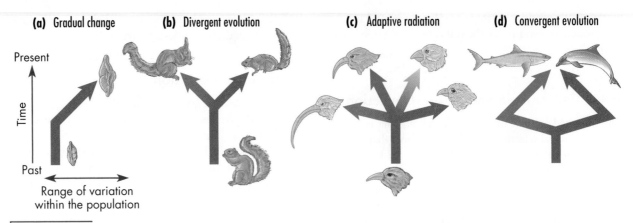

(a) Gradual change   **(b)** Divergent evolution   **(c)** Adaptive radiation   **(d)** Convergent evolution

Present

Time

Past

Range of variation within the population

Figure 9.16 **Patterns of Descent in Evolution.** (a) A population's gene pool may slowly change in a certain direction, as when one lineage of foraminiferans gradually became larger. (b) The population may split into two, each population changing in a different way as time passes. This happened when the Grand Canyon formed and blocked reproduction between two squirrel populations. (c) Adaptive radiation results in several new species stemming from a single common ancestor, as with the small honeycreepers of Hawaiian forests. (d) In convergent evolution, separate lines of distantly related organisms evolve similar forms and exploit similar environments, as with sharks and dolphins.

occupied the area. As the gash deepened into the Grand Canyon, it became an uncrossable barrier and the squirrels were divided into two separate populations. Genetic changes accumulated and the two squirrel populations diverged into two separate species that have inhabited opposite rims of the Grand Canyon ever since (Fig. 9.16b).

Adaptive Radiation

Sometimes several populations of a single species diverge simultaneously into a variety of species (Fig. 9.16c). This process, **adaptive radiation,** may take place when one ancestral species invades new territories that allow it to exploit a variety of environments and different ways of life. Hawaiian honeycreepers are a good example. One ancestral species of the colorful birds colonized the

Hawaiian Islands and radiated into about 20 different species, each adapted to survive on different foods.

Convergent Evolution

Two or more dissimilar and distantly related lineages can evolve in ways that make the organisms resemble each other superficially the way the streamlined bodies of sharks and porpoises do (Fig. 9.16d). A good example of this process, **convergent evolution,** involves the squirrel-like sugar gliders of Australian forests and the flying squirrels of Europe, Asia, and North America. Both kinds of animals soar through forests on thin folds of skin that extend along both sides of the body from wrist to ankle (Fig. 9.17). Like all of Australia's original mammals, sugar gliders are marsupials that grow in the mother's pouch (mar-

Figure 9.17 **Flying Squirrels and Convergent Evolution.** The sugar glider (a) and the flying squirrel (b) belong to two very different groups of mammals, but they have both evolved bushy tails and flaps of skin that help them float through their forest environments and exploit them more efficiently.

(a) Sugar glider

(b) Flying squirrel

supium), while flying squirrels are placental mammals. These two lineages diverged more than 150 million years ago. Geologists agree that the Australian continent separated from the rest of Earth's continents more than 50 million years ago, and had few, if any, placental mammals at the time of the separation. Without competition from placental mammals, Australian marsupials radiated into a great number of species, many of which look and act like placental mammals on other continents that inhabit similar environments. In this case, sugar gliders resemble flying squirrels closely in ways associated with the gliding life style, but their most recent common ancestor lived over 150 million years ago. Convergent evolution can also happen when different bacterial species inhabit "environments" (including hospital patients) with high concentrations of antibiotics. The different bacteria—say, different staphylococcal and streptococcal species—can independently undergo mutations that allow them to survive one or more of the drugs. Thus they can converge, independently, on an antibiotic-resistant phenotype such as the ability to survive methicillin or vancomycin.

The Tempo of Evolution

How fast do old species converge, diverge, or radiate into new ones? One idea suggests that all lines change at about the same constant rate over time. More recent analyses, however, imply that structural changes often occur in fits and starts.

Phyletic Gradualism

Traditionally, evolutionary biologists thought that after a population splits into two, natural selection and random genetic changes would cause the new subgroups to diverge from each other at about equal and constant rates. In this model, termed **phyletic gradualism,** the giraffe's short-necked ancestor would have slowly acquired giraffe-like qualities as genetic changes gradually accumulated.

Punctuated Equilibrium

Starting in the 1970s, some evolutionary biologists pointed out that if new species usually form gradually, then the fossil record should show numerous intermediate species. This certainly does occur, but paleontologists often find surprisingly little evidence of continually, gradually transforming lineages with a complete series of intermediate forms. Instead, fossils often reveal that species exist for millions of years with little change and that very rarely these long periods of

equilibrium are interrupted, or "punctuated," by great phenotypic changes, resulting in new species.

The **punctuated equilibrium** model has three main tenets:

1. Changes in body form evolve rapidly in geologic time.
2. During **speciation** (the formation of a new species), changes in body form occur almost exclusively in small populations, and new species that result are quite different from their ancestral species.
3. After the burst of change that brings about speciation, species keep basically the same form until they become extinct, perhaps millions of years later.

Models Compared

The family trees for the okapi and giraffe can help you compare the two hypotheses. Phyletic gradualism predicts that the lines eventually becoming okapis and giraffes began to diverge slowly (Fig. 9.18a). Gradually, changes accumulated that eventually prevented the two groups from interbreeding, thus making two separate

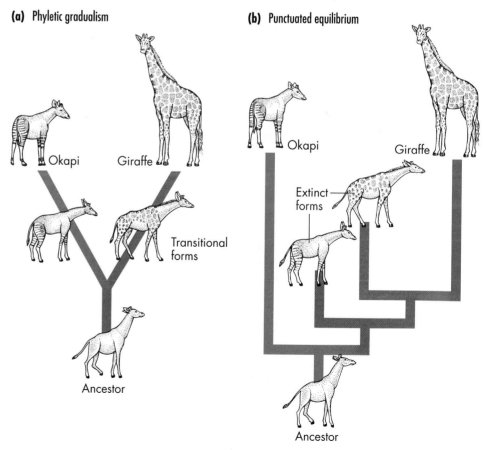

Figure 9.18 Okapis and Giraffes: Two Models of Descent. (a) In the phyletic gradualism model, okapi and giraffe evolved as two lineages slowly diverged from a common ancestor. (b) In the punctuated equilibrium model, a small population budding off from an ancestral okapi predecessor group would lead to modern okapis, while a number of more giraffelike species would arise and some would become extinct as time passed.

(a) Phyletic gradualism

Okapi   Giraffe

Transitional forms

Ancestor

(b) Punctuated equilibrium

Okapi   Giraffe

Extinct forms

Ancestor

species. Changes continued to accumulate gradually in both species until the okapi and giraffe emerged as we know them. In contrast, punctuated equilibrium predicts that offshoots of the initial evolutionary line split off repeatedly (Fig. 9.18b). These split-offs took place as small, isolated populations splintered from the main group, underwent bursts of evolutionary change, and then maintained new features for long periods (usually) before the line died out.

In fact, it seems quite likely that evolution proceeds in *both* ways. Evolution may often take place when a small population becomes isolated geographically or reproductively from other members of its species. Gradually, over many generations (but nonetheless a very short stretch of geologic time), the isolated population accumulates genetic changes. Once enough change has occurred and the new species is fitted to its environment, the species may stay essentially the same for millions of years.

Whether evolution is slow and gradual or fast and punctuated in any particular case, the fundamental raw material giving rise to the change is the same: genetic variation.

9.4 Genetic Variation: The Raw Material of Evolution

The speed of a gazelle outrunning a hungry cheetah. The tendency of a flower to attract more pollen-disseminating bees than the flower growing next to it. The ability of a *Staphylococcus* cell to destroy an antibiotic like methicillin and entrench itself in a man's foot bones. In every case, **genetic variation,** or genetic differences among individuals of the same species, cause individuals' traits to vary. Darwin was particularly impressed with the enormous variation in chickens, dogs, and other domestic animals (Fig. 9.19). He noticed on his voyages that organisms vary widely in nature, too. Darwin was unaware of genes, however, and had little or no idea how variation arises and can be passed from parent to offspring. Gregor Mendel published his genetics experiments (see Chapter 5) six years after Darwin and Wallace first published their work. If either of the two naturalists knew about Mendel's work, they didn't understand its importance enough to include it in their writings on evolution.

In the 1930s and 1940s, biologists began to combine evolutionary theory with genetics in the so-called **synthetic theory of evolution.** It suggests that: (1) Gene mutations occur in reproductive cells at high

Figure 9.19 **Big Chicken: The Picture of Genetic Variation.** The varied physical traits within a single species fascinated Darwin. Poultry breeders produced this modern chicken breed to look like Big Bird from Sesame Street by selecting from the genetic variation inherent in domestic chickens.

enough frequencies to impact evolution. (2) Gene mutations occur in random directions unrelated to the organism's survival needs in its environment. In *Staphylococcus* bacteria, for example, one mutation may give the cells more resistance to methicillin while another makes them more susceptible. The resistant cells survive, while the susceptible ones die out. (3) Natural selection acts on the genetic diversity brought about by such random mutations.

To understand how modern biologists see evolution, we need to understand how genetic variation occurs in the first place and how it is maintained in a population.

Sources of Genetic Variation

Mutations, new base pair sequences, and new combinations of genes (all of which we encountered in previous chapters) can lead to the varied genetic raw material that evolution acts upon.

Single-Gene Mutation

Mistakes or alterations in the DNA sequence of a single gene can (but don't always) alter how that gene functions (Fig. 9.20a). Some of these so-called **single-gene**

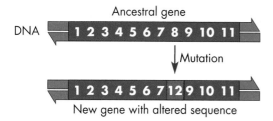

(a) Single gene mutation

Ancestral gene

DNA 1 2 3 4 5 6 7 8 9 10 11

↓ Mutation

1 2 3 4 5 6 7 12 9 10 11

New gene with altered sequence

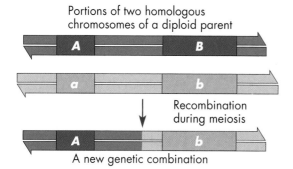

(b) Recombination

Portions of two homologous chromosomes of a diploid parent

A B

a b

↓ Recombination during meiosis

A b

A new genetic combination

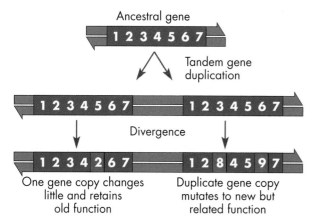

(c) Duplication and divergence

Ancestral gene

1 2 3 4 5 6 7

Tandem gene duplication

1 2 3 4 5 6 7   1 2 3 4 5 6 7

Divergence

1 2 3 4 2 6 7   1 2 8 4 5 9 7

One gene copy changes little and retains old function

Duplicate gene copy mutates to new but related function

mutations are **neutral mutations:** they leave the gene's function basically intact, and neither harm nor help the organism. The exchange of one amino acid for another with a similar property might not, for example, change the function of the cap-forming enzyme in Figure 9.5. Other mutations produce favorable effects that benefit the organism. A favorable mutation in *Staphylococcus aureus,* for example, might decrease the affinity of cell-wall–forming strands for vancomycin. Likewise, a favorable mutation might increase the number of kernels a corn plant produces. Many mutations, however, are either harmful or lethal; they reduce, modify, or destroy the function of a gene necessary for survival. A mutation that destroyed a staph cell's ability to make a cell wall would be harmful in nearly all environments.

New Genes with New Functions

Mutation can alter old genes. But how do new genes arise with new functions? One way is through **gene duplication.** Sometimes, a chance error in DNA replication or recombination creates two identical copies of a gene, and mutation in one or both of these copies can change their nucleotide sequences such that one gene maintains the original gene function, while the other acquires a related, new function. This kind of favorable gene duplication has resulted, for example, in several different genes for the oxygen-carrying protein, hemoglobin, with each gene becoming active during specific phases in human development: the embryo, fetus, child, or adult (Fig. 9.20c). In another example, the gene encoding the capping enzyme in *Staphylococcus* (see Fig. 9.5a) has duplicated and moved onto a plasmid, a circle of DNA that replicates independently in bacterial cells like an autonomous minichromosome (see Figure 7.16). This duplicate gene mutated and became *vanA,* one of the main genes that confers resistance to vancomycin.

Figure 9.20 **The Sources of Genetic Variation.**
(a) In a simple mutation, a nucleotide (here labeled 8) in a gene can be replaced with an entirely different one (here labeled 12). (b) In recombination, crossing-over leads to new combinations of alleles. (c) In duplication and divergence, an ancestral gene doubles, resulting in two copies of the gene, often nearby on the same chromosome. One copy can remain little changed (bottom left), while the other can diverge in sequence by simple mutation (bottom right) and perhaps eventually assume a new function.

Recombination

Mutation can create new versions (new alleles) of old genes. The process of **genetic recombination,** the shuffling of existing genetic material, can rearrange those alleles further (Fig. 9.20b; also review Fig. 4.16). Recombination might bring together advantageous alleles of different genes into the same cell. For example, a bacterium might have one gene with two alleles—one allele that is sensitive to vancomycin and the other allele resistant to it. The same bacterium could also have a second gene with two alleles, one sensitive to penicillin and the other resistant to it. A recombination event that randomly shuffled genetic material so that a single cell has the resistance alleles of both genes would create an organism that is strongly selected for when a physician prescribes both antibiotics for the same patient.

It is important to keep in mind that recombination merely shuffles existing alleles into new combinations but keeps the frequency the same (in the absence of selection), while mutation can change the frequency (commonness) of alleles in a population's **gene pool** (all the genes in a population at any given time) by changing one allele into another. By analogy, you could shuffle and deal the 52 cards in a deck into millions of new combinations (hands), but the deck would always have the same frequencies of cards—four aces, four queens, and so on. Only if a mutation changed an ace into a queen, for example, or a four into a seven would the frequencies change.

How Much Genetic Variation Exists?

Do mutations, recombinations, and gene duplications happen often enough to fuel evolution? In other words, do these chance occurrences really generate all the genetic variation that exists within a natural population?

Recall from Chapter 6 that researchers can now create genetic fingerprints that reveal how similar (or different) individuals are at the level of DNA. At the end of the 20th century, geneticists for the first time compared corresponding large portions of a chimpanzee and a human chromosome to see how different and variable our DNA sequences are. The biologists determined the exact sequence of about 10,000 base pairs from a region of the DNA in the middle of the X chromosome, using as subjects 30 chimpanzees and 69 humans who represented all the world's major language groups. They found that, on average, any 2 chimpanzees differed at about 13 places in the stretch of 10,000 base pairs, while any 2 humans differed only at

about 3.7 places. These results show two things. First, there is a high rate of genetic variation within a species. For example, you differ from an unrelated person by about a million base pairs over your entire genome. This amount of genetic variation is probably enough to provide the material for natural selection. Second, different species can harbor different amounts of genetic variation. Chimps are about four times as variable as humans. Assuming equal mutation rates, this might reflect a more ancient origin of the chimpanzee populations than of the human populations.

Next, biologists compared the genetic variation *within* a given human ethnic group to the variation among all groups. They were amazed to find that about 90 percent of the genetic variation that exists in the *entire human species* can be found among the individuals *within each single racial group* (Fig. 9.21). If a giant meteor destroyed everyone on Earth except a few thousand Native Alaskans, Bora Bora Islanders, or African Bantus, those few survivors would still harbor about 85 percent of our species's presently existing genetic variation.

In contrast, all the racial differences that exist between European, African, Indian, East Asian, New World, and Oceanic peoples can be explained by only about 10 percent of our species's total genetic variation. For example, just 2 to 5 of our 100,000 genes are responsible for the differences in skin color between Africans and Europeans. This means that, except for superficial traits like skin color, facial features, hair type, and body shape, numerous people belonging to different races are more similar to you genetically than many individuals of your own race. Clearly, no one can invoke significant genetic differences among groups of human beings to support their racial and ethnic prejudice.

Some animals have very little genetic variation. For example, cheetahs, the handsomely spotted cats of the African savannah that are the world's fastest runners, have less genetic variation than most other mammals, about 100 times less than humans do. This genetic uniformity may someday propel cheetahs toward extinction. Because each individual is so similar genetically, a single new virus, for example, to which all cheetahs were vulnerable could wipe out their entire population. Furthermore, because genetic variation is required for evolution to occur, cheetahs have much less chance for evolutionary change than organisms with abundant genetic diversity like chimps or dogs. If their environment changes dramatically—say, from global warming or the rise of new diseases to which all living cheetahs are susceptible—they could be doomed.

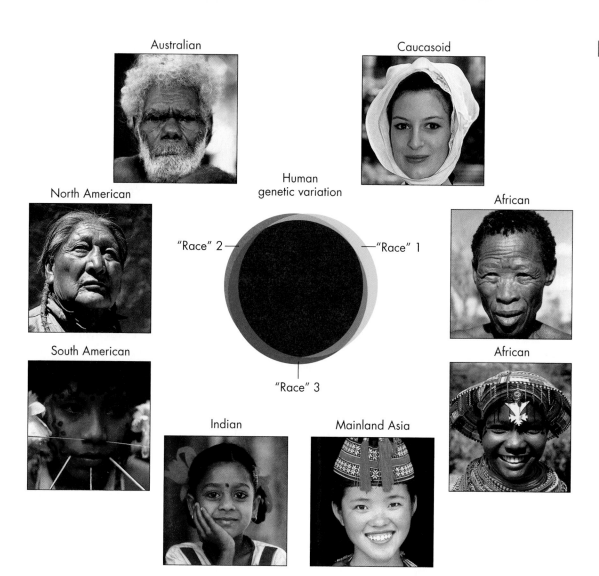

Australian

North American

South American

Caucasoid

Human
genetic variation

"Race" 2 — — "Race" 1

"Race" 3

Indian

Mainland Asia

African

African

Figure 9.21 **Genetic Variation and the Human Species**. Genetic variation between individuals within each human population is far greater than the variation between populations themselves. Each colored circle represents the genetic variation encompassed by what some people consider a separate human race. The variations overlap broadly, with few or no genetic differences confined to a single race. Because of this fact, most human geneticists think that there is no genetic basis for the concept of race.

9.5 How Is Genetic Variation Inherited in Populations?

In 1908, a young geneticist named R.C. Punnett was mulling over a puzzling genetic problem: Punnett was aware that some people are born with a mutation causing short, stubby fingers, a condition known as **brachydactyly** (Fig. 9.22). He also knew that the mutant allele is dominant over the allele for normal fingers. So, he wondered, because the mutation is dominant, why didn't everyone have brachydactyly within a few generations of the time the mutation first arose? This is another way to ask the more fundamental question of what maintains genetic diversity over time?

One day, Punnett presented the puzzle to his friend G.H. Hardy, a famous British mathematician. Hardy thought about it for a moment and then wrote two simple mathematical equations on a napkin to illustrate what he thought was happening genetically in the case of brachydactyly. To simplify the problem, Hardy assumed that there would be no significant outside influences on the population of organisms (in this case, people). He also reasoned that freedom from external pressures would create two consequences for the population: (1) The frequencies of *alleles* in the population would stay the same over the generations, and (2) the relative frequency of various *genotypes* (heterozygotes and homozygotes) in the population would stay the same after the first generation. For a century, biologists

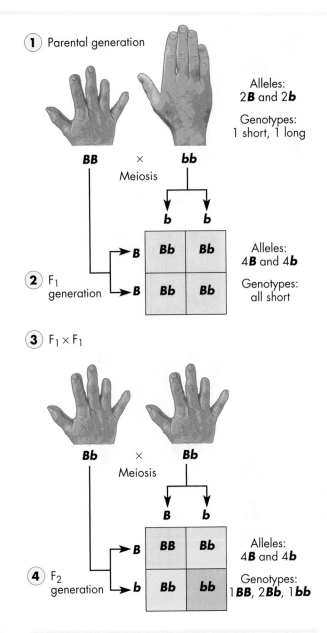

Figure 9.22 Finger Length and Hardy-Weinberg Equilibrium. The dominant allele of the brachydactyly gene causes the end bone of the fingers to be short or missing, while the recessive allele causes fingers of normal length, with three joints. Since it's dominant, why don't most people have the allele for brachydactyly? Follow the steps here and in the text to find the answer.

1 Parental generation

Alleles: 2**B** and 2**b**

Genotypes: 1 short, 1 long

BB × **bb**

Meiosis

b   b

Bb	**Bb**
Bb	**Bb**

2 F₁ generation

B
B

Alleles: 4**B** and 4**b**

Genotypes: all short

3 F₁ × F₁

Bb × **Bb**

Meiosis

B   b

BB	**Bb**
Bb	**bb**

4 F₂ generation

B
b

Alleles: 4**B** and 4**b**

Genotypes: 1**BB**, 2**Bb**, 1**bb**

Conclusion: In the absence of outside influence, the frequencies of alleles do not change over the generations, and the frequencies of different genotypes do not change after the first generation.

SOLVE IT

Explorer 9.3

The Hardy-Weinberg Principle

have called the mathematical model describing how genes behave in populations the **Hardy-Weinberg principle,** in honor of G.H. Hardy and of E. Weinberg, a German physician who discovered it independently.

Punnett and other geneticists of his day were stumped by the brachydactyly problem because they confused the concepts of genotype and phenotype. They were more concerned with the number of stubby-finger phenotypes in the population than with the frequency of the allele that caused the condition. Students sometimes confuse these today when considering a situation like this: If a person homozygous for the dominant trait of brachydactyly *(BB)* marries a person with the recessive trait for normal fingers *(bb)*, then all their first generation offspring will be heterozygous *(Bb)* and all will show the dominant stubby-finger phenotype (Fig. 9.22, Steps 1 and 2).

By considering only phenotype, we might conclude that the dominant trait (allele) has increased, and the recessive trait (allele) has disappeared. But by considering the genotype, we can see that the parents *(BB × bb)* have four alleles, half of which are *B* and half of which are *b*. Since the children's genotypes are all *Bb* (again, half *B* and half *b)*, we can see that the allele frequency has stayed the same. This was Hardy's first consequence: Allele frequencies remain constant in the absence of outside influences.

The second consequence begins to show up in the second generation. Suppose that two people from the first generation, each one possessing the genotype *Bb,* mated with each other (Step 3). Their offspring (the F₂ generation) would have the genotypes 1*BB*:2*Bb*:1*bb* and the phenotypes 3 short fingers to 1 normal fingers, just as Mendel would have predicted (Step 4). Hardy's equations (see Appendix C) show that this 1:2:1 genotype ratio would remain stable throughout additional generations as long as there were no outside perturbing influences. The above analysis reflected the special case wherein the numbers of alleles are equal. That, however, is rarely true. Biologists say that a population is in **Hardy-Weinberg equilibrium** when, in the absence of external pressure, it has both stable allele frequencies and stable genotype frequencies over many generations. Appendix C presents simple equations that allow the analysis of any ratios in the starting population. **E**xplorer 9.3 presents a mystery to solve using the Hardy-Weinberg principle.

How Do Biologists Use the Hardy-Weinberg Principle?

Using the Hardy-Weinberg principle, a biologist can predict allele frequencies in populations, and, in turn, can determine whether or not a population is evolving. The allele frequencies stay the same, and hence the population won't evolve, as long as it is free of outside influences. These "outside influences" are obviously important and a key to evolution. So what are they?

Five conditions must hold in order for allele and genotype frequencies to remain constant over many generations: (1) no mutation, (2) no migration into or out of the population, (3) large population size, (4) random mating, and (5) no natural selection at work. How likely is it that all five conditions will be met and that allele frequencies in a population will remain unchanged (nonevolving) generation after generation? Zero. Populations in nature *do* evolve, and allele frequencies do change within populations over time. The Hardy-Weinberg principle is useful only as a theoretical standard—an unchanging baseline—to compare against real populations in real environments that have outside influences. A patient like Wayne Chedwick, for example, taking multiple antibiotics for his bone infections, changes the environment of the bacteria in his body and inadvertently creates pressure that selects for drug-resistant bacteria. Evolutionary biologists study real life cases such as these, in which the five conditions listed above are *not* met, and can therefore measure the actual agents of evolution at work.

9.6   The Agents of Evolution

Although the Hardy-Weinberg equations only apply to sexually reproducing diploid eukaryotes, the basic logic is similar for a population of *Staphylococcus* bacteria lurking on your skin, in your throat, or deep in your nasal cavities. These cells can have plenty of genetic variation and the potential to show many different traits, but that variation doesn't work alone to bring about change in a population of cells evolving toward methicillin or vancomycin resistance. First something has to *act* on the genetic raw materials, the heritable variation. That something can be any of the outside factors we just saw—mutation, migration, chance in small population size, nonrandom mating, or selection. Working alone or together, these can alter allele frequencies and upset the Hardy-Weinberg equilibrium. Let's see how.

Mutation as an Agent of Evolution

By changing an original allele into a new one, mutation can alter allele frequencies in a population. The most common eye tumor in children, retinoblastoma, is an example of mutations and evolution. In about 30 percent of all cases of the disease, it seems that a new dominant mutation has arisen, probably in one parent's egg or sperm. With each new mutation, one normal allele is

removed from the human gene pool and replaced with the tumor-causing allele. This means the allele frequency changes by a tiny amount (and the Hardy-Weinberg equilibrium is disturbed slightly). The main importance of mutation to evolution, however, is not tiny changes in allele frequencies. Instead, it is the new phenotype that the new mutation may cause, and how natural selection acts on it. Obviously, eye tumors are not going to be favored by natural selection because, untreated, they usually blind or kill the child. The slightly higher allele frequencies therefore don't automatically mean humans are evolving toward more retinoblastomas.

Gene Flow: Migration and Alleles

Few populations are so isolated that they escape the arrival or departure of members. When organisms—butterflies, people, or milkweed plants—migrate from one population to another one nearby, they may take alleles away from the first group and introduce them into the second. Biologists call this change in allele frequencies due to migration in or out **gene flow.**

Archaeologists have applied the gene flow concept to a mystery of human history. Several decades ago, archaeologists began to dig up a certain type of decorated beaker at widely separated sites in Europe and the Middle East where humans lived about 9000 years ago. It seems that ancient peoples originally made the vessels in the Middle East. Over several thousand years, however, their production and use spread across Europe. Archaeologists wondered what exactly spread: Was it simply the knowledge of how to make and decorate this new pottery? Or did the Middle Eastern potters themselves migrate slowly across the continent? To answer this question, human geneticists sampled allele frequencies for 95 genes in various populations now living across Europe. They found gradients of allele frequencies as one travels from southeast to northwest. These gradients suggest that the potters themselves migrated towards the northwest, intermarrying with the local populations as they went. A single holdout tribe, the Basques living at the border of what is now Spain and France, resisted intermarrying with the migrating Middle Easterners. Today they alone display the language and genetic makeup of the prehistoric, premigration Europeans!

Gene flow is especially important in the evolution of bacterial antibiotic resistance. Many antibiotic resistance genes are present on the small circles of DNA called plasmids. These DNA circles can escape from one species of bacterium infecting a hospital patient, for ex-

How is gene flow involved in antibiotic resistance?

ample, then encounter another bacterial population in a dirty hospital drain and enter a second species by the process of transformation (Chapter 6). If another patient or staff member touches a towel, glass, or bedpan from that sink, he or she could pick up the second type of bacterium now carrying a plasmid bearing the antibiotic resistance genes. These genes, in effect, will have entered a new population and changed allele frequencies. As we've seen, they could also endanger people's lives.

Genetic Changes Due to Chance

Genetic drift is a term that refers to changes in allele frequencies that happen by chance and can't be predicted. Drift occurs most dramatically in small populations. To understand why, picture two populations of cherry trees, one large and one small. Now suppose that an allele *r* exists in one-tenth of the individuals in both populations. If a chance occurrence like a severe spring ice storm struck a large population with 1 million cherry trees and one-half died, 500,000 would still survive. It is quite likely that one-tenth of the survivors, or about 50,000, would still bear the *r* allele. But if that same storm hit a small population of just 10 trees, only one tenth of which (just one tree) had the *r* allele, and only one half (5 trees) survived at random, then there is a 50 percent chance that the single tree bearing the *r* allele would be killed. If that happened, the small tree population would lose the *r* allele completely, and the allele frequency would have changed unpredictably. Two types of genetic drift are the bottleneck effect and the founder effect. Let's look at each one.

Bottleneck Effect

Genetic drift can affect real-world organisms through a mechanism called a population bottleneck. Before we define it formally, let's consider the cheetah again. We said earlier that cheetahs have very little genetic diversity; it was actually Drs. Marilyn Menotti-Raymond and Steven O'Brien (our cat DNA researchers from Chapter 6) who discovered this fact. They studied the cheetah's mitochondrial DNA because it accumulates mutations more rapidly than DNA inside the cell nucleus. Also, they knew the rate at which mutations usually accumulate in the mitochondrial DNA of mammals, and they used this as a sort of clock from which to work backwards and calculate how long it took for the cheetah's genetic variation to shrink so dramatically. What they found coincides with a reconstructed history of the now rare spotted cats.

Until about 10,000 years ago, cheetahs had a large population inhabiting all of Africa and the Middle East, and stretching into Asia. About ten millennia ago, however, the population apparently crashed due to disease, drought, or overhunting by humans, leaving only a few thousand of the beautiful felines. Since then, cheetah populations have rebounded (Fig. 9.23), but genetic diversity remains low (despite the few mutations that have accumulated since the crash) because all of the living cheetahs derive from the same few survivors. Biologists call a situation like this, in which a large population is slashed and then recovers from a few survivors, a **population bottleneck.** They call the reduced genetic diversity based on the few surviving original alleles the **bottleneck effect.** If the chance survivors have allele

Figure 9.23 **Bottleneck Effect.** Cheetahs experienced a population bottleneck about 10,000 years ago. At that time, disease, drought, and/or overhunting by humans decreased the population size to only a few thousand individuals, which are the ancestors of today's entire cheetah population. The alleles present in today's cheetahs are primarily those that were present by chance in the survivors of that ancient bottleneck.

combinations that leave them susceptible to certain diseases, then the population's long-term future can be in doubt. Cheetahs, in fact, are highly susceptible to cat distemper, the males' sperm tend to be defective, and the cubs are sickly. Elephant seals are another example of the bottleneck effect. Hunted to near extinction in the late 1800s, a remaining population of just 20 animals has now recovered to tens of thousands, but their genetic diversity is extremely limited.

More and more species are experiencing genetic bottlenecks due to human activities. People are rapidly logging coniferous forests in America's Pacific Northwest and tropical forests in Brazil and Borneo, and stripping grasslands in North Africa, devastating these natural habitats that support millions of species. Some biologists, including Harvard ecologist E.O. Wilson, warn that the majority of species on Earth will become extinct during the next 50 to 100 years—an unprecedented extinction crisis. To help protect endangered species, biologists are using the principles of evolution and population genetics to estimate the **minimum viable population,** the smallest "bottleneck" a population can pass through and still have enough genetic variability to ensure the species' survival. With information like this, conservationists can then help citizens and governments set aside adequate natural habitats and plan other strategies to help save species from extinction.

Founder Effect

Another type of genetic drift stems from the long-term isolation of a population founded by a few individuals. Consider the 14,000 Amish people who live in thriving communities in eastern Pennsylvania. All the Amish are descendants of about 200 individuals who started emigrating from Switzerland in 1720. One of these founders, either Mr. or Mrs. Samuel King, had a recessive allele that, when present in two copies (homozygous), causes short forearms and lower legs as well as extra fingers and extra toes (Fig. 9.24). In the Amish living around Lancaster, Pennsylvania, 1 in every 14 people carries this allele instead of the 1 per 1000 found in other populations.

This kind of genetic drift is called a **founder effect,** because a few individuals split off from a large population and founded a new, isolated one. Nevertheless, both populations continue to exist. Because the small group of "founders" bears such a small fraction of the larger population's alleles, it may be a skewed sample, genetically speaking.

Dr. Fred Tenover and other bacteriologists who study the emergence of antibiotic resistance sometimes

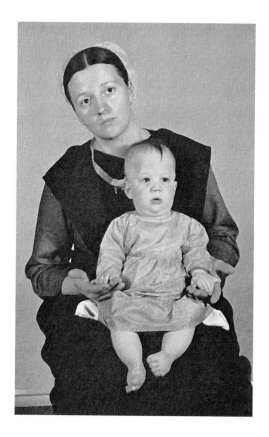

Figure 9.24   The Founder Effect. One of the original members of the Amish community in Pennsylvania happened by chance to carry a recessive allele for a rare kind of dwarfism that causes shortened forearms and lower legs, an altered upper lip, and often extra fingers and toes. Note the six fingers on this child's right hand and six toes on her right foot. Because there were few founders, this allele occurs more frequently among the Amish than in the general population.

invoke the founder effect to help explain the appearance of resistance in different places. For example, there are several ways that penicillinase—an enzyme that breaks down penicillin—can mutate, allowing bacteria to destroy a variety of other antibiotics. Often, researchers will find only one of these mutations in a given hospital, whereas in another hospital, a mutation in a different part of the same gene may confer the same resistance. The founder effect is most likely behind this kind of situation. Two different, independently occurring mutations probably arose in individual bacteria in two different hospitals, and descendants of these founder bacteria made up the resistant populations in the each of the hospitals.

Nonrandom Mating

When G.H. Hardy scratched his mathematical equations on a restaurant napkin, he made an assumption about mating: that every individual in a population has an equal chance of mating with any other member of that population. Oftentimes, however, reality is far different, and a population's mating tendencies—who mates with whom—can alter the ratio of heterozygotes and homozygotes (the genotype frequencies that Hardy predicted).

One type of nonrandom mating occurs when relatives are more likely to mate with each other than with unrelated individuals, a situation we call **inbreeding.** In many species, one member tends to mate with another that was born nearby, for example, frogs born in the same pond or people living in the same valley. Chances are good, therefore, that the two are related to some degree and have more identical alleles than nonrelatives.

Figure 9.25 **Natural Selection and the Medium Ground Finch.** In a year with normal rainfall, large and small seeds provide plentiful food for Galápagos finches with either large or small beaks. In a dry year, however, small seeds are more scarce than large seeds. Birds with beaks that are too small to crack large seeds usually die, leaving birds with larger beaks such as this medium ground finch as parents of the next generation. The frequency of alleles causing larger beaks therefore increases due to natural selection.

As a result, homozygotes become more common and heterozygotes become less common than expected by chance. In this way, inbreeding changes genotype frequencies, while leaving allele frequencies unchanged.

Although inbreeding does not change allele frequencies, it does affect the evolution of populations in sometimes insidious ways. Recall that inbreeding increases the frequency of homozygotes above what would be expected by random mating. If these individuals are homozygous for deleterious recessive alleles, then the population will have many more individuals that are less fit than in a normally breeding population, a situation called **inbreeding depression.** (In heterozygotes, these alleles would remain hidden by the dominant alleles.)

The rarest large mammal on Earth, the Javan rhinoceros, provides an example of inbreeding depression. This huge grazer once ranged throughout all of Southeast Asia, but the current population has been whittled down to about 60 individuals and split into several separated groups. Because this small population is split up, related Javan rhinos are more likely to breed with each other than if there were more of the rhinos living in a totally joined range. The inbreeding produces more homozygotes than if mating were truly random, and the result is inbreeding depression. Geneticists calculate that it would take a population of about 500 of the rhinos, and these would have to breed randomly, to prevent the deleterious effects of inbreeding depression and to keep about 90 percent of the rhino's current genetic variation throughout this century. What should we do to preserve Javan rhinos? Capture the last remaining wild rhinos and encourage breeding in captivity—but only with distant or nonrelatives? Or should we protect the survivors in the wild more carefully? We will face more and more choices like this as human activities continue to fragment the ranges and whittle down the population sizes of the world's species.

9.7 Natural Selection Revisited: How Populations Become Better Adapted to Their Environments

In natural selection, the environment does the selecting. In **artificial selection,** people choose individuals with certain phenotypes to become the parents of succeeding generations. Swiss dog breeders, for example, selected and mated the largest, strongest animals for many generations in a row to create the St. Bernard, a mammoth dog breed that has helped save skiers lost around

Greater St. Bernard Pass in the Swiss Alps. Whether selection is natural or artificial, however, an individual's ability to survive and reproduce in a particular environment relative to other members of its species is its **evolutionary fitness.**

We've seen that antibiotics like methicillin can act as agents of selection upon bacteria such as *Staphylococcus aureus*. The drugs kill the most susceptible individuals and leave the fittest cells—those that can destroy or survive the antibiotic. Similar changes have been noted in larger, more familiar organisms. Darwin noticed that the dry, barren Galápagos Islands are home to several finch species, each with its own body size, bill shape, choice of foods, and food-gathering habits. More than a century after Darwin's visit, a team led by Peter Grant of Princeton University carried out some direct experiments on these birds. They studied the medium ground finch, *Geospiza fortis* (Fig. 9.25), for several characteristics, including body size, beak size and shape, number of eggs laid, and genetic diversity. Grant's team found considerable variation in the birds' body and beak sizes, and determined through breeding experiments that genetic variation plays a large role in this physical variation.

In 1977, the medium ground finch lost most of its major food source, seeds, due to a severe drought on the Galápagos Islands. That year, none of the birds bred successfully, and 85 percent of the individuals disappeared. Which 15 percent survived? Was death random? Or did birds with certain characteristics fare better than others? It turns out that birds with larger bodies and beaks were better survivors. The team found that in a year with normal rainfall, herbs and grasses grow abundantly and produce a bumper crop of small, soft seeds. In years like that, only a few kinds of plants, growing in limited numbers, produce large seeds. Apparently, birds eat the small seeds first, and then turn to the large seeds, but only the bigger birds with stouter beaks can actually crack large seeds and get at the nutritious kernels inside. A drought year most drastically cuts the number of small plants that produce small, soft seeds. Therefore, the smaller birds, unable to crack larger seeds, are most likely to starve and die in drought years, while the larger birds with bigger beaks can switch to the harder, more sizeable seeds and survive. Peter Grant's team was able to document that in 1978, when these larger birds reproduced, the entire surviving population and their offspring had shifted to a larger average body size.

In this case, the drought was the agent of natural selection, acting upon preexisting genetic variation within the finch population, and allowing only the largest birds to survive and to reproduce. The result? A change in gene frequencies—another way of saying evolution by natural selection.

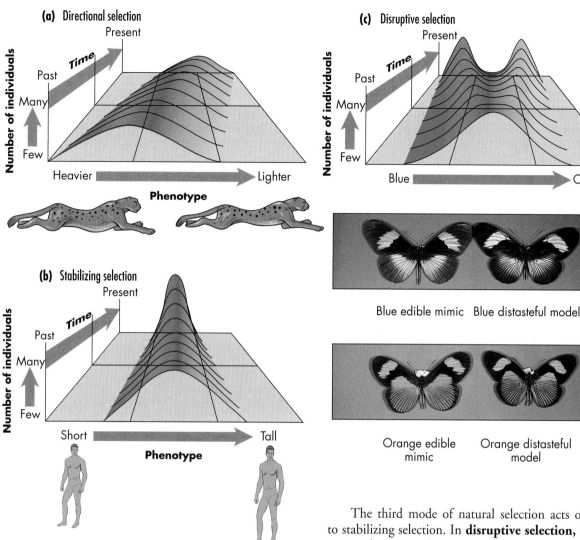

(a) Directional selection

Number of individuals
Many
Few

Past
Present
Time

Heavier ➡ Lighter

Phenotype

(b) Stabilizing selection

Number of individuals
Many
Few

Past
Present
Time

Short ➡ Tall

Phenotype

(c) Disruptive selection

Number of individuals
Many
Few

Past
Present
Time

Blue ➡ Orange

Blue edible mimic   Blue distasteful model

Orange edible
mimic   Orange distasteful
model

Figure 9.26 **Three Modes of Natural Selection.** (a) Smaller size in cheetahs is an example of directional selection. (b) Intermediate height in humans is an example of stabilizing selection. (c) The blue and orange forms of the *Pseudacraea eurytus* and *Bematistes* butterflies demonstrate disruptive selection.

Natural selection can alter a species' evolutionary trajectory in at least three different ways. In **directional selection,** a population shifts toward one extreme form of a trait (Fig. 9.26a). In the last 4 million years, for example, cheetahs have become half their former size, presumably because natural selection favored alleles that constantly pushed cheetah weight downward. Directional selection also accounts for the resistance that insects and microorganisms acquire to pesticides and antibiotics.

A second mode of natural selection is called **stabilizing selection,** because it results in individuals with intermediate phenotypes, as extreme forms are less successful at surviving and reproducing (Fig. 9.26b). Height in adult humans is a stabilized trait, with most people falling toward the center of the bell curve. Apparently in the past, very large or very small people left fewer progeny than those with intermediate heights.

The third mode of natural selection acts opposite to stabilizing selection. In **disruptive selection,** two extreme phenotypes become *more* frequent in a population (Fig. 9.26c). For example, in parts of Africa, females of the butterfly *Pseudacraea eurytus* taste good to birds, but mimic in coloration other butterfly species that are distasteful to birds. Some of the good-tasting mimics occur in a blue form, and others in an orange form (Fig. 9.26c, left two butterflies). The blue form mimics the foul-tasting blue butterfly species model called *Bematistes epaea,* while the orange form looks like the orange distasteful species model *B. macaria* (Fig. 9.26c, right two butterflies). Mimics of either foul-tasting model are likely to escape being eaten. In contrast, intermediate forms mimic neither of the bad-tasting species and are more likely to be eaten by predators. This has caused alleles for the intermediate phenotype to decrease in frequency over time.

All five agents of evolution we've been exploring— mutation, migration, genetic drift, nonrandom mating, and selection—can nudge a population away from Hardy-Weinberg equilibrium. In recent years, however, evolutionary biologists have been particularly interested in determining what role each agent plays in *maintaining* genetic variation and therefore in contributing to evolution.

Maintenance of Genetic Variation

We've seen that most natural populations have lots of genetic variation and that cheetahs and Javan rhinos are rare exceptions. But why? Shouldn't natural selection eliminate all alleles except those that confer the highest fitness? Some biologists, so-called **selectionists,** think that high rates of genetic variation exist because natural selection *maintains* genetic diversity in most populations. They argue that the observed amount of variation is selected for because environments often rapidly change in both time and space. Some alleles are selected under one set of conditions, others under different sets of conditions. As a result, they say, most populations maintain a high level of genetic variation.

Neutralists, on the other hand, think that most variation is unlinked to an organism's survival and reproduction. They point out that even organisms in very different environments often accumulate mutations in a given DNA stretch at similar rates. The gorilla, bonobo, chimpanzee, and human, for example, are all about equally distant from the orangutan along the stretch of DNA from the *X* chromosome we discussed earlier. For the great apes, then, about one mutation accumulates each 100,000 years, on average in that part of the *X* chromosome. People have called this steady tempo of change a **molecular clock.** Neutralists think that most alleles are equivalent and most mutations are neutral (that is, they lead to slightly different proteins that work about as well as the original), giving the organism neither an advantage nor a disadvantage.

To a neutralist, most genetic change over time comes from neutral mutations, unpredictable genetic drift, and gene flow. To a selectionist, most genetic change is due to the various forms of natural selection. Many biologists take a view that combines the best of both arguments. The intermediate view is that even if many mutations are neutral, natural selection can still act on plenty of non-neutral variation in every living generation and thereby affect the course of evolution.

9.8 How Do New Species Originate?

Everyone knows that a cheetah is a different species from the bacterium *Staphylococcus aureus*. But how do we know that *Staphylococcus aureus* is a different species

from *Staphylococcus epidermidis*? Or that the bonobo and the very similar chimpanzee (see Chapter 26) are two different species and not just one? According to modern evolutionary theory, a **species** is a group of populations that interbreed with each other in nature and produce healthy and fertile offspring. In short, a species in nature is **reproductively isolated** from (fails to generate fertile progeny with) other true species. Species that to a human observer may *look* identical can still live side by side and not interbreed. Conversely, very different-looking organisms can sometimes interbreed in nature, and if they do, they are considered varieties of a single species.

The definition of a species as a shared, reproductively isolated gene pool does have some drawbacks. One is that it applies only to sexually reproducing organisms. *Staphylococcus aureus* and many other kinds of microbes are asexual organisms that reproduce only by cell division, and so this definition excludes them. Also, extinct species can't be tested for reproductive isolation, and so this definition makes it harder to assign a fossil organism to a particular species. For asexual organisms and fossils, physical traits—structural, environmental, evolutionary, biochemical—have to be used instead of the criterion of interbreeding in nature.

How Do New Species Arise?

New species arise when one group of organisms becomes reproductively isolated from another. But what isolates species in the first place? What keeps separate species like lions and tigers, for example, or sets of nearly identical-looking tropical frogs, from interbreeding in nature and producing fertile offspring?

Reproductive isolating mechanisms are biological features that keep the members of species A from successfully breeding with the members of species B. Some reproductive isolating mechanisms act before fertilization (so-called prezygotic reproductive isolating mechanisms) and others act after fertilization (so-called postzygotic reproductive isolating mechanisms). Looking at isolating mechanisms that act before fertilization, some are behavioral. For example, dozens of frog species can live in the same area, but some frogs mate only in the spring and others only in the fall. One frog species may mate only in ponds, another only in flowing streams, and a third only in shallow pools and puddles. These peculiarities of the frogs' mating habits can stop gene flow from one group to another and help ensure that the species become or stay separate.

Other mechanisms that act before fertilization can be mechanical or chemical. In some spider mites, for example, the male genitals are shaped like a "key" that opens the females' genital plates; the "key" of one species can't open the "lock" of another. Sometimes, a species's sperm can't fertilize another species' eggs because a specific chemical surrounds the egg and allows binding only by sperm from the same species.

Looking at reproductive isolating mechanisms that operate after fertilization, sometimes different species mate and offspring result, but they die or are themselves infertile. Many American frogs, for instance, can interbreed, but genetic differences between the species lead to abnormal zygotes and freakish tadpoles and adults. In contrast, the mules that result from the mating of a male donkey with a female horse are very hardy hybrids, noted for strength and endurance. They usually don't produce offspring with other mules, horses, or donkeys, however, because chromosomes can't pair normally during meiosis. This prevents a flow of genes between donkeys and horses, and keeps the species separate.

How, then, do reproductive isolating mechanisms arise and lead to new species?

The Origin of New Species

Most biologists believe that many species arise after populations are split apart geographically and then evolve based on the reproductive isolating mechanisms we just discussed.

Remember the squirrels we encountered earlier that were separated by the Grand Canyon and evolved into two separate species (see Fig. 9.16b)? Their case is not uncommon. A physical barrier, such as a river, a desert, or different zones of vegetation can separate populations of a single species and prevent gene flow between them. The split populations grow more and more distinct from each other as mutation, genetic drift, and adaptation cause different sets of characteristics to accumulate. Eventually, the differences are great enough to prevent matings even if the two populations come into contact again later. This mechanism is called geographical speciation, or **allopatric speciation** (*allo* = different + *patra* = native land) (Fig. 9.27). The Galápagos finches we saw earlier arose by allopatric speciation, too, when a mainland species got separated into several populations on the islands. If the isolated populations are small, the founder effect can come into play and so can punctuated equilibrium, because in the small, separated populations, the pace of evolution can speed up for a while before stable species emerge.

It's easy to see how species can evolve where populations are totally separate (allopatric). But species can evolve even without any geographical barriers separating the populations. This is called **sympatric speciation** (*sym* = same + *patra* = native land). Biologists discount many claims that sympatric speciation has oc-

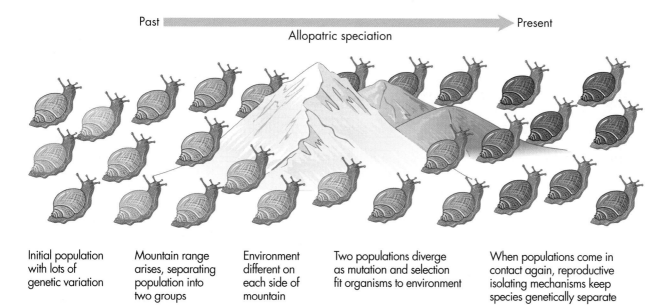

Past → Allopatric speciation → Present

| Initial population with lots of genetic variation | Mountain range arises, separating population into two groups | Environment different on each side of mountain | Two populations diverge as mutation and selection fit organisms to environment | When populations come in contact again, reproductive isolating mechanisms keep species genetically separate |

Figure 9.27 **Geographical Barriers and Allopatric Speciation.** After a mountain range or other geographical feature separates an initial population (represented here by snails of various colors) into two noninterbreeding subpopulations, natural selection and genetic drift may genetically differentiate the two groups (represented by their changing colors). If the mountain range later erodes and the two groups eventually come into contact again, they may be so dissimilar genetically that they can no longer interbreed, and they'll have become two different species (represented here by brown and blue snails).

curred, but do tend to agree that in plants living in the same geographic region, new species can sometimes arise in a single generation based on **polyploidy,** an increase in the number of chromosome sets in a cell. A genus of pretty pink wildflowers called *Clarkia* provides a good example of polyploidy. Members of the Lewis and Clark expedition (1804–1808) first collected these flowers in Idaho (Fig. 9.28). Later geneticists discovered that one species, *Clarkia concinna,* has seven pairs of chromosomes in a diploid set, while *C. virgata* has five pairs. If these two species mate, the hybrid progeny gets seven haploid chromosomes from one parent and five from the other. The diploid hybrid is sterile because the two different sets of chromosomes don't pair and move correctly during meiosis. At some time in the past, however, a **tetraploid** hybrid plant arose in which the chromosome count had doubled from 12 (7 + 5) to 24. Every chromosome in the new tetraploid hybrid, *Clarkia pulchella,* had a partner, so meiosis could occur normally. The species can't produce fertile hybrids with either diploid species *C. concinna* or *C. virgata,* but it

Figure 9.28 **Speciation Without Barriers.** The diploid western wildflower *Clarkia concinna* has seven chromosome pairs, while the related diploid species *C. virgata* has five chromosome pairs. A diploid hybrid of these two species having one set of five chromosomes and one set of seven chromosomes is sterile because of mismatched chromosomes during meiosis. But if the chromosome sets double in this diploid hybrid, then a tetraploid results with two sets of five chromosomes and two sets of seven chromosomes. Meiosis occurs normally in the tetraploid, *C. pulchella* (shown here), because each chromosome can pair with a homologous chromosome. Because the newly formed tetraploid cannot successfully mate with either parental species, and because it arose from the parental populations without a geographical barrier, it represents a case of sympatric speciation.

can fertilize itself and produce new seeds. In this case, a new species was formed in one step within one original population without a geographical barrier. Sympatric speciation is relatively common in plants, and has contributed to the development of wheat, cotton, and some of our other important crops.

9.9 More Evolutionary Pathways

Evolution can be a small-picture event or a big-picture event. When a species changes over time, as with the enlarging beak size in medium ground finches, or when one species splits into two, as in the Grand Canyon squirrels, biologists call it **microevolution,** evolution at the level of populations and species. But grander evolutionary changes can take place, too, and biologists call this evolution above the level of species **macroevolution.** Good examples of macroevolution include the sudden appearance of *Hallucigenia* in the fossil record (review Fig. 9.7) and the evolution of mammals from reptilian ancestors. Evolutionary biologists have been trying to determine whether microevolutionary mechanisms alone can account for the origin of new groups above the species level—phyla, divisions, classes, families, orders, and so on. In other words, they wondered whether macroevolution is just microevolution occurring over very long periods of time.

The origin of mammals from reptilian ancestors provides some answers (Fig. 9.29). Fossils reveal that over time, reptilian legs moved from their sideways orientation to directly under the body, the jaw muscles grew more complex, and various tooth types became more mammal-like. Fossils also reveal that ancient reptiles radiated into several evolutionary lines. Some became more like mammals, some less, and most died out without giving rise to further species. Clearly, mammals evolved as a result of many small changes in the legs, jaws, and elsewhere, and this evolution involved adaptations to environmental demands. Biologists have concluded from evidence like this that macroevolution *is* just microevolution extended over very long time periods.

How can the gradual, stepwise evolution of new features bring about whole new orders like the vertebrates? An answer comes from huge, extinct, rhinoceroslike animals called titanotheres (see Fig. 9.8). Titanotheres had large horns, which their smaller ancestors lacked. Fossil evidence shows that the huge horns arose gradually in stages from small bumps already present on the heads of earlier species. This shows that new

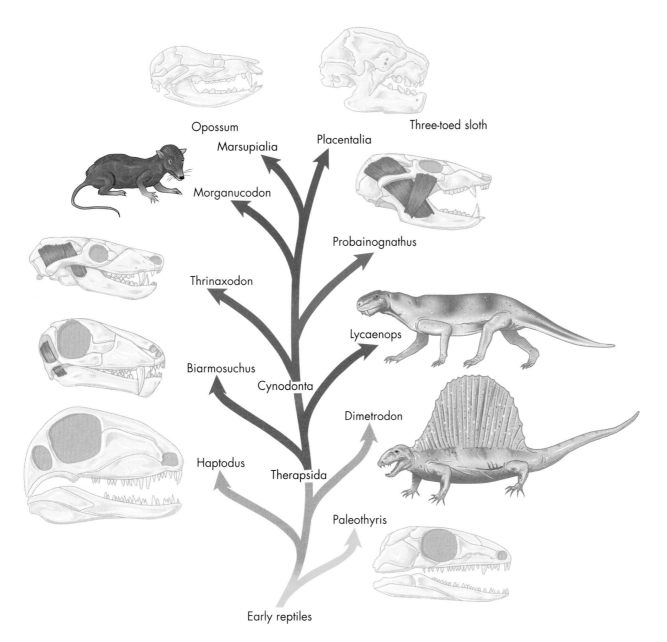

Figure 9.29 **Major Evolutionary Transitions: Reptiles to Mammal.** In the mammal-like reptiles, changes gradually occurred in many small steps. For example, the legs gradually changed from protruding sideways from the body, as in Dimetrodon, to being more underneath the body in the early mammal Morganucodon. The skull bones, muscles, and teeth developed in ways that allowed mammals to feed more efficiently and to better exploit their environments. This fossil record shows that macroevolution can occur by microevolutionary changes taking place over very long time periods.

Opossum
Marsupialia
Placentalia
Three-toed sloth
Morganucodon
Probainognathus
Thrinaxodon
Lycaenops
Biarmosuchus
Cynodonta
Dimetrodon
Haptodus
Therapsida
Paleothyris
Early reptiles

structures often evolve by a change in size, shape, or function of a pre-existing body part. The original structure may have been present for some reason totally unrelated to the future horns—the bumps didn't anticipate a future need. Once the structure was there, however, natural selection could act upon it and a new function could arise.

Oftentimes, the members of *different* species compete for the same foods and habitats, and natural selection acts on them, too. This process is **species selection** (also called **species replacement**). The evolution of the horse is a good example of species selection, as Figure 9.30 and *E*xplorer 9.4 explain.

Modern horses are large animals with long legs with a single toe at the tip; they also have high-crowned cheek teeth with complex grinding surfaces. Fossils show that an early ancestor of today's horses was the size of a German shepherd dog, had legs with four toes, and had small teeth that lacked grinding surfaces. The evolutionary changes between early and modern horses did not occur smoothly in a single, goal-seeking lineage. Instead, the changes unfolded in fits and starts, with

GET THE PICTURE

*E*xplorer 9.4

Evolution of the Horse

Figure 9.30 Species Selection, Extinction, and Evolutionary Trends. The dog-sized progenitor of modern horses, Hyracotherium, had four toes, but the modern horse, Equus, has one toe. The lineage of horses is a dense bush with many branches, most of which became extinct.

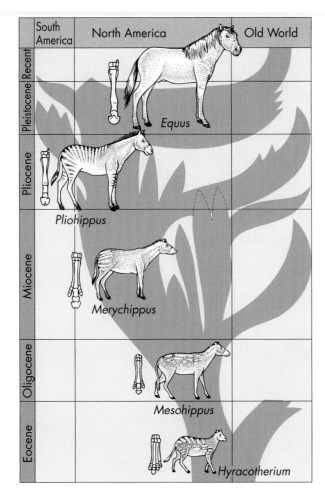

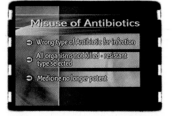

How can physicians and patients help prevent the evolution of antibiotic resistance?

different species displaying various combinations of toe number, body size, and tooth shape, and with smaller species occasionally replacing larger ones. Each species was subject to the environment at the time, including competition with other similar animals.

Species selection can explain in two different ways how horses evolved: (1) smaller-sized early horses may have tended to become extinct more quickly than larger horse species, or (2) larger horse species may have tended to form new species more quickly than small-bodied species. Perhaps both mechanisms contributed to horse evolution. Ultimately, though, evolution in all lineages leads to the great equalizer—extinction.

Adaptive Radiation, Extinction, and Modern Species

Modern organisms tend to fall into a few large taxonomic groups called phyla (singular, *phylum)* for animals, and divisions for fungi and plants (see Chapters 1 and 10). For example, mollusks (such as snails, clams, and squid) make up one of the many animal phyla and arthropods (such as crabs, spiders, and insects) make up another. Could the mechanisms of microevolution explain how major groups like these arose? Part of the answer may come from sediments laid down over 500 million years ago, as the Cambrian period began. At this time, the major animal phyla appeared in the fossil record in just a few million years, and in the hundreds of millions of years since, no new body plans have appeared. For this reason, paleontologists call the events of this time the **Cambrian explosion.**

Fossil evidence confirms that during the early Cambrian, the range of body plans was far greater than it is today. As time passed, extinction pruned many branches of the Cambrian's bushy family tree, and only organisms with the basic body plans we know today survived. After these lines were well established, they diversified, but no large enough differences have ever appeared to create any new animal phyla or plant and fungal divisions.

Evolutionary Pathways of Antibiotic Resistance

As we've explored evolution in this chapter, we've seen how natural selection can bring about new species and even higher orders of organisms. We've seen that the heavy use of antibiotics in hospitals, for example, can act as an agent of natural selection, "choosing" as parents of the next generation potentially dangerous cells like variants of *Staphylococcus aureus* that can't be killed with penicillin, methicillin, or normal levels of vancomycin. What, then, does our exploration of evolution suggest for dealing with the life-threatening problem of antibiotic-resistant bacteria?

Researchers have shown that under normal conditions (an environment free of antibiotics), antibiotic-resistant bacteria are poor competitors (review Fig. 9.5). Mutant cells use more energy to make the molecular machinery for antibiotic resistance, for example, or do other cellular tasks more poorly than normal as a trade-off for surviving antibiotics. They're actually at a disadvantage when they're *not* challenged with antibiotics because normal bacterial cells can outcompete them. In many cases, they'll die out, just like cavefish with fully developed eyes. What's the evolutionary implication here? It's this: One logical way to fight resistance is to cut the load of antibiotics in our everyday environments. Patients often demand antibiotics for colds and flu—even though these are viral infections—and doctors often comply to keep their patients happy and to ward off the possibility of secondary infections. By some estimates, half of all prescribed antibiotics are unnecessary, and medical consumers need to cut back on their demands. Farmers also use large amounts of antibiotics to stimulate the growth of farm animals and to protect fruit trees from bacterial diseases. Society could cut down on its exposure to antibiotics if farmers improved the hygiene in their feedlots and if the public consumed more organic produce or washed nonorganic fruits and vegetables more thoroughly to remove resistant bacteria and residue of antibiotics. The same principle applies to cleanliness in hospitals and around the kitchen and bathroom. Mutations occur in any given gene about 1 in every 100,000 to 1,000,000 times the gene replicates. In a bacterial population of 1,000,000 cells, therefore, between 1 and 10 of them might have a mutation in an antibiotic resistance gene. If bacterial populations are kept smaller through vigilant hygiene, it is that much less likely that a mutation will arise. Practices as simple as isolating infected patients and wearing masks and gloves when treating them can significantly reduce antibiotic resistance in clinics and hospitals, according to Dr. Fred Tenover. When applied in this way, the knowledge of evolutionary principles such as natural selection and gene flow by immigration can be translated into day-to-day solutions.

For patients like Wayne Chedwick, with serious bone infections that threaten to destroy toes and feet, effective antibiotics are a lifeline. Although methicillin stopped working for him, Dr. Joshua Fierer had an alternative. Researchers continue to develop new drugs for which resistance has not yet evolved and this is crucial, because our current arsenal of alternatives is nearly used up. Unfortunately, bacteria will eventually evolve resistance to any new antibiotics; we can only slow that process by using the principles of natural selection to our own advantage. By applying those principles, medical researchers, doctors, patients, farmers, medical consumers, and other members of society can take the necessary measures to prevent a return to the bad old days when bacterial diseases took a heavy human toll.

Chapter Summary and Selected Key Terms

Introduction Wayne Chedwick has lost four toes to a bacterial infection that destroyed soft tissue and bone, a complication of diabetes. That same infection, lodged in an ankle bone of his right foot, has threatened the entire appendage. The infection is especially dangerous because the *Staphylococcus* bacteria are resistant to the common antibiotic methicillin. The emergence of antibiotic-resistant species like this one represents **evolution** (p. 216) in action.

Emergence of Evolutionary Thought
Most 19th-century scientists believed that species were created individually and remained unchanged throughout time. Jean Baptiste Lamarck suggested a variation upon this with "created" species able to change over time to better compete in their environments. Lamarck's hypothesis, the **inheritance of acquired characteristics** (p. 217), was wrong. The notion of change over time, however, was correct. Charles Darwin and Alfred Wallace posed a third theory after observing and collecting plants and animals on extensive voyages. They both postulated **descent with modification** (p. 218) from common ancestors, by means of **natural selection** (p. 218), during which selection pressure from the environment affects individuals and their inheritable variations. Through this, nature "selects" favorable traits or **adaptations** (p. 219) that are passed along to new generations. Evolution acts on **populations** (p. 219). Traits that confer **selective advantage** (p. 219) increase an individual's chances of survival within a population in a given environment.

In modern hospitals, antibiotics are agents of selection pressure upon bacteria. Bacterial DNA can mutate and by chance confer a new ability to survive in the presence of an antibiotic. These mutants are at a disadvantage when the antibiotic is absent from the environment, but when it is present, they can outcompete normal cells, divide and redivide, and leave the next generation of bacteria with the new trait. For example, the gene that allows vancomycin resistance arose this way.

Evidence for Evolution **Fossils** (p. 222) provide evidence that organisms with body forms different from modern organisms lived earlier in Earth's history. Fossils of intermediate form such as the ancient bird/reptile *Archaeopteryx* show points along evolutionary pathways. **Homologous** elements (p. 224) derive from a single element in a common ancestor. The forelimbs of whales, people, cheetahs, and bats are homologous and support the idea of descent with modification. **Vestigial organs** (p. 224) are evolutionary "leftovers" that are now useless but resemble an ancestor's useful structures. The fishlike early embryos of vertebrates reflect a common ancestor 430 million years ago. By comparing the base pair sequences of genes and the amino acid sequences of proteins from different species, geneticists can create molecular family trees that are remarkably similar to trees based on fossil data. Other evidence for evolution comes from **biogeography** (p. 227), the geographic distribution of organisms. Darwin, for example, observed similarities between large land birds on the southern continents. Complex organs like the "camera" eye could have evolved through a series of small genetic changes but almost certainly not in a single mutation event.

Pathways of Descent Marine foraminifera are good examples of gradual evolutionary change. The squirrels living on opposite rims of the Grand Canyon were split into two populations historically and are a good example of **divergent evolution** (p. 229). The 20 species of Hawaiian honeycreepers adapted to surviving on different foods are good examples of **adaptive radiation** (p. 230). Flying squirrels and sugar gliders are unrelated but resemble each other closely; they are good examples of **convergent evolution** (p. 230). How fast do existing species diverge, converge, or radiate into new species? The model of **phyletic gradualism** (p. 231) holds that all lines change at about the same rate over time, while the model of **punctuated equilibrium** (p. 231) holds that structural changes occur in fits and starts. The formation of a new species is called **speciation** (p. 231).

Genetic Variation: Evolutionary Raw Material Traits vary between individuals of the same species because of **genetic variation** (p. 232). Where does this variation come from? Single-gene mutations or changes in the DNA sequence of a single gene can alter gene function. Some of these are **neutral**

mutations (p. 233) that don't help or harm the organism, while others can be harmful or lethal mutations. **Genetic recombination** (p. 234) can shuffle existing alleles (but the frequency in the population stays the same) while mutation can change the frequency in a population's **gene pool** (p. 234). **Gene duplication** (p. 233) can lead to new genes with new functions. Some species have very little genetic variation. Humans, for example, have only one-quarter as much genetic variation as chimpanzees.

How Is Genetic Variation Inherited?

Brachydactyly (p. 235), a finger mutation, helped G.H. Hardy develop the **Hardy-Weinberg principle** (p. 236) for how genes behave in populations. A population without perturbing influences is in **Hardy-Weinberg equilibrium** (p. 236), with stable allele and genotype frequencies over many generations. External pressures include mutations, migrations, small populations, nonrandom mating, and natural selection.

Agents of Evolution

Mutations can change allele frequencies in a population. So can **gene flow** (p. 237), which is based on the arrival or departure of members. **Genetic drift** (p. 238), or unpredictable changes in allele frequencies, often happens in small populations. A **population bottleneck** (p. 238) can occur when a large population is cut dramatically by disease, famine, or other factors, and recovers its numbers from a few survivors—a **bottleneck effect** (p. 238). Genetic drift can also occur when a few individuals split off from a large population and establish an isolated one, termed the **founder effect** (p. 239). Mating is often nonrandom. **Inbreeding** (p. 239) brings about changes in allele frequency.

Natural Selection: Adaptations to Environments

In **artificial selection** (p. 240), people (for example dog breeders) choose specific traits to emphasize in new generations. This can influence an individual's **evolutionary fitness** (p. 240). The medium ground finches of the Galápagos Islands demonstrate ongoing evolution, with the environment selecting for the trait of large beak size. Researchers have calculated a steady tempo of change in some species, which they call a **molecular clock** (p. 242).

How Do New Species Arise?

In nature, every true **species** (p. 242) is **reproductively isolated** (p. 242) from every other species. Mechanisms that bring about this isolation can be behavioral, physical, chemical, or biological. In **allopatric speciation** (p. 243), species arise when a barrier separates their ranges into distinct areas. In **sympatric speciation** (p. 243), new species arise without geographic barriers. Tetraploidy in plants is a good example of the latter.

More Evolutionary Pathways Evolution can occur on a small scale or on a large scale. Evolutionists call it **microevolution** (p. 244) when one species splits into two or radiates. **Macroevolution** (p. 244) involves evolution above the level of species, such as the appearance of new genera, families, classes, and so on. Different species sometimes compete for the same resources in a process called **species selection** (p. 245). Many new species appeared in the fossil record in just a few million years, during what is called the **Cambrian explosion** (p. 246), and most of those new lineages became extinct. Understanding the evolution of antibiotic resistance can help researchers and public health workers slow the emergence of bacteria with immunity to new antibiotics.

> All of the following question sets also appear in the Explore Life *E* electronic component, where you will find a variety of additional questions as well.

Test Yourself on Vocabulary and Concepts

In each question set below, match the description with the appropriate term. A term may be used once, more than once, or not at all.

SET I

(a) convergent evolution (b) divergent evolution (c) adaptation (d) species selection (e) adaptive radiation

1. Divergence of an ancestral species into many species that are well-adapted to slightly different parts of a new territory
2. A result of competition between members of different species for the same habitat
3. Two or more groups of organisms that live in similar environments are superficially similar but are descended from different ancestors
4. Body parts, behavior, or physiological processes that improve an organism's chance of survival and/or reproduction
5. Two or more species that have recently split from a common ancestral species that may now be extinct

SET II

(a) homologous organ (b) vestigial organ (c) phyletic gradualism (d) punctuated equilibrium

6. Rapid changes in small, offshoot populations, followed by long periods of stability
7. Slow but steady descent with modifications consistent with natural selection in ever-changing environments
8. Two structures that may have different functions but are descended from a common ancestral structure
9. A form of homology in which a rudimentary structure in one species is believed to share a common evolutionary ancestry with a functional structure in another species

SET III

(a) gradual change (b) divergent evolution (c) convergent evolution (d) adaptive radiation (e) species selection

10. Honeycreeper birds of Hawaii
11. The horse
12. Squirrels on the north and south rims of the Grand Canyon
13. Foraminiferans in the Indian Ocean
14. Flying squirrels (placental mammals) and marsupial sugar gliders, which both glide from tree to tree

SET IV

(a) population (b) gene pool (c) gene flow (d) evolutionary fitness (e) natural selection

15. The relative ability to survive and reproduce in a particular environment
16. A local group of interbreeding members of a species
17. Sum of all the alleles in a population
18. Change in allele frequency caused by individuals entering or leaving a population

Integrate and Apply What You've Learned

19. What acts as an agent of natural selection in new instances of antibiotic resistance? How do the new traits arise?
20. What is the role of the environment in the evolutionary theory of Lamarck as compared with Darwin and Wallace's?
21. Despite the similar ecology of the Galápagos Islands and the Cape Verde Islands, different plants and animals live on them. How did these observations influence Darwin?

22. What sort of fossil evidence would it take to contradict the major predictions of evolutionary theory?

23. Echidnas and anteaters both eat ants, and have long snouts and tongues, and sharp, powerful claws. These features evolved independently in the two groups. Which patterns of evolution does this exemplify and why?

24. Why is extinction the likely fate of any species?

25. Hardy-Weinberg equilibrium is relevant under only such special conditions that it does not apply to most natural populations. What, then, is its value?

26. The northern spotted owl is threatened with extinction in the Pacific Northwest. Some people argue that a minimum of several hundred reproducing individuals would be necessary in order to maintain a viable population. Preserving the habitat for this number of breeding pairs would prevent logging in much of the remaining old-growth forest. Other people argue that we can log the rest of the old growth while keeping a few breeding pairs of owls in captivity, where they could reproduce and avoid extinction. Compare the underlying assumptions of these two arguments, give your opinion, and support it.

Analyze and Evaluate the Concepts

27. Suppose you are observing a group of five species of organisms called "smurfs," each of which lives in the tree canopy in a jungle. Each species has a grasping, prehensile tail. They have numerous small physical differences, eat different diets, and prefer to remain at different levels in the canopy. The molecular data strongly suggest that they evolved from a common ancestor after the ancestral species populated the jungle. Which of the following terms best describes this evolutionary situation and why?
    (a) Parallel evolution   (b) Convergent evolution
    (c) Adaptive radiation   (d) Divergent evolution

28. The current cheetah population has little genetic variation, but Bengal tigers, with a population size roughly the same as the cheetah's, has considerable genetic variation. Which population has a greater capacity to adjust to environmental changes brought about largely through human activities in their range? Explain.

29. If you were an avid fossil hunter, would you rather search inside an active volcano, in sedimentary rock that had once been a lake bottom, or in land that had always been on the top of a mountain range?

30. Chihuahuas and Great Dane dogs do not interbreed. Why do we consider them to be members of the same species? How do the concepts of biological species, reproductive isolating mechanisms, and gene flow enter into the answer?

31. How can modern industrialized societies slow down new cases of antibiotic resistance? How can they prevent them entirely?

32. Which of the following best explains why population bottlenecks are the major cause of seriously endangered species:
    (a) The population size is reduced below effective size.
    (b) The chance survivors may not be representative of the original population in terms of variability.
    (c) The chance survivors may be more homozygous than their ancestors.
    (d) The chance survivors may harbor a higher frequency of harmful genes than their ancestors did.
    (e) All of the choices above are correct.

Hawaii's Disappearing Crown Jewels

Saving Silverswords: Volcanic land formation; Dr. Robicheaux; endangered silversword rosette

Every year, millions of visitors travel to Hawaii to swim, ride mountain bikes, play golf and tennis, or hike through the islands' fragrant forests. Let's say that this year, you're one of those lucky vacationers. There's a chance you might drive or walk to the top of the dramatic 10,000-foot volcano called Haleakala on the island of Maui. There's a smaller chance you might see one of the world's strangest plants, the Haleakala silversword, jutting from the side of a steep cliff like a spiky green brush topped with an outrageous maroon flower stalk. And there's even a slim chance you might see Dr. Robert Robichaux jumping from boulder to craggy rock, ready to plant a smaller, flowerless silversword near that same precipitous rock face, where it will be safe from hungry sheep and goats. You would be seeing one of America's most dedicated conservationists helping to preserve one of the country's most famous endangered plants. But even if you

don't make it to Maui this year, the story of Hawaii's silverswords has important answers for the mysteries surrounding life's origins and diversity.

The story starts almost 6 million years ago at the bottom of the Pacific Ocean in a "hot spot" some 18,000 feet below the waves. There, molten rock from Earth's core shot upward through a massive vent and piled lava higher and higher into the sea. Eventually, it reached tens of thousand of feet into the air, then collapsed under its own weight to below the ocean surface. By 5.6 million years ago, it built up once more to become the future island of Kauai. The newborn island was a lifeless cone with molten rock still pouring up through volcanic shafts and oozing back down to the sea in wide lava flows. It looked in most respects like every other piece of land that has swelled above our planet's oceans, starting 4 billion years ago.

At the time the first continents emerged, the sun was much younger and therefore dimmer. Earth's atmosphere was also very different chemically and together, these factors resulted in a sky that was a pale gray-blue. Somewhere on Earth in a landscape as barren and volcanic as the forming Kauai, life began. That origin event might have taken place in wet clay, in a warm pool along a rocky shore, or submerged near a hot ocean vent. But biologists think that single cells formed in that ancient terrain and gave rise to more complex cells. Those complex cells led to many-celled aquatic life-forms. Eventually, the astonishing diversity of Earth's plants, animals, fungi, and microbes arose from those aquatic ancestors.

After its formation, the island of Kauai was a prime piece of uninhabited real estate. The land began filling in with plants and animals that flew, swam, or rafted over from older islands and continents. At some point, a bird in what is now California, about 2400 miles (3900 km) to the east, accidentally picked up a sticky seed on its feathers. The seed was from a nondescript herb called a tarweed (a member of the sunflower, or aster, family). The bird flew west, eventually landing on the nearly new island, and the tarweed seed plopped off into a lava crack. There it survived, grew, and founded the silversword lineage, which includes the Haleakala silversword.

The story of Kauai continues as the massive rock plate on which the island rides kept sliding to the northwest. Lava cones kept pumping up through the deep-ocean hot spot as the plant moved above it and five more islands, including Maui, eventually formed. The last and largest, Hawaii, appeared only 700,000 years ago. The tallest of the lava cones was probably Haleakala, the highest point on what is now the island of Maui. That cone pushed upward to an astounding 50,000 feet before collapsing to a current height of more than 28,000 feet—18,000 feet below water and 10,000 feet above. Over time, wind, rain, landslides, and volcanic eruptions sculpted all the islands and produced an unusually wide collection of habitats, from snowy peaks to swamps and bogs, scrublands, wet and dry forests, and crusty lava flows. As these varied habitats opened, the tarweeds spread into them, eventually radiating into 28 closely related species, the so-called **silversword alliance**. These relatives have wildly varying forms, depending on their adopted habitats. Some are trees, some vines, some bushes, some low weeds, and some—including the Haleakala silverswords—have a circular base or "rosette" of long, hairy grayish swords. These swords grow for 40 to 60 years, finally send up a single stalk bearing 25,000 or more purplish flowers, and then the whole plant dies.

These strangely spectacular rosettes once dotted Hawaii's volcanic craters by the thousands. But voyagers and settlers released sheep, goats, and pigs onto the islands and the animals eventually ate all but a few dozen plants clinging to the highest, steepest crater walls. One quarter of the silversword alliance species are now in danger of extinction.

Botanist Robert Robichaux has devoted more than a decade to saving Hawaiian silverswords by raising seedlings and replanting them in the Haleakala crater and elsewhere. They are "the crown jewels of the Hawaiian flora," he says, and they "offer us a unique window into the evolution of life's diversity. If we lose them, we lose that window." Robichaux and colleagues have also been studying what he calls "the Holy Grail of evolution—the genetic basis of adaptive radiation," and have discovered more about the silverswords' evolutionary history than is known about any other plant. We'll trace that history in this chapter and we'll see how the silverswords' story helps illuminate our current subjects: life's origins and splendid variety, and the imminent threats to that diversity. Along the way, you'll find the current answers to some of biology's most fascinating puzzles and our questions for this chapter:

1. How did Earth form and how did life emerge here?
2. How did the first cells evolve?
3. How did multicellular life arise from early microbes?
4. How did life's grand diversity emerge from those early ancestors?
5. How do biologists classify the millions of living and extinct species?
6. What is today's biodiversity crisis?

10.1 Earth's Formation and the Emergence of Life

In many ways, the birth of an island in the Hawaiian archipelago mimics the origin of dry land on our planet billions of years ago. That primal land formation was just one step, however, in a vastly longer process that included the generation of the planet itself, and even before that the inception of the universe.

A space probe launched in 1989, the Cosmic Background Explorer (COBE), provided supporting evi-

Big bang

Clouds of dust, ice, gases form

Solar system forms

| 15 | 14 | 13 | 12 | 11 | 10 | 9 | 8 | 7 |

Billion years ago

Figure 10.1 **The Solar System Forms.** Billions of years after the Big Bang, a monstrous cloud of gases and dust condensed to form the sun and planets of our solar system. Earth formed about 4.6 billion years ago. The first signs of life and the earliest known prokaryotic and eukaryotic cells evolved after the meteor bombardment ended. Land animals were relative newcomers at 0.3 billion years, and in the sweep of time, today's organisms have just arrived.

Did life emerge in the oceans or in shallow pools on newly formed land mass?

dence for a widely accepted cosmological theory about how the universe began, the so-called Big Bang. This is the idea that a cataclysmic event took place about 13 to 15 billion years ago, during which all matter in the universe—concentrated into a single mass by unknown processes—underwent an enormous explosion and began to expand rapidly (Fig. 10.1). The COBE probe confirmed that the expansion continues today throughout the universe.

Between 5 and 7 billion years after the Big Bang, a giant disk made up primarily of hydrogen and helium, the two lightest elements, spun in our corner of the universe. Gravitational attraction among atoms led to an accumulation of matter at the disk's center. The resulting compaction created a new star, our sun. At various distances from the center of the spinning disk, planets condensed from the cold gas and interstellar dust orbiting far from the pale young sun. Earth, the third planet from the sun, formed about 4.6 billion years ago.

The Birth of Earth

The newly formed Earth was nothing but a huge, frigid ball of ice and rock, devoid of oceans and atmosphere. The planet began to warm as radioactive elements in the rock decayed, releasing heat deep within Earth. The enormous, crushing pressure of gravity and the fiery impact of meteors bombarding the planet also warmed Earth into a molten mass that did not cool for several hundred million years (Fig. 10.2a). In 2001, geologists reported the discovery of a zircon crystal they dated to 4.3 to 4.4 billion years old. This find suggests that some areas may have begun cooling faster and earlier than previously believed.

As Earth cooled, a thin crust of rock formed at the surface like the rubbery skin on a bowl of pudding. Molten rock, or lava, frequently erupted through the crust, and clouds of gases issued from the planet's interior. This same process continues today from volcanoes in Hawaii, the Pacific Northwest, Japan, and other parts of the world (Fig. 10.2b). Vapor clouds from volcanoes formed Earth's first atmosphere, and atmospheric chemists think it probably consisted of these compounds: carbon dioxide (CO_2), nitrogen (N_2), water vapor (H_2O), hydrogen sulfide (H_2S), and traces of ammonia (NH_3), and methane (CH_4). (Recall from Chapter 2 that carbon, hydrogen, oxygen, nitrogen, sulfur, and phosphorus are the major elements in living things.)

As Earth's surface and surrounding atmosphere cooled, water vapor in the atmosphere condensed and fell to the surface as rain, eventually cooling the surface further and creating a huge shallow ocean dotted with sharp volcanic cones that spewed out more lava and gases. The sun was about one quarter its current intensity, and looked pale in the sky. It is in this watery environment, with its air lacking molecular oxygen (O_2), that biologists believe life arose (Fig. 10.2c).

Life as we know it—based on carbon compounds and liquid water—could have arisen only on a planet with enough of both. Mercury and Venus, the sun's closest satellites, are blisteringly hot, with daytime surface temperatures of about 325 °C and 480 °C (520 °F and 900 °F), respectively. Under such ovenlike conditions, water and carbon compounds occur as gases. Mars may once have had flowing water and living things, but now its surface is barren, with most—if not all—of its water frozen and most of its carbon trapped

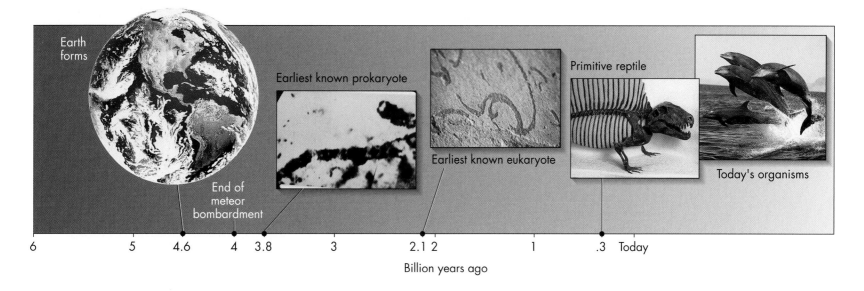

Earth forms

End of meteor bombardment

Earliest known prokaryote

Earliest known eukaryote

Primitive reptile

Today's organisms

| 6 | | 5 | 4.6 | | 4 | 3.8 | | 3 | | 2.1 2 | | 1 | | .3 | Today |

Billion years ago

in rocks. As we saw in Chapter 1, NASA and other space agencies are probing the solar system in search of evidence that life exists outside of Earth. Could there be traces on Mars? On the moons of Jupiter or Saturn? Only time will tell. Of all the planets in our solar system, we can only be certain at this point that Earth had a favorable composition, geological activity, size, and distance from the sun for life to arise and flourish.

Life's Emergence

With its liquid water and its atmosphere containing life's basic elements, early Earth was primed for living things to emerge. But how did these elements become organized into biological molecules and how did those, in turn, organize into living systems? Biologists have many competing theories and no definite answers, but they do agree on one thing: Life originated in a series of small steps, each following the laws of chemistry and physics. The origin of life is a unique branch of biology because biologists can't directly observe how life formed nearly 5 billion years ago on our planet, which has changed so dramatically. They can conduct experiments to recreate the assumed conditions of early Earth in the lab. But they can still show only how life *might* have arisen, not how it *did* arise for sure, even with the laws of physics and chemistry as a guide.

Why not look for direct physical evidence of the earliest life-forms and how they emerged in very ancient fossils? A record of the telltale chemical events may well have been laid down over 4 billion years ago in sediments that then became fossilized. Unfortunately, meteors left over from the formation of our solar system frequently flamed into Earth's atmosphere until about 3.8 billion years ago, and these no doubt melted most or all of the earliest fossils and the chemical evidence locked inside them. One of the meteors

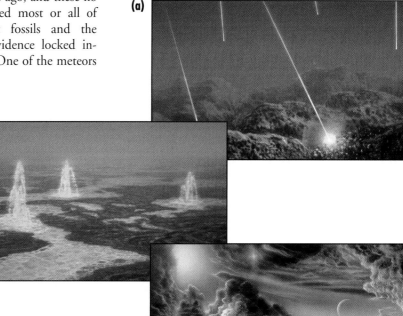

(a)

(b)

(c)

Figure 10.2 **Primordial Landscapes.** (a) Meteorites bombed Earth's surface soon after its formation and for millions of years thereafter. (b) A molten scene about 4 billion years ago, as the forces of gravity and radioactive decay melted the initial ball of rock and ice and created a red inferno of lava that boiled and smoked for millions of years. (c) By about 3.5 billion years ago, the planet had cooled, and the landscape was nothing but blue ocean, pale blue sky, and black volcanic cones.

Figure 10.3 **Stromatolites: Primitive Colonies.**
These pillow-shaped mounds are currently inhabited by large colonies of cyanobacteria growing along Australia's west coast. Strikingly similar fossil stromatolites from North Pole, Australia, are among life's earliest traces on Earth. The inset photo shows *Schizothrix* cyanobacteria growing in modern stromatolites at Exuma Cay, Bahamas.

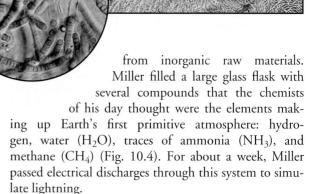

was so huge—about the size of Mars—that it knocked loose the material that formed our moon. The meteor showers pelting Earth also pounded the newly formed moon, leaving thousands of craters you can see on a clear night through binoculars.

Biologists think that the ancestors of today's living organisms evolved shortly after the intense meteor bombardment stopped. What's their evidence? Paleontologists have found fossils about 3.6 billion years old at a place called North Pole, Australia, and the preserved remains closely resemble living, stump-like stromatolites: mounds of minerals deposited layer upon layer by colorful prokaryotic cells called cyanobacteria, or blue-green algae (Fig. 10.3). These fossils show that living things must have evolved quickly from nonliving building blocks, when our planet was extremely young. If we represent the time span from Earth's formation to the present as a single 24-hour day, then Earth formed just after midnight, the heavy meteor bombardment stopped about 3:55 a.m., and cells had already evolved well before 5:30 a.m.

Biologists, then, can be pretty sure about the *when*. So the next question concerns the *what*: What happened during those first few hundred million years after the streams of meteors stopped falling? We mentioned nonliving building blocks; the first steps would have been the forming of biological molecules from inorganic compounds (Table 10.1).

How Simple Organic Molecules May Have Formed

The story of prebiotic ("before life") chemistry has a first act: the origin of small molecular building blocks such as simple sugars and amino acids and their assemblage into the large molecules (macromolecules) that characterize living organisms. No one was sure this type of spontaneous formation was even possible until the 1950s when a 23-year-old chemistry student named Stanley Miller came up with an idea. He tried to simulate in the lab the process of organic molecule formation from inorganic raw materials. Miller filled a large glass flask with several compounds that the chemists of his day thought were the elements making up Earth's first primitive atmosphere: hydrogen, water (H_2O), traces of ammonia (NH_3), and methane (CH_4) (Fig. 10.4). For about a week, Miller passed electrical discharges through this system to simulate lightning.

Amazingly, a pinkish sludge collected in the apparatus and contained large quantities of several amino acids; some lactic acid, formic acid, and urea; and traces of other biologically significant compounds. Atmospheric chemists now think that the early atmosphere was different from the one Miller recreated and might have included carbon dioxide (CO_2). Nevertheless, this classic experiment and later variations on it still proved that some organic molecules can be produced from nonbiological raw materials and physical processes like simulated lightning.

In the decades since Miller's historic test, space scientists have collected a different kind of evidence from outer space that proves the same point: Organic molecules can form in the absence of Earthly life. Astronomers have discovered enormous interstellar clouds containing organic molecules. They've also collected hundreds of **carbonaceous chondrites**, a class of meteorites containing various kinds of organic molecules. In 1969, one of these dense, charred meteorites fell in Australia. Chemists immediately examined it using careful test methods that protected the specimen from contami-

Table 10.1   Steps in the Evolution of Life

Event		Billions of Years Ago	Consequences
Full diversity of life forms present		0.5	Complete ozone screen; atmosphere like today's; 20% oxygen; large, active fishes, land plants
Shelled animals and early land plants appear		0.55	Diversity evident in fossil record; 10% oxygen in atmosphere
Many-celled organisms appear		0.67	Fossils and tracks made; oxygen and ozone accumulate in atmosphere to about 7%
First eukaryotic cells appear		2.1	Mitosis, meiosis, genetic recombination, and aerobic respiration occur; 2% oxygen in atmosphere
Oxygen-tolerating blue-green algae appear		2.0	Ozone screen begins to form; iron deposits appear on earth's surface
Strong evidence of photosynthetic organisms		2.8	Oxygen is given off into atmosphere, but still is less than 1%
Autotrophs, methane-generating bacteria, and sulfur bacteria appear; suggestive evidence of photosynthesis		3.6	Little change in atmosphere
The origin of life		3.8	Primordial atmosphere lacks oxygen

nation by Earthly substances. Intriguingly, this object from deep space contained organic compounds that closely matched the ones in the pinkish sludge in Stanley Miller's flask. This evidence from space shows that organic molecules can form from inorganic raw materials acted upon according to the universal laws of physics and chemistry. In fact, scientists think that carbonaceous chondrites and other meteorites may even have carried huge amounts of organic material to Earth's surface during the long period of intense meteor bombardment.

How Macromolecules May Have Formed

In the scientific story of life's origins, the next step after small organic molecules would have been macromolecules. Biologists think that small organic subunits would have joined, forming larger polymers such as proteins and nucleic acids. Long polymers like these tend to be fragile, especially in water. But chemists have shown that when they freeze solutions of amino acids dissolved in water or heat them to drive off water

molecules, small polypeptides (chains of amino acids) form spontaneously. Picture the shore of a black volcanic island (like those in the Hawaiian Islands) but billions of years ago when Earth was newly formed. Perhaps a "primordial soup" made up of seawater and dissolved organic molecules collected in the ancient tide pools along that lava rock shore. The sun's heat could have concentrated those precursors via evaporation, or they could have frozen at night. Drying or freezing could have forced out water and may have provided the energy to form amino acid chains (polypeptides) and other macromolecules. Experimenters have also shown that under certain conditions, organic molecules can become trapped on the surfaces of clay particles. With the input of light or heat energy, these molecules can then react and become concentrated and organized. Organic molecules trapped on iron pyrites ("fool's gold") might also act this way, leading to the same end: the formation of large organic molecules such as polypeptides, polynucleotides, polysaccharides, and perhaps lipids—the complex raw materials of life (Explorer 10.1).

GET THE PICTURE

*E*xplorer 10.1

Steps in the Evolution of Life

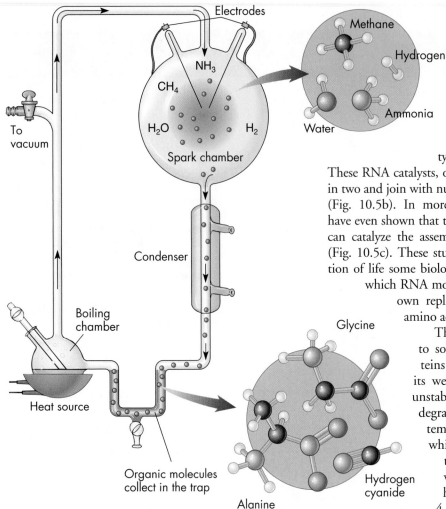

Figure 10.4 **Spark Chamber Experiments: Attempts to Simulate Earth's Primordial Conditions.** In 1955, Stanley Miller and Harold Urey designed a series of spark chamber experiments to recreate the supposed conditions on Earth before life evolved. When supplied with energy from electric sparks, raw materials in the upper chamber (methane, hydrogen gas, ammonia, and water) reacted. Organic building blocks of proteins, including glycine and alanine, collected in the lower trap along with hydrogen cyanide and other compounds.

Electrodes

Methane

Hydrogen

NH₃

CH₄

Ammonia

H₂O   H₂

Water

To vacuum

Spark chamber

Condenser

Boiling chamber

Glycine

Heat source

Organic molecules collect in the trap

Hydrogen cyanide

Alanine

Self-Replicating Systems

It's not much of a stretch to explain how small organic molecules might have accumulated and linked up into proteins and nucleic acids (although their stability was probably poor and their existence fleeting). As we saw in Chapter 1, however, self-replication is another fundamental feature of life's chemistry. So researchers studying the origins of life also need to explain how primitive organic chains could have begun making copies of themselves. Here is the most popular explanation.

RNA Catalysts

In today's cells, nucleic acids can't replicate without the help of protein enzymes. Proteins, however, require nucleic acids for *their* synthesis. This makes the evolution of self-copying molecules a circular problem. So which did come first, proteins or nucleic acids, and how did

they start copying themselves and/or each other?

Many biologists are convinced that the first systems of self-replicating molecules were based on the nucleic acid RNA (Fig. 10.5a). Experiments conducted in the 1980s showed that certain types of RNA can act as catalysts. These RNA catalysts, or **ribozymes,** can snip themselves in two and join with nucleotides or other RNA molecules (Fig. 10.5b). In more recent experiments, researchers have even shown that the ribosomal RNA in today's cells can catalyze the assembly of amino acids into protein (Fig. 10.5c). These studies suggest a stage in the evolution of life some biologists call the **RNA world,** during which RNA molecules would have catalyzed their own replication as well as the joining of amino acids into proteins.

The RNA world hypothesis seems to solve the circular problem of proteins and nucleic acids, but it still has its weak points. First, RNA is a very unstable compound, and it would have degraded quickly at the high surface temperatures of early Earth. Second, while RNA molecules can replicate themselves, they don't do it very well in a beaker simulating the harsh conditions on Earth nearly 4 billion years ago. By adding certain types of energy-transfer molecules to the system, researchers can apparently improve RNA replication. For this reason, one Nobel Prize-winning biologist, Christian de Duve, suggests that sulfur-containing compounds (called *thioesters)* may have helped transfer energy in the RNA world.

Self-Copying Molecules and Natural Selection

For life to evolve, natural selection must have come into play, even before living things existed. Biologists think that natural selection would have acted upon little sets of self-replicating RNA molecules and proteins, allowing some to survive better than others. Replication would have been inaccurate, and so the daughter molecules would have had variations in the sequences of bases or amino acids. Some of these variant molecules might have been more stable or more easily copied than others, and these would have survived longer and/or replicated faster. Such RNAs would then have accumu-

(a) RNA forms

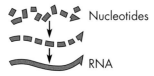

Nucleotides

RNA

(b) Ribozymes catalyze RNA replication

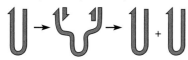

(c) RNA catalyzes protein synthesis

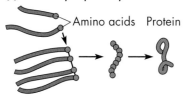

Amino acids Protein

(d) RNA encodes both DNA and proteins

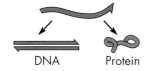

DNA   Protein

(e) Proteins catalyze cell activities

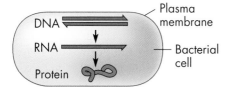

Plasma membrane

DNA

RNA — Bacterial cell

Protein

Figure 10.5 **The Origins of Life: The "RNA World" Hypothesis.** (a) According to this idea, organic subunits (nucleotides) combined spontaneously and formed RNA molecules. (b) RNA molecules acted as ribozymes, catalyzing their own replication. (c) RNA molecules also catalyzed the synthesis of protein, which in turn stabilized RNA molecules. (d) DNA molecules were copied from RNA molecules. (e) At some point, proteins began to act as true enzymes, helping build more proteins from information in RNA (red arrows). DNA assumed the role of stable information storage, while RNA continued to take part in protein synthesis; a plasma membrane also appeared from lipid molecules in the environment.

genes that encode proteins, they are surrounded by a plasma membrane, and they have complex metabolic pathways. So how did all of these features emerge from systems of self-replicating molecules in an RNA world or some other scenario?

Like the other "chapters" in life's origins, the story of how membranes, metabolism, and the translation of proteins arose is still guesswork. However, around and within every living cell, biologists can find persuasive evidence that the evolutionary steps *did* take place, and followed the laws of nature.

Membranes Evolve

Every living cell is enclosed within a plasma membrane-a flexible, semi-permeable lipid wrapper that confers many advantages and would have benefitted emerging life systems, as well. Membranes can keep harmful substances out and useful chemicals in. A membrane, for example, could keep a protein that speeds RNA replication close to that particular RNA molecule (Fig. 10.5e). Without that boundary, the protein could diffuse away, and not be concentrated enough to replicate the RNA.

In living cells, membranes arise from the preexisting membrane of a parent cell. So how could membranes have formed before life started? Researchers have heated together polypeptides and phospholipids, and shown that membranelike sheets can form spontaneously. These can fold back on themselves to form membrane-enclosed compartments and under certain conditions, can even divide. Substances in the surrounding medium can become trapped inside these membranes and then apportioned when the compartments divide. Suppose that self-replicating RNAs were trapped in such a membrane: Those compartments might then have the advantages of producing more RNAs, dividing faster, and leaving more compartments behind with protective membranes. If some RNA molecules kept membranes around themselves better

lated more quickly than others in the primordial soup. The sets of molecules were still not living cells, but over eons of this natural selection, RNA molecules would have become tuned by natural selection for stability and rapid replication. Eventually, the more stable form of nucleic acid, DNA, took over the information storage task of RNA (Fig. 10.5d), and proteins became the common catalysts within cells (Fig. 10.5e).

10.2 The First Cells

In our story of evolving life, we're not yet at the point of living cells but only of self-replicating organic molecules. As we've seen again and again, true cells have

than others, then a primitive natural selection mechanism would promote the improvement of membranes (see *E* 10.1).

Origin of Metabolic Pathways

The same kinds of selective processes that favored membranes could have led to metabolic pathways. Biologists suspect that initially, primitive self-replicating compartments could have gotten all needed materials from the primordial soup: nucleotides, amino acids, lipids, and so on. But eventually, some substances may have started running out. Let's say, then, that a membrane-enclosed RNA by chance encoded a protein that could act upon the abundant material Y and change it into the needed but scarcer material X. This little membrane-enclosed unit would have an obvious selective advantage. When substance Y became depleted, another round of genetic change and natural selection could favor a protein that could change substance Z into substance Y. Through this kind of selective pressure, metabolic pathways could have evolved: $Z \rightarrow Y \rightarrow X$, and so on (see *E* 10.1). Somewhere, at some point, replication, protein synthesis, and metabolism became so closely integrated inside membrane-bound compartments that the resulting units were indistinguishable from living cells. At that point, life had begun on Earth.

The pathway from the formation of organic molecules to the probable first cells in their organic soup was a long and complicated one—a pathway that surely involved millions of small evolutionary steps. Modern experiments can never demonstrate exactly how life did evolve, but they can show that life's origin adhered to the laws of chemistry and physics and the environmental conditions existing on Earth 4 billion years ago. Many biologists, in fact, think that once these rare conditions came into existence, life was bound to emerge. Cell biologist Christian de Duve said it this way: "The universe was pregnant with life."

Life Alters Earth

Early living organisms appear to have dramatically altered the physical conditions on Earth itself and paved the way for later life-forms.

Geologists exploring southern Greenland have discovered ancient rocks containing organic carbon deposits that were probably substances from the earliest fossilized cells. These rocks formed in ocean sediments about 3.8 billion years ago, shortly after meteorites stopped obliterating all traces of earlier events. The

metabolism in these early cells would have had to be anaerobic because, as we've seen, the primordial atmosphere lacked free oxygen. Also, these earliest living systems were probably **heterotrophs.** That is, instead of generating their own food molecules, they obtained nourishment from external sources, in this case from carbon-containing organic molecules that fell to Earth in meteors or appeared when lightning shot through Earth's first atmosphere.

Eventually, there would have been so many primitive heterotrophs removing organic molecules from their watery environment that the supply would have run out: All the free organic compounds would have been incorporated into biological molecules. For the first time—but far from the last—organisms would have changed their environment (see *E* 10.1).

Evolution of Autotrophs

As heterotrophs depleted the primordial oceans of organic materials, they would have run short of carbon and energy supplies. Natural selection would then have favored **autotrophs,** organisms that could produce their own organic molecules from simpler inorganic precursors, such as methane and carbon dioxide.

An exciting goal for biologists has been to search Earth for a living organism with traits as close to the earliest autotrophic cells as possible. To this end, researchers at the University of Hawaii and Michigan State University in 1998 reported making a dangerous dive in the submersible *Pisces V.* Their target was Pele's Vent, a hydrothermal opening that spews sulfurous, acidic hot water into the ocean near the summit of Loihi, the ever-growing but still-submerged Hawaiian island (Fig. 10.6). The team collected prokaryotic lifeforms thriving in the hellish conditions around the vent then returned them to the lab (Fig. 10.6). These organisms are a subset of autotrophs called **chemoautotrophs,** because they can generate organic molecules and also trap energy using inorganic compounds such as hydrogen sulfide and iron.

From these primitive chemoautotrophs, the researchers isolated ribosomal RNA genes and investigated how closely the cell's base-pair sequences for those genes resembled the sequences of other kinds of living cells. They found that cells living in Pele's Vent inhabit evolutionary branches near the very base of the tree of all life. Many chemoautotrophs like the ones from Pele's Vent have names that hint at the environments in which they evolved and still thrive: *Thermotoga* (*thermo* = heat); *Pyrodictium* (*pyro* = fire); *Haloferax* (*halo* = salt); and *Sulfolobus* (*sulfo* = sulfur).

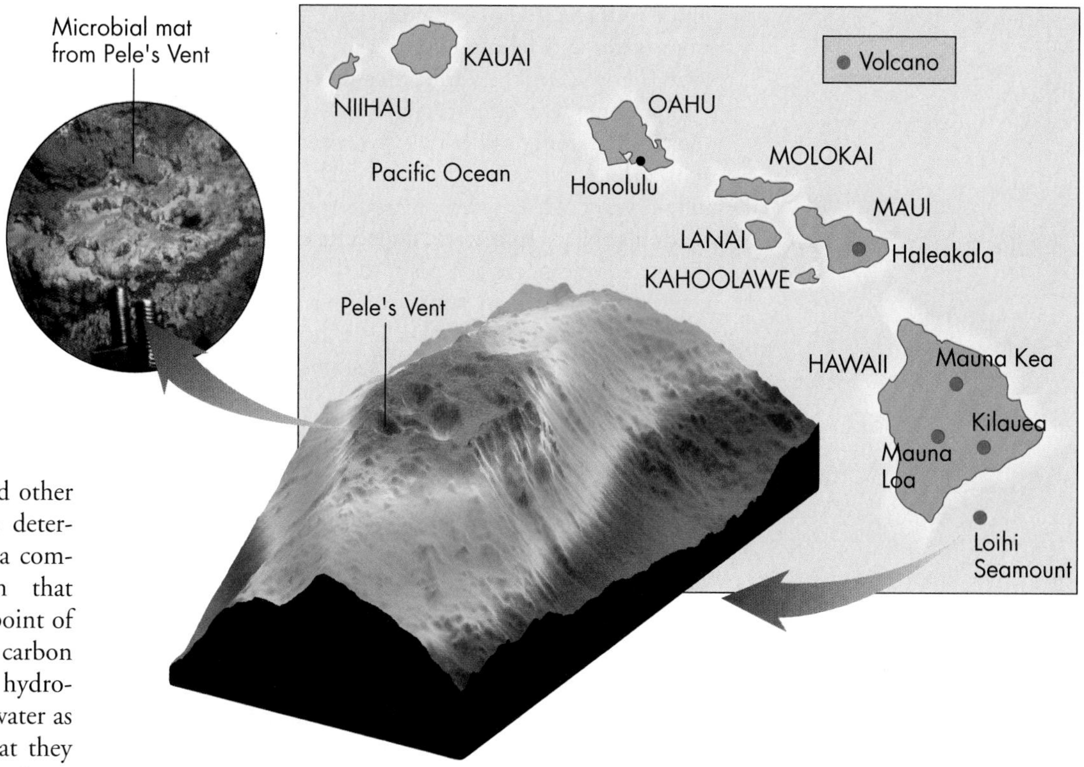

Figure 10.6 Searching for Relatives of the Oldest Living Cells.

Conditions in thermal vents at the bottom of the sea mimic conditions on early Earth in many ways. Researchers descended in a small submarine to Pele's Vent near the summit of Loihi Seamount, the newest but still-submerged Hawaiian island 20 miles off the coast of the Big Island of Hawaii. They collected primitive, heat-loving bacteria like those pictured in the inset and brought them to the lab. Genetic analysis showed that these cells are among the earliest diverging lineages of life and thus may be similar to the earliest cells.

Microbial mat from Pele's Vent

Pele's Vent

KAUAI

NIIHAU

OAHU

Pacific Ocean

Honolulu

MOLOKAI

LANAI

MAUI

Haleakala

KAHOOLAWE

HAWAII

Mauna Kea

Kilauea

Mauna Loa

Loihi Seamount

● Volcano

By studying organisms from Pele's Vent and other deep-sea hydrothermal settings, biologists have determined that all currently living organisms share a common ancestor, a primitive chemoautotroph that probably lived at temperatures near the boiling point of water. These ancient progenitors probably used carbon dioxide as a carbon source and used sulfur and hydrogen gas as energy sources (rather than light and water as do photosynthetic cells). Biologists conclude that they could survive only in oxygen-free (anaerobic) environments because all the lowest branches on the tree of life are anaerobic. And they apparently lived deep enough below the surface of oceans, lakes, and hot springs so that the water layer above them screened out the highly energetic ultraviolet light that can destroy complex biological molecules.

Although chemoautotrophs do well in some environments today, the sunlight at Earth's surface provides an abundant and available source of energy. Very early in the course of life's history, autotrophic organisms began to tap this rich resource. Evidence from the fossilized remains of stromatolites (see Fig. 10.3) show that photosynthetic cells had already evolved by 3.6 billion years ago—amazingly, just a few hundred thousand years after the end of the meteor bombardment period. Photosynthetic cyanobacteria are among today's simplest photosynthetic cells, and scientists have uncovered the earliest remains of distinct cells in 3.5-billion-year-old sedimentary rocks near the sites containing fossil stromatolites (Fig. 10.7).

Photosynthesizers Change the Atmosphere

The early photosynthesizers were still very, very simple cells that probably used hydrogen sulfide and hydrogen

gas as electron donors in their energy metabolism (review Chapter 3). These raw materials would have run out at some point, too, but water was essentially unlimited and any organism that could extract electrons from water could thrive in many more environments. During

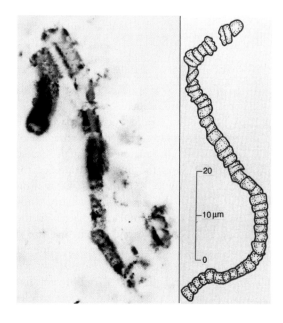

Figure 10.7 Fossils of Ancient Prokaryotic Cells.

These microscopic fossilized cells occur in a rock dated at about 3.5 billion years old. These ancient cells were similar to cyanobacteria such as those living in stromatolites (see Fig. 10.3).

photosynthesis, today's cyanobacteria pick up electrons from water and release oxygen as a waste product (review Fig. 3.25). The earliest fossils of cells closely resembling modern cyanobacteria (and that use water as the electron donor rather than hydrogen sulfide or other substance) come from Australian rocks dated about 2.8 billion years old. Based on these fossils, biologists think that cells able to split water molecules during photosynthesis and release oxygen gas into the environment had evolved by 2.8 billion years ago. This metabolic development had dramatic implications: it would allow living things to modify the planet's atmosphere in a major way (see **E** 10.1).

Oxygen Tolerance and Aerobic Cells

For over a billion years after oxygen-producing, photosynthetic cells evolved (nearly 4:10 a.m. in our day-long analogy), the atmosphere would have accumulated little or no oxygen. There is proof, instead, that the oceans literally rusted: iron compounds in seawater reacted with the free oxygen given off by the photosynthesizing cells and this formed iron oxides which fell to the ocean floor by the ton. Geologists have found iron deposits dating from 2.5 billion years ago in deep rock strata in the Hamersly Range in Australia and at other sites including the iron mines of Michigan and Minnesota. Only after most of the iron in seawater was bound to oxygen in this way would O_2 have built up in the atmosphere. Ironically, since oxygen is poisonous to anaerobic cells, this initial buildup would have harmed both the early heterotrophs and the photosynthesizers themselves.

About 2 billion years ago (1:30 p.m. in our analogy), as oxygen levels continued to build, thick-walled cells that could probably resist damage from oxygen lived and died in the Lake Superior area. Fossils of these cells tell us that mechanisms had begun to evolve (in this case, thick walls) that could ward off the toxic effects of oxygen given off by cyanobacteria and accumulating in the oceans and atmosphere.

This oxygen buildup would have had two more consequences. First, it would have acted as an agent of natural selection favoring the energy pathway of aerobic respiration, with oxygen as the final electron acceptor (review Fig. 3.14). Early cells with this metabolism might have resembled the common bacterium *Escherichia coli*. Second, the accumulating oxygen in the upper atmosphere formed an ozone screen that had a major impact on the global environment. Sunlight striking gaseous oxygen (O_2) creates ozone (O_3), which in turn absorbs ultraviolet light. This screen protects living things from some of the damage that ultraviolet light inflicts on DNA and allowed cells to start living closer to the planet's surface rather than hidden around deep ocean vents or the like (see **E** 10.1).

10.3 Evolution of Eukaryotic Cells and Multicelled Life

We haven't arrived yet at the eukaryotes—the basic organismal groups that include silverswords—and their researchers. On our 24-hour time line, living things cropped up before 5:30 a.m. (3.6 billion years ago) and have inhabited Earth for 80 percent of its history to the present (midnight). For more than half that time, the most complex life-forms were prokaryotic cells, such as the anaerobes that inhabit hot springs on the newly forming Loihi Seamount in Hawaii. Not until about 2.1 billion years ago, or 12:55 p.m. on our time scale, did eukaryotic cells appear. Multicellular organisms—animals, plants, and most fungi—evolved only about 900 million years ago (6:55 p.m.) and it was 38 seconds before midnight that silverswords appeared. We humans came along just 4 seconds before midnight!

Biologists have carefully examined the stepwise bursts of evolution that led to the first eukaryotic cells, to their diversification into single-celled groups, and then to their descendants, the multicellular eukaryotic organisms. Some biologists hypothesize that the earliest eukaryotic cells lost the ability to make tough cell walls but got toughness and rigidity from different sources, namely the nuclear envelopes and cytoskeletons (see Fig. 2.25). These new features also would have allowed them to engulf food particles in pockets of the plasma membrane (review Fig. 2.27). Nuclear envelopes may have originated as invaginations of the plasma membrane—in-pockets that came to surround and protect the host cell's naked DNA. Fossils show these envelopes starting about 2.6 billion years ago (Fig. 10.8a,b).

Eukaryotes probably engulfed something else at least 2.5 billion years ago (about noon on our time line): the precursors to mitochondria and chloroplasts (Fig. 10.8c–e). University of Massachusetts biologist Lynn Margulis and others have proposed the **endosymbiont hypothesis** (*endo* = within + *symbiont* = live together). This is the idea that mitochondria and chloroplasts originated as free-living bacterial cells. A generalized ancestral cell type then "swallowed" them. Together, the host and guest organisms formed cellular communes, viable single organisms with each member adapted to the group arrangement and benefitting from it.

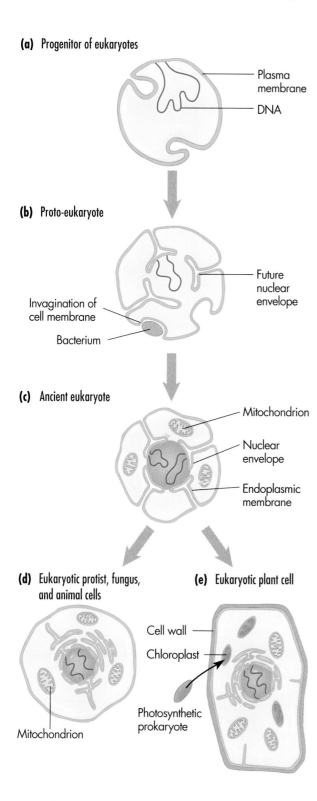

(a) Progenitor of eukaryotes

Plasma membrane

DNA

(b) Proto-eukaryote

Future nuclear envelope

Invagination of cell membrane

Bacterium

(c) Ancient eukaryote

Mitochondrion

Nuclear envelope

Endoplasmic membrane

(d) Eukaryotic protist, fungus, and animal cells

Mitochondrion

(e) Eukaryotic plant cell

Cell wall

Chloroplast

Photosynthetic prokaryote

Mitochondrion

Figure 10.8 **The Endosymbiont Hypothesis.** How did eukaryotic cells get their specific internal organelles? The best-accepted theory holds that the ancestor was a prokaryotic host cell lacking a cell wall and organelles, but capable of invaginating its cell membrane (a). This ancestral type may have merged with symbiotic bacteria that could carry out respiration in the presence of oxygen (b). These bacteria could then begin to survive inside the host cell's cytoplasm. Invaginating membranes may have given rise to the endoplasmic reticulum and the nuclear envelope (c). The nonphotosynthetic prokaryotes, once engulfed, could have evolved into mitochondria (d), while the photosynthetic ones could have given rise to chloroplasts (e).

the cell's chromosomes and contains genes similar to genes in aerobic bacteria. Chloroplasts, too, are similar in size and shape to certain photosynthetic prokaryotes, and, like mitochondria, they have separate, self-replicating DNA with bacterialike genes. Oxygen-tolerant cells large enough to accommodate them would have gotten a big boost in ATP-generating capacity and/or a source of sugars from these "hitchhiking" energy-processing organelles. Genetic analysis shows that eukaryotes probably acquired mitochondria only once from one specific group of bacteria (see Fig. 10.8). The same kind of analysis, however, shows that several different eukaryotic lineages acquired chloroplasts independently.

By 1:50 p.m. in our analogy (about 1.9 billion years ago) cells contained mitochondria and chloroplasts. Their radiation into numerous groups of single-celled organisms, however, happened far later, about 1 billion years ago (6:50 p.m.). Some biologists suggest that the evolution of sex led to the burst in diversity. The recombining of genetic material during meiosis and fertilization permitted much more genetic variability. Hence, sexual reproduction could have sped the pace of evolution considerably.

Simple Animals Emerge

About 580 million years ago (8:55 p.m. on our time line), simple animals first appeared in the fossil record. Paleontologists suspect that this surge in diversity coincided with higher oxygen levels in the atmosphere. The first fossilized animals were soft-bodied and paleontologists dubbed them the Ediacaran fauna (Fig. 10.9). Some of these strange, primitive-looking animals resembled jellyfish, and others looked more like annelid worms (even though they were probably unrelated to modern jellyfish and annelid worms). It's somewhat mysterious that soft-bodied animals became fossilized 580 million years ago,

What's the evidence for this intriguing hypothesis? Mitochondria are remarkably similar in size and shape to today's aerobic bacteria. What's more, mitochondria have their own DNA, which replicates independently of

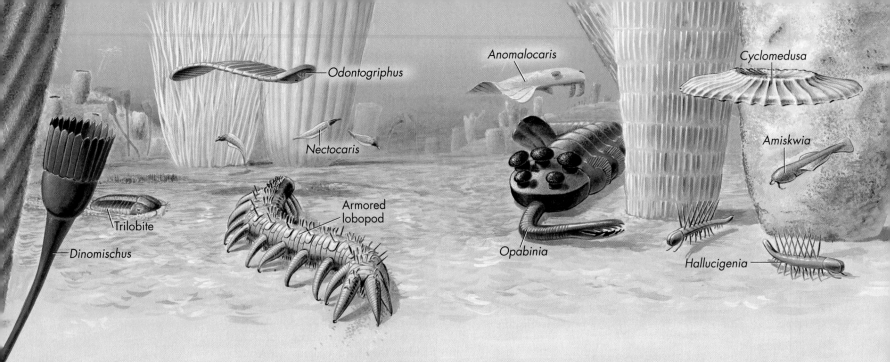

Odontogriphus

Anomalocaris

Cyclomedusa

Amiskwia

Nectocaris

Armored
lobopod

Trilobite

Dinomischus

Opabinia

Hallucigenia

Figure 10.9 **Animals of the Burgess Shale.** An artist's recreation of strange sea life from the Burgess Shale about 580 million years ago. Fossils show that most of the animals were soft-bodied, and this, in turn, suggests that few predators roamed the seas.

considering that the soft body parts of later organisms are seldom fossilized. Some paleontologists suspect that there were no scavenging animals during Ediacaran times, and so dead animals lay scattered on the ocean floor much longer than they would today—and thus were more likely to become fossilized. About 540 million years ago, this all changed: In a process dubbed the **Cambrian Explosion,** animals with hard body parts emerged and left many more fossils. The armoring of Cambrian animals such as trilobites and the prickly spined *Hallucigenia* (see Fig. 10.9) suggests that predators had begun to roam the seas. Individuals with tougher skins or shells would have been devoured less often by predators than soft-bodied neighbors. Armored animals could therefore have reproduced more effectively and left more offspring bearing their genes. Holes punctured in the bodies of some fossil trilobites offers silent evidence that survival had indeed become much more tenuous.

Wiwaxia

Large Plants and Animals Arrive

By 400 million years ago (9:52 p.m. on the time line), the atmosphere was essentially like today's, with about 20 percent oxygen content. The ozone screen was in place, and large, complex life-forms were appearing in profusion. Large fishes swam in the ancient seas, and primitive land plants grew along moist shores. Within another 200 million years, arthropods, amphibians, reptiles, birds, and mammals would move about the continents, and large stands of conifer trees and flow-

ering plants would grow abundantly, including the California tarweeds, the ancestors of Hawaiian silverswords.

Geologic Change Affects Life

The evolution of living things changed the atmosphere, rusted the oceans, and produced the ozone shield. Simultaneously, though, Earth was changing from its own geologic processes and these influenced the evolution and diversification of living things. Geologists describing these events have divided Earth's history into eons, eras, periods, and epochs, based on sudden changes in the fossil record. For our purposes, however, the names of just five time periods are useful to remember:

1. The Archean ("ancient") reaches back to the formation of the oldest known rocks.
2. The Proterozoic ("earlier life")
3. The Paleozoic ("old life")
4. The Mesozoic ("middle life")
5. The Cenozoic ("recent life")

The next few sections describe each period, as does Table 10.2.

Archean and Proterozoic Times

In the 500 million years or so between the origin of Earth (just after midnight on our time line) and the beginning of the **Archean Era** (2:55 a.m.), meteor

Table 10.2 History of the Earth

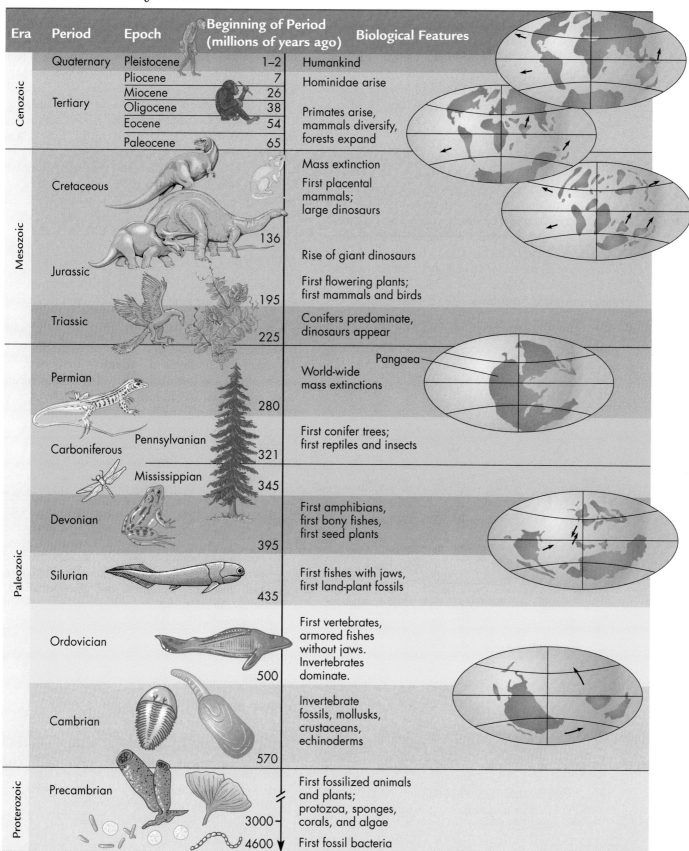

Era	Period	Epoch	Beginning of Period (millions of years ago)	Biological Features
Cenozoic	Quaternary	Pleistocene	1–2	Humankind
	Tertiary	Pliocene	7	Hominidae arise
		Miocene	26	
		Oligocene	38	Primates arise, mammals diversify, forests expand
		Eocene	54	
		Paleocene	65	
Mesozoic	Cretaceous			Mass extinction
				First placental mammals; large dinosaurs
			136	Rise of giant dinosaurs
	Jurassic		195	First flowering plants; first mammals and birds
	Triassic		225	Conifers predominate, dinosaurs appear
Paleozoic	Permian		280	World-wide mass extinctions
	Carboniferous	Pennsylvanian	321	First conifer trees; first reptiles and insects
		Mississippian	345	
	Devonian		395	First amphibians, first bony fishes, first seed plants
	Silurian		435	First fishes with jaws, first land-plant fossils
	Ordovician		500	First vertebrates, armored fishes without jaws. Invertebrates dominate.
	Cambrian		570	Invertebrate fossils, mollusks, crustaceans, echinoderms
Proterozoic	Precambrian		3000	First fossilized animals and plants; protozoa, sponges, corals, and algae
			4600	First fossil bacteria

Pangaea

impacts and volcanoes probably prevented life from emerging. Some scientists, in fact, refer to that hellish early environment as the "Hadean" (from "Hades"). But compared to that, the Archean was a peaceful time when the first continents rose from the sea, much as Hawaii did in recent times (Fig. 10.2b). During the **Proterozoic era,** beginning about 2.5 billion years ago, a phase of mountain building began and lasted for nearly two billion years. The organisms that had evolved earlier began to diversify.

Paleozoic Era

About 580 million years ago, with just one hour and 50 minutes left on our 24-hour time line, the **Paleozoic Era** began suddenly with the **Cambrian Explosion.** This was the nearly simultaneous appearance in the fossil record of all the major animal phyla that are alive today. Sometime before this sudden appearance of life-forms, **plate tectonics,** the process that moves about the landmasses of the Earth's crust by seafloor spreading and subduction, began (Fig. 10.10). The plate on which the Hawaiian Islands sit, for example, moves northwest by about 9 cm (3.5 in.) each year.

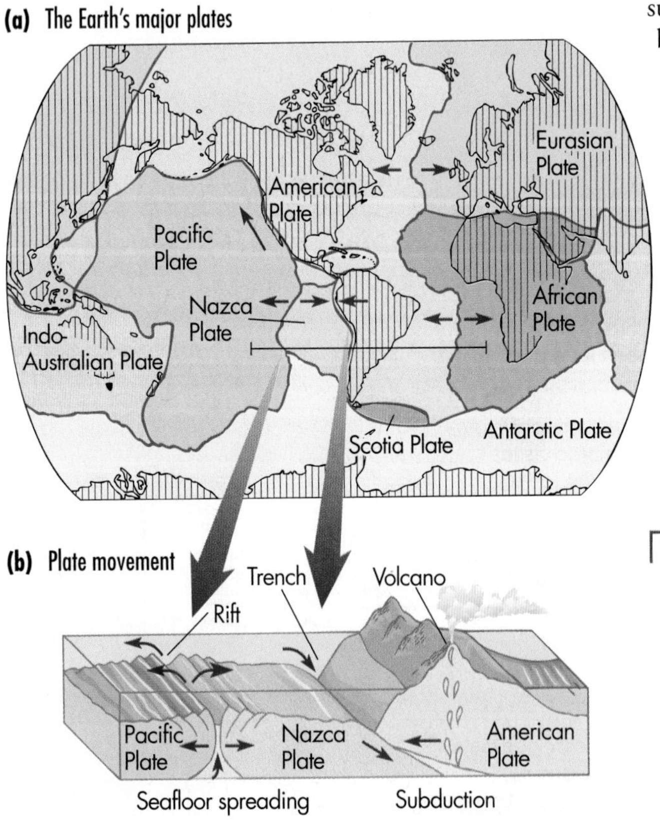

About 245 million years ago (10:42 p.m.), Earth's landmasses converged into a single supercontinent called **Pangaea** ("all Earth")(See Table 10.2). As the plates collided, landmasses buckled and tilted, lifting up huge mountain ranges; draining warm, shallow inland seas; and altering the size and position of ocean basins. The resulting climate changes were a calamity for both sea and land animals. About 75 percent of the families of amphibians (salamanders and relatives), 80 percent of the reptiles (lizards, snakes, and relatives), and 96 percent of all marine species became extinct. Paleontologists call this cataclysm the **end-Permian extinction,** because it marked the end of the Permian period in the Paleozoic Era, and they consider it history's greatest mass extinction.

Mesozoic Era

Pangaea broke apart in the **Mesozoic Era** (between 245 and 65 million years ago; about 11:20 p.m.), separating populations of organisms and providing great opportunities for evolutionary change. When Pangaea existed, for example, mammals had not yet evolved the placenta, the organ that nourishes the young in the uterus. Instead, offspring were born at a very early developmental stage, found their way to their mother's pouch, or marsupium (Latin for "purse"), where they were nourished by milk. Small primitive pouched mammals or marsupials thrived all across Pangaea, as well as on the giant landmass of Australia after it broke apart and drifted away. Thousands of species of mammals with placentas evolved on other continents, but not on Australia. Without competition from placental mammals, the marsupials radiated into a rich collection of unique animals, including kangaroos, koalas, and wombats (see Chapter 13). In most parts of the world, marsupials lost the competition with placental mammals and disappeared. In North America, for example, opossums are the only surviving descendant marsupials.

(a) The Earth's major plates

(b) Plate movement

Seafloor spreading   Subduction

Figure 10.10   Plate Tectonics: Continents Colliding. (a) The continents and oceans ride on massive crustal plates. (b) At the junctions (rifts) between plates, new crustal material oozes up from Earth's core and pushes the plates apart. Where the leading edges of plates collide, one pushes below the other, causing a deep trench, in a process called subduction. Here, the Nazca Plate dives beneath the American Plate. Plate tectonics causes most of Earth's violent geologic activity.

Along with the breakup of Pangaea came the emergence and diversification of the largest land animals ever to live, the dinosaurs. A warming climate engendered great swampy forests containing tropical plants and trees all around the globe. In those forests lived flowering plants, birds, and mammals.

Cenozoic Era

A catastrophe ended the Mesozoic and ushered in the current era, the **Cenozoic Era,** about 65 million years ago (11:20 p.m.). In history's second-greatest mass-extinction event, one half of the world's species, including the dinosaurs, died out. Current research suggests that at this time, a meteor crashed into the gulf coast of Mexico. The resulting airborne debris darkened the sky around the globe, and could have drastically cooled Earth's climates, leading to the extinctions. When the skies cleared, birds, mammals, and flowering plants continued diverging into large and varied groups of species. Plate tectonics continued in the Cenozoic, bringing the continents to their present positions (Table 10.2). These crustal shifts continued to separate groups of organisms, rafting them to entirely new locations and conditions, and further encouraging the emergence of new plant and animal groups.

10.4   Organizing Life's Diversity

The Hawaiian Islands formed relatively late in the Cenozoic Era. Kauai, the oldest of the main islands arose about 5 million years ago (1 minute and 36 seconds before midnight). Maui erupted from the sea about 2 million years ago. Hawaii surfaced 0.5 million years ago (10 seconds before midnight). And Loihi is still forming, 950 m below sea level. Despite this very brief time span in Earth's long history, the Hawaiian islands have accumulated a remarkably diverse group of plants and animals. The silversword alliance alone represents tremendous variation of plant body forms and habitats. But add to that the insects called plant-hoppers that feed on the silversword leaves; the birds that eat the planthoppers; the bats that fly zigzag paths at sunset; the fragrant blossoming plumeria trees; the goats and other hoofed animals that have brought silverswords to near extinction; and the prokaryotes thriving in the hydrothermal vents atop Loihi. Together, they show an even bigger slice of life's diversity, even on an island state as small and remote as Hawaii.

People have long recognized patterns of similarity among life's diverse organisms and have lumped similar organisms together to form logical groupings. For scientists, life's diversity must be cataloged systematically in ways that make obvious the relationship of one organism group to another. **Systematics** is the scientific study of the diversity of organisms, and how they are related in an evolutionary context. In contrast, **taxonomy** is the science of identifying organisms, naming them, and classifying them into groups such as kingdom, genus, and species.

Why Study Diversity?

Modern biologists studying and categorizing life's diversity are doing much more than simply making lists for their own sake. There are excellent reasons to push for a complete catalog of all of Earth's species (up to 50 million, some think) as soon as scientifically possible:

1. Much of Earth's diversity will likely disappear in the next 50 years due mostly to habitat destruction and other harmful human activity. Species could be lost before they are ever known to science, and that would impoverish biology, not to mention the rest of the survivors. Also, we need to understand individual species in order to learn about their roles in ecosystems, and in turn, to prevent their collapse.
2. It is crucial for us to understand organisms and their life habits in order to identify the causes of some diseases. Is an outbreak of intestinal disease in a community due to a bacterium or a protist? Which particular species is the culprit? And which antibiotic will be effective against the pathogen?
3. Many of our current and future drugs and raw materials come from nature, and explorers find new compounds regularly. Researchers, for example, found a very useful compound for treating childhood leukemias in the Madagascar periwinkle. Future investigations of little-known life-forms in remote forests, grasslands, tide pools, and other wild places will surely uncover more useful compounds for saving lives, feeding people, maintaining livestock, and developing new industrial materials.

Classification According to Linnaeus

The plants shown in Figure 10.11 are similar to each other in some ways, but very different in others. Why do biologists classify all of them in the silversword alliance when they look so distinct from each other? What does this classification mean? More than 200

When did pioneering species first inhabit Hawaii?

(a) Tree   **(b)** Mat   **(c)** Vine

Figure 10.11   **Astonishing Diversity in the Silversword Alliance.** Among the diverse Hawaiian silverswords are (a) *Dubautia reticulata,* a large shrub or tree; (b) *Dubautia scabra,* a low mat-like, ground-covering shrub; and (c) *Dubautia latifalia,* a heavily branched vine.

years ago, a Swedish botanist named Carolus Linnaeus developed a classification system that assigned every organism then known to science to a series of increasingly larger, more general and all-inclusive groups. He and his contemporaries did the assigning by observing similarities in organisms' sizes, shapes, colors, and other tangible traits.

Linnaeus also began consistently to use a system for naming each type of organism—the system of **binomial nomenclature** *(binomial* = two names). This system assigns each species a two-word name, such as *Homo sapiens,* our own species' name, or *Argyroxiphium sandwicense,* the name of the Haleakala silversword (Fig. 10.12). The first part of an organism's name is the

Figure 10.12   **The Hierarchy of Classification.**

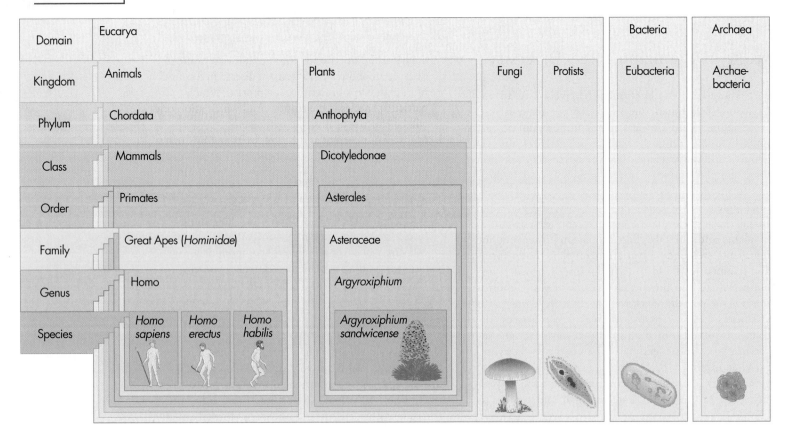

genus (pl., *genera),* a term that may include several similar species. The Haleakala silversword's genus is *Argyroxiphium.* The second part of the name, the **species,** works along with the genus term to designate only one type of organism. The species part of the Haleakala silversword's name is *sandwicense,* referring to the Sandwich Islands, an early European name for the Hawaiian Islands. As used today, the genus refers to a group of very similar organisms related by common descent from a recent ancestor and sharing similar physical traits. For example, *A. sandwicense* grows in Haleakala Crater on Maui; another genus member, *A. kauense,* grows only on Hawaii's Mauna Loa and Mauna Kea volcanoes. A related genus, *Dubautia,* also contains several species more closely allied to each other than to the members of the genus *Argyroxiphium* (Fig. 10.12).

With its unique two-part name, a species is also a unique group within a genus whose members share the same set of structural traits and can successfully reproduce in nature only with members of the same species. (Recall from Chapter 9 that it can be difficult to assign organisms to species. We can't test the mating potentials between extinct organisms represented only by fossils, for example, or between members of species that don't generally reproduce sexually. Taxonomists must use the best available evidence from each group's particular traits and life history.) All genus names begin with a capital letter, and both genus and species names are written in italics or underlined. Biologists usually write or say the full scientific name the first time they refer to it in a discussion. After that, though, they often abbreviate the genus and simply use *A. sandwicense,* for example, or *H. sapiens,* or *E. coli.*

The Hierarchy of Classification

Each organism fits into a series of categories, or *taxonomic groups,* based on shared characteristics. These are arranged in a hierarchy of broader and broader shared traits, and assigned to the species, genus, **family, order, class, phylum (or division), kingdom,** and **domain.** To see how this works, check out all the taxonomic groups for *A. sandwicense* and *H. sapiens* in Figure 10.12.

Criteria for Classification

A taxonomist's job might seem easy—just group organisms according to their shared characteristics. But which characteristics should they use? Is flower color a good measure? Should they group together all plants with yellow flowers—yellow silversword with dandelions and

yellow roses, for example? Or how about flower size: Should they group silverswords and sunflowers because both have large flower heads made of small, individual flowers? The species is an *objective* category, because biologists can often test whether two different species can interbreed successfully in nature. But the higher levels—genus, family, and so on—are a matter of opinion, albeit the well-educated opinions of taxonomists. So how do they decide what to include in different genera or families?

The answer is that taxonomists usually want to lump together organisms in a way that reflects their evolutionary history—their position on a family tree of organisms, or a **phylogeny**. This method, in turn, is called the phylogenetic approach.

Cladistics: Classification by Shared Derived Traits

To make a family tree or phylogeny, it makes sense to put the most closely related species on the tips of nearby branches and more distantly related organisms at the tips of branches diverging further down on the tree. So biologists need to identify groups sharing recent common ancestry, and to do this, often use a procedure called **cladistics.** Cladistics is the study of **clades,** taxonomic groups made up of an ancestral organism and all the descendents derived from it.

Let's look at several animals—a frog, a small bird called a Hawaiian honeycreeper; a goat of the type that eats silverswords, and a bat, Hawaii's only land mammal—and work through a cladistic approach to constructing a family tree (Fig. 10.13). A cladist looks among the traits exhibited by such a group of organims to find shared, derived traits. All four of these organisms share the trait of limbs, including legs and forelimbs (Fig. 10.13, Step 1). Thus, to a cladist, the presence or absence of limbs would not be a good character for determining relationships among the frog, honeycreeper, bat, and goat. Possession of limbs, in this case, represents an **ancestral trait,** a characteristic displayed by an ancient ancestor common to all the species in the group. We can infer ancestral traits by looking at an **outgroup,** a closely related species whose lineage diverged before all the members of the group in question. In this example, extinct fossil amphibians could serve as an outgroup; because they had limbs, we can conclude that limbs are ancestral.

In contrast to an ancestral trait, cladists look for a **derived trait,** an evolutionary novelty, a newly originated inherited change. For example, the honeycreeper, bat, and goat all develop from embryos enclosed in an amnion, a protective membrane (see Fig. 8.24). Frogs,

How does classification aid the study of biodiversity?

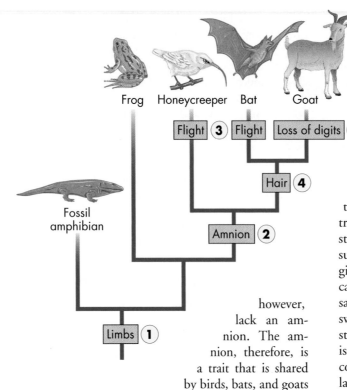

Figure 10.13 **Using Cladistics: Identifying Shared Derived Traits.** The living organisms shown here—honeycreepers, bats, goats, and frogs—share the trait of limbs. We know that limbs are an ancestral trait, based on fossil evidence. Honeycreepers, bats, and goats share the derived trait of a protective amnion membrane around the embryos, and bats and goats share hair, a newly evolved trait. Flight in the honeycreeper and the bat is an analogous trait. The goat's loss of digits is a derived trait. These patterns of shared, derived traits support the phylogenetic tree drawn here.

(Fig. 10.13, Step 5). Usually taxonomists use as many traits as they can identify among the species under consideration, always identifying shared, derived traits, to draw phylogenetic trees.

Molecular Phylogenies

Toes, hair, and other distinctive physical traits are very handy for classifying organisms, but most living things don't have limbs, let alone toes. All, however, have DNA and/or proteins, so biologists have recently been creating family trees by comparing DNA or protein structures. And instead of counting toes, they look at molecular traits such as the orders of nucleotides or amino acids. Biologists still look for shared, derived orders of subunits. Recall Dr. Robert Robichaux, the silversword expert we saw bounding over volcanic boulders to plant silverswords and help restore decimated populations of the strange and gaudy plants. He and his collaborators have isolated a silversword gene called *apetala-3* that helps control petal formation in the flowers. Once they isolated the gene, they looked at the protein it encodes and compared part of its amino acid sequence with those of petal-producing proteins in other plants (Fig. 10.14a). They found mostly identical sequences (shaded in the figure), but a few amino acids did differ from plant to plant. The silversword's petal protein was closer in amino acid sequence to the aster protein, so they grouped these together on the tree. They placed snapdragons on a lower branch, and placed rice, with the least similar sequence, diverging near the base of the tree.

Gene phylogenies like these reveal that the closest relative to members of the Hawaiian silversword alliance is the tarweed, a flowering plant in the aster family living in California and the Pacific Northwest. The degree of change between the sequences also helped Robichaux and colleagues reconstruct the timing of the silversword's evolutionary events in Hawaii. They studied the amount of genetic divergence among *apetala* genes in various members of the silversword alliance then compared them with tarweed petal genes. From this, they determined that about 5 million years ago, a tarweed seed was transported by a bird or perhaps floated from the west coast of North America to Hawaii, and took root. Kauai and the small island of Niihau had emerged from the sea at the time the tarweed seed arrived, but Maui, Oahu, and the Big Island had not. From the genetic data, Robichaux's group reconstructed this scenario: The transplanted tarweed population

however, lack an amnion. The amnion, therefore, is a trait that is shared by birds, bats, and goats and was derived (evolved) after they diverged from the frog lineage. The shared derived trait of the amnion groups the honeycreeper, bat, and goat in the same clade (Fig. 10.13, Step 2).

Next, the cladist looks among the species for additional shared, derived traits, but questions arise. Should we group bats with birds because they both fly? Or should we group bats with goats because they both have hair? The concept of shared, derived traits helps us out. Fossil evidence shows that flying evolved independently in ancestors of the bird and the bat. Flight in birds and bats is thus an **analogous trait,** a similar trait possessed by the members of two different groups, but not by their last common ancestor (Fig. 10.13, Step 3). The yellow flower color shared by certain silverswords and certain roses is another analogous trait, and so is the streamlined body shape of porpoises and sharks.

In contrast to the trait of flying, the trait of hair is a good trait for building a family tree by cladistics (Fig. 10.13, Step 4). Body hair is is a **homologous trait,** a similarity that two species share because they inherited it from a common ancestor. Furthermore, hair is a derived trait that originated in the lineage shared by bats and goats. A cladist, therefore, would use hair, a shared, derived trait, to group goats and bats in the same classification category. Finally, additional derived traits appeared in the goat lineage, as, for example, three of the original five digits present in each limb were lost

(a)

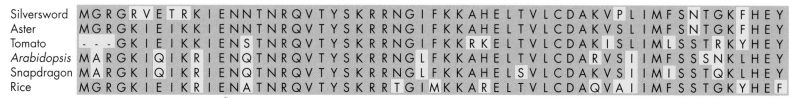

Silversword	M G R G R V E T R K I E N N T N R Q V T Y S K R R N G I F K K A H E L T V L C D A K V P L I M F S N T G K F H E Y
Aster	M G R G K I E I K K I E N N T N R Q V T Y S K R R N G I F K K A H E L T V L C D A K V S L I M F S N T G K F H E Y
Tomato	- - - G K I E I K K I E N S T N R Q V T Y S K R R N G I F K K R K E L T V L C D A K I S L I M L S S T R K Y H E Y
Arabidopsis	M A R G K I Q I K R I E N Q T N R Q V T Y S K R R N G L F K K A H E L T V L C D A R V S I I M F S S S N K L H E Y
Snapdragon	M A R G K I Q I K R I E N Q T N R Q V T Y S K R R N G L F K K A H E L S V L C D A K V S I I M I S S T Q K L H E Y
Rice	M G R G K I E I K R I E N A T N R Q V T Y S K R R T G I M K K A R E L T V L C D A Q V A I I M F S S T G K Y H E F

(b)

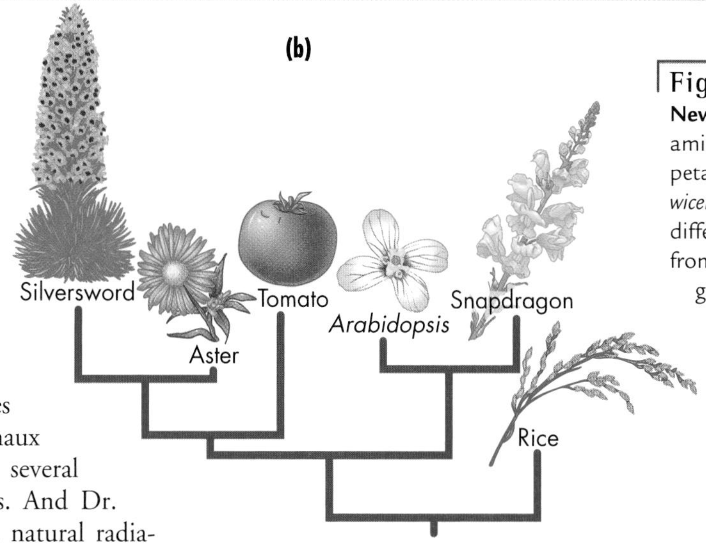

grew, spread, and underwent various genetic changes. These changes, tuned by natural selection, suited different sub-populations to different habitats. As new islands arose and their physical features changed, silversword seeds floated or hitchhiked on birds from one island to the next, radiating and forming new species on each island in turn. Robichaux and colleagues have identified several dozen silversword alliance species. And Dr. Robichaux would like to see this natural radiation continue. Part of his goal for replanting silverswords on the volcanic craters of Maui and Hawaii is to create sources for the colonization of Loihi, the next Hawaiian island to emerge about 2000 years from now.

The Tree of Life

Our understanding of the tree of all life has been revolutionized by the building of family trees using the same principles Robert Robichaux's group applied to silversword classification. Two centuries ago, the great classifier Carolus Linnaeus considered all organisms to be either plants, like the silversword, or animals, like the goats that threaten to eat certain silverswords into extinction. The more biologists observed living organisms, however, and the more they learned about their cellular and genetic structures, the more species they discovered that fit outside the plant and animal kingdoms. One of the biggest breakthroughs in understanding the relationships of all living organisms to each other came with the sequence comparisons of ribosomal RNA. The sequences of ribosomal RNA base pairs (see Fig. 7.9) are more useful for constructing family trees than the amino acid sequences of proteins. This is because rRNA is found in all organisms, and evolves rather slowly—so slowly that researchers can track the sequence changes and use them to determine the branchings that occurred in the distant past. This, in turn, allows them to con-

struct more accurate family trees. Studies of rRNA confirm that all organisms alive today occupy the tips of branches stemming from three main trunks on the tree of life (Fig. 10.15). These three great lineages are called **domains**, and they consist of the **Bacteria** (or Eubacteria), the **Archaea** (or Archaebacteria), and the **Eucarya** (or eukaryotes) (*E* xplorer 10.2).

The Bacteria and Archaea are both single-celled prokaryotes—cells lacking membrane-enclosed nuclei. The Bacteria include familiar cells like the species that cause teeth to decay (*Streptococcus*) and a major inhabitant of the human gut (*Escherichia coli*). The Archaea are also single-celled prokaryotes, but they have distinctive cell membranes and other unique biochemical and genetic traits, as well. Many *Archaea* live in harsh environments we would find hellish or disgusting. These include acidic, sulfurous hot vents and springs like Pele's Vent at the top of Loihi Seamount; very salty lakes, such as Mono Lake in California (*Haloferax*); and the insides of cows' intestines (*Methanococcus*)! Taxonomists used to lump the Bacteria and the Archaea together in a single kingdom Monera, creating five kingdoms of life (Monera, Protista, Plantae, Fungi, and Animalia) as first proposed by R.H. Whittaker in 1969. Today, Monera is replaced by the two domains we just considered. And Whittaker's four remaining

Figure 10.14 **Molecular Phylogenies: A Powerful New Approach.** (a) The top row of letters represents an amino acid sequence from the protein encoded by the petal-forming gene *apetala-3* in the silversword (*A. sandwicense*). (Each of the 20 amino acids is represented by a different letter.) The subsequent rows show sequences from other plants. The amino acids that differ between organisms are highlighted in yellow. (b) Cladistic analysis of these data forms the basis of a family tree for this protein; rice is a very distant silversword relative and aster species are close relatives.

READ ON

E xplorer **10.2**

The Tree of Life

kingdoms, the protists, plants, animals, and fungi, all fit into the domain Eucarya because they all contain eukaryotic cells.

Our current understanding merges the old notions of kingdoms with the three domains. Within the Eucarya, the kingdoms of plants, animals, fungi, and protists still work rather well, despite a few exceptions we'll see in later chapters. Within the Archaea, there are two large lineages, the Crenarchaeota and the Euryarchaeota (Fig. 10.15), which some biologists consider kingdoms and which may be more broadly accepted in the future. Within the Bacteria, there are many lineages, all in a single kingdom that we'll discuss in Chapter 11.

Notice in Figure 10.15 the somewhat surprising fact that Archaea and Eucarya are more closely related to each other than either is to Bacteria. This reflects the belief that some ancient Archaean cell probably gave rise to the eukaryotic cell nucleus, or at least donated the nuclear genes that control genetic information processing (review Fig. 10.8). Another curious fact is that the familiar organisms we see every day— birds, trees, grass, flowers, dogs, mushrooms, flies— are just tiny branches at the periphery of the tree of life. The vast expanse of evolutionary diversity, determined through molecular analyses, lies in the single-celled organisms within the domains Bacteria and Archaea, and in the kingdom Protista. Those simple living things, in other words, are far more diverse genetically from each other and from larger organ-isms than are all the familiar plants, animals, and fungi.

The base of the Eucarya contains a diverse group called the **Protista** (sometimes called Protoctista). These are generally single-celled species with nuclei such as *Euglena*, paramecia, and amoebas. The **Fungi** include multicellular heterotrophs, such as mushrooms and molds, as well as unicellular yeasts. Fungi decompose other biological materials, develop from spores, and lack flagella. The **Plantae,** which are mostly multi-cellular autotrophs, include some algae and mosses, ferns, and flowering plants like silversword; plants generally develop from embryos. Finally, the **Animalia** are multicellular heterotrophs, which lack cellulose, usually exhibit movement, and develop from embryos. Animals include the groups encompassing sponges, insects, clams, birds, and mammals. In the next three chapters of the book we explore the kingdoms and domains in greater detail. Here, we look at the very serious problem of species loss. Will our own huge populations and actions cause the next major extinction crisis?

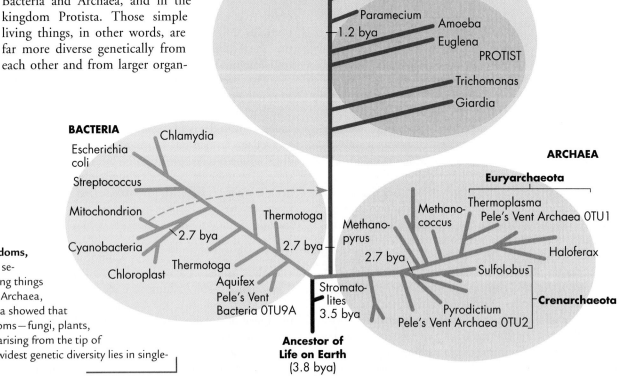

Figure 10.15 **Domains, Kingdoms, and Life's Family Tree.** Based on the sequences of ribosomal RNA genes, living things fall into three domains: the Bacteria, Archaea, and Eucarya. These revolutionary data showed that the most visible, most familiar kingdoms—fungi, plants, and animals—are just tiny branches arising from the tip of one of the three branches, while the widest genetic diversity lies in single-celled organisms.

10.5 The Biodiversity Crisis

Silverswords and their wonderfully diverse relatives shaped like vines, trees, bushes, weeds, and silvery rosettes are unusual plants on remote islands. But their endangerment and loss still represent a worldwide threat to biodiversity—a crisis that could claim millions of species in the next 50 years. Let's explore biodiversity and extinction in more detail to better understand this potential disaster.

How Much Biodiversity?

Over the past 230 years, biologists have discovered, named, and recorded 1.5 million species of living organisms.

This has been basically a hit-or-miss prospect, based on where naturalists have gone, what they've been interested in collecting, and how much time (and funding) they've had for studying new and unknown organisms. Of the 1.5 million known species, about 5 percent are single-celled prokaryotes and eukaryotes (although people are just beginning to appreciate the huge diversity in Archaea and Bacteria and most of those are still unnamed). About 22 percent of the known species are fungi and plants. And about 70 percent are animals (Fig. 10.16). Most of the animals are invertebrates (they lack backbones). The sea harbors the most varied groups of invertebrates, with 90 percent of all the higher level groups. But these thousands and thousands of taxonomic categories contain only 20 percent of the individual animal species. By far the largest number of species are land-dwelling animals, and of these, over a million named so far are insects. Ours is a planet of terrestrial insects, and

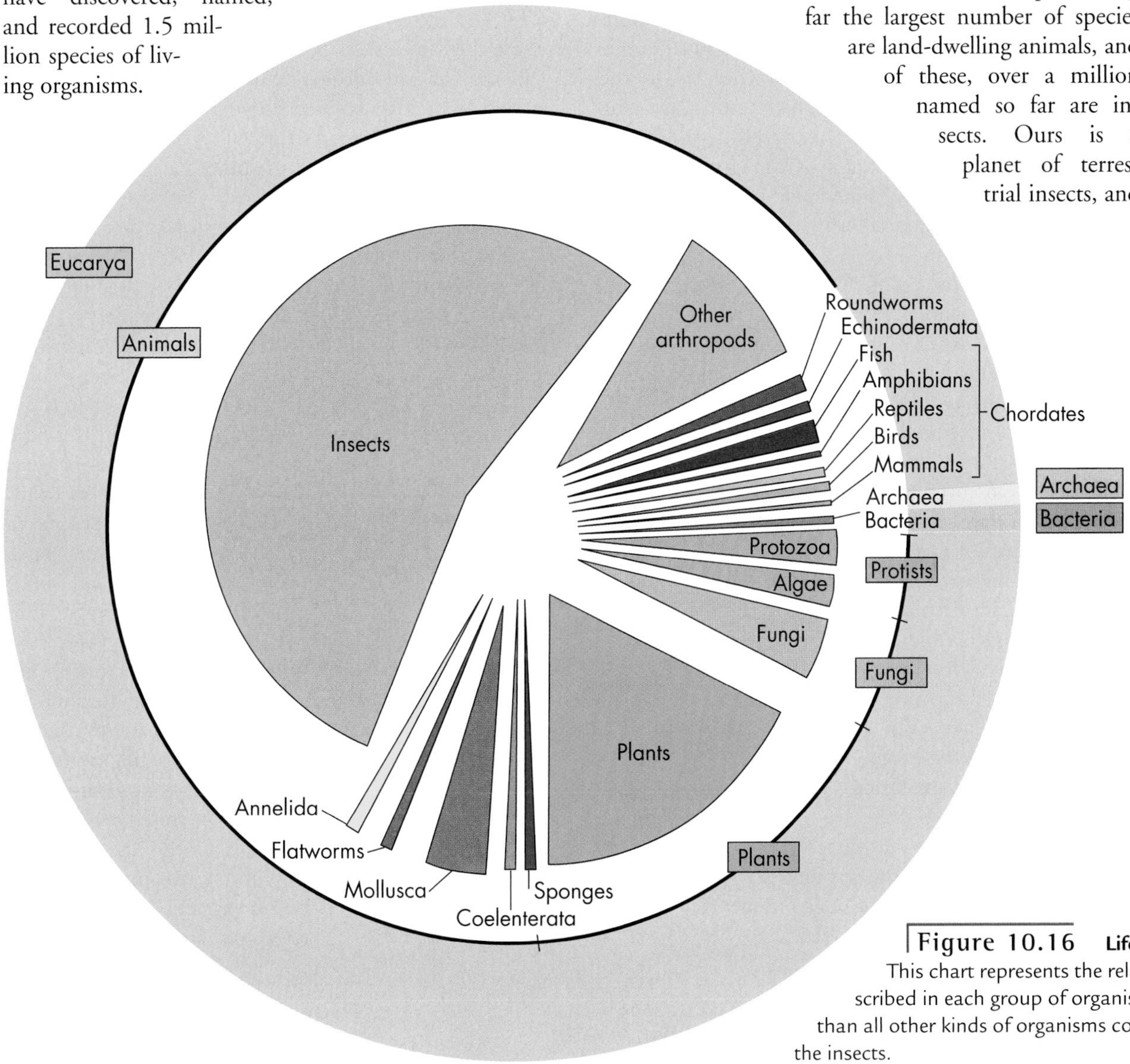

Figure 10.16 **Life's Diversity: Planet of the Insects.** This chart represents the relative number of species currently described in each group of organism. Insects account for more species than all other kinds of organisms combined. Earth is truly the planet of the insects.

Figure 10.17 **The Hawaiian Rain Forest.** Researchers study the hundreds of insect and other species inhabiting small sections of the tropical rain forest around Nanue Falls on the Hamakua coast of Hawaii. Orange flowers blaze on African tulip trees.

READ ON

*E*xplorer *10.3*

Biodiversity Hot Spots

there are probably millions of species still to be discovered and classified.

For decades, biologists estimated that an accurate number of total species would be about 3 to 5 million, based on models of the likely diversity in temperate zones (where, coincidentally, almost all the scientists have lived) plus a multiplier of 2 for the number of tropical regions. This estimate lost favor, though, when biologists went out to study the "last biotic frontiers" in the 1970s and 1980s, including the deep ocean floor and the treetops of tropical rain forests (Fig. 10.17). Over and over during this period, researchers would gather plants and animals clinging to tropical treetops or collect mud from a previously untested spot of ocean bottom and find that 90 percent of the organisms they collected were unknown to science! For example, Terry Erwin, a researcher working in the Panamanian rain forest, set off insecticidal fog bombs in individual jungle trees and collected all the organisms that fell to the ground. Focusing only on beetles, he found an amazing 163 beetle species from *each tree species*, on average. Multiplying that by the number of known tropical tree species, he calculated that there could easily be *30 million* species of insects and other invertebrates to describe in the rain forests alone—the vast majority undescribed and unnamed at this point. Adding in tropical plants, seabed organisms, and all the other undiscovered species in nontropical climatic zones, the figure of 50 million new species begins to look possible—even conservative.

Why should we care if Earth has up to 50 million species but only 1.5 million have been named? A minor reason is we're going to need a lot more biologists to discover and catalog these species before they become extinct as people rapidly disrupt and destroy large sections of tropical rain forests. Second, we could lose an irreplaceable genetic reservoir, including sources of new foods, drugs, and fibers for paper and cloth before we even find them. Third, we could get a tremendous ecological shock in the future, because we don't know how important this great degree of diversity is to the stability of all interdependent life. Is a huge number of species important for ecosystems to function? Are ecosystems with large numbers of species more stable and long-lived than those with fewer species? We think the answers are yes, but we don't

know how many species are needed and how many can be lost before an ecosystem fails. We'll explore the questions and answers to this ecological puzzle in Chapters 24 and 25. But it's clear in the meantime that human activity, such as the many modifications we've made in Hawaii, for example, is now driving unknown numbers of species to extinction, including one quarter of the plants in the silversword alliance.

The Threat of Extinction

The silversword is probably the most diverse plant group in Hawaii, but there are diverse animal groups there, as well. For example, there are at least 511 named and 300 unnamed species of the family Drosophilidae (the geneticist's friend, the fruit fly *Drosophila melanogaster,* is just one of them). All but about 20 of these delicate insects occur nowhere else in the world. Hawaii has had another remarkable radiation, as well: about 30 brightly colored honeycreepers that use different kinds of flowers and fruit (Fig. 10.18). Biologists, in fact, consider Hawaii and the other Pacific Islands to be among the world's 25 biodiversity "hot spots," small regions together containing the greatest percentage of Earth's biodiversity (*E*xplorer 10.3). Some other hotspots are the tropical Andes, the Mediterranean basin, and Indonesia and surrounding islands.

Just as many silversword species are threatened by people's farming and land development and by introduced grazers like pigs and goats, human activities are threatening biodiversity in the other 24 hot spots. In all these areas—tropical rain forests, coral reefs, drylands, and other life zones—people are destroying habitats faster than scientists can survey and describe them. Habitat destruction takes away interrelated plant, fungal, and bacterial systems as well as animals' feeding and nesting sites. When the last plant or the last breeding pair of a species disappears, the species becomes extinct. Since large areas contain more species than small areas, we can directly translate the loss of massive tracts of habitat into staggering species extinction. Biologists Peter Raven and E.O. Wilson es-

Crested honeycreeper
(Palmeria dolci)

Kauai akialoa
(Hemignathus procerus)

Grosbeak finch
(Psittirostra kona)

Figure 10.18 **Radiation in Hawaiian Honeycreepers.**

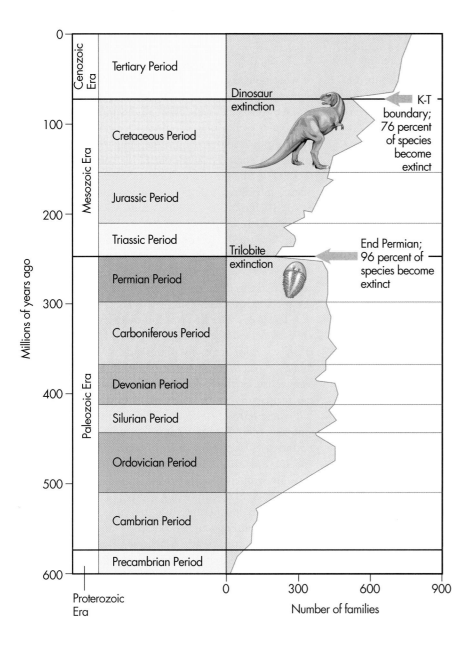

Figure 10.19 Extinction Events and Radiation. Major and minor extinctions punctuate the history of life. How will our human influence affect life's diversity in the next 50 years?

mans were on the scene. But the past events were unstoppable and this time, *we* are making drastic changes to land, sea, and atmosphere that could be avoided.

We can, of course, learn from historic **mass extinctions,** the extinction of many groups of organisms over relatively short spans of geologic time. They can serve as models for what we can expect during and after the current biodiversity crisis. Figure 10.19 shows a series of mass extinction events that have punctuated the history of life. Following each episode, a burst of evolution brought about many new forms, replacing the extinct ones. A good example is the radiation of the mammals following the disappearance of the dinosaurs. Past extinctions also teach us that during a massive die-off, species don't disappear entirely at random. In the extinctions that took place

timate that up to 20 percent of the world's species could disappear over the next 30 years due to the doubling of human populations in tropical and semitropical zones. Some biologists are skeptical about the 20 percent figure but few doubt that species are disappearing at an alarming rate (more details in Chapter 24). The extent of the loss, in fact, may begin to approach some of prehistory's great extinction episodes. The main difference is that *we* are causing this one, and we're killing things off at a much faster rate of loss than in past extinction events. We're not the only agents of extinction; meteors and global climate shifts caused thousands of dinosaur species to die out more than 64 million years before hu-

about 10,000 years ago after the last ice age, land animals with large bodies—for example, mammoths, mastodons, and saber-toothed tigers—tended to die out faster than smaller land animals. This suggests that today's extinction crisis will claim many of our favorite animals first, including rhinoceroses, giraffes, pandas, and a number of whale species.

Another lesson from the past is that genera living in many large regions tend to survive mass extinctions, while genera localized to particular small regions are especially vulnerable. If a genus is confined to one small corner of the world, like the silverswords native to Haleakala volcano, then altering this small region will eliminate critical habitat and could easily endanger or wipe out the organisms in the genus. We can assume from this that over the next centuries, our world will be enriched in widespread, fast-growing "weedy" species like rats, cockroaches, starlings, and ragweed, and be impoverished in rare, locally adapted species such as silverswords. Ironically, these rare species have tended to be our most useful sources for new medicines and foods.

A final lesson we can take from the fossil record of extinctions is that the permanent losses we cause in the next 50 years or so will not be recouped for millions of years. The evolution of new species takes a very long time. As long as our species continues to overpopulate the planet as we are doing today, species diversity is unlikely to recover because of the dramatic changes we make to their environments. Clearly, even mild extinction events have far-reaching biological repercussions. Many biologists conclude that the cost of continuing to disrupt habitats and cause extinctions is so dire we must act immediately or suffer tremendous consequences.

The Importance of Taxonomy

Along with the changes we've been wreaking upon ecosystems through pollution, habitat destruction, and the killing of organisms for food, hides, trophies, and wood, we've given new importance to the science of taxonomy. Once the domain of dusty collections and hushed libraries, taxonomy is now an exciting, high-technology field. Modern taxonomists can easily find themselves suspended from tropical tree tops, investigating ocean vents in deep-sea submersibles, or working in state-of-the-art genetics laboratories. Our survival, and that of the millions of species we are currently endangering, depends on our understanding how ecosystems change as species die out and are replaced. As we just saw, though, we're missing even the roughest estimate of how many species exist, let alone a complete list of what they are and how they live, die, and interact in the environment.

As we've also mentioned, discovering, categorizing, and investigating new species has practical significance beyond pure biological understanding. Researchers have developed many cancer-controlling drugs and other modern medicines from chemical compounds in rare plants, including breast cancer drugs from the yew tree of the Pacific Northwest and leukemia medicines from the Madagascar periwinkle flower. Nutritious fruits, roots, and seeds are no doubt growing in obscurity around the globe right now, and some of those could well become staple foods of the future and an important adjunct to the three or four dozen staple crops our entire species depends for survival.

The work Dr. Robert Robichaux is doing to conserve and restore Hawaiian silverswords is the kind of effort we must make for thousands of species in hundreds of places around the globe. If he and his colleagues succeed, Haleakala crater will have 20,000 spectacular silverswords jutting from the stark volcanic soil where only a few dozen once remained. The Nature Conservancy, World Wildlife Fund, National Science Foundation, and many other organizations are directing and funding efforts like these in biodiversity hot spots. But governments and private groups must commit more resources to combating overpopulation, pollution, and habitat destruction. And as citizens we must approve the funding, inform ourselves, and participate in direct and indirect ways. Maybe some of the readers of *Explore Life* will even become motivated to study taxonomy. It is truly a science with a deadline.

Chapter Summary and Selected Key Terms

Introduction   The birth of the Hawaiian islands helps us learn about land formation on early Earth. The arrival of a tarweed seed on the newly formed island of Kauai shows how barren land can become colonized. The ancestral tarweed's radiation into the silversword alliance is a splendid example of evolutionary radiations. Dr. Robert Robichaux is helping to conserve and replant endangered silverswords on the islands as part of an important movement to save our world's rapidly diminishing diversity.

Earth Forms, Life Emerges   A few billion years after the universe originated in the Big Bang, our solar system formed. Earth formed about 4.6 billion years ago. Earth was an inorganic world of pale blue sky, blue ocean, and black volcanic cones with an early atmosphere containing CO_2, N_2, H_2O, H_2S, NH_3, and CH_4. Nevertheless, Earth had the right combination of composition, temperature, size, and distance from the sun to harbor life.

Biologists have no ultimate proof, but they do have many models and forms of indirect evidence for how life may have originated. Meteor bombardment until about 3.8 billion years ago melted all rocks that might have contained chemical traces of the earliest organic processes or cell formations. But in rocks dated to 3.6 billion years ago, not long after the bombardment ended, there are fossils that resemble stromatolites or large mounds built by cyanobacteria. Three kinds of important indirect evidence support the idea that the organic building blocks of life were present on early Earth: spark chamber molecules, carbonaceous chondrites, or meteorites containing organic molecules.

Biologists picture a primordial soup of seawater and dissolved organic molecules that received energy from the sun hitting shallow pools or perhaps from heat near hydrothermal vents. Energy inputs could have facilitated the forming of organic precursors into amino acid chains, lipid chains, and other large organic molecules. In one well-respected model, RNA molecules formed spontaneously and acted as ribozymes (p. 256) or catalysts of further RNA formation or protein synthesis. In this **RNA world** (p. 256), natural selection favored certain RNA and protein variants more than others, and these would have accumulated more quickly. Eventually, the more stable information storage form DNA evolved, but the relationship between DNA, RNA, and proteins remained.

The First Cells
Living cells have an outer membrane, metabolism, and gene translation. Researchers have heated together polypeptides and phospholipids, which can form membrane-like sheets that in turn form compartments, trapping materials within. They can even divide under certain conditions. Self-replicating RNAs with membrane compartments would have had selective advantages. A compartment like this that could catalyze the change of an abundant material into a needed scarce one would also have had a selective advantage. Strings of such acting catalysts could have given rise to the first metabolic pathways.

Very early **heterotrophs** (p. 258) left traces of biological materials in rocks from southern Greenland dated to over 3.8 billion years old. Competing heterotrophs eventually would have used up most of the environmental supply of organic raw materials for carbon and energy supplies. Natural selection would have favored **autotrophs** (p. 258). **Chemoautotrophs** (p. 258) generate organic molecules and trap energy from inorganic compounds like hydrogen sulfide and iron. Biologists think that all currently living organisms share a common ancestor, a primitive chemoautotroph that probably lived at temperatures near the boiling point of water similar to the cells living near Pele's Vent. Fossilized autotrophic cells somewhat like today's simplest photosynthetic cells, the cyanobacteria, appear in 3.5-billion-year-old rocks. Fossils dated at 2.8 billion years old contain cells very much like modern cyanobacteria.

Eurkaryotic Cells and Multicellular Life
Eukaryotic cells containing DNA inside a nuclear membrane appeared about 2.6 billion years ago. About that time, according to the **endosymbiont hypothesis** (p. 260), eukaryotes also engulfed free-living aerobic and photosynthetic bacteria, and these led to the internal organelles mitochondria and chloroplasts. Cells containing these started appearing in the fossil record about 1.9 billion years ago, and about 1 billion years ago, radiated into numerous single-celled groups.

The first soft-bodied animals show up in the Ediacaran fossils, dated to 580 million years old. After hard parts (shells and tough skins) evolved 40 million years later, many more kinds of animals were preserved as fossils. By 400 million years ago, primitive land plants had evolved, and large fishes swam in the seas. Within the last 200 million years, the amphibians, reptiles, birds, mammals, conifer trees, and flowering plants appeared.

Just as living things affected Earth's atmosphere and oceans, spontaneous geologic changes affected life. Geologists call most of Earth's 4.6 billion year history the **Archean** (p. 262) and the **Proterozoic** eras (p. 264). During these eras, the continents formed, mountains arose, and single-celled life-forms appeared. The **Paleozoic Era** (p. 264) began far later, about 580 million years ago, with the **Cambrian Explosion** (p. 262), or the appearance of hard parts in the animal fossil record. **Plate tectonics** (p. 264) began to move about the landmasses through seafloor spreading and subduction of giant plates. The single supercontinent **Pangaea** (p. 264) formed, then broke up. The Paleozoic ended with the **end-Permian extinction** (p. 264).

The **Mesozoic Era** (p. 264) (between 245 and 65 million years ago) saw continued plate tectonics and organismal evolution, including the rise of the dinosaurs. This era ended with another massive extinction event that wiped out the dinosaurs. The current era, the **Cenozoic Era** (p. 265), has brought the continents to their present locations and the three domains of living organisms to their modern members.

Organizing Life's Diversity
The silversword alliance represents tremendous variation in plant body forms and habitats. Life's diversity is inconceivably broader. **Systematics** (p. 265) is the study of diversity and evolutionary interrelationships. **Taxonomy** (p. 265) is the science of identifying and naming organisms and classifying them into groups. Carolus Linnaeus invented the system of **binomial nomenclature** (p. 266), assigning a **genus** (p. 267) and **species** (p. 267) name to each separate type of organism. Higher order taxonomic groups include the **family, order, class, phylum (or division), kingdom,** and **domain** (p. 267). Taxonomists usually lump organisms together in a way that reflects their evolutionary history—their position on a family tree or **phylogeny** (p. 267). To create phylogenies, they often use **cladistics** (p. 267) to reconstruct patterns of descent using traits that evolved only within a given **clade** (p. 267) or group (shared, derived traits). We can infer ancestral traits by looking at an **outgroup** (p. 267). Flight in birds and bats is an **analogous trait** (p. 268), a similar trait possessed by members of two groups but not their ancestors. Body hair is a **homologous trait** (p. 268), shared by bats and goats because they inherited it from a common ancestor. An **ancestral trait** (p. 267) is shown by an ancient ancestor as well as living descendants—for example, five digits per limb. A **derived trait** (p. 267) is an inherited change of a preexisting trait. A goat's two toes are derived from the ancestral condition of five digits.

Biologists also use molecular data to create phylogenies, including the orders of nucleotides in DNA or RNA, and the orders of amino acids in proteins.

The family tree of all life includes three **domains** (p. 269): the **Bacteria**, **Archaea**, and **Eucarya** (p. 269). All the varied bacteria form a single kingdom. Archaea has two large lineages that may be considered kingdoms in the future. The Eucarya contain four kingdoms, **Protista**, **Fungi**, **Plantae**, and **Animalia** (p. 270).

The Biodiversity Crisis
Biologists have recorded 1.5 million separate species, mostly insects. But there are very likely 30 to 50 million species, still undiscovered and unnamed, and at risk of being lost before they are ever known to science. Some biologists predict that we will lose up to 20 percent of that diversity in the next 30 years due to pollution, human population growth, and habitat destruction. The **mass extinctions** (p. 273) of history teach us that radiations of new species usually follow but take millions of years. They also teach us that large, specialized organisms tend to die off first, and that "weedy" species tend to survive. Taxonomy has become a dynamic field concerned with finding and organizing species before they are lost, including those that can serve as new sources of drugs, foods, and fibers.

> All of the following question sets also appear in the Explore Life *E* electronic component, where you will find a variety of additional questions as well.

Test Yourself on Vocabulary and Concepts

In each question set below, match the description with the appropriate term. A term may be used once, more than once, or not at all.

SET I

(a) Archean (b) Proterozoic (c) Paleozoic
(d) Mesozoic (e) Cenozoic

1. Appearance of first heterotrophic cells
2. First flowering plants appeared.
3. Pangaea formed.
4. Pangaea broke apart.
5. Cambrian explosion of life-forms
6. Continents moved to present locations.
7. Adaptive radiation of birds and mammals

8. Ended with a mass extinction that eliminated the dinosaurs

9. Age of *Homo sapiens*

10. Appearance of autotrophic bacteria

SET II

(a) species (b) phylum (c) kingdom (d) class
(e) domain (f) genus

11. A taxonomic group composed of similar orders

12. A taxonomic grouping that is used as part of an organism's binomial name

13. The most limited group that includes related classes

14. The group that is composed of one or more phyla

15. The most inclusive taxonomic category

SET III

(a) cladistics (b) phylogenetic taxonomy
(c) homology (d) analogy (e) natural selection

16. A system of classification based on family trees.

17. An anatomical similarity of two species based only on adaptation to similar environments.

18. An anatomical similarity based on inheritance from a common ancestor.

19. A method of classifying organisms based upon their shared derived characteristics.

SET IV

(a) Archaebacteria (b) Fungi (c) Protista
(d) Animalia (e) Plantae (f) Bacteria

20. Multicellular autotrophs

21. Found mainly in harsh environments such as hot, deep-sea vents; sulfurous springs; and salty lakes

22. Mostly single-celled eukaryotes

23. Multicellular heterotrophs such as molds and yeasts

24. Multicellular heterotrophs that lack cellulose; most are motile

Integrate and Apply What You've Learned

25. Describe how the Hawaiian Islands formed and the similarities between their formation and the formation of the earliest land masses.

26. Researchers discovered bacteria in deep hydrothermal vents near the Loihi Seamount, the beginnings of a new Hawaiian island. What were those organisms like and what is their significance?

27. You are engaged in a debate with someone who argues that if cells arose spontaneously from nonliving matter once in the history of Earth, the same or similar events should be able to occur again at any time. What is your response?

28. How might silverswords have arrived on Hawaii and radiated into such a diverse alliance of species?

29. What is the binomial system of nomenclature?

30. Why do biologists think there could be up to 50 million species, even though they have discovered, described, and named fewer than 2 million?

Analyze and Evaluate the Concepts

31. If Earth's orbit were to shift suddenly toward that of Mars, what would be the likely effect on life on Earth? What if Earth's orbit shifted toward Venus? Cite evidence to support your answers.

32. What would the physical condition of today's Earth be if life had never evolved?

33. Why are stromatolites considered to be so important?

34. Using cladistics, how could a biologist determine whether the flippers of a whale and a seal are shared, derived traits; ancestral traits; or merely analogous traits?

The Wily Scourge: Malaria

It was one of those interminable meetings, and Mark Stover was getting a headache. The recent college graduate was sitting in the cafeteria of a small Christian seminary in east central Nigeria. The Lutheran mission's simple wood and tin buildings were clustered beneath tropical hardwood trees and palms; surrounded by rice, corn, and peanut fields; and situated just across the road from Mbamba, population 100. As Mark's fellow volunteer teachers continued debating some now-forgotten issue, Mark slipped away to his small hut, dived below the gauzy mosquito netting draped around his bed, and fell asleep.

"In the middle of the night," he recalls, "I woke up with a pounding headache. I felt like someone was jackhammering inside my skull. It was painful to move—in fact, to exist! My forehead was burning up, but I felt chilled off and on, too. I started looking back at my life," he says, convinced that he was dying. "If this is it," he thought at the time, "at least I've had this adventure in Africa and a chance to give something back to other people."

Mark dragged himself to a nearby hut, banged on the screen door, and awakened his sleeping friend, a nurse named Linda. After hearing Mark's symptoms, she drowsily pronounced them "just malaria." She gave him strong aspirin and, in the morning, drove him to the nearby city of Yola, where a blood test confirmed the presence of malaria parasites in his blood. For a week he took the classic malaria treatment, quinine, which he recalls as "nasty, bitter stuff that makes you urinate and makes your ears ring." Eventually, the infection ended. But another began just before Mark finished his two-year commitment.

"I was grading student essays," he says, and when he developed a headache, he thought, jokingly, "Well, that's just a natural response!" As the headache persisted and worsened, though, he recognized it as malaria and this time drove himself to town for medicine. Within five days, he was feeling better and was able to catch his flight for home.

Today, Mark is a medical student at Washington University in St. Louis, and has an additional perspective on his experience. He knew the risks of malaria he faced in rural Nigeria—the same, in fact, as throughout much of Africa and parts of South America and Asia. "Every week or two, some student or teacher came down with malaria," he recalls, and one student's infant daughter died from complications. "What surprised

Focus on a Tropical Scourge. Mark Stover in Nigeria (back row, center); culturing *Plasmodium falciparum*; Dr. Daniel Goldberg.

me, though," he says, was that he caught malaria twice, even though he was taking methacrine every week. This drug "was supposed to be 90 percent effective, plus I slept under a mosquito net every night and my house had good screens—all items that most Africans can't afford." Yet still he caught two cases of malaria.

These failed preventions are less surprising to Dr. Daniel Goldberg, one of Mark's medical school professors at Washington University and a well-respected malaria researcher. "Malaria prophylaxis is getting harder and harder," he says, because the parasites that cause malaria are evolving drug resistance. Millions of people like Stover travel from industrialized countries to regions where malaria is prevalent, and each year, says Goldberg, "there are a number of deaths in perfectly healthy people who come back and die from malaria." These cases, however, are a tiny fraction of the suffering malaria causes in developing countries themselves. "The disease is a major catastrophe for Africa," says Goldberg. "It afflicts nearly 500 million people a year there and on other continents and kills about 2 million a year, mostly children."

Daniel Goldberg and his colleagues have unraveled the multiple steps by which the malarial parasite attacks human blood: The single-celled protozoan that causes malaria, *Plasmodium falciparum,* enters human red blood cells and chews up millions of molecules of the protein hemoglobin, causing severe anemia and other problems. Goldberg's discoveries also helped show how standard malaria drugs attack and starve out the protozoan, and some details of how the parasite fights back through drug resistance.

Drug resistance is just one item in *P. falciparum's* arsenal of survival tricks. An evolutionary battle has raged between this infectious agent and its human hosts, says Goldberg. The organism's complicated

life cycle—alternating between the *Anopheles* mosquito and the human liver and blood—helps bamboozle our natural defenses. So does the parasite's ability to disguise itself by changing surface proteins and to send up "smoke screens" and "decoys" that confuse immune system efforts to attack and destroy. The wily protist can even cling to the insides of human blood vessels to avoid flowing through the spleen, which, says Goldberg, is "poised to recognize and remove infected blood cells."

Researchers like Goldberg believe that the elusive malaria parasite has caused half of all human deaths throughout history, making it our species' master scourge. For centuries, it has rendered large areas of damp, low-lying land uninhabitable, and it has sapped people's vitality when it didn't outright kill them. How can a single-celled organism slay and debilitate with such impunity? You'll find out as you explore this chapter's topic: the evolution and diversity of single-celled life forms. You'll see that *Plasmodium falciparum* is but one fascinating species among millions of microbes, and you'll learn about its interrelationships with and differences from other protists, prokaryotes, and viruses. By touring this largely microscopic realm, you'll find answers to these questions:

1. Why are prokaryotes so important to life on Earth?

2. Where do viruses fit in to the evolutionary tree of single-celled organisms?

3. How is the malaria parasite related to prokaryotes, to other protists, and to larger organisms?

4. How can a knowledge of prokaryotes and protists help researchers like Daniel Goldberg and future doctors like Mark Stover fight humankind's greatest killer?

11.1 Prokaryotes and Their Importance to Life on Earth

Malarial parasites are protists, eukaryotic single-celled organisms that are members of a stunningly diverse kingdom. To fully appreciate the protists—especially

complex ones like *Plasmodium falciparum*—we need to set the stage first with the simpler single-celled organisms called prokaryotes. This group is even more diverse than the protists in terms of their genetics, biochemistry, and habitats—so much so that it takes two domains, Bacteria and Archaea, as well as several kingdoms to encompass them (Fig. 11.1). Many

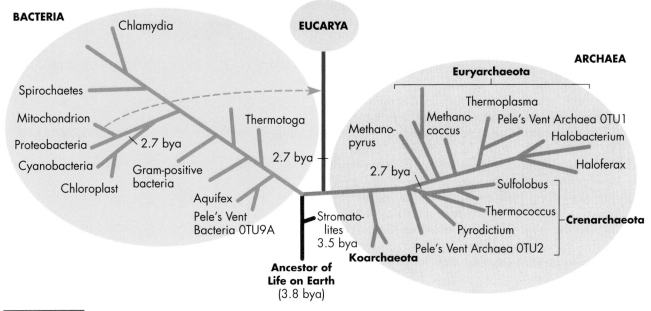

Figure 11.1 **Family Tree of the Prokaryotes.** During evolution, the prokaryotes diverged into two distinctly related groups: the Bacteria and the Archaea. The Bacteria radiated into many lines of descent, while Archaea have three main branches.

prokaryotic organisms can cause diseases, but far more prokaryotes benefit the environment and our bodies. One type may even help save people from malaria!

Prokaryotes, of course, are organisms whose genetic material is not contained within a nuclear envelope (see Fig. 2.29). Prokaryotic cells are usually much smaller than eukaryotic cells. For example, an intestinal prokaryote such as *Escherichia coli* is only about one one-thousandth the size and volume of a malarial parasite. Despite their small size, though, prokaryotes are profoundly important to the environment, medicine, and industry. Let's start by exploring their significance, then look at their general biological properties and diversity. As you read along, note that in the past, biologists used the word bacteria (little "b") to refer to all prokaryotes. Today, however, they use "prokaryote" for members of both domains Bacteria and Archaea, and use the capital "B" to designate the domain name, Bacteria (or as some biologists say, Eubacteria).

Importance of Prokaryotes

Prokaryotes have an immense ecological impact. Billions of prokaryotic cells in nearly every square meter of soil, water, and air release oxygen into the atmosphere and recycle carbon, nitrogen, and other elements. Prokaryotes consume enormous quantities of dead animals, fungi, plant matter, and human and animal wastes. They can also consume pesticides and pollutants that would otherwise poison our environment.

For decades, biologists have documented the prokaryote's enormous natural recycling role. More recently, field researchers have developed ways to encourage more bacterial breakdown of environmental pollutants (a process they call **bioremediation**). After oil spills, cleanup crews can seed the contaminated shoreline with a "fertilizer" containing nitrogen and phosphorus (Fig. 11.2). These nutrients, in combination with the organic compounds in the shoreline sludge, promote the growth of local prokaryotes with a natural appetite for greasy hydrocarbons, and within a couple of weeks, the sprayed areas dramatically improve.

Prokaryotes have great medical importance, both negative and positive, in addition to their immense environmental role. They cause hundreds of human diseases, including staph and strep infections (see Chapter 10), blood poisoning, tetanus, venereal diseases, and dental caries, as well as thousands of diseases in other animals and plants.

Despite this tremendous negative medical impact, harmful or pathogenic species are a small fraction of all prokaryotes, and many of the single-

Figure 11.2 **Bacteria Help Clean the Environment.** In 1989, millions of barrels of oil spilled from the tanker Exxon Valdez, despoiling hundreds of miles of Alaskan coastline. Cleanup crews sprayed nitrogen and phosphorus fertilizer on the fouled beaches, which encouraged the growth of naturally occurring oil-consuming bacteria.

celled organisms supply materials crucial to our health and survival. For example, *Escherichia coli* bacteria, along with other inhabitants of the human gut, produce vitamins K and B$_{12}$, riboflavin, biotin, and various co-factors that we probably absorb and use. Residents like *E. coli* may also blanket the intestinal walls so heavily that they block other dangerous microbes from passing into the bloodstream. Most plant-eating animals—cattle, sheep, rabbits, koalas, termites, and others—couldn't digest grasses, leaves, or wood without prokaryotes in their intestines actually breaking down the cellulose. You've probably demonstrated the importance of your own flora (as doctors call your internal microbes) if you've ever taken large amounts of antibiotics for an in-fection. The drugs probably killed off the beneficial or-ganisms in your intestines and reproductive system and left you, temporarily, with digestive difficulties, diarrhea, and/or yeast infections (oral, vaginal, etc.) until the nor-mal inhabitants returned and re-established a balance.

In the food and chemical industries, single-celled organisms help to produce cheese, yogurt, pickles, soy sauce, chocolate, and other foods, and to generate sup-plies of chemical compounds, such as butanol, fructose, and lysine. Commercial applications aside, much of what we know about how cells function comes from studying prokaryotes. They have small genomes and fast division rates, and are easy to grow in the laboratory. These traits have made the tiny cells favorites for inves-tigating biochemistry, molecular biology, and genetic engineering—all fields that rely on enzymes, chromo-somes, plasmids, and other components harvested from bacteria (see Chapter 7). Without their microscopic re-search subjects, biologists would be decades behind their current state of research.

Let's see, now, what all prokaryotes have in com-mon. Let's also see what makes them so important to the environment and to other organisms.

General Structure of Prokaryotic Cells

Biologists recognize thousands of prokaryotic species and assign them to the domains Bacteria and Archaea (Fig. 11.1). Bacteria and Archaea differ in basic ways, but the traits they share define the prokaryotic organism.

Cell Structure

Prokaryotic cells have an outer cell wall that surrounds the plasma membrane (Fig. 11.3). This membrane, in turn, surrounds a noncompartmentalized cytoplasm dotted with ribosomes. Prokaryotic cells lack a nuclear

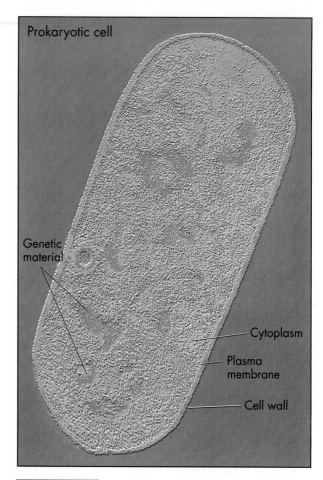

Prokaryotic cell

Genetic material

Cytoplasm

Plasma membrane

Cell wall

Figure 11.3 **Prokaryotic Cell.** A greatly magnified view of a *Clostridium perfringens* cell reveals the relatively sim-ple structure of a prokaryotic cell. This organism can cause food and blood poisoning.

envelope and other membrane-enclosed organelles. A prokaryote's chromosome is a circular strand of DNA that is loose, not tightly complexed with spool-forming proteins like eukaryotic chromosomes (see Fig. 6.8).

Plasma Membrane

In prokaryotic cells, regions of the plasma membrane carry out roles accomplished by specific organelles in eukaryotic cells. Cellular respiration, for example, takes place on a part of the plasma membrane that folds in-ward toward the cell's interior, rather than in a mito-chondrion. Likewise, photosynthesis takes place on infolded membranes instead of in separate chloroplasts. The membranes themselves are different, too; in Ar-chaea, the membrane lipids link differently to the three-carbon head of the molecule (review Fig. 2.18).

Cell Wall

In prokaryotes, the cell wall performs functions achieved by the eukaryote's cytoskeleton (see Fig. 2.26). In species within the domain Bacteria, the prokaryotic cell wall is a strong but flexible covering, made primarily of sugar-protein complexes called **peptidoglycans.** In so-called **gram-positive** organisms, these complexes occur in a single broad layer, while in **gram-negative** organisms, the peptidoglycan layer is covered by an outer sheet containing proteins and lipopolysaccharides (fat-sugar complexes). Species in the domain Archaea may have cell walls made of polysaccharide, protein, or glycoprotein, but they never contain true peptidoglycans.

Based on these cell wall variations and the differences they cause in permeability, gram-positive cells turn purple when exposed to a special stain called **Gram's stain,** while gram-negative cells turn reddish (Fig. 11.4). This staining technique is one of the first procedures a microbiologist will perform to help identify a newly found or unfamiliar prokaryote. Gram staining is important because it helps microbiologists classify prokaryotic organisms into groups of species. It's also important because gram-positive and gram-negative organisms are usually sensitive to different types of antibiotics (chemicals made by one microorganism that can slow the growth of or kill another microorganism). For example, one summer, dozens of delegates at an American Legion convention in Philadelphia developed fever, chills, and pneumonia. Biologists isolated a new species of bacterium from the sick Legionnaires and found that it was gram negative. The doctors knew that

Figure 11.4 **Gram Staining and Prokaryotic Cell Walls.** Danish bacteriologist Christian Gram developed staining procedures that differentiate prokaryotic cells into two classes based on their cell wall structures, and the walls' tendencies to absorb certain dyes. Gram-positive cells, such as the *Staphylococcus aureus* shown here, stain purple; gram-negative cells such as the *E. coli* shown here, stain red. The two species also differ in their sensitivities to various kinds of antibiotics.

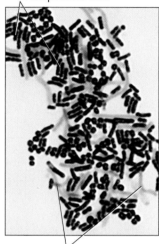

Gram-positive cells

Gram-negative cells

penicillin-type antibiotics kill gram-positive cells by blocking the growth of the cell wall but the drugs don't work on gram negatives. Fortunately, erythromycin-type antibiotics do kill gram negatives, again by blocking cell wall growth. Erythromycin turned out to be especially effective against Legionnaires' disease, and saved some patients' lives. The organism that causes anthrax — *Bacillus anthracis* — is also a gram positive bacterium that has figured prominently in news reports of bioterrorism.

Forms of Bacterial Cells

Prokaryotes are usually spheres (also called **cocci**), rods (called **bacilli**), spirals (**spirilla**), or curved, comma-shaped rods (called **vibrios**) (Fig. 11.5). A prokaryote's name will often give a clue as to its shape. For example, *Bacillus anthracis,* the cause of anthrax, is rod-shaped, while *Streptococcus pyogenes,* the pathogen that causes strep throat, is spherical, as is *Enterococcus faecalis,* which can cause infections of wounds and the urinary tract.

Figure 11.5 **Diversity of Shapes among Bacteria and Archaea.** Bacteria come in a wide range of shapes and sizes, including (a) round cocci, such as enterococci, and rod-shaped bacilli, such as *Bacillus subtilis*; (b) spiral-shaped spirilla, such as the spirochetes shown here that cause leptospirosis; and (c) comma-shaped vibrios, such as *Vibrio cholerae*, the agent that causes cholera. Note the whiplike flagella at the end of the cell in (c).

(a)

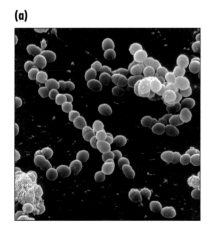

(b)

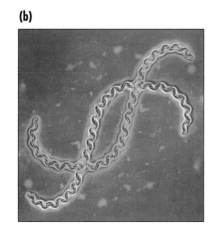

(c)

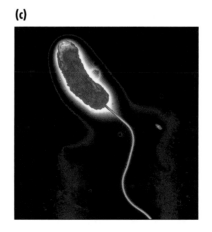

Nutrition in Prokaryotes

As a group, prokaryotes have a hugely diverse appetite. Prokaryotes can be **heterotrophs,** which consume a wide variety of organic nutrients made by other organisms, or **autotrophs,** which make their own organic nutrients from inorganic materials in the environment. Together, these two nutritional modes give prokaryotes great ecological versatility, enabling them to consume an enormous range of energy sources, including biological materials and organic substances such as methane; inorganic substances, such as hydrogen sulfide; industrial chemicals such as hydraulic fluids and toxic herbicides; and cancer-causing wastes, including vinyl chloride and polychlorinated biphenyls (PCBs). They can also live in air, soil, water, and ice; inside rocks miles under the surface of the Earth; and in the bodies of virtually all other life forms.

Most prokaryotes are heterotrophs, obtaining organic nutrients from the environment. Many are also **saprobes,** that is, they live on dead or dying organisms and act as decomposers. Without the waste-recycling activity of these prokaryotic decomposers (along with the fungi, which are multicellular saprobes), Earth would quickly accumulate a thick layer of fallen leaves, dead animals, and other organic matter that could choke out living things.

Thousands of prokaryotic species are **symbionts** (literally, "organisms that live together"), and reside on or inside other living organisms. Most pathogenic (disease-causing) bacteria are **parasites** that do harm to their hosts, but other species are neutral or beneficial to the organisms with whom they associate. The bacteria residing normally in your intestines are beneficial symbionts.

A few prokaryotes are autotrophic and produce their own organic compounds from inorganic precursors. These include **photoautotrophs,** which capture energy from light. Green and purple photosynthetic bacteria and cyanobacteria are photoautotrophs. Others are **chemoautotrophs,** which capture energy from certain chemicals. Sulfur bacteria and methane bacteria are chemoautotrophs that can obtain their carbon from carbon dioxide and their energy from the bonds of inorganic compounds, such as hydrogen gas, hydrogen sulfide, sulfur, iron, and certain nitrogen compounds. As we saw in Chapter 10, chemoautotrophs living around deep ocean hydrothermal vents may resemble some of the earliest living cells.

Reproduction in Prokaryotes

Prokaryotes usually reproduce by **binary fission,** simply dividing in half into two identical offspring. If the dividing cells remain connected, small clusters or long filaments can form (see Fig. 11.5a,b). Binary fission can be incredibly fast and efficient, with division occurring as often as every 20 minutes. If a single *E. coli* bacterium and all its progeny continued dividing every 20 minutes, after just 48 hours there would be a mass of bacteria 4000 times the weight of Earth. This doesn't happen, of course, because the availability of food, oxygen, and other resources limits their growth.

Although prokaryotes usually divide in half, they have several means of genetic recombination, including **conjugation,** the direct exchange of DNA through a strand of cytoplasm that bridges two cells, and indirect exchanges, such as transduction and transformation. **Transduction** is the transfer of genes from one bacterium to another via a virus, while **transformation** is the transfer of genes by the uptake of DNA directly from the surrounding medium. In Chapter 7, we saw how biologists use transformation to introduce recombinant DNA into foreign host cells (see Fig. 7.17). Just as sexual reproduction in plants and animals produces new genetic combinations, the above-mentioned prokaryotic processes produce new combinations that might give the cell a survival advantage when environmental conditions change. Transformation doesn't necessarily exchange DNA between members of the same species. In fact, as biologists have looked at whole genomes in several species of Bacteria, they've seen more and more cases in which a gene in one particular species is more closely related to an Archaeon gene or even a eukaryotic gene than to another kind of Bacteria. These situations are apparently cases of gene transfer between domains of life.

Many prokaryotes have yet another reproductive trick that enables them to survive unfavorable conditions. Some species such as those that cause botulism, gas gangrene, and anthrax form small, tough-walled resting cells called **endospores** (Fig. 11.6). Endospore formation is triggered by worsening environmental conditions. Endospores can survive extremes of heat, cold, drought, and even radiation for long periods, then when conditions improve, they grow into new bacterial cells. Endospores are the reason surgical instruments must be sterilized with high heat and pressure, and also why home-canned foods must be processed exactly the right way.

How Prokaryotic Cells Behave

Prokaryotes can do some amazing things. Although some can't move under their own power and must sim-

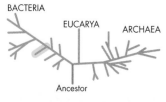

Figure 11.6 **Endospores: Survival Capsules.** Certain bacterial species can withstand extreme environmental conditions by forming tough-walled endospores, or resting cells. An endospore (circle at the right of this cell) has formed within this *Clostridium botulinum* cell—the cause of botulism food poisoning. Home food canners must apply the right amount of heat and pressure for enough time in order to kill all such endospores.

ply float along in the surrounding currents of water, blood, plant sap, or other fluid, many prokaryotes can propel themselves along gradients. The gradients can be of attractants (such as nutrients and light) or repellents (such as noxious chemicals), but they allow the cells to find and exploit appropriate food sources or avoid being eaten. Tests show that an *E. coli* cell can detect a change in nutrient concentration of just 1 part per 10,000. That's the equivalent of a person detecting the difference between a jar containing 10,000 marbles and a jar containing 10,001. Some prokaryotes, such as *E. coli,* have flagella (structurally different from eukaryotic flagella) that rotate like a propeller and push the cell toward a higher concentration of a nutrient. Some prokaryotes live in puddles where the nutrients sink to the bottom. A few of these organisms contain small crystals of iron oxide (magnetite) that act like little compasses and enable the organism to detect up from down and to move downward into the nutrient-rich sediments.

Let's look now at the many branches on the prokaryotic family tree.

The Domain Archaea

The organisms most similar to the earliest living cells are members of the domain Archaea. Most Archaea are **anaerobes,** cells that can't survive in the presence of oxygen. Many also inhabit Earth's most hostile environments and have pushed to the extreme the limits of a "livable" habitat. Biologists call such organisms **extremophiles**—extreme-loving organisms. Extremophiles can live in near-boiling water in hot springs, extremely salty lakes, and highly acidic or alkaline water and soils. Some Archaea also have weird biochemical properties. For instance, **methanogens** produce methane (CH_4, natural gas) as an essential component of their metabolism. Methanogens include the species *Methanobacterium ruminantium,* which inhabits a cow's digestive tract. Cows belch into the atmosphere the methane gas these prokaryotes produce and it actually contributes a huge volume of a greenhouse gas (see Fig. 25.27), adding to our global warming problem.

Diversity Among the Archaea

The domain Archaea consists of three main branches, which microbiologists call kingdoms. A lineage at the base (brown in Fig. 11.1, the Korarchaeota) is remarkable because biologists have yet to see a single living organism in this kingdom! We know them solely because biologists have obtained fragments of their DNA from samples of boiling hot springs in Yellowstone National

Park while looking for ribosomal RNA genes. As you can see from Figure 11.1, Korarchaeotans reside on some of the lowest branches on the tree of life and may be similar to the most recent ancestor of all surviving life forms on Earth.

Another kingdom within the domain Archaea (the Crenarchaeota) is a group of **thermophiles** (heat lovers) that use sulfur as an electron acceptor in their energy metabolism rather than oxygen (as most species do). The third kingdom of Archaea (the Euryarchaeota) includes several types of methane-generating organisms, such as the cow denizens just mentioned, and extreme salt-loving organisms or **halophiles.** The salt lovers, such as *Halobacterium halobium,* live in places like the Great Salt Lake and the Dead Sea, and in the salt flats that planes pass over when landing at the San Francisco Airport (Fig. 11.7). Interestingly, the red color of the salt pans is due to a red protein in the organisms that is similar to the light-detecting rhodopsin protein in our eyes (see Fig. 19.18).

While the Archaea are a diverse and ancient domain of life, species in the domain Bacteria rival their variety and are much better known.

The Domain Bacteria

By analyzing ribosomal RNA genes in many prokaryotes, biologists have determined that there are about 12 major evolutionary branches among the Bacteria (not all shown on Fig. 11.1). Microbiologists often call all 12 of these branches kingdoms. Standard reference books such as *Bergey's Manual of Systematic Bacteriology* categorize Bacteria by means of their form, physiology, and ecology, and thus have different numbers and sets of branches. Let's explore the most important groups of Bacteria looking at some of the 12 major branches based on RNA genes.

Primitive Bacteria

The lowest branches on the Bacterial tree of life—and thus the species that probably most resemble the last common ancestor of all Bacteria—are members of the *Aquifex* group. Like the most primitive Archaeans, Aquifex species are extremophiles. Several oxidize hydrogen or reduce sulfur in their metabolism. One genus, *Thermotoga,* lives in nearly boiling or merely scalding conditions. And another, *Deinococcus,* is resistant to high doses of irradiation. Since all these conditions were similar to those on early Earth, *Aquifex* species are probably close in many ways to very early life forms.

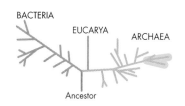

Figure 11.7 **Salt Pans and Salt-Tolerating Bacteria.** These salt pans south of San Francisco airport look ruddy because of the trillions of individual salt-loving (halophilic) archaebacteria that thrive in such highly saline water.

How do researchers use Gram staining?

Gram-Positive Bacteria

Emerging on a higher branch are the gram-positive Bacteria (all other branches are gram negatives). Gram positives include many medically important species, such as *Streptococcus mutans,* which causes tooth decay, and *Streptococcus pyogenes,* a cause of strep throat. The gram-positive *Staphylococcus aureus* (see Fig. 11.4) causes staph infections, such as toxic shock syndrome, a sometimes fatal vaginal infection that occurs when the bacteria multiply in and around a tampon. Fortunately, certain other gram-positive bacteria—the **actinomycetes**—produce more than 500 different antibiotics, natural substances that fend off competing microorganisms. Antibiotics produced by these cells include streptomycin, tetracycline, and neomycin.

Among the gram-positive Bacteria are the smallest free-living cells, the **mycoplasmas.** These simplified members of the domain Bacteria lack cell walls and live inside animals, plants, and sometimes other single-celled organisms. Only 0.2 to 0.3 μm long—20 would fit across the diameter of a red blood cell—mycoplasmas cause a mild form of pneumonia called "walking pneumonia" that usually affects young, otherwise healthy adults.

One type of gram-positive Bacteria, *Bacillus thuringiensis,* has been a major tool of plant researchers and may also provide a new means of controlling malaria. Belgian plant researchers discovered in the 1980s that *B. thuringiensis* contains a gene for a protein called Bt—a toxin that accumulates in tiny crystals and kills insects that consume them. By splicing the *Bt* gene into tobacco and corn plants with recombinant DNA methods (see Chapters 7 and 22), biologists were able to create plants with immunity to attack by hornworms and certain other insects.

Now, a Peruvian researcher has another potentially important application—fighting malaria—for a variety of the same organism, *Bacillus thuringiensis israelensis* H-14 (or Bti). Mosquitos carry the malarial protist in their digestive tracts (as we'll see in detail later). When a mosquito ingests Bti cells, the toxin they make binds to receptors on the cells that line the insect's gut, and pokes holes in them. Subsequent leakage causes gut cells to die, and the bacilli invade the mosquito's body cavity and kill it. The Bti variety does not harm people or other animals or even most insects (excepting mosquitos and other flies), because their gut cells lack the particular receptor. Dr. Palmira Ventosilla of Peru has developed a novel but simple way that people in small villages can culture Bti themselves and spread it on ponds where mosquito larvae grow (Fig. 11.8). She dis-

covered that coconut milk provides a terrific medium for culturing Bti, and she has taught villagers how to put a cotton swab soaked in living Bti cells through a hole drilled into a coconut, then reseal the hole with candle wax. Within a few hot tropical days, the bacilli have multiplied so rapidly that when the villagers sprinkle the contents of a few coconuts into a pond, most of the mosquito larvae developing in the stagnant water will die. Because mosquitos rarely fly more than two miles, this simple technique may help prevent malaria quite effectively in a local area.

Spirochetes

Bacteria with a distinctive spiral shape, the **spirochetes,** include the agents that cause Lyme disease and the sexually transmitted disease syphilis.

Figure 11.8   Bti and Coconuts: A Way to Stop Malaria? Cells of *Bacillus thuringiensis israelensis* (Bti) develop spores and produce crystals of a toxin that kills mosquitos and other flies. Here the Salitral-Piura family of Peru is innoculating coconuts with Bti. They will later spread the bacteria-laden coconut milk on ponds infested with the larvae of malaria mosquitos. This can decimate the local mosquito population and help prevent new cases of malaria.

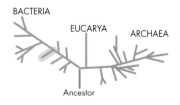

Culturing Bti in coconuts to control malaria

Chlamydia

Bacterial species of the genus *Chlamydia* cause the most frequent sexually transmitted disease in North America, called by the same name. **Chlamydia** live inside animal cells and are smaller than some viruses. They have no way of making their own ATP, and instead they must obtain energy from the cell they infect.

Cyanobacteria

The **cyanobacteria,** formerly called blue-green algae, are an environmentally important group. Like plants, cyanobacteria contain chlorophyll *a*, which absorbs light in photosynthesis. Many cyanobacteria can also fix nitrogen, that is, convert nitrogen gas (N_2) from the air (which is unusable for most organisms) to organic nitrogen compounds like ammonia (NH_3) for their own protein synthesis. This ability means that many cyanobacteria require only water, sunlight, a few inorganic nutrients, and carbon dioxide and nitrogen gases readily available in air. As a result, cyanobacteria can live almost anywhere: in fresh or salt water; on damp rocks; in soil; on tree trunks; and in hot, cold, or dry climates. Because cyanobacteria are so widespread, their ecological role is tremendous; they contribute millions of tons of oxygen to the atmosphere and biologically usable nitrogen and carbon to the environments and organisms around them.

One nitrogen-fixing species of cyanobacteria (an *Anabaena* species) grows symbiotically with a water fern (Fig. 11.9). Before planting rice, a farmer will allow dense blooms of the water fern to grow in the paddy. When the rice plants grow large, they crowd out the water fern as well as its nitrogen-fixing cyanobacterial ally. Nitrogen released from the dead ferns acts as a fertilizer for the developing rice crop and helps the farmer produce a high yield per acre without potentially toxic chemical fertilizers.

Molecular evidence shows that the chloroplasts in green plant cells have genes closely related to cyanobacterial genes (see Fig. 11.1). This, plus the fact that both have chlorophyll *a*, convinces many biologists that an ancient cyanobacterium probably entered into a symbiotic relationship with a eukaryotic cell hundreds of millions of years ago and evolved into chloroplasts in green algae (see Fig. 10.10). These green algae, in turn, eventually evolved into plants.

Proteobacteria

The largest and most diverse group in the domain Bacteria are the **proteobacteria.** These include purple bacteria such as *Rhodospirillum.* Many of these bacteria are actually purple in color, and they carry out photosynthesis with chlorophyll pigments very different from those in plant cells. Other members of the proteobacteria are heterotrophs, including the common intestinal bacterium *E. coli*; the *Rhizobium* species that fix nitrogen in peas and beans; and the bacterium that causes Legionnaires' disease. This group also includes **rickettsias,** tiny, rod-shaped parasitic bacteria transmitted by ticks, fleas, and lice and the cause of typhus fever, Rocky Mountain spotted fever, and other serious infections.

Some ancient purple bacterium probably gave rise to mitochondria in eukaryotes (review Fig. 10.15). Researchers have exploited this fact to help people with malaria. The antibiotic tetracycline blocks ribosome activity in members of the Bacterial domain, but not in eukaryotes. Doctors have found that tetracycline helps malaria patients when given along with quinine or other malarial drugs, even though the malarial parasites are protists and therefore eukaryotes. Tetracycline probably interferes with ribosomes inside the parasite's mitochondria—those ribosomes descended from some original, ancient purple bacterium.

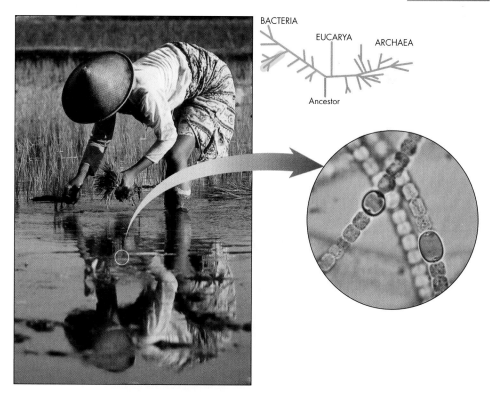

Figure 11.9 **Cyanobacteria.** Cyanobacteria often grow as filaments, with an occasional thick-walled cell (a heterocyst) that shuts out oxygen and allows nitrogen fixation to take place. Asian rice farmers fertilize their crops naturally with cyanobacteria that grow symbiotically with a water fern. When the plant dies, it returns nitrogen to the fields.

11.2 Noncellular Biological Entities

Most of the organisms we've discussed so far consist of cells. In Chapter 2, however, we encountered the HIV virus, a noncellular agent capable of causing AIDS. This section explores biological entities, including viruses, virions, and prions, that aren't organized as cells and aren't alive by the characteristics of the living state we described Chapter 1. Instead, they are nonliving parasites on and in living cells that probably evolved from them.

Viruses: Infectious Genes with a Coat

Viruses are geometric packages of genes that are 1,000 to 10,000 times smaller than the cells of *B. thuringiensis* or other prokaryotes. A virus particle is, in effect, a minute package of DNA or RNA surrounded by a protein coat, or **capsid,** and occasionally by other materials (Fig. 11.10 and Fig. 2.1). Viruses infect cells by attaching part of the capsid to the cell's exterior wall or plasma membrane and then either entering the cell or simply injecting their DNA or RNA into the cell, leaving the capsid outside. Once inside, the viral genes commandeer the cell's protein-synthesizing machinery, sometimes stopping production of cellular proteins entirely and pre-empting the machinery for the production of new virus particles. Eventually, the cell may burst and die, or may survive and gradually release thousands of viruses that can then infect other cells. Most biologists consider viruses to be nonliving because these agents lack the machinery for self-reproduction or metabolism.

There are hundreds of kinds of viruses, many causing plant and animal diseases such as gastroenteritis, her-

pes, mononucleosis, smallpox, measles, rubella, yellow fever, the common cold, influenza, poliomyelitis, mumps, rabies, some cancers, and, as we saw, AIDS— all told, a scary list. Some viruses, like the rhinoviruses that cause colds and the influenza viruses that cause flu, are temporary parasites. We encounter them in sneeze droplets or mucus sprayed into the air by cold or flu sufferers, or we pick them up from contaminated hands or surfaces, and they begin to multiply in our body cells. Eventually, though, our immune system destroys them.

Other types of viruses, however, like herpes simplex I, which causes cold sores, and herpes simplex II, which causes genital herpes, insert their DNA (or DNA copies of their RNA) into the chromosomes of nerve cells or other body cells and take up permanent residence. There they lie dormant, erupting and causing symptoms only occasionally when triggered by fever, sunlight, or other environmental stimuli.

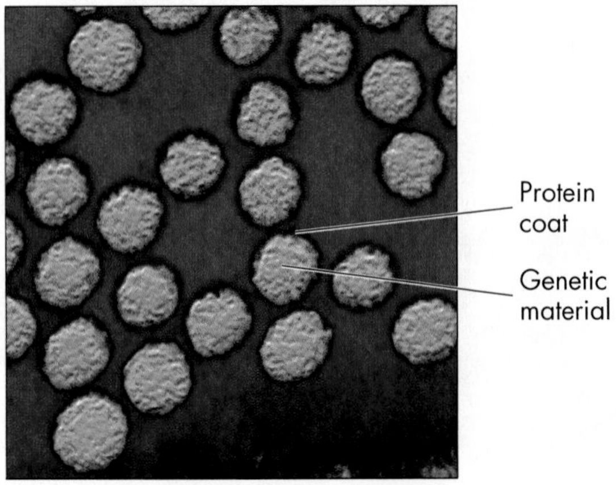

Treating Viral Disease

Because viruses aren't cells, they aren't killed by the antibiotics that control most disease-causing prokaryotes. Right now, in fact, there are really no medical cures for any viral diseases. One of the biggest challenges in modern medicine is to develop drugs that can fight viruses. One drug, called **acyclovir,** helps reduce the severity of herpes viral outbreaks. By blocking the activity of certain herpes viral enzymes, acyclovir can help control shingles, cold sores, and genital herpes. The latter, which is sexually transmitted, affects about 20 million Americans. Researchers hope that, within a decade, they'll have many drugs like acyclovir to control viruses.

The Origin of Viral Diseases

Where do new viral diseases come from? Researchers think many new human viruses arise in animals. The AIDS virus, for example, probably attacked primarily African monkeys until the continual advance of our species into formerly untouched or little-used wilderness areas brought about increased contact with monkeys and their blood. Long-distance air travel was a factor, too; once the virus infected humans, it spread fairly quickly around the world.

There are other examples of this "new virus" phenomenon. For example, Asian influenza A virus killed 20 to 30 million people around the world in 1918. The virus lives in ducks and other birds, and was apparently transmitted from birds to barnyard pigs and then to people. In another case, several people in mainly southwestern states died of a mysterious respiratory virus in

Figure 11.10 **The Polio Virus.** A protein coat surrounds the hereditary material—an RNA molecule—of the polio virus. This virus attacks nerve cells that normally stimulate muscle contraction. In poliomyelitis, a flulike disease progresses into muscle paralysis as the virus infects motor nerves.

Protein coat

Genetic material

(a)
Normal cellular
prion protein

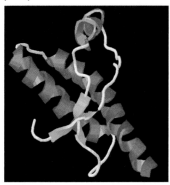

(b)
Disease-causing form
of prion protein

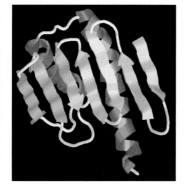

(c)

Figure 11.11 **Infectious Particles: Viroids and Prions.** Viroids are parasites made up of nothing but short RNA molecules that infect crop plants. (a,b) Prions are the proteins that cause mad cow disease and related neurological conditions. (c) Because the beef from affected cows can pass along prions to meat eaters, many European consumers have become concerned about the safety of their meat supply.

the summer of 1993. Researchers eventually identified it as the deadly Hanta virus, which is borne by rodents, is now a human pathogen, and is spread by contact with deer mouse feces.

Viroids: Infectious RNAs

The **viroids** are parasites that live inside plant cells and consist of an RNA molecule only about 300 nucleotides long, far smaller than the smallest virus. Viroid RNAs lack a protein coat. They bind to RNAs in the host cell's ribosomes and disrupt the cell's protein synthesis. Some diseases of potatoes, cucumbers, citrus trees, and artichokes stem from viroid infection.

Prions: Infectious Proteins

While the viroids consist of a single molecule (an RNA), **prions** also consist of a single molecule, but one of protein (Fig. 11.11a,b). Prions are the smallest and strangest disease-causing agents that can be transmitted from one animal to another, and they aren't well understood. They appear to lack genetic material entirely, consisting of nothing but a specific protein molecule. Yet prions are implicated in serious nerve and brain diseases, including scrapie in sheep and goats; mad cow disease in cattle (also called bovine spongiform encephalopathy); and in people, the very similar diseases kuru, Creutzfeldt-Jakob disease, and a human version of mad cow called sporadic Creutzfeldt-Jakob or variant Creutzfeldt-Jakob. Biologists are not sure how prions reproduce or cause disease, and the main protection against them now is to avoid eating animal matter that might harbor prions (Fig. 11.11c).

11.3 The Protists: Single-Celled Eukaryotes

We've been exploring single-celled organisms and their place on the universal tree of life. So where do the malarial parasites and their relatives fit in? An electron micrograph of a malarial parasite living inside a human red blood cell shows that this interloper, too, is made up of a single cell (Fig. 11.12). Furthermore, that individual parasite contains a number of complex or-

EUCARYA

BACTERIA

ARCHAEA

Ancestor

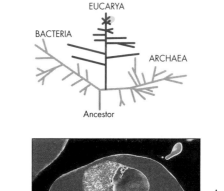

Red blood
cell
cytoplasm

Apical
complex

Nucleus

Figure 11.12 **The Malarial Parasite: Which Kingdom?** This electron micrograph of two merozoites (blue and pink) of the malarial parasite *Plasmodium falciparum* inside a human red blood cell shows that the cause of malaria is a single-celled organism with complex intracellular organelles, including an apical complex and a nucleus. These characteristics place it in the Kingdom Protista, Domain Eucarya.

Figure 11.13 **Protists and the Tree of Life.** The tree of life based on ribosomal RNA sequences shows that protists and other organisms in the Domain Eucarya are more closely related to prokaryotes in the Domain Archaea than they are to cells in the Domain Bacteria. Note the relationships of various protists to each other and to plants, animals, and fungi.

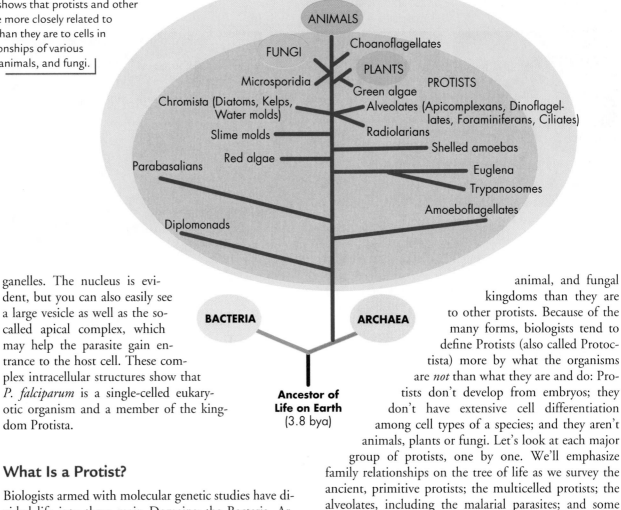

ganelles. The nucleus is evident, but you can also easily see a large vesicle as well as the so-called apical complex, which may help the parasite gain entrance to the host cell. These complex intracellular structures show that *P. falciparum* is a single-celled eukaryotic organism and a member of the kingdom Protista.

What Is a Protist?

Biologists armed with molecular genetic studies have divided life into three main Domains: the Bacteria, Archaea, and the Eucarya (Fig. 11.13 and Chapter 10). The most familiar organisms—the lively animals, green plants, and fungi—occupy just a small area near the top of life's tree. The vast majority of living species and most of life's genetic diversity lie with the microscopic, single-celled groups. We've just seen the prokaryotes in Domains Bacteria and Archaea. The remaining microbes are protists in the Kingdom Protista, Domain Eucarya. The structure of the malarial parasite and its molecular characteristics place it squarely among the protists. So let's first explore that kingdom and all its variety, then the specific organisms behind malaria.

The plants, animals, and fungi occupy individual and distinct branches on the tree of life. The protists, however, occupy numerous branches on the tree, and are beautifully complex—but for the most part invisible—organisms (Explorer 11.1). Some protists, however, are more closely related to members of the plant,

animal, and fungal kingdoms than they are to other protists. Because of the many forms, biologists tend to define Protists (also called Protoctista) more by what the organisms are *not* than what they are and do: Protists don't develop from embryos; they don't have extensive cell differentiation among cell types of a species; and they aren't animals, plants or fungi. Let's look at each major group of protists, one by one. We'll emphasize family relationships on the tree of life as we survey the ancient, primitive protists; the multicelled protists; the alveolates, including the malarial parasites; and some complex but unrelated groups. Explorer 11.1 illustrates them. But for each type, we'll also discuss the way they obtain their nourishment and how they impact natural environments.

Ancient Protistan Lineages

Recall the endosymbiont hypothesis from Chapter 10. This is the idea that the first eukaryotic cells may have arisen after the nucleus evolved (probably from an infolding of the cell membrane) and after a simple bacterium took up residence and led to mitochondria. Not long after that, ancestral eukaryotes began to live in the primordial seas and a number of lineages of protists diverged from them and diversified. Because the ancestral eukaryotes occupy very low branches on the tree of life, taxonomists call them "basally diverging" protists. A number of these single-celled organisms have living members that lack mitochondria, but recent work shows that they descended from predecessors that did possess the powerhouse organelles. These primitive protists include some important causes of human and animal disease, and perform important ecological services.

A common human parasite, now a scourge to campers and others who drink untreated water, is *Giardia lamblia. Giardia* can multiply quickly and blanket the inner surfaces of the intestines, causing severe diarrhea. Notice *Giardia's* long flagella (Fig. 11.14c), which help propel the cell through water or body fluid, and its one flat side, which helps it stay stuck to the intestinal wall.

Diverging from the tree a bit higher than the **Diplomonads,** the group that includes the *Giardia* lineage, are the **amoeboflagellates.** Members of this group live in water and soil, and usually display **pseudopodia** (literally, "false feet"), limblike cellular extensions that help the organisms move and feed. Pseudopodia can project outward in any direction, pulling the cell along.

These extensions can also stab prey so that the amoeboflagellate can then engulf the hapless cell by phagocytosis (see Fig. 2.20). Some amoeboflagellates, such as *Naegeria,* can also sprout whiplike flagella at certain stages of the life cycle, which emerge like the wheels a plane lowers at landing.

Another line of ancient, primitive protists, the **Parabasalians,** includes members that give termites their ability to digest wood. Termites chew wood into small pieces that move down the digestive tract and form a slurry with the liquid inside the insect's gut. If you gently press a termite's abdomen, the insect will extrude some of this slurry, and if you look at it under the microscope you will see very large, complex, single-celled protists such as *Trichonympha* (Fig. 11.14a). Each such cell gliding within the slurry takes in tiny wood particles via phagocytosis. But like nested Russian dolls, one inside the other, *Trichonympha* cells themselves harbor even smaller cells: bacteria that produce and secrete cellulase, an enzyme that breaks down the cellulose in

What makes Plasmodium falriparum *such a dangerous pathogen?*

(a) *Trichonympha, a Parabasalian*

(b) *Trypanosome, a Kinetoplastid*

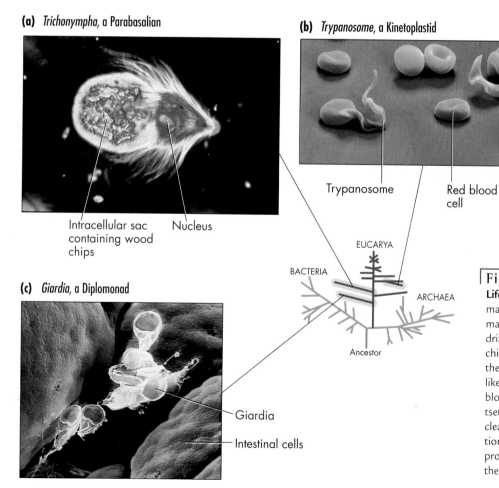

Intracellular sac containing wood chips

Nucleus

Trypanosome

Red blood cell

(c) *Giardia, a Diplomonad*

EUCARYA

BACTERIA

ARCHAEA

Ancestor

Giardia

Intestinal cells

Figure 11.14 **Protists That Diverged on "Low Branches" of Life's Tree.** (a) A *Trichonympha* cell (magnification 133✕), with its mane of flagella. The central, spherical, envelope-enclosed nucleus marks this organism as a eukaryote, even though it lacks mitochondria. An intracellular sac in the cell (left) is filled with tiny wood chips that the termite host rasped from a dead tree. Bacteria within the sac secrete digestive enzymes that break down cellulose. (b) Eel-like trypanosomes (magnified 4000✕) wriggle among a human's red blood cells. The parasite enters the bloodstream with the bite of a tsetse fly, then invades the brain and spinal cord. (c) A drink from a clear-looking but contaminated mountain stream can lead to infection by *Giardia lamblia* (magnified about 1000✕). This flagellated protist causes severe abdominal gas, cramping, and diarrhea. Notice the small size of intestinal lining cells by comparison.

wood into glucose subunits (see Fig. 2.8). The sugar molecules fuel the synthesis of amino acids and other compounds in both the bacteria and the host *Trichonympha.* The termites absorb some of the sugars, too, and that's how the insects digest and live off of wood, even though they don't have their own cellulose-digesting enzymes. Some other parabasalians are pathogens, including *Trichomonas vaginalis,* cause of a common sexually transmitted disease.

Two related groups of primitive protists are **Kinetoplastids** and **Euglenoids,** each with long whiplike flagella that move the cells. The protistan flagellum has the same basic internal structure and the same protein (tubulin) as in a human sperm cell's lashing tail (see Fig. 8.7). A kinetoplastid called *Trypanosoma brucei gambiense* causes African sleeping sickness (Fig. 11.14b). Trypanosomes are spread through the bite of the tsetse fly and can inflict lethal damage on domesticated animals as well as on human beings. They are dangerous pathogens because they can evade the host's immune system by changing their surface coats—a trait we'll encounter again with the malarial parasite.

Euglenoids are named after *Euglena,* from the Greek words for "good eye." Indeed, each of the many green, spindle-shaped euglenoids has an eyespot, or **stigma,** as well as chloroplasts that carry out photosynthesis. The stigma shades a light receptor and allows the aquatic cell to swim toward light; this maximizes its photosynthesis. Genetic evidence suggests that during evolution, the photosynthetic bacteria that gave rise to *Euglena's* chloroplasts were different from the bacteria that gave rise to plant cell chloroplasts. Some biologists see this multiple evolution of chloroplasts as evidence that symbiotic unions took place often during evolution.

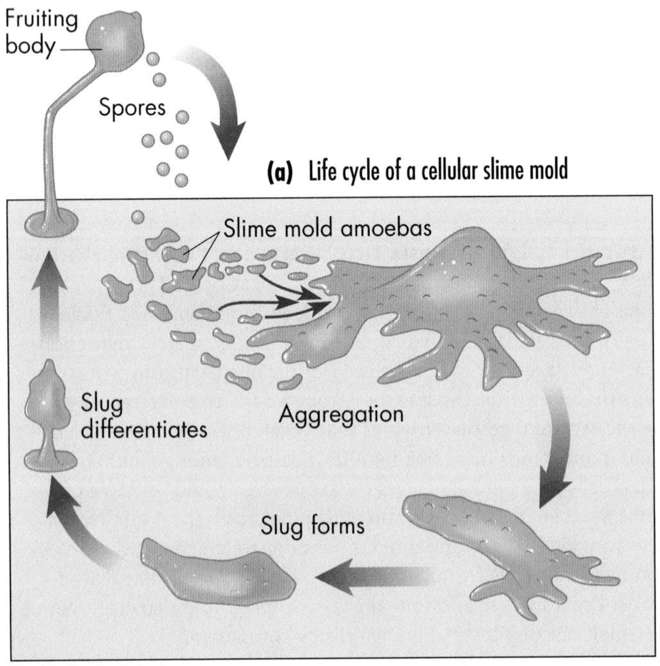

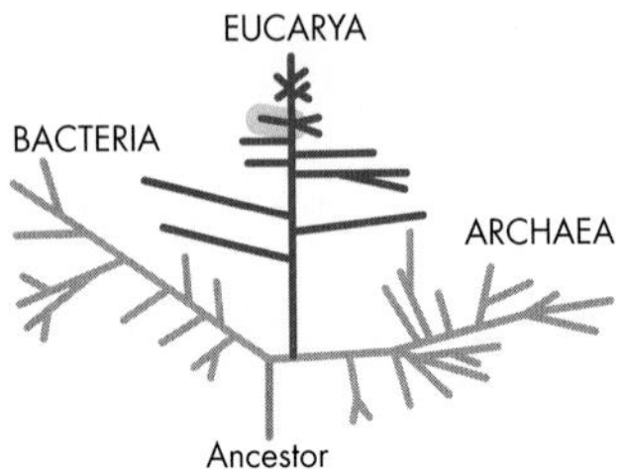

Multicellular Protistans Evolve

Most protists are single-celled organisms—but not all. Some protists stemming from the first primitive eukaryotes evolved a multicellular body form (or aspects of it) completely independently of the many-celled plants, animals, and fungi. The most common are the slime molds and the red algae.

Slime Molds

The glistening slime molds harvest food and energy by secreting digestive enzymes onto organic matter and then absorbing the digested material back into the cell. This dining habit is very similar to the way mushrooms and other fungi "eat." Slime molds, however, are not closely related genetically to mushrooms or other fungi.

A **true slime mold** exists as a flat, often huge, fan-shaped mass of cells up to 30 cm (12 in) in diameter. This mass moves about slowly like a giant amoeba in damp soil, leaf litter, and fallen trees. The mass is called a **plasmodium** (which you'll notice is the same name as the malarial parasite's genus, even though the two organisms have little in common). The true slime molds have just one enormous plasma membrane surrounding thousands of diploid cell nuclei. Amazingly, all of these nuclei divide simultaneously during mitosis.

The **cellular slime molds** are also amoebalike organisms that glide along as individual cells. They actively engulf and consume bacteria rather than secreting digestive enzymes onto organic matter. When food is scarce, the individual cells of a cellular slime mold join together and send up brightly colored, multicellular **fruiting bodies,** reproductive structures that can look like golf balls on bent tees (Fig. 11.15). Spores are produced in the balls, and these are scattered by wind or rain and germinate in new locations, dispersing the species. Ironically, recent molecular evidence suggests that cellular slime molds, including the much-studied *Dictyostelium discoidum,* are more closely related to the fungi than they are to many other protists.

Figure 11.15 **Slime Molds.** True slime molds exist as huge cells with thousands of nuclei, but form multicellular fruiting bodies. Cellular slime molds are normally amoebalike single cells. (a) Under poor environmental conditions, cellular slime molds aggregate and differentiate into multicellular fruiting bodies with stalks and spore chambers that eventually release new, genetically recombined spores. These, in turn, germinate into new amoebalike cells that individually crawl off in search of bacteria to consume. (b) Closeup of a *Dictyostelium* slime mold's single flamelike fruiting body.

(a) Life cycle of a cellular slime mold

Fruiting body

Spores

Slime mold amoebas

Slug differentiates

Aggregation

Slug forms

(b) Fruiting body

EUCARYA

BACTERIA

ARCHAEA

Ancestor

Rhodophyta: Red Algae

You may be surprised to find algae in a chapter about microbes rather than one on plants. But the term **alga** (pl., *algae)*, comes from the Latin word for "seaweed," and it has been used to describe simple chlorophyll-containing organisms that live in water, whether the organisms are true plants or not. Most **red algae** are small, delicate organisms that occur as thin filaments or flat sheets of many cells with an ornate, fanlike appearance (Fig. 11.16); some, however, are single-celled or grow in colonies of many cells. You can see from the branches of the eukaryotic tree of life that red algae are only distantly related to plants. Red algae generally live in shallow, tropical ocean waters, but a few species survive at depths of about 270 m (880 ft). These deep-sea denizens have reddish pigments in their cells that absorb light in the blue-green range—the only wavelengths that can penetrate to such great ocean depths. The red pigments absorb light energy and pass some of it to chlorophyll for use in photosynthesis (see Chapter 3). Red algae have several commercial uses. Nori, a seaweed, is a common ingredient in Asian cooking. Red algae produce a gel-like protein that forms the rubbery base of the laboratory growth medium called agar. Red algae also produce the starchy thickener called carrageenan, which chemists include in some cosmetics and paints, and which you might have noticed in the ingredient lists for some ice creams and puddings.

Complex Eukaryotes: Top Branches in the Eukaryotic Tree

We return, now, to the protists that cause malaria, as well as to a number of other complex, beautiful protists that branched off later on the tree of life than the groups we just explored.

Alveolates

Malarial parasites are **alveolates,** a recently recognized group of protists that also includes armored marine cells, the dinoflagellates and foraminiferans, and miniature predators, the ciliates. Alveolates are named for a common cell structure: a system of membranous sacs (or **alveoli**) under their plasma membranes. The cause of malaria, *Plasmodium falciparum*, is part of a subgroup of alveolates, the Apicomplexa.

Apicomplexa The malarial parasite, along with other species in the group called Apicomplexa, has a complex of rings and tubules at the cell's apex. You can see these

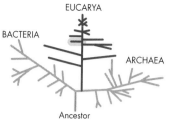

Figure 11.16 **Red Algae: Deep Dwellers.** Some red algae live in cold, inky waters at depths of 268 m (884 ft) below the ocean surface. This is due in part to the ability of their pigments to absorb the wavelengths of light that penetrate to these depths. This red alga is in the genus *Botryoglossum*.

structures in the malarial parasite in Figure 11.12. Apicomplexans also have a sporelike stage in which the cells lack any means of locomotion. You might think this would reduce a protist's ability to get around or infect hosts, but in fact, this stage helps *P. falciparum*. You can follow this protist's complex life cycle in Figure 11.17 and see an animation of it in **E**xplorer 11.2. It is truly remarkable how many different forms this single-celled organism takes as it moves from a mosquito's midgut to its salivary gland, then into a person's bloodstream. From there it enters liver cells, multiplies, then enters red blood cells and consumes the hemoglobin, usually destroying the red cell. Malaria is the most infamous apicomplexan, but other members of this group also cause serious human and animal diseases, including toxoplasmosis, the "cat box" infection that can cause a pregnant woman to deliver a malformed baby.

Dinoflagellates The apicomplexans are the most significant alveolates because malaria claims more human victims than any other infectious disease. But a second group of alveolates, including the **dinoflagellates,** occasionally has a large economic impact. Dinoflagellate means "spinning cell with flagella," and indeed, these marine protists have two flagella that cause them to spin as they swim. One flagellum winds about the middle of the cell like a belt in a groove and causes the spinning motion, while another projects backward in a second groove and propels the cell forward (Fig. 11.18a). Some dinoflagellates have internal plates of "armor," elaborately embossed vesicles filled with tough cellulose. Dinoflagellates can cause **red**

GET THE PICTURE

Explorer 11.2

Complex Life Cycle of the Malarial Parasite

Figure 11.17   **The Complex Life Cycle of Malarial Protists.** (Step 1) A tropical mosquito can inject the agent of malaria, *Plasmodium falciparum,* into a person's blood. (Step 2) The protist enters a person's liver cells, where they change to a different, dangerous phase (Step 3). (Step 4) Then the parasites reinvade the bloodstream and quickly enter red blood cells, where they multiply (Step 5). Every 48 hours, parasites exit red blood cells and infect new cells. (Step 6) Eventually, the parasite makes gamete-forming cells, which a mosquito may ingest when she sucks blood from a person with malaria (Step 7). The gamete-forming cells enter the mosquito's gut (Step 8), and form zygotes (Step 9), which develop further (Step 10), and once again enter the salivary glands (Step 11). The next time this mosquito feeds on a person, she may inject the new crop of sporozoites, infecting this person with malaria (Step 1).

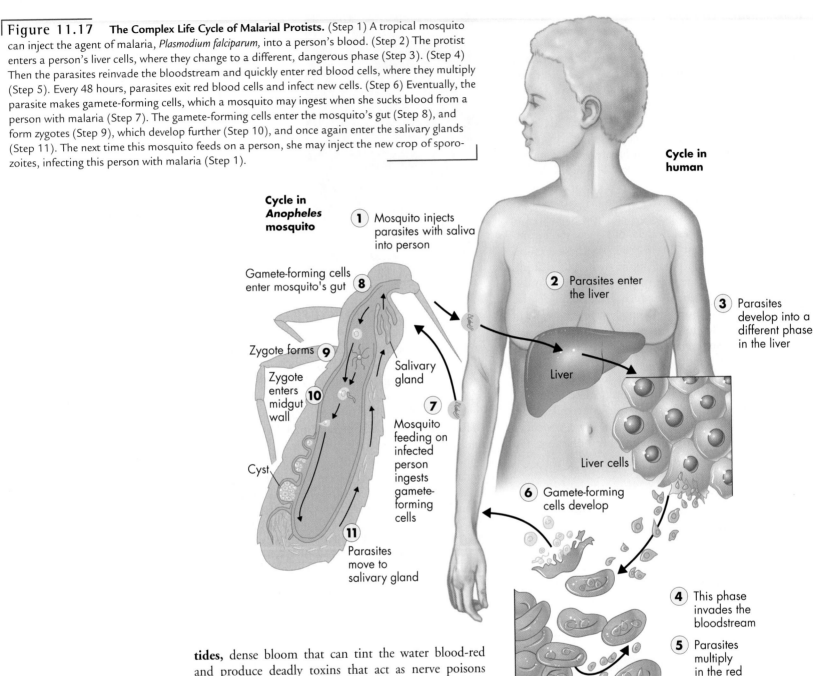

Cycle in human

Cycle in *Anopheles* mosquito

1 Mosquito injects parasites with saliva into person

2 Parasites enter the liver

3 Parasites develop into a different phase in the liver

Gamete-forming cells enter mosquito's gut 8

Zygote forms 9

Zygote enters midgut wall 10

Salivary gland

7 Mosquito feeding on infected person ingests gamete-forming cells

Cyst

Liver

Liver cells

6 Gamete-forming cells develop

11 Parasites move to salivary gland

4 This phase invades the bloodstream

5 Parasites multiply in the red blood cells

tides, dense bloom that can tint the water blood-red and produce deadly toxins that act as nerve poisons (Fig. 11.18a). The poisons build up in fish and shellfish, killing them and poisoning people who gather and eat them. Government agencies in coastal areas of North America often ban collection and consumption of shellfish during May through August, when red tides can occur.

Foraminiferans   The third group of alveolates are the **foraminiferans,** delicately shaped protists that live in the oceans and secrete whitish, calcium-based shells (or "tests") that can look like spiral seashells or cham-

bered nautiluses (Fig. 11.18b). These protozoa are so plentiful that when the organisms die, their shells pile up and contribute to forming thick limestone and chert deposits, including England's famed White Cliffs of Dover. The miniature shells often appear as fossils in rock layers and have contributed to our knowledge of Ice Ages and even of the current global warming: Geol-

(a) Dinoflagellate cells

(b) A foraminiferan shell

Figure 11.18 **Alveolate Diversity.** (a) Dinoflagellates have flagella in grooves along the long "arms"; each cell forms a coat of cellulose armor. Some dinoflagellates cause dangerous red tides, dense population explosions that tint the water red and produce deadly toxins. (b) Fossilized microscopic foraminiferan shells reveal details about past climates. (c) A single-celled hunter, the juglike ciliate *Didinium* (*left*), swallowing its prey, a slipper-shaped *Paramecium* cell (magnification 600×).

(c) Ciliates – the hunter and the hunted

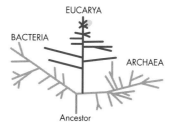

ogists have measured oxygen isotopes in foraminiferan shells and used them to reconstruct global temperature changes over the last 100,000 years.

Ciliates The fourth group of alveolates is the **ciliates,** some of Earth's most complex single-celled organisms. Ciliates are covered by hundreds of short, eyelash-like organs of locomotion. Cilia bend with coordinated, oarlike movements, propelling the cell or helping to sweep food particles into its gaping "gullet" (Fig. 11.18c). Ciliates have several organelles with functions analogous to an animal's organs: Anal pores discharge wastes. Microfilaments and microtubules support and contract like bones and muscles. Tiny toxic darts (**trichocysts**) spear and paralyze prey. Food vacuoles are filled with enzymes and function as digestive organs. Each ciliate also has a giant nucleus (biologists call it a polyploid **macronucleus**), containing many sets of chromosomes, that directs cell activities. One or more small diploid nuclei (**micronuclei**) undergo meiosis and are exchanged during sexual reproduction.

Radiolarians and Chromista

We've explored the alveolates, including the wily malarial parasites. However there are two more lineages of protists that branched off near the top of the eukaryotic family tree, but did not lead to multicelled plants, animals, or fungi as did the other groups we'll explore shortly. **Radiolarians** are single-celled protists that produce beautiful silicon-based shells that look like miniature glass ornaments (Fig. 11.19). The many shapes of radiolarian shells provide clues that paleontologists have learned to read when judging rock ages.

As their name suggests, the **Chromista** (also called Stramenopila) are mainly colored protists: most contain golden, brownish, and greenish pigment

molecules that help them to photosynthesize. Diatoms, brown algae, and downy mildew are all Chromista.

Diatoms **Diatoms** are the most common Chromista and perhaps the most exquisitely beautiful of the **phytoplankton** (literally, "floating plants"). Along with their chlorophyll pigments, most diatoms also have carotenoid pigment that gives them a golden color. Their cell walls contain silica (the main component of window glass) instead of cellulose (the main compound in plant cell walls). Diatoms also store oils rather than starchy compounds. Many people recognize diatoms because of their jewel-like, glassy shells (Fig. 11.20a). These shells fall to the ocean floor and accumulate in crumbly white sediments called **diatomaceous earth,** which manufacturers use in toothpaste, swimming pool filters, and other applications. Living diatoms are so

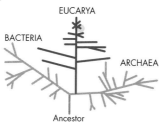

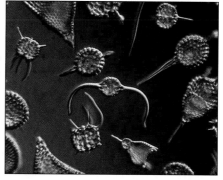

Figure 11.19 **Radiolarians.** Radiolarians, such as this mixed group of species, have intricately patterned, glassy shells.

(a) A diatom shell

(b) A kelp forest

(c) Water molds caused the Irish potato famine

Figure 11.20 **Chromista: Colored Protists.** Chromista is a recently recognized group of protists from high branches on the tree of life. (a) Diatoms, prominent components of phytoplankton, form glassy shells (magnification 4600×). (b) Here, sunlight slants below the ocean surface at Catalina Island, off the California coast near Los Angeles. The light illuminates tall, fluttering kelps, which tower like forest trees. (c) This Irish potato farmer lives in the Connemara region of County Galway. He probably had ancestors who survived the great potato famine of the mid-19th century, caused by a water mold.

abundant and ubiquitous that some biologists estimate they may contribute more oxygen to the atmosphere than all land plants combined.

Brown Algae Another type of Chromista are the **brown algae,** multicellular inhabitants of cool, offshore waters. Most members of the group are small, however the **kelps** are brown algae that can grow 100 m (over 325 ft) long and float vertically like tall trees (Fig. 11.20b). Huge floating masses of the brown alga *Sargassum* thrive in the Sargasso Sea, a mass of still water in the Atlantic Ocean north of the Caribbean Sea. These kelp forests provide important breeding grounds and hiding places for many fish and shellfish.

Brown algae range from golden brown to dark brown to black. They also contain green chlorophylls *a* and *c*, but the golden-brown carotenoid pigment colors the cells. Carotenoids also allow them to collect the blue and violet wavelengths of light that penetrate medium-deep water and thus to exploit an ocean environment too dim for many other organisms.

Kelps and other brown algae often look surprisingly like plants because they have body parts analogous to parts of land plants: Leaflike **fronds** collect sunlight and produce sugars; the stemlike **stipe** supports the organism vertically; and the rootlike **holdfast** anchors the organism to submerged rocks. Special tubelike conducting cells carry sugars produced in the fronds to the deeper

parts of the kelp. The tubes function like the specialized tissues that transport materials inside land plants, but they're not related. In fact, molecular genetic analyses show, surprisingly, that plants are more closely related to animals than they are to brown algae!

Water Molds Some chromista, such as the **water molds,** or members of the *Oomycota* (which means "egg mold"), lack colorful pigments and are not photosynthetic. Studies of their genetic material nevertheless show that they're closely related to other chromista. Water molds inhabit soil and water; some are single-celled, while others form fuzzy, branching filaments called **hyphae.** Like the true slime molds, water molds have several nuclei within a common cytoplasm and form large, immobile egglike cells, hence their Latin name. After fertilization, they form spores that disperse by swimming. One parasitic water mold struck Ireland in 1845–1847. This species causes **potato blight,** a disease that rapidly rots and kills growing potato vines (Fig. 11.20c). The Irish depended so heavily on potatoes for their daily diet that when potato blight struck, more than a million people starved, and 1.5 million more emigrated, mostly to North America.

Protists Closely Related to Plants, Animals, and Fungi

The plants, animals, and fungi, the three most familiar kingdoms, grace the top of the eukaryotic tree, being complex and multicellular and relatively recently evolved compared to many of the simpler forms we just looked at. But branching off from the same lineages that eventually gave rise to the plants, animals, and fungi are closely related protists with many similarities. The **green algae,** or Chlorophyta, are protists closely related to plants; the microsporidia are protists apparently closely related to fungi; and the choanoflagellates are protists closely related to animals. We'll consider the green algae with the plants in Chapter 12, because they appear to form a single branch on the tree of life, and

Figure 11.21 **Protists Related to Plants, Fungi, and Animals.** Ulva, a green alga undulating in a shallow tide pool, is in the protist group from which plants evolved (see Fig. 12.12). Choanoflagellates are the protist group most closely related to animals (see Chapter 13).

many botany courses cover them along with plants. The protists closely related to fungi and animals, however, we'll look at here.

Microsporidia

Microsporidia are among the simplest of eukaryotic cells. These tiny organisms live only inside animal cells. Microsporidia have among the smallest of eukaryotic genomes (smaller than many bacterial genomes, in fact), and they lack several organelles including mitochondria, stacked Golgi apparatus, and peroxisomes (small vesicles containing enzymes that break down peroxide, a byproduct of metabolism). Some biologists thought that microsporidia were eukaryotic relics, representing the primitive state before eukaryotes acquired mitochondria (review Fig. 10.10). However, by studying proteins from *Nosema locustae,* a microsporidium that lives in locusts, biologists have found the organisms to be closely related to fungi. Perhaps microsporidia lost many genes and cell organelles as an adaptation to the parasitic life style.

Choanoflagellates

The **choanoflagellates,** or "collar flagellates," are single-celled or colonial protists living in fresh water and in the oceans (see Chapter 13). The collar is formed by a ring of closely packed microvilli, small fingerlike cell

projections that encircle the flagellum. The wagging of the flagellum creates water currents that wisk prokaryotic cells into contact with the sticky collar; the cell then devours the ensnared bacteria by phagocytosis. A choanoflagellate looks very similar to the collar cells of sponges, the simplest animals (see Fig. 13.4). This, along with molecular genetic evidence, suggests strongly that choanoflagellatelike ancestors gave rise to all of the animals. Animals, of course, include humans subject to malaria—humankind's greatest medical scourge.

11.4 Fighting Humankind's Greatest Killer

Malaria is not only our species' biggest killer, it is one of our oldest recorded maladies, appearing in Hippocrates's medical journals in the 5th century B.C. Efforts to fight malaria have a long history, too (*E*xplorer 11.3). In the mid-1600s, the natives of Lima, Peru, were already using bark from a tree in the Andean cloud forests to treat the disease. Spanish diplomats returned the bark to Europe, where it became known as quinine and remained the principle treatment until World War II, when researchers developed chloroquine and other synthetic substitutes. Unfortunately, resistance to multiple malaria drugs has evolved in many populations of the deadly protist *Plasmodium falciparum* (Fig. 11.22a). Their major hosts, *Anopheles* mosquitos, also evolved resistance to DDT and other insecticides (Fig. 11.22b). As a result, the incidence of malaria began rising dramatically in the late 20th century.

We've seen how a detailed knowledge of the prokaryotes—specifically, *Bacillus thuringiensis*—has led to the simple coconut method of controlling mosquito larvae. Do you suppose there's a virus somewhere that could specifically kill either

READ ON

*E*xplorer 11.3

The Impact of Microbial Diseases on History

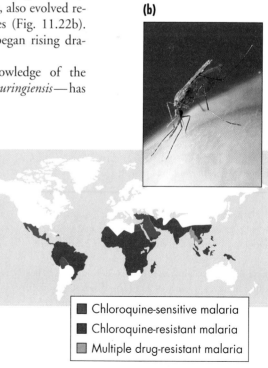

(b)

(a)

Figure 11.22 **Malaria, Mosquitos, and Hope Based on Research.** (a) A map showing the widespread status of drug-resistant malarial parasites. (b) An *Anopheles* mosquito, its gut filled with blood.

■ Chloroquine-sensitive malaria
■ Chloroquine-resistant malaria
■ Multiple drug-resistant malaria

the malarial parasite or malaria-transmitting mosquitos? How could biologists go about looking for those viruses ?

Researchers at laboratories and field stations around the world are traveling this and many other avenues in their search for drugs and preventatives. They're studying:

- How the malarial parasite disguises itself and evades our human immune system
- How it absorbs nutrients
- What the entire genomes look like in the parasites and mosquitos
- What kinds of vaccines could stymie such a genetically changeable protist
- Whether geneticists could engineer specialized mosquitos that can't transmit the protists

Dr. Daniel Goldberg has taken a different and very creative approach to the problem by studying aspects of the malarial parasite's metabolism. He found, for example, that *P. falciparum* can't make its own A and G bases for its DNA, and instead makes a transport protein that brings these bases into the cell. Importantly, the protist's transporters will carry other materials in, as well, and this may provide a "Trojan horse"—a clever way to sneak drugs into the parasite.

Goldberg has investigated another pathway in *P. falciparum,* as well. The protist dines almost exclusively on hemoglobin, gobbling the polypeptide chains and then detoxifying the leftover, potentially poisonous heme portion of the hemoglobin (containing the central iron group) in a special organelle. Dr. Goldberg discovered that antimalarial agents such as chloroquine actually control the parasites by blocking their detoxification of heme. Future drugs can target this same process—perhaps more effectively. Dr. Goldberg also discovered that the parasites have specific protein-digesting enzymes that have to act in a certain order to dismantle hemoglobin; without that precise order, they can't provide amino acids for the parasite's energy metabolism. These enzymes, therefore, could provide additional targets for new drugs: if you can stop the parasite's voracious digestion of hemoglobin, you can stop its damage to human red blood cells. Malaria research is one of the most active areas of biology and a subject to watch now that you understand the players, both eukaryotic and prokaryotic.

Chapter Summary and Selected Key Terms

Introduction
Mark Stover contracted malaria twice while serving as a volunteer teacher in rural Nigeria. He took preventative drugs and slept beneath netting and behind screens, but they didn't help. Mark's professor at Washington University, Dr. Daniel Goldberg, is an expert on malaria and has discovered how the parasite *Plasmodium falciparum* destroys human hemoglobin yet evades our nature immune defenses. This chapter describes the single-celled domains and kingdoms of life, and how knowledge of them can help fight malaria and other diseases.

Prokaryotes
These single-celled organisms lack organelles with membrane boundaries and occur in the Domains Bacteria and Archaea. They are assigned to many kingdoms within these domains. Prokaryotes have immense ecological impact as decomposers. They cause hundreds of animal and plant diseases, but harmless species also inhabit digestive tracts, produce vitamins and cofactors, and help prevent infections by pathogens. The food and chemical industries use many prokaryotes in their processes, and field researchers use some species for **bioremediation** (p. 279).

Prokaryotes have an outer cell wall surrounding a cell membrane and a cytoplasm that lacks organelles surrounded by membranes. Their single chromosome is a circular DNA molecule. Infolded regions of the plasma membrane carry out respiration and photosynthesis. The cell wall is made up of **peptidoglycans** (p. 281). **Gram-positive** cells (p. 281) have a single broad layer of this membrane; **gram-negative cells** (p. 281) have an additional outer layer of proteins and fat-sugar complexes. A procedure using **Gram's stain** (p. 281) differentiates between the two types and helps medical workers to identify species and to choose appropriate antibiotics.

Prokaryotes can be shaped as spheres (**cocci**); rods (**bacilli**); spirals (**spirilla**); or curved rods (**vibrios**) (p. 281). They can be nutritional **heterotrophs** (p. 282) or **autotrophs** (p. 282), **saprobes** (p. 282), **symbionts** (p. 282), or **parasites** (p. 282). **Photoautotrophs** (p. 282) capture and use light energy; **chemoautotrophs** (p. 282) capture energy from chemical such as CO_2, H_2, H_2S, and others.

Prokaryotes usually reproduce by **binary fission** (p. 282), but they can exchange DNA directly, through **conjugation** (p. 282), or indirectly, through **transduction** (p. 282) or **transformation** (p. 282). Some species like the cells that cause botulism can form **endospores** (p. 282) when environmental conditions worsen, and these spores can survive heat, drying, and radiation and then reform cells when conditions improve.

Members of the Domain Archaea are mostly **anaerobes** (p. 283) and **extremophiles** (p. 283) that evolved when Earth's environments were extreme. Some **methanogens** (p. 283), for example, can inhabit cows' guts and produce methane gas.

Three kingdoms within Archaea contain as yet undiscovered cells, known only by their DNA fragments. Some occur only in boiling hot springs, the **thermophiles** (p. 283), and use sulfur in energy metabolism; some are methanogens or **halophiles** (p. 283), which live in salty environments.

The Domain Bacteria has up to 12 main kingdoms. Primitive bacterial species in the *Aquifex* group are extremophiles, similar to early life forms. Gram-positive bacteria include *Streptococcus* and *Staphylococcus* species; **actinomycetes** (p. 284), which produce antibiotics; the smallest free living cells or **mycoplasmas** (p. 284); and "Bti," a killer of insect larvae that is helping control the spread of malaria.

Spirochetes (p. 284) include the spiral-shaped agents of syphilis.

Chlamydia (p. 285) live inside animal cells, can't make their own ATP, and cause a common sexually transmitted disease.

Cyanobacteria (p. 285) photosynthesize and fix nitrogen and thus can live almost anywhere with few resources. An ancient cyanobacterium probably entered into symbiosis with a primitive eukaryotic cell and evolved into chloroplasts in green algae, which, in turn, gave rise to plants.

Proteobacteria (p. 285) include purple bacteria such as *Rhodospirillum;* the intestinal bacterium *Escherichia coli;* nitrogen-fixing *Rhizobium;* and the **rickettsias** (p. 285), which cause typhus, Rocky Mountain spotted fever, and other serious diseases.

Non-Cellular Agents of Disease

Viruses (p. 286) contain DNA or RNA in a **capsid** (p. 286) and commandeer the protein-making machinery of cells. Biologists consider them to be nonliving and possibly devolved from ancient cells, and hundreds of kinds cause plant and animal diseases, including AIDS, herpes, colds, and influenza. Doctors have only a few antiviral drugs like **acyclovir** (p. 286) but are developing others.

Viroids (p. 287) are nothing but RNA molecules that live inside plant cells and cause disease.

Prions (p. 287) are nothing but protein molecules that cause scrapie, mad cow disease, and other nerve and brain diseases.

The Protists

Protists like *Plasmodium falciparum* are single-celled eukaryotes in the Domain Eucarya and the Kingdom Protista. Because protists are so varied in form, biologists define them by what they aren't: They are eukaryotes that are neither animals, plants, nor fungi; they don't develop from embryos; and they don't have extensively differentiated cells.

The ancient, primitive protists arising early, that is, on low branches of the tree of life, include *Giardia,* a **Diplomonad** (p. 289), which hikers sometimes contract by drinking untreated water, and **amoeboflagellates** (p. 289), with their limblike extensions or **pseudopodia** (p. 289). The primitive protists that give termites the ability to digest wood, *Trichonympha,* are **Parabasalians** (p. 289). Two more primitive groups include the **Kinetoplastids** (p. 290) such as the trypanosomes that cause African sleeping sickness, and the **Euglenoids** (p. 290) like the mobile green *Euglena* that have an eyespot or **stigma** (p. 290) directing them toward light for photosynthesis.

Some protists are multicellular, but this body form arose completely independently of animals, plants, and fungi. **True slime molds** (p. 290) are fan-shaped and move about like a giant amoeba with one huge plasma membrane encompassing an enormous cytoplasm and thousands of diploid nuclei. The mass is a **plasmodium** (p. 290), and it secretes enzymes that digest organic matter. An example is *Physarum.* **Cellular slime molds** (p. 290) are amoebalike, too, but move

along engulfing bacteria. An example is *Dictyostelium.* Both types send up reproductive structures called **fruiting bodies** (p. 290).

Some **red algae** (sing., **alga**; p. 291) are single celled and some are colonial but all are protists that can live at great ocean depths. Red pigments absorb wavelengths that penetrate deeply. Agar and carageenan come from red alga.

Several kinds of complex protists branched off later (higher) on the tree of life. **Alveolates** (p. 291) are protists with **alveoli** (p. 291) or membranous sacs under their plasma membrane. Alveolates include *P. falciparum* and other members of the **Apicomplexa** (p. 291), species with an apical complex of rings and tubules. Toxoplasmosis ("cat box" disease) is also caused by an apicomplexan.

Other alveolates are (1) the **dinoflagellates** (p. 291), flagellated, armored cells that cause **red tides**; (2) **foraminiferans** (p. 292), with their calcium-based shells that appear in fossils and have helped geologists reconstruct climate records; and (3) **ciliates** (p. 293), hunters covered with beating cilia and possessing toxic darts. A giant **macronucleus** (p. 293) directs cell activity, while several **micronuclei** (p. 293) undergo meiosis and are exchanged during reproduction.

Two other complex protistan groups branched off late and independently of others, the **Radiolarians** (p. 293) and the **Chromista** (p. 293). Radiolarians have delicate silicon-based shells that look like glass ornaments and pile up in thick layers, as seen in the famous White Cliffs of Dover. Chromista contain colored pigments. **Diatoms** (p. 293) are golden-brown **phytoplankton** (p. 293) with yellow and green pigments and silica shells instead of cellulose cell walls, as in plants. Sediments called **diatomaceous earth** (p. 293) are used commercially in toothpaste and swimming pool filters.

Brown algae (p. 294) are multicellular Chromista that include towering **kelps** (p. 294). Carotenoids absorb light at medium-deep ocean depths. Brown algae are plantlike with their leaflike **fronds** (p. 294), stemlike **stipes** (p. 294), and rootlike **holdfasts** (p. 294).

Water molds (*Oomycota;* p. 294) such as the cause of **potato blight** (p. 294) are also Chromista that form branching filaments or **hyphae** (p. 294) and live in wet places.

One last set of protists evolved relatively late and occupies high branches on the tree of life—branches that eventually gave rise to the plants, animals, and fungi. The **green algae** or Chlorophyta (p. 294) live in fresh water and shallow oceans, and contain the same pigments found in green plants. **Microsporidia** (p. 295) live only inside animal cells and lack several organelles. By studying *Nosema* (an internal parasite of locust cells) researchers have found microsporidia to be closely related to fungi. **Choanoflagellates** (p. 295) have cells with a collar and flagella like the collar cells

of sponges and ancestors like them probably gave rise to all animals.

Fighting Malaria Understanding the single-celled life forms is essential to fighting malaria. The disease is ancient and so are treatments such as quinine. Researchers around the world are trying numerous avenues of attack against humankind's greatest scourge.

All of the following question sets also appear in the Explore Life *E* electronic component, where you will find a variety of additional questions as well.

Test Yourself on Vocabulary and Concepts

In each question set below, match the descriptions with the most appropriate term. A term may be used once, more than once, or not at all.

SET I

(a) member of Domain Bacteria (b) member of Domain Archaea (c) viruses (d) prions (e) member of Kingdom Protista

1. Genetic material is circular DNA; gram-positive or gram-negative
2. Genetic material is unknown; made of protein only
3. Genetic material is DNA enclosed in a nucleus
4. Genetic material is DNA or RNA; capsid present
5. Genetic material is circular DNA; cellular; not pathogenic to humans

SET II

(a) spirochete (b) cyanobacteria (c) actinomycetes (d) mycoplasmas (e) extremophiles

6. Source of some antibiotics
7. Some are nitrogen-fixing autotrophs.
8. Parasitic; smallest free-living cells
9. Include bacteria that cause syphilis and Lyme disease
10. Include methanogens that live in a cow's stomach

SET III

(a) parasitism (b) saprobe (c) symbiosis (d) heterotroph (e) pathogen

11. The general term that includes all disease-causing biological entities

12. A decomposer that lives on dead or dying organisms

13. The general term that refers to an organism that must obtain organic food molecules from its environment

14. A close association of two or more living species in which each species is required for the survival of the other(s)

15. A nonreciprocal relationship in which one organism lives in or on another living organism and derives benefit (such as nourishment or shelter) from it to the provider's detriment.

Integrate and Apply What You've Learned

16. Identify the organism depicted in Figure 11.17 and explain the sequence of stages in its life cycle.

17. Give two examples of autotrophic prokaryotes and two examples of autotrophic protists.

18. Why are antibiotics effective against bacteria but not against viruses?

19. Explain why cyanobacteria are among the most self-sufficient organisms.

Analyze and Evaluate the Concepts

20. How would your life be different if the Domain Bacteria were suddenly and specifically wiped off the face of Earth? How would your life be different if members of the Domain Archaea lived inside your body?

21. Medical researchers have been less successful at identifying antiviral agents than at finding antibac-terial substances. What properties of the two types of infectious agents do you think has contributed to this situation?

22. Suppose that doctors in your town suddenly started noticing many patients with severe intestinal distress and suspect that the municipal water supply is contaminated by a microorganism. You are asked to help determine which type of microorganism may be involved. How would you investigate the organism to determine whether the culprit belonged to the Domain Bacteria, Archaea, or Eucarya? If the culprit was eukaryotic, how would you determine which group of protists was involved?

23. Which of the following organisms has mitochondria, ribosomes, a cell wall, and chlorophyll?
    (a) *Chlamydia*   (b) microsporidia   (c) diatoms
    (d) cyanobacteria   (e) viruses

24. Explain, in general, why malaria so easily eludes the human immune system.

25. Explain what Bti is and how it's being used to fight malaria.

A Fiery Fascination

Some people marry into money or titles, but Paul Bosland married into chiles—at least in a manner of speaking. Meeting and wedding a woman from New Mexico who cooked with chiles every day first kindled Bosland's interest in the fiery pods. Today, he is the top expert on nature's hottest plant genus, and a significant player in the ongoing evolution of this diverse group.

A native of San Diego, Bosland grew up in a family that was largely chileless. After training in plant breeding and genetics, he took a job at New Mexico State University in 1986, and began working with cabbage, cauliflower, and broccoli. While important, these lacked the pizazz of chiles, and Bosland soon determined that peppers would stoke his academic interest and lead to a spicier—and more productive—research career. Soon he had co-founded the Chile Pepper Institute and was editing their quarterly journal. One jovial student nick-named him "Chileman" and the name stuck, right down to the personalized license plate that still embellishes his truck. People started giving the good-natured professor chile posters, chile neckties, chile canisters, and even a brass chile door-knocker. This research topic *does* lend itself to levity, but chile peppers have their serious side, too: Bosland and his colleagues have helped New Mexico develop a $150 million annual business in fresh, dried, canned, frozen, brined, and mashed peppers. And the plant species behind this red-hot enterprise are among botany's most fascinating subjects.

All chile peppers—from the mildest green and red bell peppers to the hottest habaneras (peppers one author refers to as "thermonuclear")—belong to the same genus, *Capsicum*. Over 20 species in the genus *Capsicum* grow in the wild, but none produces fruits larger than pea-sized (peppers are technically fruits, not vegetables). Five separate groups of prehistoric South and Central Americans domesticated chile species starting 10,000 years ago. Today, plant breeders grow hundreds of varieties of chile plants, virtually all belonging to just one species, *Capsicum annuum*, but producing an amazing array of fruit and plant sizes. "Some chile plants grow 6 to 7 feet high," says Bosland, "while others may be no more than 4 to 6 inches tall." Some, he continues, have smooth leaves, while others are hairy. Some of the fruits are dull and round, while others are waxy and over a foot long. And pepper colors can range from light green to yellows, oranges, reds, and browns. Strange as it may sound, *Capsicum annuum* "is the most widely spread plant species in the world," says Bosland, because chile peppers are the most widely used cooking spice.

There's a great irony in the evolution and popularity of *Capsicum,* with its mouth-searing pungency.

▲ **A Passion for Chiles.** Paul Bosland; his Chileman license plate; and part of the pepper garden at the University of New Mexico.

This genus is part of the "deadly nightshade" family, Solanaceae, whose members—including tobacco, potatoes, and tomatoes—are known for producing toxic compounds in their leaves and/or fruits. Chile peppers make none of these toxins. Instead, they generate spicy-hot alkaloids called *capsaicins* (cap-SAY-shins) in the pepper fruit's ovary tissue, which surrounds the seeds. Birds can't taste this pungency, and so they eat the fruits and seeds with gusto, then deposit the seeds unharmed in their droppings. In this way, birds act as beneficial seed-dispersal agents for pepper plants. Mammals, on the other hand, *can* taste the heat and virtually always avoid hot peppers after one nibble. This, too, benefits *Capsicum* plants, because unlike a bird's digestive tract, a mammal's stomach contains acids that will kill the seeds. Plant scientists like Bosland believe that the hotness of peppers evolved as an adaptation to attract birds and repel mammals. How ironic, then, that we humans seek out not just the smoldering sensation of peppers—from jalapeño peppers, which are quite hot, all the way to mouth-blasting habanera peppers, which are over 40 times hotter! As you read on, you'll find out why so many humans love—sometimes even *crave*—a daily fix of hot chile flavor.

Chiles and chile researchers like Dr. Paul Bosland are a good case history for this chapter on plant diversity and evolution. Our goal here is to explore the major groups of fungi (sing., *fungus*) including the bread molds, yeasts, and mushrooms; green algae, such as the many types that undulate in tidepools; and plants, including mosses, ferns, pines, and flowering plants like chiles. We'll talk about their evolution from ancient, single-celled progenitors, their relationships to each other, and

their great ecological roles as decomposers and producers of organic matter in vast quantities. Even though chiles are just one small group in this varied sweep of fungal and plant species, they are an apt focus because:

- They are associated with fungi in several ways, both beneficial and harmful;
- Plants evolved from water dwellers to land dwellers, and chiles have most of the major adaptations to life on land;
- Many plants coevolved with animals, and the reliance on birds as chile-seed dispersers is a good example.

Humans have long influenced fungal and plant evolution, and the many chile varieties are a fine illustration. Dr. Paul Bosland, himself, has helped develop 11 unique chile varieties, including, most recently, the "NuMex Piñata," a jalapeño that changes color four times as it ripens. And he and his colleagues are continuing to use standard and biotechnological means to improve the nutritional value of cultivated chiles as well as their resistance to insects and diseases.

As you explore chiles and their place in the spectrum of fungi, green algae, and plants, you'll find the answers to these questions:

1. How do fungi make a living, and how do they reproduce?

2. What are the characteristics of green plants that have allowed them to evolve and dominate most landscapes?

3. What makes flowering plants—including the chiles—so successful?

12.1 The Lives of Fungi

Sometimes, for his research, Paul Bosland has to uproot and examine a pepper plant. He might, for example, walk to a row of plants and yank from the sandy soil a NuMex Piñata jalapeño chile plant with colorful yellow, orange, or red peppers. From the narrow branching roots he would probably see small filaments or threads extending (Fig. 12.1). Closer examination would reveal that these threads are not a living part of

the plant but some other living entity that grows in the soil and attaches to the plant roots. These fine threads are actually the bodies of organisms belonging to the biological group called **fungi,** all of which occupy a single branch on the tree of life (Fig. 12.2). The group includes mushrooms, puffballs, yeasts, molds, and certain other organisms with a similar body plan, and the members are more closely related to animals than plants. So what are the filaments doing on and sometimes inside the pepper plant root cells? Do they help or

Figure 12.1 **Plant Roots and Fungi: Helpful or Harmful?** Biologists grew lemon tree seedlings with and without mycorrhizal fungi. After 18 weeks, those with mycorrhizal (right) grew taller and stronger than those without (left).

hinder the plant's growth and its production of fiery fruits? Biologists have carried out experiments to answer these questions. But before we can fully understand the tests, we need a general background in the fungal body plan and life style. In this section, then, we'll concentrate first on the body form of fungi, then see how they obtain energy, reproduce, and finally how they interact with chiles and other plants.

The Body Plan of Fungi

The long slender fungal threads Paul Bosland unearths on the roots of a chile plant are probably of the species *Glomus intraradices*

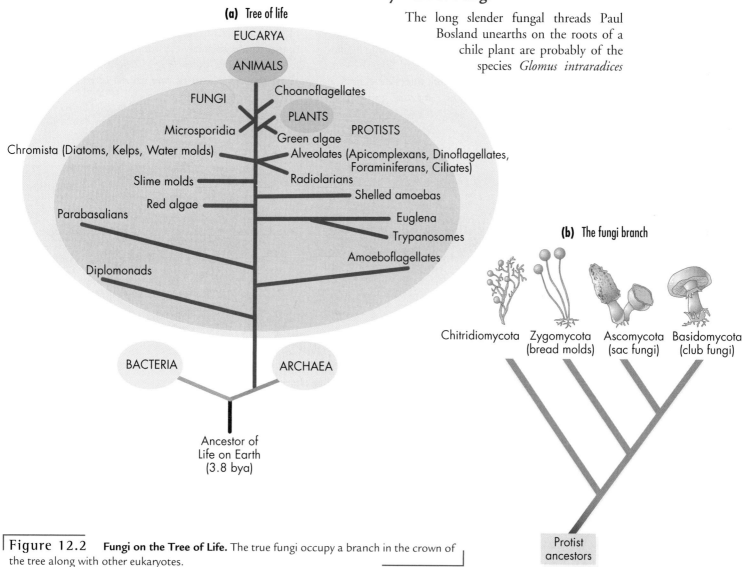

(a) Tree of life

EUCARYA

ANIMALS

FUNGI

Choanoflagellates

PLANTS

Microsporidia

Green algae

PROTISTS

Chromista (Diatoms, Kelps, Water molds)

Alveolates (Apicomplexans, Dinoflagellates, Foraminiferans, Ciliates)

Slime molds

Radiolarians

Red algae

Shelled amoebas

Parabasalians

Euglena

Trypanosomes

Amoeboflagellates

Diplomonads

BACTERIA

ARCHAEA

Ancestor of
Life on Earth
(3.8 bya)

(b) The fungi branch

Chitridiomycota

Zygomycota
(bread molds)

Ascomycota
(sac fungi)

Basidomycota
(club fungi)

Protist
ancestors

Figure 12.2 **Fungi on the Tree of Life.** The true fungi occupy a branch in the crown of the tree along with other eukaryotes.

and are typical examples of the fungal body plan. The filaments have these chief characteristics:

- They are eukaryotes (made up of eukaryotic cells).
- They are heterotrophs (they obtain their nourishment from other organisms).
- They develop a diffuse, branched, and tubular body.
- They reproduce by compact structures called spores.

Mycologists (fungal biologists) study organisms with some or all of these characteristics. As we saw in Chapter 11, mycologists also study water fungi and slime molds—organisms that apparently arose on separate lineages within the protists and therefore belong to the eukaryotic Kingdom Protista and have only distant kinship with other fungi (see Fig. 12.2). The more frequent subjects of mycology—mushrooms, toadstools, puffballs, and other more familiar fungi—occupy a single branch on the tree of life called the **Eumycota** or true fungi. We'll consider only true fungi in this chapter.

In general, fungi consist of filaments called **hyphae** (HIGH-fee; sing., *hypha;* Fig. 12.3a,b). A fungal filament is made up of cells joined end to end, like the cars of a passenger train. Hyphae often interweave into a loose mat called a **mycelium** (pl., *mycelia*), rather like

steel wool (Fig. 12.3a). You can see such a mat growing as a widening green or gray circle on the heel of a loaf of bread forgotten at the back of a refrigerator. Researchers have recently found that even the **yeasts,** which generally grow as single-celled fungi and reproduce by budding, will form hyphae under conditions of starvation.

Cell structure distinguishes fungi from prokaryotes (see Chapter 11) as well as from plants and animals (Explorer 12.1). The cells of a fungus have membrane-enclosed nuclei, unlike the cells in the prokaryotic domains Bacteria and Archaea. Fungal cells also have cell walls, which distinguishes them from animal cells. But the cell walls of fungi also distinguish them from plants, because fungal cell walls consist mostly of **chitin** (KI-tin), a nitrogen-containing polysaccharide, while plant cell walls are mainly cellulose. The tough chitin in fungal cell walls helps make fungi among the most resilient organisms of all the eukaryotes, able to withstand heat and drought. (The hard outer skeletons of insects are also made of tough chitin.) The cell walls that separate adjacent cells in a fungal filament are perforated by little holes (Fig. 12.3c). These perforations allow the cytoplasm of one cell to directly contact the cytoplasm of neighboring cells.

GET THE PICTURE

Explorer 12.1

Fungi and Plants on the Tree of Life

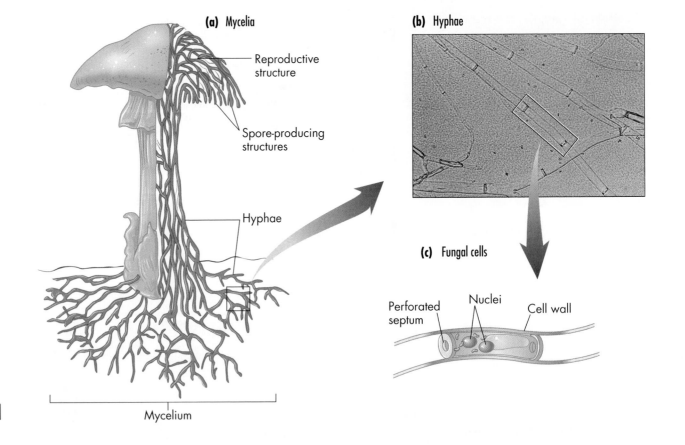

Figure 12.3   The Fungal Body. (a) A familiar mushroom is the reproductive structure in certain fungal species, and it has reproductive cells called spores. The mushroom consists of tightly packed, threadlike cells or hyphae. In the soil, these fungal threads grow as loose mats or mycelia. (b) Hyphae consist of many cells arranged end to end. (c) Some species have two nuclei per cell, and at the ends of each cell are septa, or cell walls. A hole in the septum allows cytoplasm to flow through from one cell to the next.

(a) Mycelia

Reproductive structure

Spore-producing structures

Hyphae

Mycelium

(b) Hyphae

(c) Fungal cells

Perforated septum   Nuclei   Cell wall

Figure 12.4 Is a Fungus the Largest Living Thing?
Some mycologists propose that the largest organism alive is an *Armillaria* fungus, a parasite on tree roots. This photograph taken from an airplane over Montana shows a virgin forest with two centers of root disease, which has caused the trees to look a paler green. The lower center started hundreds of years ago from a single haploid spore cell, and has grown to encompass 20 acres (8 hectares). Many mycologists think that even bigger fungi are devouring trees in Michigan and Washington State.

Even with their many interlinked cells, the bodies of most fungi are simple in structure and lack specialized tissues or organs such as the roots, stems, and leaves in plants. Fungi do, however, often form specialized reproductive structures. An example of such a reproductive structure, or **fruiting body,** is the familiar button mushroom cooks use in salads or on pizzas. The body of a button mushroom looks solid, but it actually consists of many fungal filaments packed tightly together like hair bound together into a pony tail. Below ground, an extensive, loose mycelium spreads beneath the mushroom, penetrating the soil for many square meters.

Fungal filaments or hyphae can grow very quickly: If you added together the growth at all the hyphal tips in a tangled fungal mat or mycelium, just one day's new growth could easily exceed 1 km (0.62 mi). This explains how mushrooms can literally spring up overnight on damp logs or soil. This meshwork of filaments gives a fungus a large surface-to-volume ratio, which enables even a very large mushroom to digest and absorb enough nutrients to grow rapidly.

A single individual fungus can be enormous. Figure 12.4 shows a view from an airplane over Montana. The pale green circle in the foreground contains pine trees damaged by a single fungus individual. One reproductive cell happened to land at the middle of the circle, and then the mycelia grew outward. Mycologists near

Mount Adams in Washington State have found a fungus that covers 1500 acres of forest (about 600 hectares), weighs well over 100 tons (9000 kilograms), and is between 400 and 1000 years old. In the early 1990s, some fungal biologists suggested that this one individual fungus could be considered the largest living organism in the world. As we'll see later, plant biologists were quick to challenge this claim.

Decomposing Mozart: How Fungi Make a Living

Although fungi often grow in the ground like plants, plants are autotrophs, while fungi are heterotrophs. Plants can extract energy and molecular building blocks directly from the physical environment and so can many kinds of prokaryotes. Fungi, however, must obtain their energy and materials by decomposing molecules first made by other organisms. Like animals, fungi obtain their nourishment by digesting the molecules of plants, or animals that eat plants. And like the cells of your digestive system, fungal cells secrete powerful enzymes into their surroundings that break down large organic molecules into smaller ones (Fig. 12.5). Again, like some of your intestinal cells, fungal

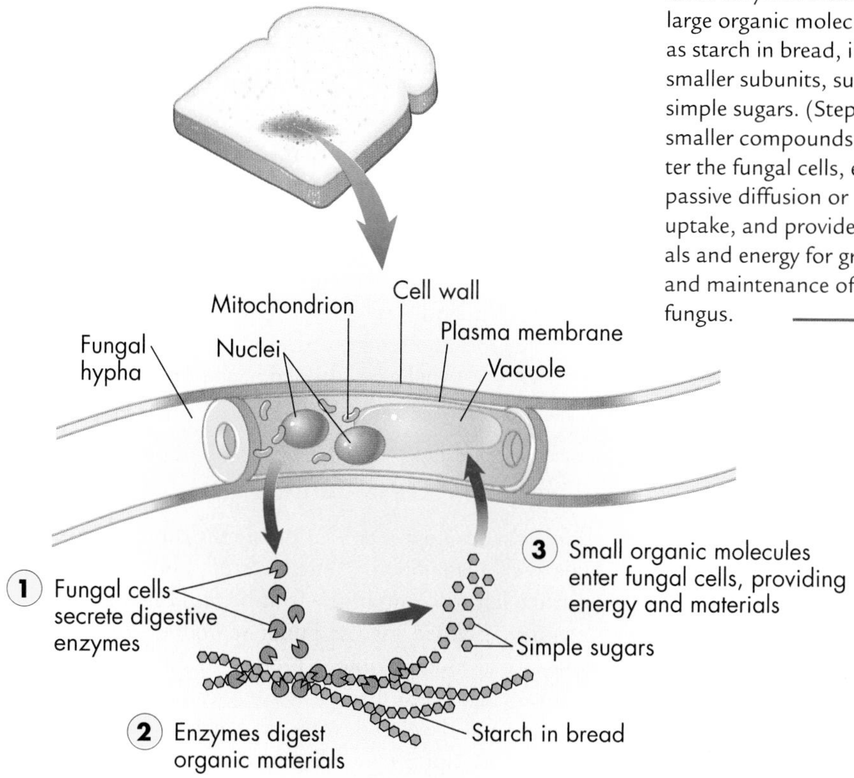

Figure 12.5 How a Fungus Makes a Living.
Fungi, such as a colony of black mold growing on a slice of bread, obtain nutrients by extracellular digestion. (Step 1) They secrete digestive enzymes into their environments, and (Step 2) these enzymes break down large organic molecules, such as starch in bread, into smaller subunits, such as simple sugars. (Step 3) These smaller compounds then enter the fungal cells, either by passive diffusion or by active uptake, and provide materials and energy for growth and maintenance of the fungus.

Cell wall
Mitochondrion
Plasma membrane
Nuclei
Vacuole
Fungal hypha

1 Fungal cells secrete digestive enzymes

3 Small organic molecules enter fungal cells, providing energy and materials

Simple sugars

2 Enzymes digest organic materials

Starch in bread

cells absorb the smaller nutrient molecules through their cell membranes. This fungal feeding strategy seems so . . . animal. So how is it different? The answer is simple and it involves *eating*. Most animals bring chunks of a plant or other organism into the mouth then *inside* the cavity of the gut before digesting it. A fungus, by contrast, digests nutrient matter *outside* of its own fungal body.

Most fungi decompose a wide range of nonliving organic matter. Fungi consume everything from leather and cloth to paper, wood, paint, and other materials, slowly reducing old buildings, books, and shoes to crumbled ruin. In the process, they are inadvertent masters of recycling. During the long evolution of the fungi, however, some **saprobes,** or decomposers of nonliving tissue, evolved into **parasites,** or organisms that live on or in other living things, without contributing to the host's survival. Today, parasitic fungi are the main cause of plant diseases, with about 5000 different fungal species inhabiting crops in fields, gardens, and orchards. This includes some major diseases of chiles, which Paul Bosland and his colleagues are working to combat. As you'll see later, other types of parasitic fungi attack people and domestic animals, causing athlete's foot, ringworm, and vaginal yeast infections, to name a few.

Clearly, the dietary habits of fungi are far from fastidious. Their activities, however, are crucial for nature's balanced ecological cycles. Plants and photosynthetic bacteria, for example, remove carbon from the air and trap it in carbohydrates. If these were the Earth's only organisms, all carbon would soon be locked up in their cells and in their remains. This doesn't happen, however, because other organisms—largely fungi and bacteria—decompose plant matter and organic wastes as well as the remains of dead organisms and your toenail clippings. From these, they release into the environment small compounds containing carbon, nitrogen, phosphorus, and other elements required for life and they re-enter the nonliving portion of the ecosystem, where they can eventually be recycled back into living cells.

The Sex Life of a Fungus

It's amazing how a colony of fungus can crop up on a slice of bread, on the underside of a board, or on an orange left too long in the fruit bowl. How did they get to these isolated places? Fungi reproduce and disperse by special microscopic reproductive structures called **spores** (Fig. 12.6). Each tiny spore can germinate into a hypha and develop into a new colony of fungus (a mycelium). Spores are adaptations for survival; they are

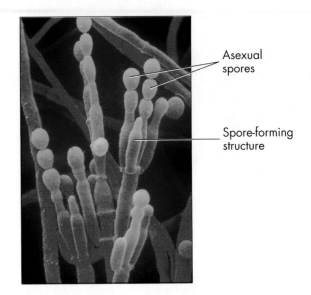

Asexual spores

Spore-forming structure

Figure 12.6 **Fungi Reproduce by Spores.** Fungi reproduce and disperse by spores-hardy, compact groups of cells that can disperse far and wide by wind and water. Here, a *Penicillium* colony is producing one particular kind of asexual spore called conidia.

able to withstand extreme conditions of dryness, heat, or cold, they can disperse by wind or water to distant locations—like a board or a fruit bowl—and then produce a new fungus when conditions are favorable.

Some spores are produced without sex. These asexual spores are genetically identical to the parent and form by modifications of cells in the parent hypha. Conidia are one common type. Fungi can also produce sexual spores (Fig. 12.7). Fungi are neither male nor female, but each haploid individual is one of two mating types, **plus** or **minus** (Fig. 12.7, Step 1). The haploid cells of each mating type can grow into hyphae and reproduce by making asexual spores (Step 2). Eventually, a haploid cell of the plus mating type can fuse with a haploid cell of the minus mating type, forming a single cell containing two haploid nuclei in one cytoplasm (a **dikaryon** (*di* = two + *karyon* = nucleus; Step 3). (Note that a dikaryon has two sets of haploid chromosomes in two separate nuclei, while a typical diploid cell has two sets of haploid chromosomes in a single nucleus.) Double-nucleated dikaryon cells can divide and form hyphae, and these threadlike structures can then join and give rise to a reproductive organ, the fruiting body (Step 4). The typical mushroom we recognize growing in a forest or field is an aboveground fruiting body. In special structures within the fruiting body, haploid nuclei can fuse, forming a diploid cell or zygote (a process called **sexual fusion** or **nuclear fusion,** Step 5). The diploid zygote usually undergoes meiosis, producing sexual spores, and the cycle is complete (Step 6). There are many variations of the fungal life cycle, and different types of fungi have different types of reproduction. But here are the essential features: The haploid stage dominates; the diploid phase is often relegated to a single cell, the zygote; and reproduction, both sexual and asexual, occurs by spores.

Fungi and Plant Roots

Fungi interact closely with other organisms, and the fungal threads on the roots of chile plants are just one example. Sometimes fungal interactions are beneficial;

other times, they are harmful, causing diseases of plants and animals. What about the fungi Paul Bosland finds on chile roots? Are they helpful or harmful? To find out, some of Dr. Bosland's Australian colleagues sterilized one plot of soil to kill all the fungi present but left an adjacent plot untreated (Fig. 12.8a). Next, they planted chiles in the treated plot as well as in the adjacent control plot (which still contained fungi in the soil). At this point, they subdivided each of the two plots and spread phosphorus-containing fertilizer on some portions and left other portions unfertilized (Fig. 12.8b). When the chile plants from all four plots matured, the researchers dried and weighed the plants.

They found that when they provided phosphorus fertilizer, the chile plants grown in fungus-infected soil weighed approximately the same as those grown in sterilized soil (Fig. 12.8c). In the plots without added fertilizer, however, they saw a dramatic difference. The chile plants grown in fungus-infected soil were 91 times heavier than the plants grown in sterilized soil! What would you conclude from this experiment? The researchers deduced that the fungi helped the chile plants take up phosphorus, an element necessary for plant growth, when quantities of the element in the soil were limited to naturally occurring levels. When they added

Do mycorrhizae boost chile plant growth?

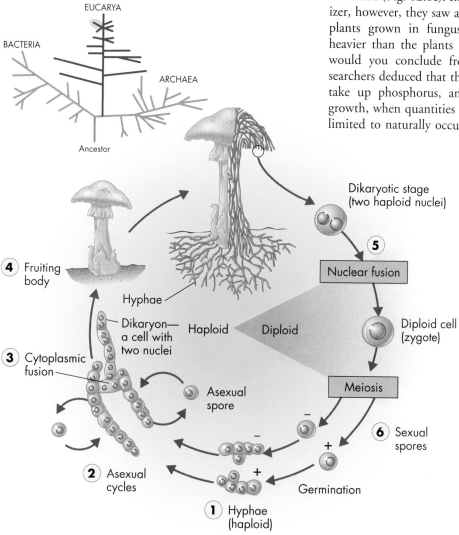

Figure 12.7 **Generalized Life Cycle of Fungi.** The haploid phase dominates the life cycles of fungi. Spores of either the minus or plus mating type (Step 1) can be produced sexually and asexually, and they grow into hyphae. Hyphae can produce spores asexually as reproductive cells (Step 2). These hardy spores can disperse to new places and grow into more hyphae. The cytoplasm of two cells from hyphae of opposite mating types can fuse and produce a cell with two nuclei (Step 3). This cell, now with two nuclei, can give rise to more hyphae, which may form a fruiting body (Step 4). Within the fruiting body, the two nuclei of opposite mating types may fuse, giving a diploid zygote (Step 5). The zygote undergoes meiosis and once more produces haploid spores (Step 6), which can again divide by mitosis and form hyphae.

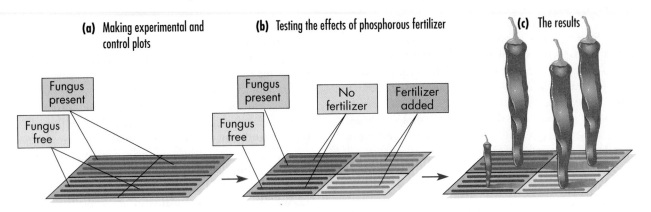

(a) Making experimental and control plots

Fungus present

Fungus free

(b) Testing the effects of phosphorous fertilizer

Fungus present

Fungus free

No fertilizer

Fertilizer added

(c) The results

Figure 12.8 **Do Root-Associated Fungi Help or Harm Chile Plants?** (a) Biologists sprayed one plot of ground with a chemical that killed all fungi, but left an adjacent plot untouched as a control. (b) Next they fertilized half of each plot with phosphorus-rich fertilizer and planted chili seeds in all four plot sections. (c) At the end of the growing season, they weighed the chile plants grown under the four conditions and found large chiles in three sectors but small ones in just the fungus-free plot without added fertilizer. Obviously, even when phosphorous concentrations were low from too little fertilizer, the presence of root-associated fungi allowed the plants to thrive and produce large chiles.

phosphorus fertilizer, however, this element was plentiful and no longer a limiting factor to growth, so the presence or absence of the fungi made no difference.

These experiments involved the fungal-root combination called **mycorrhizae** (myco-RYE-zee; sing., *mycorrhiza*, literally "fungus roots"), commonly associated with many plants. With this study, the Australian researchers proved that mycorrhizae are highly beneficial for chile plant growth. The interaction between root

fungi and the chile plants is a type of **symbiosis** in which two species have a mutually beneficial relationship. As we've seen, the plants benefit, but the fungus benefits, too, by obtaining sugars from the plant roots for its own growth. In Figure 12.9 note the close physical association of fungus and root cell: a portion of the chile root mycorrhizal fungus actually invades the plant cells and obtains sugars and other carbon-containing molecules directly from the plant. In other plant roots,

Figure 12.9 **How Root-Fungi Help Plants Grow.** Chiles usually have mycorrhizae or special fungi associated with their roots (a). The fungus grows and branches inside root cells. This photo shows mycorrhizae in a leek root cell (b). This branching produces a large amount of surface area for exchange of materials between the plant cell and the fungus. This fungus helps the plant obtain water, phosphorus, and other minerals, while the plant provides the fungus with sugars.

(a) Fungus growing inside root

(b) Fungus inside plant cell

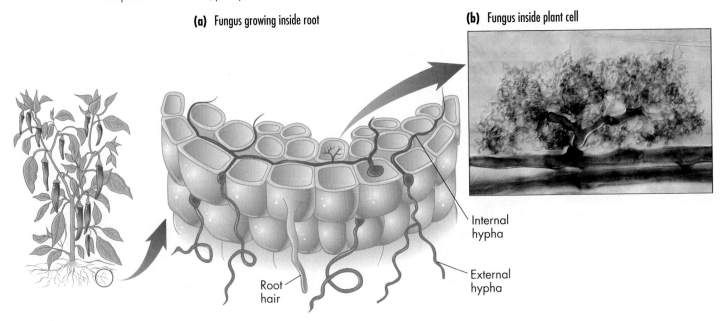

Internal hypha

Root hair

External hypha

the fungi may remain outside the plant cells. Regardless, the plant helps the mycorrhizal fungus obtain organic nutrients, and the fungus helps the plant obtain phosphorus and probably water, as well. Biologists have discovered and named several hundred species of mycorrhizal fungi associated with the roots of perhaps 90 percent of all land plants. Because fossils of the oldest land plants have fossilized mycorrhizae among the roots, botanists think that the fungus–plant interactions may have helped the transition of plants from water to land.

Lichens: More Mutual Benefit

Have you ever looked carefully at tombstones in an old cemetery? The stones are often encrusted with gray, orange, or greenish layers like those shown in Figure 12.10a. These formations are **lichens** (LIE-kenz), and are actually symbiotic associations between a fungus and either a green alga (a eukaryote; see Fig. 12.12), or a cyanobacterium (a prokaryote; see Fig. 11.9). Many botanists credit

Beatrix Potter, the author of the Peter Rabbit books, with being one of the first (perhaps the very first) to notice this association between fungi and algae.

In a colorful lichen, the fungus forms a dense hyphal mat around the photosynthetic algal or cyanobacterial cells (Fig. 12.10b). The fungal mat provides the photosynthetic cells with some protection and a supply of water, and in turn, the mat filaments are able to obtain some of the photosynthetic cells' newly made carbohydrates. Lichens take almost no nutrients from the substrate on which they live, and they are often the first organisms to inhabit lava flows or other newly exposed rocks. However, they grow extremely slowly. You can estimate the rate of lichen growth by measuring the diameter of the largest lichens found on tombstones of various ages and comparing these to the dates inscribed on the stones. While lichens can survive in bleak environments, they are particularly sensitive to the industrial pollutant sulfur dioxide. Therefore, biologists have found the disappearance of lichens to be an early indicator of air pollution. Since both species in the lichen need to reproduce at the same time, one might expect to see clever evolutionary solutions, and indeed, some exist. Many lichens reproduce by releasing little balls that contain at least one algal cell embedded in a small network of hyphae. Wherever the reproductive structure lands, it has the potential to grow into a new lichen.

(a)

(b)

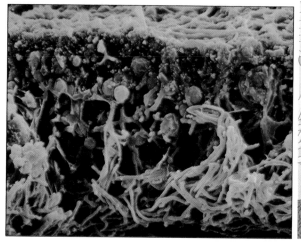

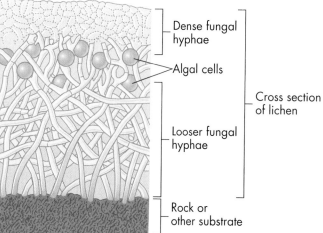

Figure 12.10 **Lichens: A Collaboration of Fungi and Algae.** (a) Colorful lichens can colonize natural and hand-carved rock surfaces. By measuring the diameter of the largest lichens on tombstones of various ages, you can make a graph of lichen growth versus time. These lichens on tombstones in Pemaquid Point, Maine, are hundreds of years old. The photograph and sketch in (b) show how fungal hyphae can form a water-trapping mesh that holds water for algal cells. The algae can undergo photosynthesis and contribute sugars for the entire lichen.

Dense fungal hyphae

Algal cells

Looser fungal hyphae

Cross section of lichen

Rock or other substrate

Recent molecular genetic studies suggest that lichenlike associations of fungi and photosynthetic algae existed 1 billion years ago, and colonized Earth several hundred million years earlier than botanists have long accepted based on fossil evidence. These associations may have altered Earth's atmosphere and climate in ways that spurred the explosion of animal evolution in the Precambrian period.

Medical Mycology

Fungi are important in medicine, both in sickness and in healing. AIDS patients are susceptible to thrush, a yeast infection of the mucous membranes, and *Pneumocystis carinii*, which is now thought to be a fungus. As we saw earlier, vaginal yeast infections, athlete's foot, and many other skin diseases are common even in people with healthy immune systems. On the plus side, many bacteria-killing antibiotics, such as penicillin, were originally isolated from fungi, as was cyclosporin, which suppresses the rejection of tissue and organ transplants. Medical researchers are continuing to focus on the many kinds of fungi in the hopes of turning up more such medically useful substances.

The Diversity of Fungi

Biologists usually classify fungi by the shapes of their sexual spore-producing structures. Spores are the only means a fungus has of dispersing to new areas, and thus

Table 12.1 Fungi: The Great Decomposers

Division	Examples	Characteristics and Significance
Lichens	Red, gray, and yellow species found on rocks	A combination of a fungus (ascomycota or basidiomycota) and an alga that live intimately together; degrades rock and helps make new soil; sensitive indicators of pollution
Deuteromycota ("imperfect fungi"); 25,000 species	Species that produce penicillin	Have lost the ability to reproduce sexually, thus their relationship to other fungi is unclear; reproduce by asexual spores; used in making drugs, cheeses, and soy sauce
Basidiomycota (club fungi; 25,000 species)	Common field mushrooms, giant puffballs, bracket fungi, toadstools, smuts	Produce spores in club-shaped basidia; the fruiting body is the familiar mushroom or toadstool which can be extremely poisonous
Ascomycota (sac fungi; 30,000 species)	Pink bread mold, brewer's yeast, morels, truffles	Produce spores in an ascus, or sac, borne in a cup-shaped body; because they include the yeasts, they are the most economically useful fungal group; powdery mildews harm fruit trees and grain crops
Zygomycota (bread molds; 600 species)	Common black bread mold	Produce diploid spores and a cottony mat of hyphae on breads, grains, or other foods and organic materials
Chitridiomycota	Species that causes potato wart disease	Sister group to other fungi; mostly aquatic; gametes have a flagellum, allowing them to swim (other fungi have lost flagellae)

they are of prime importance. Table 12.1 describes several major divisions of fungi, plus the imperfect fungi and lichens. The table focuses on how they form spores and includes other significant characteristics as well. (In the fungal and plant kingdoms, a **division** is the equivalent of a phylum in other kingdoms.) By reading a field guide to common fungi, you can learn more details about the various groups of fungi, including the bread molds; the yeast, morels, and truffles; the true mushrooms and other club fungi; and the so-called fungi imperfecti.

12.2 Plants: The Green Kingdom

Plants—chiles, ferns, mosses, and thousands of other types—occupy the branching crown of the eukaryotic tree of life along with fungi, animals, and the Chromista (see Chapter 11). **Plants** are multicellular eukaryotes that capture energy by photosynthesis and develop from embryos. As we'll soon see, plants also have an alternation of generations, with distinct multicellular haploid and diploid body forms. All plants reside on a single branch of the tree of life. Preceding that branch were the green algae—organisms that are not true plants but that have some similarities to them. We consider both plants and green algae in this chapter because both groups originated from a single common ancestor that did not give rise to other forms of life (see Figs. 12.11 and 12.2). Let's start with the green algae and then move to the plants, which are more complicated, including the chiles.

Green Algae: Ancestors of Land Plants

If you've walked by a lake or stream, or poked around in a tidepool, you've probably seen several species of green algae, although you may not have recognized them. Unlike plants, which are multicellular, **green algae,** or Chlorophyta, often occur as single cells (Fig. 12.12). Some, however, do occur as multicellular, threadlike filaments; hollow balls; or wide, flat sheets. Most species of green algae live in shallow freshwater environments or on moist rocks, trees, and soil. A few inhabit shallow ocean waters, undulating gracefully in tidepools alongside other kinds of algae. Green algae have the same types of chlorophyll (called *a* and *b*) that occur in plants, together with orange carotenoids;

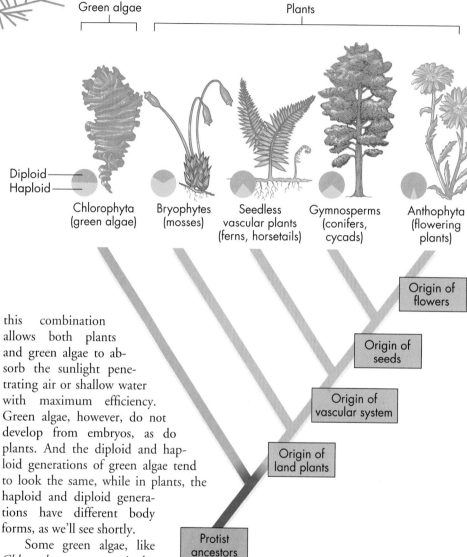

Figure 12.11 **Trends and Milestones in the Evolution of Green Algae and Plants.** The ancestors of today's green algae probably gave rise to the earliest land plants. The origins of the vascular system, seeds, and flowers were further milestones in plant evolution. The increasing dominance of the diploid sporophyte generation, indicated by the circles in the top row of figures, was a general evolutionary trend.

this combination allows both plants and green algae to absorb the sunlight penetrating air or shallow water with maximum efficiency. Green algae, however, do not develop from embryos, as do plants. And the diploid and haploid generations of green algae tend to look the same, while in plants, the haploid and diploid generations have different body forms, as we'll see shortly.

Some green algae, like *Chlamydomonas,* are single-celled. *Chlamydomonas* is a favorite laboratory organism for investigating how genes control mating, how the lashing movements of flagella can propel a cell, and what are the biochemical details of photosynthesis and aerobic respiration. *Chlamydomonas* is an oval cell propelled through freshwater pools and moist soil by two flagellae (Fig. 12.12a). Other green algae, like the delicate *Volvox,* grow in colonies. And still others, like *Ulva,* the sea lettuce, inhabit tidepools and grow a delicate, leaflike body, just two cells thick, that resembles sheets of green

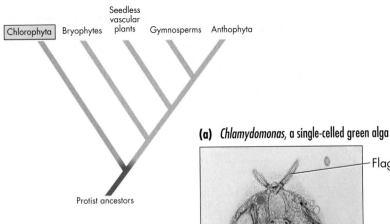

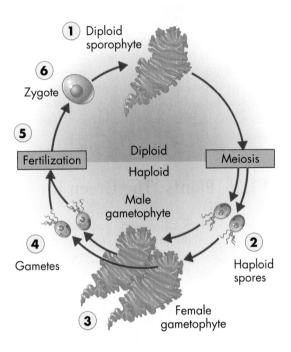

(b) Alternation of generations of *ulva*

1 Diploid sporophyte
6 Zygote
5 Fertilization
Diploid
Haploid
Meiosis
Male gametophyte
4 Gametes
3 Female gametophyte
2 Haploid spores

(a) *Chlamydomonas,* a single-celled green alga

Flagellum
Nucleus
Eyespot
Chloroplast

Figure 12.12 **Green Algae: Plant Ancestors.** Green algae, which are protists not plants, can occur in several body forms. (a) The single-celled *Chlamydomonas* is a common resident of ponds and ditches. These cells have flagellae, large plastids, numerous chloroplasts, a large eyespot, and a dark-staining nucleus. (b) *Ulva* is a multicellular green alga that usually grows in shallow tidepools. The haploid phase produces spores that look similar to individual *Chlamydomonas* cells. *Ulva*'s life cycle shows an alternation of generations between multicellular haploid and diploid body forms.

cellophane (Fig. 12.12b). Reproductive spores of *Ulva* bear a striking resemblance to a single individual of *Chlamydomonas,* suggesting close evolutionary ties.

The life cycle of the green alga *Ulva* shows an **alternation of generations,** a regular alternation of multicellular haploid and diploid body forms. *Ulva*'s life cycle shows the basic features of this alternation (Fig. 12.12b). The diploid phase of the life cycle consists of a leafy sheet of many cells (Step 1). Special cells within the sheet undergo meiosis and produce haploid spores (Step 2). Because the diploid phase produces spores, it is called the **sporophyte** generation (*sporo* = spore + *phyte* = plant, or in this case, plantlike). The haploid spores divide by mitosis into multicellular sheets that have *the same body form* as the diploid generation (Step 3). The haploid organism is called the **gametophyte,** because special cells within the sheets produce haploid gametes (Step 4). In *Ulva,* the two kinds of gametes look the same. This contrasts with the sperm and eggs of plants and animals, which have different forms. When *Ulva*'s haploid gametes fuse (Step 5), they produce a diploid zygote (Step 6), which then grows by mitosis into another leafy sporophyte (Step 1).

The alternation of generations shown by some green algae occurs in a more elaborate form in plants. Botanists think that ancient aquatic green algae probably gave rise to the simplest land plants, which still rely on water for reproduction. Let's move on, now, to explore the physical trends that emerged over time and allowed life to inhabit the land.

12.3 Plants: Green Embryo Makers

Plants can be small and colorful, like chiles, with their brightly hued pepper fruits (Fig. 12.13a). They can be gigantic, like the redwoods of the Pacific coast. They can be minute, like the shimmering green duckweed growing on a pond's surface. Or they can be drably camouflaged, like stone plants (Fig. 12.13b). Regardless of differences, however, plants have the common physical characteristics we listed earlier, including this: A plant generally develops from an **embryo,** an immature

(a) Colorful chiles

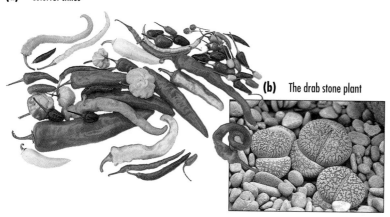

(b) The drab stone plant

Figure 12.13 **Plants: Colorful or Drab.** (a) The diversity of colors, shapes, and sizes of chile fruits is spectacular. (b) In contrast, the South African stone plant *Lithops aucampiae ssp. eunicaea* blends in among the pebbles, which may decrease the likelihood it will be seen and eaten by a hungry animal.

form of an organism undergoing the early stages of development. The diploid multicellular embryo is supported by a haploid maternal plant. Plant embryos result from sexual fusion, and most plants can reproduce sexually. A brilliantly colored chile pepper, for example, contains within its seeds embryos resulting from sexual reproduction. Plant cells also generally contain chloroplasts (review Fig. 3.21), green cytoplasmic organelles that transform the energy of sunlight into sugars. Fungi and animals lack chloroplasts or related organelles. The Chromista (the pigmented organisms we discussed in Chapter 11) have chloroplasts whose evolutionary origin may have differed from the chloroplasts in green algae. Plant cells have cell walls made chiefly of cellulose. Recall that although fungal cells have walls, they differ in composition and are usually chitin.

Plants play ecologically crucial roles. They dominate most terrestrial landscapes: just picture a forest or prairie without plants. They sustain the lives of most animals, including humans, through the production of food, fibers, fuel, and shelter. In addition, plants release the oxygen that aerobic organisms, including animals, utilize during aerobic respiration. Finally, plants display an amazing array of substances called **secondary metabolites** that are not directly needed for the organism's energy gathering and reproduction, but defend the plant against animals, fungi, and even other plants. The capsaicins in chiles are secondary metabolites, and so are the caffeine in coffee beans and the nicotine in tobacco leaves. As we saw, capsaicins probably protect chile plants from mammals with their seed-killing digestive tracts. Caffeine and nicotine also protect against certain types of animal predators.

We will investigate plant adaptations such as these in greater detail in Chapters 21 and 22, along with other details of their anatomy and physiology.

Plant Life Cycles

As in some green algae, the life cycles of plants have a regular alternation of multicellular haploid and diploid generations (***E***xplorer 12.2). This alternation of generations is a key to understanding plant biology. In many plants, the haploid and diploid generations are separate, and each phase is a free-living, multicellular form. In other plants, the gametophyte may consist of just a few cells and be inconspicuous, while the sporophyte is a larger multicellular organism. Figure 12.14b shows the alternating generations in the moss life cycle.

The familiar short, fluffy, green moss plant that covers the ground in moist habitats is the haploid gametophyte (gamete-producing) phase of the moss life cycle (Step 1). Female gametophytes produce haploid eggs, and male gametophytes produce haploid sperm (Step 2). Sexual fusion, or fertilization, produces a diploid zygote (Step 3), which grows by mitosis into a diploid embryo protected in the tissues of the maternal gametophyte plant. The embryo grows into a diploid sporophyte, a spore-producing portion of the life cycle (Step 4). The moss sporophyte is often a brown structure that looks like a miniature light post and grows up from the gametophyte. As in Ulva, specialized cells in the moss sporophyte produce haploid spores by meiosis (Step 5). These spores then germinate and once again grow by mitosis into the gametophyte generation (Step 1). Notice a key feature that distinguishes the life cycle of a moss and other plants from that of your own: In humans and other animals, meiosis produces haploid cells that themselves become gametes, eggs or sperm. In contrast, in plants, the immediate cellular products of meiosis do not become gametes directly, but grow into separate multicellular haploid individuals, which eventually produce the gametes.

As plants evolved from ancestral green algae, the sporophyte generation gradually took on more and more prominence in plant life cycles (see Fig. 12.11). As we saw in Ulva, a living green algae, sporophytes and gametophytes look much the same (see Fig. 12.12b). In plants, however—moss, for example—the two phases always look different: The moss gametophyte is green,

GET THE PICTURE

Explorer 12.2

Alternation of Generations

What is the chile sporophyte?

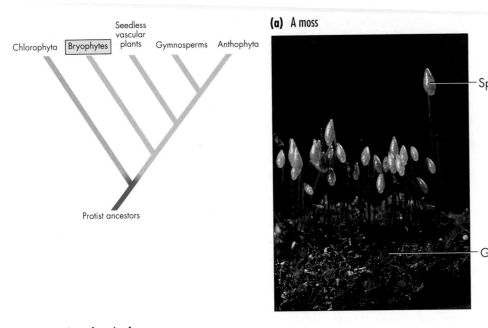

(a) A moss

Sporophyte

Gametophyte

(b) Life cycle of a moss

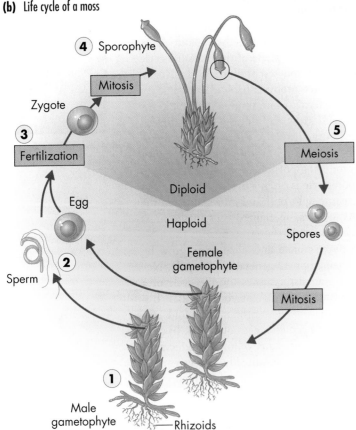

4 Sporophyte

Mitosis

Zygote

3 Fertilization

Diploid

5 Meiosis

Haploid

Egg

Female
gametophyte

2

Sperm

Spores

Mitosis

1

Male
gametophyte

Rhizoids

Figure 12.14 **The Plant Life Cycle: Alternation of Generations.** Plant life cycles alternate between a multicellular haploid generation (a gametophyte, or gamete-making plant) and a multicellular diploid generation (a sporophyte, or spore-making plant). This figure shows a fire moss (*Funaria hygrometrica*) (a) and its life cycle (b).

longer lived, and more complex, and consists of many more cells than the sporophyte. The sporophyte tends to be brown and has a "head" raised on a slender stalk that facilitates the dispersal of haploid spores (Fig. 12.14). In more complex land plants such as chiles, coconut palms, and hibiscus, the diploid sporophyte phase dominates and is the conspicuous plant we recognize.

Trends in Plant Evolution

Evidence from fossils as well as from plant genes suggests that plants probably originated from ancient aquatic green algae (Fig. 12.11 and review Fig. 11.13). The first plants to make the transition from water to land probably survived on the damp fringes of ancient oceans over 450 million years ago (see Table 10.1). At that time, the continents were relatively flat. Water levels probably changed dramatically as the seasons changed and when climatic cycles lengthened. Some of the first transitional land plants had lucky combinations of alleles that resulted in adaptations to withstand desiccation (drying out). These drought-resistant individuals would have tended to reproduce more successfully in fluctuating environments than would individuals without the allele combinations. Adaptations to drying include a waxy outer layer perforated by holes that allow gas exchange, and reproductive structures that resist desiccation. Adaptations like these opened up spacious new frontiers in damp areas near oceans. Sunlight was plentiful, the soil held rich mineral nutrients, and initially there were no plant-eating animals on the land.

Biologists think that other evolutionary milestones occurred after plants made that first transition to land (Fig. 12.15). Fluid-conduction tubes or a **vascular system** evolved and enabled water and nutrients to move efficiently throughout the plant. (We'll see details of this later in the chapter.) Seeds then originated in certain plant types, followed by flowers in other types. Seeds and fruits helped improve the chances of young plants being dispersed to new areas, and flowers helped the chances of successful fertilization of eggs by pollen grains. We already saw that plants underwent a general trend toward the dominance of the diploid generation (the sporophyte). This trend, along with the innovations for resisting drought, for conducting water and nutrients, and for dispersal and fertilization, led to the vast diversity of plants we see today. These form a spectrum from simple, low-growing mosses that lack vascular tissue to showy chile plants with their vascular systems, flowers, seeds, and specialized hot pep-

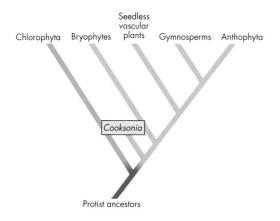

Cooksonia

Sporangia

Figure 12.15   **A Pioneer on the Land.** A 400-million-year-old fossil of early plants in the genus *Cooksonia* reveals that these ancient plants had several adaptations for terrestrial life. These included gas-exchange ports (stomata), thick-walled, water-preserving spores, and a vascular system. Branching stems produce numerous spore-forming organs.

pers. **E** 12.1 illustrates the great diversity of land plants, including the cycads, ginkgoes, gnetophytes, conifers, and flowering plants.

Simple Land Plants: Still Tied to Water

Walking through a forest today you can still see plants similar to the earliest survivors on land. All you have to do is look for **bryophytes:** the mosses, liverworts, and

hornworts. These plants grow low to the ground and require droplets of water for reproduction. In contrast, the **seedless vascular plants,** including ferns, horsetails, and club mosses, are stiffened by an internal transport system that allows them to grow taller than the bryophytes. Even with their superior height, though, the seedless vascular plants still require water to reproduce (for example, fern sperm must swim to get to the egg). In time, the seedless vascular plants were surpassed by the **seed-forming vascular plants,** the **gymnosperms** (such as pine trees) and **angiosperms** (*Anthophyta* or flowering plants). Both groups evolved hardy seeds and, for the first time, freedom from environmental water for reproduction. The final major evolutionary innovation was **flowers,** showy but sterile reproductive structures that attract animals carrying pollen, like bees, moths, and hummingbirds. The inadvertent transport of pollen to individual plant species by specific animal species has helped to rapidly diversify flowering plants into thousands of beautiful species. Genetic and fossil data suggest that an ancient green alga made the transition to land and then later diversified into mosses (bryophytes) and ferns, horsetails, and club mosses (seedless vascular plants). The seedless vascular plants, in turn, gave rise to seed-forming plants, the gymnosperms and angiosperms (Fig. 12.11).

For plants to thrive on dry land, they needed a variety of novel adaptations for the following functions:

1. Avoiding desiccation: Without surrounding water, they had to minimize evaporation while still allowing oxygen and carbon dioxide gases to move in and out of the plant.
2. Support: Without buoyancy from surrounding water, plants needed a strong internal system of support for the plant body.
3. Transport: The visible, above-ground parts of the plant needed structures that absorb water and minerals from the soil, mud, or sand, and a set of tubelike structures that distribute them throughout the plant body.
4. Reproduction: Unable to shed gametes directly into an ocean, lake, or pond, land plants had to develop ways to reproduce in air. As you'll see, the relative success of each group of land plants depends, in part, on how its particular set of adaptations met these challenges.

Bryophytes: Pioneers on Land

The modern bryophytes include mosses, liverworts ("lobed plants"), and hornworts ("horn-shaped plants").

Second only to the flowering plants in number of species, bryophytes are small organisms that generally stand less than about 3 cm (1 1/4 in) tall (see Fig. 12.14, Table 12.2, and *E* 12.1). They are often among the first plants to colonize a new area, and in most of the species, the sporophyte (which is diploid) grows like a miniature street lamp from the leafy gametophyte (which is haploid). These small, simple plants have waterproof coatings that help prevent them from drying out. Bryophytes have fairly rigid tissues that help

Table 12.2 Green Algae and Major Plant Divisions

Angiosperms (Flowering plants)

Monocots (50,000 species)	Lilies, corn, onions, palms, daffodils	Leaves with parallel veins; seedlings have just one "seed leaf" or cotyledon; flower parts usually occur in multiples of three; seed stores much endosperm 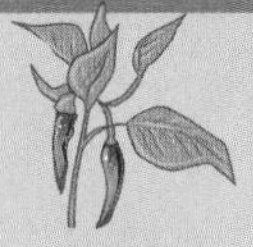
Dicots (225,000 species)	Roses, apples, beans, daisies	Leaves with netlike veins; seedlings have two cotyledons; flower parts in multiples of four or five; seed stores little endosperm 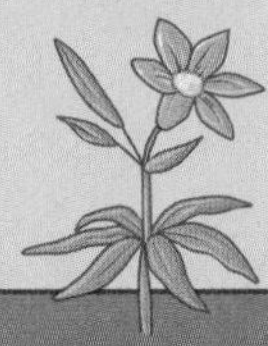

Gymnosperms

Gnetophyta (70 species)	*Welwitschia*	Low-growing, cone-producing native of southwestern Africa, with long, flat twisting leaves 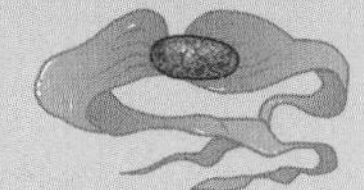
Coniferophyta (Pinophyta) (600 species)	Pines, sequoias, firs	Naked seeds produced in cones; usually have needlelike leaves or scales; produce pollen; well-developed vascular system; true roots, stems, and leaves; sporophyte is dominant and supports gametophyte; conifers harvested in great numbers for wood products 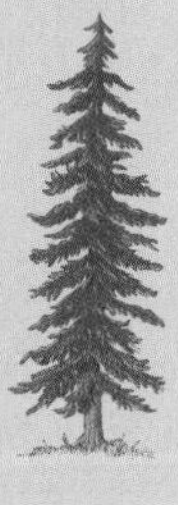
Ginkgophyta (1 species)	Ginkgos	Round fleshy cones on female trees, leaves turn golden and fall each year 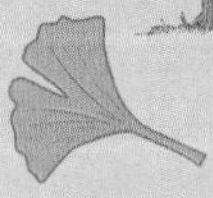
Cycadophyta (100 species)	Cycads	Flourished 200 million years ago, seeds carried on open reproductive surfaces on cones 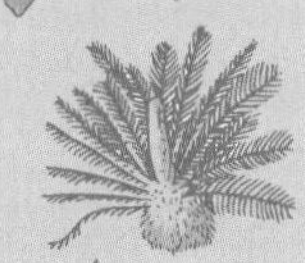
Seedless vascular plants (Sphenophyta, Pterophyta, and Lycophyta; 13,000 species)	Ferns, horsetails, club mosses	Vascular pipelines; rhizomes, stems, and fronds; gametophyte can be tiny, independent plant or grows from sporophyte; grow on shady forest floors in low-lying damp areas 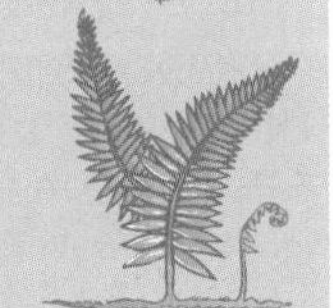
Bryophytes (Bryophyta; 24,000 species)	Mosses, liverworts, hornworts	Waterproof coatings, rigid tissues for upright growth on land, rootlike rhizoids; haploid generation (gametophyte) is dominant; often the first plants to colonize an area 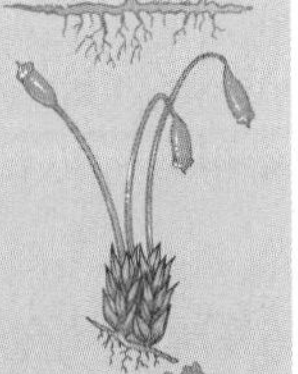
Green algae (Chlorophyta; 7000 species)	*Ulva, Chlamydomonas*	Produce carotene, chlorophyll, like the land plants; many have conspicuous haploid and diploid generations; ancestors to the land plants; found in fresh water 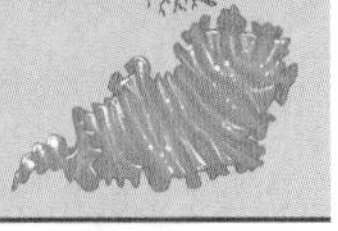

(Bracket labels at left: Embryo-Producing plants; Vascular plants; Seed plants)

keep the low plants upright. Mosses, liverworts, and hornworts, however, are small enough that most don't have the kind of internal transport system that absorbs or moves water around inside the plant and lends rigidity. Instead of true roots, most bryophytes have hairlike **rhizoids** that act only as anchors, usually absorbing neither water nor minerals. Water reaches the individual plant cells by slowly diffusing through the entire organism, like a paper towel soaking up a spill. Because of this dependence on diffusion, the plants are limited in size, and are often able to grow only in shady, moist places. In addition, bryophytes can reproduce sexually only where it's at least seasonally wet. This is because their sperm swim about with flagella as do the gametes and certain body cells of so many green algae. Bryophyte sperm can reach and fertilize an egg only when the plant is drenched in water.

The life cycles of most moss species alternate between a conspicuous gametophyte generation—the green, "mossy" plant—and a slightly less obvious sporophyte generation—the red, yellow or brown "lamp post" in many common mosses. (See Ⓔ 12.2 for more details.) While many bryophyte species have survived in moist habitats, this pioneering group did not itself give rise to other more complex land plants.

Vascular Plants

Bryophytes are generally quite successful in moist shady habitats, but home gardeners know that many kinds of vegetables and herbs "prefer" to grow in raised beds because they "hate wet feet." This is true, for example, for chile plants. Somehow, chiles have to get water up into their leaves, flowers and fruits, knee-high off the ground, and this goes for any plant with a significant vertical dimension. Upright plants clearly require some method for moving water against gravity, and this method—a system of internal tubes (the vascular system)—evolved a bit more than 400 million years ago, about the time the first insects appeared on land.

The **vascular plants,** which include the ferns, the pine trees (gymnosperms), and chiles (angiosperms) have specialized fluid conduction systems that solve the problem of how to transport water, minerals, and sugars (the products of photosynthesis) throughout the plant (Fig. 12.16a). These transport tubes are strengthened by long cells containing **lignin,** a substance that makes individual plant cell walls rigid enough to collectively support trees as huge as the General Sherman sequoia tree with a trunk wider than a good-sized truck (see Fig. 12.19). In addition to the internal transport and support system, early vascular plants had **stomata,** as well: tiny openings in the plant surface that admit carbon dioxide into the plant's interior. The 400-million-year-old fossils of a plant called *Cooksonia* found in England show these gas-exchange pores or stomata (see Fig. 12.15). These adaptations enabled vascular plants to tolerate drier habitats more efficiently than bryophytes and thus to spread farther inland into unoccupied areas. The sporophytes of vascular plants also evolved the ability to grow from the tips of the plants instead of below the tip, as in bryophytes. This allowed the diploid plant to assume a branching growth habit and with it, the production of many spore-producing organs on a single sporophyte and, in turn, many more spores and greater reproductive potential.

The Seedless Vascular Plants: Horsetails and Ferns

Ferns are probably the most familiar of the **seedless vascular plants,** plants that have a vascular transport system but don't form seeds. Seedless vascular plants include the horsetails, club mosses, and ferns (see Table 12.2). In the tropical climates of the Paleozoic era, a bit less than 400 million years ago, these first vascular plants grew as vast primordial forests, and reached sizes much larger than today's tree ferns (Fig. 12.16b). However, with the dawning of the Mesozoic era about 245 million years ago, the climate gradually grew colder and drier, and these giants disappeared, leaving behind smaller representatives of each group (Fig. 12.16c), such as the modern horsetails. In a different legacy, the bodies of the fallen giants were only partially decomposed by fungi and bacteria; this left organic compounds that became transformed over millions of years into the enormous coal deposits people mine for fuel.

Some seedless vascular plants have horizontal stems called **rhizomes** that grow on or just beneath the ground surface and survive from year to year (Fig. 12.16a). Growing up from each rhizome are many erect, leafy stalks. In horsetails and lycopods, the erect parts are **true stems;** in ferns, they are the central stalks of large leaves called **fronds.** At the end of each growing season, the stems and leaves die, and are replaced the following season by new aerial parts from new places on the slowly expanding rhizomes. This replacement is the way seedless vascular plants reproduce asexually. Figure 12.16d shows the main stages in a fern's sexual reproduction.

Horsetails, lycopods, and ferns show a continuation of the trend toward dominance by the diploid generation (review Fig. 12.11): In these seedless vascular

How can you tell chiles are seed-forming vascular plants?

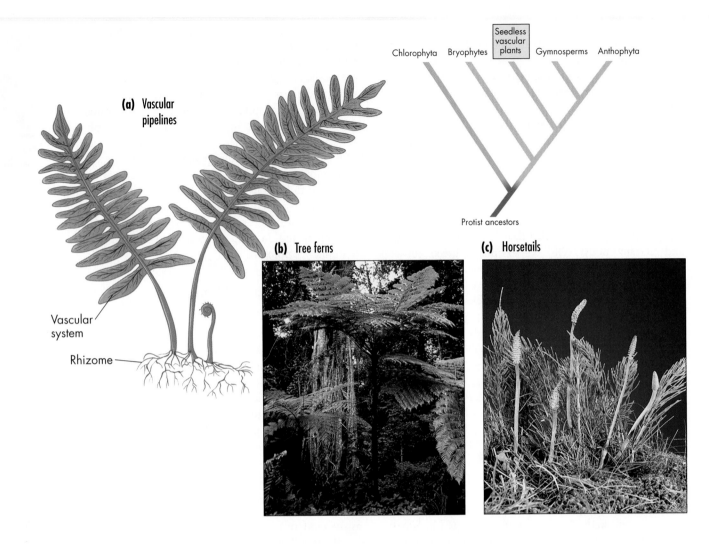

(a) Vascular pipelines

Vascular system

Rhizome

Chlorophyta   Bryophytes   Seedless vascular plants   Gymnosperms   Anthophyta

Protist ancestors

(b) Tree ferns

(c) Horsetails

plants, the sporophyte is the conspicuous adult, and the gametophyte is a small, free-living green plant. Like the mosses, ferns, horsetails, and lycopods retain swimming sperm and require water for sexual reproduction. In a sense, the seedless vascular plants are the botanical equivalents of frogs, salamanders, and other amphibians, which can live on dry land but must have water to lay their eggs. Interestingly, giant amphibians and seedless vascular plants were the dominant land organisms during the warm, swampy Paleozoic.

The Invention of Seeds

If you want to grow chiles in your garden or in a clay pot on your patio, you simply visit a gardening store in the spring, pick an envelope off the rack, and at home, shake out the smooth yellowish discs—the chile seeds. We take seeds for granted, but they are amazing little

devices. Seeds can survive for several years just sitting on a shelf, yet when they are moistened, they can burst open, allowing the little embryo housed inside—protected by the seed coat and surrounded by stored food—to grow into a new plant (Fig. 12.17a). Think of the possibilities if you could do this with a prized breed of dog or an endangered subspecies of tiger! The arrival of seeds spurred a biological revolution that dramatically changed Earth's landscapes.

We've already encountered the two major groups of seed plants, the gymnosperms and angiosperms. In the gymnosperms or naked-seed plants, including yew and pine trees, the seeds are exposed on the surface of the plant's spore-forming organ (Fig. 12.17b). In the angiosperms or *Anthophyta*, including chiles and other flowering plants, maternal tissues surround the seeds forming organs called fruits (Fig. 12.17c). As you might suspect, gymnosperms were the first to evolve.

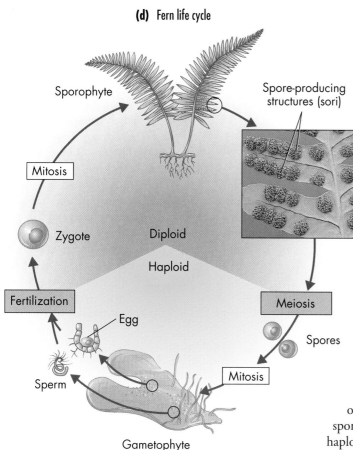

(d) Fern life cycle

Sporophyte

Mitosis

Zygote

Diploid

Haploid

Fertilization

Egg

Sperm

Gametophyte

Spore-producing
structures (sori)

Meiosis

Spores

Mitosis

Gymnosperms: Plants with Naked Seeds

As the Mesozoic Era began about 245 million years ago, the great megacontinent Pangaea was breaking apart, the continental seas had retreated, and new masses of high, dry land became open to plants and animals (see Table 10.2). While the seedless vascular plants were tied to swampy lowlands, riverbanks, and coastal lagoons, lines of plants that had already evolved with reproductive innovations—seeds—were able now to dominate the land. These organisms were the gymnosperms ("naked-seed" plants; in ancient Greece, the *gymnasium* was a place where athletic contests were held in the nude). Gymnosperms included the now-extinct seed ferns as well as the cycads, ginkgoes, and conifers, such as yews, pines, firs, sequoias, and redwoods (see Table 12.2 and (E) 12.1). The reproductive innovations of this group include (1) the pollen grain, (2) the seed, and

(3) a further shift toward the dominance of the diploid (sporophyte) generation.

The large, familiar pine tree is an example of the sporophyte generation of a gymnosperm. The sporophyte makes small, male cones (Fig. 12.18a), which contain hundreds of compartments, each of which contains hundreds of spore-forming cells. In the early spring, these spore-forming cells undergo meiosis and produce small haploid spores (Fig. 12.18b). Cell division takes place in each spore and generates a small gametophyte consisting of just four cells; this is the **pollen grain** (Fig. 12.18d). Inside the pollen grain, one of the cells will develop into the sperm. The wind can then airlift the pollen, with the sperm inside, far from the parental plant, where it can meet an egg. Notice how much smaller this gametophyte is—the thickness of a book page—compared to the size of the sporophyte—the huge pine tree. Contrast this with the sporophyte and gametophyte of a moss, wherein the gametophyte is the larger plant (see Fig. 12.14).

A gymnosperm's egg forms in a way roughly similar to the sperm formation in male cones: Cells in the female cone (Fig. 12.18c) undergo meiosis and produce haploid spores, but these are larger than the ones produced by the male cones. Botanists call the two sizes of spores **microspores,** which form the male gametophyte, and **megaspores,** which form the female gametophyte. The production of these different-shaped spores occurs not only in the gymnosperms but carries through in the flowering plants, including the chiles.

(a) Cross section of a gymnosperm seed

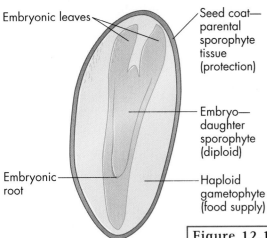

Embryonic leaves

Seed coat—parental sporophyte tissue (protection)

Embryo—daughter sporophyte (diploid)

Embryonic root

Haploid gametophyte (food supply)

(b) Pine cone

Seed

(c) Chili, angiosperm

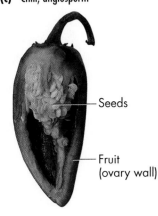

Seeds

Fruit (ovary wall)

Figure 12.17 **The Invention of the Seed: A Three-Generation Family.** (a) A seed consists of a protective seed coat, provided by the diploid sporophyte parent; a food supply, provided in gymnosperms by the gametophyte (haploid) parent; and the diploid embryo, the next sporophyte generation. (b) The seed of a pine tree is not covered by maternal tissues. (c) In contrast, the seeds of a chile are surrounded by the bright red ovary tissue of the sporophyte (maternal plant).

The female gametophyte of a pine tree remains microscopic and unable to undergo photosynthesis, but it produces the egg cell. The egg cell—surrounded by cells of the gametophyte mother, which are in turn surrounded by cells of the sporophyte mother—is called the **ovule.** The ovule becomes the seed when the egg is fertilized (Fig. 12.18d). The meeting of the egg and sperm produces the zygote, the start of the next diploid, sporophyte generation. A seed therefore contains cells from three plant generations: the outer protective layers of the parental diploid sporophyte, the nutritive cells of the haploid gametophyte, and the diploid zygote of the next sporophyte generation, the next new pine tree.

It is curious that the dominant land plants and animals in the Mesozoic, the gymnosperms and reptiles, evolved similar reproductive strategies for life on land. In both groups, water for fertilization comes from moisture in the tissues or reproductive tract of the parent rather than in splashing raindrops or standing water. Furthermore, with their tough seed coats and enclosed nutrients for the embryo, gymnosperm seeds are analogous to reptilian eggs. These too have tough shells and an internal food supply (yolk), and they could be laid and hatched on land. Seeds and shelled eggs freed both plants and animals from the dependence on water that the more ancient bryophytes and amphibians had—and still have today.

Diversity Among the Gymnosperms

Gymnosperms share certain important traits, primary among them that their seeds are not enclosed in an ovary. But the members of this group don't occupy a single branch on the tree of life (Table 12.2). Instead, they are a collection of four living divisions (or phyla): the cycads (SIGH-cads), the ginkgo, the gnetophytes (KNEE-toe-fites), and the conifers. Cycads look roughly palmlike and are found in tropical areas; only one species is native to the United States and was once common but is now rare in Florida's sandy woods. Ginkgo, or maidenhair tree, is an ornamental with fan-shaped leaves that turn a brilliant gold in the fall. Many herbal medicine producers use ginkgo extracts in their preparations for "memory improvement" and other claims. Gnetophytes are a group of naked-seed plants that display some features in common with flowering plants, including some characteristics of their cones and vascular system. Table 12.2 and **E** 12.1 describe and picture these groups in greater detail.

Conifers: Familiar Evergreens

The cone-bearing trees called **conifers** are the most familiar and largest remaining group of gymnosperms, and they include many ecologically and economically important species. Millions of acres in mountainous regions are dominated by pine, spruce, fir, and cedar, much of which is harvested for wood, paper, and resins. Smaller conifers, such as yew, hemlock, juniper, and larch, are often used for the graceful landscaping of buildings and parks. Finally, the massive sequoias and redwood trees are conifers as well as living reminders of the Mesozoic Era, an age of giants (Fig. 12.19).

Conifers have two distinctive characteristics: their narrow leaves and familiar woody cones. Their needle-shaped leaves are covered by a waterproof **cuticle,** or waxy layer. The needle shape resists drying, having little surface area relative to volume. In spring, at the start of the conifer life cycle, a large pine tree sprouts thousands of soft male cones on lower branches and the more familiar large, hard female cones on upper branches. Cells

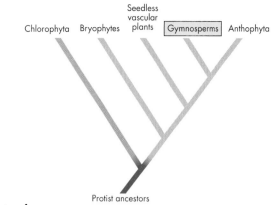

Chlorophyta   Bryophytes   Seedless vascular plants   Gymnosperms   Anthophyta

Protist ancestors

(a) Lodgepole pine

(b) Cross section of pollen-producing cone

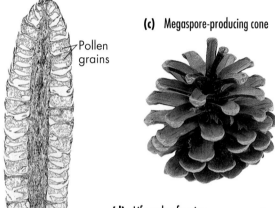

Pollen grains

(c) Megaspore-producing cone

Figure 12.18 **Innovations of the Gymnosperms.** The sporophyte of a lodgepole pine (a) produces pollen-producing, or male, cones as well as female cones. A cross section of a pollen-producing cone (b) reveals compartments in which cells are undergoing meiosis and producing haploid spores. The haploid spores develop into gametophytes, which are pollen grains, including a sperm cell. The pollen grains waft through the air to ovule-producing (female) pine cones (c), which contain the ovules, the female gametophyte and egg. (d) Life cycle of a gymnosperm, a pine tree. The tree (the diploid sporophyte) produces male and female cones, in which meiosis occurs to produce haploid spores. These spores undergo mitosis and make the male (pollen) and female gametophytes (both haploid). Fusion of sperm and egg results in the diploid zygote, which undergoes mitosis and develops again into the huge sporophyte, the pine tree.

(d) Life cycle of a pine

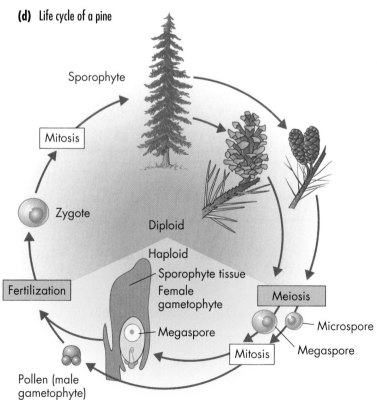

Sporophyte

Mitosis

Zygote

Fertilization

Pollen (male gametophyte)

Diploid

Haploid

Sporophyte tissue

Female gametophyte

Megaspore

Mitosis

Meiosis

Microspore

Megaspore

in these cones produce pollen and eggs and form the pine seeds.

Conifers could colonize higher and drier reaches of the continents during the Mesozoic because of their numerous attributes:

• Drought-resistant leaves
• Protective seed coats
• Haploid male gametophytes reduced to a few cells that become airborne pollen grains
• Female gametophytes also reduced to a few cells that produce eggs protected in the ovule
• A well-developed vascular system that produces wood and stiffens the trunk, branches, and roots
• The tendency to form the kinds of mycorrhizal associations with fungi we saw earlier
• The ability to survive and produce new generations, sometimes for several centuries

Because of these modifications for life on land, conifers are still a successful group, with more than 600

What are the roles of the chile flower?

Figure 12.19 **A Giant Gymnosperm.** The General Sherman Tree is certainly one of Earth's largest individual organisms. It's 275 feet (84.6 meters) tall, has a trunk that's 36.5 feet (11.1 meters) in diameter, and it weighs 1,385 tons—as much as 250 adult African elephants or 11 blue whales! The General dwarfs both neighboring trees in Sequoia National Park and the visitor at its base.

modern species. Nevertheless, one final set of evolutionary changes gave flowering plants—including our case history plants, the chiles—their still greater success.

12.4 Flowering Plants: A Modern Success Story

When you think of a chile, you usually think not of the plant but of the red, hot pepper. That structure, the chile plant's fruit, is one of the important innovations of the most recent group of plants to evolve, the *Anthophyta,* or flowering plants. Naturally, the flower is the

other! These plants (also called angiosperms, or *Magnoliophyta*) appeared as the continents rose, and drier, colder conditions became more commonplace on Earth's landmasses about 165 million years ago. During the modern, or Cenozoic, era the flowering plants became—and remain—the dominant land plants (review Fig. 12.11).

As flowering plants came to dominate the landscape, mammals became the dominant land animals. Both groups of organisms evolved with reproductive structures that protect and nourish developing embryos. Mammals evolved the placenta and the womb, while the flowering plants evolved a new reproductive structure, the **flower,** consisting of a female flower part, the **carpel** (containing at least one **ovary,** the maternal tissue that houses and protects the ovules) and at least one **stamen,** the part of the flower that produces the male gametophyte, the pollen (Fig. 12.20). Flowering plants also evolved **fruits,** the mature, ripened ovary containing the **seeds.** Flowers help the process of pollination and fruits promote seed dispersal. Flowering plants also usually have broad leaves that collect light efficiently. Flowers, fruits, and broad leaves along with a rapid life cycle together allowed flowering plants to radiate into more species than the combined numbers of all other plant groups. Figure 12.20 gives a brief overview of the flowering plant life cycle. You'll find a more detailed description of the flowering plant life cycle in Chapter 21.

Today, flowering plants are the most common and conspicuous species in Earth's tropical and temperate regions. It is not surprising, therefore, that when some biologists claimed that the largest living thing is a fungal clone (review Fig. 12.4), some botanists, not to be outdone, claimed that a flowering plant is actually the champion. They located a huge stand of quaking aspens in Utah. This 106-acre, 6000-ton stand arose from a single original seed, and each shoot arises from a single immense root system (Fig. 12.21). This giant clone weighs more than the fungal clones pictured in Figure 12.4. But the rivalry continues.

From the monumentally successful flowering plants of tropical and temperate regions come virtually all of our crop plants (wheat, rice, corn, soybeans, fruits, and vegetables such as green peppers) and beverages (coffee, tea, colas, and fermented drinks), as well as spices (including paprika and chile powder), cloth, medicines, hardwoods, ornamental plantings, and, of course, flowers—symbols of beauty, affection, and renewal throughout human history. Table 12.2 shows the two major groups of flowering plants—the **monocots,**

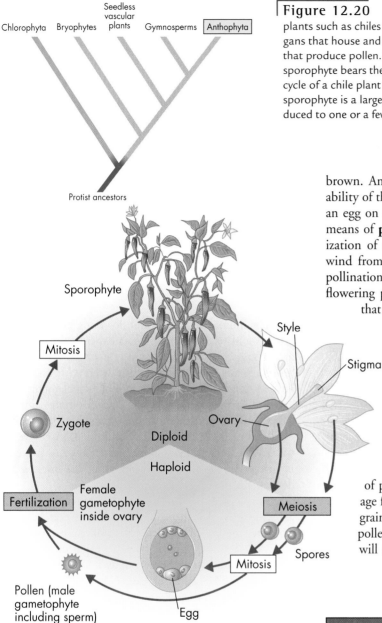

Chlorophyta Bryophytes Seedless vascular plants Gymnosperms [Anthophyta]

Protist ancestors

Sporophyte

Mitosis

Zygote

Diploid

Ovary

Style

Stigma

Haploid

Fertilization

Female gametophyte inside ovary

Meiosis

Mitosis

Spores

Pollen (male gametophyte including sperm)

Egg

Figure 12.20 **Innovations of Flowering Plants.** Flowering plants such as chiles evolved flowers. These contain ovaries, or organs that house and protect ovules, as well as stamen, or organs that produce pollen. The ovary develops into the fruit. The diploid sporophyte bears the flowers, fruit, and (often) broad leaves. The life cycle of a chile plant is typical of flowering plants, in which the sporophyte is a large, free-living plant and the gametophyte is reduced to one or a few cells dependent on the sporophyte.

brown. Another common trait is **self-pollination,** the ability of the pollen from a single chile flower to fertilize an egg on the same flower. But chile plants have other means of **pollination,** or pollen dispersal for the fertilization of eggs. Chile pollen can also be dispersed by wind from one individual to the next, allowing cross-pollination between different plants. And like so many flowering plants, chile flowers can also attract animals that inadvertently bring about cross-pollination in a way unavailable to naked-seed gymnosperms.

Chile flowers are usually small but they produce **nectar,** a sugary fluid that attracts butterflies, aphids, ants, and other animals. In general, when you see a plant with bright, showy flowers, you can usually assume that the blossoms attract insect, bird, or mammal pollinators. These "envoys" accidentally carry loads of pollen from one plant to another as they forage for sweet nectar or for the protein-rich pollen grains. A flying or walking animal dusted with pollen and scouting for more of its favorite food will substantially increase the chances of pollina-

which have a single embryonic leaf, and the **dicots,** which have two embryonic leaves—and their main characteristics.

Flowers and Fruits and the Coevolution of Plants and Animals

Chiles are typical flowering plants in many ways, with their flowers of various sizes and hues and their often large, showy fruits in green, gold, orange, red, or

Figure 12.21 **Is a Flowering Plant the Largest Living Thing?** Biologists from the University of Colorado claim that a stand of quaking aspen in Utah is the largest living individual organism. These aspen are the genetically identical offspring of single seed and share a common, interconnected root system. The current unresolved controversy over what is the biggest organism is a standard and healthy aspect of the scientific process. What's the largest, in your opinion?

(a) Bat pollinating a flower   **(b)** Cedar waxwing eating a berry   **(c)** Seeds dispersed by wind   **(d)** Dispersed seeds trapped in fur

Figure 12.22 **Flowers, Fruits, Dispersal, and Coevolution.** (a) Adaptations evolved in certain bat species for collecting nectar from particular kinds of flowers, and adaptations evolved in the plants for attracting the bats, which inadvertently assist in pollinating the flowers. (b) A cedar waxwing feeds berries to its young. Adult birds, following digestion, deposits the seeds, along with some fertilizer, in new locations. (c) Many plants use the wind for dispersal, such as milkweed (*Asclepias synaca*) seeds, equipped with sails blowing in a summer breeze. (d) Other plants can hitch a ride clinging to a mammal's fur, such as these prickly seeds of the common cocklebur on the dog's ear and neck.

How do chile plants exemplify coevolution?

tion, whether in a chile flower, a wild rose, or an apple tree (Fig. 12.22a).

After pollination and fertilization, the seeds begin to develop, and the wall of the ovary eventually matures into a fruit. Fruits not only surround the seeds but also aid in dispersal by attracting hungry animals. Oftentimes, animals attracted to a fruit's colorful skin or pleasing flavor eat the fruit, and then excrete the seeds along with feces in some new location (Fig. 12.22b). Paul Bosland suspects that chile fruits evolved their bright colors as a means of attracting birds. At the same time, though, he thinks they probably evolved their fiery capsaicin compounds as repellents for mammals. When a chile seed passes through the digestive tract of a mammal, the seed is killed, but the molten taste of wild and most domesticated chiles tends to keep mammals away (excluding ourselves!). Why do we like the pungency of chiles? The production of pain is often followed by the release of natural brain opiates called endorphins, which mask pain and leave a sense of pleasure in its place. Just as people can become addicted to opium, they can become addicted to their own endorphins and will do seemingly irrational things on a regular basis—like eating hot peppers or running long distances—to get it! Birds have a much simpler reason for eating chiles: They can't taste the hot capsaicins due to subtle differences in their pain receptors. And the chile fruits are perfectly good sources of food calories and vitamins. When birds eat pepper fruits and seeds, the little life-units pass through their digestive tracts ready to germinate and are often dropped in new regions suitable for the growth of new chile plants.

The mutual dependence of flowering plants on pollinators and seed dispersers and of these animal species on the same plants for nutrition is an example of **coevolution.** Plants that had genes for flowers with sweet nectar, a strong fragrance, or a bright color might have survived in greater numbers because more animals would have visited them in their constant foraging for food and carried away pollen. The animals that were attracted to the new food sources probably also survived in greater numbers, thus perpetuating the interdependence.

This coevolution has resulted in specialized physical structures in both plants and animals that help ensure the tradeoff of nutrients for pollination and seed dispersal. The mouthparts of many types of bees, butterflies, moths, birds, and even bats are the right shapes for tapping the nectar or pollen of the flowers they visit. Flowers pollinated by hummingbirds, which see best in the red spectrum and have long, narrow beaks and a poor sense of smell, for example, have bright red flowers, such as fuchsia, with a long, slender shape but little fragrance. Conversely, flowers pollinated by bees tend to be yellow or blue and very fragrant, corresponding to the bees' vision and ability to perceive odors.

Plants have evolved other mechanisms of seed dispersal, including "parachutes" that loft the seeds into the wind and hooks for hitchhiking onto animals passing by (Fig. 12.22c,d). The many kinds of seed dispersing mechanisms are almost as diverse as the thousands of color, size, and heat combinations of chile peppers.

Chapter Summary and Selected Key Terms

Introduction

Paul Bosland is one of our nation's experts on chile plants. Members of this group, including wild and domesticated species, produce unique, protective alkaloids called capsaicins that give chile peppers their pungency. Chiles are associated with fungi in several ways. They have most major adaptations to life on land, and demonstrate the characteristics of flowering plants. They coevolved with animals, and humans have long influenced their traits.

Lives of Fungi

Fungi (p. 300) are eukaryotes and heterotrophs, and have a tubular body that produces spores. Fungi include the mushrooms, puffballs, yeasts, and molds, and have a similar body plan: filaments called **hyphae** (p. 302) packed into a mat called a **mycelium** (p. 302). Even **yeasts** (p. 302), which are primarily single-celled fungi that bud, form hyphae if starving for nutrients. Fungi have cell walls made of **chitin** (p. 302) perforated by little holes that connect the cytoplasms of adjacent cells. Fungi form **fruiting bodies** (p. 303), such as the familiar button mushroom used in cooking. A fungus digests nutrient matter outside of its own fungal body. Nutrients include a wide range of nonliving organic matter from wood to leather to dead and living tissues of plants and animals. Most fungi are **saprobes** (p. 304), but some are **parasites** (p. 304). Fungi reproduce and disperse by means of **spores** (p. 304) produced either asexually or sexually via the typical fungal life cycle. Fungal associations with plant roots, or **mycorrhizae** (p. 306), provide phosphorous in exchange for a place to live inside root cells. These fungus–plant interactions, a type of **symbiosis** (p. 306), may have helped the transition of plants from water to land. **Lichens** (p. 307) are symbiotic associations between a fungus and a green alga or cyanobacterium. Fungi cause many significant diseases, including vaginal yeast infections, athlete's foot, and thrush. Many kinds of antibiotics are derived from fungi, as well. There are four major divisions of fungi.

Plants: The Green Kingdom

Green algae (p. 309) were the ancestors of land plants, and share the same kinds of carotenoids and green chlorophyll. *Chlamydomonas*, *Volvox*, and *Ulva* are typical algae.

Plants: Green Embryo Makers

A **plant** (p. 309) is a multicellular eukaryote that generally develops from an embryo and generally contains chloroplasts. Plants are major ecological producers. Plants have an **alternation of generations** (p. 310), a regular alternation between multicellular haploid and diploid generations. The **sporophyte** (p. 310) is the spore-generating plant and the **gametophyte** (p. 310) is the gamete-generating plant.

Trends in plant evolution include adaptations to prevent drying; adaptations for fluid conduction and vertical support; and adaptations for reproduction and dispersal including seeds, flowers, and fruit.

The **bryophytes** (p. 313), including the mosses, liverworts, and hornworts, are small, simple, and low to the ground. The haploid gametophyte is green and leafy, while the diploid sporophyte looks like a miniature street lamp. Bryophytes have waterproof coatings and **stomata** (p. 315). Most also produce anchorlike **rhizomes** (p. 315) but no roots.

All more complex plants are **vascular plants** (p. 315), with transport tubes stiffened by **lignin** (p. 315). **Seedless vascular plants** (p. 313, 315) including ferns, horsetails, and lycopods, grow taller than bryophytes. Most have **true stems** (p. 315), large leafy **fronds** (p. 315) and horizontal stems called **rhizomes** (p. 315). **Seed-forming vascular plants** (p. 313) include the **gymnosperms** (p. 313, 317) and **angiosperms** (p. 313). Gymnosperms have naked seeds or seeds that develop in cones unenclosed by maternal tissue. Reproductive innovations include the pollen grain, the seed, male and female spores of unequal sizes (**microspores** and **megaspores;** p. 317), and a major shift toward dominance by the sporophyte generation. The male gametophyte consists of just a few cells, the **pollen grain** (p. 317). The female gametophyte is microscopic and produces the **ovule** (p. 318). Gymnosperms, unlike all earlier plants, need no water for reproduction. **Conifers** (p. 318) are the familiar evergreen trees we use so heavily for lumber and landscaping. They have needlelike leaves covered by a waxy waterproof **cuticle** (p. 318).

Flowering Plants: Modern Success Story

Chiles are flowering plants, the Anthophyta (also called angiosperms). Innovations include the **flower** (p. 320), with a **carpel** (p. 320) producing at least one **ovary** (p. 320) or maternal tissue that protects the ovules, and at least one **stamen** (p. 320), or pollen-producing structure. **Fruits** (p. 320) are the mature, ripened ovary containing the **seeds** (p. 320). Flowers, fruits, and seeds together allowed flowering plants to radiate into more species than all other plant groups and become Earth's most widespread, conspicuous species. Two major groups include **monocots** (p. 320) and **dicots** (p. 321). In chiles, the capsaicin is an evolutionary innovation that discourages predation by mammals but not birds. Chile plants have several means of **pollination** (p. 321), or pollen dispersal allowing for the fertilization of eggs.

All of the following question sets also appear in the Explore Life *E* electronic component, where you will find a variety of additional questions as well.

Test Yourself on Vocabulary and Concepts

In each question set below, match the description with the appropriate term. A term may be used once, more than once, or not at all.

SET I

(a) hyphae (b) mycelium (c) mycorrhizae
(d) chitin (e) spore

1. Seedlike reproductive structure
2. Tough nitrogen-containing polysaccharide in cell walls
3. Filaments made up of cells joined end to end
4. A fungal-root association
5. A loose mat, like steel wool

SET II

(a) gametophyte (b) sporophyte (c) microspore
(d) megaspore (e) sporangium

6. The dominant, diploid plant in the life cycle of flowering plants
7. In mosses, meiosis occurs within this structure, resulting in haploid cells that grow into gametophytes
8. In flowering plants, these cells are produced within the ovary
9. The haploid portion of a moss that is anchored to the soil by rhizoids
10. In flowering plants, these cells develop into pollen grains

SET III

(a) green algae (b) bryophytes (c) cycads (d) ferns
(e) angiosperms

11. Contain vascular tissue, but do not produce seeds
12. The group that probably gave rise to plants
13. Looks like a palm tree, but produces seed-bearing cones
14. Its life cycle is fairly evenly divided between haploid and diploid phases, although the diploid plant depends on the haploid plant for support and nutrition

(a) pollen grain (b) cone (c) seed (d) ovule (e) ovary
(f) petal

15. Found only in angiosperms, where it protects the embryo and often develops into a fruit

16. A protective structure of gymnosperms and angiosperms that develops from an ovule and enclosed embryo

17. Protects the embryo in both gymnosperms and angiosperms, eventually becoming part of the seed

18. Immature male gametophytes in the seed-producing plants

Integrate and Apply What You've Learned

19. What are the main structures of fungi and how do these organisms obtain nutrients?

20. Recent evidence from molecular genetic studies suggests that fungi may be genetically more similar to animal cells than to plant cells. Which aspects of fungal biology support this hypothesis? Which contradict it?

21. Outline the major trend that is apparent when comparing the life cycles of bryophytes, ferns, and seed-producing plants.

22. In surveying the representative plants in the plant kingdom, what are the primary evolutionary trends that are evident?

Analyze and Evaluate the Concepts

23. Which adaptations identify chile plants as members of the Anthophyta (flower plant group)? Which are unique to chiles? What advantages do these unique adaptations confer?

24. Outline two arguments, one for placing green algae in the Plant Kingdom, and one for placing green algae with the protists. Evaluate the merits of the two placements.

25. Discuss the differences between the fern life cycle and the conifer life cycle.

26. How does the presence of mycorrhizae, lichens, and pollinators contribute to the success of plants? Discuss each separately.

27. Name adaptations of the flowering plants that might account for their greater diversity than any other plant group.

Not Your Average Hobby

▲
Deconstructing Decomposers. Bruce Gill; dung beetle rolling a dung ball; dung beetle collection.

B ruce Gill has the kind of hobby he can't discuss over dinner: Dung beetles. Without dung beetles, the bearded Canadian biologist explains, the world would be "a pretty smelly, foul place." But Gill's interest goes beyond their ecological significance as animals that decompose environmental waste materials. He's fascinated by their amazing behavior and diversity, as well: Some species roll dung into orbs the size of croquet balls. Others, called "kleptos," steal dung from honest, hard-working comrades. And some specialize in eating only sloth dung, millipede dung, or tortoise dung. Gill's hobby is clearly not your stamp-collecting, model-building, or muffler-knitting kind of pastime, and it's certainly not for the faint-hearted. But through his intense personal interest, Gill has discovered a dozen new dung beetle species and expanded the world's knowledge and appreciation of what *Smithsonian* magazine called "nature's own pooper scoopers."

As a boy growing up in Vancouver, Gill became fascinated with the big striped June bugs that crawl under streetlights and smash into window screens. In time, this simple curiosity would lead to a doctoral dissertation on Panamanian dung beetles, and then to an important day job in entomology (the study of insects). Dr. Gill is the co-manager of a Canadian government agency that, he says, "identifies insect hitchhikers that move around the globe in international commerce." His job is to keep these "hitchhikers" from "getting established in Canada (where they could) destroy our forests, farms, or

livelihoods." In his free time, though, Gill returns to his passion, often paying his own way to sites in Canada, the U.S., and various parts of Latin America to study new kinds of dung beetles.

Gill is best known for discovering evidence of dung beetles among fossilized blocks of feces left by giant plant-eating dinosaurs. He published his study with geologist Karen Chin of the University of California at Santa Barbara in 1996. Before their study came out, scientists had traced the evolution of the beetle lineage back 300 million years, but dated beetles that eat dung to only 40 million years ago. In that relatively recent geological era, mammals were already depositing their droppings as they roamed the continents. Gill and Chin, however, pushed back the age of the oldest dung beetles to 76 million years and established them as the custodians of dinosaur detritis, as well.

How much detritis did dinosaurs leave? If you've ever seen an elephant pat—a pile of elephant dung—at the zoo, picture a collection two to four times bigger beneath an enormous, herbivorous dinosaur. Big as it is, an elephant pat on the African savanna disappears overnight as thousands of dung beetles chop up the rich resource and drag or roll it away. So too, says Gill, did the dinosaur dung beetles clean up their companions' monumental messes.

Oh sure, other decomposers feast on animal droppings, too: earthworms, flies, molds, bacteria. All are crucial to ecological recycling. But none can compare to the dung beetles' speed and flair. One Oklahoma rancher who corrals his cattle in one paddock then shifts them to another made an interesting observation: 48 hours after moving out the animals, every speck of manure is gone—the work of 11 native prairie dung beetle species. By contrast, Australian ranchers tried raising cattle on arid grasslands where only kangaroos had grazed. Australian dung beetles, comments Bruce Gill, "were adapted to little marsupial pellets and couldn't handle a great, big, wet cow pie." Soon the pies were piling up and plastering the ground into a hard dry pavement that suppressed grass growth. Only by importing and releasing bigger, hungrier dung beetles were the ranchers able to raise cattle in Australia. (So far, the imported dung beetles haven't caused other ecological problems, as so often happens with introduced exotics.)

The dung beetle gets its flair from elaborate anatomy and behavior. These insects, says Gill, evolved "all kinds of 'cutlery' to cut up, package, and move dung quickly," including sharp pointed teeth and forelegs "shaped like forks and knives." Some dung beetles just locate feces, eat, lay eggs, and run. But others tunnel up from underneath and pull the protein-rich material down into elaborate burrows where their eggs can hatch and feed. Still others fashion large dung balls, which they roll away and bury. "Klepto" species have developed the trick of tunneling vertically to cart off stored dung from the larders of harder working neighbors. The great diversity of dung beetles is impressive in itself: There are at least 10,000 species, and perhaps as many as all the birds, mammals, fishes, and other vertebrates combined—upwards of 50,000. Toss in the specializations some beetles show for eating sloth dung or millipede dung, for instance, and you have a flamboyant animal group that could well attract scholars like Gill. In fact, so many generations of naturalists were fascinated before Gill that dung beetles inspired a classic Egyptian icon: Their famous scarab beetle rolling the sun across the sky was based on observations of dung beetles rolling dung balls!

Dung beetles are a good case history for this chapter because they remind us that ours is a planet of terrestrial insects: Two out of every three living species so far discovered and named by biologists are "bugs," and nearly half of all species are beetles. From this group, we can learn a great deal about the animal body, animal evolution, animal diversity, relationships between the major animal groups, and the many threats to Earth's beetles—not to mention other species and ourselves. This chapter will explore animal life, and along the way, you'll find answers to these questions:

1. What are the general characteristics that all animals share?

2. What are the many different types of animal body plans?

3. How did the chordates (animals with notochords) evolve and what are their innovative features?

4. Where do primates, including humans, fit into the spectrum of animal species?

13.1 The General Characteristics of Animals

Picture a dung beetle rolling a big ball of dung, then think about the animal's characteristics. What might lead you to classify it as an animal and not, say, a plant or a fungus?

1. **Animals are multicellular.** Dung beetles have millions of cells and whales have trillions, unlike most prokaryotes and protists, which are usually single-celled organisms.
2. **Animals are heterotrophs.** Unlike plants, animals cannot manufacture their own food and instead obtain nourishment by consuming other organisms or their byproducts, including the wastes devoured by dung beetle larvae.
3. **Animals are self-propelled.** Unlike most fungi, which are also multicellular heterotrophs, animals avoid danger or search for mates and food by moving about on their own at some point in their life cycle (usually throughout the entire cycle). A dung beetle rapidly rolling a ball of cattle manure is typical of such self-propulsion. Fungi and plants often produce spores that disperse the species, but these tend to move passively rather than under their own power.
4. **Animal bodies are diploid and develop from embryos.** Animals reproduce primarily by sexual reproduction, and their bodies consist of diploid cells. Most plants develop from embryos, too, but all plants have a multicellular haploid phase in addition to a multicellular diploid phase (see Fig. 12.14). Every animal grows from a zygote formed when egg and sperm unite (two unequally sized gametes). The animal zygote becomes a ball of cells (see Figure 8.10) and passes through various distinct stages of development before becoming an adult.

Innovations in Animal Evolution

One of the fascinating things about studying animals is that as we explore the major groups—including the dung beetles and other insects—we can see how various animal features arose: heads, tails, wings, legs, backbones, internal organs, and so on. Fundamental to virtually all evolved animal traits—from simple to complex, and from ancient to recent—were four major anatomical and physiological innovations: multicellularity, the origin of tissues, the origin of three tissue layers in a bilateral body, and the origin of specific patterns of embryonic development (Fig. 13.1).

(1) First, multicellularity arose. Animals evolved from single-celled protists, and very early on, such cells gained the ability to stick together in colonies. (2) Next, cells within colonies became specialized into inner and outer tissue layers (recall the endoderm and ectoderm we encountered in Chapter 8), as well as dedicated nerve cells. This early specialization was associated with an increasingly organized body that was often radially symmetrical (round and often flattened, like a tuna can rather than a ball). (3) A third set of innovations included an additional middle tissue layer between the inner and outer layers (recall the mesoderm from Chapter 8); formation of a body cavity in that middle layer; and **bilateral** (two-sided) **symmetry** with an anterior (front end), a posterior (rear end), and a left and right side, rather than a radial body plan wherein all sides are similar. (4) A fourth set of innovations brought about new forms of embryonic development. These involved the pattern of cell divisions in the very early embryo (recall cleavage divisions from Chapter 8); the way the middle tissue layer relates to a cavity inside the body; and the way the mouth and anus arise in the embryo (recall gastrulation from Chapter 8). Together, these four innovations enabled evolving animals to exploit their environments efficiently and to develop an enormous array of body plans, appendages, and life histories. Today, the innovations serve as a framework that helps us categorize animals, from simple sponges to highly organized dung beetles and millions of others.

When Did Animals Arise?

As a group, animals have a long and interesting evolutionary history. Paleontologists have found the earliest fossils of soft-bodied animals in rocks from the Edicara region of southern Australia, and dated them to about 560 to 600 million years ago (Fig. 13.2). Fossils of burrows and trails laid down in ancient mud, however, suggest that animals already existed 800 million to 1 billion years ago. Another interesting finding also helps date the origin of animals. Recall from Chapter 10 that more than 3 billion years ago, stromatolites—free-form "footstools" covered with greenish glaze—evolved along the edges of warm seas. These stromatolites were built of layer upon layer of cyanobacteria (simple photosynthetic cells) and other prokaryotes. Significantly, about 1 billion years ago, the diversity of stromatolites began to fade. Paleontologists think they were disappearing because very early animals, already established at that time, were feeding on the stool-shaped microbial communities.

What distinguishes the animal kingdom from the plant and fungal kingdoms?

Single choanoflagellate

Colonial choanoflagellate

Figure 13.1
Animal Groups and Evolutionary Innovations.

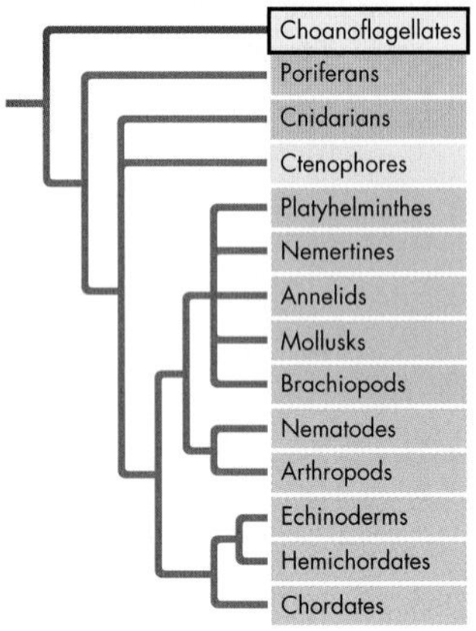

Choanoflagellates

Protist ancestor

Poriferans

Choanoflagellates
Poriferans
Cnidarians
Ctenophores
Platyhelminthes
Nemertines
Annelids
Mollusks
Brachiopods
Nematodes
Arthropods
Echinoderms
Hemichordates
Chordates

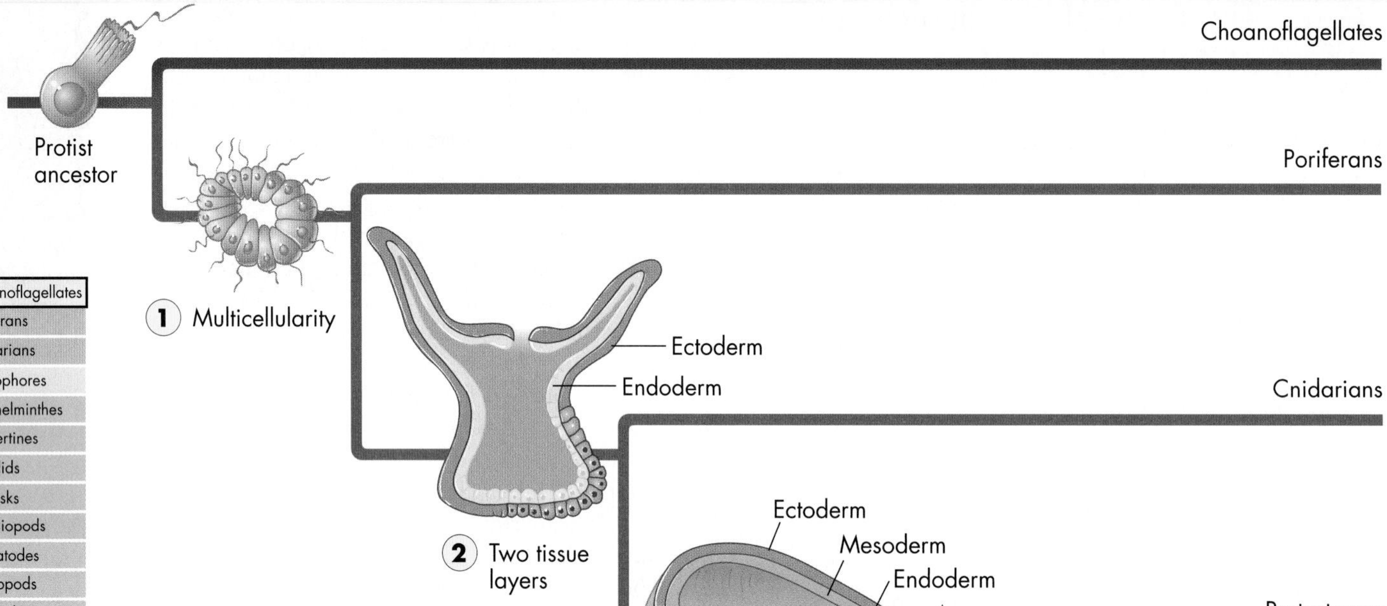

1 Multicellularity

Ectoderm
Endoderm

Cnidarians

2 Two tissue layers

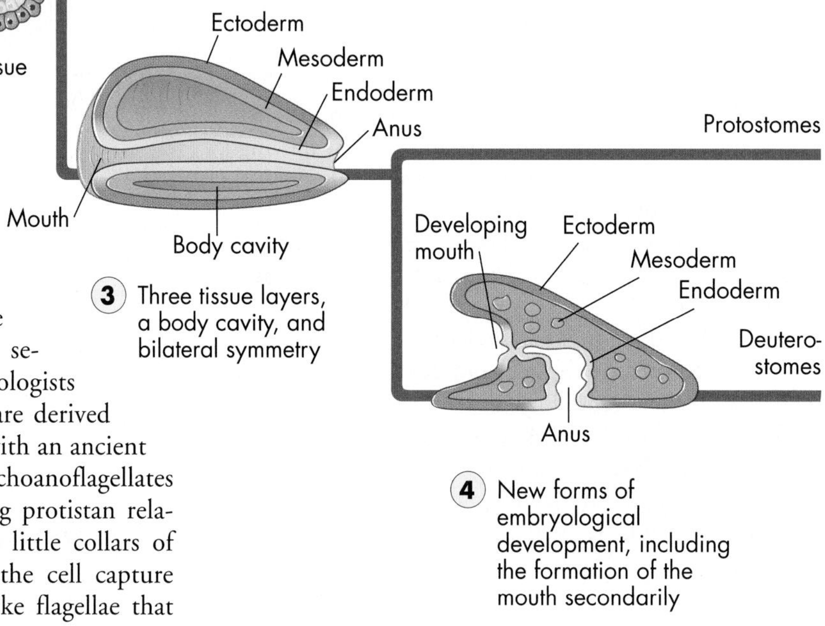

Ectoderm
Mesoderm
Endoderm
Anus

Mouth
Body cavity

3 Three tissue layers, a body cavity, and bilateral symmetry

Protostomes

Developing mouth
Ectoderm
Mesoderm
Endoderm
Deutero-stomes

Anus

4 New forms of embryological development, including the formation of the mouth secondarily

Animal Origins

In the next few pages, we'll explore dozens of very distinctive animal groups. Despite their vast diversity of shapes and complexities, all animals share similarities in cell structure and gene sequences that convince **zoologists** (biologists who study animals) that all animals are derived from a single lineage that originated with an ancient ancestral protist. The protists called choanoflagellates (see Chapter 11) are the closest living protistan relatives of animals. These protists have little collars of microvilli, cell extensions that help the cell capture small food particles and long, whiplike flagellae that

sweep small prokaryotes toward the microvilli (see margin p. 327). As we'll see, sponges (Poriferans), the simplest animals, have food-capturing cells that are virtually identical to these choanoflagellates. In addition, some choanoflagellate species grow in small colonies consisting of several cells. Ancient protists

Figure 13.2 **Fossils Show the Antiquity of Animal Life.** This Ediacaran fossil dated to 600 million years ago shows the unusual shape of a long extinct worm (*Dickinsonia costata*).

with this sort of colonial life habit may have foreshadowed the multicellularity of animals. Many zoologists think that animals may have arisen from ancient choanoflagellate protists, and that sponges may have branched off early from the rest of the animal lineage.

Some time after the first animals evolved, the activities of photosynthetic organisms caused the amount of oxygen in the atmosphere to rise rapidly (see Chapter 10). This permitted a great evolutionary radiation of the animal kingdom—the rapid origin of large numbers of animal species with a great variety of body forms. All the animals that ever existed were products of this radiation, including the first spiderlike animals to crawl on land, the giant dragonflies of primordial forests, the lumbering sauropods (herbivorous dinosaurs—plus the dung beetles that cleaned up after them), and the towering mastodons and ground sloths of the last ice age. At least 35 different major body plans evolved, which we categorize into distinct **phyla** (sing., **phylum**). Most of these still have living members. Together, the animals number more than a million individual species—and that's a conservative estimate. The total could be ten times higher or even more.

Animal Relationships

Biologists investigate the evolutionary relationships between animals by (1) comparing body forms or (2) comparing gene sequences. Comparing body forms is useful, but has a complication: Animals from separate phyla can be so different that it is often difficult to know what to compare. For example, is the "forehead" of a dung beetle with its antenna more similar to the forehead of a person or the "forehead" of an earthworm? Comparing nucleic acid and protein sequences is a more precise and generally applicable tool, and modern biologists use it frequently to help reconstruct the relationships of various animal groups.

One of the favorite molecular tools is ribosomal RNA (see Chapter 7), because these genes generally change slowly during evolution. Chapter 9 discussed the way biologists draw evolutionary trees based on molecular data. Recent genetic analyses have revolutionized our understanding of animal relationships. The evolutionary tree shown in Figure 13.3 and in *E*xplorer 13.1 presents our best current understanding of animal relation-

GET THE PICTURE

*E*xplorer *13.1*

Animals on the Tree of Life

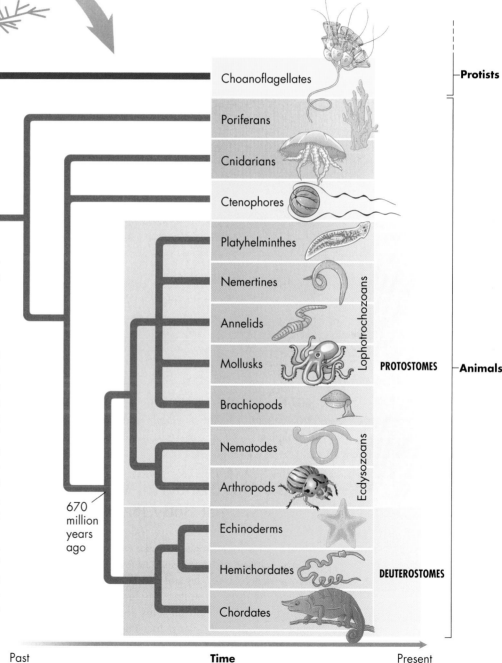

Figure 13.3   Evolutionary Relationships of the Major Animal Phyla. Animals arose from protistan ancestors similar to today's choanoflagellates. The sponges were an early branch off the main line of animal evolution. Evolutionary innovations arose as lineages diverged. Although there is still some argument between molecular biologists and paleontologists, the protostomes and deuterostomes probably split about 670 million years ago, and the Cambrian explosion of animals with outer shells that left fossils started 544 million years ago.

Table 13.1 Some Key Animal Phyla

Phylum	Examples	Number of Species	Notable Features
Porifera	Sponges	10,000	Generally asymmetrical saclike bodies; central opening; body wall perforations; spicules; no evolutionary descendants; no true tissues
Cnidaria	Jellyfish, hydras, corals, sea anemones	10,000	Radial symmetry; grow as polyps or medusae; two tissue layers; gastrovascular cavity; nerve cells, nematocysts, planula larvae; no body cavity
Ctenophores	Comb jellies	100	Radial symmetry; rows of cilia and tentacles; two tissue layers; common plankton species
Platyhelminthes	Flatworms, including tapeworms, flukes	15,000	Bilaterally symmetrical with a head; three tissue layers; true organs and organ systems; life cycles usually complex and include two or more hosts; no body cavity
Nemertea	Ribbon worms	900	Bilateral symmetry; cephalized; organ systems, unique proboscis; regeneration; lives on ocean bottom
Annelida	Segmented worms, including earthworms, polychaete worms, leeches	18,000	Bilaterally symmetrical with a head and gut tube; lined body cavity; segmentation; hydroskeleton; move with bristles pushing against ground; crop, gizzard; closed circulatory system; earthworms occur widely and help aerate soils
Molluska	Snails, clams, octopuses, squid, slugs	120,000	Bilaterally symmetrical with a head and a gut tube; three tissue layers with a lined body cavity; gills, mantle; open circulatory system; very common marine and freshwater organisms

ships based both on body form *and* molecular data. These figures provide a visual guide for the rest of the chapter.

13.2 The Wonderful Variety of Animal Body Plans

Dung beetles are animals, just like we are, but they are covered with a hard outer shell, while we have a bony internal skeleton. Beetles and people represent just two of the many kinds of animal "architectures" we see in the different animal phyla. This section explores the major kinds of body plans, focusing on evolutionary relationships between the phyla and how new shapes and functions arose. The orderly and linear organiza-

tion of our story should help you learn the different body plans, but it can leave a misimpression: that one type of living organism gives rise to another. In fact, no organism alive today is an ancestor of any other living organism. All modern organisms, no matter how simple, occupy the *tips* of branches on the tree of animal life—branches that split off (diverged evolutionarily) lower on the trunk (see Fig. 13.3). In fact, most animals on most branches of the animal family tree arose and then became extinct after existing as distinct species for only hundreds of millions of years. Another potential misconception is that evolution is a single directed path toward "bigger and better" organisms. In truth, there are many instances wherein animal lineages lost features present in ancestral predecessors. Today's surviving species may be better adapted to today's environmental conditions, but that doesn't make

Table 13.1 Continued

Phylum	Examples	Number of Species	Notable Features
Brachiopoda	Lamp shells	300	Two shells, resemble clams; lophophore feeding apparatus plus coelom; clearly protostomes
Nematoda	Roundworms	20,000	Bilaterally symmetrical with a head and gut tube; three tissue layers with unlined (false) body cavity; hydroskeleton; extremely common in soils and as parasites on other animals and plants
Arthropoda	Spiders, mites, ticks, scorpions, millipedes, centipedes, insects, lobsters, shrimp	1–30 million	Bilaterally symmetrical with a head and gut tube; lined body cavity; segmentation; exoskeleton for support and protection; specialized segments and appendages; jointed legs; tracheae or gills; acute senses; most diverse phylum in living world
Echinodermata	Sea stars, sea urchins, sea cucumbers	7000	Gut tube and lined body cavity; a head in some larvae and adults; no segmentation; first endoskeleton; unique water vascular system for locomotion; separate evolutionary line from Molluska, Annelida, and Arthropoda
Hemichordata	Acorn worms	85	Bilateral symmetry; embryos develop as echinoderms do; complete gut; gill slits; dorsal nerve cord
Chordata	Invertebrates like sea squirts, and vertebrates like fish, amphibians, birds, reptiles, and mammals	50,000	Bilaterally symmetrical with head, gut tube, lined body cavity and segmentation; stiff rod of cartilage (notochord); dorsal tubular nerve cord; gill slits

them "better" overall. Finally, evolutionary innovations have no "plan" or "goal" or "direction." At each point in time, natural selection works on populations of organisms in ways that allow individuals with certain inherited traits to leave more offspring. We can identify successful innovations in various animal groups, but they usually arose through genetic accidents that happened to benefit survival. With all this in mind, let's begin by looking at the simplest living animals, the sponges (Table 13.1).

The Sponges: Irregular Bodies Without Tissues

Sponges are models of simplicity. Their bodies perforated with many tiny openings or pores, sponges belong to the phylum **Porifera** (pore-bearing). They are generally asym-
metrical, aquatic organisms lacking distinct tissues. For centuries, people thought sponges were plants because the adults are **sessile** (stationary) and are permanently attached to rocks, pilings, sticks, plants, or other animals. Sponges, however, are multicellular animals with swimming larvae like the one in Figure 13.1 (1), and they filter and consume fine food particles from the water. Figure 13.4 summarizes the important features of sponge biology.

Lacking true tissues, sponges are essentially collections of cooperating cells. Pores lined with special cells (**choanocytes** or collar cells) perforate the body wall of the sacklike animal. These cells have a flagellum, which draws a current of water through the pore, and a sticky collar, which catches the sponge's food: bacteria, protists, and other small organic particles suspended in the water. The water then exits an opening at the top of the sponge. A gelatinous layer inside the body wall contains

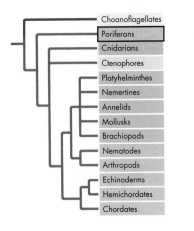

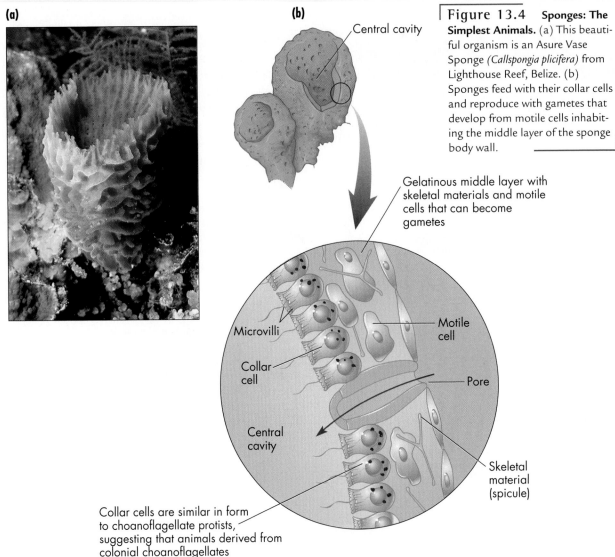

Central cavity

Gelatinous middle layer with skeletal materials and motile cells that can become gametes

Microvilli

Motile cell

Collar cell

Pore

Central cavity

Skeletal material (spicule)

Collar cells are similar in form to choanoflagellate protists, suggesting that animals derived from colonial choanoflagellates

several different kinds of motile cells, some of which differentiate into gametes. The middle layer also contains skeletal material, including fibers related to hair protein or tiny pointed **spicules** made of silica or calcium carbonate. The irregularly shaped, tan bath sponges sold commercially are the tough, fibrous skeletons left after the animal dies. As we mentioned earlier, collar cells closely resemble choanoflagellates.

Cnidarians and Ctenophores: The Origin of Tissues and Radial Body Plans

Animals "invented" multicellularity [see Fig. 13.1 (Step 1)]. This made a second set of innovations possible: radial symmetry and tissue layers. Two representative phyla, the cnidaria and the ctenophores, show this body plan.

Cnidarians: The Stinging Nettle Animals

Some of the most ephemeral and beautiful animals are members of the ancient phylum **Cnidaria** (nih-DARE-ee-ah; Greek *knide* = stinging nettle). This phylum diverged deep in the tree of life and includes the translucent hydras, the gossamer jellyfish, the sea anemones, and the colorful corals (Fig. 13.5a–d and *E* 13.1). Most live in the oceans, but a few, such as the hydras, inhabit fresh water.

Cnidarian Body Plan Cnidarians have a **radial body plan** consisting of a central axis with structures

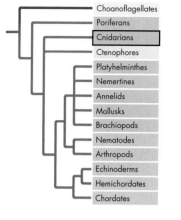

Choanoflagellates
Poriferans
Cnidarians
Ctenophores
Platyhelminthes
Nemertines
Annelids
Mollusks
Brachiopods
Nematodes
Arthropods
Echinoderms
Hemichordates
Chordates

(a) Hydra (polyp)

(b) Jellyfish (medusa)

Figure 13.5 **The Cnidarians: Stinging Nettle Animals.** Cnidarians diverged deep in the animal tree of life. The group includes (a) hydras, (b) jellyfish like this Blue Bottle (*Physalia utriculus*), (c) sea anemones such as this Whitespotted Rose anemone *(Tealia lofotensis)*, and (d) corals. Each member has a (e) radial body plan and (f) two tissue layers and a few specialized cell types.

(c) Sea anemone (polyp)

(d) Coral

(e) Radial body plan

Jellyfish, from top

(f)

Tentacle

Mouth

Epidermis

Gastrodermis

Mesoglea

Tentacle

Gastrovascular cavity

Polyp form

Mouth

Medusa form

Nerve cell

Nematocyst discharged

Nematocyst-bearing cell

Epidermis

Mesoglea

Muscle cells

Digestive enzyme–secreting cell

radiating outward like the spokes of a wheel. A cross section of a cnidarian's body wall reveals two tissue layers, the **epidermis** (derived from the embryonic ectoderm) on the outside, and the gastrodermis (derived from the embryonic endoderm) on the inside, lining the central gastrovascular (digestive/vascular) cavity. Sandwiched between the two tissues is a jellylike substance called **mesoglea** (literally, "middle glue"). Jellyfish resemble jelly because of the extreme thickness of the mesoglea. The mass of one monstrous North Atlantic jellyfish that is nearly 3 m (10 ft) across and weighs a ton is virtually all mesoglea. The gastrovascular cavity has a single opening that serves as both mouth and anus. Tentacles move prey into the cavity, where the food is digested.

Cnidarian bodies have two basic forms: (1) The **polyp,** a hollow, vaselike body that stands erect on a base and has a whorl of tentacles surrounding a mouth near the top (Fig. 13.5a,c,f), and (2), the **medusa** (pl., *medusae),* an inverted umbrella-shaped version of the polyp, with tentacles and mouth pointing downward (Fig. 13.5b,f). Sea anemones, corals, and most hydras are polyps as adults, while jellyfish are medusae.

Cnidarian Cell Types

Cnidarians have several specialized cell types not found in sponges. Embedded in the epidermis of cnidarian tentacles are cells that contain remarkable tubelike organelles called **nematocysts,** or stinging capsules. The slightest contact triggers the tube to evert like a sock turned inside out. The sharply pointed end of this nematocyst can penetrate prey and release a paralytic toxin. Several human deaths have been attributed to nematocysts from *Chironex,* a genus containing many large tropical jellyfish species.

Cnidaria and all animals that diverged more recently from this early animal lineage also have contractile cells equivalent to muscle cells and nerve cells—elongated cells that can conduct electrical signals. These cell types help cnidarians to detect prey, coordinate body movements, and capture victims and move them into the gastrovascular cavity. In cnidaria, nerve cells are arranged in an uncentralized, loose network that conducts information in all directions, and coordinates body activities.

Other specialized cells line the gastrovascular cavity and produce enzymes that help break down food extracellularly. Extracellular digestion allows animals to digest larger food pieces, and thus expand the diet compared with the intracellular digestion of very tiny food bits found in most protists and sponges.

Reproduction in Cnidarians

The cnidarian life cycle often has a polyplike generation that reproduces asexually alternating with a sexually reproducing medusalike generation (Fig. 13.6). Sexual reproduction produces a so-called **planula larva** covered by cilia that helps the new organism move around.

While sea anemones, hydras, and jellyfish usually live as independent individuals, thousands of coral polyps live together in huge colonies. Each individual

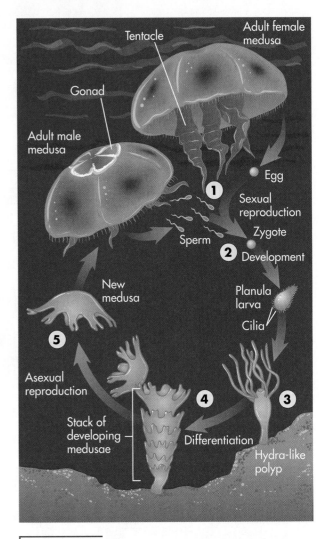

Figure 13.6 **Life Cycle of a Jellyfish.** Adult jellyfish live as medusae and release sperm or eggs into the water (Step 1). The fertilized egg develops into a small, flat, solid planula larva, which is propelled by cilia (Step 2). The planula attaches to a rock and develops into a stationary polyp (Step 3). The polyp can undergo a type of division that results in a colony of polyps resembling a stack of saucers (Step 4). The stack can reproduce asexually as each saucer develops tentacles and swims away as a new medusa (Step 5). At maturity, the medusa develops male or female gonads, and the cycle continues.

Figure 13.7 **Coral Bleaching.** Pollution and temperature changes in the oceans are causing coral colonies to expel their photosynthetic algae. "Bleaching" starves the coral polyps because they get much of their nourishment from their symbiotic algae. This bleached coral *(Diplora strigo*s) lives in the Gulf of Mexico.

polyp in the colony is encased in its own cup-shaped skeleton of calcium carbonate. These skeletons can form giant reefs and atolls, including the only biological structure visible from space, the 1900-km-long (1200-mi-long) Great Barrier Reef off Australia's eastern coast. Reef-building corals get their spectacular colors and obtain much of their energy from microscopic photosynthetic algae that live symbiotically inside their cells. Unfortunately, changes in seawater purity and temperature, as well as human activities around coral reefs, are causing an epidemic of coral "bleaching," the expulsion of the algae from the coral cells. Bleaching often leads to death of the coral (Fig. 13.7). Researchers fear that pollution and global warming will accelerate this trend and perhaps destroy the largest and most impressive structures ever made by living organisms.

Ctenophores: The Comb Jellies

Tow a bottle just below the ocean's surface half a mile from most seashores and you'll probably capture a **ctenophore** (TEEN-o-for; *cten* = comb + *phore* = bearer), a small, transparent, usually globe-shaped animal with two tentacles and eight rows of cilia that resemble combs (*E* 13.1). The tentacles contain sticky cells that capture food particles, and the moving cilia send streams of water past the tentacles. At night, these animals emit a bluish glow. Like the cnidarians, which they closely resemble, they have radial symmetry and two tissue layers. But ctenophores have more extensive

digestive and muscular systems, and lack nematocysts and the polyp/medusa body alternatives. Although there are relatively few species of ctenophores, they can make up a significant portion of the plankton in some regions of the sea.

The Origin of Bilateral Symmetry

Animal evolution saw a third set of basic innovations: the "invention" of three tissue layers as well as of bilateral symmetry (right and left sides). Animals with these characteristics have been amazingly successful and diverse, including the tapeworms, dung beetles, clams, starfish, and, of course, you. Figure 13.3 shows that bilateral animals form two major evolutionary branches, the protostomes and the deuterostomes, distinguished by two different patterns of early embryonic development. Recall from Chapter 8 that early animal embryos generally form a hollow ball (the blastula). This ball indents, with the infolding cells becoming the digestive tract (see Fig. 8.10). In the evolutionary branch that contains only invertebrates (animals without backbones) such as dung beetles, the initial indentation of the embryonic ball becomes the mouth. **Protostomes** (PRO-toe-stomes; "first mouth") are the animals in this evolutionary line (Fig. 13.8a). In the other evolutionary line—the one that yielded the vertebrates—the initial indentation (blastopore) becomes the anus, but a second opening becomes the mouth (Fig. 13.8b). These animals are the **deuterostomes** (DUE-ter-oh-stomes; "second mouth").

Besides showing differences in the embryonic origin of the mouth, the major lines of animal descent have two additional differences: First, protostome eggs often have a spiral cleavage pattern, while deuterostome eggs cleave in a radial fashion. Second, the two groups show differences in how the body cavity forms; this cavity is also called the **coelom** (se-LOM) and is an internal space lined at least in part with mesoderm. In protostomes, solid blocks of mesoderm split to form a coelom, but in deuterostomes, the mesoderm forms a pocket from the developmental precursor of the gut, and the enclosed space becomes the coelomic cavity. Radial animals, the cnidaria and ctenophores, lack a coelom. The animals that have one, however, reap several advantages. The space within the coelomic cavity allows the reproductive and digestive organs to evolve more complex shapes and functions than a totally solid body plan. This fluid-filled chamber cushions the gut tube and other organs and thus protects them. And finally, because the gut is suspended in a cavity, digestion can take place undisturbed by the activity or inactivity of the animal's outer body wall.

Figure 13.8 **Two Lineages of Bilateral Animals: The Protostomes and Deuterostomes.** The two main lineages of bilateral animals, represented here by fruit flies (a) and sea urchins (b), differ in three significant ways: embryonic origin of the mouth; the origin of the main body cavity, the coelom; and the way the embryos cleave. The 8-cell stage photos here are the mud snail, with the same radial cleavage as an insect embryo's, and the sand dollar embryo, a close relative of the sea urchin with matching spiral cleavage.

(a) Protostomes

Fruit fly embryo, surface view

Fruit fly larva, cross section

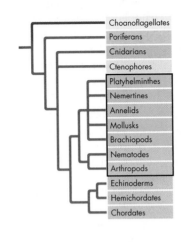

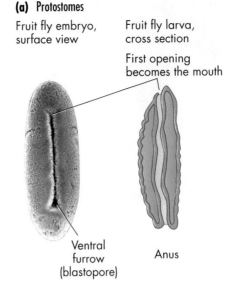

First opening becomes the mouth

Ventral furrow (blastopore)

Anus

Mud snail embryo, 8-cell stage

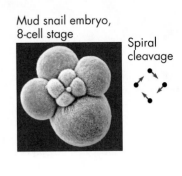

Spiral cleavage

(b) Deuterostomes

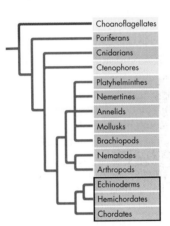

Sea urchin embryo, surface view

Blastopore

Sea urchin larva, cross section

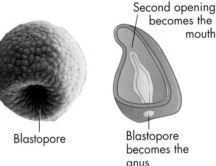

Second opening becomes the mouth

Blastopore becomes the anus

Sand dollar embryo, (related to sea urchin), 8-cell stage

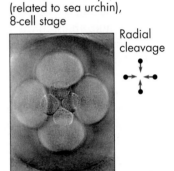

Radial cleavage

The rest of this section looks at the two great branches of animal evolution, the protostomes and then the deuterostomes.

Protostomes: The Diverse Invertebrates

Earth's thousands of dung beetle species, its millions of other insect types, and most of its other immensely varied animal phyla are protostomes. Throughout most of the history of **zoology** (the study of animals), scholars had to guess at the relationships between these diverse phyla by comparing gross physical traits. Recent evidence from molecular genetics, however, has allowed zoologists to construct careful geneologies based on genes that code for ribosomal RNA, that control development, or that govern parts of the cytoskeleton. They now think that one huge set of protostomes includes all animals with a specialized feeding apparatus called a **lophophore** or a specialized larval form called a **trochophore.** For this reason, they are calling the group the **Lophotrochozoans** (LO-fo-TRO-ca-zoans), and include in it several kinds of worms and all the mollusks (such as snails, clams, and squid). A second set of protostomes includes animals that form a tough external skeleton, which they shed periodically by molting. This set of protostomes they call the **Ecdysozoa** (eck-DI-so-ZO-ah) or molting animals, and include in it the arthropods (such as dung beetles and lobsters), the nematodes (or roundworms), and several smaller phyla.

(a) Planaria, a free-living flatworm

(b) The flatworm body plan

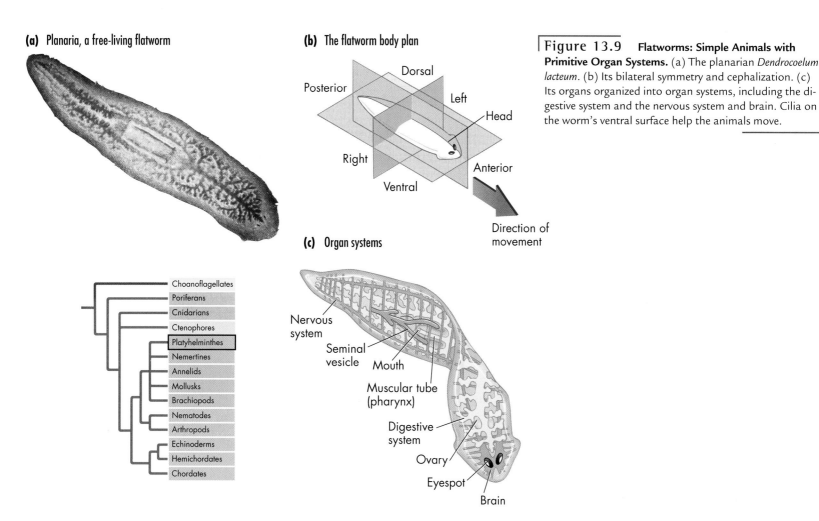

(c) Organ systems

Lophotrochozoans: Specialized Feeding Tentacles or Specialized Larvae

Lophotrochozoans are protostomes in several different, varied phyla that often have either the special feeding tentacles called lophophores or the special ciliated larva called trochophores. The feeding tentacles contain an extension of the body cavity (the coelom) as well as cilia that help collect small food particles and bring them to the animal's mouth. The special trochophore larva has bands of cilia in characteristic rows. Zoologists are still trying to understand how the different phyla within the Lophotrochozoans are related to each other. But we can explore what they do know about the groups and their relatedness.

Flatworms: The Platyhelminthes

The phylum Platyhelminthes (PLAT-ee-HEL-minthes; Greek, "flat worms") contains guess what?—**flatworms,** with a self-descriptive body plan (Fig. 13.9 and

E 13.1). Some flatworms are free-living. The **planarians**, for example, inhabit freshwater lakes, rivers, or bodies of salt water (Fig. 13.9a). Most flatworms, however, including the flukes and tapeworms, are parasites that live within the bodies of their hosts (Fig. 13.10).

Figure 13.10 **Tapeworms: Parasitic Flatworms.** The beef tapeworm infects about 60 million people worldwide. It can reach 60 feet in length inside a person's intestines. The life cycle starts with consumption of larvae. These develop and attach to the host's gut, where it can sicken the host. Body segments develop reproductive organs and embryos, then the segments detach and are released in the person's feces. If a cow eats food contaminated by human feces bearing embryos in detached body segments, the young grow in the new host and penetrate the gut wall, pass through the bloodstream, and lodge in the cow's muscles, forming cysts. If the cysts are not killed by cooking, they can start another cycle.

The blood fluke of Southeast Asia, for example, is responsible each year for 200 million cases of a human disease called schistosomiasis, which is second only to malaria in claiming human victims. Flatworms have a specialized larva that some regard as a modified trochophore. Like many Lophotrochophorans, flatworms have radially cleaving embryos.

Bilateral Symmetry and Cephalization As in other bilateral phyla, the flatworms have right and left halves that are mirror images (Fig. 13.9b). The anterior and posterior (front and back) ends, however, are different, and so are their dorsal and ventral (top and bottom) surfaces. In addition, flatworms display **cephalization** (Greek, "head"); one end generally leads during locomotion and contains both a nerve mass that serves as a brain, and specialized regions that can sense light, chemicals, and pressure. When a bilateral animal moves forward, the head, with its sensors and brain, encounters a new environment first. Depending on the data the head collects, an animal can continue forward or back up and try a different direction. This evolutionary adaptation is so successful that heads occur in almost all animals more complex than flatworms.

Tissues, Organs, and Organ Systems Flatworms display two additional advances seen in all bilateral animals: three distinct tissue layers and true organs and organ systems. In flatworms and other bilateral animals, the middle tissue layer, or **mesoderm,** is made up of living cells, not jelly as in cnidarians, and lies between an outer cell layer, the ectoderm, and an inner cell layer, the endoderm. The mesoderm is important because it gives rise to muscles, blood, and other tissues. Flatworms also have organs and organ systems, described in Figure 13.9c. These multicellular structures help the animal obtain and digest food, dispose of wastes, and reproduce. Flatworms do not have a coelom, a mesoderm-lined body cavity, so instead are termed **acoelomate.** Biologists think, however, that flatworms may have lost a coelom that was present in the group's ancestors. The reproductive system has both testes and ovaries (making most flatworms hermaphroditic). The worms pair up and exchange both sperm and eggs with another individual and so both are fertilized.

Flatness is a key to a flatworm's success. Because it is so thin, oxygen and carbon dioxide diffuse directly through the tissue layers to every cell, nutrients diffuse outward from the branching intestine to all body cells, and the animal needs no extra internal or external skeleton for support.

Parasitic Flatworms

In contrast to free-living planarians, parasitic flatworms, such as flukes and tapeworms, live a sheltered life, with most of their needs provided for by one or more different hosts. A tapeworm's head is called a **scolex,** and it is little more than a knob with ghoulish hooks or adhesive suckers around the mouth that attach to host tissues (Fig. 13.10). Their bodies consist mostly of hundreds of individual reproductive units. The tapeworms that infect a cow's intestines shed embryos that bore into its muscles and form protective cysts. If a person consumes undercooked beef bearing these cysts, the cysts can grow into new tapeworms in the person's intestines (the second host). Both flukes and tapeworms shed thousands of eggs and larvae, and these are their only means of reaching new hosts and continuing their life cycle.

Nemertea: The Ribbon Worms

Ribbon worms are unfamiliar to most people because they live mainly at the bottom of the oceans (**E** 13.1). Ribbon worms can be less than one centimeter long or up to 60 m (195 ft) long! Ribbon worms are similar to flatworms in general shape, having bilateral symmetry, cephalization, and organ systems. Nemertines have in addition, however, a unique proboscis ("nose") that distinguishes this phylum. A ribbon worm stores its proboscis in a body cavity that represents the animal's coelom, and then shoots it out at prey. Little nail-shaped structures (stylets) can pierce the body of the prey and then the proboscis pulls the prey to the mouth. In addition to this special feeding apparatus, the ribbon worms have a circulatory system enclosed in vessels, and a two-way gut. This innovation helps the ribbon worms process some types of food more efficiently than flatworms. Ribbon worm embryos cleave spirally, like flatworms and many other Lophotrochophorates, and they have what experts categorize as a modified trochophore larva. Nemertines also have the ability to regenerate to almost nightmarish proportions: If you cut a ribbon worm into 20 pieces, each fragment will grow a new, complete worm.

Annelids: Segmented Worms

Marine sandworms, common earthworms, and leeches are all members of the phylum **Annelida,** the **segmented worms** (**E** 13.1). The term annelid means "tiny rings" and refers to the external segments visible on members of this phylum. Annelids belong to one of three classes. The largest class includes colorful marine worms that burrow in the mud or sand and bear such common names as fireworms and feather dusters (class Polychaeta, "many-bristled") (Fig. 13.11a). Familiar red earthworms are in the class Oligochaeta ("few bristles"). These ubiquitous inhabitants of moist soils, can number 50,000 per acre and literally eat their way through dense, compacted earth, excreting the displaced material in small, dark piles, aerating and mixing the soil. Like dung beetles, earthworms benefit soil ecology, but they function much more slowly.

Leeches, which live mainly in fresh water, are in the class Hirudinea. Leeches parasitize other animals by sucking their blood. Physicians sometimes use a substance (called hirudin) they obtain from leeches to block undesirable blood clotting in their patients. A leech can consume three times its weight in blood and go for as long as nine months between meals (Fig. 13.11b).

Annelid reproduction is generally sexual. Many marine annelids shed sperm and eggs into seawater, where the gametes unite and the egg cleaves in a spiral fashion. Each embryo then develops into a trochophore larva.

Earthworms and leeches are hermaphrodites; each individual produces both sperm and eggs. When two individuals couple, they exchange sperm so that each fertilizes the other's eggs.

Characteristics of Annelids Annelids not only display bilateral symmetry, cephalization, and a gut that is totally tubular, but also have a coelom, a fully lined body cavity. Their additional trait of segmentation offers certain advantages. Consider an earthworm, with 100 or more body segments (Fig. 13.11c,d). Each segment is separated from the next by an internal partition, or **septum** (pl., *septa*), and each segment contains a set of internal structures. Most earthworm segments contain two excretory units called **nephridia** (Fig. 13.11d). Each nephridium removes excess water and wastes from the body fluids by means of a ciliated funnel and excretes them through a pore in the body wall.

Each segment also contains a fluid-filled compartment of the coelom and is surrounded by circular and longitudinal muscles in the body wall. As the circular muscles squeeze against the incompressible fluids in the

(a) Bearded fireworm (class Polychaeta)

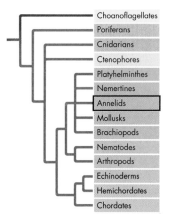

(b) Leech (class Hirudinea)

Figure 13.11 **Segmented Worms: Annelids.**
Segmented worms range from the delicate, including (a) this bearded fireworm, to the despised such as (b) this black leech. (c) Annelids like this common earthworm have a closed circulatory system. Each segment (d) is a repeated module containing organs of the excretory, nervous, and locomotory systems.

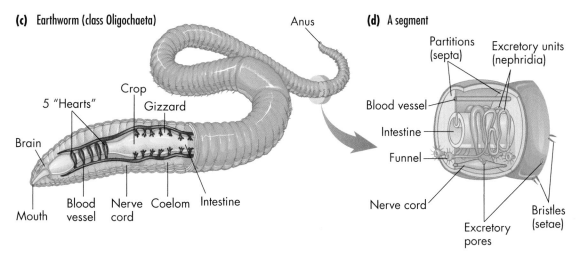

(c) Earthworm (class Oligochaeta)

(d) A segment

coelom and gut, the fluid force is transmitted to adjacent segments, creating a **hydroskeleton,** an internal skeleton made of fluid. Contractions in sequential segments produce waves of force that propel the animal forward. **Setae,** which are pairs of bristles on each segment, push against the ground with each contraction and help the animal move. In addition to nephridia, muscles, and coelom compartments, each segment contains clusters of nerve cells connected to the brain by nerve cords.

Annelids also have a **closed circulatory system**. Blood is carried entirely in tubes, or vessels, and at no point bathes the body tissues directly. This adaptation allows the animal to pump blood at higher pressure through its body. The result is a more regular and constant delivery of oxygen and nutrients, and a more efficient removal of wastes from cells at times of high activity.

Mollusks: Soft-Bodied Animals

In your mind, picture snails, slugs, scallops, and squid. They seem to be very different types of animals, but because of their structurally similar body plan, they are all **mollusks**, members of the phylum **Mollusca** (Fig. 13.12a,b and *E* 13.1). The name of this phylum comes from the Latin for "soft," and if you have ever eaten a snail or oyster, you remember that—except for the hard shell—the main sensation they produce is squishy. Mollusks are bilateral protostomes: they produce a trochophore larva, a tiny, fringed stage, as do several other phyla in the Lophotrochophore group of protostome invertebrates (see Fig. 13.3).

Characteristics of Mollusks The wide variety of mollusks share a distinct body plan. As Figure 13.12c shows, each mollusk has (1) a **head,** housing the mouth, brain, and sense organs; (2) a **foot,** a muscular

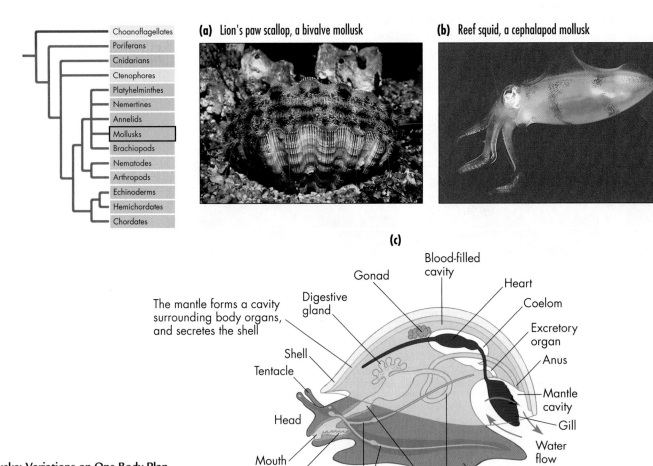

(a) Lion's paw scallop, a bivalve mollusk

(b) Reef squid, a cephalopod mollusk

Choanoflagellates
Poriferans
Cnidarians
Ctenophores
Platyhelminthes
Nemertines
Annelids
Mollusks
Brachiopods
Nematodes
Arthropods
Echinoderms
Hemichordates
Chordates

(c)

Gonad
Blood-filled cavity
Heart
Coelom
Digestive gland
Excretory organ
The mantle forms a cavity surrounding body organs, and secretes the shell
Anus
Shell
Tentacle
Mantle cavity
Gill
Head
Water flow
Mouth
The muscular foot aids in locomotion
The rasp-like radula helps obtain food
Nerve cords
Gut

Figure 13.12 **Mollusks: Variations on One Body Plan.**
(a) This lion's paw scallop *(Lyropecten nodusus)* is a colorful bivalve.
(b) The reef squid *(Sepioteuthis lessoniana)* of Indonesia is a strangely beautiful cephalopod. (c) A generalized molluscan body plan.

organ used for gripping or creeping over surfaces or through sand; (3) a **mantle**, a thick fold of tissue that secretes the calcium carbonate that makes up the hard molluscan shell; and (4) a **visceral mass** containing all the internal organs (labeled separately on Fig. 13.12c). Most members of the phylum have a special feeding organ called the **radula.** This strap-shaped structure bears rows of tiny teeth and works like a cheese grater, rasping off successive layers of food. You can see the scrape marks of the radula on the bottom of a pumpkin if a slug has grazed on the bright orange exterior. Food passes through a complete gut, from mouth through to anus.

Branches of the mollusk's circulatory system flow through the **gills,** the animal's gas exchange surface, and then to the pumping chambers of the heart, and to a **blood-filled cavity,** where internal organs and tissues are bathed with blood (see Fig. 13.12c). Most mollusks have an **open circulatory system** because the blood is confined within vessels in only certain parts of the body and flows through open spaces, where it bathes body tissues directly, in other regions. Specialized organs and organ systems for respiration and for circulation of blood represent evolutionary advances not found in less complex animals.

Some mollusks have well-developed nervous systems with large brains, acute senses, and a capacity for learning. For example, Italian researchers trained octopuses to attack a red ball, but to leave a white ball unmolested. They allowed a second group of octopuses to watch the exercise from an adjoining tank, then tossed red and white balls in with these observer animals. The experimenters found that the second group also attacked the red balls and left the white balls alone. Other sets of octopuses easily learned the reversed task—to attack white balls and leave red balls untouched. These experiments showed that mammals are not the only animals with the ability to learn a task merely by watching another individual perform it.

Some Classes of Mollusks

Let's look at the three main classes of mollusks: gastropods, bivalves, and cephalopods. **Gastropods** (class Gastropoda, "belly foot") include snails, garden slugs, and sea slugs, or nudibranchs (NEW-di-brancks; see E 13.1). Most snails are identifiable by their coiled shells, as well as by **torsion**—an internal twisting of the body mass during embryonic development. Nudibranchs often produce bitter or toxic substances that repel attackers. Their bright colors serve as a warning that a bad meal lies ahead for the predator.

Bivalves (class Bivalvia, "two valves") have two half shells that enclose bodies of oysters, clams, mussels, scallops, and relatives (Fig. 13.12a and E 13.1). Strong muscles close the valves and stretch an elastic ligament at the hinge. Closed shells protect the animals from most predators. When the muscles relax, the shell snaps open. Rapid closing of the shell in scallops produces a water jet that propels the animal. When a mollusk's shells are open, particles can enter and irritate the soft tissue of the mantle, which responds by secreting layers of nacre (mother-of-pearl) around the offending object. Pearls result from reddish, whitish, or black nacre deposits.

Bivalves are **filter feeders;** they have gills that, in addition to collecting oxygen and releasing carbon dioxide, can strain out and collect tiny food particles suspended in water. Beating cilia draw water across the gills, and a mucous layer traps the particles, which are then passed into the mouth. One biologist has estimated that each week, all the water in San Francisco Bay is drawn through the bivalves that lie half-buried in the silt and sand at the bottom of the bay.

Cephalopods (class Cephalopoda, "head foot") include squids, octopuses, and the chambered nautilus (Fig. 13.12b and E 13.1). Cephalopods are the most complex mollusks and evolved as fast-swimming predators of the deep sea. The cephalopod foot bears a circle of 8 or 10 arms, each studded with suckers, and it terminates in a funnel, or **siphon.** Thus, a single organ, the foot, has become specialized for land travel in the gastropods and for hunting, swimming, and feeding in the cephalopods. The cephalopod mantle forms a muscular enclosure, which can expand and draw water into the mantle cavity or contract and force it out of the siphon, jet-propelling the mollusk backward. These explosive bursts can carry the animal to safety or bring its suckered tentacles within reach of prey. Cephalopods have a largely closed circulatory system that pumps blood faster and helps support the active hunter life habit.

The coordination for hunting and feeding in cephalopods depends on acute senses, a large brain, and the most complex nervous system among the invertebrates. The largest cephalopods, the giant deep-sea squid, can weigh 450 kg (1000 lb) and reach nearly 18 m (60 ft) in length. Giant squid have the largest eyes in the animal kingdom (they can grow larger than a car's headlights), and these highly sensitive organs form images like our own. Because of experiments like the ones with the colored balls, biologists suggest that octopuses may be the most intelligent invertebrates.

How does classification contribute to preservation?

Brachiopods: The Lamp Shells

Animals in the phylum **Brachiopoda** ("arm-foot") live as solitary creatures at the bottoms of oceans. Brachiopods have two shells that enclose the body (hence the common name, "lamp shells"). These two shells superficially resemble clam and other bivalve mollusk shells, but the two brachiopod shells are usually shaped differently from each other, and their hinge is very different from a bivalve's. The brachiopods feed with a lophophore, the feeding apparatus consisting of tentacles with cilia on their outside and an extension of the coelom on the inside. Although there are only 300 or so brachiopod species alive today, they have a rich fossil record and were among the most abundant animals in the Paleozoic Era. Until recently, biologists generally classified brachiopods as deuterostomes, but were really not very sure where they belonged; recent molecular genetic data show that they belong clearly with the protostomes.

While biologists now understand that the Lophotrochozoans occupy a single branch on the tree of life (Fig. 13.3), they are still far from understanding the evolutionary relationships of the various Lophotrochophorans to each other.

Let's switch our focus, now, to the other group of protostomes, the animals that molt.

Ecdysozoans: The Molting Animals

We arrive now at the group that includes dung beetles, the Ecdysozoa, the animals that periodically shed an external skeleton or cuticle. Two very successful phyla molt this way, the roundworms and the arthropods (jointed-leg invertebrates, including dung beetles). Some small phyla are ecdysozoans, too, such as the **rotifers,** small common pond organisms that have a cuticle formed within the epidermis.

Roundworms: The Nematodes

In sheer numbers of individuals, the most abundant animals on Earth are **roundworms,** members of the phylum **Nematoda** (*nema* = thread). Roundworms, as the name implies, are round in cross-section and most are very small (Fig. 13.13a, and *E* 13.1). One cubic meter of rich soil can contain 3 billion nematodes. One study counted 90,000 individuals in a single rotten apple! Biologist A.M. Cobb once said that if our planet's lands and seas disappeared but the nematodes somehow

stayed in place, a clear outline of Earth and its geological features would remain. When the environment grows harsh, nematodes can curl up, dry out, and shut down their metabolism for up to 30 years. Then, when conditions improve and water returns, the animals rehydrate and revive: instant nematode! While many roundworms are free-living, hundreds are plant and animal parasites that damage crops and cause diseases.

Characteristics of Roundworms   Roundworms are unsegmented, bilateral worms with three tissue layers and a **pseudocoelom** (false body cavity), a body cavity only partially covered with mesoderm. This feature was once thought to be primitive, but modern molecular genetic data (the first animal genome to be sequenced was a nematode) clearly shows that nematodes diverged higher, not lower, on the tree of life. The roundworm's simplified body plan therefore is likely to be based on

Figure 13.13   **The Roundworms: Phylum Nematoda.** (a) A common soil dwelling nematode *(Pristionchus pacificus).* Grotesque limb enlargement (b) can occur when parasitic roundworms block lymph vessels and the tissues accumulate fluid. This victim of elephantiasis has one hugely swollen leg and one normal one.

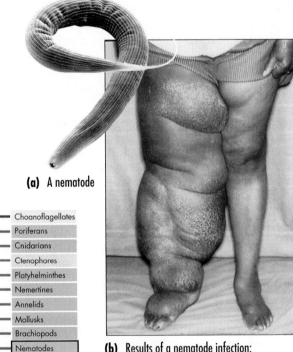

(a) A nematode

Choanoflagellates
Poriferans
Cnidarians
Ctenophores
Platyhelminthes
Nemertines
Annelids
Mollusks
Brachiopods
Nematodes
Arthropods
Echinoderms
Hemichordates
Chordates

(b) Results of a nematode infection: elephantiasis

a loss of features present in its ancestors. The nematode's false coelom acts as a hydroskeleton, providing support and rigidity for the soft animal, as its incompressible liquid contents are squeezed by the body wall muscles. The nematode body surface is covered by a cuticle made of protein fibers secreted by the epidermis. Periodically during development, the animal sheds this cuticle and replaces it with a new one, like other Ecdysozoans. Nematodes have a complete gut with mouth and anus, and dorsal and ventral nerves. Strangely, the nematode muscle cells send long processes to the nerves rather than the other way around, as in most animals.

The nematode *Caenorhabditis elegans* was the first animal whose DNA sequence was determined completely. It has about 19,000 genes, and 969 cells. Because most individuals of this species are hermaphroditic (a single individual makes both eggs and sperm), it is easy to perform the matings necessary to analyze nematode genetics. This, plus its small, transparent body and rapid life cycle, has made *C. elegans* a favorite research subject for looking at the genetic basis of embryonic development.

Roundworms and the Environment Dozens of types of free-living roundworms help consume rotting plant and animal matter. That makes them ecologically important decomposers like the bacteria and fungi and the more specialized dung beetles. However, the parasitic roundworms get most of the fame—or infamy. At least 1000 nematode species parasitize plants, and some ecologists estimate that roundworms consume fully 10 percent of all crops annually. Nearly 50 species parasitize humans, entering in food or contaminated water or through bare skin. They cause many diseases, including trichinosis (from eating undercooked worm-infested pork), ascariasis (a common disease in tropical regions, characterized by lung infections and intestinal blockages due to masses of worms), hookworm (an infestation in the internal organs, also common in the tropics), and elephantiasis (Fig. 13.13b). Clearly, much of the nematode's evolutionary success comes at the expense of other organisms, including ourselves.

Arthropods: Jointed Legged Animals

Dung beetles are members of Earth's largest phylum (not just of animals but of *all* organisms): **Arthropoda** (*arthro* = joint + *poda* = leg). Arthropoda is another major phylum in the Ecdysozoa, and its members, the so-called **arthropods** include fossil trilobites, spiders, mites, ticks, scorpions, centipedes, millipedes, lobsters, crabs, beetles, and other insects (**E** 13.1). The insects, numbering at least 1 million species, make up the vast majority of all animal species (Fig. 13.14). As Chapter 10 explained, there may be many, many more insect species as yet undiscovered—perhaps 10 to 30 million more! The insects' astounding diversity and success are based on their body plan: They are bilateral, with a head, a one-way gut, and a fully lined body cavity. Arthropods have several other shared characteristics, including an external skeleton (which they periodically shed), specialized body segments, acute sensory systems, and rapid movement and metabolism due to special respiratory structures. All of these contribute to their astonishing success.

Part of that success was the arthropod's initial invasion of the land. Fossils of the earliest land-living animals yet discovered were tiny spi-

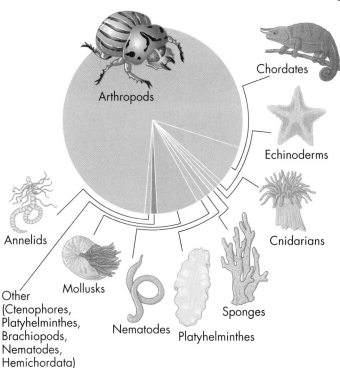

Chordates

Arthropods

Echinoderms

Annelids

Cnidarians

Other (Ctenophores, Platyhelminthes, Brachiopods, Nematodes, Hemichordata)

Mollusks

Nematodes

Platyhelminthes

Sponges

Figure 13.14 **Arthropods: The Phylum Richest in Species.** Compared to the insects and other arthropods, all other species make up a small slice of the animal kingdom. If the projections of 10 to 30 million total tropical insect species are correct, the rest of the animal kingdom is, in fact, a minuscule slice! Our era could be called the Age of the Arthropods.

derlike organisms that lived among the stems of *Cooksonia* plants (review Fig. 12.15) more than 405 million years ago. These arthropods, called "trigs" (trigonotarbids), had jointed legs, lungs, mouthparts, armor plates on the abdomen, and eyes with ten or so loosely packed lenses—all traits that resemble modern predatory spiders.

The Arthropod Exoskeleton Like the nematodes and the other Ecdysozoa, arthropods have an outer protective cuticle, which they molt. The arthropod **exoskeleton,** or external skeleton, completely surrounds the animal and provides protection, strong support, and rigid surfaces that muscles can pull against (Fig. 13.15a). This thick, hard cuticle contains the polysaccharide **chitin** (see Chapter 2), in addition to sugar-protein complexes, waxes, and lipids that make the body covering waterproof. In crustaceans, such as crabs, lobsters, and shrimp, the exoskeleton also contains calcium carbonate crystals, which make it a hard, inflexible armor. Besides providing shieldlike protection from enemies and resistance to general wear and tear, the exoskeleton prevents internal tissues from drying out. This is extremely important for arthropods that live on land.

Exoskeletons do, however, have a major disadvantage: The animal cannot grow larger unless it periodically sheds its constricting armor and produces a larger exoskeleton; this is the *molting* process. During the period between the shedding of one exoskeleton and the hardening of the next, the animal is soft and vulnerable. The arthropod exoskeleton remains thin and flexible at the **joints,** the hingelike areas of the legs and body (Fig. 13.15b). The presence of jointed appendages allows arthropods to move quickly and efficiently above the ground or sea floor instead of dragging the body directly along the ground on stubby legs or bristles.

Specialized Arthropod Segments Arthropods have body and leg segments but their segmentation differs

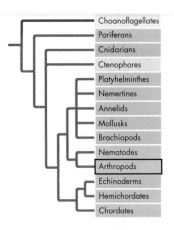

Choanoflagellates
Poriferans
Cnidarians
Ctenophores
Platyhelminthes
Nemertines
Annelids
Mollusks
Brachiopods
Nematodes
Arthropods
Echinoderms
Hemichordates
Chordates

Figure 13.15 **Diversity Among the Arthropods.** (a) A giant centipede from Thailand, with many similar segments and poison claws. (b) A molting shrimp from Malaysian waters. (c) A South African dung beetle rolling a dung ball. The hard, confining exoskeleton is molted and replaced as the animal grows.

(a) Chilipoda: a centipede

The arthropod body is segmented and originally each segment probably had a pair of jointed appendages

(b) Crustacea

Arthropods periodically molt their exoskeletons

(c) Insecta

Arthropods have regional body specializations like the head, thorax, and abdomen of insects

from an annelid's. Each annelid segment is basically the same as any other, whereas arthropod segments are different and highly modified, giving the animals a far greater repertoire of activities. The body segments are usually fused into a few major regions; dung beetles and other insects, for example, have three regions: **head,**

thorax, and **abdomen** (Fig. 13.15c). In spiders and crustaceans, by contrast, the fused body segments are reduced to an abdomen plus one region with fused head and thorax, the **cephalothorax.**

During evolution, arthropods have evolved a veritable Swiss Army knife of tool-like appendages specialized for walking, swimming, and flying, with each species having its own modifications. Some arthropods also developed pincers or palps (feelers), which facilitate hunting and feeding. From the head region grew other appendages: *mouthparts* that allow chewing and sucking and *antennae* that sense odors and vibrations. The shrimp in Fig. 13.15b, for example, has appendages modified for eating (mouthparts), grasping (pincers), walking (legs), mating (swimmerets), and fast swimming (tail). The dung beetle in Figure 13.15c has specialized cutlerylike mouthparts for carving up dung; long back legs for dung ball rolling; and horns for fighting off competitors. The millions of arthropod modifications like these enabled the animals to exploit a wide variety of environments.

Organ Systems Arthropods metabolize and move rapidly, and, in fact, some insects do both at the highest rates in the animal kingdom. Some tiny flies can generate enough metabolic energy to beat their wings 1000 times per second. High metabolic rates require rapid oxygen delivery, and the arthropods' efficient respiratory organs provide a large surface area for collecting oxygen and releasing carbon dioxide quickly. Insects have **tracheae** (TRAY-kee; sing., *trachea* [TRAY-kee-a]), branching networks of hollow air passages; spiders and relatives have **book lungs,** chambers with leaflike plates for exchanging gases; and finally, aquatic arthropods have gills, flat tissue plates that act as gas-exchange surfaces. High metabolic rates and rapid movements allow arthropods to fly, run, swim, and roll dung balls faster than any group we've discussed until now. Speed, in turn, allows them to escape predators with agility and to disperse over wider ranges. Arthropods also have a complete gut and an open circulatory system.

Sense Organs Most arthropods have antennae capable of detecting movement, sound, or chemicals with great sensitivity. The antennae of several male moth species, for instance, can detect the **pheromones,** or odor signals, given off by adult female moths at distances of over 11 km (about 7 mi). Arthropods also have various types of organs on the head and body that detect sound and taste chemicals, and many have special compound eyes made up of 2500 or more six-sided segments called **facets** (see Fig. 8.18). Many compound eyes are capable of color vision and can detect the slightest movements of prey, mates, or predators.

Major Arthropod Classes The successful and highly divergent phylum Arthropoda is divided into a number of taxonomic classes, most of which will sound quite familiar.

Centipedes (class Chilopoda) have a series of flattened body segments, each bearing a pair of jointed legs that move the centipede swiftly in search of insects, worms, small mollusks, or other prey. Centipedes kill their prey with poison claws, which are modified legs on the first body segment (Fig. 13.15a). Some huge tropical centipedes are dangerous to humans, but the common varieties that lurk in damp basements are harmless and consume insects.

Millipedes (class Diplopoda) are slow-moving counterparts to the centipedes. They have, as the term Diplopoda suggests, two pairs of legs per segment, a round body, and a preference for decaying vegetable matter instead of live prey.

Crustaceans (class Crustacea) include crabs, shrimp, lobsters, barnacles, crayfish, and sow bugs (Fig. 13.15b and *E* 13.1). These animals are so different from each other that only two generalizations apply: (1) Almost all have an exoskeleton hardened with calcium salts that covers most of the animal as a protective shell, or **carapace,** and (2) all have two pairs of antennae. The lobster, a familiar crustacean, has walking legs and pincers sprouting from its cephalothorax, while its abdomen bears feathery swimming appendages.

Crustaceans are so numerous and so diverse that they are found in most bodies of water. There are even a few terrestrial species, like the familiar sow bugs (also called pill bugs or armadillo bugs) that scurry away when you lift a rock or flower pot sitting on the ground. Many small aquatic species, such as fairy shrimp, brine shrimp, and copepods, serve as the primary food for various fish, whales, and other species. Barnacles cling tenaciously to tidal-zone rocks; the natural cement they secrete is strong enough to keep the animals from washing away in pounding ocean surf. Biologists are studying barnacle glue in hopes of developing medical cements that work when wet for fastening dentures and rejoining broken bones.

Arachnids (class Arachnida) include spiders, ticks, mites, and scorpions (see *E* 13.1). Arachnids belong to the subphylum Chelicerata, which also includes the extinct trilobites, the sea spiders, and the primitive-looking horseshoe crabs often seen on Atlantic coastal beaches. Arachnids lack antennae; have a cephalothorax and segmented abdomen; and usually have six pairs of appendages. The first pair of appendages is modified into **chelicerae,** or poison fangs, used for killing prey or for self-defense. The second pair holds the prey while the spider injects poison or enzymes. The other four pairs are the spider's eight walking legs. Spiders also have organs called **spinnerets** at the rear of the

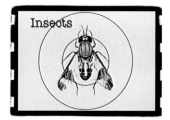

What is the significance of segmentation?

abdomen that reel out silk threads that are, size-for-size, stronger than steel. Despite their sinister reputation, most spiders are capable of poisoning only very small animals. The potent poisons of the black widow and brown recluse spiders, however, can be dangerous to humans. Present-day scorpions have a body form very similar to the huge water scorpions that lived 500 million years ago and that probably gave rise to the land scorpions and spiders. Mites and ticks are, for the most part, merely irritating biters, but some ticks carry serious diseases, including Lyme disease and Rocky Mountain spotted fever, with its fever, rash, and joint pain.

Insects (class Insecta) are the largest—and, by that measure, most successful—class of animals on Earth (see **E** 13.1). Like the dung beetles, most insects live on land in habitats from the tropics to the poles. Many zoologists attribute insect success to the general arthropod traits we've already considered, but also to the insect's small size and (in many) the ability to fly. Small size enables insects to exploit a vast array of microhabitats—the bark of a tree, the backs of leaves, the dense thickets of an animal's fur, a pile of animal dung, or the universe of midair.

Specific body parts also help account for the insects' success. Many insects have organs for smelling, touch reception, tasting, seeing, and hearing in various parts of the body. An insect's head bears one pair of antennae; its thorax bears three pairs of legs and usually one or two pairs of wings; and its abdomen is usually free of appendages. A series of modified segments, the mouthparts, enables the insect to feed efficiently. We've seen the dung beetles' all-purpose "cutlery" mouthparts (Fig. 13.15c). Mosquitoes have pointed stylets that pierce and suck. Others, like locusts and grasshoppers, have chewing mouthparts that can quickly decimate foliage.

Insects have evolved various ways to grow, despite the confining exoskeleton, and various ways to thrive, despite the changing seasons. In some insects, like grasshoppers and cockroaches, the embryo emerges as a miniature version of the adult, but without wings or mature reproductive organs. This organism feeds, grows, and molts five or six times, gradually attaining adult size and characteristics. In most insects, however, including flies, butterflies, and dung beetles, the embryo develops into an immature form, or **larva,** eats voraciously, then forms a transitional stage, or **pupa,** sometimes inside a cocoon. A complete change in form, or **metamorphosis,** takes place in the body within the pupal exoskeleton—a kind of supermolt. Finally, a nonmolting, reproductively mature adult emerges (Fig. 13.16). In insects that metamorphose, the larvae and adults can be adapted to very different foods and environmental conditions. This successful evolutionary solution allows larvae to specialize in feeding and obtaining energy, and the adult to specialize in reproduction.

Insects have tremendous environmental significance. To assess the relative importance of insects and humans, ask yourself how many types of organisms would become extinct if people were suddenly wiped off the face of the Earth. As you can probably guess, very few would disappear and many would rebound. In contrast, if insects became extinct, Earth would change dramatically. As we saw in the introduction, without dung beetles, animal dung would pile up and cover large expanses of land. In addition, without the help of

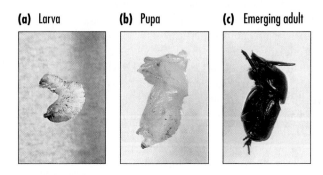

(a) Larva   **(b) Pupa**   **(c) Emerging adult**

(d) Hormonal control

Periodic releases of ecdysone (molting hormone)

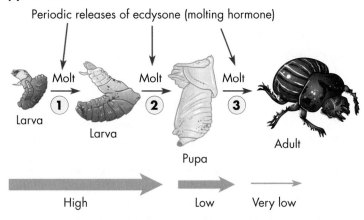

Figure 13.16   **From Grub to Beetle: Hormones Provoke Metamorphosis.** (a) A grub, the larval form of a dung beetle, sheds its skin, or molts, several times as it grows larger. (b) Eventually it changes into a pupa, and then an adult. (c) Finally, the adult emerges from the pupa case. (d) These changes in form are directed by hormones. As the larva grows, surges of ecdysone regulate the timing of each larval molt. Throughout these early molt cycles, levels of juvenile hormone remain high (Step 1) and maintain the animal's larval form. Eventually, levels of juvenile hormone begin to drop, and with the next-to-last surge of ecdysone, the animal molts into a pupa (Step 2). With the final surge of ecdysone, juvenile hormone levels are very low and the animal is transformed into an adult beetle (Step 3).

their insect pollinators, thousands of plant species would die off and the animals that depend on those plants would also disappear quickly.

Insects have additional significance for medicine, agriculture, and science. Insects are vectors of many diseases, including malaria and yellow fever. Insects are major agricultural pests, consuming a substantial portion of grains and other crops each year. On the other hand, insects—the fruit fly *Drosophila melanogaster* in particular—have contributed greatly to our understanding of genetics, cell biology, and developmental biology. *Drosophila's* was the second animal genome to be fully sequenced. Initial analysis suggests that the insect has 13,600 genes—about 5,000 fewer than the nematode worm.

Perhaps the most fascinating of the arthropods are the **social insects:** termites, ants, wasps, and bees. Most species of these insects live in large colonies with labor divided among **castes,** or subgroups, that differ in appearance and behavior. Such insect colonies are highly successful. Ants, for example, may make up one-third the weight of all the animals on Earth, with 200,000 ants for every person now alive! Insect colonies are highly evolved, functioning, in a sense, as a single well-coordinated organism with the capacity to simultaneously build homes and cities, defend their own, harvest food, and reproduce.

Now that we've investigated the marvelous diversity of protostomes, let's turn to the deuterostomes, the animals whose embryos make a first opening for the anus and a second opening for the mouth.

Deuterostomes: The "Second-Mouth" Animals

Another glance at the animal family tree in Figure 13.3 shows the relationships of the deuterostomes to other groups. After the sponges and radial animals diverged from the trunk of the tree, the bilateral animals evolved and formed the subdivision we just explored, the protostomes, as well as the deuterostome phyla: the Echinoderms, the Hemichordates, and our own phylum, the Chordates.

Echinoderms: The First Endoskeletons

An expert on invertebrates once called the echinoderms "a noble group especially designed to puzzle the zoologist." The phylum **Echinodermata** includes sea stars, brittle stars, sea urchins, sea cucumbers, and sea lilies (Fig. 13.17 and **E** 13.1). Echinoderms may constitute 90 percent of

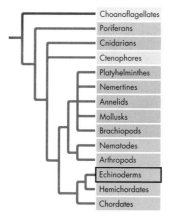

(a) Sea urchin

(b) Sea cucumber

(c) Anatomy of a sea star

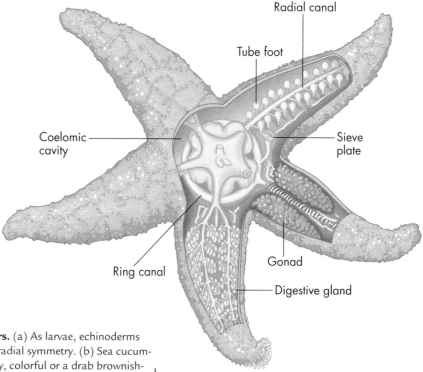

Radial canal

Tube foot

Coelomic cavity

Sieve plate

Ring canal

Gonad

Digestive gland

Figure 13.17 **Echinoderms: Spiny-Skinned Ocean Dwellers.** (a) As larvae, echinoderms (such as the sea urchin) are bilateral, but the adults have fivefold radial symmetry. (b) Sea cucumbers are sometimes spiny, like this one, but can be smooth, bumpy, colorful or a drab brownish-green. (c) Sea star anatomy.

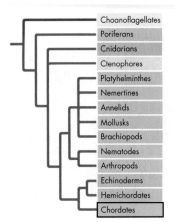

the biomass near the ocean floor. The word echinoderm means "spiny-skinned," and virtually all members have spines, bumps, spikes, or unappetizing projections that help to protect these slow-moving marine creatures from predators. These spines are the calcium-based extensions of the animal's **endoskeleton,** its internal support system. Besides their innovative endoskeleton, echinoderms also have a unique hydraulic pressure system, called the **water vascular system,** that derives from the coelom and that facilitates locomotion. Echinoderms have an odd mixture of other traits. Some, including the absence of excretory

and respiratory systems, may represent a secondary loss from more complex ancestors.

Echinoderm larvae are bilaterally symmetrical, but adults of many species are headless, brainless, and have a fivefold radial symmetry due to the loss of cephalization and bilateral symmetry. Nerve trunks run along each of the adult's arms and unite in a ring around the mouth. This simple system allows coordinated, but slow, movement of the limbs that can generate enough force to open a tightly shut clam shell. Capable of eating more than ten clams or oysters in a day, sea stars can be serious predators on commercial oyster beds.

Hemichordates: Wormy Deuterostomes

The small phylum of **Hemichordata** ("half chordates") consists of a few hundred species of worms that burrow in marine muds, and whose embryos develop in much the same way as echinoderm embryos do (**E** 13.1). The most common hemichordates are acorn worms. Based on recent molecular genetic evidence, zoologists now group echinoderms and hemichordates more closely to each other than to our phylum, the chordates (see Fig. 13.3).

Hemichordates have a complete gut, and several other similarities to chordates. Hemichordates have gill slits that open from the pharynx to the water, similar to the gill openings of a fish or the gill arches of a human embryo (Fig. 8.19). Furthermore, they have a dorsal nerve cord (which chordates share), as well as a ventral nerve cord (which chordates lack). If hemichordates have a structure related to the Chordate notochord, no one has found it yet. You may never see a hemichordate, but they are significant to bi-

Figure 13.18 **The Phylum Chordata.** Chordates arose within the deuterostomes, and then diversified widely.

ology: Future investigation should help us understanding vertebrate evolution, especially the origin of the chordates.

13.3   Evolution of the Chordates

Chordates, or members of the phylum **Chordata,** are the most familiar and complex animals on Earth (Fig. 13.18 and ⒠ 13.14). The group includes ourselves, our pets, our livestock, and the subjects of the most popular nature films. Like the sea stars and acorn worms, chordates are deuterostomes. Unlike those simpler animals, however, the chordates have a stiff but flexible cord, the notochord, running down their backs (hence the names *Chord*ata and *chord*ate). In some chordates this cord is replaced during development by a series of interlocking bones called the backbone, or **vertebral column,** that provides internal support for the body. We call chordates with backbones **vertebrates.** And while there are only about 50,000 species of vertebrates compared with millions of invertebrate species, the vertebral column has a winning evolutionary design that led to Earth's fastest runners, highest fliers, deepest divers, most agile climbers, and best problem solvers, and to the largest animals that ever lived.

Characteristics of Chordates

The graceful, powerful, and fleet chordates of land, sea, and air seem much more complex than radial animals, protostomes, or the other deuterostomes. A few chordates, however—especially the sea squirts—seem simple and rather like squishy blobs at first glance. Nevertheless, all chordates, sea squirts included, demonstrate the major animal innovations: bilateral symmetry, cephalization, a fully lined body cavity, and a one-way gut tube with an opening at each end. On top of that, chordates have a few novelties of their own that allowed for their radiation into so many active, adroit groups (Figs. 13.18 and 13.19).

Notochords and Nerve Cords

Chordates have a new and novel structure, the **notochord,** a solid, flexible rod of cartilage that provides internal support and generally runs from the brain to the tip of the tail, at least in the embryonic stage (Fig. 13.19). Chordates also have a second innovative structure, the dorsal, hollow nerve cord, or **spinal cord,** which is a hollow tube of nerve tissue that runs the length of the animal, just above (dorsal to) the notochord. (Recall from Chapter 8 and Fig. 8.11 that during development, the notochord helps induce the nerve cord to form above it.) The nerve cord acts like a central "trunk line" carrying impulses that help integrate the body's movements and sensations. In most chordates, the nerve cord is present throughout embryonic and adult life.

Gill Slits

Chordates also have a third innovation, **gill slits,** pairs of openings through the digestive tract's anterior region, the **pharynx,** or throat. If you still had gill slits, they would be openings from the back of your throat, to the outside of your neck. Simple chordates such as sea squirts use gill slits to filter food particles from the surrounding water. Fish use gill slits for gas exchange. In most reptiles, birds, and mammals, gill slits are either vestigial structures found only in the embryo or they develop into other structures. For example, your eustachian tube is a vestige of one of the gill slits that runs from the back of your throat to your middle ear. This tube allows you to equalize the pressure in your ears with that of the environment. This is particularly important when you drive up a mountain, fly in an airplane, or dive in deep water.

Myomeres

A fourth major chordate trait is usually evident in the larva or embryo: blocks of tissue called **myomeres** flank the notochord and nerve cord and generate the muscles and bones. Myomeres are the major signs of segmentation in chordates. When you eat fish, you can easily see muscle layers that derive from the myomeres in the stacked units of white or pink meat that flake off under your fork. Or feel your own ribs; each is the product of a single myomere that arose when you were an embryo.

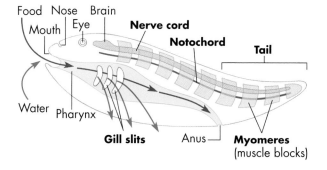

Figure 13.19   Chordate Innovations in the Animal Body Plan. The five distinguishing chordate traits—notochord, nerve cord, gill slits, myomeres or muscle blocks, and tail—are shown in this generalized chordate body.

Tails

The notochord, nerve cord, and myomeres extend into the **tail,** the fifth chordate characteristic. In chordates—including humans—the tail protrudes beyond the anus at some point during development. Most chordates have a tail throughout life. But in chordates like ourselves that are tailless as adults, the tail appears only briefly in the embryo.

Importance of Chordate Characteristics

The new chordate characteristics—the notochord, spinal cord, gill slits, tail, and muscle blocks—had dramatic evolutionary implications (Fig. 13.20). The physical support of a notochord, and later of a vertebral column, allowed chordates to grow as large as dinosaurs and whales. The dorsal, hollow nerve cord allowed centralized nerve coordination and the evolution of acute senses. From these grew a wide range of behaviors, including intelligence, agility, and the vocal communications of a wolf pack hunting a moose. The gill slits could take part in filter feeding or in exchange of respiratory gases. The tail, in its many manifestations, became a major organ of locomotion (as in a goldfish), balance (as in a kangaroo or monkey), and even communication (as in a male peacock or black tail deer).

Three Major Branches in the Chordate Lineage

Biologists divide all 50,000 or so chordate species into three subphyla: the **Urochordata** ("tail chordates"), such as sea squirts; the **Cephalochordata** ("head chordates"), such as lancelets; and the **Vertebrata** (the chordates with backbones), such as frogs, snakes, fish, birds, and mammals.

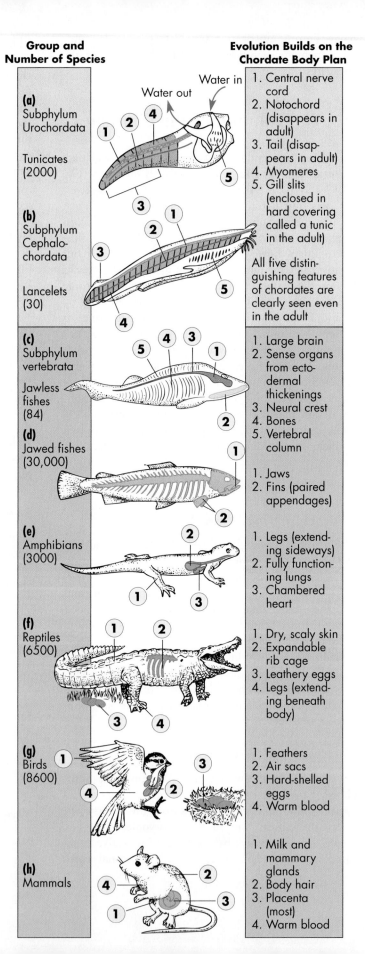

Group and Number of Species

(a) Subphylum Urochordata

Tunicates (2000)

(b) Subphylum Cephalochordata

Lancelets (30)

(c) Subphylum vertebrata

Jawless fishes (84)

(d) Jawed fishes (30,000)

(e) Amphibians (3000)

(f) Reptiles (6500)

(g) Birds (8600)

(h) Mammals

Evolution Builds on the Chordate Body Plan

1. Central nerve cord
2. Notochord (disappears in adult)
3. Tail (disappears in adult)
4. Myomeres
5. Gill slits (enclosed in hard covering called a tunic in the adult)

All five distinguishing features of chordates are clearly seen even in the adult

1. Large brain
2. Sense organs from ectodermal thickenings
3. Neural crest
4. Bones
5. Vertebral column

1. Jaws
2. Fins (paired appendages)

1. Legs (extending sideways)
2. Fully functioning lungs
3. Chambered heart

1. Dry, scaly skin
2. Expandable rib cage
3. Leathery eggs
4. Legs (extending beneath body)

1. Feathers
2. Air sacs
3. Hard-shelled eggs
4. Warm blood

1. Milk and mammary glands
2. Body hair
3. Placenta (most)
4. Warm blood

Figure 13.20 **Evolution Builds on the Chordate Body Plan.** Each group of chordates shows a slightly different set of evolutionarily new characteristics, each modifying the adaptations of earlier species.

Urochordata: The Tunicates

The urochordates, also called tunicates, diverged early in the chordate lineage (see Fig. 13.21 and ⓔ 13.1). Sea squirts (or ascidians) are the squishy tunicates we mentioned earlier; they live as stationary, sacklike adults that attach to submerged rocks or harbor pilings. Sea squirt larvae are free-living, however, and look like miniature tadpoles. Muscle blocks in the trunk and tail power their swimming, and the tail itself is stiffened by the notochord. The nerve cord coordinates their movements, and they filter food from seawater via gill slits (Fig. 13.20a,b). The larvae eventually settle down, attach to a submerged rock, and undergo a dramatic metamorphosis to the stationary adult, resorbing its tail along with notochord and nerve cord. The adult develops an outer **tunic** enclosing a large basket-shaped pharynx perforated by hundreds of gill slits. (This tunic gives the common name tunicates to urochordates.) Cilia line the gill slits, and their beating draws water in through the mouth or incurrent siphon, and sweeps it out again through the excurrent siphon. The pharynx captures food particles suspended in the feeding current and these pass to the stomach. Sometimes they forcibly squirt water out—to the surprise of many a novice student in zoology lab. Hence the popular name sea squirt.

In contrast to the sea squirts, another class of urochordates, the larvaceans, retain the five major chordate features even as adults. These urochordates are important to our study of chordates because the ancestor of the fishes probably evolved from a urochordate-like larva. One of the ways biologists study how new traits arise during evolution is to compare the genes of urochordates and fish.

Cephalochordata: Lancelets

In the second group of chordates, the notochord runs all the way to the very front of the head, hence the name *cephalo*chordate (ⓔ 13.1). The cephalochordates called **lancelets,** or **amphioxus,** are small, streamlined marine animals about 5 cm (2 in) long. They live half-buried, tail first, in the sandy bottoms of shallow saltwater bays and inlets (Fig. 13.21b). While sea squirts lose their notochord, nerve cord, and tail as adults, lancelets retain the five major chordate traits throughout life. In common with sea squirts, however, lancelets are filter feeders; they have cilialike structures around their mouths and sticky, food-trapping mucous regions in the pharynx. If food becomes scarce, lancelets can pull up anchor and move to more fertile waters. The ability to move as adults gives lancelets access to food supplies

that are unavailable to the stationary sea squirts. Note from the tree in Figure 13.18 that cephalochordates are the sister group of the vertebrates.

The Vertebrates: Animals with Backbones

About 550 million years ago, at the dawn of the Paleozoic, ferocious relatives of the cephalochordates began to swim the ancient seas and leave a fossil record. These new animals became fierce and cunning predators, with enhanced ways to detect food and to snare and devour it. These animals were fishes, the first vertebrates.

The Fishes The earliest fish were streamlined filter feeders that lived in the muddy bottoms of ancient seas. About 30 cm (1 ft) long, they had fixed, circular, jawless mouths that could filter sediments by muscular pumping rather than solely by ciliary action. These fishes are called **agnathans** (*a* = not + *gnath* = jaw) meaning "jawless."

Jawless fish had four main features in addition to the usual chordate characteristics (Fig. 13.20c): (1) a large, three-part brain; (2) thickenings of the embryonic ectoderm that developed into pairs of highly specialized sense organs (eye lenses, ears, and nose); (3) the neural crest (review Figs. 8.11 and 8.15), a group of migratory cells in the developing embryo that form the gill bars,

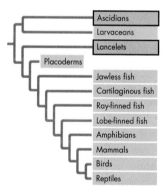

Figure 13.21 at left shows a cladogram listing: Ascidians, Larvaceans, Lancelets, Placoderms, Jawless fish, Cartilaginous fish, Ray-finned fish, Lobe-finned fish, Amphibians, Mammals, Birds, Reptiles.

Figure 13.21 **Chordates without Vertebrae.** Urochordate larvae have a streamlined body shape that resembles the generalized chordate. (a) Adult ascidians like this blue and gold Solitary tunicate living off the Philippines have lost many chordate features during their metamorphosis from larvae. (b) Lancelets such as *Amphioxus* (a cephalochordate) seem fishlike superficially. However, lancelets rest vertically with transparent bodies planted tail first in the sand; just the anterior tip is exposed and filters tiny food morsels from seawater.

(a) An adult ascidian

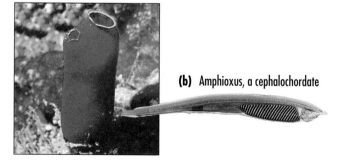

(b) Amphioxus, a cephalochordate

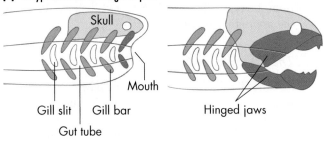

(a) A hypothesis for the origin of jaws

Skull

Gill slit   Gill bar

Gut tube

Mouth

Hinged jaws

(b) A placoderm fish

(c) A cartilaginous fish

(d) A lobe-finned fish

(e) A ray-finned fish

Figure 13.22   **Fishes: The Most Diverse Vertebrates.** (a) The gill support bones (gill bars) of early jawless fishes may have evolved into the powerful jaws of the first jawed fishes, (b) the placoderms. (c) Cartilaginous fishes, including sharks, have a skeleton of cartilage that is lightweight and flexible. (d) Lobe-finned fishes, such as coelacanths, today inhabit the deep waters off the Comoro Islands east of India. (e) The bony, jawed fishes include ray-finned fishes, which, in turn, include teleosts such as the large anglerfish shown here.

some hormone-producing cells, and some nerve cells that make the guts work more efficiently; and (4) bone, a meshwork of cells and proteins hardened by calcium, phosphate, and other minerals. The tough bony plates encased and protected the cunning brains of early fish, like the placoderm fossil shown in Figure 13.22b, and served as coats of armor that shielded the ancient fish from dangerous invertebrates such as giant sea scorpions. There are still jawless fish species alive today, like the lamprey, but they have lost the bony plates. Neural crest cells gave rise to bars of bone in the arches between the gill slits, and this support, in concert with the action of strong muscles, helped the fish draw water more forcefully through the gills and into the pharynx than did the cilia of earlier animals. This more powerful current allowed

the agnathans to consume greater quantities of food than their earlier cousins did and to grow 6 to 30 times larger. Modern jawless fishes, including the lamprey and hagfish, have very flexible internal skeletons of cartilage. Lampreys are parasites that attach to their prey by suction. Once attached, a lamprey rasps through the victim's body wall with a sharp tongue, digests its tissues, and then consumes the liquified remains. Many lamprey species can be serious pests of commercially fished species in Lake Michigan. In some lamprey species, however, the adult is not a parasite and, in fact, reproduces but does not feed at all.

Four Groups of Fishes Evolve

The ancient jawless fishes gave rise to all modern fishes, but they also gave rise to amphibians, reptiles, birds, and mammals. We'll explore those descendant groups

Ascidians
Larvaceans
Lancelets
Placoderms
Jawless fish
Cartilaginous fish
Ray-finned fish
Lobe-finned fish
Amphibians
Mammals
Birds
Reptiles

later. Figure 13.22 shows the four main lines of fish evolution. These include (1) the jawless fishes (such as lampreys) that we just discussed; (2) the cartilaginous fishes (such as sharks); (3) the ray-finned fishes (such as trout and goldfish); and (4) the lobe-finned fishes (such as lungfish and coelacanths). Fig. 13.22 and **E** 13.1 will help you visualize these groups and their evolutionary relationships.

Jawed fishes appeared in the fossil record about 470 million years ago as the **placoderms** (*placo* = flat + *derm* = skin) (see Fig. 13.22). The placoderms had three basic new characteristics that were so useful they appeared in nearly all the vertebrates that followed. These innovations include hinged jaws, vertebrae, and paired fins. Jaws come from the bones (gill bars) that first appeared in the jawless fishes and support the gills (Fig. 13.22a). Jaws allowed placoderms to consume large chunks of food—kelp fronds, clams, other fish. This was a big competitive advantage and some became physically huge; a person can stand upright in the fossilized jaws of one placoderm species. The second innovation, the vertebral column, is a series of bones that largely replace the notochord during development and arch over and protect the spinal cord.

The resulting vertebral column provides a site of attachment for muscles, resulting in more powerful propulsion through the water. The third innovation, paired, lateral fins, provides more control over swimming direction, speed, and depth.

Placoderms (Fig. 13.22b) gave rise to more modern fishes, and the Devonian period (about 345 to 395 million years ago) is called the Age of Fishes because fishes dominated Earth's lakes and seas. Most of today's fishes fall into two main groups, the cartilaginous fishes and the bony fishes.

Cartilaginous fishes have lost the heavy bony plates of the placoderms, and their skull, vertebrae, and the rest of their skeletons are made entirely of cartilage, a matrix of fibrous proteins in a flexible ground substance. Cartilaginous fishes, or **Chondrichthyes** (con-DRICK-thees; *chondro* = cartilage + *ichthys* = fish) include sharks, skates, and rays. In skates and rays, large fins provide lift, much like graceful underwater wings, while in sharks, stiff fins knife through the water, helping these large and common oceanic hunters to maneuver easily toward their victims (Fig. 13.22c).

Bony fishes include sturgeon, gar, bass, bluegill, trout, tuna, and virtually all of today's familiar fresh- and saltwater fishes. Also called **Osteichthyes** (os-tee-ICK-thees; *osteo* = bone + *ichthys* = fish), bony fishes have two major groups. The **lobe-finned fishes,** which arose about 400 million years ago, had large, muscular, lobe-shaped fins supported by bones that allowed them to "walk" across the bottoms of shallow bays. They also had **lungs,** or air sacs, that allowed them to breathe air. When water levels fell, these adaptations may well have allowed them to survive by breathing air and migrating overland to other pools or bays. Present day lobe-finned fishes include the **lungfishes** and the **coelacanths** (Fig. 13.22d).

A second group of bony, jawed fishes are the **ray-finned fishes,** including **teleosts** such as perch, zebrafish, and angelfish (Fig. 13.22e). Teleosts lost their ancestors' fleshy fins, but developed much more versatile spiny fins with webs of skin over delicate rays of bone. Their lung sac lost its connection to the exterior and evolved into a swim bladder, an internal balloon below the backbone that can change volume and allow the animal to adjust its swimming depth. These fishes radiated into a huge group, and with about 30,000 recorded species, teleosts are the largest group of vertebrates. Most fish, however, are relegated by their anatomy to life in water. The great landmasses were to be dominated by other vertebrates—descendants of the early lobe-finned fishes. Nevertheless, fish evolved important new adaptations, including the skull, bone, hinged jaws, paired fins, and vertebrae.

Vertebrate Descendants of the Ancient Fishes

Fossil hunters have found strong evidence that during the Devonian, a new group of animals branched off from the lineage that included air-breathing lobe-finned fishes: this new group included the **amphibians**—vertebrates that can live both on land and in water (*amphi* = both + *bios* = life) (**E** 13.1).

Amphibians About 375 million years ago, the first amphibians overcame the formidable problems of life on land, including more efficient means of walking, breathing, and staying moist. Modern amphibians include approximately 3000 species of frogs, toads, salamanders, and wormlike apodans.

In early amphibians, the fleshy fins of lobe-finned fish had transformed into front and hind **legs** containing strong bones and powerful muscles. Extending sideways from the body, the limbs could support the animal's weight far better than lobed fins; this allowed the animal to move about on land for greater distances, even without water's buoyant support (Fig. 13.23). Salamander bones still show this characteristic pattern that emerged in 380-million-year-old lobe-finned fishes.

Laborious walking, however, required a great deal of energy from food. Amphibians needed this energy,

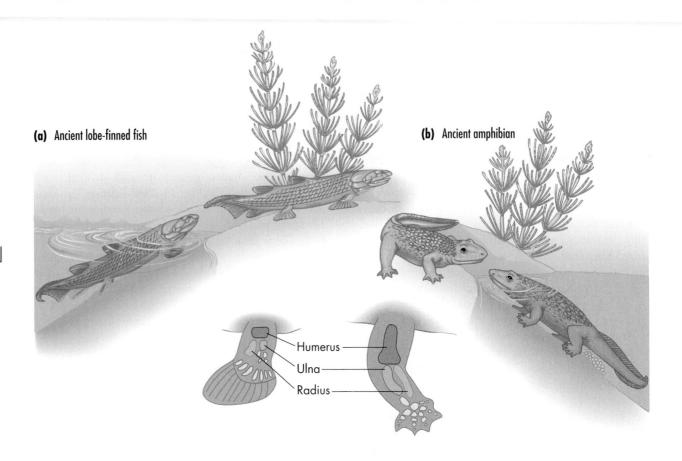

Figure 13.23   The Transition to Land. (a) Ancient lobe-finned fishes, like today's lungfish, would have risen to the surface to gulp air when the stagnant water became depleted of oxygen. From time to time they may have ventured into shallower water and to land in search of invertebrate food. Natural selection would have favored alleles that modified the limbs, over millions of years, into those of early amphibians (b).

(a) Ancient lobe-finned fish

(b) Ancient amphibian

Humerus
Ulna
Radius

Figure 13.24   Modern Amphibians. The brilliant gold and red pigments in the skin of (a) the tropical play-actor frog *(Dendrobates histrionicus)* and (b) the Chinese salamander *(Tylototriton verrucosus)* warn would-be predators of the amphibians' poison skin glands. Frogs and salamanders are by far the most common amphibians. The rarer amphibians called apodans burrow in the soft, moist soils of tropical rain forests in search of insects and have lost all trace of limbs.

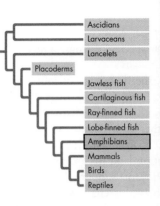

(a) Frogs

Ascidians
Larvaceans
Lancelets
Placoderms
Jawless fish
Cartilaginous fish
Ray-finned fish
Lobe-finned fish
Amphibians
Mammals
Birds
Reptiles

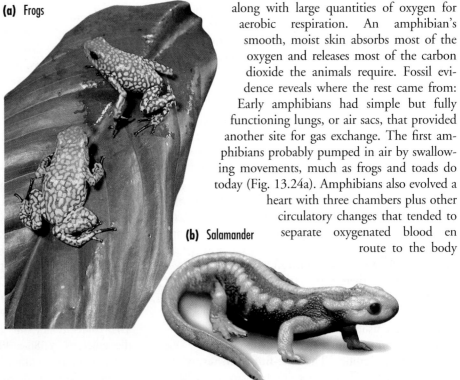

(b) Salamander

along with large quantities of oxygen for aerobic respiration. An amphibian's smooth, moist skin absorbs most of the oxygen and releases most of the carbon dioxide the animals require. Fossil evidence reveals where the rest came from: Early amphibians had simple but fully functioning lungs, or air sacs, that provided another site for gas exchange. The first amphibians probably pumped in air by swallowing movements, much as frogs and toads do today (Fig. 13.24a). Amphibians also evolved a heart with three chambers plus other circulatory changes that tended to separate oxygenated blood en route to the body tissues from deoxygenated blood bound for the lungs. This separation became complete in crocodilians, mammals, and birds and allowed more and more vigorous activity levels to emerge.

Because an amphibian's skin must remain moist for gas exchange, the animals have always been restricted to life in damp places or near the water's edge. In addition, amphibians lay eggs with a clear, jellylike coating that must also stay moist, or the embryos will die before the fishlike tadpoles emerge and wriggle away. Most amphibians can move away from water at times, but must always return to reproduce. In this sense, they are analogous to the mosses, ferns, and other ancient land plants whose sperm must swim to an egg.

Today, frogs, toads, salamanders, and a legless group called the **apodans** are the only remaining members of the class Amphibia. Alarmingly, amphibian populations all over the world apparently began declining in the late 20th century. Some scientists think that viruses, parasites, acid rain, global warming, or pollutants that mimic amphibian hormones and thus interfere with growth may be contributing to the decline or extinction of many species.

Reptiles   In the vast, steamy Carboniferous swamps, amphibians crawled about in profusion, but they were not alone. Another kind of four-legged land animal—**reptiles**—lumbered about, too (Fig. 13.25). The early reptilians superficially resembled crocodiles and evolved four important innovations for life on land that freed them from dependence on wet environments, much as the early seed plants were free to inhabit areas further from standing water:

1. Their dry, **scaly skin** provided a barrier to evaporation and sealed in body moisture. This eliminated the skin surface as a major site for gas exchange.
2. Their expandable **rib cage** could draw in large quantities of air like a bellows. This extra air could be dis-

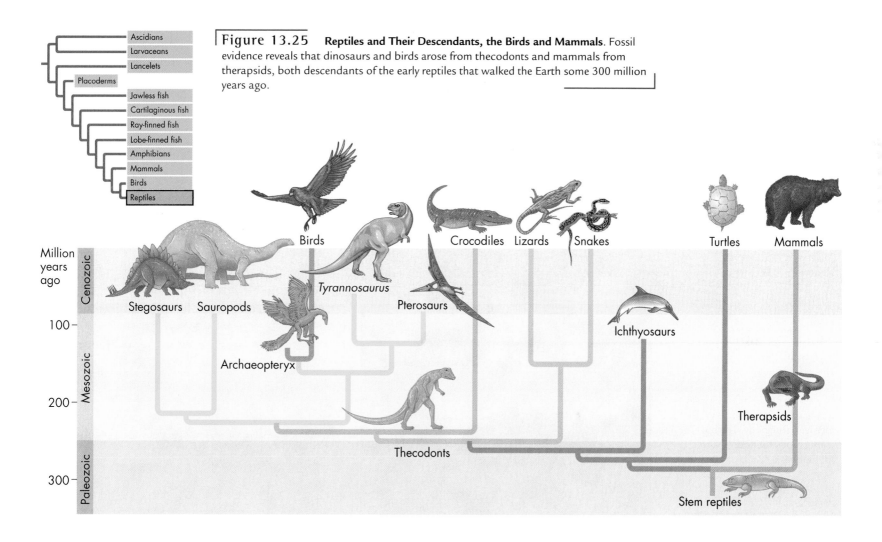

Figure 13.25   **Reptiles and Their Descendants, the Birds and Mammals**. Fossil evidence reveals that dinosaurs and birds arose from thecodonts and mammals from therapsids, both descendants of the early reptiles that walked the Earth some 300 million years ago.

tributed well because modifications in the heart and circulatory system separated oxygenated and deoxygenated blood more fully than in the amphibians.

3. Innovations in reproduction eliminated reliance on open water for reproduction. Females produced so-called **amniote eggs.** These eggs essentially encased the developing embryo in a pool of fluid (the amniotic sac). They also provided a source of food (the yolk) and surrounded both embryo and food with membranes and a leathery shell that prevented crushing and drying out. Males had copulatory organs that could deliver sperm directly into the female's body (internal fertilization) rather than into the surrounding water (external fertilization).

4. Finally, the legs of the early reptiles extended directly beneath the body rather than out to the side (as amphibians' legs did then and continue to do today). This arrangement provided better support and made walking and running easier.

These new characteristics were crucial to the evolution of all the modern reptiles—the crocodiles, turtles, lizards, tortoises, snakes, and the ancient lizardlike tuatara from islands off New Zealand (*E* 13.1). They were also, however, pivotal to rise of the birds and mammals, as we'll see shortly.

The biggest, fiercest reptiles that ever lived were **dinosaurs** ("terrible lizards"). The earliest crocodile-like reptiles gave rise to two lineages: the **thecodonts,** or small lizards that ran on two legs and led to the dinosaurs, and the **therapsids,** small, heavyset, fiercely toothed animals that led to the mammals.

The warm Cretaceous period (144 to 65 million years ago) was the heyday of dinosaurs including the 15-m-long (50-ft-long) meat-eating *Tyrannosaurus* and the 24-m-long (80-foot-long) plant-eating *Apatosaurus*

(formerly *Brontosaurus*) (Fig. 13.26a). Dung beetles, as we saw earlier, had already evolved and were making use of giant dinosaur pats by this time. The grand radiation of reptiles into forms that swam, flew, and lumbered across Earth lasted nearly 150 million years, and inspired the common name for the entire Mesozoic Era: the Age of Reptiles. Most of the great and diverse reptiles died out, however, in a massive extinction event at the end of the Cretaceous.

A few types of small reptiles weighing less than 20 kg (44 lb) survived the extinctions and radiated once again during the Cenozoic (the current geologic era), which began about 65 million years ago. Today, there are only about 6500 species of reptiles. Alligators, caimans, and crocodiles are all streamlined carnivores that inhabit warm climates. The tortoises and turtles have a tough and unique protective structure, the shell, with its upper curved carapace and its lower flat plastron. The lizards and snakes are elongated reptiles that inhabit wet, dry, or hot environments and sometimes reach great size. The heaviest lizards today are the Komodo dragons of Indonesia, which can weigh up to 115 kg (255 lb) (Fig. 13.26b), and the longest snakes are the pythons of the same region, which can grow to 8 m (27 ft).

Most biologists no longer consider the reptiles as a formal taxonomic group because reptiles do not constitute *all* the animals on a single branch of the tree of life; the branch that includes the reptiles also includes the birds. Some zoologists think of the animals as feathered and nonfeathered reptiles.

Birds   Except for domesticated animals such as dogs, cats, and cows, the most common chordates we see around us are **birds** (*E* 13.1). Curiously, some biologists have recently suggested that, from a zoological per-

(a)   Apatosaurus

(b)   Komodo dragon

Figure 13.26   **Extinct and Living Reptiles.** While ancient amphibians had legs directed out to the sides, reptiles, such as the dinosaurs (a) had legs more directly underneath the body. The huge, carnivorous Komodo dragon of Indonesia *(Varanus komodoensis)* (b) can grow up to 3 m (10 ft) long and weigh 115 kg (255 lb).

spective, birds are actually a kind of dinosaur. The first winged vertebrates—the giant, soaring pterosaurs (Fig. 13.25)—might seem like logical ancestors to the birds, but pterosaurs died out long before birds evolved. Instead, small, two-legged, lizardlike thecodonts appear to be the real forerunners of the birds. Paleontologists have unearthed six fossil skeletons of one of the oldest birds, the crow-sized *Archaeopteryx,* at different sites in rocks dated back to the Upper Jurassic period (150 million years ago). The fossil imprints suggest that this animal was a true intermediate: It had scaly skin, curving claws, a long, jointed tail, and sharp teeth like a reptile, but it had birdlike feathers on its forelimbs.

Most of the birds' evolutionary adaptations prepared them for efficient flight (Fig. 13.25). **Feathers** are marvelously lightweight structures made of dead cells containing the protein keratin (see Chapter 2). Birds also have lightweight, hollow bones and a breastbone, or keel, enlarged into a blade-shaped anchor for the powerful pectoral muscles that raise and lower the wings. The legs are reduced to skin, bone, and tendons and can be folded up like an airplane's landing gear, reducing drag during flight.

Flight is a strenuous activity that requires plenty of oxygen for the aerobic respiration of muscle and other tissues, and birds have a third set of modifications that ensures an adequate supply of oxygen. First, birds are **homeothermic** (meaning "constant-temperature"). They maintain a constant internal temperature slightly warmer than our own body temperature regardless of changes in the outside air or water temperatures. This constancy helps with the steady production of ATP energy during cellular respiration, which in turn fuels the activities of wing and leg muscles. In fact, the earliest feathers may have been more useful as insulation than as aids to flight. Most fish, amphibians, and reptiles have body temperatures that vary; as a result, they tend to be active in warm environments but sluggish in cold ones. (See Chapter 14 for more about warm- and cold-bloodedness.) Second, birds have lungs connected to a series of **air sacs** that exchange oxygen and carbon dioxide in an efficient one-way flow. And third, birds have a **four-chambered heart** that completely separates oxygenated and deoxygenated blood so only the former reaches body tissues (see Chapter 15).

Finally, in addition to the evolutionary features of feathers, constant body temperature, and air sacs, birds produce amniote eggs with hard shells rather than leathery coverings. These innovations free birds from dependence on water for reproduction, just like reptiles' leathery eggs, but the shells are more impact resistant and thus more protective for the embryo inside.

Birds' evolutionary innovations were so successful that in the last few million years of the Cenozoic, birds radiated into a highly diverse class with more than 8600 species specialized for life in the trees, at the sea shore, in freshwater lakes, in the desert—even on icebergs!

Mammals   While the second lineage of reptiles—the fierce, heavyset therapsids—gave rise to the mammals (Fig. 13.27), they developed their own distinct body traits at least 180 million years ago. The very early mammals resembled shrews, and until the end of the Cretaceous, they probably scurried around the great dinosaurs' feet along with the dung beetles. These progenitor mammals survived the mass extinctions of dinosaurs and other animals about 65 million years ago and radiated into 5000 modern species. Our current geological era, the Cenozoic, is in fact the Age of Mammals (*E* 13.1).

Mammals (Latin, *mamma* = breast) are warm-blooded and have a four-chambered heart, just like birds. These two traits probably evolved independently in the two groups, however. Mammals have two additional traits unique to their class: milk and body hair or fur. A fluid rich in fats and proteins, **milk** is produced in **mammary glands** and is used to nourish newborns. Many mammals have dense carpets of hair called **fur** covering the entire body. Others, such as certain monkeys and humans, have sparse body hair. A few, including the whales and porpoises, have only a few very sparse hairs. These marine mammals rely instead on thick layers of fat called **blubber** for insulation. Significant changes occurred in the mammalian skull, as well.

In addition to milk and to hair or fur, most mammals also have a unique reproductive structure, the **placenta** (Fig. 13.20h). This spongy organ supports the growth of the embryo to a fairly complete stage of development before birth. Placental mammals descended from one branch of the earliest mammals, while two other branches led to nonplacental mammals that nurture embryos in different ways. The **monotremes** lay leathery eggs and warm them until the young hatch, then nurse the young with milk. There are only three species of monotremes: the duck-billed platypus and the two spiny anteaters (or echidnas) of Australia and New Guinea (see Fig. 13.28a). The **marsupials** give birth to immature live young no bigger than a kidney bean. When the newborn emerges from the birth canal, it crawls upward into an elastic pouch of skin on the mother's ventral surface (the **marsupium**), attaches to a teat, starts to consume milk, and continues to develop inside the pouch for several months (Fig. 13.28b). There are dozens of marsupial species,

Three Toed Sloth

What are the common characteristics of mammals?

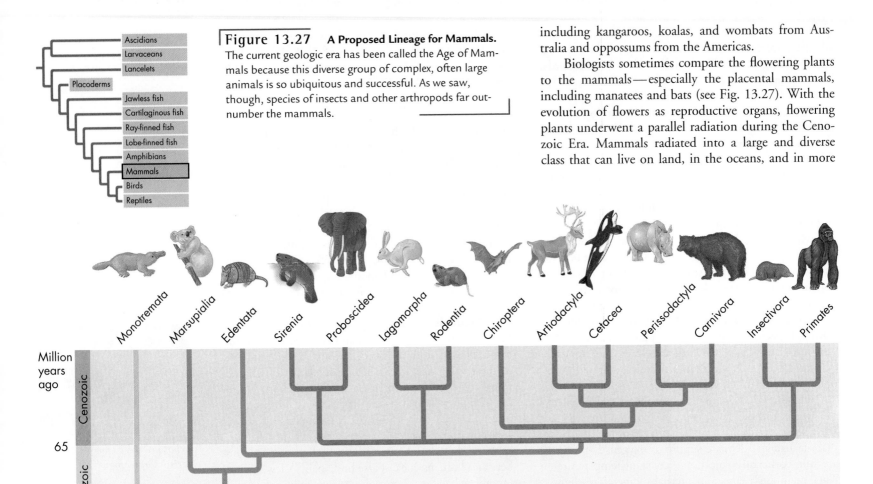

Figure 13.27   A Proposed Lineage for Mammals.
The current geologic era has been called the Age of Mammals because this diverse group of complex, often large animals is so ubiquitous and successful. As we saw, though, species of insects and other arthropods far outnumber the mammals.

including kangaroos, koalas, and wombats from Australia and oppossums from the Americas.

Biologists sometimes compare the flowering plants to the mammals—especially the placental mammals, including manatees and bats (see Fig. 13.27). With the evolution of flowers as reproductive organs, flowering plants underwent a parallel radiation during the Cenozoic Era. Mammals radiated into a large and diverse class that can live on land, in the oceans, and in more

Figure 13.28   Modern Mammals. Mammals are remarkably diverse in body size and appearance. The echidna (a) (here, the Australian short-nosed spiny anteater, *Tachyglossus aculeatus*) is an egg-laying monotreme from Australia and New Zealand that nourishes its young with milk. (b) This black-faced gray kangaroo, an Australian marsupial, gives birth to tiny young that develop in a pouch. Here, the young is large enough to graze along with its mother. Placental mammals include all the familiar types you can see in Figure 13.27.

(a) A monotreme

(b) A marsupial mammal

specialized niches than any other class of animals except perhaps the birds. Living right alongside the land mammals—or more accurately, under—are the thousands of species of dung beetles adapted to consuming their solid wastes.

13.4   Human Origins

One branch of the mammalian family tree gave rise to the **primates,** the order that includes *Homo sapiens,* the "wise man." We humans are clearly mammals with our body hair, warm blood, mammary glands, and mode of giving birth. But we also have a unique combination of behavioral abilities—including spoken and written language, agriculture, and extensive tool use—that has allowed us to dominate the environment like no other animals before us. One small example is our ability to study and categorize dung beetles. Despite our unique abilities, scientists have ample evidence, both fossil and genetic, that humans represent one recent branch of the Primate order within the mammals. Moreover, our branch separated only about 6 million years ago from the lineage leading to chimpanzees. Ethologist Desmond Morris called humans "naked apes," and so, it seems, we are. We are subject to the same kinds of evolutionary forces we discussed in earlier chapters, including natural selection and genetic drift. Because human beings are a branch of the primate evolutionary tree, the shape and size of that tree help us understand ourselves (Fig. 13.29).

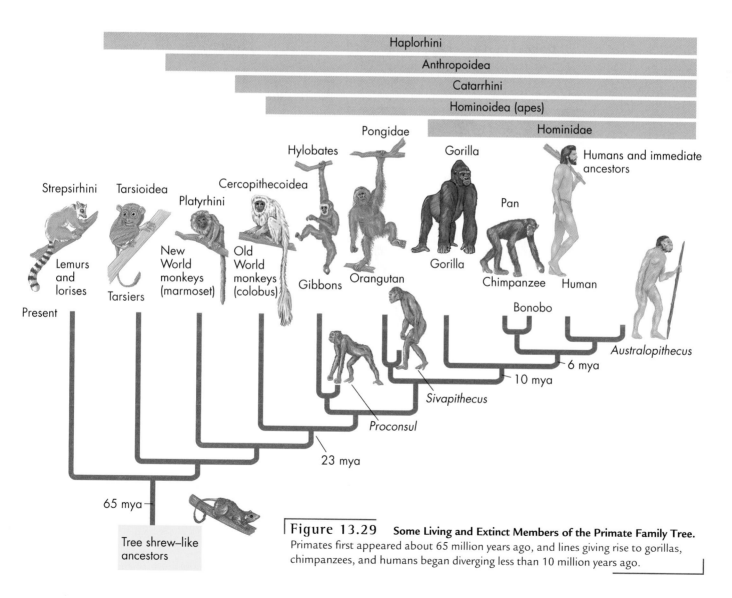

Figure 13.29   **Some Living and Extinct Members of the Primate Family Tree.** Primates first appeared about 65 million years ago, and lines giving rise to gorillas, chimpanzees, and humans began diverging less than 10 million years ago.

Figure 13.30 **Living Primates.** (a) The lineage leading to lemurs, such as this red-fronted lemur mother and its infant, diverged from other primates early in primate phylogeny. They have limited color vision and rely more strongly on the sense of smell than other primates. (b) Tarsiers live in the tropical forests of the Philippines, Sumatra, Java, and Borneo, where they harvest fruit and hunt insects, lizards, and birds by night. (c) This scolding white-faced capuchin *(Cebus capucinus)* is a New World monkey of Costa Rican rain forests. Old World monkeys include the colobus pictured in Figure 13.29.

(a) A lemur

(b) A tarsier

(c) A New World monkey

The Primate Family Tree

Taxonomists divide the order Primates, with its 150 or so currently living species, into two suborders: the **Strepsirhini** (which means "moist nosed"), and the **Haplorhini** (which means "hairy nosed") (see Fig. 13.29 and *E* 13.1).

Strepsirhini

Strepsirhini are small, mostly arboreal (tree-dwelling) primates such as lemurs and lorises, with the claws, long snout, and side-facing eyes common in early mammals (Fig. 13.30a). Lemurs are fuzzy animals found only on Madagascar, with foxlike muzzles, long bushy tails, and long front and back limbs (Fig. 13.30a). Most species live in trees and feed on nectars, fruits, leaves, and sometimes insects. Many species of lemurs became extinct shortly after the large primate *H. sapiens* came to inhabit Madagascar less than 2000 years ago. Continued destruction of forests on the island threatens the remaining lemurs' survival.

Haplorhini

The Haplorhini include the tarsiers and the **anthropoids.** The tarsiers (Fig. 13.30b) have huge, forward pointing eyes, which help them see in their treetop habitat. Uniquely, their legs are also elongated due to long tarsal (ankle) bones. The Anthropoidea (see Fig. 13.29) consist of the New World monkeys or Platyrhini ("flat-nosed"); the Old World monkeys or Catarrhini ("downward-pointing nose"); and the apes and humans.

New World monkeys, which inhabit the forests of southern Mexico and Central and South America, have flatish noses with widely separated nostrils oriented somewhat laterally (Fig. 13.30c). Some of the larger bodied New World monkeys, such as spider monkeys and howler monkeys, have grasping, or **prehensile**, tails. These tails, which have a naked, touch-sensitive pad at the tip, serve as a type of fifth hand, aiding the monkeys in dangerous crossings between branches of adjacent trees.

Old World monkeys, which live in tropical forests and savanna regions of the Eastern Hemisphere from Africa to India and Southeast Asia, have closely set, downward-pointing nostrils and lack prehensile tails (Fig. 13.29). Many have buttock calluses that help pad and protect their posterior. The rhesus monkeys used in medical research are Old World primates, as are the langurs of Asia and the baboons and mandrills of Africa.

Hominoids include the apes and humans: Apes are the largest primates and our closest relatives. Apes are large, tail-less animals with long arms, large brains, and complex social behavior (see Fig. 13.31). The apes (the nonhuman hominoids) include the gibbons (Hylobatidae), the orangutans (Pongidae); and the Hominidae, including the gorillas and chimpanzees of Africa and humans.

Trends in Primate Evolution

Fossil evidence suggests that the earliest primates lived high in the canopy of the tropical forest. These early primates evolved several important specializations that lead to their success as a group and that form the background for human physical features. These traits included manual dexterity, upright posture, acute vision, large brain, extended infant care, and modified teeth (see Fig. 13.31).

Dexterity of the Hands

Natural selection tends to eliminate clumsy individuals who bumble awkwardly along high tree branches. Primates evolved an **opposable thumb,** the ability to spread their digits widely and to touch their "thumb" with the tips of each of the other four fingers. This allows them to curl their digits around branches in a power grip, as well as to grasp objects with precision—whether a piece of fruit or a pencil.

Upright Posture

Hands cannot manipulate objects if they are supporting the weight of the body. We humans are the only fully upright primates that walk consistently on two legs (we are **bipedal**). Nevertheless, many monkeys and apes spend large amounts of time in vertical rather than horizontal positions. Even though gorillas and chimpanzees walk on all fours by touching the ground with their feet and the knuckles of their hands, their arms are so long that the angle of the body is still fairly erect. **Upright posture** improves an animal's ability to see and can leave the hands free for other activities.

Acute Vision

Life in the trees is rich with visual information, such as fluttering leaves, dappled spots of sunlight, and tangled tree limbs. It can take keen vision to focus on fruits and

Dexterity: Thumbs can touch each of other four digits; fingers can curl around objects; items can be held with precision between two digits.

Infant care: Primate babies are rather helpless and require prolonged periods of parental care.

Upright posture: Changes in the skeleton allowed primates to spend more time with their backbone in a nearly upright position, clinging to tree trunks, or hanging from branches.

Brain: Primates have very large brains for animals their size. Large brains are required for dexterity, for interpretation of complex visual information, and for social skills involved in primate infant care.

Vision: Acute vision facilitates accurate decisions during rapid movements in the trees, and color vision enhances detection of both prey and predators.

Teeth: Teeth became broader as primate diets shifted from mainly insects to tough plant material; further changes that could accommodate an omnivorous diet occurred in lines giving rise to humans.

Figure 13.31 Six Trends in Primate Evolution.

insect foods amidst this backdrop. Furthermore, an error in judging the precise location of a food object on a particular tree limb could result in a fatal fall. Thus, natural selection strongly favored depth perception, or **stereoscopic vision,** in primates. Depth perception occurs because an animal's left and right eyes see the same object from slightly different angles. Try looking at one of your fingers while holding it at arm's length; look first with one eye closed and then the other. Notice how your finger seems to shift position. The brain interprets this difference and creates a realistic three-dimensional picture of the environment. A primate's color vision also helps it recognize members of its species, identify ripe fruits, find edible young leaves, and locate cryptic prey and predators lying in ambush.

A Very Large Brain

The evolution of life in the trees was accompanied by the enlargement of specific parts of the brain. The brain of a rhesus monkey, for example, is much larger than the brain of a similar-sized dog. Per gram of body weight, apes and humans have the most complex brains of any mammals—in terms of brain size, number of nerve cells, and numbers of interconnections.

Zoologists have posed several hypotheses to explain the increase in brain size and intelligence during primate evolution. The political hypothesis holds that primates' complex social interactions were the main driving force behind increased brain size. Larger, more nimble brains might help animals relate better to others in the troop, and perhaps achieve higher social rank and thus leave more offspring. The dietary hypothesis suggests that a larger brain helped primates obtain enough food in the forest canopy, sustain better physical condition, and reproduce more effectively. More agility with grasping and manipulating foods would have helped as well, according to this theory. Gorillas, for example, eat plants defended by nettles, spines, hooks, and thick husks, and have to make as many as eight different manipulations to eat a single food, like using leaves to protect their hand from nettles, peeling the food, and then twisting tough fibers apart. The ability to recognize nutritious foods, harvest them effectively, remember where they grow, and return to them during the right seasons would require large brains as well as learning and would be subject to natural selection.

Both the political and dietary hypotheses make sense and, in fact, could well have played equal roles in increasing primate intelligence.

Infant Care

Along with their mobile life in the trees, primates evolved new reproductive strategies. Compared with most other mammals, primates have smaller litters of young, often just one baby at a time. The infants tend to be helpless and depend on their parents for complete physical care long after birth. In general, the higher a species' intelligence, the more time the parent devotes to the care and nurturing of the young, and the more complex skills the young must master to succeed as an adult. The primates' greater parental investment pays off in their offspring's higher survival rates.

Teeth

Primates have several tooth types, modified for an omnivorous diet of both plant and animal matter. The earliest mammals had pointed teeth, adapted for slicing up the bodies of insects. Primate ancestors may have had similar teeth, but as early primates began to eat more plant foods, individuals with broader, flatter back teeth (molars) could chew these new foods more efficiently. In the line giving rise to humans, the jawline became U-shaped, while in most other primates, it is more V-shaped.

Together, these primate characteristics—manual dexterity, upright posture, stereoscopic color vision, unusually large brain, extended infant care, and various tooth types—helped the primate lineages succeed.

Emergence of Modern Primates

Evidence from genetic studies and from fossilized bones suggests that monkeys, apes, and humans (the anthropoids) arose from early primate ancestors starting about 40 to 35 million years ago (review Fig. 13.29). Monkeys diverged into New and Old World lineages, while the apes and humans (the hominoids) radiated into various niches throughout the Old World. The evolutionary history of the apes is particularly interesting; the early part of it is also our own.

Some 20 million years ago, a line of primitive Old World apes radiated into many species. These ape species had many different body sizes and lived in a diverse array of habitats from tropical forests to more open grasslands and woodlands. One of these apes was *Proconsul africanus* (see Fig. 13.29), a tree-living, fruit-eating African ape that walked about on all fours. Adaptive radiation within this group led to a lineage that eventually included all of today's so-called **great apes**: the orangutans, the gorillas, the humans, and two species of chimpanzees, the common chimpanzee *(Pan troglodytes)* (see Fig. 13.32) and the smaller pygmy chimp, or bonobo *(Pan paniscus)* (see Chapter 26).

Biologists have discovered the close relationships between the great apes by comparing their DNAs. The genetic distance between humans and chimpanzees is about the same as, or a little less than, the genetic distance between chimps and gorillas. This means that a chimp is genetically as close or closer to *you* as it is to a gorilla!

The human and chimp lineages split about 6 million years ago—an extremely short span in geologic history. Using a 24-hour analogy, Earth formed at midnight, life appeared at 4:10 a.m., the first vertebrates appeared at 9:35 p.m., and the first humans arose only 38 seconds before the clock struck midnight. Let's explore those last 38 seconds.

(a) A gorilla

(b) A chimpanzee

Figure 13.32 **The Great Apes.** Orangutans *(Pongo pygmaeus)* such as those pictured in Figure 13.31 live in remote forests of Sumatra and Borneo. Their arms are longer than a chimpanzee's or gorilla's, and they move about rather clumsily in search of fruit and tender shoots. These animals are intelligent, though less social than gorillas or chimpanzees. (a) Gorillas are the largest and most powerful nonhuman primates; most, including this lowland gorilla *(Gorilla gorilla)* inhabit misty lowland forests of western and central Africa. Male gorillas can be nearly 2 m (6.5 ft) tall and weigh over 200 kg (450 lb), and much of their feeding occurs on the ground. Gorillas live in mixed-sex, mixed-age social groups and travel together through the forest in search of food. (b) Chimpanzees *(Pan troglodytes)* are our closest living relatives and live in the forests of western and central Africa in large social communities. They eat mostly fruits and leafy matter as well as termites and other insects. Males occasionally hunt mammals, including monkeys, and then share the meat among themselves and with females and young. In nature they use tools (sticks and grass) and in captivity can be trained to carry out complex tasks, including how to communicate in American Sign Language.

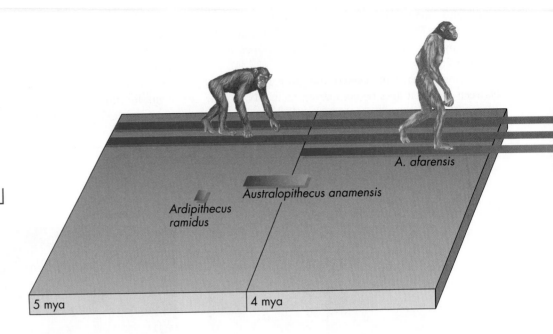

Figure 13.33 **A Family Album of Human Ancestors.**
The apelike *Australopithecus afarensis* lived about 3.75 million years ago and was the first fully upright hominid. Even older hominids include *Ardipithecus ramidus* and *Australopithecus anamensis*. The large-boned *Australopithecus robustus* lived about 2 million years ago, and recent evidence shows that its hand was adapted for precision grasping and it may have used tools. *Homo habilis* was an active tool user who lived about 1.5 million years ago and preceded *H. erectus*. Then *H. sapiens* emerged from *H. erectus* about 500,000 years ago. These species did not lead to humans in an advancing ladder, but occupied ends of branches in a family tree whose precise arrangement is still under investigation.

A. afarensis

Australopithecus anamensis

Ardipithecus ramidus

5 mya

4 mya

The Appearance of *Homo sapiens*

The gene sequences in people and chimps are 98.5 percent the same. The 1.5 percent of genetic differences has a huge impact on our respective capabilities. Paleontologists do not yet understand entirely how our apelike ancestors became fully human. It is clear, though, that the human family tree had many branches and every few months, new fossil and genetic discoveries are helping to illuminate the branches and how they arose. It is also very clear that human evolution did not follow a ladder from ape to human, and instead, that the human lineage is the sole surviving branch of an evolutionary bush. To understand our evolution, we need to explore several questions: What changes led to modern humans? How long ago did those changes occur? And what forces brought about those changes?

Innovations in Human Evolution

At least four critical innovations arose as apelike ancestors evolved into humans. First was the origin of bipedalism, walking on two feet. This happened about 6 million to 4 million years ago, when our lineage diverged from the one leading to modern apes. Second was the invention of toolmaking, the manufacture of stone implements to cut flesh from the bones of game animals and to crush their bones for the nutritious marrow inside. Recent discoveries show that toolmaking began about 2.5 million years ago. The third innovation was the dramatic expansion of the human brain far beyond the relative brain sizes of chimps and gorillas. And the fourth change, occurring a few tens of thousands of years ago, was the use of brain power for abstract thought, including artistic, musical, linguistic, technological, and other skills that have led to our current domination of the world.

The Root of the Bush of Human Evolution

In the mid-1990s, paleontologists discovered 4.4-million-year-old fossils in East Africa belonging to a species they call *Ardipithecus ramidu* (*ardi* = floor + *pithicus* = ape + *ramidus* + root). These fossils—the bones and teeth of 17 individuals—show a mix of chimpanzeelike and humanlike traits, including large canine teeth but relatively small molars, suggesting a diet of soft fruits and vegetables. It's not clear whether *Ardipithecus ramidus* walked upright, but recently discovered fossils may answer the question.

Soon after scientists found *A. ramidus*, a member of the famous Leakey family of British paleontologists, Maeve Leakey, found a slightly younger fossil (4.2 million years old) with bones that show upright walking. These animals, called *Australopithecus anamensis* (*austral* = southern + *pithecus* = ape + *anam* = lake) were still not as fully upright as we are today and their legs were short, so the gait would have been different as well. This hominid was significant, though, because it lived and walked on two legs 500,000 years earlier than the oldest bipedal animal found before it.

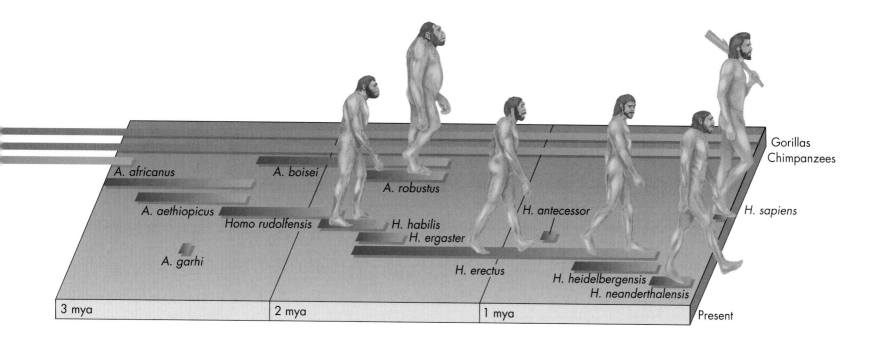

A. africanus

A. aethiopicus

A. boisei

A. robustus

Homo rudolfensis

A. garhi

H. habilis

H. ergaster

H. antecessor

H. erectus

H. heidelbergensis

H. neanderthalensis

Gorillas
Chimpanzees

H. sapiens

3 mya | 2 mya | 1 mya | Present

Were *A. ramidus* and *A. anamensis* on our branch of the evolutionary bush (Fig. 13.33)? No one knows yet. *A. ramidus* may be an offshoot from our lineage, but its discovery does bring us closer to knowing how our lineage split off from that of chimpanzees and bonobos. Maeve Leakey hypothesizes that the other early hominid, *A. anamensis,* may be a direct ancestor to the famous "Lucy" fossil.

Lucy and Her Kin

In the mid-1970s, British anthropologist Mary Leakey, her son Richard (Maeve's husband), and numerous colleagues found fossils of a species they named *Australopithecus afarensis* (after the Afar region of East Africa). One very complete skeleton recovered by Donald Johanson was named "Lucy" after the hit song by the Beatles, "Lucy in the Sky with Diamonds." *A. afarensis* had a very apelike skull and teeth, a brain only slightly larger than a chimpanzee's (450 ml), and long arms but short legs. Nevertheless, its head sat on top of the backbone like ours does rather than projecting forward like an ape's, and the hands were distinctly humanlike.

At a separate site called Laetoli in East Africa, the Leakey team found remarkable evidence of hominid behavior—probably among *A. afarensis* individuals—living 3.5 million years ago. They found small humanlike footprints made in soft ash in a string nearly 40 footprints long (Fig. 13.34). The size and depth of the tracks suggest that a male and female may have walked side by side through newly fallen ash from a nearby volcano. The closeness of the two tracks suggests that the male may have been touching the female as they walked. A smaller individual (probably a child) apparently trailed along behind them, stepping in the male's footprints, perhaps to conceal his own, or perhaps because the ash was hot, or maybe just as a game.

This momentous find helped show that bipedalism—walking erect on two feet—evolved a million years before human ancestors had the enlarged brains or stone tools of later fossil evidence. It also showed that ancestral humans were living in at least small social groups and walking on the open savanna 3.5 million years ago, and not just living in forests, as did most other primates.

The genus *Australopithecus* radiated into a multibranched human family tree with many species. Such species include the large-boned *Australopithecus robustus,* which lived about 2 million years ago and had hands adapted for precision grasping, maybe even allowing it to use tools. Some of these may have been living simultaneously or in overlapping time periods (see Fig. 13.33). Figuring out how they are related to each other and to later species is difficult because of gaps in the fossil record. In 1999, however, a series of exciting finds helped to fill in part of this fossil gap.

Who Were the First Toolmakers?

On an Ethiopian plain overlooking a shallow lake, a humanlike individual bent over an antelope carcass about

Figure 13.34 **Making Tracks at Laetoli.** Mary Leakey's team discovered these footprints in 1978. They dated them to 3.6 million years old and think the tracks were made by an *Australopithecus afarensis* couple walking through loose ash in what is now Tanzania. Smaller imprints inside the male's footprints may have been a child's.

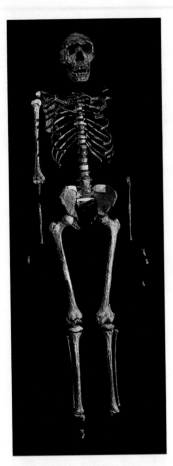

Figure 13.35 *Homo erectus*: **A 1.6-Million-Year-Old Skeleton.** This nearly complete skeleton is that of a 12-year-old boy who would have been 6 ft tall as an adult. Notice the heavy skull and large, broad teeth.

2.5 million years ago. Grasping the hammerstone he had made and carried with him, he struck the antelope's lower leg bone twice, crushing the bone and exposing the fatty, nutritious marrow, which he devoured. He also cut out the animal's tongue, leaving curved parallel tool marks on the jawbone—and perhaps carried the prize off to family members. Paleontologists found the crushed leg bone and scarred jawbone less than a meter from fossils of a new species. The bones were clearly more advanced than Lucy and presumably belong to the humanlike butcherer of antelope. This new species, *Australopithecus garhi* (*garhi* means "surprise" in the Afar language), was clearly not itself a human. It is a candidate, however, for the ancestor of our own genus, *Homo*. It also shows that tool use had emerged by 2.5 million years ago and perhaps was invented first by the now-extinct genus *Australopithecus* and not by the genus *Homo*.

Until *Australopithecus garhi* surfaced, *Homo habilis*, the "handy human," was considered the earliest tool user starting about 2 million years ago. These ancient members of our own genus, *Homo*, had a large face, big teeth, and trunks and limbs fully adapted for walking upright. At 1.5 m (5 ft) tall, however, *Homo habilis* individuals were larger than members of the genus *Australopithecus*, and their brains, at about 700 ml, were 50 percent larger. The "handy man's" brain was still only half the size of a modern human's. Its increased size, however, suggests a real increase in intelligence and probably a heightened degree of manual dexterity and social skills along with it. *H. habilis* individuals appear to have made crude stone tools by cracking rocks and flaking off chips, using the sharp chips to harvest meat from tough animal hides. Access to more meat, to the highly nutritious bone marrow, and maybe even to turnips, yams, and other root vegetables cooked over campfires may have allowed *Homo habilis* to outeat and outcompete the *Australopithecus* species that lived about the same time. The ability to make tools—the first emergence of technology—has dominated the rest of human evolution.

Humans Disperse

Tool-using *H. habilis* appears to have remained in African savannas and woodlands, but fossil evidence shows that a new species of human arose in Africa then migrated away. **Homo erectus** ("erect human") had evolved by 1.8 million years ago and spread into northern Africa. From there, *H. erectus* moved to southern Asia, Indonesia, and probably into southern Europe. *H. erectus* fossils are intermingled with a new kind of stone tool, the Acheulean hand ax. This implement, while still simple, required finer technological skill to manufacture than the hammers, choppers and scrapers of the earlier hominids.

A fairly complete *H. erectus* skeleton of a boy (Fig. 13.35) gives us important clues about human evolution. The shape of the boy's large skull suggests that brain's frontal lobes may have changed and allowed for greater abstract thought and reasoning. The boy's narrow pelvis tells us he had yet to reach adult size and sexual maturity. Both of these factors suggest that the human development we undergo today may have already emerged 1.6 million years ago: We are born in a physically helpless state, but our growth and maturation processes continue for 15 to 20 more years. This pattern insures that our growing brains will have sufficient time to store enormous amounts of survival information and that our lengthy physical dependence will keep us in contact with potential teachers—parents, grandparents, siblings, and neighbors—throughout childhood.

With its expanded geographical range and effective use of new tools, *H. erectus* appears better able than its predecessors to deal with extremes of climate and varied food resources. Fossil evidence suggests that these humans hunted elephants, bears, antelope, and other large game in cooperative bands and transported, butchered, and cooked the meat over warmth-providing campfires. *H. erectus* probably first used fire about 500,000 years ago as the species spread to the temperate zones of Europe and Asia. Cooperative hunting, cooking, and food sharing were probably developed and maintained through **cultural transmission;** each new generation learned from their elders.

How and When Did Our Species Supplant *Homo erectus*?

By about 500,000 years ago, *Homo sapiens* had emerged with a smaller face, smaller teeth, and a larger brain than *Homo erectus*. The populations that *H. erectus* established in Africa, Europe, and Asia were eventually supplanted everywhere by *H. sapiens*. Anthropologists are still debating, however, whether *H. sapiens* arose directly from *H. erectus* at the sites where the different groups of humans live today or whether *H. sapiens* arose from *H. erectus* in one place and then migrated throughout the world and developed the regional features of modern Asians, Africans, and Europeans.

Biologists have used mitochondrial DNA to test the two hypotheses. As we said in Chapter 8, all the DNA in your cells' mitochondria comes from your mother's egg and none comes from your father's sperm. As a result, you can trace your mitochondrial DNA to your mother, your mother's mother, and so on back hundreds of thousands of years. By examining the genetic variation among mitochondrial DNAs from people around the world, and estimating rates of mutation, some evolutionary geneticists are convinced that all people alive today are descended through the female line from a small group of people who lived about 200,000 years ago, probably in Africa. (This is how the term "Mitochondrial Eve" was coined, as we discussed in Chapter 2.) Most anthropologists currently accept this idea.

Neanderthals

The classic movie images of "cave men" come from a separate species, *Homo neanderthalensis,* the Neanderthals, who shared the same general regions of Europe and the Middle East with *Homo sapiens* for tens of thousands of years (see Fig. 13.33). The first Neanderthal fossils turned up in Germany's Neander Valley in 1856. Neanderthals have been much maligned as brutish, ignorant cave dwellers, but the facts suggest a very different picture. Living in harsh, cold environmental conditions, Neanderthals were short, stocky, powerfully built people with a large, protruding face; an extremely large, prominent nose; and a characteristic projecting brow ridge. Their brain was similar in organization but *larger* (1400 ml) on average than a modern human's (1300 ml). The tools and other artifacts they left behind suggest an ability to deal with the environment through sophisticated cultural behavior rather than brute physical force. They made spears and spearheads for hunting large game and scrapers for cleaning animal hides. They routinely built shelters, and their large front teeth were often worn, perhaps from chewing hides to soften leather for clothing. The skeletons of old and disabled Neanderthals show that others cared for the elderly and infirm, and some Neanderthal groups appear to have buried their dead. There's no hint, however, that Neanderthals had language or believed in an afterlife.

Although *H. sapiens* and *H. neanderthalensis* lived in the same general areas at the same time, there were so few of each that they probably seldom met. Furthermore, modern researchers have been able to isolate DNA from a Neanderthal fossil, and it is different enough from our own that it rules out substantial interbreeding between the two

species. Neanderthals appear to have occupied a branch of the human tree that became extinct some 34,000 years ago, perhaps due to competition from their human neighbors—us.

The Explosion of Culture

About 40,000 years ago, anatomically modern humans arose and used their brainpower in dramatically new ways. Their faces were smaller, flatter, and less projecting than the Neanderthals'; the heavy brow ridges had all but disappeared; and their skulls were higher and rounder, with thinner cranial bones. Their limbs were more slender, but still stoutly athletic compared to our own; their teeth were smaller; and most important, their cultural remains were vastly more complex. They created stone knives, chisels, scrapers, spearheads, and axes for shaping other tools of rock, bone, and ivory. Most significant of all, they apparently had developed **symbolic thought.** They left hundreds of sophisticated cave paintings, engravings, and sculptures, suggesting a major development of symbolic forms of communication, probably accompanied by music, increased language abilities, and a sense of mystery (Fig. 13.36). Their living sites and shelters became larger and more complex, and human burials became more common

(a) Cave paintings, France

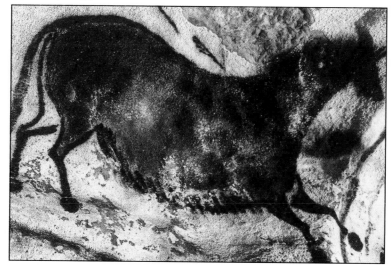

(b) The dung beetle sun god

Figure 13.36 **Art and the Evolution of Symbolic Thought.** People left numerous cave paintings throughout Europe and other continents beginning about 29,000 years ago. (a) This famous cave painting from Lascaux Cave in southern France depicts a running horse. Art forms and techniques evolved along with human civilizations. (b) This ancient Egyptian wall painting was discovered in the tomb of Nefertari in the Valley of the Queens, Thebes. The ancient Egyptian god Khepri is represented here as a man with a scarab head. Khepri was a sun god to the Egyptians, who believed he pushed the sun across the sky just as a dung beetle (scarab) pushes a ball of dung across the ground.

and elaborate, with indications that they held spiritual and supernatural beliefs.

These cultural developments allowed people to move into regions they couldn't inhabit before. They followed the herds of mammoths, woolly rhinoceroses, reindeer, and other game into the arctic regions of Eurasia and across the Bering Sea into the Americas. They also built boats to carry them across uncharted waters to New Guinea and Australia. By 25,000 to 15,000 years ago, people had occupied virtually all the habitable regions of Earth.

Anthropologists use the term **cultural evolution** for the accumulation of useful skills and knowledge, and the discarding of harmful practices, passed down through thousands of human generations. Cultural evolution included the domestication of animals and the development of agriculture by about 10,000 years ago. And this brings us back to our case study, the dung beetle. Dung beetles evolved with or before the dinosaurs, and so they accompanied all the game animals people hunted and later domesticated (not to mention we hominids ourselves). By the time people were recording their human thoughts with words and symbols, they already had an appreciation for "nature's pooper scoopers": Ancient Egyptians symbolized the force that makes the sun move across the sky as a scarab beetle (a dung beetle), rolling a bright yellow dung ball over Earth!

Will Humans Continue to Evolve?

How does the evolution of symbolic thought alter the course of evolution itself? Because our brains have invented ways to transport humans virtually instantaneously anywhere on the globe, our ethnically diverse species is now one vast, interbreeding population. We saw in Chapter 9 that evolution proceeds fastest in small, isolated populations. Ironically, inventions like the steam engine, train, automobile, and airplane dramatically slowed the propensity of humans to evolve new physical innovations by increasing our mobility and thus speeding gene flow. Our ability to control food supplies, reduce predators, and eliminate many diseases also dramatically blunts the effects of natural selection on ourselves and our domesticated animals and plants. Even if we continue to alter the natural world for the worst, with pollution, global warming, and increased radiation, our technology is likely to spare at least some of us for the future. This is not to say that we are immune to extinction. Our species could die out if a large asteroid collided with Earth, or the world's ecosystems totally collapsed from global warming or a nuclear winter. We remain, however, the only animals—among all the cnidarians, protostomes, lophotrochozoans, hemichordates, and other vertebrates—with control over most aspects of our existence. That control may encompass not only our own outcome but the future of all species on our planet, as well.

Chapter Summary and Selected Key Terms

Introduction
Dung beetles are members of the largest order of arthropods and the largest phylum of animals. They are ecologically significant decomposers of animal wastes, and their biology helps us understand the animal body, animal evolution, animal diversity, and relationships between the major animal groups.

The General Characteristics of Animals
Animals are multicellular. Animals are heterotrophs. Animals are self-propelled at some point in their life cycle. And most animals are diploid and develop from embryos. Animals' evolutionary innovations include multicellularity, cell specialization, tissue layers, and **bilateral symmetry** (p. 327), plus new forms of embryonic development. Fossil animal traces and other evidence suggest that animals existed about 1 billion years ago. Based on morphologic and genetic evidence, **zoologists** (p. 328) (biologists who study animals) think animals make up a single lineage derived from an extremely ancient protistan ancestor. They categorize

animals with distinctly different body plans into about 35 **phyla** (p. 329).

Variety of Animal Body Plans
Sponges belong to the phylum **Porifera** (p. 331). They are multicellular; lack true tissues; and have swimming larvae and stationary (sessile) adults that filter fine particles from the water. **Cnidaria** (p. 332) include jellyfish, sea anemones, and corals. They have a **radial body plan** (p. 332) and have two basic forms, the stationary **polyp** (p. 334) and the swimming **medusa** (p. 334). Cnidarians have tentacles with stinging capsules or **nematocysts** (p. 334). **Ctenophores** (p. 335), or comb jellies, have rows of cilia and tentacles. They are radial and have two tissue layers, but lack nematocysts and an alternation between medusae and polyps.

Protostomes (p. 335) are animals in which the first indentation of the embryonic ball becomes the mouth. **Deuterostomes** (p. 335) are animals in which the first indentation becomes the anus and a second becomes the mouth. The two groups also have differences in how the body cavity or **coelom** (p. 335) forms. Protostomes are divided into **Lophotrochozoans** (p. 336), or animals with a lophophore feeding apparatus and/or

a trochophore larva, and **Ecdysozoans** (p. 336), or molting invertebrate animals.

All the following animals are lophotrochozoans:

- **Flatworms** (p. 337) have bilateral symmetry and **cephalization** (p. 338); three tissue layers (mesoderm, endoderm, ectoderm); organ systems; and some are free living while others are parasites.
- **Annelida** (p. 338) or **segmented worms** (p. 338). Feather dusters, earthworms, and leeches are all annelids. Annelid worms have a **coelom** (p. 335), a fully lined body cavity. They also have segments and a **closed circulatory system** (p. 340).
- **Mollusks** (p. 340) are members of the phylum **Mollusca** (p. 340). They include slugs, scallops, and squid. Mollusks have a **head** (p. 340), a muscular **foot** (p. 340), a **visceral mass** (p. 341), a **mantle** (p. 341), and a **radula** (p. 341). They also have an **open circulatory system** (p. 341) with blood bathing some tissues from pools in open spaces. **Gastropods** (p. 341), **bivalves** (p. 341) (which are filter feeders), and **cephalopods** (p. 341) are all mollusks.
- Lamp shells or animals in the phylum **Brachiopoda** (p. 342) have two distinctive shell sizes and feed with a lophophore.

All of the following animals are ecdysozoans:

- **Roundworms** (p. 342) are animals in the phylum **Nematoda** (p. 342). Also called nematodes, they have a **pseudocoelom** (p. 342) that contains fluids and provides support and rigidity for the animal.
- **Arthropods** are members of the world's largest phylum, **Arthropoda** (p. 343). These animals have legs with **joints** (p. 344), and an **exoskeleton** (p. 344) made of **chitin** (p. 344). Arthropods have specialized body segments including a **head** (p. 344), **thorax** (p. 345), and **abdomen** (p. 345), or a fused **cephalothorax** (p. 345). Organ systems include the **tracheae** (p. 345) in insects, the **book lungs** (p. 345) in spiders, and the **gills** (p. 341) in aquatic animals such as shrimp. Arthropod classes include the **centipedes** (p. 345), **millipedes** (p. 345), **crustaceans** (p. 345), **arachnids** (p. 345), and **insects** (p. 346). Some arthropods molt many times. Some have a **larva** (p. 346) that forms a **pupa** (p. 346) then undergo a complete change or **metamorphosis** (p. 346). Ants, wasps, bees, and termites are **social insects** (p. 347).
- Deuterostomes include the **Echinodermata** (p. 347) or sea stars, brittle stars, sea urchins, and relatives. These marine animals have an **endoskeleton** (p. 348) and a **water vascular system** (p. 348) that assists in locomotion. A second deuterostome phylum is **Hemichordata** (p. 348), or mud-burrowing worms, such as acorn worms, with gill slits and a dorsal nerve cord. The last deuterostome phylum is **Chordata** (p. 349). Most members have a **notochord** (p. 349); a dorsal nerve cord or **spinal cord** (p. 349); **gill slits** (p. 349); **myomeres** (p. 349); and a **tail** (p. 350) at some point in the life cycle.

The subphylum **Urochordata** (p. 350) includes the tunicates such as sea squirts. "Tunicates" come from the structure called the **tunic** (p. 351). The free-living larvae have a notochord, nerve cord, and tail, but the stationary adults lose all three. The subphylum **Cephalochordata** (p. 350) includes **lancelets** (p. 351), or **amphioxus** (p. 351), which resemble minnows and keep all five chordate traits throughout life, even though they are filter feeders. The subphylum **Vertebrata** (p. 350) includes all the animals with a bony **vertebral column** (p. 349). Chordates with backbones are called **vertebrates** (p. 349). Bone first evolved in the **agnathans** (p. 351), or **jawless fishes** (p. 351).

Modern fish include a few jawless fish species, but most are **jawed fishes** (p. 353) in two main groups: **cartilaginous fishes** (p. 353) or **Chondrichthyes** (p. 353); and **bony fishes** or **Osteichthyes** (p. 353), which incorporate the **ray-finned fishes** (p. 353) and the **lobe-finned fishes** (p. 353). Hinged jaws arose in the **placoderms** (p. 353), or armored fishes, living 470 million years ago. They also had the

first vertebrae and paired fins. The first lobe-finned fish 400 million years ago had **lungs** (p. 353) and are related to the modern day **lungfish** (p. 353) and **coelacanths** (p. 353). Ray-finned fish include **teleosts** (p. 353) such as perch and angelfish.

Amphibians (p. 353) were the first land animals, with **legs** (p. 353) instead of fins, and breathing lungs. **Reptiles** (p. 355) had additional innovations for life on land: an expandable **rib cage** (p. 355), dry **scaly skin** (p. 355); **amniote eggs** (p. 356) with a leathery shell and a yolk; and legs extending beneath the body. Ancient reptiles called **thecodonts** (p. 356) gave rise to the biggest, fiercest reptiles that ever lived, the **dinosaurs** (p. 356). Living descendants of the dinosaurs are the **birds** (p. 356), with innovations for flight that include **feathers** (p. 357) and lightweight, hollow bones. Birds are **homeothermic** (p. 357) or have a constant internal temperature. They also have **air sacs** (p. 357) connected to the lungs, and they have a **four-chambered heart** (p. 357). Their amniote eggs have a hard shell.

Ancient reptiles called **therapsids** (p. 356) gave rise to the mammals. **Mammals** (p. 357) are warm-blooded and have a four-chambered heart, like birds. But they also have **mammary glands** that produce **milk** (p. 357), and they have **fur** (p. 357). **Monotremes** (p. 357) like the echidnas and platypus lay eggs but nurse their young with milk. **Marsupials** (p. 357) give birth to very immature young, which finish developing in a pouch or **marsupium** (p. 357). Placental mammals have a **placenta** (p. 357).

Primates (p. 359) are an order of mammals divided into two suborders, the **Strepsirhini** (p. 360) and the **Haplorhini** (p. 360). Strepsirhini include lemurs and lorises. Haplorhini include the tarsiers and **anthropoids** (p. 360), which include the **New World monkeys** (p. 360), **Old World monkeys** (p. 361), and **hominoids** (p. 361): the apes—gibbons, orangutans, gorillas, chimpanzees—and humans. Primates have an **opposable thumb** (p. 361); partially or fully **upright posture** (p. 361); stereoscopic vision (p. 362); a large brain; extended infant care; and various kinds of teeth.

Modern monkeys, apes, and humans arose from ancestors starting 40 to 35 million years ago. Our hominid lineage diverged from the one leading to modern apes about 6 million years to 4 million years ago. They were **bipedal** (p. 361); they walked on two feet. Tool-making began about 2.5 million years ago. The human brain expanded and the power of abstract thought appeared less than 1 million years ago.

Australopithecus afarensis (p. 365) is the oldest hominid yet discovered that definitely walked upright 3.5 million years ago. *Australopithecus garhi* (p. 366) was a toolmaker 2.5 million years ago. It preceded *Homo habilis* (p. 366), which made tools 2 million years ago. *Homo erectus* made more sophisticated tools by 1.8 million years ago and migrated out of Africa. Skills were passed to others through **cultural transmission** (p. 366).

Homo sapiens (p. 366) emerged by 500,000 years ago. A separate species lived at the same time, *Homo neanderthalensis* (p. 367), but died out some 34,000 years ago. Anatomically modern *Homo sapiens* arose about 40,000 years ago and left evidence of **symbolic thought** (p. 367). Humans also started accumulating and passing on skills and knowledge through the process of **cultural evolution** (p. 368). We now control most aspects of our environment and have ever-expanding populations that threaten the balance of nature.

All of the following question sets also appear in the Explore Life **E** electronic component, where you will find a variety of additional questions as well.

Test Yourself on Vocabulary and Concepts

In each question set below, match the descriptions with the appropriate term. A term may be used once, more than once, or not at all.

SET I

(a) sponges (b) cnidarians (c) flatworms
(d) roundworms (e) none of the above

1. Radial symmetry; true tissues; sessile and motile forms; beginning of coordination by means of electrical signals
2. Extra cellular digestion; parasitic and free-living forms; three tissue layers
3. Neither radially nor bilaterally symmetrical; intracellular digestion of food molecules; beginning of cell specialization
4. Bilaterally symmetrical; beginning of cephalization; beginning of organs organized into organ systems
5. Extracellular digestion; one-way gut; parasitic and free-living forms

SET II

(a) flatworms (b) roundworms (c) segmented worms (d) flatworms and roundworms (e) roundworms and segmented worms (f) flatworms, roundworms, and segmented worms

6. Both free-living and parasitic forms
7. First phylum to exhibit bilateral symmetry, cephalization, a tubular gut, a coelom, and segmentation
8. One-way gut and either a coelom or a pseudocoelom

9. Setae, nephridia, and a hydroskeleton

10. Tapeworms, liver flukes, and the organisms that cause schistosomiasis are members of this phylum.

SET III

(a) arthropods (b) echinoderms (c) annelids
(d) mollusks (e) none of the above

11. Deuterostomes

12. Protostomes

13. Complex brain capable of observational learning; image-forming eyes; unsegmented

14. Includes clams, crabs, and lobsters

15. Specialization of body segments; exoskeleton made of chitin; social groups showing division of labor

SET IV

(a) spinal cord (b) vertebral column (c) gill slits
(d) notochord (e) myomeres (f) tube feet

16. A hollow, dorsal tube or nerve tissue found at some stage in all chordates

17. A solid, flexible rod that is present in all chordate embryos

18. Found in vertebrates but not in nonvertebrate chordates

19. Bilateral, segmented blocks of tissue surrounding the vertebral column

20. A structure originally used to filter food from seawater

SET V

(a) agnathans (b) placoderms (c) ray-finned fishes
(d) Chondrichthyes (e) lobe-finned fishes
(f) urochordates

21. Bony fishes with jaws that are not believed to be ancestral to land vertebrates

22. Bony fishes with jaws and lungs; an extinct group of these fishes is believed to be ancestral to land vertebrates

23. Modern cartilaginous fishes including sharks and rays

24. An extinct group of fishes but the first to exhibit jaws and paired appendages

25. Jawless fishes; an extinct group of these fishes is believed to be the first of the vertebrates

SET VI

(a) amphibians (b) reptiles (c) birds (d) mammals
(e) agnathans

26. Vertebrates that have lungs and a three-chambered heart but do not produce amniote eggs

27. Vertebrates that exhibit internal fertilization and produce amniote eggs

28. Vertebrates that gave rise to both birds and mammals

29. Warm-blooded vertebrates that include egg layers as well as bearers of live young

30. Homeothermic, with air sacs and a four-chambered heart

SET VII

(a) New World monkeys (b) Old World monkeys
(c) lemurs (d) gorillas (e) humans (f) neanderthals

31. True monkeys without a prehensile tail

32. True monkeys with a prehensile tail

33. The only surviving hominid

34. Moist-nosed primates from Madagascar

35. The largest of the apes

SET VIII

(a) *Australopithecus afarensis* (b) *Homo habilis*
(c) *Homo erectus* (d) *Homo neanderthalensis*
(e) *Homo sapiens*

36. Neanderthal

37. The first human toolmakers

38. A fully bipedal, small-brained link between apes and humans

39. Associated with tools, such as hand axes that were more advanced than scrapers and choppers, and with cultural transmission of information

Integrate and Apply What You've Learned

40. What is a coelom and why is it important?

41. Why is segmentation considered to be an important evolutionary advance?

42. What are some of the dung beetle's adaptations for their particular food source and way of life?

43. Although cephalopod mollusks appear to be the "smartest" of the invertebrates, most zoologists consider the arthropods such as dung beetles to be the most successful invertebrates. What characteristics of the arthropods account for their spectacular success?

44. In what ways are echinoderms similar to some of the more primitive invertebrates, in what ways are they similar to the chordates, and in what ways are they unique?

45. Compare chordate characteristics with vertebrate characteristics. Are there more animals classified as chordates or as vertebrates?

46. In what ways are reptiles more suited to life on land than amphibians?

47. Trace the phylogenetic lineage of birds and mammals from the first vertebrates.

48. Which physical and behavioral traits do we share with the other primates, and which are currently thought to be particularly "human"? Are the particularly human traits unique to human beings?

49. What evidence do we have that bipedalism preceded large brain size in the evolution of humans?

Analyze and Evaluate the Concepts

50. Which invertebrate phyla are lophotrochozoans? Which are ecdysozoans? Which characteristics set the two groups apart?

51. What arguments support the contention that the sponges split from the main line of animal evolution before the cnidarians arose? Consider symmetry, cell structure, and tissues.

52. Consider the animal phylogenetic tree in Figure 13.3. If you were a researcher interested in the diversity of animal genomes, which animal genomes would you choose to study? Include nematodes and fruit flies in your answer, among other animals, and explain why you would or wouldn't study them.

53. Some zoologists think that Earth should be called the Planet of the Insects. Why?

54. What are some reasons why the evolution of bilateral symmetry was fundamental to the radiation of animals?

55. Summarize the new evolutionary features that appeared in the amphibians. What aspects of amphibian biology inhibit them from being the dominant land vertebrates?

56. Some zoologists characterize birds as "flying dinosaurs." What do they really mean?

57. There are more species of fishes than all other vertebrates combined. What might explain this observation?

58. Child birth is more hazardous in humans than any other primate. The infant's head is large in relation to its body, and the mother's pelvic bones are arranged for upright posture. Design a hypothesis for the role of natural selection in determining infant head size, maternal pelvic shape, and extended infant care. List at least one prediction of your hypothesis that could be tested by examining the fossil record.

59. Which of the following traits seems to be the most necessary precondition for the spread of humans from the savanna to temperate climates? (a) bipedalism (b) loss of body hair (c) complex toolmaking skills (d) slender stature and less massive skulls (e) cultural practices such as burying the dead.

PART 3 The Dynamic Body

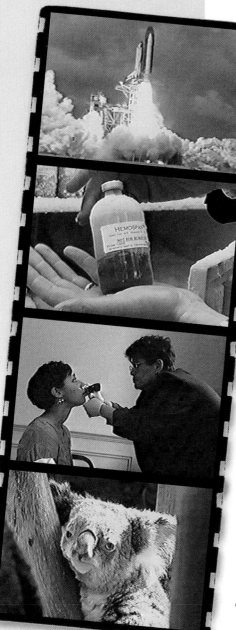

CHAPTER 14
Body Function,
Survival, and the
Steady State

*"You feel every pound of that
thrust going through your
body."*

CHAPTER 15
Circulation and
Respiration

*At first, the search for a
blood substitute "was an unbelievable failure."*

CHAPTER 16
Immune System

*"Since I've been in college,
I get a cold with every
change of season."*

CHAPTER 17
Nutrition and
Digestion

*"She's the human
equivalent of 140 years
old."*

CHAPTER 18
Hormones:
Messengers of Change

*"If I miss playing, I get a
little bit moody or grumpy."*

CHAPTER 19
Nervous System

*"I can see a car coming from
the side quite easily."*

CHAPTER 20
Muscles and Skeleton:
The Body in Motion

*"It's just like stepping on a patch
of ice on a cold day."*

A Medical Pioneer in Orbit

▲
Space Travel and Homeostasis. Bernard Harris in a space lab; space shuttle lifting off; Harris testing a fellow astronaut on the shuttle.

On April 26, 1993, Dr. Bernard Harris strode across a steel gangplank with six fellow astronauts, entered the Space Shuttle *Columbia*, and buckled himself into a deeply padded recliner seat, head low, feet high. For two hours, he and the other crew members waited in this heady position. Meanwhile, engineers and technicians readied the booster rocket, rumbling below them like a 20-story building with mammoth engines. Eventually, the rocket engines began to roar thunderous blasts and as the final countdown began, the 5-million-pound space shuttle started vibrating in every direction. *Columbia* lifted off amidst a fireball visible for 20 miles. The shuttle's main engines produced over 7 million pounds of thrust that day, or 15 times more than a commercial airliner at takeoff. "And," recalls Harris, "you feel every pound of that thrust going through your body."

The sensation of triple gravity (three "Gs") caused by the powerful engines pressed him deep into his seat but it lasted only eight and a half minutes. By then, *Columbia* had arched upwards and soared safely out of Earth's atmosphere into an orbit 240 miles above the planet's blue and white surface. At that altitude, says Harris, the engines cut off and "you went from three 'Gs' to zero gravity." Everything in the shuttle started floating, from the dirt on the floor to the astronaut's hair and clothing, and as soon as Harris unbuckled himself, he, too, began to float upward. "There was a fuzziness in my head," he recalls, which he characterizes as "a cloudiness, like you just woke up." The blood that normally resides in the legs "has no pressure to keep it down," he explains, "so it ends up in your chest and head" and produces the drowsy, headachy sensation.

This fluid shift to the head was only one of many physical changes Harris would experience along with *Columbia*'s other crew members for that mission, and every other astronaut on one of NASA's many space flights. In a sense, they have all been guinea pigs in a historic experiment on **homeostasis**—how the body adapts to changes in the external environment and maintains constant internal conditions.

The environment of deep space is about as hostile and lethal as environments come—a vacuum devoid of oxygen and blasted by subfreezing temperatures, lethal radiation levels, and a near or total absence of gravity, depending on the proximity of stars and planets. Today's space vehicles provide breathable air, liveable temperatures, and shielding from most radiation, but they can't reproduce Earth's gravity. Therefore, the human body, through the delicate balancing act of homeostasis, adjusts to the sudden weightlessness of space in various ways. Ironically, the responses are "jerryrigged" solutions because the human body had not experienced the conditions of space before the 1950s and so has no specifically evolved mechanisms to deal with weightlessness. The adjustments work well enough, however—at least in the short term.

The fluid shift away from the lower limbs, for example, produces what astronauts jokingly call "chicken legs." Another related adjustment is an increase in urination as the body tries to eliminate some of the fluid that has pooled in the head and chest. This helps reduce the overloaded feeling, but it also causes the blood volume to drop, and the red blood cells to become more concentrated. This concentration, says Harris, signals the bone marrow to stop making new red cells, and astronauts then tend to become somewhat anemic (a condition of too few red blood cells).

The bones receive another message and make another adjustment, too: Bearing no body weight in space, they start releasing some of their calcium and slowing the formation of new bone cells. In his 9 days and nearly 3 million miles in Earth orbit, Harris lost almost 1 percent of his total bone. It is not clear, yet, whether astronauts on long flights will be able to recover from the substantial bone loss they accrue. His muscles, too, began to disassemble themselves in the absence of gravity and normal exercise. "I probably lost somewhere in the neighborhood of 10 to 15 percent of my muscle mass," he says, but this can be "rebuilt" back on Earth.

An average citizen might find changes like these distressing. But to Harris, the homeostatic adjustments were an expected part of the astronaut calling—a calling that came when he was just nine years old in the early 1960s, living on a Navajo reservation, and gazing nightly at an intensely starry and intriguing sky. Harris's mother had accepted a government teaching post in the Southwest, and "it took me out of what we'd infamously call 'the

ghetto' in Houston to a different environment." This new place had quality science education and challenged Harris "to learn at an early age how to deal with people."

Every choice he made after that took him closer to the space program—from medical school after college back in Texas, to pilot training, to studying the loss of bone and muscle with a NASA researcher. Harris became an astronaut in 1991, and two years later became the first African American to walk in space as his first shuttle flight docked with the Russian Space Station *Mir*. Nervousness over the physical dangers he would face was no match, he remarks, for his "excitement, exhilaration, and sense of accomplishment after dreaming of space flight for 20-some odd years."

The courage of Harris and the other astronauts has helped biologists learn much more about the body's remarkable homeostasis and its constant adjustments to change. This change can be the radical shifts that occur in extreme environments like deep space, deserts, high mountains, or polar caps. It can be the subtler variations in the temperate climates where most of us reside. Or it can involve alterations brought on by injury, disease, childbirth, or other processes. Regardless, homeostasis allows dozens of aspects of our blood and fluid chemistry and body functioning to stay within narrow *survivable* limits, day after day.

As we explore body function, survival, and homeostasis in this chapter, we'll discuss how cells and tissues are organized in an animal's body. We'll also see how that structural order allows needed materials to reach—and wastes to exit—all parts of the body. As we go along, you'll find answers to these questions:

1. How do tubes and transport mechanisms help supply the body with its daily needs?

2. How is the complex animal's body organized into cells, tissues, and organs?

3. How does an animal's body maintain the steady state of homeostasis?

4. What mechanisms keep our salt and water content balanced?

OVERVIEW

Explorer 14.0

Body Functions, Survival, and the Steady State

14.1 Staying Alive: Problems and Solutions

The changes Bernard Harris experienced in space were specific to the state of near-weightlessness called **microgravity.** (True weightlessness occurs in deep space; the shuttle and its passengers still receive a tiny tug of gravity while in orbit 240 miles above Earth.) The changes, nevertheless, are nearly identical to true weightlessness, and they represent a class of responses to environmental change that we will encounter again and again as we explore animal anatomy and physiology. **Anatomy** is the study of biological structures, such as bones, skin, and kidneys; **physiology** is the study of how such structures work—how bones grow, how skin cells make a tight body cover, how the kidneys cleanse the blood, and so on. Living organisms and their cells can be subjected to wide fluctuations in their immediate environments: changes in temperature, light, acidity, salinity, and availability of water, minerals, and nutrients. Changing environmental factors like these, including microgravity while in Earth's orbit, create an external setting that is constantly shifting. Most living cells, however, can survive only within a narrow range of temperature, acidity, and other parameters. The central problem for a living thing, therefore, is to maintain a steady state internally despite an often harsh and fluctuating external environment. And the solution is homeostasis (Explorer 14.1).

GET THE PICTURE

Explorer 14.1

Strategies for Body Function

Maintaining Homeostasis

Living organisms are made up of cells, and essential life processes must go on within individual cells as well as within entire, many-celled animals. Essential life processes include (1) capturing energy and essential materials from the environment, (2) exchanging gases such as oxygen for energy harvest and the waste carbon dioxide, and (3) maintaining a body of a particular shape, fluid composition, and in some cases temperature that stays within a given range. Single-celled organisms, such as an amoeba, exchange materials with their environment by means of diffusion and active transport directly across the plasma membrane (see Chapter 4). Even some multicellular animals such as the flatworm (review Figure 13.9) can rely on direct diffusion because their bodies are so thin and flat that each cell lies very close to the outside world. Larger, thicker animals, however, like bumblebees, peregrine falcons, or astronauts require other solutions.

Large bodies are generally too thick for each of their cells to exchange materials directly with the environment. For example, Bernard Harris, a muscular man about six feet tall, has a pancreas lying about four inches deep in his body and an equivalent distance from the outside, oxygenated air. Your pancreas is probably about the same depth, give or take a bit. A number of special systems evolved to overcome this distance problem, both for the cells lying deep inside the pancreas and all the other cells and organs within multicellular organisms.

Three systems service the cells of the pancreas. Their general function serves as a good overview of all the physiological systems in Dr. Harris's body. Each system has a tube running from the outside environment to a location deep inside the body, capable of bringing supplies in to the cells or carrying away wastes. Each tube has special regions where substances inside the tube can be exchanged with body fluids bathing the tube. This exchange takes place over short distances via diffusion and by active transport. The body fluids surrounding each cell exchange oxygen, nutrients, wastes, and other substances with the circulatory system, which carries materials to every cell in the body in a moving stream, the blood. The substances can leave this stream by diffusion and can enter every body cell, again via diffusion and active transport, or wastes can be released by cells, enter the tube, and be carried away. **E** 14.2 animates this strategy of tubes and diffusion.

Let's see how this general strategy works in one particular physiological system, Dr. Harris's respiratory system (Fig. 14.1): Oxygen from the air in the space station enters his body through a tube (the windpipe; Step 1) and diffuses into the bloodstream in a special exchange region, the lungs (Step 2). The oxygen diffuses into the blood (Step 3), and the flowing stream of blood then carries the dissolved oxygen throughout his body (Step 4). Finally, the oxygen passes out of his blood and into individual body cells by diffusion (Step 5), where the cell uses the gas in aerobic respiration (Step 6).

Tubes and transport mechanisms give a many-celled organism some control over the immediate environment surrounding most of its cells. Still, each organism functions as a whole and directly contacts the physical environment with its skin, scales, fur, or other outermost surface and so remains subject to environmental fluctuations. We'll see in the next sections how a large array of structural, functional, and behavioral mechanisms come into play to keep both the organism and its cells on an even keel.

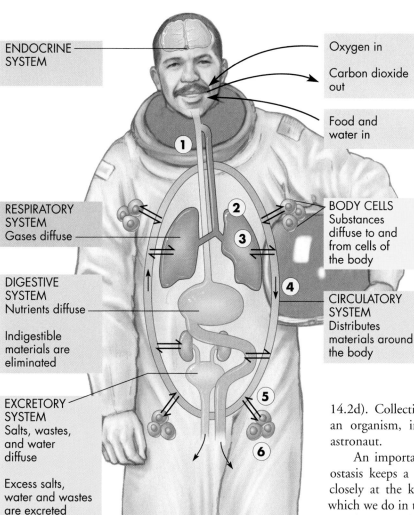

ENDOCRINE
SYSTEM

Oxygen in

Carbon dioxide
out

Food and
water in

RESPIRATORY
SYSTEM
Gases diffuse

BODY CELLS
Substances
diffuse to and
from cells of
the body

DIGESTIVE
SYSTEM
Nutrients diffuse

Indigestible
materials are
eliminated

CIRCULATORY
SYSTEM
Distributes
materials around
the body

EXCRETORY
SYSTEM
Salts, wastes,
and water
diffuse

Excess salts,
water and wastes
are excreted

Figure 14.1 **Tubes and Diffusion: A Strategy for Exchanging Materials with the Environment.** Animals—including astronauts—obtain nutrients, exchange gases, and get rid of wastes by tubes that penetrate deep into the body. The tubes of the digestive system, respiratory system, and excretory system exchange substances with the blood by diffusion. Substances diffuse to and from the blood and the individual cells.

stomach, for example, is an organ containing four tissue types cooperating in the function of food storage and digestion (Fig. 14.2c). Several organs can form an **organ system,** which is two or more interrelated organs that work together, serving a common function; the digestive system, for example, contains your mouth, stomach, liver, and other organs functioning together in food processing (Fig. 14.2d). Collectively, all the organ systems make up an organism, in the case of our central topic, an astronaut.

An important part of understanding how homeostasis keeps a steady internal state is looking more closely at the kinds of tissues that make up organs, which we do in this chapter. Each of the next six chapters then introduces the functioning of organs within particular organ systems, such as the circulatory system, immune system, or nervous system.

14.2 Body Organization: A Hierarchy

If an animal's cells were crammed together haphazardly, gases, fluids, and other materials could not be transported around the body in a useful way. Instead, cells are organized into tissues, organs, and organ systems (Fig. 14.2).

A **tissue** is an integrated group of cells of similar structure performing a common function within the body. For example, cells that line the inside of your mouth or your stomach form an epithelial tissue (Fig. 14.2a,b). An **organ** is a unit composed of two or more tissues that together perform a certain function. Your

The Four Types of Tissues

A person's body is a marvelously complex and integrated whole containing over 200 cell types that carry out myriad specialized roles like secreting stomach acid or building hair proteins. Despite all this diversity of cell types, there are just four types of tissues: epithelial, connective, muscle, and nervous tissues. These, in turn, make up the organs, most of which contain all four tissue types. Consider a major player in your digestion of food, your small intestine. The organ's interior is lined with epithelial tissue. Connective tissue binds this epithelium to a layer of muscle tissue surrounding the organ. Finally, nervous tissue relays signals that can stimulate muscle contraction and speed the passage of materials through the small intestine.

Figure 14.2 **Hierarchy of Organization in the Animal Body.** The billions of cells (a) in an animal's body are neatly organized into tissues (b), organs (c), and organ systems (d) that carry out body functions efficiently. Organ systems working together in a coordinated fashion help an animal stay alive. Here we show a mucus-secreting goblet cell in the tissue that lines the stomach, an organ of the digestive system.

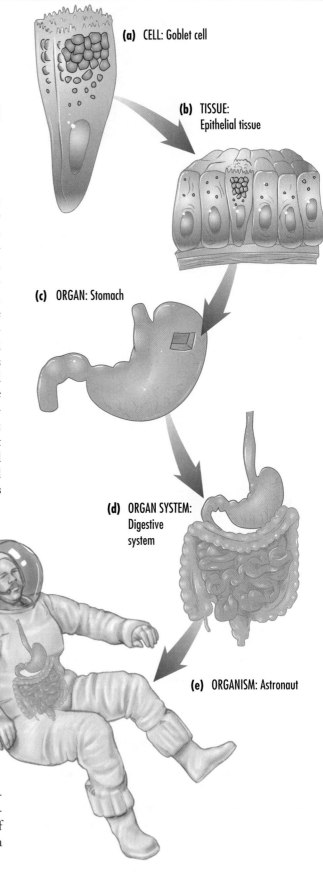

(a) CELL: Goblet cell

(b) TISSUE: Epithelial tissue

(c) ORGAN: Stomach

(d) ORGAN SYSTEM: Digestive system

(e) ORGANISM: Astronaut

Epithelial Tissue

Like a raincoat on a stormy night, **epithelial tissue** lines or covers body surfaces with sheets of cells. We just saw that an epithelium lines the inner surface of the stomach, and it also covers us with the outer layers of our skin (Fig. 14.3a). Because epithelial sheets are often subject to abrasion, epithelial cells tend to divide continuously. Epithelial tissue has several crucial functions. It (1) receives environmental signals (e.g., cells of the inner ear), (2) protects the body from foreign substances (e.g., skin) (3) secretes sweat, milk, wax, and other materials (e.g., sweat glands), (4) excretes wastes (in the kidney) and (5) absorbs nutrients, drugs, and other substances (in the intestines) (Fig. 14.3b). Since glands are organs that secrete, it's no surprise that epithelial tissue makes up the two major kinds of glands: the **exocrine glands,** which secrete substances like sweat onto the skin and digestive juices into the gut, and some **endocrine glands,** which secrete the internal chemical messengers within the body called **hormones** (see Chapters 8 and 18). As we'll see there, hormones such as insulin and testosterone help regulate and control many body functions. Epithelia have so many important roles, it's also no surprise that damage to epithelial tissue can block certain normal body functioning. For example, cigarette smoke damages the epithelium lining the breathing tubes and as a result, the epithelial lining removes less of the harmful debris that enters lungs and passageways, including dirt, pollen, molds, and, of course, cigarette smoke.

Epithelial sheets typically have one surface facing a space, like the air, in the case of the skin, or the urine-storing cavity of the urinary bladder. Cell junctions (review Figure 2.29) bind adjacent epithelial cells tightly together and stop fluid (such as urine) from leaking across the lining and into the body. Junctions also link the inner surface of the epithelium to a thick underlying meshwork of protein and polysaccharide fibers, the basement lamina

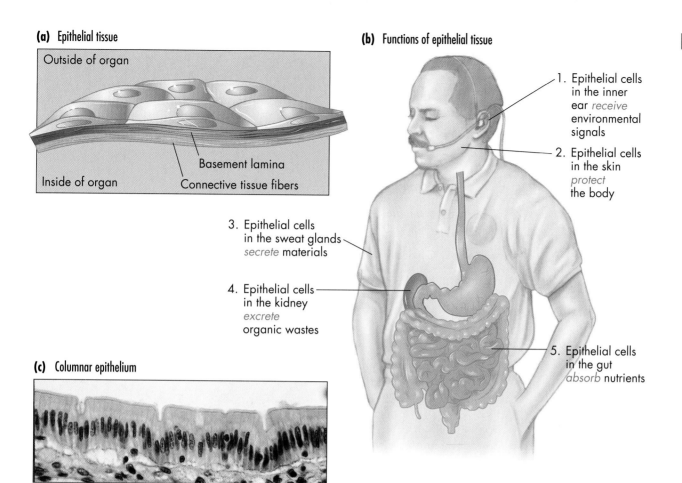

(a) Epithelial tissue

Outside of organ

Inside of organ

Basement lamina

Connective tissue fibers

(b) Functions of epithelial tissue

1. Epithelial cells in the inner ear *receive* environmental signals

2. Epithelial cells in the skin *protect* the body

3. Epithelial cells in the sweat glands *secrete* materials

4. Epithelial cells in the kidney *excrete* organic wastes

5. Epithelial cells in the gut *absorb* nutrients

(c) Columnar epithelium

Figure 14.3 **Roles of Epithelial Tissue.** (a) Epithelial tissues are sheets of cells that cover body surfaces. (b) Epithelia receive signals, protect the body, secrete materials, excrete wastes, and absorb nutrients. (c) Micrograph of the epithelium that lines a mammal's intestine (magnified 160 times).

(Fig. 14.3a). Without this firm bond, the skin would peel right off your body. Just beneath the basement lamina lies the second tissue type, connective tissue.

Connective Tissue

Reach up and gently squeeze the upper part of your outer ear—the stiff material is cartilage, a type of connective tissue that gives the ear shape, strength, and flexibility. **Connective tissue** has relatively few cells, which are dispersed within a large amount of extracellular material. Connective tissue binds other tissues and supports flexible body parts and connective tissue cells produce the extracellular material that forms a matrix, or framework, for other structures (Fig. 14.4a). This matrix usually includes fibers of the proteins collagen and elastin (Fig. 14.4c). Collagen is the most abundant protein in the body; weight for weight it is as strong as steel. Ear cartilage contains large amounts of collagen. Elastin, also part of the ear, is stretchy (as the name implies) and helps structures keep their shape, even,

for example, when pierced and supporting rings or posts.

Connective tissue (1) fills spaces between muscles; (2) links epithelial layers to underlying organs; (3) stores fat; (4) makes up the tendons that attach muscles to bones and the ligaments that attach bones to other bones; and (5) forms cartilage (Fig. 14.4b). Blood and bone are also types of connective tissue, even though blood is a liquid containing red and white cells, and bone is a hardened matrix housing bone-forming cells.

In space, Bernard Harris experienced bone loss. Why would homeostasis, a system evolved to keep the body functioning on an even keel, allow such a seemingly self-destructive process? Bone forms a rigid framework that withstands gravity, and it is a complex of proteins, phosphorus, and calcium. (The latter is an element needed for muscles to contract, for nerves to transmit information, and for blood to clot.) In space, Harris's bones sensed "mechanical unloading" or a sudden lack of body weight due to the absence of gravity.

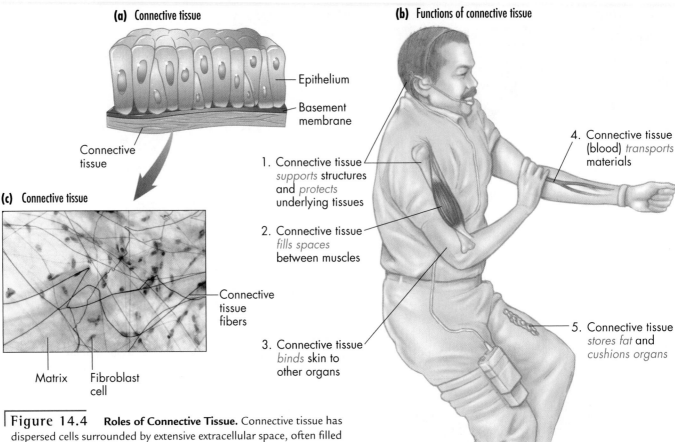

(a) Connective tissue

Epithelium

Basement membrane

Connective tissue

(c) Connective tissue

Connective tissue fibers

Matrix   Fibroblast cell

(b) Functions of connective tissue

1. Connective tissue *supports* structures and *protects* underlying tissues

2. Connective tissue *fills spaces* between muscles

3. Connective tissue *binds* skin to other organs

4. Connective tissue (blood) *transports* materials

5. Connective tissue *stores fat* and *cushions organs*

Figure 14.4   Roles of Connective Tissue. Connective tissue has dispersed cells surrounded by extensive extracellular space, often filled with fibers. It binds other tissues together and supports flexible body parts. (a) Connective tissue lies just below the basement membrane (lamina) to which epithelial cells are attached, and it produces extracellular material that forms a matrix, or scaffold, for other structures. The matrix includes fibers of the proteins collagen and elastin. (b) Connective tissue supports structures, fills spaces, binds organs together, transports materials, and stores fat. (c) Micrograph of human breast tissue shows collagen, elastin, and other kinds of connective tissue fibers.

The homeostatic mechanisms of the body work to constantly remodel bone, depositing bone at points of heavy stress, such as the spots where very strong muscles attach, and removing bone where there is less stress. Because microgravity reduces stress virtually everywhere in the skeleton, hormonal signals told Dr. Harris's bones to start "remodeling" themselves by removing calcium and releasing it into the blood. Signals also slowed the formation of new bone. The same process goes on in a bedridden patient: The absence of muscular and mechanical stress leads to the removal of calcium and the slowdown of new bone production.

Bone remodeling is a good example of the general strategy of homeostasis: When external conditions change (in Bernard Harris's case, when gravity all but disappeared), natural mechanisms act to bring about a proper body response. In Harris's situation, the proper response would be dumping calcium, because keeping up a robust bone structure would "cost" more in energy expenditure than it would be "worth," since there is essentially no gravity pulling down on the body and strong structural support is less important. The bone remodeling and calcium dumping led to a thinning and weakening of Harris's bones and to an increase of calcium in the urine and in the blood. In postmenopausal female astronauts, the situation becomes even more pronounced. Shannon Lucid, for example, spent 188 days in space in 1996 at the age of 55. Even before her space flights, her bones were receiving a signal to release some stored calcium since, after menopause, the element would no longer need to be heavily "stockpiled" for building future fetal bones or producing milk. A woman who doesn't take estrogen supplements typically loses 1 or 2 percent of the calcium in her bones each year after menopause. This is why women may be concerned about osteoporosis, literally porous bones, due to calcium loss. An astronaut in space typically loses 1 or 2 percent of bone calcium per *month*. Unless postmenopausal women astronauts take estrogen, space flight could have a longer lasting effect on them than on their male colleagues in terms of rebuilding weakened bones back on Earth.

Muscle Tissue

Thanks to a third tissue type, most animals are capable of movement: **Muscle tissue** contracts, or shortens, and thereby applies force against objects (Fig. 14.5a). (In Chapter 20, we'll see three types of muscle tissue: the striped or striated type that moves limbs and other body parts; the smooth type that propels internal organs; and the heart muscle that keeps this vital pump squeezing and moving blood through the body.) Grab your upper right arm with the left hand and raise your right hand up to your chin, bending your arm at the elbow. The hardening bulk you feel is muscle tissue as it changes from long, thin tubes to short, thick tubes to move the arm. Muscle cells are generally quite long and thin and are filled with protein fibers. The fibers slide past each other like two parts of a telescope, and cause the cell to shorten. During space flight, Harris's muscles atrophied—particularly those muscles that maintain the body upright against gravity, such as the calf muscle, the back portion of the thigh, the buttocks, and the muscles that keep the back straight and the head upright. Long periods of bed rest or disease can cause similar muscle wasting, as when you wear a cast on a broken limb. People are often shocked at the skinny arm or leg they see when a cast comes off after six or eight weeks. Atrophy of unused muscles is another good example of homeostasis: It prevents the body from spending valuable resources to build muscle tissue that is not being used as much as before.

Nervous Tissue

A fourth type of tissue, **nervous tissue,** helps control and coordinate the actions of organs, organ systems, and the whole body. Nervous tissue, in turn, makes up a system, the **nervous system,** encompassing the nerves, sense organs, and brain. Nervous tissue transmits electrochemical impulses; for example, you feel a pin prick on your fingertip because the pin deforms nervous tissue and this triggers a nerve impulse (Fig. 14.5b; see also Chapter 19). The essential feature of nervous tissue is its ability to transmit signals and thereby communicate. Nerve cells, called **neurons,** sense changes in their environment. They then process that information and command muscles to contract or glands to secrete; these

(a) Muscle tissue

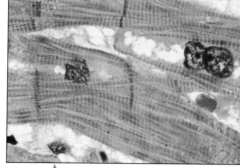

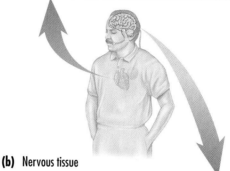

(b) Nervous tissue

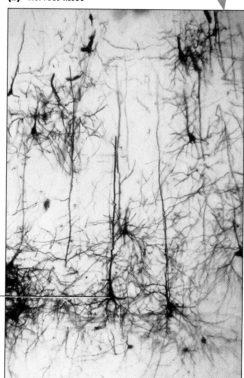

Neuron

Figure 14.5 **Muscle Tissue and Nervous Tissue.** (a) Muscle tissue can contract and shorten. Heart muscle (here magnified about 1000 times) pumps blood through the body. Smooth muscle squeezes fluid through the digestive tract, and skeletal muscles move body parts. (b) Nervous tissue, like this section from the human cerebrum, transmits information from one body part to another.

actions then help the animal adjust to the initial change. Nerve cells have very long, thin processes that act like telephone wires carrying messages from one place to another in the body. The longest cells of the body are nerve cells.

Nervous tissue accomplishes a key feature of homeostasis: the communication of various body parts with one another. Large animals need sophisticated avenues of coordination and communication, which integrate far-flung body parts so the left hand (or paw or tentacle) knows what the right hand is doing—at least most of the time—and the organism functions efficiently as a single entity. The electrical signals of nervous tissue and the nervous system rapidly integrate and regulate body functions. At the same time, blood-borne chemical signals (hormones) given off by the endocrine system cause slower, longer lasting body reactions and changes. We'll see more about the nerves and hormones in this and later chapters, and how they help organisms to maintain homeostasis—the internal constancy they need to survive.

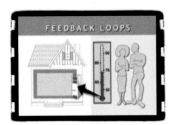

What role do feedback loops play in homeostasis?

14.3 How Do Our Bodies Maintain a Steady State?

When Dr. Harris performs routine checkups on patients, he can get a fairly accurate reading of how healthy they are by taking their temperature and blood pressure, listening to their heart and lungs, and analyzing the chemical content of their blood and urine. Since medical researchers have established a normal range of values for each measurement, chances are good that if the so-called vital signs fall within the correct ranges, the body is keeping the internal environment at a constant healthy level that, in turn, promotes the cells' efficient functioning.

On the space shuttle *Columbia*, Bernard Harris was both a physician and, in a sense, a medical guinea pig, providing information on astronauts' vital signs while subjected to microgravity. Dr. Harris outfitted himself and the other crew members with various types of electronic sensors. He collected data from those sensors, took blood and urine samples, and conducted tests before, during, and after the space flight. His goal was to determine baselines for body functioning on Earth and then to compare them with changes astronauts experience during space flight, including increased urination, mild anemia, and bone and muscle wasting. The core of our story is not precisely what he and colleagues found but rather *how* an astronaut's body (and yours, as well) establishes and maintains an internal even keel whether on a college campus on a warm, sunny day, or in the frigid darkness of space. The answers involve a set of marvelously integrated strategies based on feedback loops.

Feedback Loops: Mechanisms for Homeostasis

The thermostat in the space capsule, and the one in your house employ the same general strategy for maintaining a steady condition with minor fluctuations. In your house, the thermostat *senses* the temperature, *evaluates* it (is it above or below the temperature range you set?), then *acts* by turning the furnace (or perhaps the air conditioner) on or off. The thermostat's activity requires: (1) a **receptor** to sense the environmental conditions (a thermostat contains a thermometer for that); (2) an **integrator** to evaluate the situation and make decisions (the internal workings of the thermostat signal the furnace or air conditioner to turn on or off), and (3) an **effector** to execute the commands (in your house, the furnace or air conditioner is the effector). Again and again in our discussion of animal homeostasis, you'll encounter these common elements: receptors, integrators, and effectors.

Now, these elements often interact in a **feedback loop,** a series of steps that *sense* a change in the external or internal environment, *evaluate* the new situation, and *react* in a way that modifies the original change (Fig. 14.6). In a house, the feedback loop involves the thermostat's recognition of a temperature drop in the house; the triggering of the furnace to pour out more heat; and the thermostat's recognition of the new, higher temperature leading to a temporary shutdown of the furnace.

Feedback loops can be negative or positive. Here's what that means: A **negative feedback loop** resists change by sensing a deviation from the baseline condition, then turning on mechanisms that oppose that trend and bring things back toward baseline. For example, your steady body temperature depends on a biological thermostat located in a small region of the brain called the **hypothalamus** (see Chapter 8) lying above the roof of your mouth. The thermostat in the hypothalamus is set at about 37.8 °C (100 °F) , which we measure as about 37 °C (98.6 °F) in the somewhat cooler mouth. Now suppose your body suddenly becomes too cold—think, for example, of the hundreds of people tossed into the icy North Atlantic as the *Titanic* sank in 1912. Specialized nerves in the hypothalamus detect the temperature change in blood flowing through the brain, while other nerves analyze information arriving from temperature receptors in the skin. The hypothalamus integrates the information and then causes muscles to begin to contract randomly (shiver), which generates heat; this heat counteracts the original change—the sudden cooling. The main point is that negative feedback loops tend to act to keep conditions constant—in this case, a constant body temperature. (E xplorer 14.2 allows you to test some negative feedback loops in action.)

Let's look at what happens when you have a fever. When virus particles or bacterial cells invade the body and begin multiplying, certain white blood cells release substances that can reset the brain's natural thermostat, the hypothalamus, to a value higher than normal (say, 38.8 °C or 102 °F rather than 37.8 °C or 100 °F). The thermostat-resetting substances cause the local release in the hypothalamus of prostaglandins, molecules that turn up the body's thermostat. The body's "furnace," the muscles, start working overtime, contracting in uncontrollable shivers, even under heavy blankets, and the body temperature soars. This heat helps inhibit the bacterial or viral growth, perhaps by allowing more activity of the infection-fighting white blood cells. Aspirin brings down body temperature by interfering with the synthesis of prostaglandins. When the fever "breaks," the thermostat is reset once again to a lower tempera-

ture. Since the body is still hot, however, it initiates cooling behavior, largely by sweating.

Yet another example of a negative feedback loop involves the bone loss Dr. Harris experienced on the space shuttle. When bones are under a load—for example, the weight of gravity or the pulling of active muscles upon them (Fig. 14.7)—sensors in the bone detect the stress. Bone cells integrate the information, and respond by depositing calcium in the stressed location. When the restructured bone supplies appropriate support to meet the stress level, the bone-forming cells cease depositing calcium. In microgravity, Harris's muscles essentially were not stressed by gravity at all. The bone cells, not sensing or detecting this normal stress, ceased depositing calcium. With normal turnover, calcium levels began to drop within a day of entry into space. When Harris returned to Earth after nine days, the stress on his bones from gravity returned, and the negative

SOLVE IT

E xplorer 14.2

Negative Feedback Loops and Homeostasis

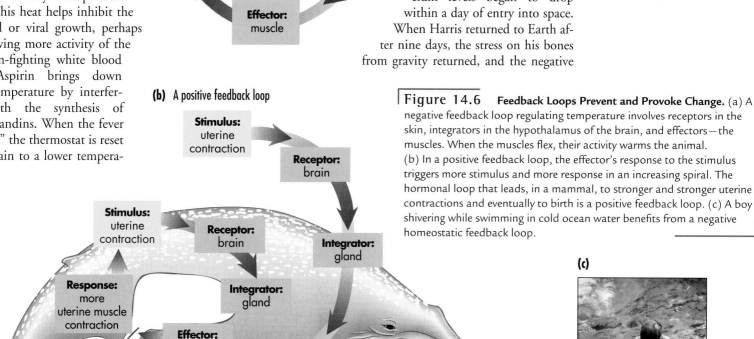

(a) A negative feedback loop

(b) A positive feedback loop

Figure 14.6 **Feedback Loops Prevent and Provoke Change.** (a) A negative feedback loop regulating temperature involves receptors in the skin, integrators in the hypothalamus of the brain, and effectors—the muscles. When the muscles flex, their activity warms the animal. (b) In a positive feedback loop, the effector's response to the stimulus triggers more stimulus and more response in an increasing spiral. The hormonal loop that leads, in a mammal, to stronger and stronger uterine contractions and eventually to birth is a positive feedback loop. (c) A boy shivering while swimming in cold ocean water benefits from a negative homeostatic feedback loop.

(c)

Figure 14.7 Bone Loss in Space: A Negative Feedback Loop in Action. Astronauts must exercise several hours a day to help prevent muscle atrophy and bone loss. With stress (including gravity) these tissues respond by growth and maintenance. In the weightlessness of space, the negative loop shuts down these cellular activities and the astronaut experiences muscle and bone wasting.

feedback loop began operating once again to deposit calcium in his bones. Researchers are not certain, however, whether his body or that of other astronauts can ever fully restore the bone lost during weightlessness. The negative feedback loop for calcium deposition is mediated by hormones, and Chapter 18 discusses these chemical signals in more detail.

Negative feedback loops can help an animal's body (including a person's body) resist internal change away from a baseline. But homeostasis employs another strategy: positive feedback loops which bring about rapid change. In a **positive feedback loop,** an initial change of external or internal environment in one direction is amplified further and further in the same direction. Probably the most familiar example of positive feedback is the terrible screech you sometimes hear from a public-address system. A microphone picking up sound from the speakers sends the sound back, where it is further amplified. The microphone picks up that louder rebroadcast sound, the speakers reamplify it once more, and soon the sound mounts to an ear-piercing squeal.

In animals, positive feedback loops that disrupt the steady state are rarer than negative ones that maintain the steady state, and the positive loops usually act for a specific purpose. An example is the positive feedback loop that drives labor and delivery during birth (Fig. 14.6b). The hormone oxytocin is a chemical signal that causes the uterus to contract; these strong muscular actions cause more oxytocin to be released, more contractions to occur, and so on. The feedback produces stronger and longer contractions spaced closer and closer together until, at the climax of the cycle, a baby is born (review Fig. 8.22). This explosive action stops the loop and restores homeostasis—in this case, a noncontracting uterus—is restored. A colloquial way of describing a positive feedback loop is as a "vicious cycle."

As you might suspect, positive feedback loops can sometimes lead to disease. For example, a certain type of white blood cell (called a neutrophil) is involved in a positive feedback loop associated with inflammation. Inflammation brings about the red, itchy, swollen bumps that can occur during a bacterial infection. These white blood cells release an enzyme that converts an inactive protein in the blood into a smaller, active protein called kinin that can cause fluid to leak out of blood vessels (and lead to the swelling). Kinin also attracts more of these white blood cells to the site of the

infection, which leads to the release of more kinin, and so on, in a vicious cycle. If the body does not moderate this cycle, swelling may eventually become so intense that the breathing tubes become blocked or the blood vessels leak so much fluid that the blood pressure drops to a dangerous or lethal level. Luckily, this type of runaway positive feedback is rare.

14.4 Homeostasis, Circulation, and Behavior

Despite the bitter cold outside the orbiting space shuttle and *Mir* space station, Bernard Harris's body remained at about 98.6 °F. The vehicles were heated, of course, but his own homeostatic mechanisms were responsible for the precise control over his body temperature. With our steady body temperature, we humans are considered **homeotherms** (*homeo* = same + *therm* = heat). This is also true of other mammals and the birds: All are homeotherms or, colloquially, "warm-blooded" animals. Homeothermic animals have a large reservoir of warm, moving fluid whose distribution in the body can be controlled to a certain degree when the environment is too hot or cold.

Homeotherms maintain body temperature via feedback loops like those we just described, as well as changes in blood circulation and in behavior. To see how these mechanisms work, think again about the victims of a boating accident—anything from a rowboat in a lake to the *Titanic* in the North Atlantic. A person plunged into cold water may start shivering quickly, but the body does some other things automatically, as well: It shunts blood from one region to another, and it behaves in certain ways triggered by a survival instinct.

Shunting of Body Fluids

A warm-blooded animal's skin is crisscrossed by blood vessels that contain part of this blood "reservoir." When a mammal, say, a passenger on the *Titanic*, falls into cold water, tiny muscles divert blood from these surface vessels so that the warm blood and hence the person's body heat, stays in the body core instead of circulating in the periphery, where it would dissipate more easily into the cold water. The same would be true if you went for an icy swim on New Years' Day (some people in "Polar Bear Clubs" do this!) or if the heater on the space shuttle malfunctioned and an astronaut like

Bernard Harris became chilled. The shunting of blood from the surface toward the core requires **vasoconstriction,** literally, blood vessel constriction, a rapid closing down of millions of tiny vessels near the body's surface and in extremities such as the nose, ears, limbs, fingers, and toes that sends a greater volume of blood toward the body core (Fig. 14.8a). If, on the other hand, an animal—a galloping horse, let's say, or an active tennis player—becomes overheated a on a hot day, blood can be shunted into these same surface vessels. This involves **vasodilation,** or blood vessel opening, during which millions of peripheral vessels quickly open wider and so blood from the body's core rushes to skin, head, and limbs (Fig. 14.8b). This conveys heat toward the body surface, where the surrounding lower temperature air or water can absorb it. Vasodilation also explains why your own skin may feel flushed and look glowing or even red in hot weather or during heavy exercise. Another mechanism called **countercurrent flow** allows an animal's body to conserve the heat from exposed extremities—flippers, hands, feet, and so on.

Behavioral Adaptations

Let's go back once again to a person who has fallen overboard into cold water. A typical survival response would be to thrash around, because, for the reasons we just saw, exercise does tend to warm you up. A victim might also try to grab another person or a piece of floating debris and desperately try to climb up and away from the freezing water. Ironically, a victim who moves around in cold water will lose body heat 35 percent faster than one who stays motionless. A person could survive up to three hours in water that's 50 °F (10 °C), but only by remaining still and keeping the head—which radiates fully 50 percent of your body heat—out of the water. The best behavioral strategy is to climb out onto floating debris, to cover the head, and never to flail or allow the head to remain in the water. A motionless person would warm up a thin film of water around the body. This layer would decrease heat loss compared to someone thrashing around, constantly recirculating cold water to the body surface.

In a sense, because *Homo sapiens* evolved in warm equatorial regions, we humans are still tropical organisms. We need special gear—warm and/or waterproof clothing, hats, boots, gloves, wetsuits, spacesuits—just to survive for extended periods in a cold environment. Even with all its mechanisms for maintaining a steady internal temperature, the body may fall behind in its efforts to keep up with heat loss during a long period in

(a) Vasoconstriction shunting of blood away from body surface if chilled

(b) Vasodilation shunting of blood to body surface if overheated

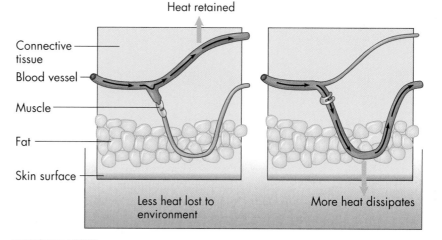

Figure 14.8 **Temperature Regulation.** The body helps regulate temperature by shunting blood to and from the skin. Tiny muscles form a ring around blood vessels near the body surface. (a) When the ring of muscles contracts in the process of vasoconstriction, the blood vessels close, preventing blood flow to the surface. This allows animals—say, "Polar Bear" swimmers in Lake Michigan—to retain body heat deep in their core. (b) When the circular muscles relax in vasodilation, they allow warm blood to flow near the cool environment. This allows the body—say, of a tennis player on a hot day—to lose excess heat.

the cold, especially if not covered in the right protective gear. A state called **hypothermia** sets in if the body's core temperature drops below 35 °C (90 °F). (The inverse of this, or **hyperthermia,** can set in when environmental temperatures are too high and exposure is too prolonged. Children or pets left in overheated cars sometimes suffer from hyperthermia, as can marathon runners on a hot day.)

During a lengthy exposure to cold air or water, rapid vasoconstriction shunts blood to the body core, leaving the extremities vulnerable to frostbite. Because vasoconstriction restricts blood flood to the brain, victims of hypothermia can also display confusion, poor judgement, drowsiness, and strange behavior. Activity slows in the nerves of the arms, legs, and other peripheral areas, and this helps explain the stumbling, clumsiness, and lethargy that are additional symptoms of hypothermia. In the blockbuster movie *Titanic,* the heroine, played by Kate Winslet, can barely think or respond by the time she sees the rescue boat but after a painfully long time does manage to blow a whistle and attract help to the makeshift raft on which she is floating. The sketch artist (played by Leonardo DiCaprio), having remained immersed in near freezing water, dies of hypothermia.

The thousands of warm-blooded bird and mammal species display a huge range of survival behaviors when confronted by overly cold or hot conditions. Hibernating in a tree trunk or cave is one familiar example. Another is seasonal migration to a more temperate area to ride out the winter months. Others include huddling together to preserve body heat, as do Emperor Penguins in the dead of the Antarctic winter; building or seeking shelters, like Arctic foxes; and restricting activity to warmer or cooler parts of the day, as do desert mice.

Variable-Heat Animals

With the elaborate mechanisms we've just seen—negative feedback, circulatory shunting, and behavioral adaptations—homeotherms have a marked advantage: their activity can remain high (including escaping from predators and capturing prey) regardless of environmental temperatures. The majority of animal species, however, are "cold-blooded," including some animals with backbones (amphibians, reptiles, and fish) and all animals without backbones (insects, jellyfish, worms, clams, crabs, and so on). How do they manage? "Cold-blooded" animals obtain most of their heat from the environment and their body temperatures can fluctuate widely. Biologists call animals with variable internal temperatures **poikilotherms** (POY-kill-oh-therms; *poikilo* = various + *therm* = heat) and have learned that many of their mechanisms for maintaining a liveable temperature range are behavioral. A lizard or alligator, for ex-

ample, instinctively moves in or out of the sun, does heat-generating "push-ups" or basks with its jaws gaping open (Fig. 14.9)—all strategies that keep its body temperature between 35 °C and 40 °C (95 °F and 104 °F). A sick lizard will actually induce its own artificial fever by baking itself in the sun until its body temperature soars and the bacterial or viral infection dissipates.

Some kinds of poikilotherms, however, have physiological adaptations in addition to behavioral ones that help regulate body heat. In some oceangoing fishes, such as tunas and sharks, hardworking red-colored muscles enable the animals to swim rapidly and at the same time generate enough excess heat to warm the body core. Other poikilotherms have mechanisms that help them cope with variable body temperatures. Some freshwater fishes, for example, have two sets of cellular enzymes, one that functions best at cool temperatures and another that functions best at warm temperatures.

Body temperature is important but it's just one feature kept in check by homeostasis. To maintain an animal's total internal steady state, despite the vagaries of the external world, dozens of systems work simultaneously and under elaborate coordination and control. Let's return to Bernard Harris on the space shuttle and examine another homeostatic system that maintains constant internal conditions, but in this case, the composition of the body fluids: the excretory system and the kidney.

14.5 Maintaining Salt and Water Balance

Facing the strange stress of microgravity, astronauts become keenly aware of their bodies' natural water balance and the role it plays in their lives. Picture a 2-liter bottle of soda pop or juice; that's how much body fluid shifted from Bernard Harris's legs and lower torso into his upper chest, neck, and face in the first few hours after entering the shuttle *Columbia*. The fluid shift made his face look round and smooth, and it caused an uncomfortable pressure in his head and sinuses. Over the next few days, however, his body eliminated some of the fluid by means of a homeostatic process that attempted to reestablish a steady state for fluid balance. Specialized receptors in the blood vessels leading to Harris's brain and heart started registering "Overfilled!" This triggered the release of chemical signals (particular hormones) that caused his body to excrete extra sodium and water in a higher-than-normal volume of urine. This removed some of the upper body fluid and relieved

Figure 14.9 Behavior Helps Control Body Temperatures. Some animals use behavioral mechanisms to help keep temperatures within a certain range. For example, an alligator often rests with its mouth open; this allows water to evaporate from the body and cool the blood. Notice the leech on the alligator's lip.

some of the congestion in his head and chest. Unfortunately, it also eliminated some water from his blood (but no blood cells). The remaining blood cells therefore became more concentrated. This, in turn, triggered a negative feedback loop that slowed down the production of new red blood cells by his bone marrow and caused him to become slightly anemic. The key player in all of these homeostatic events is the **kidney,** an amazing living filtration device that maintains the chemical constitution of the blood and the level of water and salts in the body fluids despite enormous external changes like the sudden weightlessness of space. Obviously, the kidney and its functions evolved under terrestrial conditions as adaptations to challenges like thirst, extra salt in the diet, and waste products in the blood. But when the strange demands of space travel confront the body, the kidney's activity changes slightly to deal with the new stress.

Over a period of hours or days, the amount of water entering an animal's body must precisely equal the amount leaving—a condition called **water balance.** Animals take in water by eating and drinking, and as a product of metabolism (Chapter 3). Animals lose water by evaporation across the skin and respiratory surfaces, such as the lungs; in sweat and feces; and in other wastes, such as urine. **Urine** is a fluid that carries from the body the nitrogen-containing waste products formed by the breakdown of protein and nucleic acid molecules. The body must equalize water gained and lost each day so that it doesn't dry out or become waterlogged. The body must also rid itself of wastes. An animal's digestive system rids the body of undigested solid wastes, but the **excretory system** does two things: it (1) cleanses the blood of organic waste molecules and (2) carries wastes out of the body as urine or its equivalent through a special set of excretory tubes. As we will see, the kidneys are central to both excreting dissolved organic wastes and balancing water content by removing the right amount of water in the urine.

In complex animals, including astronauts, blood vessels carry blood adjacent to or very near every cell. The blood's fluid and salt content is normally maintained within narrow limits. What's more, inside each body cell, the pH, salt content, and waste concentration cannot vary too much, or individual cells will cease to function, and the animal will die. The fluid inside cells, the **intracellular fluid,** makes up about two-thirds of the total body water. This intracellular fluid undergoes a continual exchange of materials with the **extracellular fluid**—fluid in the spaces between cells. Figure 14.10 illustrates this exchange. Extracellular fluid, in turn, is in equilibrium with the watery portion of the blood—

the **plasma,** and the plasma composition is maintained by the excretory system. This explains why we call the kidney an amazing organ. By maintaining the contents of the blood, the kidneys indirectly maintain the intracellular fluid within each of the body's billions of cells. The kidney's crucial cleansing and fluid balancing activities, however, bear a high cost: The kidneys are metabolically more active than the heart! Without a continuous energy flow to these vital organs, the kidneys would fail and body fluids would become too salty or too watery for an animal to survive.

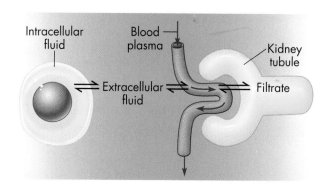

Figure 14.10 **Water Compartments in the Body.** Homeostasis involves a continuous exchange of water and materials, and establishes an equilibrium between intracellular fluid, extracellular fluid, blood plasma, and filtrate, leading to urine.

Nitrogen-Containing Wastes

If you've ever built a bookcase, then you know that short pieces of boards remain after you saw and hammer wood to build the furniture. In a similar way, residues of molecules remain unused after cells process nutrients and build from them new molecules and cell parts; these residues are wastes that must be removed. **Excretion** is the process that removes the byproducts of metabolism and rids the organism of excess water and salts. Excretion differs from elimination, the expulsion of solid wastes as feces (see Chapter 17). Undigested solid materials never leave the interior of the digestive tube; they never, in other words, actually enter the body or its cells. By contrast, metabolism within cells is the source of most waste products excreted in urine, and these wastes must be collected from every cell and floated away. Let's look more closely, now, at the excretion process and what it removes.

Three Types Of Waste

When it comes to using energy, living cells are very flexible: they can burn fats, carbohydrates, and/or proteins for energy, depending on what's available. Cells burn fats and carbohydrates "cleanly," changing them to acetyl-CoA. This molecule, you may recall, feeds directly into the Krebs cycle portion of aerobic respiration (see Fig. 3.13), and leads only to the waste products water and carbon dioxide. These small H_2O and CO_2 molecules diffuse out of cells and eventually may be expelled from the body. In contrast, before cells can use proteins for the production of ATP in aerobic respiration, they must first break the proteins down into amino acid subunits, and then chop each amino acid into a nitrogen-containing portion (the —NH_2, or amine group) and the carbon-containing remainder of the molecule (review Fig. 2.10). Cells transform the carbon-containing portion of the amino acid into acetyl-CoA, which enters the Krebs cycle, but the amine becomes ammonia (NH_3), a strong-smelling alkaline ingredient we recognize from cleaning solutions, or the ion ammonium (NH_4^+). Even at low concentrations, ammonia is highly toxic to cells, so organisms must get rid of it quickly. Many animals that live in water excrete ammonia directly into the surrounding pond, stream, or seawater, where it is diluted so much that it becomes nontoxic. Land animals and certain other aquatic animals, however, turn the ammonia into less caustic, less damaging compounds before excreting it.

Birds, reptiles, insects, and land snails convert ammonia into the less toxic **uric acid** through special enzyme activities in the liver. The relatively insoluble substance continues to circulate suspended in the blood, like silt carried along in a river, and the kidneys gradually remove it. Producing uric acid as an excretory product helps land animals conserve water, because this relatively insoluble nitrogenous waste can pass out of the body as a paste containing little moisture. The white portion of the ubiquitous bird droppings that decorate window ledges, park statues, and the occasional human head is mainly uric acid. Uric acid can also be stored in almost solid form in a special part of a chick or turtle egg, away from embryo. Without such an effective form of waste disposal, hard-shelled eggs (and the animals that hatch from them) might never have evolved.

Mammals and a few kinds of fishes turn ammonia into **urea,** a combination of ammonia and carbon dioxide. This conversion also takes place in the liver. Just as dirty dishwater goes through sewer pipes to a central water treatment facility for final disposition, mammal and fish livers secrete urea, which dissolves in water, into the blood. The bloodstream then carries the urea to the kidneys, where they remove it from the blood and excrete it in urine, the fluid wastes excreted by the kidney.

GET THE PICTURE

Explorer 14.3

The Human Excretory System

How Do Kidneys Work to Maintain a Steady State?

Kidneys are not just amazing, they are amazingly *fast.* Just how fast does your kidney filter something you have consumed? Most people can perform an experiment—although it sounds a bit indelicate—to answer this question. For many diners, eating tender, pale

green shoots of asparagus is a pleasurable mealtime event, but the gastronomic experience has a peculiar sequel: afterwards, they can smell the characteristic scent of asparagus in their urine. (Some people lack the ability to perceive this smell.) A sulfur-containing biochemical in the food crosses the gut, enters the bloodstream, is filtered out by the kidneys, and appears in the urine. How fast does it happen? To find out, you could eat asparagus and then urinate at various times thereafter, each time performing an olfactory assay for the sulfur-containing compound. For many people, the essence of asparagus can show up within 20 minutes. The key to the excretory system's rapid functioning lies in the kidney's complicated internal structure and in its efficient plumbing system.

The Human Excretory System

If you've ever been hungry, tired, and cold and had a very full urinary bladder simultaneously, then you'll probably understand why urination is sometimes called our most compelling bodily function. And imagine having to do it in the space shuttle or a space station with nearly zero gravity where drains don't work! (In case you've ever wondered, Dr. Harris reveals that astronauts visit small lavatory compartments and use devices that vacuum the droplets so they don't wind up floating around the cabin interior for the rest of the flight!) This urgency associated with excretion—whether assisted by gravity or not—has to do with the necessary waste-removal role of the excretory system and with its particular anatomy.

In humans, as in other vertebrates, the main organ of the excretory system is the **kidney.** A person's two plump, dark red, crescent-shaped kidneys, each about the size of an adult's fist, are located just below and behind the liver, and they receive a large, steady flow of blood (Fig. 14.11 and **E**xplorer 14.3). From the blood, the kidneys filter and remove excess water and waste substances, and these materials collect as concentrated urine in a central cavity in each kidney. Wastes leave the central cavity and then flow down a long tube called the **ureter** (yer-EET-er). Sometimes the ureter can become painfully blocked by **kidney stones,** crystals usually formed of calcium and organic compounds. Drinking plenty of water decreases the likelihood of developing this excruciating condition. The two ureters dump urine into a single storage sac, the **urinary bladder,** which in an adult can hold about 500 ml (1 pt) of fluid. The bladder can distend to hold a bit more, but when it is full, it stretches local nerves, causing a feel-

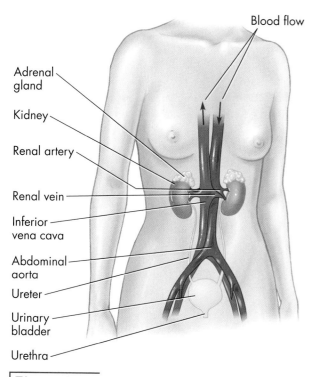

Adrenal gland

Kidney

Renal artery

Renal vein

Inferior vena cava

Abdominal aorta

Ureter

Urinary bladder

Urethra

Blood flow

Figure 14.11   The Human Excretory System.

proximated the kidney's general sorting and cleansing action and its three basic processes—filtration, reabsorption, and secretion (Fig. 14.13).

Filtration occurs at the bulbous tip of the nephron. During this first stage, blood pressure forces small molecules, such as glucose, amino acids, ions, water, and urea, out of the blood plasma through a filter, and into the cavity of the nephron. Large protein molecules and blood cells are left behind in the blood because they are too large to fit through the pores in the filter.

During **tubular reabsorption,** the nephron sorts the small molecules, shunting useful ones back to the blood and retaining the wastes in the nephron.

Finally, during **tubular secretion,** the nephron checks the blood supply one last time and removes from circulation excess ions, drugs, or other wastes that still remain and secretes them into the forming urine.

The way the kidney performs the three basic functions of filtration, reabsorption, and secretion is truly a wonder of natural engineering that can be appreciated by viewing the animation in **E** 14.3. This "engineering" is based largely on the delicate loops within the nephrons and the flow of blood around them.

ing of urgency that can temporarily eclipse all other concerns. During urination, bladder contents drain through a single tube, the **urethra** (yoo-REE-thruh), which carries urine out of the body.

A lamb kidney bought from the butcher and cut in half lengthwise looks a lot like a human kidney. Both have three distinct visible zones (Fig. 14.12a): (1) There is an outer **renal cortex,** which is a bit like a thick orange peel; this is where initial blood filtering takes place. (2) The central zone, or **renal medulla,** is divided into a number of pyramid-shaped regions (Fig. 14.12b). The medulla helps conserve water and valuable dissolved materials (solutes). The functional units of the kidney, twisted tubules called **nephrons,** reach from the cortex down into the medulla (Fig. 14.12c). (3) Inside these two zones lies the funnel-shaped, hollow inner compartment, or **renal pelvis,** where urine is stored before it passes into the ureter and then collects in the bladder.

An Overview of the Nephron's Activity

Have you ever cleaned out a desk drawer—sorting, discarding, saving certain items for future use, and then rechecking the drawer one last time? If so, you have ap-

Structure of the Nephron

Each nephron is a long, twisted, looping tubule that carries out filtration, absorption, and secretion in the healthy kidney with efficiency and ease (Fig. 14.12c). Each human kidney contains roughly 1 million nephrons, each extending from the outer cortex through the medulla and draining into the renal pelvis. If all the nephrons in an adult's kidney were straightened out and placed end to end, they would form a microscopically slender tube about 80 km (50 mi) long.

Near the outer surface of the kidney, one end of the nephron, called **Bowman's capsule,** is cup-shaped like a punched-in balloon (Fig. 14.12c,d). The other end of the nephron drains away urine towards the bladder. Between these two ends, the tubule meanders and tangles like a long strand of spaghetti. The capsule leads into the convoluted **proximal** ("near") **tubule,** which then dips sharply down into the medulla. There, it makes a hairpin curve and heads back up into the cortex. The U-shaped section of the nephron, known as the **loop of Henle,** is chiefly responsible for reabsorption of water from the tubule contents back into the blood. The loop of Henle connects to the part of the nephron farthest from Bowman's capsule, the **distal** ("far") **tubule.** Distal tubules from various nephrons join, and form large **collecting ducts,** which receive liquid

Figure 14.12

Anatomy of the Human Kidney: A Blood-Cleansing Organ. (a) A cross section of the kidney reveals its basic anatomy. (b) A pyramid consists of many nephrons. (c) Each nephron is a twisted tubule that filters the blood and conveys the filtrate to the collecting duct. (d) Nephrons are amply supplied with blood from nearby vessels. (e) Blood filtration takes place in Bowman's capsules through holes in the capillary wall and the special coffee-filter anatomy of podocyte cells (f) or the capsule lining cells with their fingerlike extensions encircling the capillaries.

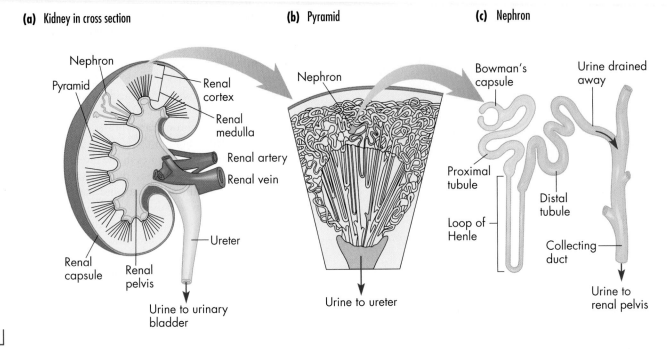

(a) Kidney in cross section

Nephron
Pyramid
Renal cortex
Renal medulla
Renal artery
Renal vein
Ureter
Renal capsule
Renal pelvis
Urine to urinary bladder

(b) Pyramid

Nephron
Urine to ureter

(c) Nephron

Bowman's capsule
Urine drained away
Proximal tubule
Distal tubule
Loop of Henle
Collecting duct
Urine to renal pelvis

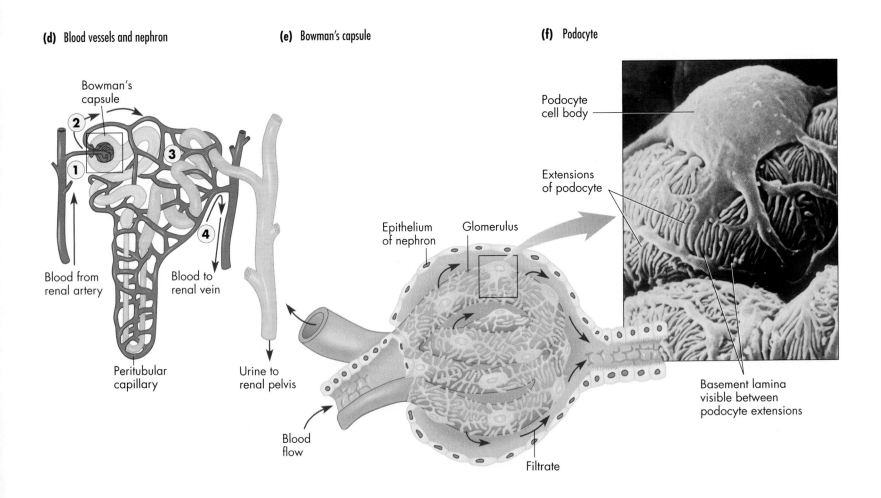

(d) Blood vessels and nephron

Bowman's capsule
2
1
3
4
Blood from renal artery
Blood to renal vein
Peritubular capillary

(e) Bowman's capsule

Epithelium of nephron
Glomerulus
Urine to renal pelvis
Blood flow
Filtrate

(f) Podocyte

Podocyte cell body
Extensions of podocyte
Basement lamina visible between podocyte extensions

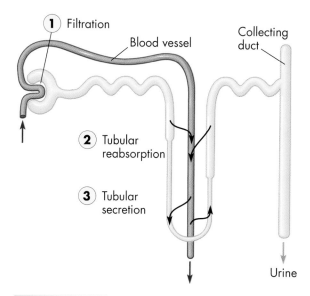

① Filtration

Blood vessel

Collecting duct

② Tubular reabsorption

③ Tubular secretion

Urine

Figure 14.13 **Filtration, Reabsorption, and Secretion: The Roles of the Nephron.** In filtration (Step 1) water and other small molecules are filtered from the blood and enter the nephron. In tubular reabsorption (Step 2), water, salt, and nutrients are returned to the blood. In tubular secretion (Step 3), some ions and drugs are secreted from the blood back into the nephron; from there they move with the urine into the collecting duct, renal pelvis, and ureter and pass into the bladder for elimination.

wastes from several distal tubules, drain into the renal pelvis and carry away the urine. About every 15 seconds, a small amount of urine collects in this pelvis or hollow inner compartment, and waves of muscle contraction in the ureter sweep the urine through the ureter to the urinary bladder for storage and eventually for excretion through the urethra. (Consider, for a minute, what would happen to an astronaut if gravity rather than these peristaltic waves of contraction caused urine to move from the kidney into the bladder.)

Blood Flow Around Nephrons

At any given time, about one-fifth of the total blood in our bodies is passing through the blood vessels to and from the kidneys. The pathway of blood flow around each nephron is critical to kidney function and is similar to the action of an incredibly delicate coffee filter. Blood makes its closest contact with the nephron in the Bowman's capsule, as a small artery branches and brings blood to the **glomerulus,** a wad of capillaries, which are fine, thin-walled blood vessels (Fig. 14.12e). Minute

holes pierce the cells that make up the capillary walls (Fig. 14.12e,f), and fluid from the blood percolates through the delicate filter and into Bowman's capsule. The fineness of these capillary pores is ultimately responsible for the nephron's blood filtration activities.

When capillaries leave the capsule, they do not merge directly into veins, but instead rejoin to form another small artery or arteriole, which exits the capsule (Fig. 14.12d, Step 2). This arteriole carries blood into a second network of capillaries that surrounds the looped portion of the nephron (Step 3). These capillaries finally merge into a small vein that connects to the large vein that leaves the kidney (Step 4). These veins carry blood away from the kidney and toward the heart.

Nephrons at Work

During his first few hours in microgravity, Dr. Harris's kidneys were working overtime as his rate of urine production increased to counteract the rush of blood and tissue fluid to his head and chest. In times of water overload, it's almost embarrassing how well the kidneys work to remove fluid and chemical wastes while at the same time conserving mineral ions, glucose, and other needed materials. Each day in an adult human, about 180 L (48.5 gal) of fluid—enough to fill a bathtub—passes from the blood into the cavities of the kidneys' Bowman's capsules. We do not, of course, urinate 180 L a day, but instead the much more reasonable amount of about 1.5 L (3 pt) per day. Clearly, the nephrons accomplish a great deal of water conservation before they produce that smaller quantity of urine. *E* 14.3 provides an overview of the main stages and details of kidney function.

Filtration

This first stage of urine formation occurs in Bowman's capsule (Fig. 14.12e). In the cup of the capsule, the walls of the capillaries are built of thin cells perforated by pores, with the blood on one side of the cells and a thick basement membrane or lamina on the other. Across the basement membrane lie the epithelial cells that form Bowman's capsule. On the other side of those cells is the cavity (lumen) of Bowman's capsule. The epithelial cells of each Bowman's capsule are called **podocytes** (literally, "foot cells"). Each podocyte has thousands of fingerlike projections that enclasp each capillary completely like fingers wrapping around a hose (Fig. 14.12f) Blood pressure forces some of the plasma, the yellowish fluid portion of blood, through tiny pores in the walls of the capillaries (Fig. 14.14, Step 1), just as water pressure forces water out of a perforated garden hose or lawn sprinkler. The filtered liquid, containing various dissolved substances but lacking blood proteins, then passes through the fingers of the podocytes into the cavity of Bowman's capsule. The transport of water; sodium, potassium, and chloride ions; and sugars, amino acids, and urea out of the capillaries and into the cavity of Bowman's capsule is passive and does not require any special output of energy other than blood pressure. The fluid, or **filtrate,** in the capsule is still very much like blood plasma, except that it contains no large proteins.

Tubular Reabsorption

As soon as the filtrate enters the nephron's twisted proximal tubule, reabsorption begins and returns to the blood most of the water, sodium, potassium, and chloride ions, sugars, and amino acids that were just filtered out of the blood into Bowman's capsule (Step 2). Cells move these materials out through the walls of the proximal tubule and into the extracellular space, and finally the materials pass back into the blood plasma of the capillaries entwining the proximal tubule. Instead of being driven passively by blood pressure like the process of filtration, tubular reabsorption depends on active transport of solutes (at a continuous cost of ATP energy) across the plasma membranes of nephron cells. As the tubule reabsorbs ions in this way, 80 to 85 percent of the water in the original filtrate passively follows the ions back into the capillaries via the process of osmosis, the movement of water from a solution with a few dissolved materials across a membrane into a concentrated solution (Step 3).

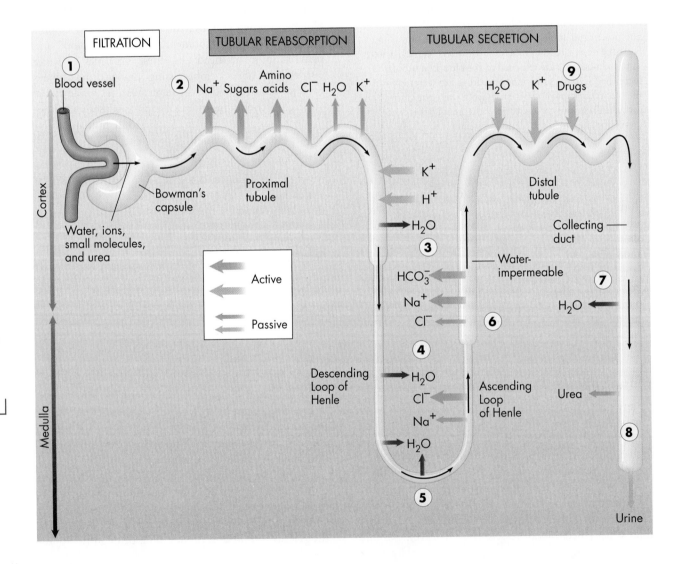

Figure 14.14 How the Nephron Makes Urine. As a nephron extends through the kidney's cortex and medulla and dumps urine into the collecting duct, various substances enter and leave the filtrate. Thin lines represent segments of the nephron wall that are permeable to water, while thick lines represent wall segments impermeable to water. Narrow arrows represent passive diffusion of materials into or out of the nephron tubule, while wide arrows represent active transport against concentration gradients. Tubular reabsorption activities in blue, and tubular secretion in red. Tubular contents are shown as yellow. The text traces nephron function and material movements step by step.

Next, the filtrate—minus most of its water but still containing urea and salts—passes through the loop of Henle. Here, the cells of the nephron use energy to pump chloride ions (Cl^-) into the extracellular fluid surrounding the tissue cells of the kidney's medulla, and sodium ions (Na^+) follow passively (Step 4). As a result of this ion pumping, the innermost part of the medulla becomes very salty compared to the upper part, as suggested by the gradient of grey in Figure 14.14. This means that the concentration of Na^+ and Cl^- is lowest towards the outside of the kidney, and grows increasingly higher toward the inner part of the medulla, the part containing the hairpin curve of the loop of Henle (see Fig. 14.12c). Because the tissue surrounding the loop is salty, water diffuses out of the loop by osmosis and reenters the blood of the nearby capillaries (Step 5). This exodus of water ends, however, as the filtrate rounds the bend and moves up the loop, because that part of the tube is not permeable to water (Step 6).

From the water-tight region of the tubule, the filtrate, still containing urea, some salt, and some water, passes toward the collecting duct (Step 7). In that duct, once again, the walls are permeable to water, and even though the filtrate passing through is already quite concentrated, the surrounding tissues are still saltier. Thus, still more water diffuses out of the collecting ducts by osmosis and enters the blood of the capillaries (Step 7). In addition, the walls of the collecting duct allow a certain amount of urea to pass back out, rather then being excreted in the urine. This urea increases the brinelike concentration in the inner medulla and causes still more water to be removed and conserved.

Strangely enough, a thirsty desert animal like a camel excretes very little urea. Instead, the kidney reabsorbs urea into the animal's blood, where it circulates throughout the body, and eventually enters the digestive tract with water following in by osmosis. Because water follows the urea back into the digestive tract, and because water can diffuse from the large intestine back into the blood, this mechanism—the retention of urea when water is scarce—helps the desert animal's body conserve precious water in a parched environment.

Now, back to human excretion. By the time the filtrate (at this point called urine) has reached the part of the collecting tubule in the innermost (and saltiest) region of the medulla, much of the water has been reabsorbed (Fig. 14.14, Step 8). In fact, about 99 percent of the water originally filtered from the blood in the Bowman's capsule has by now been returned to the body's circulation. That's why we excrete 1.5 liters of urine each day instead of 180 liters and why the urine has a high concentration of wastes relative to water content.

Tubular Secretion

The kidneys' activities—specifically filtration and reabsorption—are mainly responsible for the volume of the urine produced. (In an astronaut, that volume is initially higher than normal because of the blood and fluid shift toward the head and the body's homeostatic response to the problem.) The kidneys' third major activity, **tubular secretion,** is responsible for removing unneeded materials from the blood. The proximal and distal tubules remove many kinds of undesirable substances from the extracellular fluid surrounding the nephron and secrete them into the forming urine. These materials include hydrogen and potassium ions, ammonia, and certain drugs like the antibiotic penicillin and the sedative phenobarbital (see Fig. 14.14, Step 9). Tubular secretion, therefore, is an important blood-cleansing process. It can also help maintain an appropriate pH level in the blood, because nephron cells secrete more hydrogen ions into the urine if the blood is too acidic and fewer hydrogen ions if the blood is too alkaline.

Tubular secretion is the physiological process that makes drug testing possible—checking a person's urine to see if he or she has taken drugs. Various laboratory techniques can detect even minute traces of the metabolic breakdown products of marijuana, cocaine, heroin, sleeping pills, tranquilizers, morphine, codeine, and many kinds of prescription drugs. If a person takes a drug overdose and loses consciousness and no one is sure which drug was consumed, physicians quickly test the urine to identify the drug and determine the best treatment for saving the patient's life. Two additional uses—testing athletes and employees for drug use on the playing field or on the job—are currently used as well, but have created considerable controversy.

The Kidney and Homeostasis

During his weeks in space, Dr. Harris's kidneys helped his body adjust to the dramatic loss of gravity. These organs not only cleansed his blood of urea but also helped reestablish and maintain constant body conditions in several ways. As we already saw, the kidneys helped Harris's body get rid of the excess fluid in the blood and tissue spaces that migrated up into his head and chest after gravity stopped tugging it toward his feet.

The extra urine production decreased Harris's blood volume and lessened the puffiness in his upper body and face. However, it created a secondary problem: As water was removed from his blood in the first couple of days in space, Harris's blood cells grew more concentrated just the way draining a reservoir in the au-

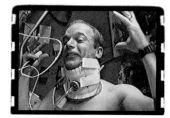

How do the kidneys help resolve an astronaut's fluid shift toward the head?

tumn concentrates the fish. The body has a feedback loop involving a hormone called **erythropoietin** (ee-RITH-ro-po-EE-tin) that deals with changes in the concentration of blood cells. Normally, if the concentration of red blood cells drops below about 45 percent or so of the total blood volume, the kidney senses the blood's decreased oxygen-carrying capacity and responds by secreting erythropoietin (Fig. 14.15). The hormone travels through the blood to the bone marrow, where it stimulates new red blood cells to form. The increased concentration of red blood cells carries more oxygen to the kidneys, and the organs respond by releasing less erythropoietin. Consider what would happen to this feedback system in Bernard Harris after his excretory system eliminated the extra water from his body in the first couple of days in space and his blood cells got concentrated as a result. His kidneys would stop secreting erythropoietin and the bone marrow would stop making new red blood cells and the concentration of red blood cells would decrease as the older cells die. Now, what would happen when Harris returned to Earth's mighty pull?

He would drink extra water to replace the fluids now draining away from his head and chest and sinking to his legs, and this would dilute his blood cells, creating a slight anemia. What would happen to the level of erythropoietin at this point? It would rise, causing the bone marrow to increase blood cell manufacturing activities and once again raise the level of red blood cells to normal.

Harris's kidneys helped him maintain a steady internal environment a third way, related to the proximal tubule's tendency to reabsorb calcium ions. Recall that when a person is weightless or nearly so, negative feedback mechanisms in the bones sense too much calcium then begin to dismantle the bones and dump the extra calcium ions into the blood. When the blood passes through Bowman's capsules in the kidneys, the calcium ions pass into the filtrate. As the filtrate passes through the tubule, the process of tubular reabsorption removes calcium ions from the filtrate, so that they can once more enter the blood. But reabsorption can bring calcium ions up only to a certain level. If that level is exceeded, as it is with massive bone degradation accompanying space travel, then the excess appears in the urine. This leads to the worry over irreversible bone loss on long space missions.

Now let's return to the actual mechanism whereby Harris's system "knows" to remove water from the blood.

How Do Hormones Control Water Excretion?

Just as you can turn a water faucet on or off to increase or decrease water flow, your body can regulate nephrons, resulting in an increase or decrease in urine flow. We've looked at several factors that influence the amount of water in the urine: The amount of water initially filtered into Bowman's capsule, which can be influenced by blood pressure; the salt concentration of the extracellular fluid in the kidney's medulla; and the degree to which the distal tubule is impermeable to water. Two hormones are involved in these processes: antidiuretic hormone regulates the permeability of the collecting duct to water, while the hormone aldosterone regulates salt reabsorption.

Antidiuretic Hormone

Have you ever heard someone say that coffee and beer are diuretics? That means that substances in these bever-

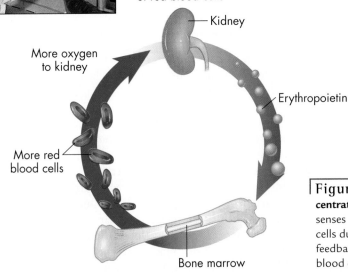

Low concentration of red blood cells

Kidney

More oxygen to kidney

Erythropoietin

More red blood cells

Bone marrow

Figure 14.15 **How the Kidney Maintains Red Blood Cell Concentrations.** A feedback loop involving the hormone erythropoietin senses a drop in blood oxygen level. This can be due to loss of red blood cells during injury or even space travel, and to anaerobic exercise. The feedback loop then stimulates the bone marrow to produce more red blood cells.

ages cause the body to produce higher than normal amounts of urine. **Antidiuretic hormone (ADH,** also called vasopressin) is aptly named because it slows down **diuresis,** or abundant urine production. ADH controls how much water the nephron reabsorbs from the filtrate and returns to the blood. Special nerve cells in the hypothalamus of the brain produce ADH, and long extensions of those cells extend to a nearby region, the pituitary gland, and store the hormone (Fig. 14.16). When the concentration of salt and other solutes rises in the body—say, after you eat a bag of salty potato chips (Fig. 14.16, Step 1)—the hypothalamus detects the changes and causes the pituitary gland to release some of its stored ADH into the blood (Step 2). Upon reaching the kidney, ADH makes the walls of the distal tubule and collecting ducts temporarily more permeable to water (Step 3) so more water is reabsorbed into the bloodstream, diluting the too-high concentration of blood solutes (Step 4). Since water is drawn from the filtrate into the blood, the blood becomes less salty, but at the same time, the urine becomes more concentrated (Step 5). Finally, less salty blood causes the brain's secretion of ADH to decrease (Step 6), thus completing a negative feedback loop.

As beer drinkers occasionally discover to their dismay, alcohol inhibits ADH secretion, leading to the excretion of sometimes embarrassing quantities of urine. This causes dehydration—a major contributor to the hangover a person can feel after drinking too much alcohol.

Aldosterone

While ADH "adjusts the tap" to control the body's reabsorption of water, another hormone, **aldosterone,** secreted by the adrenal glands (which sit on top of the kidneys) controls the absorption of salt. In brief, if either the concentration of sodium ions or the blood pressure drop, the kidney secretes the enzyme **renin.** Renin initiates a signal cascade in the blood that eventually produces **angiotensin II,** which triggers the release of aldosterone from the adrenal gland. Aldosterone causes the distal tubule and collecting duct to reabsorb sodium ions from the tubule contents into the blood. Chloride ions follow passively, and the increasingly salty blood slows the release of renin. This completes the negative feedback loop and, in turn, maintains the concentration of sodium in the blood within the narrow range that is best for the activity of nerve cells and other crucial internal activities.

Studies of astronauts in space have shown that their blood is low in sodium ion concentrations and, in fact, low in the concentrations of all particles. No one knows yet why this condition exists. Future experiments will have to be conducted on astronauts in space to explain why weightlessness seems to bring about mysterious shifts in blood chemistry.

Kidneys: Adapted to the Environment

Our planet teems with life: Organisms have evolved to survive in environments that are hot, cold, dry, wet, acidic, salty, and every imaginable combination and variation. We've seen the close association between a kidney's filtering, balancing, and secreting functions and its multilayered, nephron-packed structure. It's no surprise that biologists have found striking variations in the kidneys—specifically in the nephrons—of animals with markedly different life histories. Take the kangaroo rat, for instance, a small rodent inhabiting the

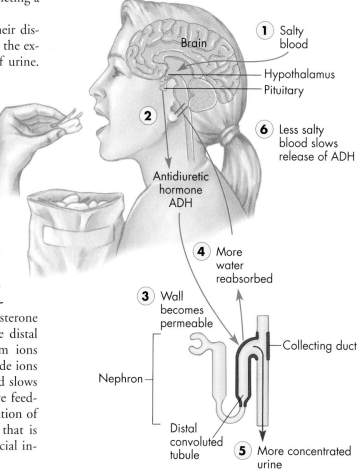

Figure 14.16 How Hormones Regulate Salt and Water Balance. Salty blood sensed by the brain's hypothalamus (Step 1) causes the release of antidiuretic hormone (ADH) (Step 2), which in turn acts on the distal convoluted tubule and collecting duct, making them more permeable to water (Step 3). Water then leaves the forming urine, is reabsorbed into the blood (Step 4), and the urine becomes more concentrated (Step 5). When levels of salt in the blood decrease, ADH release is shut down (Step 6).

deserts of the western United States. This animal conserves water to an extreme degree, and each of its nephrons has a very long loop of Henle. With this modification, its kidneys can reabsorb practically all the available water back into the animal's body as the filtrate moves through the long loop. As a result, the kangaroo rat can produce urine 25 times more concentrated than its own blood—the highest concentration of any mammal.

In contrast, look at an aquatic mammal like a beaver. It takes in a great deal of water in its food and through its skin and must get rid of the fluid, not conserve it. In a beaver's kidney, the loops of Henle are short and reabsorb less water; as a result, the animal produces great quantities of urine with only twice the concentration of its own blood. Humans have a combination of long and short loops of Henle and so produce a urine of variable concentration, depending on how much water is available in the environment at a given time.

The Kidney and Disease

An organ as central to homeostasis as the kidney is impossible to live without and difficult to replace. People with kidney diseases often spend many hours each week connected to a **dialysis** machine, a large instrument that mimics the filtering activities of a healthy kidney (Fig. 14.17).

The kidney is also involved in the health of the circulatory system. Many people, in recent years, have begun watching the amount of sodium in their diets, because medical research has shown that too much sodium can lead to elevated blood pressure and strain a heart already clogged and weakened by atherosclerosis (see Chapter 15). We just saw that if you eat a salty diet, the kidneys will homeostatically balance the situation by excreting less water. This water retention dilutes the blood and makes it less salty. At the same time, though, it also increases the total volume of fluid in the circulatory system. This, in turn, elevates the blood pressure (like forcing more water into a water balloon) and causes the heart to work harder.

Physiologists have also discovered that the heart can secrete a hormone that helps keep blood pressure in check. When the blood volume and pressure increase, heart cells can give off **atrial natriuretic factor (ANF),** and this hormone causes sodium to appear in the urine. (Here's a way to remember ANF: the Na symbol for sodium comes from its Latin name "natrium.") Water soon follows the sodium by osmosis, and the volume of urine goes up while the blood volume goes down. As a result, the blood pressure goes down as well. Biotechnologists would like to produce large quantities of ANF using recombinant DNA techniques to treat people with chronic high blood pressure.

Sometimes after an accident or allergic reaction, blood flow and blood pressure start to fall so low that blood flow to tissues is inadequate and the body goes into a medical state called **shock.** The kidneys can counteract falling blood pressure by releasing the enzyme renin, the same enzyme involved in sodium ion homeostasis. Renin converts a blood protein into the hormone angiotensin II, and in this case, the angiotensin II causes blood vessels to constrict. The constriction elevates blood pressure the way that tightening the nozzle of a garden hose increases the pressure in the length of hose behind it. Angiotensin also triggers the adrenal glands to secrete aldosterone as we just saw, which results in sodium and water reabsorption and thus even higher blood volume and pressure.

The details may sound a bit complicated, but the message from this section is simple: Hormones help reg-

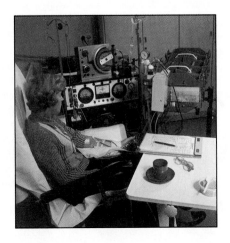

Figure 14.17 **Dialysis and Kidney Disease.** The kidneys can lose their exquisite ability to cleanse the blood in several ways. Bacteria from feces can contaminate the urinary tract, attacking the kidneys. An autoimmune attack on the kidneys by white blood cells can block and destroy glomeruli. Finally, poisoning by mercury, lead, or certain solvents can damage the kidney tissue. If the number of functioning nephrons drops below 10 to 20 percent, the person can suffer extreme tissue swelling as a result of salt and water retention, as well as a buildup of metabolic byproducts like urea. These make the blood acidic and can lead to coma or death if the pH drops below 6.9. Waste-laden blood from a patient's artery can be routed through an artificial kidney or kidney dialysis machine, containing a long porous membrane bathed by a solution much like normal blood plasma, where wastes diffuse out of the blood. While the artificial kidney can carry out filtration and help sustain the patient's life, it lacks the real kidney's reabsorption and secretion activities. Patients also must receive drugs such as AT3 (see Chapter 7) to prevent blood clotting while on dialysis, and these may subject them to infections. Finally, five- to ten-hour treatments every two or three days gobble up much of a patient's time.

ulate how much salt and water we retain, and this, in turn, affects blood content, blood pressure, and other factors vital to homeostasis and health.

Salt and Water Balance

Ever wonder why a thirsty sailor can't simply drink seawater? The answer lies with the salt and water balance we have been discussing: Human kidneys cannot make urine as salty as seawater. For every 1 L (1 qt) of seawater humans consume, we produce 1.3 L of urine. This would cause rapid dehydration. In contrast, a camel *can* drink seawater because its kidneys can make urine that is more concentrated than seawater. Likewise, marine animals such as sharks, flounder, or sea gulls drink seawater readily. Obviously, animals face different osmotic challenges depending on the saltiness, dryness, and wetness of their natural environments. Let's look at another important aspect of homeostasis on land, at sea, and in fresh water: **osmoregulation,** the regulation of osmotic balance.

Salt and Water Balance in Land Animals

We land animals are bathed in a sea of air, and so we tend to lose water through evaporation across the body and breathing surfaces as well as through urination. This brings about needs so familiar we rarely think about them: (1) The need to drink water, (2) the need to dump excess salt by sweating and urinating (or in some animals, excreting salt through special salt glands), and (3) the need to conserve water through physical and behavioral mechanisms. Our human skin surface, for example, is dry and helps seal in water—not as well, though, as an insect's waxy cuticle, a reptile's scaly skin, a bird's feathers, or a mammal's fur. In one experiment, researchers sheared a camel, and the animal's water expenditure increased by 50 percent!

The moist skin of an earthworm, snail, or amphibian allows those animals to lose much more water through evaporation, so they tend to behave in ways that help them conserve water: Amphibians, for example, tend to remain in cool, wet places, and snails tend to come out only in the cool, damp night air. People exhibit water conserving behaviors, too: Desert dwellers tend to wear clothing that minimizes evaporation and protects them from the sun, and they tend to be most active at dawn and dusk, when temperatures are lower.

Salt and Water Balance in Aquatic Animals

You might think that fishes, seals, crabs, and other aquatic animals have it made when it comes to conserving water and balancing salts, since they don't have to face the problem of constant evaporation like we land animals do. In fact, though, they simply have a *different* continual problem due to their body fluids having a salt concentration either higher or lower than the surrounding water. If an animal lives in an aquatic environment saltier than its internal fluids—in the ocean, say, or a salt lake—water tends to diffuse outward and leave the animal dehydrated. If an animal lives in an aquatic environment less salty than their blood—as in a freshwater lake or river—then water tends to diffuse into their bodies and waterlog the organisms. To overcome these challenges, various adaptations have evolved in different groups for excreting water and balancing salt.

The easiest solution to osmoregulation is to have body fluids with the same concentration as the surrounding water, and that is the situation in invertebrates such as oysters and sea stars. The salt levels of their body fluids conform to their surroundings, whether they are salty or dilute. Other marine invertebrates, such as crabs and brine shrimp, have internal mechanisms to maintain steady levels of water and salts regardless of environment.

In fishes, marine mammals such as seals, and other aquatic vertebrates, body fluids can be more or less concentrated than their surroundings, but the concentrations remain more or less steady through homeostatic mechanisms and anatomical structures. Sharks, for example, have very little net change in their internal water concentration because their body fluids have roughly the same concentration of dissolved substances as seawater. The retention of urea by the shark's kidney contributes to this high concentration of dissolved substances in the blood. Any salts that build up are excreted by a special rectal gland.

In freshwater fishes and amphibians, the blood has a higher salt concentration than the environment. Water tends to flow in across the skin by osmosis, so these animals have no need to drink; instead, their kidneys produce large amounts of dilute urine, which rids the body of excess water. Unfortunately, salt also tends to leave the body and diffuse into the water. The gills of freshwater fishes are able to actively accumulate and absorb salt from the lake or river water.

Finally, for most ocean-going fish and mammals, the seawater is saltier than their body fluids and thus the animals tend to lose needed water through their skin and gain excess salt. As an evolutionary consequence, saltwater fish *do* drink seawater and they retain most of the water. They produce very little urine, while their gills actively pump out excess salt. Sea lions, whales, and other marine mammals have a different solution: They

How do marine animals balance the blood's salt and water content?

never need to drink and instead take in salt water only with food. Their very efficient kidneys conserve water strongly and excrete excess salt.

It is clear from these descriptions that whatever an animal's environment, its excretory system works continuously to keep water and dump salt—or the reverse!—depending on what's needed to maintain a steady state in the body tissues. Homeostasis—whether of body fluid content, oxygen levels in the tissues, calcium in the blood and bones, or body temperature—is costly but necessary to survival. Understanding the many body parts and mechanisms that contribute to this steady state helps us appreciate the beauty of our own physiology and that of other animals. It also encourages us to take care of our health in intelligent ways. And it gives us a deeper understanding of the extreme environments explorers like Bernard Harris have faced and overcome in the high mountains, desert, polar regions, ocean abyss, and the radiation-blasted darkness and weightlessness of space.

Chapter Summary and Selected Key Terms

Introduction   During Dr. Bernard Harris's 9 days in space, he lost over 10 percent of his muscle mass and 1 percent of his bone. His body literally disassembled itself in a normal, natural attempt to maintain a steady internal state, **homeostasis** (p. 372), despite the very abnormal conditions of microgravity in space.

Staying Alive: Problems and Solutions
Homeostasis is a recurring theme in the study of **anatomy** (p. 374), or biological structure, and **physiology** (p. 374), or the function of biological structures. Essential life processes include capturing energy and materials, exchanging gases, and maintaining a body of a particular shape, fluid composition, and (usually) temperature range. Some body systems that help maintain homeostasis have a supply tube running from the outside environment to the inside; special regions of the tube where substances can be exchanged with body fluids; transport of essential materials by those fluids; and delivery of the materials to all cells.

Body Organization
An animal's cells are highly organized into a hierarchy of structures at increasingly complex levels: **tissues** (p. 375), or cell groups performing similar functions; **organs** (p. 375), or units of two or more tissues collectively performing a certain function; and **organ systems** (p. 375) of two or more related organs working together, serving a common function. The four tissue types include: **epithelial tissue** (p. 376), or sheets of cells that line or cover body surfaces; **connective tissue** (p. 377), or loose groups of cells embedded in a matrix that binds other tissues and forms a framework of support within body parts (Bernard Harris's bone loss involves changes to the bone's connective tissue); **muscle tissue** (p. 379), which shortens, applies force against objects, and allows animals to move; and **nervous tissue** (p. 379), which transmits information and helps control and coordinate the actions of organs, organ systems, and the whole body.

Maintaining a Steady State
Homeostasis often works via **feedback loops** (p. 380). These loops involve the action of a **receptor** (p. 380) that senses an original change in the environment; an **integrator** (p. 380) that evaluates the situation and makes decisions; and an **effector** (p. 380) that executes commands. A **negative feedback loop** (p. 380) resists change by sensing a deviation from the baseline condition and opposing it. A household thermometer and furnace work this way, and so do the human body's internal mechanisms for maintaining a steady oral temperature of about 98.6 °F. A **positive feedback loop** (p. 382) brings about rapid change by sensing an initial environmental change, responding to it, and then amplifying the response further and further. This type of response brings about mammalian birth.

Homeostasis, Circulation, and Behavior
The maintenance of body temperature in a warm-blooded animal or **homeotherm** (p. 382) is a classic example of homeostasis involving specialized circulatory mechanisms. **Vasoconstriction** (p. 383), or blood vessel constriction, shunts blood away from the periphery and toward the body core, controlling heat loss. **Vasodilation** (p. 383), the enlarging of blood vessels, shunts blood to the body surface, promoting heat loss. **Countercurrent flow** (p. 383) promotes blood flow but also conserves heat. Adaptive behaviors also help prevent **hypothermia** (p. 383), a drop in core temperature, or **hyperthermia** (p. 383), a rise in core temperature. These behaviors include hibernation, migration, huddling, building shelters, shedding heavy coats, and depositing fat layers. **Poikilotherms** (p. 384) (so-called cold-blooded animals) have more fluctuation in core temperature and tend to bask in the sun or use other behavioral mechanisms to help control heat gain and loss. Some have enzymatic or tissue modifications that help tolerate cold or generate heat.

Maintaining Salt and Water Balance
The **kidney** (p. 385) is an amazing filtration device that maintains the constitution of the blood and the level of water and salts in body fluids. An animal's body must maintain **water balance** (p. 385); it must take in and lose the same amount of water over a period of hours or days. **Urine** (p. 385) produced in the kidneys carries fluid and nitrogenous wastes from the body. The kidneys are part of the **excretory system** (p. 385), which cleanses the blood and carries wastes out of the body as urine or its equivalent. **Intracellular fluid** (p. 385) makes up about two thirds of the total body water. **Extracellular fluid** (p. 385) resides in the spaces between cells. **Plasma** (p. 385) is the fluid portion of blood. The kidneys help maintain salt and water balance in all three fluids. **Excretion** (p. 385) removes the byproducts of metabolism and rids the organism of excess water and salts. The amino portion of amino acids, the building blocks of protein, is a leftover chemical waste product that can be removed as ammonia (by many aquatic animals); as **uric acid** (p. 386) (by birds, reptiles, insects, and land snails); or as **urea** (p. 386) (by mammals and a few kinds of fishes). The human kidney filters urea and other substances from the blood and excretes it into **ureters** (p. 386), which lead to the sacklike **urinary bladder** (p. 386). Urine then leaves the body through the **urethra** (p. 387). The kidney has three zones: the **renal cortex** (p. 387), where initial blood filtering occurs; the **renal medulla** (p. 387), which helps conserve water and dissolved materials; and the **renal pelvis** (p. 387), where urine is briefly stored.

The kidney's functional units are tubules called **nephrons** (p. 387), each consisting of a **Bowman's capsule** (p. 387), a tangled **proximal tubule** (p. 387), a hairpinlike **loop of Henle** (p. 387), and a **distal tubule** (p. 387) that drains into a **collecting duct** (p. 387). Each nephron is the site of three main functions: filtration, reabsorption, and secretion. In **filtration** (p. 387), the cuplike Bowman's capsule filters water, urea, and solutes from blood moving through the tuft of capillaries (**glomerulus**; p. 389) within the capsule. Once in Bowman's capsule, the fluid is called the **filtrate** (p. 390). During **tubular reabsorption** (p. 387), most of the water, glucose, amino acid molecules, and useful ions leave the nephron tubule and reenter the bloodstream via the nearby capillaries.

During **tubular secretion** (p. 387, 391), certain drugs and other organic molecules, as well as hydrogen, potassium, and other ions, are actively removed from the blood plasma and secreted into the urine. Water continues to leave osmotically, and the urine reaching the bladder contains only 1 percent of the water originally filtered from blood plasma. The kidneys secrete the hormone **erythropoietin** (p. 392) in response to low concentrations of oxygen, and this hormone stimulates the production of new red blood cells.

In mammals, water balance is regulated by thirst, which is triggered by nerve activity in the hypothalamus of the brain, and by **antidiuretic hormone (ADH)** (p. 393) and **aldosterone** (p. 393), which help control the functioning of the nephrons and hence the amount of water they absorb. The kidneys also help regulate normal blood pressure and volume by responding to **atrial natriuretic factor** (p. 394), which is secreted by the heart and causes the kidneys to excrete more sodium ions (and hence more water) from the blood. The kidneys also secrete **renin** (p. 393), which converts a blood protein to the hormone **angiotensin II** (p. 393); this causes blood vessels to constrict, thereby raising blood pressure. People with kidney disease often need **dialysis** (p. 394).

An animal's excretory system is adapted to its environment: Animals inhabiting dry deserts, lush pastures, freshwater lakes or streams, or salty oceans each have adaptations appropriate to their surroundings that maintain the constancy of their internal fluids via **osmoregulation** (p. 395). Land animals are in constant danger of dehydration. Body coverings help conserve moisture, and so do the kidneys by forming concentrated urine. Other adaptations can be behavioral or physiological. Aquatic invertebrates can have osmotic levels in which the contents of their body fluids follow that of the external environment, or they can have a more constant internal environment, maintained despite changes in the external surroundings. In aquatic vertebrates the excretory systems help maintain body fluids at constant levels that may be the same, more, or less salty relative to the surrounding water.

All of the following question sets also appear in the Explore Life ⓔ electronic component, where you will find a variety of additional questions as well.

Test Yourself on Vocabulary and Concepts

In each question set below, match the description with the appropriate term. A term may be used once, more than once, or not at all.

SET I

(a) anatomy (b) physiology (c) homeostasis
(d) positive feedback (e) negative feedback

1. The aspect of biology that is devoted to an understanding of the structure of microscopic and visible structures of organisms

2. A dynamic steady state maintained by the organism that enables cells to survive and function

3. The biological specialty primarily concerned with the functioning of cells, organs, and organ systems

4. A process in which an initial change in one direction is amplified further and further in the same direction

5. The primary way in which cells maintain stable internal conditions despite ever-changing environmental conditions

SET II

(a) epithelial tissue (b) connective tissue (c) muscle tissue (d) nervous tissue (e) more than one of these

6. Secretion, excretion, protection, and absorption are among the primary functions of this tissue. The cilia of cells of this tissue can be damaged by cigarette smoke.

7. Composed of contractile cells that are either attached to skeletal parts or act to move the contents contained within hollow body organs

8. Cells of this tissue function primarily to integrate incoming stimuli with actions taken by effector organs.

9. Many of the cells of this tissue produce fibers that are released from the producing cells and lie in the extracellular matrix. Tendons and ligaments are examples of this tissue.

10. Found within the organs of the digestive system

SET III

(a) urine (b) ammonia (c) uric acid (d) urea

11. The most toxic of the nitrogenous wastes, but also the one that is most soluble in water

12. The nitrogenous waste that assists the most in water conservation

13. Produced from filtrate by the processes of tubular secretion and reabsorption

14. A water-soluble nitrogenous waste that is less toxic than ammonia

SET IV

(a) filtration (b) tubular reabsorption (c) tubular secretion (d) filtration and reabsorption (e) filtration and secretion (f) filtration, reabsorption, and secretion

15. Primarily the function of the glomerulus and Bowman's capsule

16. Collectively the function of the nephron

17. Compounds move from the blood into the tubule.

18. Compounds move from the tubule into the blood.

19. Podocytes are an essential element in this process.

Integrate and Apply What You've Learned

20. What common strategies for the maintenance of homeostasis are shared by the digestive, respiratory, and excretory systems?

21. What is the essential difference between a tissue and an organ?

22. Is it likely that creatures such as single-celled protists or jellyfish could attain the size of a human being and live on land? Explain.

23. Explain why negative feedback loops are more important in the maintenance of homeostasis than positive feedback loops.

24. In hot, dry climates, athletes drink water as much as once an hour, even if they are not thirsty. Why?

25. After working at a hard physical activity on a hot and dry day, a dehydrated student drinks several cans of beer. Will the student's thirst be quenched and the body rehydrated?

26. Biologists often consider the fish kidney to be less an organ of excretion than of osmoregulation. Is this true for freshwater fishes as well as saltwater (marine) fishes? Explain why.

Analyze and Evaluate the Concepts

27. How did homeostatic mechanisms attempt to stabilize Dr. Bernard Harris's bones, muscles, and red blood cells, and why were they only partially successful from the standpoint of his health upon return to Earth's gravity?

28. Which of the following situations is the best example of negative feedback?

(a) A cyclist speeds up while coasting down a steep hill and slows down when the terrain flattens out.

(b) A merchant raises the price of an item. Fewer people buy the item and the merchant lowers the price.

(c) A motorist sees a red stoplight in the distance and slows down in preparation for stopping.

(d) A motorist sees a red light in the distance and slows down until the light turns green.

(e) In order to counteract a low-grade fever, a student takes an aspirin. The student's body temperature returns to normal.

29. Which of the scenarios below is the best example of positive feedback?

(a) A cook reasons that if a cake needs to bake for 45 minutes at 350 °F, then raising the oven temperature to 400 °F will shorten the time to 30 minutes and raising it to 450 °F will shorten the time to 20 minutes.

(b) In order to counteract traffic jams from the northern suburbs of a city, the city council builds more roads. When people find out about the improved commuting conditions, more people move to the northern suburbs. The city council once again authorizes more road building.

(c) While heating up a pot of soup, a cook increasingly raises the temperature of the burner. The soup heats faster and faster until it boils over on to the surface of the stove and puts out the flame.

(d) Sighting his commuting bus, a passenger runs faster and faster to catch it; however, he then trips and the bus leaves without him.

(e) A shop keeper notices that a particular item is not selling. The price of the item is lowered, but still no one buys it.

30. Tadpoles swimming in a pond excrete nitrogenous wastes as ammonia, but adult frogs hopping about on land excrete nitrogenous wastes as urea. Explain.

31. In some forms of diabetes, large amounts of glucose are filtered from the blood into the nephron, where it accumulates in urine. Would you expect a patient with this condition to produce more urine or less urine than most other people? Explain.

32. Why is the urine of a dehydrated person dark yellow-brown, while that of a hydrated person only faintly yellow?

33. You are a doctor and your hospitalized patient has this medical situation: the region of the nephron including Bowman's capsule functions normally, but the proximal and distal tubules, as well as the loop of Henle, have ceased to function in reabsorption. Which of the following conditions would you expect to find in this patient?

(a) dehydration   (b) excessive fluid retention
(c) uremic poisoning   (d) high blood pressure

34. Which of the functions of the nephron do we rely on when testing the urine of athletes for forbidden drugs: filtration, tubular reabsorption, or tubular secretion? Explain.

A Life-Saving Substitute

- Two cars collide head on, trapping both drivers in the wreckage.
- An earthquake sends walls and ceilings crashing down on to a family in their beds.
- A man is wheeled into surgery for removal of a massive lung tumor.
- A soldier falls to the ground with bullets in his arm and shoulder.

The people in these catastrophes are going to need blood: fresh, disease-free blood of their own blood type (A, B, O, AB; see Fig. 5.11), banked and ready for transfusion. Blood is a life-giving, oxygen-bearing commodity that we tend to take for granted. We often assume that in a modern, industrialized country, it will be quickly available should we have the misfortune to need it. But blood shortages are growing more common as fewer and fewer people—less than 5 percent of the population—are willing donors. Exhaustive testing for the viruses that cause AIDS and hepatitis and for other contaminants has made our blood supply safe, but at a price: A hospital patient can now pay $1000 for a single unit (pint) of blood.

To researcher Robert Winslow, there's a straightforward answer to this dilemma: Artificial blood. Dr. Winslow has spent nearly two decades in pursuit of a blood substitute that works well and won't harm patients. Artificial blood has, in fact, been a top research goal for many drug companies and for the U.S. Army. But as simple and worthy as this search may sound, says Winslow, "it's an unbelievably complicated thing to do."

In the mid-1980s, medical workers began finding HIV in banked blood, but had no reliable way to weed out donors who carried the virus. Concern over the blood supply "was near panic at that point," says Winslow, especially in the military. Winslow led an Army research effort to find a safe blood substitute with a "shelf life" longer than human blood's 42 days. His team tried removing the oxygen-carrying molecule, hemoglobin, from outdated red blood cells, modifying it slightly, then infusing it into animals, but it was, he recalls, "an unbelievable failure." A cubic centimeter of blood contains 4.5 to 5.5 million red blood cells, and each of these tiny lozenge-shaped cells contains 350 million hemoglobin molecules. Once outside the delicate cell membrane, however, naked hemoglobin molecules cause little blood vessels throughout the body to constrict—a "horrible" property, says Winslow, that is potentially fatal.

Faced with this stumbling block, but with HIV and other screening tests making donated blood safe again, the Army abandoned its search for substitutes in 1991. Winslow, however, kept researching

▲
Searching for a blood substitute. Dr. Winslow in the lab; one new blood substitute: climbers on Mt. Everest.

the problem at the University of California, San Diego. Eventually, he founded nearby Sangart, Inc., to produce and test what he believes is the solution to the puzzle of artificial blood. His product, still under development, carries and transfers oxygen like real blood. It doesn't have a blood type, so it's universal. It can be frozen and stored indefinitely, it costs less than human blood, and so far, it doesn't seem to harm recipients.

Other researchers in other companies have created their own formulations, but Winslow is understandably more optimistic about Sangart's and at this printing is planning clinical trials in patients. If the artificial blood proves safe and effective enough, the market for it could be as high as 8 million units a year in the United States and 200 million globally. From the human perspective, "market" means automobile accident and earthquake victims, surgery patients and soldiers, diabetes sufferers who lose limbs to poor blood circulation in tiny vessels, and huge populations in underdeveloped countries with tainted or insufficient blood supplies.

The creation of artificial blood is our case study in this chapter and serves as a lead-in to the two main subjects. The first is **circulation**—how blood moves through the body carrying oxygen and nutrients to tissue cells and carrying wastes away. The second is **respiration**—how sufficient oxygen is drawn into the body and transferred to blood, and in return, how carbon dioxide is expelled. The two systems and their vital organs, the heart, lungs, and blood vessels, are delicately intertwined. This means that we'll encounter both breathing and blood circulation when we consider topics such as the fate of mountain climbers and high-altitude dwellers; the development of breathable liquids to help patients with lung failures; the causes of high blood pressure and heart disease; and the effects of tobacco smoking on the lungs and blood vessels.

While exploring these critical aspects of physiology, you'll find the answers to these questions:

1. What is blood and how does it reach tissues throughout the body?

2. How does the heart circulate the blood and what is the lymphatic system?

3. How are heartbeat and blood pressure controlled?

4. How does oxygen leave the blood and enter cells, and why do those cells need oxygen?

5. How do the lungs draw in air and expel carbon dioxide, and how does the body control breathing rate?

6. What are comparable strategies for gas exchange in other animals?

7. How did Robert Winslow and colleagues create their artificial blood product?

15.1 Blood: A Liquid Tissue for Transport

A safe human blood substitute is clearly needed. But why is making one so difficult? To understand that, let's look at what the blood does in the healthy body or when transferred into a surgery patient, let's say, and how it carries out those many roles.

Components of Blood

If you prick your finger with a sterile pin then press out a small drop of blood onto a glass slide, you'll immediately see the bright red color. If you examined the drop with a microscope, you'd see the reason for the crimson hue: Your view would be nearly filled with red blood cells, or **erythrocytes** (*erythro* = red + *cyte* = cell) (Fig. 15.1a). You'd also see a much smaller number of white blood cells, or **leukocytes** (*leuko* = white) in the same microscope field, along with some **platelets,** which are small globular cell fragments, and a watery yellow fluid surrounding all the solid elements.

Since blood is red, one might assume that the most common element is the red blood cells. But is it? To find out, you'd need not just a droplet of blood but a few milliliters of blood in a test tube. Then you'd need to put the tube in a centrifuge, an apparatus that spins a sample rapidly and separates components according to their density. What you'd see is that within the centrifuged blood, the top half or so is the pale yellow liquid or **plasma,** which is more than 91 percent water (Fig. 15.1b). Plasma's main function is transporting blood cells and dissolved substances, including salts, sugars, and fats from the foods we eat. Plasma also contains water and a storehouse of important dissolved salts and proteins, which additional laboratory procedures can separate out. One group of plasma proteins, the globulins, includes **antibodies,** defensive molecules that attack invaders (described in Chapter 16). The **albumins,** another group of large blood proteins, help maintain osmotic pressure and bind to toxic substances in the blood. A third plasma protein, **fibrinogen,** is essential for blood clotting. To make substitute blood plasma in the laboratory, one would need to add the proper amount of salts to sterile water. The liver and other organs would then provide proteins like albumin and fibrinogen.

In a test tube of centrifuged blood, the dense portion below the yellowish plasma is the part that proves hardest to reconstruct in artificial blood. This 45 to 50 percent has two bands (Fig. 15.1b): on top is a thin gray band representing less than 1 percent of blood volume and consisting of white blood cells plus cell fragments, the platelets, which help blood to clot. The lower, much wider red band contains red blood cells. A physician can use the width of this band of red blood cells to help diagnose medical problems. If the band is too narrow, the person has too few red blood cells and is probably **anemic,** perhaps suffering from fatigue and shortness of breath. Anemia can result from heavy blood loss, iron deficiency, poor vita-

min B_{12} absorption, sickle cell disease (see Chapter 5), certain infections, and other causes. The red blood cells are simple, but crucial to blood's role in carrying oxygen, and constructing them—or something that performs their functions—is the major challenge to Winslow and other biologists trying to make artificial blood.

Red Blood Cell Shape and Function

Fully one third of the approximately 75 trillion cells in your body are red blood cells. Let's look at those 25 trillion cells, see how they work, and discuss what it would take to replace them. Clues to red blood cell function

Figure 15.1 **Blood and its Components.** (a) In this colorized scanning electron micrograph, you can see (1) dark background areas representing plasma (which is actually yellowish and watery); (2) platelets (small pink clusters in a yellowish mesh); (3) white blood cells (here, blue fuzzy balls); and (4) red blood cells (dark red, biconcave disks). (b) Spinning blood in a centrifuge separates components according to their density. Yellowish plasma collects on top. Further laboratory treatment can separate plasma into water; proteins; and dissolved salts, sugars, fats, and amino acids. Fully 99 percent of the remaining material is the dense red blood that collects in the lower portion of the tube. Somewhat less dense white blood cells and platelets form a narrow grayish zone just above the packed red blood cells.

(a)

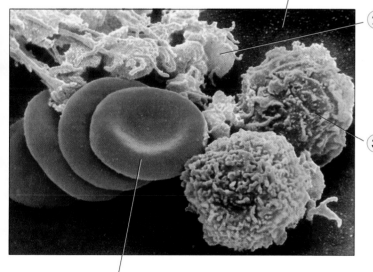

(1) Plasma: Liquid portion of blood; about 50% by volume; contains proteins involved in blood clotting, antibodies, and other proteins of the immune system, albumin, lipid-carrying proteins, hormones, plus salts, sugars, amino acids, and other small molecules.

(2) Platelets: Also called thrombocytes, cell fragments that contain no nucleus; less than 1% of blood volume, derived from break-up of certain large white blood cells; regulates blood clotting; contains serotonin, a regulator of blood vessel diameter.

(3) White blood cells: Also called leukocytes; less than 1% of blood volume; larger than red blood cells; some move like an amoeba, some engulf cell debris, help defend body from microorganisms and foreign particles; several subclasses, including neutrophils, basophils, eosinophils, macrophages, and lymphocytes.

(4) Red blood cells: Also called erythrocytes; disk-shaped with 7-μm diameter, no nucleus, about 50% of blood volume; packed with hemoglobin protein; transports oxygen.

(b) Separating blood contents by density

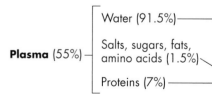

Plasma (55%) —
- Water (91.5%)
- Salts, sugars, fats, amino acids (1.5%)
- Proteins (7%)

Platelets and **White blood cells** (1%)

Red blood cells (99%)

come from the shape of the cells and their contents. Each red blood cell resembles a doughnut without a hole (Fig. 15.1a)—biologists call it a **biconcave disk** shape. In contrast, white blood cells are shaped more like golf balls, and many other body cells are shaped like ice cubes. So why does the red blood cell have that particular disklike shape? Being rounded, they move through vessels fairly easily. Their flattened shape allows more of the cell contents to be close to the cell's surface compared to a spherical cell; this suggests that cell surface exchanges are especially crucial for red blood cell function. Proteins of the cytoskeleton just inside the cell membrane help maintain the red blood cell's disk shape (see Fig. 2.27). Embedded in that membrane are the proteins and carbohydrates that give the cell its A, B, O, or AB blood type. Infusing a patient with the wrong blood type can lead to a dangerous clumping reaction in the blood vessels. Artificial blood lacking blood cell membranes, surface markers, and hence blood types would avoid this potential problem.

Although humans are eukaryotes, their mature red blood cells lack a nucleus, mitochondria, and other cell organelles. That's because these organelles are squeezed out as the red cells grow and develop, and then are not replaced. The loss of organelles leaves more space for important proteins in the cell. Red blood cells help to transport carbon dioxide to the lungs, partly because they contain the enzyme **carbonic anhydrase.** This enzyme speeds the conversion of carbon dioxide (CO_2) and water (H_2O) to bicarbonate (HCO_3^-) and hydrogen ion (H^+). When there is lots of carbon dioxide in the environment, the extra hydrogen ions turn the blood more acidic. Most of all, though, the red blood cells are chock full of the protein **hemoglobin,** which transports oxygen. The flattened shape of red blood cells makes sense in that they allow more of each cell's millions of hemoglobin molecules to lie near the outer membrane than if red blood cells were spheres. Any artificial blood substitute would therefore need some type of oxygen-transporting molecule that somehow mimics this flattened "wrapper" or at least the cell membrane's ability to allow fast, efficient gas exchange between the blood and its surroundings.

Hemoglobin: An Oxygen-Carrier Molecule

Hemoglobin molecules are far smaller than red blood cells. In fact, each red blood cell contains about *350 million* hemoglobin molecules. Hemoglobin's specialized structure allows it to act like an oxygen sponge and enables a pint of blood, let's say, to soak up 70 times more oxygen than could a pint of water. Hemoglobin has a unique structure with four polypeptide chains (called globin chains). Two of these, the alpha-globin chains, are identical, as are the other two, the beta-globin chains. Each of the four globins in a hemoglobin molecule is wrapped around a heme group, a series of chemical rings surrounding an iron atom. Heme is red in color and hence makes your blood red. Oxygen binds to the single iron atom at the center of each heme (Fig. 15.2).

Researchers like Robert Winslow have tried the most direct approach to making a blood substitute: eliminating the red blood cell "wrapper" and putting naked heme groups directly into a test animal's blood vessels. The problem, though, is that the heme binds so tightly to oxygen that it won't release the gas to cells that need it—hardworking muscle cells in the legs, let's say. This fact helps us understand the role of the hemoglobin's protein portion, the globin chains: They cause the heme to take up and release oxygen under appropriate physiological conditions. The plain heme idea didn't work. But it does help us see the complexity of blood's functioning in the body.

With its four globin chains and central heme group, the hemoglobin molecule is a wonder of adaptation. It not only picks up oxygen in the lungs and carries it to the tissues but also picks up carbon dioxide from the tissues and releases it in the lungs. Carbon dioxide binds directly to the protein chains, not to the central heme group. Furthermore, the versatile hemoglobin molecule binds more or less easily to oxygen, depending on the conditions. The blood in the body's brain and muscle tissues, for example, is more acidic than the blood in the lungs. That's because the active brain and muscle cells are giving off carbon dioxide from the burning of food molecules for energy. Carbonic anhydrase enzymes in the red blood cells (mentioned earlier) quickly convert CO_2 and water to bicarbonate and hydrogen ion, and this conversion makes the blood more acidic. Under these more acidic conditions, hemoglobin binds oxygen less readily, thus O_2 tends to break away from hemoglobin and diffuse from the bloodstream into tissue cells, which need it. This helps insure that cells will get the oxygen they need. But it also adds another level of complexity for those trying to make a successful blood substitute.

Production of New Red Blood Cells

We said that red blood cells are relatively short lived and can't divide and reproduce themselves. So how fast does the body make new blood cells, how long do they stay in circulation, and where do they arise in the body? The answers are additional considerations when design-

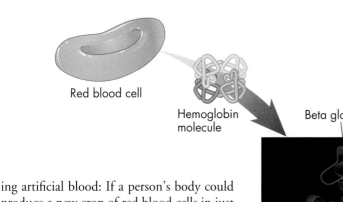

Red blood cell

Hemoglobin molecule

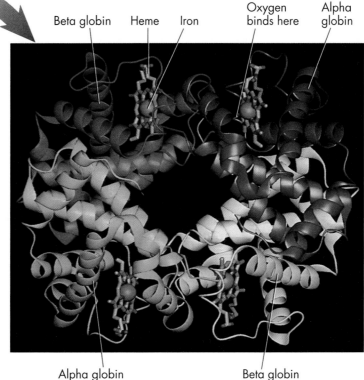

Beta globin   Heme   Iron   Oxygen binds here   Alpha globin

Alpha globin   Beta globin

Figure 15.2

Hemoglobin. A single red blood cell contains about 350 million hemoglobin molecules. Each hemoglobin molecule consists of four polypeptide chains—two beta-globin chains (here, red and pink) and two alpha-globin chains (here, blue and orange). Each chain surrounds a heme (green ball). An iron atom at the center of each heme (red) binds oxygen; thus the entire hemoglobin molecule transports four oxygen molecules from the lungs to the other tissues.

ing artificial blood: If a person's body could produce a new crop of red blood cells in just a couple of hours, then when he or she became injured, artificial blood could be added for a very short time, mainly to keep the blood volume at a normal level, and soon the missing cells would be replaced naturally. However, if red blood cells form slowly over days or weeks, then a replacement would have to last much longer. It turns out that individual red blood cells die after about 120 days of circulating in the body and are then replaced from stem cells in the bone marrow. The bone marrow stem cell populations are self-regenerating: When a stem cell divides in two, one daughter cell begins to differentiate into a mature blood cell, while the other differentiates into another stem cell and in effect replaces the original one. Some stem cells give rise to red blood cells; others generate white blood cells and the precursors to platelets.

In Chapter 14, we saw that the body has an important feedback mechanism that can sense a drop in red blood cell concentration after a sudden blood loss, as from a wound, and tells the bone marrow to begin making new cells (Fig. 15.3). A sudden blood loss can signal the bone marrow, too. The feedback mechanism starts when a decrease in the number of red cells in the blood lowers oxygen levels to the liver and kidneys. These organs then begin to produce the protein hormone **erythropoietin** (eh-RITH-ro-po-EE-tin; Greek, red + to make). This hormone travels through the bloodstream to the bone marrow, where it stimulates stem cells to divide and differentiate more quickly; as more red blood cells are produced, they carry more oxygen to tissues and organs throughout the body, including the liver and kidney. As a result, these organs slow their production of erythropoietin, and division of bone marrow stem cells slows. Because it takes about five days for a cell to pass down the "assembly line" of red blood cell differentiation, the first brand new red blood cells

won't appear in the blood in substantial numbers until about five days after the initial blood cell loss. Ideally, then, a blood substitute would last at least five days while the body generates new cells of its own to restore the original concentration.

As you might imagine, erythropoietin could be an important supplement to artificial blood to help stimulate new blood cell production after blood loss from an accident, surgery, or a battle. Indeed, there are effective erythropoietin drugs on the market that do stimulate red blood cell formation. Unfortunately, some unscrupulous athletes have tried using these drugs to gain an advantage without putting in the necessary training work. They've resorted to taking erythropoietin ("epo") to stimulate an increase in red blood cells in the hopes of increasing oxygen delivery to their muscles. For someone without blood loss, however, this may be counterproductive, because the increased concentration of red blood cells also increases blood viscosity; this in turn makes the heart work harder than normal, trying to pump the thicker, more molasses-like fluid through the vessels.

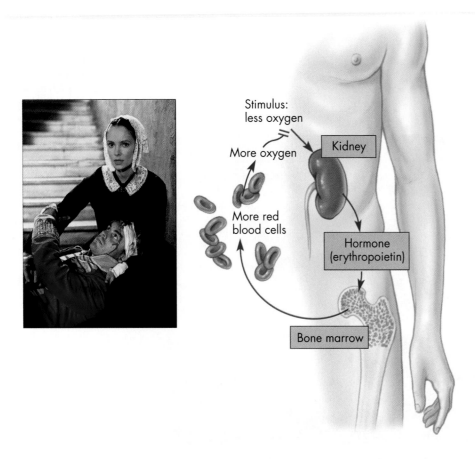

Stimulus:
less oxygen

More oxygen

Kidney

More red
blood cells

Hormone
(erythropoietin)

Bone marrow

Figure 15.3 **Replacing Lost Blood Cells: A Negative Feedback Loop.** Florence Nightingale (played by Jaclyn Smith) tends to a "wounded" soldier, complete with stage blood. When a true blood loss occurs, and the number of red blood cells drops, the kidney receives less oxygen and it releases the hormone erythropoietin. This stimulates the bone marrow to make more red blood cells. Increased numbers of red blood cells carry more oxygen to the kidneys (and other organs), and the kidneys stop releasing the hormone.

White Blood Cells: Defense of the Body

For every 500 red blood cells, there is only one white blood cell, or **leukocyte** (*leuko* = white + *cyte* = cell; see Figure 15.1). White blood cells are larger than red blood cells, and they retain their nucleus when they mature. Some leukocytes can move like an amoeba, an ability that allows them to squeeze through blood vessel walls and patrol the fluid-filled spaces between cells. Some can also engulf and take in debris (see Fig. 2.20). These characteristics help leukocytes to defend the body against invasions by microorganisms and other foreign materials, and to consume cell debris. There is so much to say about leukocytes and their role in the body's immune defense that Chapter 16 is devoted entirely to the subject.

Do you think a blood substitute would need white blood cells? Leukocytes are crucial to a person's long-term chances of surviving bacterial and viral infections. In the short term, though, after an auto accident, say, or cancer surgery, doctors can control most bacterial infections by keeping the wounds as sterile as possible and by using antibiotics (drug-resistant organisms, however, can be a problem, as we saw in Chapter 9). Since Dr.

Winslow's intended purpose for artificial blood is short-term emergencies, he doesn't try to replace the functions of white blood cells.

Platelets: Plugging Leaks in the System

We saw that blood has a third cellular component, the platelets (also called **thrombocytes;** see Fig. 15.1). Platelets are not entire cells, but are irregular fragments that have broken off from large specialized cells in the bone marrow. Platelets are crucial blood components because they help to plug small leaks in the circulatory system. Were it not for the blood-clotting action of platelets, an animal's entire blood supply might literally drain away through even a minor wound. Even if a person or other animal were to lose a significant amount of blood through an unexpected trauma, however, the remaining blood would contain enough platelets to keep up the blood's normal clotting functions. So Dr. Winslow's blood substitute has no special additive to replace platelet function.

All of the blood's components are essential to its collective functions in the body: oxygen delivery; carbon dioxide removal; blood clotting; immune defense; and helping maintain the body's proper fluid volume. Artificial blood for emergency use can be simpler: just a sterile saline solution for the right blood volume and osmotic conditions, and some red-blood-cell equivalent to carry oxygen and carbon dioxide. The latter characteristic is the real challenge, as we'll see. But first, let's explore the intricate system of vessels through which the blood flows, as well as alternative systems for both blood and circulation in other animals.

15.2 Circulatory Systems

In a town without roads, delivery trucks would be useless. Likewise, blood—whether authentic or artificial—would be of little value to an animal unless it could circulate its cargo of gases, nutrients, and other substances throughout the body. The human circulatory system has a highly branched network of vessels totalling thousands of miles in length—its "road system," if you will. A constant flow of blood passes through the vessels to within 0.1 mm (0.004 in.) of each body cell. A distance this tiny allows materials in the blood to diffuse quickly into cells and replenish nutrients and other necessary materials. Animals living in various kinds of environments have a range of different needs for gases, nutrients, and waste removal, as well as numerous kinds of restrictions upon what would be an appropriate delivery system. The range of solutions tells us something fundamental about the range of needs. So let's look at the circulatory systems "invented" by other animals.

Open Circulatory Systems

Most people hate tarantulas. They never get past their revulsion over the animal's eight thick, hairy legs. But inside of this large spider lies a circulatory system typical of most animal species on our planet. Tarantulas have an **open circulatory system** (Fig. 15.4a), in which a clear blood equivalent called **hemolymph** circulates rather freely in the body unconfined by tubes or vessels. Most other arthropods, as well as mollusks, also have open circulatory systems. The tarantula's heart is little more than an elongated, pulsating tube that runs from the rear of the body to the front (Fig. 15.4b). The heart's pumping action sends hemolymph in the direction of the animal's brain, where it leaves the heart tube and percolates toward the back of the animal through the open body cavity, bathing the tissues in the animal's gut and other internal organs. Finally, the hemolymph returns to the heart tube through special pores. An open circulation is not a very efficient transport system, but it suffices for spiders, insects, and many other invertebrates.

What can a tarantula's circulatory system tell us about designing artificial blood? The hemolymph in these spiders transports oxygen by means of a huge protein called **hemocyanin.** As its name suggests, hemocyanin is blue (cyan) rather than red, it's totally unrelated to hemoglobin, and it's not contained within blood cells. How can tarantulas get away with having their oxygen-transporting molecules free and uncontained within cells? Hemocyanin consists of 48 subunits compared to hemoglobin's 4, and each hemocyanin subunit is nearly 5 times larger than a single globin chain. Maybe Robert Winslow or other artificial blood researchers could make a human blood substitute without cells if hemoglobin could be altered to be a huge, multi-subunit protein instead of being small and having four subunits. Let's keep that evolutionary idea on the "back burner," as they say, then return to it after we investigate the other major type of circulatory system, the kind found in our bodies.

Figure 15.4 **An Open Circulatory System.** (a) In an open circulatory system, the blood moves briefly in a vessel, but is then dumped in to an open area in which it freely bathes the internal tissues before re-entering the vessel. (b) The open circulatory system of the Mexican red-legged tarantula *Brachypelma emelia*.

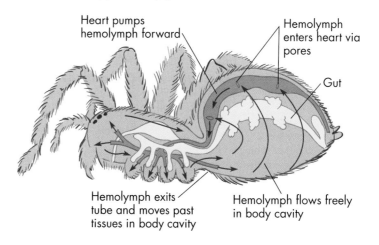

(a) An open circulation system

Pump

(b) Circulatory system of a tarantula

Heart pumps hemolymph forward

Hemolymph enters heart via pores

Gut

Hemolymph exits tube and moves past tissues in body cavity

Hemolymph flows freely in body cavity

Closed Circulatory Systems

Segmented worms like earthworms and vertebrates like humans have a **closed circulatory system,** with its blood completely contained inside a system of vessels (Fig. 15.5a).

A closed circulatory system has several advantages over an open one. First, fluid contained within a network of closed tubes can be more easily shunted to specific areas where it is needed, much as a farmer can send the water in irrigation pipes to different fields. Second, because the fluid is completely contained within pipelines, the circulatory system can exert more pressure on the blood, forcefully distributing blood to areas distant from the pump, like a giraffe's head or a whale's tail. You've no doubt discovered the results of pressurized fluid yourself: You can squirt water on an unsuspecting friend much farther away when you partially close off the opening of a garden hose with your thumb than you can by dribbling the water out of the unobstructed hose end. The features of the closed circulatory system allow it to deliver blood more efficiently

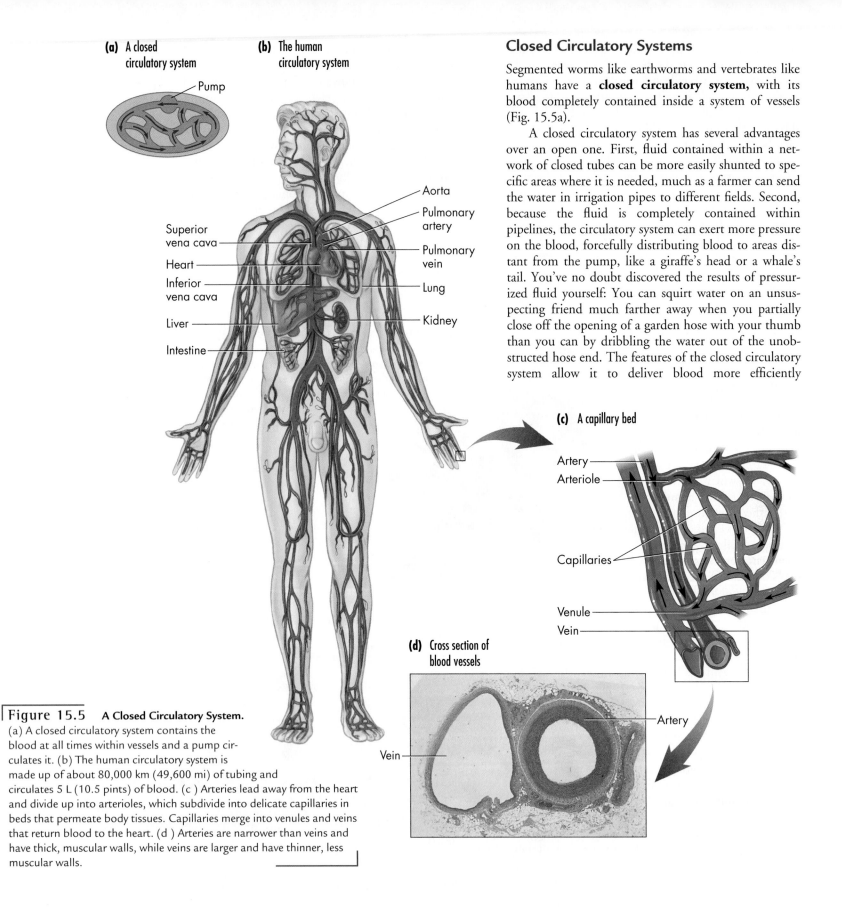

(a) A closed circulatory system

Pump

(b) The human circulatory system

Superior vena cava

Heart

Inferior vena cava

Liver

Intestine

Aorta

Pulmonary artery

Pulmonary vein

Lung

Kidney

(c) A capillary bed

Artery

Arteriole

Capillaries

Venule

Vein

(d) Cross section of blood vessels

Vein

Artery

Figure 15.5   A Closed Circulatory System.
(a) A closed circulatory system contains the blood at all times within vessels and a pump circulates it. (b) The human circulatory system is made up of about 80,000 km (49,600 mi) of tubing and circulates 5 L (10.5 pints) of blood. (c) Arteries lead away from the heart and divide up into arterioles, which subdivide into delicate capillaries in beds that permeate body tissues. Capillaries merge into venules and veins that return blood to the heart. (d) Arteries are narrower than veins and have thick, muscular walls, while veins are larger and have thinner, less muscular walls.

throughout larger, more complex, and more active organisms than an open system could. The human circulatory system is just one variation on this basic theme (Fig. 15.5b).

15.3 The Human Circulatory System

It takes an immense network of circulatory "tubing" to carry the 25 trillion red blood cells that deliver gases and materials to your body's other 50 trillion cells (Fig. 15.5b-d). If all of the blood vessels in your body were stitched end to end, the single long pipeline would circle the earth twice! To keep the blood flowing through all these vessels, your heart must beat with a regular rhythm and maintain the flow at a pressure that is high enough to force blood into your brain, nose, toes, and all the tissues in between. These enormous tasks are possible because of both the anatomy and the activity of the blood vessels and heart (*E* xplorer 15.1). This vital system is also subject to many diseases and conditions, as we'll see later.

An Overview of Circulation

Let's start by exploring the pathways that blood takes through the human circulatory system. Blood leaving the right side of the heart passes to the lungs, organs that bring blood into contact with oxygen from the environment. Oxygen diffuses from the air in the lung sacs into the blood. Once the blood has picked up oxygen, it returns to the left side of the heart for further pumping. The left side of the heart then pushes the blood with enough force to deliver oxygen to every cell, tissue, and organ in the body. On this circuit, for example, the blood moves through the gut, where it picks up nutrients (details in Chapter 17); through the kidneys, where it dumps waste materials for excretion from the body (see Chapter 14 for details); to the muscles and brain, where it unloads oxygen and picks up carbon dioxide; to endocrine organs, where it picks up hormones (see Chapter 18), and so on. Eventually, the blood returns to the right side of the heart and begins another circuit.

This arrangement can be thought of as two separate circulatory loops. In the lung loop, or **pulmonary circulation,** the right side of the heart pumps blood directly to the lungs and back to the left side of the heart.

In the second loop, a body loop called the **systemic circulation,** the left side of the heart pumps blood to all organs except the lungs, and the blood returns to the right side of the heart. Because of this double loop arrangement, oxygen-rich and oxygen-poor blood never normally mix. Furthermore, the body can control the two circulatory systems somewhat independently, with the right side of the heart using less pressure to pump blood to the nearby lungs, and the left side of the heart using high pressure to reach all body tissues quickly. Not all vertebrates have this two-loop arrangement, as biologists have learned by studying the evolutionary origin of the circulatory systems in fish, reptiles, amphibians, birds, and mammals.

Blood Vessels

"Vessels" is a catch-all phrase for thick-walled tubes wider than your thumb, delicate pipelines much finer than a hair, and everything in between. All the vessels must contain fast-moving blood cells (or whatever substitute a researcher like Dr. Winslow can put into them instead). So what are our blood vessels like?

Blood moves away from the heart in **arteries,** vessels with thick, multilayered, muscular walls ranging in size from the diameter of a garden hose to the thickness of a pencil or less (Fig. 15.5c,d). The contraction of these wall muscles, along with the beating of the heart, helps keep blood under pressure—the force we call **blood pressure.** The pulse you can feel in your neck and wrists is actually sequential spurts of blood passing through arteries. Arteries branch and form smaller vessels called **arterioles.** These vessels are too small to be seen without a microscope, and the muscle layer surrounding them is thinner than in the arteries. Arterioles branch again into the dense, weblike network of delicate **capillaries,** minute vessels that permeate the fingertips, earlobes, lungs, liver, and all the tissues of the body. Capillaries are microscopic vessels only about 8 micrometers (0.0003 in.) in diameter, and rarely more than about 0.1 mm (0.004 in.) from any body cell. A capillary's diameter is so small that red blood cells can pass through only in single file. Capillary walls are just one cell layer thick, so they are also very delicate. Capillaries are so numerous, however, that they make up almost all of a person's blood vessels; if all capillaries were filled with blood at the same time, they could contain the entire 5 L (10.5 pints) of human blood. The capillaries' ultrathin structure is one key to the efficiency of the circulatory system, because materials can readily diffuse into and out of the narrow capil-

GET THE PICTURE

E xplorer 15.1

The Human Circulatory System

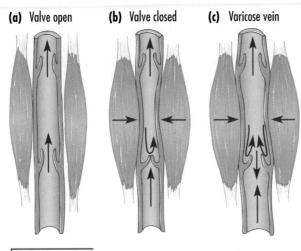

(a) Valve open **(b)** Valve closed **(c)** Varicose vein

Figure 15.6 **Valves Maintain a One-Way Blood Flow in Veins.** (a) When blood moves in one direction, it forces the valve open, and the fluid flows unimpeded. (b) If blood begins to flow in the opposite direction, the backward force pushes the valve flaps together and closes the valve. Thus, the valve in veins allows blood to flow in only one direction—toward the heart. (c) In a varicose vein, the valve is damaged and blood flows backward, causing the vein to distend and look very large and blue in a limb, usually the lower leg and foot.

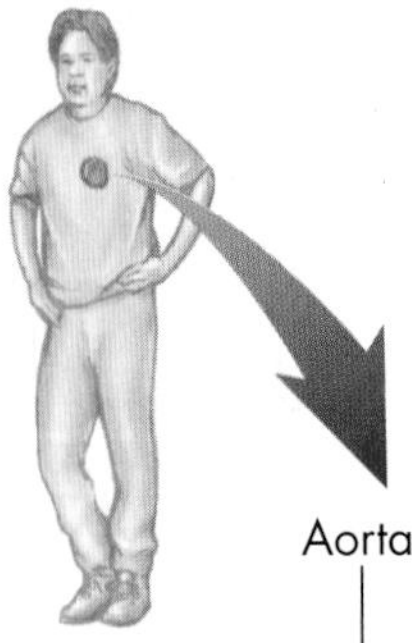

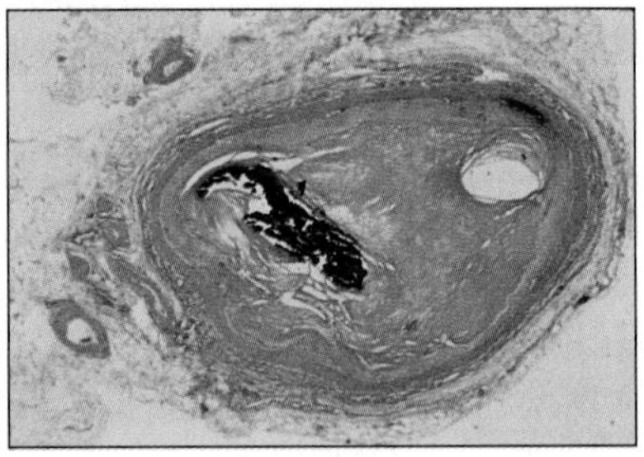

Figure 15.7 **The Pumping Heart.** (a) A human heart has large ventricles, and the coronary arteries carry a large blood supply to them. The inset (b) shows plaque build-up in a coronary artery (here it looks black), blocking almost all of the blood flow.

(a)

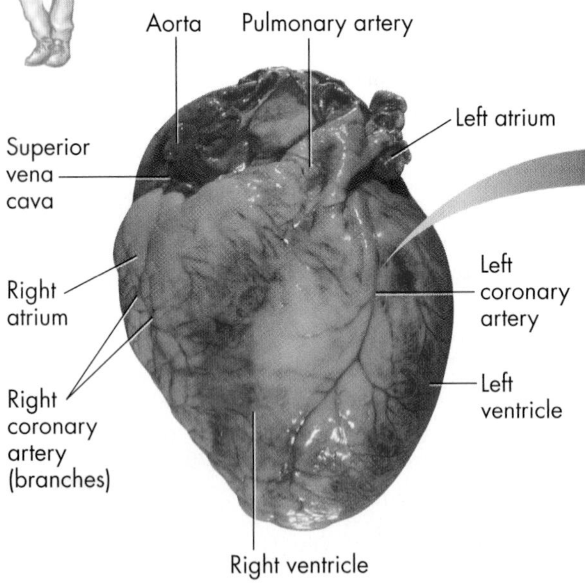

Aorta

Pulmonary artery

Left atrium

Superior vena cava

Right atrium

Left coronary artery

Right coronary artery (branches)

Left ventricle

Right ventricle

(b)

laries and the single cell layer that encompasses them. Any successful design for artificial blood must easily pass through these capillaries and leave them unharmed. Recall from the chapter opener how Dr. Winslow sent naked hemoglobin through animal's vessels without the enclosing "bags" represented by red blood cell membranes. The attempt failed horribly, in his view, because the capillaries constricted tightly around the proteins.

One of the most marvelous features of the vertebrate circulatory system is the **capillary beds,** networks that link the arterial and venous sides of the circulation (see Fig. 15.5c). The arterial side of each capillary bed conveys fresh blood (pictured in red) *away from the heart* and to the capillaries. Oxygen, nutrients, carbon dioxide, and metabolic wastes can move quickly across the capillary walls and into and out of the extracellular fluid and nearby tissue cells. The oxygen-depleted blood (shown in blue) then continues to move through the bed to the venous side; there the capillaries leading away from the tissues feed into larger vessels known as **venules,** which in turn merge and become veins. **Veins** carry blood *toward the heart,* where the circulation cycle begins again. Perhaps you can see the bluish veins close to the skin surface of your wrists. These blue vessels contain hemoglobin with less oxygen bound to it and this is not as bright red as oxygen-rich blood.

Blood in the veins is under lower pressure than blood in arteries. This is because blood in the veins has already traveled some distance from the heart and been slowed by passage through the narrow capillaries, and because veins have a larger diameter and less muscular walls which squeeze their contents less forcefully than artery walls (see Fig. 15.5d). This low-pressure fluid flows toward the heart's right side and does not flow backward or pool in the extremities owing to a system of **valves**—tonguelike flaps that extend into the internal space (or **lumen**) of the vein (Fig. 15.6). The heart has valves, too, as we'll see; these are similar to but distinct from the venous valves. Like a one-way door, when blood pushes a valve from one direction, it opens and allows blood to pass, but when blood pushes from the other direction, the valve slams shut. If venous valves become damaged, blood can flow backward, causing the vein to distend and become visible as a large blue bulb, a **varicose vein.** Whatever form artificial blood takes, it must not be unduly damaged by the slamming shut of the valves or the small eddies they create.

The Heart

At the physiological center of the circulatory system lies the **heart** (see Figs. 15.5a, and 15.7). Roughly the size of a large, lopsided apple, your heart pumps a teacupful of blood with every three beats, 5 L (5.3 qt) of blood every minute, and upwards of 7200 L of blood every day of your life. Over a lifetime, this amounts to 2.5 billion heartbeats and enough blood to fill a building six stories high and a city block long. Although medical engineers have tried valiantly, replacing the living circulatory pump has proved exceedingly difficult.

This tireless organ, like the hearts of all mammals, has four chambers divided into two pairs: the right atrium and right ventricle, and the left atrium and left ventricle. Each **atrium** (pl., *atria*) receives blood from veins. (In ancient Rome, an atrium was a roofless room with a pool to collect rainwater.) The more muscular chamber, the **ventricle,** receives blood from the atrium and pumps it through an artery to either the pulmonary or systemic circulation. Let's follow the pathway of blood through the heart, starting with the pulmonary circulation (Fig. 15.8). *E* 15.1 animates and describes this route:

Step 1. The right ventricle of the heart pumps oxygen-poor blood past a one-way valve, the **pulmonary semilunar valve,** with three half-moon-shaped flaps, into the Y-shaped **pulmonary arteries.**

Step 2. The pulmonary arteries carry oxygen-poor blood to capillaries in the lungs, where the blood picks up oxygen from inhaled air and unloads carbon dioxide into exhaled air.

Step 3. The freshly oxygenated blood flows to the **pulmonary veins** (the only veins in an adult person that carry oxygen-rich blood) and then passes to the heart's left atrium.

Step 4. The left atrium pumps the oxygenated blood through another one-way valve, the mitral or left atrioventricular valve, into the heart's left ventricle. This chamber sucks in much of the blood as it relaxes. (For a comparison, clench your fist tightly and then relax your fingers, and see how air is sucked into the space inside your fist.)

Step 5. The thick walls of the muscular left ventricle contract around this blood in a wringing motion until enough pressure develops to push open yet another one-way valve, the aortic semilunar valve, and squirt the blood into the **aorta**—the main artery leading to the systemic circulation. This body loop conveys blood to capillary beds in brain, muscles, kidney, and other distant tissues, then returns it via venules and veins that lead back to the right atrium again.

Step 6. The right atrium receives oxygen-poor blood from the body tissues via two large veins, the **superior vena cava** (Fig. 15.7) and the **inferior vena cava** (see Fig. 15.8). These vessels collect blood from the upper and lower body, respectively. Finally, the right atrium pumps its load of oxygen-poor blood into the right ventricle through a fourth valve, the petal-shaped tricuspid or right atrioventricular valve, which prevents blood from flowing back when a ventricle contracts. Again, the ventricle provides negative pressure that sucks blood into the cavity. The circuit is

Figure 15.8 **Pathways of Blood Flow.** The pulmonary circulation carries deoxygenated blood from the right ventricle (Step 1) through the pulmonary arteries (Step 2), through the lung capillaries where it picks up oxygen and turns bright red again, then back to the left atrium (Step 3). The systemic circulation takes oxygenated blood from the heart's left ventricle (Step 4), through the aorta (Step 5) to the body. In the capillaries of body tissue, oxygen is delivered and the now deoxygenated blood returns to the heart's right atrium (Step 6), ready for another circuit.

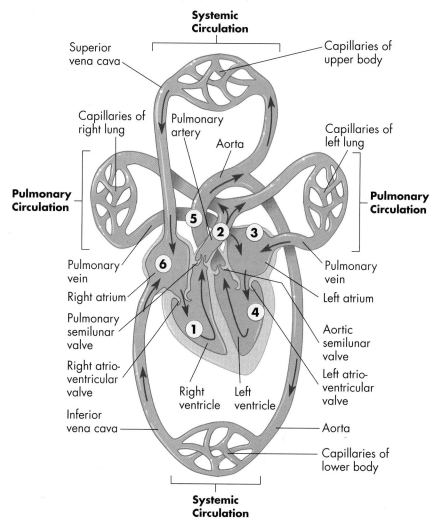

now complete, and as the right ventricle contracts, it pushes blood once more into the pulmonary circulation.

The Action of Heart Valves

Together, the four valves in the heart prevent blood from reversing its normal flow. The closing of valves makes the "lub-dub, lub-dub" sounds you can hear through a doctor's stethoscope or in a horror movie. As the ventricles begin to contract, the pressure of surging blood pushes the atrioventricular valves shut, and you hear a low-pitched "lub." Then, when the ventricles begin to relax, pressure in the aorta and pulmonary artery rapidly forces valves in both of these arteries to slam closed, producing the quicker, higher-pitched "dub." In some people, the heart's pumping strength can be diminished by the backwashing blood that escapes through weak or defective valves, and can sometimes be heard as a **heart murmur.** Heart surgeons can usually replace severely diseased valves with either valves from a pig's heart or a mechanical device.

How Does the Body Control Heartbeat?

Like the muscles in your arm, each of the heart's four chambers is made up of specialized cells organized into contractile fibers. In your arm, however, your biceps muscle can contract either weakly (e.g., when you pick up a pencil) or more strongly (when you pick up a book as heavy as this one), and the strength of contraction depends on what percentage of the muscle fibers contract at any given time. In your heart, however, all the fibers in the **cardiac** (heart) muscle contract with each heartbeat. Blood flow through the heart increases mainly by faster beating.

What accounts for this specialized "all or none system" so characteristic of heart muscle? First, heart muscle cells are linked electrically by special regions, the **intercalated** (in-TER-cal-ated) **disks.** Junctions (review Fig. 2.29) between cells in these disk regions pass electrical impulses instantaneously from muscle cell to mus-

cle cell, and these electrical signals stimulate muscle contraction. As a result, neighboring sections of the heart wall contract and relax together in a superbly coordinated pumping action that keeps blood flowing smoothly through the system.

Second, cardiac muscle contracts automatically—that is, without stimulation from the nervous system. (Nerves do help speed or slow the heart rate, however, as Chapter 19 describes.) Different subsets of heart muscle cells have different intrinsic rates of contraction. Some, called **pacemaker cells,** contract slightly faster than all the others, and because each contraction spreads quickly throughout a region of the muscle, pacemaker cells ignite contractions in the entire heart and set the rate of the heartbeat.

Pacemaker cells are located near the upper right atrium in a region called the **sinoatrial (SA) node** (Fig. 15.9a). Immediately before a beat, an electrical impulse spreads from the SA node across the walls of both right and left atria, causing the two chambers to contract in

(a) Signal from SA node: atria contract

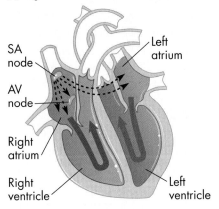

(b) Signal from AV node: ventricles contract

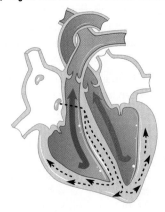

Figure 15.9 **Electrical Impulses Drive the Heartbeat.** (a) Pacemaker cells in the sinoatrial (SA) node send an impulse across the right and left atrial chambers, which respond by contracting. (b) Upon reaching the atrioventricular (AV) node, the signal is delayed for a split second, after which it reaches the ventricles and triggers their contraction, a squeeze that forces blood to the lungs and the rest of the body.

unison. A second node, the **atrioventricular (AV) node,** is another area of modified cardiac muscle cells located at the junction of the right atrium and right ventricles. When the electrical signal generated by the SA node reaches the AV node, there is a brief delay before the signal passes to the ventricles. Upon reaching the ventricles, the signal triggers a contraction of these chambers. The brief delay in the transmission of the pacemaker impulse at the AV node is necessary for an efficiently beating heart: If the atria and ventricles were to contract simultaneously, blood in the atria might flow back into the veins instead of forward into the ventricles.

People with irregular heartbeats can often be helped with an artificial pacemaker. This small electrical device, powered by a battery or by the decay of a small amount of radioactive material, sends rhythmic electrical impulses that stimulate the cardiac muscle to contract at an appropriate time.

Each time a heart's atria or ventricles contract, they are said to be in **systole** (SISS-toe-lee). The opposite, or relaxed, phase is known as **diastole** (die-ASS-toe-lee). Together, these two phases make up the **cardiac cycle**—the contraction-relaxation sequence of atria and ventricles that makes up a single heartbeat. As the ventricles relax in diastole, they suck blood from the atria into their chambers, a bit like a turkey baster drawing up juice.

Blood Pressure

When a person loses blood from a major wound, there is a disastrous decrease in blood pressure because the pressurized system has been breached. The main reason for giving blood transfusions in the battlefield or at an accident scene is to maintain blood pressure. Often, paramedics will simply infuse saline because fresh whole blood of the right blood type is not available in an ambulance. In a sense, saline was the first successful blood substitute because it helps restore blood volume and pressure. It doesn't carry much oxygen, however. One of Dr. Winslow's objectives in developing a hemoglobin-based artificial blood is to create a resource for transfusions in the field that will not only raise blood pressure but also carry needed oxygen. Normal blood pressure, generated by muscles in the walls of the heart and arteries, is substantial: blood spurts out more than 2 m (6.5 ft) high from a severed aortic artery! To deliver blood quickly and continuously to cells far from the heart, blood pressure must remain adequate; an extreme blood pressure drop can send a person into shock and actually be fatal in itself.

Blood Pressure and Disease

A healthy young adult male at rest has a typical blood pressure of about 120 mm Hg during systole and about 80 mm Hg during diastole. This is written as 120/80 and stated as "120 over 80." A nurse or doctor measures your blood pressure in millimeters of mercury (mm Hg); a pressure of 120 mm Hg could literally lift a column of mercury 120 mm high (4.5 in.). Young women and people who exercise regularly tend to have blood pressure readings lower than 120/80. Patients are considered to have high blood pressure or hypertension if their readings are persistently higher than 140/90. Hypertension is often called "the silent killer" because it can cause considerable damage to blood vessels long before it causes pain or disability.

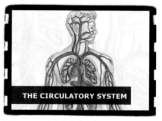

How can high blood pressure lead to even higher blood pressure?

High blood pressure causes local thickening in the smooth muscle cells surrounding blood vessels. You've probably heard of **arteriosclerosis**, a disease characterized by thickened artery walls (*sclero* = hard). In one form of this disease, called **atherosclerosis**, cholesterol accumulations lead to waxy deposits or **plaques** that build up inside the arteries and gradually obstruct blood flow (see Fig. 15.7b). Obstructing blood flow increases blood pressure the way tightening a nozzle boosts water pressure inside a running hose. To overcome the mechanical obstruction and resistance of plaque buildups, the heart's ventricles must work harder. If plaques accumulate in the heart's own arteries (the **coronary arteries**), the muscle can become oxygen-deprived and this can lead to **angina pectoris,** a condition of squeezing chest pressure and pain. Plaques in coronary arteries can attract platelets and these, in turn, sometimes cause blood clots to form. If a clot breaks off, it can lodge further down the artery's interior and completely cut off blood flow to a section of heart muscle. This clotting and sudden blockage is called a **heart attack** or **myocardial infarction,** and it can produce disability or even cause death. If the blood clot forms in a brain artery and cuts off blood flow to brain tissue, the event is called a **stroke,** and this can result in deficits of speech and/or movement.

People can prevent or at least manage hypertension by maintaining ideal weight, exercising several times per week, avoiding tobacco, reducing salt and alcohol consumption, and controlling stress. Certain drugs can also help, including diuretics to reduce blood volume (and hence pressure) and beta blockers to reduce heart contraction and blood pressure.

Blood pressure is obviously a key to cardiovascular health. So what do numbers like 120/80 actually mean? When a medical worker wraps an inflatable cuff around your upper arm and vigorously pumps in air, it increases the pressure in the cuff to, say, 200 mm Hg, and this constriction temporarily prevents blood from flowing into the main artery in your arm. The worker will listen through a stethoscope placed on the artery inside your elbow, but won't hear anything yet because the artery is closed by the pressure. The medical worker then releases air from the cuff until the pressure drops below that of the beating heart ventricle (say, 120 mm Hg). He or she can hear the artery open when the ventricle contracts, but then slaps shut from the force of the cuff when the ventricle relaxes. Next the worker releases more air from the cuff, dropping its pressure to below that of the ventricles, even at rest (say 80 mm Hg). Once again, he or she will hear no sound because the artery is permanently open. In this example, the patient's blood pressure is 120 over 80.

Blood traveling away from the heart has its high pressure (for example, 120) because of the heart's muscular pumping. As blood moves through the arteries, however, and into the voluminous capillary beds, the pressure dissipates (e.g. to about 40). This lowered pressure allows veins to serve as the body's blood reservoir, capable of holding as much as 80 percent of total blood volume at any given time. The movement of muscles in the arms, legs, and rib cage gently "milks" blood through individual veins, while the one-way valves in vessels in your extremities prevent backflow and keep the blood moving toward the heart. If no muscular milking takes place over a long period of time—for example, while you're sitting throughout a long plane ride—blood and other fluids accumulate in the extremities. Some scholars suggest that the death of crucifixion victims was due in large part to blood pooling in the legs.

Regulation of Blood Flow

Remember the big bluish protein molecules in tarantula blood? The huge bristly spiders have hemocyanin, an oxygen-carrying molecule dissolved directly in the blood and not contained within cells. Recall, too, that Robert Winslow and other researchers tried extracting hemoglobin from outdated blood and injecting it directly into test animals needing a blood transfusion. Remember his comment, too? "An unbelievable failure." As we saw, hemocyanin is a huge, multichain molecule; when the much smaller hemoglobin is removed from red blood cells and circulated directly in the plasma, blood vessels constrict and carry less blood. If blood can't circulate freely into the capillaries, then cells

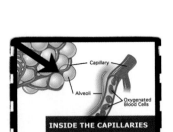

Why do some people's lips turn bluish on a cold day?

downstream starve for oxygen. Hence, the "failure." Let's look briefly at the body's strategies for regulating blood flow as a way to understand how free hemoglobin molecules derail the circulatory system.

From moment to moment, the distribution of blood varies in an animal's arteries, veins, and capillaries, depending on oxygen consumption. During vigorous exercise, for example, a higher volume of blood moves to your oxygen-starved muscles. Later, if you eat an energy bar, your intestines absorb nutrient molecules which, in turn, move into a suddenly increased blood supply in the digestive tract (see Chapter 17). When one region of the body, say, the leg muscles, uses oxygen more rapidly, then the walls of arterioles in other body regions contract and cut down blood flow. As we saw in Chapter 14, contraction of the vessel walls is called **vasoconstriction** ("vessel constriction"); it causes a decrease in the diameter of blood vessels and restricts blood flow while more blood moves to other regions. In the body regions requiring a greater blood flow, a reverse process, **vasodilation,** takes place. The diameter of the vessels increases, and more blood can move through.

Local blood flow is also regulated by **precapillary sphincters,** or rings of smooth muscle around the capillary's upstream origin (Fig. 15.10). When we get extremely cold, vasoconstriction and the closing of precapillary sphincters can decrease blood flow in capillary beds near the skin surface in the fingers or toes. The decrease in bright red oxygenated blood in these tissues can make the skin take on a yellowish or bluish cast (Fig. 15.10a). Likewise, eyedrops that "get the red out" contain chemicals that constrict the eyeball's arterioles so they appear less prominent. When we are very warm, say, during active exercise, vasodilation allows the capillary beds to become engorged, and the skin can feel hot and look red (Fig. 15.10b).

Several factors regulate the tension of arteriole muscles and so modulate vessel diameters. Among these are neurotransmitters (see Chapter 19); hormones, including the gas nitric oxide (see Chapter 18); and finally, oxygen itself. In artificial blood experiments, it became clear that hemoglobin circulating freely in the bloodstream must disrupt one or more of these controls. But which? Current evidence, says Robert Winslow, points to both nitric oxide and the oxygen supply.

Nitric oxide (NO) is a gaseous molecule containing one nitrogen atom and one oxygen atom. NO is produced locally by the cells that line the blood vessels (endothelial cells), and causes the smooth muscle fibers surrounding arterioles to relax and thus increases local blood flow. During sexual arousal, for example, cells

(a) Vasoconstriction

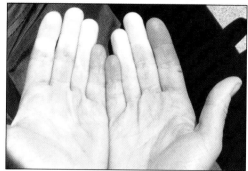

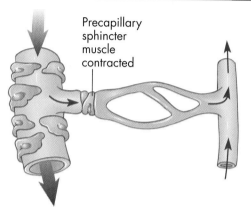

Precapillary sphincter muscle contracted

(b) Vasodilation

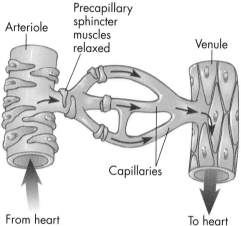

Arteriole

Precapillary sphincter muscles relaxed

Venule

Capillaries

From heart

To heart

Figure 15.10 **Evidence of Blood Flow.** (a) In the fingers of a person with Raynaud's syndrome, whitish, mottled regions are a sign of restricted blood flow or vasoconstriction. Raynaud's syndrome involves the excessive tightening of smooth muscle cells around arterioles. The condition is inherited, but researchers don't know many details about its mechanism. When those smooth muscles constrict—as they tend to do in cold weather—they diminish blood flow to the capillaries. (b) The full flush of heavy exercise comes from relaxed arteriole muscles, dilated vessels (or vasodilation), and increased blood flow to the skin, where excess heat can disperse to the environment.

lining blood vessels in the penis release NO, extra blood flows into the penis as a result, and this causes an erection. Experiments have shown that free hemoglobin mops up NO, and thus decreases its local concentration. Without it, the arteriole muscles can't relax and in fact tighten up further, decreasing blood flow to the tissue cells.

On top of this NO mechanism, free hemoglobin molecules carry more oxygen than the same molecules would inside red blood cells. Experiments show that an increased oxygen supply can trigger vasoconstriction. This prevents a local tissue area from getting too much oxygen delivered and not having enough carbon dioxide carried away. At this point, Dr. Winslow and fellow researchers are unsure which commodity—NO or increased oxygen—is mainly responsible for the vessel

constriction that takes place when free hemoglobin is used as a blood substitute. It is clear, though, that this approach won't work and that biologists will have to modify free hemoglobin in some way to develop a safe blood substitute.

Blood Clotting

A patient suffering drastic blood loss needs something else in addition to normalized blood pressure: a way to stop the bleeding. Fortunately, as you have probably experienced throughout your life, we have a natural system that plugs up wounds—a system biologists call the **blood clotting cascade.** In a cut vessel, several separate events take place that halt the flow of blood

Figure 15.11   The Formation of a Blood Clot: A Lifesaving Cascade. (a) When a wound severs a tiny blood vessel, numerous processes are set in motion, including vasoconstriction, the formation of a platelet plug, and then coagulation, or blood clotting. (b) A cascade of proteins acting on each other ultimately converts prothrombin to the enzyme thrombin. Thrombin cleaves a small piece from the fibrinogen protein, turning it into fibrin, and fibrin molecules line up and form a fiber. Fibrin threads then trap red blood cells in a blood clot.

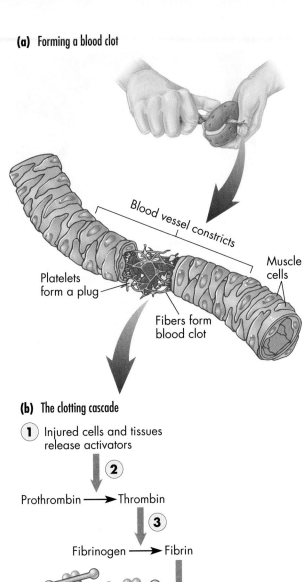

(a) Forming a blood clot

Blood vessel constricts

Muscle cells

Platelets form a plug

Fibers form blood clot

(b) The clotting cascade

① Injured cells and tissues release activators

②

Prothrombin ⟶ Thrombin

③

Fibrinogen ⟶ Fibrin

Fibrin monomer

Peptides

④

Cross-linked fibrin strand

(Fig. 15.11a): (1) Smooth muscles in the walls of the damaged vessel contract and partly close off the vessel. (2) Platelets, the circulating cell fragments we saw earlier, react by releasing a hormone, **serotonin,** that keeps the muscles contracting. (3) Platelets at the injured site begin sticking to each other and to rough surfaces at torn edges of the wound, forming a plug. (4) The next stage is **coagulation,** the actual formation of a blood clot.

The formation of a blood clot (coagulation) involves a series of steplike changes in blood proteins (Fig. 15.11b). First, injured cells release activators (Step 1), which, in a cascade of reactions, stimulate the conversion of an inactive protein (**prothrombin**) into an enzyme (**thrombin**) (Step 2). This enzyme transforms **fibrinogen** (a protein that normally circulates dissolved in the plasma) into tough, insoluble threads of a related protein, **fibrin** (Step 3). Fibrin threads in turn become woven into a strong, wiry mesh that traps red blood cells, creating a blood clot (Step 4).

People with **hemophilia** lack a functional copy of a gene encoding a protein in the clotting cascade. As a result, they do not form normal blood clots. Unless a person with hemophilia receives periodic intravenous injections of solutions containing the normal protein, even a minor wound can lead to major blood loss and possibly death. Clotting is another factor Robert Winslow is investigating in his research designs and products. A successful blood substitute must allow normal clotting but not stimulate unwanted clots. These could clog blood vessels in vital organs such as the heart, causing a heart attack (infarct), or in the brain, causing a stroke. In fact, quickly administering clot-dissolving enzymes to heart attack or stroke victims can dramatically reduce damage to the heart or brain.

The Lymphatic System

Our human circulatory system and that of many other vertebrates is so highly pressurized that it forces fluid from the blood plasma through the thin walls of capil-laries. This fluid—water plus dissolved plasma components—then accumulates in the spaces between cells. A second system of fluid-containing vessels, the **lymphatic system,** drains this squeezed-out fluid from extracellular spaces and returns it to the bloodstream (Fig.

(a) The lymphatic system

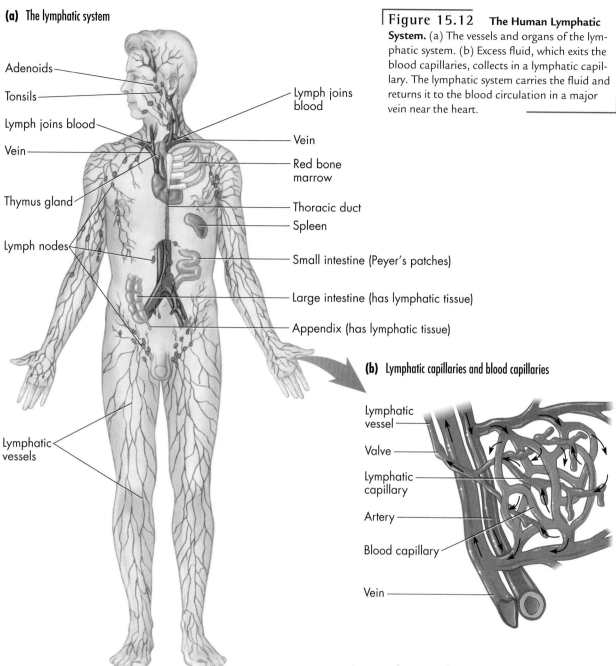

Adenoids

Tonsils

Lymph joins blood

Vein

Thymus gland

Lymph nodes

Lymphatic vessels

Lymph joins blood

Vein

Red bone marrow

Thoracic duct

Spleen

Small intestine (Peyer's patches)

Large intestine (has lymphatic tissue)

Appendix (has lymphatic tissue)

Figure 15.12 **The Human Lymphatic System.** (a) The vessels and organs of the lymphatic system. (b) Excess fluid, which exits the blood capillaries, collects in a lymphatic capillary. The lymphatic system carries the fluid and returns it to the blood circulation in a major vein near the heart.

(b) Lymphatic capillaries and blood capillaries

Lymphatic vessel

Valve

Lymphatic capillary

Artery

Blood capillary

Vein

15.12a). Lymphatic vessels also play a major role in the immune system (see Chapter 16).

Unlike the blood vessels, the lymphatic vessel system consists of capillaries and larger vessels that do *not* form a totally enclosed circulatory loop. Instead, lymphatic capillaries are minute tubes that are closed at one end and siphon off excess fluid from the body tissues. After seeping through the thin walls into the lymphatic capillaries, the fluid is called the **lymph.** Lymph contains mostly water, but also white blood cells, salts, proteins, dead cells, and sometimes invasive microorganisms, such as bacteria and viruses. Lymphatic capillaries merge, like creeks joining to form a stream, and streams combining to form a river (Fig. 15.12b), and finally, two major lymphatic vessels emit a surge of lymph into veins near the heart. The lymph fluid is then merged once again with the plasma portion of circulating blood.

Bean-shaped filtering organs called **lymph nodes** prevent debris in the lymph from mixing into the blood (Fig. 15.12a). As lymph moves along a lymph vessel, it percolates through the nodes, like oil passing through an oil filter, and the node filters out dead cells and other debris. The nodes also harbor large numbers of infection-fighting **lymphocytes** (a type of white blood cell), which help protect the body from bacteria and other foreign materials. You've probably experienced painful, swollen lymph glands in the neck, underarms, or groin area during a bout of infection; the swelling indicates that these organs have increased their filtering activities. Unfortunately, just as lymph vessels can carry dead cells away from an infection site, they can also transport migrating cancer cells to new sites. This is why doctors often remove lymph nodes and vessels along with cancerous tissue in cancer patients.

Like veins, lymphatic vessels transport fluids under low pressure. Valves keep the lymph flowing in one direction, and the squeezing force of contractions in muscle tissue surrounding the lymph vessels propels the fluid. A person with inactive muscles, say, an injured skier in a hospital bed, can suffer lymph accumulation, which results in swelling or **edema** (eh-DEE-muh).

The lymphatic system also has two organs that are somewhat separate from the network of vessels. The soft, V-shaped **thymus** located at the base of the neck in front of the aorta plays a role in the body's immunity (Chapter 16). The **spleen,** about the size and shape of a banana in an adult, lies just behind the stomach. The spleen filters blood, produces phagocytic (debris-gobbling) white blood cells, and stores platelets and some blood cells. Despite these functions, surgeons can remove a person's injured spleen without causing any severe consequences to overall health because the bone marrow and liver can perform essential spleen functions. We saw in Chapter 11 that the spleen will collect red blood cells infected by the malaria parasite. The spleen also collects and destroys the hemoglobin in most blood substitutes so far developed. A major effort of researchers like Robert Winslow has been to increase the time that a blood substitute remains active in the body before the spleen takes it out. We'll see later how he dealt with this issue.

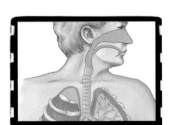

How can you tell tracheae from trachea?

15.4 The Respiratory System

Blood transports many important substances around the body, with oxygen being the most "time sensitive." Any successful blood substitute would have to do the same. Nature has provided us with an elegant system for picking up needed oxygen and eliminating carbon dioxide wastes. Let's explore the structure and activity of the respiratory system to examine the challenges of gas exchange and the evolutionary solutions that have arisen in different animal groups.

Why Do Animals Have a Respiratory System?

Think back to Chapter 3 for a moment and to the cell's harvest of energy from food molecules. In the process of cellular respiration, enzymes in an organism's mitochondria remove energized electrons from food molecules, passing them to electron carriers and releasing carbon dioxide as a waste product (review Fig. 3.15). The carbon dioxide produced this way diffuses from the cell and the electron carriers pass electrons to mitochondrial proteins. These proteins eventually pass the electrons to oxygen gas (along with hydrogen ions) to oxygen gas, and water molecules form. Cells use the energy released by electron transport to make ATP, the cell's energy currency (see Fig. 15.13). Cells in a large complex animal like a human take up oxygen from the watery fluid that surrounds the cell and also dump carbon dioxide there. Where does the extracellular fluid get its oxygen? From a group of organs specialized to extract oxygen from the environment: the respiratory system.

Biologists use "respiration" in two ways. We just reviewed cellular respiration. In contrast, **organismal respiration** is the way a whole animal exchanges carbon dioxide and oxygen with the atmosphere. We mammals respire when we breathe, but other animals have other solutions.

Respiration and Diffusion

We saw in Chapter 14 that large animals rely on systems of tubes to bring oxygen into the complex multicelled body and dump carbon dioxide out. They rely on different tubes—the circulatory system—to transport the gases to and from each body cell. And the two systems interact (review Fig. 14.1). Less complicated animals have the same needs but simpler solutions. Take a caterpillar (a moth larva), for example (Fig. 15.13a). Atmospheric gases enter and leave the body through **spiracles,** portholes that open onto branching tubes called **tracheae** (TRAY-key-ee). At the farthest tips of those tubes deep inside the insect's body, oxygen directly enters the surrounding fluid, and from there diffuses to surrounding tissue cells (Fig. 15.13b). The oxygen passes across the plasma membranes, enters the cells' cytoplasm, and moves to the mitochondria, where oxygen takes part in cellular respiration (Fig. 15.13c). In each step, the oxygen passes by diffusion, the spontaneous

Figure 15.13 **Gas Exchange at the Organismal, Cellular, and Molecular Levels.** (a) In a caterpillar, little openings called spiracles allow air to enter the body. (b) Air tubes called tracheae penetrate deep into the body. Oxygen from the air they carry can then diffuse into tissue fluids. At the cellular level, oxygen diffuses into the cell and through the cytoplasm, and carbon dioxide diffuses in the other direction, each moving down their concentration gradients. (c) Mitochondria produce carbon dioxide as they dismantle organic molecules and utilize oxygen as the final acceptor of electrons in the electron transport chain.

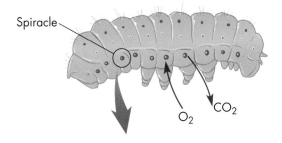

(a) A silkmoth caterpillar's spiracles

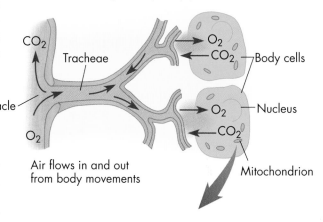

(b) The silkmoth caterpillar's respiratory system

Air flows in and out from body movements

migration of a substance from a region of higher concentration to a region of lower concentration.

Diffusion underlies gas exchange at the subcellular, cellular, and sometimes organismal levels. The need for gas exchange via diffusion has powerfully influenced how the shapes and functions of respiratory structures evolved:

1. Because water generally surrounds cells, gases must reach cells by passing through a layer of liquid.
2. Diffusion is much slower in water than in air—an amazing 300,000 times slower. As a consequence, each animal cell must be close to a source of oxygen and a sink for carbon dioxide so that gases can diffuse to and from each cell quickly. Flatworms don't have a respiratory system because their bodies are so thin and flat (review Fig. 13.9) that the "deepest" interior cells are never more than about 0.5 mm from the surface. Oxygen and carbon dioxide therefore can diffuse quickly and directly between cells and the animal's moist surroundings.
3. The greater the exposed surface for exchange, the more gas can diffuse between a liquid and the atmosphere, or between two liquids. If you hang wet laundry on a line, for example, and expose a large surface area to the atmosphere, the water will evaporate faster than if you leave the laundry in a heap with a small surface area-to-volume ratio.
4. Diffusion rates speed up or slow down depending on the gas concentrations and gas pressures in the liquid and in the surrounding atmosphere. Returning to the laundry analogy, water leaves the liquid phase in the clothes and enters the atmosphere much faster if the air is dry than if the humidity is high.

These four principles of diffusion have influenced the shape and activities of animal bodies, in the flatworm and tarantula as well as in aquatic and land vertebrates. Let's look at solutions for gas exchange in fish, frogs, birds, and humans.

Strategies for Gas Exchange

Fish, tadpoles, and many other animals that live in the water have **gills:** organs specialized for gas exchange that develop as outgrowths of the body surface. A developing fish has several tunnels connecting the inside of the mouth with the external environment. A fish's gills form from bars of tissue in between these tunnels (Fig. 15.14a). Each gill bar is subdivided into hundreds of flexible **gill filaments,** which, in turn, support many thin, platelike structures. These delicate plates, or **lamellae,** hold the key to how a large, active, aquatic animal can respire efficiently in water, even though most lake, river, or ocean water holds only one twentieth as much oxygen as air. Embedded within each lamella is a lacy meshwork of capillaries lying just one cell layer away from the water that passes through the gill filament. This proximity of blood to oxygen-bearing water means that oxygen readily diffuses across the cells into capillaries, while carbon dioxide easily diffuses outward. The pumping heart then circulates the oxygen-rich blood throughout the animal's body.

The steady opening and closing of a fish's mouth pumps a constant stream of water through the mouth to the gills, over the gills, then out through a flap (the operculum) that protects the gills (Fig. 15.14a). This water flow moves in the opposite direction to blood flow within the capillaries inside the gill's lamellae (Fig. 15.14b). The result is a **countercurrent exchange,** in which two fluids with

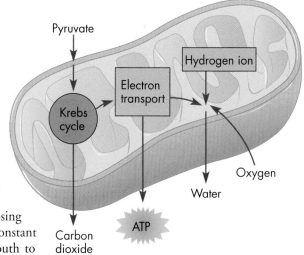

(c) Mitochondrion: the source of carbon dioxide and the sink of oxygen

(a) Water flow

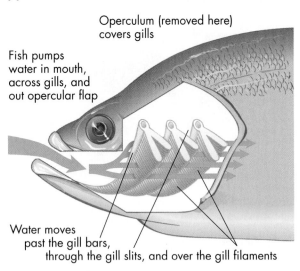

Operculum (removed here) covers gills

Fish pumps water in mouth, across gills, and out opercular flap

Water moves past the gill bars, through the gill slits, and over the gill filaments

Figure 15.14 Fish Gills, Water Flow, and Countercurrent Exchange Allow Oxygen Harvest from the Water. Water flows past delicate gill structures (a). The functioning of gills (b) allows oxygen to diffuse from water into the circulating blood. (c) A countercurrent exchange explains the diffusion of oxygen from water to air.

(b) Gill function

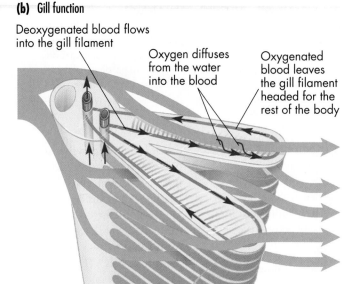

Deoxygenated blood flows into the gill filament

Oxygen diffuses from the water into the blood

Oxygenated blood leaves the gill filament headed for the rest of the body

(c) Countercurrent exchange

Water and blood flow in opposite directions

Blood

Oxygen

Lamellae

Water

Thus, along the entire route, the concentration of oxygen (red dots) in the water is always a bit higher than in the blood

different characteristics (here, different oxygen contents) flow in opposite directions. As they flow, materials are exchanged along all their points of contact owing to diffusion from high to low concentrations. As a result, the animal can collect enough oxygen for aerobic cellular respiration and an active lifestyle even though water holds so much less oxygen than air.

We land animals have access to the higher oxygen content of air, but we face a different threat: the drying out of our respiratory surfaces, which would bring gas exchange to a lethal halt.

Respiratory Adaptations in Land Animals

The only animals that live surrounded entirely by dry air are terrestrial arthropods and mollusks, such as spiders and snails, and land vertebrates, including reptiles, birds, and mammals. Not coincidentally, the arthropods and vertebrates have evolved specialized internal respiratory channels. We saw how a spider's trachaeae bring oxygen deep into its body. The standard solution in air-breathing vertebrates is, of course, lungs.

A few species of fish that live in the warm, stagnant, oxygen-poor water of swamps have evolved lungs, blind-ended internal pouches that connect to the outside by a hollow tube (Fig. 15.15a). These swamp fishes get a meager amount of oxygen through their gills, and the lungs supplement it. In many fishes, the lung pouch loses its connection to the exterior and serves not as a true lung, but as a **swim bladder,** a bag of gas that

helps the fish maintain its depth in the water without sinking or floating.

In an amphibian, such as a frog (Fig. 15.15b), the lungs are simple sacs with walls that are richly endowed with a dense lacework of blood capillaries. The uncomplicated baglike lungs of amphibians have far less surface area for gas diffusion than the convoluted lungs of reptiles, birds, and mammals; however, gases can also diffuse through an amphibian's moist skin, which supplements the lungs so that the animals can obtain enough oxygen to support their way of life.

Figure 15.15 **Pouches and Air Exchange.** (a) Fish develop outpocketings from the digestive tract. If these lose their connection to the exterior, they become swim bladders, organs that hold gases and control the animal's buoyancy in the water. If the outpocketings maintain their connection to the exterior, they become lungs that can help the animal exchange gases between the atmosphere and the blood. (b) Amphibians have baglike lungs with relatively little interior surface area for gas exchange. (c) Birds have air sacs that interconnect with the lungs and lighten the animal for flight. A bird takes two breaths to move air completely through its respiratory system. The first breath draws air into the posterior air sacs and then into the lungs. The second breath draws in more air, and this pushes air from the lungs, into the anterior air sacs, and then out of the body. This one-way flow provides a high level of oxygen to the bird's actively contracting wing muscles.

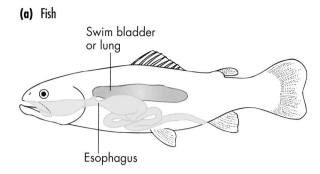

(a) Fish

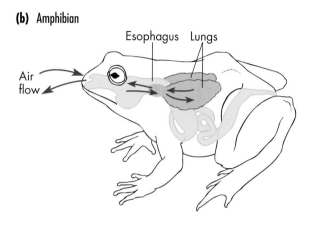

(b) Amphibian

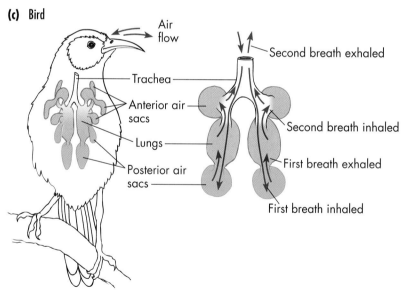

(c) Bird

In contrast to amphibians, air flows through a bird's more complicated set of air sacs and tubes in a one-way path (Fig. 15.15c). Because this one-way flow through the lungs prevents the mixing of fresh and "stale" air, birds can sustain extremely high levels of activity for much longer periods than we mammals can manage. Birds can even flap actively over the top of Mt. Everest while most humans standing still at the peak need oxygen tanks to avoid passing out.

The Human Respiratory System

Most mountain climbers attempting to scale Mt. Everest carry supplemental oxygen (and never reach the summit, anyway). A handful of people, however, have climbed to the top breathing on their own. Swedish climber Göran Krupp, for example, rode a bicycle from his home more than 7,000 miles to Kathmandu, Nepal. He proceeded to Everest base camp with 150 pounds of gear in his backpack and with no assistance from hired Sherpas. He then trudged to the summit without an oxygen tank, descended once again to Kathmandu, then pedaled his bike back home to Sweden. The human respiratory and circulatory systems can obviously adapt to low-oxygen environments—even extreme ones. In working on blood substitutes, Dr. Winslow wondered how the respiratory system could manage such feats of adaptation. Being a climber himself, he travelled to Mt. Everest with a group in 1981 and performed a number of experiments on his companions while on or near the mountain's summit to find answers (Fig. 15.16). He learned that the amount of hemoglobin in the blood is crucial but that the way the molecule picks up and re-

leases oxygen is also critical to survival at high altitude and under other extreme conditions. This background helped him in his efforts to design a better blood substitute, as we'll see later.

Birds and fish have the benefit of a unidirectional current of air or water moving past their gas exchange surfaces, but we don't and neither do other mammals.

Figure 15.16 **High Mountain Feats.** Robert Winslow conducted physiological studies on Mt. Everest climbers in 1981. Virtually all such climbers required supplemental oxygen.

Instead, our lungs operate more like an amphibian's with an in-and-out air flow or **tidal ventilation,** a bit like the ebb and flow of the tides. The air we breathe in travels through an inverted tree of hollow tubes leading into the lungs. There, gases are exchanged across a thin, moist membrane before the air moves out again through the same branching set of tubes. Usually, a quantity of air (about 500 ml, or a pint, in humans) is inhaled and exhaled in a regular rhythm. This tidal pattern means that fresh air enters the lungs only during half of the respiratory cycle and that a quantity of unexpelled, stale dead air (the so-called residual volume) filled with carbon dioxide remains in the lungs and airways at all times, mixing with the fresh air that enters from outside. (Imagine the amount of dead air a giraffe has to clear from its six foot long windpipe!) Nevertheless, we have special adaptations that increase the rate of gas exchange and compensate for an in-and-out ventilation system, as **E**xplorer 15.2 shows.

GET THE PICTURE

E*xplorer 15.2*

The Human Respiratory System

Respiratory Passageways for Air Flow

When a mammal breathes in, air enters the respiratory system through the nose and sometimes through the mouth. The moist cavities of the mouth or nose warm and humidify the air, and open toward the back (posteriorly) into the **pharynx** (FAIR-inks) or throat (Fig. 15.17a). The pharynx branches into a pair of tubes; one, the **esophagus** (ih-SOFF-uh-gus; Greek, gullet), leads to the stomach. The other, the windpipe, or **trachea** (TRAY-key-uh) (not the *tracheae* of spiders and insects), is the airway leading into the lungs. At the forward (anterior) end of the trachea lies the **larynx** (LAIR-inks), or voice box, housing the vocal cords. Just above the opening to the larynx is a flap of tissue called the **epiglottis** (epi, above, glot-, language), which normally shields the larynx during swallowing and prevents food from accidentally entering the lungs.

A few centimeters below the human larynx, the trachea branches into two hollow passageways called **bronchi** (sing., *bronchus*), each of which enters a lung. Finer and finer branchings of these tubes create an inverted tree with thousands of narrowed airways, or **bronchioles.** These eventually lead to millions of tiny, bubble-shaped sacs called **alveoli** (al-VEE-oh-lie; sing., *alveolus*) where gas exchange actually takes place (Fig. 15.17b). A moist layer of epithelial cells lines each alveolus, and blood capillaries surround this layer. Human lungs contain roughly 300 million alveoli, and if the linings of all these delicate bubbles were stretched out simultaneously, they would occupy about 70 square meters—enough surface area to cover a badminton court, or 20 times the body's entire skin surface! Oxygen diffuses readily out of the alveolus and into red blood cells that squeeze down through the center of the nearby capillaries (Fig. 15.17c). Meanwhile, carbon dioxide leaves the blood, diffusing out of the capillary and entering the alveolus. From the alveolus, carbon dioxide is expelled to the outside with the next exhalation. In the medical condition called **asthma** (AZ-ma), normal exhalation and inhalation can be inhibited. This blockage can be due to inappropriate contractions of the smooth muscle cells that wrap the smaller bronchi and bronchioles; to increased mucus secretion; and to swelling of the tissues surrounding the airways. Allergies to pollen, molds, or house mites can also trigger asthma attacks. Drugs that reduce inflammation (see Chapter 16) can help open the airways during such attacks.

Cells lining the larger airways produce a sticky mucus ideally suited to capturing any dirt particles or microorganisms we accidentally inhale. Brushlike cilia on cells lining the airways continuously clear this mucus from the bronchi and sweep any trapped debris up toward the pharynx, where they can be swallowed or expelled (Fig. 15.17d). Inhaled tobacco smoke damages the cells lining the alveoli and paralyzes the cilia in the airways. The smoke from a single cigarette can immobilize the cilia for hours and lead fairly quickly to a hacking smoker's cough—the respiratory system's attempt to rid itself of airborne garbage that accumulates because the cilia no longer sweep it out (Fig. 15.17e). Long-term abuse of the airways can lead to a near-total breakdown of the system. As cilia break down and other natural defenses are overwhelmed, it becomes increasingly likely that genetic changes in lung cells will go unrepaired and that the lungs will develop cancer or **emphysema,** a degenerative disease in which the alveoli

(a) Human respiratory system

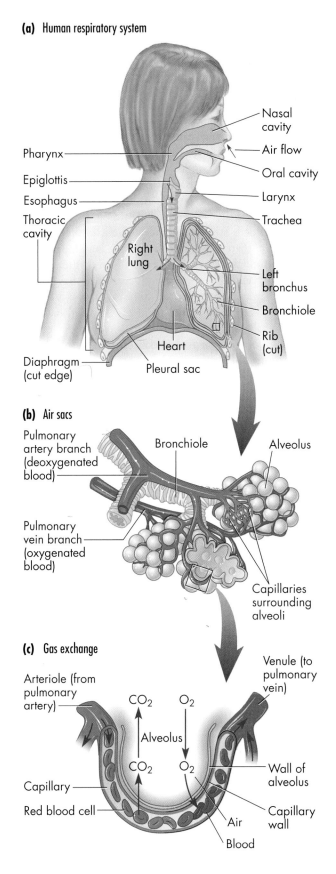

Nasal cavity

Pharynx

Air flow

Epiglottis

Oral cavity

Esophagus

Larynx

Thoracic cavity

Trachea

Right lung

Left bronchus

Bronchiole

Heart

Rib (cut)

Diaphragm (cut edge)

Pleural sac

(b) Air sacs

Pulmonary artery branch (deoxygenated blood)

Bronchiole

Alveolus

Pulmonary vein branch (oxygenated blood)

Capillaries surrounding alveoli

(c) Gas exchange

Venule (to pulmonary vein)

Arteriole (from pulmonary artery)

CO_2 O_2

Alveolus

CO_2 O_2

Wall of alveolus

Capillary

Red blood cell

Air

Capillary wall

Blood

Figure 15.17 **The Human Respiratory System.** (a) Anatomy of the airways and lungs. (b) Close-up view of the branched passageways, or bronchioles, and the cluster of tiny air sacs, or alveoli. (c) The diffusion of oxygen and carbon dioxide between the alveoli and the blood passing in a nearby vessel. (d) Hairlike cilia protruding from the cells that line the trachea and bronchioles sweep mucus and debris from the respiratory passageways. (e) In a smoker's lungs, the cilia are paralyzed by cigarette smoke and the lung turns black from accumulated tar. These lungs also show signs of emphysema.

(d) Cilia line the airways

(e) Smoker's lung

steadily deteriorate. Also, the tars and gases in cigarette smoke damage blood vessel walls and increase the chances of heart attack or stroke. Studies show that 4 out of every 10 smokers—approximately 400,000 Americans each year—die as a direct result of their smoking habit. This makes smoking by far the leading cause of preventable death each year. Even breathing the smoke from someone else's cigarettes can have serious health consequences. In fact, passive smoking kills about 53,000 Americans each year, making it the third leading cause of preventable death after direct smoking and alcohol abuse. Children raised around smokers are more likely to have respiratory and blood-vessel problems. *E*xplorer 15.3 delves further into the link between smoking, lung cancer, and heart disease.

Ventilation: Moving Air Into and Out of Lungs

Healthy lungs look like pink, spongy, deflated balloons. A fluid-filled **pleural sac** surrounds each lung, which

READ ON

*E*xplorer 15.3

Heart Disease, Smoking, and Other Assaults on Circulation and Respiration

hangs in the **thoracic cavity,** the region within the rib cage, over a domed sheet of muscle, the **diaphragm** (Fig. 15.18a). The fluid surrounding the lungs is under lower pressure than the air inside, and this pressure difference enables the lungs to remain slightly expanded even when no air is being taken in. A collapsed lung results when a pleural sac is punctured and the fluid drains away or air enters.

Several special adaptations we mammals have give us a powerful bellowslike air intake. Part of the intake depends on anatomy. The **ventilation** of the lungs is possible because the dome-shaped diaphragm muscle pulls down towards the stomach during **inhalation** or the drawing in of fresh air (Fig. 15.18b). You can feel this if you place both hands on your chest with a couple of fingers on ribs and the other on your abdomen, then breathe in. **Exhalation,** the passive release of air from the lungs, occurs when these steps are reversed (Fig. 15.18c).

Control of Ventilation by the Brain

Let's return to Dr. Winslow's experiments on Mt. Everest and the question of how climbers like Göran Kropp can scale the highest peaks without oxygen tanks. Dr. Winslow found that climbers like Kropp can extract oxygen more efficiently from the air than most people. They have what's called an enhanced hyperventilatory response: When operating in an oxygen-poor atmosphere, they tend to breathe much more deeply and rapidly than normal people. This is due partly to differences in the way their bodies control breathing.

In each of us, respiratory control centers in the brain help determine when and how we breathe. For example, the diaphragm and the muscles between ribs (the **intercostal muscles**) respond to nerve impulses generated by the **medulla oblongata.** This region in the brain connects with the spinal cord (see Fig. 15.18a). When one set of nerve cells in the medulla oblongata fires, the breathing muscles contract and we inhale. During normal, relaxed breathing, these inspiratory nerve cells stop firing after about two seconds, allowing the diaphragm and intercostal muscles to relax and us to exhale. During panting, however, the inspiratory nerve cells signal expiratory nerve cells, they trigger stronger muscle contraction, and exhalation is more forceful. We can, of course, regulate our breathing consciously to some extent, and that's because nerve cells located in other parts of the brain can exert control over the medulla oblongata's respiratory centers.

Try holding your breath for the count of ten seconds. Cessation of breathing has two effects on the body: it decreases the blood level of oxygen, and at the same time increases the blood level of carbon dioxide. Which gas does the body monitor to control breathing? Biologists have known the answer ever since 1875 when three French physiologists found out through a dramatic and disastrous experiment on themselves. The three ascended to a high altitude in a hot-air balloon equipped with bags of pure oxygen to be used as needed. Their balloons passed the 7500 m (24,600 ft) level (about 1000 m lower than the summit of Mt. Everest), but they felt no need to use the oxy-

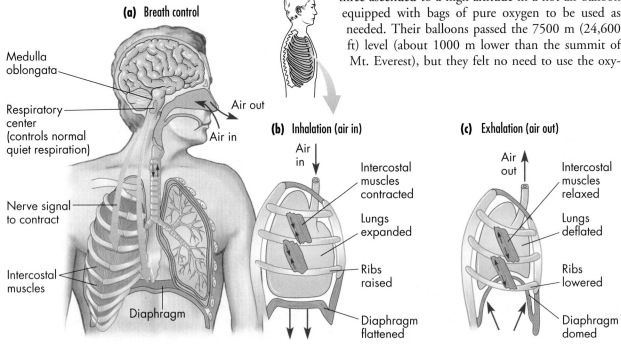
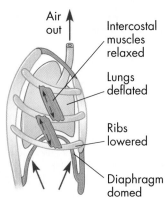

(a) Breath control

Medulla oblongata

Respiratory center (controls normal quiet respiration)

Air out

Air in

Nerve signal to contract

Intercostal muscles

Diaphragm

(b) Inhalation (air in)

Air in

Intercostal muscles contracted

Lungs expanded

Ribs raised

Diaphragm flattened

(c) Exhalation (air out)

Air out

Intercostal muscles relaxed

Lungs deflated

Ribs lowered

Diaphragm domed

Figure 15.18 **Control of Breathing.** (a) Nerve cells in the medulla, a region in the brainstem, stimulate the rib muscles to contract. (b) As a result, the diaphragm flattens out, the lungs fill with air, and we inhale. (c) During exhalation, the intercostal muscles relax, the diaphragm relaxes into a dome shape, the ribs move closer together, the chest capacity drops, and air is expelled.

gen. One of the scientists recorded—in an oddly scrawled handwriting—that they were feeling no ill effects at all. When the balloon finally descended, two of the three were dead. The survivor's sad conclusion: The human body is very poor at sensing its own need for oxygen when deprived of it at high altitude.

In contrast, the brain readily senses an increase in carbon dioxide. If a person again and again rebreathes a small volume of air in a stuffy room, the concentration of carbon dioxide gradually rises, and this is detected by sensors in the brain's medulla oblongata and in chemical receptors in the aorta and the carotid arteries. (These receptors also respond to low oxygen levels.) The sensors notify the brain's respiratory center that carbon dioxide levels have increased, and the brain deepens and speeds the person's ventilation, reestablishing the blood's optimal oxygen-carbon dioxide balance. Sensors in the muscles and joints can also detect body movements, and they trigger the brain to increase breathing rates even before carbon dioxide levels in the blood begin to change. The Göran Kropps of the world obviously have very sensitive carbon dioxide sensors and an exaggerated response to the "low oxygen" signal.

Most of the rest of us tend to show periodic breathing when traveling above 9000 ft: our breathing grows rapid and increasingly deep then is followed by shallower breathing that eventually stops completely for up to 10 seconds. This pattern, repeated over and over, is apparently based on the body's attempt to balance two conflicting situations. Rapid breathing at high altitude brings more oxygen into the blood, but expels huge quantities of carbon dioxide. The same is true when you blow on the coals of a campfire or inflate an air mattress with very rapid, deep breathing or **hyperventilation**. The decrease in carbon dioxide causes the blood to become more alkaline. Recall from Chapter 4 that carbon dioxide and water combine and form carbonic acid, which breaks down to bicarbonate and hydrogen ions that acidify the blood. When carbon dioxide concentrations are low, on the other hand, including after hyperventilation, the blood becomes more alkaline. Apparently, when you temporarily stop inhaling at high altitude, carbon dioxide builds up and this helps restore the blood's normal pH.

Oftentimes, people who normally live and work at low altitude ascend too rapidly to 10,000 feet or more—whether by foot or by vehicle—and can get **mountain sickness.** The heart and lungs suddenly have to work extra hard because of the "thin air" (low oxygen pressure) at high altitude. The result can be a headache, lethargy, dizziness, difficulty sleeping and loss of appetite. In severe cases, the traveler can become uncoordinated, off-balance, confused, drowsy, nauseated, and irrational due to swelling around the brain, or can suffer fluid buildup in the lungs and fall into a coma. Studies have shown that a drug called acetazolamide (Diamox) eases the symptoms of mountain sickness. This drug inhibits the action of carbonic anhydrase, the enzyme that helps convert the carbon dioxide in blood into bicarbonate. With the enzyme blocked by the drug, blood levels of carbon dioxide stay higher, and people breathe faster. Robert Winslow tested the effectiveness of Diamox on four climbers at the 20,400 foot level of Mt. Everest. He found that the climbers did breathe faster on Diamox and this helped confirm the body's mechanism for controlling breathing rate. Winslow even dragged a stationary bicycle up Mt. Everest to test the climber's physical performance while on Diamox, but found that it was no better than without the drug. He concluded that the drug can help prevent acute mountain sickness, but it doesn't improve a person's climbing performance.

Engineering a Blood Substitute

Robert Winslow has a unique perspective on both respiration and circulation; his high altitude studies have expanded and informed his decades-long attempt to create successful blood substitutes. The high altitude studies are a great training ground for the artificial blood studies, he says, "because they told us it's not enough to just have some kind of hemoglobin. It has to pick up oxygen and deliver it in the correct ways, and if you understand those mechanisms, it's possible to design" workable substitutes.

Winslow and colleagues knew that hemoglobin would be the best molecule for carrying and delivering oxygen in artificial blood (Table 15.1). They knew that it had to be delivered without its natural "wrapper," the red blood cell membrane, because these delicate bags are the reason blood gets outdated on blood bank shelves, as well as the reason patients have A, B, O and other blood types. But they also knew that free raw hemoglobin molecules create an osmotic imbalance in the blood. Furthermore, free hemoglobin molecules are rapidly filtered out and excreted by the kidneys (Table 15.1). And those hemoglobin molecules that aren't tossed out cause the blood vessels to constrict. So how to proceed?

Winslow's group tested many kinds of blood substitutes on mice, rats, monkeys, and other lab animals. They also took a clue from the blood of the tarantula with its unenclosed but very large hemocyanin molecules. His group tried collecting and purifying hemoglobin from outdated human blood then treating it with long, stringlike organic molecules called polyethylene glycol (Table 15.1). This acts like an-

Why can Dr. Winslow freeze Hemospan but not real blood?

Table 15.1 Making a Successful Blood Substitute

	Hemoglobin in red blood cell*	Free hemoglobin†	Modified hemoglobin‡
Spleen	Only old red blood cells destroyed	Rapid destruction	Not destroyed as fast as free hemoglobin
Kidney	Does not pass into urine	Leaks into kidneys and urine, destroys kidney function	Does not leak into kidney
Hemoglobin	Releases oxygen when it is needed most	Does not easily release oxygen	Releases adequate oxygen
Arterioles	Healthy control of vasoconstriction	Abnormal vasoconstriction and elevated blood pressure	Does not cause abnormal vasoconstriction
Durability in body	Lasts 120 days	Lasts only 12 hours in circulation	Lasts 2–3 days in circulation
Shelf life	Shelf life of 42 days when refrigerated	Long shelf life	Long shelf life

(Polyethylene glycol)

* Does everything well but sit on a shelf.
† Does everything poorly but accept oxygen and last for long periods.
‡ Has numerous benefits and appears to be safe.

hemoglobin molecules are not quickly filtered out by the kidneys. Even more importantly, the treated hemoglobin doesn't mop up nitric oxide, so it doesn't cause tiny blood vessels to constrict and harm recipients. The Sangart company's blood substitute, which Winslow calls Hemospan, can be frozen, even freeze-dried, and stored indefinitely. It carries and transfers oxygen in the ranges of real blood. "The 'span' part [of Hemospan] keeps the blood vessels open," says Winslow, "and the 'hemo' part delivers the oxygen." It has no blood types so it's universal. And it's cheaper than human blood because one unit of outdated blood can be processed into three units of Hemospan.

Winslow sees many critical uses for artificial blood:

- Emergencies like earthquakes and tornadoes where local blood supplies can run out quickly.
- Battlefields, accident scenes, and medical surgeries in remote rural areas.
- A replacement for whole blood in places like Africa and Mexico where HIV, hepatitis, and other contaminants have tainted some banked blood supplies.
- Routine scheduled surgeries where patients need a unit or two of blood but can't donate and stockpile their own weeks before surgery. Instead, the patient could donate blood in the operating room immediately before the start of the surgery, says Winslow, then he or she could receive a pint or two of blood substitute as a blood volume stabilizer. During the surgery, the patient's own blood could then be transfused back with less chance of contamination, mislabeling, or wrong blood typing that costs lives every year in American hospitals.
- Blood banks for explorers in deep sea submersibles, polar research stations, and even space stations or other planets.

Sangart has been testing Hemospan and hopes to release the product by 2004. Other companies are also testing artificial blood products designed in other ways. It seems likely that our biological knowledge of blood, blood circulation, and respiration will soon lead to a safe, effective substitute that many of us will someday need.

tifreeze, explains Winslow, and attracts a cloud of water around each hemoglobin molecule. The cloud, in turn, creates a "shell" that effectively makes each molecule larger. Because of their large size, these altered

Chapter Summary and Selected Key Terms

Introduction We each have the life-giving, oxygen-bearing resource called blood, but extra supplies of it are short because people are becoming too busy or too reluctant to donate it to blood banks. Artificial blood is a major research goal—and a difficult one for reasons we see in this chapter. To understand them takes a knowledge of blood as a tissue; of how blood moves through the body delivering oxygen and removing wastes via the **circulatory system** (p. 400), and how air is drawn into the body as an oxygen source and how carbon dioxide wastes are exhaled via the **respiratory system** (p. 400).

Blood: Liquid Transport Tissue Blood is made up of **erythrocytes** (p. 400) or red blood cells, **leukocytes** (p. 400) or white blood cells; small globular **platelets** (p. 400); and the pale yellow fluid **plasma** (p. 400) surrounding the blood cells and platelets. Plasma contains **antibodies** (p. 401), **albumin** (p. 401), **fibrinogen** (p. 401), and other important dissolved proteins. One third of all your cells—25 trillion out of 75 trillion—are red blood cells. Each has a biconcave disk shape and contains **carbonic anhydrase** (p. 402) which speeds the conversion of CO_2 and H_2O to bicarbonate, HCO_3^-, and thus the removal of CO_2 from tissue cells. Red blood cells contain **hemoglobin** (p. 402) molecules, 350 million per cell. Each of the four globin chains is wrapped around a central iron-containing heme group. A hemoglobin molecule has two alpha chains and two beta chains. Each hemoglobin molecule picks up four oxygen molecules (which bind to the heme groups) in the lungs and carries them to the tissues. There, a lower pH (brought about by CO_2 conversion to bicarbonate and H^+ ions) causes oxygen to break away from hemoglobin and diffuse out of the bloodstream. Some of the CO_2 then enters the red blood cell and binds to the globin chains.

A red blood cell lives for about 120 days, then is replaced by stem cells in the bone marrow. When blood cell numbers drop due to blood loss, the kidneys and liver sense lowered oxygen delivery and produce the hormone erythropoietin. This stimulates the bone marrow to produce more red blood cells and as they carry more oxygen to the liver and kidneys, erythropoietin production drops once again in a negative feedback loop. For every 500 red blood cells there is just one white blood cell or leukocyte. The leukocytes help defend the body against invaders and also clean up debris. Artificial blood, intended for short term emergencies, would not require white blood cell equivalents. Platelets or **thrombocytes** (p. 404) are cell fragments that help plug small leaks in the circulatory system.

Circulatory Systems
Tarantulas are typical of most invertebrates in that they have a clear blood equivalent called **hemolymph** (p. 405) that moves through an **open circulatory system** (p. 405). A heart tube pumps hemolymph toward the brain, then the fluid percolates through the body cavity outside of tubes, bathing the internal tissues. Hemolymph contains a huge bluish protein, **hemocyanin** (p. 405) with 48 subunits; it is not contained within cell membranes.

Some invertebrates (like earthworms) plus all vertebrates have **closed circulatory systems** (p. 406) in which blood is completely contained inside vessels. These can efficiently shunt blood to needed areas and are pressurized.

Human Circulatory System
An immense, intricate system of tubules channels blood to all tissues of the body. The heart pumps at a regular rhythm and helps maintain a fairly high blood pressure. The right side of the heart pumps blood directly to the lungs and back to the left side of the heart. This "lung loop" is the **pulmonary circulation** (p. 407). The left side of the heart pumps blood to all organs except the lungs and the blood returns to the right side of the heart; this "body loop" is the **systemic circulation** (p. 407).

These two loops prevent oxygen-rich blood from mixing with oxygen poor blood.

There are several types of blood vessels. **Arteries** (p. 407) are large, hose-like vessels with muscular walls that help keep up the blood pressure. Arteries branch into smaller **arterioles** (p. 407), and these branch into dense web-like networks of **capillaries** (p. 407). Naked hemoglobin molecules cause arterioles to constrict. Each capillary bed has an arterial side that conveys oxygenated blood away from the heart. Oxygen diffuses out of the capillaries and into surrounding tissue cells. The capillary bed also has a venous side which conveys oxygen-poor blood and carbon dioxide wastes back to the heart and lungs. The venous capillaries merge into **venules** (p. 408) which merge into **veins** (p. 408). **Valves** (p. 408) or one-way flaps keep venous blood moving toward the heart, even though veins don't have muscular, pulsating walls and the blood pressure in them is lower.

The **heart** (p. 409) pumps a third of a teacup full of blood with each heart beat—2.5 billion beats in a lifetime. Two heart chambers called **atria** (sing. *atrium*) (p. 409) receive blood from veins. Two chambers called **ventricles** (p. 409) receive blood from the atria. The heart's right ventricle pumps oxygen-poor blood through a one way valve into the **pulmonary artery** (p. 409), which carries the blood to capillaries in the lung. Here oxygen is picked up and CO_2 is unloaded. The oxygen-rich blood now flows to the **pulmonary veins** (p. 409) then into the heart's left atrium. From the left atrium, oxygenated blood flows through a valve into the heart's left ventricle. Then the ventricle contracts strongly and squirts the blood past a valve into the **aorta** (p. 409), the main throughway to the body loop. After traveling through the tissue capillary beds and back in the vessels and veins, the oxygen-poor blood enters the **superior vena cava** (p. 408) (upper body) and **inferior vena cava** (p. 409) (lower body) and is channeled to the heart's right atrium. From the right atrium, blood flows through a fourth valve into the right ventricle and into the lung loop.

The heart beat is coordinated and controlled. **Intercalated disks** (p. 410) link heart muscle cells electrically. **Pacemaker cells** (p. 410) set the pace for the contraction of other heart muscle cells. Pacemaker cells are located in the **sinoatrial (SA) node** (p. 410) in the right atrium, and before a heart beat, an electrical impulse signals from the SA node across both atrial walls. A second node, the **atrioventricular (AV) node** (p. 411) relays the signal and coordinates the contraction of the ventricles. Contraction of the atria or ventricles is called **systole** (p. 411). Relaxation is called **diastole** (p. 411). Together, they form the **cardiac cycle** (p. 411). When you have your blood pressure taken, it measures the pressure during systole (e.g. 120 mm Hg) and diastole (e.g. 80 mm Hg).

Blood flow is regulated to various parts of the body. During **vasoconstriction** (p. 412), blood vessel diameters narrow and blood moves to other regions. During **vasodilation** (p. 412) diameters expand in an area and more blood flows to it. **Nitric oxide** (p. 412) is involved in control of blood flow and because free hemoglobin molecules mop up NO and cause vasoconstriction, they are a poor blood substitute by themselves.

When the circulatory system is breached by a wound, vasoconstriction helps reduce blood flow. Platelets release serotonin, which maintains the constriction. Platelets begin to form a plug. Then **coagulation** (p. 414) begins: Injured cells release activators and these stimulate **prothrombin** (p. 414) conversion to **thrombin** (p. 414). Thrombin is an enzyme that converts dissolved **fibrinogen** (p. 414) proteins into tough threads of **fibrin** (p. 414) that create a blood clot.

Mammals have a second circulatory system, the **lymphatic system** (p. 414) that drains excess tissue fluids or **lymph** (p. 415) and returns it to the blood plasma. **Lymph nodes** (p. 415) are filtering organs active in sustaining immunity. The lymphatic system has two additional organs, the **thymus** (p. 416) which produces certain white blood cells and the **spleen** (p. 416) which stores platelets and blood cells and filters out and destroys aging red blood cells.

The Respiratory System
Because animal cells undergo cellular respiration, each cell needs a supply of oxygen and a way to dump carbon dioxide. This includes the trillions of cells in interior areas far from the surface. The respiratory system is a set of tubes and organs that facilitate the exchange of oxygen and carbon dioxide at the level of the whole organism. Many arthropods, including caterpillars, have branching tubes called **tracheae** (p. 416) that permeate the body, are washed by free hemolymph, and serve as sites of gas exchange by diffusion.

Fish, as well as many invertebrates, have **gills** (p. 417) or gas exchange organs that develop as outgrowths of the body surface. In fish gills, water passes in a one way flow across delicate plates (lamellae) encompassing capillaries. The resulting **countercurrent exchange** (p. 417) in which water moves opposite to blood flow allows an efficient gas exchange, even though water holds far less oxygen than air. Amphibians have simple sac-like lungs and their moist skin contributes to gas diffusion. Birds have a complicated set of air sacs as well as a one way air flow that prevents the mixing of stale and fresh air and can sustain birds' high activity levels even at high altitudes where oxygen levels are low.

Humans have an in-and-out air flow or **tidal ventilation** (p. 420), meaning that some stale air remains in the lungs. Air flows through a branching treelike set of passages. It moves in the nose and/or mouth and into the throat or **pharynx** (p. 420). One branch from the pharynx, the **esophagus** (p. 420), leads to the stomach. A second branch off, the windpipe or **trachea** (p. 420) leads into the lungs. The voice box or **larynx** (p. 420) and tissue flap or **epiglottis** (p. 420) lie at the forward end of the trachea. The trachea branches into **bronchi** (p. 420), one per lung, and these branch into **bronchioles** (p. 420) that lead to tiny bubble-like sacs or **alveoli** (p. 420) where gas is actually exchanged. The lungs are surrounded by a **pleural sac** (p. 421) and hang in the **thoracic cavity** (p. 422) above the **diaphragm** (p. 422).

Lung ventilation is possible because the diaphragm pulls down during **inhalation** (p. 422) and relaxes during **exhalation** (p. 422). A region of the brain stem sends nerve impulses to the diaphragm and rib muscles to contract, leading to inhalation. Levels of carbon dioxide inform this brain center and it increases or decreases inhalation rates accordingly.

Robert Winslow used his knowledge of blood, the circulatory system, and the respiratory system to design a blood substitute. He coated free hemoglobin with polyethylene glycol (a type of antifreeze) to form a shell that enlarges each molecule. This treatment allows the hemoglobin to transport gases and yet not cause vasoconstriction and other harmful side effects.

All of the following question sets also appear in the Explore Life *(E)* electronic component, where you will find a variety of additional questions as well.

Test Yourself on Vocabulary and Concepts

In each question set below, match the description with the appropriate term. A term may be used once, more than once, or not at all.

SET I

(a) plasma (b) erythrocytes (c) leukocytes
(d) erythropoietin (e) platelets (f) lymph (g) renin

1. A hormone that stimulates the rate of division of stem cells in the bone marrow
2. Blood cells that contain hemoglobin and carbonic anhydrase
3. A category that includes cells of the immune system
4. A fluid that is initially derived from tissue fluids and is returned to the plasma
5. Cell fragments that are important in the initiation of blood clotting

SET II

(a) artery (b) vein (c) capillary (d) arteriole
(e) lymphatic

6. The only category of blood vessel with walls thin enough for nutrient exchange with the tissue fluid
7. Blood vessel with valves that prevent back flow
8. Large, thick-walled vessel that carries oxygenated blood
9. Blood leaving this kind of vessel next enters a capillary bed
10. Vessels of this kind carry blood under the highest pressure in the body

SET III

(a) swim bladder (b) bronchiole (c) alveolus
(d) larynx (e) epiglottis (f) bronchus

11. The region in a mammalian lung where gas exchange occurs
12. A flap of tissue that prevents swallowed food from entering the larynx
13. A lunglike structure found in fishes
14. Site of the vocal cords; the voice box
15. Small airway that is contained within the substance of the lung

SET IV

(a) thoracic cavity (b) diaphragm (c) pleural sac
(d) intercostal muscles

16. A dome-shaped muscle that assists in breathing
17. Action of these muscles enlarges the thoracic cavity
18. Fluid-filled cavity in which the lungs lie
19. Relaxation of these muscles results in lowering of the rib cage
20. A puncture in this structure results in a collapsed lung

Integrate and Apply What You've Learned

21. Compare the level of oxygen in the blood that leaves the heart of a fish with that leaving the heart of a mammal.
22. State the constituents of blood and the primary functions of each.
23. What keeps the blood moving in a single direction in the heart, the arteries, and the veins?
24. What organs are included in the lymphatic system? Why do we need this second circulatory route?
25. Contrast the terms respiration and cellular respiration.
26. Contrast the flow of air through the respiratory system of a bird with that of a mammal.
27. Snorkelers often hyperventilate before a deep dive. Why might this be dangerous?

Analyze and Evaluate the Concepts

28. Why couldn't researchers just put naked hemoglobin molecules into a person's body as a blood substitute?
29. What does the red blood cell membrane do in the body?
30. Suppose that you had to select members for a track team that was to compete in a championship meet in Atlanta, Georgia. The team members will arrive in Atlanta individually from their homes the day before the meet. You must choose between two runners who have previously shown themselves to be equally fast many times in the past. Runner A lives in New York City; runner B lives in Denver. Both runners have been home continuously for the past two months. All other things being equal, whom would you choose and why?
31. Suppose that you leave your home near sea level, spend a week skiing in the mountains of Colorado, then return home. Compare your body's production of the hormone erythropoietin two days before leaving home to its production two days after arriving in the mountains, and its production two days after returning home again.
32. On a planet with high humidity and an atmosphere containing 30 percent oxygen instead of 21 percent like ours, you discover large, cow-sized organisms with cellular respiration and many other metabolic details similar to Earth's animals. Can you predict how the organism's respiratory system might differ from that of a cow?
33. A person who hyperventilates while blowing on a campfire might black out and become unconscious. During this situation how would the control centers of the brain function in restoring homeostasis?
34. Why is a bird's lung more efficient than an amphibian's lung?
35. How did Robert Winslow's team create a potentially commercial blood substitute, and what are some of the ways it could be used?

The Uncommonly Unpleasant Cold

Adrienne Wilson is feeling miserable—again. She is slumped in a waiting room at the University of Virginia Medical Center in Charlottesville, eyes red, nose clogged, throat aching. "How did I get *another* cold?" she mumbles to no one in particular. "Besides, what *is* a cold?" Adrienne closes her eyes suddenly, inhales twice through parted lips, and sets off a monumental sneeze. "And why," she wonders out loud, "aren't there better drugs?"

Ms. Wilson is having a bad day, but at least she's having it in an ideal place: The University of Virginia has an active research program on the common cold. Several experts, led by Dr. Jack Gwaltney, have collectively examined tens of thousands of cold sufferers. Ms. Wilson, a student at the University, is waiting to see Dr. Gwaltney as part of an experiment to test a new cold medicine. The ultimate aim of the study is to understand in more detail the intricacies of disease resistance based on our bodies' defense against foreign invaders, including the viruses that cause colds.

As one of North America's top authorities on the common cold, Dr. Gwaltney can already explain why Adrienne got "*another*" cold." Very few people, he says—"maybe 10 percent of the population"—never experience cold symptoms. The average person, he says, reports having two and a half colds every year. Adrienne's second cold of the season simply puts her in the normal category. Dr. Gwaltney also has an in-depth knowledge of what a cold *is*, and perhaps how Adrienne caught this latest one.

A cold, says Gwaltney, is the manifestation of a viral infection of the respiratory tract. It involves nasal symptoms like a stuffed-up or runny nose and throat symptoms such as cough or soreness. Most of the time, it also affects the sinuses and inner ear. The culprit, a virus (see Chapter 2), is typically one of 100 different strains of **rhinovirus** (literally, "nose virus"). But, says Gwaltney, very similar symptoms can result from five or six other types of viruses, including influenza virus and respiratory syncytial (sin-SISH-al) virus. A cold virus usually enters through the nose and then defeats the body's first line of defense—the skin lining the nose and sinus passages, the mucus given off by those skin cells, and the beating cilia that also line the sinuses. As the cilia sweep the viral invaders toward the back of the throat, they encounter perfect protein entry ports on the nasal cell membranes. Through these ports (ICAM1 receptors,

▲
Focus on the Common Cold.
Adrienne Wilson with a cold; Dr. Jack Gwaltney; Adrienne gets examined in the clinic.

pronounced eye-cam-one; Fig. 16.1, Step 2), the virus gains entrance to the nasal cells. Once inside, the virus inserts its nucleic acid into the nasal membrane cell and commandeers the cell's growth mechanism to make new virus particles. These, in turn, burst out, kill the cell, and reinfect new cells nearby.

Surprisingly, this cellular takeover and destruction do not cause the cold symptoms we're all familiar with. Instead, Gwaltney explains, the runny nose, sore throat, and so on result from our body's second line of defense against intruders, **inflammation,** with its redness, heat, and swelling. The nasal passages' copious production of mucus is part of this response and is an early attempt to "wash" viral particles away from the cells and tissues. But as we'll see later, the defensive strategy itself gets carried away and contributes to the miseries of a cold.

The body also has a third, slower but more specific line of defense that involves the entire **immune system,** with its network of protective organs, tissues, and cells. White blood cells (see Chapter 2), for instance, produce targeted proteins called **antibodies** that attack only very specific invaders—in this case, a particular strain of rhinovirus. Unfortunately, it takes *two weeks* for the white blood cells to gear up and mass produce these molecular weapons if needed, and in the meantime, most cold sufferers will have turned to over-the-counter drugs for relief.

Dr. Gwaltney tells patients like Adrienne Wilson that "at the very earliest sign of a cold," they should take a 12-hour antihistamine tablet plus a generic pain reliever to fight the side effects of inflammation. Right now, however, there are no truly effective drugs against cold viruses themselves. Later in the chapter, we'll see why Gwaltney recommends these medicines, and why more effective drugs are still a thing of the future.

There are also social factors involved in catching a cold. Behavioral research has shown that the stress of losing a job, for instance, leads to a higher rate of illness in people exposed to rhinovirus. So does having few social ties. And so does the constant stress of being a college student, like Adrienne Wilson. Says Gwaltney, "If you put a rhinovirus in people's noses (experimentally), 95 percent will become infected . . . but of those, only three quarters will develop an illness." Someday cold researchers will unearth the secret of symptom-free infection in that lucky one quarter, and employ it to help the rest of us. In the meantime, we can use the common cold as a way to explore the immune system and our bodies' defenses against disease-causing agents like rhinovirus.

In this chapter, we'll see how our disease-resistance mechanisms enable us to recognize invaders and eliminate them. We'll also see how, once aroused, natural immune protection can form a life-long defense against particular threats. Along the way, you'll find answers to these questions:

1. What is disease resistance?

2. What are the body's nonspecific responses to invasion by microbes or viruses?

3. How does the immune system form targeted antibodies and provide specific resistance against invaders?

4. How does the body regulate immunity and what goes awry in diseases like arthritis, in which the defense system starts attacking a person's own cells and tissues?

5. How do medical workers use vaccines to deliberately transfer immunity to people, pets, and other animals?

16.1 Disease Defense and Nonspecific Resistance

Our body's ability to fight off disease-causing agents (or **pathogens**) such as rhinovirus is called **resistance.** As we saw in the introduction, we have two types of resistance. **Nonspecific resistance** to diseases includes a first line of defense in the form of physical barriers to entry and a second line of defense based on nonspecific cell activities—cellular functions that fight disease in the same way regardless of the invader's characteristics. These two lines of defense protect us from a wide variety of pathogens, and they work quickly. Because of their general nature, however, they can fail to combat many important disease-causing organisms. The second type of resistance (and our third line of defense) is **specific resistance** to disease, or **immunity.** Immunity has the advantage of combatting particular species or strains of pathogens and other invaders. Its disadvantage is that it works more slowly than nonspecific resistance.

Disease resistance is based on three main functions: recognizing the invader, communicating between defensive cells, and eliminating the invader. In this section, let's investigate how the physical barriers and cellular activities of nonspecific resistance protect the body. Then in the next section, we'll explore the more complicated mechanisms of specific resistance to disease.

Nonspecific Resistance to Disease

Adrienne's cold symptoms actually resulted from her body's nonspecific defenses (**E**xplorer 16.1). Regardless of the type of invader—virus versus bacteria, for example—nonspecific defenses can involve:

1. *Physical barriers* such as the skin and mucous membranes.
2. *Nonspecific substances* (listed below) that fight many different sorts of viruses and bacteria. (In contrast, specific substances, including antibodies, are highly

targeted to recognize and stop only particular species and strains of bacteria or viruses.)
3. The *cellular activities* of natural killer cells, an attacking type of white blood cell; of phagocytes, patrolling cells that engulf particles; and of many cell types that produce inflammation, with its redness and swelling.

The First Line of Defense: Physical Barriers

If you are holding a tack and your finger slips, you might puncture your skin. The skin and the mucous membranes that line the respiratory and digestive systems are the body's first line of defense against unwanted intruders (Fig. 16.1, Step 1). They form a physical barrier that keeps out most microorganisms capable of multiplying within the body and causing infection. A puncture, tear, slash, abrasion, burn, or other break in this smooth, elastic barrier can allow dirt and microorganisms to enter the bloodstream. Fluids secreted by the skin or mucous membranes can help the body resist disease by washing away pathogens in a nonspecific way. Furthermore, some of these secretions, such as sweat, saliva, and tears, contain lysozyme (review Fig. 3.8), an enzyme that can dissolve certain kinds of bacterial cell walls and hence kill bacterial cells.

The Second Line of Defense: Nonspecific Substances and Cellular Activities

Returning to the tack in your finger, the metal point would probably introduce microbes into your system and these would quickly come into contact with growth-inhibiting antimicrobial proteins in body fluids. The **interferons** are an important class of antimicrobial proteins released by certain virus-infected cells. Interferons diffuse to other cells and, in turn, cause them to produce proteins that inhibit viral replication

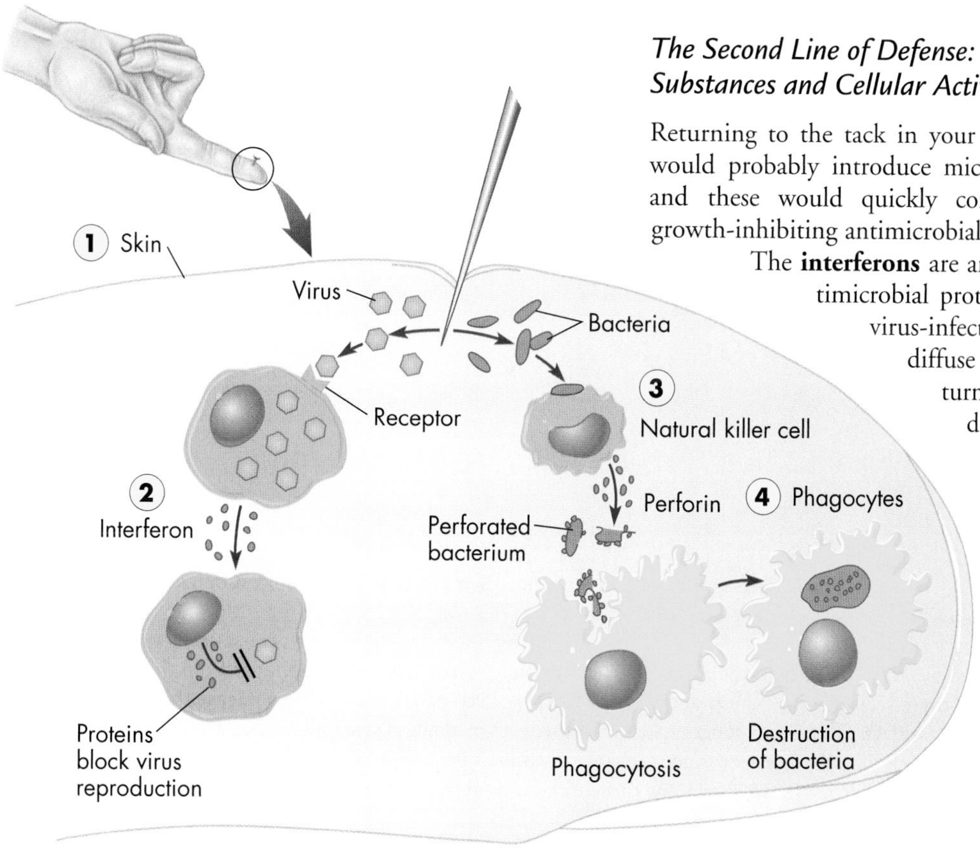

> **GET THE PICTURE**
>
> **E**xplorer 16.1
>
> Nonspecific Immune Responses

1 Skin

Virus

Receptor

2 Interferon

Proteins block virus reproduction

Bacteria

3 Natural killer cell

Perforin **4** Phagocytes

Perforated bacterium

Phagocytosis

Destruction of bacteria

Figure 16.1 **Nonspecific Defenses.** Nonspecific defenses work the same way toward each invader; i.e., they are not *specific*. They include (1) physical barriers; (2) nonspecific substances; and nonspecific cellular activities of natural killer cells (3) and phagocytes (4).

(Step 2). Another type of antimicrobial proteins is the so-called **complement system,** a group of about 20 plasma proteins that attack and destroy microbes and stimulate inflammation. These proteins can become activated by either antibody binding to its target or by components in a bacterium's cell wall. This activation then intensifies (or complements) other disease-fighting mechanisms, both nonspecific and specific.

Bacteria entering with the tack may encounter **natural killer cells,** which are a type of white blood cell that can kill many kinds of infectious microbes and some types of tumor cells in a nonspecific fashion. Natural killer cells can recognize some cell surface molecules and release proteins called **perforins** that perforate or poke a hole in bacterial plasma membranes (Step 3).

An invading bacterium might also encounter **phagocytes** (*phago* = eat + *cyte* = cell), or specialized scavenger cells that devour debris (Step 4; review Fig. 2.20). Once consumed by a phagocyte, the microbe is

killed by enzymes stored in the little bags called lysosomes that we discussed in Chapter 2 (Fig. 2.17).

Finally, a tack puncture—a mechanical breech—can induce a series of nonspecific internal resistance reactions called the **inflammatory response,** with its redness, pain, heat, and swelling. Inflammation captures invaders at the site of the injury and prepares the tissue for repair. An even better example of inflammation is a cold like Adrienne Wilson's. A rhinovirus attack upon and entry into the nasal membrane cells is a biological breech (Fig. 16.2, Step 1). The virus particles infect the nasal cells, take over their information-processing machinery, and cause them to start churning out new virus particles. This eventually causes some cells in the nasal membrane to burst and die (Step 2). We said earlier that Adrienne's cold symptoms came not from the virus particles themselves, and now we can add that they aren't the result of her killed membrane cells, either. Instead, they stem from her body's inflammatory response, a reaction to the virus.

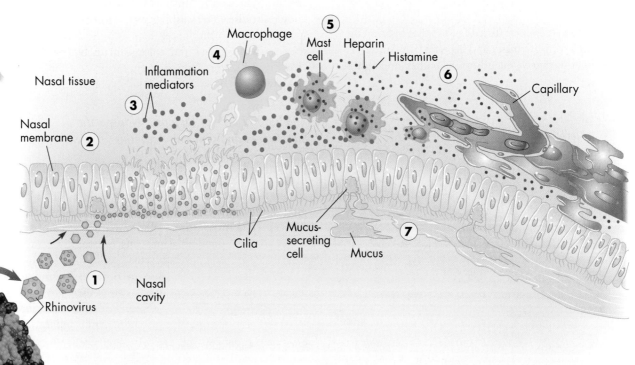

Figure 16.2 **Nonspecific Defenses and a Cold Sufferer's Symptoms.** (1) In a cold sufferer like Adrienne Wilson, rhinovirus invades cells lining the nasal cavities. In this rhinovirus model, the reds, blues, and greens represent different polypeptide chains. (2) Infected cells die, releasing debris. (3) The debris causes cells to release inflammation mediators. (4) Inflammation mediators attract phagocytes called macrophages from the blood circulation, and these scavengers clean up debris. (5) Mediators also cause mast cells to release their granules of histamine and heparin. (6) These substances cause blood vessels to dilate and capillaries to leak fluid. (7) Mediators also cause mucous glands to secrete mucus. Eventually, inflammation inhibitors limit the response by blocking the action of the mediators. Insert shows the three-dimensional structure of a rhinovirus. **E** 16.1 animates this inflammatory response.

Damaged tissues such as virus-infected membrane cells release **inflammation mediators,** or chemical signals that provoke inflammation reactions (Step 3). Inflammation mediators include the large proteins called complement; small hormones called **prostaglandins;** and short proteins called **cytokines.** These mediators cause scavenging white blood cells (phagocytes) to leave the bloodstream and converge on the injured area in search of invaders. The phagocytes devour the intruders and clean up debris by phagocytosis (Step 4).

Inflammation mediators also act on **mast cells,** round cells that occupy connective tissue in the skin and mucous membranes and are filled with little packets of chemicals called **histamine** and **heparin** (Step 5). Mast cells stimulated by inflammation mediators such as prostaglandins or cytokines release the chemicals from these packets into the space surrounding the cell. Released histamine binds to surface receptors on the cells of nearby capillaries and on smooth muscle cells. This binding, in turn, causes capillaries to dilate and become leaky (Step 6). Dilation brings more blood to the infected area (in Adrienne's case, the nose and sinuses), and allows fluid to leak into tissue spaces, causing redness, nasal congestion, and swelling around the nose and eyes. (We take the drugs called antihistamines to combat these symptoms.) The heparin released by mast cells binds to the clot-inhibiting protein AT3 (see Fig. 7.2) and lessens the chance that blood vessel leaking will lead to a blood clot. Finally, inflammation mediators also stimulate mucous glands to secrete more of the slippery coating; sneezing, nose blowing, and coughing up this mucus flushes away debris from the nasal passages (Step 7).

The small hormones called prostaglandins also stimulate mucus secretion, and a mucus buildup in the throat probably stimulated Adrienne's hacking cough. Prostaglandins produced by mast cells cause smooth muscle cells in the bronchioles to constrict, as well, and this probably made Adrienne's breathing more difficult. Finally, prostaglandins also participate in resetting the body's thermostat, and this contributed to her fever (see Chapter 14). To round off Ms. Wilson's list of cold symptoms, the short proteins called cytokines probably triggered her sore throat. Most of us are familiar with the red, swollen inflammation of a finger punctured by a rose thorn or tack. But as you can see from this discussion of inflammation mediators, Adrienne's cold is also classic inflammation at work.

Dr. Jack Gwaltney bases his advice to cold sufferers like Adrienne on a detailed understanding of how inflammation causes cold symptoms. Dr. Gwaltney recommends that at the first sign of cold symptoms, one take several kinds of over-the-counter drugs:

1. A 12-hour antihistamine tablet containing chlorpheniramine or clemastine. This will block histamine receptors on blood capillaries and counteract the effects of histamines, thus decreasing tissue swelling in the nose and eyes.
2. A nonsteroidal anti-inflammatory drug such as ibuprofen or naproxen. This will mimic the action of cortisol, the body's natural inhibitor of inflammation. Ibuprofen also blocks prostaglandin production and thus slows mucus secretion and helps bring down a fever.
3. If you have a cough, Gwaltney recommends taking a cough suppressant to control irritation to the throat and chest.

Together, these medications diminish a cold's miserable symptoms, although they don't fight the cold viruses themselves. A few drugs actually block viral replication (we saw some HIV-fighting drugs in Chapter 2, for example), and researchers like Dr. Gwaltney are working on various kinds. So far, however, they tend to be only partially effective and to have bothersome side effects. Perhaps future drugs will be able to quickly and specifically target the particular viral culprits behind each person's cold. In the meantime, our body's own specific resistance does eventually gear up—albeit slowly—to fight colds and flu and return us to health.

Why has the search for effective cold drugs been so difficult?

16.2 Specific Resistance to Disease

Our first and second lines of defense are a fast-acting "one size fits all" approach to combatting disease that, as we've seen, can also cause the "overreactions" of inflammation. On the other hand, specific resistance—our third line of defense—acts more slowly, but tailors a defensive response to the invading agent—for example, the exact strain of rhinovirus that attacked Adrienne's nasal passages. Specific resistance involves your body's complete immune system, a large network of organs, cells, and molecules (Table 16.1). The organs include your bone marrow, spleen, lymph nodes and thymus. The cells include several classes of white blood cells that constantly circulate in the bloodstream and lymph. Finally, the molecules include special cell surface molecules called MHC proteins; signaling chemicals secreted by white blood cells; and the uniquely targeted attack proteins we mentioned earlier, the antibodies. A remarkable and

Table 16.1 Players in the Immune Response

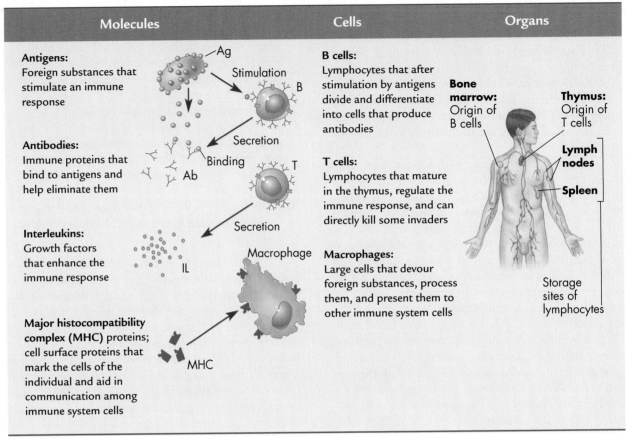

Molecules	Cells	Organs
Antigens: Foreign substances that stimulate an immune response	**B cells:** Lymphocytes that after stimulation by antigens divide and differentiate into cells that produce antibodies	**Bone marrow:** Origin of B cells
Antibodies: Immune proteins that bind to antigens and help eliminate them	**T cells:** Lymphocytes that mature in the thymus, regulate the immune response, and can directly kill some invaders	**Thymus:** Origin of T cells
Interleukins: Growth factors that enhance the immune response	**Macrophages:** Large cells that devour foreign substances, process them, and present them to other immune system cells	**Lymph nodes** **Spleen** Storage sites of lymphocytes
Major histocompatibility complex (MHC) proteins; cell surface proteins that mark the cells of the individual and aid in communication among immune system cells		

unique feature of the immune system is that it features **immunologic memory:** it can remember an invader to which it has already been exposed, such as last year's strain of cold virus. This memory feature is one of the reasons people get on average only two or three colds each year. To understand how this wonderful and specific defense system works, we need to investigate the cells and molecules of the immune system in more depth.

Molecules of the Immune System

The immune system's silent battles are fought at the level of molecules. These can be molecules on the surfaces of invading viruses or microbes, for instance, acting as antigens (defined below). And they can be antibodies, regulatory proteins, and other examples of the body's defensive molecules.

Antigens

Antigens are chemicals that generate an immune response. You can start to understand antigens by exam-

ining the structure of the rhinovirus in Figure 16.2. Notice that the viral surface is bumpy, with individual proteins projecting by a few amino acids here and a few there. Each stretch of three to six amino acids has a specific shape characteristic for that particular amino acid sequence. The immune system can recognize each specific molecular shape of about this size when attached to a larger molecule and initiate a response to it. Any substance that can stimulate an immune response is called an antigen, and the class includes the surface proteins of a rhinovirus. Many antigens are proteins, but carbohydrates and lipids can also act as antigens. Even small molecules, when attached to a larger protein, can act as antigens. Invaders can have many recognizable surface proteins; the outer coats of most viruses, bacteria, fungal cells, and larger parasites (such as the malaria agent we encountered in Chapter 11) contain many different proteins and carbohydrates functioning as antigens.

Antibodies

The immune system responds to antigens by producing **antibodies,** protective proteins made by special types of white blood cells and secreted into the blood and lymph (Table 16.1 and Fig. 16.3). (One way to keep these molecules straight is to remember that an anti*gen gen*erates an immune response, and an anti*body* is made by your *body*.) An antibody acts by binding to an antigen projecting from the surface of an infecting rhinovirus, bacterium, or other foreign agent. The antibody is therefore recognizing the invader and at the same time marking it for elimination.

Regulatory Proteins

Besides antigens and antibodies, specialized white blood cells make an additional group of proteins that work together in regulating the immune response. We saw that some of these, the cytokines, are actors in the nonspecific response to invaders such as rhinoviruses. Some immune cells secrete proteins called **interleukins** (Table 16.1) that speed or slow the dividing and maturing of other immune cells. In this way, interleukins are involved in the immune system's communication activities. Other immune system molecules stud the surfaces of white blood cells, especially the proteins in the **major histocompatibility complex,** or **MHC.** These MHC proteins are specific for each individual and are the factors that cause the body to reject transplanted tissues such as skin grafts or organs such as donated kidneys. Their main function is to aid communication among immune system cells; the tissue "labelling" leading to transplant rejection is just a side effect.

Cells of the Immune System

Several types of white blood cells cooperate in generating immunity (Table 16.1), but among them, lymphocytes are the driving force. **Lymphocyte** means "cell of the lymph system" and indeed, these small, round, colorless white blood cells spend much of their time inside lymphatic tissues (review Fig. 15.12). One major type of lymphocyte called a **B cell** (short for **B lymphocyte**) is formed in the *b*one marrow. B cells are vitally important to our health because these cells make and secrete the antibody proteins that coat free-floating invaders such as rhinovirus, other viruses, and bacteria, and mark them for destruction.

The other major type of lymphocyte, called a **T cell** (short for *T* **lymphocyte**), originates in the bone marrow but matures in the *t*hymus. T cells kill foreign cells, often eukaryotic invaders like malarial parasites or body cells no longer recognized as "self" because they have become cancerous or infected with virus. Certain T cells also help regulate the activities of other lymphocytes. A third kind of white blood cell is the **macrophage** (liter-

ally, "big eater"). Macrophages are large, specialized phagocytes that, along with some other cell types, help lymphocytes recognize antigens; this stimulates the lymphocytes to attack invaders. Then the macrophages help clean up by consuming debris from the dismantled intruders.

The Specific Immune Response in Action

Let's return to Adrienne Wilson sitting in Dr. Gwaltney's waiting room, sneezing and blowing her nose. Even as her nonspecific defense mechanisms are generating cold symptoms through inflammation, her specific defenses have begun an additional series of steps that specifically targets the particular strain of rhinovirus inflicting her misery. Her body's specific resistance involves two main types of immunity. One is the **antibody-mediated immune response,** in which antibody proteins circulate freely throughout the body, bind to the invaders, and help to eliminate them. The other is the **cell-mediated immune response,** in which specific T lymphocytes circulate through the body and the immune cells themselves directly attack the invader. Because the body's attack on rhinovirus involves mainly the antibody-mediated immune response, we'll discuss it first. You can follow these steps in Figure 16.3 and *E*xplorer 16.2.

> **GET THE PICTURE**
> *E*xplorer 16.2
> Specific Immune Reponse

1. The virus attacks nasal lining cells and reproduces itself inside them, and cells begin dying as the virus particles break out and spread. Cellular debris, including rhinovirus proteins, begins to accumulate and may leak into the bloodstream and lymphatic system.
2. As the circulating debris floats past B cells, Y-shaped antibodies anchored in the plasma membranes of some of these stationary B cells bind to antigens—the bumps on the rhinovirus (see Fig. 16.2). In addition, phagocytes that have engulfed virus particles can pass (or "present") viral antigens to the membrane-bound antibody on B cells, where they bind. This binding represents recognition of the invader and leads to a specific immune response tailored for this strain of rhinovirus.
3. Certain T cells that are members of a class called **helper T cells** also become activated by the viral antigen and then interact with B cells that have bound themselves to the rhinovirus antigen.
4. The helper T cells secrete interleukins.
5. These protein growth factors communicate with B cells and stimulate them to divide into 2, 4, 8, 16, and eventually hundreds of B cells identical to the initial cells.

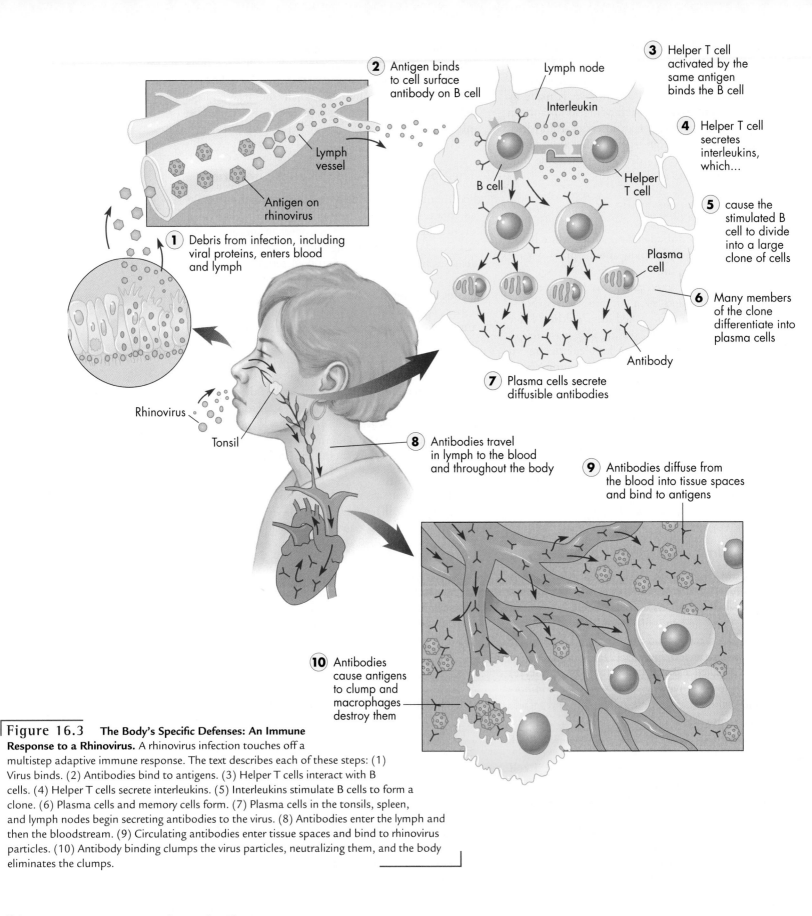

2 Antigen binds to cell surface antibody on B cell

3 Helper T cell activated by the same antigen binds the B cell

Lymph node

Interleukin

4 Helper T cell secretes interleukins, which...

Lymph vessel

B cell

Helper T cell

5 cause the stimulated B cell to divide into a large clone of cells

Antigen on rhinovirus

1 Debris from infection, including viral proteins, enters blood and lymph

Plasma cell

6 Many members of the clone differentiate into plasma cells

Antibody

7 Plasma cells secrete diffusible antibodies

Rhinovirus

Tonsil

8 Antibodies travel in lymph to the blood and throughout the body

9 Antibodies diffuse from the blood into tissue spaces and bind to antigens

10 Antibodies cause antigens to clump and macrophages destroy them

Figure 16.3 **The Body's Specific Defenses: An Immune Response to a Rhinovirus.** A rhinovirus infection touches off a multistep adaptive immune response. The text describes each of these steps: (1) Virus binds. (2) Antibodies bind to antigens. (3) Helper T cells interact with B cells. (4) Helper T cells secrete interleukins. (5) Interleukins stimulate B cells to form a clone. (6) Plasma cells and memory cells form. (7) Plasma cells in the tonsils, spleen, and lymph nodes begin secreting antibodies to the virus. (8) Antibodies enter the lymph and then the bloodstream. (9) Circulating antibodies enter tissue spaces and bind to rhinovirus particles. (10) Antibody binding clumps the virus particles, neutralizing them, and the body eliminates the clumps.

6. Many members of the clone differentiate into antibody-secreting "factories" known as **plasma cells.** Other members of each clone of cells may become **memory cells** (which we'll discuss shortly). Because each cell division cycle can take many hours, this step can take several days and helps explain why Adrienne Wilson continues to suffer cold symptoms for more than a week.

7. Once mature, each of the plasma cells (nestled in the tonsils, spleen, and lymph nodes) secretes into the lymph fluid over 1000 copies of a free-floating antibody molecule. This molecule can bind specifically to the antigen on the surface of the rhinovirus and to it alone.

8. The secreted antibodies now leave the tonsils, spleen, and lymph nodes carried in the lymphatic vessels. The lymph drains back into the bloodstream near the heart, carrying the defense molecules with it into the blood, which then distributes them throughout the body.

9. The antibodies pass from blood capillaries into spaces between cells all over the body, and wherever they encounter their target—a rhinovirus—they zero in and bind to it like a dart in a bull's eye.

10. This binding of antibodies to rhinovirus proteins neutralizes the virus by causing virus particles to clump together. The body eliminates these clumps because the antibody coating attracts macrophages like chocolate frosting attracts ants, encouraging the scavenging macrophages to devour, digest, and destroy the antibody–virus complexes. This activity eventually overwhelms the rhinovirus, and Adrienne Wilson is finally able to recover from her cold.

It's a marvel of biological research that we can understand in such detail how the body fights an infection through the rapid nonspecific resistance response followed by the slower-acting but specific immune response. In fact, biologists and physicians know much, much more about the activities of antibodies, B cells, and T cells. So let's explore a little deeper.

How Antibodies Work

The antibodies Adrienne made to her particular infecting rhinovirus strain will protect her from another cold caused by the same strain, but she's likely to catch another cold within several months anyway. Why? The answer requires a better understanding of what antibodies are, how they perform their functions, and the limits to their powers.

Antibody Structure

Antibodies are **immunoglobulins** (im-uno-GLOB-ulins), or globular proteins of the immune system. Each antibody consists of four polypeptides, or chains of amino acids (Fig. 16.4). Two of these chains are longer and are called **heavy chains,** and two are shorter and are called **light chains.** In any given antibody molecule, the two heavy chains are identical to each other, and the two light chains are identical to each other. Linked together, the four chains make a molecule shaped like the letter Y, with a stem made only of two heavy chains and two arms, each made of a light chain and part of a heavy chain (Fig. 16.4).

Antibodies' two main functions are related to this Y-shaped structure. (1) They recognize antigens by binding to them. The tips of the two upper arms each bind to an antigen of complementary shape like a lock fits a key. (2) They mark antigens for elimination from the body. Once the antibody's Y arms bind to an antigen, the lower stem of the Y stimulates components of the immune system such as macrophages to attack and destroy whole antibody–antigen complexes (see Fig. 16.3, Step 10).

To understand how antibodies do their jobs, researchers have compared the exact amino acid sequences of different antibody proteins directed against different antigens. By doing this, they've discovered that in the main class of antibody circulating in the blood, the stem

Figure 16.4 **Y-Shaped Antibodies: Variable "Arms" Bind Different Antigens.** Antibodies are Y-shaped molecules with specific antigen-binding sites located at the tips of the arms. Antigen molecules, wedge-shaped in (a) and rectangular in (b), fit into the binding sites of antibody molecules like keys in locks. The elimination of an antigen bound to an antibody depends on the stem of the Y. (c) A computer graphic reveals how an antigen, in this case the bird's egg enzyme lysozyme (blue), binds to an antibody (the red heavy chain, and the light yellow chain).

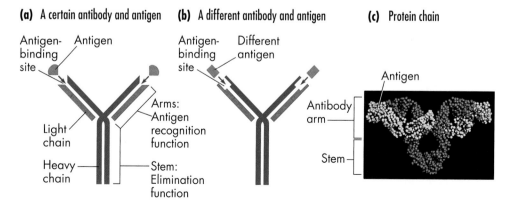

(a) A certain antibody and antigen

Antigen-binding site
Antigen
Light chain
Heavy chain
Arms: Antigen recognition function
Stem: Elimination function

(b) A different antibody and antigen

Antigen-binding site
Different antigen

(c) Protein chain

Antigen
Antibody arm
Stem

section always has the same amino acid sequence, and hence the same shape. This is true whether the antibody is directed against a rhinovirus, a strain of *E. coli* bacteria, a tetanus toxin, or some other antigen. In contrast, the tips of the two arms have a huge variety of different amino acid sequences and thus a huge range of different shapes. These many shapes can bind in complementary key-in-lock fashion to an enormous array of antigens—each antibody to an antigen of a particular shape (see Fig. 16.4). But because antibodies are so specific, an antibody against one rhinovirus strain may not protect against any of the hundred or so other strains. And, says Dr. Jack Gwaltney, a shifting pattern of different cold and flu viruses will come through a community every year.

Antigen-Binding Sites: Specificity and Diversity

The antibody molecules made by a single clone of B cells will bind only to an antigen of a single shape. Luckily, though, a person can make more than 100 million distinct types of antibody molecules, each with a unique binding site. By adding antibodies from all the clones of B cells together, an individual's antibodies can recognize and combine with 100 million different antigens, including pollen from a wide spectrum of different flowers, the outer coats of thousands of different viruses and bacteria, and the cell walls of many different molds.

Doctors have harnessed the huge spectrum of antibody shapes and activities as a powerful tool for diagnosing illnesses. One example of this is an accurate test for specific antibodies to the AIDS virus, HIV. When HIV first enters a person's body, he or she begins to make antibodies to it according to the scheme outlined in Figure 16.4. Blood tests can identify these antibodies against HIV. Anyone who has these antibodies in their blood is then called HIV-positive, a term that indicates exposure to the virus. Unfortunately, these antibodies do not protect against the virus because the virus wreaks its havoc *inside* cells, where it is hidden from the antibodies. Nevertheless, identification of HIV-positive individuals can be used to help slow the spread of the virus. By identifying HIV-contaminated blood from exposed people, medical workers can avoid accidental transfusions to other patients and help prevent new cases of AIDS.

A person who is HIV-positive makes many different types of antibody molecules to different "bumps" on the HIV particle. Biologists, however, can isolate a single type of antibody reacting to a single type of "bump" on a virus, bacterium, or cancerous cell. Biologists have learned to grow a single clone of B cells in the lab and isolate the antibodies. These so-called **monoclonal antibodies** are then highly useful in medical and industrial applications for targeting and binding specific antigens. For example, they have created a monoclonal antibody called Herceptin, which is directed against *HER2,* a growth factor receptor gene overabundantly produced in many women with invasive breast cancer. When the doctor injects these monoclonal antibodies into a woman with breast cancer, they selectively seek out and bind to the receptor and prevent it from triggering the cancer cell to divide. Physicians also use monoclonal antibodies to detect specific molecules, including human chorionic gonadotropin (see Chapter 8); this is the basis of the "dipstick" home pregnancy tests.

Figure 16.5   **Different Antibody Classes Have Different Destinations.** (a) Some antibodies can cross the placenta and confer immunity on the fetus. (b) Some antibodies are secreted in saliva, tears, and milk and can help a breast-feeding infant resist infection. (c) Other antibody molecules bind to mast cells in the skin and mucous membranes lining the nose, throat, airways, and intestines and are involved in allergic responses.

(a) Some antibodies can cross the placenta

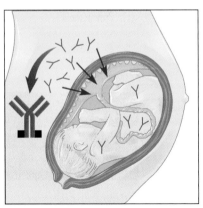

(b) Other antibodies are secreted into body fluids

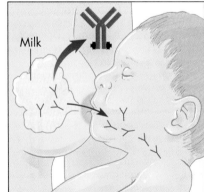

Milk

(c) A third class of antibodies is involved in allergies

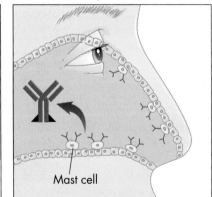

Mast cell

How Antibodies Trigger Elimination of an Invader

We've seen that the arms of a Y-shaped antibody bind to an antigen, while the stem of the Y triggers the elimination of an invader. We've also seen that in the main class of antibody circulating in the blood, all molecules have the same amino acid sequence of their stem section, while the sequences differ in their arms, allowing them to bind differently shaped antigens. In addition to this main class, biologists have identified other classes of antibodies with different elimination activities. Each class varies slightly from other classes in the amino acid sequence of the stem (and hence the stem shape) (Fig. 16.4).

One stem class circulates in the blood and efficiently eliminates rhinovirus and other viruses and bacteria. These antibodies can also cross the placenta from a pregnant woman into her baby, providing the baby with some protection for a few weeks after it's born (Fig. 16.5a). The blood protein fraction called **gamma globulin** is rich in this class of antibody. Doctors sometimes inject gamma globulin into a patient who has just had surgery, or a patient headed for a country with hepatitis, typhoid, or other endemic diseases, to provide that traveler with immediate protection.

Another class of antibodies has a stem shape that allows these molecules to enter bodily secretions such as saliva, tears, milk, and the mucus from a runny nose or the intestinal lining. These molecules can pass to a nursing infant through the mother's milk (Fig. 16.5b) and can temporarily help protect the child from diseases to which the mother is immune—for instance, the strain of her last cold bug. Passing along antibodies is a strong rationale for breast-feeding.

A third antibody class binds to mast cells in the skin and intestinal lining. Recall that mast cells contain bags of chemicals that cause the swelling associated with inflammation (see Fig. 16.2, Step 5). Cells bound to this antibody class help fight parasites, but also produce the irritating condition we call **allergy** (Fig. 16.5c), which acts much like the body's nonspecific responses to a cold. Sometimes allergies can be dangerous: If the mast cells throughout the body react all at once to an antigen— let's say toxin from a bee sting—then vessels dilate throughout the entire body. This can lead to a life-threatening drop in blood pressure known as **anaphylaxis.**

B Cells in Action

Antibodies are obviously talented with their single basic Y-form and their millions of uniquely shaped recognition sites. But why does it take them so long to help a cold sufferer like Adrienne? This time lag between infection and elimination stems from the way the immune system produces antibodies. Every B cell is a small, colorless, rough-surfaced sphere (Fig. 16.6) that carries in its membrane many copies of the kind of antibody it will secrete when an antigen of the appropriate shape comes along and stimulates it. Immunologists have discovered that each individual B cell carries and secretes antibodies with binding sites of *only one shape.* "One cell, one anti-

body" is the shorthand way to remember this important phenomenon.

We're protected from most pathogens because our bodies contain millions of different B cells and each one is able to make an antibody of a different shape. These cells are lodged in lymph nodes, including Adrienne's, and tonsils or are present in blood and lymph. Each B cell responds to a different antigen, and the response involves the initiation of **cell proliferation,** as you can see in Figure 16.7.

1. Each B cell has on its surface copies of an antibody with a unique shape based on the amino acid sequences of the Y-shaped antibody molecule.
2. When rhinovirus antigens entered Adrienne's body, some bound to antibodies on the B cells that fit them best.
3. This binding of antigen to antibody stimulated the B cell to divide rapidly and thus *selected* it to proliferate into a **clone,** a group of cells all descended from a single parent cell.
4. In about 10 days, each single B cell generated a clone of 1000 daughter cells. The combined process of selection by antigen and proliferation into a clone is called the **clonal selection mechanism** of antibody

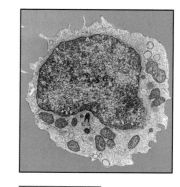

Figure 16.6 **B Cell: Colorless, Bumpy, Essential.** Every B cell secretes antibodies of only one shape: one cell, one antibody. In this artificially colored photo, the blue area is the cytoplasm, the red area is the mitochondria, and the yellow area is the nucleus.

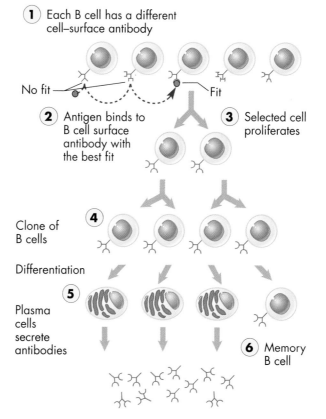

① Each B cell has a different cell–surface antibody

No fit — Fit

② Antigen binds to B cell surface antibody with the best fit

③ Selected cell proliferates

Clone of B cells **④**

Differentiation

Plasma cells secrete antibodies **⑤**

⑥ Memory B cell

Figure 16.7 **B Cells: Clonal Selection and Antibody Production.** Each of the millions of types of B cells can recognize an antigen of specific shape (1). Antigen binding selects a B cell (2) to proliferate (3) into a clone of identical B cells (4). These differentiate into plasma cells (5), which can secrete antibodies to the specific antigen, or memory cells (6), which can respond months or years later.

formation. The time it takes for cell division contributed to Adrienne's long-lasting cold.

5. Some of the daughter cells in each clone in Adrienne's tonsils stopped dividing and differentiated into plasma cells, living antibody factories. Plasma cells focus their energy on antibody production and survive only a few days.

6. Other cells of the clone did not differentiate immediately into antibody-secreting factories. Instead, they became memory B cells, or **memory cells.** Like the original B cell, memory B cells advertise their antibody on their surface and can proliferate when stimulated by the same type of antigen that triggered the original B cell (in Adrienne's case, the specific rhinovirus strain she caught). With this second round of division, the memory B cells form their own clones of both plasma and memory cells, all with the ability to make antibodies of the same specificity as those carried by the original B cell.

The first time a specific antigen stimulates an animal, only a few cells may respond. This first response is known as a **primary response.** But the next time the animal encounters the same antigen, say, the same strain of rhinovirus, it is better prepared because it has a large pool of memory cells partway along the road to becoming differentiated plasma cells. The net result is a stronger, swifter reaction to the invader—a **secondary response.** Memory B cells can thus make the difference between resisting disease and succumbing to it.

By the time healthy newborns grow to adolescence, they will have developed many sets of protective memory cells, providing immunity to many kinds of infections. Memory cells can survive for several decades, and this explains why most of us rarely contract measles or chicken pox a second time. If a grandparent had measles as a second-grader and is then exposed to measles while babysitting a sick grandchild, a large pool of memory cells in the older person is stimulated to produce antibodies that help eliminate the viruses before they have a chance to cause disease.

The message about cells bears repeating: Each type of antigen stimulates specific individual B cells to form a clone of daughter cells. Some of these identical daughter cells will secrete antibodies that target and destroy invaders immediately like guided missiles. Others will remain in reserve as memory cells in case of later exposure to the same antigen.

The Antibody Diversity Puzzle

Your body contains B cells that can form an antibody to virtually any antigen you could ever encounter: a rare virus from Borneo, a pollen grain from a Mongolian steppe grass, and even molecules of an organic chemical just invented by a chemist in the laboratory! *But how?* How can your cells contain the huge number of genes necessary to code for millions of different antibodies, including antibodies against molecules you will probably never encounter and some that haven't even been invented yet? Recall from Chapter 7 the "central dogma" of biology: that each polypeptide is encoded by a single gene. Does that mean there is a different gene for each different antibody? If that were true, we'd need more genes for making individually shaped antibodies than for all our other body functions combined. In a sense, the immune system contradicts this dogma. To see how, consider two familiar kinds of restaurants. In one, a particular main dish comes with certain set side dishes—no substitutions. That's how most genes work. In another, you can build a unique dinner by selecting one appetizer from six choices, one salad from four selections, one entree from twenty alternatives, and one dessert from five. In a similar way, a mature antibody gene is built from gene parts put together from three or four subparts, each with many different possible base pair sequences. By this special mechanism of gene recombination, a relatively small number of genes can create an almost infinite number of antibody protein shapes.

T Cells in Action

Although Adrienne Wilson was still feeling sick when she came to Dr. Gwaltney's clinic, her fast nonspecific resistance was already fighting the rhinovirus infection and so was her slower antibody-mediated immune response—her antibodies and B cells. Yet another group of cells, the T cells, were also involved. T cells help regulate immune responses and are involved in the **cell-mediated immune response.** During this, T cells directly attack and eliminate foreign intruders, usually large invaders like parasites, cancer cells, some virus-infected cells, or tissue transplants rather than small invaders like rhinovirus or bacteria.

Types of T Cells

While all B cells have the same function—to produce antibodies—T cells play at least three different roles in the immune response. (1) Half of all T cells are **cytotoxic T cells,** which destroy foreign cells, including cells transplanted to a person from another individual, or body cells transformed by virus infection or cancer (Fig. 16.8). (2) **Helper T cells** communicate via chemical signals with B cells and with the other types of T cells; this "dialogue" can initiate an immune response or aug-

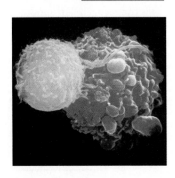

Figure 16.8 T Lymphocytes Destroying a Cancer Cell. This false-colored scanning electron micrograph captures a rounded T cell (yellow) in the process of attacking a single, larger cancer cell (purple).

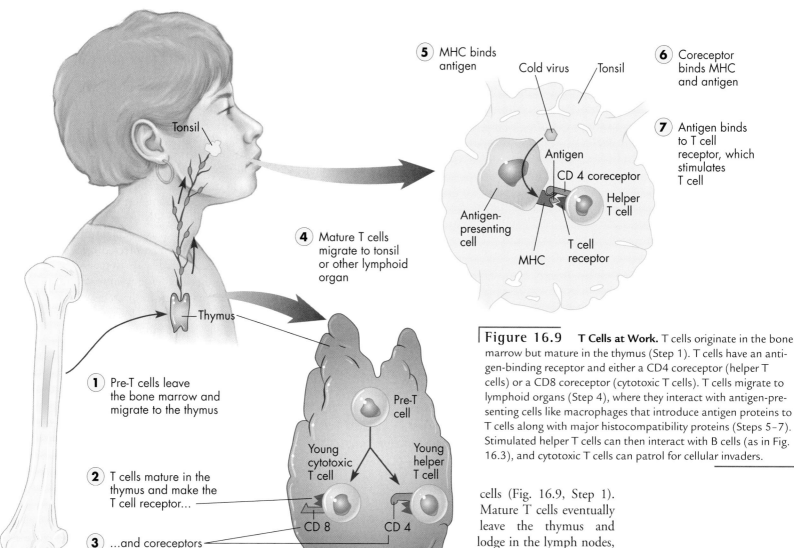

⑤ MHC binds antigen

⑥ Coreceptor binds MHC and antigen

⑦ Antigen binds to T cell receptor, which stimulates T cell

Cold virus
Tonsil
Antigen
CD 4 coreceptor
Helper T cell
Antigen-presenting cell
MHC
T cell receptor

Tonsil

Thymus

④ Mature T cells migrate to tonsil or other lymphoid organ

Pre-T cell

Young cytotoxic T cell

Young helper T cell

CD 8

CD 4

① Pre-T cells leave the bone marrow and migrate to the thymus

② T cells mature in the thymus and make the T cell receptor...

③ ...and coreceptors

Figure 16.9 **T Cells at Work.** T cells originate in the bone marrow but mature in the thymus (Step 1). T cells have an antigen-binding receptor and either a CD4 coreceptor (helper T cells) or a CD8 coreceptor (cytotoxic T cells). T cells migrate to lymphoid organs (Step 4), where they interact with antigen-presenting cells like macrophages that introduce antigen proteins to T cells along with major histocompatibility proteins (Steps 5–7). Stimulated helper T cells can then interact with B cells (as in Fig. 16.3), and cytotoxic T cells can patrol for cellular invaders.

cells (Fig. 16.9, Step 1). Mature T cells eventually leave the thymus and lodge in the lymph nodes, spleen, and skin. The thymus shrinks as a person matures, and the immune system functions less effectively. This helps explain why people over 50 are urged to get flu shots at the start of each winter's flu season.

ment an existing one. (3) **Suppressor T cells** slow or stop ongoing immune responses.

How T Cells Work

Viewed through a microscope, T cells are practically indistinguishable from B cells: both are small, spherical, and colorless. But the two types of lymphocytes differ in their pathways of development. B cells arise in the bone marrow and mature there before eventually migrating to either the spleen, tonsils, or lymph nodes. T cells arise in the bone marrow, too, but pre-T cells then migrate to the thymus, where they acquire their definitive capabilities as one of the several types of mature T

T Cell Receptors: Key to T Cell Function

While in the thymus, pre-T cells develop the ability to bind specific antigens. Antigen-binding ability comes from antigen-binding proteins embedded in their cell surface membrane, just as in B cells. In B cells, however, the receptor protein is an antibody, while in T cells it's a special **T cell receptor** (Fig. 16.9, Step 2). Another protein, called the **coreceptor,** is associated with the T cell receptor, and researchers picture it as a cane-shaped molecule like the ones in Figure 16.9.

Different types of T cells display different coreceptors on their surfaces. Helper T cells have a coreceptor called CD4, and young cytotoxic T cells have CD8 coreceptors (Step 3). After T cells mature in the thymus, they migrate to lymph nodes (Step 4), where they sometimes encounter invaders.

Self and Nonself: Major Histocompatibility Proteins

The T cell receptor binds antigen only when a macrophage or other immune cell presents antigen to it along with the self-marking membrane molecules called **major histocompatibility proteins (MHC)** (*histos* = web of a loom, or tissue). MHC molecules are a key to the T cell's two main functions: the ability to directly kill cells bearing antigens and the ability to regulate the immune response.

Each person's cells have a unique array of MHC proteins serving as cellular fingerprints that distinguish self from nonself. Biologists first recognized MHC proteins for their role in determining whether an animal will accept or reject a tissue graft, such as skin transplanted from a donor onto a burn victim. That transplant can be rejected if the MHC "fingerprints" of the graft's cells differ from the recipient's. You've probably heard of desperate searches by doctors and families to find compatible donors of bone marrow, livers, kidneys, or other organs to save a gravely ill loved one. What physicians are searching for in cases like these are organs with MHC proteins very similar or identical to the patient's. If the MHC proteins are different, the patient's cytotoxic T cells will recognize the grafted tissue as nonself and attack it, usually causing rejection.

During an immune response to a foreign invader, MHC proteins and coreceptors work in concert with the T cells. As Figure 16.9 shows, phagocytes attach foreign antigens to their MHC proteins (Step 5). The coreceptor on a T cell then binds to this MHC–antigen complex (Step 6), bringing the two cells very close together. The antigen then binds to the T cell receptor (Step 7), which activates the T cell and causes it to divide into a clone. The clone of activated T cells can then do its job: helping B cells produce antibodies (Fig. 16.3, Step 3), killing virus-infected or cancerous host cells, or attacking and destroying invaders.

Even in cases of mild illness like Adrienne's cold, T cells are doubly important. First, helper T cells stimulate B cells to divide rapidly in the cold sufferer or other patient (see Fig. 16.3). Second, cytotoxic T cells can directly target and kill virus-infected cells, such as nasal membrane cells infected with rhinovirus. Cytotoxic T cells can kill other cells in a manner similar to natural killer cells: releasing perforin proteins, which poke holes in the target cell (see Fig. 16.1, Step 3). Both of these T cell actions help a cold sufferer or other patient regain her health.

T Cells and AIDS

Sooner or later, Adrienne will get rid of her rhinovirus infection. People infected with HIV, the human immunodeficiency virus, however, generally don't eliminate the infection from their bodies. Why not? The

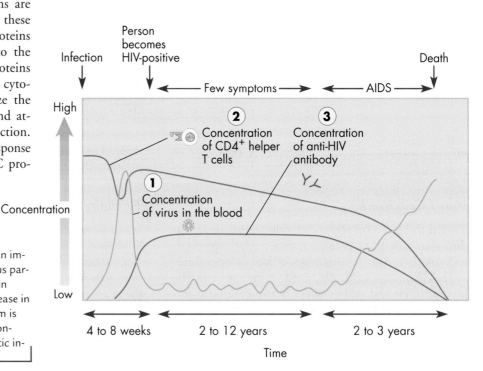

Figure 16.10 **AIDS: A Sharp Decline in Helper T Cells.**
A single act of unprotected sexual contact can introduce HIV (human immunodeficiency virus) into a person's body. Within a few weeks, virus particles multiply (violet line) (Step 1) and CD4$^+$ helper T cells decline in numbers (blue line) (Step 2). Recovery of helper T cells and the increase in anti-HIV antibody (red line) (Step 3) suggest that the immune system is mounting a defense against the invader. Nevertheless, helper cells continue to die and finally the classic signs of AIDS emerge: opportunistic infections, wasting of the body, and dementia.

difference is simple but profound: rhinovirus infects nasal epithelia, but HIV *infects T cells themselves* by binding to the CD4 coreceptor. Because, as we just saw, helper T cells are central regulators of the immune response, this infection damages or destroys the body's specific defense mechanism.

Since researchers first studied and described AIDS about 20 years ago, medical workers have come to diagnose AIDS partly by the decrease of T cells bearing the CD4 coreceptor ($CD4^+$ cells). Figure 16.10 shows what can happen to these cells after even a single act of unprotected sexual contact with a person already carrying the virus. After a few weeks, the concentration of HIV particles rises in the blood (Step 1), and the concentration of $CD4^+$ cells dips (Step 2). At about two months after infection, antibodies to the virus first appear (Step 3), $CD4^+$ cells begin to recover, and as a consequence, the amount of virus circulating in the blood drops. The anti-HIV antibodies signal that a person has become HIV-positive. Over the course of the next 2 to 12 years, the concentration of the person's $CD4^+$ cells gradually falls, eventually approaching zero. (You may remember from the Chapter 2 case study that Dan's T cell count fell to nearly zero at one point.) Simultaneously, the circulating antibody continues to decrease, the AIDS virus multiplies steadily, and the secondary infections that characterize full-blown AIDS set in, including Dan's purple spots from Kaposi's sarcoma and his rampant yeast and bacterial infections. Sometimes, an AIDS patient dies within a matter of months from pneumonia or other complications that could have been defeated by a healthy immune system. The decline in $CD4^+$ cells leads to AIDS by the series of steps pictured in Figure 16.11 and discussed in *E* 16.2.

16.3 Regulating the Immune System

Whether a person has a devastating illness like Dan's AIDS or a mild infection like Adrienne's cold, our immune system, with its ability to recognize and eliminate specific molecules, is crucial to our daily lives. Its precision is so pinpoint, in fact, you have to wonder why it can produce antibodies and cytotoxic T cells against foreign cells but doesn't usually attack and destroy its own proteins. The answer is **self-tolerance,** the lack of an immune response to one's own molecules. Biologists think that self-tolerance develops because the B cells capable of responding to our own antigens are selectively

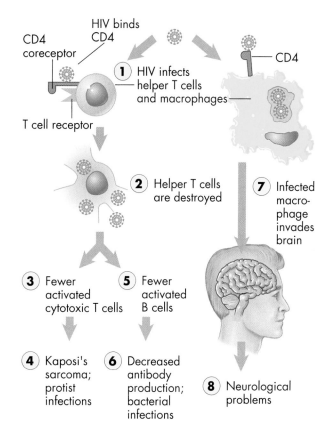

Figure 16.11 How the AIDS Virus Wreaks Havoc in the Human Body. (1) The virus that causes AIDS infects helper T cells by binding to the CD4 coreceptor. (2) The virus destroys the infected helper T cells. (3) Too few helper T cells remain to induce cytotoxic T cells to form. (4) Cytotoxic T cells are unavailable to kill cancer cells, such as those that cause the tumors of Kaposi's sarcoma or to ward off other opportunistic infections. (5) Without helper T cells, B cells are not stimulated to produce antibodies. (6) The victim's body can't combat viruses in the bloodstream, or bacterial invaders. (7) CD4 coreceptors also exist on macrophages, which take in HIV by phagocytosis. Infected macrophages can enter the brain, causing neurological problems, such as dementia, in some AIDS patients (8).

killed off when we are fetuses and newborns. The few self-reactive cells that manage to survive are apparently held in check by interactions with special suppressor T cells in a way that biologists don't yet fully understand.

Unfortunately, sometimes tolerance goes awry, leading to **autoimmune diseases,** attacks on certain cells or tissues by the person's own immune system. This may be the result of misdirected helper T cell activity. **Rheumatoid arthritis** is an autoimmune disease that afflicts millions of people in the United States

Figure 16.12   Autoimmune Disease: A Breakdown of Self-Tolerance.
In a person with rheumatoid arthritis, overly active immune cells spark inflammation and joint destruction. Some immunologists think that excessive, misdirected helper T cell activity is the major culprit. Researchers are still looking for more specific drugs with few side-effects to fight autoimmune diseases.

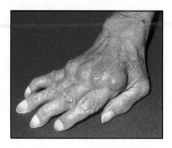

ther's histocompatibility proteins, burrows into the uterus. So why doesn't the mother's body reject the half-foreign fetus? Studies have shown that the uterus is a special immunological zone during pregnancy. During these nine months, a woman's body rejects a graft from the fetus if it is placed anywhere but the uterus. Researchers are still studying how embryos are protected in the womb, but it must be quite complicated, precise, and efficient in order to defend the tiny cluster of tissues from the mother's powerful network of immune cells and organs.

Despite this protection, the fetus sometimes does come under attack. About one couple in 15 have **Rh incompatibility,** a situation in which their newborn infant can contract a serious anemia called **erythroblastosis fetalis.** Many people have Rh antigen on the membranes of their red blood cells and are thus said to be Rh-positive. Others, however, lack the Rh antigens and are Rh-negative. When an Rh-positive man and an Rh-negative woman produce a baby, the infant may be Rh-positive (Fig. 16.13, Steps 1–2). If the baby's blood cells mingle with the mother's during delivery (Step 3), and its Rh antigens enter her bloodstream, they may stimulate her immune system to produce antibodies

alone. In people with this condition, the joints swell painfully, the fingers can become gnarled and twisted, and everyday movements, like buttoning a shirt, can be painful or even impossible (Fig. 16.12). **Multiple sclerosis** is another autoimmune disease, in which a person's immune system reacts to substances (myelin) that surround and protect nerve cells. In some types of diabetes, a patient's immune system destroys some of the pancreatic cells that normally produce insulin. This hormone helps regulate the levels of sugar in the blood, and without it, metabolism is disrupted.

Pregnancy is an interesting exception to the immune system's recognition and elimination of foreign antigens. The human fetus, containing some of the fa-

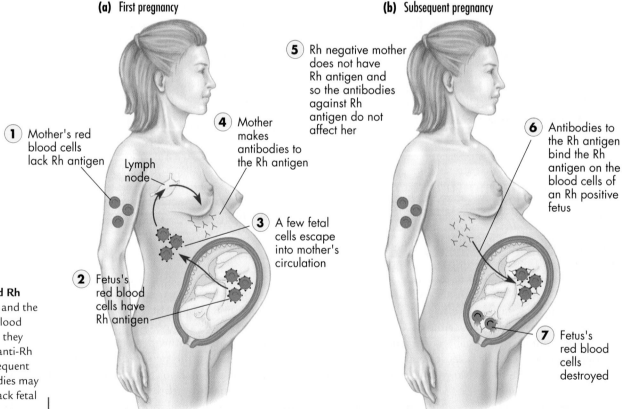

(a) First pregnancy

① Mother's red blood cells lack Rh antigen

Lymph node

④ Mother makes antibodies to the Rh antigen

③ A few fetal cells escape into mother's circulation

② Fetus's red blood cells have Rh antigen

(b) Subsequent pregnancy

⑤ Rh negative mother does not have Rh antigen and so the antibodies against Rh antigen do not affect her

⑥ Antibodies to the Rh antigen bind the Rh antigen on the blood cells of an Rh positive fetus

⑦ Fetus's red blood cells destroyed

Figure 16.13   Pregnancy and Rh Disease.
(a) If a fetus is Rh-positive and the mother is Rh-negative, and if fetal blood cells escape into the mother's body, they may trigger the mother to produce anti-Rh antibodies (Steps 1–4). (b) In subsequent pregnancies, these maternal antibodies may enter the fetal bloodstream and attack fetal blood cells (Steps 5–7).

against the Rh antigen (Step 4). Because it takes several days for antibodies to form, the newborn is safe from harm. The antibodies against the Rh antigen don't affect the mother, either, because her cells don't produce Rh antigen (if they did, she'd be Rh-positive, not Rh-negative) (Step 5). During a subsequent pregnancy, however, the antibodies against the Rh antigen already present in the mother's blood from the first delivery can cross the placenta and attack the new fetus's red blood cells if they carry Rh antigen (Steps 6 and 7). Debris from the attacked cells can lead to anemia, brain damage, or even death of the fetus.

Physicians prevent this dangerous situation by injecting an Rh-negative mother with antibodies against the Rh antigen at the birth of her first Rh-positive child. These antibodies (called *Rhogam*) bind to the Rh antigens on fetal blood cells that may have entered the mother's circulation during delivery. Covered by these injected antibodies, the Rh-positive antigen fails to stimulate the mother's immune system, and the next fetus will be safe from attack.

16.4 Immunization

In the last 200 years, medical practitioners have learned to manipulate the immune system to help patients in two ways, termed passive and active immunity.

Passive Immunity with Borrowed Antibodies

Let's say you were out hiking and suddenly saw a rattlesnake, which reared up and bit you on the leg. Not only would you be shocked and frightened, but there would be no time to spare. The venom contains toxic proteins that stop nerves from functioning and that damage blood vessels. You would need a quick-acting remedy to keep these poisons from harming or even killing you. The best treatment would be **passive immunization:** injecting antibodies made by one individual into another individual (Fig. 16.14). Technicians prepare the anti-snake-venom antibodies by injecting a horse or rabbit with inactivated snake venom, which induces the mammal to form antibodies against the venom proteins. Workers then collect these antibodies, now called **antivenin.** When you rush into a clinic after a snakebite, a doctor injects this antivenin into you immediately and it quickly circulates through your bloodstream. There it can combine with (neutralize) the snake toxin in a typical antigen–antibody fashion be-

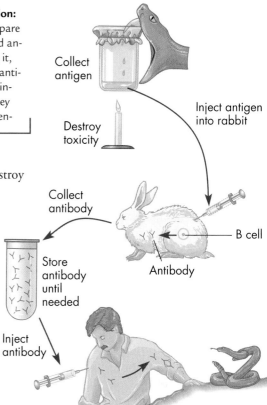

Figure 16.14 **Passive Immunization: Transferred Antibodies.** Technicians prepare commercial snakebite medicine (so-called antivenin) by collecting venom, inactivating it, and then injecting it into a rabbit. Later, antibodies are collected from the rabbit and injected into the snakebite victim, where they bind to and inactivate the toxic venom, rendering it harmless.

Collect antigen

Destroy toxicity

Inject antigen into rabbit

Collect antibody

Store antibody until needed

B cell

Antibody

Inject antibody

Rabbit antibody combines with antigen and deactivates it

fore, one hopes, the venom can destroy any of your nerve cell function.

Passive immunization has one great advantage: It works very fast. But it also has the disadvantage of acting for only a short time. Your immune system would soon recognize the borrowed antibody molecules as foreign and eliminate them. With the antibody molecules would go your passive protection, leaving you vulnerable to snake venom if bitten again.

Active Immunity with Altered Antigen

You've probably received many vaccines in your lifetime for diseases like polio, diphtheria, and/or measles. Unlike antivenin, which provides short-term passive immunity, vaccines like these provide long-term protection by stimulating **active immunity,** the production of antibodies or antigen-specific T cells by the individual's own immune system. Safer and more effective than most drugs, vaccines provoke a specific response aimed at one microbe or toxin, and nothing else in the body.

Edward Jenner, an 18th-century English country physician, developed the first vaccine almost 200 years ago. To help protect people from smallpox, a severely disfiguring disease that killed one out of every four people in Jenner's time, he injected people with a small amount of cowpox virus (the word **vaccination,** in fact, comes from the Latin, *vacca* = cow). Jenner based his technique on the observation that milkmaids who contracted cowpox from the cows they milked always recovered and almost never got smallpox. Indeed, the cowpox virus he injected caused mild sickness and discomfort, but it still protected people against smallpox. By 1978,

Figure 16.15 Active Immunization: Stimulating B Cells With Vaccines. Modern medicine has 20 types of safe, effective vaccines in its arsenal. This 4-year-old child is receiving a vaccination. Smallpox was once a major scourge but vaccines have successfully eradicated it. Programs are underway to develop new vaccines against HIV, malarial parasites, rhinovirus, and other invaders, but the obstacles are large.

modern vaccines based on Jenner's original ones had successfully eradicated smallpox worldwide (Fig. 16.15).

Modern vaccines are made of microbes and toxins that have been killed or otherwise modified in the laboratory so they can't cause disease. The first shot with altered germs stimulates the production of antibody-producing cells and memory cells (review Fig. 16.7). A **booster shot,** which is simply a second dose of the same vaccine, can then induce the memory cells to differentiate and form still more antibody-producing and memory cells. If a vaccinated person later comes in contact with live bacteria, virus, or toxin carrying those antigens, his or her body will already contain many antibodies and memory cells that can quickly eliminate the dangerous agents and prevent disease. Vaccination is a slow-acting process, requiring a booster shot and several weeks for the development of adequate protection. Nevertheless, the active immunity it stimulates lasts a long time, sometimes a lifetime.

Many medical researchers see new vaccines as the best hope for combating infectious diseases that can't be treated or cured by other approaches. Unfortunately, many viruses, including rhinovirus, come in many different strains. And other invaders like the malarial parasites we encountered in Chapter 11 can change their surface properties so often that they evade active immunity. Although it's been two centuries since Jenner's first vaccine, medical workers still have developed only about 20 safe, effective vaccines. Contemporary researchers are using recombinant DNA and other forms of biotechnology to create synthetic vaccines against influenza, HIV, and other viruses that can "outwit" the immune system.

Ironically, Dr. Gwaltney's research has shown that, for unknown reasons, a person makes circulating antibodies to the rhinovirus only about half the time they catch a cold. And because there are over 100 different strains of rhinovirus, it's going to be hard—maybe impossible—to make safe, effective vaccines against each one. In the meantime, Dr. Gwaltney's research group is aiming for new antiviral drugs that will directly eliminate cold viruses or keep them from spreading. And they continue to recommend over-the-counter measures (antihistamines, pain killers, and cough suppressants) to block the side effects from our nonspecific responses, while they continue to search for new and better drugs that will actually fight the rhinovirus and other viral invaders.

Chapter Summary and Selected Key Terms

Introduction A cold can be miserable and Adrienne Wilson knows it, because she has another one. Researchers like Dr. Jack Gwaltney of the University of Virginia are probing the secrets of the common cold and how our immune systems both contribute to the misery but also eventually end it. **Resistance** to disease (p. 428) involves a network of organs, cells, and molecules that defend the body from invaders of all kinds, including cold viruses such as **rhinovirus** (p. 427). The most familiar part of this network is the **immune system** (p. 428), and the specifically targeted proteins or **antibodies** (p. 428) it produces.

Nonspecific Resistance **Nonspecific resistance** (p. 428) involves physical barriers such as skin, a first line of defense against intruders such as **pathogens** (p. 428). Nonspecific responses also include the **in-**

flammatory response (p. 430), with its generation of **inflammation** (p. 428). The second line of defense includes nonspecific substances and cell activities. The molecules include **interferons** (p. 429) and proteins of the **complement system** (p. 430). **Inflammation mediators** (p. 431) include **prostaglandins** (p. 431), and **cytokines** (p. 431). **Natural killer cells** (p. 430) can kill many types of microbes, using **perforins** (p. 430). An invading bacterium might encounter **phagocytes** (p. 430). The nonspecific molecules also (1) cause mast cells (p. 431) to release **histamine** (p. 431) and **heparin** (p. 431), which cause tissue fluid leaking; (2) stimulate more mucus formation to float away virus and debris; (3) cause a hacking cough (prostaglandins) and/or a sore throat (cytokinins). Dr. Gwaltney recommends taking an antihistamine, a generic pain killer, and a cough suppressant (if you have a cough) to lessen these symptoms of inflammation as soon as they set in.

Specific Resistance **Specific resistance** (p. 429) or **immunity** (p. 429) is a slower series of reac-

tions than the nonspecific responses, but is more targeted to particular invaders or damaged cells (malignant or viral-infected) within the organism. Any substance that can stimulate an immune response is an **antigen** (p. 432), including viral and bacterial surface proteins. The **antibody-mediated immune response** (p. 433) responds to some invaders by producing protective proteins called **antibodies** (p. 432). It has the remarkable feature of **immunologic memory** (p. 432). **Interleukins** (p. 433) speed or slow the dividing and maturing of other immune cells and so are involved in the system's communication activities. **Major histocompatibility complex (MHC)** (p. 433) proteins stud the surface of white blood cells, label most body cells as "self," and aid in immune system communication.

Several types of white blood cells, or **lymphocytes** (p. 433), are part of the **cell-mediated immune response** (p. 433). **B cells** or **B lymphocytes** (p. 433) are formed in bone marrow; these secrete antibody proteins and mark invaders for destruction. **T cells** or **T lymphocytes** (p. 433) come from the bone marrow but mature in the thymus and kill virus-infected or cancerous cells. **Macrophages** (p. 433) consume debris from intruders and present antigen to B and T cells.

Antibodies are able to mark and eliminate intruders because they are **immunoglobulins** (p. 435), or globular proteins with four polypeptide chains that form a Y shape. The tips of the two upper branches of the Y have variable amino acid sequences and thus shapes, and bind to an antigen of complementary shape like a lock fits a key. This marks the antigen-bearer for elimination. The lower stem of the Y stimulates the immune system to attack and destroy the entire antibody–antigen complex. A person can make more than 100 million distinct types of antibody molecules, each with a unique binding site. Biologists prepare individual clones of particular B cells and collect the targeted antibodies they give off, also called **monoclonal antibodies** (p. 436) for use in medicine and industry, including AIDS diagnosis. Some classes of antibodies can cross the placenta, and doctors can inject **gamma globulin** (p. 437) to protect travelers and surgery patients. Another class enters bodily secretions, including mother's milk, and can pass into a nursing baby. A third class binds to mast cells in the skin and intestinal lining and inadvertently causes **allergies** (p. 437) and, rarely, the dangerous condition called **anaphylaxis** (p. 437).

Each B cell has surface copies of a unique antibody. In a **clonal selection mechanism** (p. 437), a B cell encounters and binds to an antigen, let's say, a rhinovirus protein; this selects the bound B cell to divide and proliferate into a **clone** (p. 437) of identical B cells. Within 10 days, a single B cell can generate a clone of 1000 daughter cells that mature into either **plasma cells** (p. 438), which start generating antibody molecules, or into **memory B cells** (p. 438), which wait to be triggered by future antigen. An animal has a **primary response** (p. 438) the first time it encounters an antigen, but memory cells turn a second encounter into a much faster **secondary response** (p. 438).

In the cell-mediated immune response, T cells directly attack and eliminate foreign intruders. **Cytotoxic T cells** (p. 438) destroy intruders, mutated cells, or transplanted foreign cells directly. **Helper T cells** (p. 438) communicate with B cells and other T cells to build up an immune response. **Suppressor T cells** (p. 439) slow or stop immune responses. T cells have a **T cell receptor** (p. 439) and a **coreceptor** (p. 439) protein. These bind to antigens only when the antigen is presented by a macrophage (or other immune cell) along with a self-marking **MHC protein** (p. 440). The latter label our cells like fingerprints and are the crucial elements to match during organ or tissue transplantation. If these labels are too different, an organ recipient's cytotoxic T cells will recognize the trans-

plant as foreign and destroy it. During an HIV infection, the virus infects the T cells themselves and as their numbers decline, lethal opportunistic infections can arise and claim the patient's life.

Regulating the Immune System
B and T cells don't attack our own tissues because of **self-tolerance** (p. 441), which develops when we are fetuses or newborns. If this tolerance goes awry, an **autoimmune disease,** (p. 441) such as multiple sclerosis or rheumatoid arthritis, can result. During pregnancy, a woman's immune system leaves the fetus—which is one-half foreign—undisturbed. An exception involves **Rh incompatibility** (p. 442).

Immunization
Doctors can confer fast-acting immunity of short duration or slow-acting but longer-term immunity. **Passive immunization** (p. 443) uses an **antivenin** (p. 443) against snakebite, for instance, or gamma globulin for a traveler. **Active immunity** (p. 443) involves vaccines and **vaccinations** (p. 443) that safely stimulate the immune system to create antibodies against virus (such as smallpox) or bacteria to protect against potential future exposure. Researchers would like to create vaccines against the common cold but it is proving difficult because there are so many strains of rhinovirus and other cold viruses.

All of the following question sets also appear in the Explore Life *E* electronic component, where you will find a variety of additional questions as well.

Test Yourself on Vocabulary and Concepts

In each question set below, match the description with the appropriate term. A term may be used once, more than once, or not at all.

SET I
(a) lymphocyte (b) macrophage (c) interferon
(d) complement (e) inflammation

1. A type of scavenging white blood cell
2. Localized increase in blood flow, redness, and swelling
3. A protein that prevents viral multiplication
4. A group of proteins that enhance the inflammatory and immune responses

5. A kind of white blood cell that can mature into B or T cells

SET II
(a) antigen (b) antibody (c) interleukin (d) MHC proteins (e) B cells (f) T cells

6. Protein growth factors that stimulate the development of antibody-producing cells
7. Any molecule that triggers an immune response
8. Cell-surface proteins that mark the identity of cells as "self" or "non-self"
9. Develop into cells that produce circulating antibodies
10. Major regulatory cells of the immune system that mature in the thymus

SET III
(a) immunoglobulins (b) helper T cells
(c) memory cells

11. CD4 cells that are attacked by HIV are one kind of this cell.
12. Can cross the placenta and confer immunity in a fetus against bacteria and viruses to which the mother also has protection
13. A long-lasting, active immunity occurs because a small clone of these cells is formed at the time of immunization and is activated to produce secondary responses

Integrate and Apply What You've Learned

14. List the body's nonspecific resistance responses and contrast their effectiveness as a group with that of the specific immune responses.
15. Relate the stages of a cold to the body's nonspecific and specific responses.
16. How is one antibody different from another and what results from this difference?
17. What characteristics of antibodies make them superb diagnostic reagents—chemicals that can test whether or not another substance is present?
18. Although antibodies are proteins and proteins are coded by genes, we can manufacture more kinds of antibodies than we have genes. Explain how this paradox can occur.
19. Contrast self-tolerance with autoimmunity. Which cells are believed to be active in each process?

Analyze and Evaluate the Concepts

20. What is the function of the antibodies found within the blood of people infected with the HIV virus?

21. Most colleges require entering students to show proof of vaccination against measles, but they do not require such proof of anyone over the age of 35. Why?

22. While on a hike, a companion is bitten by a poisonous black widow spider. One member of the group suggests the bitten person should be rushed to the hospital to be protected by active immunity. Another says that passive immunity would be better in this case. Which do you think would be more helpful to the bitten friend and why?

23. You are the head of a rhinovirus research lab, and one of your graduate students suggests that because the rhinovirus changes its antigens so often, passive immunity against rhinovirus would be a better protective strategy for older people than active immunity. Evaluate the pros and cons of this idea, and decide whether your lab should invest in further investigation of this idea.

24. What are some things you can do the next time you catch a cold and why would they help? How might your answer be different 10 years from now?

Up a Tree

She's so fuzzy and adorable that people come from hundreds of miles away just to watch her sit in a tree and sleep. Born in 1980, she's the human equivalent of 140 years old. She never drinks water and her favorite food is her only food. She rests or sleeps 80 percent of the time, eats during half of her waking hours, and moves about for just 4 minutes a day. Meet Point Blank ("P.B." for short), the oldest koala in captivity.

P.B. lives in the Small Mammal House at the Lincoln Park Zoo in Chicago. She's been a favorite attraction there since she arrived in 1985 and has outlived an elderly companion who passed away at age 19. Hers is an upbeat story of long life, health, and successful zoo husbandry. And it's an ironic story, because although P.B. has the simplest diet out of 226 species at the zoo, it requires a biweekly airlift of hand-picked eucalyptus stems and leaves—80 pounds of them in a box the size of a refrigerator—to keep this 14-pound (9-kg) marsupial happy!

Koalas eat nothing but eucalyptus leaves. No vitamins. No special treats for good behavior. Nothing. You might think that would make feeding time a breeze once the leaves are procured. But "these animals are *real* picky," says Dr. Sue Crissey, director of nutritional services for the Brookfield Zoo, also in Chicago, and the consultant responsible for planning P.B.'s diet at Lincoln Park. "There are probably 600 species of eucalyptus trees," Crissey says, "but koalas will only touch 35 of them."

The head veterinarian at Lincoln Park, Dr. Robyn Barbiers, is responsible for P.B.'s general health and works closely with Dr. Crissey, with a staff of koala keepers, and with a eucalyptus grower in Florida to insure a constant supply of leaves and twigs. "Every day," says Dr. Barbiers, "we offer at least four different species to the koala. Depending on the season, we may get four species shipped or ten. Throughout the year, it's 15 species total. But our koala is sporadic in what she likes . . . there's no predicting it." Beyond that, P.B. will "nibble a little bit of every leaf and then drop the rest." Now *that's* picky. And it helps explain how a 14-pound animal can run through nearly 5 times its weight in expensively shipped leaves every week.

But not entirely. The explanation rests with the koala's unique diet and digestive system. Here is a mammal that lives without shelter in the branches of eucalyptus trees in its native Australia, rarely comes down to the ground, virtually never drinks water, and exists for decades on nothing but the moisture and nutrients in leathery, gray-green eucalyptus leaves. If you tried this regimen—we'll call it the All-New Revolutionary Koala Diet—you'd lose weight fast! But that's because the leaves contain

▲
Feeding, Digestion, Nutrition. Sue Crissey and Robyn Barbiers; P.B. the Koala; Feeding time at the zoo.

few nutrients and have a bitter, nauseating, turpentine flavor due to their toxic essential terpene oils and tannins. The fact that your body could never absorb enough nutrients from these nasty-tasting leaves is only the beginning. You'd also dry out from water loss, have a continuously full digestive tract from the bulky, fibrous leaves, and be hungering for your old foods within a matter of hours.

How can P.B. live so long and contentedly on the Koala Diet when you couldn't do it for a day? For that matter, how can a robin eat mostly worms? A hyena raw meat, hide, and bones? A cow mostly grass and hay? A sea lion mostly fish? You'll discover the answers in this chapter as we explore the subjects of digestion and nutrition: How do animals, which are heterotrophs, obtain energy and materials—including amino acids, sugars, fatty acids, vitamins, and minerals—from the foods they ingest? And how do the energy and materials, in turn, fuel activity, growth, and maintenance in these multicellular organisms?

You'll see how the koala's unique digestive organs allow it to utilize eucalyptus leaves, with a little help from intestinal bacteria. You'll see, too, how for every animal, digestive anatomy and physiology are closely tied to what the animal eats and what its body needs. You'll see how the koala's digestive system breaks down food in a series of mechanical and chemical steps that release nutrients slowly and dependably. And you'll see how, in overall action, the koala's digestive system is similar to a dog's, a horse's, a cow's, and a person's.

Along the way, you'll find out why eucalyptus leaves poison virtually all other mammals but not the koala. And you'll find answers to these questions:

1. What are nutrients?
2. How do animals digest food?
3. What are the parts of the digestive tract?
4. What is the form and function of our own human digestive system?
5. How can we make sense of the voluminous but conflicting and frequently changing information on healthy diets and proper nutrition?

17.1 Nutrients: Sources of Energy and Elements

To plan out the right diet for P.B.—even a diet as simple as straight eucalyptus leaves—Drs. Crissey and Barbiers rely on **nutrition**, the study of precisely how much protein, carbohydrates, lipids, vitamins, and minerals, and how many calories of food energy an animal must consume to stay alive and healthy. Without this knowledge, zoo workers could never be sure their koalas, camels, cobras, and hippopotami were getting enough of the right foods to fend off disease. Without scientific investigation of human nutrition, parents might wonder if their children were getting the elements essential for full mental and physical development. And adults would wonder if they were eating the right foods to prevent cancer, heart disease, and obesity, yet have enough energy for their daily activities and exercise.

Even with their highly specialized diet, koalas—as heterotrophs—still need a supply of organic and inorganic nutrients, including carbohydrates, lipids, proteins, vitamins, minerals, and total calories, just as do other zoo animals and zoo visitors. We encountered these molecules in Chapters 2 and 3. Here we explore their roles in the animal body.

Carbohydrates: Sources of Carbon and Energy

The sugars in fruit; the starches in potatoes, rice, bread, and pasta; and the cellulose in the cell walls of eucalyptus leaves are rich sources of energy and of carbon atoms. Together, these and certain other foods provide the nutrients we call carbohydrates. As Chapter 2 explained, carbohydrates include monosaccharides, such as the glucose found in honey; disaccharides, such as sucrose, or table sugar; and polysaccharides, such as potato starch and cellulose (see Fig. 2.8).

Our human digestive system can derive glucose from the sugars in grapes or peaches or from the starch in rice, wheat, corn, or potatoes. After a meal, the simple sugars pass into the bloodstream, and the circulatory system carries them to cells throughout the body, where they supply most of the energy for glycolysis and aerobic respiration (Fig. 17.1a). Nerve cells in the brain and throughout the body are particularly sensitive to fluctuations in blood glucose levels. If starving, the body will break down its fat stores first, then move on to its own muscle tissues, converting the subunits to glucose and providing the sensitive nervous system cells with what they need to stay fully active.

The cellulose fibers in plant cell walls, including the cells of leaf blades, and the "strings" in celery stalks, are also complex carbohydrates. We can't digest these cellulose fibers and neither can other animals. So how can strict vegetarians like koalas, cows, termites, and zebras get what they need from their all-plant diets? The answer is they have cellulose-digesting microbes in their digestive tracts, and these microbes break down the cellulose into glucose subunits, providing the animals with energy. Even with its cellulose-digesting microbes, the koala still has a challenging diet because eucalyptus leaves contain a large amount of lignin, an indigestible material that strengthens plant cell walls (see Chapter 21). The koala can digest a small part of the lignin, although experts aren't sure how. But eucalyptus leaves are 36 to 56 percent cell walls—more fibrous than "All-Bran" high-fiber cereal, at 32 percent. Analysis shows that because of the high percentage of cellulose and lignin in the leaves, koalas get most of their nutrients (87 percent) from the fluid inside the eucalyptus leaf cells. We get some cellulose in our own diets, in vegetables, fruits, and whole-grain cereals, but because we can't digest it at all, it provides the **dietary fiber** or **roughage** we need to help propel wastes through the large intestine.

Soluble fiber comes from pectin, a polysaccharide that helps glue adjacent cell walls together in plants. Pectin is particularly abundant in apples, citrus fruits, and strawberries. Soluble fiber forms a gel when mixed with water, and nutrition researchers think that dietary soluble fiber helps lower a person's cholesterol (and thus decreases the risk of arterial disease) by binding to and eliminating from the body a substance made from cholesterol, called bile. Insoluble fiber won't dissolve in water, and it's plentiful in bran.

This fiber type helps move material through the intestines faster, and researchers think that because of this characteristic, insoluble fiber may help decrease the risk of colon cancer.

The koala, of course, gets a huge amount of fiber—so much that it poses a problem: How can a small animal have a big enough digestive tract to hold all this bulky material while bacteria ferment some of it and release energy compounds? We'll see the answer later.

When it comes to carbohydrates, we often hear, "Avoid table sugar, desserts, and other sugary foods." The fact is, sugar provides calories of food energy, our largest nutritional need. Table sugar doesn't provide many vitamins or minerals, but neither do honey,

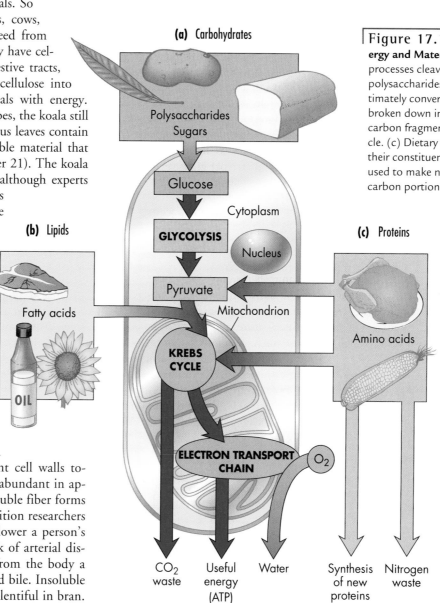

(a) Carbohydrates

Polysaccharides
Sugars

Glucose

Cytoplasm

GLYCOLYSIS

Nucleus

Pyruvate

Mitochondrion

(b) Lipids

Fatty acids

OIL

KREBS CYCLE

(c) Proteins

Amino acids

ELECTRON TRANSPORT CHAIN O_2

CO_2 waste

Useful energy (ATP)

Water

Synthesis of new proteins

Nitrogen waste

Figure 17.1 **How the Body Gets Energy and Materials from Foods.** (a) Digestive processes cleave carbohydrates in foods to polysaccharides and sugars. The sugars are ultimately converted into glucose. (b) Lipids are broken down into fatty acids and then two-carbon fragments that can enter the Krebs cycle. (c) Dietary proteins are broken down into their constituent amino acids, which can be used to make new proteins. In addition, the carbon portion of amino acids can be modified to intermediates in the Krebs cycle, and the amino part can be excreted as nitrogen waste.

brown sugar, or raw sugar. The issue is whether a person substitutes sugar-laden foods for more nutritious ones—a candy bar, say, instead of an apple. Bacteria in a person's mouth also prefer sucrose as a fuel, and the sugar allows the microbes to grow rapidly and produce acid that can cause cavities in your teeth. Some people crave sugars and other carbohydrates to the point of overeating, and this has led to the promotion of the low-carbohydrate, high-protein diets many now follow. We'll revisit that later, too.

Lipids: Energy-Storage Nutrients

Most people have only to visualize the solid white fat of a bacon slice or the slippery golden oil in salad dressing to know what a lipid is. Eucalyptus leaves are also filled with toxic oils that are lipid based. Like carbohydrates, lipids supply food energy, but animals generally store much more energy in the form of lipids than of carbohydrates, as our waistlines often reveal. Even a lean person, though, with no visible "spare tire," stores about seven times as much lipid as glycogen (a storage form of glucose). Why is lipid a better form for storing energy? First, lipids provide about twice as much energy per gram as carbohydrates (9 Calories per gram versus 4.5). After a meal of oily sunflower seeds or fatty meat, an animal breaks down lipids to products that move directly into the Krebs cycle (Fig. 17.1b). The Krebs cycle harvests energy from these products, and their carbon atoms become intermediates for building amino acids and other substances the body needs.

A second reason animals often store fat rather than carbohydrate is that fats and water don't mix, while carbohydrates tie up twice their weight in water. In fact, if you stored carbohydrates instead of fats, your body would weigh twice as much. Without fat stores, walruses couldn't live through an Arctic winter (Fig. 17.2), hummingbirds couldn't migrate across the Gulf of Mexico, and wolves would have more trouble surviving between the rabbits and other prey they catch as irregular meals.

Finally, certain essential nutrients, including vitamins A, D, E, and K, are fat soluble, and lipids help the digestive tract absorb these vitamins and deliver them to cells. The human body can't generate one key fatty acid—linoleic acid—from component parts, so it becomes an **essential fatty acid** that we must eat in our diets in order to build

parts of the cell membrane involved in importing materials. Lipids, then, are necessary nutrients and important storage molecules. But many of us eat a diet too rich in fats and oils, and this increases one's risk of obesity, colon cancer, and heart disease. In fact, according to a study of 88,000 nurses, a person who eats red meat (beef, pork, or lamb), with its relatively high fat content, every day has a risk of colon cancer three times higher than someone who ingests animal fat from those sources just two times a week or less.

Protein: Source of Amino Acids

Both a koala's thick, silvery fur and its sharp claws curved for tree-climbing are made almost totally of a single type of protein, keratin (Fig. 17.3). In marsupials and in all other mammals, skin, cartilage, tendons, bone, muscle tissue, and even the cornea that covers the eye are largely protein. So are enzymes, antibodies, hemoglobin, and some hormones, all necessary for life to continue and for cells to divide or recycle their constituents. The ubiquity of protein and its constant turnover means that animals need a continuous supply of the nutrient; from it, their digestive systems can extract amino acids and build new proteins (see Fig. 17.1c).

For a koala, getting enough protein is a real challenge and it helps explain why the animal eats so much.

Figure 17.2 Lipids: High-Energy Storage. Fats are a compact way to store calories. Active animals, such as this plump walrus at Chicago's Brookfield Zoo, have a thick fat layer that acts as superb insulation as well as a source of stored energy that the animal can use to generate heat and power activities. Human inactivity can lead to excess lipid storage.

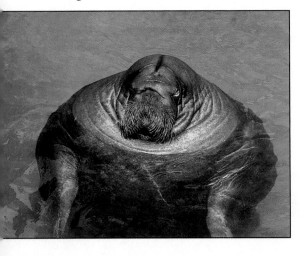

Figure 17.3 Claws and Fur Are Purely Protein. The koala's thick fur is made up almost entirely of the protein keratin, as are the claws on its fingers and toes. On the koala's back paws, the "index" and middle toes are fused together and the claws make a small comb the animal uses for grooming. On the koala's front paws shown here, the first two digits oppose the other three, making it look like it has two thumbs. These help the animal strongly grasp branches.

The protein concentration in eucalyptus leaves (8 percent by weight, excluding water) is very low compared with other plant foods such as oats (13 percent) or almonds (21 percent). Furthermore, the poisonous tannins and terpenes in eucalyptus leaves inhibit the animal's body from digesting and absorbing protein.

A 150-pound (68-kg) human is much larger and much more active than a 20-pound (9-kg) koala, and requires about 1 g (0.04 oz) of protein per kilogram (2.2 lb) of body weight per day. A college student of average height and weight needs at least one sixth of a pound (about 75 g) of pure protein each day to replace daily losses. This is about what you would obtain from a single cooked chicken breast. Some experiments with lab animals have suggested that eating too much protein can lead to kidney damage, because those organs must work overtime to excrete the nitrogen wastes derived from the amino acids in a high-protein diet. This, however, is in dispute; the advocates of low-carbohydrate, high-protein diets see these kidney-damage studies as flawed.

The amounts of protein needed are one current nutritional issue; the types of protein needed are another, and this has important implications for vegetarians. The body can synthesize many of the 20 amino acids it needs as long as the amino (nitrogen-containing) portion of the molecule is available (review Fig. 2.10). Just as your body can't make linoleic acid for lipids, however, it can't make eight amino acids: lysine, leucine, phenylalanine, isoleucine, tryptophan, valine, threonine, and methionine. For us, these are **essential amino acids** (Fig. 17.4) that we must eat in our foods every day, since our bodies don't store free amino acids. Children also need extra supplies of the amino acids histidine and arginine, because their bodies make enough for maintenance but not enough for growth.

Without one or more of the essential amino acids, the body can't build the full spectrum of proteins it needs to replace worn-out parts or to generate new cells. We saw in Chapter 7 that cells manufacture proteins by adding one amino acid at a time to a growing polypeptide chain. If even one of these amino acids is missing, the chain stops elongating, a bit like running out of red yarn while knitting a blue and red striped sweater; the project stops until you can get more. That's why lacking even one essential amino acid can stop protein synthesis in the body.

Most animal proteins in foods such as meat, cheese, eggs, and milk contain the eight essential amino acids as well as the other twelve. Many plant proteins, however, don't have the eight essentials. This is why vegans, strict vegetarians who avoid all animal products, including eggs and dairy products, must combine particular vegetable foods each day to ensure that they'll get enough of the right amino acids. Rice, for example, contains little lysine but plenty of methionine, so they'll eat rice together with beans, which are deficient in methionine but contain enough lysine (see Fig. 17.4). Some people in developing nations show symptoms of protein deprivation— swollen belly, skin and hair loss, and low levels of energy—because they eat a diet consisting mostly of starchy cereals and/or root crops that can lack essential amino acids. As a result, their bodies draw proteins from their own tissues, dismantling them and using their essential amino acids for new protein synthesis. Excess amino acids from their foods are simply excreted and wasted. When a human being starves, it's usually the lack of protein, rather than the lack of food energy, that leads to death.

Vitamins and Minerals: Important Nutrients

An animal's body, whether human or koala, needs relatively large amounts of protein and carbohydrate each day for building, repair, and energy. It needs another set of nutritive substances too, the vitamins and minerals, but only in minute amounts.

Vitamins

Vitamins are organic compounds needed in small amounts for normal growth and metabolism. Animals can't synthesize or, in some cases, store vitamins in large quantities, though, so they must take in a set of specific vitamins every day in their food.

Nutritionists like Sue Crissey doing research on human diets have found that people require 14 vitamins. Vitamins A, D, E, and K are soluble in fat, while the B vitamins and vitamin C are water soluble. Fat-soluble vitamins tend to be stored in the body's fat tissues; because accumulations can produce serious side effects, nutritionists warn against taking high doses of vitamins A, D, E, or K.

Rather than being stored in fat tissues, water-soluble vitamins move from the digestive system directly

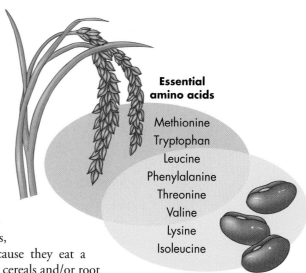

Essential amino acids

Methionine
Tryptophan
Leucine
Phenylalanine
Threonine
Valine
Lysine
Isoleucine

Figure 17.4 **Essential Amino Acids.** A person must eat the eight essential amino acids on this list in his or her daily diet. If one or more are missing, the body begins to break down proteins in its own tissues and use the amino acids they contain for critical functions, such as maintaining the lungs, heart, and brain. Meat contains all eight essential amino acids. Rice contains only six, and beans a slightly different set of six, but a diet that includes both rice and beans provides all eight.

into the bloodstream and are then picked up, as needed, by the tissue cells. Our kidneys filter out and eliminate any amounts beyond the cells' immediate needs. That's why nutritionists say it's pointless to take huge amounts of vitamin C, for example. Koalas need vitamins, too; a few of these are supplied by the leaves, and the rest by bacteria that ferment leaves in the animal's digestive tract. Bacteria in our digestive tracts also provide us with some vitamins.

Minerals

An animal's body also needs small amounts of another type of nutrient: **minerals,** specific inorganic chemical elements. **Major minerals** are those elements we need in amounts greater than 0.1 g each day; **minor minerals** are those we need in amounts less than 0.01 g daily. An adult person has about 2 kg (4.5 lb) of minerals in his or her body—mostly the calcium and phosphorus in bones and teeth. Our tears, blood, and sweat taste salty because of the sodium, potassium, and chlorine in the fluids. Sulfur is found in many proteins, and magnesium in many enzymes; the bones also hold a reservoir of magnesium along with calcium. The human body contains less than 1 teaspoon of minor minerals, but these elements are still critical to survival. Perhaps most important is iron, lying at the center of the hemoglobin molecule in our blood and essential to the transport of oxygen (see

Chapter 15). Thyroid hormones contain the element iodine; zinc is an important component of some enzymes and gene-regulating proteins; and we need fluorine for healthy bones and teeth. Studies of koalas in the wild reveal that while they tend to get most of the minerals they need from eucalyptus leaves and dirt, they occasionally have mineral deficiencies such as too little copper in the bloodstream and liver.

A person's need for specific minerals can change over time. For example, prior to menopause, a woman needs more iron every month because of periodic blood loss, and she requires extra iron throughout pregnancy to supply nutrients for the fetus's blood. She also needs extra calcium while pregnant for building fetal bones, and even more calcium while nursing. After menopause, natural declines in estrogen cause a loss of calcium that can lead to osteoporosis, so postmenopausal women, too, need to consume enough calcium.

Because vitamins and minerals are required in such small quantities, most people in affluent countries get more than enough of both simply by eating a well-balanced diet. A poor diet due to poverty or poor eating habits can indeed lead to vitamin or mineral deficiencies, however, with potentially serious consequences. A long-term absence of vitamin A, for instance, can lead to blindness, while a prolonged lack of B vitamins may lead to convulsions and other neurological disorders.

Food as Fuel: Calories Count

Most Americans have a far greater problem with overeating and obesity than with malnutrition. While most children and half of adults are in a healthy weight range (Fig. 17.5), fully 25 percent of teenagers and 50 percent of adults are either **overweight** or **obese** (a body weight more than 10 or 20 percent over ideal). Research shows that obesity contributes to cardiovascular diseases, some forms of diabetes, joint problems, generally poor health, and an overall decrease in the quality of life, including reduced earning power due to illness and prejudice against their appearance.

Height

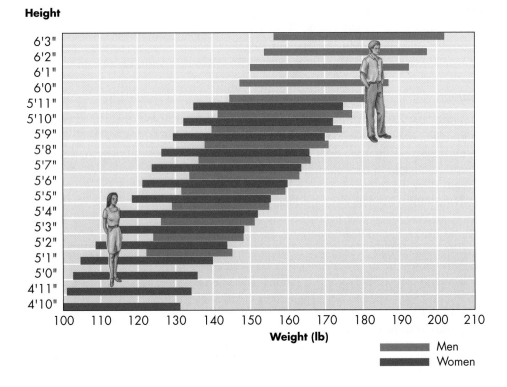

Weight (lb)

■ Men
■ Women

Figure 17.5 **Ideal-Weight Chart.** Ideal weight depends on frame size, physical condition, and age. At 6 ft and 195 lb, a heavy-boned, muscular football tackle would not be overweight, while a sedentary office worker with a medium frame probably would be. Height and weight are given for people not wearing shoes or other clothing.

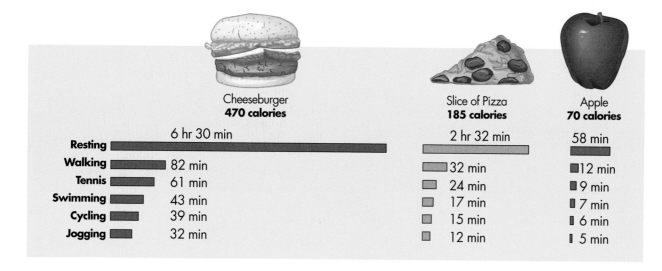

Cheeseburger
470 calories

Slice of Pizza
185 calories

Apple
70 calories

	Cheeseburger	Slice of Pizza	Apple
Resting	6 hr 30 min	2 hr 32 min	58 min
Walking	82 min	32 min	12 min
Tennis	61 min	24 min	9 min
Swimming	43 min	17 min	7 min
Cycling	39 min	15 min	6 min
Jogging	32 min	12 min	5 min

Figure 17.6   Food and Exercise Equivalents. As this chart shows, you can swim off a cheeseburger in 43 minutes or jog it off in 32 minutes, but at rest, your body will take 6 hours and 30 minutes to burn those same 470 Cal.

While obesity is common in North America, some people become so obsessed with body weight and thinness that it seriously threatens their health. In the disorder called **anorexia nervosa,** a person (typically a middle-class teenaged girl) restricts her food intake severely, becomes cadaverously thin, and yet still sees herself as overweight. In a separate disorder called **bulimia,** a person secretly binges on huge helpings of cake, ice cream, cookies, bread, or other high-calorie foods, then purges with self-induced vomiting, laxatives, diuretics, fasting, or vigorous exercise. Both disorders have social, psychological, and probably neurochemical roots and can lead to organ damage and even death if untreated.

Food energy is usually measured in kilocalories (kcal), also called Calories (Cal). A calorie (lower case "c") is the amount of energy needed to raise the temperature of 1 mL of water 1 °C; 1000 cal (1 kilocalorie or 1 Cal) is therefore a thousand times that energy amount. A large apple contains about 100 Cal worth of energy-producing compounds; and jogging 1.6 km (1 mi) burns about 100 Cal of stored energy.

How much energy do we need? That varies with age, sex, body size, and activity level. A normally active female college student needs around 1800 to 2000 Cal a day to fuel her total metabolic needs; a male college student needs about 2200 to 2500 Cal. Carbohydrates and protein each provide about 4.5 Cal/g, while fat provides 9 Cal/g. A female college student, then, would typically need the equivalent of about 400 g (14 oz) of pure sugar to provide the necessary calories. One of the strategies koalas have evolved to help overcome the small amount of calories in their spartan diet is to minimize their **basal metabolic rate,** the rate at which their bodies use energy while resting. Their rate is about half of our rate and is probably among the lowest of all mammals. Koalas

require only about 500 Cal each day, about the amount of energy in 200 g (a large bowl) of breakfast cereal or 500 g (a large pile) of eucalyptus leaves.

Figure 17.6 provides a revealing look at calories and energy, and a simple way to understand weight gain and loss. It shows the calories in three common snack foods and the amount of energy you'd have to expend in various physical activities to work off that food. For example, it takes 30 minutes to run off the calories in a cheeseburger, or an hour to burn them off playing tennis, but 6 1/2 hours to use them up sitting in a chair watching television! When an animal's food intake exceeds its energy needs, the inevitable result is storage of the leftover energy as body fat. The secret of weight control is basically this: take in only as many kilocalories as the body needs for fuel. Sound weight-loss programs combine calorie reductions, primarily through the decreased intake of sugar and fat, with increased physical activity. The result is less energy in and more energy out, with the differences made up gradually from the body's fat reserves. The koala eats such a low-calorie, high-fiber food source that it has to take a reverse strategy with regard to exercise: To conserve calories and hide from predators, it spends only four minutes a day moving about (Fig. 17.7)!

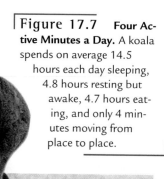

Figure 17.7   Four Active Minutes a Day. A koala spends on average 14.5 hours each day sleeping, 4.8 hours resting but awake, 4.7 hours eating, and only 4 minutes moving from place to place.

17.2 How Animals Digest Food

A koala chewing tough, leathery leaves; a polar bear devouring a seal; and a college student eating a chef's salad in the school cafeteria are all consuming food in forms that their cells can't use directly. Animal digestive systems must break down foods into usable small molecules and absorb them into the bloodstream, where they can then be distributed to body cells.

Intracellular and Extracellular Digestion

In the simplest animals—the sponges—some body cells take in tiny whole food particles directly from the water and break them down enzymatically to release usable nutrients. This strategy is called **intracellular digestion** (Fig. 17.8a), and it circumvents the need for the mechanical breakdown of food in a mouth or for the chemical digestion of food in a gut or other cavity. Intracellular digestion is clearly a simpler strategy, but it

puts an upper limit on the size of food particles an animal can take in, and, in turn, limits its dietary choices.

Evolutionary changes brought about a way animals could exploit larger food particles. **Extracellular digestion** is the enzymatic breakdown of larger pieces of food into constituent molecules outside of cells, but usually within a special body organ or cavity (Fig. 17. 8b). Nutrients from the broken-down foods pass into body cells lining the organ or cavity, and take part in body metabolism. Because the vast majority of animals rely on it, let's explore extracellular digestion in more detail.

Patterns of Extracellular Digestion

In rather simple animals, such as a cnidarian or flatworm (see Chapter 13), extracellular digestion takes place in the **gastrovascular cavity,** an internal sac with a single opening through which whole particles of food enter and undigested wastes leave. This system works well for small, thin organisms, but has important limitations: The animal doesn't break food pieces apart mechanically. As a result, digestive enzymes in the saclike gut can only break down the outer portions of ingested food chunks.

More complex animals have digestive tracts with two openings (Fig. 17.9). Food enters the **mouth** at one end of the digestive tract, moves in a single direction through the **gut,** and wastes exit at the other end of the tube, the **anus.** Between the two ends, a variety of specialized regions perform particular digestive roles. The overall result is the absorption of nutrients into the circulatory system. The circulatory system then distributes nutrients to each cell in the body.

(a) Intracellular digestion

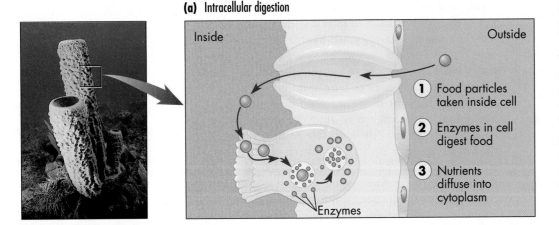

Inside / Outside

1. Food particles taken inside cell
2. Enzymes in cell digest food
3. Nutrients diffuse into cytoplasm

Enzymes

(b) Extracellular digestion

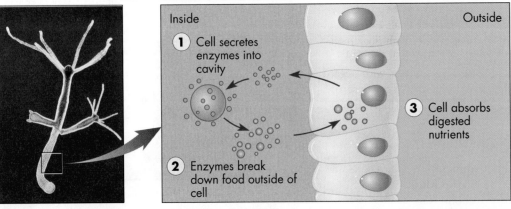

Inside / Outside

1. Cell secretes enzymes into cavity
2. Enzymes break down food outside of cell
3. Cell absorbs digested nutrients

Figure 17.8   **Strategies for Digestion.** (a) A simple animal like a sponge has no gut or similar organ and instead carries out intracellular digestion. Tiny food particles are taken into cells (Step 1), and digestive enzymes within the cell break the small particles into constituent molecules (Step 2). Nutrients then diffuse into the cell cytoplasm, where they can fuel cellular activities (Step 3). (b) A hydra has a central digestive cavity that can take in and digest relatively large pieces of food. In a process called extracellular digestion, cells lining the digestive cavity secrete enzymes into that central space (Step 1). There, the enzymes break down food into constituent nutrients (Step 2), and the nearby cells then absorb these nutrients (Step 3).

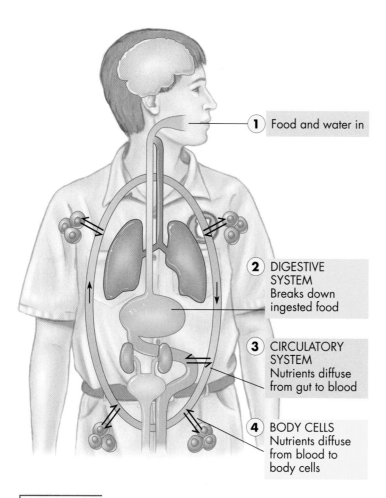

Figure 17.9 **Basic Strategy of the Digestive System.** A tube conveys food and water from the environment into the body (Step 1). The digestive system breaks down the food into nutrient molecules (Step 2). Then the nutrients diffuse through the gut-lining cells into the circulating blood (Step 3). Finally, nutrients diffuse from the blood into the cells of the body's tissues (Step 4).

1 Food and water in
2 DIGESTIVE SYSTEM Breaks down ingested food
3 CIRCULATORY SYSTEM Nutrients diffuse from gut to blood
4 BODY CELLS Nutrients diffuse from blood to body cells

17.3 The Digestive Tract

Let's return to the koala and its diet of nothing but eucalyptus leaves. This diet poses three main problems: (1) it has few nutrients; (2) it has a high concentration of bulky, indigestible dietary fiber; and (3) it has a large number of compounds that are toxic to most other mammals. Somehow, the koala's digestive tract must solve these problems without creating an animal as big as a buffalo. Let's look at digestive tracts, and compare the koala's to our own to see how that fuzzy marsupial solves its problems.

In people, koalas, and other vertebrates, the digestive tract is divided into five main regions: mouth, esophagus, stomach, small intestine, and large intestine (Fig. 17.10). Together, these form the **alimentary canal,** or **gastrointestinal tract** (also called the gut). In addition, nearby **accessory organs,** such as the salivary glands, liver, gallbladder, and pancreas, produce enzymes, bile, and other materials and deliver them into the tract at appropriate times, which aids digestion. Together, the alimentary canal and accessory organs accomplish the step-by-step conversion of foods into nutrients circulating in the bloodstream and able to supply all the body's cells.

Now let's contrast the human digestive system to a koala's, displayed diagrammatically (Fig. 17.10 right and *E*xplorer 17.1). The most remarkable feature of the koala gut is its huge **caecum** (SEE-cum), a dead-end sac with a 2 L capacity that is six times as long as the animal itself! As you can see from its position, the koala's caecum corresponds anatomically to the human appendix, which is vestigial and no bigger than your little finger. The koala's caecum is the largest of any mammal's, and about 40 percent of a koala's weight is the material inside this sac. Imagine if you had a 50- or 60-pound appendix! Comparing again, the koala's proximal colon is huge relative to our colon, but its distal colon is very narrow, containing lumps of compacted waste leaf material. What are the functions of the koala's caecum and colon? Other mammals that eat mostly leaves, like cattle, have large stomachlike sacs that act as fermentation chambers. Bacteria growing in the cow's sac (or rumen) break down the cellulose and provide energy to the cow. But as we've seen, a koala gets only 13 percent of its energy from cellulose in cell walls and the rest from the contents of eucalyptus leaf cells. So what is the caecum doing if it is not fermenting cell walls? To understand the specialized gut of the koala, we first need to understand the function of a generalized gut.

Functions of the Gastrointestinal Tract

Within the alimentary canal, a four-step digestive process takes place that extracts nutrients from food. The steps include ingestion and mastication, digestion, absorption, and elimination (Fig. 17.10 center). First, the animal brings food into its mouth (**ingestion,** Step 1).

GET THE PICTURE

*E*xplorer 17.1

Human and Koala Digestive Systems: Structure and Function

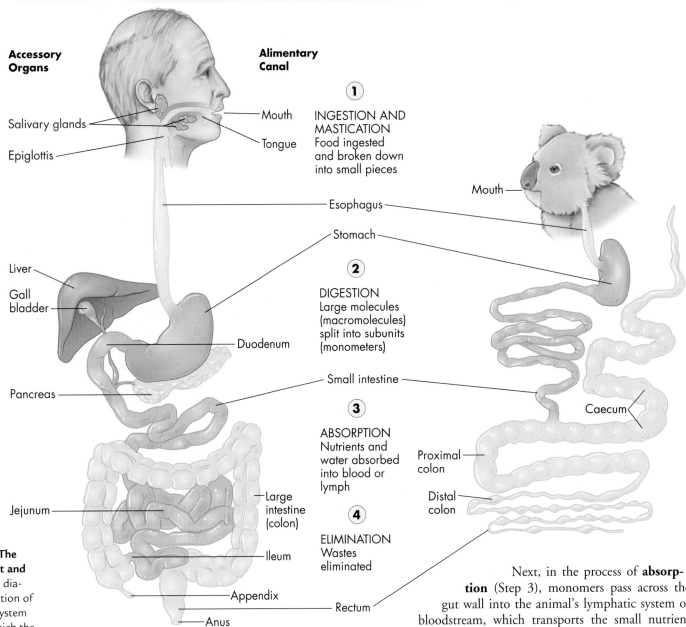

Accessory Organs

- Salivary glands
- Epiglottis
- Liver
- Gall bladder
- Pancreas
- Jejunum

Alimentary Canal

- Mouth
- Tongue
- Esophagus
- Stomach
- Duodenum
- Small intestine
- Large intestine (colon)
- Ileum
- Appendix
- Rectum
- Anus

1 INGESTION AND MASTICATION
Food ingested and broken down into small pieces

2 DIGESTION
Large molecules (macromolecules) split into subunits (monometers)

3 ABSORPTION
Nutrients and water absorbed into blood or lymph

4 ELIMINATION
Wastes eliminated

- Mouth
- Caecum
- Proximal colon
- Distal colon

Figure 17.10 **The Gastrointestinal Tract and Steps of Digestion.** A diagrammatic representation of the human digestive system (left). The steps by which the digestive system processes foods (center): ingestion and mastication, digestion, absorption, and elimination. The koala's digestive system (right). The koala has a highly specialized alimentary canal, with the largest caecum (homologous to our appendix) of any mammal relative to its size. This organ helps the woolly marsupial cope with its fibrous, toxic-laden diet.

It then breaks the food into small pieces (**mastication**). In many mammals and other vertebrates, this mastication involves slicing or grinding by teeth. In the case of many birds, mastication occurs in a muscular sac called the **gizzard,** which usually contains stones that grind against each other and pulverize food. (Researchers have also found huge gizzard stones in some dinosaur fossils.) Enzymes secreted into the gut tube break down the large macromolecules of food into smaller molecules, or monomers (**digestion,** Step 2).

Next, in the process of **absorption** (Step 3), monomers pass across the gut wall into the animal's lymphatic system or bloodstream, which transports the small nutrient molecules to each cell in the body. Finally, most of the water in the food is absorbed into the circulation, and undigested residues are eventually eliminated (**elimination,** Step 4).

Tissues of the Gastrointestinal Tract

As we saw in Chapter 13, the gut is a long tube, and these alimentary functions are facilitated by the tube's four tissue layers. The innermost layer, which faces the cavity of the digestive tract, is a mucous membrane, or **mucosa** (Fig. 17.11). Some mucosa cells produce digestive enzymes, while others secrete **mucus,** a slimy coating that lubricates food passing through and keeps the gut from digesting itself.

Surrounding the mucosa is the **submucosa,** a connective tissue that is richly supplied with blood and lymph vessels and with nerve cells. Next is the **muscularis**, a double layer made of muscle fibers, some encircling the gut (circular muscles) and others running along its length (longitudinal muscles). The contraction and relaxation of these fibers, or **peristalsis,** knead the food, mix it with digestive juices, and propel it along with rhythmic sequential contractions like toothpaste squeezed through a tube. Finally, a thin outermost layer, the **serosa,** forms a band around the other tissues and joins to the sheet (the **mesentery**) that attaches the gastrointestinal tract to the inner wall of the body cavity.

17.4 The Human Digestive System

Most vertebrates share these strategies of digestion, and we're no exception. Because we have such a varied diet, however, the human digestive system (Fig. 17.10 and *E* 17.1) makes a better general model than the koala's highly specialized digestive system or that of many other animals. You might be surprised to learn that our gut, from mouth to anus, is approximately 8 m long, or about the height of a two-story house. All the sandwiches, brownies, apples, milk, and other foods a person eats pass through the central cavity, or **lumen,** of this tube and undergo one digestive process after another. Let's follow a turkey sandwich through its digestive journey to see how the nutrients are liberated.

The Mouth, Pharynx, and Esophagus

Your teeth are superbly adapted to the job of cutting, tearing, and grinding the plant and animal tissues that omnivorous mammals (like ourselves) eat (Fig. 17.12a). Teeth can stand up to regular wear and tear because the **enamel** covering them and made up of calcium salts and protein is the hardest substance in the body. The enamel on a person's permanent teeth will actually generate sparks if struck against steel! Working along with the teeth is a muscular **tongue,** the principal organ of

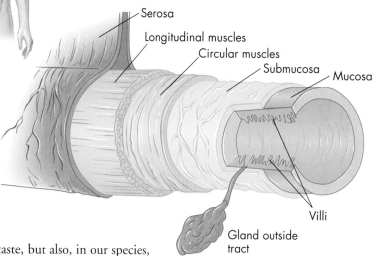

Figure 17.11 Anatomy of the Alimentary Canal. A cross section of an intestine reveals four main layers: the mucosa, submucosa, the circular and longitudinal muscles, and the serosa. Fingerlike villi extend the inner surface area of some portions of the gut.

Serosa
Longitudinal muscles
Circular muscles
Submucosa
Mucosa
Villi
Gland outside tract

taste, but also, in our species, an organ that along with the **larynx,** or voice box, forms the sounds of spoken language (Fig. 17.13a). When you take a bite of a sandwich, your tongue moves some of the food toward the molars for grinding and some to the incisors for cutting, and shapes each small bite into a soft, moist lump, or **bolus,** that you can easily swallow.

Koalas have the same basic types of teeth as you do, but each of its molars have four **V**-shaped ridges that leave characteristic marks on a eucalyptus leaf as they break up the leaf blade into cells and crush the cells to release their nutritive contents (Fig. 17.12b). When these ridges wear down after years of cutting up tough leaves, a koala (especially one in the wild) can't access the leaf's contents, and the animal often dies of diseases and/or malnutrition. (Despite her age, Point Blank's teeth are still in surprisingly good shape, according to Drs. Crissey and Barbiers.)

As a mammal chews (Fig. 17.13a), three large pairs of **salivary glands** that lie in the tissues surrounding the oral cavity secrete clear, watery **saliva,** which mixes with the food. Saliva contains primarily mucus and water, and these moisten food particles and help them cling together in a bolus. Saliva also contains small amounts

(a) Human teeth

(b) Koala teeth

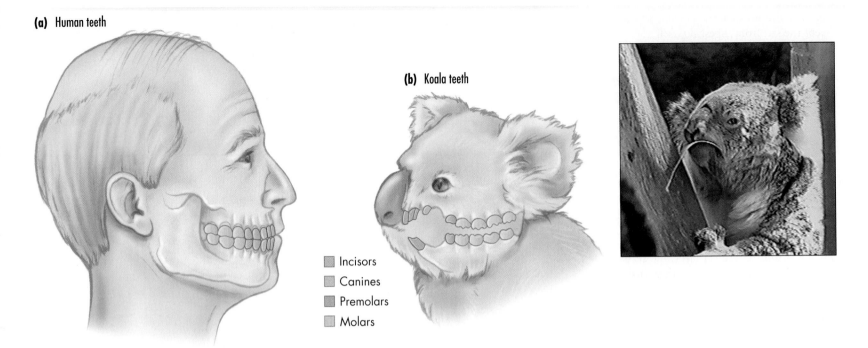

■ Incisors
■ Canines
■ Premolars
■ Molars

Figure 17.12   Teeth: Unspecialized and Specialized. Our human teeth (a) are not specialized for particular foods and can tear, slice, and crush many kinds of foods. In contrast, a koala's mouth (b) has sharp cutting teeth (the incisors) in the front, and somewhat sharp grinding teeth (the premolars and molars) in the back. A large gap separates the front and back teeth. The animal inserts a eucalyptus leaf through the gap on one side of the mouth to the molars on the other side, and grinds sideways, leaving little chevron-shaped crush marks on the leaves and liberating the fluid cell contents.

of **amylase,** a digestive enzyme that begins the breakdown of carbohydrates (such as the starch in slices of bread or in the lettuce and tomato of the sandwich). Just before swallowing begins (Fig. 17.13b), the tongue pushes the bolus of food up and back against the **hard palate,** which forms the floor of the nasal cavity and most of the roof of the mouth. Next, the food encounters the **soft palate** at the back of the mouth above the throat. This flexible muscular sheet rises as swallowing starts, preventing food from entering the nasal cavity. The flaplike **epiglottis** moves backward and downward, closing off the opening to the trachea, or windpipe.

Figure 17.13   Anatomy and the Swallowing Reflex. The anatomy of our mouth and throat area allows us to chew and swallow food. (a) Once we have chewed food, the tongue shifts a bolus of moistened food to the back of the mouth (b, Step 1); the soft palate rises, preventing food from entering either nasal cavity (Step 2); and the flaplike epiglottis moves backward and downward, closing off the glottis or opening to the trachea (windpipe) (Step 3). The larynx rises (Step 4), the food bolus passes through the pharynx (Step 5), triggering a reflexive swallowing action that sends food into the esophagus and toward the stomach. If you place your fingers lightly on either side of your throat just below your chin as you swallow, you can feel the larynx move up and forward, opening the esophagus, which receives food.

(a) Chewing

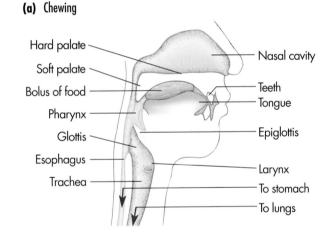

Hard palate
Soft palate
Bolus of food
Pharynx
Glottis
Esophagus
Trachea
Nasal cavity
Teeth
Tongue
Epiglottis
Larynx
To stomach
To lungs

(b) Swallowing

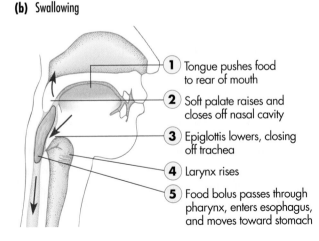

1 Tongue pushes food to rear of mouth
2 Soft palate raises and closes off nasal cavity
3 Epiglottis lowers, closing off trachea
4 Larynx rises
5 Food bolus passes through pharynx, enters esophagus, and moves toward stomach

This opening is called the **glottis.** Next the food bolus moves through the **pharynx,** the passageway for food between the mouth and esophagus and for air between the nose and throat. Food entering stimulates a reflexive swallowing action in the pharynx.

Swallowing then delivers the food bolus to the **esophagus** (eh-SOF-eh-gus), the pipeline to the stomach. To enter the stomach, the bolus passes through the **cardiac sphincter** (SFINGK-ter), a ring of muscle located at the junction of the stomach and esophagus (Fig. 17.14a,b). This muscular ring usually stays tightly contracted, like the drawstring on a purse, preventing the stomach's contents from backing up into the esophagus. People often experience "heartburn" when the stomach is very full; this burning sensation is due to small amounts of stomach acid seeping out past the cardiac sphincter into the esophagus, which lacks the stomach's heavy mucous lining. Despite their names, neither heartburn nor the cardiac sphincter is related to the heart.

The Stomach

Next, the food bolus (here, of chewed turkey sandwich) passes into the elastic, J-shaped bag called the **stomach** (Fig. 17.14b). The stomach stores food for later processing—an adaptation that enables large animals to eat larger meals and hence feed less often. The average human stomach can comfortably hold about 1 L (about a quart; recall that the little koala's caecum holds twice as much!). When the stomach is full, waves of peristalsis in the muscular stomach wall churn and mix the contents with an acid bath of gastric juice secreted by glands in the stomach wall.

Gastric juice is a mixture of water, hydrochloric acid (HCl), mucus, and **pepsinogen,** a precursor to the protein-cleaving enzyme, **pepsin.** The hydrochloric acid makes gastric juice acidic enough to kill off most bacteria or fungi contaminating foods. This acid contributes to the breakdown of food pieces into constituent protein fibers, fat globules, and so on, and it also converts pepsinogen into pepsin. Pepsin breaks the peptide bonds that link amino acids in proteins. Therefore, during the time that protein-containing foods (such as the turkey in a sandwich) are in the stomach, they are partially digested to short polypeptide segments. Although digestion of the starch in the bread of the sandwich begins in the mouth, once food enters the stomach, very little additional digestion of starches and other carbohydrates (or of fats, like those in mayonnaise) takes place. This takes place later, in the small intestine.

The result of the chemical activity and mixing waves in the stomach is a pasty, milky, and highly acidic soup called **chyme** (KIME) (Fig. 17.14c). Chyme passes through another sphincter, the **pyloric sphincter,** into the small intestine at a rate of about a teaspoonful every three seconds after a meal (Fig. 17.14d). It usually takes one to four hours for the stomach to process a meal and for the chyme to pass, spoonful by spoonful, into the small intestine—less time for a high-carbohydrate meal, more time for a fatty meal.

Pancreas, Liver, and Gallbladder

The small intestine carries out most chemical digestion of food, as well as most absorption of nutrients. This portion of the gut tube, however, does not accomplish these tasks alone. Three accessory organs assist this por-

Figure 17.14 **Function of the Stomach.** Before swallowing food, the cardiac sphincter is closed (a). A food bolus pushes on the cardiac sphincter, which opens (b), allowing the food to enter the stomach. (c) Muscle contractions moving in peristaltic waves across the stomach mix the food with digestive enzymes and stomach acid, forming chyme. (d) The pyloric sphincter then opens in repeated brief bursts that allow small amounts of chyme to squirt into the duodenum (upper part of the small intestine), where it is digested further.

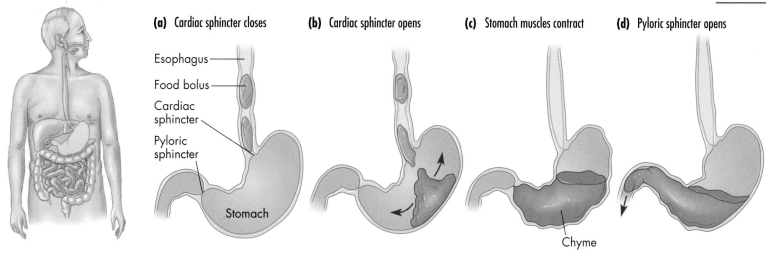

(a) Cardiac sphincter closes **(b)** Cardiac sphincter opens **(c)** Stomach muscles contract **(d)** Pyloric sphincter opens

Esophagus
Food bolus
Cardiac sphincter
Pyloric sphincter
Stomach
Chyme

tion of the gut tube—the pancreas, the liver, and the gallbladder—by dumping in substances that aid digestion and absorption.

The Pancreas and Its Digestive Enzymes

The **pancreas** is a narrow, lumpy organ situated close to where the stomach joins the small intestine (see Fig. 17.10). The pancreas produces a host of digestive enzymes and secretes them into the small intestine. These pancreatic enzymes include **proteases,** which break down proteins; **lipases,** which digest fats; and enzymes such as amylase, which complete the digestion of carbohydrates. The pancreas also secretes bicarbonate ions (the main ingredient in "Tums" and other antacids), which buffer or neutralize stomach acid entering the small intestine. This buffering is vital because, unlike stomach enzymes, pancreatic enzymes can't function in an acidic environment, and the acid could damage the small intestine.

Some pancreatic cells secrete the hormones **insulin** and **glucagon** directly into the bloodstream. These hormones help keep the levels of blood glucose within a certain range (see Chapter 18); when that homeostasis fails, a person can develop diabetes or other blood sugar disorders.

The Multipurpose Liver and the Gallbladder

The smooth, lobed, rather hemispherical **liver** is the largest gland in the human body (see Fig. 17.10). It weighs about 2 kg (4.5 lb) in an adult and performs biological tasks as diverse as destroying aging red blood cells, storing glycogen, and dispersing glucose to the bloodstream as circulating levels of the sugar drop.

One of the liver's most important functions is to produce **bile salts,** molecules of modified cholesterol. Bile salts are stored as the yellow-green liquid **bile** in the **gallbladder,** a pear-shaped sac on the underside of the liver. Bile salts act like detergents and break up fat droplets in the small intestine (Fig. 17.15). Cholesterol-lowering drugs bind to bile salts and cause them to pass down the intestinal tract without being absorbed. This prevents the recycling of cholesterol from bile salts and thus decreases the amount of cholesterol in the blood.

Besides synthesizing bile, the liver picks up, stores, and sometimes synthesizes amino acids, glucose, and glycogen, and stores some vitamins and other compounds that cells need to function normally. Certain liver cells also contain special enzymes that can detoxify poisons. For example, the liver transforms molecules of ammonia—a toxic, nitrogen-containing waste created by the breakdown of amino acids—into urea, which is less toxic and is excreted in urine. The liver also detoxifies alcohol; but if the

Figure 17.15 **Fat Digestion.** Bile salts travel from the liver through the bile duct to the gallbladder (Step 1). The presence of fats in the small intestine stimulates the gallbladder to squeeze drops of bile down a duct into the small intestine (Step 2). Like a detergent mixed with cooking oil and water, bile disperses dietary fat globules into tiny droplets (Step 3), providing a larger surface area for the action of fat-digesting enzymes, and a more rapid breakdown of lipids. Bile salts also aid absorption of fully digested fat molecules across the lining of the small intestine (Step 4). Bile salts travel in the bloodstream back to the liver (Step 5), where they eventually are recycled. Fats enter the lymphatic system (Step 6).

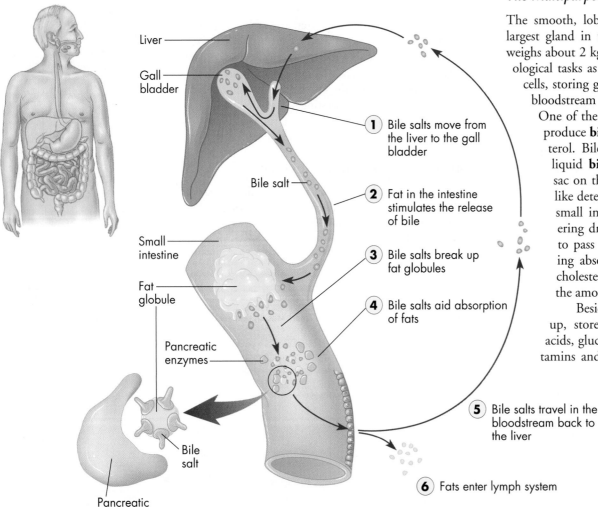

Liver

Gall bladder

Bile salt

Small intestine

Fat globule

Pancreatic enzymes

Bile salt

Pancreatic enzyme

1. Bile salts move from the liver to the gall bladder
2. Fat in the intestine stimulates the release of bile
3. Bile salts break up fat globules
4. Bile salts aid absorption of fats
5. Bile salts travel in the bloodstream back to the liver
6. Fats enter lymph system

liver is overloaded with the drug year after year, it can become irreversibly scarred, producing a potentially fatal condition called **cirrhosis** (sih-ROH-siss).

The Small Intestine: Digestion and Absorption

The **small intestine** is a remarkable coiled tube about 6 m (20 ft) long. It's the main site of carbohydrate and fat digestion, and is also the place where protein digestion is completed so that nutrients can be absorbed into the blood. The small intestine begins just below the stomach and has three main regions along its length: the upper section, or **duodenum** (doo-oh-DEE-nuhm), the central **jejunum**, and the remainder, the **ileum** (see Fig. 17.10 and *E* 17.1).

The small intestine is a marvel of compact biological engineering. If the surface of its inner lining were a smooth tube like a garden hose, there would be a relatively small surface area for digestion and absorption. Instead, that inner lining is so convoluted that it houses a huge absorptive surface. The intestinal lining is pleated into large numbers of folds (Fig. 17.16a,b), and each fold is covered with fingerlike extensions known as **villi** (sing., **villus**), which project into the lumen and come into contact with chyme (Fig. 17.16c). Further, cells on the outer layer of each villus are carpeted with **microvilli,** microscopic brushlike projections of the plasma membrane (Fig. 17.16d). The combination of folds, villi, and microvilli creates a total surface area in the human small intestine the size of a tennis court.

By the time a turkey sandwich you ate for lunch is reduced to chyme and reaches the small intestine, it contains partially digested carbohydrates and proteins, as well as undigested fat. Here, enzymes secreted by the intestine and the pancreas interact with bicarbonate ions and bile salts. Together, they gradually complete the digestion of the carbohydrates (from bread, lettuce, and tomato) into simple sugars; the proteins (from

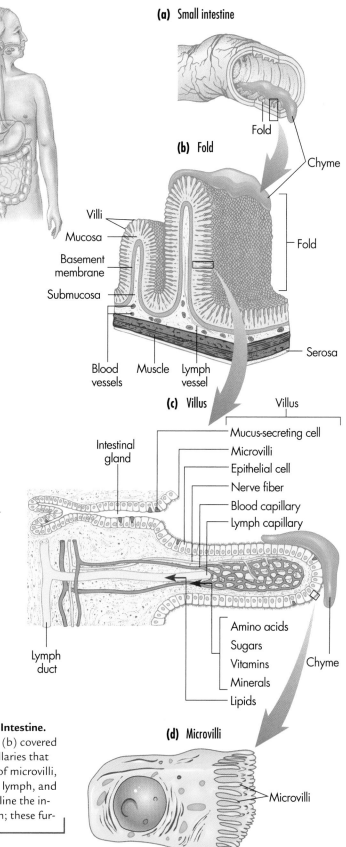

(a) Small intestine

Fold

Chyme

(b) Fold

Villi

Mucosa

Basement membrane

Submucosa

Blood vessels   Muscle   Lymph vessel

Fold

Serosa

(c) Villus

Villus

Intestinal gland

Mucus-secreting cell

Microvilli

Epithelial cell

Nerve fiber

Blood capillary

Lymph capillary

Lymph duct

Amino acids

Sugars

Vitamins

Minerals

Lipids

Chyme

(d) Microvilli

Microvilli

Figure 17.16   **Highly Absorptive Lining of the Small Intestine.**
The small intestine (a) is pleated into large numbers of folds (b) covered with villi. Each villus (c) encompasses blood and lymph capillaries that carry away nutrients absorbed through the brushlike carpet of microvilli, with their enormous combined surface area. Lipids enter the lymph, and other nutrients enter the blood. (d) The surface of cells that line the intestinal cavity contain microvilli, like the hairs of a little brush; these further increase the surface area of the gut.

turkey meat) into amino acids; and the fats (from mayonnaise) into fatty acids and glycerol. These nutrient molecules are small enough to move across the plasma membranes of the microvilli and enter the intestinal cells. The amino acids and sugars then pass into blood capillaries, along with most of the available vitamins and minerals from the tomato slice and lettuce, while lipids enter tiny lymph capillaries (lacteals) (see Fig. 17.16c). The remaining contents of the small intestine, including the small amount of indigestible roughage in the bread, lettuce and tomato, pass into the large intestine.

Why would we starve on a diet of dry grass?

The Large Intestine: Site of Water Absorption

The last 1.2 m (4 ft) of the human alimentary canal is the **large intestine,** or **colon.** This section of the gut tube ascends on the right side of the body cavity, cuts across just below the stomach, then descends on the left side and ends in a short tube called the **rectum** (see Fig. 17.10 and ⓔ 17.1). Extending from the colon near its junction with the small intestine is the small caecum with its extension, the **appendix.** This, as we saw, is a miniature version of the koala's giant caecum. A person's finger-shaped appendix plays no known role in digestion, but it may help fight infections in the gut (as may the koala's caecum). An inflamed appendix, most often based on blockage by hardened waste matter, can lead to the medical emergency called **appendicitis.**

The colon absorbs water, ions, and vitamins from the chyme. The rectum stores the semisolid undigested wastes, or **feces,** and helps excrete them. The colon is twice as wide as the small intestine, but its walls lack the many folds and villi of the small intestine. The smoother surface has a lower surface area for absorption, but it also presents less resistance to the movement of chyme through the tube.

Each day, about 0.5 L (about 1 pt) of chyme (now minus the nutrients absorbed in the small intestine) reaches the colon, along with 2 to 3 L (2 to 3 qt) of water, some from food and the rest secreted by the stomach and intestines themselves. The body can't afford to lose this much water, however, and much of the water is reabsorbed as the chyme moves slowly through the colon over a period of 12 to 36 hours. This absorption gradually transforms the chyme from fluid to semisolid, and the wastes (including undigestible cellulose fiber from the bread, lettuce, and tomato of the turkey sandwich we started with) are stored until their pressure against the colon wall triggers a **bowel movement** (defecation), the muscular expulsion of wastes through the rectum and out the anus.

Digestion is never 100 percent efficient, and some nutrients invariably pass into the colon from the small intestine. A variety of bacteria, including *Escherichia coli* and *Lactobacillus* and *Streptococcus* species, reside in the human intestine and live on these remaining nutrients; in the process, they produce a number of vitamins, including thiamine (vitamin B_1), riboflavin (vitamin B_2), vitamin B_{12}, and vitamin K. The colon absorbs these vitamins along with fluids.

Coordination of Digestion

As your turkey sandwich moves through your digestive system, the food is mechanically and then chemically broken down and most of its nutrients are absorbed. This complex process can't take place willy nilly, and in fact, it's closely controlled by your nervous system and your endocrine (hormonal) system. Nerves throughout the digestive tract communicate between "upstream" and "downstream" regions, speeding or slowing the propulsion of food appropriately. Nervous activity in the brain and spinal cord can also speed up or slow down digestion. For example, the sight, taste, smell, or even thought of food (Fig. 17.17, Step 1) can cause signals from higher brain centers to travel via nerves to the salivary glands (Step 2), causing them to secrete saliva, and to secretory glands in the stomach lining, causing them to secrete gastric juices—hydrochloric acid and pepsinogen—into the stomach. Food arriving in the stomach and pushing against the stomach wall can trigger the same response.

Food in the stomach also lowers the acidity of the contents (Step 3), and this triggers stomach-lining cells to secrete a hormone called **gastrin** (Step 4). Gastrin acts on other nearby stomach cells, causing them to secrete more hydrochloric acid (Step 5). This raises the acidity (lowers the pH) of gastric juice, and helps speed the breakdown of the food (Step 6). When the pH drops to about two again, the stomach stops secreting gastrin and in turn, the extra acid. Food literally helps to stimulate its own digestion through a negative feedback loop for regulating stomach acid. Long periods of emotional stress can stimulate acid production in some people (and in lab animals) and lead to **ulcers,** craterlike sores in the mucosa of the stomach or small intestine. Research has shown that a bacterial species called *Helicobacter pylori* can be the causative agents in certain ulcers and can be treated with antibiotics.

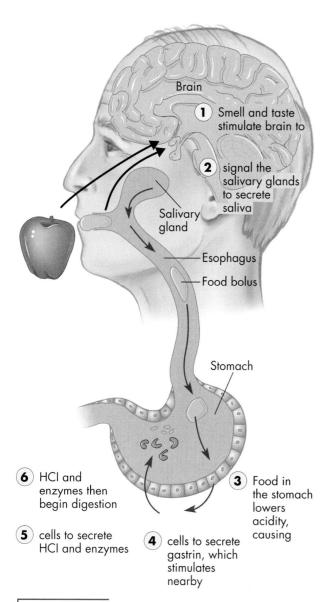

Figure 17.17 **Nerves and Hormones Coordinate Digestion.** Your body begins the digestive process when food is available—not when the stomach is empty. The control of digestion involves the senses (Step 1); signals from the brain to the salivary glands (Step 2); and lowered stomach acidity (Step 3), which induces gastrin secretion (Step 4) and then the secretion of HCl and digestive enzymes (Step 5). These then begin digestion (Step 6).

Several other hormones help coordinate the timing and amount of enzyme secretion after you eat. For example, the peptide **secretin** from your small intestine causes your pancreas to secrete bicarbonate ion, which then enters your small intestine and neutralizes the stomach acid. Partially digested proteins cause the small intestine to release a second hormone, **cholecystokinin** (CCK; KOH-luh-siss-toe-kine-in). This hormone triggers your pancreas to release protein- and fat-digesting enzymes. CCK also works on regulatory centers in the brain and produces the sensation of being full. This is why nutritionists often advise dieters to eat the protein foods in a meal first, so they'll feel satisfied sooner.

Working together, nerves and hormones—the body's rapid- and slow-acting control agents—fine-tune the secretion of digestive juices, making enzymes and ions instantly available to break down food but only when food is present and the powerful agents are needed.

The Koala's Survival Strategy

We've seen, so far, that koalas stay alive by eating only one food, a food that few other animals can stomach. We've also seen that koalas have the animal kingdom's biggest caecum, and that while it's filled with fermenting bacteria that can break down cellulose, koalas still get 87 percent of their food energy from the contents of eucalyptus leaf cells and only 13 percent by fermenting the cell walls themselves. Looking at these facts, you've got to wonder what the huge caecum and swollen proximal colon are doing if they're not serving as fermentation chambers for bacteria that digest cellulose. And you've also got to wonder if a koala relies primarily on cell contents for energy, how it can ever obtain enough food calories given the enormous mass of indigestible cell walls filling up the gut (in Point Blank's case, 80 pounds of it a week)?

Koala researchers solved these puzzles by studying how solid fibers and soluble components move through the koala's gut. It turns out that it takes about nine days for small particles and dissolved nutrient compounds to pass through the digestive system, but only about a day for large particles such as leaf fragments to move through. The retention time for small particles is one of the longest among all the mammals. Apparently, the colon's back-and-forth squeezing movements wash dissolved substances and small particles into the caecum. There, bacteria can ferment them and help detoxify some of the tannins and other poisonous compounds from the leaves. (Enzymes in the liver detoxify the terpenes and other essential oils, and the "disarmed" compounds are excreted in the urine.) Meanwhile, larger leaf fragments from the animal's diet settle in the distal colon, where water can be recovered and most of the bulk quickly eliminated. Through these digestive sorting processes, the furry animals can quickly move through and excrete large quantities of tough, fibrous leaves while retaining the most easily digested parts— the cell contents—for longer periods to extract more nutrients from them. This sorting strategy, coupled with the animal's low basal metabolic rate, sleepy demeanor, and minimal energy expenditure, insures that the koala can survive on a meager, otherwise toxic diet that other animals shun. Until the late 1800s, when hunters took huge numbers of koalas for their thick, soft pelts, Australia's eucalyptus forests were filled with millions of the marsupials. Today, the living teddy bears are rare.

What happens when leptin levels drop?

17.5 Healthy Diets and Proper Nutrition

You might be surprised to learn that the koala's feeding strategy has lessons for our own nutrition. If a person wants to lose weight, he or she needs to eat food with low energy content and high roughage content, while still maintaining an adequate intake of amino acids, vitamins, and minerals, and couple that with increased exercise and stepped up basal metabolic rate. That's hard for most overweight people to do but it's an equation that works and recent animal research shows why.

Enormously obese mice such as the one shown in Figure 17.18a led researchers to discover how hormones regulate body weight. Mice like these have a mutation that blocks a feeling of fullness and so they keep on eating even when their energy intake levels exceed their energy expenditures. The mutation, in other words, seems to block the normal "stop eating" signal, and it was this that allowed biologists to understand how weight-regulating hormones function in both mice and humans. The hormone called **leptin** is produced by fat cells. A second hormone involved in fat regulation is the insulin secreted by the pancreas. Figure 17.18b shows how these two hormones influence human fat formation and **E**xplorer 17.2 illustrates their action. Fat cells secrete leptin (Step 1) and also cause the pancreas to liberate insulin (Step 2). These hormones travel in the blood to the brain (Step 3), where they can activate neural circuits that speed energy use (Step 4) and repress neural circuits that control the desire to eat (Step 5). During a meal, when the "I'm full" factors we discussed earlier, like cholecystokinin, act on the brain, they interact with leptin/insulin-sensitive nerve cells, and cause the individual to stop eating (Step 6). These interactions result

in energy balance (Step 7), by modulating the size of the fat cells and their production of leptin (Step 8). Obviously, if the "stop eating" signal is broken as in the mutant mouse, the animal will keep eating and become obese.

Investigation of a few rare human families whose members are extremely obese demonstrated that, with slight differences, the basic regulatory pathway works the same way in people. And just as in mutant mice, giving leptin to obese humans that have a mutation in the leptin gene reduces weight markedly. While most obese people do not have a mutation in leptin, they may have genetic variation in other elements of the pathway, such as the activity of brain hormones (Fig. 17.18, Steps 3–6). This gives real hope that treatments for obesity could become available in the future, as **E** 17.2 discusses.

Look again at Figure 17.18b and think of the way you feel when you go on a diet. You slow down your intake of food, and your fat cells start to shrink. You are going to start to feel slim and trim and full of energy, right? Wrong! You feel hungry all the time, constantly fatigued, and can't think of anything but food. As you can see from Figure 17.18b, while dieting, your fat cells start to shrink, produce less leptin, and signal the pancreas to make less insulin. The drop in leptin revs up your brain circuits that stimulate eating and conserving energy and now you feel like eating larger meals and sitting around and doing less exercise. What a cruel trick! This is what makes dieting and maintaining weight loss so difficult. This plus the fact that American consumers find fat- and sugar-laden foods in every restaurant, cafeteria, supermarket, convenience store, and vending machine they pass.

Given this discouraging picture, how can one turn on fat-burning enzymes, reduce the size of fat cells, and cut body weight to an ideal level? Nutritionists recommend cutting the amount of fats and oils we eat, because human cells convert dietary fats to body fat more easily than other nutrients. They also recommend cutting sugars, because in an already overweight person, when the blood sugar levels are high, glucose is often rapidly converted to fat and stored in fat cells rather than entering muscle cells, where the sugar can be burned. Some people have found that high-protein, low-carbohydrate diets are tasty and reduce food cravings, helping them to eat the right number of calories each day to lose weight or maintain weight loss. But the jury is still out on how healthy these diets are in the long run.

Most importantly, dieting must be accompanied by increased daily physical activity. Exercising most or all

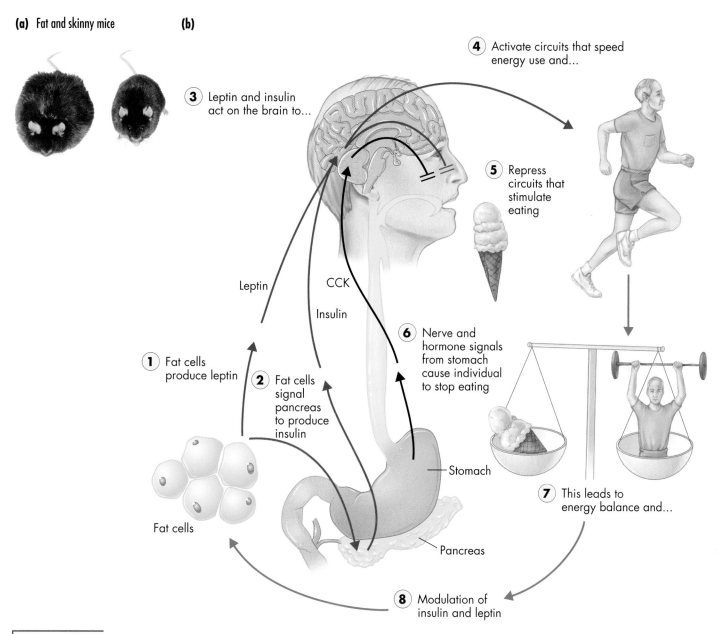

(a) Fat and skinny mice

(b)

3 Leptin and insulin act on the brain to...

4 Activate circuits that speed energy use and...

5 Repress circuits that stimulate eating

Leptin

CCK

Insulin

1 Fat cells produce leptin

2 Fat cells signal pancreas to produce insulin

6 Nerve and hormone signals from stomach cause individual to stop eating

Stomach

Fat cells

Pancreas

7 This leads to energy balance and...

8 Modulation of insulin and leptin

Figure 17.18 **Hormones Regulate Body Weight.** (a) Geneticists discovered the extremely obese mouse on the left, and found that it has a mutation in a gene necessary to maintain a normal body weight. The mutation alters the protein hormone leptin in such a way that it can't block overeating. The text and *E* 17.2 explain the current understanding of how hormones control body weight.

days of the week seems to turn up the metabolic rate so that the body burns more fat—not just during exercise sessions, but for many hours at rest, as well. Exercise decreases fat tissue and increases muscle mass, making the body look and feel trimmer. Finally, moderate daily exercise reduces the appetite, whereas both fasting (such as observing very-low-calorie diets) and inactivity increase appetite.

The moral of the story is, eat low-calorie, high-fiber foods like a koala, but don't exercise like one! How's your diet?

Chapter Summary and Selected Key Terms

Introduction

Point Blank is the oldest koala in captivity and like all koalas, has a diet limited to tough, bitter eucalyptus leaves. Its anatomy and physiology, however, allow the animal to absorb all the water, nutrients, vitamins, and minerals it needs from this single source. That makes the koala an interesting case history for studying nutrition and digestion.

Nutrients: Sources of Energy and Elements

Nutrition (p. 448) is the study of what animals need to consume to stay alive and healthy. The basic biological molecules we encountered in Chapter 2 are part of the nutrition story. Carbohydrates, including sugars, starches, and cellulose, are made up of glucose subunits and form the main source of energy and carbon atoms for most animal species. Leaf- and plant-eating animals like the koala can't actually digest cellulose themselves, but instead harbor bacteria with enzymes to break down cellulose. Cellulose serves as **dietary fiber** (p. 449) or **roughage** (p. 449) in our diet that helps propel wastes through the large intestine.

Lipids, solid like bacon fat or liquid like salad oil, are our sources of fatty acids and provide twice as much energy per gram (9 C/g) as carbohydrates (4.5 C/g). They are a more compact energy storage form in animals' bodies. Lipids dissolve vitamins A, D, E, and K and help the digestive tract absorb and deliver them to cells. Linoleic acid is an **essential fatty acid** (p. 450) or lipid component we can't generate and must eat in our diets.

Proteins are our sources of amino acids. Animal bodies contain many kinds of proteins in hair, claws, skin, cartilage, tendons, bone, muscle, enzymes, antibodies, hemoglobin, and so on. We need a continual supply of amino acids for cell division, growth, and repair. A 68-kg (150-pound) human needs 75 g of pure protein per day. We have eight **essential amino acids** (p. 451). Since we can't make them, we need to eat them daily in foods, or protein synthesis will be interrupted. Most animal proteins contain all 8 plus the other 12 amino acids. But many plant proteins don't, so vegetarians must combine plant foods like rice and beans to insure enough of all eight.

Vitamins (p. 451) are organic compounds needed in small amounts for normal growth and metabolism, but that an organism can't synthesize or store and must get in foods. We require 14 vitamins, some fat soluble, some water soluble. **Minerals** (p. 452) are the specific inorganic chemicals that we need. We need **major minerals** (p. 452) such as sodium, potassium, and phosphorous in amounts greater than 0.1 g/day. We need **minor minerals** (p. 452) such as iron, iodine, and zinc in amounts less than 0.01 g/day.

Food consumption can often get out of balance with energy expenditure. One quarter of all teenagers and one half of all adults are **overweight** or **obese** (p. 452). At the same time, **anorexia nervosa** (p. 453) and **bulimia** (p. 453) are not uncommon.

A typical active female college student needs 1800 to 2000 Calories/day; an active male needs 2200 to 2500. The **basal metabolic rate** (p. 453) is the rate at which an animal uses calories of food energy while resting. Koalas have one of the lowest of all mammals, and this helps them survive on their diet of eucalyptus leaves.

How Animals Digest Food

The simplest animals, including the sponges, use **intracellular digestion** (p. 454), taking particles into cells and breaking them down enzymatically. Most animals use **extracellular digestion** (p. 454), the breakdown of larger pieces of food in a special body cavity or organ. Cnidarians and flatworms have a **gastrovascular cavity** (p. 454), but no way to break foods apart mechanically, so the pieces must be small. More complex animals have a **mouth** (p. 454) at one end, an **anus** (p. 454) at the other, and a **gut** (p. 454) in between with specialized regions for different aspects of digestion.

The Digestive Tract

The five regions of the vertebrate digestive tract, the mouth, esophagus, stomach, small intestine, and large intestine, make up the **alimentary canal** or **gastrointestinal tract** (p. 455) (also called the gut). **Accessory organs** (p. 455), such as the salivary glands, liver, gallbladder, and pancreas, aid digestion, helping in the step-by-step conversion of foods into nutrients circulating in the bloodstream. The human's system is more typical than the koala's, which has a giant **caecum** (p. 455) involved in fermenting cellulose and a huge proximal colon. A cow has a stomach-like sac or **rumen** (p. 455) that acts as a fermentation chamber.

The alimentary canal carries out **ingestion** (p. 455), or taking in food; **mastication** (p. 456), or breaking up food (in birds, this can involve the **gizzard** [p. 456] instead of a mouth and teeth); **digestion** (p. 456), or the chemical breakdown of food into smaller molecules; **absorption** (p. 456) of the molecules into the bloodstream; and **elimination** (p. 456), or expulsion of undigested residues.

The gut has four layers: the **mucosa** (p. 456), which produces digestive enzymes and **mucus** (p. 456); the **submucosa** (p. 457), a connective tissue supplied with blood vessels; the **muscularis** (p. 457), or muscle layer that carries out **peristalsis** (p. 457); and the thin outer **serosa** (p. 457).

The Human Digestive System

Your mouth and teeth, coated by hard **enamel** (p. 457) are well suited to chewing food. The muscular **tongue** (p. 457) is the main organ of taste and moving food about in the mouth. The **larynx** (p. 457), or voice box, forms speech, along with the teeth, tongue, and lips. Food is swallowed as a moistened lump or **bolus** (p. 457). **Salivary glands** (p. 457) secrete watery **saliva** (p. 457), which moistens the food, and saliva contains a starch-breaking enzyme, **amylase** (p. 458). The bolus moves past the **hard** and **soft palates** (p. 458) and the **epiglottis** (p. 458), closing off the **glottis** (p. 459). The bolus then moves through the **pharynx** (p. 459) then down the **esophagus** (p. 459), main pipeline to the stomach, passing through a ring of muscle, the **cardiac sphincter** (p. 459). The bolus enters the J-shaped bag or **stomach** (p. 459), containing gastric juice made up of water, hydrochloric acid, mucus, and **pepsinogen** (p. 459)—a precursor to the protein-cleaving enzyme **pepsin** (p. 459). The milky, acidic **chyme** (p. 459) passes through the **pyloric sphincter** (p. 459) to the small intestine. The **pancreas** (p. 460) secretes digestive enzymes, including **proteases** (p. 460) and **lipases** (p. 460). It also secretes the hormones **insulin** (p. 460) and **glucagon** (p. 460), which regulate blood glucose levels.

The **liver** (p. 460) destroys old blood cells, stores glycogen, and adds glucose to the bloodstream when levels drop. It produces **bile salts** (p. 460) which are stored as **bile** (p. 460) in the **gallbladder** (p. 460); bile breaks up fat droplets in the small intestine. The liver also stores and synthesizes certain nutrients and detoxifies alcohol and other poisons.

The **small intestine** (p. 461) is 6 m (20 ft) long, the main site of carbohydrate and fat digestion, and completes protein digestion. The upper section is the **duodenum** (p. 461), the central section is the **jejunum** (p. 461), and the rest is the **ileum** (p. 461). The inner surface is pleated into folds topped with fingers or **villi** (sing., *villus*) (p. 461) carpeted with **microvilli** (p. 461) that together create a huge surface area for absorbing nutrients. The **large intestine** or **colon** (p. 462) ends in the **rectum** (p. 462). The human **appendix** (p. 462) is tiny, vestigial, and homologous to the koala's caecum, although both may help fight infection in the gut. The colon stores and then excretes **feces** (p. 462). Bacteria residing in the human intestines break down nutrients and produce vitamins B_1, B_2, B_{12}, and K.

The nervous and hormonal systems coordinate digestion. Food in the stomach triggers release of gastric juices and the hormone **gastrin** (p. 462), which causes nearby cells to release more hydrochloric acid. Food literally stimulates its own digestion. Too much acid (as well as particular bacteria) can lead to **ulcers** (p. 462). Other hormones, including **secretin** (p. 463) and **cholecystokinin** (p. 463) from the small intestine, help neutralize acid in the intestine, help digest proteins, and produce the sensation of fullness.

We saw that the koala has a huge caecum full of cellulose fermenting bacteria but gets most of its food energy from cell contents. Research reveals that small particles and dissolved nutrients get washed into the koala's caecum where bacteria act on them for up to nine days, fermenting cellulose and other carbohydrates, and helping detoxify tannins. Larger leaf fragments settle into the distal colon and are quickly eliminated, reducing the dietary bulk in the animal's intestines.

Healthy Diets and Proper Nutrition

If you want to lose weight, eat foods with enough nutrients but with lower energy content and higher roughage content, then increase exercise to step up basal metabolic rate. New research on the action of the hormones **leptin** (p. 464) and insulin helps explain normal fat storage and obesity. Cutting fats and sugars helps, but exercising is the key to burning fat and turning up the metabolic level.

All of the following question sets also appear in the Explore Life *E* electronic component, where you will find a variety of additional questions as well.

Test Yourself on Vocabulary and Concepts

In each question set below, match the descriptions with the appropriate term. A term may be used once, more than once, or not at all.

SET I

(a) carbohydrates (b) lipids (c) proteins
(d) vitamins (e) minerals

1. Energy-dense nutrient that is required for the absorption of fat-soluble vitamins
2. This class of biochemicals includes enzymes and some hormones

3. In nutrition, functions almost exclusively as an energy source
4. Organic enzyme precursors and activators that we require but cannot synthesize
5. Our cells can synthesize all that we need from its subunits.

SET II

(a) gastrovascular cavity (b) accessory organ
(c) gizzard (d) gastrointestinal tract

6. A bird's substitute for teeth
7. Site of extracellular digestion in a saclike, nontubular gut
8. A one-way gut with mouth and anus
9. Glandular outpocketings of a gut tube, such as the liver and pancreas

SET III

(a) mucosa (b) muscularis (c) cardiac sphincter
(d) pyloric sphincter (e) duodenum (f) colon

10. Pulses out chyme into the small intestine after a meal by alternately contracting and relaxing
11. Lines the lumen of the gut and contains cells that produce digestive enzymes
12. Guardian of the entry to the stomach
13. Primary site of water reabsorption from the gut
14. First section of the small intestine where secretions from the liver and pancreas enter the lumen of the gut

SET IV

(a) gastrin (b) secretin (c) cholecystokinin
(d) pepsin (e) bile salts (f) proteases

15. A protein-digesting enzyme that acts in an acid environment
16. Made from cholesterol, acts as a detergent that breaks up and surrounds droplets of fat
17. A hormone produced by the stomach that regulates the secretion of the stomach's main enzyme
18. A hormone produced by the small intestine that acts to produce a feeling of fullness after a meal
19. A small-intestine hormone that causes the pancreas to secrete an alkaline compound that neutralizes stomach acid

Integrate and Apply What You've Learned

20. What is so unusual about the koala's diet, what problems does that diet pose, and what problems would it pose for you if you were stuck in its favorite kind of tree for a week?
21. What structures in the digestive system act to break down ingested food physically? Why is this breakdown essential for digestion? What would happen to your digestive process without it?
22. Which organs of the human digestive system are the primary sites of digestion and absorption? What makes them primary?
23. What vital functions occur in the liver and pancreas?
24. Explain how the nervous and endocrine systems coordinate digestion.
25. Explain the model for how leptin and insulin regulate body fat.
26. Why are digestive enzymes like pepsin secreted in an inactive form?

Analyze and Evaluate the Concepts

27. You are a biologist trying to develop a weight-control therapy and decide to formulate a pill that causes people to feel full early in a meal so they eat less. What type of substance might you try, and why would you try it?
28. If a person had a mutation that blocked the function of cholecystokinin, what would be the resulting condition?
29. Explain how someone could obtain enough calories and still starve to death.
30. Why does the existence of essential amino acids pose potential problems for strict vegetarians and how can they overcome this problem?

Stress on Campus

Stressors and Relaxers. Wade Yandell talking about stress; Dr. Plotsky in the lab; Wade playing basketball.

Ask Wade Yandell about stress. He's a pre-med student at the University of Oregon. That alone is worth five or six points on a stress scale of ten. But Yandell is also majoring in biology and must pass a long list of tough prerequisites in chemistry, physics, and math before he can even register for most biology classes. On top of that, he studies 4 to 5 hours per day, then puts in another 16 hours a week in lectures and labs. And just to round things out, Wade also works for a genetics professor on campus to earn spending money and research experience: Count 10 to 12 hours more doing zebrafish experiments.

Not convinced yet? Wade is also sending out numerous expensive applications to medical schools, while worrying about getting top grades for his transcript. And here's the clincher: This hard-working college senior endures the gray skies and cold drizzle of western Oregon after leaving his home and family on the warm, breezy Hawaiian island of Oahu.

Wade is facing what biologists call **stressors,** activities or events that cause **stress,** the body's response to fear, pain, or other disturbances of our normal physiological equilibrium. How does he bear all those stressors? To start with, Wade has an innate resistance to Oregon weather. "Coming here freshman year, a lot of people from Hawaii couldn't handle it and wound up going back home or transferring to another school on the mainland. For me, though, it's never been that bad." And then there's basketball and weightlifting. "I like to go to the gym everyday," he says. "It's a good way to relieve stress." When he's too busy to exercise or misses his ride to the gym, "I get a little bit moody or grumpy." Before and during finals week, when time for pumping iron or shooting hoops is even tighter, that mood can escalate to "cranky and irritable. It feels like you never have enough time," he says, "and like you haven't had enough sleep, even if you have."

After each academic quarter ends, Wade recharges with rest and extra recreation. Then summer brings a major regeneration, with a return to Oahu to "sleep a lot, eat a lot, and go out with friends."

On the surface, Wade's college experience is pretty typical, and his remedies—rest, exercise, and socializing—are recommended stress-busters. Internally, his responses are probably perfectly normal, too. And that's where an expert on hormones like Paul Plotsky comes in.

A professor at the Emory University School of Medicine in Atlanta, Georgia, Dr. Plotsky studies the complex chain of signals, secretions, and reactions initiated when the body is under stress. Stressors can be real, he says—"life and death sorts of physical threats like sprinting away from a predator so you don't become lunch." But they can also be just the sorts of daily anxieties a college student faces and, behind them, the generalized "fuzzy sense of not being in control or knowing what's going to happen next."

Whether stalked by a hungry lion or an impending chemistry midterm, alarm signals in the brain tell the small adrenal glands located on the kidneys to secrete chemical messengers called stress hormones. These move through the bloodstream and, explains Plotsky, "activate a whole bunch of systems in the body." They shunt blood away from your hands and feet and toward your main internal organs. They increase blood pressure and heart rate. They cause a flood of stored energy to reach your heart, brain, and large muscles. They temporarily slow down your digestive, reproductive, and immune functions. And they dull your ability to feel pain while sharpening your readiness to act.

This is all appropriate and adaptive, says Plotsky, and if the stressor stops, "things will more or less return to some semblance of normalcy." But if stress becomes chronic, as in a couple on the verge of divorce, a person facing bankruptcy, or a student flunking out of college, "then the activation becomes repeated or continual and starts to have very detrimental effects." These can include susceptibility to infection, cardiovascular disease, digestive problems, sexual dysfunction, and perhaps impaired memory. Chronic stress, says Plotsky, calls for regular sleep, meals, and exercise, as well as diversions, hobbies, and socializing. Anything is helpful, he concludes, "if it returns some semblance of control and predictability to the environment . . . and gives you a sense of mastery and making progress."

Stress, stress hormones, and healthy ways of coping are all part of our topic in this chapter: hormones. **Hormones** are molecular messengers in the body that help trigger and coordinate our physiological responses to change, whether that change is imposed on us from the external world or arises spontaneously from within.

In this chapter, we'll talk a lot more about how stress hormones work for and against us. We'll also discuss other kinds of chemical messengers that regulate growth, digestion, pregnancy and nursing, and other body functions. Along the way, you'll learn how hormones determine a pygmy's height, why most sports have outlawed anabolic steroids, and what causes menstrual cramps. And by the end, you'll also find the answers to these questions:

1. What kinds of molecular messengers move through an animal's body and what do they do?

2. How does the body's endocrine system do its job and where does stress fit in?

3. How do molecular messengers help animals maintain a steady internal state but also adjust to change?

18.1 Molecular Messengers at Work in the Animal's Body

An overworked college student trying to get into medical school. A mother with a toddler who badly needs a diaper change and an infant who is crying to nurse *now*. An employee with a difficult and uncompromising boss who wants that report *immediately*. In all three situations, the people feel stress. And in each case, the stress response is mediated by **molecular messengers,** substances that are produced by one group of cells and secreted into those cells' environment, and then provoke some change in another set of cells. Molecular messengers permeate your body and regulate your life. They help control the amount you eat each day; how quickly you grew during childhood and adolescence; whether you are happy, depressed, or tense; how much salt and sugar flows in your blood; and a thousand other aspects of your physiology and behavior. Molecular messengers are clearly central to survival and normal functioning. So what are they, and how do they work?

Figure 18.1   How Molecular Messengers Work. (a) Molecular messengers share a general strategy of operation. Regulator cells sense a change in their environment (Steps 1 and 2), and release a molecular messenger (Step 3). This travels through air, water, or body fluid (Step 4); binds to a receptor on or in a target cell (Step 5); and alters the cell's activity to respond to the initial change. (b) In the specific instance of Wade Yandell worrying about an upcoming exam (Step 1), his brain signals his adrenal glands (Step 2), and they release cortisol (Step 3). This stress hormone binds to receptors on liver cells (Step 4) and these release more glucose (Step 5), fuel for the body.

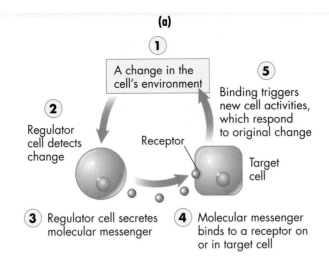

(a)

1 A change in the cell's environment

2 Regulator cell detects change

Receptor

5 Binding triggers new cell activities, which respond to original change

Target cell

3 Regulator cell secretes molecular messenger

4 Molecular messenger binds to a receptor on or in target cell

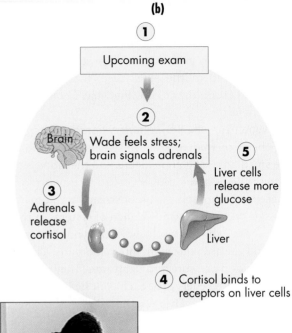

(b)

1 Upcoming exam

2 Wade feels stress; brain signals adrenals

Brain

3 Adrenals release cortisol

5 Liver cells release more glucose

Liver

4 Cortisol binds to receptors on liver cells

GET THE PICTURE

E*xplorer* 18.1

Stress Hormones

The General Strategy of Molecular Messengers

Although various molecular messengers mediate stressful events like the ones just mentioned, they all share a general strategy of operation. First, a change occurs in the immediate environment of a **regulator cell,** a cell that monitors local conditions. The regulator cell detects this change (Fig. 18.1a, Step 1), and in response, secretes a molecular messenger (Step 2). The molecular messenger then moves through the fluid surrounding the cell—air, water, or fluid between cells—and may become dispersed locally or at a distance (Step 3). Eventually, the molecular messenger contacts **target cells,** cells that can detect and respond to the messenger. The molecular messengers generally bind to **receptors** inside target cells or on their surfaces (Step 4). Receptors are proteins that bind only to specific molecular messengers and then relay a signal to the cell that the messenger is present in the cell's environment. After binding the messenger molecules, the receptors change shape, and trigger new activities in the cells that adjust them to the original perturbation (Step 5).

In this chapter, we'll see many examples of this general strategy for molecular messengers. For now, let's take one example from Wade Yandell's situation: worrying about an upcoming exam. This worry eventually alters the environment of regulator cells in a specific part of the adrenal gland sitting on top of each of his kidneys (Fig. 18.1b, Steps 1 and 2). These adrenal cells detect the change and release a specific molecular messenger, a stress hormone called **cortisol** (Step 3), which diffuses into the blood. The blood carries cortisol throughout the body, and some reaches the liver. There, target cells contain a re-

ceptor for the messenger (Step 4). The receptor binds the cortisol stress hormone, and this binding can stimulate a change in the liver cells: the synthesis of glucose from amino acids, thus providing more fuel for the body (Step 5). Glucose can then race through Wade's bloodstream and supply energy that may help him cope better with the upcoming exam. E*xplorer* 18.1 animates the general model of molecular messengers.

Receptors

A molecular messenger like cortisol moving through the bloodstream can bathe many cells, but only certain cells—the target cells—respond. Why does a molecular messenger act on target cells but not other cells? The answer lies with the receptors each target cell has on its plasma membrane or in its cytoplasm. Cells that are *not* targets for molecular messengers lack receptors for the specific messenger in question. The signal from a radio station is a good analogy. Only people with a radio tuned specifically to the station's frequency will receive the signal, while others without the tuned radio receiver will be totally unaware of the station's radio waves constantly in their environment. Likewise, liver cells have many receptors for the stress hormone cortisol, while other cell types have few cortisol receptors or none and so are unaffected by its presence.

There is specificity between a receptor and its particular molecular messenger, because their three-dimensional shapes fit together, just as a key fits into a lock. The shape of the receptor (the lock) is complementary to the shape of the messenger (the key), and so they can bind together. And just as with a lock receiving a key, the arrival of the molecular messenger causes the receptor to change shape subtly after binding. It is this shape change that heralds the messenger's arrival and triggers the change in cellular activity (opens the lock). Returning to Wade's case, then, the binding of cortisol to receptors in liver cells causes a change in receptor shape that triggers a new activity in the cell: it starts the cell making enzymes that will, in turn, convert stored amino acids to the ready fuel glucose.

Second Messengers

When molecular messengers such as hormones bind to receptors on cell surfaces and subtly change their shape, it usually initiates one of two processes. For some hormones, including cortisol, the receptor originally resides *in the target cell's cytoplasm,* and hormone binding causes the receptor to move from the cytoplasm into the cell's nucleus. There it binds to DNA and alters the activities of specific genes. For other hormones, including insulin, which we discussed in Chapter 14, the receptors are em-

bedded *in the target cell's surface.* The immediate effect of binding to one of these surface receptors is often the production *inside the cell* of one of a number of small molecules. The small molecule, called a **second messenger** (the hormone was the "first messenger"), then initiates a series of other reactions within the cell that culminate in the cell's response to the hormone. One important second messenger is called *cyclic AMP,* and it's chemically related to one of the nucleotide subunits (adenine) in DNA. Once triggered by the arrival of the hormone (the first messenger) cyclic AMP can cause channels to open in the cell's membrane; it can trigger the cell to build enzymes and other specific molecules; or it can cause the cell to secrete substances like sweat or milk.

An Overview of Molecular Messengers

During finals week, Wade's body made lots of the stress hormone cortisol, which, as we've seen, is a type of molecular messenger. But there are five different kinds of molecular messengers, each playing markedly different roles in the physiology of animals (Fig. 18.2).

Hormones are molecules produced by one set of cells and transported in body fluids, usually the blood, to other parts of the body. These parts can be near or far, and upon arrival, the hormones bind to a target cell and cause a distinct change in cell activity (Fig. 18.2a). For example, vigorous exercise like playing basketball or weightlifting stimulates the **pituitary gland,** an important secretor of several hormones that is located at the

Figure 18.2   **Molecular Messengers: Targets and Transport.** (a) Hormones are secreted into the bloodstream, travel some distance, diffuse out again, and act on a target cell. (b) Paracrine hormones diffuse a short distance and act on an adjacent cell. (c) Some nerve cells secrete neurohormones, which pass into a blood vessel, travel some distance, then diffuse out again and reach a target cell. (d) Pheromones are secreted by regulatory cells in glands with ducts, leave the body, and stimulate target cells in another organism's body. (e) Many nerve cells secrete neurotransmitters that diffuse across a narrow cleft to a target cell.

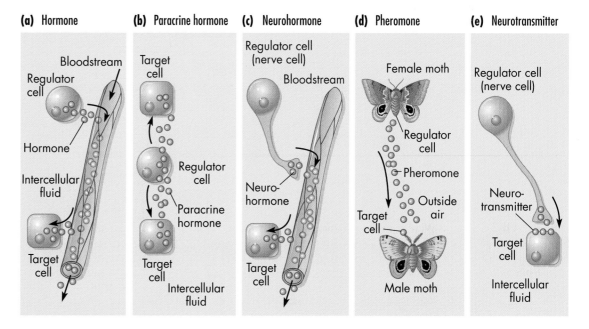

(a) Hormone
(b) Paracrine hormone
(c) Neurohormone
(d) Pheromone
(e) Neurotransmitter

base of the brain, to release growth hormone. Growth hormone circulates in the blood, binds to receptors on bone and muscle cells, and stimulates their growth, helping athletes to build stronger bones and bigger muscles.

Hormones are products of the **endocrine system,** a group of cells and organs at various locations in the body that secrete these molecular messengers into body fluids (see Fig. 18.4 and ⒠ 18.1).

In general, hormones are produced by **glands,** groups of cells organized into secretory organs. Some glands secrete hormones directly into the extracellular fluid and blood rather than into a duct or a tube; these are called **endocrine glands** (Fig. 18.3a). The pituitary, adrenal, and thyroid glands are all endocrine glands, and, along with about ten other endocrine glands, they produce most of the body's hormones.

In contrast, some glands secrete materials into ducts (tubes) that generally lead out of the body or into the gut; these are called **exocrine glands** (see Fig. 18.3b). Exocrine glands produce and release body exudates such as sweat, saliva, milk, digestive enzymes, and the silken threads that silkworms and spiders spin into cocoons and webs.

Paracrine hormones are hormones secreted by individual regulator cells; they diffuse only a short way through the extracellular fluid and act only on adjacent target cells (Fig. 18.2b). Regulator cells in our intestines, for example, detect protein molecules from food and

secrete a paracrine hormone that causes nearby target cells to release protein-digesting enzymes (different from those just mentioned) into the gut. Without this paracrine secretion, your body would extract fewer nutrients from the food you eat.

Neurohormones are molecular messengers secreted by nerve cells, and these messengers travel through the bloodstream to distant target cells elsewhere in the body (Fig. 18.2c). For example, just before Wade takes a midterm physics exam, nerve cells in his brain release a neurohormone called CRH into the blood (details later). Through a couple of intervening steps, this neurohormone stimulates the release of cortisol. Cortisol then brings about some of the physical signs of stress—sweating, shaky hands, dry mouth, rapid heartbeat, and weak knees. Another neurohormone, *oxytocin,* released from the pituitary gland, travels to a woman's uterus and stimulates contractions during birth.

Pheromones are compounds secreted by one individual that affect another individual's behavior or physiology. As we just saw, in many insects and mammals, pheromones are secreted by exocrine glands. They

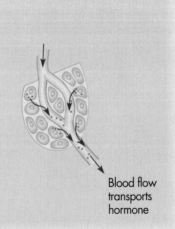

(a) Endocrine gland secretes hormones directly into cell surroundings

Blood flow transports hormone

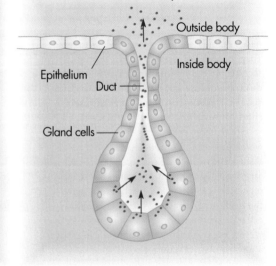

(b) Exocrine gland secretes substances such as sweat into a duct that leads out of body

Outside body

Inside body

Epithelium

Duct

Gland cells

Figure 18.3 **Glands With and Without Ducts.** (a) An endocrine gland such as the adrenal secretes hormones into the extracellular fluid that surrounds cells. From there, the chemicals pass into blood vessels and travel to distant sites in the body. (b) An exocrine gland such as a sweat gland secretes sweat into a duct, which leads the substance out of the body or into the digestive tract.

leave the body via ducts, travel in air or water, and stimulate target cells in other animals of the same species located up to several kilometers away (Fig. 18.2d). Take the sexually mature female silk moth, for instance. A scent gland in her abdomen secretes a pheromone that wafts on air currents for distances of up to 11 km (7 mi). Pheromone molecules that happen to hit target cells in a male silk moth's antennae can trigger wild sexual excitation. This leads the male to fly long distances toward the origin of the pheromone and a potential mate.

Neurotransmitters are a special type of molecular messenger and are nerve-signaling compounds released by individual nerve cells. Like the other four molecular messengers mentioned, neurotransmitters are substances secreted by one cell (a nerve cell) that have their effects on other cells. But neurotransmitters are much more local in their action than the other four molecular messengers: Pheromones can travel miles to their target cell. Hormones can travel five feet or more—say, from your pituitary to your toes. And paracrine hormones can diffuse ten micrometers to their targets. Neurotransmitters, however, diffuse only 0.03 micrometers to their target cells. Neurotransmitters usually diffuse a short distance from a nerve cell to a nerve or muscle cell, but sometimes diffuse to a gland cell (Fig. 18.2e). For example, when Wade is playing basketball and wants to make a jump shot, specific nerve cells running from his lower back down both legs release a neurotransmitter. This substance diffuses across a short space to muscle cells in the calves and causes his lower leg muscles to contract and propel his body upward toward the basket. We'll learn much more about neurotransmitters in Chapter 19.

For the rest of our discussion here, we're going to focus on hormones and neurohormones, and explore their role in stress and other familiar responses in an animal's body.

18.2 Hormones, the Endocrine System, and Stress

The Main Types of Hormones

Although animal hormones perform an incredible variety of tasks and control a huge spectrum of responses, most are either polypeptide hormones, steroid hormones, fatty acid hormones, or amino acid-derived hormones. Hormones of each type play a role in the body's response to physical or emotional stress.

Peptide Hormones

Peptide hormones are made of strings of amino acids—either peptides (just a few amino acids long) or polypeptides (many amino acids linked together). For example, oxytocin, a peptide hormone that increases the strength of uterine contractions in childbirth, is just nine amino acids long. The polypeptide hormone CRH (corticotropin-releasing hormone), which is 41 amino acids long, is released by cells in your brain when it detects a stressful event, as we'll see later. Peptide and polypeptide hormones usually act by stimulating second messengers in target cells.

Steroid Hormones

The major stress hormone, cortisol, is a **steroid hormone.** Steroids are lipids made from cholesterol, and they contain rings of carbon atoms with various attached atoms. Steroid hormones also include the estrogen and testosterone generated in the human ovaries and testes, respectively (see Chapter 8). And another steroid hormone, **ecdysone,** causes insects to shed their old exoskeletons. Steroid hormones often act by binding receptors that bind DNA and alter gene activity.

How are stress hormones related to the fight or flight response?

Fatty Acid Hormones

Some hormones have structures related to fatty acids. **Prostaglandins** are an example; these fatty acid-derived messengers act as paracrine hormones. They can be secreted by most cells of the body and have many activities. They play a role in inflammation (review Fig. 16.2), for example, and in intensifying pain. Some prostaglandins assist childbirth by softening the mother's cervix, the tight ring that surrounds the birth canal, thereby allowing the baby to pass through. Prostaglandins can cause headaches and menstrual cramps by altering blood vessel diameters and triggering uterine muscle contractions. Drugs such as aspirin and ibuprofen relieve pain by blocking the production of prostaglandins. Prostaglandins often act by stimulating the formation of second messengers in target cells.

Amino Acid-Derived Hormones

The adrenalin "rush" you feel as you stand at the top of a steep ski slope, or as an instructor calls on you the morning after you skipped doing your homework, is caused by the stress hormone called **epinephrine** (or sometimes **adrenaline**). Epinephrine is derived from an amino acid, as is the thyroid hormone that helps determine your metabolic rate. Both are small molecules produced when amino acids are chemically modified, and act via second messengers.

These four types of hormones regulate a multitude of activities in your body and arise from a variety of endocrine glands, as we will see in the next section.

The Endocrine System

The **endocrine system** includes a number of hormone-secreting glands in various body locations. Figure 18.4 displays these glands in the body and briefly summarizes

a few of their functions, as does E xplorer 18.2. Rather than just listing and describing these organs one by one, let's see how their activities are integrated in a single important physiological network, the stress response.

The Endocrinology of Stress

When an animal faces a physical threat, such as a hungry lion, an angry boss, or a surprise exam in a college class, the endocrine and nervous systems react with a coherent series of activities, collectively called the stress response. Paul Plotsky of Emory University, as one of

Figure 18.4   Glands in the Human Endocrine System.

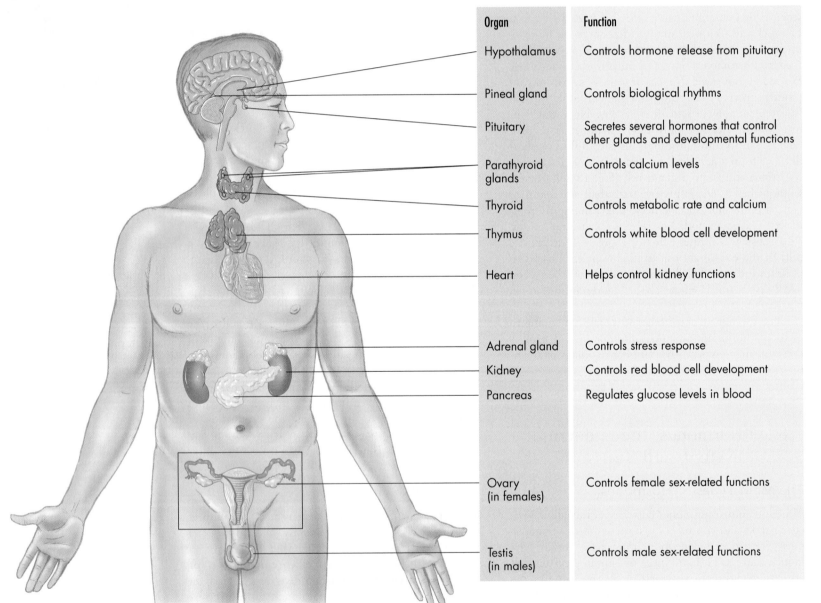

Organ	Function
Hypothalamus	Controls hormone release from pituitary
Pineal gland	Controls biological rhythms
Pituitary	Secretes several hormones that control other glands and developmental functions
Parathyroid glands	Controls calcium levels
Thyroid	Controls metabolic rate and calcium
Thymus	Controls white blood cell development
Heart	Helps control kidney functions
Adrenal gland	Controls stress response
Kidney	Controls red blood cell development
Pancreas	Regulates glucose levels in blood
Ovary (in females)	Controls female sex-related functions
Testis (in males)	Controls male sex-related functions

the nation's top stress researchers, studies the stress response in laboratory animals (specifically newborn rats) as a model for human stress. By using Wade Yandell as our example, we can follow the stress pathway in the human body and see the normal roles of regulatory cells and target cells. Let's begin with two glands at the base of the brain, the hypothalamus and pituitary, where the stress response is integrated.

Stress Integration: Hypothalamus and Pituitary

Wade is about to start his physics exam and he feels very nervous. He sees the exam questions stacked on a desk at the front of the auditorium. He hears the instruction to sit in an even-numbered seat. He becomes aware of a heavy, wooden sensation from studying half the night instead of sleeping. His stomach churns from strong coffee and a wolfed-down donut. Wade's brain perceives all these sensations and feelings, and consolidates the signals into a global sense of being out of control and threatened, and signals the **hypothalamus,** a collection of nerve cells at the base of the brain (Figs. 18.4 and 18.5a, Step 1,

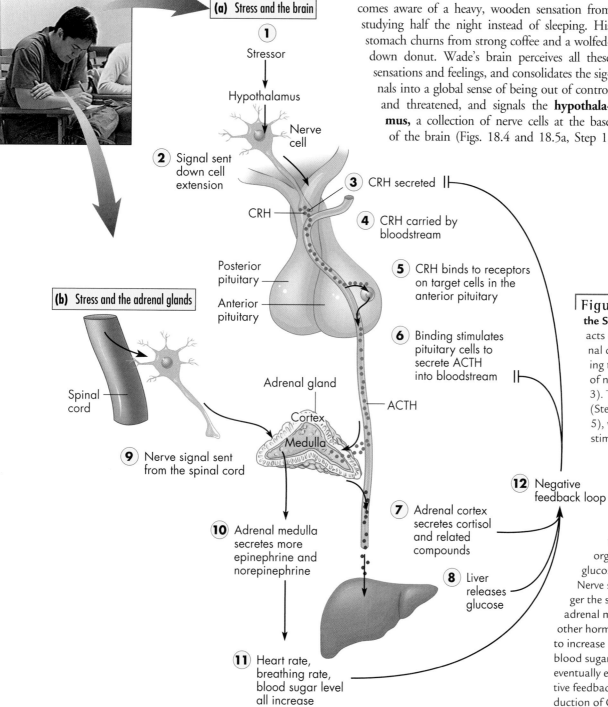

(a) Stress and the brain

1 Stressor

Hypothalamus

Nerve cell

2 Signal sent down cell extension

3 CRH secreted

CRH

4 CRH carried by bloodstream

5 CRH binds to receptors on target cells in the anterior pituitary

Posterior pituitary

Anterior pituitary

6 Binding stimulates pituitary cells to secrete ACTH into bloodstream

(b) Stress and the adrenal glands

Spinal cord

Adrenal gland

Cortex

Medulla

ACTH

9 Nerve signal sent from the spinal cord

10 Adrenal medulla secretes more epinephrine and norepinephrine

7 Adrenal cortex secretes cortisol and related compounds

8 Liver releases glucose

11 Heart rate, breathing rate, blood sugar level all increase

12 Negative feedback loop

Figure 18.5 **Anterior Pituitary and the Stress Response.** (a) Stress (Step 1) acts on the hypothalamus, sending a signal down a cell extension (Step 2), causing the release of CRH from the endings of nerve cells into the bloodstream (Step 3). The CRH travels in the bloodstream (Step 4) to the anterior pituitary (Step 5), where it binds to target cells and stimulates them to secrete ACTH into the bloodstream (Step 6). This, in turn, travels throughout the body. (b) ACTH eventually reaches the adrenal cortex and causes it to release cortisol and related compounds (Step 7); these stimulate the liver and other organs to change fats and proteins into glucose and release the sugar (Step 8). Nerve signals from the hypothalamus trigger the spinal cord (Step 9) to stimulate the adrenal medulla to release epinephrine and other hormones (Step 10) that allow the body to increase its heart rate, breathing rate, and blood sugar (Step 11). The stress response is eventually ended and normal returns by a negative feedback loop (Step 12) that slows the production of CRH and ACTH.

and  18.1). These important regulatory nerve cells integrate the incoming signals and, given the stressful situation, quickly send electric signals down cell extensions (axons). These signals rapidly reach the ends of the nerve cells' axons in the upper portion of a stalk that projects downward from the hypothalamus (Step 2). After receiving these electric signals, the tips of the axons secrete a neurohormone, **CRH** (corticotropin-releasing hormone) (Step 3). We'll see shortly why it's called this.

CRH secreted from the hypothalamus diffuses into special blood vessels (Step 4), which carry it directly to a fingertip-sized bulb hanging from the brain just above the roof of both the nasal cavities, the pituitary gland (Step 5). The mature pituitary in an adult is shaped like an upside down light bulb and has two lobes, a posterior (rear) lobe, which develops from the brain, and an anterior (front) lobe, which develops from the roof of the embryonic mouth. Together, these two lobes secrete nine major kinds of hormones, many of which control other endocrine glands. For this reason, the pituitary has been called by some the "master gland." The stress response illustrates well the main features of the interaction between hypothalamus and anterior pituitary and how they control other glands.

When CRH diffuses out of the special blood vessels leading from the hypothalamus and into the anterior pituitary, it encounters target cells with CRH receptors on their surface (see Step 5). CRH binds to these receptors and this "docking" stimulates pituitary cells to synthesize and release another peptide hormone, called **ACTH,** or adrenocorticotropic hormone (Step 6). Adrenocorticotropic hormone gets its name from its activity on turning on (*trop* = to turn on) the cortex (outer layer) of the adrenal gland and causing it to release stress hormones or cortisol. (Now the name CRH also makes sense; it is called corticotropin-*releasing* hormone because it causes the pituitary to release ACTH.) The ACTH enters the blood, which distributes this hormone throughout the body, including the adrenal glands on top of the kidneys.

Let's review the stress pathway so far: Wade's eyes, ears, stomach, and so on send messages to his hypothalamus, which interprets the signals and responds by secreting the hormone CRH. CRH travels in the blood quickly to the pituitary gland, which secretes another hormone, ACTH. ACTH enters the bloodstream and reaches another set of the body's key glands, the **adrenal glands,** which sit atop the two kidneys (*ad* = towards + *renal* = kidney) (Figs. 18.4 and 18.5b).

Adrenal Glands: Makers of Stress Hormones

The adrenals are major producers of stress hormones. Each adrenal gland has a middle portion, or **medulla,** and an outer layer, or **cortex** (from the Latin for *bark*). Each portion releases a different hormone with a different role in the stress pathway.

The Adrenal Cortex   When stimulated by ACTH, the adrenal glands' outer covering makes the major stress hormone **cortisol,** as well as other hormones not related to stress (Fig. 18.5b, Step 7; cortisol gets its name from the adrenal *cort*ex). Acting on target cells in the liver, fat tissue, and most other organs, cortisol causes stored proteins, lipids, and carbohydrates to be broken down and glucose to be rapidly generated; this simple fuel, in turn, provides stressed cells with a source of quick energy (Step 8). CRH causes the pituitary to release the "turn on" factor, ACTH, into the blood.

The Adrenal Medulla   The middle portion of the adrenal gland makes the other important stress hormones. When activated by the sudden stress of a hungry lion or a pop quiz, the body initiates the alarm reaction or **fight-or-flight response.** In this rapid response to a sudden threat, nerve signals from the spinal cord (Fig. 18.5b, Step 9) cause the adrenal medulla to increase its secretion rate for epinephrine and **norepinephrine,** which are chemically similar and derived from amino acids (Step 10). Carried through the blood, these hormones reach target cells in the heart, lungs, intestines, and elsewhere, and accelerate the stress response that Wade experienced and that we all know from time to time (Step 11): The hormones cause the heart to beat faster, blood sugar levels to increase, the breathing rate to speed up, and blood to be shunted away from organs such as the stomach and intestines. The result of all this is that the muscles receive more blood, oxygen, and sugar, and thus the animal is better able to defend itself or move away to safety. This is what Paul Plotsky means when he says stress hormones help in "life and death sorts of physical threats" like fleeing a predator so "you don't become lunch."

Stress and Negative Feedback Control

Once the hungry lion leaves, or the exam is over, the body must calm down, because it can't carry on its everyday activities like digesting food, making reproductive cells, and protecting itself through immune surveillance when it's in a state of readiness for "fight or flight." Luckily, there is such a mechanism for restoring prestress equilibrium: The endocrine system can regulate the levels of hormones in the blood through negative feedback loops (review Fig. 14.6 and  18.1). Just as a thermostat senses rising temperatures and sends a signal to shut down the furnace, the end-product of a biological pathway feeds back information to the pathway's initiator to decrease its action and turn down the whole system.

What is the main stress hormone?

In the stress response we've been exploring, the stress hormone cortisol acts on receptors in two brain regions, the hypothalamus and pituitary, to stop their release of CRH and ACTH (Fig. 18.5, Step 12, and *E* 18.1). With the decrease in levels of these activator molecules, the stress response itself diminishes and the body returns to normal. A second negative feedback pathway helps this return to normalcy, as well: the surge of glucose fuel allows faster running, clearer thinking, or whatever activities are needed to cope with the stressful situation. The escape and subsequent sense of safety again decrease the secretion of CRH and, in turn, of cortisol. For example, as Wade starts his physics exam and realizes that he studied the right material and is able to solve the problems easily, his CRH levels decrease, and his physiology can move toward normal again.

But what happens if the stress does not disappear quickly? One part of Dr. Plotsky's research is trying to discover whether prolonged, high-level stress early in life can permanently alter the body's feedback response to stress, rendering it less able to control the stress pathway once it's initiated. He and his colleagues indeed found that rats stressed heavily as pups (by their mother's absence for long periods) grew up with too few cortisol receptors in their brains (Fig. 18.6). As a result, in adulthood, the rodents could not decrease their CRH levels and turn down cortisol release normally: Their stress thermostats, in a sense, were permanently set on "high." We don't yet know the degree to which these results on rats apply to humans.

Additional Roles of the Hypothalamus, Pituitary, and Adrenals

By exploring the stress response in a pre-med student like Wade, we've touched on the major features of the endocrine system and hormones: the role of the regulatory glands, the hypothalamus and pituitary; the role of important organs such as the adrenal glands; the functioning of peptide and steroid hormones in negative feedback loops; and the role of laboratory studies. The hypothalamus and pituitary however, play other crucial roles in the body besides responding to stress, including the regulation of reproductive activities and growth. The adrenals are also involved in regulating water excretion and can affect sexual development.

Other Roles of the Anterior Pituitary and Hypothalamus

The anterior lobe of the pituitary synthesizes and secretes the ACTH that turns on stress hormones, as we saw. But it also makes and releases about half a dozen other hormones, including growth hormone, thyroid-stimulating hormone, the LH and FSH involved in sperm and egg production (see Chapter 8), and **endorphins.** Endorphins can act as natural pain killers and are the source of "runner's high." In each case, the hypothalamus gives off releasing hormones or release-inhibiting hormones that control the pituitary's secretion of these hormones. Nursing in mammals is a good example of these positive and negative factors. A specific release-inhibiting hormone from the hypothalamus usually suppresses the release of the hormone prolactin by the anterior pituitary. During pregnancy, the hypothalamus secretes prolactin-releasing hormone, and prolactin levels rise. The suckling of an infant reduces the amount of prolactin-inhibitory hormone from the hypothalamus, and these dual controls—more stimulation and less inhibition—result in more prolactin release from the anterior pituitary. Prolactin, in turn, causes a woman's breasts to produce milk (Fig. 18.7a).

Roles of the Posterior Pituitary

The pituitary's posterior lobe—a second rounded protrusion behind the first—is also controlled by the hypothalamus, but communicates with the hypothalamus in a different way. Recall that the hypothalamus signals the anterior pituitary by secreting neurohormones into blood vessels that travel directly to the anterior pituitary (review Fig. 18.5a). In contrast, the hypothalamus signals the posterior pituitary by sending extensions of nerve cells called axons all the way from the hypothalamus to the posterior pituitary (Fig. 18.7b, Step 1). Hormones made in hypothalamus cells travel down the long nerve cell extensions and accumulate at each tip, in the posterior pituitary. The hormones taking this path include the water-balancing hormone ADH, which acts on the kidney (review Fig. 14.16) and oxytocin, with its role in nursing.

Let's continue considering a nursing mother to see how the posterior pituitary works. As her baby begins to suckle, sensitive nerves in the woman's nipples send signals to the hypothalamus (Fig. 18.7b, Step 2). This, in turn, relays nerve impulses to the fiber tips and causes the posterior pituitary to release oxytocin into the bloodstream (Step 3). When oxytocin reaches target cells in the woman's milk ducts, it causes those muscle cells to contract, forcing milk into the ducts and out through the nipple, where the baby can suck and swallow it.

Other Roles of the Adrenal Glands

We saw that during Wade's physics exam, his adrenal cortex secreted the stress hormone cortisol. This outer

Figure 18.6 **Rat Studies, Stress, and Human Depression.** Dr. Plotsky tends to the lab rats in one of his stress studies.

Figure 18.7 **Posterior Pituitary and Milk Release.** The sucking action of this 3-month-old baby boy (a) stimulates milk release (b). Nerve cells in his mother's hypothalamus have axon tips extending into the posterior pituitary; these store oxytocin (Step 1). A nerve impulse from the suckling (Step 2) causes oxytocin to be released into the bloodstream (Step 3) and to stimulate milk production for the suckling baby.

(a) Nursing and the pituitary

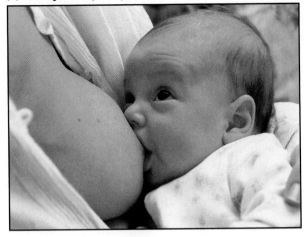

(b) Control of oxytocin production

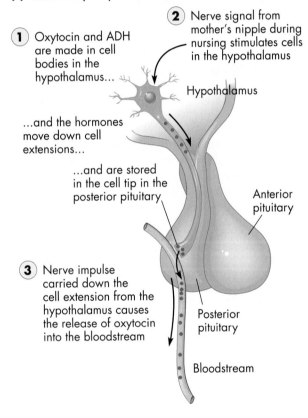

① Oxytocin and ADH are made in cell bodies in the hypothalamus...

...and the hormones move down cell extensions...

...and are stored in the cell tip in the posterior pituitary

② Nerve signal from mother's nipple during nursing stimulates cells in the hypothalamus

Hypothalamus

Anterior pituitary

③ Nerve impulse carried down the cell extension from the hypothalamus causes the release of oxytocin into the bloodstream

Posterior pituitary

Bloodstream

adrenal layer also liberates two other types of hormones—one so vital that in its absence, death comes quickly. That secretion is the steroid hormone **aldosterone** and it causes the kidney to conserve sodium and water and to excrete potassium (review Figure 14.14). Without aldosterone, your body would excrete too much sodium and water in the urine, and conserve too much potassium. Your blood volume would then drop precipitously, your blood potassium would skyrocket, and your life would end within days.

The adrenal cortex also secretes male steroid sex hormones, or **androgens,** that help stimulate the early development of male sex organs and also stimulate pubic and underarm hair growth in human females (see Chapter 8).

18.3 Molecular Messengers: Agents of Homeostasis and Change

We watched Wade's hypothalamus, pituitary, and adrenal glands in full operation as he faced a tough exam and then saw how negative feedback control brought his stress responses down again as he answered the test questions easily. His hormones helped prepare him to meet a perceived threat and survive it. (Good study habits didn't hurt, either.) But animals encounter multiple survival challenges each day, and several endocrine glands and several kinds of hormones help prevent or provoke internal change in response to those environmental fluctuations. Some hormones help maintain processes or substances within a very narrow range of concentrations compatible with life. We talked about homeostasis and the steady internal state in Chapter 14, and in this section, we'll see how hormones contribute to controlling the levels of glucose and calcium in the blood. Sometimes, of course, great changes are required for life: Sexual maturation is one. We'll also see how hormones help usher in those transformations.

Hormones of Homeostasis

Two glands in your neck and one nestled near your intestines carry out the important jobs of regulating your metabolic rate, controlling your body's level of calcium, and balancing your blood sugar.

The Thyroid Gland

The **thyroid,** a bumpy pink gland at the base of the neck (Fig. 18.8a), acts as the body's metabolic thermostat, regulating its use of energy as well as its growth. The thyroid's main hormone is **thyroxine,** which contains iodine and is derived from amino acids. The cascade of events leading to thyroxine release is similar to the stress pathway: the hypothalamus sends a releasing hormone to the anterior pituitary, which sends a thyroid-stimulating hormone to the thyroid gland, which releases thyroxine.

What happens if the thyroid gland is severely underactive? In a child with this condition (referred to as **cretinism**), the body cannot break down carbohydrates or build proteins fast enough, which leads to low body temperature, sluggishness, stunted growth, and mental retardation. In an adult with an underactive thyroid gland (a condition called **myxedema**), weight accumulates easily and mental efforts are slowed. Fortunately, both conditions respond to drugs containing thyroxine.

A person deprived of iodine in the diet can develop a **goiter**, an enlarged thyroid visible as a large lump in the neck (Fig. 18.8b). It may seem strange that an iodine *deficiency* could lead to a larger thyroid rather than a smaller one, but it's based on another negative feedback loop (*E* 18.1). When the body has made enough thyroxine, the hormone itself feeds back and blocks the brain from making more of the signal that turns on the thyroid gland (to make thyroxine). When iodine is missing from the diet, the thyroid can't form thyroxine, so it can't feed back and turn off the brain signal. It's a bit like a furnace with a broken thermostat. The pituitary pumps out the thyroid-stimulating hormone full throttle, trying to get the thyroid to grow larger to do its job. The thyroid grows larger and larger but still can't produce thyroxine due to the lack of iodine. The result is often **hypothyroidism,** a condition of low thyroid activity leading to sluggishness, dry skin, weight gain, and other symptoms. Fortunately, goiters are rare in this country because the iodine in ordinary iodized table salt supplies enough for our thyroid glands to make thyroxine. The thyroid also works with another set of glands, the parathyroids, to control calcium levels.

The Parathyroid Glands

The **parathyroid glands** are a set of four small, dark patches of cells embedded in the posterior side of the paler thyroid (Fig. 18.8a). Collectively, these neck glands steady the level of calcium in the blood.

Let's say Wade has a muffin and a glass of milk for breakfast. Milk is rich in calcium, so calcium levels would

(a) Normal thyroid and parathyroid glands

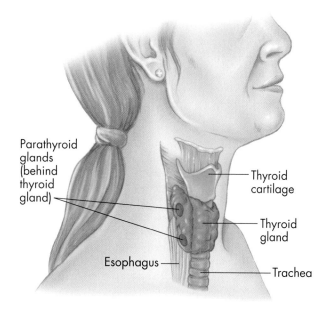

Parathyroid glands (behind thyroid gland)

Thyroid cartilage

Thyroid gland

Esophagus

Trachea

(b) Goiter, an enlarged thyroid

Figure 18.8 **The Thyroid Gland and Goiter.** Normally, thyroxine, produced by the thyroid (a), blocks the release of thyroid-stimulating hormone from the pituitary in a negative feedback loop. (b) Without iodine, however, the thyroid can't make thyroxine, and so the pituitary makes thyroid-stimulating hormone continuously, causing the thyroid gland to enlarge into a goiter. This Burmese woman from the village of Peinnenpin has a large, prominent goiter.

start to rise in his blood. His thyroid responds to this by releasing the hormone calcitonin, which causes the bones to deposit excess calcium. Later that day, after a pick-up basketball game, the calcium levels in Wade's blood start to fall and his parathyroid glands release parathyroid hormone. This molecular messenger causes his bones to release more calcium back into the blood and his body to pick up more calcium from food—say, from the broccoli in his dinner salad. Without enough circulating calcium ions, nerves and muscles malfunction and a person or other animal will suffer tingling extremities, hyperactive reflexes, muscle contractions, and perhaps death. That makes parathyroid hormone the second hormone (besides aldosterone) that we absolutely need for survival.

The Pancreas

A serious reminder of our need to maintain a steady internal state through hormones is **diabetes mellitus,** a disease of high blood sugar and excessive sugar excretion, and our most common hormonal disorder. Diabetics produce large amounts of dilute, sugary urine and often have increased thirst. Without treatment, they risk **coma** (unconsciousness) due to dehydration or poisons accumulating in the blood. Diabetes mellitus stems from an inability to make or to respond to **insulin,** a hormone made by the pancreas, an elongated organ nestled near the intestines (Fig. 18.9a). The **pancreas,** a secretory organ near the stomach, is an interesting organ because it serves as both an exocrine gland and an endocrine gland (review Fig. 18.3). Its exocrine role is making and secreting digestive juices for the small intestine (review Fig. 17.10). Its endocrine role is secreting the peptide hormones **glucagon** and insulin from cells called the **islets of Langerhans.**

Let's return to Wade Yandell for a moment to see how insulin and glucagon act to regulate glucose levels in the blood. In Wade and most other people, a feedback loop involving glucagon and insulin keeps blood sugar levels steady. After a breakfast of cereal and toast, nutrients enter his bloodstream, and his blood sugar levels rise (Fig. 18.9, Step 1). In response, one type of secretory cells in the islets in his pancreas secrete insulin (Step 2). This hormone causes target cells in the liver and elsewhere to remove glucose from the blood and store it as glycogen (Step 3). Then, as Wade goes to class, rides his bike, and works in the lab, his blood sugar levels fall (Fig. 18.9b, Step 4). A second type of pancreatic islet cells detect the drop and release glucagon (Step 5). This causes liver cells to break down glycogen and release glucose, so blood levels once again rise (Step 6). This dual control acts like a car's brake and accelerator to keep blood sugar levels within a narrow, healthy range.

Now let's take a person with diabetes of the type in which the pancreas fails to make insulin (called Type I diabetes). If this person eats the same breakfast of cereal and toast, the blood sugar levels rise, but islet cells in the pancreas fail to make insulin. Without this hormone, body cells fail to remove glucose from the blood and high

Figure 18.9   **The Pancreas and Blood Sugar Levels.** The pancreas regulates levels of glucose in the blood. (a) After a healthy person consumes carbohydrates, say, in a sports drink, glucose levels rise in the blood (Step 1). Special pancreatic cells (Islets of Langerhans cells) then release insulin (Step 2); this binds to its receptors on liver and other cells, causing them to remove glucose from the blood and store it as glycogen (Step 3). (b) When blood levels of glucose drop from bike riding or other heavy exercise, for example (Step 4), the pancreas secretes glucagon (Step 5); this binds to specific receptors on liver cells, causing them to break down glycogen and release glucose into the blood (Step 6). This push-pull relationship of glucagon and insulin helps maintain blood glucose at a fairly constant level despite the nutrient content of our meals and the amount of fuel we consume during physical activity.

(a) Function of insulin

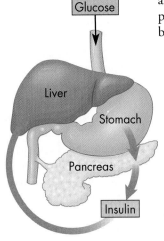

1 Glucose levels rise in blood

3 Cells take glucose from blood and store as glycogen

2 Islet cells in pancreas detect high glucose and secrete insulin

(b) Function of glucagon

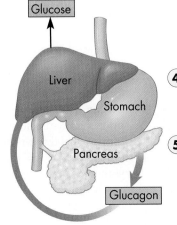

4 Glucose levels fall in blood

6 Cells dismantle glycogen and release glucose into the blood

5 Specific cells in pancreas detect low glucose and secrete glucagon

levels keep circulating. The kidneys remove the sugar along with large amounts of water (for the reasons we discussed in Chapter 14). This explains a diabetic's voluminous urine and frequent thirst, as well as the danger of dehydration. Because body cells cannot remove glucose from the blood, they "starve" in the midst of plenty.

Fortunately, there is treatment for people with this type of diabetes (which usually begins in people before they are 20): daily insulin injections by needle or by small automated pump devices. As some diabetics grow older, they risk long-term damage to fingers, toes, and retina of the eye due to poor circulation. Their levels of HDL ("good cholesterol") are too low, and this leads to damaged arteries and poor circulation in the extremities. Recent tests showed that careful blood sugar monitoring and frequent injections can diminish these harmful side effects.

There is a second, more common type of diabetes (Type II) that usually affects overweight people over 35. These people fail to make enough insulin receptor molecules on their target cells, and hence they can't respond fully to insulin. Fortunately, a regimen of diet, exercise, and weight loss usually helps this type of diabetes.

Hormones and Change

All the hormones we've talked about so far help keep key body substances at a fairly steady level. Sometimes, though, an animal's body needs to undergo daily changes (waking and sleeping, for example) or permanent physical changes (such as during puberty). A different set of hormones brings about this necessary change.

The Pineal Gland and Melatonin

In general, humans are diurnal animals—most active during the day—and our stress hormones peak about the time we wake up, and then fall to a low point in the early evening. This natural surge of cortisol helps prime your system with a surge of glucose that literally breaks your fast even before morning food and coffee. Nocturnal animals like rats and owls have the opposite schedule of hormone activity.

These fluctuating daily patterns are called **circadian rhythms** (sir-KAY-dee-uhn; Latin, *circa* = about + *dies* = day). Circadian rhythms can reflect the changing length of daylight during the four seasons, and research shows that a small brain structure, the **pineal gland** (PIHN-ee-uhl) (review Figure 18.4), is involved in measuring day length and controlling responses to it—moods, activity cycles, reproductive cycles, and perhaps even puberty.

The pineal gland is a bumpy, pea-sized knob that resembles a small pine cone (hence *pine*al) lying deep in the brain. In some animals, including lizards, the pineal gland is called the "third eye" because it is structured like the retina and directly perceives light. In humans and other mammals, light receptors in the eyes send signals to the pineal. Regardless of species, the knobby little gland secretes a hormone called **melatonin** in response to darkness, and this hormone promotes sleep and inhibits gonadal activity (egg and sperm production). People sometimes take melatonin to help them sleep and adjust to jet lag when they travel. We need further studies, however, to confirm whether this is truly safe and effective. Melatonin may also be involved in **seasonal affective disorder (SAD).** Some people experience depression, sleepiness, weight gain, and lethargy when the days grow short in winter, then feel happier and livelier in spring and summer. Exposure to bright lights for several hours per day in winter helps some people with SAD. It is not clear whether melatonin actually helps, and many tests show no benefit.

Puberty and Steroid Hormones

In humans, puberty begins at about age 10 to 14. Once begun, surges of testosterone and estrogen stimulate the development of the mature male and female secondary sexual traits—enlargement of female breasts, deepening of the male voice, development of larger muscles, and so on (see Chapter 8)—as well as the production and delivery of eggs and sperm.

Synthetic versions of these steroid hormones called **anabolic steroids** have found their way into the world's locker rooms, gyms, and health clubs. The pressure to look muscular and win competitions is so strong that as many as 20 percent of professional football players, 5 percent of college athletes, and 7 percent of high school boys have taken synthetic testosterones such as stanozolol. Anabolic steroids can mimic the adolescent upsurge of real male hormones that promote rapid muscle bulking in 15- to 16-year-old boys. Unfortunately, the drugs also confer the other masculinizing effects of testosterone in exaggerated proportions: heavy hair growth, acne, and premature baldness in men; and deepening of the voice, growth of facial hair, increased body hair, and enlargement of the clitoris in women. These changes in women appear to be permanent. Anabolic steroids also suppress a male's own hormones, and this can lead to shrunken testes and enlarged breasts. Ominously, research also shows that damage to the kidneys, liver, heart, and possibly the prostate gland, which produces some of the seminal fluid, are all too

When the pancreas secretes insulin, is it acting as an exocrine gland or an endocrine gland?

common in users of anabolic steroids, as are depression, anxiety, hallucinations, paranoia, and other psychological symptoms. Coaches who care about their players' long-term health make sure they don't take anabolic steroids.

Thyroxine and Metamorphosis

Human puberty brings about major body changes but the transformation can be even more dramatic in other species. Many insects, amphibians, and other animals undergo **metamorphosis,** a revolutionary change in body form from the larval stage to the adult (see Chapter 13, p. 346). In tadpoles, metamorphosis is triggered by **thyroxine** from the animal's thyroid gland. Thyroxine acts on target cells in the tail and flank; this causes the long tail to be resorbed and legs to sprout from the formerly smooth sides. The tail-less, four-legged frog can now move about on land as well as in water. Interestingly, stress can accelerate metamorphosis.

The Western Spadefoot Toad lives in western states and lays its eggs in desert pools that accumulate during winter rains but dry out and disappear in the spring. In especially dry years, the pools disappear early, the tadpoles mature prematurely, and small toads hop away over land. In wet years, metamorphosis is delayed, and larger toads emerge later in the spring. Researchers studying the timing of toad metamorphosis found that thyroxine brings it about but not in response to the releasing hormone that normally triggers thyroxine release. Instead—and unexpectedly—the releasing hormone used by the toad is our old friend, CRH, the releasing hormone associated with the stress response. Researchers were intrigued to learn that just before birth, the human fetus and placenta begin secreting CRH. This initiates the softening of the cervix and leads to the uterine contractions that begin labor. Researchers have found a link between stress in the fetus, the early secretion of CRH, and premature births, a leading cause of infant mortality. What appear to be unrelated events—toad metamorphosis and premature human births—are actually outcomes of a single, ancient hormonal pathway that allows the immature organism to escape an unfavorable, highly stressful habitat—either a dry pond or an inhospitable womb.

The body's many responses to physical or psychological stressors—mediated by the endocrine system— are marvels of adaptation. As we saw, however, with Wade Yandell and others, when stress becomes chronic, those often-triggered fight-or-flight responses can lead to long-term health problems. Wade's solutions, including regular exercise, extra sleep, and relaxing summers in Hawaii, are themselves highly adaptive and help both his hormones and his health.

Chapter Summary and Selected Key Terms

Introduction   A pre-med major with a part-time job, Wade Yandell knows all about **stress** (p. 468) and **stressors** (p. 468). Dr. Paul Plotsky of Emory University studies stress and the role of **hormones** (p. 469): molecular messengers that help trigger and coordinate our physiological responses to external or internal change.

Molecular Messengers at Work   **Molecular messengers** (p. 469) work via a general strategy. A **regulator cell** (p. 470) detects changes in its environment, and in response secretes a molecular messenger, which moves through air, water, intra- and intercellular fluids, and the bloodstream. The messenger contacts **target cells** (p. 470), which detect the message and respond by carrying out an activity that adjusts to the original change.

Receptors (p. 470) on the target cell's surface, or inside the cell cytoplasm or nucleus, are proteins that bind to only specific molecular messengers and relay to the cell the information that the messenger is present. Often, the binding of the messenger to the receptor causes the production of a **second messenger** (p. 471) inside the cell such as cyclic AMP that triggers cellular change such as the binding of an enzyme or the secretion of sweat or milk. Other receptors bind DNA and alter gene activity after binding to hormone molecules.

Hormones are produced by one set of cells and are transported (usually in blood) to other parts of the body. An example is growth hormone. The **endocrine system** (p. 472) is a group of organs at various body locations that secrete molecular messengers into body fluids. Organs of the endocrine system are often groups of secreting cells or **glands** (p. 472). **Endocrine glands** (p. 472) dump molecular messengers directly into extracellular fluid and blood. Most hormones travel this way.

Paracrine hormones (p. 472) are secreted by individual cells and act on adjacent target cells. Paracrine hormones, for example, aid in digestion. Nerve cells can release **neurohormones** (p. 472) that travel through the bloodstream to distinct target cells. Neurohormones like CRH trigger the release of stress hormones. **Pheromones** (p. 472) are molecular messengers secreted by one individual that affect another individual's behavior or physiology. Silk moth sex attractants are examples. Pheromones are released by **exocrine glands** (p. 472), which release molecular messengers or other substances into ducts that usually lead out of the body. Sweat, saliva, and milk are released by exocrine glands. **Neurotransmitters** (p. 473) are nerve-signaling compounds released by individual nerve cells that diffuse to a nearby neuron, muscle, or gland cell. Neurotransmitters, for example, help trigger muscle contraction.

Hormones, the Endocrine System, and Stress   There are many types of hormones. **Peptide hormones** (p. 473) are chains of amino acids. CRH is a peptide hormone. **Steroid hormones** (p. 473) like estrogen, testosterone, and **cortisol**

(p. 476), the stress hormone, contain rings of carbon and attached atoms. **Fatty acid hormones** (p. 473) include **prostaglandins** (p. 473). **Amino acid–derived hormones** (p. 473) include **epinephrine** (p. 473). The stress pathway starts when the **hypothalamus** (p. 475) receives information about a stressful situation in the environment and sends a releasing hormone to the anterior **pituitary gland** (p. 471), stimulating it to release the peptide hormone **ACTH** (p. 476). ACTH, carried by the blood, triggers the release of cortisol from the outer **cortex** (p. 476) of each **adrenal gland** (p. 476). Nerves control the release of **epinephrine** (p. 473) by the gland's central **medulla** (p. 476). Negative feedback loops regulate the stress pathway by decreasing the release of CRH and in turn **cortisol** (p. 476). Paul Plotsky discovered that stress early in development can cause too few receptors to form on brain cells, which interferes with the feedback mechanism and causes a permanently increased level of reaction to stress.

The pituitary's anterior lobe makes and secretes several hormones in addition to the ACTH that turns on cortisol, including growth hormone, LH and FSH, thyroid-stimulating hormone, and **endorphins** (p. 477). The pituitary's posterior lobe secretes ADH (p. 477), which regulates water retention, and oxytocin, which regulates milk release.

The adrenal cortex secretes **aldosterone** (p. 478), which also participates in sodium, potassium, and water regulation in the blood and urine. The adrenal medulla secretes epinephrine and **norepinephrine** (p. 476) in reaction to the **fight or flight** response (p. 476). The adrenals also secrete small amounts of male sex hormones or androgens. The **thyroid** (p. 478) and **parathyroid glands** (p. 479) help regulate metabolic rate, calcium levels, and blood sugar. Imbalances can lead to **goiter** (p. 479) and other conditions.

The **pancreas** (p. 479) helps regulate blood sugar levels and, when healthy, prevents high blood sugar or **diabetes mellitus** (p. 479), chronic high blood sugar. Groups of cells called **islets of Langerhans** (p. 480) secrete the hormones **glucagon** and **insulin** (p. 480), which regulate blood sugar.

Fluctuating daily patterns called **circadian rhythms** (p. 481) govern our sleep/wake cycles. The **pineal gland** (p. 481) in the brain interprets daily light length and intensity, and secretes **melatonin** (p. 481) in response to darkness. Rising levels of melatonin promotes sleep and inhibits reproductive activity.

In some animals including insects and amphibians, **thyroxine** (p. 482) from the thyroid gland triggers

metamorphosis (p. 482), or a change in body form such as from a tadpole to a frog.

> All of the following question sets also appear in the Explore Life *E* electronic component, where you will find a variety of additional questions as well.

Test Yourself on Vocabulary and Concepts

In each question set below, match the description with the appropriate term. A term may be used once, more than once, or not at all.

(a) CRH (corticotropin releasing hormone) (b) hormone of the anterior pituitary gland (c) negative feedback with a metabolite such as calcium or glucose (d) stimulated by the nervous system (e) neurohormone (f) stimulated by another hormone

1. Cortisol
2. ACTH in the stress response
3. Parathyroid hormone
4. Epinephrine
5. Oxytocin
6. Glucagon
7. Melatonin
8. Insulin
9. Thyroxine
10. Growth hormone (*Hint*—See *E*xplorer 18.2)

Integrate and Apply What You've Learned

11. What happens inside Wade Yandell's body, step by step, when he shows up one hour late for his job at the laboratory?
12. Men who take anabolic steroids have small testicles. Why might that be true? Consider two facts: (1) a hormone from the brain stimulates the function of the testes, which in turn produce testosterone; and (2) high levels of testosterone or anabolic steroids block the release of the brain hormone in a negative feedback loop.

13. What would happen in an animal's body if a hormone were produced and circulated in the blood, but no cells were present bearing receptors for that hormone?
14. Explain how insulin and glucagon work to regulate the level of blood sugar.
15. Contrast the role of the adrenal medulla and adrenal cortex in the stress response.
16. Some endocrine glands are more essential than others for survival. Excluding the hypothalamus and pituitary, name three glands that are among the most essential and one that is least essential, and defend your choices.
17. Rarely, grown women develop mustaches and other masculine traits and are found to have a tumor or other abnormality in their adrenal glands. Can you reason out why this might occur and which part of the adrenal glands would be affected?

Analyze and Evaluate the Concepts

18. Which types of hormones are more important: those that inhibit change or those that stimulate change? Support your answer.
19. Mrs. A shows signs of a thyroid hormone deficiency, but she has above-normal levels of thyroid hormone. What might be the problem?
20. Diabetes that develops during childhood is generally treated with insulin. Adult-onset diabetes is usually less severe and can often be treated by dietary restrictions. One of these forms of diabetes is caused by destruction of the beta cells in the pancreas; the other is generally caused by a lowered number of insulin receptors on body cells. Which kind of diabetes is the result of lack of beta cell production of insulin? Explain your reasoning.
21. In France, truffles are a culinary treat. In order to find these wild underground fungi, farmers use female pigs to detect the truffles, because pigs are able to detect and are strongly attracted by the scent emitted by the fungus. Scientists have recently found that male pigs make a compound chemically very similar to the odor-producing material of truffles. How does this help to explain the truffle-hunting ability of pigs?

Silence on the Mind

▲ **Signs and Signals.** Sherry Greer wearing red EEG cap; Jo and Sherry signing in ASL; Helen Neville in her office.

Sherry Greer has lived in silence for a quarter century. She inherited a genetic condition from her two deaf parents and was herself born deaf. Each snail-shaped structure or **cochlea** in Sherry's inner ears formed abnormally when she was a fetus, and as a result, they could never receive sounds and transmit aural patterns to her brain. But Sherry's years have been rich in language communication, because on the day she was born, her parents began speaking to their infant daughter through the gestures of American Sign Language (ASL). By 8 months of age, Sherry was babbling back to them with strings of tiny hand signs, including her first word, the sign for "Mama." At 2, she was forming phrases like any typical toddler—but with gestures. By 4, she was totally fluent in her native ASL but went on to learn English as a second language at home and in grade school. After high school, she attended college at

the National Technical Institute for the Deaf and, since graduating, has worked as a sales associate for a department store.

As a deaf member of the hearing world, Sherry's life has been different—and notable—in several ways. For example, she relies on special devices in her house, like a phone, doorbell, and fire alarm that flash lights instead of ringing. Her phone also has a keypad for typing messages that are transmitted to another party's phone screen. Sherry's deafness has also given her special capabilities. For instance, her peripheral vision is especially acute and she can easily catch all the ASL signs a friend sitting beside her might make as Sherry faces fully forward to drive or watch television. "I can also see a car coming from the side quite easily," she says through her ASL interpreter, Jo Larson-Muhr, "or a deer moving on the side of the road." This capacity, says Larson-Muhr, "makes deaf people better drivers in general." And, Larson-Muhr adds, most can also quickly and accurately "read" a stranger's honesty and intentions by using heightened visual awareness of facial expressions and nuances of behavior.

There's another fascinating difference: When Sherry is communicating in ASL her brain is more fully active than is a hearing person's brain while reading, talking, or listening to spoken language. How does Sherry know what's going on in her own brain? She doesn't officially, but Dr. Helen Neville does, because she has tested Sherry in her lab at the University of Oregon. For two decades, Neville has studied a somewhat surprising subject: how experience inscribes the brain, shapes its nerve circuits, and molds its activity. Her work is part of **neuro-**

science, the study of the brain and nervous system, and more specifically, the part relating to brain "plasticity." More than any other organ, the brain is plastic: Its functioning is both flexible and resilient. Neville and colleagues in her field around the world have discovered that a range of factors can shape how the brain adapts to changes and rebounds from injuries; how it senses the world; how it reacts to various stimuli; and how it regulates the body's activities. The factors that shape the brain can include mental stimulation (sights, sounds, smells, movements, and so on), deprivation (the absence of sound or some other type of sensation), and disease or trauma that physically alters the brain. For many kinds of brain activity, this plasticity is greatest in the developing child. But at some point, a window of time closes—a so-called **critical period**—after which experience no longer has as powerful a stimulating and shaping effect.

While studying brain plasticity and critical periods in deaf volunteers like Sherry Greer, Helen Neville has used two modern tools and has made several important discoveries. One tool is called **EEG (electroencephalography)** and involves a cap, electrodes, and a sophisticated monitoring and recording device. With EEG, Dr. Neville can determine exactly *when* and *in what order* sets of brain cells give off bursts of electrical activity in response to an incoming stimulus such as a sight or sound. For example, she and her technician can place a red elastic cap dotted with white electrodes on Sherry's head, show her a video of a woman signing in ASL, then record Sherry's brain waves as she takes in the visual messages. The other tool is called **fMRI (functional magnetic resonance imaging),** and it creates high-resolution images of brain areas as they work. With fMRI, Neville can pinpoint exactly *where* the electrical waves are arising in Sherry's brain as she reads or interprets ASL. This combination of tools allowed Neville to draw an interesting conclusion about people like Sherry who learned ASL in infancy: They show activity in the same areas of the brain's left side that "light up" in hearing people when they listen to or speak a language. ASL users, however, also have areas of activity on the brain's *right* side—the hemisphere that decodes spatial patterns (such as ASL's hand and facial movements.)

Neville has learned that in people deaf since birth or early childhood, the peripheral vision is sharper. This is because the brain's hearing center (the auditory cortex) is unused and some of its circuits get reassigned to the brain's visual center (the visual cortex). But Neville found that the phenomenon only works if the hearing loss occurred before age four. This suggests that much of the brain's auditory wiring becomes permanent by that age. A window thus closes on this aspect of brain plasticity, and people who lose their hearing after age four don't get the benefit of enhanced side vision.

Neville studied yet another such window for the learning of new vocabulary words and grammar. Her research shows that the brain can absorb new vocabulary throughout life, but loses the capacity to learn grammar if not introduced to it by grade school age. People not taught a *first* language—whether English, ASL, or some other—in childhood can still learn individual words later, but they can never order them into meaningful sentences. If, however, the child learns a first language and its grammar early enough, then the brain can learn second or multiple languages. Those additional languages are best absorbed before puberty too, however, because of critical periods for language. Neville draws several conclusions from her work on the brain: "People should talk to their kids. Read to them . . . . Kids with more language experience show more specialized brain systems and wind up learning larger, more productive vocabularies earlier." The parents of a deaf child, she says, should never delay his or her exposure to a gestural language like ASL, even though they might prefer their child to learn English. And finally, "Second language programs usually start in middle school, but the time to add French, Spanish, or German is *before* puberty, not after."

It's remarkable that biologists have come so far in understanding the nervous system and brain that they can confidently predict the best time to learn American Sign Language and the precise costs to the child in delaying it. This chapter explores the basic biology of the nerves and their electrical impulses. It covers the sense organs and how they take in stimuli from the environment. And finally, this chapter discusses the brain, with its global processing and control of reactions, movements, sensations, body systems, thoughts, language, and imagination. Nervous systems can be as simple as the networks of cells that allow flatworms to withdraw when poked. Or they can be as complex as the brain of an Einstein or a Madame Curie. But regardless, all nervous systems are built on the same principles and they help animals survive by responding to the environment and controlling internal processes.

You'll encounter many examples of neural activity in this chapter, from the owl's ability to hunt at night to more about Sherry's keen peripheral vision and special right brain activity. You'll also discover the answers to these questions:

1. How do individual nerve cells receive and transmit information?

2. How do nerve cells communicate with each other in networks?

3. How does the brain function and how is its activity coordinated with the nervous system, allowing an animal to sense, react, and behave in ways that promote survival?

4. How do sense organs work?

19.1 The Structure and Function of Nerve Cells

When Sherry communicates through ASL, her brain directs the movement in her hands and arms. When she "listens" in ASL, she perceives language signs with her eyes, and her brain decodes these visual signals as words and phrases. All these events are possible because of her **nervous system,** a network of specialized cells in humans and virtually all other animals. These specialized cells send messages—usually directly—to other specific cells in the body. The nervous system has the crucial task of controlling and coordinating the activities of an animal's many body systems. It *receives* information from the environment; it *integrates* this information (in other words, it makes sense of it for the animal), and it *affects,* or causes a change in, the way the body functions. All these roles require *communication*. This communication can take place at the microscopic level between different parts of an individual nerve cell or between neighboring nerve cells. And it can take place at the macroscopic level between different individuals in a community, allowing animals to eat, flee from, or mate with other animals at appropriate times. We'll follow these micro and macro levels of communication in this chapter, first examining the way single nerve cells "talk" to each other, then how networks of nerve cells communicate, and finally, how an animal's sense organs detect what's happening in its environment and allow it to react and behave appropriately.

Nerve Cell Structure

Sherry Greer's nervous system—and in fact, every person's—contains over 1000 different kinds of cells. Some, like the nerve cells that control hand and finger movements, are about two feet long. (Equivalent cells extending from a giraffe's spinal cord to its legs and hooves can be 3 m (10 ft) long!) Others, like the nerve cells that link one brain region to another, may be only a few millimeters long or less. All 1000 cell types in the nervous system, however, can be classified as either **neurons**, the nerve cells that accomplish the actual communication tasks, or **glial cells** (GLEE-al; Greek, *glia* = glue), support cells that surround, protect, and provide nutrients to neurons and may influence them in as yet unknown ways. The human brain contains 100 billion neurons supported by 10 times as many glial cells, or 1 trillion! Figure 19.1 shows an ultrathin slice of brain tissue with darkly stained individual neurons branching against a golden background. In fact, brain tissue is so densely packed with neurons and glial cells that the whole slice would look solid black if it weren't for the special type of stain used here, which darkens only 1 of every 1,000 to 10,000 neurons. Revealed this way, let's zoom in on one neuron and explore its unique structure.

Anatomy of a Nerve Cell

All neurons—including the billions of brain neurons that help Sherry Greer understand and speak American

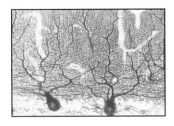

Figure 19.1 **Nerve Cells Are Organized into Highly Elaborate Networks.** Neurons from a cat's cerebellum interconnect via delicate extensions, establishing a complex cellular network with point-to-point contact of nerve cell to nerve cell. This photo of neurons called Purkinje cells shows two distinct cell bodies and hundreds of branches to other neurons.

Figure 19.2 **Neurons Transmit Information from One Cell to Another.** A neuron consists of four main parts: the dendrites, cell body, axon, and axon terminals. At the terminal, the sending cell transmits a signal across the synapse to the receiving cell.

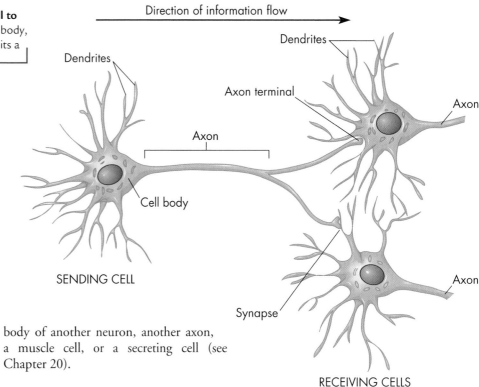

Direction of information flow

Dendrites

Dendrites

Axon terminal

Axon

Axon

Cell body

SENDING CELL

Synapse

Axon

RECEIVING CELLS

Sign Language—function by *collecting* information and *relaying* it to other cells in the body. Specific portions of each neuron perform these collecting and relaying functions (Fig. 19.2 and **E**xplorer 19.1).

Neurons gather information by means of fine, branching cell processes called **dendrites** (from the Greek for *little tree*). Dendrites pick up signals from other nerve cells and pass these impulses in one direction, toward the **cell body,** the major portion of the neuron. The cell body houses the usual complement of intracellular organelles, including nucleus, mitochondria, endoplasmic reticulum, Golgi apparatus, and ribosomes. With this cellular machinery, the cell body produces the proteins that make up the rest of the nerve cell and the enzymes that assist the cell's activity. The cell body takes up oxygen from the blood in the course of its protein-building activities. An active neuron uses more oxygen than one at rest. Therefore, Dr. Helen Neville is able to detect patterns of brain activity with an fMRI device by indirectly measuring how vigorously the cell body absorbs oxygen.

After passing down the dendrites and across the cell body, the nerve impulse—still traveling in a single direction—enters the neuron's long, thin cell extension, its **axon** (Greek for *axle*). In some of your spinal nerves, the axon extends all the way from your spinal cord to your toe muscles, while the cell body and dendrites of those same neurons are located back in the spinal cord itself. A neuron may have dozens of dendrites with up to thousands of branches, but it usually has just one axon. A **nerve**—the sort a dentist might accidentally hit with a drill, for example—is actually a group of axons and/or dendrites from many different neurons gathered in a bundle like a telephone cable.

Also like a telephone wire, the terminals or tips of an axon (it can have a few branches, too) stop at the receiver: they are the site of communication between cells. When a nerve impulse moves from dendrites to the axon and reaches the tips, it enters bulblike processes or axon terminals located very close to the cell membranes of other cells (Fig. 19.2). The two nearly contacting membranes plus the minute gap or cleft between them is a unit known as a **synapse** (SIN-apps), and a neural impulse can cross this synapse quickly. A neuron's axonal terminals may form synapses with a dendrite, a cell

body of another neuron, another axon, a muscle cell, or a secreting cell (see Chapter 20).

Sending and Receiving Cells

The two cells on either side of a synapse have different functions. The *sending cell* (or **presynaptic cell**) transmits a message down its axon and across a synapse to the other cell, the *receiving cell* (or **postsynaptic cell**) (see Fig. 19.2). This cell, in turn, may propagate the same message down its axon. Just as information flows in one direction within a neuron, it flows in one direction between neurons, from sending cell to receiving cell.

The axon of a sending cell may branch and rebranch, forming synaptic junctions with up to 1000 other cells. Conversely, 1000 other neurons might form synapses with the dendrites and the cell body of a single neuron, like hundreds of hands reaching out to touch a central object simultaneously. With these multiple links, an animal's communication network is literally a net, as Figure 19.1 reveals. The 100 billion neurons of the human brain, for example, make 1000 trillion synaptic contact points—more than all the known stars and planets in the universe. This most intricate lacework of interconnected neurons allows for all the simple and complex behavior we observe among animals. At one end of the continuum, it lets a sea slug withdraw its siphon when touched. At the other end, it lets a human being speak and understand the complicated symbols of ASL, such as those for "I'm going to class now."

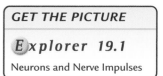

GET THE PICTURE

Explorer 19.1

Neurons and Nerve Impulses

Recall that an animal's nervous system receives and integrates data, and this affects a change in body function. Within a given neuron, the dendrites act as the receptors, the cell body acts as the integrator, and the axon and synapse act as the effectors of a nerve signal. Let's see what a nerve signal is, how it moves along a cell, and then later, how it can contribute to complex behavior like language.

A Nerve Cell at Rest

We said in the introduction that Dr. Helen Neville sometimes monitors a subject's brain waves with a special stretchy red EEG cap that looks like a geodesic dome crossed with a hairnet. The EEG technique detects the electrical activity of nerve cells, specifically the **nerve impulse,** an electrochemical reaction that allows specific electrically charged atoms to rush into and out of the nerve cell.

The Resting Potential

The best way to see how neurons generate a nerve impulse is to focus on a single patch of a resting neuron's plasma membrane. Like all cells, nerve cells have an electrical difference across their plasma membranes, even when unstimulated.

That is, the inside of the cell is negatively charged with respect to the outside (Fig. 19.3a and **E** 19.1). This difference in electric charge, which can be measured in volts like a battery, represents an amount of potential energy called the **resting potential.** A typical flashlight battery has only about 20 times the voltage in one of your neurons. Think about the literal brain power in 100 billion neurons!

A cell's resting potential is due to differences in the ions (electrically charged atoms) inside and outside the cell. The most important ions for nerve signaling are sodium (Na^+), potassium (K^+), and chloride (Cl^-).

Cells are bathed in a fluid that has a high concentration of sodium ions and a relatively low concentration of potassium ions (Fig. 19.3b). In contrast, the fluid inside the neuron is relatively rich in potassium ions and low in sodium ions. The high concentration of sodium ions outside the plasma membrane is like the thousands of music fans pushing and straining to get into an amphitheater for a rock concert. In contrast, the musicians and their crew inside the stadium warming up before the concert are like the potassium ions inside the cell. The musicians and crew can occasionally leave the stadium through the stage doors, but to keep the fans out requires strong gates, and also security officers to throw out the occasional fan that sneaks inside. Neurons have proteins in their membranes that act like these doors, gates, and security guards.

Protein Channels and a Pump

Three proteins are primarily responsible for maintaining a neuron's resting potential. The **potassium channel** is a protein pore through the membrane. The potassium channel pore contains a single gate that can open or close; thus the channel acts a bit like the stage door that permits the music crew to pass out of the stadium in that it allows potassium ions to pass slowly through the membrane and leak down their concentration gradient from the inside of the cell to the outside (Fig. 19.3b). Recall that a cell's cytoplasm is a gel-like substance containing many proteins. These proteins are mostly negatively charged and are too bulky to pass through the membrane. The slow leak

Figure 19.3 **A Neuron's Resting Potential.** (a) A peek inside an axon shows that the fluid just inside the cell membrane is more concentrated in negative charge relative to the fluid outside the cell. (b) Potassium ions (blue circles) leak slowly through the potassium channel from the inside of the cell to the outside (Step 1). In contrast, closed sodium channels prevent an influx of sodium ions (orange circles) (Step 2). The sodium-potassium pump, a protein embedded in the membrane, pumps potassium ions into the cell and sodium ions out (Step 3). This requires substantial energy because the pump moves ions against their concentration gradient, almost like making water flow uphill.

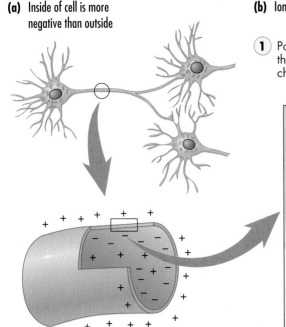

(a) Inside of cell is more negative than outside

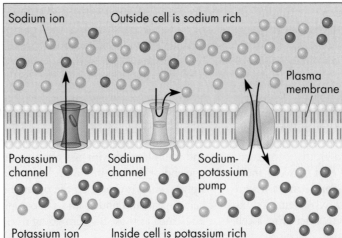

(b) Ion differences across the membrane

1 Potassium ions leak through potassium channel

2 Sodium channel is tightly closed

3 Sodium-potassium pump pushes sodium ions out of the cell and potassium ions into the cell

Sodium ion   Outside cell is sodium rich

Plasma membrane

Potassium channel   Sodium channel   Sodium-potassium pump

Potassium ion   Inside cell is potassium rich

of positively charged potassium ions out of the cell thus leaves the inside of the cell with a slight negative charge.

In contrast to the potassium channel, a second protein pore, the **sodium channel,** acts more like the secured gates before the concert. It closes tightly and allows almost none of the sodium ions outside the cell to leak inward. The sodium channel has two gates: an activation gate and an inactivation gate. These act in a coordinated way, finely controlling the length of time the channel is open. At rest, one gate is open and the other closed.

The third protein that helps maintain the neuron's resting potential is the **sodium-potassium pump.** Like the security guards, the pump opposes the inappropriate leakage of ions through the channels by using energy to force sodium ions out of the cell. The pump also escorts potassium back into the cell. Note that the pump keeps the potassium and sodium ions from falling down their concentration gradients. *This takes work.* Similar pumps act continuously in every animal cell and collectively use more ATP than any other body activity (see Chapter 3).

The actions of these three proteins in a neuron's plasma membrane maintain a steady voltage differential between the cell's outside and its inside. *E* 19.1 animates this process. In its resting state, the neuron is said to be **polarized;** it has an imbalance of electrical charges, negative on the inside and positive on the outside.

The Action Potential: A Nerve Impulse

If a resting neuron already has an electrical charge—the resting potential—then what does Dr. Neville's red polka-dotted cap detect? A nerve impulse, or **action potential,** is the *reversal* of the charge on a resting cell; it's a bit like what happens when the gates finally open and all the fans suddenly rush into the stadium.

How, then, does this reversal, this action potential, come about? It all starts with a stimulus, such as a light flash, the sizzling taste of a chili pepper, the sight of an ASL word being signed, or an impulse from another nerve cell. The stimulus tweaks a resting neuron (Fig. 19.4a), causing the second of the two gates in some of the sodium channels to open (Fig. 19.4b). With both gates open, sodium ions begin to leak into the cell. As more sodium ions enter the cell, the difference in charge between the inside and outside begins to decrease. The cell begins to **depolarize,** or lose its state of electrical polarization.

When the original difference in charge between the outside and inside decreases enough, a threshold is passed and something quite dramatic occurs: In the patch of plasma membrane where the stimulus first arrives, most of the sodium ion channels open wide (Fig. 19.4b). Sodium ions can now rush into the cell, and the inside of the neuron becomes positively charged with respect to the outside. This reversed polarity is the action potential. You can see it animated on *E* 19.1.

Sodium channels in the area of the action potential remain open for only about $\frac{1}{1000}$ of a second, and then inactivate as the inactivation gates close (Fig. 19.4c). For a few thousandths of a second after closing, the sodium channels cannot open again. This temporary state of nonresponsiveness limits the number of action potentials a neuron can fire each second.

Figure 19.4 **How a Neuron Generates an Action Potential.** (a) The inside of a cell at rest is electrically negative with respect to the outside. Recall that potassium ions (blue) slowly leak through potassium channels in the membrane. (b) After a stimulus, sodium ions begin to leak across the membrane, which decreases the electrical differences between the inside and outside of the cell, and the cell comes less negative. When the cell is depolarized past a threshold, the sodium gates open widely, sodium rushes into the cell, and the charge across the membrane reverses, becoming positive on the inside with respect to the outside. This leads to the action potential. (c) Shortly after, the sodium channels close and the potassium channels open; potassium rushes out, returning the cell membrane to negativity again. In fact, it's now more negative than the resting state. (d) The potassium channels close and the sodium-potassium pump can re-establish the cell's resting potential.

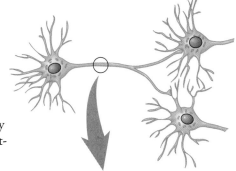

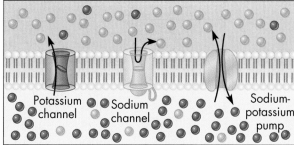

(a) Resting potential

Potassium channel • Sodium channel • Sodium-potassium pump

(b) Sodium channels open; potassium channels close

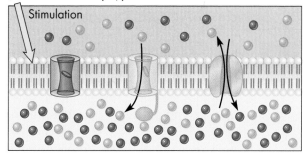

Stimulation

(c) Sodium channels close; potassium channels open

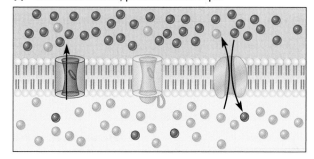

(d) Potassium channels close; pump restores resting potential

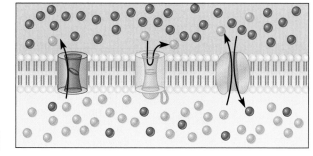

About the time the sodium channels close, the potassium channels open fully, allowing potassium ions to rush out of the cell (Fig. 19.4c). With this outpouring, the electrical potential falls to a level below the original resting potential.

Eventually, the potassium channels close. The "bailing" action of the sodium-potassium pump restores the neuron to its original resting potential (Fig. 19.4d). The neuron is now ready for another stimulus to trigger a new action potential. And the entire process of stimulus, action potential, and recovery has taken only a few milliseconds.

What would happen to an action potential if the gates in the ion channels were stuck shut? You have probably experienced the answer to this in your dentist's office. He or she will often inject novocaine as a pain-killing anesthetic before a procedure. Novocaine works by preventing the opening of the ion channels in nerve endings in the gums, tongue, and lips. Because the gates are closed, ions cannot pass through and there can be no action potentials in the neurons that relay pain signals to your brain. Without action potentials, you feel little pain, even if the dentist's drill touches an exposed nerve.

Action potentials have a peculiar property: They are **all-or-none responses;** that is, either they don't occur at all, or they do occur and are always the same strength for a given neuron, regardless of how powerful the stimulus may be. A jab to the ribs is more intense than a tender caress, but it's not because the jab causes larger or faster action potentials. Instead, the jab simply causes more cells to fire more impulses more frequently than the caress. The action potential generated by the mechanism we just discussed—when they occur in thousands of neurons at once—create the signal Dr. Neville detects with her EEG electrode cap. The question is, how do these impulses allow nerve cells to communicate?

How Does a Nerve Impulse Travel Down a Cell?

Sherry's arms, hands, and fingers move when forming an ASL word because action potentials move from cell bodies in her spinal cord down axons in her arms to muscles that move the limb and its extremities. How does an action potential travel down an axon? The process is rather like the Wave of rising and cheering fans that travels around a sports stadium. Each individual in the stadium represents a sodium channel, and standing up, throwing arms in the air, and sitting down again represents the opening and closing of sodium channels.

Just as each individual stands because the person sitting nearest has just started to rise, so too does each sodium channel open because the channel next to it has just opened. The open gates allow sodium ions to rush into the cell and generate an action potential that itself propagates along the axon—in this case down Sherry's arm.

Direction of Impulse Travel

Interestingly, within the body, the impulses in a given neuron travel in one direction only (from spine to finger, for example, but not from finger to spine), because of the short nonresponsive period. For a few thousandths of a second after an action potential, a membrane patch cannot experience another action potential because the sodium channel is temporarily inactivated. It's a bit like the Wave that passes around a football stadium; once you've stood, raised your arms, then sat back down, you don't want to do it all again too soon. The result of the temporary inactivation in the neuron is that the signal moves forward to the next patch but never moves backward toward the previous one.

The Speed of Impulse Travel

In many invertebrates, a nerve impulse traveling down a neuron moves only about 2 m (6.5 ft) per second. Think how sluggish a tennis match would be if players had to wait a full second for an impulse to travel from their brains to their toes! One way to speed the travel of nerve signals along an axon is for the diameter of the axon to increase. Just as a wider pipe has less resistance to water flow than a narrower one, a larger axon has less resistance to the flow of ions. A few invertebrates, such as the fast-moving squid, have in fact evolved huge axons over 1 mm in diameter that carry impulses rapidly. A person, however, could never make do with giant axons: To contain the 100 billion neurons and 1000 trillion interconnections we need to produce our own complex actions, thoughts, and language, our heads would have to be 10 times bigger! Instead, vertebrates have evolved a means of rapid impulse propagation without giant neurons.

Insulating Sheaths Speed Impulse Travel

We vertebrates have greatly speeded nerve impulses because many of our neurons are insulated with fatty sheaths. We said earlier that the nervous system has ten times more glial, or supportive, cells than actual neurons. Some specialized glial cells situate themselves along an axon like fatty sausages on a string and wrap their plasma membranes around the axon surface like an electrician wraps a wire with black tape (Fig. 19.5a). Together, these special glial cells called Schwann cells form a lipid-rich layer, the **myelin sheath,** that insulates the axon from the fluid outside the cell (see Fig. 19.5a,b). In the tiny gaps between these insulating "sausages" lie bare regions, and ions can flow across the axon membrane only in these uninsulated nodes. Between the nodes, electrical current flows in the cytoplasm and extracellular fluid, and when this current reaches the next node, it causes sodium channels to open there. The result of this glial cell arrangement is that the nerve impulse moves rapidly from one node to the next, bouncing along the axon 20 times faster than if the axon lacked the myelin sheath. This rapid method of propagation, taking place in insulated neurons, enables Sherry Greer to make the rapid, flowing hand movements of sign language, allows Andre Agassi to respond quickly to his opponent's overhead smash in tennis, and lets Andre Watts play a Rachmaninoff piano concerto with lightning-fast fingers. Even at their fastest, however, in neurons carrying impulses associated with touch, for example, or that conduct impulses to skeletal muscles, nerve impulses are transmitted only about 130 m (400 ft) per second, 2000 times slower than the speed of electricity in a wire. Although most of our neurons have myelin sheaths, many neurons that communicate from one neuron to the next don't.

Mutant mice whose axons lack insulating sheaths, such as the one in Figure 19.5c, shiver uncontrollably. In the absence of the sheaths, nerve impulses can move erroneously between the axons from side to side as well as from end to end. Likewise, the human disease multiple sclerosis (MS), whose symptoms usually first appear in early adulthood, causes some nerve cells to lose their insulating sheath. Without this insulating layer, nerve impulses can't move rapidly from node to node along the cell, and transmission slows, leading to double vision, weakness, and wobbly limbs.

19.2 Communication Between Nerve Cells

Just as ASL helps one person communicate with another, action at the special cell junctions called synapses helps some neurons communicate with other cells. The nerve impulse can pass information across a synapse from one neuron to another neuron, or from a neuron to a muscle cell. In either case, the signal moves in just one direction from the axon part of the neuron to the neighboring cell. Some neurons send impulses to other cells through direct electrical contact, like plugging in an electrical cord (see **E**xplorer 19.2). Most neurons in the human body, however, talk to other cells chemically.

(a) An insulated axon

Insulating sheath

Glial cells

(b) Cross section

Schwann cell making a myelin sheath

Uninsulated node

No myelin sheath

Axon   Normal   Mutant

(c) Normal mouse and *shiverer* mouse

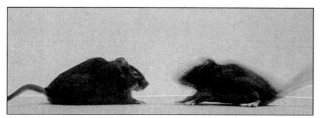

GET THE PICTURE

Explorer 19.2

Transmission Across the Synapse

Figure 19.5 **An Insulated Axon Propagates Impulses More Efficiently.** (a) Special glial (support) cells wrap the axons of many vertebrate neurons, thus insulating them. The action potentials propagate from the uninsulated nodes between the glial cell wraps, and so skip rapidly down the axon. (b) Cross section of a neuron from a normal mouse (left) shows a glial cell called a Schwann cell wrapped around an axon and a fatty layer, the myelin sheath. A cross section of neurons from a mutant *shiverer* mouse (right) reveals no encircling glial cells or myelin sheath. (c) A photograph taken with a long exposure time shows that the normal mouse holds perfectly still, while the mutant *shiverer* mouse lives up to its name.

Figure 19.6 **At a Chemical Synapse, Neurons Transmit Information Across a Cleft.** (a) A schematic diagram of a synapse. (b) An electron micrograph of a synapse reveals that a synaptic cleft separates the sending cell, with its packets of neurotransmitter, from the receiving (in this case a muscle cell). (c) How a chemical synapse functions, in five steps, from the arrival of an action potential (Step 1) to the diffusion of neurotransmitter molecules (Step 2), receptor binding (Step 3), sodium ion passage (Step 4), and the re-entry or removal of neurotransmitter (Step 5).

Chemical Synapses: Messages Spanning Gaps

We saw earlier that in a synapse, the bulblike axon terminal end is separated from a bulging structure on a receiving cell by the narrow synaptic cleft (Fig. 19.6a,b). A neural impulse can pass from the sending cell when a tiny amount of a chemical moves across the cleft and touches off a new impulse in the receiving cell. But what is this chemical and how does it work? You can see an animation in *E* 19.2, as well as reading what follows.

Within the axon's tip, enzymes synthesize a chemical signal called a **neurotransmitter.** The neurotransmitter enters small round vesicles, and in an axon at rest, these accumulate at the axon terminal. When an action potential moves along a neuron and reaches the end of the axon, special channels open and allow calcium ions to rush into the bulbous terminal (Fig. 19.6a, Step 1). The increase in calcium ion concentration causes the little bags of neurotransmitter

in that sending cell to fuse with the cell's outer membrane. This releases thousands of neurotransmitter molecules into the synaptic cleft that separates the sending and receiving cells. The neurotransmitter molecules diffuse across the cleft (Step 2) in less than a millionth of a second and bind to receptor proteins embedded in a trough-shaped plasma membrane of the receiving cell's (usually) knoblike process (Step 3). This binding can cause the receiving cell to generate an action potential (Step 4). If the receiving cell is a muscle cell, it will contract; if it is another neuron, an action potential might be generated in it.

As Sherry Greer forms an ASL phrase like "I'm going to class now," millions of neurons in the brain and spinal cord take part. The idea to form the phrase leads to signals running down axons in the spinal cord and crossing synapses to the dendrites of other neurons. Those neurons send action potentials to their axon terminals, where neurotransmitter is released, moves across synapses to muscle cells in the arms, hands, and fingers, and causes them to contract and form appropriate shapes.

A medical condition called **myasthenia gravis** demonstrates how important it is for receiving cells to have proper receptor proteins. In people with this disease, muscle cells have few working receptors on their cell surfaces for one particular neurotransmitter; as a result, the muscle cells cannot respond normally to nerve signals, and affected people experience muscular weakness and fatigue.

Cocaine Addiction: Action at the Synapse

Just after a neurotransmitter enters the synaptic cleft and triggers an action potential in the receiving cell, enzymes break down the chemical signal or the sending cell reabsorbs it (Fig. 19.6a, Step 5). Without this rapid cleanup, the messenger molecules would remain in the cleft and would continuously stimulate the receiving cell.

A faulty cleanup operation is central to the high people feel when taking cocaine. Cocaine blocks the cleanup in brain cells of a specific neurotransmitter called **dopamine.** The blocked cleanup occurs in the cells of the so-called central reward system (Fig. 19.7). Because the neurotransmitter remains for a longer time at these pleasure synapses, the person feels euphoria and excitement. The body begins to compensate by stepping

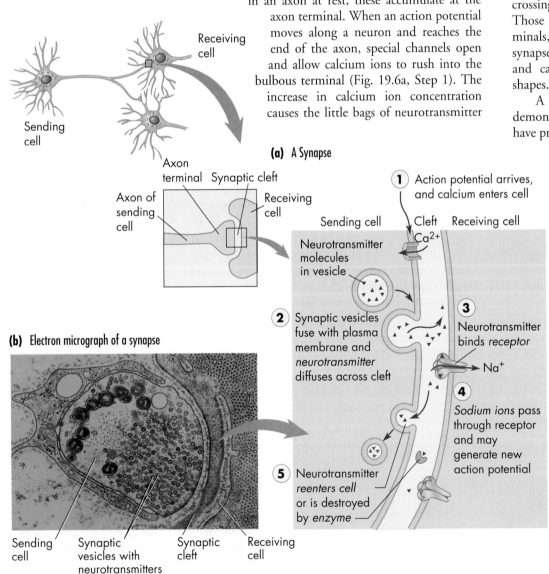

(a) A Synapse

Receiving cell

Sending cell

Axon terminal   Synaptic cleft

Axon of sending cell

Receiving cell

① Action potential arrives, and calcium enters cell

Sending cell   Cleft Ca^{2+} Receiving cell

Neurotransmitter molecules in vesicle

② Synaptic vesicles fuse with plasma membrane and *neurotransmitter* diffuses across cleft

③ Neurotransmitter binds *receptor*

Na^+

④ *Sodium ions* pass through receptor and may generate new action potential

⑤ Neurotransmitter *reenters cell* or is destroyed by *enzyme*

(b) Electron micrograph of a synapse

Sending cell

Synaptic vesicles with neurotransmitters

Synaptic cleft

Receiving cell

up its ability to destroy the neurotransmitter dopamine. This jazzed-up destruction mechanism, however, works faster and more efficiently whether cocaine is present or not and brings about drug tolerance: When the narcotic is gone, the reward center is starved of stimulation, and the user can no longer enjoy normally pleasurable activities such as good food and good sex. Instead, the addict's body craves more cocaine, and this causes the destructive cycle of drug-seeking behavior called **addiction.** Dopamine is so powerful in the brain's pleasure center that addiction to cocaine (especially the form called "crack") takes hold faster than addiction to other drugs, including heroin and tobacco.

Numerous animal species, from mice to monkeys, quickly learn to push a lever to obtain cocaine. Laboratory mice will experience their first "high," crave the drug, become addicted, and die from a self-delivered overdose, all within one week's time. The drug can also cause brain lesions and damage. Len Bias, a promising young basketball star drafted by the Boston Celtics in 1986, decided to celebrate his good fortune by snorting cocaine for the first time. Within a few hours, he was dead from brain seizures and cardiac arrest. *E*xplorer 19.3 summarizes the known effects on the human body of many illicit, addictive drugs, as well as numerous legal prescription drugs.

Types of Neurotransmitters

The dopamine involved in cocaine addiction is just one of more than 50 different known chemicals that can serve as neurotransmitters in the nervous systems of various animals. In fact, dopamine itself can have other dramatic effects. While increased dopamine can bring on addiction, decreased quantities are associated with **Parkinson's disease** and contribute to its symptoms of rigidity, weakness, and shaking. Some patients are helped by the drug levodopa (L-dopa), which brain cells can convert into dopamine. Medical researchers have also gotten promising results by transplanting into Parkinson's patients fetal brain cells capable of generating dopamine.

Of the many other neurotransmitters biologists have studied, the two they understand best are **acetylcholine** and **norepinephrine.** Like most neurotransmitters, acetylcholine transmits nerve impulses within the brain. But it also relays nerve signals to the skeletal muscles that help maintain posture, breathing, and limb movements, including the gestures of ASL, and in addition, it is released by the vagal nerve, slowing the heart rate. Norepinephrine is found in synapses throughout the brain (and elsewhere in the nervous system), and biologists agree that it helps keep our moods and behavior

Figure 19.7   Cocaine Acts at Synapses in the Pleasure Center of the Brain. Just 15 minutes after taking cocaine, a PET scan of a living brain reveals the accumulation of the drug in the reward center of the brain. There it blocks the destruction of dopamine. The persistence of dopamine at the synapse stimulates "pleasure" neurons more vigorously than normal. Unfortunately, the body compensates by improving its neurotransmitter destruction mechanisms, and so without the drug, the person feels no pleasure. This leads to addiction. This photo shows that glucose metabolism, a measure of brain activity, has dropped by half in the brain on cocaine.

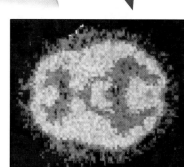

on an even keel. People suffering clinical depression, with its pessimistic moods, disturbed sleep, altered appetite, and diminished energy levels may actually have norepinephrine deficiencies within the brain. Conversely, the condition called mania, characterized by overactivity, irritable moods, and recklessness, may stem from too much norepinephrine in the brain.

In the early 1990s, researchers discovered that the gas nitric oxide (NO) is secreted by some neurons and may dilate blood vessels, control blood pressure, and influence memory. More than 2 percent of neurons in the brain produce nitric oxide, and neurons that cause smooth muscle in the gut to relax also use this gas as a neurotransmitter. Recall from Chapter 8 that nitric oxide also mediates the erection of the penis.

Fine Control at the Synapse

The chemical communication between neurons can either encourage or discourage the receiving cell from firing, and this lends control to nervous system activity. Since most neurons form synapses with many other cells, the ultimate activity of any neuron depends on the cumulative impact of all the impulses it receives—some exciting neural firing, some inhibiting it. The receiving cell simply sums up all the information impinging on it from the sending neurons over a short period of time. If chemical encouragements outnumber discouragements, the threshold is reached and the neuron fires an impulse in an all-or-none fashion. This allows the precise control in animal nervous systems that leads to simple behaviors—blinking, scratching, or yawning—and complex ones—fleeing a predator, recognizing a face, or creating a series of ASL signs to introduce one friend to another.

READ ON

*E*xplorer 19.3

Drugs and the Nervous System

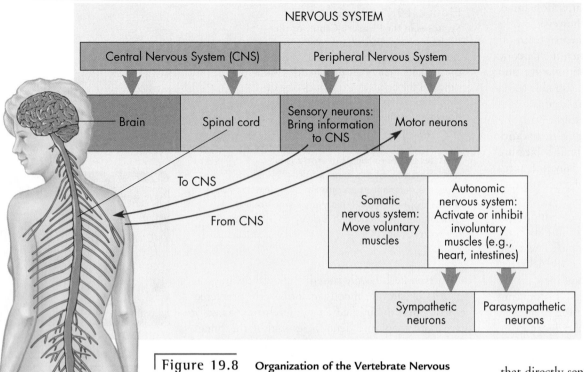

NERVOUS SYSTEM

Central Nervous System (CNS) | Peripheral Nervous System

Brain | Spinal cord | Sensory neurons: Bring information to CNS | Motor neurons

To CNS

From CNS

Somatic nervous system: Move voluntary muscles

Autonomic nervous system: Activate or inhibit involuntary muscles (e.g., heart, intestines)

Sympathetic neurons | Parasympathetic neurons

Figure 19.8 **Organization of the Vertebrate Nervous System.** Your nervous system has two main branches, the central nervous system (CNS), with brain and spinal cord that control the processing and integration of information; and the peripheral nervous system (PNS), whose sensory neurons carry information to the CNS and whose motor neurons carry information from the CNS. Thirty-one pairs of spinal nerves branch off the spinal cord with sensory and motor neurons. The somatic nervous system runs your voluntary muscles. The autonomic nervous system activates or inhibits involuntary muscles. The sympathetic and parasympathetic neurons of the autonomic nervous system run to many of the same organs but often act in opposition, turning the organ's activity on or off, up or down.

19.3 Animal Nervous Systems

The learning of human language depends on nervous system activity. It depends on *sensing* signals—sounds or sights—from other people, *integrating* the signals to understand them, and *reacting* with a coordinated response to the communicated information. This is true whether the communication is spoken, written, signed, postural, or chemical (as in many other animal species). Regardless, sensing, integrating, and reacting require highly organized networks of neurons. Let's look, now, at how nervous systems are constructed and function, focusing mainly on the nervous systems of vertebrates like ourselves.

The Vertebrate Nervous System

The grouse's elaborate mating dance, the cobra's hypnotic weaving, the vigorous kicks and turns of an aerobic dancer: All of these depend on a highly organized nervous system, and in vertebrates, that system is organized into two main units. The **central nervous system (CNS)** (Fig. 19.8) is the director, thinker, and information processor. The CNS consists of the **brain,** which performs complex neural integration, and the **spinal cord,** which carries nerve impulses to and from the brain. The CNS also allows **reflexes,** or involuntary but predictable responses to stimuli. The **peripheral nervous system (PNS)** is the "go-between" or "middleman"; it includes **sensory neurons** that directly sense the environment and the **motor neurons** that directly contact muscles and glands and make them work. The peripheral nervous system connects the central nervous system with the sense organs, muscles, and glands of the body. The interplay between the central and peripheral nervous systems allows an animal to sense environmental stimuli, integrate the information, and respond appropriately, providing all the fascinating behaviors we can see in ourselves and other animals.

A Reflex Arc

Have you ever stepped on a tack or sharp piece of gravel with your bare toe? If so, the muscles in your leg lifted your toe immediately, even before you realized what had happened (Fig. 19.9). That muscle response was driven by a simple nerve circuit called the **reflex arc,** an uncomplicated neural loop that links a stimulus to a response in a direct way. A reflex arc shows the clear relationship between the central and peripheral nervous systems. Reflex arcs require no conscious input from the brain and bring about behaviors that are generally rapid, involuntary, and nearly identical each time the stimulus is repeated—a knee jerk, for example, or the pullback of a toe from a tack or a finger from a hot stove.

Reflex arcs usually involve:

1. A sensory neuron in the peripheral nervous system that receives information from the external or inter-

nal environment—a tack stabbing a toe is this kind of information—and then transmits the message along a spinal nerve into the spinal cord (Fig. 19.9, Steps 1, 2).

2. **Interneurons** in the central nervous system that relay messages between nerve cells and integrate and coordinate incoming and outgoing messages (Step 3). Most of the brain is made up of interneurons.

3. Motor neurons, which relay messages from interneurons in the CNS out to muscles or secretory glands (Step 4). Motor neurons allow an organism to *act*—to respond to the information the sensory neuron brings in from the environment. In this case, the organism is you, and the action is pulling your toe back from the tack (Step 5).

The sensory and motor neurons enter or leave the spinal cord in pairs of nerves called **spinal nerves,** shown in Figures 19.8 and 19.9. Let's look more closely now at the motor neurons in these pairs of spinal nerves.

Motor Neurons

Some of your motor neurons are under voluntary control and activate muscles on command. Examples are Sherry Greer's motor neurons that allow her to make specific ASL signs during a conversation. These voluntarily controlled motor neurons form the **somatic nervous system.** A different set of motor neurons in the **autonomic nervous system** acts subconsciously and automatically to regulate the body's internal environment, controlling glands, the heart muscle, and smooth muscles in the digestive and circulatory systems.

The Autonomic System of Motor Neurons

A person's autonomic nervous system performs many of its duties through the spinal cord and lower centers of the brain (in or near the neck). There are two sets of autonomic neurons, called parasympathetic and sympathetic neurons, that function in opposition, a bit like the accelerator and brakes of a car.

In general, the **parasympathetic** nerves function as a housekeeping system for the body, stimulating the stomach to churn, the bladder to empty, and the heart to beat at a slow and even pace during most daily and nightly activities. During exercise or embarrassment, or when an emergency arises and generates intense anger, fear, or excitement, the **sympathetic** nerves dominate, increasing heart rate, dilating the pupils of the eyes (which lets in more light), expanding the bronchioles of

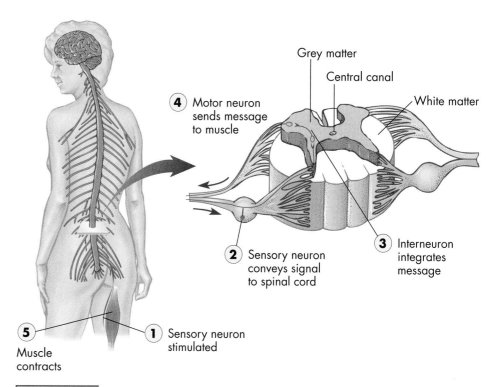

Figure 19.9 **Reflex Arcs and Spinal Nerves.** Follow the five steps to see how a reflex arc connects the peripheral and central nervous systems.

the lungs (which improves O_2 and CO_2 exchange), and slowing down nonessential digestive activities until the emergency is past. Sympathetic and parasympathetic neurons often control the same organ and help to speed up or slow down its activity as needed. For example, most sympathetic neurons can speed and strengthen the heartbeat by secreting the neurotransmitter norepinephrine, while parasympathetic neurons can slow it by secreting acetylcholine.

In Chapter 18, we saw how common it is to experience stressful changes that make us feel trapped and yet be unable to respond with satisfying physical activity (like fleeing a traffic jam or putting a pie in your boss's face). The result can be a flood of stress hormones as well as higher concentrations of epinephrine and norepinephrine and with them, stress-related health problems such as high blood pressure. Some blood pressure medications (so-called beta-blockers) specifically inhibit the sympathetic synapses that stimulate the heart and blood vessels, thereby keeping the heart rate slower and steadier.

Both branches of the autonomic system—the sympathetic and parasympathetic neurons—act primarily through reflex loops. Consider the loop that controls the emptying of the urinary bladder. Sympathetic motor neurons cause the bladder muscle to relax and the muscles that close the sphincter (through which the bladder empties) to shut tight. When the bladder fills, the bladder wall stretches. This activates stretch receptors, which signal the spinal cord via sensory neurons. These stimulate parasympathetic motor neurons that feed back to the bladder muscle and cause the urine-storing organ to contract and the sphincter to relax, and thus urine to be expelled. In infants, this reflex loop is the only control over bladder function, so the organ empties whenever it fills. With toilet training, the child's brain gradually learns to exert conscious control over the pelvic muscles that regulate urination.

Some cultures have explored conscious control over the autonomic nervous system—with intriguing results. Yogis in India, for example, sometimes claim that they can survive being sealed in an underground chamber for many days by consciously lowering their metabolic rate and using far less oxygen than is normally required. Some of these claims are surely exaggerations. But by measuring the yogis' breathing, pulse, and oxygen utilization with reliable instruments, scientists have concluded that some yogis actually can consciously lower their metabolic rate below normal resting levels. **Biofeedback** training, more common in this hemisphere, uses electronic devices to indicate changes in blood pressure, blood flow to extremities, or other external states. By making the trainee aware when these changes start to occur, the device often reinforces the internal states that lead to lowered blood pressure, warmed hands and feet, reduced pain, deep relaxation, and so on. Eventually biofeedback training can allow conscious control over parts of the autonomic nervous system.

The Central Nervous System

The sense receptors of the peripheral nervous system serve as windows on the environment. Without the central nervous system to interpret the data and coordinate the responses, however, an animal couldn't act appropriately. Important to this coordinating and responding is the neural thoroughfare that runs through the spinal cord.

The Spinal Cord: A Neural Highway

You've probably heard of "gray matter" and "white matter." But what are they? If you took a cross section from a human spinal cord, you'd see a butterfly-shaped core of grey material surrounded by an oval field of white (see Fig. 19.9). The gray matter contains nerve cell bodies and is a zone of many synapses, where local neural traffic occurs (such as the reflex arcs we discussed earlier). The white matter, on the other hand, is like an interstate freeway; it consists mainly of thin, mostly myelin-covered axons that transport information long distances up and down the spinal cord to and from the brain. Because many of these long axons are insulated with white, fatty myelin sheaths, (review Fig. 19.5), the entire white matter has its whitish color.

The integration of signals from sensory neurons and interneurons begins in the spinal cord. Most major data processing, however, occurs in the brain.

The Brain: The Ultimate Processor

We all know that the brain embodies our feelings and strivings, our knowledge and memories, our musical and verbal abilities, and our sense of the future. Biologists still do not fully understand, however, how it performs these tasks. Over many decades, studies of patients with strokes or injuries to particular, often small parts of the brain have revealed the seats of specific complex behaviors and specific emotions in individual brain regions. Recent tools dramatically reinforce this concept, including the fMRI that Dr. Helen Neville uses to locate the brain regions involved in language. One of the big themes of modern neuroscience has been that brain anatomy can open a map to behavior.

The human brain consists of four interconnected parts (Fig. 19.10): (1) the brainstem, the "stalk" of the brain, which relays messages between the brain and spinal cord; (2) the highly rippled cerebellum attached to the brainstem and posterior to it; (3) the diencephalon, which sits just above the brainstem; and (4) the large, folded cerebrum, whose right and left halves sit atop the brainstem and span the inside of the head from just behind the eyes to the bony bump at the back of the skull.

The Brainstem: Fundamental Regulation

The **brainstem** plays three crucial roles: It helps integrate sensory and motor systems, it regulates body homeostasis, and it controls arousal. The lowest part of the brainstem, the **medulla oblongata,** lies inside the skull above the level of the mouth. The medulla oblongata helps keep body conditions constant (see Chapter 14) by receiving data about activities in various physiological systems and by regulating respiratory rate, heart rate, blood pressure, and many other subconscious body activities. Above the medulla lie the **pons,** a relay center that helps control breathing, and the **midbrain,** which contains fiber tracts running between the anterior and posterior brain, as well as structures involved in sight and hearing.

The Cerebellum: Muscle Coordinator

Attached to the pons, at the middle of the brainstem, is the **cerebellum** (Latin, *little brain*), a convoluted bulb that serves as a complex computer, comparing outgoing commands with incoming information about the status of muscles, tendons, joints, and the position of the body in space. The result is a fine-tuning of motor commands. Learning how to make the specific, rapid movements of sign language or how to run an offensive pattern in basketball involves the cerebellum.

The Diencephalon

The **diencephalon** (*di* = second + *cephalon* = brain) (Fig. 19.10a) is an extension of the brainstem above the cerebellum, and it includes the thalamus, hypothalamus, and pineal gland (review Fig. 18.4). The **thalamus** makes up most of the diencephalon, and it is the main relay station for sensory signals moving between the cerebellum and other parts of the brain to another. It also plays a role in awareness and in the acquisition of knowledge.

Lying below the thalamus is the **hypothalamus,** which regulates the pituitary gland and provides the link we saw in Chapter 18 between the nervous and endocrine systems. The hypothalamus also has important roles in maintaining homeostasis: It helps to regulate body temperature, water balance, hunger, and the digestive system. By electrically stimulating different regions of the hypothalamus, researchers can make an animal behave as if it feels alternately hot and cold, hungry and satisfied, angry and content. Some regions of the hypothalamus seem to be pleasure centers: A rat with electrodes permanently implanted in those areas will continue pressing a foot pedal to stimulate its pleasure centers until the animal drops from exhaustion. The third part of the diencephalon is the **pineal gland,** shaped like a pea-sized pine cone. The pineal secretes

(a) Major regions of the brain

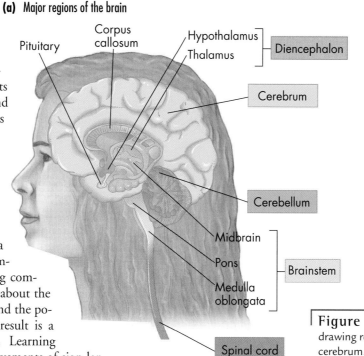

(b) MRI of the brain

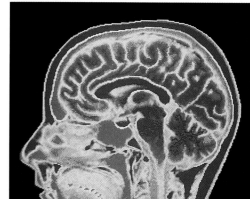

Figure 19.10 **Major Regions of the Brain.** (a) The drawing reveals the brainstem, diencephalon, cerebellum, and cerebrum, as well as major parts of each section. (b) A slice through the brain as viewed in magnetic resonance imaging (MRI) reveals the living tissues in remarkable detail. These images are from a normal 42-year-old woman and clearly show the cerebral cortex, corpus callosum, pons, medulla, spinal cord, and other structures of the brain and head.

melatonin and may be involved in the operation of the body's biological clock (see Chapter 17).

Extending through the brainstem from the diencephalon down to the spinal cord is the **reticular formation,** a network of tracts that reaches into the cerebellum and cerebrum. When you're awake, specific neurons of the reticular formation are actively firing. When you're asleep or unconscious, those neurons fire more slowly. This selective shaping of neural input by the reticular formation and the diencephalon helps a person to concentrate on a dinner partner's conversation in a noisy restaurant, and plays a part in the groggy incoherence of an early-morning phone conversation.

The Cerebrum: Perception, Thought, Humanness

The two side-by-side hemispheres of the brain's **cerebrum** (Latin, *brain*) fit like a cap over most of the other brain regions. This mass of tissue embodies not only the attributes we consider human, such as self-awareness, speech, and artistic ability, but also traits that we share with other vertebrates, such as sensory perception, motor output, and memory. In humans, the **cerebral**

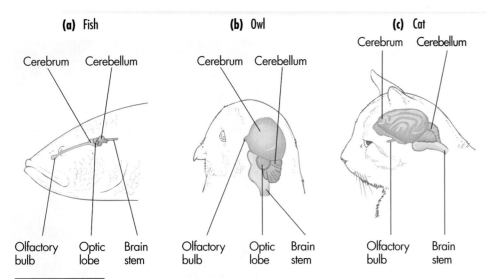

(a) Fish   **(b)** Owl   **(c)** Cat

Cerebrum Cerebellum   Cerebrum Cerebellum   Cerebrum Cerebellum

Olfactory Optic Brain   Olfactory Optic Brain   Olfactory Brain
bulb   lobe   stem   bulb   lobe   stem   bulb   stem

Figure 19.11 **Brain Differences in Various Vertebrates.** Compare the cerebrum of a fish, owl, and cat to that of the human in Figure 19.10a. The increasing size and surface convolutions correspond to increased intelligence and an enlarged behavioral repertoire.

cortex, the cerebrum's highly creased and infolded surface layers, contains about 90 percent of the brain's neuron cell bodies. By looking at the brains of fish, birds, and cats (Fig. 19.11), you can see how this cap ballooned in size relative to other brain regions during vertebrate evolution and became increasingly convoluted, with a massively enlarged surface area folded and fissured so it fits into the skull.

One of the great unsolved questions of modern biology is precisely *how* the cerebrum can control complicated traits. Biologists have many ideas but no single answer. One principle has been well established, however: Basic cerebral functions, including sensations (like

seeing or hearing), motor ability (movement), cognitive functions (like language and perception), affective traits (emotions), and even some character traits (like friendliness or shyness), can be mapped to specific regions of the cerebrum (Fig. 19.12).

Motor and Sensory Centers

Before the advent of modern brain imaging methods like fMRI, researchers were limited to mapping the cerebral cortex primarily during surgery. Despite its billions of neurons, brain tissue has no pain receptors, so a surgeon can operate on this highly complex organ with only a local anesthetic for the scalp and skull incision. By inserting extremely fine, low-current electric probes into the exposed brain, neurosurgeons can stimulate specific regions and watch to see which muscles the patient moves or ask the alert patient to report the sensations he or she feels. Studies like these allowed neuroscientists to build maps of the brain regions that control feeling and movement in various parts of the body. If you run your right index finger from your left ear straight up to the top of your head, the arc traces the surface area devoted to the left hemisphere's **motor cortex,** the part of your brain that controls muscles on the right side of your body, including the ones that move your right hand (Fig. 19.12b). Likewise, the corresponding arc on the surface of the brain's right hemisphere moves the left side of your body.

Just behind the motor cortex lies the **sensory cortex,** which registers and integrates sensations from body

Figure 19.12 **Mapping the Human Brain: Specific Functions Reside in Specific Regions.** (a) A map of the cerebral cortex reveals the mental functions associated with various regions of the brain's convoluted surface. Association cortex is shown in tan. (b) If a person's body parts grew in proportion to the size of the corresponding motor cotex region for each part, he or she would resemble the strangely distorted human shown stretched across the brain's surface and in three dimensions on the next page. (c) A similar view of the sensory cortex.

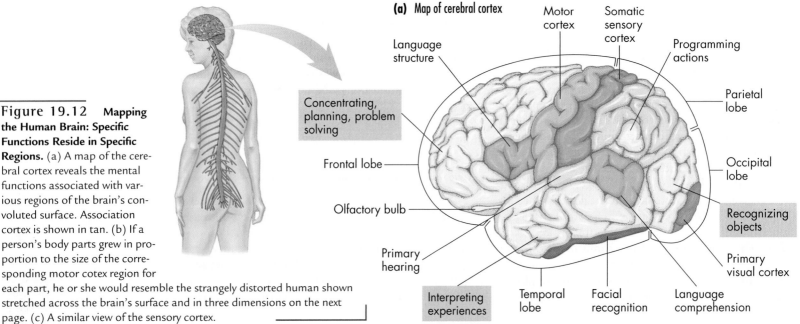

(a) Map of cerebral cortex

Language structure

Concentrating, planning, problem solving

Frontal lobe

Olfactory bulb

Primary hearing

Interpreting experiences

Motor cortex

Somatic sensory cortex

Programming actions

Parietal lobe

Occipital lobe

Recognizing objects

Primary visual cortex

Temporal lobe

Facial recognition

Language comprehension

parts (Fig. 19.12c). Again, the brain's left side receives sensations primarily from the body's right side, and vice versa.

On the maps of both the motor cortex and the sensory cortex, the body parts seem to have exaggerated proportions: Lips, face, and fingers appear extremely large, and the trunk too small. This distortion reflects the large number of delicate touch receptors in face and fingers and the large number of muscles we need for speech and manual dexterity compared to the small numbers of neurons we need for sensation and movement in the trunk. Other specific brain regions are dedicated to vision, hearing, and smell. Recent studies employing intense stimulation of a monkey's fingertips show that brain maps can change with use. People who lose their vision, for example, and then learn Braille, the raised-dot writing system for the blind, would have more cortical area devoted to fingertip sensation after Braille training than before.

Many higher cerebral functions map to specific regions, too. One area of the cerebral cortex, for example,

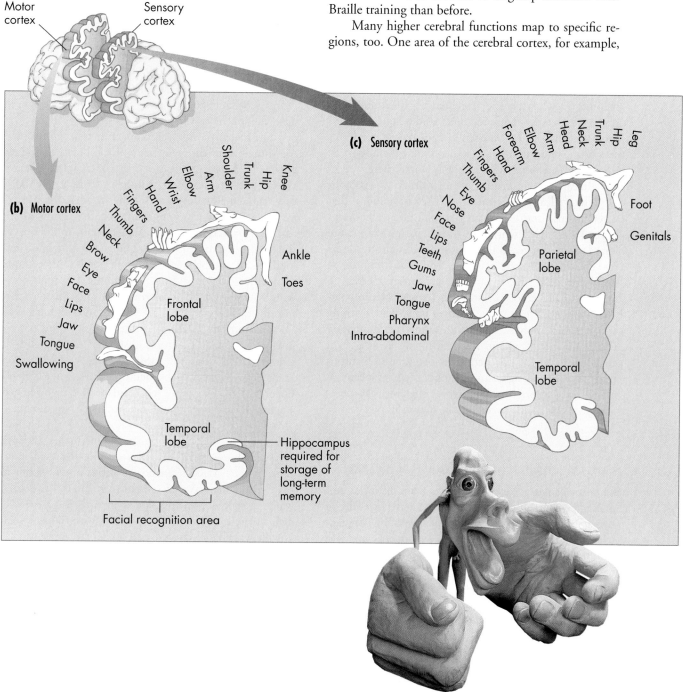

Motor cortex

Sensory cortex

(b) Motor cortex

Fingers
Hand
Wrist
Elbow
Arm
Shoulder
Trunk
Hip
Knee
Thumb
Neck
Brow
Eye
Face
Lips
Jaw
Tongue
Swallowing

Frontal lobe

Ankle

Toes

Temporal lobe

Hippocampus required for storage of long-term memory

Facial recognition area

(c) Sensory cortex

Hand
Fingers
Thumb
Forearm
Elbow
Arm
Head
Neck
Trunk
Hip
Leg
Eye
Nose
Face
Lips
Teeth
Gums
Jaw
Tongue
Pharynx
Intra-abdominal

Parietal lobe

Foot

Genitals

Temporal lobe

Which side of the brain processes American Sign Language?

is dedicated to recognizing faces (Fig. 19.12a). A person who suffers an injury to this part of the brain can no longer recognize loved ones' faces, even though he or she can still recognize their voices!

Memory: Storage of Past and Present

Deep in each cerebral hemisphere lies a small bilateral structure called the **hippocampus,** which plays a crucial role in the formation of long-term memory. Have you ever been interrupted while trying to memorize a new phone number? You probably forgot the number because your hippocampus failed to fix it in long-term memory before the interruption occurred. A patient with a damaged hippocampus can recall events that happened even decades before the injury, but fails to remember new facts and experiences (including the shapes of objects, phone numbers, and people's names) for more than a few moments. Clearly, we need this area of the brain to help us lay down new memories, although the memories aren't actually stored there.

Researchers are making great strides in understanding how we learn and remember. They have found that when intense and/or repeated stimuli alter a neuron's normal impulse activity, the altered pattern can cause long-lasting changes in the expression of neurotransmitter genes. These changes, in turn, can alter communication between cells by revamping the activity needed to trigger impulses across specific synapses. Neurobiologists sometimes refer to the changes underlying learning and memory as "hard wiring," meaning that a memory is laid down or an item is learned only as long-term changes are physically inscribed in the way the neurons interact. These long-lasting changes seem to occur in many different brain regions, which suggests that memory involves far more than just the hippocampus.

Language: The Human Trait

Complex language distinguishes us from all other animals. The control of this most human activity resides in two areas of the cerebral cortex, one governing language structure (or generation) and the other language comprehension (see Fig. 19.12a). If a **stroke** (paralysis or numbness caused by destruction of brain tissue due to a blood clot or blood vessel break) disrupts the area of language generation, the patient might understand spoken and written language but be unable to speak well. One such patient, talking about a dental appointment, used halting phrases but could still convey meaning to listeners: "Monday-Dad and Dick-Wednesday nine o'clock-ten o'clock-doctors-and-teeth." Conversely,

damage to the site of language comprehension allows the person to speak fluently in sentences, but often to convey no meaning to others. When describing a picture of two boys stealing cookies behind a woman's back, one such patient stated, "Mother is away here working her work to get her better, but when she's looking the two boys looking in the other part."

Mapping shows that the brain is asymmetrical. In most people, language abilities reside mainly on the brain's left side (remember, *l*anguage, *left),* as do the underpinnings of analytical thought and fine motor control. The right hemisphere is mainly responsible for intuitive thought, musical aptitude, the recognition of complex visual patterns, and the expression and recognition of emotion. Almost all right-handers and 75 percent of left-handers have left-brain language areas. Many left-handers, however, have language areas on *both* sides of the cerebral cortex.

Dr. Helen Neville is a pioneer in trying to unravel the functions of different parts of the brain. She knew about the language areas on the brain's left side, of course, and that early exposure is necessary if a child is to learn language fluently. But she wondered if and exactly how the two were related: Does something about early language training lead the left brain to develop as the language center? Is it the rapidly changing pitch and loudness of audible speech sounds? Or is it the unique stimulation provided by the grammar of language—word order, verb tenses, pronouns, and so on—that imprint the left brain, regardless of sound or silence?

To find out, Neville recruited three groups of volunteers. One included hearing subjects with English as a native language and no ASL training. A second group included people with genetic deafness, like Sherry Greer, who learned ASL as an infant and then learned English later (usually in grade school). In the third group were people like Sherry's translator Jo Larson-Muhr—born to deaf parents and instructed in both ASL and English from infancy on. Neville used functional MRI (fMRI) to map their active brain regions as each group silently read sentences in English and then watched sentences being signed in ASL. Figure 19.13 depicts the brain maps for the first two groups, which Neville was able to create based on her research.

In brief, here's what she found:

- Native speakers of English only showed strong activation in the classic left-brain regions when reading English, but showed weak activation or none at all when viewing ASL (which they couldn't understand).
- Deaf subjects showed strong activation in both left- and right-brain regions when viewing sentences being signed in ASL, but showed only weak activity in the

right brain (and none in the left) when reading English (which they tended to learn late in childhood).

- Hearing subjects fluent in both ASL and English showed strong activity in both left- and right-brain areas when viewing ASL signing, but when reading English showed strong activity only in the classical left-brain areas.

A more important question is: What do Dr. Neville's findings mean? She explains that when a person is using the language they learned from birth, whether English, ASL, or both, the standard language regions of the left hemisphere "light up" (our term) on the fMRI. This, she says, implies that experience with any grammatic language—and not simply one with spoken sounds received by the ears—causes the brain to develop its specialized language areas in the left. Native ASL speakers have a unique situation in that they're exposed as infants to a grammatic language that also has strong visual/spatial aspects. That's why both left and right sides of the brain learn to respond. Finally, deaf subjects who learned English in grade school—sometimes years after a critical period for language acquisition in early childhood—experience weak right brain activity but no left when reading English as adults. This, says Neville, suggests that delayed or imperfect learning of a language may lead to altered brain organization for that language. This finding is important for parents to know so that they are careful to expose their children—especially if the youngsters are hearing impaired—to languages early in childhood. Furthermore, the study focuses attention on the brain's amazing adaptability and plasticity, as well as on very precise brain regions and their central role in specific behaviors. Neville's work is helping neuroscience answer perhaps its biggest underlying mystery: How do these highly individualized regions arise within the brain, and once dedicated to specific behaviors, how do they work smoothly and seamlessly together in one conscious, coordinated mind?

(a) Viewing written English

Left hemisphere · Hearing subjects · Right hemisphere

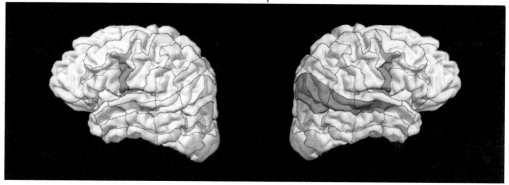

Deaf subjects

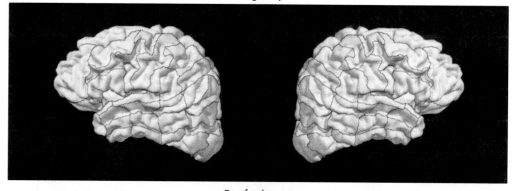

(b) Viewing American Sign Language

Hearing subjects

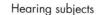

Deaf subjects

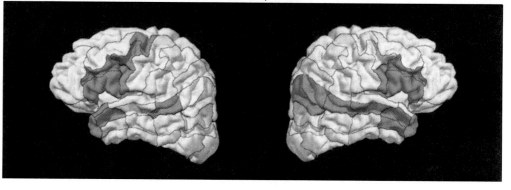

Figure 19.13 **Do Language Areas Require Sound or Will Any Structured Language Activate Them?** Dr. Neville mapped the brains of hearing and deaf subjects by fMRI as they viewed sentences in English (a) or in American Sign Language (b). (The figure shows left and right hemispheres.) Compare the brain areas used by hearing and deaf subjects when reading English. What do you conclude? Make the same comparison when the subjects are viewing sign language. What conclusions can you draw?

Complete answers lie in the future. But one part of the story involves communication between the brain's hemispheres.

Right Brain Versus Left Brain

If separate functions are encoded in each side of the brain, how do the two sides "talk"? A likely answer involves the **corpus callosum,** a tissue bridge between the left and right cerebral hemispheres. Neurosurgeons discovered some answers while treating patients with **epilepsy,** or recurring nervous system attacks that often result in seizures and/or unconsciousness and can affect body movements and sensations. In epileptics, nerves fire back and forth across the brain in a crescendo of positive electrical feedback. This firing can continue to escalate, equivalent to the screeching feedback from a microphone, until a seizure results. Surgeons thought that they could interrupt the loop by "splitting the brain," or severing the corpus callosum. These split-brain patients no longer had seizures (or had dramatically fewer) and seemed completely normal to family and friends. But when neurobiologists tested them in controlled experiments, they found strange anomalies (Fig. 19.14). For example, consider a split-brain subject who was shown a key so that only the visual centers on the right side of the brain could perceive it. When asked what he saw, the subject could point to a key with his left hand, which, recall, is controlled by the right brain. The patient couldn't *name* the object, however, because only the right brain saw the key, not the left brain. Since the patient knew what he saw but couldn't say it, surgeons concluded that they had cut off the avenue of communication between the right visual cortex and the left brain's region of language ability. Neurosurgeons use the split-brain procedure much less frequently today to control epilepsy, but its prior use dramatically underscored the highly localized nature of specific brain regions and the specific abilities tied to them and, at the same time, the need for integration and communication between them.

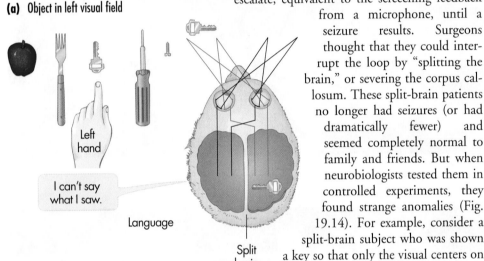

(a) Object in left visual field

Left hand

I can't say what I saw.

Language

Split brain

(b) Object in right visual field

Right hand

I saw a key!

Language

Split brain

Association Cortex

Despite mapping many functions to different parts of the brain, neurobiologists know little about their coordinated action. Each of the cerebral hemispheres, for example, contains regions they vaguely label **association cortex.** These areas appear to integrate information from other areas and reconstruct it at different levels of consciousness. To a large extent, association cortex contains the physiological underpinnings of emotions, memory, personality, reasoning, judgment, and the conglomeration of traits we call intelligence, but exactly *how* they reassemble information and coordinate it is basically a mystery.

The Diseased Brain

The human brain is so remarkable and so central to all of our activity that malfunctions can lead to devastating diseases like Alzheimer's disease and schizophrenia.

Alzheimer's Disease

Consider an engineer who forgets that he has already determined the costs of a project and needlessly recalculates all the estimates. Or a homemaker who finds it hard to remember the names of familiar objects around the house and how to balance the checkbook. These people may be showing signs of a relentlessly advancing condition called **Alzheimer's disease.** Within three to ten years of the first symptoms, Alzheimer's sufferers are often unable to dress themselves, to care for themselves, and eventually even to walk, sit up, or smile.

Currently, Alzheimer's affects as many as four million Americans and accounts for over half of the

Figure 19.14 **Two Minds in One Brain.** (a) A patient in which the corpus callosum has been surgically severed can perceive a key with the right visual cortex and point to it with the left hand. But since the patient's right brain cannot communicate with the language region in the left brain, the person knows what he or she saw but can't say it. (b) When the patient perceives the key with the left visual cortex, there is no communication block with the language region of the same hemisphere, and thus the person knows and can say what he or she saw.

recorded cases of dementia (mental deterioration). It attacks otherwise healthy middle-aged and older people and causes a progressive loss of mental function. While forgetfulness, confusion, and neglect of daily tasks are characteristic, only the examination of brain tissue after a patient's death can confirm Alzheimer's. The brains of Alzheimer's victims have usually lost neurons in areas necessary for thinking and remembering, and have accumulated twisted filaments within the brain cells and aggregations of protein around blood vessels that cut blood flow to specific regions of the cerebrum (Fig. 19.15a,b). Neurobiologists are beginning to understand the mechanisms that underlie Alzheimer's disease, but are still far from finding a cure for this devastating condition.

Schizophrenia

Consider a young man who watches strangers talking and assumes they are gossiping about him. The same man sees people walking behind him and concludes that the FBI is after him, or hears a radio commercial and believes it's directed specifically at him. Or consider a young woman who begins to experience disordered thoughts and can't keep irrelevant ideas from creeping in when answering a question. She withdraws from contact with other people, and sits motionless, crouched on the floor for hours at a time. These people have **schizophrenia** ("split mind"), a condition characterized by psychosis, or false beliefs that can't be changed by evidence, and hallucinations, or phantom experiences perceived to have happened.

Schizophrenia tends to run in families: Only about 1 out of every 100 people worldwide has schizophrenia, but among the relatives of schizophrenics, the figure is 15 out of 100. Biologists think that genetic factors, environmental factors, or both predispose the members of certain families to schizophrenia. Stress can also trigger the disease in genetically susceptible people.

Physicians have found that many patients with schizophrenia have enlarged **ventricles,** the fluid-filled spaces within the brain, and a corresponding loss of brain tissue (Fig. 19.15c,d). It's not yet clear whether these changes are causes or effects, and why some sufferers of schizophrenia lack the brain changes altogether.

Certain drugs alleviate the delusions, hallucinations, and disordered thinking of schizophrenia, and these drugs work by blocking one type of receptor for the neurotransmitter dopamine. This has led some neuroscientists to speculate that too much dopamine in certain parts of the brain may bring on schizophrenia's symptoms and thus to look for other treatments related to this possible underlying cause.

19.4 How Do Sense Organs Perceive the World?

Sherry Greer, the young woman in our case study, moves through a silent world, devoid of sounds, yet she has extraordinarily acute

Figure 19.15 **Diseased Minds and Brains.** PET scans comparing a normal brain (a) and the brain of an Alzheimer's patient (b). The dark areas in (b) reveal diminished blood flow to critical brain regions and degradation of tissue. Additional PET scans compare a normal brain (c) to that of a person with schizophrenia (d). The extensive red zones highlight areas of unusual brain activity as well as enlarged ventricles with consequent loss of brain tissue.

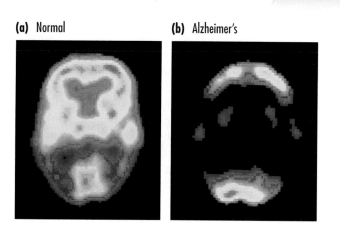

(a) Normal   **(b)** Alzheimer's

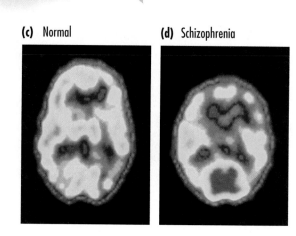

(c) Normal   **(d)** Schizophrenia

vision and touch. For most of us, sense organs are keys to knowing and interacting with the surrounding environment. But what *are* senses and how do they work? The sense organs are groups of specialized cells that receive stimulus energy (such as light, sound, odors or flavors, pressure, or heat) and convert it into another kind of energy—the kind that can trigger a neural impulse. Sensory neurons then relay the impulses to the brain, which, in turn, interprets (1) *what kind* of stimulus the sense organ received (what is the pitch of the sound, for example, the color of the light, or the nature of the chemical) and (2) *how strong* is the stimulus (for example, is it loud or soft, bright or dim, concentrated or dilute)? Our day-to-day survival depends on accurate answers to these questions.

Your eyes are generally tuned to visual stimuli, your ears to aural input, and so on. A strong enough mechanical, chemical, or electromagnetic stimulus, however, can often cause any sense organ to generate an action potential. Perhaps you've been bumped on the head or eye and "saw stars." You saw these flashes because cells in the visual pathway are sensitive to pressure as well as to light.

Some animals are masters at detecting sensations we can only pick up with highly sophisticated instruments. Rattlesnakes, for example, have a pair of **pit organs** beneath the eyes that enable the reptiles to "visualize" objects in the dark by detecting the heat patterns they give off (Fig. 19.16). Certain fishes have sense organs that act like volt meters, detecting even the tiniest motion as distortions in the electric field.

The two types of language Dr. Neville studies— spoken and sign language—depend on our most complicated and significant senses, hearing and vision. Let's focus on them in particular, and see how each works.

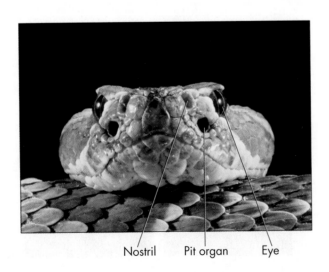

Nostril   Pit organ   Eye

The Ear: Collector of Vibrations

When you clap your hands together or say a word out loud, you generate an aural message that, like all sounds, is really a wave of compressed air. How does the normal ear detect such a wave, analogous to the oscillations we create when we plunge a stone through the still surface of a pool of water? Sound waves first strike the flexible, sculptured outer ear (Fig. 19.17). This, in turn, funnels sound waves to the **eardrum,** a taut membrane that stretches like a drumskin across each ear canal (Fig. 19.17). The eardrum vibrates in time with the sound waves, causing a chain of three tiny bones in each middle ear to vibrate. The last bone in the series causes a fluid to vibrate in the coiled **cochlea** (Greek, *kokhlos* = snail), the most complex mechanical structure in the human body, with over a million moving parts!

How Does the Ear Hear?

The cochlea is a transducer: it changes one type of energy (sound vibrations) into another (nerve impulses). Here's how. The cochlea is shaped like three parallel tubes coiled like a snail shell (Fig. 19.17). Attached to a partition between two of the tubes are rows of box-shaped neurons called **hair cells,** each crowned by an elegant bundle of tiny, brushy extensions called **stereocilia** that stand erect like a Mohawk haircut (Fig. 19.17). In most of the 18 million deaf Americans, the hair cells don't function properly. Researchers have recently discovered that under certain conditions, hair cells can regenerate, and this is giving new hope for deafness due to defective hair cells. When Sherry Greer was born, tests confirmed that she could not perceive any sounds because her cochlea had developed abnormally. Instead of hearing her mother's voice, she saw her facial expressions, watched her hands moving in the intricate patterns of ASL, and sometimes felt the signs against her skin.

Vibrations in the fluid inside the cochlea cause the hair cells to move relative to a membrane that lies on top of them (Fig. 19.17, Step 4). The movement bends the brushy stereocilia, which are attached to their neighbors by thin, springlike links. The latest theory for how we hear says that sound vibrations affect these links and cause ion channels to open in the hair cell membrane (which, recall, is a specialized type of neuron). Ions enter the hair cell, trigger an impulse, and thereby convert the mechanical stimulation of the sound wave into an electrochemical signal (Fig. 19.17, Step 6). Finally, the hair cell transmits the signal across one or more

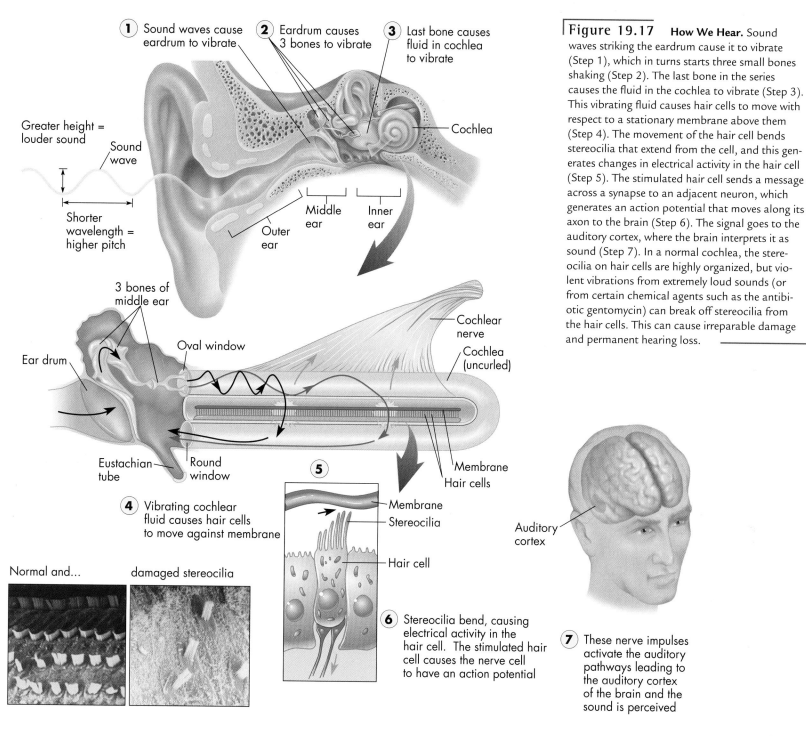

1 Sound waves cause eardrum to vibrate

2 Eardrum causes 3 bones to vibrate

3 Last bone causes fluid in cochlea to vibrate

Cochlea

Greater height = louder sound

Sound wave

Shorter wavelength = higher pitch

Outer ear

Middle ear

Inner ear

3 bones of middle ear

Oval window

Ear drum

Eustachian tube

Round window

Cochlear nerve

Cochlea (uncurled)

Membrane

Hair cells

4 Vibrating cochlear fluid causes hair cells to move against membrane

5

Membrane

Stereocilia

Hair cell

Normal and...

damaged stereocilia

Auditory cortex

6 Stereocilia bend, causing electrical activity in the hair cell. The stimulated hair cell causes the nerve cell to have an action potential

7 These nerve impulses activate the auditory pathways leading to the auditory cortex of the brain and the sound is perceived

Figure 19.17 **How We Hear.** Sound waves striking the eardrum cause it to vibrate (Step 1), which in turns starts three small bones shaking (Step 2). The last bone in the series causes the fluid in the cochlea to vibrate (Step 3). This vibrating fluid causes hair cells to move with respect to a stationary membrane above them (Step 4). The movement of the hair cell bends stereocilia that extend from the cell, and this generates changes in electrical activity in the hair cell (Step 5). The stimulated hair cell sends a message across a synapse to an adjacent neuron, which generates an action potential that moves along its axon to the brain (Step 6). The signal goes to the auditory cortex, where the brain interprets it as sound (Step 7). In a normal cochlea, the stereocilia on hair cells are highly organized, but violent vibrations from extremely loud sounds (or from certain chemical agents such as the antibiotic gentomycin) can break off stereocilia from the hair cells. This can cause irreparable damage and permanent hearing loss.

synapses to a sensory neuron in the cochlear nerve that leads to the brain.

Receiving the signal, how does the brain know whether the incoming sound is the high-pitched whistle of a lovely song bird or the low growl of an angry grizzly bear? The answer involves the mechanical properties of the cochlea. Unlike a guitar string, which vibrates equally along its entire length when stimulated by the appropriate pitch, different parts of the cochlea vibrate maximally when stimulated by different pitches. High pitches excite motion in the membrane near the wide part of the cochlea, while low pitches cause movement near the cochlea's narrow tip. Because each region of membrane connects to a different part of the brain, different pitches stimulate different brain cells. It's as if a musical staff is inscribed across a certain part of the brain. A baby can distinguish the characteristic tones of its mother's voice because that voice stimulates hair cells at specific places in the cochlea, which in turn correspond to specific brain areas. Soft, soothing sounds like a mother's voice cause the stereocilia on the hair cells to sway gently, and fewer neurons to fire action potentials. Loud sounds from a jackhammer, say, or a heavy metal rock band, vibrate the stereocilia more vigorously and cause more neural firing; very loud noises can actually vibrate the stereocilia until they break off. This can permanently damage the hair cells and along with them the sense of hearing.

The ear has jobs beyond hearing: It also helps us keep our balance and detects the rapid acceleration of an elevator as it drops or climbs or of a car speeding away from a green light. Most people rely heavily on their sense of hearing. People with hearing loss, however, can often get along quite well, thanks to extra compensation by their sense of sight.

The Eye: An Outpost of the Brain

Sherry's ability to communicate with others depends on her gestures and above all, her eyes. Visual sensations are possible because eyes convert light energy into neural energy.

How the Eye Detects Light

Picture a rose bush with huge, fragrant yellow blossoms. On a pitch-dark night, you might be able to smell the roses and feel the sharp thorns. But seeing it takes at least a low level of light. Light bouncing off the plant strikes the human eye and passes through the protective transparent outer layer, or **cornea** (Fig. 19.18a). If you gently place your finger over a closed eyelid and turn your eye from left to right, you can feel the bulge of your cornea. After penetrating the clear cornea, light passes through a transparent fluid and then through the **pupil,** which is a black, circular, shutterlike opening in the **iris** (the eye's colored portion). Next, light crosses the **lens,** a circular, crystalline structure that, together with the cornea, focuses light through another fluid onto the retina. The **retina** is a multilayered sheet that lines the back of the eyeball and contains light-sensitive **photoreceptor cells;** these begin converting light energy to electrochemical energy (nerve impulses) that can form brain patterns and create a visual image of a rose or other object. Because the lens is curved, it bends the

light rays. As a result, the image of the object it projects on the retina is backwards (left and right are reversed) and upside down (Fig. 19.18a). Once the neural signal reaches the brain, however, it is integrated and sorted out so that we perceive the image in its actual orientation.

Within the three-layered retina, the rear layer (closest to the back of the eye and just inside the eye's tough outer covering) consists of jet-black pigment cells that protect the photoreceptors from extraneous light. In the second layer, nestled just in front of the pigment cells, are two types of photoreceptor cells called **rods** and **cones** because of their distinctive shapes (Fig. 19.18a). Rods are very sensitive to low levels of light, but can't distinguish color, whereas cones can detect color but need more light. (Remember *c*one for *color.*) Cones explain the trouble you have seeing colors in dim light. Night hunters like owls and cats have a retina jam-packed with rods, allowing them to pick up the dimmest reflection, but so few cones that their color vision is probably very poor.

In the retina's third layer, in front of the rods and cones, lie sensory neurons. These synapse with the rods and cones and with other neurons that send axons toward the brain. These axons collect into a bundle called the **optic nerve,** which exits the eye and leads to the brain.

The Visual Pigment

To understand how the eye works, let's concentrate on the rods and how they receive light. The front part of each rod contains the nucleus and forms synapses with other neurons. The rear part detects light by means of stacks of membranous disks housing millions of molecules of the light-sensitive pigment molecule **rhodopsin** (see Fig. 19.18b). A protein part of each rhodopsin molecule snakes back and forth across the membrane and cradles the actual pigment portion, a small ring-and-chain molecule called **retinal.** (Eyes synthesize retinal from the compound that gives carrots their bright orange color; the old saying that carrots are good for your vision turns out to be true!) The structure of retinal is really the key to vision, since it changes shape when light hits it. This shape change unleashes a cascade of reactions that we eventually perceive as light.

We can tell bright light from dim because brighter light stimulates more neurons. These generate more action potentials in a shorter period of time; the action potentials travel down the optic nerve to the optic centers in the brain, and the brain then interprets these patterns as stronger light. We see the position of a light source because photoreceptors in different parts of the retina receive light from different parts of the visual field and link up with different parts of the brain. We see color because different cones are sensitive to light of different wavelengths.

The eyes and ears seem very different but are actually similar in several ways. They can both locate objects in space because they contain cells that can detect physical forces (sources of sound or light). They each create a map in the brain with a one-to-one correspondence between the position in space of the sound or light source and the location on the brain map. Together, hearing and seeing help animals interpret the environment around them in a way that improves their chance of survival. And even if one sense is lost, the brain can partially compensate through the other, as in Sherry Greer's acute peripheral vision and her ability to face forward and drive a car while simultaneously seeing and understanding the ASL signs of a friend in the front passenger's seat.

Taste and Smell: Our Chemical Senses at Work

We humans are visual beings and sight seems to dominate our senses. Nevertheless, taste and smell probably play a far greater role in your daily life than you think. The tongue and nose receive and detect flavor and odor molecules—actual chemical tidbits of the environment. That's why biologists consider taste and smell to be our **chemical senses.**

People sometimes think of the tongue as a perceptual genius and the nose as a sensory dullard. However, precisely the opposite is true. The tongue is studded with **taste buds,** each of which consists of a pore leading to a nerve cell. The nerve cell is surrounded by accessory cells arranged in an overlapping pattern that resembles an artichoke or onion bulb (Fig. 19.19a). The nerve cells in taste buds have receptors capable of receiving flavor molecules, but they can distinguish only four general classes of flavors: sweet, salty, bitter, and sour. We can tell similar foods apart—beef from pork, beets from turnips—and sense the subtle difference in their flavors partly because these foods stimulate different receptor types to different degrees, but mostly because aroma molecules from the food bind to receptors in the nose.

The **olfactory epithelium** consists of button-sized patches of yellowish mucous membrane high in the nasal passages. Olfactory neurons and their processes lie embedded in these patches. One end of the nerve cell,

(a) Anatomy of the eye

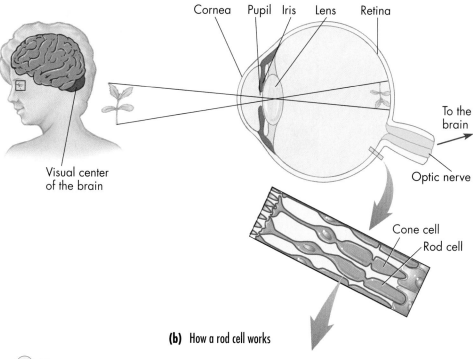

(b) How a rod cell works

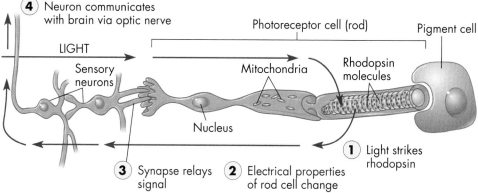

Figure 19.18   How Eyes Work. (a) A cross section of the eye reveals how light lands on the retina. (b) A view of neurons, a rod cell, a pigment cell of the retina, and how a rod cell works. Light passes through several cell layers before hitting rhodopsin (Step 1), the light-sensitive pigment inside rod cells. When light strikes rhodopsin, a portion of the molecule changes shape. This molecular change alters the electrical potential of the rod cell (Step 2). Synapses relay a signal to adjacent neurons (Step 3), and eventually, a stimulated neuron sends a message along the optic nerve to the visual centers of the brain (Step 4). The brain interprets the action potential as a flash of light.

shaped somewhat like a flower, binds odor molecules (Fig. 19.19b). The other end is a long, spindly axon that carries signals to the olfactory bulb, or smell region, on the underside of the brain (Fig. 19.12a). If you take a bite of a sandwich and simultaneously pinch your nose shut, you can still perceive the four primary tastes,

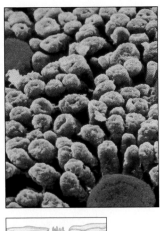

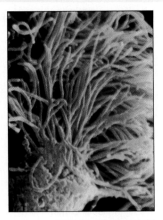

Figure 19.19   **The Nose Knows and So Does the Tongue.** Taste buds on the tongue (colorized red) (a) and odor-detecting cells in the nasal lining (colorized yellow) (b) consist of receptor cells that generate a nerve impulse when stimulated by a specific chemical.

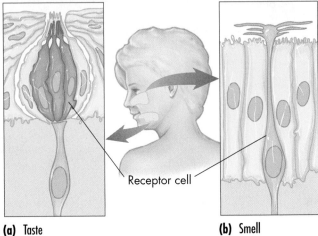

Receptor cell

(a) Taste

(b) Smell

whiff of a Christmas tree or a brand of sunscreen used on a particular beach in particular company. The explanation for such odor-stimulated *deja vu* lies in the anatomy of the nose and brain. The olfactory lobes are closely connected to the brain's **limbic system,** a series of small structures (including the hippocampus and hypothalamus) largely responsible for generating fear, rage, aggression, and pleasure and for regulating sex drives and reproductive cycles. A smell, therefore, directly stimulates the brain's centers of memory, emotion, and sexuality.

Smells can also affect the endocrine system. Women roommates, for example, often have synchronized menstrual cycles, and women with irregular periods often grow more regular when in a man's company on a routine basis. In both cases, the effects seem to be largely olfactory and due, perhaps, to pheromonelike molecules in the sweat, which act on the brain and its hormonal feedback loops with the body.

Our main message about sense organs is straightforward: An animal's brain can decipher the stimulus it receives as light, sound, taste, or smell because each sense organ is tuned to a different physical or chemical stimulus and sends a signal to a different part of the brain. And as we saw earlier, the brain's marvelous plasticity allows it to heighten one set of sensations if another set is dimmed. That explains why Sherry Greer can detect light and motion at the edges of her visual field with exceptional acuity, an ability that helps compensate for the absence of sound.

but not primary odors—and thus very little of the food's complex flavor.

For centuries, people have noticed the peculiarly intimate connections between smells, memories, and emotions. Who hasn't experienced a flood of memories, complete with appropriate emotions, when catching a

Chapter Summary and Selected Key Terms

Introduction

Sherry Greer was born deaf and her brain never received aural information during a critical early imprinting period. Sherry did learn languages (ASL from infancy and English in childhood) and went on to have a successful academic and business career. Dr. Helen Neville studies the effect of sensory input or deprivation on brain development. Her studies are part of the investigation of nervous system functioning, of brain plasticity, and of our species's unique ability to use language.

Nerve Cells

Most animals have a **nervous system** (p. 486), a network of specialized cells that receives information, integrates it, and allows an effect—a change—to take place in the way the body functions. There can be over 1000 types of nervous system cells, but each is either a **neuron** (p. 486) that participates in cell-to-cell communication, or a **glial cell** (p. 486) that supports, surrounds, protects, and/or provides nutrients to neurons. A neuron has signal-receiving branches called **dendrites** (p. 487), a **cell body** enclosing the nucleus (p. 487), and a long thin **axon** (p. 487) that transmits an impulse. A **nerve** (p. 487) is a group of axons from many different neurons bundled together. A **synapse** (p. 487) is a junction between neurons, including a minute gap, the synaptic cleft, where a neural impulse crosses between two adjacent neurons. The **nerve impulse** (p. 488) is an electrochemical reaction that allows specific electrically charged atoms to rush into and out of the nerve cell. The difference in electric charge (based on charged sodium, potassium, and chloride ions) between the inside and outside of a neuron generates potential energy called the **resting potential** (p. 488).

Protein-lined pores called **potassium channels** (p. 488) and **sodium channels** (p. 489) allow the ions to move in or out of the cell. A **sodium-potassium pump** (p. 489), another specialized protein, forces sodium ions out of the cell and escorts potassium ions back in. In its resting state, the neuron is polarized, that is, it has an imbalance of electrical charges across the membrane. An **action potential** (p. 489), or nerve impulse, involves the reversal of the charge on the resting cell and a series of electrochemical changes propagated along the neuron. A stimulus (such as a sound or light flash) causes sodium to enter the cell, and it depolarizes, or loses its state of polarization. A threshold is reached, sodium channels open wide, sodium ions rush in, and the inside of the neuron becomes positively charged. Sodium channels inactivate then close, potassium channels open, then the neuron is restored to its original resting potential. The action potential is propagated along the axon and moves in one direction only. An insulating sheath of fatty material (a **myelin sheath;** p. 491) greatly speeds nerve impulses.

Communication Between Nerve Cells

When a nerve impulse reaches the terminal end of an axon, it triggers the release of a chemical signal or **neurotransmitter** (p. 492). These molecules move into the narrow space separating sending and receiving cells, the neurotransmitter molecules diffuse across the cleft, and they bind to receptors in the receiving cell's plasma membrane. This can cause an action potential in the receiving cell. Enzymes clean up the neurotransmitter so a new round of reception and stimulation can occur. Cocaine blocks the cleanup of **dopamine** (p. 492) and can lead to **addiction** (p. 493). Other neurotransmitters include acetylcholine and norepinephrine. Some neurons transmit signals directly through specialized gap junctions in electrical synapses.

Animal Nervous Systems

Vertebrates like ourselves have a **central nervous system (CNS)** (p. 494) or information processor that includes the **brain** (p. 494) and **spinal cord** (p. 494). We also have a **peripheral nervous system (PNS)** (p. 494) or "middleman" that includes sensory neurons that sense the environment and motor neurons that control muscle contraction. We have simple nerve circuits called **reflex arcs** (p. 494) that link a stimulus like stepping on a tack directly to a rapid, involuntary response. Reflex arcs usually involve a **sensory neuron** (p. 494) that receives information, **interneurons** (p. 495) that relay the incoming messages in the central nervous system,

and **motor neurons** (p. 494) that relay outgoing messages of response to muscles or glands.

Voluntarily controlled motor neurons form the **somatic nervous system** (p. 495). A different set of subconscious, automatic motor neurons makes up the **autonomic nervous system** (p. 495) and controls the body's internal environment. Within the autonomic system, **parasympathetic** (p. 495) nerves help maintain normal functions like digestion and resting heartbeat. **Sympathetic** (p. 495) nerves help regulate normal functions like blood pressure as well as responding to emergencies with the fight-or-flight response. The spinal cord within the central nervous system includes gray matter, or cell bodies and nerve fibers without thick insulation, and white matter or axons insulated with white fatty sheaths. Local neural traffic occurs in the gray matter; long-distance communication moves up and down the white matter.

The brain has a **brainstem** (p. 496), including the **medulla oblongata, pons,** and **midbrain** (p. 496), which function mainly in homeostasis, control, sensory and motor integration, and arousal. The brain has a **diencephalon** (p. 497), which includes the **thalamus** (p. 497), **hypothalamus** (p. 497), **pituitary** (p. 497), and **pineal gland** (p. 497). It also has a **cerebellum** (p. 497), which monitors body position and times motor commands. And the brain has a **cerebrum** (p. 497), a convoluted mass of tissue responsible for all our higher functions, including self-awareness and speech, as well as memory, sensation, and motor output. Various regions of the outer layer or **cerebral cortex** (p. 497, 498) have specific functions in generating movement (the **motor cortex;** p. 498), interpreting sensory input (the **sensory cortex;** p. 498), creating language, and so on. The **hippocampus** (p. 500) is involved in forming both short- and long-term memories. Parts of both the left and right cerebral cortex are involved in language. A **stroke** (p. 500) can disrupt language comprehension or generation. Most English speakers show activity in the left brain when using language. Native ASL speakers show activity in both left and right brain, based on both the grammar and the spatial patterns of the gestural language.

The **corpus callosum** (p. 502) is a tissue bridge between the right and left cerebral hemispheres that allows a person to both see an object and name it. Doctors sometimes sever this bridge in patients with **epilepsy** (p. 502). **Association cortex** (p. 502) is the name for regions of the cerebral cortex that integrate information from other areas and participate in emotion, memory, personality, reasoning, judgment, and intelligence. In **Alzheimer's disease** (p. 502), brain tissue deteriorates as filaments and protein globs accumulate, leading to dementia. In **schizophrenia** (p. 503), the mind is disturbed by psychosis and hallucinations. Both diseases are at least partially inherited and show distinctive brain changes.

Sense Organs   We hear because sound waves vibrate the **eardrum** (p. 504), then a chain of tiny bones, then fluid in the coiled **cochlea** (p. 484, 504), a highly complex structure with over one million moving parts. The cochlea changes the energy from sound into nerve impulses. **Hair cells** (p. 504) crowned by **stereocilia** (p. 504) move in response to sound vibrations and trigger nerve impulses that the brain interprets as high, low, grating, melodious, and so on. The inner ear is also involved in maintaining balance.

We see because light passes through the transparent outer layer or **cornea** (p. 506); penetrates the **pupil** (p. 506), or opening in the colored **iris** (p. 506); traverses the crystalline **lens** (p. 506); and is focused onto the **retina** (p. 506). Photoreceptor cells — **rods** (p. 506) and **cones** (p. 506) — in the retina detect light and color, respectively, and send information to the brain via the **optic nerve** (p. 506). Rods house molecules of the light-sensitive pigment **rhodopsin** (p. 506), with its actual pigment portion called **retinal** (p. 506).

Taste and smell are our chemical senses because sensory cells in the **olfactory epithelium** (nose lining) (p. 507) and in the tongue's **taste buds** (p. 507) receive and detect flavor and odor molecules directly from the environment.

All of the following question sets also appear in the Explore Life ⒺE electronic component, where you will find a variety of additional questions as well.

Test Yourself on Vocabulary and Concepts

In each question set below, match the description with the appropriate term. A term may be used once, more than once, or not at all.

SET I

(a) dendrite (b) cell body (c) axon
(d) axon terminal (e) synapse (f) sodium channel

1. A neuron's metabolic center

2. The receiving end of a neuron

3. Region of an axon where synaptic vesicles open and liberate a neurotransmitter substance

4. A small space across which neurotransmitters diffuse

5. Region along which action potential travels

SET II

(a) resting potential (b) action potential
(c) myelin sheath (d) nerve impulse
(e) potassium channel

6. A series of action potentials that occur sequentially along an axon

7. The rapid transmission of an action potential from node to node along an axon requires this

8. The rapid entry of sodium ions into one site on an axon followed by the rapid outflow of potassium ions

9. A condition in which the inside of a neuron is maintained at a negative charge relative to the outside by means of the transport of ions through pumps and channels

SET III

(a) CNS (b) PNS (c) sensory neuron (d) motor neuron (e) interneuron (f) autonomic nervous system
(g) synapse

10. Includes sympathetic and parasympathetic nerves that regulate the internal organs of the body (g) synapse

11. Composed of the brain and spinal cord

12. Consists of neurons that carry information from sense organs to the brain as well as from the brain to muscles

13. Neurons that do not synapse directly with either sense organs or with muscles, but only with other neurons

14. The neuron category that comprises most of the brain

SET IV

(a) sympathetic nerves (b) parasympathetic nerves
(c) somatic motor nerves (d) gray matter
(e) white matter

15. Motor neurons of the autonomic nervous system that take charge of regulating internal organs when the body is calm and quiet

16. Motor neurons of the autonomic nervous system that take charge of regulating internal organs during times of emergency or stress

17. Motor neurons that regulate the contraction of muscles attached to the skeleton

18. The axons traveling up and down within the spinal cord

19. Found within the CNS; consists mostly of cell bodies and synapses

Integrate and Apply What You've Learned

20. Compare the relative advantages and disadvantages conferred by electrical synapses and chemical synapses. Could large, active, intelligent animals have evolved with only electrical synapses?

21. Despite our large brains and capacity for learning, we also exhibit many reflexive behaviors. Name three different human reflex activities, stating their common features.

22. Let's say you discover a particular mutation in mice that prevents the formation of axon terminals at the ends of axons. What would be the physiological result of such a mutation?

23. Sometimes mussels and clams consume microbes that contain saxitoxin. In vertebrates, this compound blocks the flow of sodium ions through sodium channels. One mussel can contain enough saxitoxin to kill 50 people! How would saxitoxin affect an action potential in an individual neuron? What would be the symptoms of saxitoxin poisoning?

24. Refer to the discussion of a reflex arc. Which elements of the reflex arc are in the central nervous system and which are in the peripheral nervous system?

25. Explain how each of the receptor cells for vision and hearing responds to a stimulus.

26. You discover that one species of flatworm has nerve centers that are substantially larger than those of another species of flatworm. Which type of neuron would you expect the larger ganglia to contain more of: sensory neurons, interneurons, or motor neurons?

Analyze and Evaluate the Concepts

27. How is Sherry Greer's excellent peripheral vision an example of brain plasticity?

28. Which features of a neuron prevent an action potential in a postsynaptic cell from immediately moving backward and restimulating the presynaptic cell?

29. Many pesticides prevent the enzymatic inactivation or removal of the neurotransmitter acetylcholine, which stimulates the contraction of skeletal muscles. As a result, acetylcholine remains active within the synapse between a neuron and its muscle cell for an extended period of time. What effect would this have on the postsynaptic cell (which in this case is a muscle cell)?

30. Which characteristics would be helpful in enabling an animal to localize the source of a sound? More than one answer may be correct. Support your choice(s).
    (a) ears close together on the front of the head
    (b) immovable ear flaps
    (c) ears far apart on the sides of the head
    (d) movable ear flaps

31. Hearing aids work by amplifying and transmitting sound waves through the inner ear to set up fluid motion in the cochlea. Can a hearing aid help a person whose hair cells are destroyed? Why or why not?

An Almost Fall

D r. Marjorie Woollacott is buckling Mary Walrod into a harness and is about to cause the 77-year-old Oregonian to slip on an "electric banana peel." That's what Woollacott and her students affectionately call the unique equipment in her equally unique laboratory at the University of Oregon in Eugene. Woollacott studies balance and how so many of us start to lose it through age, inactivity, or lopsided exercise programs. What she is learning could have a major impact on college students, their parents, their grandparents, and the rest of society.

Once snugly supported within the harness, Mary is in no danger, even when her feet slip out from beneath her as she steps gingerly along Woollacott's 20-meter (65-foot) long walkway. The test equipment has a hidden slip plate that moves forward just as a person's heel strikes it in a normal step. The effect, says Woollacott, is "just like stepping on a patch of ice on a cold day" or a banana peel on the sidewalk. As the harness prevents a tumble, the laboratory equipment simultaneously captures the "almost fall" on videotape and records electrical signals from the person's muscles.

Mary is a fairly active retiree and takes walks in the park near her home, although, she admits, "Maybe not as much as I should!" For her age, her balance is pretty good, but when the slip plate moves beneath her right heel, she has a characteristic response for a person in her 70s: Mary swings her arms up, wobbles back and forth at the hips, and steps forward with her other leg.

"People used to think that your hip and trunk muscles mainly controlled your balance," says Dr. Woollacott. But her studies have shown that the strength, speed, and coordination of the ankles and legs is, in fact, the deciding factor in whether people have good balance or are unsteady on their feet. Moreover, by testing two kinds of runners, she and her students were able to trace the roots of balance to particular muscle fibers within the leg muscles.

When Dr. Woollacott tested sprinters in her lab—people who train for short, fast running events like the 50-meter dash—she found that they exhibit superior balance. Canadian sprinter Nicole Commissione, for example, was scarcely affected by the slip plate track and merely took one small step to regain firm footing. This superior balance, says Woollacott, is based on a sprinter's high proportion of so-called fast-twitch muscle fibers, or special cells within the muscles that deliver quick power. However, when Woollacott tested marathoners like Kenyan Billy K. Harper on the slip plate, these runners lost their balance much more easily, and staggered and wavered more than did

▲ **A Question of Balance.** Marjorie Woollacott and co-worker hooking up Mary Walrod to an experimental harness; Nicole Commissione sprinting; Woollacott doing a yoga pose to improve her balance.

the faster, stronger sprinters on the track. This, Woollacott explains, is due to the marathoner's high proportion of slow-twitch muscle fibers, or cells within the muscles that provide endurance but respond more slowly to a slip. The two kinds of runners would probably show big differences in reaction time tasks, as well, says Woollacott, like pressing a button quickly or catching a fast-moving ball. This, she says, is because sprinters also tend to build up the fast twitch muscle fibers in their upper bodies through weightlifting, while for many long-distance runners, she says, lower-body exercise is "pretty much it," and they lose the muscle fibers in their arms, shoulders, and trunk that provide quick power.

There is a surprising link between long-distance runners like Billy Harper and senior citizens like Mary Walrod, and it's this: All of us naturally lose fast twitch fibers as we age. But the decline can be faster if we are generally sedentary or if we are selectively inactive—that is, build up just one set of muscles the way marathoners usually do with their legs. "Use it or lose it" is an appropriate motto for the musculoskeletal system that generates the body's physical support and movement. It helps explain why slips and falls are the seventh leading cause of death in older Americans. And it underlies a number of Marjorie Woollacott's personal choices and opinions.

At age 52, the Oregon researcher does hatha yoga each morning then walks three miles to work. Woollacott's father helped inspire these good habits: Laurence Hines, now 83, still hikes 45 minutes every day in a national forest near his home in Sedona, Arizona. His daughter's lab work has been an equal inspiration for him, too: She has tested him many times on the slip plate and confirmed that his continued, vigorous activity is keeping his legs strong and his balance sound, and this, in turn, helps keep him on the hiking trail.

Woollacott's work shows the importance of staying active as we age. "This is going to be really important as the Baby Boom generation gets into their 70s and 80s," she says. "They will be at greater risk for falls and ending up in nursing homes," and that could swamp our medical system. Most people in nursing homes are there because they are no longer steady on their feet, and in fact, a fall and resulting broken hip or other bone often sends an older person to bed, never to live independently again. Dr. Woollacott's studies also help much younger people, including college-age athletes, plan training programs. "Their exercise regimens need to be focused," she says, "to help them build up the types of muscle fibers that are best for their particular sport." That's true whether their sport is swimming, ballet, rowing, tennis, basketball, sprinting, or ping pong, and whether they need quick power or greater endurance.

As we explore the musculoskeletal system in this chapter, we'll see how muscles and bones make exercise possible and how movement and physical work help most animals survive day-to-day. We'll see how bones grow and act as scaffolds and levers. We'll see how muscle fibers contract by means of organic motors. And we'll see why Americans—even at young ages—are growing more and more sedentary and how that lack of activity can strongly affect health.

As we go through the chapter, you'll discover the answers to these questions:

1. How do skeletons work?

2. How do muscles move skeletons?

3. How does training for a sport affect the body's muscles and skeleton?

20.1 The Skeleton: The Body's Framework

Take a look at Figure 20.1, which includes some of Dr. Woollacott's data from testing a young subject and an older subject on the slip plate. Can you see differences in how the two people responded? What happened to the arm? The knee? The angle of the upper body with re-

spect to upright stance? How did their stride lengths compare? Our task in this chapter is to explore the body's system of muscles and bones in enough detail to be able to analyze the kinds of foot slips Dr. Woollacott studies and records. Our goal is to understand which muscles pull against which bones and in roughly what sequence to restore the balance we all need to walk upright.

Here's a simple introduction to the muscles and bones: If you're sitting in a chair right now, swing your

Figure 20.1 Tracings from an "Almost Fall."

(a)

(a) Mary Walrod and Dr. Marjorie Woollacott. Notice the reflective markers the researcher has applied to Mary's joints. (b) A stick figure traced from the reflective markings glued to a young person moving on the laboratory walkway and slip plate. (c) A stick figure of an older person experiencing an experimental slip trial. The red arrow marks the location and movement direction of the slip plate in each trial. Thick black lines represent the right side of the body; thin blue lines represent the left side.

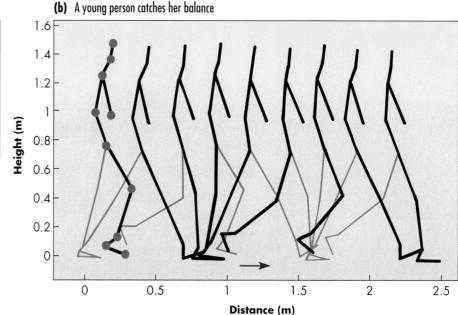

(b) A young person catches her balance

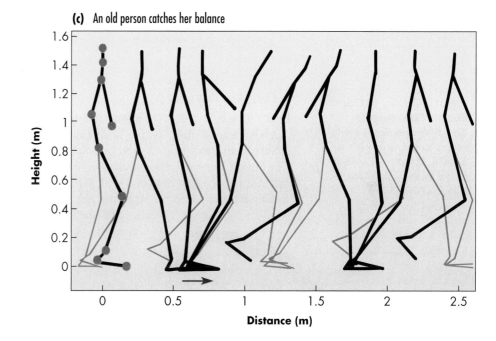

(c) An old person catches her balance

lower leg out in front of you, pivoting it about the knee, until your entire leg becomes straight. This is like the forward kicking movement in Figure 20.1b (red arrow). How does a small, natural movement like that take place? It involves a stiff rod (your upper leg bone) that has a stationary object at one end (your hips), a movable object at the other (your lower leg), and muscles pulling between them like a stretched rubber band. Crucial to this contraction is the **skeleton,** the rigid body support to which muscles attach and apply force (*E* xplorer 20.1). Most people think of the skeleton as a solid framework like a shrimp's outer hard shell or the rigid rods of bone in your arms and legs. And these *are* skeletons. But there's another less familiar type of skeleton, too, that depends on a liquid, like the hydraulic brake system in a car, rather than on a solid rod or shell to transmit force. Let's look for a moment at this simple type of skeleton.

Water as a Skeletal Support

Sea anemones in an ocean tidepool, earthworms in the backyard soil, and snails in a river bed all have the same kind of internal support based on a liquid core (made up of water or body fluid) wrapped in a muscular sheath. The two sea anemones in Figure 20.2 show how a water skeleton or **hydroskeleton** (*hydro* = water) works. This type of fluid internal skeleton is actually much like a water balloon: If you squeeze on one end, the fluid transmits force to the other end. The sea anemone's central digestive cavity is filled with seawater and its surrounding body wall contains two layers of muscles: One layer extends lengthwise from the animal's base to its tentacles; and the other encircles the wall. These two muscle layers act as **antagonistic mus-**

cle pairs—groups of muscles that move the same object in opposite directions. When the lengthwise muscles contract, the animal becomes shorter and wider (as when you push a water-filled balloon from both ends simultaneously). When the circular muscles contract, the animal becomes longer and narrower (as when you wrap your hands around a balloon and squeeze).

Hydroskeletons are common among invertebrates, with their lack of bone, but even people rely on hydroskeletons in some situations. When we want to lift a

---

GET THE PICTURE

E **xplorer 20.1**

Skeletons, Muscles, and Movement

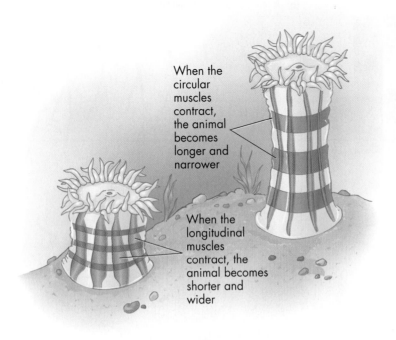

When the circular muscles contract, the animal becomes longer and narrower

When the longitudinal muscles contract, the animal becomes shorter and wider

Figure 20.2 Sea Anemones and Water Skeletons.

heavy lamp or stack of books off the floor, we tend to hold our breath and tighten our abdominal muscles. The compressed fluid of the abdominal cavity contributes to the body's rigidity and helps support the load. The human penis and those of many other mammals are stiffened by a hydroskeleton, as well: valves allow blood to flow forcefully into the organ but not to flow out as freely, and the trapped fluid causes the penis to expand and become rigid.

Skeletons: Stiff Rods and Curved Plates

A simple hydroskeleton is fine for sea anemones living with the extra buoyant support of ocean water and for earthworms slithering inch by inch through the soil. A rapidly moving land animal, however, needs a solid framework made up of inter-

locked proteins, minerals, and polysaccharides. It can be an **exoskeleton**—a stiff external covering that surrounds the body, as in shrimp, beetles, scorpions, and other arthropods (Fig. 20.3), or it can be an **endoskeleton**—like our own internal set of rigid rods and plates and those of other vertebrates (Fig. 20.4; see **E** 20.1).

A mammal's endoskeleton is made of **bone,** a living tissue that contains a meshwork of the long, stringy protein **collagen** hardened by calcium and phosphate. The human skeleton has 206 individual bones, and these are found in either the body's main axis, or in the limbs (appendages) that branch off that axis (**E** 20.1).

The Skeleton of the Body Axis

Mary Walrod, the sprinter Nicole Commissione, the marathon runner Billy Harper, and virtually every other human has a body axis made up of the skull, vertebral column, ribs, and **coccyx** (tailbone). Part of the **skull** is the **cranium,** which encloses and protects the brain, and part is a collection of face and middle ear bones (Fig. 20.4 and see Fig. 19.17). The skull also contains the **hyoid** bone, the only human bone not jointed with other bones, which is suspended under the back of the mouth by ligaments and muscles.

Since humans are upright animals, our head sits on top of our backbone, or **vertebral column,** rather than in front like a dog's or lizard's. The human backbone is made up of 33 **vertebrae** (sing., *vertebra*) (Fig. 20.5). Each vertebra is a rigid bone shaped like a tunafish can that fits together with other vertebrae in a stack and supports the trunk. An arch extends from the round box and encases and protects the spinal cord. **Processes** or protuberances provide attachment sites for muscles (Fig. 20.5). Rubbery cartilage discs separate the vertebrae and allow the bones to move without grinding or pinching each other. This movable stack of vertebrae and discs gives flexibility to the body axis, allowing a shortstop to bend forward and scoop up a line drive, a gymnast to do a back walkover, or a person to flex to catch his or her balance. The vertebral column is a bit fragile, however, and requires special precautions as you lift, stand, or sit to avoid developing lower back problems. Curving forward from the vertebrae are 12 pairs of ribs, which meet in the front and attach at the breastbone (**sternum**) and form a protective compartment around the heart and lungs.

The Skeleton of the Body's Branches

Our human skeleton has appendages attached to the main axis. The shoulder (**pectoral girdle**) and hip (**pelvic girdle**) are branching-off points that support

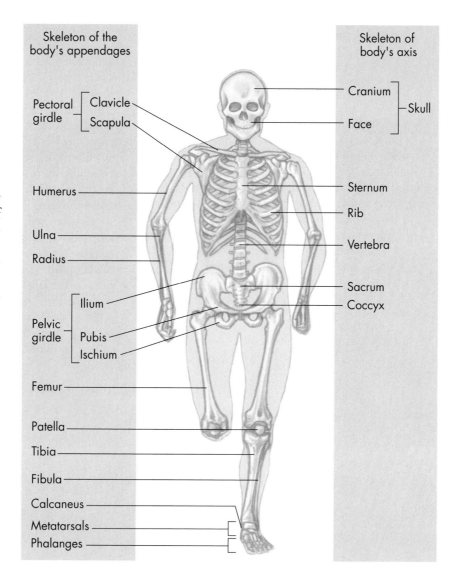

Skeleton of the body's appendages

Skeleton of body's axis

Pectoral girdle — Clavicle, Scapula

Cranium, Face — Skull

Humerus

Ulna

Radius

Pelvic girdle — Ilium, Pubis, Ischium

Femur

Patella

Tibia

Fibula

Calcaneus

Metatarsals

Phalanges

Sternum

Rib

Vertebra

Sacrum

Coccyx

the arm and leg bones and allow them to swing (see Fig. 20.4). If we compare our arm bones to the bones of a horse's front leg, it's clear that they share a common anatomical theme of extension and support (Fig. 20.6 and **E** 20.1). The pressures of natural selection acting over thousands of generations have molded the horse's anterior appendages for superb running ability, just as entirely different pressures have led to the adaptations in our arms and hands for grasping pencils, throwing baseballs, and turning book pages.

Bones: A Living Scaffold

As you walk, ride, drive, or bicycle today, the bones of your skeleton are supporting your body and anchoring your muscles. They also encase your vital organs, form your blood cells, and store your body's excess calcium and phosphate ions. You might think of your bones as nonliving mineral sticks and rods, but they're alive! A closer look at a large bone such as the **femur,** extending between the hip and knee joints, shows how bones support, anchor, encase, and store—as well as how they self-repair when broken.

The femur's long shaft (Fig. 20.7a) allows it to support the body and transmit your weight to the ground. Bones aren't smooth pipes; they have projections, or processes, that serve as attachment sites for ligaments and tendons. **Ligaments** are connective bands linking bone to bone, while **tendons** connect bone to muscle. You can easily find and feel examples of these skeletal structures in your body. There is a large muscle attachment site on your skull bone where your head joins the

back of your neck; this process (which feels like a bump) links to muscles that keep your head erect. If you tighten your **quadriceps** (thigh muscles), you can feel the ligament that attaches your **patella** (kneecap) to your **tibia** (shinbone) just below your knee. And you can easily feel tendons between the muscles in your upper arm and the bone in your forearm by putting your hand beneath the edge of a desk, lifting, and, with the other hand, touching the "cables" that project on the inside of your elbow.

Now let's zoom in still closer on the femur to see how it—and other bones—can perform so many important functions (Fig. 20.7b). A cross section through the hip joint reveals that bone has many small spaces between the hard parts. The spaces store fat or contain blood vessels. Regions with few, small spaces form **compact bone.**

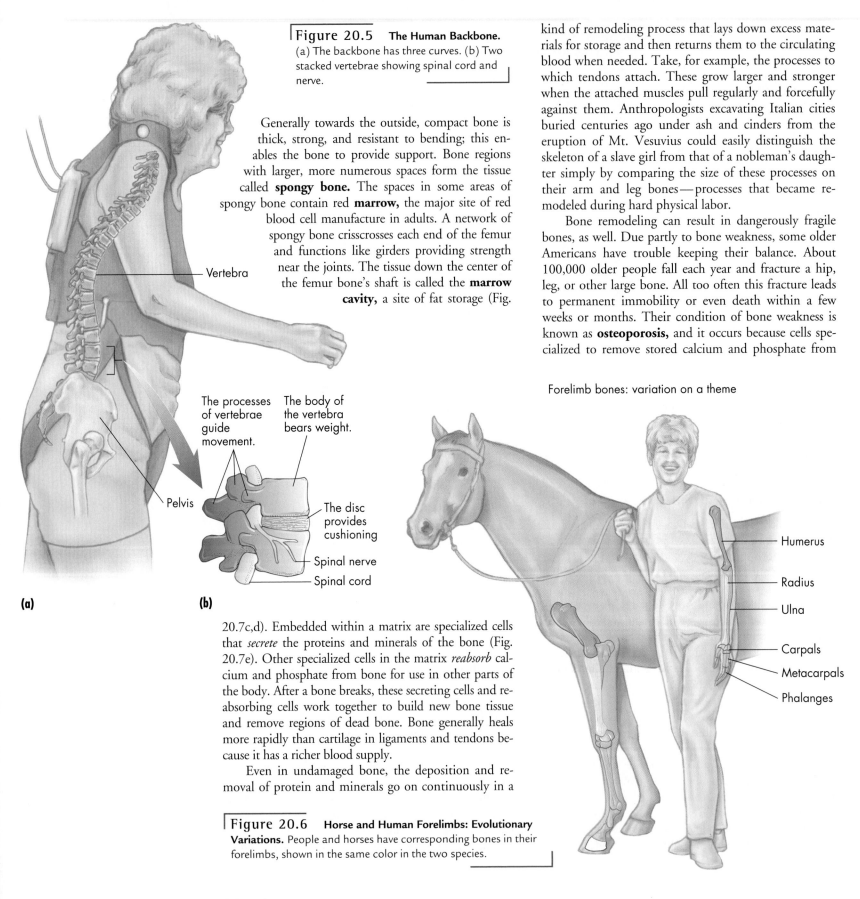

Figure 20.5 **The Human Backbone.** (a) The backbone has three curves. (b) Two stacked vertebrae showing spinal cord and nerve.

Generally towards the outside, compact bone is thick, strong, and resistant to bending; this enables the bone to provide support. Bone regions with larger, more numerous spaces form the tissue called **spongy bone.** The spaces in some areas of spongy bone contain red **marrow,** the major site of red blood cell manufacture in adults. A network of spongy bone crisscrosses each end of the femur and functions like girders providing strength near the joints. The tissue down the center of the femur bone's shaft is called the **marrow cavity,** a site of fat storage (Fig.

— Vertebra

The processes of vertebrae guide movement.

The body of the vertebra bears weight.

Pelvis

The disc provides cushioning

Spinal nerve

Spinal cord

(a)

(b)

kind of remodeling process that lays down excess materials for storage and then returns them to the circulating blood when needed. Take, for example, the processes to which tendons attach. These grow larger and stronger when the attached muscles pull regularly and forcefully against them. Anthropologists excavating Italian cities buried centuries ago under ash and cinders from the eruption of Mt. Vesuvius could easily distinguish the skeleton of a slave girl from that of a nobleman's daughter simply by comparing the size of these processes on their arm and leg bones—processes that became remodeled during hard physical labor.

Bone remodeling can result in dangerously fragile bones, as well. Due partly to bone weakness, some older Americans have trouble keeping their balance. About 100,000 older people fall each year and fracture a hip, leg, or other large bone. All too often this fracture leads to permanent immobility or even death within a few weeks or months. Their condition of bone weakness is known as **osteoporosis,** and it occurs because cells specialized to remove stored calcium and phosphate from

Forelimb bones: variation on a theme

Humerus

Radius

Ulna

Carpals

Metacarpals

Phalanges

20.7c,d). Embedded within a matrix are specialized cells that *secrete* the proteins and minerals of the bone (Fig. 20.7e). Other specialized cells in the matrix *reabsorb* calcium and phosphate from bone for use in other parts of the body. After a bone breaks, these secreting cells and reabsorbing cells work together to build new bone tissue and remove regions of dead bone. Bone generally heals more rapidly than cartilage in ligaments and tendons because it has a richer blood supply.

Even in undamaged bone, the deposition and removal of protein and minerals go on continuously in a

Figure 20.6 **Horse and Human Forelimbs: Evolutionary Variations.** People and horses have corresponding bones in their forelimbs, shown in the same color in the two species.

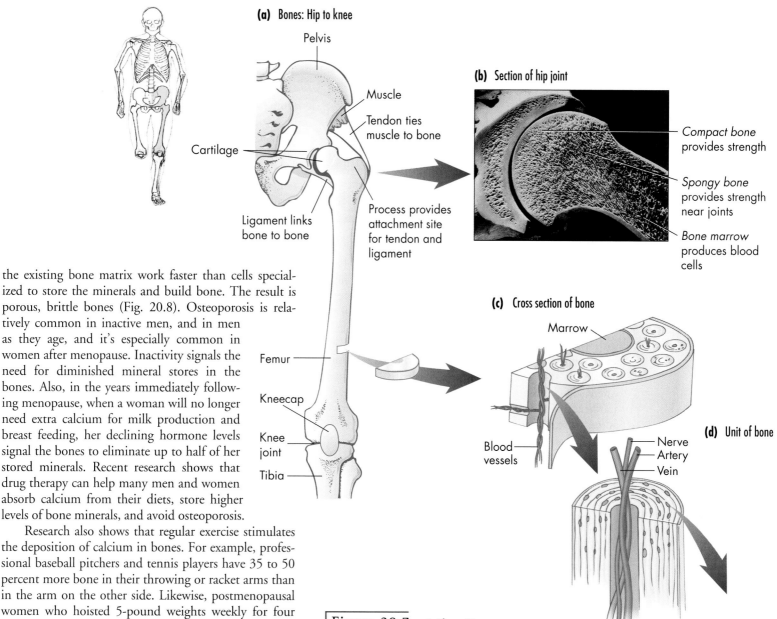

(a) Bones: Hip to knee

Pelvis

Muscle

Tendon ties muscle to bone

Cartilage

Ligament links bone to bone

Process provides attachment site for tendon and ligament

Femur

Kneecap

Knee joint

Tibia

(b) Section of hip joint

Compact bone provides strength

Spongy bone provides strength near joints

Bone marrow produces blood cells

(c) Cross section of bone

Marrow

Blood vessels

(d) Unit of bone

Nerve
Artery
Vein

(e) Bone cell embedded in bone matrix

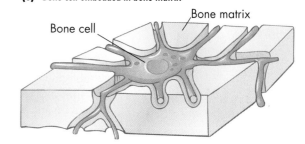

Bone matrix

Bone cell

Figure 20.7 **A Short Tour of a Long Bone: The Femur.** The femur helps support the body, anchor muscles and tendons, form blood, and store minerals. (a) The main parts of the femur. (b) The compact and spongy bone layers and bone marrow in a cross section of a human femur where it fits into the hip. (c,d) A cross section through the femur shows how rings of bone cells encircle blood vessels. (e) Individual bone cells secrete a matrix of protein and salt around themselves.

the existing bone matrix work faster than cells specialized to store the minerals and build bone. The result is porous, brittle bones (Fig. 20.8). Osteoporosis is relatively common in inactive men, and in men as they age, and it's especially common in women after menopause. Inactivity signals the need for diminished mineral stores in the bones. Also, in the years immediately following menopause, when a woman will no longer need extra calcium for milk production and breast feeding, her declining hormone levels signal the bones to eliminate up to half of her stored minerals. Recent research shows that drug therapy can help many men and women absorb calcium from their diets, store higher levels of bone minerals, and avoid osteoporosis.

Research also shows that regular exercise stimulates the deposition of calcium in bones. For example, professional baseball pitchers and tennis players have 35 to 50 percent more bone in their throwing or racket arms than in the arm on the other side. Likewise, postmenopausal women who hoisted 5-pound weights weekly for four years lost only half as much bone mass in their arms as did more sedentary women in the same age group. Luckily, with proper diet, exercise, and perhaps drug treatments, osteoporosis is not an inevitable result of aging.

Joints: Where Bone Meets Bone

The swinging of arms and legs as a person walks or falls on Dr. Woollacott's slip plate depends on joints. In fact, the movement of any bone with respect to an adjacent stationary bone requires a **joint,** or articulation, an expanded portion where the adjacent bones nearly touch. Slightly movable joints, such as those between adjacent vertebrae in the spine, have the pads of cartilage that

Figure 20.8 Osteo-
porosis: A Thinning of the
Bone Matrix. (a) Normal bone
shows a thicketlike meshwork
of fibers. (b) The decrease in
estrogen that accompanies
menopause can lead to osteoporosis, a thinning and weakening of the bones. Both of
these bone samples come from
a person's lower back.

(a) Normal bone **(b)** Osteoporosis

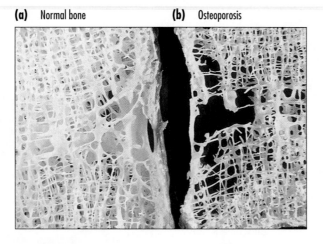

absorb shock but allow limited mobility (review Fig. 20.5). Freely movable joints, such as the shoulder, hip, and knee (see Fig. 20.7a), have pads of cartilage at the ends of the two adjoining bones. They also have a flattened sac called the synovial membrane, which is filled with a fluid (the synovial fluid) that cushions the joint and eases the gliding of bones across each other. Inflammation of this joint sac is called **arthritis.** In places where skin, tendons, muscles, and bone rub against one another, small sacs filled with fluid or **bursae** (*bursa* = purse; sing., *bursa*) cushion the movement. Inflammation of these sacs from overuse, such as in a weekend basketball game or dance session, can cause **bursitis.**

20.2 Muscles: The Body's Movers

Bones provide support, but without muscles to move them, the body would be a static pile of sticks. This section investigates how muscles work at the level of the whole organism to recover balance. Then it explores the cellular construction of muscles and finally their molecular mechanism of action.

How Bones Act as Levers

When Marjorie Woollacott begins to test an experimental subject like Mary Walrod to see how her muscles and bones interact to maintain balance, the researcher places reflective markers on various parts of Mary's body. Then Woollacott trains a video camera on Mary's movements and records how they respond after stepping on the simulated "icy patch" or "banana peel" section of the laboratory walkway. The cameras and associated computers

follow the positions of joints in space. Next, Dr. Woollacott makes stick figures of her subjects catching their balance based on the reflector patches she can track in her videotapes. These figures help her understand part of the story of why older people have a more difficult time keeping their balance than younger people. But there is more to it, and to understand the rest, she had to investigate the way bones pivot at joints, as well as how different people use their muscles, building on the basis of muscle structure and function.

Bones pivot at joints because of the way muscles pull on them. Animals with hard skeletons have muscles arranged in **antagonistic pairs.** For example, contraction (shortening) of the shin muscle (tibialis anterior) flexes your foot toward your body (Fig. 20.9a), while its antagonist, the calf muscle (gastrocnemius), extends your foot to point away from the body (Fig. 20.9b and *E* 20.1). Often, when a muscle contracts, its antagonist relaxes. Dr. Woollacott found, however, that when older people slip on her laboratory plate, they often contract both groups of muscles at the same time, stiffening the leg and preventing the small compensating movements it takes to restore balance.

We can produce the ordered movements of walking, lifting, throwing, and so on because of our **skeletal muscles,** muscles attached to bones that move the skeleton. One end of a skeletal muscle (called the **origin**) attaches to a bone that generally remains stationary during a contraction, like an anchor, while the other end (the **insertion**) attaches across a joint to a bone that moves. For example, reach down and feel your calf muscles at the back of your lower leg. These muscles attach to your femur just above the back of your knee, and at the opposite end, to your heel bone. When the calf muscles contract, the shortening forces the foot to rotate around a pivot, the ankle joint, exactly as a lever (here, the foot) rotates around a fulcrum (the ankle joint). Because of this arrangement, a small contraction of the muscle transmits a large movement to the bone. Figure 20.1b shows this change in angle during walking.

The Body's Muscle System

Each muscle's arrangement and position is crucial to its function. Figure 20.10 shows many of the body's major skeletal muscles. Dr. Woollacott has found that several muscles are important for regaining balance after a slip, including parts of the "quads" (**quadriceps**) in the front of the thigh; three muscles running down the back of the thigh (the semitendinosus, semimembranosus, and biceps femoris) that form what we generally refer to as the "hamstrings"; a shin muscle in the front of the lower leg

Figure 20.9 **Muscles Move Bones to Action.** (a) Simultaneous contraction of the shin muscle and relaxation of the calf muscles flex the foot (flexion), while (b) contraction of the calf muscles and relaxation of the shin muscle extend the foot (extension). Because the ankle joint acts as a fulcrum, people with longer heel bones (levers) can often jump higher. (c) Muscle contractions and lever action drive New York Yankee's Bernie Williams to soar upwards for a fly ball at Seattle Mariner's Stadium in October, 2000.

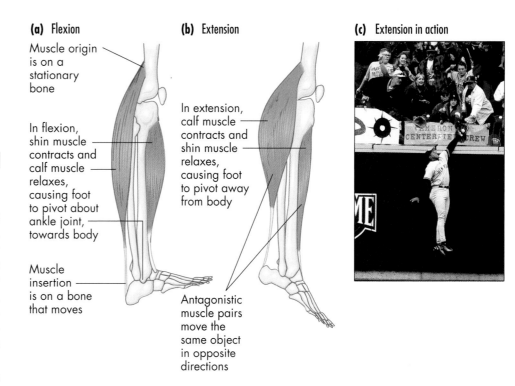

(a) Flexion
Muscle origin is on a stationary bone
In flexion, shin muscle contracts and calf muscle relaxes, causing foot to pivot about ankle joint, towards body
Muscle insertion is on a bone that moves

(b) Extension
In extension, calf muscle contracts and shin muscle relaxes, causing foot to pivot away from body
Antagonistic muscle pairs move the same object in opposite directions

(c) Extension in action

(tibialis); and the "abs" (abdominals). Dr. Woollacott knew that electrical events stimulate muscle contraction and that she could record the electrical activity in her subject's muscles by placing electrodes on the skin above the muscle in question. Figure 20.11 shows a runner with some of these attached electrodes, as well as data recorded from young and older adults during an unexpected slip. The signal is measured in volts and is displayed as a graph. When the "trace" or graphed line goes up, it means that the muscle has responded to the slip.

The technique for recording electrical changes in the muscles and plotting the changes on paper is called **electromyograph,** or **EMG** (Fig. 20.11). To learn how an EMG works, we need to back up first, and look at the structure of individual muscle cells, since these allow entire muscles to shorten and hence to move the skeleton.

Muscle Cells, Muscle Proteins, and the Mechanism of Contraction

The muscles of your arms, legs, and trunk—indeed all of your skeletal muscles—are made up of **muscle fibers:** giant cells that have many nuclei and may extend the full length of the muscle, up to several centimeters (Fig. 20.12a). If we took a small portion of a single muscle cell and enlarged it, we could see that the plasma membrane encloses several long bundles called **fibrils** (Fig. 20.12b). Each fibril is organized into units called **sarcomeres** (*sarco* = flesh). The sarcomeres in various fibrils line up and give the muscle cell a striped pattern leading to an alternate name for skeletal muscle: striated muscle. Surrounding each fibril is a network of modified endoplasmic reticulum, called the sarcoplasmic reticulum in muscle cells. Within each fibril are bundles of protein filaments (Fig. 20.12c). Mitochondria appear regularly between the fibrils. As we have seen, muscles work by shortening. This contraction happens because the sarcoplasmic reticulum acts like a car's electric system; the bundles of protein filaments act like the car's engine; and the mitochondria supply the energy system, the fuel for action.

Protein Filaments in Muscle Cells

The muscle protein filaments shown in Figure 20.12c actually do the job of contracting the muscle. From the dark line at the end of each sarcomere, numerous thin protein filaments project towards the middle of the sarcomere. These thin filaments are made mainly of many copies of the round protein **actin.** Towards the middle of the sarcomere lies a second type of filament, thicker than the first. These thick filaments are made of the golf club–shaped protein **myosin.** If you grab a bundle of seven pencils and push the eraser end of the middle pencil out about half way, then the middle pencil would be like the thick filament in the middle of the cell, and the surrounding six pencils would be like the thin filaments attached to the end plate. Figure 20.12c shows one thick and one thin filament.

The Mechanism of Muscle Contraction

How do these neat arrays of actin and myosin filaments allow muscle cells to shorten and help stop you from falling after slipping on a banana peel? Biologists observing skeletal muscle tissue with an electron microscope noticed that as the muscle contracts, each sarcomere within the muscle cell becomes shorter. This causes the whole fiber to shorten. The *filaments* within each muscle cell, however, do *not* shorten during contraction. Instead, as each sarcomere shortens, the filaments *slide past each*

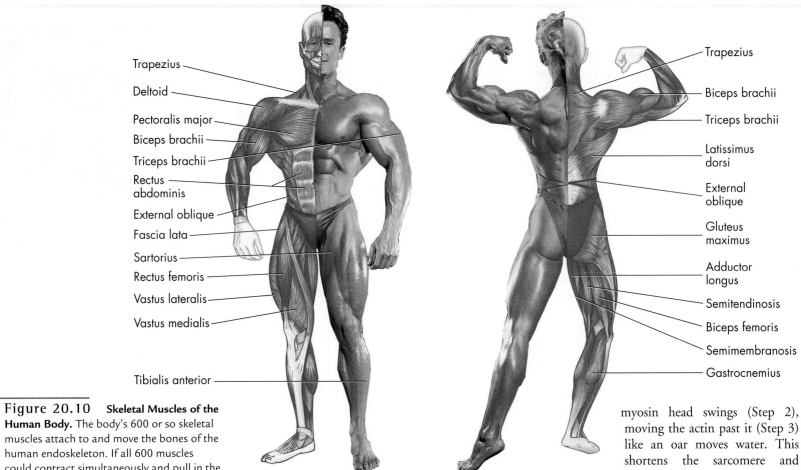

Trapezius

Deltoid

Pectoralis major

Biceps brachii

Triceps brachii

Rectus abdominis

External oblique

Fascia lata

Sartorius

Rectus femoris

Vastus lateralis

Vastus medialis

Tibialis anterior

Trapezius

Biceps brachii

Triceps brachii

Latissimus dorsi

External oblique

Gluteus maximus

Adductor longus

Semitendinosis

Biceps femoris

Semimembranosis

Gastrocnemius

Figure 20.10 **Skeletal Muscles of the Human Body.** The body's 600 or so skeletal muscles attach to and move the bones of the human endoskeleton. If all 600 muscles could contract simultaneously and pull in the same direction, they could lift 25 tons. In her slip plate experiments, Marjorie Woollacott monitors movement in the abdominals (rectus abdominis); the quadriceps (rectus femoris portion in the front of the thigh), the hamstrings (biceps femoris muscle at the back of the thigh, plus the semitendinosis and semimembranosus muscles), and the shin muscle (tibialis anterior).

other. To picture this, try thinking of a woman putting her hands into a furry muff on a cold day; her hands slide past the insides of the muff, her elbows come nearer together, but neither the length of her forearms nor the length of the muff changes. This process, going on right now in the muscles of your eyelids as they blink, your chest cavity as it expands with each breath, your hands as they hold this book, is called the **sliding filament mechanism** of muscle contraction (*E* xplorer 20.2).

This filament sliding is made possible by a unique arrangement and behavior of the proteins in the thin and thick filaments. Within each thick filament, many individual myosin protein molecules bind together, each with a long, straight tail at one end and a head at the other (Fig. 20.12c,d). The myosin head of the thick filament reaches out to a thin filament (made of actin) like an oar dipping into the water, and the two filaments attach temporarily (see Fig. 20.12d, Step 1). The

GET THE PICTURE

E xplorer 20.2

How a Muscle Contracts

myosin head swings (Step 2), moving the actin past it (Step 3) like an oar moves water. This shortens the sarcomere and hence the entire muscle a tiny bit. The myosin head then releases the thin filament, swings back to its original position, and binds the thin filament again, this time in a different place, like a rower's oar taking another stroke. The process of reaching, attaching, swinging, and detaching happens over and over for each myosin molecule as the muscle contracts, literally sliding the actin thin filament past it.

Energy for Muscle Contraction

Not surprisingly, just as a rowing crew must eat heartily the night before a regatta, the reaching and swinging of myosin "oars" also requires energy. In muscle fibers at rest (Fig. 20.12d, Step 1), the myosin head is bound to ADP and a phosphate. After the power stroke (Step 2), the myosin head releases the ADP and phosphate (Step 4), and an energy packet, ATP, binds to the myosin head (Step 5). The binding of ATP causes the myosin head to release the thin filament and cleave the ATP to ADP and phosphate (Step 6). This delivers a burst of energy which powers the return of the myosin head to its original position, where it can bind to the thin filament once more (Step 7).

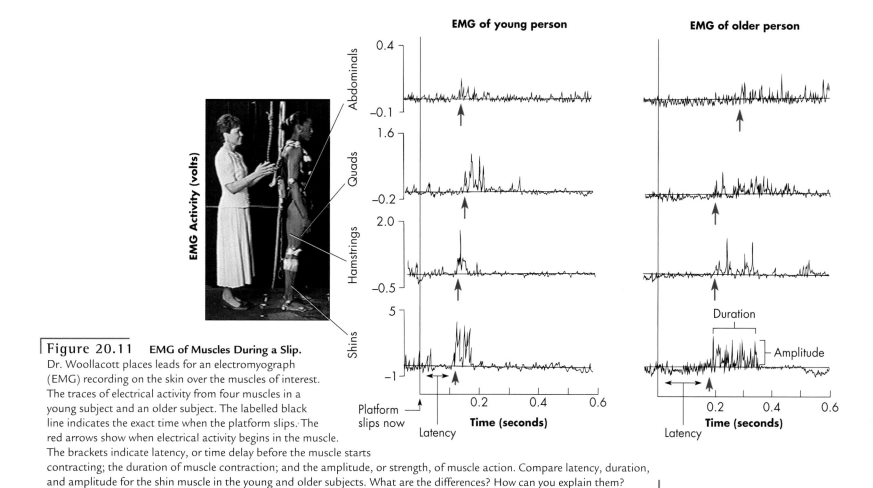

Figure 20.11 **EMG of Muscles During a Slip.**
Dr. Woollacott places leads for an electromyograph
(EMG) recording on the skin over the muscles of interest.
The traces of electrical activity from four muscles in a
young subject and an older subject. The labelled black
line indicates the exact time when the platform slips. The
red arrows show when electrical activity begins in the muscle.
The brackets indicate latency, or time delay before the muscle starts
contracting; the duration of muscle contraction; and the amplitude, or strength, of muscle action. Compare latency, duration,
and amplitude for the shin muscle in the young and older subjects. What are the differences? How can you explain them?

The mechanism of sliding protein filaments explains how muscles shorten, and muscle shortening explains how Mary Walrod can catch her balance on the
laboratory slip plate. The sliding mechanism does not,
however, explain how her muscles "know" *when* to contract and *how much* to contract to prevent a fall. That
requires signals from cell membranes.

Controlling Muscle Contraction

Muscle contraction enhances an animal's survival only
if that contraction is controlled and takes place at the
appropriate time. If Mary Walrod's arms and legs
started to react and contract *before* she got to the slip
plate, she could fall—or at least "almost fall"—into the
harness. The normal control over muscle contraction is
exerted by electrical events—in fact, the very events Dr.
Woollacott detects with her electrodes (see Fig. 20.11).
Motor neurons from the spinal cord run to and connect
with all our skeletal muscles (Fig. 20.13a). The motor
neuron releases a neurotransmitter substance onto the

cell membrane of the muscle cell. This generates an action potential in the muscle cell that is similar to the
one in a nerve cell (review Fig. 19.4). This signal causes
sarcoplasmic reticulum to release calcium into the cell's
cytoplasm. The calcium causes a blocking protein to
"get out of the way," thus allowing the myosin head to
associate with the thin filaments and hence start the
contraction cycle shown in Figure 20.12d. When the
muscle's action potential passes, ATP-fueled pumps
bring the calcium ions back into the sarcoplasmic reticulum; without calcium, the myosin can no longer bind
and slide past the thin filaments, and so contraction
stops and the muscle relaxes.

Understanding the roles of ATP and calcium in
muscle contraction helps to explain rigor mortis, the
muscle stiffening that occurs from 3 to 24 hours after
an animal dies. Because cellular membranes start to leak
after death, calcium ions seep from the sarcoplasmic
reticulum. The increase of calcium ions in the cytoplasm allows the myosin head to bind actin. Because
ATP synthesis ceases after death, there is no ATP to

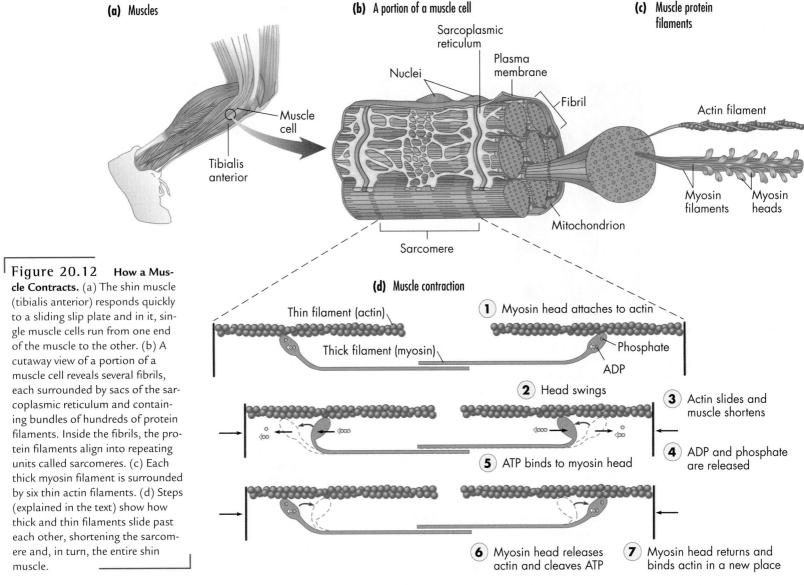

(a) Muscles

Muscle cell

Tibialis anterior

(b) A portion of a muscle cell

Sarcoplasmic reticulum

Nuclei

Plasma membrane

Fibril

Mitochondrion

Sarcomere

(c) Muscle protein filaments

Actin filament

Myosin filaments

Myosin heads

(d) Muscle contraction

Thin filament (actin)

Thick filament (myosin)

(1) Myosin head attaches to actin

Phosphate

ADP

(2) Head swings

(3) Actin slides and muscle shortens

(4) ADP and phosphate are released

(5) ATP binds to myosin head

(6) Myosin head releases actin and cleaves ATP

(7) Myosin head returns and binds actin in a new place

Figure 20.12   How a Muscle Contracts. (a) The shin muscle (tibialis anterior) responds quickly to a sliding slip plate and in it, single muscle cells run from one end of the muscle to the other. (b) A cutaway view of a portion of a muscle cell reveals several fibrils, each surrounded by sacs of the sarcoplasmic reticulum and containing bundles of hundreds of protein filaments. Inside the fibrils, the protein filaments align into repeating units called sarcomeres. (c) Each thick myosin filament is surrounded by six thin actin filaments. (d) Steps (explained in the text) show how thick and thin filaments slide past each other, shortening the sarcomere and, in turn, the entire shin muscle.

cause the myosin head to release the thin filament (Fig. 20.12, Steps 5 and 6) so the "oars" get "stuck in the water." As a result, myosin can't glide past actin, so muscles can neither contract nor relax. Instead, they remain stiff until the protein filaments themselves are degraded by enzymatic action.

How Muscles Work to Help Control Balance

We started the chapter with Mary Walrod, the sprinter Nicole Commissione, and the marathoner Billy Harper, each trying to restore their balance on Dr. Woollacott's slip plate device. We've seen that their balance depends on muscle contraction; that filaments sliding within muscle cells bring about this contraction; that motor

neurons control the start and end of a contraction cycle; and that an EMG can detect electrical activity in the muscles. Specifically, the electrodes glued to a subject's limbs and trunk can detect the huge electrical signal given off when an action potential ripples across a large muscle. As we saw, the signal is measured in volts, and when the trace goes up, it means that the muscle has responded to the slip. Looking at the EMG signals from various subjects, we can now understand the different strategies that older people like Mary use to catch themselves after a slip compared to younger people like Nicole or Billy.

On the graphs in Figure 20.11, you can see the time it takes for a person's shin muscle to begin to respond to Dr. Woollacott's slip plate. This time delay is called the **latency.** What is the difference in latency be-

Figure 20.13 — Skeletal Muscle, Cardiac Muscle, and Smooth Muscle Compared.

Figure 20.13 **Skeletal Muscle, Cardiac Muscle, and Smooth Muscle Compared.** (a) Several skeletal muscle fibers can contact together (a motor unit) when branches of the same motor neuron (shown in purple) contract them all. The more motor units that fire, the stronger the contraction. (b) Cardiac muscle cells have junctions (intercalated disks) that allow electrical communication between adjacent cardiac muscle cells, coordinating their contraction. (c) Smooth muscle cells often have electrical connections (gap junctions) that allow communication and coordination; they tend to contract in sequential waves.

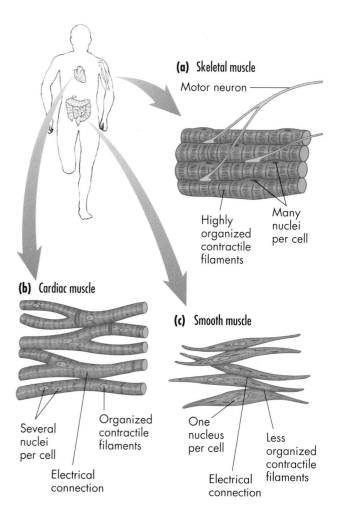

(a) Skeletal muscle
Motor neuron
Highly organized contractile filaments
Many nuclei per cell

(b) Cardiac muscle
Several nuclei per cell
Organized contractile filaments
Electrical connection

(c) Smooth muscle
One nucleus per cell
Less organized contractile filaments
Electrical connection

tween the younger and older person's shin muscles? How about the strength (or amplitude) of the response, which is revealed by the height of the tracing? The results showed that the older person responds later and with less strength than the younger person. To make up for this slower, weaker response, what happens to the duration of the muscle electric signal—the length of time it continues to stimulate the shin muscle? The graph shows that the older person compensates for the late, weak response by contracting for a longer period of time.

You can use this same graph to compare the order in which people's muscles in different parts of the body come into play during a slip (Fig. 20.11). Do the muscles start contracting high in the body and move down, or do they start low in the body and then move up? Do the younger and older subjects use their muscles in different orders during the slip plate experiments? In both age groups, people tend to start low and move up. However, infants who are learning to walk and are at the pull-up-and-stand stage use a very different balance-keeping strategy from the same children a few months later in the toddling stage.

Cardiac Muscle and Smooth Muscle

The mechanisms we just discussed for contracting, powering, and controlling apply to skeletal muscles that attach to bones and move the skeleton (Fig. 20.13a). Animals with backbones (vertebrates), however, have two additional types of muscles that move fluids rather than bones. Cardiac muscle, found only in the heart, drives blood through the circulatory system, while smooth muscle pushes food through the digestive tract and causes the urinary bladder, uterus, and blood vessel walls to contract.

Cardiac muscle consists of a network of heart muscle cells that are electrically connected to each other (review Fig. 15.7). These electrical connections allow adjacent cells to communicate back and forth (Fig. 20.13b). An impulse initiated any place on the heart muscle propagates through the entire organ, causing the whole heart to contract as a unit. The impulse usually begins inside the heart itself, in its pace-

maker (review Fig. 15.9), not in nerves running to the heart. This quality allows the heart to keep beating during a heart transplant even though the vital organ is temporarily removed from the body.

In contrast to cardiac muscle, skeletal muscle fibers do not communicate with each other directly. Several skeletal muscle cells may contract at the same time because they are innervated by the same motor neuron coming from the spinal cord. This contractile group is a **motor unit** (Fig. 20.13a). The more motor units that fire, the more forceful the muscle contraction. Like skeletal muscle, cardiac muscle is striated, or striped, because of the way its thin and thick filaments are organized.

The **smooth muscle** cells of the digestive tract, urinary bladder, blood vessels, and other hollow body organs lack striations because their thick and thin filaments are not as well ordered as those in skeletal and cardiac muscle (see Fig. 20.13c). Smooth muscle cells usually have a single nucleus, and many communicate with other smooth muscle cells electrically via gap junctions, as do cardiac muscle cells. This communication allows the rhythmic pushing of food down the digestive tract, the voiding of urine, the pushing of a baby through the birth canal, or the maintenance of blood pressure.

We've seen now how muscles and skeleton act together to keep balance and how smooth muscle and heart muscle propel other organs. Now let's see how major muscle groups act together when we do something fun, like play baseball or tennis or swim. At the same time, we'll see how the type of sport activity you choose can ultimately affect your body tone and your ability to stay in balance.

20.3 The Physiology of Sport and Exercise

Sport is big business—in colleges, in city stadiums, and on radio and television. Although owners spend billions of dollars promoting their teams, comparatively little funding is available for rigorously controlled experiments to discover how to improve athletic performance in people and other animals. The research that has been done, however, is beginning to present a clearer picture of how our muscles are fueled, about the body changes we can expect from training, and about the epidemic of diseases related to inactivity in our society. Let's explore some of those results now.

Why would a marathon runner have a slip plate response like an older adult's?

Getting Fuel to the Motor

While thick and thin filaments cause muscle cells to contract, we saw that the fuel for muscle cell contraction is ATP. We saw in Chapter 3 that mitochondria and certain enzymes in the cytoplasm are the powerhouses that make ATP and therefore provide cells with fuel. These cell parts cooperate in muscle cells to establish three energy systems: (1) the immediate system, (2) the glycolytic system, and (3) the oxidative system.

The duration of physical activity dictates which energy system the body uses at any given time. The **immediate energy system** is instantly available for a brief explosive action, such as one heave by a shot-putter. It depends on a muscle cell's stores of ATP plus the high-energy molecule **creatine phosphate,** and it can fuel muscle contraction for several seconds. The second energy system, the **glycolytic energy system,** is based on the splitting of glucose by glycolysis in the muscles. The glycolytic system can sustain heavy exercise for a few minutes, as in a 200-m swim. After that, the **oxidative** (or **aerobic**) **energy system** takes over for long periods of exercise. When the distance runner Billy Harper runs a 10-kilometer race, his muscles use the oxidative system throughout most of the race, until the final sprint. If enough oxygen is present, this aerobic energy system can produce energy by breaking down carbohydrates, fatty acids, and amino acids mobilized from other parts of the body. Clearly, anyone interested in melting away body fat should engage in aerobic activities like rapid walking, swimming, bicycling, cross-country skiing, or jogging, which rely primarily on the oxidative energy system and its ability to use fats as fuel.

Slow-Twitch Muscle Fibers

Not all muscle fibers use the three energy systems equally. We saw in the chapter introduction that long-distance runners have more slow-twitch muscle fibers while sprinters have more fast-twitch fibers. Here's what all of that means. **Slow-twitch muscle fibers** (also called slow oxidative muscle fibers) obtain most of their ATP from the oxidative system (Fig. 20.14a). Slow-twitch fibers require about one-tenth of a second to contract fully, they are packed with mitochondria, they receive a rich supply of blood, and they have large quantities of the red protein **myoglobin,** which stores oxygen in muscle cells. These characteristics make slow-twitch fibers deep red, like the dark meat of a chicken (see Fig. 20.14c). Slow-twitch fibers are resistant to fatigue and thus are able to contract for long periods of time. Athletes, such as Billy Harper, who trained for endurance sports like running marathons have a large proportion of slow-twitch muscles. Slow-twitch muscles also provide functions critical for survival, such as maintaining posture. Without the contraction of slow-twitch muscles in your jaw, your mouth would be wide open as you read this sentence.

Fast-Twitch Muscle Fibers

Whereas slow-twitch fibers bestow endurance, **fast-twitch fibers,** or fast glycolytic muscle fibers, provide quick power—the kind of power needed for a 50-yard dash, for example, or one clean and jerk of a heavy barbell (see Fig. 20.14b). Fast-twitch fibers derive most of their ATP from glycolysis, and they reach maximum contraction twice as quickly as slow-twitch fibers. They soon grow fatigued, however, since they quickly run through their limited stores of ATP. Fast-twitch fibers are white (see Fig. 20.14c), because they are packed with white actin and myosin filaments (which maximize contractile force), and they contain very little myoglobin for oxygen storage. The white meat in the breast of a chicken is the fast-twitch muscle that powers the wings and enables the chicken to suddenly burst away from a fox and fly to safety in a tree. By contrast, a duck's breast is dark meat, consisting of red, slow-twitch fibers that can sustain the beating of the wings without fatigue during long migratory flights. A third kind of muscle fiber, **fast oxidative-glycolytic muscle fiber,** has characteristics midway between fast glycolytic and slow oxidative fibers; these fibers are moderately powerful and moderately resistant to fatigue.

Although fiber types are genetically determined to a large degree, training can partly change how the muscle

Figure 20.14 **Muscle Fibers and Sports Performance: Slow-Twitch and Fast-Twitch.** (a) A person trained for aerobic endurance events like cross-country skiing or long-distance running is more likely to have a high proportion of slow-twitch fibers. (b) A person highly trained for, and naturally good at, weightlifting, shot-putting, or sprinting is more likely to have a high proportion of fast-twitch muscle fibers. (c) Biologists use a stain that darkens tissue containing the ATP-splitting myosin head; this makes the myosin-rich fast-twitch fibers in a cross section of muscle tissue appear light-colored, while the slow-twitch fibers appear as dark patches.

(a) Cross country skier: slow-twitch muscles, oxidative energy system

(b) Weight lifter: fast-twitch muscles, glycolytic energy system

(c) Fiber types differ in their concentration of myosin

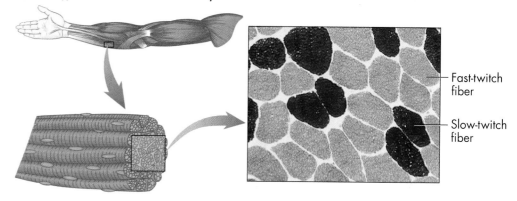

Fast-twitch fiber

Slow-twitch fiber

fibers function. For example, endurance training can cause both fast and slow fibers to develop an increased oxidative capacity but a reduced explosive strength. Clearly, a sprinter would want to avoid that type of training. In contrast, strength training improves immediate energy supply systems and suppresses oxidative capacity. Although it frustrates some body builders, the total *number* of muscle fibers is apparently genetically determined and is not affected by weight training. Lifting can, however, increase fiber *thickness* by increasing the number of thick and thin filaments in muscle cells.

One of Dr. Woollacott's graduate students hypothesized that two kinds of athletes with different fiber types—mainly fast-twitch versus mainly slow-twitch—should have balance recovery strategies that look more like young people and older people, respectively. So she tested college-age athletes like Nicole and Billy, trained for either sprinting or distance running. She found that indeed, the sprinters (with more fast-twitch fibers) were like most young people in her tests, but that the distance runners (with more slow-twitch fibers) were more similar to older subjects, although not quite as extreme. She concluded that muscle fiber type plays a role in balance recovery strategy, but that other factors must come into play, as well. This information has been useful to physical therapists planning exercise programs to strengthen older people and help them prevent falls. What kinds of exercise, in general, do you suppose they recommend?

Athletic Training and the Stress Response

For most animals, survival is linked to locomotion, and in nature, locomotion often must be rapid and coordinated if the animal is to survive. Heavy exercise, strenuous work, and the intensity of a basketball game or sprint workout stimulates the same fight-or-flight response that electrifies the escape of a field mouse from a diving hawk.

Recall our discussion of stress in Chapter 18. Stress initiates a cascade of hormone signals that raises the amount of vital fuels like glucose and fatty acids in the blood; elevates heart rate and blood pressure; dilates air passages and increases breathing rate; and diverts blood from the skin and digestive organs to the skeletal muscles, supplying food and oxygen where they are needed the most (review Fig. 18.5). A strenuous athletic training session does the same things. As a result, the system adapts. Consider Miguel Indurain, the Spanish winner of the grueling Tour de France bicycle race for several years in the early 1990s. At rest, his heart beat at a remarkably low 28 beats per minute. To put this rate in perspective, compare it to your own by counting the pulse in your wrist for 10 seconds and multiplying by 6.

Chances are, your resting heart beats at least twice as fast as Indurain's heart. During a time trial—an all-out race against the clock—Indurain's heart rate soared to 195, a rate faster than most of us could possibly achieve. Exercise mimics the fight-or-flight response. It calls into action all the major physiological systems of the body.

How Athletic Training Alters Physiology

Since heavy exercise evokes the fight-or-flight response, people who exercise several times a week enter this state of stress repeatedly. What effects do such repeated periodic challenges have on the body's homeostatic mechanisms? Training works by causing "breakdown" followed by "overshoot"—a breakdown of stored fuel, for example, followed by the increased deposition of fuel molecules; or a slight breakdown of muscle tissues, with an overall strengthening as the tissues are repaired. To increase fitness without injury, therefore, one must carefully, gradually, and *progressively* augment the *intensity, frequency,* and *duration* of workouts.

One goal of human athletic training is to boost the amount of oxygen a person can deliver to working muscles (the so-called maximal oxygen uptake). What changes occur after training that increase one's ability to take up oxygen? After training, hearts pump more blood per minute than before. This is due to an increase in the amount of blood pumped per beat, not an increase in the heart rate. Well-trained endurance athletes have enlarged heart cavities and thicker heart walls.

In addition, physically fit people can better use the increased oxygen pumped by their larger, stronger hearts because their muscles extract more oxygen from the blood. The muscle cell mitochondria grow larger and produce more ATP; the muscles build more myoglobin, which stores more oxygen; and there is an increase in the number of blood capillaries in the muscles. Together these phenomena provide the muscles with a greater oxygen supply.

Evolution has brought about permanent physiological changes in various animals that are similar to those in exercising people. Take, for example, the pronghorn antelope—the animal world's champion long-distance runner and a denizen of high-altitude prairies in Wyoming, Idaho, Nevada, and Oregon (Fig. 20.15). Pronghorn antelopes can run at an average speed of 64 kilometers per hour (40 mph) for half an hour or longer. To understand what sorts of physiological tricks may have evolved in the pronghorns, physiologists trained the animals to run on a treadmill wearing face masks connected to a gas analyzer. They found that the pronghorn's maximal oxygen uptake was five times that of goats and other related mammals of comparable size. Instead of uncovering some unique physiological attributes, however, the researchers found that pronghorns simply have somewhat larger lungs, slightly more of the oxygen-binding protein hemoglobin in their blood, and a bit more mitochondria in their muscle cells. In other words, natural selection—probably in the form of preying wolf packs—has fine-tuned the pronghorn into a world-class endurance athlete. In Chapter 3, we saw the same phenomenon in hummingbirds, with their highly packed mitochondria in flight muscle cells.

Anyone still needing a nudge to begin or continue a regular exercise program can expect yet another benefit from it. Exercise stimulates the release of **endorphins,** naturally occurring morphines made in the pituitary gland and other brain regions. These peptide hormones reduce pain and enhance a feeling of well-being. The release of endorphins during strenuous exercise may explain, in part, why many people experience feelings of relaxation and contentment after a workout. Add to this the dramatic positive effects on the heart and on a person's chances of living a longer, more vigorous life, and it becomes more and more difficult to justify a sedentary life style.

Exercise and Diseases of Inactivity

Despite publicity about the health benefits of exercise, many people in industrial nations have adopted sedentary life styles: riding to work in a car, sitting at a desk or workbench all day, and reading the paper or watching television at night. Our softer life styles have led to

Figure 20.15 **The Pronghorn: The Animal World's Champion Long-Distance Runner.** Antelopes have no special anatomical tricks, despite their ability to run up to 20 miles in half an hour. Instead, natural selection has simply resulted in the optimization of their oxygen utilization system, their muscular system, and their fuel delivery. This pronghorn antelope (*Antilocapra americana*) is running across its home in Montana's National Bison Range.

an increase in **hypokinetic disease,** maladies caused by "too-little motion." Low back pain, obesity, and heart disease can all result from sedentary life styles, although family heredity, injuries, and infections can also contribute to these ills. In general, however, the incidence and severity of these problems can be lessened by activating muscles, bones, the heart, lungs, nerves, and hormones with regular aerobic exercise.

Available evidence suggests that a physically inactive person is about twice as likely as an active one to die of heart disease. It is ironic that many people recognize the importance of diet and exercise to their pets' health, but fail to apply the same principles to their own human bodies. This inspired Swedish exercise physiologist Per Olof Astrand to advise sedentary adults to "Walk your dog whether you have one or not."

Chapter Summary and Selected Key Terms

Introduction   Dr. Marjorie Woollacott studies the muscles and bones involved in restoring balance after a slip on an "electric banana peel." Researchers like her help people prevent falls—a leading cause of death in the elderly—and plan for better exercise programs throughout life.

The Skeleton   The **skeleton** (p. 513) is a rigid body support to which muscles attach and apply force. Many invertebrates have **hydroskeletons** (p. 513) with a liquid core wrapped in a muscular sheath. A sea anemone has a typical water skeleton with two layers of muscle, one extending the animal's length and the other encircling it. The layers act as **antagonistic muscle pairs** (p. 513) moving the same object (such as the animal with its enclosed liquid core) in opposite directions.

Arthropods have **exoskeletons** (p. 514), or stiff external coverings surrounding the body. Vertebrates have **endoskeletons** (p. 514), or rigid internal rods and plates surrounded by muscles and skin. A mammal's endoskeleton is made of **bone** (p. 514), with its meshwork of collagen hardened by calcium and phosphate. The bones of the **skull** (p. 514) enclose and protect the brain and sense organs. The backbone (**vertebral column**; p. 514) is made up of **vertebrae** (p. 514), or boxlike units that are stacked and flexible and support the trunk and encase the spinal cord. **Shoulder (pectoral)** and **hip (pelvic girdles)** (p. 514) are branch-off points that support arm and leg bones and allow them to swing. **Processes** (p. 514) provide attachment sites for muscles.

A bone like the **femur** (p. 515) forms a joint, or articulation with other bones and can have regions of **compact bone** (p. 515) and **spongy bone** (p. 516). Attached to bones are **ligaments** (p. 515), or connective tissue bands linking bone to bone. **Tendons** (p. 515) connect bone to muscle. **Marrow** (p. 516) in the bone's center is the site of blood cell production. Some specialized bone cells secrete proteins and minerals, while others reabsorb the minerals for use elsewhere in the body when needed. Bone remodeling goes on continuously. When mineral-reabsorbing cells work faster than mineral-storing cells, bone thinning and weakness, called **osteoporosis** (p.516), can occur.

Bone movement requires a **joint** (p. 517), where adjacent bones nearly touch but are separated by shock-absorbing cartilage pads. Bones pivot at joints because muscles pull on them. **Antagonistic pairs** (p. 518) of muscle relax and contract to move a body part as in pointing or flexing the foot. Walking, for example, requires many sequential muscle actions.

Muscles   **Skeletal muscle** (p. 518) connects to and moves the skeleton, with an **origin** (p. 518) on one bone and an **insertion** (p. 518) on another. Several muscles are needed to maintain and restore balance and to carry out physical activities. Skeletal muscles are made up of **muscle fibers** (p. 519), giant cells with many nuclei that extend the length of the muscle. A contractile mechanism within each fiber allows it to shorten. Thin filaments made of the protein **actin** (p. 519) slide past thicker filaments made of the protein **myosin** (p. 519). The plasma membrane encloses bundles called **fibrils** (p. 519), organized into units called **sarcomeres** (p. 519). The sliding depends on the structure of the actin and myosin proteins and involves reaching, attaching, swinging, and detaching again and again. This **sliding filament mechanism** (p. 520) of muscle contraction requires ATP energy. Motor neurons running from the spinal cord to all our skeletal muscles release neurotransmitters that stimulate an action potential in the muscle cell and the start of a contraction cycle. When a person slips, as on an "electric banana peel," there is a delay (called latency) before muscles, bones, and limbs can respond and restore balance. The same motor neuron from the spinal cord innervates a motor unit or contractile group. Skeletal muscle appears striated or striped because of the organization of thick and thin filaments.

Cardiac muscle (p. 523) contains networks of heart muscle cells electrically connected to each other. An impulse to contract propagates through the entire organ because of these connections. Cardiac muscle is also striated. **Smooth muscle** (p. 523) surrounds hollow organs like the intestines or urinary bladder and lacks striations. The cells communicate with each other electrically, like cardiac muscle cells, and contract rhythmically, pushing food down the digestive tract, for example. Several skeletal muscle cells may contract together as a **motor unit** (p. 523).

Sports and Exercise Physiology   The **immediate energy system** (p. 524) provides power for one brief explosive action based on ATP and **creatine phosphate** (p. 524). The **glycolytic energy system** (p. 524) can power heavy exercise for a few minutes. The **oxidative energy system** (p. 524) sustains an exerciser for long periods. **Slow-twitch muscle fibers** (p. 524) get most of their ATP from the oxidative system. Marathoners have a large proportion of slow-twitch fibers. **Fast-twitch muscle fibers** (p. 524) provide quick power and get most of their ATP from glycolysis. **Fast oxidative-glycolytic muscle fibers** (p. 524) have in-between characteristics and are moderately powerful and moderately enduring. Dr. Woollacott's research group found that older people slip and fall more because the proportion of fast-twitch fibers decreases with aging and they react more slowly to unstable footing. Marathoners have a similar pattern, but sprinters recover very quickly and tend not to lose balance.

Exercise is a beneficial form of stress, taxing your body's major physiological systems and helping to build strength and endurance. Appropriate training (regulating the intensity, frequency, and duration of the stress) increases fitness. Inactivity can lead to low back pain, obesity, heart disease, and other health problems.

All of the following question sets also appear in the Explore Life **E** electronic component, where you will find a variety of additional questions as well.

Test Yourself on Vocabulary and Concepts

In each question set below, match the description with the appropriate term. A term may be used once, more than once, or not at all.

SET I

(a) exoskeleton (b) endoskeleton
(c) hydroskeleton (d) more than one of the above

1. Muscles attach to a rigid tissue.
2. Acts like a water balloon
3. The vertebral column, ribs, sternum, and skull
4. Most commonly made of chitin and found in arthropods
5. General skeletal category that describes a braced framework made of bone

SET II

(a) muscle fiber (b) fibril (c) sarcomere
(d) myosin (e) actin (f) more than one of the above

6. The fundamental unit of contraction in a fibril
7. Muscle filaments made of protein
8. Another name for a whole muscle cell
9. An organized array of thin and thick filaments in a muscle cell
10. Muscle filaments that have a movable head region

Integrate and Apply What You've Learned

11. What are several functions of bone tissue?
12. How can an earthworm crawl when it has no bones on which muscles can attach?
13. What are antagonistic muscles and why are they important?
14. Contrast actin filaments with myosin filaments. Explain their roles in muscle contraction.
15. One might think that ATP would be spent in the contraction step that causes the myosin heads to squeeze the actin fibers closer together, but he or she would be wrong. In which step is ATP spent? Present an hypothesis for why the mechanism might work the way it does.
16. Contrast the structure and location of skeletal, cardiac, and smooth muscle fibers.
17. Why is a marathoner's experience on Dr. Woollacott's slip plate more like an older subject's such as Mary Walrod than like a sprinter's?
18. Why is it important to stay active and keep exercising throughout life?

Analyze and Evaluate the Concepts

19. Explain what happens to the muscles and bones of your leg when you sit on a stool or chair and lift your calf and foot until they are parallel with the floor.
20. A skeptical student says it's impossible for the sarcomere to contract without the actin and myosin filaments becoming shorter. Build a toothpick model to demonstrate that this actually happens.
21. An ice skater fell and damaged connective tissue that joins his hip bone to his femur. His teammates told him he had tendonitis, but he was positive he didn't. Why was he so sure?
22. An agricultural worker accidentally swallows an insecticide that acts by blocking the transfer of a signal from nerves to muscles. Her skeletal muscles become limp, but her heart muscle continues to contract rhythmically. How can you explain those symptoms?

PART 4 The Living Plant

A Tangled Yarn

Hoping for a New Cash Crop. Gordon Scheifele; Geof Kime; *Cannabis sativa,* low THC variety.

Will hemp be an important crop for the 21st century? A small group of Canadian farmers, researchers, and plant processors think so. And they've already convinced some icons of the fashion and auto industries that they're right.

Most people know hemp only as the illegal marijuana plant. However, hemp *(Cannabis sativa)* has two forms: **marijuana,** with its high levels of a mind-altering chemical called **THC** (tetrahydrocannabinol), and industrial hemp, with its extremely low levels of THC but an abundance of strong plant fibers. Most people are also unaware that farmers have been growing hemp for nearly 5000 years. The tall, lanky stalks, in fact, have been a major source of strong plant fibers for twine, rope, nets, soft fabrics, canvas, and paper all over the world. For instance, Rembrandt and Van Gogh painted on hemp canvas. Thomas Jefferson, George Washington, and Henry Clay grew fields of hemp. Benjamin Franklin milled paper from

it. And the first two drafts of the Declaration of Independence were written with quill pens on hemp paper.

For centuries, people on several continents ate oily hemp seeds in rough breads and cakes and pressed the seeds for cooking oil and paint solvents. Herbalists used drug compounds from the seeds, flowers, and leaves to deaden pain and induce sleep. And in modern times, millions of illegal drug users have smoked or eaten these same plant parts trying to get a euphoric or hallucinogenic effect. Many of them have gotten something entirely different: impaired memory, reduced driving ability, and declining academic performance, as well as psychological drug dependency.

Biochemists in the 1960s and 1970s figured out that *Cannabis* users were actually seeking the plant's own protective alkaloid, THC, which it produces in response to insect and bacterial attacks, drought, heat, and other stresses. Ironically, THC has been the undoing of the North American hemp fiber industry, at least until recently. In the 1930s, American and Canadian lawmakers banned the growing of *Cannabis sativa,* and for over 60 years, the hemp industry has been, in the words of Canadian plant researcher Gordon Scheifele, a "sleeping giant." Some plant breeders, working entirely illegally, produced marijuana hemp with very high levels of THC. Other legitimate plant breeders in Europe and Asia, however, continued during those decades to develop new hemp fiber varieties that make almost no THC (less than 0.3 percent). Government agronomists like Gordon Scheifele kept an eye on this work because Canadian farmers are always in need of alternative crops with potentially large export markets. Soybeans, long grown in Asia but introduced to North America in the mid-20th century, became just such a "new" crop. It seemed to Scheifele and several colleagues that, given the right political climate, low-THC hemp might be the soybean of this new millennium.

Enter two new players: Joe Strobel, a retired schoolteacher and farmer in southern Ontario, searching for a replacement for his acres of tobacco, and his friend Geof Kime, an engineer looking for a technological challenge. In 1994, with help from Scheifele and other government researchers, the two petitioned a Canadian agency to let them grow a few acres of very-low-THC hemp as a test. They also urged the Canadian legislature to take the mature plant stalks and fibers off the list of controlled substances, and in 1998, this became law. Strobel and Kime incorporated a business, Hempline, to grow, pick, and process hemp stalks. And Kime invented machinery that could handle the tough hemp fibers without the clogging and tangling that usually brings harvesters, balers, and other traditional farm machinery to a dead stop.

In 1998, Hempline started filling orders for paper and textile mills. These included the fabric sources for fashion magnates like Calvin Klein and Ralph Lauren, with their newly introduced hemp clothing lines. Hempline has supplied producers of garden compost and animal bedding, and, notes Scheifele, "Queen Elizabeth's horse stables are now using the bedding material." Hempline has even exported hemp fibers to major automakers in Detroit. Pressed and molded hemp fibers turn out to make a cheap, strong, ecologically friendly substitute for fiberglass and plastic in car parts like dashboards, door panels, and sound-deadening mats.

Hemp growing is an apt case history for this chapter's two topics: **plant anatomy**—the main parts and tissues of the plant body—and general **plant function**—how the parts work together in the living organism. In typical land plants, we will encounter an outer protective layer; then a system of internal tubes that carry water and nutrients; and finally a central system of ground tissue that stores materials and provides support. Which tissue types make up the hemp plant's strong fibers and pulpy stem tissue? Where are they and what do they do for the plant? What are they made of? How can a farmer get better quality fibers? Which hemp tissues generate THC and why? How can biologists decrease natural levels of THC and increase fiber production? And how do hemp plants grow up to 20 feet tall in a single summer? You'll find information about all these issues here, as well as answers to these basic questions:

1. What are the main parts of vascular plants?
2. How do plants grow?
3. What are the unique features of plant roots?
4. What are the features of plant shoots?
5. What are the structures, functions, and uses for various parts of the hemp plant?
6. How likely is it that plant breeders will be able to breed the drug effect completely out of *Cannabis sativa* and produce hemp with more and better fibers?

21.1   Vascular Plants

Cannabis sativa is just one species among nearly half a million vascular plants (Fig. 21.1). In Chapter 12, we discussed both nonvascular plants (such as mosses and liverworts) and vascular plants (such as pines, chiles, and hemp). Here we concentrate on the latter group, which is Earth's biggest and most dominant collection of autotrophs. As a group, vascular plants are distinguished by their **vascular tissue,** or internal transport tubes, an important adaptation for life on land. One type of these transport tubes, for example, is supported by the soft, flexible hemp fibers that Gordon Scheifele hopes will become a major Canadian agricultural export. Hundreds of thousands of vascular plant species dominate the terrain on most continents and include seedless vascular plants (e.g., ferns, cycads, and ginkgoes) and the seed plants, the conifers and flowering plants (Fig. 21.1). Seed plants have two adaptations in addition to their internal transport tubes: they have sexual reproduction with internal fertilization (some nonvascular plants have this, as well) and they produce a tough, protective seed coat around the developing embryo. We saw very similar innovations in the

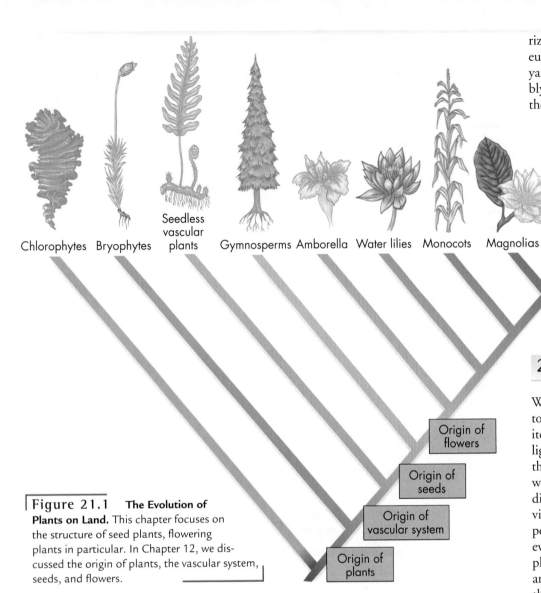

rizes some of the differences between monocots and eudicots, the plant types most of us see in our homes, yards, parks, fields, and woods every day. You'll probably refer back to this figure as you read about the various plant structures in this chapter. In each case, the organisms have the same general architecture. We start with a general orientation on the plant body, followed by an overview of how plants grow, since the two topics are closely connected. Then we'll look in more detail at two main parts of a plant, the shoot and the root, discussing the specific plant parts in each body region and how they grow.

21.2 The Plant Body

We saw in Chapter 11 that green algae were the ancestors of land plants, and these simple organisms inhabited shallow ocean waters flooded with sunlight. The light supplied a power source for internal reactions, and the sea supplied the algae with other needed resources: water, dissolved mineral nutrients, dissolved carbon dioxide (from which algae generated organic molecules via photosynthesis), dissolved oxygen, and buoyant support for the sometimes large body. The land plants that evolved from these aquatic ancestors have the same physical requirements for light, water, nutrients, gases, and support. However, they obtain these resources from the soil and air, not seawater. A plant's roots generally extend into the soil, and from it they take in water and mineral nutrients. Soil provides no direct access to light, however, and thus root cells have no direct way of manufacturing their own food by photosynthesis. In contrast, a plant's stem and leaves invade the air, and absorb carbon dioxide and light, allowing photosynthesis to occur. Air, however, provides no mineral nutrients, very little water, and no physical support. The various structures typical of land plants supply these conflicting needs.

The Plant Axis: Shoot and Root

The realities of the physical environment on dry land dictate the form and function of a typical vascular land plant with its shoot in the air and its root in

Figure 21.1 **The Evolution of Plants on Land.** This chapter focuses on the structure of seed plants, flowering plants in particular. In Chapter 12, we discussed the origin of plants, the vascular system, seeds, and flowers.

internal fertilization and leathery eggshells that evolved in reptiles. The nonflowering seed plants (such as pine and fir trees) are the gymnosperms, and the flowering plants, such as maples, chiles, and hemp) are the angiosperms.

Recent molecular genetic analysis reveals that flowering plants include several quite distinct lineages: Some species arose very early in plant evolution like *Amborella* (a shrub from New Caldonia), and the many kinds of water lilies. Angiosperms also include **monocots** (or Monocotyledons, MON-oh-cot-il-EE-dons), the magnolia-related plants; and the **eudicots** (YOU-die-cots or "true dicots"). Monocots include corn, bamboo, Bermuda grass, true lilies, and orchids. Eudicots (formerly called "dicots") include hemp, oak trees, peach trees, daisies, roses, and broccoli. Figure 21.2 summa-

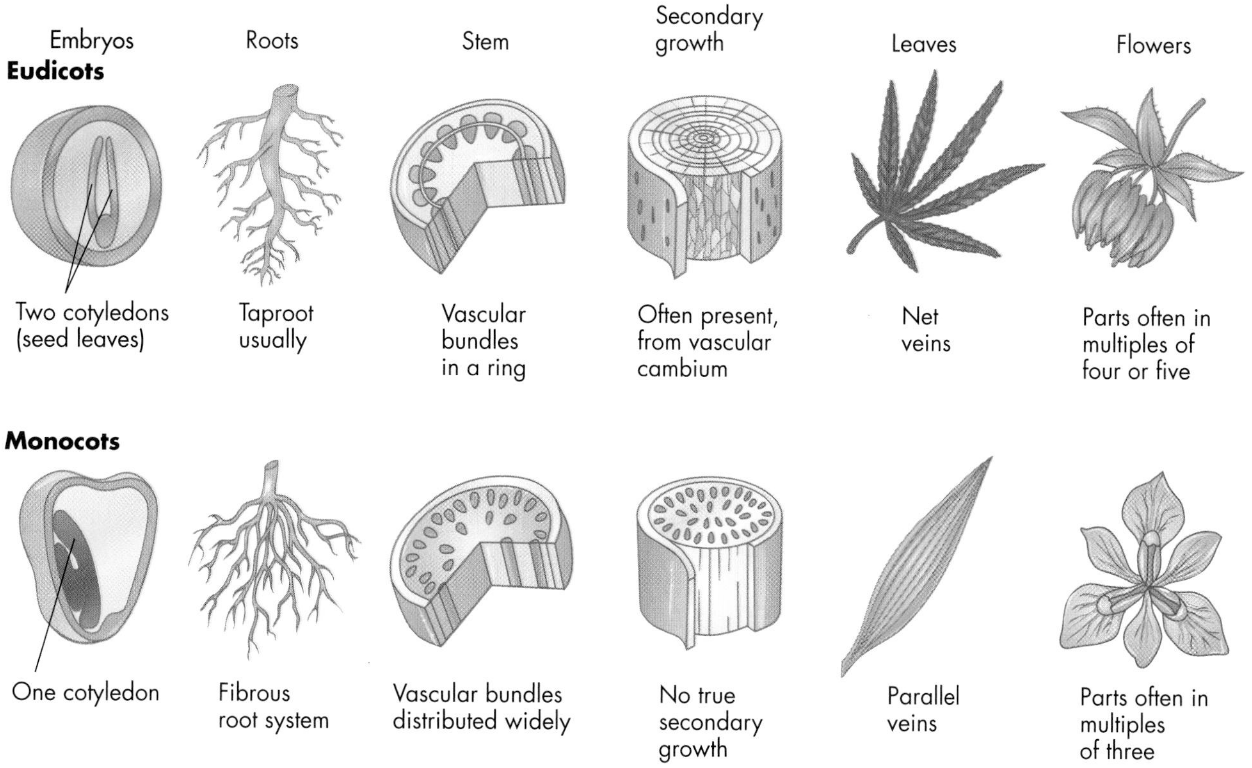

	Embryos	Roots	Stem	Secondary growth	Leaves	Flowers
Eudicots	Two cotyledons (seed leaves)	Taproot usually	Vascular bundles in a ring	Often present, from vascular cambium	Net veins	Parts often in multiples of four or five
Monocots	One cotyledon	Fibrous root system	Vascular bundles distributed widely	No true secondary growth	Parallel veins	Parts often in multiples of three

Figure 21.2 The Plant Structures of Monocots and Eudicots Compared.

the ground. Together, the shoot and the root make up the plant **axis,** its main—usually vertical—regions (Explorer 21.1).

The Shoot

The **shoot** includes the stem, branches, leaves, flowers, and fruits (Fig. 21.3). The shoot of a hemp plant, for example, can be 10 to 20 feet (3 to 6 m) tall. The **stem,** a stiff bundle of internal transport tubes plus surrounding tissues, supports the rest of the shoot. The stem is enclosed within an external waterproof coating that minimizes water loss. Hemp stems are only about an inch in diameter despite their very tall height. In trees and shrubs, the main stem matures into a large, woody, bark-covered trunk with many side branches supporting thousands of leaves and flowers.

The Root

Roots are branching organs that grow downward into the soil. Roots help support the plant physically by spreading out through the soil and anchoring the plant in it, and nutritionally by absorbing and transporting water and mineral nutrients that the plant cannot take in from the air. Hemp plants have a thick, trunklike **taproot** system that develops directly from the embryonic root and grows straight down into the soil. Lateral roots branch off the sides of the main taproot (Fig. 21.4a). Pear trees and carrot plants also have tap roots. Taproots store water, as well as food in the form of starch. In contrast, in grasses, including corn, and a few other kinds of plants, the embryonic root dies back early in the organism's life. In these plants, the roots that will support the adult plant grow downward and outward from the plant's stem, branching repeatedly and forming a **fibrous root** system (Fig. 21.4b). Fibrous roots store less starch than taproots, but efficiently anchor the plant and absorb water and nutrients. A different type of root emerges from above-ground stem tissue and grows through the air before reaching the soil. Called **adventitious roots,** these aerial roots can grow to enormous proportions (Fig. 21.4c). If you have visited Honolulu, Hawaii, you may have shopped in the International Market. This entire shopping mall is contained within the adventitious roots of a single enormous banyan tree! Other adventitious roots include the thick prop roots of a corn plant and the narrow structures that allow ivy plants to cling to walls and fences.

GET THE PICTURE

Explorer **21.1**

The Plant Axis: Cells and Tissues

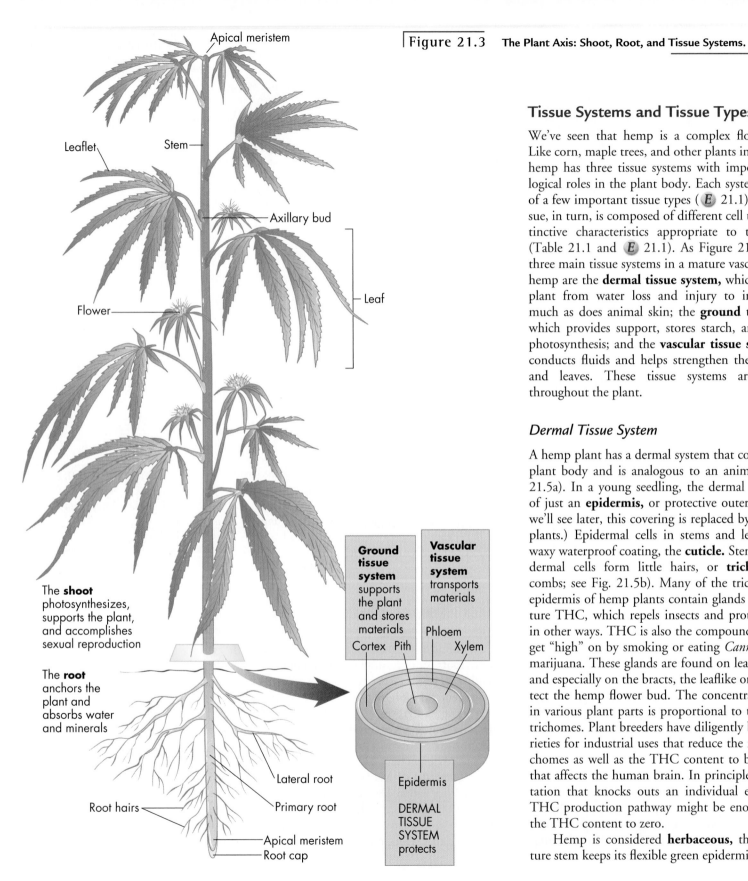

Apical meristem

Leaflet

Stem

Axillary bud

Leaf

Flower

The **shoot** photosynthesizes, supports the plant, and accomplishes sexual reproduction

The **root** anchors the plant and absorbs water and minerals

Lateral root

Root hairs

Primary root

Apical meristem

Root cap

Ground tissue system supports the plant and stores materials

Cortex Pith

Vascular tissue system transports materials

Phloem Xylem

Epidermis

DERMAL TISSUE SYSTEM protects

Figure 21.3   The Plant Axis: Shoot, Root, and Tissue Systems.

Tissue Systems and Tissue Types

We've seen that hemp is a complex flowering plant. Like corn, maple trees, and other plants in this category, hemp has three tissue systems with important physiological roles in the plant body. Each system is made up of a few important tissue types (**E** 21.1), and each tissue, in turn, is composed of different cell types with distinctive characteristics appropriate to their function (Table 21.1 and **E** 21.1). As Figure 21.3 shows, the three main tissue systems in a mature vascular plant like hemp are the **dermal tissue system,** which protects the plant from water loss and injury to internal tissues much as does animal skin; the **ground tissue system,** which provides support, stores starch, and carries out photosynthesis; and the **vascular tissue system,** which conducts fluids and helps strengthen the roots, stems, and leaves. These tissue systems are continuous throughout the plant.

Dermal Tissue System

A hemp plant has a dermal system that covers the entire plant body and is analogous to an animal's skin (Fig. 21.5a). In a young seedling, the dermal tissue consists of just an **epidermis,** or protective outer covering. (As we'll see later, this covering is replaced by bark in some plants.) Epidermal cells in stems and leaves secrete a waxy waterproof coating, the **cuticle.** Stem and leaf epidermal cells form little hairs, or **trichomes** (TRY-combs; see Fig. 21.5b). Many of the trichomes on the epidermis of hemp plants contain glands that manufacture THC, which repels insects and protects the plant in other ways. THC is also the compound people try to get "high" on by smoking or eating *Cannabis* grown as marijuana. These glands are found on leaves and stems, and especially on the bracts, the leaflike organs that protect the hemp flower bud. The concentration of THC in various plant parts is proportional to the number of trichomes. Plant breeders have diligently bred hemp varieties for industrial uses that reduce the number of trichomes as well as the THC content to below the level that affects the human brain. In principle, a single mutation that knocks out an individual enzyme in the THC production pathway might be enough to reduce the THC content to zero.

Hemp is considered **herbaceous,** that is, the mature stem keeps its flexible green epidermis and does not

Figure 21.4 **Patterns of Root Growth.** (a) Dock has a central taproot with fine lateral roots. (b) Grass has numerous fibrous roots that anchor the plant firmly, as does this marigold plant. (c) Banyan trees have dozens of adventitious roots that help support the plant and absorb water and minerals.

(a) Tap root

(b) Fibrous roots

(c) Adventitious roots

produce wood (which we'll talk more about later). In trees and woody shrubs (whose stems obviously do generate wood), the epidermis is replaced: the outer areas of the bark or **periderm** substitute for a thinner "skin."

Ground Tissue System

A plant's root and shoot contain **ground tissue** packed around the tubes of the vascular system. The ground tissue system makes up the bulk of most plant organs; it stores the starchy products of photosynthesis, helps keep the plant from collapsing into a formless heap, and has other, more specialized roles. There are three types of ground tissue, which vary in strength and flexibility: parenchyma, collenchyma, and sclerenchyma.

Parenchyma (pair-EN-keema) makes up the majority of ground tissue in both root and shoot, and consists of loosely packed, thin-walled, rounded cells such as those that store starch in a potato (Fig. 21.6a). In a hemp plant, much of the central, pithy core consists of parenchyma. Hemp farmers call this the "hurd," and once processed, sell it as bedding for horse stalls. Most parenchyma cells have only a thin primary cell wall (a flexible porous layer immediately surrounding the plasma membrane) laid down as the cell grows. Parenchyma cells are often unspecialized and can give rise to new cells and cell types in a wounded adult plant. Some parenchyma cells, however, *are* specialized. In the leaves, for example, parenchyma cells photosynthesize but store little starch, while in the roots, they often store large quantities of starch.

Collenchyma (coll-EN-keema) tissue is tougher than parenchyma because individual collenchyma cells often have a cylindrical shape and thick walls deposited as the cells enlarge. Collenchyma cells are often connected end to end in long, stringy fibers such as those just beneath the epidermis of a celery stalk or the central rib of a hemp leaf (Fig. 21.6b). The word *collenchyma* is based on the Greek word for glue, and in fact, these shiny cells do help hold the plant body together. In cross section, collenchyma cells have uneven, thickened corners (see Table 21.1).

Sclerenchyma (skler-EN-keema) tissue (Greek, *hard*) is a third tissue type in the ground tissue system. Because of its hardness, sclerenchyma often surrounds and reinforces the tubes of the vascular system. Sclerenchyma cells die at maturity and leave behind a thick cell wall that is deposited after the cell stops growing and is hardened with a tough complex polymer called **lignin.** Sclerenchyma tissue includes stone cells, or **sclereids** (SKLER-ids), which are hard, crystalline cells (Fig. 21.6c). Sclerenchyma cells give pears their grittiness and walnut shells their hardness. And sclerenchyma explains how hemp fibers can form strong ropes and supple fabrics. While in most flowering plants long sclerenchyma cells called

(a) A petal's dermal tissue

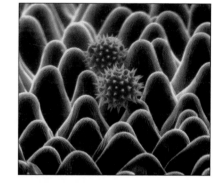

(b) A leaf's dermal tissue

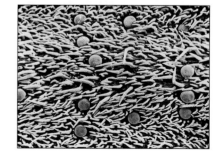

Figure 21.5 **The Dermal System.** These scanning electron micrographs show (a) the bumpy epidermis covering a sunflower petal and (b) the waxy cuticle and epidermis of a thyme leaf, with its hairlike, protective trichomes. Thyme trichomes create surface "hairiness" that retards water loss; other trichomes secrete the herb's aromatic oils. Both physically and chemically, trichomes also discourage certain predators from eating stems and leaves. Hemp trichomes secrete the alkaloid THC, which naturally poisons insects.

Table 21.1   Characteristics of Cell Types in Flowering Plants

Tissues and Cell Types	Structure	Function	Tissues and Cell Types	Structure	Function
Epidermis			**Xylem**		
Guard cells	Pairs of large cells in epidermis surrounding a pore	Regulate gas and water exchange in leaves	Tracheids	Elongated tapering cells; primary and secondary cell walls with pits; dead at maturity	Conduct water in all vascular plants
Trichomes	Hairs or scales jutting from epidermis	Protection; retard water loss	Vessel elements	Long and narrow, but generally shorter than tracheids; primary and secondary cell walls with both pits and perforations; stacked end-to-end in a vessel; dead at maturity	Main water-conducting cell in xylem of flowering plants
Parenchyma					
Parenchyma cells	Many-sided, often thin primary wall; alive at maturity; found throughout plant	Photosynthesis, storage, local conduction of materials, wound healing			
Collenchyma			**Phloem**		
Collenchyma cells	Rectangular; unevenly thickened primary cell wall at corners; alive at maturity	Support of young stems and leaf ribs	Sieve tube member	Elongated cell; primary cell wall with sieve plates; alive at maturity, but lacks functional nucleus; stacked end-to-end in sieve tube	Main food-conducting cell in phloem of flowering plants
Sclerenchyma					
Fibers	Very long and narrow; primary and very thick secondary cell wall; often dead at maturity	Support in stem cortex and vascular system			
Sclereids	Often long, but shorter than fibers; primary and thick secondary cell walls; living or dead at maturity	Support and protection throughout plant	Companion cells	Somewhat elongated; primary cell wall; alive at maturity and closely connected to sieve tube member	Help regulate activities of sieve tube member

fibers surround and support the internal transport vessels, the fibers in hemp plants can be especially strong and long, with individual cells reaching up to 2.5 inches (64 mm) in length.

All three types of plant tissues, parenchyma, collenchyma, and sclerenchyma, can provide the dietary fiber that is so important to the health of our digestive and circulatory systems. **Soluble dietary fiber** dissolves

(a) Most ground tissue cells are parenchyma cells, such as these large hemp storage cells. (b) Hemp stems contain collenchyma cells, with their irregularly thickened corners. Here, they appear white surrounded by reddish stain. (c) Hemp stem cortex contains sclerenchyma cells such as those shown here. These cells have thick walls but lack cytoplasm when mature. (d) A cross section of a hemp stem reveals the thick-walled sclerenchyma fibers in the phloem.

(a) Parenchyma

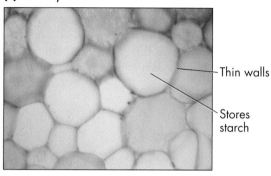

—Thin walls

—Stores starch

(b) Collenchyma

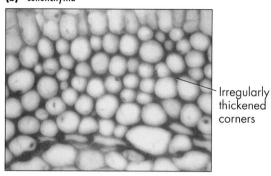

—Irregularly thickened corners

(c) Sclerenchyma

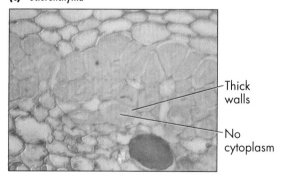

—Thick walls

—No cytoplasm

(d) Sclerenchyma fibers

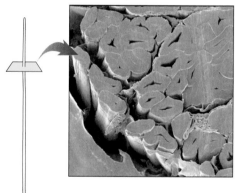

in water and is found naturally in plant gums and in oats, peas, and beans. Food producers use soluble dietary fibers as thickening agents in various foods. Thickeners in some ice creams and salad dressings, for example, come from the pulverized seeds of the carob tree. The soluble fiber found in oats and psyllium (SILLY-um) may help reduce a person's risk of heart disease by binding to cholesterol in the blood and blocking its absorption in the digestive tract and from there into body tissues. **Insoluble dietary fiber** such as wheat bran does not dissolve in water, but it binds to water molecules and helps substances pass through the digestive system more rapidly. The cellulose

in the cell walls of plants that make up soluble and insoluble fiber often taste good and add to our diet. However, some fibrous plant materials—including fibers in the vascular systems of woody plants—are so tough and hardened that they are inedible to us and are appetizing only to termites.

Vascular Tissue System

Within a vascular plant, water and materials travel in two kinds of tubular tissues, xylem and phloem, each of which forms a continuous pipeline extending from root tips to leaves. In hemp, those thin pipes can be 20 feet long; in tall trees they can be 300 feet long! In general, after roots absorb water and minerals from the soil, those substances then move to the stems and leaves through rigid water pipes called **xylem.** The material we know as wood consists mostly of xylem. In contrast, **phloem** transports dissolved sugars and proteins from "source to sink"—that is, from cells that produce or store sugars to cells that use sugars rapidly.

In most flowering plants, xylem is composed of several kinds of cells, but the main components are **tracheids** and **vessel elements** (Fig. 21.7a,b). Both tracheids and vessel members transport water only after they have died: Each cell type becomes hollow at maturity, when the cell contents disintegrate and leave behind empty cell walls. Table 21.1 provides more details about the shapes and functions of tracheids and vessel elements. A look at one, the vessel elements, explains in general how these tubes conduct water. Vessel elements have **pits,** or tiny

gaps in their side walls that let water move through (Fig. 21.7), and the elements are stacked directly end to end like the clay pipe sections of a water main. These "pipe sections" have holes in their end walls (larger than pits) called **perforation plates,** and these allow water to flow through the stacked vessel members unimpeded (see Table 21.1). A researcher at Harvard University reported in 2001 that tangles of polymer molecules located between the stacked ends of xylem vessel elements, may respond to calcium, potassium, or other ions and speed water uptake through the xylem "pipes." Even dead xylem cells, in other words, appear to help regulate water flow in plants, perhaps in response to brighter light or other influences on ion concentrations.

Like xylem, phloem is also a vital transport tissue composed of cells arranged end to end (Fig. 21.7c). While xylem cells are thick-walled and dead, mature phloem cells are thin-walled and contain living cytoplasm. While xylem transports minerals and large quan-

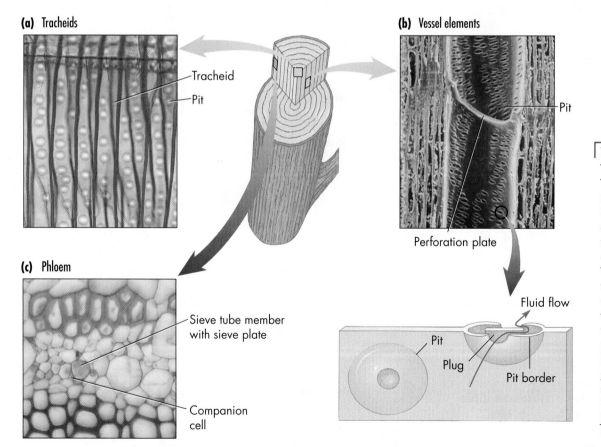

(a) Tracheids

Tracheid

Pit

(b) Vessel elements

Pit

Perforation plate

Fluid flow

Pit

Plug

Pit border

(c) Phloem

Sieve tube member with sieve plate

Companion cell

Figure 21.7 **Vascular Tissue System: Material Transport and Support of Plant Structures.** (a) Pitted, hollow xylem cells called tracheids transport water in all types of vascular plants, including pine, shown here. (b) Vessel elements are also pitted, such as these from an oak tree. Stacked end to end, vessel elements transport water but only in flowering plants. (c) Phloem in this hemp stem transports organic substances through sieve tube elements, which lack a nucleus, but are adjoined by nucleated companion cells.

tities of water, phloem transports sugars and amino acids dissolved in small quantities of water. Finally, while xylem transports materials in one direction, from roots to leaves, phloem can move sugars from photosynthesizing leaves to storage sites in the roots in summer or from roots to developing leaves in the spring.

The conducting cells of phloem tissue are called **sieve tube members** because their end walls, called **sieve plates,** usually contain large pores (Fig. 21.7c) and allow fluids and solutes to pass in and out, as through a sieve. Sieve tube members remain alive but lack a functioning nucleus. A small **companion cell** with a nucleus directs the activities of each nearby sieve tube member. Chapter 23 discusses how plants transport water, sugars, and minerals in more detail.

Thanks to the vascular system, the root and shoot function as a partnership; the root provides water and nutrients, which travel to the shoot via the xylem vessels, and the photosynthesizing shoot provides energy and organic molecules in the form of sugars and other substances, which travel to the root via the phloem. This body plan, plus the rather inflexible cell walls surrounding each plant cell, constrain the way a plant can grow. Hemp, for example, could never grow so tall and fast in a single summer without this root/shoot partnership.

What identifies hemp as an annual?

21.3   An Overview of Plant Growth

Picture, for a moment, a favorite pet dog or cat. Now imagine the chaos that would ensue if this pet were to continue growing throughout life like a plant, producing new eyes, ears, legs, feet, livers, and other organs. The result could be a bizarre creature with three tails, seven legs, and four eyes. Conversely, think what would happen if a hemp plant grew like a dog or cat, keeping its original root, stem, and single initial leaf, and each organ simply grew bigger and bigger as the plant matured. Clearly, the growth pattern that suits an animal would never work for a plant and vice versa!

Indeterminate Growth

Throughout life, a hemp plant adds new organs, such as branches, leaves, and roots. Cells at the tips of the root and shoot give rise to the new plant parts. Because a plant's cell walls are rigid, the organisms can't move to-

ward favorable conditions or away from harm the way animals can. A plant can usually overcome this rigidity and immobility, however, by means of growth: plants can *grow toward* light, water, and mineral nutrients and *away* from harmful situations. Like virtually all other land plants, hemp displays this special survival growth pattern called **indeterminate growth.**

One bizarre example of indeterminate growth involves the common houseplant philodendron in its native habitat, the tropical forest floor. Over a period of days or weeks, the living stem of a philodendron seedling lying on the ground appears to move through the forest like a green snake. In fact, no leaf or other part of the plant actually moves forward. Instead, the stem grows in one direction along the ground, forming leaves and roots in each segment as its habit of indeterminate growth dictates. The trailing section then dies. If you made a time-lapse movie, you could see this snakelike movement.

A plant's continual, indeterminate growth is based on **meristems,** tissues that remain perpetually embryonic and that give rise to new cells and cell types throughout the plant's life. Meristems give rise not only to vegetative parts of the plant, such as new leaves and stem, but also to the cells that produce reproductive parts—the flowers and the eggs and sperm they contain. Unlike animals, then, plants don't establish a special line of reproductive cells (germ cells) during early embryonic development. This means that any of its cells with this perpetual potential to divide can make reproductive cells, given the right environment.

Types of Meristems

In just a few summer months, a hemp plant grows 2 to 4 meters (6.5 to 13 ft) vertically into the air. An oak tree at the edge of the hemp field, however, would grow much slower but it would expand in two dimensions, getting both taller and thicker. Two types of meristems cause these two types of growth: the apical meristems and the lateral meristems.

Apical Meristems

Plants increase in length due to **apical meristems,** perpetual growth zones at the tips (apices; sing., *apex*) of roots and stems (Fig. 21.8a). Apical meristems allow shoots to grow upward toward the light and roots to grow ever deeper into the soil toward water. Growth arising from apical meristems is called **primary growth,** as in a tall, herbaceous hemp plant with its slender, flexible, green stem (**E**xplorer 21.2). The xylem and phloem produced by apical meristems are called **primary xylem** and **primary phloem.**

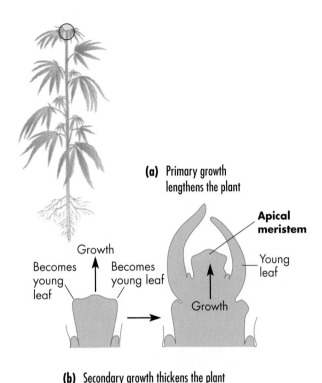

(a) Primary growth lengthens the plant

Growth

Becomes young leaf

Becomes young leaf

Apical meristem

Young leaf

Growth

(b) Secondary growth thickens the plant

Epidermis
Cortex
Primary phloem
Procambium
Primary xylem
Pith

Lateral meristems
Ruptured epidermis
Cork
Cork cambium
Cortex
Primary phloem
Secondary phloem
Vascular cambium
Secondary xylem
Primary xylem

Growth

Growth

Pith

Figure 21.8 **Primary and Secondary Growth.** (a) Primary growth from apical meristems at the tips of root and shoot causes the plant to lengthen. (b) Secondary growth from lateral meristems causes the shoot and root to increase in diameter.

Lateral Meristems

The second type of meristem, **lateral meristems,** are cylinders of dividing cells in the stems and roots of some plants that cause these parts to become thicker (Fig. 21.8b). This increase in diameter (girth) as a result of cell divisions in the lateral meristems is called **secondary growth** (**E** 21.2). This contrasts with primary growth, an increase in length due to cell divisions at the apical meristems. As you might expect, lateral meristems are found in plants that become quite thick and strong, like most trees and shrubs, but they are lacking in corn and other monocots, as well as in many eudicots, such as hemp and dandelions, that survive for just a single growing season. Herbaceous plants such as hemp lack secondary growth, while woody plants such as oak or maple have secondary growth.

Vascular cambium and cork cambium are types of lateral meristems that give rise to secondary xylem and phloem and hence to stem or trunk enlargement. **Vascular cambium** is a type of lateral meristem that produces **secondary phloem** to the outside and **secondary xylem** to the inside; this inside tissue is the material we know as **wood.** Another lateral meristem, **cork cambium,** generates **cork,** the waterproof outer part of the bark of trees and shrubs.

By observing trees growing in your neighborhood, you can more easily understand the distinction between primary growth from the apical meristems and secondary growth from the lateral meristems. Locate a tree that someone drove nails into years ago to attach a sign or put up a tree house. Did the nails move higher as the tree grew? No. Instead, the tree grew around the nails. Primary growth, which lengthens the tree, occurs only at the apical meristems, which lie at the ends of branches far above the nail. The nail, however, will eventually be buried in bark and wood as secondary growth from lateral meristems adds new tissue to the sides of the plant.

Plant Growth and Seasonal Cycles

Flowering plants have an indeterminate growth pattern based on meristems. But as any gardener knows, some plants live only one season, whereas others live from 2 to 5,000 years or more. In a single season, an **annual plant** sprouts from a seed, matures, produces fruits and new seeds, and dies (Fig. 21.9a). Hemp is an annual, and a male hemp plant dies after shedding pollen about five months after planting, and a female dies when the seeds are mature, about a month after the male. Marigolds, zinnias, petunias, poppies, and soybeans are

(a) Annual (hemp)

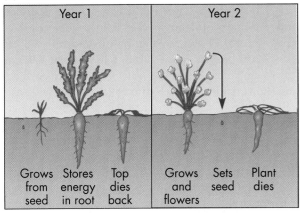

Year 1 — Grows from seed, Sets seed, Dies

Year 2 — Grows from seed, Sets seed, Dies

(b) Biennial (carrot)

Year 1 — Grows from seed, Stores energy in root, Top dies back

Year 2 — Grows and flowers, Sets seed, Plant dies

(c) Perennial (lupine)

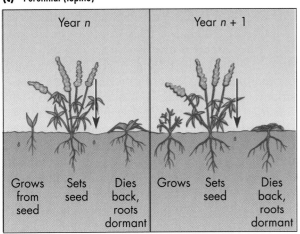

Year n — Grows from seed, Sets seed, Dies back, roots dormant

Year n + 1 — Grows, Sets seed, Dies back, roots dormant

Figure 21.9   Annuals, Biennials, and Perennials. (a) Annuals, such as zinnias or hemp, have one-year life cycles and pass the winter in seed form. (b) Biennials, like carrot or Queen Anne's lace, grow the first year, reproduce the second year, and then die. (c) Perennials, such as lupine or rose bushes, survive and grow year after year.

familiar annuals that gardeners and farmers plant from seeds every year. **Biennials** are plants that have a two-year life cycle: They grow from seeds to adults in the first year; in the second year, the adults produce flowers, fruits, and seeds and then die (Fig. 21.9b). Celery, cabbage, carrots, and Queen Anne's lace are all biennials. Many biennials and most annuals are herbaceous, lacking secondary growth. Finally, **perennials** live for many years and typically bloom and set seeds several times before the adult plant dies (Fig. 21.9c). Some perennials, like tulips, lupine, and dahlias, are herbaceous; others, like rose bushes and oak trees, are woody.

21.4   Roots

A hemp plant's taproot system snakes down through the soil and provides the tall, skinny plant with a number of important services. Roots anchor the organism firmly to one spot, penetrate the soil and absorb water and minerals, and often store starch. Our discussion of root anatomy begins deep in the soil at the very tip of the root.

The Root: From Tip to Base

Root Cap

At the tip of a every root is a dome-shaped **root cap** (Fig. 21.10). As the root grows downward, rough soil particles damage and scrape off cells at the surface of the root cap. These damaged cells slough off and cover the rest of the root with a slime that eases penetration through the soil.

Apical Meristem

Damaged root cap cells are replaced by the apical meristem, a little disk of cells just above the root cap that produces new cells from both its surfaces. The cells from the lower surface of the meristem become root cap cells, while those derived from the upper surface elongate, adding length to the root. After cells elongate, they differentiate and develop their specialized roles as members of the dermal, ground, or vascular tissue systems. These sequential activities—division, elongation, and maturation—are displayed in order along a growing root from the tip backwards toward the plant (Fig. 21.10a).

The Root: From Outside to Inside

The Dermal System: Epidermis and Root Hairs

If you were to pull up a hemp plant, knock the dirt from the roots, and cut a thin slice of root with a sharp blade (or cut a carrot in half crossways) you'd see that the root consists of several concentric cylinders, like drinking glasses of different diameters nested one inside the other. The outer cylinder is a protective layer just one cell thick. This epidermis arises from the outer cells in the zone where cells mature (Fig. 21.10a). The root epidermis absorbs water and minerals from the soil, and just as microvilli enlarge the absorptive surface area of an animal's intestinal cells (see Fig. 17.16), tiny extensions of the root epidermis called **root hairs** extend the root's absorptive capacity (Fig. 21.10b). Each root hair is an extension of a single epidermal cell, but their numbers can be immense, and their collective surface area amazingly huge: A single 4-foot-tall ryegrass plant can have about 14 billion root hairs, with a combined surface area the size of a tennis court!

Root hairs are delicate and short-lived, with new ones continually arising. Since most water absorption occurs in the root hairs near root tips, gardeners must be careful not to tear off young root tips when transplanting flowers or shrubs, and they often get the best results by fertilizing fruit trees several feet out from the trunk, nearest the root tips, where root hairs are numerous.

The Ground System

Just inside the cylinder of epidermis tissue lies a thick layer of cells called the **cortex** (Fig. 21.11a). You can see this in a slice from a hemp taproot or a bright orange carrot root. The cortex makes up most of the root's bulk and often stores excess starch.

The innermost layer of the cortex is the **endodermis,** a cylinder of tightly packed cells just one cell thick (Fig. 21.11b). Water moving inward from the soil can pass freely between most cortex cells without entering their cytoplasm. When the water reaches the endodermis, however, it must pass through the living cytoplasm of endodermal cells before reaching groupings of xylem and phloem called vascular bundles and hence the rest of the plant. Thus, these cells function like gatekeepers for the water and dissolved minerals passing into the root's vascular system and then the rest of the plant. Each cell in the endodermis is a bit like a brick in a circular brick wall (Fig. 21.11b–d). The outside surface of the "bricks" faces the rest of the cortex, and the inside surface faces the vascular tissue in the root's center. The side walls of these cells are impregnated with a waxy, water-resistant substance. This narrow, water-resistant belt—the **Casparian strip**—encircles each cell and lies between it and the next endodermal cell like the mortar between bricks (Fig. 21.11d). Because of the Casparian strip, water can't pass between endodermal cells or diffuse from one endodermal cell to the next. Instead, water can pass from the cortex to the vascular tissue only by moving through the cytoplasm of the endodermal cells. This has an important consequence for the plant: Water and minerals seeping from the soil through the epidermis and into the cortex can reach the central vascular tissue (and from there move throughout the rest of the plant) only by first passing through the living cytoplasm of the endodermal cells. Because plant cells contain a different mineral composition in their cytoplasm than occurs in the soil, plants take up minerals against a concentration gradient. This requires energy

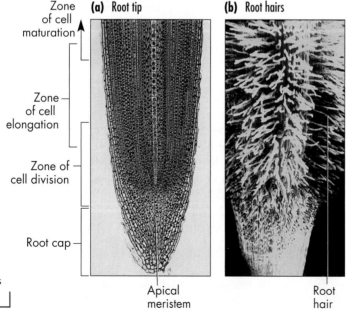

Zone of cell maturation

(a) Root tip

(b) Root hairs

Figure 21.10 The Growing Root Tip. (a) A longitudinal section of a root reveals, from the tip upward, the protective root cap just below the apical meristem; the small new cells of the zone of cell division; and the lengthening cells of the zone of cell elongation. (b) Epidermal cells and root hairs in the zone of maturation. Hemp roots have similar root tip zones and root hairs.

Zone of cell elongation

Zone of cell division

Root cap

Apical meristem

Root hair

expenditure and is a major job of the endodermis.

The Vascular System

The endodermis in a taproot from hemp or carrot surrounds yet another cylinder—a central zone of vascular tissue that carries water and other materials between the roots and the stems and leaves. Moving inward through the vascular system, one encounters the pericycle, phloem, xylem, and, in some plants, pith (see Fig. 21.11c,e).

The **pericycle** lies just inside the endodermis and consists of one or more layers of parenchyma cells. The pericycle gives rise to the lateral roots, which grow out

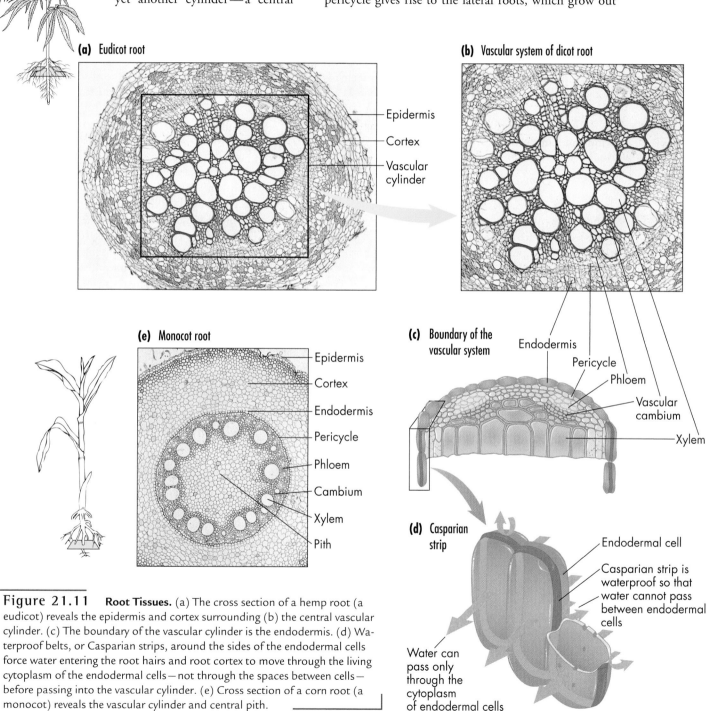

(a) Eudicot root

- Epidermis
- Cortex
- Vascular cylinder

(b) Vascular system of dicot root

(e) Monocot root

- Epidermis
- Cortex
- Endodermis
- Pericycle
- Phloem
- Cambium
- Xylem
- Pith

(c) Boundary of the vascular system

- Endodermis
- Pericycle
- Phloem
- Vascular cambium
- Xylem

(d) Casparian strip

- Endodermal cell
- Casparian strip is waterproof so that water cannot pass between endodermal cells
- Water can pass only through the cytoplasm of endodermal cells

Figure 21.11 **Root Tissues.** (a) The cross section of a hemp root (a eudicot) reveals the epidermis and cortex surrounding (b) the central vascular cylinder. (c) The boundary of the vascular cylinder is the endodermis. (d) Waterproof belts, or Casparian strips, around the sides of the endodermal cells force water entering the root hairs and root cortex to move through the living cytoplasm of the endodermal cells—not through the spaces between cells—before passing into the vascular cylinder. (e) Cross section of a corn root (a monocot) reveals the vascular cylinder and central pith.

What are the roles of a plant's shoot?

horizontally through the cortex parallel to the surface of the ground and then, often, downward into the ground.

Encircled by the pericycle are the xylem and the phloem, which are arrayed in various ways in plant roots, depending on the plant species. In the roots of most eudicots, including hemp, the central vascular system contains a star-shaped core of xylem vessels, with phloem nestled in the angles between the points of the star (see Fig. 21.11a,b). In contrast, most monocots have a ring of phloem inside the pericycle that surrounds a ring of xylem (Fig. 21.11e), which in turn encloses a central core of large, thin-walled parenchyma storage cells called **pith.** The pith helps thicken hemp roots, and this, in turn, allows them to anchor the tall plants in the soil more firmly, so they stay erect despite wind and rain storms.

Secondary Growth in the Roots

In perennial plants, such as shrubs and trees, roots experience both primary growth from the apical meristems and secondary growth from lateral meristems. The root tips continually probe downward or outward (primary growth) as the older parts of the root system become woody and thus stronger and more capable of anchoring a large plant such as a maple tree (secondary growth). *E* 21.2 describes secondary growth in roots in more detail.

21.5 The Shoot

Hemp plants are amazingly tall for their girth, reaching up to 20 feet tall yet growing no thicker than your thumb. By contrast, the tallest California redwood trees can be wide enough to drive a truck through but can also reach 100 m (328 ft) into the sky. Regardless of height and thickness, vertical growth is based on the stem, and stems have two fundamental tasks: (1) supporting the plant's leaves, flowers and fruit, if present, and (2) acting as a central corridor for the transport of water, minerals, sugars, and other substances. The fibrous "corridor" surrounding the phloem inside a hemp stem provides the soft material useful for making paper and fabrics.

The modified stems of some plants can store starch and allow plants like ivy to adhere to vertical surfaces. Some plants have modified stems called **rhizomes** that grow horizontally underground and help anchor the

plant and promote vegetative reproduction. In this section, we discuss stems as well as another important part of the shoot, the leaves. Flowers are part of the shoot, too, of course, but their architecture and activities make the most sense in the context of plant reproduction, our subject in Chapter 22.

Stem Structures and Primary Growth

Shoot Apical Meristem

The shoot lengthens as specialized cells at the tip of the plant divide and redivide. The shoot tip, or apex, contains the apical meristem. As the apical meristem at the tip of a branch (the **terminal bud**) continues to grow upward, it leaves behind groups of cells like smoke billowing behind a steam locomotive or dirt piling up behind a burrowing earthworm. Some of these cells, called **leaf primordia,** continue to divide and form leaves. Other cells produced by the apical meristem form buds of embryonic tissue in the space between a leaf and the stem. These buds (or **lateral buds**) can give rise to branches or flowers, and are important for determining a plant's overall shape. Dr. Gordon Scheifele explains that if farmers are planting hemp mainly to harvest the seed for food or oil, they sow only about four plants per square meter (Fig. 21.12a). This thin density allows the lateral buds to form branches so the hemp plants grow bushier and develop more leaves and flowers. In contrast, a farmer wanting to harvest hemp fiber sows hemp seed at about a hundred plants per square meter. This high population density suppresses the growth of lateral buds, and the terminal bud continues to grow straight and very tall, resulting in plants like the one in Figure 21.12b.

Besides meristems in the terminal and lateral buds, additional meristems give rise to the epidermis, to the ground tissues, and to the primary xylem and phloem (Fig. 21.13).

Let's look one by one at the specialized shoot tissues laid down by the apical meristem, and how their growth helps push the plant skyward. (*E* 21.2 animates this topic.)

Dermal System: Epidermis

As in the root, the dermal system of the shoot consists of a layer of epidermal tissue one cell thick. While root epidermis absorbs water, shoot epidermis resists water loss. Just as a plastic bag keeps a sandwich from drying out, stem epidermal cells have a water-resistant waxy

(a) **(b)**

Figure 21.12 **Plant Form: A Consequence of Lateral Buds.** (a) When farmers plant hemp seeds sparsely, the lateral buds generate long, leafy side branches. (b) When farmers plant hemp seeds densely, the stalk grows straight and tall with few, short branches.

coating, the **cuticle,** that keeps the plant from desiccating (Fig. 21.14).

Ground System: Cortex and Pith

As in the root, the ground system of the stem is primarily cortex (Fig. 21.14a) made up of parenchyma cells. Stems, however, which support the weight of the upper parts of the plant and keep the plant vertical in the air, need more structural reinforcement than roots, which are supported by the soil. In many stems, a strand of collenchyma (see Table 21.1) around the outer edge of the cortex, just inside the epidermis, helps provide this support. Many plants, including hemp, also have a strong central core of pith that contributes extra support, even though the root's ground system may lack it. Gordon Scheifele explains that the pith of a hemp plant is part of the pulpy material from the plant's central core that can be used to make rough, low-quality paper or horse stall bedding but is unsuitable for high-quality paper or fabrics.

Regardless of other support tissues, it's the shoot's vascular system that usually provides the strongest structural support.

Vascular System

In the stem of a herbaceous annual such as hemp or a young perennial, such as an apple seedling, xylem and phloem are organized in groupings called **vascular bundles.** Within each bundle, a sheath of parenchyma or sclerenchyma cells surrounds strands of xylem and phloem, with xylem lying toward the center of the stem. In the stems of most eudicots, vascular bundles form a ring around the stem's central core of pith, and a layer of cambium lies between the xylem and phloem (Fig. 21.14a). Cambium from each vascular bundle extends

Figure 21.13 **The Apical Meristem.** (a) A longitudinal section of the growing tip of a coleus plant reveals the primordia that will give rise to new leaves, flowers, and branches. (b) A cross section diagram shows the areas responsible for primary growth.

(a) Longitudinal section of shoot tip

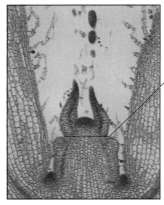

(b) Cross section of shoot tip

A shoot tip cut cross-ways at this level reveals meristem giving rise to the epidermis, to the xylem, phloem, and cambium, and to the ground tissue

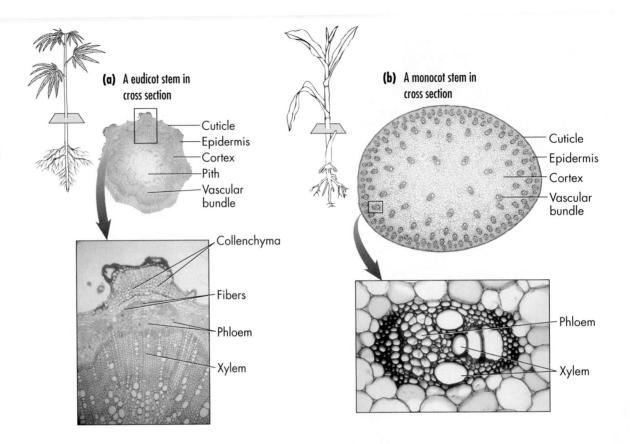

(a) A eudicot stem in cross section

- Cuticle
- Epidermis
- Cortex
- Pith
- Vascular bundle

- Collenchyma
- Fibers
- Phloem
- Xylem

(b) A monocot stem in cross section

- Cuticle
- Epidermis
- Cortex
- Vascular bundle

- Phloem
- Xylem

to the adjacent bundles, forming a complete ring of cambium around the pith. In conifers and woody eudicots, the cambium layer, which is sandwiched between xylem and phloem, produces the secondary growth of wood and bark, resulting in the enlargement of the stem. In monocots and a few eudicots, by contrast, the vascular bundles are scattered throughout the cortex, there is no vascular cambium, and the stems are incapable of true secondary growth (Fig. 21.14b).

Stem Structures and Secondary Growth

We saw in Figure 21.13 that the apical meristem produces cells that divide and form the primary xylem and primary phloem, as well as a ring of meristematic cells, the vascular cambium. Late in the first summer of a tree seedling's life, lateral meristem cells in the vascular cambium begin dividing and produce additional xylem and phloem cells (review Fig. 21.8b). Botanists call the new conducting vessels made by the vascular cambium secondary xylem and secondary phloem. Like their primary counterparts (which, recall, are laid down at the apical meristem), these secondary conducting vessels transport water and other materials. Because the cambium makes far more secondary xylem than secondary

phloem, most of the stem tissue in a sapling, a bush, or an older tree is made up of dead xylem cells (or **wood**) in a thick interior rod (Fig. 21.15).

Wood

A tree cut down in a temperate region—for instance, a fir tree in Southern Canada—will have concentric

Figure 21.15 **Wood: A Result of Secondary Growth in Stems.** A cross section of a two-year-old basswood tree (*Tilia* sp.) stem reveals several years of secondary growth in rings.

rings of lighter and darker wood. If the chair you're sitting in is made of wood, you can probably see evidence of these rings as the grain. These **growth rings** (Fig. 21.15) occur because the cambium produces larger cells in the spring and summer (lighter areas), when water and sunlight are plentiful, and smaller cells in the fall and winter, when water tends to be more scarce. In moist, tropical regions, on the other hand, where a tree can grow year round, the wood may have no obvious growth rings. Because trees produce larger rings in years with good weather, scientists can use the width of growth rings in ancient trees to reconstruct historical weather patterns. Scientists have dated a massive volcanic eruption on the Greek island of Thera, which must have darkened skies around the world with ash, to approximately 1625 B.C. They did this by identifying very narrow growth rings in 5000-year-old bristle-cone pines of eastern California, already producing wood at the time of the eruption and still alive today.

As a tree grows older, the xylem at the center of the trunk gradually stops conducting water and minerals, and newer xylem closer to the bark takes over. The nonconducting central wood, or **heartwood,** becomes infiltrated with oils, gums, resins, and tannins, all of which make it dark, aromatic, and resistant to rot. The **sapwood** around the outside continues to transport water in the plant.

Bark

When the stem expands during secondary growth, it ruptures the seedling's original cortex and epidermal layers. The small tree still needs a protective waterproof covering, however, and the periderm, with its cork cambium and layer of cork cells, develops here. Together, the periderm, the underlying phloem, and many layers of cork cells that die but whose cell walls remain in place make up the outer protective **bark**. Sheets and plugs of cork for flooring, bulletin boards, and bottle stoppers come from the periderm of cork oak trees native to Spain, Portugal, and Algeria. Cork cutters are careful to leave the sugar-conducting phloem intact. Without a functioning phloem all the way around the trunk, sugars made in the leaves via photosynthesis can't move downward to the roots, and the tree eventually starves and dies.

Bark does more than keep water in; it helps keep plant-eating insects, fungi, viruses, and other parasites out. Healthy trees also secrete resins that ooze out of holes bored by insects, often forcing the intruders back out. The famous gifts of the "Three Wise Men," frankincense and myrrh, are fragrant tree resins. The precious yellowish "jewel" called amber is nothing but fossilized resin.

Structure and Development of Leaves

Hemp plants have a characteristic radiating leaf shape that became a counterculture symbol in the 1960s. Today, it's just a good example of a highly evolved solar collector. A hemp plant's first pair of true leaves each have just a single leaflet. The plant's second pair of leaves each have 3 leaflets, the third 5, and so on, until each leaf has 11 or 13 leaflets (Fig. 21.16a). This highly divided form helps wind pass through the tall plant's thousands of leaves without bending and breaking the slender stalk. The radiating leaflets also stretch like fingers to collect every possible photon of light energy.

Figure 21.16 **Leaf Architecture: Eudicots, Monocots, Conifers.** Eudicot leaves such as those from the hemp plant (a) are often broad, net-veined, and each leaf has a petiole and a blade. Maple leaves such as this Norway Maple (b) reveal colored plant pigments when chlorophyll breaks down in autumn. Monocots such as this corn plant (c) usually have parallel veins and a slender blade attached to the stem via a sheath, not a petiole. The leaves of conifers like this white pine (d) are needlelike and grow in clusters.

(a) Hemp: a compound leaf

(b) Maple leaf

(c) Monocot leaves

(d) Gymnosperm leaves

This leaves the ground below the hemp plant heavily shaded and discourages weed growth—so much so that hemp farmers can usually avoid herbicides altogether. This ecological attribute compares very favorably with a fiber crop like cotton, which requires heavy pesticide and herbicide spraying.

Most plants have fewer, simpler leaves than hemp. Regardless of form, however, nearly all leaves share the same basic functions: (1) exposing a photosynthetic surface, usually one that is large and flat and oriented to catch the maximum amount of sunlight; (2) taking in carbon dioxide from the atmosphere as a carbon source for photosynthesis, while at the same time conserving as much water vapor as possible; and (3) helping to draw water and nutrients up through the vascular system.

Leaf Blade and Petiole

Picture a typical elm, oak, or maple leaf (Fig 21.16b). Most leaves have a broad, flat portion, as do these familiar leaves, called the **blade.** Most also have a **petiole** (a slender stalk) or a **sheath** (an expanded portion at the base of the leaf encircling the stem) that connects the blade to the plant stem at an attachment site called a **node.** Leaf blades are often flat and thin, which maximizes the available surface area for absorbing light and carbon dioxide. Leaf shapes are frequently modified, however, and this helps balance the conflicting needs of taking in carbon dioxide while conserving water. Desert plants such as jojoba and creosote have small leaves covered with a thick, waterproof cuticle and hairy surfaces that together reflect light and help retard water loss. Cactus leaves are modified into small, dry, needlelike spines that defend the plant from hungry, thirsty animals rather than contributing to photosynthesis. Many plants have leaves modified for clinging and climbing, for capturing little pools of water, and for attracting animals toward inconspicuous blossoms. In a few species, including pitcher plants and sundews, the leaves actually help trap visiting insects, which are then digested for their proteins.

As we've seen, the leaves of eudicot species can also be **compound,** like hemp leaves, consisting of many small leaflets, or **simple,** like oak or maple leaves, consisting of a single blade.

Leaf Tissues

A leaf's internal anatomy is as intimately associated with its many functions as its outer form.

Dermal Tissue

A leaf's upper epidermis is protected and sealed by a waxy coating, the cuticle. Tiny openings in the lower epidermis called **stomata** (sing., *stoma*), let in carbon dioxide for photosynthesis and let out water vapor and oxygen. Stomata are surrounded by **guard cells** that cause the stoma (Greek, *mouth*) to open and close (Fig. 21.17; also see Fig. 23.6).

Ground Tissue

Sandwiched between the upper and lower epidermal layers is the leaf's **mesophyll** layer, made up of two types of parenchyma (Fig. 21.17). **Palisade parenchyma,** the main photosynthetic tissue, lies just beneath the upper epidermis. It usually consists of one or more rows of vertically oriented, column-shaped cells, each enclosing dozens of chloroplasts. A layer of rounder cells, the **spongy parenchyma,** lies between the palisade

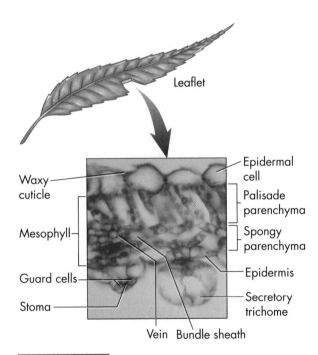

Leaflet

Waxy cuticle

Mesophyll

Guard cells

Stoma

Epidermal cell

Palisade parenchyma

Spongy parenchyma

Epidermis

Secretory trichome

Vein Bundle sheath

Figure 21.17 **Leaves: An Inside View.** A small section taken from a leaf of a hemp plant reveals the cuticle-covered upper epidermis, layers of palisade and spongy parenchyma cells, a vein, or bundle of vascular tissue surrounded by a sheath, and a lower epidermis ventilated by an opening—a stoma—that are flanked by guard cells.

parenchyma and the lower epidermis. The loosely packed spongy parenchyma cells provide a huge surface area (analogous to an animal's lung) that absorbs carbon dioxide from the air that enters the stomata. Carbon dioxide entering the stomata can move rapidly through the spongy parenchyma to the palisade parenchyma above, where most of the photosynthesis takes place.

Vascular Tissue

Leaves have an elaborate "plumbing" system that distributes the products of photosynthesis and brings in water and minerals. A plant's vascular system is continuous between leaves, stem, and roots. Monocot leaves have bundles of xylem and phloem called **veins** running parallel to each other along the long axis of the leaf (see Fig. 21.16c). Eudicot leaves, including hemp leaves, have net veins; that is, the veins form a branching pattern. Surrounding individual veins in some plants are **bundle-sheath cells,** which protect the xylem and phloem from direct exposure to air (see Fig. 21.17). Collenchyma or sclerenchyma fibers may also stiffen the veins and leaf margins, and help each leaf remain flat, which improves solar collection.

Hemp for the Future

We posed several questions in the chapter introduction. For example, how can a hemp plant reach 10 to 20 feet tall in a single growing season? The answer is that all summer long, billions of mesophyll cells in the plant's thousands of leaves trap solar energy and produce carbohydrates. These molecules, among other things, form the strong cell walls of the fibers in the plant's slender stem, allowing the plant to grow taller and taller from its apical meristems. The result is a plant that rapidly reaches up to six meters in height, fortified by tough fibers perfect for making high-quality paper, cloth, and even automobile parts.

Will hemp be the gigantic new crop for the 21st century that soybeans were for the 20th century? No one can really say. But as plant research becomes more and more sophisticated, the chances grow better that geneticists can breed hemp with little or no THC. This could further reduce the risk of people using and abusing hemp's protective alkaloid. And it could allow farmers to grow huge fields of the tall, gangly plants with their wealth of usable fibers, pulp, seeds, oil, and other reflections of plant architecture.

Chapter Summary and Selected Key Terms

Introduction   Canadian plant researchers and farmers would like to see hemp varieties very low in THC become the basis of a major export industry. The plants produce strong fibers, nutritious seeds, oil, and pulp that can be used in products from designer clothing to automobile dashboards. Hemp is a good case study for **plant anatomy** (p. 531) and **plant functions** (p. 530). Its industrial form is rich in plant fibers. Its **marijuana** (p. 530) form has high levels of **THC** (p. 530).

Vascular Plants   Nearly 500,000 vascular plant species have been identified, each having internal transport tubes or **vascular tissue** (p. 531). The biggest group are the flowering plants or angiosperms and include, among others, the **monocots** (p. 532; corn, bamboo, grasses) and the **eudicots** (p. 532; hemp, peach trees, roses).

The Plant Body   Land plants must get all of their needed resources from soil, water, and air. The plant **axis** (p. 533) includes the root and shoot. The **shoot** (p. 533) consists of **stem,** branches, leaves, flow-

ers, and fruit. The **root** (p. 533) is usually below ground. A **taproot** system (p. 533) grows straight down and usually stores water and starch. **Fibrous root** systems (p. 533) branch and form a mass that anchors and absorbs water and nutrients exceptionally well. **Adventitious roots** (p. 533) are above ground.

The plant body has three tissue systems: The **dermal tissue system** (p. 534) includes the skinlike **epidermis** (p. 534) and waxy **cuticle** (p. 534) that cover and protect the plant from invaders, injury, and water loss. The **ground tissue system** (p. 534) makes up the bulk of most plant organs (leaf, stem, etc.). Thin-walled, rounded **parenchyma** (p. 534) cells store water and starch, often photosynthesize, and sometimes give rise to new cells. **Collenchyma** (p. 534) cells are tough, thick-walled, shiny, and gluelike, helping hold the plant body together. Stacks of these cells, end to end, form celery ribs and the central ribs of hemp leaves. **Sclerenchyma** (p. 534) cells are hardened and give nut shells their solidity, pears their grittiness, and hemp fibers their strength.

The **vascular tissue system** (p. 538) includes the water-conducting **xylem** (p. 538) and the **phloem** (p. 538), which conducts sugars and proteins dissolved in plant fluids. Xylem contains **tracheids** (p. 538) and **vessel members** (p. 538), which grow, die, hollow out, and conduct water when stacked end to end. **Pits** (p. 538) allow water to pass between overlapping xylem

cells; **perforation plates** (p. 538) allow water to flow from one cell to the stacked cell above. Phloem contains **sieve tube members** (p. 539) or thick-walled living pipe sections with porous end walls or **sieve plates** (p. 539). These members lack nuclei, but each has a nucleated **companion cell** (p. 539).

How Plants Grow   Plants have **indeterminate growth** (p. 539), meaning they add new organs throughout life, enlarging from the tips of root and shoot. While animals can usually flee danger or stress or move toward needed resources, plants must grow toward light, water, and nutrients and away from harm. Indeterminate growth is based on **meristems** (p. 539), or tissues that stay perpetually embryonic and capable of giving rise to new cells and new organs throughout life.

Apical meristems (p. 539) increase shoot or root length at the tips (apices). Growth from apical meristems is **primary growth** (p. 539). Apical meristems produce **primary xylem** (p. 540) and **primary phloem** (p. 540). Lateral meristems produce **secondary xylem** (p. 540) and **secondary phloem** (p. 540). **Lateral meristems** (p. 540) increase girth; this is **secondary growth** (p. 540). Two types of lateral meristems are (1) **vascular cambium** (p. 540), which produces the secondary phloem and secondary xylem tissue we call **wood** (p. 540); and (2) **cork cambium** (p. 540), which generates **cork** (p. 540) or the waterproof outer part of bark. **Annual plants** (p. 540) live and die in a single season, sprouting from a seed, maturing, producing fruit and new seeds, and then dying. **Biennials** (p. 541) have a two-year growth cycle, growing from seeds to adults in year one, then producing seeds and dying in year two. **Perennials** (p. 541) live for many years, producing seed several times (often annually) before the adult plant dies.

Roots

Roots anchor the plant firmly, absorb water and minerals, and often store starch. A **root cap** (p. 541) covers and protects the growing apical meristem, with its zones of cell division, elongation, and maturation (pp. 541, 542). The root has the three tissue systems mentioned earlier. The dermal system includes the root epidermis cells and their extensions, the **root hairs** (p. 542), which absorb water and minerals across a huge surface area. The root ground tissue system has a thick layer of cells, the **cortex** (p. 542), that often stores starch. The inside edge of the cortex is a cylinder one cell thick, the **endodermis** (p. 542), that regulates water passage by means of a waxy water-resistant substance (p. 542) and water-resistant belts, the **Casparian strips** (p. 542).

The roots vascular tissue system lies in a central zone that carries water and other materials between roots, stem, and leaves. The **pericycle** (p. 543) lies inside the endodermis cylinder and gives rise to lateral roots and encircles the xylem and phloem. In most monocots, xylem and phloem are scattered bundles surrounded by **pith** (p. 544).

The Shoot

Some plants have modified stems called **rhizomes** (p. 544). Hemp plants have a very tall slender vertical stem that supports the rest of the shoot (leaves, flowers, fruits, and seeds) and houses the vascular tissue. The shoot lengthens at apical meristems. The one at the tip of a branch, the **terminal bud** (p. 544), grows upward and leaves behind **leaf primordia** (p. 544) that form leaves or **lateral buds** (p. 544), which in turn give rise to leaves or flowers.

The shoot has a dermal system made up of epidermis and cuticle. The shoot has a ground system like the root, with cortex and pith. Finally, the shoot has a vascular system containing xylem and phloem. In non-woody stems, these conducting pipelines are organized into **vascular bundles** (p. 545). These can form a ring joined by cambium around the pith (as in most eudicots) or scattered throughout the cortex (as in most monocots). In woody stems, the vascular cambium gives rise to secondary xylem and secondary phloem. Dead secondary xylem forms **wood** (p. 546). These new conducting cells tend to be larger in spring and summer, smaller in fall and winter, and this creates **growth rings** (p. 547) in the wood. **Heartwood** (p. 547) is nonconducting; **sapwood** (p. 547) contin-

ues to transport water. Stem expansion during secondary growth ruptures the original cortex and epidermal layers. **Bark** (p. 547) forms, made up of periderm, underlying phloem, and layers of cork cells.

Leaves are part of the shoot and usually specialize in solar collection and carbohydrate production. Hemp has **compound leaves** (p. 548), divided into many leaflets. Oak and elm leaves have **simple leaves** (p. 548) in form. All leaves collect sunlight; take in CO_2 but regulate water vapor loss; and help draw water and nutrients up through the vascular system. Most leaves have a broad, flat **blade** (p. 548) and a **petiole** (p. 548) or **sheath** (p. 548) connecting blade to stem at a **node** (p. 548). Leaves are sometimes modified to defend the plant, collect water, or attract and/or trap insects. Leaves, too, have the three tissue systems. Their dermal tissue has a cuticle-covered epidermis perforated by openings called **stomata** (p. 548), which are surrounded by **guard cells** (p. 548) that regulate water loss and CO_2 entry. The ground tissue consists of a **mesophyll** (p. 548) layer with photosynthetic **palisade parenchyma** (p. 548) cells near the upper leaf surface and CO_2-absorbing **spongy parenchyma** (p. 548) cells near the lower leaf surface and stomata. Bundles of xylem and phloem called **veins** (p. 549) carry water, nutrients, and the products of photosynthesis. In some plants, **bundle sheath cells** (p. 549) surround veins. Monocots have **parallel veins** (p. 547) while eudicots have **net veins** (p. 547). Collenchyma and sclerenchyma may also stiffen the veins and leaf margins to help flatten and strengthen the solar-collecting leaf surface.

All of the following question sets also appear in the Explore Life *E* electronic component, where you will find a variety of additional questions as well.

Test Yourself on Vocabulary and Concepts

In each question set below, match the description with the appropriate term. A term may be used once, more than once, or not at all.

SET I

(a) vascular plants (b) monocots (c) eudicots
(d) woody plant (e) herbaceous plants

 1. Angiosperms that have secondary growth

 2. Angiosperms that do not grow in girth; stems are generally green

 3. Flowering plants that include all woody plants as well as many herbaceous types

 4. Flowering plants that include grasses, lilies, and palm trees but not roses, daisies, and apple trees

 5. The most inclusive category in the list; includes angiosperms and gymnosperms

SET II

(a) root (b) shoot (c) meristematic tissue
(d) dermal tissue (e) ground tissue (f) vascular tissue

 6. Includes stem, branches, leaves, flowers, and fruit; usually the predominant above-ground part of a plant

 7. Found either as epidermis with or without cuticle, or as periderm

 8. Primary and secondary xylem and phloem constitute that tissue

 9. Region of absorption of water and minerals; usually nonphotosynthetic and anchored in soil

 10. The major storage tissue; helps to support the plant

SET III

(a) pericycle (b) endodermis (c) root hair
(d) cuticle (e) pith (f) vascular cambium

 11. Increases surface area for absorption

 12. Characterized by the Casparian strip

 13. Found in root and stem; gives rise to secondary xylem and phloem

 14. Waterproofing exterior layer of stems

 15. Lateral roots grow from this layer

Integrate and Apply What You've Learned

 16. Which tissues and structures from the hemp plant give rise to marketable products?

 17. Describe the plant structures and tissue types you are consuming when you eat a baked potato and a salad with lettuce, tomatoes, celery, and carrots.

 18. What are the closest analogies you can make between the types of plant tissues and the tissues of animals?

 19. Explain the tissue sources for primary and secondary growth in roots and stems.

 20. Contrast the arrangement of vascular tissues in monocot and dicot roots.

 21. Contrast the arrangement of vascular tissues in monocot and dicot stems.

Analyze and Evaluate the Concepts

22. A hemp plant can grow as tall as a young maple tree. What tissues allow each to reach four to five meters tall?

23. Which of the structural features of roots help them anchor the plant, and which help them take up water and mineral nutrients?

24. Where would you look in a poppy and a rose to locate cells dividing by mitosis in the plant's shoot?

25. Why can peeling off circular rings of bark from birch trees (girdling the tree) kill the entire tree?

26. Look in the produce section of a grocery store for plants in the mustard family: broccoli, cauliflower, brussels sprouts, mustard green, kale, kohlrabi, and cabbage. What organs and tissues have farmers modified through standard breeding techniques to produce these different vegetables?

The Rice Revolution

▲
Efforts to Engineer Rice. Ray Rodriguez; tiny plants growing from cell tissue; recombinant rice plants growing at Applied Phytogenetics.

Raymond Rodriguez embodies two very different versions of the American dream. He founded his own high-tech company, complete with investors, stock options, new products, and the promise of spectacular financial rewards. But he's also the son of Mexican immigrants who moved to California in the 1930s, became citizens, and worked hard to buy a home and send their children to college. Throughout high school, Ray worked along side his parents picking grapes near the agricultural town of Kerman, California. As he went off to college in October 1965, Ray made "a pledge to myself that I would never harvest grapes again. And I was correct. I never did." Twelve years later, he had a professorship at the University of California at Davis, one of the nation's top plant research centers. With grape harvesting well behind him and with two radically different perspectives on agriculture and American enterprise, he became a pioneer in the genetic improvement of rice.

"People devote more time to rice growing, selling, and consuming than to any other single human activity except sleep!" says Dr. Rodriguez. One third of the world's population, or two billion people, get most of their daily calories from rice. The so-called "Green Revolution" of the 1950s and 1960s brought standard plant-breeding techniques and improved fertilizers to the task of increasing rice production and that has helped many developing countries enormously. Nevertheless, the Green Revolution's huge jumps in production leveled off long ago, and now rice harvests are "just creeping up incrementally every year. At some time in the next 20 to 30 years, there's going to be a crossover point where there will be way too many people to feed and not enough productivity. What," Rodriguez asks rhetorically, "is going to stimulate our next green revolution?" He is answering his own question with new genetic research to improve the yields and nutritional qualities of rice.

For a cereal grain, rice has some oddities: The seed needs to be buried beneath six to ten inches of muddy water, and the little embryo inside the seed is unusually "patient." Most cereal seeds (wheat, rice, barley, and so on) germinate within a couple of days: They soak up moisture, start to grow, then burst upward through a thin soil layer and unfurl their first leaves in the sun. Rice seedlings, however, must grow up and up through the flooded rice paddy for a week or more before they can start gathering sunlight and making their own carbohydrates.

The food stored in the rice seed is obviously critical to the rice seedling's survival during this "patience." And many years ago, Ray Rodriguez started studying exactly how the rice embryo gears up to use the stored materials during those first crucial days.

Rodriguez brought unusual tools to the task. He had already been one of the pioneers of genetic engineering, the biotechnology that allows researchers to cut genes from one organism then splice them into another. To modify plants, investigators use a variation on the techniques we saw in Chapter 7. Dr. Rodriguez helped create some of the basic approaches still used to splice animal genes into bacteria, but started applying them to plant biotechnology because agriculture is so central to human well-being. Genetic engineering tools allowed him and his co-workers to probe a series of genes that control germination in rice seeds. Some of these genes instruct the building of plant hormones; the hormones then activate other genes and these, in turn, code for enzymes that break down starch, protein, and lipids in the seed. Together, says Rodriguez, this series drives the growth of the embryo's root and shoot until the little rice seedling can pop up out of the muddy water, gather sunlight, and start photosynthesizing.

Once he knew how this "relay team" of rice genes worked, Rodriguez could use other biotechnology methods to hook a genetic control sequence from an active rice gene to nonrice genes of interest and re-splice them into the rice genome. This gene grouping could then drive the building of other highly useful proteins such as lysozyme or lactoferrin from human milk. These two proteins help protect a breast-fed baby from disease, but infants who are fed formula don't get them. Rice starch is a common ingredient in formula, however, so "If you put those two proteins in rice," says Rodriguez, "you now have a baby formula that can combat disease." His company, Applied Phytogenetics, is engineering rice to carry these and other proteins to improve people's diets and general health. "I think," he concludes, "that genetically engineered cereal grains are a prime example of what people really had in their hearts and minds 20 years ago" when they started developing biotechnology.

How does Dr. Rodriguez put a human breast milk gene into a rice plant and get it to make human proteins in rice seeds? You'll find out in this chapter as we explore three important subjects: Plant reproduction and development; the main kinds of plant hormones and their control over plant growth and development; and the merging of development and hormones in the enterprise of plant biotechnology, including improvements to the world's most important crop and food source, rice. Along the way, you'll discover answers to these questions:

1. How does a plant's life cycle differ from an animal's?
2. How do flowers develop and function to allow plants to reproduce sexually?
3. How do plant embryos, seeds, and fruits develop and grow?
4. What are the structures and activities of plant hormones?
5. How do hormones control development, flowering, and fruit formation?
6. How do plant biotechnologists genetically modify plants?
7. What are the promises and potential pitfalls of plant engineering?

22.1 An Overview of the Plant Life Cycle

To engineer rice to produce novel proteins for human health, Ray Rodriguez must build upon a basic understanding of how rice plants grow flowers and seeds, produce embryos, and develop into adult plants. A key to reproduction and development in plants is the alternation of generations we talked about in Chapter 12: A plant's life cycle involves two different generations of multicellular individuals, one haploid generation and one diploid generation. The diploid generation is the plant we usually see and are familiar with—the green stalk of the rice plant (Fig. 22.1) or the thick trunk and spreading branches of an apple tree. The haploid generation is often much smaller, and, in the case of rice and

Figure 22.1   Life Cycle
of a Rice Plant.

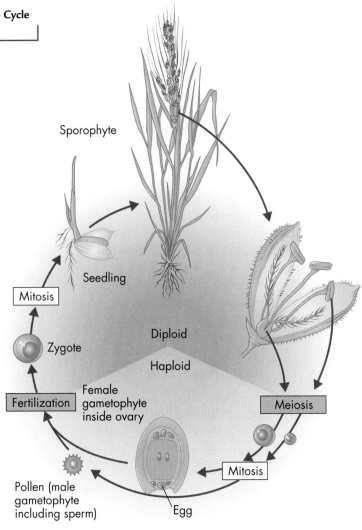

Sporophyte

Seedling

Mitosis

Diploid

Zygote

Haploid

Fertilization

Female
gametophyte
inside ovary

Meiosis

Mitosis

Pollen (male
gametophyte
including sperm)

Egg

GET THE PICTURE

Explorer 22.1

The Life Cycle of the
Rice Plant

other flowering plants, is usually enclosed within the flower and is too small to see without a microscope. An animal—a mouse, a house cat, or a mosquito, for example—has a conspicuous multicellular diploid generation in the young and adult animals. But the haploid phase is a single cell (the egg or sperm) rather than a multicellular individual, as in plants. In certain land plants, including mosses and ferns, both multicellular generations within the plant life cycle are easily visible (review Fig. 12.14).

As flowering plants, rice shows this same alternation of generations, with a diploid sporophyte producing spores and the haploid gametophytes producing gametes (*Explorer 22.1). By understanding the rice plant's life cycle, Ray Rodriguez could

think about which plant parts to modify to produce lysozyme or lactoferrin proteins for infant formula. Should he choose the rice sporophyte or the rice gametophyte?

22.2   Flowers: Sex Organs of Flowering Plants

Dr. Rodriguez decided that because people eat rice grains, he should try to modify the grain itself. So he turned his attention to the sporophytes, the flowers they produce, and the rice grains (seeds) that derive from those flowers.

Flower Structue: The Four Whorls

The flowers of dicot plants like roses are much more familiar than the flowers of most monocots, including rice. So to start our look at flower structure, let's begin with the parts of a rose. Dicot flowers generally consist of four rings of structures; from the outside in, these are the sepals, petals, stamens, and carpels (Fig. 22.2). Only the structures in the inner two rings produce gametes. The outer two rings, while often quite beautiful, are sterile. The **sepals** in the outermost ring and the **petals** in the second ring can have shapes, fragrances, and showy colors that attract animal pollinators.

In a rose or other dicot, structures in the two inner rings of the flower produce eggs and sperm. The male reproductive structures, the pin-shaped **stamens** (STAY-menz) produce the pollen. Within each stamen, the **anther** (the "pin head") contains two pollen sacs and sits high on a **filament,** (the "pin shaft"). This long filament aids pollen dispersal, especially in plants where wind—instead of insects—carries away the pollen. Making up the innermost ring of structures is one or more **carpels** (Greek, *karpo* = fruit); these female parts are often shaped like a wine bottle and may fuse together, as do the carpels in a rose blossom. The base of the carpel is the **ovary** and it houses the **ovules,** the structures that contain the egg and that later mature into the seed. The "neck" of the ovary, or **style,** supports a sticky surface or **stigma** on which pollen grains germinate. Pollen landing on the stigma grows through the style toward the ovules, and a sperm in the pollen tube fertilizes the egg.

The numbers and sizes of flower parts vary from species to species. Roses and apples (also dicots) have flower parts in multiples of five. You can confirm that by turning a fresh apple upside down and counting the little brown sepals (and often stamens) at the bottom, left over from the former flower. In monocot flowers, like the lily, a monocot, sepals and petals are both colorful. Rice is a monocot, too, and a form of grass; grass flowers lack sepals and petals, but they do have similar protective blades surrounding the gamete-producing organs called the palea and lemma (Fig. 22.2b and *E* 22.1). Lilies and irises, both monocots, have flower parts in multiples of three. In many kinds of flowers, both dicots and monocots, each flower contains male and female parts. In other species, such as oaks, the flowers contain either male or female parts, but not both.

Some species, like corn and oaks, have separate male and female flowers on the same plant. These plants are termed **monoecious** (Greek, "one house"). In contrast, willows and hemp (our case study in Chapter 21) have two types of individual plants, some with only male flowers and others with only female flowers. Such plants are called **dioecious** ("two houses").

When most people think of flowers they think of the large, showy blossoms in a florist's shop. Flowers like these do play an important role in the sex lives of the species, as does a peacock's splendid tail. They don't, however, directly attract members of the opposite sex; instead, they attract animal pollinators. For example, bee-pollinated flowers are often bright blue or yellow, and have distinctive patterns called "honey guides" that direct the bee to the nectar. We can seldom see the patterns on bee-pollinated flowers, but bees can, with their vision in the ultraviolet range of the light spectrum. Flowers pollinated by birds and certain moths are often bright red—a color these animals see well. Flowers pollinated by night-flying animals such as bats and particular types of moths are often white and very fragrant, and in many cases, open only at night.

Most plants don't need animals for pollination because their pollen is carried by wind or simply sifts down to other parts of the same plant. Logically enough, self-pollinated flowers are generally small, dull, and lack fragrance. The flowers of ragweed and of birch and hazelnut trees grow on branches where the wind can lift and disperse the pollen on air currents. Rice is self-pollinating and so were Mendel's pea plants (review Fig. 5.3), requiring neither wind nor animals. Rice and pea flowers are inconspicuous and enclosed in little "boxes." Ray Rodriguez points out that this self-pollination trait made it harder for rice

seed companies to develop new vigorous hybrids the equivalent of the many kinds of hybrid corn. Researchers like Dr. Rodriguez can open the closed rice flowers by hand and cross-pollinate them with other rice individuals. It's been much more difficult to perform large-scale matings of one variety to another in an open field, however, and this has slowed rice research over the years. Let's go on, now, to see what happens within a rice or other flower type as it makes gametes.

How Flowers Make Sperm and Eggs

Ray Rodriguez wanted to help prevent infant deaths due to bacterial infection in the developing world. So he decided to splice into rice plants the gene for lysozyme, a protein usually secreted in human milk, because rice flour is a common ingredient in infant formula for bottle-fed babies. Rice flour is made by milling rice grains (seeds), which develop from the rice flower. Thus, Dr. Rodriguez based his experiments on an understanding of how seeds develop, beginning with the formation of sperm and eggs in flowers. You can follow the steps in Figure 22.3.

Pollen Grains and Sperm

Pollen grains form in the anthers of the flower (Fig. 22.3a). Initially, all cells in an anther are diploid. Special cells undergo meiosis and produce four haploid microspores, which are usually very small (as their name suggests). Each haploid microspore develops a tough and often highly complex cell wall. Inside the cell wall of the microspore, the haploid nucleus divides by mitosis, yielding two cells. One cell eventually becomes the pollen tube during pollination, and the other divides again by mitosis, and becomes two sperm, the gametes. The pollen grain is the male's multicellular haploid generation (gametophyte; Fig. 22.3a). In many people, airborne microspores cause the seasonal annoyance of runny noses and itchy red eyes we call hay fever.

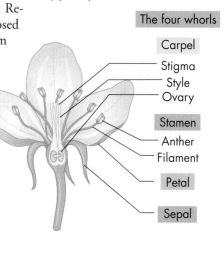

(a) Rose, a dicot

The four whorls
Carpel
Stigma
Style
Ovary
Stamen
Anther
Filament
Petal
Sepal

(b) Rice, a monocot

Anther
Palea
Filament
Lemma
Stigma
Ovary

Figure 22.2 **Flowers: Beauty and Variety.** The flowers of (a) a rose, a dicot; and (b) rice, a monocot.

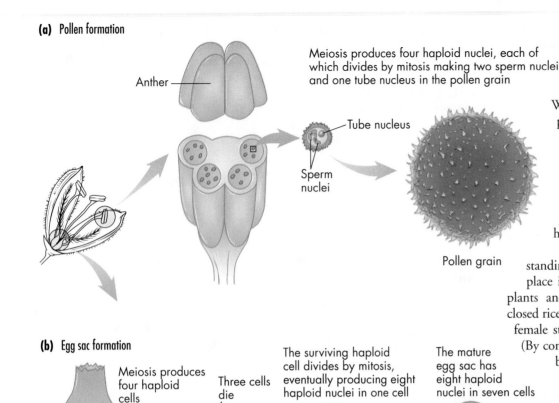

(a) Pollen formation

Anther

Meiosis produces four haploid nuclei, each of which divides by mitosis making two sperm nuclei and one tube nucleus in the pollen grain

Tube nucleus

Sperm nuclei

Pollen grain

(b) Egg sac formation

Meiosis produces four haploid cells

Three cells die

The surviving haploid cell divides by mitosis, eventually producing eight haploid nuclei in one cell

The mature egg sac has eight haploid nuclei in seven cells

Carpel

Ovary   Ovule

Megaspore

The haploid egg cell combines with one sperm nucleus and makes the embryo

The large central cell with two nuclei combines with one sperm nucleus and makes the endosperm

Figure 22.3 Development of Pollen Grains and Female Gametophytes.

Egg Sacs and Eggs

Eggs form in the flower's carpels (Fig. 22.3b). In a carpel, the inner wall of each ovary begins to form small mounds of cells. These develop into ovules, which, if fertilized, eventually become the **seeds** (in rice, the rice grain). In an ovule, certain diploid cells undergo meiosis and produce four large haploid megaspores. Three of these four haploid megaspores die. The surviving cell undergoes mitosis, but the cell's cytoplasm doesn't divide. This eventually produces eight haploid nuclei, which migrate to special positions and the cytoplasm divides, producing a group of seven cells with eight nuclei called the **egg sac;** this is the

female's multicelled haploid generation (gametophyte).

Pollination and Fertilization

We've seen how the diploid spore-producing plant generates the haploid spores; how these produce the several haploid gamete-producing plants; and how these, in turn, form sperm and egg. The plant is now ready for **pollination,** the transfer of pollen to female flower parts. This process may lead to *fertilization,* the fusion of male and female haploid cells.

Ray Rodriguez needed a thorough understanding of how pollination and fertilization take place in rice in order to engineer new types of rice plants and seeds. Pollination takes place within the closed rice flower as pollen from the anther falls onto the female stigma of the same flower (Fig. 22.4, Step 1). (By contrast, in an insect-pollinated flower like a pear blossom, bees carry pollen from the anthers of one pear blossom to the stigma of another.) Once a pollen grain has become stuck on a stigma, it breaks open and the pollen tube grows down the length of the style to the ovary, bringing along the sperm. The pollen tube eventually reaches the ovule and embryo sac with its eight haploid nuclei (Fig. 22.4, Step 2).

Flowering plants undergo a **double fertilization:** One haploid sperm nucleus fuses with two haploid nuclei in the large central cell, forming the triploid nutritive tissue, or **endosperm,** containing the starch that will nourish the embryo within the seed and that can be ground into rice flour and other flours from wheat, oats, and other grains (Fig. 22.4, Step 3). The other haploid sperm nucleus fuses with the haploid egg nucleus, and forms the diploid zygote (Step 4). The zygote divides mitotically and forms the plant embryo, which matures into the diploid plant, the familiar pear tree or rice plant (Step 5). The ovule wall surrounding the zygote and endosperm becomes the seed coat; the entire ovule becomes the seed; and the ovary wall surrounding the seed becomes the fruit (Step 5). You can see this animated in *E* 22.1.

If double fertilization stopped today, most humans would die of starvation in a few months. That's because the greatest source of calories for most of Earth's people is endosperm from rice or wheat. Endosperm supplies calories and some vitamins but, unfortunately, it lacks many other nutrients. In Southeast Asia alone, a quarter of a million children become blind every year because rice forms almost their whole diet yet it lacks the vitamin A they need to develop proper vision. An additional 124 million children around the world have vitamin A levels above the range that leads to blindness but still so low it compromises their health.

Early in 2000, European researchers announced that they had genetically engineered rice endosperm to produce carotene, or provitamin A, the orange substance in carrots, pumpkins, and other plants. Our bodies use provitamin A from foods to form the vitamin A we need for proper vision. The production of provitamin A turns the rice grains a golden color (Fig. 22.5). More work is needed to transfer the yellow rice trait

Figure 22.5 **Golden Rice: Potential Help for Ailing Vision.** Using the methods of genetic engineering, European researchers turned rice endosperm from white to gold. Rice endosperm normally has very little carotene, a precursor to vitamin A. The genetically modified rice, however, produces this golden substance in abundance. Biologists hope that they can introduce the genes for provitamin A into the local varieties of rice eaten by children at risk for vitamin A deficiency. This alteration could provide a low-cost, effective way to improve the vision and general health of millions of people living with vitamin A deficiency.

into a hardy commercial crop, but this genetically modified food does have enormous potential to save millions of people from impaired vision.

Flowers are shaped by the forces of evolution, and one of the more interesting aspects of plant reproduction is the evolutionary interactions of flowers with their animal pollinators. Among the products of this co-evolution are flowers that open at night and are pollinated only by nocturnal bats, and the largest, foulest-smelling flower on Earth, which attracts its fly pollinators by giving off a stench like rotting flesh (Fig. 22.6).

Figure 22.6 **Co-evolution of Flowers and Their Pollinators.** The world's largest recorded flower, *Rafflesia arnoldii*, grows in the Sumatran rain forest and smells like rotten meat, attracting fly pollinators.

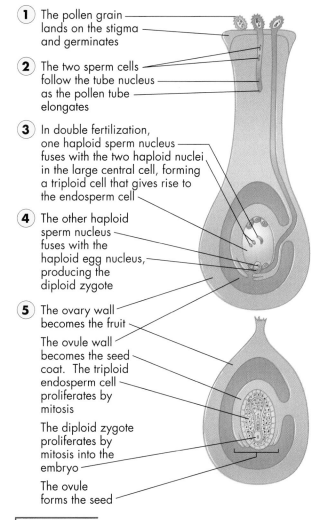

① The pollen grain lands on the stigma and germinates

② The two sperm cells follow the tube nucleus as the pollen tube elongates

③ In double fertilization, one haploid sperm nucleus fuses with the two haploid nuclei in the large central cell, forming a triploid cell that gives rise to the endosperm cell

④ The other haploid sperm nucleus fuses with the haploid egg nucleus, producing the diploid zygote

⑤ The ovary wall becomes the fruit

The ovule wall becomes the seed coat. The triploid endosperm cell proliferates by mitosis

The diploid zygote proliferates by mitosis into the embryo

The ovule forms the seed

Figure 22.4 **Pollination and Fertilization.**

22.3 Development of Embryos, Seeds, and Fruits

We've seen that most of a rice grain is filled with starchy endosperm and that European researchers have modified it to produce vitamin A. In his quest to produce human lysozyme, Dr. Rodriguez could have chosen to alter rice endosperm. Instead, however, he decided to capitalize on a different system: the way the tiny embryo in the rice grain breaks down stored endosperm for its own rapid growth. To understand why, we need to see how a seed forms, how it stores endosperm, then how the embryo uses it.

Inside young ovules, the endosperm receives additional nutrients from the parent tissues and enlarges. In many plants, such as pears and peaches, the developing embryo inside the seed uses up the endosperm as it grows, and that nutritive tissue shrinks to a thin sheet. In rice and other grains, however, the endosperm continues to fill up most of the seed.

How a Plant Embryo Develops

To focus on how the embryo develops within the seed, let's start with the familiar peanut; it houses a plant embryo that you've no doubt seen and can easily dissect for

this discussion. Then we'll return to Dr. Rodriguez's work with rice seeds and embryos.

The peanut plant, a member of the pea family, has delicate yellow flowers that bloom around the plant's lower branches and leaves (Fig. 22.7a). After peanut flowers self-fertilize (as did Mendel's pea plants), the petals drop off. Curiously, the ovary of each flower then grows downwards and pushes into the soil, where the embryo starts to grow surrounded by maternal tissues. Embryo cells first divide by mitosis into a heart-shaped mass (Fig. 22.7b). The lobes of the heart become the cotyledons (the first leaves) with a little mass between them that will become the shoot apical meristem. The rest of the embryo becomes the embryonic stem and embryonic root. From that point on, growth in the meristems takes over and eventually a recognizable plant forms.

You can see some basic embryonic structures by buying a bag of peanuts in the shell, peeling the woody outer covering from one, then closely examining the ordinary peanut inside. The tough peanut shell is actually the **fruit** formed from the enlarged ovary wall, while the reddish papery coating around each peanut is the **seed coat,** the wall of the ovule. The two oval halves of the peanut are the dried remains of the **cotyledons,** the embryo's fleshy seed leaves. Peanut embryos, having two cotyledons, are easily recognized as dicotyledons (dicots). Plants with just one cotyledon—rice, lilies, and palms, for example—are called monocotyledons (monocots), as we saw in Chapter 12.

In the peanut, the tiny fleck between the two large fleshy cotyledons is the dried embryo; if you look closely enough you'll see a pair of tiny leaves attached to the **epicotyl** (*epi* = upon), the embryonic stem above the cotyledon. The short length of stem below the attachment of the cotyledons is called the **hypocotyl** (*hypo* = below). Below the hypocotyl is the embryonic root.

Inside the peanut seed, this little embryonic plant becomes dormant. It stops growing; it dramatically slows its use of oxygen and production and use of ATP; it loses water to ten percent or less of its composition; and it waits for the proper conditions before it resumes development and

(a) The peanut plant and flower

(b) Embryo growth

Petals

Stigma

Anther

Style

Ovary   Ovule

Embryo

Cotyledons

The embryo divides by mitosis

Attachment to maternal tissues

Future shoot

Future root

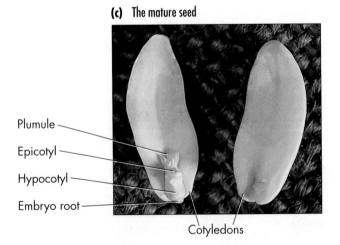

(c) The mature seed

Plumule

Epicotyl

Hypocotyl

Embryo root

Cotyledons

Figure 22.7   **Development of a Peanut Embryo.**
(a) Yellow flowers grow on the lower branches of the peanut plant. After fertilization, the flower petals fall off, and the ovary and style bury themselves into the ground. (b) Within the ovule inside the ovary, cells of the zygote divide by mitosis and produce a globe-shaped and then a heart-shaped embryo. (c) The lobes of the heart develop into the cotyledons, and between them the plant axis forms. This axis includes the future shoot—consisting of epicotyl and hypocotyl—and the embryonic root.

becomes a plant. That resumption of growth in the seed after a period of dormancy is called **germination,** and the dormancy leading up to it can be amazingly long: Peanuts can stay dormant for years—if they aren't roasted for a snack first! Rice seeds can wait decades, and some seeds, including lotus, can last hundreds of years, dormant but alive.

Germination: The New Plant Emerges

Ray Rodriguez decided to focus on the germination stage of rice development in order to engineer rice to produce human lysozyme. Now we can see why. He discovered that germinating rice seeds turn on genes that encode immense quantities of the enzyme amylase. This, in turn, rapidly digests the starch stored in the endosperm. (Your salivary glands also produce amylase as one of the initial steps of digestion; review Fig. 17.10.) Dr. Rodriguez reasoned that if he could induce the germinating rice seeds to make large quantities of human lysozyme in place of amylase, then technicians could harvest this human protein and add it to infant formula, where it could help protect nursing babies from bacterial disease.

In grains like rice and barley, germination is called **malting.** As malting takes place, cells in the seed pro-

duce amylase (Fig. 22.8, Step 1). Amylase breaks down starch into the sugar maltose (Step 2), the sweet substance used in making beers, malt liquors, and malted milk. Other enzymes convert maltose to sucrose and eventually glucose, which fuels the growth of the sprouting plant (Step 3).

Now Ray Rodriguez needed to figure out the control mechanisms that regulate when and how germinating rice grains produce amylase and by extension, the spliced-in lysozyme. It turns out that plant hormones and other plant growth regulators control germination. So that's our next topic.

22.4 Plant Growth Regulators

Plant researchers have known for half a century that plant hormones control a plant's stages of development, growth, and reproduction. One early area of research was a peculiar disease of the rice plant known to Japanese farmers as "foolish seedling disease" because afflicted rice plants grow far taller and faster than normal after germination, and in the process become spindly and blow over in the wind and die. It turns out that the disease is caused by a chemical secreted by a fungus that

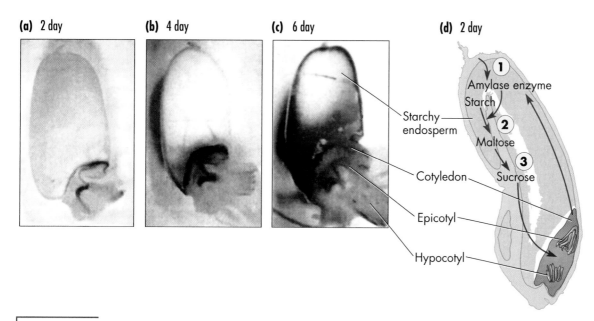

(a) 2 day **(b)** 4 day **(c)** 6 day **(d)** 2 day

Amylase enzyme
Starch
Maltose
Sucrose
Starchy endosperm
Cotyledon
Epicotyl
Hypocotyl

Figure 22.8 **Germination.** To study the regulatory mechanisms at work during germination, Dr. Rodriguez bioengineered rice grains to produce a marker enzyme. The blue stain indicates the amount of enzyme activity at 2, 4, or 6 days after germination and in various parts of the seed. Notice how much the embryo grows during this period.

invades and infects rice plants. This chemical is a plant hormone, or growth regulator. Let's look at the whole spectrum of plant growth regulators and see how these important chemicals control not only germination and shoot growth, but also the budding of flowers, the falling of leaves, and many other aspects of plant biology. Then we can return to Dr. Rodriguez's rice experiments in our last section and see how he used plant growth regulators to help him succeed in engineering a new, more nutritious kind of food.

Types of Plant Growth Regulators

There are five major classes of plant hormones—three that promote and regulate growth, and two that inhibit growth or promote maturation (Table 22.1 and *E*xplorer 22.2).

READ ON

*E*xplorer 22.2

Plant Growth Regulators

Gibberellins

The culprit behind foolish seedling disease belongs to the class of plant growth regulators called **gibberellins** (JIB-er-ELL-ins) and demonstrates what a global effect plant hormones have on a plant. Gibberellins are made in a variety of organs, such as young leaves, embryos, and roots. They can move through the plant's vascular system and help regulate plant height. Too little gibberellin results in dwarf plants, but too much results in long, pale, "foolish" stems. Excess gibberellins can turn a rosette of cabbage leaves into a 15-foot giant (Fig. 22.9). Gibberellins induce the seeds of rice, barley, and other grasses to germinate. Plant scientists have found that gibberellins are also involved in flowering, fertilization, and the growth of fruits, new leaves, and young branches.

Auxins

Like gibberellins, **auxins** (AWKS-ins) generally promote growth. The name, in fact, comes from the Greek *auxein,* meaning "to increase or augment." The chemical structure of natural auxin is related to the amino acid tryptophan, a building block found in most proteins. Instead of being produced in various tissues and traveling through the vascular system like gibberellins, auxins are generally made in shoot

Table 22.1 Internal Regulators of Plant Growth and Development

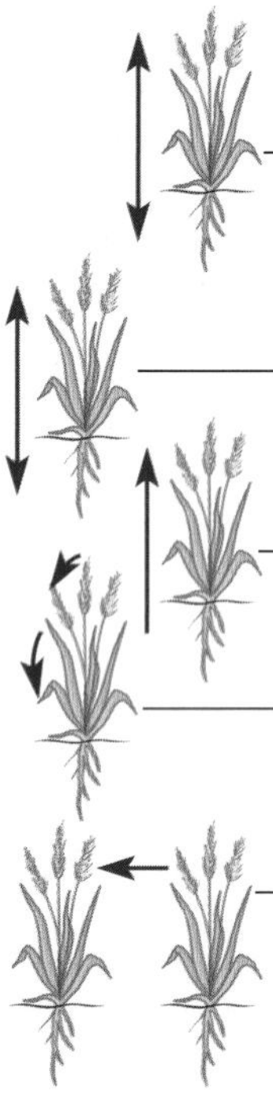

Regulator	Transport	Action
Hormones		
Gibberellins	Upward and downward in vascular system	*Promote growth:* Promote stem lengthening in dwarf and rosette plants and seed germination in grasses; involved in flowering and fertilization; involved in growth of new leaves, young branches, and fruits
Auxins	From tip of shoot downward in vascular system	*Promote growth:* Augment growth by cell elongation; inhibit growth of lateral buds; foster growth of ovary wall; prevent leaf and fruit drop; orient root and shoot growth
Cytokinins	From root upward in vascular system	*Promote growth:* Stimulate cell division; kindle growth in lateral buds; block leaf senescence
Abscisic acid	Short distances in leaf and fruit	*Inhibits growth:* Opposes the three growth-promoting hormones; induces and maintains dormancy
Ethylene	Through air, as a gas	*Promotes maturation:* Enhances fruit ripening; promotes dropping of leaves, flowers, and fruits
Pigment		
Phytochrome	Not transported; remains in cell that produces it	*Detects light:* Changes form in response to light; mediates flowering, germination, growth, and plant form

apical meristems and developing leaves. They diffuse from the site of production downward toward the roots.

Auxins help bring about growth by promoting cell elongation. Auxins also act in other ways that prevent leaves, fruits, and flowers from falling off prematurely and that inhibit growth in lateral buds and roots in favor of growth at the apical meristem. As we saw in Chapter 21, this controls whether a plant, including hemp (see Fig. 21.12), is tall and narrow or short and bushy.

In high concentrations, auxins can cause uncontrolled growth and plant death. In fact, some herbicides are based on this principle. The synthetic auxin 2,4-D, for example, is an active ingredient in Agent Orange, a powerful herbicide the military sprayed to defoliate jungle vegetation during the Vietnam War. Such herbicides appear to turn on such rapid tissue growth that the phloem becomes plugged with nonconducting cells. Without a functioning phloem, the plant literally grows itself to death.

Cytokinins

The third class of plant growth regulators are the **cytokinins** (SIGH-toe-KY-nins), and they generally stimulate cell division, including cytokinesis or division of the cytoplasm (review Fig. 4.5). Cytokinins are chemically related to adenine, a base in DNA. Cytokinins are less mobile than auxins (and far less so than gibberellins) and appear to move in opposition to auxins—from root upward to shoot, not shoot downward to root. Cytokinins promote the growth of lateral buds, not shoot tips, as do auxins. Like most plant hormones, cytokinins also have other effects, including the prevention of leaf aging, or **senescence.**

Abscisic Acid

If plants had only gibberellins, auxins, and cytokinins, they would grow constantly. Sometimes, however, it's more advantageous for the plant to stop growing—to close its stomata, decrease the level of photosynthesis, drop aging leaves, or become dormant like a rice grain waiting to germinate. **Abscisic acid** (ABA) is a growth inhibitor that blocks protein synthesis and new growth. ABA moves only short distances from where it's produced. The name "abscisic acid" comes from ABA's role in accelerating **abscission,** the dropping (literally, "cutting away") of leaves and fruits. Despite its name, plant scientists now think that ABA's main role is to induce and maintain dormancy, or metabolic slowdown, espe-cially in buds, and to promote the closing of stomata in leaves, which prevents excess water loss.

Ethylene

The fifth major plant growth regulator, **ethylene** (ETH-ih-lean), is a simple two-carbon molecule that exists as a gas at normal temperatures and disperses from one plant or plant part to another by air. Ethylene is behind the old adage about one rotten apple spoiling the whole barrel. The hormone is produced by ripening fruits, and it stimulates ripening in nearby fruits. Ethylene also stimulates the aging and dropping of leaves and fruits, and it may have an important role in plant self-protection.

<table>
<tr><td>**22.5**</td><td>**How Growth Regulators Work in the Plant Life Cycle**</td></tr>
</table>

Dr. Rodriguez knew the five possible players responsible for causing the rice plant to move through its life cycle of germinating, flowering, setting seed, and aging. He was especially interested, of course, in which of the five hormone types turns on the production of starch-digesting amylase in germinating rice, and what, before that, turns on the hormone itself.

Hormonal and Environmental Control of Germination

Seeds of some plants, like willows and poplars, germinate shortly after they reach a new location and take in enough water to soften the seed coat so that the tiny plantlet can burst out. But an annual such as a daisy, for example, that sprouted in late summer might not grow to full size and produce its own seeds before the cold temperatures and dry conditions of winter set in. Many temperate plants, therefore, germinate in spring or early summer after surviving winter as a dormant embryo in a seed. Then in spring, rising temperatures, lengthening hours of daylight, and rain or melting snow trigger hormonal activities within the embryo that, in turn, trigger germination.

What about in rice? It turns out that moisture is the environmental trigger that stimulates the release of gibberellin in rice. This hormone then turns on the genes for the amylase enzyme, and once built, the enzymes start breaking down the starch fuel, which the embryo's growth (review Fig. 22.8).

Figure 22.9 **Hormones and Tall Cabbages.** Gibberellin placed on a cabbage plant (*Brassica oleracea*) causes the shoot to elongate and flower (right) instead of remaining in the normal rosette form (left). Bolting, or the growth of the shoot to form a flowering stalk, is controlled by gibberellin and takes place naturally in a biennial before flowering in the second year.

For other kinds of seeds, light, not moisture, controls germination. The seeds of some desert plants, for example, germinate only in deep, dark, cooler soil. Presumably, desert plant seeds that germinated in the light (near the soil surface) would tend to dry out and die.

Lettuce and many weeds have an opposite situation: Their seeds are so tiny that emerging seedlings buried deeper than a few millimeters run out of food reserves before reaching the sunlit surface, where they can begin to photosynthesize. Their germination depends on light. Knowing this, biologists at Oregon State University tried plowing and planting fields at night instead of during the day. They found a dramatic reduction in the subsequent growth of weeds whose seeds require light to germinate compared to identical fields plowed in daylight. The biologists think that some farmers could avoid using weed-killing herbicides simply by plowing at night.

Environmental Cues for Plant Growth and Orientation

If you plant a rice seed upside down, does the root grow into the air and the shoot into the soil? No, roots grow down and shoots grow up no matter how the seed falls into the dirt. What controls this behavior? A plant's orientation is based on **tropisms** (TRO-piz-ims), bending or directional growth in response to environmental stimuli. (Greek, *tropism* = turning toward.) A tropism is *positive* if the organism's orientation or growth is *toward* the stimulus, and *negative* if the organism's orientation or growth is *away from* the stimulus. Botanists have named a number of tropisms by the external stimulus that causes them. For example, root growth toward water is called **hydrotropism;** stem growth (as in vine tendrils) toward the touch pressure of a supporting object like a wall is **thigmotropism;** the orientation of a plant part toward the ground is **gravitropism** (turning toward gravity); and the orientation toward light is

phototropism. These last two tropisms help explain how a rice seedling's roots grow down and its shoot grows up.

Gravity and Plant Growth

A root grows down and a shoot up because plants are sensitive to gravity. The root grows down toward the pull of gravity, and the shoot grows up against the pull of gravity (Fig. 22.10a). In a plant laid on its side, cells on the upper side of the root become longer, while cells on the underside remain the same length. As a result of this differential growth, the root bends downward. What causes this differential growth? Experiments have revealed that root cells contain dense starch granules (called **amyloplasts**) that sink to the bottom of the cells. Some plant physiologists suspect that amyloplasts may respond to the pull of gravity and somehow cause auxin to move to the lower side of a root and block the elongation of lower cells. As the upper cells continue to elongate, the roots bend and grow downward, thrusting toward a more likely source of moisture and minerals.

Studies of mutant tomato plants that grow horizontally and not vertically suggest the mutants lack cell receptors for auxins. Auxins and perhaps other plant hormones may bind to receptors in cell membranes, as do many hormones in animals.

Light and Plant Growth

Plants need light for photosynthesis, but sometimes they can't get access to it. For example, a wild rice seedling might, by chance, germinate beneath the bushes at the edge of a forest. The new plant can grow away from the shade and toward light, however, and this will maximize its exposure to the sun.

The first experiments to probe how plants orient toward light appeared in a book that Charles Darwin and his son Francis published in 1881. The Darwins knew that shining light on the side of a grass plant causes it to bend toward the light source a short distance back from the growing tip (Fig. 22.11a). They performed a series of experiments on oat and canary grass seedlings to find out where the controller of that bending response resides: in the curving portion of the seedling or in its tip. They shielded the section of the seedling that actually bends toward the light with a piece of foil to keep the light out and observed that the seedlings bent toward the light anyway (Fig. 22.11b). Then they covered just the tip of a seedling with a foil cap and saw that this time, the seedling didn't bend to-

Figure 22.10 **Gravitropism: Growth Toward or Away from Gravity.** (a) Due to gravitropism, a plant placed sideways grows upward. (b) Differential cell growth leads to root building. (c) Starch grains (amyloplasts) sinking to the bottom of starch-storing cells may help signal "up" from "down."

(a) In a plant laid on its side, the shoot bends upward, and the root bends downward

(b) The root tip bends because cells on the upper root surface elongate relative to cells on the lower root surface

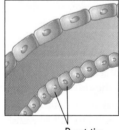

Root tip cells

(c) Root tip cells

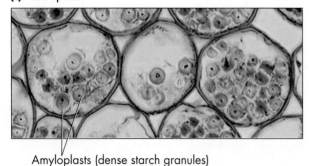

Amyloplasts (dense starch granules)

Figure 22.11   A Plant's Tip Controls Phototropic
Bending. (a) By testing grass seedling, Charles and Francis Darwin noted that grass seedlings bend toward a source of light. (b) By shading the bending portion, they deduced that the controller is not located there. (c) Shading the tip revealed that this portion of the plant responds to light by sending a signal to the stem portion, which then bends.

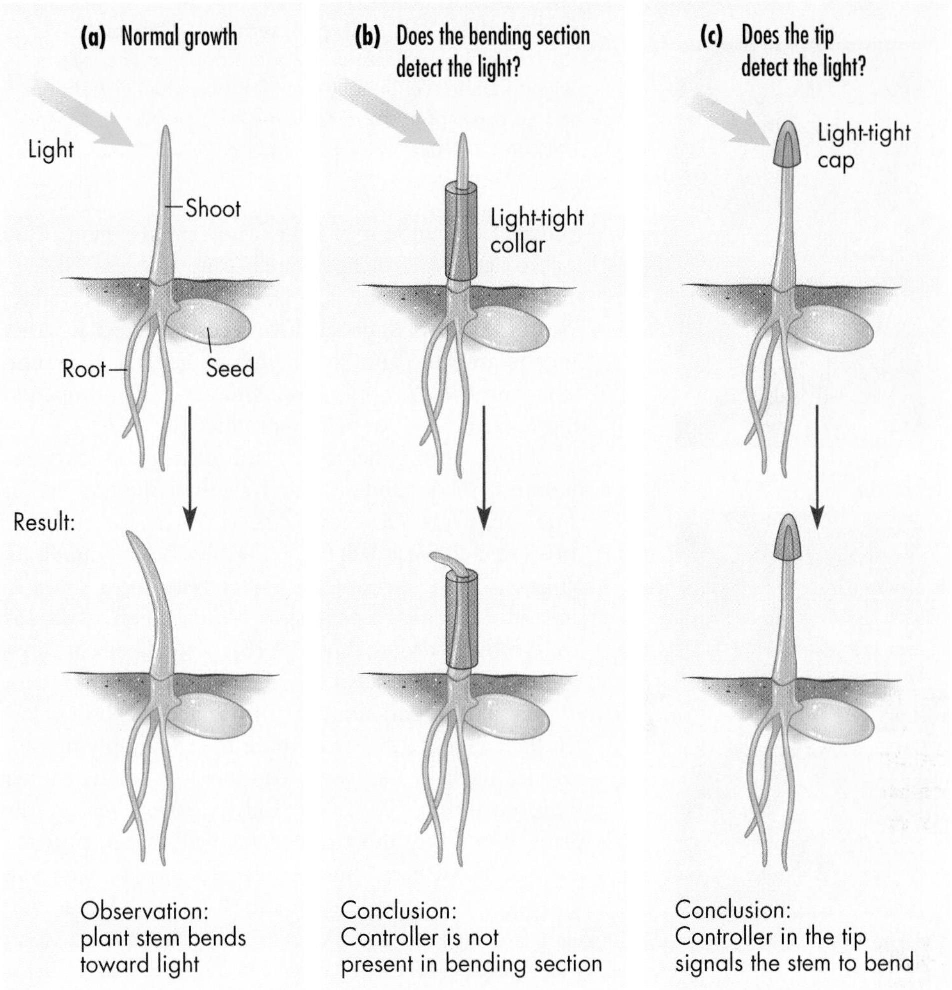

(a) Normal growth

Light
Shoot
Root
Seed
Result:

Observation:
plant stem bends
toward light

(b) Does the bending section detect the light?

Light-tight collar

Conclusion:
Controller is not
present in bending section

(c) Does the tip detect the light?

Light-tight cap

Conclusion:
Controller in the tip
signals the stem to bend

ward the light. Their conclusion: The seedling's tip sends a signal to its lower portion, causing it to bend in the appropriate direction (Fig. 22.11c).

Later research showed that the growth regulator, auxin, comes from the seedling's tip and accumulates on the side away from the light. Evidence shows that the accumulated auxin causes the pH inside cells to drop and the acidic environment causes an enzyme to become active. The enzyme weakens cellulose in the cell walls, allowing these walls to stretch and the entire cell to elongate. This cell stretching and lengthening, in turn, causes the seedling to bend toward the light.

Other Plant Movements

Plants don't always move toward or away from a stimulus. Leaves of the sensitive plant *Mimosa pudica,* for example, quickly droop if touched anywhere along their length (Fig. 22.12). The leaves of the Venus flytrap also snap shut if a small frog or insect triggers any of the little hairs inside the trap. Movements not toward or away from a stimulus are called **nastic responses,** and they're often not slow nor based on the growth of cells, as tropisms tend to be. For example, the Venus flytrap snaps shut within seconds and then slowly reopens.

Another category of nastic responses involves so-called sleep movements, wherein plants' leaves droop at night but lift in the daytime. Even if you move the plants into complete darkness, these sleep movements will continue for a few days, drooping at night, lifting during the day (Fig. 22.13). Sleep movements are strong evidence that plants, like animals, have internal biological clocks that meter the passage of time and trigger regular responses.

Shape Changes in Response to Light

Light can profoundly affect a plant's shape. A good example is a crabgrass plant growing in a dark spot, say, under a board—the plant becomes long, spindly and

pale yellow, not healthy and green. Deprived of direct light, the plant expends little energy making chlorophyll and instead grows longer and longer until it grows out from under the board and reaches the light. Even in less extreme shade, plant survival depends on the plant's ability to detect light levels and adapt by growing toward it.

How Plants Detect Light

Plants have a light-sensitive pigment called **phytochrome** (literally, "plant color") that, like a light meter, allows them

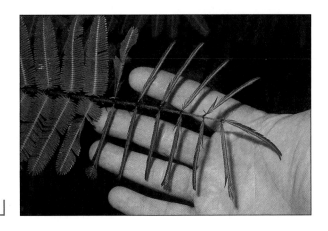

Figure 22.12   Nastic Response.
A sensitive plant's leaves are normally openly spread and pointed upward. When disturbed, however, the leaves quickly close up and droop. This occurs because a large cell in the axil beneath the leaf's connection to the stem rapidly loses water when disturbed, and deflates, allowing the whole leaf to droop.

3:00 P.M.

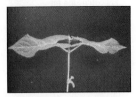

9:00 P.M.

11:00 P.M.

2:00 A.M.

6:00 A.M.

12:00 NOON

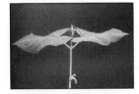

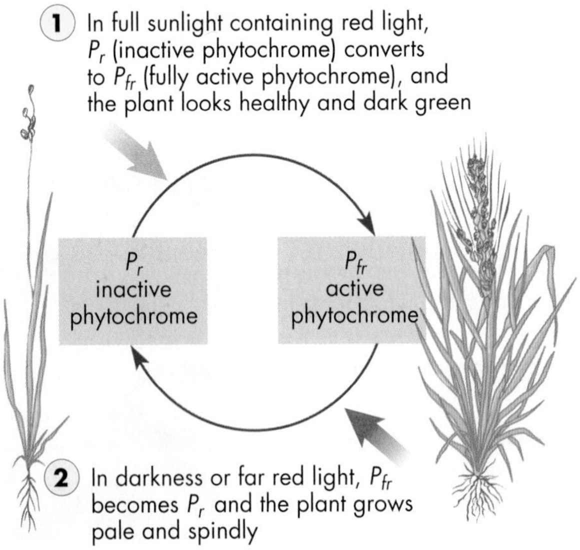

Figure 22.14 Phytochromes and Plant Growth Patterns.

Figure 22.13 **Sleep Movements in Plants.** Leaves of the bean plant *Phaseolus vulgaris* droop at night, extend by noon, and remain fully open for more than eight hours even when kept in the dark. These "sleep movements" are evidence of a biological clock in plants.

to detect the amount of light in the environment. Phytochrome isn't a true hormone because it doesn't diffuse from cell to cell, but it is an internal regulator. Phytochrome allows light-sensitive seeds to detect whether light is available and so whether to germinate or not. Phytochrome also spurs flowers to develop and open at appropriate times for plant reproduction.

Phytochrome helps a plant distinguish between darkness, shade, and sun, and also tell the amount of time spent in each. How? The chemical has different forms that change, depending on the amount and wavelength of light striking it. Phytochrome responds to light of a specific color, just as do the pigments in our eye's retina (review Fig. 19.18). The form of phytochrome called P_r absorbs red light, which is abundant in normal daylight. Another form called P_{fr} absorbs far-red light, but it's easiest to think of it as simply not absorbing daylight very well. Another way to remember P_{fr} is to think of the "fr" as "fully reactive," because the plant responds to this form of phytochrome by growing greener and lusher. Thus, in full daylight, P_r changes to P_{fr} (Fig. 22.14, Step 1) and the plant grows green and lush. But in the dark, P_{fr} spontaneously changes to P_r, the nonreactive form (Step 2), and the plant grows long, pale, and spindly. Phytochrome is an important growth regulator affecting plant shape, but it's not

1 In full sunlight containing red light, P_r (inactive phytochrome) converts to P_{fr} (fully active phytochrome), and the plant looks healthy and dark green

P_r inactive phytochrome

P_{fr} active phytochrome

2 In darkness or far red light, P_{fr} becomes P_r and the plant grows pale and spindly

the only one. Substances at the tips of the plants also help control how they grow in response to their environments.

Growth Regulators and Plant Shape

Plants come in all different shapes and sizes. Spruce trees, for example, are usually conical, with a single main trunk and numerous side branches that are long at the base and shorter at the top. Maple trees, in contrast, are roughly spherical, with several main branches but no central trunk, and branches that roughly form a sphere, not a cone. These cone and sphere shapes come from **apical dominance.** Apical dominance is, literally, domination by the tip of the plant and is based on auxin from the apical bud moving down the shoot and inhibiting lateral buds. Auxin promotes growth in the stem, but blocks growth of lateral buds. Because auxin is continuously broken down as it moves down the stem, its concentration drops off. That's why buds closest to the tip of the main stem are most inhibited (and smallest), while branches farther from the tip of the plant are the least inhibited (and oldest and largest), resulting in a conical shape. Cytokinin also comes upward from the roots, counteracts the effects of auxin, and causes the lateral buds and the lowest branches to grow more strongly. In a plant with a more spherical shape, the apical bud may produce less auxin, and lateral buds and branches may be less sensitive to inhibition by the hormone; thus, the branches both high and low on the stem can grow strongly. If a gardener or a browsing animal picks off a plant's apical bud, the buds farther down the stem are released from apical dominance and can start growing, producing a bushy shape. Commercial Christmas tree growers take advantage of apical dominance by pruning back the tips of branches. This allows buds on inside branches to grow and yields a fuller, more desirable tree.

By now, we've seen that plants usually respond to hazards or opportunities in the environment by growing and changing in form under the control of plant growth regulators. Now let's see how this responsiveness also includes life-cycle events such as flowering, fruit formation, dropping leaves, and dying back before winter.

Control of Flowering and Fruit Formation

Plants tend to flower at specific times of year: crocuses in early spring; petunias in mid-summer; poinsettias at Christmas, and so on. How do plants "know" when to flower? A plant's germination and growth are synchronized to the seasons when that species has the greatest potential for success, and the same is true for flowering and generating fruit. Environmental factors change with the seasons—temperature, moisture, and the number of hours of sunlight in a day—and these are the outside cues, the external regulators, that control the internal hormonal triggers of events in the plant's life cycle. Seasonal timing is critically important for annual plants such as corn, which flower and produce seeds just once in a single season. But it's also crucial for perennials such as roses and oak trees, which also must produce flowers and fruits (rose hips and acorns) at the right times.

Temperature and Flowering

Some plants, including beets, carrots, turnips, and other biennials, flower only after they're exposed to cold. Exposure to cold followed by warming signifies that winter has come and gone. Because flowering comes early in the spring, biennials produce seeds early in their second summer, and the seedlings have plenty of time to grow into hardy plants that can withstand the following winter.

Light and Flowering

For many plants, temperature alone is too risky a trigger for flowering. A cold September followed by a warm October and a harsh November could find a plant covered with flowers but buried in snow. Many plants track the seasons by **photoperiod,** the length of light and dark periods each day, rather than by daily temperature. Researchers noticed that some plants such as rice require days *shorter than* a particular length to bloom (for rice, it's days with fewer than 12 hours of daylight); they called these plants **short-day plants.** Other plants like spinach require days longer than a certain length (for spinach, longer than 14 hours), and these they called **long-day plants.** Plants that flower without regard to photoperiod are **day-neutral plants.** Generally, long-day plants blossom in spring, when the days are becoming longer, and short-day plants flower in the late summer, when the days are becoming shorter.

Do plants actually measure the length of daylight or the length of the night? (In other words, is a short-day plant really a long-night plant?) To find out, botanists exposed ragweed (which, like rice, is a short-day plant), to a short day and long night but interrupted the period of darkness with a short flash of light. The result: no flowers formed. Because interrupting the night blocks flowering, plants apparently measure night length rather than day length. Thus even a quick peek at a short-day plant like a poinsettia stored in a dark closet before Christmas may interrupt the plant's "long night" and prevent it from setting flowers.

The phytochrome "hourglass" is the agent that allows such accurate timing of photoperiod, and it works so accurately that all the individuals of a species can flower within a day or two of each other. This provides a ready source of pollen that wind or animals can convey to other blossoms of the same species. And a plant's photoperiod often coincides with the most active food-seeking portion of an insect's (or other pollinator's) life cycle—evidence of the coevolution between plant and pollinator.

Hormones and Flowering

Plant scientists wondered whether phytochrome in the cells of each flower bud measures the day-night cycle for that particular bud or whether the signal to flower comes to the bud from other parts of the plant. To test these possibilities, experimenters exposed chrysanthemum buds to one light cycle and the chrysanthemum leaves to another and checked the plants for flowers. The results showed that leaves, not the bud, were responsible for sensing photoperiod. If the leaves control flowering, however, then they must send a flower-generating signal to the bud to open into a flower. Plant physiologists, however, are not yet sure what the flower-inducing hormone is.

Triggers to Fruit Development

Fruits have to develop at the right times of year, too, so that they can attract animals or allow other agents to disperse the seeds within the fruits to new locations. Although fruits help protect the seeds and disperse them, they are energy-expensive organs to produce—especially large, juicy ones such as pears, plums, and oranges. Not surprisingly, some plants ensure that fruits develop only around viable seeds.

The dozens of tiny white seeds on the outside of a red strawberry, for example, help control the development of a normal berry. If all the seeds are removed when the berry is still small and green, it never enlarges and ripens. If, however, one removes young seeds from the strawberry and treats the berry with auxin, it will grow normally. This suggests that the seeds promote berry growth by secreting auxin.

Once fruits have developed, the airborne hormone ethylene causes them to ripen. When a peach ripens, it gets softer through the partial breakdown of cell walls, sweeter through the conversion of organic acids and starches to sugars, more fragrant through the production of aromatic molecules; and it turns from green to yellow-red by the formation of carotenes and other pigments. These changes enhance the fruit's appeal to animals that eat it and distribute its seeds, and the plant growth regulator ethylene is behind most of these changes. Since most fruits release ethylene gas as they begin to ripen, green fruits ripen faster if you place them in a plastic bag along with ripe fruit.

How Plants Age and Prepare for Winter

Dropping leaves and dying are natural parts of the plant life cycle. Dr. Rodriguez notes that rice plants will produce two or three crops of seeds if trimmed back after each harvest, but after one or two seasons they eventually die. Trees and other perennials often drop their leaves and become dormant during the cold months. These changes are related and, like the other parts of the life cycle, are controlled by plant growth regulators.

Senescence and Abscission

Senescence, or growing old and dying, and **abscission,** or "scissoring" away mature fruits and dying leaves, are technical terms for the events of autumn we all know and enjoy: the turning of green leaves to glorious shades of red, gold, orange, and purple, followed by the fading and finally the falling of the leaves (Fig. 22.15). Several external and internal events control this sequence. In many annual plants, flowering itself is a key to senescence, because when the apex of each shoot converts into a flower bud, it ends the open growth potential of that branch. When the last branch tip has converted to a flower, the plant stops growing and soon begins to senesce.

Figure 22.15 **Senescence Unmasks the Flaming Colors of Fall.** The breakdown of chlorophyll allows yellow, red, and orange light-gathering pigments to show. These vibrant trees are growing along the Kancamagus Highway (Route 112) in New Hampshire. Eventually, the colors will fade and the leaves will drop off.

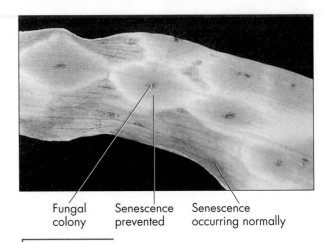

Fungal colony   Senescence prevented   Senescence occurring normally

Figure 22.16 **Cytokinins Block the Essence of Senescence.** A fungal infection dotting a browned, senescent corn leaf causes islands of green to remain. The fungus releases a plant hormone—in this case, cytokinin, which blocks ABA, the hormone causing senescence.

In perennials, the plant conserves resources over the dry, cold months of winter, and the **deciduous** growth habit (the dropping of leaves) is very common. In the cold and dark of winter, leaves are much less efficient photosynthetic organs and provide an avenue for too much water loss. In fall, a deciduous tree such as a sugar maple withdraws valuable nutrients from the leaves, including sugars and amino acids, and transports them to the trunk and roots, where they're stored. When the chlorophyll breaks down, other brilliantly colored leaf pigments still intact in the leaves can show through, including red or purple anthocyanins, yellow xanthophylls, and red and orange carotenes. Eventually, the leaves fall, and the plant survives winter in a dormant state.

Which plant hormones control senescence and leaf drop? Researchers noticed that a certain kind of fungus will cause a green spot to remain on a yellowing corn leaf, and they determined that the fungus releases cytokinin (Fig. 22.16). These results make them suspect that autumn conditions probably influence cytokinin levels in the plant, and these, in turn, influence when leaves senesce.

Environmental cues and plant hormones also govern when leaves fall. During the long, sunny days of summer, each leaf produces large quantities of auxin, and this hormone prevents the leaf from completing an **abscission zone** (Fig. 22.17)—an area of relatively weak cells at the base of the stem, where the leaf will eventually separate from the branch. In the fall, auxin levels drop, enzymes begin to break down the cell walls

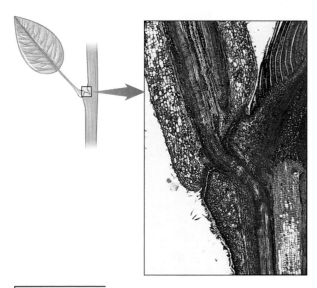

Figure 22.17 **Abscission Zone Allows Fruit Detachment and Leaf Fall.** Growth and maturation result in an abscission zone, where fruit or leaf will eventually detach. This light micrograph of an elm leaf section stained with dye shows the weakened zone, a diagonal stripe of purplish cells across the leaf attachment.

one by one in the abscission zone, and eventually, a gust of wind or a few pelting drops of rain will send the leaf fluttering to the ground.

Once the leaves have fallen, the plant survives winter in a state of greatly reduced metabolic activity known as **dormancy** (Latin, *dormire* = to sleep). A dormant plant uses much less energy and builds proteins much more slowly for cell division and maintenance. Phytochrome may time the changing light cycles, and other hormones relay the message — "slow down metabolism!" — to individual cells, helping maintain the state of dormancy.

22.6   Plant Biotechnology

Plants, including rice, are complicated organisms, and Ray Rodriguez had to know a great deal about the anatomy, reproduction, development, genetics, hormonal control, and biochemistry of rice in order to engineer a new kind of rice plant that can provide calories of food energy for people's diets but also make special proteins, like human lysozyme, that can help protect babies from disease. With his knowledge of plant biology, why couldn't Dr. Rodriguez use traditional ways of altering crops for human benefit instead of turning to

biotechnology? After all, for thousands of years, Asian farmers have used simple breeding techniques, selecting seed from the strongest, most productive rice plants to sow for the next generation. This selection process gradually changed the genetic constitution of wild rice into today's high-yielding, rapidly growing, disease-resistant cereal crop. But, explains Dr. Rodriguez, traditional methods are too slow to keep pace with the dramatic growth of human populations and our ever-expanding demand for food. Furthermore, he says, traditional breeding relies on natural genetic variation that pre-exists in the rice genome, and this would never include a human gene — no matter how beneficial — without the deliberate manipulation of recombinant DNA methods.

Recombinant DNA Methods

Dr. Rodriguez's first step was to engineer a useful, desirable piece of foreign (nonrice) DNA with the potential to be expressed in a germinating rice grain. To do this, he isolated from rice DNA a gene for making amylase that is normally transcribed into RNA in the germinating embryo. He wasn't interested in the amylase gene itself but rather the DNA near the gene that contains the genetic "on switch" (the regulatory element) that turns on amylase production when triggered by plant hormones (Fig. 22.18 and *E*xplorer 22.3). He then spliced the human DNA that codes for lysozyme next to the control sequence for rice amylase. It was then time to insert the recombinant DNA into rice chromosomes inside rice cells. For that, Dr. Rodriguez used the techniques of plant tissue culture.

Tissue Culture

A researcher can grow plant cells in a laboratory vessel much as they grow animal or bacterial cells. The investigator begins this **tissue culture** process by taking a bit of plant tissue from some part of a plant. Ray Rodriguez takes a rice embryo from a rice grain, but a piece of a leaf or stem can also work (Fig. 22.19a, Step 1). (*E* 22.3 animates this process.) Next, the biologist puts the tissue in a nutritious culture medium in a warm, well-lighted room (Step 2). Soon the cells of the tissue lose the distinctive characters that mark them as embryo, leaf, or stem and form an unorganized mass of cells called a **callus** (pl., *calli*) (Step 3). Dr. Rodriguez breaks up rice callus and places bits of it on the surface of a solid culture medium. Now he's ready to insert the recombinant DNA into the rice cells.

GET THE PICTURE

*E*xplorer 22.3

Plant Biotechnology

Figure 22.18 Making Recombinant Rice DNA.

(Step 1) Experimenters extract DNA from rice cells. (Step 2) Next, they isolate DNA containing an amylase gene and then separate the protein-coding portion of the gene from the regulatory regions—the DNA that turns on amylase production during germination. (Step 3) Now the experimenters extract the lysozyme gene from human DNA, and (Step 4) splice it to the regulatory region of the amylase gene. This results in a recombinant DNA molecule that, when inserted into a rice chromosome, should result in a rice plant that produces human lysozyme.

Rice cells

Human cells

1

Rice DNA

Human DNA

Regulatory element   Amylase gene

Regulatory element   Lysozyme gene

2

3

4

Regulatory element for transcription in germinating rice

Protein coding for lysozyme

new gene he grows to adulthood. Then he cross-pollinates them, thus establishing a line of plants homozygous for the new trait. If all has gone well, when these recombinant plants produce rice grains, they will randomly manufacture human lysozyme along with natural amylase at the time of seed germination. The germinated rice can then be used as an ingredient for infant formula, and the milk substitute will be fortified with the bacteria-killing enzyme lysozyme. Dr. Rodriguez says that fortified rice flour should be for sale from his company Applied Phytogenetics by the time this book is published.

Promises and Problems of Plant Biotechnology

Plant researchers see two separate promises for recombinant DNA: (1) The addition of novel genes to improve the food value of crops directly, and (2), the addition of genes that will increase crop yields.

The engineering of rice, for example, promises to help improve the health of millions of people around the world by incorporating provitamin A in the starchy endosperm for improved vision, lysozyme for disease-fighting infant formula, or iron to help prevent anemia. Most people get enough calories daily, but not enough of specific required nutrients, and modifications like these would help provide better health and nutrition. Dr. Rodriguez is also working to engineer rice strains that produce vaccines against diseases like cholera. Introducing small amounts of a vaccine through the diet over a long period of time is an especially effective way to immunize people, Dr. Rodriguez says, and this type of delivery makes sense in developing countries that lack enough doctors and clinics to inoculate local populations by injection.

As the human population continues to grow faster than the food supply, agriculture will fall short and starvation will increase. This is why plant scientists are also adding genes to improve crop yields. Ideally, they would develop wheat, rice, or tomato plants with the drought resistance of cacti, the salt tolerance of marsh grass, and the nitrogen-fixing ability of legumes, as well as the ability to resist insect pests, diseases, and herbicides. This would allow increasingly larger harvests and the use of land that is currently unplantable. Early attempts have included:

- The Flavr Savr tomato, engineered so that the production of an enzyme that speeds ripening is inhibited, and hence the tomatoes have a longer shelf-life. Some professional chefs have questioned the need for a more "durable" tomato, and some consumers com-

Rodriguez coats tiny particles of gold with recombinant DNA and loads these little DNA-coated bullets into a **gene gun** (Fig. 22.19a Step 4 and Fig. 22.19b). This apparatus shoots the gold particles at high speed into the callus tissue he has grown. Inside the cells, the DNA detaches from the gold (which is inert inside the cell), and it moves into the nucleus, occasionally integrating into the chromosomes. The introduced DNA doesn't usually replace the amylase gene, but instead inserts itself into some other (probably random) spot in a chromosome.

Now Rodriguez adds to the tissue culture medium a mixture of plant hormones that promote cell differentiation (Fig. 22.19a, Step 5). Some of the callus cells respond by growing and developing into tiny plantlets. An individual cell from the embryo, root, stem, or leaf can then proliferate into an entirely new plant with its own roots, stems, leaves, and flowers (Step 6). This fact suggests that each plant cell contains a complete set of genetic information.

When these recombinant plants have grown large enough, Rodriguez transfers them to containers and grows them in the greenhouse, checking through a separate set of procedures to see which ones are harboring the human lysozyme genes. Those rice plants with the

(a)

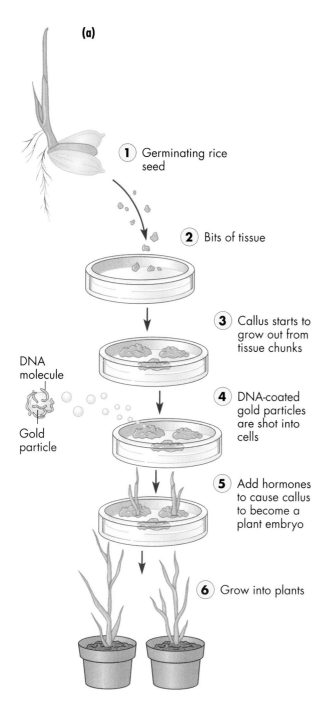

① Germinating rice seed

② Bits of tissue

③ Callus starts to grow out from tissue chunks

DNA molecule

Gold particle

④ DNA-coated gold particles are shot into cells

⑤ Add hormones to cause callus to become a plant embryo

⑥ Grow into plants

(b)

Figure 22.19 **Plant Tissue Culture and Transformation.** Ray Rodriguez uses tissue culture and a gene gun to insert recombinant DNA including the human lysozyme gene into rice cells. (a) He puts bits of embryonic rice tissue into a petri dish in suitable medium and these proliferate into callus—undifferentiated groups of cells (see text for steps). He shoots callus with a gene gun (b) loaded with gold particles coated with the recombinant DNA. In some of the cells, the DNA leaches off the inert gold pieces and enters the chromosomes inside the callus cells. Dr. Rodriguez adds the proper plant hormones and the callus differentiates into a little plant, with root and shoot. He determines which individuals have been genetically transformed, then grows the recombinant rice plants in the greenhouse.

the pollen. Experiments have shown that very large quantities of this Bt pollen will kill monarch butterflies. Subsequent research showed that the life cycle of corn and monarch butterflies are offset in nature so that corn pollen isn't present when the butterfly larvae hatch and feed. Nevertheless, some observers now fear that the Bt gene might escape from corn into weeds and decimate native populations of helpful insects.

- Crops engineered to resist herbicides, including plant-killing chemicals like Roundup. With "Roundup Ready" soybeans, for example, a farmer can spray a field containing both soybeans and weeds with the herbicide but only the weeds die. The goal was to decrease the amount of herbicide farmers use on their fields, but many consumers, especially in Europe, object to corn, soybeans, and other foods modified with herbicide resistance, and have boycotted stores selling the products as whole foods or as ingredients. In the United States, more than one third of the corn and soybean crops already contain recombinant genes.

- *Arabidopsis,* a wild mustard plant related to cabbage and broccoli, engineered to grow in saltier soil. Researchers hope that by perfecting these tests in wild mustard then applying the same techniques to crop species, they could enable farmers to cultivate previously unusable acreage.

One of the most common objections to genetically modified foods is that up until now, plant researchers have been inserting antibiotic-resistance genes along with genes for desirable traits so they can locate recombinant individuals. For example, Dr. Rodriguez inserted a gene for resistance to the antibiotic kanamycin along with the lysozyme gene. He did this so that he could treat the rice plantlets with kanamycin to kill plants that lacked the

plain about the flavor, while other chefs have welcomed the sturdier produce and some consumers prefer the taste! At this printing, the controversy has led the Flavr Savr's producers to remove the engineered tomato from the market.

- Corn engineered to produce *Bt,* a bacterial toxin that kills insect pests. The first generation of Bt corn does discourage certain insects, but it was designed to express the toxin in every cell of the plant, including

new gene. Many people worry that these marker genes will contribute to society's problems with growing antibiotic resistance in pathogenic organisms. Because of these fears, the USDA conducted tests and determined that in at least one case, the kanomycin-resistance gene used in the Flavr Savr tomato, this technique poses no risk to human health. Nevertheless, to allay people's concerns, Dr. Rodriguez points out that plant researchers should devise better ways than antibiotic resistance markers of finding recombinant "needles" in the "haystack" of individuals they use in their research.

Another major objection stems from the widespread use of Bt genes and the realities of evolution by natural selection. Growing an insect-killing gene in millions of corn plants over thousands of acres is bound to encourage the evolution of resistance in insect pests. Since Bt toxin is a common insecticide that organic farmers sprinkle on their crops, they worry that widespread resistance among insects would eliminate one of their most powerful tools. People in many countries have begun organizing to block the sale of genetically modified foods, while others are convinced that we will need the fruits of biotechnology, as well as of traditional plant breeding, in order to feed the burgeoning human race in the decades to come. We'll return to this controversy in Chapter 23.

Chapter Summary and Selected Key Terms

Introduction   Two billion people get most of their daily calories from rice. Professor Ray Rodriguez is pioneering genetic engineering techniques to modify rice with genes for human lysozyme and other proteins. To do this, he needed an extensive understanding of plant reproduction, development, growth regulators, and genetics.

The Plant Life Cycle   Plants have an alternation of generations, or two different generations of multicellular individuals, one haploid, one diploid. A rice seedling is a multicellular diploid sporophyte that produces haploid reproductive cells or spores. Haploid spores divide by mitosis to produce tiny haploid plants or gametophytes. The female gametophyte is just seven cells, including the egg; the male gametophyte or pollen grain contains both a cell that grows toward the egg and two sperm. Fertilization of egg and sperm produces the zygote, which divides into the embryo inside the seed. This then grows into a new sporophyte.

Flowers   Dicot flowers like roses have outer rings of **sepals** (p. 554) and **petals** (p. 554), and inner rings of **stamens** (p. 554) and **carpels** (p. 554). Within each stamen, the **anther** (p. 554) contains pollen sacs and sits high on the **filament** (p. 554). Within each carpel, the base is the **ovary** (p. 554), which houses **ovules** (p. 554); these form part of the future seeds. The ovary neck or **style** (p. 554) supports a sticky **stigma** (p. 554), on which pollen grains germinate. Dicot flowers often have parts in multiples of five, while many monocots have parts in multiples of three. **Monoecious** (p. 555) plants have separate male and female flowers on the same plant; **dioecious**

plants (p. 555) have male and female flowers on separate plants.

Pollen grains form in anthers of stamens on the diploid sporophyte. Special cells undergo meiosis and produce microspores. In each microspore, mitosis produces two sperm cells and a future pollen tube cell. Together, these three cells compose the pollen grain.

In the carpel, special cells in the inner wall of the ovary undergo meiosis and produce four megaspores. Three die. One undergoes mitosis and produces the **egg sac** (p. 556), with seven cells including one egg, and one large cell with two nuclei. When combined with a sperm nucleus, this double cell generates triploid **endosperm** (p. 556), which stores energy and nutrients in the **seed** (p. 556).

The transfer of pollen by wind, gravity, or animals leads to **pollination** (p. 556) and fertilization. Flowering plants undergo **double fertilization** (p. 556): one of the two sperm in the pollen fuses with the two nuclei in the egg sac and gives rise to endosperm. The other sperm fuses with the egg cell, producing the zygote. The zygote divides by mitosis into the embryo. The ovule wall becomes the **seed coat** (p. 558). The entire ovule becomes the seed. The ovary wall becomes the **fruit** (p. 558). European researchers modified rice endosperm to produce provitamin A, capable of helping prevent vision problems in people who get most of their food calories from rice.

Embryos, Seeds, Fruits   The **ovary wall** (p. 558) houses the seed within its tough **seed coat** (p. 558). The zygote inside the ovule undergoes mitosis, becoming a globe then a heart-shaped mass. The lobes become the **cotyledons** (p. 558), with the future apical meristem in-between. In a peanut, the woody shell is the fruit, the paper coating on each peanut is the seed coat, and the two halves of the peanut are the cotyledons, or seed leaves. Dicots have two initial "seed leaves," while monocots have one. A short stem or

hypocotyl (p. 558) projects below the cotyledons; the stem above is the **epicotyl** (p. 558), which becomes the plant's entire shoot.

Germination (p. 559) is resumption of growth in the seed after a period of dormancy. In grains like rice or barley, cells in the single cotyledon generate and release enzymes like amylase that break down starch for the growing embryo.

Plant Growth Regulators   Japanese farmers and researchers discovered gibberellins as the cause of foolish seedling disease. **Gibberellins** (p. 560) regulate height, play a role in rapid stem growth, induce seeds to germinate, and are involved in flowering, fertilization, and growth.

Auxins (p. 560) promote growth. They are usually made in shoot apical meristems and developing leaves, and diffuse downward. They promote cell elongation. **Cytokinins** (p. 560) usually stimulate cell division and move from root upward to shoot. **Abscisic acid (ABA)** (p. 560) is a growth inhibitor. It's involved in **abscission** (p. 560), or leaf and fruit drop. But it also induces and maintains dormancy. **Ethylene** (p. 560) is involved in fruit ripening.

How Growth Regulators Work   Germinating seeds need control so, for example, seedlings don't wind up growing during a temporary warm spell in late fall and then freeze in early winter. Moisture and/or light can trigger the release of gibberellins, which, in rice, turn on the genes for amylase production and fuel release. Roots grow down and shoots up; this plant orientation is based on **tropisms** (p. 562), bending or directional growth in response to environmental stimuli. **Hydrotropism** (p. 562) is growth toward water. **Thigmotropism** (p. 562) is turning or growth toward touch pressure as from a supporting structure. **Gravitropism** (p. 562) is growth toward gravity. **Phototropism** (p. 562) is growth toward light. The root tends to grow downward with **gravity**

(p. 562), while the shoot grows upwards against gravity. Charles Darwin and son Francis figured out that a seedling's shoot tip sends a signal to its lower portion, causing it to bend toward light, gravity, and so on. Later researchers showed that the signal is auxin that leads to pH changes, cell wall stretching, and cell elongation.

Plants sometimes display relatively quick movements or **nastic responses** (p. 563) not based on growth toward or away from a stimulus. The sensitive plant *Mimosa pudica* and Venus flytrap are examples, as are sleep movements.

Plants detect light through the pigment **phytochrome** (p. 563). Two interconvertible forms, P_{fr} and P_r act like a chemical hourglass, timing darkness and light.

Plant shape is governed by growth regulators. **Apical dominance** (p. 564) based on auxin from the apical bud and cytokinin from the roots establishes a conifer tree's cone shape.

Flowering and fruit formation can be based on temperature (exposure to cold followed by warming) and/or light. Many plants track the seasons by **photoperiod** (p. 564), or length of light and dark periods. **Short-day plants** (p. 564) need a long night; **long-day plants** need a short night. Photoperiod is irrelevant to flowering in **day-neutral plants** (p. 564). The phytochrome "hourglass" times the length of light and dark periods.

Flowering is based on a flower-generating substance that may be a combination of other hormones and phytochrome. Auxin and ethylene are involved in the timing of fruit development.

Senescence (p. 566), or the process of aging and dying, is also based on plant growth regulators, as is **abscission** (p. 566), or the falling of mature fruits or dying leaves. A **deciduous** (p. 566) growth habit is common in perennials. Cytokinins influence when leaves senesce. Auxin prevents the completion of an **abscission zone** (p. 566), but when auxin levels drop, the zone is completed, and a leaf or fruit can fall. Many plants overwinter in a state of **dormancy** (p. 567).

Plant Biotechnology
Standard plant-breeding techniques are too slow to keep up with the growing world demand for more food and more nutrients in foods. They are also limited to existing genomes, whereas plant researchers using recombinant DNA techniques can introduce new genes, such as human lysozyme, into rice. Dr. Raymond Rodriguez spliced the lysozyme gene next to control region DNA for rice amylase. He used **tissue culture** (p. 567) techniques to grow **callus** (p. 567) tissue (pl., *calli*) from bits of rice embryos. Then he used a **gene gun** (p. 568) to send in gold particles coated with the recombinant DNA. He promoted cell differentiation with hormones so seedlings would grow from the callus, selected those

seedlings with the recombinant DNA, then grew the rice plants in a greenhouse. The adult rice plants produce grain containing lysozyme.

Plant biotechnology can allow the addition of new genes to improve the food value of crops and to improve crop yields. Examples of genetically modified foods include the Flavr Savr tomato, Bt corn, Roundup-ready corn and soybeans, and salt-tolerant *Arabidopsis*. A common objection involves the antibiotic-resistance genes inserted to help locate recombinant individuals. Another objection involves worry over the evolution of resistance to Bt in insect pests.

> All of the following question sets also appear in the Explore Life E electronic component, where you will find a variety of additional questions as well.

Test Yourself on Vocabulary and Concepts

For each question set, complete or match the statements with the most appropriate term from the list. Any answer may be used once, more than once, or not at all.

SET I
(a) sporophyte (b) megaspore (c) microspore
(d) gametophyte (e) gametes (f) zygote

1. Each haploid _____ develops into a seven-celled structure in corn, pear, and rice.
2. Each _____ develops into a pollen grain.
3. The pollen grain is one kind of _____.
4. Rice leaves and stems are part of the _____.
5. Each pollen grain includes two _____.

SET II
(a) sepals (b) carpel (c) ovary (d) ovule (e) anther
(f) monoecious plant

6. Has both male and female flowers
7. The structure that contains one or more pollen sacs
8. Contains the egg and later matures into the seed
9. Includes the ovary, style, and stigma
10. May aid reproduction but does not actually participate in reproduction

SET III
(a) gibberellins (b) auxins (c) cytokinin
(d) abscisic acid (e) ethylene

11. Travels through the air from one fruit to another, causing maturation of flowers and fruits
12. Induces and maintains dormancy by inhibiting protein synthesis
13. A growth-promoting hormone that maintains apical dominance
14. Promotes leaf, flower, and fruit drop; produced by ripening fruit
15. Stimulates cell division, especially in lateral buds
16. Cause of foolish seedling disease; important regulator of plant height and seed germination
17. Growth promoter that diffuses from the root upward

SET IV
(a) tropism (b) phytochrome (c) apical dominance
(d) photoperiod (e) long day plant (f) short day plant

18. A plant that flowers when the night is less than a certain length.
19. A light-sensitive pigment that regulates seed sprouting and flowering
20. Directional plant growth in response to external stimuli such as light or internal hormonal stimuli
21. The environmental cue by which short- and long-day plants regulate their life cycles
22. The process of inhibition of lateral bud growth by auxins

Integrate and Apply What You've Learned

23. Corn plants are monoecious. What does that mean and how does it compare to a dioecious plant?
24. Which plant parts do you eat when eating an apple? A green pea? Rice?
25. If mammals such as dogs, cats, and humans are analogous to angiosperm sporophytes, what structures (if any) in mammals are analogous to the plant ovary and stamen? Explain.
26. What is meant by double fertilization? What is the significance of this term? Where would you find the product(s) of double fertilization in rice?
27. The Christmas cactus is a short-day plant that normally blooms in the late fall. How would phytochrome regulate this cycle?
28. How do plant hormones interact and regulate the conical and spherical shapes of many kinds of trees?
29. Which hormones are the primary regulators of leaf senescence, seed germination, fruit growth and ripening, and dormancy?

Analyze and Evaluate the Concepts

30. After many years of work, plant geneticists have developed new varieties of corn in which the enzyme that converts newly photosynthesized sugar to starch has been inhibited. How would this alteration affect the product sold to consumers?

31. Is a tomato a fruit? Why?

32. Suppose travelers from Earth to a distant planet discovered humanoids (peoplelike organisms) there whose reproductive cycle resembled the rice plant. What might the sex lives of these aliens be like?

33. Consider a garden variety pea plant with its outer green pods and inner peas. What parts correspond to the ovary wall, the ovule wall, and the cotyledons? About how many ovules are there in each ovary in a green bean?

34. Let's assume that you have become an agricultural biologist and must choose a crop and decide what traits would make it more useful. Describe the crop, the new traits, and, in general, how you might achieve the desired modification using recombinant DNA techniques.

35. Fruit growers would like to have genetic strains of peaches or pears that do not ripen until they are treated with a natural hormone. This treatment could be done just before the grocers put the fruit out on their shelves. Which plant hormone might a plant physiologist try to manipulate to achieve this goal?

36. Suppose that during an investigation of a plant's phytochrome gene you discover a mutation that causes the plant to grow pale and spindly. How do you think the mutation might achieve this effect?

Weeds and Super Weeds

If you've ever done yard work and maintained a grassy backyard, then you've probably learned firsthand why people use the expression "growing like a weed." You pull a dandelion from your lawn this week and it's back next week just as big, tough, ugly, and hard to pull out as before. What makes weeds so vigorous, so expansive, so—exasperating? Ask Joe Di Tomaso. He has the biggest weed problem in America, or close to it.

Dr. Joseph Di Tomaso is an ecologist who studies weeds at the University of California at Davis, and a major part of his job is fighting yellow star thistle. This hairy, spindly, gray-green weed grows to nearly a meter tall and produces a dozen or more bright yellow flowers, each guarded by a whorl of vicious spines. Finding it an attractive ornamental, California settlers first imported yellow star thistle from Europe or Asia before 1870. Today, says Di Tomaso, "it's probably the single most common plant in California, including all the native species."

The Golden State, with its vast tracts of rangeland, wildland, fallow fields, roadsides, and forests has its own indigenous thistles. These, Di Tomaso explains, "grow one or two plants per acre." But when yellow star thistle takes over an area, "you see one or two plants every *inch*." The yellow stars growing in one square meter can produce 29,000 seeds in a summer, 95 percent of them capable of growing into new plants even if they have to wait a decade to germinate. Each year, yellow star thistle infests 15 to 20 million acres of California's nonagricultural land. That's a collective area equivalent to the entire states of Connecticut, Maryland, and Delaware combined!

Dr. Di Tomaso has other weed responsibilities, as well, including wild blackberry vines, which he considers "pretty close to the perfect weed" because of its aggressive growth and protective spines; poison oak, to which three quarters of us have skin allergies; and the low-growing green plant called salt cedar, which draws minerals out of the soil and sterilizes the surrounding ground to all but its own seedlings and other salt-tolerant plants. How does Di Tomaso explain the rambunctious nature of weeds in general? And knowing that, how does he then deal with California's colossal overgrowth of yellow star thistle?

The secret of any weed's success, he says, "is an age-old question that has been asked a thousand times but never answered." Weeds, however, do tend to share certain traits, he adds: Rapid growth. Abundant seed production. The ability to spread through stems and roots as well as seeds. And, tremendous genetic variability, allowing them

▲
Fighting Problem Weeds. Joe Di Tomaso in the test field; yellow star thistle; star thistles by the acre.

to occupy different kinds of habitats. "So there are morphological and physiological characteristics that help a weed's success," says Di Tomaso. But the environment must also be right for these advantages of form and function to win out.

Yellow star thistle, for example, is an imported "alien" that has no natural insect enemies or diseases in California. When it's mature and the spines are glistening around every flower, foragers such as cows and deer shy away from it and eat nearby competitors, instead, allowing the thistles to spread further. What's more, says Di Tomaso, yellow star thistle tends to grow where people long ago cleared native perennial grasses and replaced them with annual grasses that their livestock found tastier. During California's long dry season, those replacement grasses dry up and disappear, leaving the alien thistle to take over. What allows yellow star to be so aggressive? Because it happens to have a very long taproot that can pull up needed water even when the soil surface layers are dusty dry.

Joe Di Tomaso uses a broad knowledge of plant physiology in his work along with specific expertise on ways to mow down, burn out, pull up, and poison different kinds of weeds in the correct sequences and combinations. One area of plant physiology that he finds quite helpful is plant transport: how plants move the products of photosynthesis around internally, and how they transfer water, minerals, and other needed substances in a dynamic pattern that sustains life. Plant transport—our focus in the current chapter—is important to understanding why yellow star thistle can survive California's dry sea-

son. Plant transport explains why salt cedar can "salt out" its competitors. And it helps weed specialists like Dr. Di Tomaso find ways to eliminate unwanted plants. Finally, this and the other aspects of plant physiology we explored in Chapter 22 help him address the public's concerns about "super weeds." Some observers of plant bioengineering are concerned about the transfer of genes for, say, herbicide resistance into crop plants. They fear that a genetic transfer into artichokes, let's say, could allow herbicide resistance to hop into relatives like yellow star thistle growing nearby and create dangerous, indomitable super weeds that could never be controlled. You'll learn in this chapter how likely Dr. Di Tomaso thinks this scenario really is, and what the fear of super weeds is teaching plant researchers.

Along the way, you'll find answers to these central questions:

1. How do plant roots draw water and minerals up through special tubes?

2. How does water reach the top of tall trees?

3. How does a plant avoid losing too much water on a hot day?

4. How does a plant move sugars through a second set of internal tubules?

5. Which nutrients must a plant have to grow?

6. How do weeds maximize their uptake and use of needed resources?

7. What is the worry over super weeds all about?

23.1 Water Balance in Plants

When Dr. Di Tomaso pulls up a yellow star thistle, he is invariably impressed with the length of its taproot (Fig. 23.1). This structure is ten times longer than the short, fibrous roots of the native grasses that occupied California prairies before pioneers introduced the noxious yellow weed in the 1800s. In late spring, a star thistle plant may stand only three inches tall, but its taproot already extends three feet down into the soil! These long roots are a key to yellow star thistle's success, because for nearly every land plant, water is the biggest need.

Over the course of one growing season, a single tall plant—corn or sunflower, let's say—consumes 17 times more water than a person! But the water essentially passes right through; all but two percent of the imbibed moisture evaporates from the plant's leaves.

A plant's insatiable "thirst" is based on a fundamental difference in the way plant and animal bodies work. The fluids in most animal bodies travel in a recirculating pathway, maintained in a circulatory system that recycles water by and through organs with little water loss and relatively little input. In contrast, plant fluids travel in a flow-through pathway, a generally

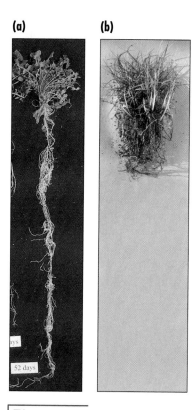

(a)   **(b)**

52 days

Figure 23.1 **Yellow Star Thistle: Root Cause.** One secret of the yellow star thistle's success is its long taproot (a), which allows the plant to access water throughout the dry California summer, when the native grasses with their short, fibrous roots, are parched and brown. (b) This Berkeley sedge grass has extensive fibrous roots.

one-way path from roots through stems to leaves, then out into the environment (Fig. 23.2). A steady supply of water enables a plant to carry out photosynthesis (discussed in Chapter 3). Water also stops the plant from wilting, allowing it to remain erect with turgid or swollen cells. Furthermore, water helps a plant to transport substances internally and to stay cool despite a baking sun on outstretched leaves and stems. In California, in the desiccating heat of late summer, when shallow-rooted annual plants are withering and browning, the star thistle continues to suck moisture from the deep soil—and grow like a weed. Star thistle, in fact, draws up water so efficiently through its taproot and grows so luxuriantly that it can starve its competitors of light—even those that don't die off from drought alone.

Let's trace the one-way flow of water through a plant and see how the physical properties of water and plant architecture propel this life-sustaining movement.

How Roots Draw Water from the Soil

Rains come mostly in the winter and early spring in California, and a yellow star thistle absorbs this rain water from the soil into its roots (Fig. 23.3). Millions of root hairs project from the cells of lateral roots into the spaces between soil particles. These hairs have a huge combined surface area for absorption. In many plants, root-associated fungi called mycorrhizae contribute to water uptake (review Fig. 12.1). After the water enters the roots, the plant stores some of it in special parenchyma cells in the root's innermost cylinder of tissue.

Water normally enters roots by simple diffusion, with no energy expenditure by the plant. Water

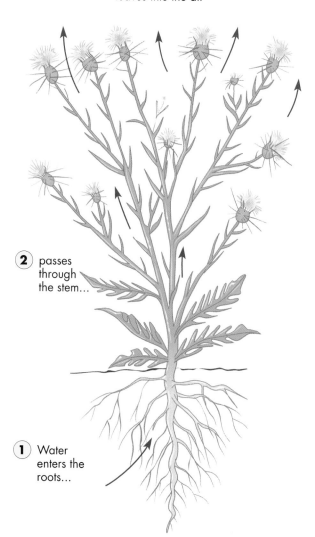

(3) and exits from the leaves into the air

(2) passes through the stem...

(1) Water enters the roots...

Figure 23.2 **Water Flows Through Plants in a One-Way Path.** Sunlight powers a one-way route for the flow of water and inorganic nutrients through the plant's xylem cells.

diffuses into root hairs and then passes through the root cortex. Some of the fluid moves through the cytoplasm of root cells, but most of the water passes within the spaces between cell wall fibers (Fig. 23.3, Steps 1 and 2). Next, water reaches the endodermis, the cell layer that separates the root cortex from the central cylinder, containing the plant's fluid conducting tubes, the xylem and phloem (review Figs. 21.7 and 21.8). Water can no longer move along and through the cell walls because the Casparian strip, the waxy belt surrounding each endodermal cell, acts like a gasket that prevents water from flowing in the spaces between cells. The Casparian strip forces water to pass through the cytoplasm of endodermal cells before it enters the hollow, water-conducting tubes of the xylem (Fig. 23.3, Step 3). Some water enters the xylem by diffusion, and some is drawn in as water leaves the root xylem and moves upward toward the stems and leaves.

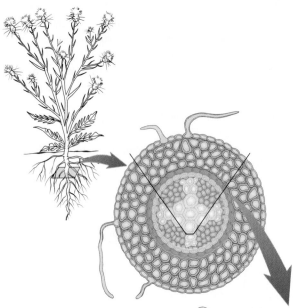

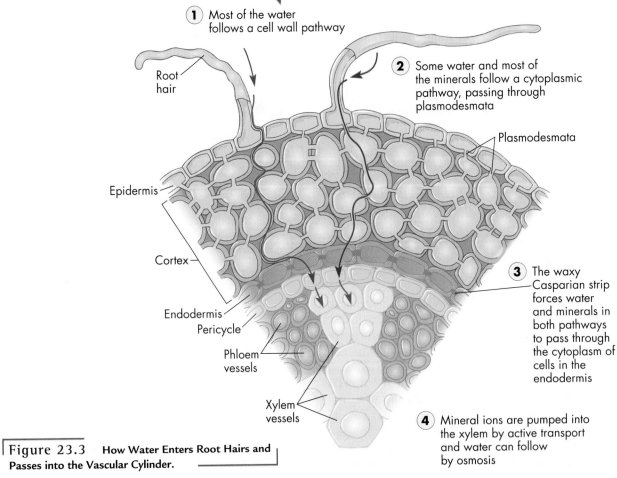

1 Most of the water follows a cell wall pathway

Root hair

2 Some water and most of the minerals follow a cytoplasmic pathway, passing through plasmodesmata

Plasmodesmata

Epidermis

Cortex

3 The waxy Casparian strip forces water and minerals in both pathways to pass through the cytoplasm of cells in the endodermis

Endodermis
Pericycle

Phloem vessels

Xylem vessels

4 Mineral ions are pumped into the xylem by active transport and water can follow by osmosis

Figure 23.3 **How Water Enters Root Hairs and Passes into the Vascular Cylinder.**

Mineral Transport

Water enters the roots of a thistle plant and moves through the endodermis passively, without energy cost to the plant. The transport of mineral ions, however, is another story (Fig. 23.3, Step 4). Plant researchers demonstrated that a rutabaga, a member of the cabbage family, can contain a concentration of potassium ions 10,000 times higher than the watery solution surrounding their roots. Ions can't move up a concentration gradient without energy input, and plant cells use energy to accumulate minerals in two ways. First, root epidermal cells have plasma membrane "transporters" or special proteins that actively transport minerals across the membrane barrier. Second, the root's parenchyma cells also have membrane proteins that use ATP energy to actively pump ions into the adjacent xylem. Water then enters the xylem via osmosis and ions move throughout the plant dissolved in the water.

Water Transport: Root to Leaf

So far, we've seen that water moves into a plant's roots by simple diffusion and ions enter the xylem by active transport. But how does water reach the crown of a 1-meter-tall yellow star thistle, or more impressively, the top of a California redwood tree, Earth's tallest living thing, towering more than 100 m (328 ft) above the forest floor? By the same simple mechanisms?

Plant scientists have proposed four possible hypotheses about water transport to the tops of plants:

1. The plant root might *push* water upward.
2. Pumps distributed along the xylem might convey water upward like a bucket brigade.
3. Water might move upward by capillary action in the xylem.
4. Evaporation of water from the leaves (as water changes from liquid to gas and moves into the atmosphere) might *pull* water up from below. But which is correct?

Root Pressure, Xylem Pumps, and Capillary Action

Farmers have long noticed that sometimes, in the early morning, the leaves of grass, tomatoes, strawberries, and numerous other plants are rimmed with tiny droplets of water. Plant researchers call this **guttation** and it results from upward pressure from the roots (Fig. 23.4). Could guttation force water all the way to the tops of trees?

Guttation droplets appear at special openings on leaf edges when the soil is nearly saturated with water and the leaves are not losing much through evaporation—in early morning, for example. Water and minerals leave the water-soaked soil and enter the vascular cylinder. Membrane proteins in the root's parenchyma cells expend energy pumping ions into the xylem; water then follows by the process of osmosis. The volume of fluid in the xylem increases, and this fluid must move somewhere. It tends not to flow back between the endodermal cells because the waxy Casparian strips prevent both the solutes and the water from leaking back out of the vascular cylinder (see Fig. 23.3). The fluid *can* move up the xylem, however, generating a force called **root pressure.** Root pressure can force water out of the leaves, where it can accumulate in tiny droplets.

After measuring the force of root pressure, botanists determined that it's much lower than the force needed to move water to the tops of tall trees. Such a force would have to be about 150 pounds per square inch (psi), roughly equivalent to the force you would exert on the floor by balancing all your weight on the tip of one big toe. What's more, many plants, including pines, don't develop root pressure at all, but still grow hundreds of feet tall. This rules out hypothesis 1.

As for hypothesis 2, you can perform a simple test of it yourself. To determine whether water is pushed along via little pumps in xylem tubes, you can pierce a plant stem with a needle. If the fluid was under pressure like the water in a garden hose or the blood in an artery, it would spew out from the hole you made. In fact, no fluid comes out and instead, air is sucked in.

Hypothesis 3, **capillary action,** is the tendency of water to move upward in a thin tube, and indeed, studies show that capillary action does occur in narrow xylem vessels. However, water can cling to and creep up the inside of narrow tubes only to a height of about 0.5 m (20 in). While this would be enough for young yellow star thistle plants, it's clearly not enough for mature thistle plants and far too little for redwood trees. Plant physiologists were left with hypothesis 4—that water is somehow *pulled* upward to the tops of tall plants. And this explanation turned out to be correct.

Transpiration

The major explanation for water transport in plants lies in a fascinating bit of natural engineering called evapo-

Figure 23.4   Root Pressure and Guttation. Root pressure can force water out of special openings in some small plants like the strawberry plant leaflets shown here. The pressure is never high enough, however, to push water all the way to the tops of tall trees.

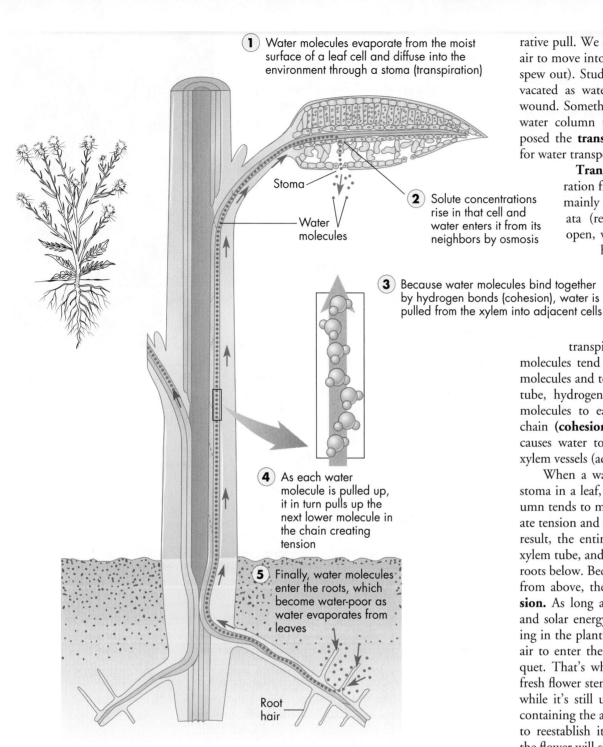

(1) Water molecules evaporate from the moist surface of a leaf cell and diffuse into the environment through a stoma (transpiration)

Stoma

Water molecules

(2) Solute concentrations rise in that cell and water enters it from its neighbors by osmosis

(3) Because water molecules bind together by hydrogen bonds (cohesion), water is pulled from the xylem into adjacent cells

(4) As each water molecule is pulled up, it in turn pulls up the next lower molecule in the chain creating tension

(5) Finally, water molecules enter the roots, which become water-poor as water evaporates from leaves

Root hair

Figure 23.5 **Transpiration-Cohesion-Tension Mechanism of Water Transport.** Transpiration, the evaporation of water from a plant surface, and the hydrogen bonds that cause water molecules to adhere to each other provide the energy and tension that pull water to the tree tops.

rative pull. We saw that puncturing a plant stem allows air to move into a xylem vessel (rather than for water to spew out). Studies show that air, in fact, fills the space vacated as water moves up the stem away from the wound. Something from above is obviously pulling the water column up, and to explain it, plant scientists posed the **transpiration-cohesion-tension** mechanism for water transport in plants.

Transpiration is the loss of water by evaporation from stems and leaves, with water exiting mainly through the little openings called stomata (review Fig. 21.17). When stomata are open, water molecules move from a region of high concentration inside the leaf cells to a region of lower concentration in the air surrounding the plant (Fig. 23.5). Water's basic physical properties allow it to form a strong, "sticky" chain that helps explain transpiration from roots to leaves. Water molecules tend both to *cohere* strongly to other water molecules and to *adhere* to unlike molecules. In a xylem tube, hydrogen bonds (review Fig. 2.4b) link water molecules to each other in a long, unbroken liquid chain **(cohesion),** while additional hydrogen bonding causes water to adhere to the cellulose lining of the xylem vessels (adhesion).

When a water molecule evaporates from an open stoma in a leaf, the water molecule below it in the column tends to move up and replace it, and, in turn, create tension and pull along the next water molecule. As a result, the entire liquid column moves upward in the xylem tube, and a new water molecule is drawn into the roots below. Because water molecules pull each other up from above, the entire column is under constant **tension.** As long as the water column remains unbroken and solar energy causes evaporation, water keeps moving in the plant's vascular tissues. This force also causes air to enter the xylem vessels of cut flowers in a bouquet. That's why florists recommend you submerge a fresh flower stem and remove an additional inch or two while it's still underwater. This eliminates the section containing the air bubbles, and allows the water column to reestablish itself so transpiration can continue and the flower will stay fresh longer.

Water Stress: A Break in the Chain

On a hot, dry day in California's Central Valley surrounding U.C. Davis, evaporation from the leaves of a native live oak tree can outstrip water absorption by the roots, and tension on the chain of water molecules in each of the millions of xylem tubes becomes greater and greater. When this occurs, the adhesive forces of water molecules pulling inward against the xylem walls can grow so strong that the diameter of a tree trunk literally shrinks, a bit like a soda straw sucked flat. In summer, the upward tension on the chain of water molecules in a plant can become so intense that the column can snap. In fact, plant physiologists with sensitive microphones can actually hear the snapping and popping inside plants on a hot, dry day!

Once the water column breaks, an air bubble forms inside the xylem tube, and transpiration can no longer pull water from above. If the temperature drops, evaporation slows, or the soil becomes damp again fairly quickly, root pressure from below, as well as capillary action within the narrow xylem tubes, can rejoin the ends of the broken water column. If conditions aren't reversed soon enough, however, wilting can destroy the plant.

Stomata and Regulation of Water Loss

Transpiration rate is crucial to a plant's daily survival, and plants carefully regulate their water loss. In the leaf epidermis, each stoma is enclosed by two kidney-shaped guard cells (Fig. 23.6, Step 1). Like curved water balloons, guard cells arch away from each other when they swell with water, increasing the size of the opening between them (Step 2). When guard cells lose water and become flaccid, they slump together and close off the stoma, blocking water loss via transpiration (Step 3). In a plant like the yellow star thistle, the leaves are covered with gray, cottony hairs and these may also help decrease evaporation from the stomata.

What causes the guard cells flanking the stoma to swell or deflate and thus open or close the plant's "portholes"? Experiments show that when a plant begins to dry out, guard cells pump out potassium ions and water follows, leaving passively by osmosis. Loss of water from guard cells allows them to slump together, close the pore, and stop further water loss. After a rain, when the plant has plenty of water, potassium ions are pumped into guard cells, water follows by osmosis, guard cells swell, the stoma opens, and carbon dioxide can enter the leaf once more, allowing photosynthesis to continue.

In addition to drying, several other environmental agents affect guard cells. These include carbon dioxide, light, temperature, and daily rhythms in the plant. Together, they cause stomata to stay open, in general, when gas exchange can take place without threat of excess water loss, but to stay closed when there's a danger of dehydration. Plants like the yellow star thistle that live in arid environments often close their stomata for much of the day. Instead, the stomata open and the plant absorbs carbon dioxide only at night, when the air is cooler. This prevents the plant from losing too much water. Photosynthesis is then completed during the day when light is available, but it takes place with the stomata closed.

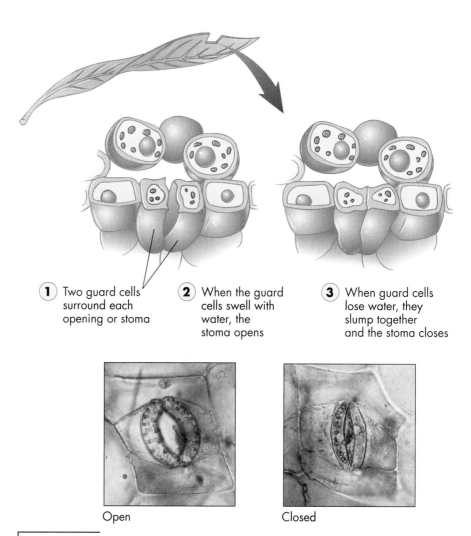

① Two guard cells surround each opening or stoma

② When the guard cells swell with water, the stoma opens

③ When guard cells lose water, they slump together and the stoma closes

Open   Closed

Figure 23.6 **Stomata: Regulators of Water Loss.** The epidermis on a leaf's lower side (here, of a common houseplant, a *Tradescantia*) is ventilated by stomata. When the leaf's cells have plenty of internal moisture, calcium ions and the plant growth regulator ethylene cause potassium ions to enter the two guard cells that surround each opening. Water follows by osmosis, the guard cells swell, and the stoma opens (Steps 1 and 2), allowing carbon dioxide to enter the leaf and photosynthesis to resume. When leaf cells lose turgor pressure because of excess evaporation, they release the growth regulator abscisic acid. This affects the concentration of calcium ions and, in turn, potassium ion concentrations inside guard cells. Water then exits the guard cells by osmosis, allowing the cells to grow flaccid and the stoma to close (Step 3).

GET THE PICTURE

Explorer 23.1

Water Pathways in Plants

The combined mechanisms of transpiration and guard cell swelling help maintain a constant supply of water to plant cells—a major aspect of homeostasis. **E**xplorer 23.1 summarizes and animates the pathways of water movement through plants. These basic principles hold for gymnosperms as well as flowering plants. But for these mechanisms to work in any plant, the cells must absorb the right quantities of ions and other mineral nutrients—our next topic.

23.2 Movement of Nutrients Throughout the Plant

Star thistle has a built-in distribution problem: In the rainy spring, it generates sugars in its leaves but stores organic nutrients in its roots. Then in the late summer, when photosynthesis is difficult because stomata must stay closed to conserve water, the thistle must move nutrients stored in its roots upward to power the construction of flowers and seeds (Fig. 23.7). How does a plant distribute and redistribute photosynthetic products without a pump and circulatory system like an animal's heart and blood vessels?

Translocation: Movement of Sugar

Plant scientists devised some clever methods for studying **translocation,** the movement of nutrients in the phloem of plants. Phloem is so delicate that puncturing the tubes even with a fine glass needle stops the flow of solutes. Tiny insects called aphids, however, can suck plant sap from phloem cells without disrupting them. Plant physiologists found that if they allow an aphid to insert its sharp feeding tube into the phloem, then cut away the insect's body, the phloem's contents ooze out of the little tube, and they can study this plant sap. Using the aphid technique, they discovered that the plant sap in phloem tubes contains about 30 percent sugars (mostly sucrose) and 70 percent water.

Plant scientists also discovered that they can expose a plant's leaves to radioactive carbon dioxide, and once the carbon is incorporated into sugars by photosynthesis, they can trace its path through the phloem. With this radioactive procedure, they found that sugar often flows downward from leaves to other plant parts and this flow can be 10,000 times faster than normal diffusion. This convinces them that translocation involves a mass movement of phloem fluid based on bulk flow.

4 and are carried upward, where they are used for growth of flowers and seeds

1 In the spring leaves make carbohydrates, which...

2 are translocated to the root where they are stored as starch

3 In the fall, sugars are cleaved from starch...

Figure 23.7   **Seasonal Sugar Transport.**

The best current hypothesis for phloem transport is the **pressure flow hypothesis,** the idea that phloem transports substances from "sources," or areas of production, to "sinks," or areas of use, down a pressure gradient set up by osmosis. Let's see how this works in a flowering plant like the yellow star thistle. Photosynthesizing cells in the leaves (the source) produce sugars and the companion cells load the sugars into the phloem's sieve tube elements by an active transport process (Fig. 23.8, Step 1). This energy costly transport increases the solute concentration locally in the phloem, and water follows by osmosis from the xylem's nearby conducting elements (Step 2). The influx of water into the phloem cells plumps them up, increasing their internal pressure like little water balloons. This force is called turgor pressure (Latin, *turgere* = to be swollen). This pressure in the phloem forces the sugary solution through sieve plates at the ends of each sieve tube member and carries it away from the leaf (Step 3).

Meanwhile, cells in the root and other parts of the plant (the sinks) remove sugary solutes from the phloem (Step 4), and the solute content drops, decreasing the osmotic pressure in those regions. Now water flows out of the phloem and back into the xylem tubes, where transpirational pull carries it upward again (Step 5). Ordinary water pressure, the active loading activities of companion cells, and the unloading of sugars by root cells are thus behind the movement of sugars, amino acids, and a few mineral ions from one part of a yellow star thistle plant to another. *E*xplorer 23.2 animates these concepts.

In the early spring, a similar mechanism probably causes plant sap to rise in the stems of biennials and perennials. As the weather warms and the days lengthen, the roots (now the sources) begin breaking down stored starch into sugar and loading it into the phloem's sieve tube members. Water from the xylem follows by osmosis, and as the roots bring in more water from the soil, enough pressure is created to push the sap up the phloem and into the shoot, where the emerging leaves (now the sink) take sugar from the phloem and use it for energy and raw materials. Ever hear the phrase, "the sap's rising"? New Englanders carefully tap into the phloem of maple trees in early spring, drain off some of the sweet rising maple sap, and boil it down to make maple syrup.

In a sense, sugar translocates itself, since its production and use create osmotic pressure and mass flow. Other nutrients, however, move passively by the flow of fluid in both the xylem and the phloem.

Transport of Inorganic Substances in Xylem and Phloem

Plants transport a long list of nutrients (in addition to carbon, oxygen, and hydrogen in sugars) (Table 23.1) and they do it in either of two ways: (1) Some nutrients, such as calcium, move only in the xylem; they're carried along in the column of water that moves upward from the roots to the leaves via transpiration. These nutrients are deposited permanently wherever they leave the xylem and enter a living cell, and they can't be redistributed to the other parts of the plant. (2) Other nutrients, such as sulfur or phosphorus, can be redistributed to new tissues as needed, carried along passively in the phloem with the sugars and other phloem contents. Figure 23.9 illustrates how plant physiologists used radioactive tracers and photographic film to learn that calcium moves only in the xylem, while sulfur can move through the xylem and then into the phloem to reach growing leaves.

Plants can have mineral deficiencies just like animals can, and the two different modes of transport explain why different mineral deficiencies affect plants in different ways. Calcium

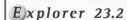

GET THE PICTURE

*E*xplorer 23.2

The Movement of Sugars in Plants

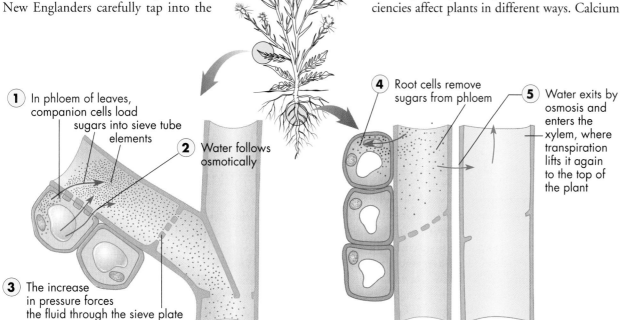

1. In phloem of leaves, companion cells load sugars into sieve tube elements

2. Water follows osmotically

3. The increase in pressure forces the fluid through the sieve plate toward the root

4. Root cells remove sugars from phloem

5. Water exits by osmosis and enters the xylem, where transpiration lifts it again to the top of the plant

Figure 23.8 Translocation: How Plants Transport Sugars.

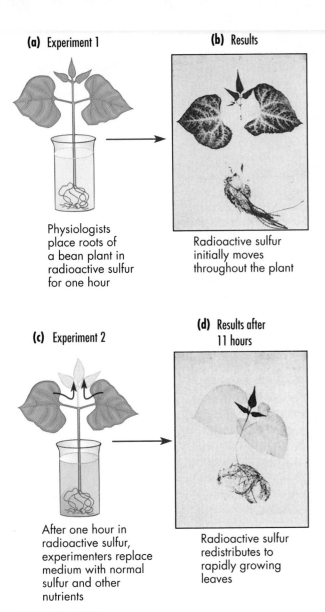

(a) Experiment 1

Physiologists place roots of a bean plant in radioactive sulfur for one hour

(b) Results

Radioactive sulfur initially moves throughout the plant

(c) Experiment 2

After one hour in radioactive sulfur, experimenters replace medium with normal sulfur and other nutrients

(d) Results after 11 hours

Radioactive sulfur redistributes to rapidly growing leaves

Figure 23.9 Movement of Labeled Sulfate Reveals Transport of an Inorganic Ion.

Table 23.1 Plant Nutrients and Their Functions

Nutients (Percent of Dry Weight)*	Location/Function
Macronutrients	
Carbon (45.0)	In all organic compounds
Oxygen (45.0)	In most organic compounds, including all sugars and carbohydrates
Hydrogen (6.0)	In all organic compounds
Nitrogen (1.0–4.0)	In proteins, nucleic acids, chlorophyll, and coenzymes
Potassium (1.0)	Involved in activating enzymes, protein synthesis, and regulation of osmotic balance
Calcium (0.5)	In cell walls and starch-digesting enzymes; regulates cell membrane permeability
Magnesium (0.2)	In chlorophyll and many cofactors
Phosphorus (0.2)	In nucleic acids, some coenzymes, ATP, and some lipids
Sulfur (0.1)	In proteins, some lipids, and coenzyme A
Micronutrients	
Iron (0.01)	In electron transport molecules; involved in synthesis of chlorophyll
Chlorine (0.01)	Essential to photosynthesis
Manganese (0.005)	Essential to photosynthesis; activates many enzymes
Boron (0.002)	Unknown
Zink (0.002)	In some enzymes; involved in protein synthesis
Copper (0.0006)	In chloroplasts and some enzymes
Molybdenum (<0.0001)	Essential to nitrogen fixation and assimilation
Elements Essential to Only Some Plants	
Silicon (0.25–2.0)	In the cell walls of grasses and horsetails
Sodium (trace)	Essential to a few desert and salt-marsh species
Cobalt (trace)	Essential to nitrogen fixation

* Concentration of nutrients expressed as percentage of dry weight in a typical plant. Actual proportions vary greatly from species to species.

deficiency, for example, always produces symptoms first in new leaves. The old leaves contain enough calcium, but the mineral can't move from these old leaves to the new leaves. In contrast, phosphorus deficiency always produces symptoms first in old leaves; as the plant's environment becomes depleted of the element, the plant mobilizes phosphorus in older leaves and transports it through the phloem to areas of new growth. Plants with phosphorus deficiency therefore have the characteristic yellow veins in their old leaves long before new leaves show this sign.

23.3 Plant Nutrients

Wild daisies can share a meadow with dozens of other plant species, but when yellow star thistle spreads through a field, it chokes out virtually all the other plants. Why? We've already seen that in late summer, the aggressive, spiky weed can take in water through its deep taproot, while shallower-rooted plants dry up and die. But what about the role of nutrients? Another possibility is that yellow star thistle is exceptionally good at absorbing some plant nutrient and effectively starves the neighboring plants for that mineral. To see if this is true, we have to understand more about plant nutrients. Curiously, what may have been the very first quantitative experiment in biology probed the mechanisms of plant nutrition.

Around 1600, the Dutch physician Jan Baptista van Helmont questioned an assertion made by Aristotle centuries earlier that a plant's body derives most of its substance from soil. To test this hypothesis, van Helmont filled a container with 91 kg (200 lb) of dry soil and planted in it a 2.2-kg (5-lb) willow tree. For five years, he watered the tree with rain water. At the end of that time, he dug the tree out of its pot, removed the soil and dried it, weighed both soil and tree, and discovered that while the tree had gained 75 kg (165 lb), the soil had lost a mere 27.5 g (about an ounce). With this, van Helmont succeeded in showing quite clearly that the main mass of the willow tree came not from the soil, as Aristotle had suggested, but from some other source.

Today, plant scientists know that plants synthesize organic compounds from the carbon and oxygen in air, from the hydrogen in water, and from small amounts of minerals in soil (review Fig. 3.22). Fully 96 percent of a plant's dry weight is made up of carbon, oxygen, and hydrogen. Nevertheless, a plant can't live without the dozen or so chemical elements that make up the remaining four percent.

Plants need at least 16 different chemical elements, the so-called essential elements, which different species require in different amounts. All plants require the **macronutrients** in relatively large amounts, and the **micronutrients** in much smaller amounts. By drying many types of plants, weighing the dried plant matter, and analyzing its chemical content, plant physiologists were able to tabulate information on the nine macronutrients—carbon, oxygen, hydrogen, nitrogen, potassium, calcium, magnesium, phosphorus, and sulfur—and the seven micronutrients—iron, chlorine, manganese, boron, zinc, copper, and molybdenum. Plants use both macro- and micronutrients in amounts roughly comparable to the levels animals need. Just as animals suffer vitamin and mineral deficiency diseases if they lack essential nutrients, plants also show specific symptoms relating to specific deficiencies. *E*xplorer 23.3 lets you diagnose plant deficiencies.

Macronutrients

In Chapters 2 and 3, we discussed the roles of carbon, oxygen, and hydrogen in biological molecules and reactions. Here, we'll concentrate on the other six macronutrients that make up most of the rest of the atoms in carbohydrates, proteins, lipids, and nucleic acids.

Nitrogen Fixation

After carbon, oxygen, and hydrogen, nitrogen is the most important of the macronutrients because it's an essential component of amino acids and proteins, chlorophyll, coenzymes, and nucleic acids. Nitrogen is often the most important nutrient that limits growth—the less nitrogen available, the slower the plant grows. Although gaseous nitrogen (N_2) makes up 78 percent of Earth's atmosphere, plants can't use it directly. Instead, the nitrogen must be fixed, or converted from the simple gas N_2 into some other form, such as ammonia (NH_3) or nitrate ions (NO_3^-), by a process called **nitrogen fixation.** Plant cells can then modify these fixed nitrogen compounds and incorporate them into plant proteins.

Farmers and home gardeners often add fixed nitrogen to the soil in the form of industrially produced nitrate-containing fertilizers. Wild-growing plants like star thistle, of course, don't have the benefit of added fertilizers and must depend instead on nitrogen-fixing bacteria, including cyanobacteria, which can convert molecular nitrogen to usable forms. Many of these microorganisms live independently in the soil and reduce nitrogen gas to ammonia. Some of the most interesting ones, though, live in the root cells of certain vascular plants (Fig. 23.10). Peas, beans, alfalfa, clover, lupine, and other **legumes,** or members of the pea family, have swellings, or **nodules,** on their roots that house nitrogen-fixing bacteria. Certain nonlegumes such as alder trees also form associations with nitrogen-fixing bacteria. Most other plants, including star thistle, don't form nitrogen-fixing nodules and get their nitrogen from soil-living bacteria, instead. Figure 23.10 shows how two kinds of plants, a soybean (a legume) and a yellow star thistle (a nonlegume) get enough nitrogen.

Nature is full of exceptions, and it turns out that soil-living, nitrogen-fixing bacteria can't survive in the acidic, waterlogged environment of bogs. Carnivorous plants such as the Venus flytrap and the pitcher plant have evolved the ability to gain fixed nitrogen from a different source: the proteins of the insects and other small animals they trap and digest.

Root nodules in legumes are a classic example of symbiosis, a close association of two different species: The plant supplies the bacteria with high-energy carbohydrates, while the bacteria provide biologically useful nitrogen. For centuries, farmers have rotated crops to

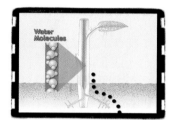

How would nitrogen be transported in a yellow star thistle?

SOLVE IT

*E*xplorer 23.3

What Is Wrong with this Plant?

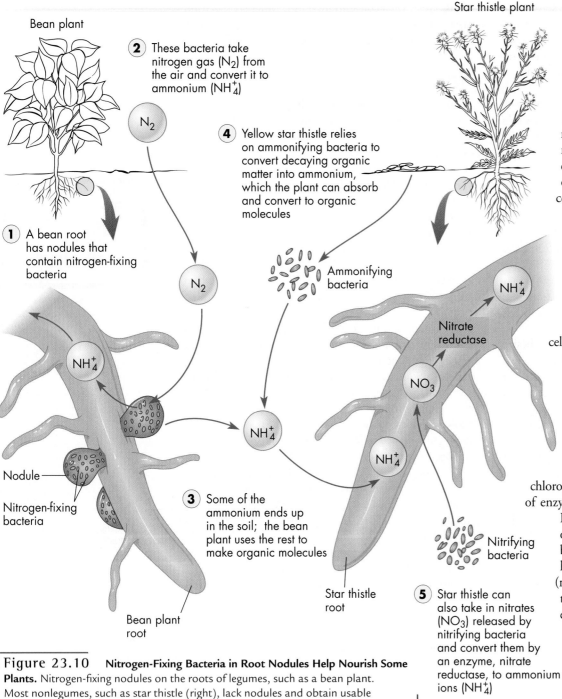

Figure 23.10 **Nitrogen-Fixing Bacteria in Root Nodules Help Nourish Some Plants.** Nitrogen-fixing nodules on the roots of legumes, such as a bean plant. Most nonlegumes, such as star thistle (right), lack nodules and obtain usable nitrogen directly from the soil.

growth of water ferns in their flooded rice paddies; cyanobacteria living symbiotically in the ferns fix nitrogen and this enriches the soil for the rice plants.

Other Macronutrients

Plants can often require 4 to 40 times more nitrogen than they do the other 5 macronutrients, but these other elements are still essential to normal plant growth and development. Calcium is an element that controls cell membrane permeability. Because of this, it plays a role in the opening and closing of stomata, directional growth in plant cells, responses to changing day lengths, and the tendency of roots to grow down and shoots to grow up. Calcium is also an important component of pectin, a substance that glues adjacent cells together and helps young plants make strong cell walls. Cooks use commercial pectin in jams and jellies to make them jell.

Potassium regulates osmosis in plant cells such as guard cells and also helps activate enzymes, including some involved in protein synthesis. Potassium deficiency causes mottled or burnt-edged leaves. Magnesium is a macronutrient because magnesium atoms are components of chlorophyll molecules and cofactors of many kinds of enzymes. Phosphorus occurs in the backbone of DNA and RNA molecules, in ATP and other high-energy compounds, and in membrane phospholipids. Because phosphorus, like sulfur, is a mobile nutrient in the plant (review Fig. 23.9), it moves from older leaves to newly developing leaves and shoots. This explains why phosphorus deficiency is first evident in the plant's oldest leaves (for example, in phosphorus-deficient tomato seedlings, whose older leaves turn purple). Because sulfur is an important component of two amino acids, plants require it for building most proteins, as well as for manufacturing some fats and coenzymes.

Micronutrients

Plants require only small amounts of micronutrients for healthy growth and so deficiencies of the micronutrients are rare. Iron, for example, occurs in several energy-

take advantage of this symbiosis. Early farmers may have been unaware of microbes, but they observed that if they raised clover or alfalfa one year, the following year's crop of corn or wheat would grow more luxuriantly. Likewise, rice farmers have encouraged the

harvesting proteins within the mitochondria and is also involved in making chlorophyll. A tomato plant with an iron deficiency makes too little chlorophyll to mask the yellow pigments in leaves. A deficiency of copper (needed in chloroplasts and certain enzymes) can cause deformed stems, leaves, and fruits in many plant species. Chlorine deficiency stunts growing tomato roots and fruits and causes the entire plant to wilt. Because zinc plays a role in protein synthesis, zinc-deficient apple and peach trees become stunted and grow miniature leaves.

Now, what about the question of whether yellow star thistle steals nutrients from other plants? The answer is no one knows, and researchers like Joe Di Tomaso will have to perform experiments to find out. They do, however, know details like this about salt cedar, a weed that grows along stream banks and in other wet, low-lying areas. Salt cedar is an alien species from the Mediterranean region that settlers imported as an ornamental (Fig. 23.11). The low-growing green bush, explains Dr. Di Tomaso, has a tremendous capacity to absorb the salts of sodium, calcium, magnesium, boron, and other minerals from the deep soil, move them up the xylem, then secrete them from salt glands onto the leaves. When it rains, these salts then wash down onto the surrounding soil surface and effectively poison the soil for most of salt cedar's former neighbors, which lack its strong salt tolerance. Over many years, says Di Tomaso, salt cedar actually changes an area's entire ecosystem (see Chapter 25).

Figure 23.11 **Salt Cedar: Winning Through Mineral Relocation.** Salt cedar (Tamarix chinensis) lifts mineral salts from deep in the soil, secretes them from salt glands, and rain carries them to the soil surface. Only salt-tolerant neighbors (if any) can survive.

Soil: The Primary Source of Minerals

Plants get their carbon, oxygen, and hydrogen from air and water but obtain all the other macro- and micronutrients from the soil. For this reason, soil composition has a major influence on the kinds of plants and, indirectly, the kinds of animals that can grow in a particular region of our planet.

Soil is a mixture of organic and inorganic material that starts with bedrock—relatively unbroken, unweathered rock. Over time, water, wind, heat, and cold disintegrate the bedrock and produce the inorganic parts of soil, from coarse sand and silt to small clay particles (Fig. 23.12). Bacteria, fungi, algae, lichens, and plants extract minerals from rocks, sand, silt, and clay and use them to synthesize organic materials. When the organisms die, the decomposing organic matter in the soil becomes **humus.**

Air spaces between soil particles are crucial to plant life. In a soil with a mixture of particle sizes, the spaces contain about half water and half air, with the water forming a continuous film over the soil particles. The size of soil particles, however, determines how much water the soil can hold: Soils made up primarily of coarse sand tend to hold water poorly, while soils made up mostly of fine clay particles hold water tenaciously. Plants such as cacti, which grow in rapidly draining sandy soils, absorb water quickly and store it, while plants that grow well in clay soils have roots resistant to dense, soggy, soil deprived of air. The best agricultural soils are deep layers of **loam,** with a high mineral and humus content and a mixture of particle sizes.

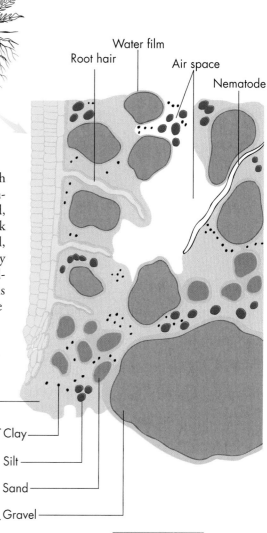

Figure 23.12 **Soil: A Source of Air, Water, and Nutrients for Plants.** Soil has inorganic and organic particles, as well as pore spaces filled with air and lined with water. Root hairs probe those spaces and take in water and minerals that dissolve from soil particles.

23.4 Plants Defend Themselves Against External Threats

Yellow star thistle plants have wicked spines surrounding every flower and spaced along the stems. The energy it takes to produce all these spines is apparently well spent: Deer, cows, and most other herbivores shun the nasty weeds in favor of plants that are easier to eat. Plants, after all, can't run away from hungry pests but they aren't defenseless. Yellow star thistle spines are just one kind of protective barrier. Others include spiky leaf hairs and sticky sap; various toxic compounds; and walling off injured areas.

Chemical Protection

The star thistle spines are an excellent deterrent but a starving animal may well persevere despite the pain in its mouth. The spines (or their equivalents on other plants) are therefore only a first line of defense. A second line of defense includes chemical warfare. Horses that eat star thistle develop a nervous disorder called "chewing disease" that is fatal once symptoms appear—drowsiness, unsteady gait, lack of coordination, and restlessness. This condition is brought about by a **secondary compound,** a molecule that helps ensure plant survival by repelling, killing, or interfering with the normal activities of plant-eating organisms. (Plant primary compounds are biological molecules necessary for normal growth and regulation.) Weed experts don't yet know which secondary compound in yellow star thistle can kill a horse. But plant biologists have discovered that in some tree species, up to 50 percent of the plant matter is defensive secondary compounds! These are so effective that in a temperate forest in a typical year, predators (including insects) consume only about seven percent of the total leaf surface area.

Over 10,000 plant chemicals are toxic to animals, and these secondary compounds include such familiar substances as cyanide, camphor, cocaine, caffeine, nicotine, and tannins. Camphor and tannins can discourage insect pests from eating a plant's leaves or laying eggs on them. The nicotine in tobacco can poison various insect predators. Cyanide compounds in birdfoot trefoil, a member of the pea family, can kill snails that feed on its leaves. Some plants generate compounds that mimic insect hormones, disrupting insect metamorphosis. And recent evidence shows that compounds given off by microbes and injured plant cells can induce the plant to respond, making antibiotics that kill attacking fungi and bacteria.

In response to insect damage, many plants increase their production of secondary compounds, in some cases reaching ten percent of the total weight of fruits or vegetables. Since these compounds can often harm people as well as insects it's best to stay away from diseased or damaged celery, apples, peanuts, and other produce. Some, such as safrol—found in the leaves of the sassafras plant, whose roots flavor our root beer—can even cause cancer.

Many secondary compounds are harmful to animals, but a few have given rise to important human medicines. These include the pain reliever aspirin from willow bark, the heart medicine digitalis from foxglove, the main ingredient in oral contraceptives from yams, cancer treatments from periwinkle flowers and the Pacific yew tree, and the pain reliever morphine from poppies. Many scientists expect to find additional medicines in as-yet-unstudied plant species, and this is a primary reason they are fighting to preserve vast tracts of unexplored rain forests.

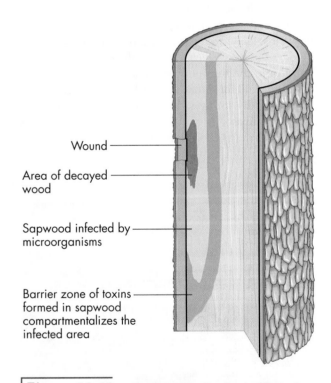

Wound

Area of decayed wood

Sapwood infected by microorganisms

Barrier zone of toxins formed in sapwood compartmentalizes the infected area

Figure 23.13   Compartmentalization in Plants: A Response to Wounding. A penetrating wound will inevitably allow microbial and/or fungal infections into the plant, but a zone of parenchymal cells (in lavender) around the infected area begins to produce tannins and other toxic secondary compounds. These kill the invaders, contain the damage, and allow growth to continue around the walled-off area.

Walling Off Injured Areas

If protective chemicals are a second line of defense, plants have a third line, as well: a growth response that protects them from viruses, bacteria, fungi, animals, or simple physical injury, such as loss of a limb in a storm. A plant's rigid cell walls prevent new cells from migrating into a wounded area and healing the damage. Plants can, however, wall off the damaged area—a response that prevents invaders from gaining access to healthy tissues. This walling-off process is called **compartmentalization.** It involves the production of toxic secondary compounds in the invaded area and the plugging of nearby xylem and phloem tubes with thick saps or resins that prevent invaders from spreading to other parts of the plant (Fig. 23.13). Once the injury is walled off, the tree can survive and grow.

23.5 Weed Control and Super Weeds

We said that plants can't run away from hungry pests, but in a sense, that's exactly what yellow star thistle has done. The prickly weed is native to dry regions of Europe and Asia, where insect predators keep the plant in check. One reason it's been able to gobble 15 million acres of open land in California is that it arrived on this continent without its natural insect predators, which include some voracious weevils. In the western United States, in other words, the yellow star thistle can see no weevil and hear none, either. Weed control experts like Joe Di Tomaso have recently tried importing weevils from Greece that consume yellow star thistle and hope that these insect predators may help reduce the star thistle population to controllable levels.

One of Dr. Di Tomaso's jobs at U.C. Davis is advising farmers and ranchers about the various methods for controlling yellow star thistle, wild blackberry, poison oak, salt cedar, and other problem weeds. These include tilling the soil and burning the weeds, but the main approach is applying **herbicides,** chemicals that kill plants. Most herbicides target biochemical processes carried out by plants but not animals. For example, the herbicide 2,4-D is an analog of the plant growth regulator auxin. The 2,4-D causes phloem to grow so fast it gets plugged and blocks substance transport, leading to death. Another herbicide was glyphosate, sold by Monsanto under the brand name Roundup.

Glyphosate is a simple molecule based on glycine, the smallest amino acid (Fig. 23.14a). In the soil, it immediately binds tightly to clay particles and breaks

(a) Glyphosate

Glyphosate

P—CH_2—Glycine

(b) Biosynthetic pathway

Enzyme

Phenylalanine

Tyrosine

Tryptophan

(c) Soybean plants treated with glyphosate

Figure 23.14 **How the Herbicide Glyphosate Works.** Glyphosate (Roundup) is a small molecule (a) that inhibits an enzyme in the pathway for building three essential amino acids (b). People and many other animals lack this enzyme; they can't make these amino acids and so are unaffected by glyphosate. Plants, such as the brown weeds in this photo, sprayed with the herbicide (c) die because their synthesis of these amino acids is blocked. The soy beans are still green because plant scientists genetically engineered them to resist glyphosate.

down. When sprayed on a plant, however, glyphosate binds to and inhibits an enzyme that helps build the amino acids phenylalanine, tyrosine, and tryptophan (Fig. 23.14b). Do these amino acids sound familiar? Look back at Figure 17.4 to find the answer: Two of them are essential amino acids; they occur in all of your proteins but you lack the enzymes necessary to make them, and so you have to acquire them in your diet. Humans don't have the enzyme that is blocked by glyphosate, and that's why Roundup can kill plants but has virtually no effect on people or other animals.

When Joe Di Tomaso sprays glyphosate on star thistle, he finds that the foliage quickly absorbs the chemical and it moves into plant tissues through the cell cytoplasm pathway (shown as the pale blue path in Fig. 23.3), and then enters the vascular system. Once in the phloem, glyphosate translocates by means of pressure flow, moving from photosynthesizing leaves to lateral meristems, areas of active growth. (This is why the label on glyphosate advises the user to apply it on a sunny morning: If the plant is not photosynthesizing rapidly, the pressure flow mechanism is not operating and the herbicide does not reach the tips of the roots and shoots before it quickly breaks down into harmless compounds in the soil.) Once in the meristems, glyphosate blocks the plant's ability to make the three amino acids it requires for survival. The plant immediately stops growing and, lacking new protein synthesis for growth and repair, it slowly begins to degrade. Soon the plant yellows and dies from dehydration (Fig. 23.14c).

The Fear of Super Weeds

Roundup is obviously a powerful tool in the fight against yellow star thistle and its runaway spread throughout California wildlands. It's such a powerful weapon against weeds, in fact, that Monsanto developed "Roundup-ready" varieties of soybeans, corn, and other crops. The splicing of genes for glyphosate resistance into crop species creates plants that can survive a spraying of the herbicide, while weeds growing in or around the fields die off.

But some plant researchers and most antibiotechnology groups have expressed concern over the potential escape of Roundup resistance genes into wild plant populations. Say, for example, that an artichoke grower planted "Roundup-ready" artichokes (which are essentially domesticated thistles) and through rare natural hybridization, the genes jumped into nearby popula-

tions of yellow star thistle weeds. Could the result be "super weeds," weeds now resistant to glyphosate and no longer controllable with the formerly powerful tool?

No, says Joe Di Tomaso. In the first place, he explains, no company would ever invest enough research funds in a small, specialty crop like artichokes to engineer it with Roundup resistance genes. And secondly, he says, even if star thistle somehow did acquire Roundup resistance, "it's not going to become some even worse, horrifically spiky monster plant with deadly fruits and leaves like poison oak. The term 'super weed' is very misleading," he remarks. "It really just means you would just have to switch to some other herbicide."

If there is a true danger of herbicide resistance genes transferring from crops to weeds, he says, it would involve crops grown widely enough to be engineered with Roundup resistance and then also planted in areas that happen to have closely related weeds. For example, sorghum is a widely grown cereal grass and is closely related to the weed called johnson grass, and the huge commercial crop canola is closely related to wild mustard. The chances are greater of a gene transfer between pairs like this, says Di Tomaso, so it's important for sorghum and canola oil farmers to be aware of the local weed populations and to avoid planting genetically modified crops if they are present.

"I think the scientific community needs to get together," he concludes, "to develop a set of guidelines for what are realistic fears and what is just misinformation."

Do you think guidelines like this would be strong enough protection against the escape of herbicide resistance from crops to weeds? Do you think Roundup resistance, Bt-toxin genes (Chapter 22), and other kinds of genetic modifications to crops pose a threat to the environment or to the food supply? How can we balance potential risks like these with the realities of human malnutrition and starvation? *E*xplorer 23.4 provides suggestions to help you investigate this important current issue in more detail and make your own decisions.

READ ON

*E*xplorer 23.4

Plant Biotechnology:
Taking a Stand

Chapter Summary and Selected Key Terms

Introduction   Dr. Joseph Di Tomaso is a weed expert who works to help control yellow star thistle, a major infester of open land in California. No one can say precisely why a weed grows so vigorously, but most weeds grow rapidly; produce seeds abundantly; spread by stems, roots, and seeds; and can adjust to a wide

range of habitats. Many, like star thistle, are also imported "aliens" with no natural predators. This plant, in addition, has a long taproot that allows it to pull water from deep soil layers during dry spells. Star thistle is a good case history for studying how plants transport water, minerals, and sugars; how they meet their nutritional needs; and how genetic engineering has led to fears of uncontrollable super weeds.

Water Balance in Plants   Water is a land plant's greatest need because it passes through the

plant in a one-way flow from roots, up the stem, and out the leaves. Water allows photosynthesis, tissue turgor, and internal transport of minerals and organic compounds like sugars. Water diffuses into root hairs then through the root cortex. Some passes inside cell wall spaces and some moves through root cell cytoplasm. When water reaches the endoderm, waxy Casparian strips prevent it from traveling between cells and force it to pass through endodermal cell cytoplasm before entering xylem tubes.

Water moves passively, without energy supplied by the sun rather than the plant. Root cells, however, actively pump in mineral ions and raise internal concentrations higher than in the fluid outside the roots.

How does water reach the tops of tall plants? Roots push some water up, squeezing little drops out of leaves. This **guttation** (p. 577) is based on **root pressure** (p. 577), but the force is too weak to push water to tree tops. If a biologist punctures xylem, air flows in rather than fluid spewing out, confirming that no pumps or pumping mechanisms transport water up xylem tubes. **Capillary action** (p. 577) can move water up thin tubes like xylem vessel elements, but only to a height of about 0.5 m (20 inches). The **transpiration-cohesion-tension** mechanism (p. 578) can explain how plants transport water to heights over 20 inches. Water molecules cohere to each other (like molecules) and adhere to the walls of xylem vessels (unlike molecules). When a water molecule evaporates from a leaf via **transpiration** (p. 578), the entire column of water molecules (linked by **cohesion** (p. 578) is pulled up the xylem tube and new water molecules are pulled into the roots. A rent in the xylem can allow air bubbles to break the chain. The rapid evaporation from leaves on a very hot, dry day can outstrip water absorption by roots, and the xylem walls can be sucked inward and/or the water chain can break, leading to plant wilting. The swelling or deflating of stomata, based on potassium ion levels and osmotic water gain or loss, controls the rate of transpiration from leaves.

Movement of Nutrients

Plants need to move around sugars and other organic nutrients, depending on their production and use levels in different plant parts. **Translocation** (p. 580) is the movement of nutrients in plant phloem. The **pressure flow hypothesis** (p. 580) is the idea that phloem transports substances from areas of production to areas of use or storage. Photosynthetic cells in leaves load sugars into sieve tube elements by active transport. Water follows osmotically, and the increased internal cell pressure sends the sugars through the sieve tubes away from the leaves. Storage cells in the root or growing plant parts needing sugar for metabolic energy requirements remove sugars from the phloem, the solute concentration drops, water flows out of the phloem into the xylem by osmosis, and it is carried upward again by transpiration.

Some nutrients, like calcium, are transported only in xylem. Others, like sulfur and phosphorous, move in the phloem. Plants can have mineral deficiencies, and these show up in various distinctive ways in the leaves and stems.

Plant Nutrients

In about 1600, Dutch researcher Jan Baptista van Helmont found that virtually all of a tree's weight comes from some source other than the soil. Modern scientists worked out that source: photosynthesis using carbon and oxygen from the air and hydrogen from water. These elements account for 96 percent of a plant's dry weight, but the plant can't live without the other 4 percent, the **macronutrients** (p.583), including N, K, Ca, Mg, K, and S (and, of course, C, O, H) and the **micronutrients** (p. 583) Fe, Cl, Mn, Bo, Zn, Cu, and Mb (see text for full names).

Nitrogen is the most important macronutrient (after C, O, and H), and gaseous nitrogen (N_2) must be fixed into ammonia (NH_3) or nitrate ions (NO_3^-) by the process of **nitrogen fixation** (p. 583). **Legumes** (p. 583) have root **nodules** (p. 583) harboring nitrogen-fixing bacteria. Other plants absorb nitrates given off by soil-living, nitrogen-fixing bacteria. Plants get most macro- and micronutrients from the **soil** (p. 585). Decomposing organic materials form **humus** (p. 585). **Loam** (p. 585) has a high mineral and humus content and mixed particle sizes.

Plant Defenses

Plants have physical defenses, like the spines on a yellow star thistle, and chemical defenses, including **secondary compounds** (p. 586) such as cyanide, nicotine, caffeine, and tannins that poison or discourage animal predators. Plants can also wall off injured areas through **compartmentalization** (p. 587).

Weeds and Super Weeds

Weeds can be pulled, burned, or tilled under, but the most common control method is spraying with **herbicides** (p. 587), including glyphosate or Roundup. Glyphosate blocks an enzyme that plants need to build three essential amino acids. Animals lack the enzyme and so remain unaffected by glyphosate. Genetic engineers have spliced glyphosate resistance genes into corn, soybeans, and other crops. This has raised major concerns over the escape of these genes into weed populations and the creation of uncontrollable super weeds. Whether these fears are justified is controversial at this point. Weed experts like Joe Di Tomaso say no; objectors to plant engineering say yes.

All of the following question sets also appear in the Explore Life **E** electronic component, where you will find a variety of additional questions as well.

Test Yourself on Vocabulary and Concepts

For each question set, complete or match the statements with the most appropriate term from the list. Any answer may be used once, more than once, or not at all.

SET I

(a) stomata (b) guard cell activity (c) Casparian strips (d) guttation (e) root pressure (f) capillary action

1. Osmotic loss of water due to potassium loss determines the diameter of the entry portals for air into the leaf.

2. Structures that form a barrier to passive flow of water into the center of the root

3. The creeping of water that results from a combination of cohesion and adhesion

4. Droplets of water that appear on leaves as a result of root pressure

5. Movement of water up from the root as a consequence of osmosis into cells of the xylem

SET II

(a) transpiration (b) adhesion (c) cohesion (d) evaporation (e) transpiration-cohesion-tension

6. The binding of water molecules together by the formation of hydrogen bonds between them

7. A theory of upward movement of water column in the xylem because of loss of water molecules from the top of the column

8. The loss of water by a plant through its leaves and stems

9. The bonding of water molecules to molecules other than water

10. The movement of molecules of liquid water into the gaseous phase

SET III

(a) soil (b) humus (c) loam (d) micronutrients (e) macronutrients

11. Nitrogen, calcium, potassium, phosphorus, and sulfur, among others

12. Iron, copper, and chlorine, among others

13. The part of the soil comprising decomposing leaves and other organic matter

14. An agriculturally rich kind of soil

15. Begins as inorganic bedrock that changes over time and acquires an organic component

SET IV

(a) nitrogen fixation (b) symbiosis (c) legumes

16. Occurs both in nodules and in soil, producing a

form of nitrogen that can be used for amino acid biosynthesis

17. An association of mutual benefit between two dissimilar organisms such as nitrogen-fixing bacteria and leguminous plants

18. Plants that house nitrogen-fixing bacteria in root nodules

19. Conversion of N_2 into ammonia or nitrate ion

Integrate and Apply What You've Learned

20. Yellow star thistle is a normal range plant in Mongolia and Eastern Europe but a catastrophic weed in California. What are some of the reasons for this?

21. "In plants, water moves only in a one-way path from roots to leaves, and it does so only in the xylem." Do you agree or disagree with the statement? Explain.

22. Do water and ions move from the soil to the plant and then throughout the plant passively by diffusion, by means of active transport, or by both means? Explain.

23. What general principle of nutrient transport in plants is illustrated by the facts that symptoms of calcium deficiency appear first in new leaves, whereas symptoms of phosphorus deficiency appear first in old leaves?

24. What is the evidence for and against root pressure, xylem pumps, and capillary action as the most essential mechanism for moving water from roots to leaves?

Analyze and Evaluate the Concepts

25. A Vermont family taps maple trees and makes syrup from the sap in the early spring. What material are they harvesting? Where does it come from in the plant? What forces the sap to flow?

26. A typical bag of fertilizer from a garden store will have a label reading "5-10-10," which means it is 5 percent nitrogen, 10 percent phosphoric acid (a source of phosphorus), and 10 percent potash (a source of potassium). According to Table 23.1, however, plants need far more carbon, oxygen, and hydrogen than they do the elements in the fertilizer. So why doesn't the fertilizer contain mostly carbon, oxygen, and hydrogen?

27. In an attempt to keep salad greens crisp, a chef puts them in a strong potassium chloride solution. What do you predict will be the result?

28. Why does Roundup kill weeds but not harm people or the environment? Under what circumstances would it not kill soybeans or corn?

PART 5 Ecology and Populations and Communities

Crashes and Clashes

Fish, Fishes, Fisheries.
Lendall Alexander; summer
flounder; Dr. Michael
Sissenwine.

Lendall Alexander is standing on the deck of his fishing boat, gazing out across the gray-blue waters off Cundy's Harbor, Maine. A fourth-generation fisherman, Alexander often rises before 4 a.m. and remains at sea for several days in a row. Like many others, he finds the fishing life a hard one, but adds, "Nothing is more fun" than to find a school of fish, scoop dozens into a strong net, and "put a bag of fish on the deck. There's nothing any better!"

Fishing has a three-century-long history in New England and for much of that history, the public viewed commercial fishers as hearty adventurers and their livelihood as noble. Today, though, many citizens see the whole fishing enterprise as tarnished with greed, and New England fishers as willing to take more fish than the region's waters can produce.

Until the mid-20th century, no one worried about sustaining or endangering fish populations. A wide continental shelf system called the Georges Bank lies offshore from Cape Cod and the coast of Maine (Fig. 24.1), and upwelling nutrients have always supported an enormous population of fish and other sea life. Fishing people traditionally used "groundfish" to refer to cod, haddock, flounder, hake, pollack, redfish, and other finned-fish living on or near the bottom of that area. By "trawling," or dragging nets across the sea floor, fishers were able to catch thousands of pounds of groundfish at a time. Because most of the fish species produce hundreds of thousands of eggs and larvae each year, the fish populations could recover fairly quickly. The marine communities off New England—including the groundfish, herring, tuna, clams, crabs, lobsters, and other organisms—were therefore quite stable. That, however, was already changing when Lendall Alexander started fishing full time.

In the 1960s, huge foreign fishing vessels—floating factories—from Japan, Norway, Taiwan, and other countries started trawling on Georges Bank alongside smaller American boats and harvesting millions of metric tons of fish annually. Around 1970, after years of such enormous catches, some of the fish populations started dwindling—an indication that overfishing was damaging the area's natural ecological balance.

In 1976, the United States extended its territorial limit to 200 miles off shore and the "factory ships" left. In their wake, the American fishing fleet grew, by then with technological improvements such as radio communication, radar, sonar, electronic navigation, and satellite weather data to help crew members find large aggregations of target fish such as Atlantic cod or yellowtail flounder. This fleet growth was also spurred by the skyrocketing demand for fish among health-conscious American consumers. As the demand rose, so did prices, along with the profitability for fishers, their families, and their towns.

This "gold rush," as New Englanders sometimes call it, continued until the early 1990s, when once again, overfishing took a drastic toll and fish stocks plummeted to record low levels. Explains Dr. Michael Sissenwine of the Northeast Fisheries Science Center in Woods Hole, Massachusetts, fishing people believed in "a legacy of inexhaustibility." However, marine environments are as vulnerable "to the fish hook and net," he says, as terrestrial ones are "to the ax and plough." For a number of fish stocks in the Georges Bank area, "it was common," Sissenwine says, "for at least half or three quarters of the fish in a population to be removed every year," leaving them at "less than ten percent of their pristine level."

For the past several years, Sissenwine has been actively involved in management programs that limit the access of commercial fishers to marine animals off New England. They have closed certain areas in certain seasons, regulated net and mesh sizes, issued restricted numbers of fishing permits, and limited each day's catch per boat to levels like 300 pounds of summer flounder or 100 pounds of cod. As a result, stocks of cod, haddock, flounder, and other groundfish are recovering—some quickly, some slowly. "The great challenge now," says a fishing contemporary of Alexander's from Cape Cod, Massachusetts, "is how do you allow an increased access to those fish without repeating the mistakes of the past?"

Lendel Alexander's experiences and the subjects of fishing, overfishing, and wildlife conservation in New England waters make an ideal case history for this chapter. Our topic here is the **ecology of populations and communities,** that is, the study of how groups of organisms are distributed in a particular area at a particular time and how they interact with other species coexisting in the same locale. Biologists have learned a great deal about these natural distributions and interactions, as well as how humans are strongly affecting them all over the world through direct harvesting of organisms and through altering or destroying their habitats. Our own burgeoning human populations are at the root of most ecological disruptions and form an important theme throughout the chapter.

As we explore population and community ecology, you'll discover the answers to these questions:

Figure 24.1 **Georges Bank: One of the World's Richest Fisheries.** A satellite image of the Cape Cod area. The relatively shallow region of Georges Bank, where Lendall Alexander and fellow fishers harvest thousands of pounds of groundfish each year, is northeast of Cape Cod, off this photo.

1. At what level do organisms interact in ecology?
2. What limits where a species lives?
3. What factors limit population size?
4. How has the human population grown?
5. Where do organisms live and how do they make a living?
6. How do species interact in communities?
7. How are communities organized in space and time?

24.1 Ecology: Levels of Interaction

Lendel Alexander's livelihood and that of his fellow fishers depends on knowing about the habits of fish: Where do yellowtail flounder or other crop fish live in the oceans? What times of year are best to harvest them? At what depths do they feed? What do they eat? In short, they need as much information as possible to locate and catch the fish efficiently. The science of **ecology** studies the distribution and abundance of organisms and how organisms interact with each other and the nonliving environment. Ecology applies research tools and the scientific method to probe the explanations behind observed phenomena. Ecologists like

Mike Sissenwine want to know what factors *cause* yellowtail flounder to prefer a certain type of ocean bottom; what other organisms, including prey and predator species, actually *regulate* the flounder's population sizes; and how physical factors such as ocean temperature and bottom characteristics *control* where the fish survives. The tenets of ecology are crucial to human survival. That's because our food, our shelter, the quality of our water, and even fundamental factors such as the air we breathe, the temperature of Earth's surface, and, to some degree, its climate depend on where organisms live and how they interact.

A key word in the definition of ecology is *interact*. Organisms interact with other living things and with the nonliving physical surroundings. For a yellowtail flounder, the living environment includes the burrowing worms and other small invertebrates it eats, the blackback and witch flounder that competes with it for food, the mackerel that prey on the larval yellowtail, and the roundworms that parasitize and weaken its body. The physical environment includes the temperature, salinity, and depth of the water; whether the bottom is sandy, rocky, or muddy; and the annual cycle of storms and ocean upwellings.

So numerous are each organism's interactions with other living things and the physical environment that biologists organize their study of ecology into a hierarchy of four levels: populations, communities, ecosystems, and the biosphere.

- A **population** is a group of interacting individuals of the same species that inhabit a defined geographical area. The yellowtail flounder of the Georges Bank, the saguaro cacti of central Arizona, and a swamp-gas–producing species of bacteria in the mud of a Louisiana bayou are all examples of populations.
- A **community** consists of two or more populations of different species occupying the same geographical area. The yellowtail flounder and the skates that prey on them, and saguaro cacti and the bats and insects that pollinate them are organisms in communities.
- An **ecosystem** consists of a community of living things interacting with the physical factors of their environment. The yellowtail flounder's ecosystem includes the muddy bottom and the cold temperatures and pressure of the water pushing down from above, the witch flounder and other fish it competes with for food, the invertebrates the yellowtail eats, and the roundworms that parasitize the flounder's flesh.
- The **biosphere** consists of all our planet's ecosystems and thus is the portion of Earth that contains living species. The biosphere includes the atmosphere, the oceans, and the soil in which living things are found, as well as global phenomena, such as climate patterns, wind currents, and nutrient cycles that affect living things.

This chapter and the next examine ecological interactions at increasingly higher levels of organization. We begin by discussing the factors that influence a population's location and size in time and space. Then we turn to community ecology and discuss how living organisms interact with each other.

What ecological levels does this represent?

24.2   What Limits Where a Species Lives?

Space on our planet is an ecological vacuum waiting to be filled by living organisms. Yet few species are scattered evenly throughout the world; they exist, instead, only in certain spots and are absent entirely from other places. What limits where an organism lives on a global scale and in individual locales?

Limits to Global Distribution

Why do yellowtail flounder occur off shore from Cape Cod, and not off the coast of Japan or Madagascar? Why does Lendel Alexander catch them at a depth of 40 to 70 m and not in shallower or deeper waters? Understanding why an organism's range extends to one part of the world but not to another is important to people when those organisms serve as food or materials, or when they cause disease in people, domestic animals, or crop plants. In general, three conditions limit the places where a specific organism might be found: physical factors, interactions with other species, and geographical barriers.

Physical Factors

Organisms may be absent from an area because the region lacks the proper sunlight, water, temperature, mineral nutrients, or any one of a host of physical or chemical requirements. For example, pine trees from the Austrian Alps photosynthesize best at a cool 15 °C (59 °F), but the *Hammada* bush of the Israeli desert photosynthesizes best at a scorching 44 °C (111 °F). Neither plant can survive in the other's environment because genetic adaptations fostered by natural selection have left each plant specialized for a particular set of physical conditions.

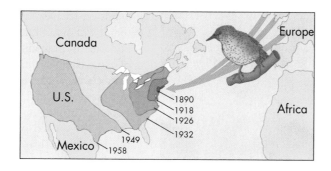

Interactions with Other Species

Other species may block survival and limit a population's distribution. If certain species are already firmly established in an area, they may prevent the incursion of new species by monopolizing food supplies or acting as predators or parasites. Large regions of Africa, for instance, are nearly uninhabitable for humans and cattle because the resident tsetse fly transmits the protist that causes sleeping sickness.

Geographical Barriers

A species may be absent from an area because a geographical barrier blocks access. Seas, deserts, and mountain ranges can be so wide or high that an organism cannot crawl, swim, fly, or float across the barrier. Europeans artificially bridged such a gap in 1890, when about 80 starlings were introduced into New York City's Central Park (Fig. 24.2). Now, millions of the dark, speckled birds chatter throughout America's cities and countrysides. North America turned out to be a congenial home for these organisms. Only their inability to cross the Atlantic had previously stopped their dispersal to North America.

24.3 What Factors Limit Population Size?

One goal of researchers like Michael Sissenwine, working with the Fishing Management Council in New England, is to maintain populations of yellowtail flounder and other crop fish at optimal levels so the Georges Bank area can sustain a healthy population of the animals. One measure of the population of yellowtail flounder is the weight of fish caught per year. Take a quick look at Figure 24.3a. What happened to the yellowtail flounder population between 1962 and 1996?

What happened to the populations of skates and spiny dogfish, a type of small shark? Mike Sissenwine wants to learn what factors caused this dramatic crash of the yellowtail population, how it might relate to the increase in skates and spiny dogfish, and how to restore populations to their original state to maintain a stable yield of marketable fish.

This practical side of ecology is concerned with a population's **density,** the number of individuals in a certain amount of space, for example, the number of metric tons of fish caught per year on Georges Bank. When reporting density, it is often helpful to take into account the species' habitat—that yellowtail flounder on Georges Bank may occupy only the sandy bottoms or that in a certain field a particular species of butterfly

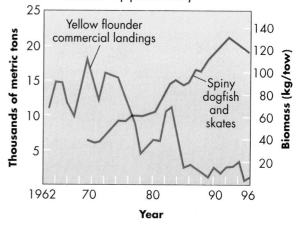

(a) The yellowtail flounder commercial catch: A reflection of population density

(b) A yellowtail flounder

Figure 24.3 **The Population Crash of Yellowtail Flounder.** (a) The total tonnage of yellowtail flounder landed by commercial fishers reflects a crash in the fish's population size. By contrast, populations of dogfish and skates have soared. (b) Yellowtail flounder (*Limanda Ferruginea*) reside on the ocean bottom.

pends on its age. For a yellowtail flounder, the chances that it will die as a newly hatched larval fish is very high, but after an individual becomes an adult, the likelihood of dying is much smaller. This is reflected in a **survivorship curve,** a plot of data representing the proportion of a population that survives to a certain age (Fig. 24.6). Look first at the human survivorship curve. The chances that a 90-year-old person will live for 1 more year are much slimmer than the chances that a 20-year-old will live for 1 additional year. In contrast, yellowtail flounder die mostly in the first month or two after hatching as they drift in the open ocean and become prey to mackerel, herring, and shrimplike krill. Once they settle on the bottom and become adults, it is less likely that a yellowtail will die in each passing month. The human follows a **late-loss** survivorship curve, whereas the flounder follows an **early-loss** survivorship curve.

A table of numbers used to generate a survivorship curve is called a life table, and it shows the **life expectancy** (average time left to live) and probability of death for individuals of each different age. Insurance companies use life tables to determine policy costs for customers of different ages. From the survivorship curve in Figure 24.6, you can understand why insurance companies charge an 80-year-old man more for insurance

might cluster mainly around yellow sulfur flowers. Human populations illustrate quite well the effect of clustering. The density of Earth's population averaged over the whole planet is less than one person per square kilometer (two per square mile). As Figure 24.4 shows, however, most people live near ocean coasts, riverbanks, and lake shores, making local densities in those areas much higher than the average—and putting pressure on fisheries the world over.

Factors Affecting Population Size

Curious biologists want to understand both how population density changes over time and the factors that cause it to change. Population size may change when individuals enter or leave the population. Clearly, if more members enter than leave, the population will grow, but if more individuals leave than enter, the population will shrink. Individuals can enter a population by either birth or immigration, and members can leave a population by either death or emigration (Fig. 24.5). If the number of individuals gained is exactly equal to the number lost, then the population shows **zero population growth.** For people, an average of 2.1 children per couple results in zero population growth.

Let's look now at how populations change as individuals leave a population by death or join a population by birth. We will assume that the effects of immigration and emigration remain constant and equal in the populations we are studying.

How Survival Varies with Age

Death and birth clearly affect population size. For many species, the likelihood that an individual will die de-

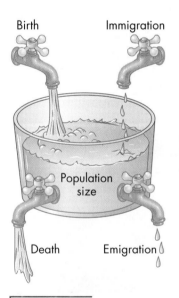

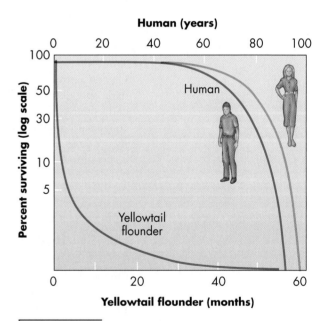

than an 80-year-old woman: He is more likely to die in the next year than she is.

How Fertility Varies with Age

Survivorship curves help predict how many individuals of each age class will leave a population through death. The major force that counteracts death is the birth of new individuals. Birth rates, like death rates, depend on age, and this fact is revealed by **fertility curves,** graphs of reproduction rate versus the age of female population members. (Since only females bear young, ecologists often view populations as females giving rise to more females.) The age of female population members is a key parameter for fisheries scientists like Dr. Sissenwine, because if fishers harvest sexually immature animals, then the population will rapidly decrease. Regulatory agencies can use fertility curves of yellowtail flounder to require that fishers use a net size with holes big enough to let animals through that are so small they have not yet reproduced.

The human fertility graph in Figure 24.7 shows that of every 100 American women between the ages of 20 and 25, on average about 12 will have a baby girl in any given year. Women younger than 20 or older than 30 are less likely to reproduce. Note the difference in the curves for 1960 and 1999. Fertility curves are significant because with them and with a knowledge of a population's age structure, one can predict future population growth. For example, a population with a high proportion of 20- to 30-year-old women is going to grow much more rapidly than a population with few women in these prime childbearing years.

How Populations Grow

Birth rates and death rates are frequently the major factors influencing changes in population size. By combining the two (but discounting immigration and emigration), population ecologists can make a model for how populations grow. Given plenty of nutrients, space, shelter, water, benign weather, and the absence of predators or disease, every population will expand infinitely, because all organisms have a high innate reproductive capacity under ideal conditions. The capacity for reproduction under idealized conditions, or **biotic potential,** is amazing. A yellowtail flounder can lay hundreds of thousands of eggs in her lifetime, and it takes 2 years for an egg to develop to a 1-kg sexually mature fish. Thus, in 12 years, with no limits, a single female could give rise to a mass of yellowtail flounder greater than the mass of Earth! Darwin made similar

(a) Human fertility curve

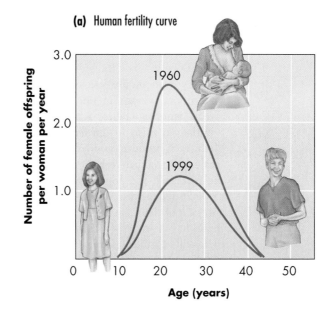

(b) Pore size in fish nets reflect the yellowtail flounder fertility curve

Figure 24.7 **Fertility Curves.** (a) Younger and older American women are less likely to reproduce than are women between 20 and 30. Contrast the childbearing habits of American women in 1960 with those of women in 1999, in terms of the number of children and the mother's age at delivery. Americans are having fewer children and are delaying reproduction to later ages. (b) The mesh size in Lendel Alexander's trawl net reflects the fertility curve for yellowtail flounder; prereproductive fish slip through the net and only larger, older fish are captured.

calculations for elephants and wondered why species do not reproduce to their potential. His solution to this puzzle was a key feature in the theory of natural selection. To fully understand the answer, we must first study patterns of population growth, then return to the question of what limits population size.

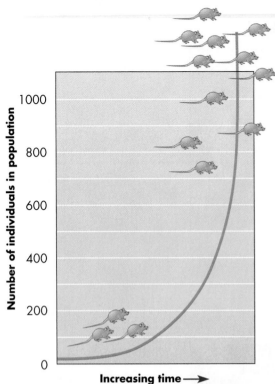

Figure 24.8 **The J-Shaped Curve of Population Growth.** In the absence of predators and disease, and in the presence of unlimited food, good weather, and pure water, a population could grow exponentially forever. The curve is steeper if individuals have more offspring per year and if the initial population size is larger.

GET THE PICTURE

Explorer 24.1

Population Curves: From Booms to Crashes

Rapid population growth is easier to visualize when plotted on graphs such as those in Figure 24.8. Let's say that individuals of a long-lived mouse species grow to reproductive age in one year and that, ignoring males, a population initially consists of ten female mice that each produce on average one female offspring per year. At the end of the first year, the population will include 20 female mice (the 10 mothers and each of their daughters). If each of those has 1 more female per year, there will be 40 female mice at the end of 2 years, 80 after 3 years, and so on. The explosive increase results in a **J-shaped curve** representing **exponential growth** (**E**xplorer 24.1).

What factors do you think might affect the shape of the J-shaped curve of Figure 24.8? It turns out that there are two factors: (1) the reproductive rate per individual and (2) the initial population size. If the reproductive rate per individual increases to two female offspring per female mouse rather than one each year, the population would grow much faster. Ecologists represent the reproductive rate per individual by the symbol r. For human populations, the reproductive rate per individual can vary widely. Women are capable of having about 30 children each, but they rarely reach that grand potential. The highest known human fertility rate was among the Hutterite communities of Canada's prairie provinces in the early 20th century, where the average family had 12 children. In contrast, in modern China, women are strongly discouraged from having more than a single child.

While the reproductive rate per individual is an important factor in the contour of the J-shaped curve, the size of the initial population is also important. If there are only five female

mice in the initial population instead of ten, as in the first example, and each produces one female offspring per year, then only five more will be added by the end of the first year rather than ten. This is why small populations add fewer individuals per year than larger ones, all other factors being the same.

Under the artificially ideal conditions we've discussed so far, any population would follow the J-shaped curve of exponential growth if the birth rate exceeded the death rate by even a small amount. Fortunately, real organisms in natural situations do not follow the J-shaped exponential growth pattern—at least not for long. Organisms in nature cannot sustain limitless growth at the full force of their reproductive potential because food supplies and living space are finite. Hence, our planet is not covered with elephants neck-deep in flounders. The realities of supply and demand explain this curb on population growth despite the organism's reproductive capabilities.

Limited Resources Limit Growth

The J-shaped curve gives ecologists an idealized standard against which to measure growth in real populations. A classic case of growth in a real population was that of sheep on the island of Tasmania, south of Australia, in the early 19th century (Fig. 24.9). When English immigrants first introduced sheep to the new environment, resources were abundant and the sheep population expanded nearly exponentially for a couple of decades, following a J-shaped curve during this time. As the density of sheep on the island rose, competition for limited resources increased, and by 1830, each sheep had a smaller share of food and living space. As a result, each individual was less likely to survive and more likely to die, and each had a smaller chance of reproducing. After 1850, the total growth rate decreased, and the

Figure 24.9 **Sheep on Tasmania: The Growth of Real Populations Often Follows an S-Shaped Curve.** Sheep populations on the island of Tasmania (blue line) initially followed a J-shaped curve, but finally they overshot the carrying capacity of the land. Eventually, their numbers oscillated within the range of the carrying capacity. This gives rise to the general S-shaped curve of logistic growth (pink line).

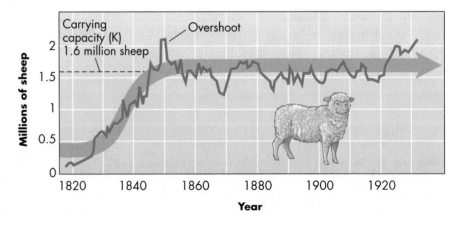

population size fluctuated around a mean of about 1.6 million sheep.

As Figure 24.9 shows, the graph representing population growth began like a J-shaped curve, but then flattened into an **S-shaped curve** representing **logistic growth,** a situation in which a large population grows more slowly than a small population would in the same area (**E** xplorer 24.2). The density at which a growing population levels off—1.6 million sheep on Tasmania in the preceding example—is called the **carrying capacity** (ecologists abbreviate this with the letter K). Carrying capacity is the number of individuals an environment can support for a prolonged period of time. At the carrying capacity, individuals are using all of the resources available to them.

Carrying capacity is not a constant number for each species and environment. You can tell that by looking at the fluctuations around the carrying capacity for the sheep shown in Figure 24.9. Because environments change with the season, with alterations in weather patterns, and with changes of species composition within the community, we should think of carrying capacity as a range of densities toward which populations tend to move from initial densities that can be higher or lower.

Population Crashes

While population growth often slows and reaches a plateau, as it did with sheep in Tasmania, in some species there can be a bust following the boom: a rapid decline following a period of intense population growth. The growth of reindeer introduced onto an island off the southwest coast of Alaska represents a frequently observed pattern (Fig. 24.10). From an initial population of 25 animals in 1891, the herd grew to about 2000 reindeer in 1938 and then crashed to 8 animals by 1950. The crash can be readily explained on the basis of carrying capacity. When the reindeer were first introduced, lichens and other food sources were plentiful, having accumulated for centuries without predation. Thus, the island's carrying capacity was high. After the deer ate the accumulated food, however, new food would appear only as the remaining lichens regrew slowly during each short summer growing season. This change in carrying capacity of that environment is what caused the reindeer population to crash (**E** 24.2).

Some organisms repeatedly experience a boom-and-crash sequence of population growth. Water flea populations, for example, ascend in a roughly J-shaped curve until they overshoot the carrying capacity of their environment, critically deplete resources, and then crash. Only after the environment has become renewed can

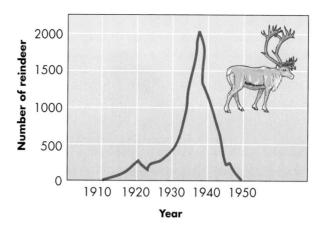

Figure 24.10 **Overexploiting Limited Resources Can Lead to a Population Crash.** A reindeer population introduced to a small island in the Aleutian chain southwest of Alaska initially had a J-shaped growth curve, since they were able to exploit centuries' worth of slow-growing lichens in a few decades. Once the food ran out, however, their population crash was rapid and spectacular.

GET THE PICTURE

E xplorer 24.2
Population and the Future

the water flea population again expand, only to crash again, in a continuing cycle.

We have seen that populations in nature usually follow either the S-shaped curve in which density slowly reaches the environment's carrying capacity and then remains relatively stable (as with the Tasmanian sheep) or the boom-and-crash pattern of rapid growth followed by an overshoot of the carrying capacity and a precipitous decline in density (as with the reindeer and water fleas). What characteristics of an organism's environment help determine the growth pattern?

How the Environment Limits Growth

The limited growth of real populations is due to such factors as limited food supplies, limited living space, and interactions with other organisms.

The Role of Population Density

Some limits to population growth depend on how dense a population is, and others do not. For instance, in a sparsely populated colony of prairie dogs, the incidence of flea-transmitted bubonic plague is also low. But when the prairie dogs are densely packed, outbreaks of plague often wipe out entire populations (Fig. 24.11). Competition among members of the same species for limited resources is another way density can regulate population size. For example, all the squirrels in a given forest might compete for the same nut crop.

Some factors limit population expansion regardless of population density. A harsh winter, for example, might kill 25 percent of a deer population regardless of population density. In practice, it can be difficult to separate the effects of population density from other factors, and this interaction leads many ecologists to

Figure 24.11 Dense Dogs and Disease. Among prairie dogs like these in South Dakota, overcrowding allows fleaborne bubonic plague to spread more easily. This is an example of how density can regulate populations.

look at whether the mechanism that limits population growth originates outside or inside the population.

Extrinsic or Intrinsic Population-Regulating Mechanisms

Extrinsic population-regulating mechanisms originate outside the population and include living factors, such as food supplies, natural enemies, and disease-causing organisms, as well as physical factors, such as weather, shelter, pollution, and habitat loss. In contrast to extrinsic factors that limit population size, intrinsic factors originate within an organism's anatomy, physiology, or behavior. For example, crowded conditions and depletion of resources can cause many marsupials, such as kangaroos and koalas, to absorb their own developing embryos; this intrinsic response lowers the rate of population growth.

Competition

The most important intrinsic population-regulating mechanism is competition among members of the same species for resources that they require but are available in limited supply. The effects of competition among members of a species depend on population density. As the population grows and resources diminish, competition for food and space becomes intense. In a coastal tidepool, for example, where a population of barnacles grows on a rock, other barnacles cannot occupy that same spot, even if food is present in abundance. An example of contest competition occurs in a wolf pack, where an aggressive alpha female ensures that she will be the only female in that pack to reproduce.

While extrinsic and intrinsic factors in the environment limit the growth of all species, they result in populations following either a roughly S-shaped curve or a boom-and-crash curve. Let's look at the factors that dictate which shape growth curve a population follows in nature.

Population Growth and Strategies for Survival

Evolution acts on organisms in ways that increase their individual genetic contributions to future generations. To survive, reproduce, and thus make a genetic contribution to the future, individuals must allocate their limited energy supplies. A fast-growing organism that expends most of its energy enlarging its body may have little energy left over for reproducing. Conversely, an individual that expends a huge amount of energy attracting a mate or producing thousands of eggs may have little energy remaining for day-to-day survival activities. The way an organism allocates its energy is its **life history strategy.**

To see how different life history strategies work, let's contrast the life history strategy of a dandelion and a rhinoceros. A dandelion reproduces rapidly, based on fast embryonic development and the production of large numbers of small seeds containing few stored nutrients. Dandelions experience an early-loss type of survivorship: most of the light, windborne seeds die shortly after germination. Because of these traits, they can quickly fill a newly plowed field before winter comes or the corn grows and shades them out (Fig. 24.12a). Our expression "to grow like a weed" reflects this type of life history strategy.

In contrast, a rhinoceros reproduces only after about 5 years of age, and the development of its embryo (usually single) is very slow (gestation takes about 15 months). Although rhinoceroses have only one calf at a time, the newborns are huge—the weight of an average male college student (Fig. 24.12b). Once born, the new individual survives for about 40 years, thus rhinoceroses experience a late-loss type of survivorship. Rhinoceroses are highly specialized to compete for resources in their environments. Adaptations include immense size, thick, armor-plated skin, and a fingerlike extension of the upper lip that pushes grasses and twigs into the mouth. Rhinoceroses are quickly approaching extinction because people slaughter them for their nasal horns. While these protuberances are made of the very same protein found in our hair and fingernails, some people have a superstition that the powdered horn can act as an aphrodisiac, and they are willing to pay more for the material than for gold, ounce for ounce. Knowing the life strategy of a rhinoceros, you can appreciate how hard it is for their populations to become re-established once decimated.

We can apply this knowledge of life history strategies to what you've learned about yellowtail flounders. Which type of strategy does it follow? Yellowtail flounders produce thousands of eggs, most of which die before becoming adults. Thus they seem to follow the first strategy. By contrast, the spiny dogfish is a "live bearer," and incubates its eggs until they hatch inside its uterus. As a result, a female produces only a few, large young each season, so their strategy seems more like the strategy of the rhinoceros. Many species have some elements of each strategy and can't be easily pigeon-holed. Which strategy—if either—do you think people follow? Let's look in more detail at the human population.

24.4 The Human Population

Human beings have achieved unparalleled mastery over their environment through improvements in agriculture, medicine, sanitation, transportation, and industrialization. Are we—with our gleaming cities, gigantic corporate farms, and shiny sewage systems—governed by ecological rules? Or have we somehow moved beyond booms, crashes, and growth curves? The answers involve a look at human history, long-term population trends, and some predictions for our future population growth.

Trends in Human Population Growth

You can see in Figure 24.13 the three historical phases of human population growth, and our current staggering rate of population growth.

Hunting and Gathering Phase

Until about 10,000 years ago, the human population grew slowly (Fig. 24.13a) as people existed by hunting animals, catching fish, and gathering roots and fruits from nature (Fig. 24.13b). The worldwide population was probably about 10 million by 8000 B.C. During this early phase, the human life history strategy emphasized slow embryonic development, long lives, large bodies, few offspring, extended parental care, and highly specialized brains that help us compete for resources with cunning efficiency.

Agricultural Phase

Population growth accelerated during a second phase of human history, beginning about 10,000 years ago, when people started planting and tending crops and domesticating animals in the **agricultural revolution.** The shift to agriculture (Fig. 24.13c) was rapid and worldwide, perhaps because people can transmit their **culture,** or ways of living, to others. As agricultural techniques spread and improved between about 8000 B.C. and 1750 A.D., world population increased from 10 million to about 800 million. Because agriculture uses some resources more efficiently, its practice increases the environment's carrying capacity for humans.

Industrial Phase

A third phase of growth began in 18th-century England with the **industrial revolution.** Inventions such as the steam engine triggered vast changes that transformed a populace living mainly as farmers and craftspeople into a population working mainly in factories and living in crowded cities (Fig. 24.13d). In the next 250 years, much of the world would follow this pattern of industrialization and social upheaval. A farmer with a steam engine attached to a tractor could accomplish the work of dozens of people in a single day and thus increase food production. A steam-driven train or ship could

(a) Dandelion: energy investment in rapid reproduction and many offspring

(b) Rhinoceros: energy investment in large bodies and few offspring

Figure 24.12 Rapid Reproduction Versus Careful Investment. (a) The dandelion develops rapidly, has a small body, and reproduces just once early in life. The emphasis is on reproductive productivity that offsets losses induced by the environment. (b) In contrast, the rhinoceros develops slowly, reproduces later in life rather than earlier, has few offspring, and has a large body. The emphasis is on efficient competition for resources.

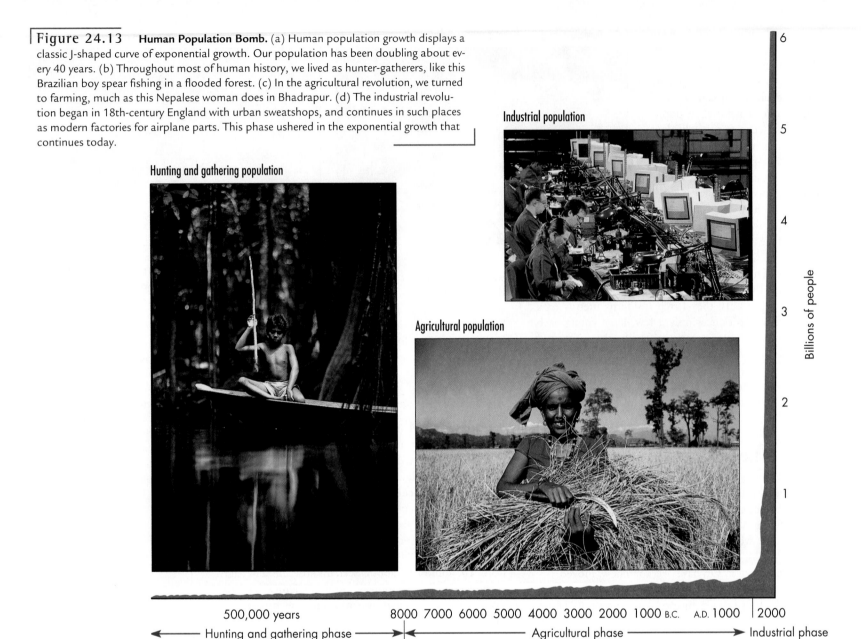

Figure 24.13 **Human Population Bomb.** (a) Human population growth displays a classic J-shaped curve of exponential growth. Our population has been doubling about every 40 years. (b) Throughout most of human history, we lived as hunter-gatherers, like this Brazilian boy spear fishing in a flooded forest. (c) In the agricultural revolution, we turned to farming, much as this Nepalese woman does in Bhadrapur. (d) The industrial revolution began in 18th-century England with urban sweatshops, and continues in such places as modern factories for airplane parts. This phase ushered in the exponential growth that continues today.

Hunting and gathering population

Industrial population

Agricultural population

Billions of people

6

5

4

3

2

1

500,000 years | 8000 7000 6000 5000 4000 3000 2000 1000 B.C. | A.D. 1000 | 2000

◄——— Hunting and gathering phase ———►|◄——— Agricultural phase ———►|◄— Industrial phase

Graph showing explosive growth of human population

rapidly distribute food and other necessities of life, and thus blunt the impact of local famine.

In recent times, the rise in human population has been staggering. While it took from the beginning of life until 1950 for the first 2.5 billion people to accumulate on Earth, it took just 40 years—a blink of evolutionary time—for a second 2.5 billion to be added. At current growth rates, by 2025 an additional 5 billion *more* people will join the planet's current population of over 6 billion. How old will you be when the 11 billion figure is reached? What might life be like when there are twice as many people as there are today?

As you look at the graph of human population growth, the towering ascension should unnerve you: It is the familiar J-shaped pattern of exponential growth, much like that of the island reindeer just before they overexploited their environment and suffered a population crash. By analyzing the causes of our own population boom, ecologists hope to learn how humans can avert a crash in the future.

Change in Human Population Size

How did the agricultural and industrial revolutions quicken the pace of the human population explosion? Did the invention of agriculture *allow* human populations to increase, or were people *forced* to invent agricultural practices to help support population densities that were already exceeding the carrying capacity of the land? Many observers believe the latter and suggest that population growth has been a constant feature of the human experience, continually forcing people to adopt new strategies for increasing the amount of food their land could produce. Carrying capacity, however, cannot be increased forever; the productivity of the land must, at some point, be reached and exceeded.

Birth Rates and Death Rates

To understand the causes of human population increase, particularly the tremendous surge after the industrial revolution, we must recall that the population growth rate equals the birth rate minus the death rate. Prior to 1775, the birth rate in developed countries like those of northern Europe was slightly higher than the death rate, and so the population enlarged at a low rate (Fig. 24.14, Step 1). After 1775, as industry expanded, people enjoyed improved nutrition, better personal and public hygiene, protection of water supplies, and the reduction of communicable diseases such as smallpox. These innovations caused a gradual decline in the death rate (Step 2). While the death rate began to decline in 1775, the birth rate did not start to drop in developed countries until about a hundred years later. Consequently, each year many more people were born than died, and this caused an increase in the rate of population growth. By the last decade of the 20th century, both the birth rates and death rates in industrialized nations had dropped to all-time lows, and the gap between them had once again narrowed (Fig. 24.14, Step 3). For example, Japan in 1999 had a birth rate of 9.5 per thousand, and a death rate of 7.3 per thousand, giving a low growth rate. A changing pattern from high birth rate and high death rate to low birth rate and low death rate is called the **demographic transition** (Fig. 24.14, Step 4).

In industrial Europe and North America, the demographic transition occurred in the first half of the 20th century and reduced overall growth rates to low levels. The populations of Africa, Asia, and Latin America, however, have continued to grow at high rates for much of this century. The concept of a demographic transition helps explain why. The death rates in those

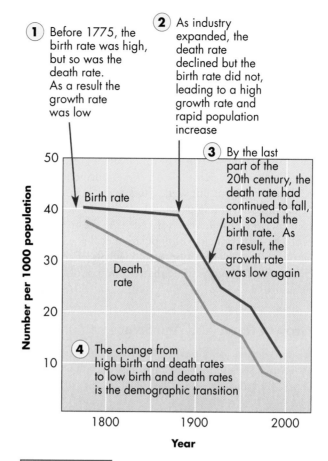

1 Before 1775, the birth rate was high, but so was the death rate. As a result the growth rate was low

2 As industry expanded, the death rate declined but the birth rate did not, leading to a high growth rate and rapid population increase

3 By the last part of the 20th century, the death rate had continued to fall, but so had the birth rate. As a result, the growth rate was low again

4 The change from high birth and death rates to low birth and death rates is the demographic transition

Figure 24.14 **The Difference Between Birth Rate and Death Rate Detonates the Population Bomb.**

countries remained high until the mid-20th century, when the spread of Western medicine and public health technology helped spare lives in record numbers. Simultaneously, however, the birth rate remained high for cultural reasons, and so the gap between the two rates widened. The huge excess in the birth rates caused an enormous net growth rate and hence an immense population boom in the 20th century.

Evidence suggests that the developing nations are now moving through the demographic transition, with a change from high birth rates and high death rates to low birth rates and low death rates. In the poverty-stricken country of Bangladesh, for example, women in 1975 had an average of 6.7 children each, but by 1999, the rate had fallen to 3.0. Still, the birth rate (27.2 per thousand) exceeds the death rate (9.3 per thousand) substantially, and so the population continues to grow. While the drop in birth and death rates provides a hopeful sign, three or four children per mother is still a houseful of kids. A family with many children in a developing country, however, has a less destructive impact on the global environment, in general, than does an average family with just two children in an industrialized society. This is because people in industrial nations consume many times more resources (for example, heating oil, gasoline, and electricity).

While people in industrial societies use up far more than their share of the world's resources and give off far more than their share of the world's pollutants, demographers point out that the most important political and ecological problem our world faces is the huge number of unemployed young people who will grow up in poor countries over the next few decades. At the root of the problem lies a principle of population ecology: A population with a large fraction of individuals of reproductive age tends to grow faster than a group of oldsters the same size.

Growth Rates and Age Structure

A sure sign of a population's growth rate is its **age structure**: the number of individuals in each age group (Fig. 24.15 and **E** 24.2). The age structure of a grow-

ing Swedish population in 1900—a time when the death rate had already declined substantially but the birth rate had yet to fall—shows a high percentage of people in the younger age classes (Fig. 24.15a). This results in a pyramid-shaped age distribution. The Swedish population's age structure in 1950 shows few teenagers, owing to a low birth rate in the years before World War II, as well as a bulge, the postwar baby boom, of children less than 10 years old (Fig. 24.15b). By 1977, after the Swedish birth rate had dropped close to the death rate, each age class was only slightly smaller than the younger one below it (Fig. 24.15c). In 2002, the Swedish population had changed to a bullet-shaped age profile with almost no difference in the size of age classes until the end of the human life span (the point of the bullet) (Fig. 24.15d). By 2050, the pyramid is likely to be slightly top-heavy with older people because Swedes are having smaller families (Fig. 24.15e).

Age structure is significant because it helps predict a population's growth potential. A pyramid-shaped population with many young people, like the African nation of Uganda in 1969, will grow rapidly (Fig. 24.15f). In contrast, a bullet-shaped population, like Sweden in 1977, will be stable or decline. With a look

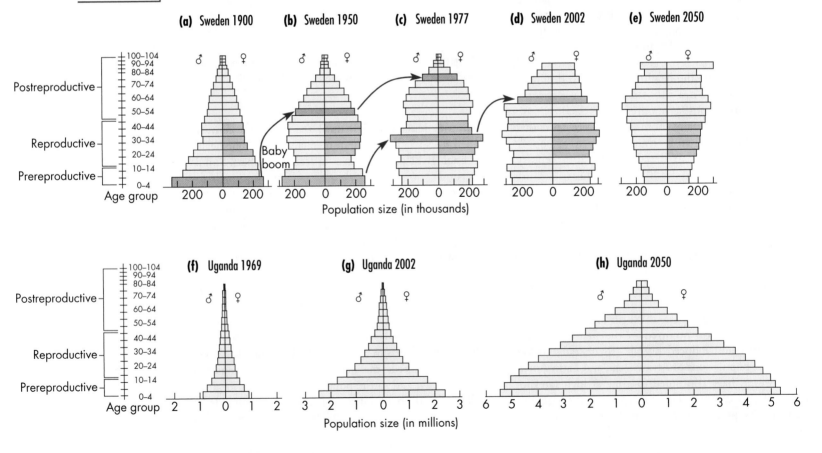

at data graphed this way, you can infer the kinds of social services needed by different populations: schools for pyramid-shaped populations, and health-care facilities for the elderly and pension plans for bullet-shaped ones.

Population of the Future

Humanity passed the 6 billion mark in 1999, and if present growth rates continue, another 3 to 5 billion will be added in the next 40 years. Even if, from this year forward, family size were reduced to the replacement rate of 2.1 children per woman, the global population would continue to grow for decades because the youthful citizens of countries like Uganda will soon attain reproductive age. Estimates for a stable human population by the year 2040 range from a low of 8 billion to a high of 14 billion, depending on the success of birth control campaigns. Regardless, our planet's human population will continue to grow in the not-so-distant future (**E** 24.2).

These astounding figures lead us, quite logically, to ask whether the planet can support billions more people at a reasonable standard of living. At present, no one can accurately predict what Earth's ultimate carrying capacity will be. Scientists do suspect that coal, oil, and some of the other resources on which we now depend are likely to become exhausted early in the 21st century. One thing is certain: Without dramatic steps taken immediately and decisively, the crush of humanity within our lifetimes will reduce or forever destroy complex and delicate biological systems such as tropical forests, as well as millions of individual species that have taken entire geological eras to evolve on our planet. The crash of the groundfish populations off the coast of Cape Cod due to overfishing is an example of what might happen to entire ecosystems upon which we depend.

24.5 Where Do Organisms Live and How Do They Make a Living?

The mushrooming human population makes it imperative to successfully manage resources such as the groundfish populations in Georges Bank. To do this, Dr. Sissenwine needs to understand not only how fish populations have changed over time, including the disastrous drops that have occurred, but also where the organisms live. For example, how many groundfish species live on the very bottom, like yellowtail flounder,

and how many cruise just above the bottom, like cod. He also needs to know how yellowtail interact with other organisms, their prey, predators, and competitors. These assemblies of different species in a particular area at a particular time are called communities.

Habitat and Niche

To understand the intricate web of relationships among organisms in a community, we must discover where the organisms live and how they make a living. The physical place in the environment where an organism resides is its **habitat.** A habitat is like an organism's address or home. In describing the general places where aquatic organisms live, for example, an ecologist might speak of an open-water habitat, a shore habitat, a muddy bottom, or a surface-film habitat. The yellowtail flounder occupies an open-water habitat as a fingerling, but then settles to an ocean-bottom habitat as an adult.

Whereas a species' physical home is its habitat, its functional role in the community is its **niche.** The niche is analogous to the organism's job—how it gets its supply of energy and materials. The yellowtail flounder's niche is that of a bottom-feeding carnivore. The niche of the single-celled algae inhabiting the surface waters of Georges Bank is that of a primary producer, using the energy of sunlight to fix carbon into biological molecules. A niche is what an organism does in and for a biological community.

Limits to Niches

The niche an organism actually occupies is often restricted by other organisms. For example, consider a warbler's niche in the forests of Cape Cod (Fig. 24.16a). The bird has the potential to eat insects wherever they occur in trees, at any height and at any distance from the trunk. The bird might also nest any time in June or July. The potential range of all biotic and abiotic conditions under which an organism can thrive is called its **fundamental niche.** If a warbler could catch insects any place in the tree, it would be operating in its fundamental niche for prey location.

A warbler in eastern forests cannot obtain insects just anywhere, however, because several species compete for food, and each species performs a slightly different role in the community. The different species obtain insects at different heights in the trees, at different distances from the trunk, and their heavy eating comes at slightly different times during the year, depending on when they nest. The myrtle (or yellow-rumped) warbler, for instance, eats insects at the base of trees, the

In what ways does human population growth threaten ecosystems?

bay-breasted warbler specializes in insects in middle branches, and the Cape May warbler seeks insects at the outer edges of the top branches (Fig. 24.16b). Thus, other community residents may force a warbler species from its broader fundamental niche into its narrower **realized niche,** the part of the fundamental niche that a species actually occupies in nature. This kind of niche

restriction is rather common. Species interactions like these can be a major factor in defining a species' distribution and abundance.

24.6 How Species Interact

The interaction of *Homo sapiens* and *Limanda ferruginea* (the yellowtail flounder) has been disastrous for the latter. Other types of species interactions also occur in nature and increase, decrease, or leave unchanged the abundance of either or both species. For example, the activities of fishers like Lendall Alexander have apparently caused the numbers of spiny dogfish to increase, while decreasing the populations of other fish species.

Ecologists have categorized interactions between species into four general types. In (1) competition and (2) predation, one or both of the species suffer. In (3) mutualism and (4) commensalism, neither species is harmed by the interaction. Let's investigate each of these types of species interactions.

Competition Between Species

There are never enough good things to go around— sunny spots in which to germinate and put down roots, sheltered places to build nests, or marine worms of a particular size to consume. Because of such restrictions, two different species often compete for the same limited resource, and this interaction restrains the abundance of both species. The key feature of **interspecific competition,** the use of the same resources by two different species, is that one or both competitors have a negative effect on the other's survival or reproduction.

A Model for Species Competition

To see how competition works, imagine populations of yellowtail flounder and witch flounder. If only one species inhabits a region of the Georges Bank, then, in the absence of fishers like Lendall Alexander, that species would increase in numbers until competition between the species' own members limited the population size, as we saw earlier with the Tasmanian sheep. But if both yellowtail and witch flounder inhabit the same area, any individual fish will compete against not only members of its own species, but against individuals from the other species as well. The outcome will depend on the relative strengths of the *inter*specific and *intra*specific competitions.

Figure 24.16 Niche: An Organism's Role in the Community. (a) A warbler in its fundamental niche might find food at any height in the tree and any distance from the trunk, and nest at any time in June or July. (b) A warbler in its realized niche might be restricted to finding food at certain positions in the tree and nesting in just a couple of weeks of summer.

(a) A warbler in its fundamental niche might find food at any height in the tree, any distance from the trunk, or nest at any time in June or July

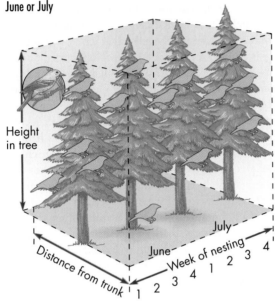

(b) A warbler in its realized niche might be restricted to finding food at certain positions in the tree, and nesting in just a couple of weeks of summer

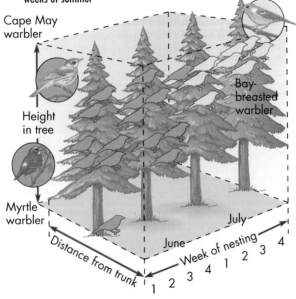

If a high density of one species (for instance, yellowtail flounder) affects the growth of the other species (witch flounder) more than it affects its own growth, then the first (yellowtail) will eliminate the second (witch flounder). A situation like this, where one species excludes another through competition, is called **competitive exclusion.** In another possible interaction, competition can affect each species' own population growth more than it affects the other species. In this case, the two species could end up coexisting. As a final possibility, each species may slow the growth of the other equally. In this case, the species with the biggest population to start with would take over the community.

These possibilities can be predicted by a mathematical model, and they certainly look good on a computer screen. But do they happen in nature with real organisms?

Natural Causes of Competitive Exclusion

Laboratory and field experiments with organisms as different as *Paramecium,* beetles, field mice, and aquatic plants typically show that competition does take place and that it can result in either competitive exclusion or in species coexisting.

For example, the Caribbean island of St. Martin has two lizard species that are slightly different in size but eat the same kinds of insects (Fig. 24.17). To find out whether the two species actually compete with each other, researchers fenced off large squares of land that contained individuals of the smaller species, the larger species, or both. They found that where both species coexisted, members of the larger species had less food in their stomachs, grew more slowly, laid fewer eggs, and were forced to perch higher in the bushes than when that species lived alone in an enclosure. Studies like these proved that strong competition does exist in natural populations and that the presence of one species can limit another species to its realized (rather than fundamental) niche.

Or take another example. In much of California's Mojave Desert, harvester ants (*Veromessor pergandei*) compete with a single other large ant species, and nearly all the harvester ants have mouthparts smaller than their competitors' (Fig. 24.18). In contrast, at a site near Ajo, Arizona, harvester ants compete with one quite large ant species and one quite small one. At this site, the harvester ants have mouthparts intermediate in size between those of their rivals. Ecologists interpret these field observations to mean that natural selection has acted on harvester ant populations to produce body

Figure 24.17 **Lizard Experiments Reveal Resource Partitioning.** Two lizard species from St. Martins Island. *Anolis gingivinus* has bulbous toes, which allow it to perch on high branches, while *Anolis pagus* has elongated toes that are better-suited to low perches in bushes and grasses.

sizes different from individuals of coexisting species; in other words, the character of body size is displaced by natural selection depending on characters in competing species. Character displacement among harvester ants apparently allows resource partitioning, since different-sized ants consume different-sized seeds. As a result, several potential competitors can coexist in the same habitat by occupying slightly different realized niches.

While competition often limits population size in both interacting species, the second major type of community interaction, predation, is beneficial for one species but harmful to the other.

Predation

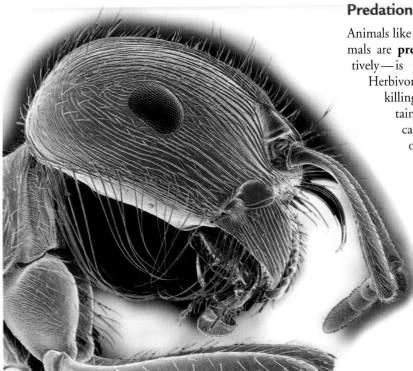

Animals like lions, yellowtail flounders, or Cape Cod fishers that kill and eat other animals are **predators,** their food—zebras, marine invertebrates, or flounders, respectively—is **prey,** and the act of procurement and consumption is **predation.** Herbivores such as harvester ants eat plant parts and often harm the plant without killing it. Parasites, like the roundworms that inhabit the livers of flounder, obtain their food from a living host organism, often without killing it. Disease-causing organisms, or pathogens, are usually fungi, bacteria, or protists that obtain nourishment from a plant or animal host and weaken or kill it. Here we focus on how predation affects the population size of both prey and predator on a short time scale, and how hunter and hunted evolve strategies to outwit each other on a longer evolutionary time scale.

Do Predator Populations Control Cycles of Prey Populations or Vice Versa?

Many populations of predator and prey rise and fall in cycles like the ripples in a pond. For example, wild populations of the snowshoe hare and lynx periodically rise and fall in phase on about a 10-year cycle (Fig. 24.19). Does the rise in the number of predators force this cycle? Or do large populations of hare make it possible for predator populations to expand? Field studies suggest that when hare populations are high, they provide an abundant resource for lynx, and this allows the lynx population to start increasing. The large numbers of hares soon overgraze their limited winter food sources, however, and so their population size decreases dramatically due to starvation (rather than to predation by lynxes). The lynx population then crashes because the cats cannot find enough to eat. It takes several years to regrow the vegetation, and when it does begin to return, the hare population once again increases steadily over a two- to three-year period, leading to a surge in the lynx population that lags a year or two behind the hares. Thus there are three factors (hare, lynx, and vegetation) involved in the arctic oscillations, not just two, and the size of the hare populations both modifies the abundance of vegetation and drives the lynx population cycle.

Recent experiments have shown that in other cases, predators can be the driving force behind the population swings of their prey. The southern pine beetle is a pest that destroys huge tracts of forest in the southern United States (Fig. 24.20). Like the hare and lynx, pine beetle populations rise and fall with a cycle of several years. Are these fluctuations due to extrinsic factors like weather? Or are they driven by intrinsic factors such as the delicious taste of the pine beetle to predators such as the clerid beetle? To find out, researchers put netting around loblolly pine trees in a Louisiana forest and fol-

Figure 24.18 **Ant Mouths and Character Displacement.** When a single large competitor is present, the harvester ant *Veromessor pergandei* has small mouthparts with a broad distribution. When a large and a small competitor are present, it has larger mouthparts and a narrower distribution. Natural selection has apparently caused hereditary shifts in mouthpart size, which hones this species of harvester ant to compete efficiently in different parts of its range, depending on its competitors.

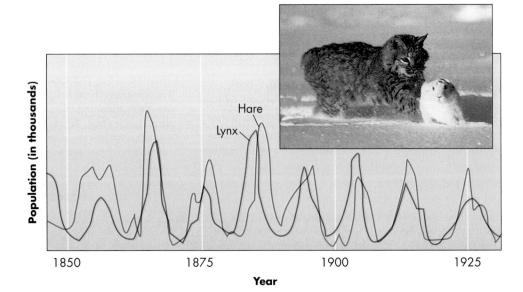

Figure 24.19 **The Hare-Lynx Cycle.** In northern Canada, hare populations occasionally boom when vegetation is abundant and crash when food becomes scarce. When hare populations rise, lynx also increase and devour more of their prey.

Figure 24.20   Pine Beetles: Predators Can Drive
Population Cycles. To test the effect of predators on populations of pine bark beetles, researchers from Pineville,
Louisiana, enclosed some trees with predator-proof cages
for five years, while nearby uncaged trees served as controls. They found that populations protected from predation grew better than those exposed to predators. They
concluded that predators can indeed help drive the cycles
of population growth.

lowed population changes for five years. The netting
kept predators from pine bark beetles, and nearby un-
netted trees served as controls. If predators in fact control pine beetle population cycles, then bark beetles in
protected areas should increase more rapidly than populations exposed to predators. Alternatively, if extrinsic
factors like the weather were responsible for population
changes, then both groups would be equally affected.
The ecologists found that the predators indeed were effective at decreasing the survival of bark beetles. These
results show that at least in some cases, predators can
drive the cycles of prey populations.

As with bark beetles and their predators, populations of predator and prey may grow or shrink over the
short term, but over the long term, genetic changes can
influence the evolutionary balance between hunter and
hunted.

The Coevolution of Predator and Prey

Predator-prey interactions lead to a grand coevolutionary race, with predators evolving more efficient ways to
catch prey, and prey evolving better ways to escape. It is
hard to tell which set of adaptations—the predator's or
the prey's—is more fascinating.

Predators need a way to catch their food, and the
two main options are pursuit and ambush. Predators
that pursue their prey are selected for speed and often for
intelligence as well. Carnivores store information about
the prey's escape strategies and make quick choices while
in pursuit. In keeping with evolutionary pressures, vertebrate predators generally have larger brains in proportion
to their body size than the prey they catch.

Some predators ambush their prey. A familiar example is the frog that zaps flying insects by snapping
out a sticky tongue. Certain mantis insects ambush
their prey by hiding in the open with a camouflaged,
plantlike appearance (Fig. 24.21a). Those mantises that
carry genes for resemblance to the plants they inhabit

Figure 24.21   A Nat-
ural Arms Race: The Coevolution of Predator and Prey.
(a) A praying mantis can
hide on a branch and am-
bush prey. (b) A moth mimics a dead leaf. (c) A flounder
can partially bury itself in the
sand or change colors to
match the bottom.

(a) A mantis hides

(b) A moth mimics a dead leaf

(c) A flounder can match its color to its surroundings

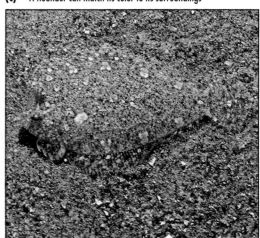

can be nearly invisible to prey and are thus more effective at ambushing their food and surviving to reproduce than mantises without those genes. An ambush can be even more effective if the predator can lure the prey, such as the anglerfish with a wormlike lure that entices prey into its gaping mouth.

Despite the stealth, athleticism, and often cunning of predators, prey species have evolved some remarkably devious tricks—in addition to speed—that help them avoid being eaten. One defense strategy is **camouflage:** the use of shapes, colors, or behaviors that enable organisms to blend in with their backgrounds and decrease the risk of predation. Many insects have evolved shapes that look like twigs, flowers, or leaves, complete with phony leaf veins (Fig. 24.21b). Still other insects, and a few amphibians, escape detection by resembling bird excrement on leaves. Behavior plays a role, too; when a flounder nestles to the bottom, it will flap a bit to cover itself with sand. Some flounders can even change their colors to match the color of the bottom on which they rest (Fig. 24.21c). In another adaptation to predation by fishers, flounders and other fish in Georges Bank have begun to become sexually mature at earlier ages and smaller sizes. This is an apparent response due to "un-natural selection": the fishers prey on larger fish, leaving only the small ones that escape through the mesh of fishing nets. If some of the small fish are genetically able to breed, then they are left to continue the stock and the population tends to reach maturity at a smaller size.

Chemical warfare is another defense strategy. Eucalyptus and creosote bushes, for example, produce distasteful oils or toxic substances that kill or harm herbivores. People sometimes plant oleanders as decorative shrubs because they resist insect pests, but the leaves are so poisonous that chewing a few can kill a child. Animals are not without their own arsenals: Toads, stinkbugs, and bombardier beetles (Fig. 24.22a)

produce highly offensive chemicals that repel attackers. Poisonous prey species often evolve brightly colored patterns, enabling the experienced predator to recognize and avoid them. This is called warning coloration. The brilliantly colored but poisonous strawberry frogs of South America (Fig. 24.22b) have evolved this strategy.

Mimicry

Many nonpoisonous prey species masquerade as poisonous species; this is one form of the process called **mimicry,** wherein one species resembles another (review the pairs of blue and orange butterflies in Fig. 9.29b and c). Mimicry could arise in the following way. Individuals of a nonpoisonous butterfly species that by chance contain alleles causing them to resemble the poisonous species even slightly may occasionally escape predation if a hungry animal mistakes them for the poisonous species. A selective pressure like this could, over time, allow the nonpoisonous species to accumulate more and more alleles for resemblance to the poisonous neighbor.

In another type of mimicry, two or more foul-tasting species come to resemble each other during evolution. An example of this is the beautiful viceroy and monarch butterflies (Fig. 24.23). Experimenters wondered whether this was a case in which only one of the species is foul-tasting or both are disgusting to bird predators. To resolve this issue, researchers removed the characteristically colored wings from seven species of butterflies and presented the now unfamiliar wingless insects to redwings (a type of blackbird). The birds took a small nibble from each, but fully consumed only a few of the species, which must have tasted fine. After sampling either a viceroy or a monarch, however, a bird would shake its head, drink excessively, and then strictly avoid *both* types of insects thereafter. Clearly, both the viceroy and monarch are foul-tasting, and birds can recognize desirable prey by taste as well as sight. Since a bird that learns to avoid one species will also avoid the other, this type of mimicry helps both kinds of insects avoid predation.

Parasites: The Intimate Predators

Parasites are insidious kinds of predators; they are usually smaller than their hosts, often live in close physical association with individual victims, and generally just sap their strength rather than killing them outright. Ectoparasites, like fleas, ticks, and leeches, live on the host's exterior, while endoparasites, like tapeworms, liver flukes, and some protozoa, inhabit internal organs or the bloodstream. In a special type of parasitic interac-

Figure 24.22 Chemical Protection. (a) The bombardier beetle (*Brachinus* species) defends itself by spewing a volatile irritant from special glands. (b) South American strawberry frogs are bright crimson, and this coloration warns predators of a poisonous meal.

(a) Jet-spraying bombardier beetle

(b) Strawberry frogs of South America

(a) Butterfly closely related to one from which viceroy evolved

(b) Viceroy

(c) Monarch

(d) Red-winged blackbird vomits after eating monarch

Figure 24.23 **Warning Coloration and Mimicry.** (a) Ancestors of the dull-colored *Limenitis arthemis* butterfly evolved into (b) the flamboyant viceroy (*Limenitis archippus*), whose coloration pattern mimics (c) the unrelated monarch butterfly (*Danaus plexippus*). Experiments show that monarchs and viceroys are both distasteful to birds, such as redwing blackbirds. The nearly identical bright coloration of the two butterfly species protects both from predation, because a bird will avoid both species after tasting only one and vomiting (d). Thus, *Limenitis* butterflies that carried alleles for resemblance to monarchs were selected for during evolution.

tion, certain insects develop inside the body of another insect, such as a caterpillar or a maggot, and inevitably kill it.

Commensalism and Mutualism

The community relationships we have discussed so far—competition and predation—have involved harm to at least one of the species. Sometimes, however, neither species is harmed by their interactions. In **commensalism,** one species benefits from the alliance, while the other is neither harmed nor helped, whereas in **mutualism,** both species are helped.

Commensalism

Commensalism is common in tropical rain forests, and the most easily observed examples are the epiphytes, or air plants, that grow on the surfaces of other plants. Epiphytes include Spanish moss, many beautiful orchids, and large and small bromeliads that festoon the

Figure 24.24 **Commensalism: One Species Benefits, the Other Is Neither Harmed nor Helped.** On coral reefs, an anemone fish can find safe haven within the folds of a sea anemone, which stings other fish.

branches or decorate the forks of trees. Using the tree merely as a base of attachment, epiphytes take no nourishment from the host and do no harm unless their numbers become excessive, their collective weight begins to stress the tree limbs, and they block light penetration and interfere with the tree's photosynthesis. Other commensal relationships include birds that nest in trees, algae that grow harmlessly on a turtle's shell, and the small fish that live among the stinging tentacles of sea anemones—unharmed and safe from predators (Fig. 24.24).

Mutualism

In a mutualistic interaction, both species benefit. An example involves the yucca, a tall, stately member of the lily family that grows in hot, dry regions of the western United States. The pointed leaves of a yucca plant jut upward from the parched ground like a sheaf of swords. Above the leaves rises a single awesome flower stalk, 4 m (13 ft) tall and loaded with a thousand white blossoms.

Each spring, white female moths flutter about the cream-colored yucca flowers (Fig. 24.25) and play out a curious set of behaviors honed by natural selection. Arriving at the younger flowers near the top of the stalk, the female yucca moth visits several blooms and collects a ball of pollen, which she carries beneath her "chin" with specially modified mouthparts. Bearing this pollen ball, she then flies to another plant and enters an older flower near the base of the stalk. Here she extends a long, sharp drill from the tip of her abdomen, pierces the older flower's ovary, and pumps her abdomen. The pumping causes eggs to pass through the drill and into the flower. The moth now executes one of nature's most remarkable sequences: She moves up to the flower's green stigma, separates a bit of pollen from the pollen ball, then rubs the pollen directly across the stigma. This pollina-

Figure 24.25 **Coevolutionary Partners.** *Yucca whipplei* grows in the Mojave Desert and has a mutualistic relationship with the yucca moth (inset). The moth lays its eggs in the ovary at the base of a yucca flower and then pollinates the flower. Moths depend on the plant for food and reproduction; the plant depends on the moth for pollination.

tion may lead to fertilization, ensuring that fruits and seeds will develop and serve as the food source for the moth's offspring as they become caterpillars.

Fortunately for the plant, the caterpillars do not eat all of the seeds from fertilized flowers as they grow. The plump, mature caterpillars eventually emerge from the fruits, drop to the ground, burrow into the soil, and metamorphose. The next season, adults emerge, mate, and repeat the cycle. In this relationship, both species benefit: the yucca moth gets a high-energy nectar reward and the plant becomes pollinated. If one species becomes extinct, so will the other.

24.7 Organization of Communities

The interactions we have considered—competition, predation, commensalism, and mutualism—affect not just pairs of species, but entire communities consisting of tens to hundreds of species. Before understanding how communities work, we must first identify these hundreds of species. Astonishingly, it was not until 1993 that the U.S. government first instituted a complete census of all living things in the United States. Called the National Biological Survey, the program aims to determine the constitution, and hence the health, of biological communities all across the country. When biologists identify the species living in communities, they help to answer an important question posed by early community ecologists: Are communities fixed groups of species unfailingly bound together? Or are they more fluid entities, consisting of one set of species at one time and another overlapping but slightly different set at another time? Research projects undertaken nearly 50 years ago

confirmed that a community is not a "superorganism," a package of highly specific groupings of plants and animals. Rather, communities consist of some species that happened to immigrate into the area and can survive under the available physical conditions, and some species that will grow only if other species are also present. Given this premise, a goal of ecologists is to learn how the addition or subtraction of a species affects the whole community in the short and long term.

Communities Change over Time

Occasionally, a cataclysm will strip an area of its original vegetation. This occurred in 1980 with the volcanic explosion of Mt. St. Helens in Washington State and again in 1988 when fires swept through much of Yellowstone National Park. A farmer clearing a field can even cause such a change. Left to nature, however, a regular progression of communities will regrow at the site in a process called **succession.**

Soon after a region is denuded, a variety of species begin to colonize the bare ground. These species make up a **pioneer community,** and they modify environmental conditions, such as soil quality, at the site. Conditions produced by the activities of the pioneer community determine which additional species can establish themselves in the area and form a **transition community.** Changes are rapid at first, as more and more species join the transition community. Eventually, a particular community of plants and animals becomes relatively stable, and changes take place more slowly over time. Such an assemblage is often called a **climax community,** but most ecologists today recognize that change is constantly occurring even in old communities, driven by such environmental variables as hurricanes, floods, and fires, as well as global climatic changes.

A well-studied example of ecological succession is in Glacier Bay, Alaska, where a glacier burying thousands of square kilometers of land has melted back about 100 km (62 mi) in the last 200 years, exposing the ground below. The most recently exposed areas are inhospitable piles of boulders, sand, and gravel lacking usable nitrogen and essentially devoid of plant or animal life. Figure 24.26 depicts the succession of plants that reclaims the rocks as habitable space for living things. The first plants to colonize this barren scene include rockrose and alder trees with nitrogen-fixing nodules on their roots that provide the plants with this growth-limiting element. Other plants follow and overgrow the early colonists, and eventually leave a **mosaic climax community** consisting of patches of spruce-hemlock forest intermixed with bogs containing water-soaked sphagnum moss.

Species Diversity in Communities

The mosaic climax forest at the foot of a glacier and the ocean bottom of Georges Bank have relatively few species compared to coral reefs and tropical rain forests, which are amazingly dense with interacting species. The total number of species found in a community is its **species richness,** or **species diversity.** In most communities, there are few common species but many rare types of organisms. For example, English ecologists captured a group of almost 7000 moths and identified individuals of 197 different species in one local area. Fully one quarter of the moths belonged to a single species and another quarter belonged to just five other species. The remaining half of the moths fell into 191 species—some represented by just one or two individuals. Paradoxically, rare species are common, and common species are rare.

What factors allow a region to have large numbers of rare species? Both *latitude* (north-south position) and *isolation* (peninsulas, island chains, or other out-of-the-way locales) influence **species richness.** Some communities in tropical latitudes, for example, have about 600 types of land birds, while an area of similar size in the arctic tundra may have only 20 to 30 species of land birds (Fig. 24.27).

A single hectare of mainland tropical forest can have 300 different species of trees and tens of thousands of insect species; on a peninsula, however—even a tropical one like Baja California or Florida—the species richness is diminished, as Figure 24.27 illustrates. This is because in an isolated area, the many rare species can easily become extinct and their replacement from the mainland is unlikely due to isolation. Species diversity on island chains is limited in such specific ways that ecologists are now applying the principles of island ecology to the design of nature preserves, which are, in fact, islands within a sea of human development.

Recent observations support the conclusion that the more resources available in an area—water and solar energy, for instance—the greater the species richness the area can support. This helps explain the

(a) The retreating glacier

(b) The rockrose: an early invader

(c) The climax community includes patches of sphagnum moss

(d) The Pattern of Succession

Time	0	10 years	20 years	50 years	80 years		100 years
	Glacier	Pioneer mosses, horsetails, rockrose	Alders invade prostrate willows, cottonwood	Alder thicket	Sitka spruce invade	Hemlock, sitka spruce forest	Hemlock, sphagnum moss bog in low spots

Soil becomes more acidic →

Soil gains nitrogen →

Figure 24.26 **Succession in the Path of an Arctic Glacier.** (a) As massive glaciers receded in Alaska's Glacier Bay, the first plants to colonize this barren scene form a pioneer community of wind-dispersed species: mosses, horsetails, fireweed, willows, cottonwood, and (b) the matlike rockrose (*Dryas ootopetala*). Most are severely stunted due to nitrogen deficiency, but rockrose has nitrogen-fixing nodules on its roots that provide the growth-limiting element. Within a few years, it crowds out other plants and forms a dense mat over the rocky soil. In areas exposed for 20 years or so, alder trees begin to invade from other sites. Like rockrose, alder roots also bear nitrogen-fixing nodules, and thus these trees can grow rapidly, add nitrogen to the soil layer, and stimulate the growth of willows and cottonwood. In areas exposed for about 50 years, dense thickets of these plants shade the pioneer species and kill them. By about 80 years, Sitka spruce invades, and nitrogen released by the alders enables spruce to form dense forests that shade out the alders and willows. Finally, after 100 years or so, shade-tolerant hemlock trees invade below the canopy of spruce branches and needles, and eventually become the most common tree in the climax community. In low places, however, sphagnum moss invades the forest floor (c), soaks up large amounts of water, and kills trees by choking off their roots' oxygen supply. This leaves a mosaic climax consisting of patches of spruce-hemlock forest intermixed with sphagnum bog.

numerous varieties of plants found in tropical forests and the richness of Georges Bank compared to other open ocean areas.

Other factors besides resources influence species richness, including competition and predation. Competition can increase by dividing niches into more and more specialized compartments. Recall the warblers and the Caribbean lizards discussed earlier. With competition forcing smaller niches, a community can accommodate more species.

Predation can also increase species richness. A sea star along the coast of Washington State, for example, preys on 15 species of barnacles, snails, clams, and mussels. When experimenters removed the sea star from sections of the shore, the diversity dropped to eight species because the mussel population increased and crowded out the other invertebrates. By eating young mussels, the sea star reduces competition for space and so preserves a higher species richness. Species like this predatory sea star, whose activities determine community structure, are called keystone species. The abundance of predators and parasites in the tropics may help explain why the lower latitudes have such a high species richness.

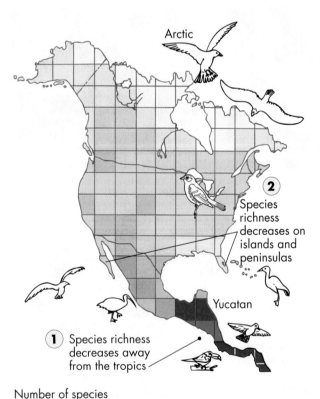

Figure 24.27 Latitude and Isolation Affect Species Richness.

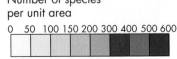

① Species richness decreases away from the tropics

② Species richness decreases on islands and peninsulas

Arctic

Yucatan

Number of species per unit area

0 50 100 150 200 300 400 500 600

Species Diversity and Community Stability

For years, biologists have warned that species richness is dwindling all over the globe. The loss of tropical forest ecosystems is one familiar example, and the cutting of old-growth (virgin) forests in the Pacific Northwest is another. Until recently, however, there was little actual data on how species richness affects the health of an ecosystem. The diversity-stability hypothesis suggests that because the species in a community encompass many different traits, a diverse community is more likely to contain at least some species that can survive environmental disturbances such as drought, fire, hailstorms, or overfishing. In contrast, the species-redundancy hypothesis predicts that within a community, various species are so similar that as long as major kinds of organisms are represented, species richness should not strongly affect community function.

In one of the first critical tests of these hypotheses, researchers in Minnesota and Quebec measured the amount of living plant material accumulated in plots of grassland (Fig. 24.28). These plots contained differing numbers of species, from undisturbed native prairies with more than 20 species per plot to abandoned farmers' fields with less than 10 species per plot. The results showed that the species-rich plots were four times more productive in drought years, and they recovered more quickly than species-poor plots. These results were predicted by the diversity-stability hypothesis; they suggest that a community's ability to resist adverse environmental conditions increases with species richness. The long-term stability of communities may thus depend on the biodiversity of their many interrelating species. These results for grasslands help underscore the urgency of pleas to conserve that biodiversity.

Species Richness in Agricultural Communities

If the diversity-stability hypothesis holds for ecosystems in addition to grasslands, then the most vulnerable biological communities are those with the fewest species, for example, huge tracts of land planted only with corn or wheat and treated chemically to kill all other plant and animal species. As you might imagine from your understanding of the principles of ecology and evolution, such a situation would be unstable. Indeed, Peruvian cotton farmers found that DDT helped control cotton boll-worms—but only temporarily, until the worms evolved resistance to the pesticide.

Their experience led to the first success of **integrated pest management**, a system based on the principles of community ecology and aimed at keeping pest populations below economically harmful levels with a minimum of chemical pesticides. Integrated pest management requires a broad view of all interactions between the crop and other species: weeds that compete for sun and nutrients; insects that eat the crop; predators that eat the insect pests; and animals that pollinate the crop. Pest managers have found that by carefully controlling the timing, spacing, and intermixing of crops, they can enhance the activities of natural predators. What's more, they can introduce biological control agents from other regions that feed exclusively on the pest, including parasitoid wasps and the insect-killing bacterium *Bacillus thuringiensis*. In addition, pest managers can release into the environment millions of sterilized adult male insects, which then greatly outnumber the natural males. When the sterile males mate with wild females, the matings produce no fertilized eggs, and the population size decreases. Finally, pest managers use chemical controls only when insects appear poised to do damage above some pre-established level of economic injury. And when spraying is necessary, they choose compounds and application methods to hit target species, leaving "innocent bystanders" alone.

Because farmers and ranchers using integrated pest management must consider the principles of community ecology, their job is more difficult than simply spraying the fields randomly with chemicals. However, because it avoids a whole host of problems—pests evolving resistance, resurgence of pest populations, and outbreaks of other pest species—integrated pest management, by increasing species diversity in croplands, offers an ecologically and economically acceptable alternative to managing crops and saving the 33 percent that are lost to pests each year.

The Future of Species Richness

Perhaps the greatest challenge in the world today is the immense increase in human population and its disastrous disturbance of Earth's biological communities. The collapse of the groundfish fishery off the coast of Cape Cod is an example. But in the species-rich communities of tropical latitudes, human population pressure is threatening the destruction of entire ecosystems. Tropical peoples have traditionally cleared land for agriculture by burning the forest, planting crops for a few years, and then moving on when the soil becomes depleted of nutrients. With enough time, the process of ecological succession can repair these small wounds. Now, however, the J-shaped curve of the human population results in vast areas of the rain forest going up in smoke to provide farmland for crops and pastureland on which to graze cattle for export to developed nations. These fires are so huge and so common that they

Figure 24.28   Investigating Species Richness and Community Resilience. These ecologists from the Missouri Conservation Department are identifying and recording all of the species in a small section of tall grass prairie in Missouri's Paintbrush Prairie Natural Area.

are clearly visible on satellite photos from space (Fig. 24.29). Many ecologists fear that the plant and animal communities of the tropics may not be able to recover from such extreme disturbances and that early in the 21st century, no tropical rain forests will remain intact. Still worse, many fear that this disruption may cause a distressingly large percentage of all living species to become extinct in our lifetimes.

Ecologists consider it an urgent research priority to learn what makes communities resilient and how they may (or may not) be able to persist in the face of human encroachment. We must solve this problem if our most diverse and interesting communities are to survive through the 21st century.

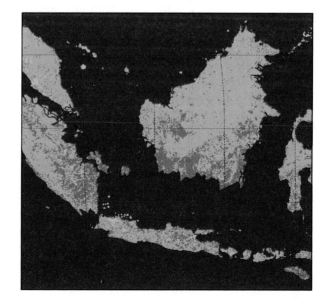

Figure 24.29   Diversity and Destruction in the Tropics. A satellite photo of Indonesia, Malaysia, and Borneo from July to December 1997 reveals huge areas of rain forest on fire. Forest clearing is intended to accommodate new agriculture. Red shows areas of fires; pale blue, electric lights; dark blue, water; and tan, land cover.

Chapter Summary and Selected Key Terms

Introduction

Lendel Alexander fishes for a living off the coast of Massachusetts. In the past three decades, fishers have decimated fish populations in the formerly rich coastal waters. Researchers like Michael Sissenwine study the **ecology of populations and communities** (p. 593), or how groups of organisms are distributed in a particular area at a particular time and interact with other species in the same locale. Their goal, using the study of **ecology** (p. 593), is to try to determine safe limits for fish catches and thus to manage marine resources to sustain healthier populations and communities in the future.

Ecological Levels

A **population** (p. 594) is a group of interacting individuals of the same species that inhabit a defined geographical area. A **community** (p. 594) consists of two or more populations of different species occupying the same geographical area. An **ecosystem** (p. 594) consists of communities of living things interacting with the physical factors of their environment. The **biosphere** (p. 594) consists of all the ecosystems of Earth, including all living species; the atmosphere, oceans, and soil; and the climate patterns, wind currents, and nutrient cycles that affect living things.

Limits to Where Species Can Live

Several factors determine where a species can or can't exist: physical factors such as temperature, water, sunlight, and nutrients; interactions with other species that may block a population's survival or distribution; and geographical barriers that prevent access such as seas, deserts, and mountains.

Limits to Population Size

Part of restoring decimated populations to healthy levels is understanding **population density** (p. 595), or the number of individuals that live in a certain amount of space. Death and birth affect population size, and are reflected in a **survivorship curve** (p. 596), a plot of data representing the proportion of a population that survives to a certain age. Humans follow a **late-loss** survivorship curve (p. 596). Flounders follow an **early-loss** survivorship curve (p. 596). **Life expectancy** (p. 596) is the average time left to live. **Fertility curves** (p. 597) graph the reproductive rate versus the age of female population members. **Biotic potential** (p. 597) is a population's capacity for reproduction under ideal conditions. **Exponential growth** (p. 598), or explosive population increase in which birth rate exceeds death rate, results in a **J-shaped curve** (p. 598) when graphed. Limited

resources impact population growth, changing J-shape growth to **logistic growth** (p. 599), indicated by an **S-shaped curve** (p. 599). When individuals are using all the resources available to them, they have reached the environment's **carrying capacity** (p. 599).

The physical environment limits population growth in several ways. Some factors depend on population density. Dense populations of prairie dogs, for example, have more flea-transmitted bubonic plague. Some factors are independent of population density, like droughts or floods. Some growth-limiting factors originate outside the population—they are extrinsic; these include food supplies and weather. Some factors originate inside the population—they are intrinsic; for example, when crowded kangaroos resorb their embryos. The most important intrinsic population regulatory mechanism is competition.

Limiting factors lead to the evolution of **life history strategies** (p. 600), the way an organism allocates its energy. Species such as dandelions display rapid rates of reproduction, often with an early-loss type of survivorship where most individuals tend to die early. Species such as rhinoceroses have a slow rate of reproduction and maintain a density close to the environment's carrying capacity. These species usually experience late-loss survivorship.

The Human Population

Our own species is governed by natural ecological principles. For tens of thousands of years during our hunting and gathering phase, human populations grew slowly to about 10 million. Beginning about 10,000 years ago, we entered an agricultural phase, and population began to grow faster, to about 800 million by 1750. We are now in an industrial phase, and human population growth is now staggering—6 billion and rising rapidly. Birth rates and death rates can increase and decrease independently. A changing pattern from high birth and death rates to low birth and death rates is called the **demographic transition** (p. 603). This change took place in the first half of the 20th century in North America and Europe, but later in Asia, Africa, and other areas. In some developing nations, it hasn't yet taken place. On all continents, the **agricultural revolution** (p. 601) and the **industrial revolution** (p. 601) accelerated human population growth.

A population's **age structure** (p. 604), the number of individuals in each age group, reflects its growth rate. Sweden currently has a bullet-shaped age distribution, with roughly equal numbers of people at most age levels up to old age, where there are smaller numbers. Uganda now has a pyramid-shaped age distribution, with many young people, fewer middle aged individuals, and even fewer old people.

Where Organisms Live and Thrive

An organism resides in its **habitat** (p. 605), its physical place in the environment. Its function role in the community is its **niche** (p. 605). The potential range of all conditions under which an organism can thrive is its **fundamental niche** (p. 605). The part of the fundamental niche that a species actually occupies in nature is its **realized niche** (p. 606).

How Species Interact

Two species often compete for limited resources. In this **interspecific competition** (p. 606), one or both competitors negatively impact the other's survival. Where one species excludes another through competition, it's called **competitive exclusion** (p. 607). Two species can coexist or merely slow each other's growth. What causes competitive exclusion? Species with similar requirements must use the same resources in different areas; this is resource partitioning. The splitting can be unequal and lead to one species's severe limitation. Hereditary change can take place in two or more species due to their competition; this is character displacement.

A second major type of community interaction is **predation** (p. 608). **Predators** (p. 608) kill and eat other animals, their **prey** (p. 608). Predators and prey influence each other's population sizes, and the interactions lead to adaptations for better ways of catching and escaping prey. Typical examples of better ways of escaping are **camouflage** (p. 610), **chemical warfare** (p. 610), and **mimicry** (p. 610). **Parasites** (p. 610) are predators that live on or in a host. In predation, only one species benefits while the other is harmed. In **commensalism** (p. 611), for example, the growth of Spanish moss on trees, one species benefits while the other (the tree) is neither harmed nor helped. In **mutualism** (p. 611) both species are helped, for example, both the yucca moth and yucca plant.

Organization of Communities

Communities change over time. After a volcanic eruption, for example, there is a **succession** (p. 612) or regular progression of communities, from **pioneer** to **transition**, **climax**, and **mosaic climax communities** (p. 612). The total number of species found in a community is its **species richness** (p. 613), or **species diversity** (p. 613). Latitude and isolation influence species richness. The warm tropics are richer in species than northern areas. An isolated peninsula like Baja California has diminished species richness. Many other factors influence species richness, including resources, competition, and predation. Species richness and diversity help stabilize communities, including agricultural communities. Many farmers now use **integrated pest management** (p. 614), utilizing ecological principles. Humankind's destruction of natural habitats and reduction of species diversity are among the greatest global problems we face.

Test Yourself on Vocabulary and Concepts

For each question set, complete or match the statements with the most appropriate term from the list. Any answer may be used once, more than once, or not at all.

SET I

(a) population (b) community (c) ecosystem
(d) biosphere (e) more than one of the above

1. Includes all the species in a habitat but not the abiotic elements
2. A shallow pond and its inhabitants
3. The most inclusive term in the list
4. The only term that is limited to one species
5. Includes biotic as well as abiotic elements

SET II

(a) survivorship curve (b) life table (c) age structure of a population (d) biotic potential (e) density

6. The probability of death in any year by individuals at particular ages constitutes a _____.
7. In a population growing logistically, the _____ is limited by the carrying capacity.
8. The number of individuals of one species in a given habitat constitutes its _____.
9. Comparisons of graphs of the _____ of different human populations can be used to predict which populations are likely to grow the most rapidly within the foreseeable future.
10. A graph showing the percentage of a given population surviving to age 5, 10, 15, 20, 25, and so on is called a _____.

SET III

(a) community (b) habitat (c) fundamental niche
(d) realized niche

11. The potential range of a species within a habitat
12. The actual range of a species within a habitat
13. The geographic location of a particular species in a community
14. A collection of populations in a particular place at a particular time

SET IV

(a) mutualism (b) commensalism (c) competition
(d) predation

15. A term that can be used to describe the behavior of pathogenic parasites as well as fast-running hunting carnivores
16. A type of interspecific interaction exemplified by tree-dwelling animals that are of no benefit to the tree
17. The usual relationship between plants and pollinating animals
18. Characterized by interspecific interactions that may lead to resource partitioning and coevolution on the one hand, or death on the other
19. Often exemplified by character displacement, limiting the realized niche

SET V

(a) succession (b) pioneer community (c) climax community (d) species richness (e) keystone species

20. An initial group of species that affect conditions for future arrivals
21. A natural series of communities that occupy a particular geographic location in turn
22. Generally the most stable and long-lasting assemblage of species in a particular location
23. Degree of diversity of populations in a habitat
24. A species, often a predator, whose activities determine the fundamental structure and richness of a community

Integrate and Apply What You've Learned

25. Explain the cause of the fluctuations in the populations of lynx and hare shown in the graph in Figure 24.19.
26. Explain the difference between a yellowtail flounder's habitat and its niche.
27. How would you classify the life history strategy of the human population? Explain your answer.
28. Which level of ecology (population, community, ecosystem, or biosphere) would address the problem of pesticide runoff into streams? Which level(s) are involved in an infectious disease like rabies, which can be carried from one raccoon to another or from a raccoon to a person?
29. Whitetail deer are found throughout the United States, except in California, Nevada, Utah, and the Four Corners area of the Southwest. Blacktail deer live throughout the western half of the United States, but not in the East or most of the Midwest. What are some hypotheses that could account for this difference in distributions?
30. Assume that in 1915, lily pads invaded a huge reservoir and their population size doubled every year thereafter. Ecologists studying the reservoir find that after 80 years of growth, just half of the reservoir is covered with lily pads. How many years will it be before lily pads cover the entire reservoir?
31. Virginia and Phil are having an argument. She thinks that having a certain number of children but having them later in life—say, when the mother is between 30 and 35 instead of between 20 and 25—will reduce a country's annual population growth rate. Phil claims that the mother's age is immaterial with respect to annual population growth rate as long as the women have the same number of children. Who is correct and why?
32. The islands of Hawaii and Iceland both encompass volcanic regions where pioneer communities pave the way for successive communities. Which region would exhibit the fastest rate of succession and the greatest species richness?

Analyze and Evaluate the Concepts

33. In order for the human population to achieve zero population growth, which of the following must occur? More than one answer may be correct.
   (a) There must be more postreproductive individuals than reproductive individuals.
   (b) There must be more prereproductive than reproductive individuals.
   (c) There must be the same number or fewer prereproductive individuals as there are reproductive individuals.
   (d) One set of parents should have only two children (approximately).
34. A model plant population that has not reached the carrying capacity is likely to:
   (a) grow exponentially
   (b) grow but not at an exponential rate
   (c) remain stable in number
   (d) decline in number
   (e) crash
35. Brown-headed cowbirds are known for their tendency to lay eggs in other birds' nests. An adult Kirtland's warbler will seldom push a cowbird egg from the nest. Apply the concept of coevolution of

parasite and host to this fact, and suggest possible future scenarios involving the interaction of Kirtland's warblers and brown-headed cowbirds.

36. The redstart is a threatened songbird with a habitat far from the forest's edge in eastern climax forests. Apply concepts of succession and the ecology of island communities to design ways to help preserve such species.

37. Which of the following results would you predict after large predatory carnivores such as wolves and mountain lions have been eliminated from a community containing both carnivores and deer? More than one answer may be correct.
   (a) The deer population would initially boom.
   (b) The number of deer would eventually crash as a result of overgrazing and starvation.
   (c) The number of deer would remain unchanged because the same number that had been killed by the predators would now die of disease.
   (d) The number of deer would initially decline because the keystone species had been removed.

38. With time, the size of the brain of predators, relative to size of the body, has grown larger. The same has also been true of prey species. However, 60 million years ago, predators' brains were significantly larger than their prey, whereas modern predators have brains only slightly larger than their prey. Why do you suppose this occurred?

39. When farmers first begin to use pesticides, crop yields often increase dramatically for a few years but then decline to levels below those obtained before pesticides were used. How could you explain this phenomenon?

A Checkered Future

▲
Warming climate, shifting ranges. Camille Parmesan on a mountain trail; close-up of Edith's checkerspot butterfly; evidence of global warming.

You've probably heard of global warming and maybe even experienced some of the severe weather spawned by it. But Texan Camille Parmesan has devoted most of her professional life to the problem and has arguably done as much to save our planet from overheating as any other individual. Her approach? Counting butterflies.

This summer, like most summers, Dr. Parmesan will toss gear into the back of a dusty pick-up truck and head for the high country of California. For eight weeks, she'll camp and hike above 10,000 feet in Sequoia and Yosemite National Parks and in the forests around Mammoth Mountain. Her object is to track down populations of a small, delicate organism called Edith's checkerspot butterfly, then make a census count of the striking black and orange insects as they flutter about in flowering alpine meadows. Her broader goal is to continue documenting the shifting home range of this "indicator species"—a tell-tale organism whose changes reveal ecological information. As summers become hotter and hotter, groups of Edith's checkerspots move farther north and higher in the mountains than the year before and disappear from some previous locations. The insect's life cycle is quite sensitive to temperature, and as we'll see, increases in heat or cold can lead to its extinction in particular locales.

Most meteorologists and climatologists agree that the average global temperature is 0.5 °C (or about 1 °F) warmer now than in 1900. During that century or so, levels of carbon dioxide (CO_2) and methane (CH_4) in the atmosphere have shot up by 25 percent, along with smaller increases in ozone (O_3) and nitrous oxide (NO). Many scientists are convinced that human activities are causing most of the build-up in atmospheric gases: forest clearing and burning, cattle grazing, operating combustion engines, and releasing industrial pollutants. They also think that those gases, in turn, produce a **greenhouse effect:** like the glass in a greenhouse, they allow light energy from the sun to pass through toward Earth, but trap infrared heat energy that bounces back from Earth's surface toward space. Many scientists predict that by 2100, this greenhouse effect will have raised global temperatures by 2 °C (3.5 °F) or more, and that the consequences will be dire.

Camille Parmesan, a trained ecologist and conservation biologist at the University of Texas, Austin, acknowledges the validity of many such predictions and can give a succinct overview of them:

• "The Midwest, our cornbelt, is expected to dry as it gets warmer," she says, and farmers may have to shift away from America's biggest crop, corn, to some other staple that tolerates heat better—or else begin history's biggest irrigation project.

- Warm-adapted tropical diseases like cholera and malaria could spread northward.

- Land set aside for nature reserves could become unsuitable for its inhabitants. "We've counted on climatic stability," says Parmesan. But as Earth heats and temperature zones shift northward, these small "ecological islands" surrounded by human development could become too hot or dry to support the animals and plants they're supposed to protect.

- As the polar ice caps melt and vast regions of permafrost thaw, newly decomposing organic matter could release huge additional quantities of CO_2 and further accelerate global warming.

- With that melting, sea level could rise by as much as 1 meter, inundating ports, industrial zones, and coastal housing. "In low-lying places like the small island states of Indonesia," Parmesan explains, a one-meter rise could cause "whole countries to disappear underwater!"

Even facing such sobering threats, many people remain unconvinced and demand more proof. Before they give up their gas-guzzling cars or shut down industries, they want hard evidence that global warming is mostly our own fault and not some natural cyclic process. They also want evidence that the warming is truly affecting ecosystems and organisms. And this is where Dr. Parmesan's studies come in. She knew that butterfly collectors traditionally keep some of the best records available on animal populations in the wild. The best-tracked of America's species has been Edith's checkerspot butterfly. Parmesan gathered data on the insect that hobbyists and professional collectors have been recording since 1900. Then she combined it with her own recent observations in the field to create a very complete range map. Where did the butterfly live in 1900? Where does it live now? Have the insect populations shifted to the north and higher up into the mountains as average global temperatures have climbed?

Parmesan generated 45,000 miles worth of greenhouse gases with her own aging four-wheel-drive vehicle as she travelled to most of the check-erspot's habitats between Baja California and Canada, and between the Pacific coast and Eastern Nevada. But the ecological information she gathered was crucial. During a single century, she writes, "the entire range of the species has shifted northward by 92 km and upward into the mountains by 124 meters." This corresponds closely to the shifting of climatic zones due to global warming: 105 km northward and 105 meters upward into the mountains.

To replicate this experiment with more species, Parmesan also collected historical records and field observations on 35 separate butterfly species in Europe. "I found that two thirds of the butterflies had shifted their ranges northward," she reports. "And it just absolutely blew me away. We're talking 50 to 200 kilometers of shift since 1900!" Parmesan enlisted most of Europe's major butterfly ecologists as co-authors on her paper, and their findings—plus her Edith's checkerspot study—are the first solid evidence that global warming is affecting nature in dramatic ways. Will it help solve the crisis before we lose the Midwestern cornbelt, the species in many nature reserves, and low-lying countries like the island states of Indonesia? Only, says Parmesan, if people shift their attitudes and consumption patterns. So far, she adds, "Americans still have this view that most of Earth is wild and we can't really do anything to hurt it. I hear this argument all the time—'we can't affect the *atmosphere*!' "

In this chapter, you'll learn how in ecosystems, communities of organisms interact with their immediate physical environments. You'll explore how the water, soil, and air near the planet's surface (a zone called the biosphere) supports life. And you'll see how human activities (like generating carbon dioxide and methane) threaten to destabilize the delicate balances and interdependencies that have evolved over millions of years. You will also learn about potential solutions to global warming, to the ozone hole, to pollution, to habitat loss, and to other major environmental issues, so that you can act as an informed voter and consumer.

As you explore this subject, you'll discover answers to these questions:

1. Through which pathways does energy flow in ecosystems?

2. How do carbon, nitrogen, water, and other materials cycle from the physical environment into organisms, and back again?

3. How do global climates affect Earth's life zones?

4. What are biomes, Earth's major communities of life?

5. What life zones occur in watery environments?

6. How are humans changing the biosphere?

25.1 Pathways of Energy and Materials

The northward shift of the Edith's checkerspot butterfly helps us clearly define an ecosystem. An **ecosystem** includes the community of organisms in an environment as well as the air, sunlight, water, soils, and other nonliving physical factors that surround them. In a high Sierra alpine meadow, for instance, the organisms might include the checkerspot butterfly, the penstemon and Indian paintbrush flowers it feeds on, and the ravens that eat the butterfly. The physical surroundings include the specific air and ground temperatures in the alpine meadow, the moisture available from melting snow and afternoon thunderstorms, and the acces-sible chemical elements in the stony mountain soil. Ecologists focus on two major interchanges between the living community of organisms and its nonliving environment: the flow of energy, and the cycling of materials.

Energy Flow and Material Cycling

Energy flows through ecosystems in a one-way path: It enters living things from the physical world, passes from one organism to another, and eventually escapes back to the physical environment (Fig. 25.1). As in most ecosystems, Edith's checkerspot butterflies get their energy from sunlight—but not directly. The leaves of wildflowers capture light energy in biological molecules. In the early part of its life cycle as a caterpillar, the checkerspot harvests some of this energy when it eats the leaves. Later, the adult butterfly takes in more energy when it sips the flower's sugary nectar. In every ecosystem, some energy eventually escapes from living organisms into the surrounding physical world, usually in the form of heat. The butterfly generates heat when it flutters over the alpine meadow, just as your body gives off extra heat when you walk or run (see Chapter 3). Heat energy passed into the environment this way is so diffuse that it can't be used by living organisms and thus is lost forever to the living world.

In contrast to the one-way flow of energy, materials *cycle* through ecosystems. For example, carbon and oxygen atoms move from the nonliving environment into organisms and then back again to the nonliving environment (Fig. 25.1). At high elevations in the Sierra Nevada mountains, for example, wildflower leaves take in carbon dioxide from the atmosphere and capture it in carbohydrates and other molecules. When caterpillars eat the leaves, some of the carbon becomes

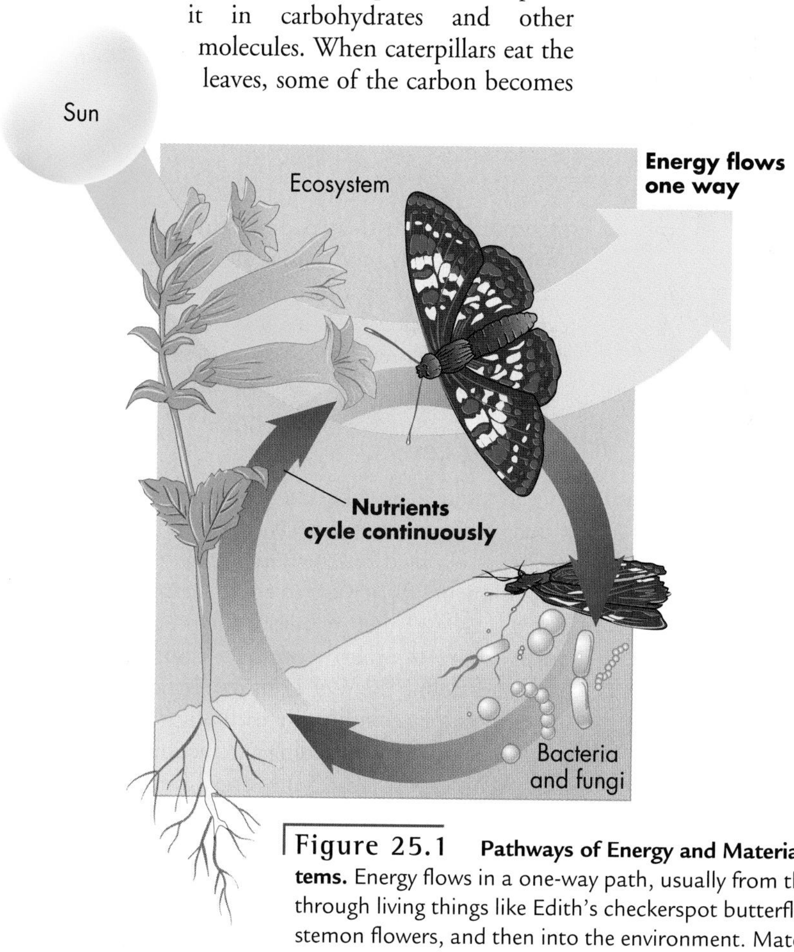

Sun

Ecosystem

Energy flows one way

Nutrients cycle continuously

Bacteria and fungi

Figure 25.1 **Pathways of Energy and Materials in Ecosystems.** Energy flows in a one-way path, usually from the sun through living things like Edith's checkerspot butterfly and penstemon flowers, and then into the environment. Materials cycle from the physical world to the living world, and back again.

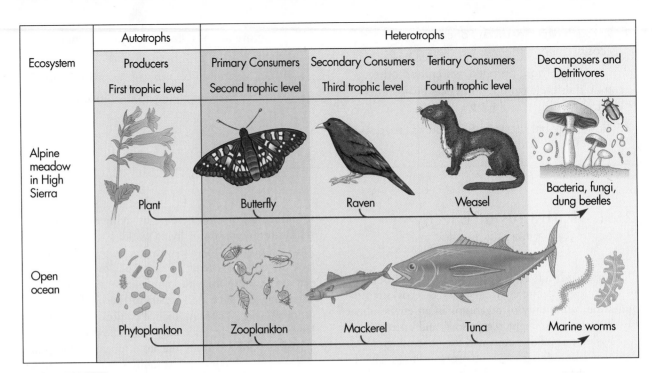

Ecosystem	Autotrophs	Heterotrophs			
	Producers	Primary Consumers	Secondary Consumers	Tertiary Consumers	Decomposers and Detritivores
	First trophic level	Second trophic level	Third trophic level	Fourth trophic level	
Alpine meadow in High Sierra	Plant	Butterfly	Raven	Weasel	Bacteria, fungi, dung beetles
Open ocean	Phytoplankton	Zooplankton	Mackerel	Tuna	Marine worms

Figure 25.2 **Feeding Levels and Food Chains.** Organisms vary among ecosystems, but within each ecosystem, producers form the first trophic level, consumers occupy the next few trophic levels, and decomposers and detritivores utilize wastes from all levels.

SOLVE IT

Explorer **25.1**

Trophic Levels

Figure 25.3 **Producer Cells Can Live In Consumer Cells.** Corals, like these golden species from Tasmania, are consumers. Within their coral cells, however, live small algal cells (in this case, with predominantly yellow pigments). Each alga is a primary producer that traps light energy by photosynthesis. The coral then obtains nourishment directly from the algal cells.

incorporated into their bodies. After the caterpillar becomes a butterfly, and this adult eventually dies, bacteria and fungi break down its tissues, and most of the carbon is recycled back into the atmosphere as carbon dioxide (**E** xplorer 25.1).

Feeding Levels

When organisms take energy directly from sunlight, water, soil, or other aspects of the nonliving environment, it is called **autotrophy** (*auto* = self + *trophy* = nourishment). When organisms take in energy first captured by some other organism, the process is **heterotrophy** (*hetero* = other + *trophy* = nourishment). Both strategies define the way a species obtains nutrients and, in turn, its place within the community. An organism's position in a web of feeding interactions boils down to this: Who eats whom? Some organisms produce all the biological molecules they need for their growth from nonliving substances taken directly from the environment; these autotrophic organisms (or autotrophs) are called the **primary producers,** and include all plants (such as the penstemon), certain protists, and some bacteria. In contrast, heterotrophic organisms (heterotrophs) obtain their energy and materials, including carbon and other atoms, by consuming the primary producers or each other. Heterotrophs, therefore, are an ecosystem's **consumers,** and include animals like checkerspot butterflies, fungi, and many kinds of microbes. Ecologists assign every organism in a community to a **trophic level,** or feeding level, depending on whether it is a producer or a consumer (Fig. 25.2 and **E** 25.1).

Producers

At the lowest trophic levels lie the primary producers—the bacteria and plants that support all other organisms directly or indirectly. In most ecosystems on land, green plants are the producers. An example is the wildflower Indian paintbrush, in the alpine meadows of the Sierra Nevada. In a lake or stream, primary producers would include algae; in hot springs, they would be mainly heat-tolerant bacteria. By collecting solar energy and carbon dioxide, primary producers build energy-rich biological molecules (review Fig. 3.22). Producers also absorb nitrogen, phosphorus, sulfur, and other necessary atoms and fix them into biological molecules. Ecologists say that producers provide both the energy-fixation base and the nutrient-concentration base for the entire ecosystem.

In most ecosystems, plants are the major producers. In the deep seas, however, as well as deep in the Earth, and in hot, acidic, or otherwise inhospitable environments, such as the hot springs we just saw, prokaryotes are usually the primary producers (review Chapter 11).

Primary Consumers

At the second trophic level are the **primary consumers,** the organisms that eat the producers. Herbivores (plant eaters), such as Edith's checkerspot butterflies and deer, efficiently digest plant matter for energy and serve as ecological links between producers and other levels. Some primary consumers are inadvertently quite ingenious: coral polyps, for example, contain algae within their cells. The algae harvest sunlight, and deliver biological molecules directly to the coral cells (Fig. 25.3).

Secondary Consumers

At the next highest level are **secondary consumers,** carnivores (meat eaters) that consume the herbivores. A raven or deermouse eating a checkerspot caterpillar is a carnivore.

Tertiary Consumers

In the next trophic level are the **tertiary consumers,** carnivores that eat other carnivores; for example, a weasel may eat a deermouse and a tuna may eat a mackerel. Finally, a few ecosystems have one more trophic level containing carnivores that eat tertiary consumers, such as a cougar eating a weasel or a shark eating a tuna.

Detritivores and Decomposers

A special class of consumers, the **detritivores,** obtain energy and materials from **detritus,** organic wastes and dead organisms that accumulate from higher levels. Earthworms, nematodes, dung beetles, and other kinds of insects, and carrion feeders like the condor in the High Sierra are specialized detritus consumers. **Decomposers** break organic molecules into inorganic subunits, which can in turn be recycled by primary producers. Bacteria, fungi, and slime molds are all decomposers critical to nutrient cycling.

The simplified diagram in Figure 25.2 suggests that each trophic level leads directly to the next in a simple chain, and indeed, **food chains** do exist in nature, with groups of organisms involved in linear transfers of energy from producer to primary, secondary, and tertiary consumers. More commonly, however, feeding relationships resemble not chains, but complex interwoven webs.

Feeding Patterns In Nature

Organisms usually consume more than one other species. Edith's checkerspot, for example, feeds on both penstemon and Indian paintbrush as well as on other wildflowers. Some animals feed at several trophic levels. A high-level carnivore like a cougar may eat herbivores, such as deermice, as well as other carnivores, such as weasels. As an omnivore ("all eater"; see Chapter 17), you yourself might eat vegetables (primary producers), ice cream (from cows, which are primary consumers), tuna fish (secondary consumers), and mushrooms (decomposers). And, surprisingly, different sexes of some species feed at different trophic levels. A horsefly male feeds on nectar, and is therefore an herbivore, while the female horsefly sucks blood and is a carnivore. Ecologists call complicated interconnected feeding relationships **food webs** (Fig. 25.4). In a High Sierra

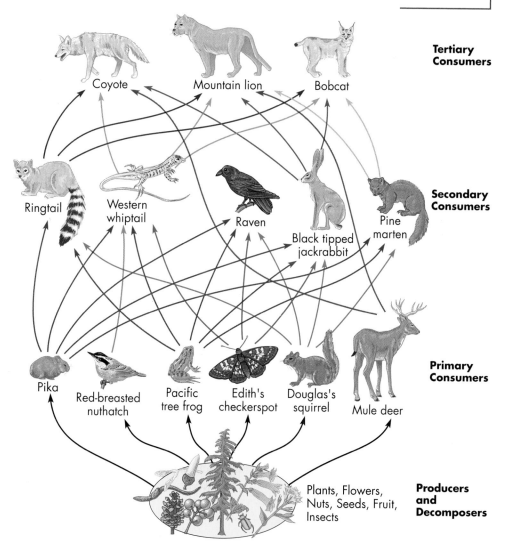

Figure 25.4 **A Food Web: Interdependencies in the Living World.** In a High Sierra ecosystem, dozens of species derive energy from each other in a complex grazing food web. Others (not shown here) participate in the detritus food web. Both webs are linked at numerous points by the activities of specific organisms. Only a few interactions are symbolized here. In an actual food web, there could be hundreds or thousands.

Tertiary Consumers

Coyote Mountain lion Bobcat

Secondary Consumers

Ringtail Western whiptail Raven Black tipped jackrabbit Pine marten

Primary Consumers

Pika Red-breasted nuthatch Pacific tree frog Edith's checkerspot Douglas's squirrel Mule deer

Plants, Flowers, Nuts, Seeds, Fruit, Insects

Producers and Decomposers

ecosystem, for example, the major producers are pine, spruce, and fir trees. In an Eastern hardwood forest, they could be oaks, maple, and hickory. These producers support two main food webs: a grazing food web that stems directly from the living plants and a decomposer/detritus food web that begins with dead plant parts and animal wastes.

Whether an organism is a producer or a consumer, it needs energy for movement, active transport of nutrients and ions, and synthesis of proteins and nucleic acids, and it needs other large molecules for growth and repair. Keep in mind that regardless of the trophic relationship, a key factor is the flow of energy from one organism to the next.

How does the greenhouse effect change an area's energy budget?

25.2 Energy Flow Through Ecosystems

Because producers obtain their energy directly from the environment, their activities set a limit for the amount of energy that can be captured and channeled throughout the entire ecosystem. Ecologists have closely studied the hardwood forest ecosystem to measure available energy and how organisms spread it.

Energy Budget for an Ecosystem

Within the White Mountains of New Hampshire lies the Hubbard Brook Experimental Forest, a fragrant zone of tree-studded rolling hills that has engaged biologists' attention for more than 30 years. By studying the input of energy, water, mineral nutrients, and organic matter into zones within the forest and tracing their incorporation into both living organisms and the physical environment, researchers have learned in detail how energy flows through an entire ecosystem.

The summer sun brings a huge amount of solar energy into the Hubbard Brook forest. However, most of this energy reflects back into the atmosphere as light or heat, or causes water to evaporate from the soil and plants. In total, only about 2 percent of the sunlight is converted by producers to chemical energy in the form of organic compounds (via photosynthesis; Fig. 25.5). Ecologists call this small fraction an ecosystem's **gross primary productivity.** The gross primary productivity limits an ecosystem's structure, including how fast birch trees will grow, for example, and how many butterflies will thrive there.

Plants use about half of the gross primary productivity to fuel their own cellular respiration, eventually losing

most as heat. The amount of energy remaining after respiration is called the **net primary productivity,** the amount of chemical energy that is actually stored in new leaves, roots, stems, flowers, and fruits. Of all the energy impinging on the ecosystem, only the net primary productivity—about 1 percent of the light energy striking the forest—is potentially available to consumers. But most of the new leaves and twigs end up as dead material on the forest floor, and this fuels the detritus food web. Amazingly, in a forest, the litter and decomposing humus stores nearly twice as much energy as found in the majestic banks of leaves and branches overhead.

Only a small fraction of the energy stored aboveground in the forest enters the grazing food web, and most of that is lost as animals radiate heat in respiration, leaving only about 0.01 percent of the total sunlight becoming assimilated in animal bodies. Only a tiny amount of energy then remains in the soil or exits the ecosystem in streams.

The information in Figure 25.5 allows ecologists to formulate three general principles about the energy budget of the Hubbard Brook forest: (1) Even in a lush, leafy green forest, plants and other producers convert only a small fraction (2 percent or less) of the solar energy that enters the ecosystem into stored chemical energy. (2) Animals ingest an even smaller amount (in this case, 0.01 percent) of the energy in the grazing food web. (3) As energy flows through the trophic levels of the ecosystem, metabolic activities (mostly aerobic respiration) release it back into the air, where it ultimately returns to space as heat, a form of energy that does little or no work.

Pyramids of Energy and Biomass

Ecologists have discovered another important fact in their experiments on energy flow in ecosystems: Food chains on land rarely have more than four or five links (excluding parasites, decomposers, and detritivores), and ocean food chains based on plankton are generally limited to about seven links. When there are hundreds or even thousands of species in an alpine meadow, a patch of forest, or a coral reef, why do food chains tend to be so short? Consider the California ground squirrel. This rodent eats seeds, leaves, or berries, assimilates some of the energy, then excretes some in liquid and solid wastes. The ground squirrel uses most of the assimilated energy in aerobic respiration, and to maintain and repair its body, storing less than 2 percent of ingested food energy in new tissues or offspring. The consequences of a huge loss through respiration and a small net increase in growth is that ground squirrels store very

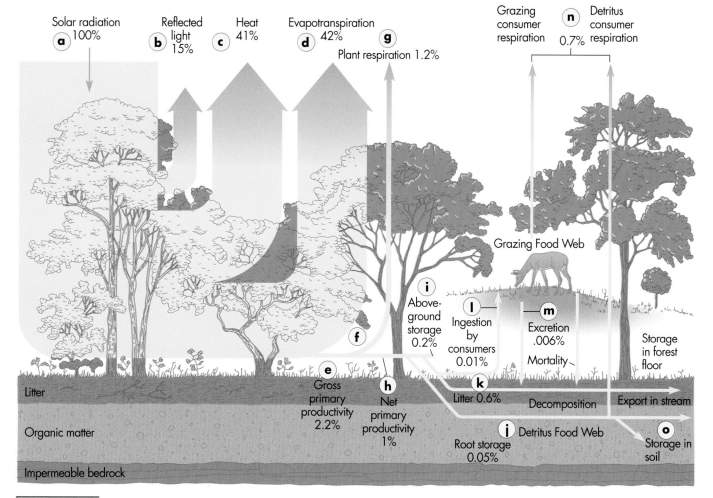

Figure 25.5   Energy Budget of a Hardwood Forest. Energy flows through an eastern hardwood ecosystem such as the Hubbard Brook Experimental Forest. If we take all the sunlight reaching the forest as 100 percent (a), then 15 percent of the energy is reflected as light (b); 41 percent is dissipated into the environment as heat (c); and 42 percent of the energy is returned during moisture evaporation and movements within plants (d). Only a small fraction of the original energy becomes fixed in organisms and their waste products (e–n).

little energy in a form that can be used at the next trophic level, say, by a coyote. That small amount of stored energy is what keeps food chains short.

Energy Pyramids

Ecologists draw **energy pyramids** with building blocks proportional in size to the amount of energy available at different trophic levels in an ecosystem. Figure 25.6 shows the energy pyramid for a river ecosystem in Silver Springs, Florida. This river system includes eelgrass and algae (primary producers); turtles, snails, and caddisflies (herbivores); and beetles, sunfish, and bass (carnivores)

(some are shown in Fig. 25.6). At each trophic level, the energy stored by the organisms is substantially less than that of the level below it. Ecologists have constructed similar energy pyramids for the wildflowers, butterflies, and ravens in alpine meadows, as well as for other ecosystems.

Biomass Pyramids

Stepwise energy decline explains why food chains usually have only four links: The small amount of energy available at the top is too difficult to collect. Ecologists have invented a handy way to measure the diminishing

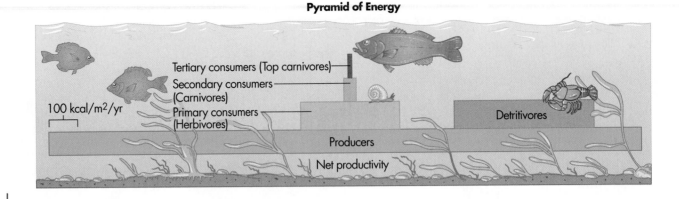

Pyramid of Energy

Tertiary consumers (Top carnivores)
Secondary consumers (Carnivores)
Primary consumers (Herbivores)

100 kcal/m²/yr

Detritivores

Producers

Net productivity

Figure 25.6   Energy Pyramid in a Florida River Ecosystem. The energy pyramid for this ecosystem shows the net primary productivity of various feeding levels: the amount of energy fixed during photosynthesis in plants minus the amount the plants use during cellular respiration. Surprisingly, the net productivity of detritivores is larger than that of herbivores.

returns through **biomass:** the dry weight of organic matter at a particular level. Biologists collected organisms in various trophic levels in the Silver Springs ecosystem, and then dried them to remove their water content and weighed the remaining material. They then displayed the data as a **pyramid of biomass,** which showed that each trophic level contained, as a rough approximation, only about 10 percent of the biomass in the level just below it (Fig. 25.7). By moving up four levels, and assuming the 10 percent carryover per link, one finds very little biomass in the top carnivores, such as large, meat-eating bass. Taking it one step further, an angler would need to catch and eat 10 kg (22 lb) of bass to put on 1 kg (2.2 lb) of human tissue. The 10 kg of bass would have come from 100 kg of smaller carnivorous fish, 1000 kg of insect herbivores, and 10,000 kg (about 10 tons) of plants. All that for just 1 kg of human being!

Ever hear the phrase, "Eat low on the food chain"? This is based on energy and biomass pyramids, and refers to the fact that it takes 10 kg of grain to build 1 kg of human tissue if the person eats the grain directly, but it takes 100 kg of grain to build 1 kg of human tissue if a cow eats the grain first, and the person eats the beef. Eating lower on the food chain—eating producers, not consumers—saves precious resources on a small planet. (Chapter 17 discusses vegetarian diets in more detail.)

Biological Magnification

Energy pyramids in ecosystems have an important ramification: **biological magnification,** the tendency for toxic substances to build up in progressively higher levels of a food chain. Many chemical insecticides, such as DDT and chlordane, resist breakdown in the environ-

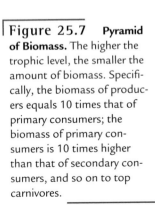

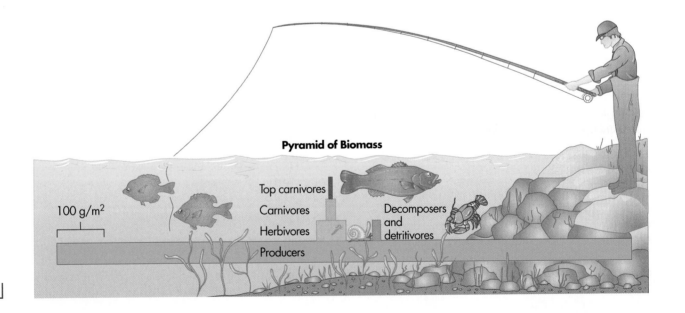

Pyramid of Biomass

Top carnivores
Carnivores
Herbivores

100 g/m²

Decomposers and detritivores

Producers

Figure 25.7   Pyramid of Biomass. The higher the trophic level, the smaller the amount of biomass. Specifically, the biomass of producers equals 10 times that of primary consumers; the biomass of primary consumers is 10 times higher than that of secondary consumers, and so on to top carnivores.

ment and, when eaten, tend to be stored in body fats. If farmers spray DDT on their cabbage plants to control cabbage looper caterpillars, some of the chemical runs off into streams and lakes. There, instead of breaking down, some of it may enter water plants later eaten by herbivorous fish (Fig. 25.8). Fish and other animals cannot break down or excrete the toxin, and it is instead stored in their body fats.

The magnification continues because consumers store all the DDT present in all the prey they have eaten. The concentration of DDT in the body of a top carnivore, say an osprey, can reach levels 10,000 times greater than in the water plants that originally took it up. This concentration of DDT can interfere with calcium metabolism and result in thin-shelled, fragile eggs. During the period DDT use was allowed in the United States, from the early 1940s until it was banned in 1968, the numbers of large birds of prey dropped significantly. This group included ospreys, falcons, hawks, eagles, and even the magnificent condor, a huge dark brown vulture with a bright orange head, scarlet neck, and a 10- to 12-foot wing span that soared above the mountains and dry foothills of California. DDT is still used in nearly two dozen developing nations in Asia, Africa, and Central and South America to control mosquitoes that spread malaria, the disease that kills more humans each year than any other disease (see Chapter 11). As a result, DDT continues to magnify in local food chains.

Organisms don't metabolize or recycle many toxic chemicals. They do, however, break down and recycle many other compounds containing nitrogen, sulfur, phosphorus, and other elements, and this recycling process is necessary to the long-term health of ecosystems.

25.3 How Materials Cycle Through Global Ecosystems

As it flutters about the open meadow, an Edith's checkerspot butterfly is a temporary repository of energy. It is also a storehouse of materials: The butterfly's body contains carbon, nitrogen, phosphorus, and other elements that originate in the environment and eventually recycle back to it. Unfortunately, human activities are causing a significant change in the global cycles of these materials. At the start of this chapter, we saw how human disruptions of the carbon cycle are leading to increasing levels of CO_2 and other greenhouse gases, and to global warming, with its direct effect on butterfly populations and home ranges. The cycles of water, nitrogen, and phosphorus are also important globally. To understand how potential changes in these cycles might affect the functioning of ecosystems, let's see how these materials cycle through the living and nonliving world.

Biologists consider materials stored temporarily in the atmosphere, in soil, or in living organisms to be in *pools,* or reservoirs. In general, the pool in organisms is much smaller than the pool in the physical environment. Like the level of water in a leaky bathtub with a leaky faucet, pools remain constant in size as long as a substance's rate of entry equals its rate of departure. Let's explore the cycling among the major "pools" of water, nitrogen, phosphorus, and carbon.

The Water Cycle

In the global **water cycle,** water moves from the atmosphere to Earth's surface as rain or snow, and back again

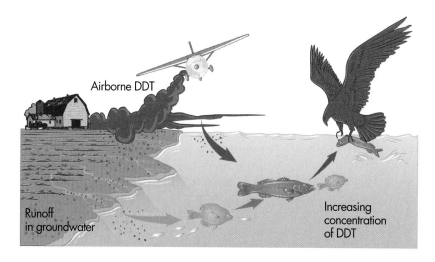

Airborne DDT

Runoff in groundwater

Increasing concentration of DDT

Figure 25.8 **Biological Magnification.** The DDT a farmer sprays may protect cabbages from caterpillars. It runs off into nearby streams, however, and is taken up and incorporated into water plants, then small fish, then larger fish, and then fish-eating birds. The pesticide accumulates in each higher level of the food chain, since living things cannot metabolize it completely. High accumulations in predatory birds such as the osprey result in fragile eggshells and in declining populations of these beautiful animals.

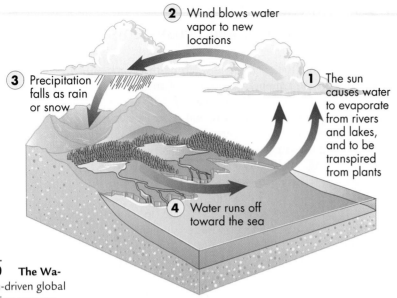

2 Wind blows water vapor to new locations

3 Precipitation falls as rain or snow

1 The sun causes water to evaporate from rivers and lakes, and to be transpired from plants

4 Water runs off toward the sea

Figure 25.9   The Water Cycle. A sun-driven global exchange involving evaporation, precipitation, runoff, and transpiration cycles water from atmosphere to Earth's surface to plants and back again.

to the atmosphere as the sun causes water to evaporate from puddles, ponds, rivers, oceans (Fig. 25.9) and from the leaves of plants in a process called **transpiration.** In some ecosystems on land, more than 90 percent of the moisture passes into plants and evaporates from their leaves, and only 10 percent evaporates directly from surfaces in the environment. In ecosystems like these, which include large tracts of tropical forests, the plants literally create their own rain: Moisture moves from plants to air to clouds and back to Earth in the form of rain wherever the clouds blow. When people cut down a forest's trees for timber or agriculture, water runs off to the sea rather than evaporating; as a consequence, clouds fail to form downwind, rainfall decreases, and climate patterns change.

The global warming we have been experiencing will have a strong effect on the water cycle. Water evaporates more rapidly from a warmer sea, and warm air holds more moisture. As winds carry moist clouds toward mountains like the Sierra Nevada range, snowfall accumulates in record levels. Changes like these in the water cycle, in turn, affect checkerspot butterfly populations. When the snow is extremely deep, adult butterflies don't emerge from their cocoons until late in the season—the end of July or early August—when all danger of freezing temperatures has past. In the warm late summer, caterpillars then rapidly develop into adult butterflies, which mate, lay eggs, and reach the overwintering (cocoon) stage before snow comes again in late September. Shallow snow levels, on the other hand, en-

courage earlier emergence (in June or early July) when unexpected cold snaps can wipe out most of the butterflies. Camille Parmesan has suggested that warm global temperatures and the resulting heavier snows in the Sierra Nevada may be contributing through this process to the recent colonization of higher elevations (up to 124 meters higher) by Edith's checkerspot butterflies. For this species, ironically, deeper snows equal later emergence, which equals better survival in the mountains.

The Nitrogen Cycle

Living organisms contain large quantities of nitrogen in their proteins and nucleic acids. In agricultural ecosystems—which we rely on for most of our food—nitrogen is often the factor that limits productivity. Nitrogen gas makes up about 79 percent of our atmosphere, but ironically, most organisms can't use the gaseous form of nitrogen. Instead, they depend on a few species of nitrogen-fixing bacteria to trap nitrogen in biologically useful forms such as ammonium (NH_4^+) and nitrates (NO_3^-). Other bacterial species return nitrogen to the atmosphere as nitrogen gas and complete the **nitrogen cycle** (Fig. 25.10).

Because available nitrogen often determines how well crop plants will grow, farmers have long fertilized their fields to increase the amount of ammonia and nitrate in the soil. The Pilgrims, for example, observed Native Americans burying fish they had caught along side corn seeds and copied the practice. Soil bacteria broke down the fish proteins and nucleic acids and produced ammonia; the corn assimilated these available nitrogen atoms and the plants grew taller and faster. Many farmers add useful nitrogen to their soils by practicing *crop rotation:* They plant *legumes*—crops such as beans, clover, or alfalfa—one year and corn, wheat, or sugar beets the next. This takes advantage of the nitrogen-fixing bacteria in the legumes' root nodules and the nitrogen released naturally into the soil.

Today, most large agribusiness farms around the world depend on nitrogen fertilizers produced through an industrial process rather than a bacterial process. In fact, nitrogen fixed in chemical factories may now represent 30 percent of the input to the global nitrogen cycle—a truly colossal shift in the natural cycle. Unfortunately, industrial nitrogen fixation requires tremendous heat and pressure, which is usually produced by burning huge quantities of fossil fuels. In some areas, people dump more energy into the soil in nitrogen fertilizers than they extract from the soil in food calories. Since reserves of fossil fuels are limited, this enormous

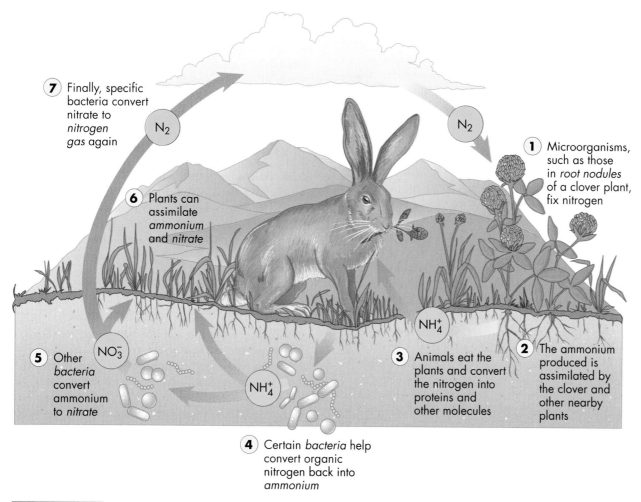

⑦ Finally, specific bacteria convert nitrate to *nitrogen gas* again

N₂

N₂

① Microorganisms, such as those in *root nodules* of a clover plant, fix nitrogen

⑥ Plants can assimilate *ammonium* and *nitrate*

NH₄⁺

② The ammonium produced is assimilated by the clover and other nearby plants

⑤ Other *bacteria* convert ammonium to *nitrate*

NO₃⁻

NH₄⁺

③ Animals eat the plants and convert the nitrogen into proteins and other molecules

④ Certain *bacteria* help convert organic nitrogen back into *ammonium*

Figure 25.10 **The Nitrogen Cycle.** (1) In the process called nitrogen fixation, nitrogen gas (N_2) from the air is changed into ammonia (NH_3), a colorless gas that dissolves readily in water. Nitrogen fixation can occur as a result of lightning or volcanoes, but much more often occurs through the action of bacteria living in soil or in root nodules of legumes such as clover and aspen trees. (2) Plants absorb dissolved ammonia (NH_4^+) from the soil and assimilate it into biological molecules. (3) When animals eat plants, they break down the plant's compounds and use the nitrogen to build new animal proteins and other cell parts. (4) After an animal excretes urea or uric acid, or after an organism dies, certain bacteria change nitrogen-containing biological molecules into ammonium again. (5) Plants can then assimilate this dissolved ammonia (NH_4^+) or still other bacteria can add oxygen atoms to the nitrogen, forming nitrate (NO_3^-). (6) Plants take in some of the nitrogen in the two forms. (7) Finally, a fourth set of bacteria acts on the remaining nitrate, changing it back to nitrogen gas, and thereby completing the cycle.

reliance on industrially fixed nitrogen fertilizers represents an energy drain that can't go on indefinitely.

The Phosphorus Cycle

Because nitrogen gas, water vapor, and carbon dioxide are airborne, they are blown anywhere by the wind. This mobility makes the nitrogen, water, and carbon cycles truly planetary. In contrast, certain other substances, such as calcium and phosphorus, lack a gaseous phase and thus cycle locally. Nevertheless, local cycles of these elements can cause great fluctuations in the populations of local organisms. Phosphorus is essential for life. It is a component of cell membranes, nucleic acids, and ATP, the energy currency of cells. The **phosphorus cycle** consists of two interlocking circuits, one that acts locally during short stretches of time, and another that operates on a more global scale over vastly longer time periods.

Eutrophication

As with other cycles, our human activities have altered the dynamics of the phosphorus cycle, especially in aquatic ecosystems like the Florida Everglades, where phosphorus is often the factor limiting primary productivity. Phosphates are major ingredients of agricultural fertilizers and until recently were also main components of detergents. For years, phosphorus-rich water from fertilized sugarcane and vegetable fields ran directly into Florida's Lake Okeechobee. The additional phosphates allowed algae and other aquatic plants to grow faster, and the ecosystem became **eutrophic** (*eu* = true or truly + *trophic* = fed), or overly supplied with nutrients that support primary production. This process, termed **eutrophication,** continued as algae and other plants "overgrew" and then died in the lake. Decomposing bacteria then fed on the dead algal cells and used up so much dissolved oxygen that fish suffocated in massive fish kills in the mid-1980s.

Ecologists dramatically demonstrated this process with an experiment in a lake in Ontario, Canada (Fig. 25.11). They wondered whether phosphorus could be the limiting factor in a lake's primary productivity. To test this hypothesis, they found a lake with a natural hourglass shape and divided it into two sections by

Figure 25.11 **The Phosphorus Cycle and a Lake Divided.** Researchers divided experimental Lake 226 in northwestern Ontario into two parts. Next, they fertilized the near basin with nitrogen and carbon, and the far basin with nitrogen, carbon, and phosphorus. After two months, the near basin was clear, with little life. But the far basin experienced an algal bloom, readily visible from the air. Their conclusion was that too much phosphorus from detergents or fertilizers can lead to a lake's eutrophication.

stretching a vinyl curtain across its narrowest part. This clever strategy separated the lake into an experimental side and a control side with equivalent conditions at the start of the experiment. The researchers fertilized both halves of the lake with nitrogen and carbon compounds, then added phosphorus to the experimental half. Without added phosphorus, the control section showed no change in organisms or ecological productivity. Within two months, however, the half with the added phosphorus developed an algal bloom that was visible from an airplane. When workers stopped fertilizing the lake with phosphorus, the algal bloom faded and the lake recovered its original condition. This spontaneous recovery suggests that people can improve the health of lakes and streams by restricting the runoff of phosphorous-containing substances such as detergents and fertilizers.

The Carbon Cycle

As with water and nitrogen, carbon atoms move globally in a vast **carbon cycle** from the physical environment through organisms and back to the nonliving world. The carbon cycle is closely linked to energy flow, because producers—including the photosynthetic plants of meadows, forests, and oceans—trap not only light energy but also carbon in sugar molecules. The trapped carbon comes from carbon dioxide in the surrounding air or water (Fig. 25.12). As the cycle proceeds, consumers ingest organic carbon compounds synthesized by producers. Then, via respiration, both consumers and producers return carbon to the nonliving environment in the form of carbon dioxide.

Carbon accumulated in wood eventually returns to the atmosphere as a result of fires or through consumption and respiration by decomposers, such as fungi and bacteria, and by detritivores. Organic carbon can leave the cycle for even longer periods of time when sediments bury organic litter, which decomposes only partially and gradually transforms into coal or oil. Carbon also leaves the cycle when the cast-off calcium carbonate shells of marine organisms sink to the ocean floor and become covered with sediments that compress them into limestone. Eventually, however, even these carbon deposits recycle into atmospheric carbon dioxide as the limestone erodes and the fossil fuels are burned in automobiles or in industry. As with the other cycles, human activities are altering the global dynamics of the carbon cycle: Fossil fuel combustion, industrial pollution, and deliberate forest clearing and burning, particularly in the tropics, are releasing more and more carbon dioxide into the atmosphere. Some of this becomes fixed by

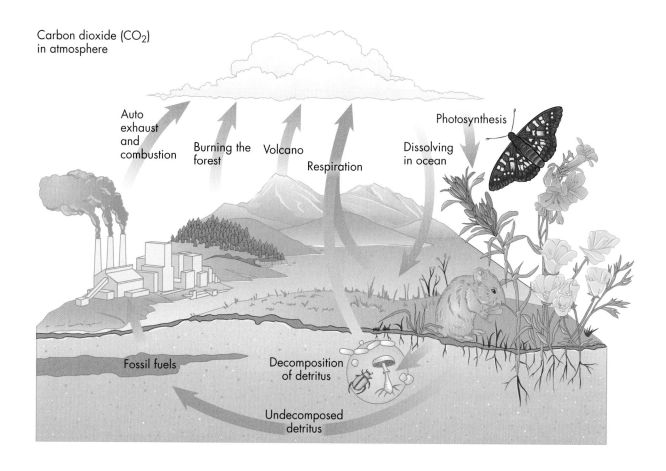

Carbon dioxide (CO_2)
in atmosphere

Auto exhaust and combustion

Burning the forest

Volcano

Respiration

Dissolving in ocean

Photosynthesis

Fossil fuels

Decomposition of detritus

Undecomposed detritus

Figure 25.12 **The Carbon Cycle.** Carbon atoms follow a cycle from atmospheric carbon dioxide, to biological molecules, to organic molecules in the soil, to geological deposits of fossil fuels, and back to carbon dioxide. Disruptions of this cycle can lead to increased atmospheric carbon dioxide. As many scientists suggest, changes in the carbon cycle can also lead to global warming and the range shifts Camille Parmesan has measured in Edith's checkerspot butterflies.

photosynthesis but some contributes to the greenhouse effect and global warming. The northward shift of Edith's checkerspot is just one consequence of this altered carbon cycle. Altered water patterns, changes in agriculture, and rising ocean levels—these are all part of the fallout from global warming. We'll return to this topic later in the chapter.

25.4 Global Climates and Earth's Life Zones

The shifting range of butterflies that Camille Parmesan discovered in the American West and in Europe confirms that global warming is affecting organisms and ecosystems directly. To understand these shifts and their possible repercussions, we must understand some background information first: What gives different parts of the world their unique combinations of temperature and rainfall? And how do these physical environments influence the types of organisms that live in Earth's many zones?

This global view of biology involves the **biosphere,** the portion of our planet that supports life. The biosphere includes every body of water; the atmosphere to a height of about 10 km (6 mi); Earth's crust to a depth of many meters; and all the living things within this collective zone. As huge as the biosphere may seem, it is actually a delicate veneer at our planet's surface. If you inflated a round balloon to a diameter of 20 cm (8 in) to represent Earth, then the thickness of its taut rubber skin would be proportional to the biosphere, the thin film of life encircling our world. This section explores how global physical factors—particularly worldwide currents of air and water—produce regional climates. The following section then discusses how climates determine the abundance and distribution of living organisms.

What Generates Earth's Climatic Regions?

A June freeze in the Sierra Nevada that follows a low-snow winter and a warm spring can wipe out local populations of Edith's checkerspot butterfly that emerge early from their cocoons. This reveals just how powerfully weather can affect biological systems. **Weather** is the condition of the atmosphere at any particular place and time, including its humidity, wind speed, temperature, and precipitation. In contrast, **climate** is the accumulation of seasonal weather events over a long period of time. Within the biosphere, weather has temporary, local effects, whereas climate is the major physical factor determining the abundance and distribution of living things. What, then, causes climate? The answers involve the uneven way that sunlight heats our planet; the behavior of water and air at different temperatures; and Earth's rotation.

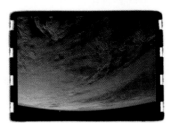

By what mechanisms could the greenhouse effect alter precipitation patterns?

The Earth Heats Unevenly

The tropics are warm and the poles are cold because the sun heats Earth's surface unevenly. Like a flashlight beam shining directly down onto a table from above versus one coming in obliquely from the side, sunlight hitting Earth directly is more intense than sunlight striking at an angle (Fig. 25.13a). The precise angle the sun makes at a point on the globe depends on the distance between the point and the equator (the *latitude*) and the season.

Temperatures fluctuate with the seasons because our planet spins on an axis tipped at a constant tilt relative to the sun. Consequently, during the summer, the Northern Hemisphere inclines maximally toward the sun, and receives more sunlight, while in winter, it tips away from the sun and receives less sunlight (Fig. 25.13b). The seasons are reversed in the Southern Hemisphere. This uneven illumination caused by Earth's tilted axis helps explain the warm temperatures and abundant growth of organisms in summer and the cold temperatures and dormancy of living things in winter.

Where Does Rain Come From?

Sunlight striking Earth's tilted sphere heats the air. Warm air has different properties than cool air, and these account for the formation of rain and snow. Cold air weighs more per unit volume than

warm air, and so tends to sink through lighter, warmer air. This is why New England farmers plant cold-sensitive fruit trees on the sides of valleys instead of on the valley floor, where cold, heavy surface air settles at night. Warm air tends to rise through cold air, causing smoke, steam, and hot-air balloons to drift upward.

Dense, cool air holds less moisture than light, warm air. This is why water droplets condense on a cold bathroom window pane when you take a hot shower. This same principle ultimately brings about rain and explains why a region like the tropics is so wet (Fig. 25.14). Powerful sunlight at the equator heats the air, which picks up moisture by evaporation from land surfaces and from plant leaves. The hot, moist air rises, and

Figure 25.13 **Earth and Sunlight.** (a) Uneven light levels cause uneven global heating. (b) Earth's tilted axis generates the seasons.

(a) Light intensities

A column of light striking the equator... is more intense than the same column of light spread over a larger area near the poles

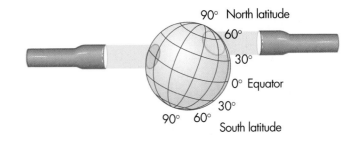

(b) Seasons

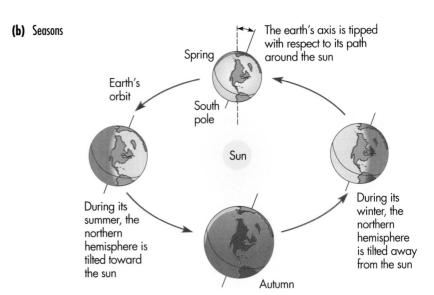

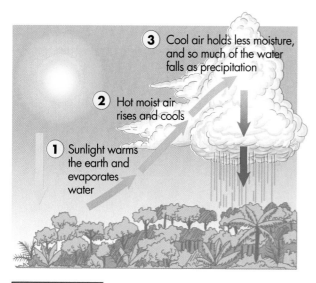

③ Cool air holds less moisture, and so much of the water falls as precipitation

② Hot moist air rises and cools

① Sunlight warms the earth and evaporates water

Figure 25.14 **Let It Rain: How Precipitation Forms.**

redirect the flow of ocean waters in slow, circular patterns in the North and South Atlantic and the North and South Pacific (Fig. 25.16). Ocean currents, like air currents, circulate heat and hence influence climate and the distribution of plants and animals. The Gulf Stream, for example, carries warm water from the tropics up the eastern coast of North America, then across the Atlantic, warming northern Europe: Compare ocean temperatures off the coast of Ireland and southern Alaska, which are about the same latitude (Fig. 25.16).

Short-term irregularities do occur in wind and ocean currents, such as the El Niño event, a temporary reversal of ocean currents in the South Pacific that occurs about every five years or so. This event leads to weather shifts on several continents. Nevertheless, in general, the sun and Earth's tilt and rotation create stable patterns of rainfall and temperatures that determine

once aloft, it cools, releasing the water it can no longer hold as rain onto the lush tropical forest.

Currents of Air and Water

Uneven heating of the globe generates circular currents of air and water that create different climates. Hot air ascending near the equator cools, and then travels north and south at high altitude (Fig. 25.15). At about 30° latitude (the latitude of Houston, Texas, and Cairo, Egypt, to the north and Cape Town, South Africa, and Sydney, Australia to the south), this cold, dry air falls to Earth. The dry, descending air creates the great deserts of Australia, North and South Africa, and North America.

The dry air from the deserts moves toward the equator, replacing the ascending hot air (Fig. 25.15). This moving air mass interacts with the rotating Earth, and causes surface winds. These predictable breezes, the **trade winds,** propelled traders' sailing ships in past centuries. Analogous currents in other circulating coils produce winds from west to east over much of North America. This flow carries moisture-laden air from the Pacific Ocean over the Coast Range and Cascade Mountains of the Pacific Northwest, where it dumps the rain and snow that encourage the lush growth of temperate rain forests.

Ocean Currents

The heating and rotation of the Earth not only affect the winds, they also drive ocean currents. Continents

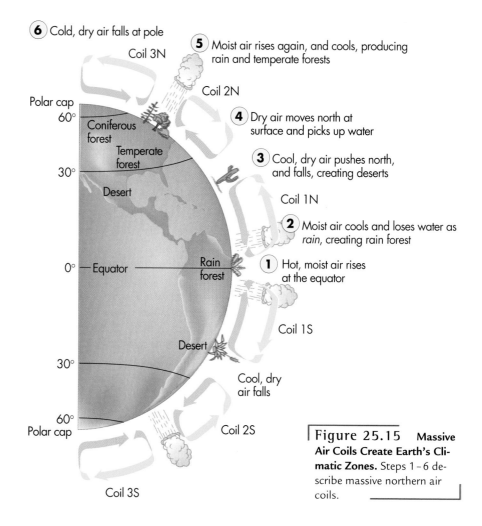

⑥ Cold, dry air falls at pole

⑤ Moist air rises again, and cools, producing rain and temperate forests

④ Dry air moves north at surface and picks up water

③ Cool, dry air pushes north, and falls, creating deserts

② Moist air cools and loses water as *rain,* creating rain forest

① Hot, moist air rises at the equator

Coil 3N

Coil 2N

Coil 1N

Coil 1S

Coil 2S

Coil 3S

Polar cap 60°

Coniferous forest

Temperate forest

30°

Desert

0° Equator

Rain forest

30°

Desert

Cool, dry air falls

60° Polar cap

Figure 25.15 **Massive Air Coils Create Earth's Climatic Zones.** Steps 1–6 describe massive northern air coils.

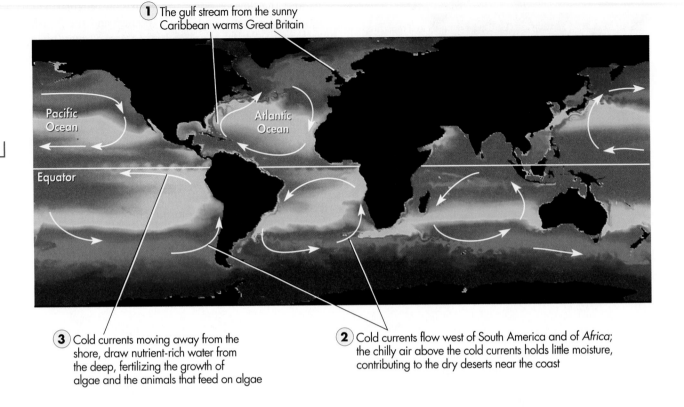

Figure 25.16 Ocean Currents Flow in Four Circular Patterns, Redistributing Heat. This satellite image charts ocean temperatures based on the color scale. If the arrows were not drawn on the map, could you determine ocean currents from water temperatures alone?

① The gulf stream from the sunny Caribbean warms Great Britain

③ Cold currents moving away from the shore, draw nutrient-rich water from the deep, fertilizing the growth of algae and the animals that feed on algae

② Cold currents flow west of South America and of *Africa*; the chilly air above the cold currents holds little moisture, contributing to the dry deserts near the coast

the general character of Earth's major communities of plants and animals.

25.5 Biomes: Earth's Major Communities of Life

Wherever similar climatic conditions exist—the deserts of Australia and Africa, the rain forests of Indonesia and Brazil—plants have evolved similar adaptations that help them exploit the climate's benefits and minimize its drawbacks. **Biomes** are large terrestrial geographic regions containing distinctive plant communities (Fig. 25.17). Major plant types characterize biomes because plants best reflect adaptations to rain, temperature, light, and wind. Furthermore, as primary producers, plants influence the consumers and decomposers that coexist in the biome.

Two major climatic features—temperature and moisture—set boundaries within which life will flourish. Tropical rain forests, for example, appear in regions with high amounts of precipitation and high tem-

peratures, while tundra appears in regions with cold and moderately dry climates. The differences in temperature and moisture cause differences in productivity in the different biomes. For example, productivity is high in temperate and tropical rain forests, and low in tundra and desert regions. In the next few pages, in Figures 25.17 through 25.25, you can "visit" eight of Earth's major biomes and learn about their characteristics. They are arranged roughly in order from equator to poles—tropical rain forests, savannas, deserts, temperate grasslands, chaparral, temperate forests, coniferous forests, and tundra. We discuss the polar ice caps, as well.

Tropical Rain Forest

Nearly half of all living species reside in the world's warm, wet tropical **rain forests** occurring near the equator (Fig. 25.18). Heavy rains tend to leach nutrients from tropical soils; thus, the biomass of the living forest itself is the biggest source of nutrients. Rain forests have an upper story, or emergent layer of tall trees that capture direct sunlight; a canopy of shorter

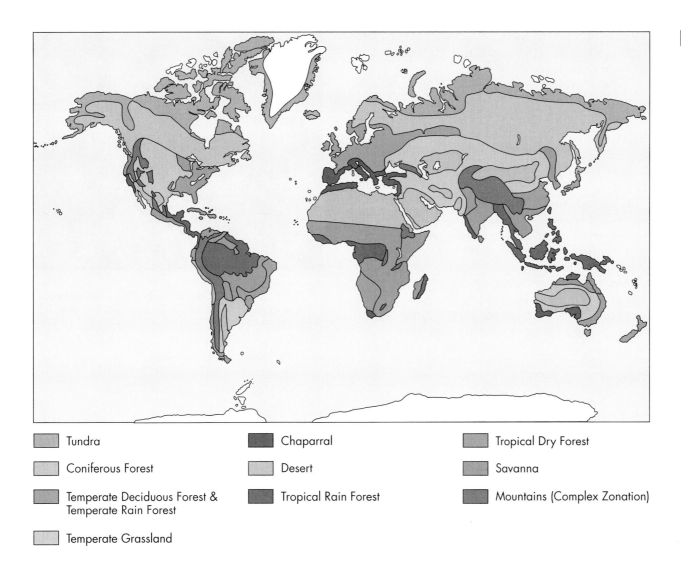

Figure 25.17 Map of the World's Major Biomes.

Tundra

Coniferous Forest

Temperate Deciduous Forest & Temperate Rain Forest

Temperate Grassland

Chaparral

Desert

Tropical Rain Forest

Tropical Dry Forest

Savanna

Mountains (Complex Zonation)

forest trees where leaves, flowers, fruits and arboreal animals abound; an understory penetrated only by dim light; and a dark forest floor where plants often have huge, deeply green leaves. Species diversity is not quite as rich in the adjacent biome called tropical dry forest, where the organisms have evolved a tolerance for longer dry seasons.

People are rapidly destroying the timeless beauty and incredible species richness of tropical rain forests in order to plant crops, mine for minerals, harvest hardwood trees, and/or graze cattle for export. This destruction threatens the survival of millions of species, including useful medicinal plants as yet undescribed.

Figure 25.18 Tropical Rain Forest: Lush Equatorial Biome. Dense, wet forest covers most of the Amazon Basin in Brazil, and much of the world's equatorial regions. Here, palms, amarillos, and other deep-green plants surround the trunk of a large fig tree.

Once gone, tropical forests are unlikely to return because most or all of the animals that pollinate and disperse the forest plants will be gone.

Savanna

Year-long warmth with an extended dry season results in a tropical **savanna,** or dry, open grasslands with sparse tree coverage (Fig. 25.19). The world's major savannas lie in Africa, South America, Australia, and parts of Southeast Asia and North America. Situated between tropical forests and deserts, savannas contain stunted, widely spaced trees with tall grasses growing in between that support grazing animals like giraffes and kangaroos. The herbivores, in turn, support lions, dingoes, and other carnivores. Many savanna species, including the rhinoceros and the elephant, are threatened with extinction because people destroy their habitats and kill them for their beautiful coats, ivory tusks, or spectacular horns.

Desert

Desert regions receive less than one-tenth of the annual rainfall of tropical rain forests; hence, desert plants are widely spaced and cover less than one-third of the ground surface (Fig. 25.20). The deserts of Mexico, Australia, and North and South Africa formed where dehydrated air—rained out in the tropics—descends back toward Earth. China's Gobi Desert formed at the center of a huge continent, far from the moist sea air.

Figure 25.19 **Savanna: Tropical Dry Land.** This tropical savanna in Kenya, sprawling below the Loldaika Mountains, is home to the giraffes and thorn trees shown here.

Figure 25.20 **Desert: Parched Zone Where Life Is Sparse.** Desert regions like this area of sandstone cliffs in Utah receive very little annual rainfall and support only widely spaced desert plants, including the evening primrose and locoweed dotting the foreground.

Some deserts form downwind of tall mountain ranges. Moist air crossing California's Sierra Nevada mountains, for example, drops its moisture as rain or snow on the western side of the peaks, leaving the Great Basin of Nevada and Utah in a rain shadow that produces a desert. Daytime temperatures can soar to 49 °C (120 °F), while at night, heat tends to radiate and leave a biting chill to the air. Many desert plants have adaptations such as fleshy stems and spines. Desert animals tend to be active only in the evening or early morning hours, avoiding the heat of the day. Human activity is expanding the world's deserts through desertification, in which overgrazing and poor irrigation practices remove grass from the grasslands that rim the deserts.

Temperate Grassland

Bordering many of the world's deserts are **grasslands,** treeless regions dominated by dozens of grass species (Fig. 25.21). Known as *prairies* in North America, *pampas* in South America, *steppes* in Asia, and *veldt* in Africa, these regions are wetter than deserts, drier than forests, and have seasonal extremes of hot and cold rather than wet and dry. Frequent wildfires in the dry grasses prevent grasslands from turning into forests. Open grasslands support animals like bison, antelope, prairie dogs, anteaters, armadillos, coyotes, snakes, and

Figure 25.21   **Temperate Grassland: Treeless Sea of Grass.** This grassland region in Montana was once home to dozens of native grasses and vast herds of elk, buffalo, and other grazing animals.

produce excellent topsoil. In early spring, wildflowers like wood sorrel, bluebells, and violets bloom on the sunny forest floor. Later, only shade-tolerant plants such as ivy and honeysuckle thrive. Many animals of the temperate forest, such as deer and squirrels, remain year-round. Some, such as bears and snakes, hibernate in winter. Others, like robins, migrate to warmer regions. People have utilized temperate forests so heavily for timber that in the United States, only 4 to 5 percent of the virgin forests that once covered much of the land remain.

Coniferous Forests

Across much of Canada, northern Europe, and Asia, at latitudes with cold, snowy winters and short summers,

hawks. Grassland soils are richer in organic matter than the soils of other biomes.

Chaparral

A biome called the **chaparral** (shap-uh-RAL; Spanish, "thicket"), or temperate scrublands, borders grasslands and deserts along the shores of the Mediterranean and along the southwest coasts of North and South America, Africa, and Australia (Fig. 25.22). Chaparral has hot, dry summers and cool, wet winters. Chaparral plants are generally less than 2 m (6.5 ft) tall and have small, leathery, often hairy leaves that stay green all year. Because grasses grow during the wet winters and then die and dry out during the hot summers, this biome, like the grasslands, experiences frequent fires. In developed areas large cataclysmic fires can spread and consume homes built within or adjacent to large areas of the tinder-dry chaparral.

Temperate Forest

Temperate forests occur in eastern North America, Europe, China, Japan, New Zealand, Australia, and the tip of South America (Fig. 25.23). **Temperate forests** are dominated by broad-leaved trees and have intermediate amounts of rainfall and fairly moderate temperatures that fluctuate between summer highs and winter lows. The dominant trees, including oak, maple, birch, and hickory, drop their leaves and become dormant until spring. The fallen leaves allow for the recycling of nutrients and

Figure 25.22   **Chaparral: A Biome of Low Bushes and Frequent Fires.** Sage, ceanothus, and other evergreen shrubs grow in the chapparal biome of California's Santa Monica Mountains.

Figure 25.23 **Temperate Forest: Deciduous Hardwoods Bordering Grasslands.** These lush woods in rural Illinois are temperate forests filled with dozens of hardwood species.

vast forests of cone-bearing pine, fir, spruce, and hemlock trees grow in the **coniferous forest** (Fig. 25.24). Conifer leaves, which are needle-shaped and have thick, waxy cuticles, combat water loss and are not shed each winter, thus they can begin collecting sunlight as soon as the short growing season begins. Coniferous forests support animals such as bark beetles, elk, porcupines, wolves, and lynx. Coniferous forests contain the world's greatest lumber reserves. Traditionally, loggers have harvested this timber by *clear-cutting,* the practice of sawing down all standing trees, clearing the land, then planting seedlings of a single tree species, usually Douglas fir. Biologists have found that the major source of fixed nitrogen for old-growth forests is probably the flat or stringy, gray-green, orange, or yellowish organisms called lichens. Since lichens don't thrive until the forest canopy is about 100 years old, young replanted fir forests grow and are recut long before this, and the forest soils often become depleted of nitrogen, jeopardizing

future tree crops. Some modern foresters are now harvesting only a few of the mature trees at a time.

Tundra

At the northern boundary of the coniferous forest, fragrant, giant conifers give way to the low vegetation of the **tundra,** a cold, treeless plain (Fig. 25.25). With annual temperatures of −5 °C (+23 °F) or less in the tundra, soil thaws to only about 1 m (39 in). The deeper, permanently frozen soil, or *permafrost,* prevents most trees from growing. Plants, including grasses, sedges, mosses, lichens, and heathers, grow low, where they can absorb warmth re-radiated from the solar-heated ground and minimize the effects of wind and desiccation. These plants support herbivores like cari-

Figure 25.24 **Coniferous Forest: Fragrant Evergreens in the Northern Latitudes.** Temperate rain forests, like the one shown here along the Nooksack River in Washington state, are found only in the Pacific Northwest and parts of New Zealand, and constitute a biome closely related to the coniferous forest.

Figure 25.25 **Tundra: Cold, Treeless Plain in the Far North.** This starkly beautiful tundra landscape lies in Alaska's Denali National Park. Caribou graze in the foreground.

bou, arctic hare, and lemmings, and they in turn are prey for foxes, weasels, and snowy owls. During summer, clouds of mosquitoes and flies fill the air, reproducing on the soggy ground. Many birds migrate to the tundra in the summer, feast on the insects, and breed during the long summer days.

Polar Caps

The **polar ice caps** are icy, treeless regions at our planet's highest latitudes. The arctic ice cap is a vast frozen ocean covering the planet's North Pole. The antarctic ice cap encompasses all of the continent Antarctica, a landmass lying beneath a great raft of ice and surrounded by frigid seas. Together the polar ice caps take up millions of square kilometers of land and ocean surface. These icy regions are not considered true biomes because they lack major plants; only animals and microbes survive in these regions. In the Arctic, polar bears hunt seals and other aquatic animals; in the Antarctic, penguins and seals inhabit the continent's ice shelf.

25.6 Life in the Water

Most of us are familiar with at least some of the biomes we just explored on dry land. When viewed from space, however, Earth is a blue planet because three-fourths of

its surface is covered by water. Most of that area is **marine**—salty oceans and seas—with less than 1 percent of the surface covered by freshwater lakes, rivers, streams, swamps, and marshes. This freshwater area is tiny compared to the oceans, but it still amounts to tens of thousands of square kilometers. Ecologists do not call marine or freshwater communities biomes. Their studies, nevertheless, show that aquatic ecosystems differ from each other in significant ways, including their productivity, with coral reefs and estuaries being far more productive than lakes or the open ocean.

Properties of Water

As we saw in Chapter 2, water has unique physical characteristics, and these properties strongly influence aquatic organisms. For example, it requires more energy exchange to heat or chill water than it does to change air temperature. Aquatic organisms therefore suffer fewer rapid temperature changes than their land-dwelling counterparts. In addition, air grows steadily denser as it cools, but water has its greatest density at 4 °C (39 °F). Because of this, ice at 0 °C (32 °F) floats. If it didn't, lakes would freeze from the bottom up, only the upper layers would thaw in summer, and bodies of water outside of the tropics would support little life.

The next few sections describe the different water habitats, beginning with freshwater ecosystems, including running and standing waters, and then various saltwater habitats, including estuaries and oceans.

Mountain Streams

A stream bubbling from a mountain spring or from melting snow is cold and clear but contains few nutrients. As the water tumbles downhill and over rocks, it picks up oxygen and more nutrients. Species that live in these turbulent zones include algae, mosses, and trout, which thrive only where oxygen is plentiful and water temperatures are low. As nutrients accumulate and the stream widens and is less shaded, the photosynthetic productivity of algae and green aquatic plants increases. A river's middle stretch has the greatest species diversity. In the lower, slower-moving section of a large river, the water becomes cloudy and enriched with nutrients. The decreased light lowers the rate of photosynthesis and primary productivity drops once again. In these areas, microbes and invertebrates live on detritus in the bottom sediments; catfish and bass replace trout. Around the world, people tend to use rivers as sewers and have dammed, diked, and channeled rivers in ways that often create significant ecological problems.

Freshwater Lakes

Lakes have different life zones that depend on light penetration and depth. In shallow areas, where light can reach the lake bottom, abundant producers like water lilies, cat-tails, and algae exist and support consumers such as insects, snails, amphibians, fishes, and birds. Farther from shore, the light-penetrated top layer of water supports huge populations of photosynthetic algae and small crustaceans. The darker bottom region supports mainly insect larvae, scavenger fishes, and decomposers. In lakes, dissolved nutrients are as important as light; excess phosphorus, for example, can cause a bloom of plant growth in a lake or pond. As the seasons change, changing water temperatures stir up the nutrients in bottom sediments.

Estuaries

Ecologists use the term **estuaries** for the areas where rivers meet oceans, fresh and salt water mingle, and temperatures and salt concentrations vary widely with the tides and seasons. Estuary organisms like brine shrimp can often tolerate wide ranges of salinity. Constant water movements stir up nutrients, making estuaries some of the Earth's most productive ecosystems and nursery grounds for many fish species. In populated areas, people often dump urban and industrial effluents into estuaries or make landfills of them to create new waterfront property. This significantly decreases estuaries' great biological productivity.

Tide Pools

At the ocean's edge lies the **intertidal zone,** a region that is underwater at high tide, exposed to air at low tide, and pounded by waves and wind. Most intertidal producers (kelp and other algae) have structures that anchor them to rocks, while most intertidal animals have tough bodies or shells as well as underwater "glues" that help fasten the organisms to the rocks. Intertidal organisms including sea anemones, sea urchins, crabs, sea stars, and certain fishes, tolerate hot, dry spells or changes in salinity as the tide rolls out each day. A group of burrowing organisms, including clams, snails, and worms, excavate beneath the extensive tidal mudflats. The ocean's most productive region extends from the intertidal zone to the edge of the *continental shelf,*

the submerged part of the continents. Huge populations of phytoplankton and zooplankton serve as a nutrient base for the rest of the ocean's consumers, including whales.

Coral Reef

In the tropics, the shallow ocean zone is home to reef-building corals and the photosynthetic microorganisms that inhabit them and give **coral reefs** their fantastic colors (Fig. 25.26). The crannies and caves in these stony colonies create sheltered nesting sites and hiding places for the most diverse and productive communities in the seas. These, too, are threatened by pollution and global warming with its resultant ocean temperature changes.

The Open Ocean

Beyond the continental shelf and covering the deep abyssal plane, the water in the **oceanic zone** or **open ocean** is nearly deserted, even in the sunniest upper regions near the surface. The open ocean lacks nutrients such as phosphate and usable nitrogen, and overall is less productive than the arctic tundra. The ocean's most productive areas occur mainly where currents cause nutrients to well up from very deep waters; a good example is the very productive cold area off the coast of Peru that supports the food chain on and around the Galápagos Islands. The open oceans help modulate climate

Figure 25.26 **Coral Reef: Fantastic Productivity Beneath Shallow Waves.** This coral reef off a small island in the Caribbean Sea is home to reef-building corals and tropical fish, most with fantastic colors.

and maintain favorable concentrations of oxygen and carbon dioxide in the atmosphere.

25.7 Change in the Biosphere

The poles, the tundra, and the oceans—all are so vast and timeless that they seem unchanging and indestructible. Change, however, is part of nature. "People need to stop looking at the world as stable," says Camille Parmesan. "It's not stable!" Climates cool down or heat up, continents drift, organisms evolve, butterflies shift their ranges northward, and each change influences the ecosystem and the larger biosphere. Recall from Chapter 10 how the evolution of photosynthesis about 2.8 billion years ago revolutionized the atmosphere, causing oxygen to accumulate in the air and the oceans to "rust." Today, human activities are causing equally pervasive modifications in the biosphere—but the modifications are coming far faster than organisms usually adapt to change via evolution. These alterations include global warming, the degradation of the Earth's ozone layer, and the loss of habitat as humans convert more of the planet to their own use. Our species must learn to use ecosystems without causing permanent damage. But how? These are issues humanity must approach and solve before it's too late.

Global Warming

Camille Parmesan's studies have helped illuminate the problem of global climate change by showing that, indeed, butterfly species in North America and Europe are shifting their home ranges northward and higher into the mountains as the land and air heat up (Fig. 25.27). Her work is extremely important, but it's just part of the evidence that human activity is altering ecosystems and the biosphere. Other studies have shown that as spring temperatures have come earlier and earlier in the British Isles (as elsewhere), 20 species of British birds are now laying their eggs weeks earlier. Eighty-three percent of British frog species are laying eggs earlier, as well. And across Europe and North America, the leaves are appearing on oak, birch, and maple trees up to 20 days earlier. What's more, meteorologists have found that the 20th century was by far the hottest 100 years in the last 1000. What's more, about half of the century's 0.5 °C (1 °F) average warming has come in just the past 30 years. After weighing all the possible causes, the United Nations Intergovernmental Panel on Climate Change deter-

(a) Shifting populations

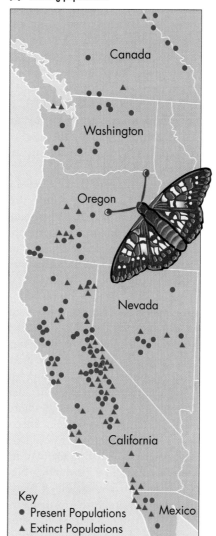

Key
● Present Populations
▲ Extinct Populations

(b) The greenhouse effect

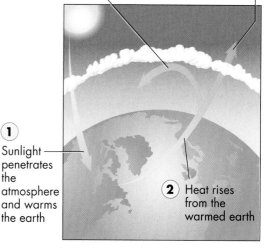

③ While some of the heat escapes to space, some is trapped by gases in the atmosphere, thus warming the earth

① Sunlight penetrates the atmosphere and warms the earth

② Heat rises from the warmed earth

Figure 25.27 **The Greenhouse Effect, Global Warming, and Shifting Butterflies.** (a) As warm zones move northward and higher into the mountains, species such as Edith's checkerspot butterfly shift their ranges, as you can see from the blue triangles representing recent locations of populations. (b) Researchers have shown that an increase in carbon dioxide and other air pollutants is causing global warming, as reflected heat becomes trapped in the atmosphere.

mined that people are mostly to blame for the greenhouse effect.

Greenhouse gases include carbon dioxide, methane, chlorofluorocarbons, and nitrous oxide. Carbon dioxide alone, however, is responsible for about half the human-induced greenhouse gases. Not only do we release much more CO_2 than the other gases, explains Dr. Parmesan, but the carbon dioxide is stable in the atmosphere for over 100 years. Methane (CH_4) is generated by decomposer bacteria in flooded rice fields and in the guts of cattle and termites, as well as from the burning of coal and natural gas. Far less methane is released than carbon dioxide, however, and CH_4 breaks down 10 times faster in the atmosphere. Chlorofluorocarbons (CFCs) are industrial chemicals that contain atoms of chlorine, fluorine, and carbon, and are used as coolants in refrigerators and air conditioners, as well as in plastic foam and insulation materials. Like carbon dioxide, CFCs remain in the atmosphere for up to 100 years, but they also break down ozone (see below). Laws restricting their use have started to help, but decades of accumulated CFCs still hang above the Earth. Nitrous oxide (N_2O_2), or laughing gas, is produced by microbes in soils, by the burning of forests and fossil fuels, and by the production of chemical fertilizers. It is a less abundant but still significant contributor to global warming.

The global production of carbon dioxide has increased by roughly 25 percent in the past century (review Fig. 3.23) as our burning of coal, oil, and gasoline has accelerated. The clearing of temperate forests and tropical rain forests and the burning of felled trees is another major source of the gas. Some of the excess carbon dioxide dissolves in the ocean. Much of it, however, enters the atmosphere—much more, unfortunately, than can be removed by plants during photosynthesis.

Many atmospheric scientists predict that sometime between the years 2025 and 2075—well within many of our lifetimes—the accumulating blanket of greenhouse gases will send global temperatures up an additional 1 to 3.5 °C (2 to 5 °F) on average (*E* xplorer 25.2). A change this big could have disastrous effects. Ice near the poles would melt, and sea levels would rise like mercury in a thermometer, inundating many of the world's most populous coastal cities, including New York, Los Angeles, London, Stockholm, Hong Kong, and Tokyo. The dust bowl conditions that plagued the United States in the 1930s could occur in the great grain-producing regions of the American Midwest, Canada, and the Soviet Union. Irrigation would probably not solve the problem because groundwater reserves would run out quickly. Ominously, even the coldest years of the last decade have been warmer than nearly every year of a century ago, and we are beginning to see widespread droughts, forest fires and crop losses as an apparent result.

Most ecologists believe that we must decrease our consumption of fossil fuels through an emphasis on energy conservation and a commitment to renewable energy sources such as wind power and solar energy. As individuals, we could, for example, lower the temperature of our homes in winter, use less air-conditioning in summer, and rely on public transportation, bicycles, and our own two feet far more often than we do now. Farmers could employ agricultural practices that better

sustain the levels of organic matter in the soil, and people in tropical regions could reduce clear-cutting and burning of forests and step up efforts to replant denuded areas. Finally, biologists could actively develop more drought-resistant and salt-tolerant plants that can substitute for our staple crops if the world's breadbaskets become dust bowls. We must all become well-enough informed to take an active role as voters and consumers in helping governments and ourselves make ecologically sound decisions (*E* xplorer 25.3).

The Ozone Hole

Chlorofluorocarbons not only act as greenhouse gases, they also attack *ozone* (O_3) molecules in Earth's protective **ozone layer,** a zone encircling the planet several miles above its surface that absorbs about 99 percent of the ultraviolet light that would otherwise penetrate and destroy many biological molecules, including DNA. Ozone forms naturally when ultraviolet light streaming from the sun strikes atmospheric oxygen gas, with its two atoms of oxygen.

Ominously, in 1985, atmospheric scientists first reported decreases in ozone concentrations in the upper atmosphere. Decreases were especially strong over much of Antarctica. The "ozone hole," which is now larger than the European continent, contains only half as much ozone as it did a decade ago (Fig. 25.28). Decreases in ozone concentrations have also been detected over the North Pole and parts of Australia, and some biologists worry that it might threaten Canada's huge coniferous forest biome.

Experts have shown that CFCs started accumulating in the atmosphere in the middle of the 20th century, proving that they are generated by humans. These long-lived compounds act as catalysts that convert ozone (O_3) to oxygen (O_2), with each CFC molecule continuing to destroy ozone molecules over and over again for decades.

Scientists predict that the thinning ozone layer will result in more skin cancers, more eye cataracts, and more problems with the human immune system. In addition, plants will suffer more mutations, stunted growth, and surface damage. Measurements showed that antarctic phytoplankton living under the ozone hole were about 10 percent less efficient at converting sunlight and CO_2 into carbohydrates, a phenomenon that, in turn, harms that region's entire food chain, from krill to fishes to birds and mammals, including whales.

By international agreement, the production of CFCs was to be phased out by the year 2000. But be-

GET THE PICTURE

E xplorer 25.2

Global Warming

READ ON

E xplorer 25.3

Personal Solutions

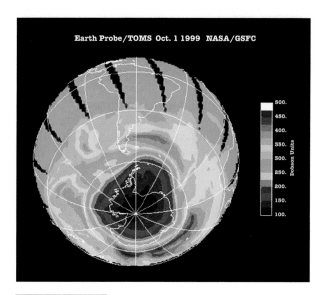

Figure 25.28 **The Ozone Shield.** Earth's ozone layer absorbs most of the sun's ultraviolet rays. This layer hangs in the stratosphere, 17 km (about 10 miles) above the planet's surface. In the last quarter century, the ozone layer has thinned dramatically over Antarctica, due primarily to human use of chlorofluorocarbons (CFCs). This computer-generated image shows the areas of most severe thinning (black, purple, and blue) in 1999.

cause the compounds are long-lived and act as catalysts, the ozone layer won't return to normal for about 100 years. More research is needed to identify the potential ecological effects of decreased ozone and increased ultraviolet light. We already know, however, that changes in the populations of especially sensitive organisms, like certain kinds of phytoplankton, can cause a collapse of entire ecosystems. Solutions based on the principle of sustainability are intended to prevent just this kind of collapse.

Sustainability

Experts predict a near doubling of world population during this century, from 6 billion in late 1999 to more than 10 billion by 2100. Supporting all these people, even at a minimum level, will require a five- to tenfold increase over current economic activity: more agriculture, shipping, building, transporting, financing, using resources, burning down forests, throwing away trash. How can we possibly keep this up? How can Earth support twice as many of us? What kind of future will we have? Will it be a global version of the boom-and-bust population cycle, with ever more hungry, impoverished people in a constantly worsening environment headed for a series of catastrophes that slash the population? Or will we learn to sustain ourselves in some kind of balance with the biosphere?

Our future depends on **sustainability**—balancing the levels of human population and economic growth against the quantities of available resources and the quality of the physical environment. In the past, economic development has come at the expense of environmental quality. High prosperity in certain countries has rested, in part, on environmental changes in other countries. To ensure a stable global environment over the long term, say experts like William Ruckelshaus, former administrator of the Environmental Protection Agency (EPA), all people must have a reasonable level of prosperity and security, and this will require a change in human attitudes and actions as great as those that brought about the agricultural and industrial revolutions of the past.

The sustainability revolution will require industries to use fewer materials and less energy, to rely on wind and solar energy instead of fossil fuels wherever possible, and to recycle their wastes into other products and processes. Sustainable agriculture will involve

- Rotating crops to increase yields
- Building up rich soil by preventing erosion and by using natural fertilizers, such as animal manure, nitrogen-fixing crops, and blue-green algae
- Controlling weeds, insects, and plant diseases through the use of integrated pest management, through genetically altered crop plants, and through the planting of traditional and nontraditional crops.

Creative economic solutions will also play a part in global sustainability. Among the many ideas now under consideration are debt-for-nature swaps and conservation easements. In the former, wealthy nations would cancel the debts of developing nations in exchange for the promise to preserve natural habitats. In the latter, rich nations would pay the equivalent of mineral rights to secure forested areas and nature preserves against destruction.

In addition to such approaches, say various experts on sustainability, all nations, rich and poor, will have to provide their citizens with information on sustainability techniques and birth control. They will have to fund international agencies to manage global environments. And they will have to be fully cooperative and committed to the sustainability revolution. The alternative—

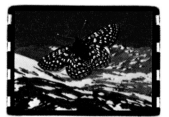

How could our efforts toward sustainability help preserve Edith's checkerspot?

the crash of the human population following its present boom—is just too dismal a prospect.

Ecological researchers like Camille Parmesan hear protests all the time such as "We can't affect the atmosphere! We're just humans!" and "We don't have to worry about global warming. Butterflies and other species will just shift around a little bit. No problem!" Our species *is* capable of changing the global environment, however, and has done so in a stunningly short time period. But we are also capable of understanding that we're mismanaging Earth's resources and of foreseeing the consequences. Our own survival and that of millions of other species depend on working out new solutions to preserving our environment while also supporting our own needs. And it starts with taking responsibility for our own actions: Some scientists, for example, have been reluctant to admit that human activity is causing global warming. But, says Dr. Parmesan, the evidence is getting so strong that even the most conservative are saying, "Well, it'd be awfully good insurance for us to start doing something now." "No one can tell you 100 percent for sure," Parmesan says, "but if you were selling me an insurance policy, I'd grab it."

Chapter Summary and Selected Key Terms

Introduction

Dr. Camille Parmesan is counting butterflies to learn about global warming. By tracking down populations of Edith's checkerspot butterfly, she has been able to document a distinct shifting in the animal's home territory. Over the course of the 20th century alone, the animal has moved 92 km northward and 124 meters higher into the mountains, spurred on by hotter and hotter summers. Shifting home ranges are just one consequence of global warming. The other predicted outcomes are loss of cropland in the Midwest to drought and heat; the northward spread of tropical diseases; the rise of sea level; the melting of polar ice caps; and the obsolescence of many nature preserves. This chapter discusses how communities of organisms interact with their physical environments in ecosystems, how the biosphere supports life, and how human activities are altering both and bringing about an increased **greenhouse effect** (p. 619) and global warming.

Pathways of Energy and Materials

An **ecosystem** (p. 621) includes the community of organisms in an environment and the physical factors that affect them. Energy flows through ecosystems in a one-way path from the physical world to organisms and back to the environment with a loss (to heat) at each exchange. Materials cycle through ecosystems, from nonliving environments to organisms and back without being diminished.

Plants and certain microbes are **autotrophs** (p. 622) ("self feeders"), while animals, fungi, and many microbes obtain nourishment by **heterotrophy** (p. 622) ("other feeders"). Autotrophs are **primary producers** (p. 622) of biological molecules. Heterotrophs are **consumers** (p. 622). Ecologists assign organisms to **trophic levels** (p. 622) or feeding levels. Producers are at the lowest trophic levels. They provide the energy fixation base (that is, they fix energy) and the nutrient concentration base (they concentrate nutrients) for the ecosystem.

Primary consumers (p. 623) eat producers; they are herbivores. **Secondary consumers** (p. 623) are carnivores that eat herbivores. **Tertiary consumers** (p. 623) eat other carnivores. **Decomposers** (p. 623) and **detritivores** (p. 623) are special classes of consumers because they decompose organic wastes or dead organisms (**detritus**; p. 623). These relationships can be seen as **food chains** (p. 623) but a more accurate depiction of all the interconnected feeding relationships in a community is the **food web** (p. 623).

Energy Flow Through Ecosystems

Ecologists have studied the energy budgets for certain forests. Only about 2 percent of the sunlight striking the forest is converted to chemical energy by producers; this is the **gross primary productivity** (p. 624). Half of this goes to fuel the producer's own cellular respiration. The rest is the **net primary productivity** (p. 624). This makes only about 1 percent of the original light energy potentially available to consumers, but most of the new leaves and twigs end up on the forest floor as detritus.

Ecologists draw **energy pyramids** (p. 625) to reflect the amounts of energy available at different trophic levels. They use the dried weight of organic matter at a particular level, or the **biomass** (p. 626), to create a **pyramid of biomass** (p. 626) to show how much energy was actually fixed. It takes 10 kg of bass to build 1 kg of human tissue. It takes 100 kg of smaller fish to build the bass; 1,000 kg of insects to build the smaller fish tissue; and 10,000 kg of plants to build the 1,000 kg of insects. This is why "eating low on the food chain" makes ecological sense. Because of food webs and energy pyramids, toxic substances tend to build up in secondary and tertiary consumers in a process called **biological magnification** (p. 626). This happened with DDT and birds of prey.

Materials Cycle Through Ecosystems

In the global **water cycle** (p. 627), water moves from the atmosphere to Earth's surface and back to the atmosphere as water vapor. In the **nitrogen cycle** (p. 628), nitrogen gas from the atmosphere is fixed in biologically useful forms such as NH_4^+ and NO_3^-, then returns to the atmosphere as nitrogen gas. Legumes have nitrogen-fixing bacteria that fix quantities of N_2. Phosphorous is exchanged between the water, soil, and organisms in the **phosphorous cycle** (p. 629). Phosphates from fertilizers and detergents can run off into lakes and cause an overgrowth of algae and aquatic plants called **eutrophication** (p. 630). Carbon atoms also move in a vast **carbon cycle** (p. 630) from CO_2 in the air to carbon compounds in producers to consumers then back to the atmosphere via respiration and detritus breakdown.

Global Climates and Earth Life Zones

The **biosphere** (p. 631) is the portion of our planet (atmosphere, water, soil) that supports life. The condition of the atmosphere at any particular place and time is **weather** (p. 632). Over long periods, those conditions create **climate** (p. 632). Climates differ around the world because the sun heats Earth's tilted surface unevenly. This also leads to rain patterns, and air and water currents. These differing climatic zones lead to **biomes** (p. 634) or major plant communities over large geographical areas. We discuss eight major biomes: **tropical rain forests** (p. 634), **savannas** (p. 636), **deserts** (p. 636), **temperate grasslands** (p. 636), **chaparral** (p. 637), **temperate forests** (p. 637), **coniferous forests** (p. 638), and **tundra** (p. 638). At Earth's poles are **polar ice caps** (p. 639). Earth is three-fourths covered by water, most of which is **marine** (p. 639), or covered by salty oceans and seas. Less than 1 percent is covered by freshwater lakes, rivers, streams, swamps, and marshes. We discuss running and standing freshwater ecosystems, **estuaries** (p. 640), coral reefs, and oceans.

Change in the Biosphere

Human activities are causing massive changes in the biosphere. These include global warming due to air pollutants and the resulting greenhouse effect; the release of chlorofluorocarbons that degrade the **ozone layer** (p. 641); and habitat destruction leading to the loss of biodiversity. Our future depends on **sustainability** (p. 642), or balancing population and economic growth with available resources and environmental health. This requires crop rotation, soil enrichment, integrated pest management, and the efforts of every citizen to reduce energy consumption and recycle materials.

> All of the following question sets also appear in the Explore Life *E* electronic component, where you will find a variety of additional questions as well.

Test Yourself on Vocabulary and Concepts

In each question set below, match the description with the appropriate term. A term may be used once, more than once, or not at all.

SET I

(a) producers (b) consumers (c) ecosystem (d) trophic level (e) detritivores

1. The term that includes a community of organisms at several trophic levels forming a food chain or web within a particular physical environment.
2. There may be one, two, three or even four trophic levels within this group of organisms.
3. A group of consumers that includes fungi, bacteria, and several invertebrate and vertebrate animals that consume dead organisms and organic wastes.
4. The trophic level that includes photosynthetic plants, bacteria, and protists.
5. A feeding level within an ecosystem.

SET II

(a) energy pyramid (b) biomass (c) biological magnification (d) material cycles (e) greenhouse effect

6. The dry weight of organic material at any trophic level.
7. A phrase descriptive of the fact that each trophic level in an ecosystem contains substantially less stored energy than the level immediately below it.
8. The tendency for increased concentration of toxic substances in the higher trophic levels of an ecosystem.
9. Global cycling of important compounds or elements, utilizing the atmosphere, Earth, oceans, and organisms as temporary reservoirs.
10. The trapping of infrared radiation near Earth's surface by carbon dioxide and other atmospheric gases.

SET III

(a) water cycle (b) nitrogen cycle (c) phosphorous cycle (d) carbon cycle

11. A global cycle in which bacteria, leguminous plants, and the atmosphere play key roles.
12. A global cycle powered by solar energy, with transpiration (evaporation from plants) as an important link.
13. Found in local cycles within ecosystems; involves calcium compounds in rocks and soil, as well as organic compounds in plants and animals.
14. A global cycle in which photosynthesis and cellular respiration are important; characterized by lengthy periods during which some of the material may temporarily leave the cycle as rock, wood, or litter.

SET IV

(a) rain forest (b) savanna (c) desert (d) grasslands (e) chaparral (f) temperate forest (g) coniferous forest (h) temperate rain forest (i) tundra (j) polar ice cap

15. Characterized by the richest agricultural soil.
16. Characterized by deciduous trees such as oak and maple and many hibernating and migratory animals.
17. Has the greatest species richness of all the terrestrial biomes.
18. Not present in the United States.
19. Bordering continental coastlines, characterized by wet winters and hot, dry summers.

Integrate and Apply What You've Learned

20. Is Edith's checkerspot butterfly a producer, or a primary, secondary, or tertiary consumer? Support your answer.
21. How much of the solar energy that strikes Earth is eventually found within the bodies of carnivorous animals? What happens to the rest?
22. In Quito, Ecuador, there are snow-capped mountains on the equator. How is that possible?
23. What is the likely fate of tropical rain forests after they are harvested and converted to pasture lands for cattle?
24. Why are some of Earth's regions much colder on average than others and why do deserts form the way they do? Use a separate piece of paper to sketch out a picture of Earth to help you answer.
25. Which terrestrial and aquatic areas of the world have the highest net primary productivity? What do these areas have in common?

Analyze and Evaluate the Concepts

26. How is a global problem like greenhouse gases and warming temperatures affecting local populations of an individual species like Edith's checkerspot butterfly?
27. A few aquatic ecosystems exhibit inverted biomass pyramids, with fewer phytoplankton producers than consumers at any one time. How could such an inverted pyramid originate? What would you predict the energy pyramid would look like? Defend your answer.
28. Choose an ecosystem near your house—a park, an abandoned lot, a corn field, a pond, or any other familiar place—and diagram the main elements of that ecosystem's food web.
29. Explain the climate where you live in terms of the angle at which the sun's rays strike Earth, the tilt of Earth's axis, the coils of air and the prevailing winds they help create, and the prevailing ocean currents nearest your community.
30. Which biome surrounds the place you live? Describe some adaptations of the local plants and animals that allow them to survive in your biome. How have humans altered the biome you live in?
31. Florida panthers in the Everglades and cougars in many other parts of the United States are the top carnivores. On the basis of pyramids of energy or biomass, explain why no larger beasts prey exclusively on panthers or cougars.

Lost Cousins Found

▲
Living Links. Frans de Waal; chimpanzees; bonobos.

• **K**akowet, a bonobo at the San Diego Zoo, started screaming and waving his arms at his keepers one day. The zoo workers had drained and cleaned the moat in the bonobo enclosure and were just leaving to turn on the water valve—unaware that several young primates had jumped into the dry ditch and would soon be in danger of drowning. Based on Kakowet's obvious insight and warning signal, the youngsters were saved.

• Laura, an infant bonobo in the San Diego Zoo nursery, played a trick one morning. After being told to "clean her plate," she did so quickly and stealthily. Only later did Laura's keeper discover that the baby had complied by hiding the offending food in her diaper.

• Loretta and Linda, two adult females, came across some sugarcane stalks that each wanted.

They turned to each other, rubbed genitals, then shared the food peacefully, all the while ignoring an onlooking male. Only after the pair left did he get the leftover sugarcane.

Dr. Frans de Waal has spent most of his research career watching bonobos do surprising and revealing things like these. Through his studies and elegant descriptions of the "forgotten" apes, de Waal is changing the way we humans see ourselves. The Dutch-born animal behaviorist is one of the world's top experts on bonobos, chimpanzees, and other great apes. He heads a group of 20 researchers at Yerkes Regional Primate Research Center in Atlanta, the nation's oldest and largest facility for studying apes and monkeys. De Waal's unit is the Living Links Center, and his team compares the brains, genes, and social lives of apes with our own.

Bonobos were among the last of the large mammals described by biologists and have been both misunderstood and misnamed. Until 1929, primate experts thought these rainforest dwellers were "pygmy chimps"—simply a smaller variety of the chimpanzee *(Pan troglodytes)*, which means roughly "cave-dwelling recluses of the forest." That year, though, anatomists described distinct physical differences between the two primates and assigned the "newcomer" its own Latin name, *Pan paniscus* ("diminutive forest spirit"). The bonobo's smaller head and longer legs give it a graceful appearance. Says de Waal with obvious predilection, "Bonobos have more style!" Biologists know that the chimp and bonobo share 98.5 percent of their DNA sequence with each other as well as with us, their human cousins. The two primate species are, by far, our closest living relatives and we are theirs. But de Waal considers bonobos to be "living links" that resemble the extinct predecessor of all three species—chimps, bonobos, and people—more closely than do modern chimps.

The public knows considerably more about chimpanzees than bonobos because chimps have long starred in circuses and zoos. Animal behaviorists know more about chimps, too, because researchers like Jane Goodall have spent nearly 40 years observing their natural behavior in the wild. She and her successors have drawn a detailed portrait of highly intelligent, highly emotional animals that create and use tools, engage in cooperative hunting, and communicate through a complex collection of sounds, postures, and facial expressions. This eyewitness observation has replaced an antiquated, misinformed view of chimpanzees as "peaceful vegetarians." Today's portrait, instead, is of fierce and powerful apes engaged in lethal combat, infanticide, and predation on other animals.

Chimpanzees, writes Frans de Waal, are the "Machiavellis of the primate world." Their communication is laced with themes of threat, aggression, political alliances, dominance, and submission. Their societies are ruled by large, aggressive males who ally themselves with other males and protect females within their territories but have limited daily interaction with the opposite sex. De Waal likens an angry male chimpanzee to an "unstoppable steam-engine" who raises his pelt, uproots small trees, and threatens all onlookers with a beating.

Contrast this with the bonobo. Dr. de Waal has carefully documented these primates' natural social behavior and finds their societies to be dominated by females. These females bond to each other, leaving subordinate males to inherit status in the community from their mothers. Bonobos are much closer to the egalitarian, peace-loving vegetarians people once thought chimps to be, and they exhibit a striking sexuality. They are ruled by a "Make Love Not War" ethic, says de Waal. Bonobo sex is "casual, almost more affectionate than erotic," de Waal writes, and occurs in every flavor: male-female, male-male, female-female, male-juvenile, female-juvenile. Sex is used to lessen conflict and resolve

power issues. So are "squeal duels." These back-and-forth vocal exchanges among bonobos are remarkably language-like and they de-escalate tensions and lead to peaceful resolutions instead of fighting, biting, and killing.

"The special relevance of bonobos and chimps," says de Waal, "is their close relation to us" and what it may reveal about our own evolution. The bonobo, in particular, has shown people that "there is a lot more flexibility within our direct lineage than you might assume." If bonobos didn't exist, "you'd get a lot of emphasis on male bonding and male dominance and hunting and warfare. Which all match very nicely with the chimpanzee," he points out. "But the bonobo is another species that is equally close to us but that does quite different things."

These fascinating observations about primates help us to see ourselves in fresh ways and are part of the study of **animal behavior,** the responses animals make to stimuli from the environment. For virtually every animal species, biologists wonder how particular behaviors aid survival, how they develop, how they evolved, and whether they are guided by genetics, learning, or a combination of the two. Some behaviors such as primate facial expressions, for example, are inherited and relatively fixed throughout life. Others, such as hunting or tool use, are more influenced by learning.

In this chapter, we'll explore various issues in animal behavior, as well as much more about the evolutionary triangle formed by chimps, bonobos, and people. We'll also encounter mother geese, dancing bees, and homing butterflies. We'll talk more about our own social and sexual behavior. And you'll discover the answers to these questions as you read along:

1. Do genes control animal behavior?
2. How can an animal's experience influence its behavior?
3. What is the role of natural selection in shaping behavior?
4. How does social behavior evolve?

26.1 The Genetic Bases of Behavior

Bonobos display a staggering range of behaviors. Some traits are carried out in virtually the same way by every individual in the species—swallowing ginger leaves, for example. Others are highly variable and highly individual—for instance, bonobo communication. Some traits are strongly determined by genes and change only over many generations, while others are modified in a single individual's lifetime as things happen to it in its environment. Most behavior, however, falls somewhere along a continuum from purely gene-based to purely learned. Let's look at some examples along this spectrum.

A Predetermined Trait: Eggshell Removal

Some of the first controlled experiments in animal behavior were the work of Dutch biologist Nikolaas Tinbergen in the 1940s and 1950s. "When studying a new species," observes Frans De Waal, "the first thing one needs to do is draw up . . . a systematic description of its behavior patterns." And that's what Tinbergen did with animals such as the black-headed gull of northern Europe. Each spring, males of this species select a position on the coastal sand dunes and begin stretching forward and calling out raucously in a stereotyped ritual. If a newly arrived male tries to claim another male's territory, the occupant usually rushes the invader, strikes him, and drives him away.

If a female approaches, the males usually react quite differently, initiating a stylized mating dance in which the pair stretch forward, strut side by side, and then, as if by prearranged choreography, turn their faces away from each other. Next the female taps on her suitor's bill with her own, and he regurgitates fish into her mouth. After dancing and dining, the pair may mate, select a nesting place, then noisily defend it from all comers. The female deposits khaki-colored eggs and the parents take turns warming and turning the eggs in the nest. After each chick hatches, the parent on duty carts off the pieces of broken eggshell and dumps them far from the nest (Fig. 26.1).

Tinbergen wondered: What makes the gulls remove broken eggshells? To find out, he placed smooth rocks, flashlight batteries, corks, lightbulbs and other rounded objects into gull nests and watched as the birds incubated them alongside the eggs. When he tried to put in bottle caps, toy soldiers, sea shells, and paper squares, however, the gulls ejected them. Tinbergen concluded that when the birds saw or felt an object with a jagged edge, they removed it from the nest. Since genes help specify how a bird's nervous system sees and feels objects, this shell-removing trait must be largely determined by genes. These results explain the immediate, or proximate, cause of the stimulus for eggshell removal. But why should natural selection have favored such genes? The answer is that the insides of eggshells are very shiny, despite their drab, camouflaged exterior. The shiny broken eggshells reflect light that is easy for hawks and other predators to spot, and attracts them to the area and a tasty meal of gull chick. Alleles of gull genes that brought about the removal behavior would thus be favored by natural selection.

Here's another example of genetic differences resulting in different behaviors. In the early 1900s, Robert Yerkes, a pioneer in ape research, carefully observed two zoo animals, Chim and Panzee. Neither Yerkes nor his contemporaries knew it at the time, but like the animals in Figure 26.2, Chim was a bonobo and Panzee was a chimpanzee. Yerkes noted substantial differences in their behaviors, despite their common environment, including the sensitivity of each ape to others and the intensity of their aggression. As these two species arose from a common ancestor millions of years ago, a widely differing set of behaviors best suited each to its environment and survival needs. Today, the two distinct species express their behavioral differences even when raised under identical circumstances. This shows how strongly genetics can influence behavior even in complex, intelligent species.

Figure 26.1 **Black-Headed Gulls Remove Broken Eggshells.** When a chick hatches, a parent grasps the broken eggshells and removes them from the nest. What advantage do they gain by this activity?

Behavior and Natural Selection

We've seen how traits like the number of toes on a horse evolve under the pressure of natural selection (Chapter 9). But how do behaviors evolve? Clearly, nat-

Figure 26.2   **Primate Cousins: Bonobos Raised with Chimpanzees.** Robert Yerkes observed a bonobo (left) and a chimpanzee (right) in the 1920s and first commented on the behavioral differences between them.

ural selection must choose among genes controlling alternative behaviors in which one leads to greater reproductive success than others. Honeybees provide a good example of clear-cut alternatives based on identifiable genes.

Researchers have noticed that some honeybees eject diseased bee pupae from their honeycombs. These "hygienic" bees uncap chambers containing the diseased young and then throw them out of the nest. Other strains of honeybees are "nonhygienic": They do not uncap infected chambers and they don't throw diseased pupae out, even though the disease can spread rapidly and devastate the hive.

To see how genes are involved in this "hygienic" behavior, researchers crossed the two kinds of bees and found that a single gene controls capping/uncapping, while a different gene controls removal/nonremoval. This work demonstrates that individual genes can govern specific acts of behavior and also that more than one gene may be needed to control each element of a complex behavioral pathway. This example is still among the best available evidence that individual genes can govern specific acts of behavior appropriate within an animal's environment.

26.2   Experience Can Influence Behavior

Bonobos are wonderfully in touch with the world that surrounds them, responding to their own experiences and to the actions of their relatives and neighbors in sometimes novel ways. Laura's quick and creative response—hiding unwanted food in her diaper—is an example of this novelty. But bonobos are among the most intelligent animals on Earth. What about the rest? Experiments by animal behaviorists show that across the wide spectrum of animal species, some acts change very little with experience, while others are more strongly influenced by the environment. Let's look at points along this spectrum.

Fixed Action Pattern

Some behaviors unfold in a fixed series of actions. A good example is egg rolling in the greylag goose. When a greylag goose is sitting on her nest and an egg rolls out of its place in the shallow depression on the ground, she reacts in a particular way: She waddles up to the egg, stretches out her graceful neck, and gently wraps her chin and neck around the egg (Fig. 26.3). Like a hockey player advancing a puck, she pushes the egg along by moving her head from side to side to keep the lopsided object rolling straight. Biologists know that experience plays a minor role in this behavior because even geese raised in isolation retrieve eggs this way the first time one rolls from the nest. If an experimenter removes the errant egg once the goose has begun rolling it, remarkably,

Figure 26.3   **Egg Rolling in the Greylag Goose.** When an egg rolls from the nest of a greylag goose, the female retrieves it by arching her neck gracefully around the stray and rolling it back with both sides of her beak. She will retrieve anything egglike near the nest, including baseballs and tin cans.

she carries on like a feathered mime, rolling and nesting a nonexistent egg. A series of stereotyped physical movements of this sort is called a **fixed action pattern.**

Fixed action patterns can be triggered by sounds, smells, tastes, or other specific environmental signals. Seeing a round object near the nest causes a greylag goose to begin the egg-rolling behavior; she will retrieve a tennis ball as tenderly as her own egg! For the black-headed gull, feeling and seeing an object in the nest with a jagged edge can trigger the eggshell removal we saw earlier.

Fixed action patterns usually proceed from start to finish in an all-or-none way and in a particular sequence. Yawning is a classic example (Fig. 26.4) and so

Figure 26.4   **Bonobo Yawn: A Fixed Action Pattern.** A yawn in a bonobo or a human is a fixed action pattern that is similar no matter who is yawning. That includes this lounging chimpanzee at Gombe Stream, Tanzania, the research station Jane Goodall made famous. Yawns last about six seconds and they are difficult to stop midway through. In addition, they act as a releaser; a yawning person spontaneously induces onlookers to yawn.

is the swallowing reflex. Have you ever tried to stop swallowing a mouthful of food midway down? It can't be done; more than a dozen muscles fire in coordinated sequence, each step unalterably triggered by the step before it. But an animal's internal state and maturation can modify these behavioral patterns. For example, a female goose will display egg-rolling behavior only after she lays her eggs and before the eggs hatch. At other times, a round object outside the nest won't trigger the response.

Behavioral Acts with Limited Flexibility

Some behavior patterns respond to cues and unfold in a set sequence, but can still be modified by experience. For example, when a seagull returns to its nest from a feeding foray, its chicks peck at the parent's wagging bill. In the herring gull, adults have a distinctive red spot on their bill and chicks peck at that spot (Fig. 26.5a). When chicks are very young they'll peck at anything resembling the parent's red bill spot. They'll actually peck more excitedly at a swinging stick with red spots painted on it than at a live gull's bill (Fig. 26.5b)! As the chicks mature, however, they beg only from their own parents: They've acquired information from the environment about their parents' physical characteristics, and this modifies the initial action pattern. Animal behaviorists lump most examples of such flexible behavior into two categories: imprinting and learning.

Imprinting: Learning with a Time Limit

In the 1930s, Austrian biologist Konrad Lorenz was studying ducks and geese, and discovered one of the most fascinating phenomena in all of animal behavior. At the time, Lorenz knew that goslings normally follow their mother from the nest to a nearby pond or lake within a day after hatching. The researcher tried walking away from a group of newly hatched goslings in the nest while he made gooselike honking sounds. To Lorenz's delight, the gaggle of goslings followed him around as readily as they would have their real mother (Fig. 26.6). He speculated that to survive, newly hatched greylag geese must quickly learn to recognize and follow their parents, and they are apparently programmed to trail after the first large moving object they see, even if it is a gray-bearded, pipe-puffing biologist!

Lorenz called this activity **imprinting:** the recognition, response, and attachment of a young animal to a particular adult or object. Imprinting can occur only

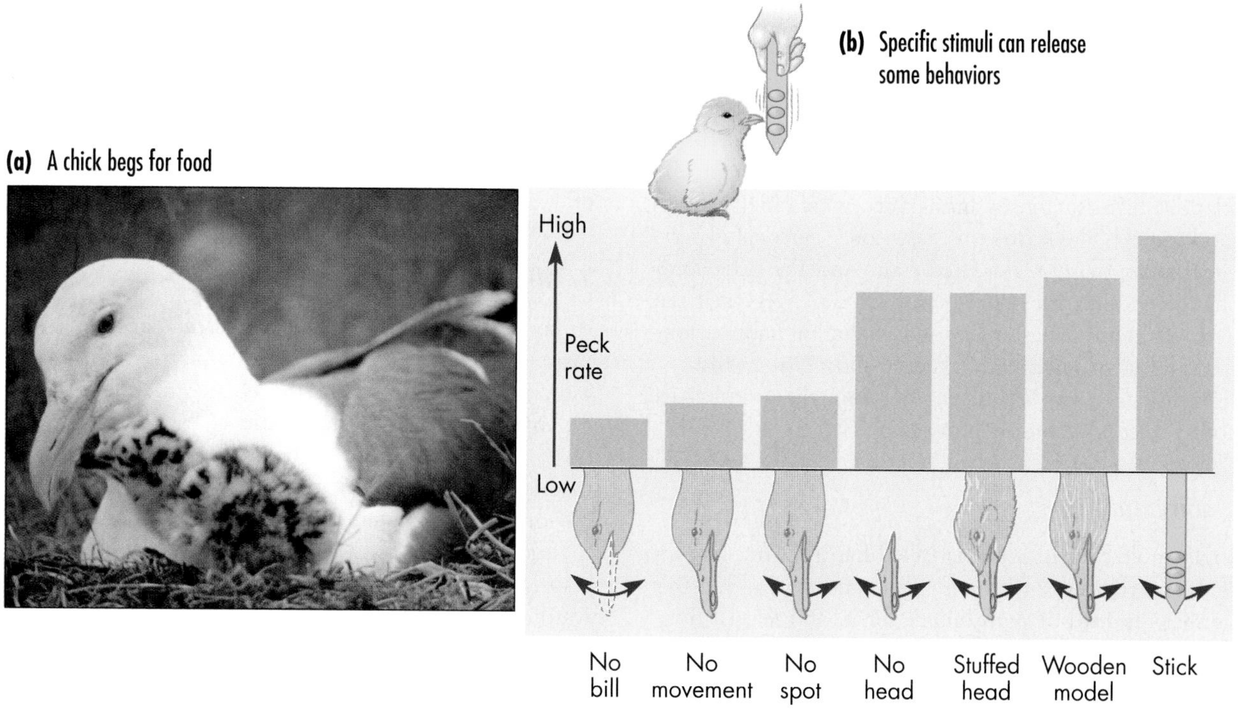

(a) A chick begs for food

(b) Specific stimuli can release some behaviors

High

Peck rate

Low

No bill | No movement | No spot | No head | Stuffed head | Wooden model | Stick

Figure 26.5 **The Red Spot and Chick Pecking.** (a) A hungry gull chick pecks at the red spot on its parent's bill, an action that triggers a regurgitated meal. (b) To identify what provokes pecking behavior, experimenters offered young chicks various artificial objects to peck at and watched for the rate of pecking in response. The results showed that movement plus a red spot on a pointed object is most effective at causing the response.

within a specific developmental stage, a **sensitive period.** In geese, the sensitive period for following lasts from the time of hatching to the second day. Hand-reared ring doves will sexually imprint on humans and as adults will actually prefer to court humans over other ring doves, but this sensitive period for sexual imprinting is later than the one for following a human (or a mother goose). The flexibility comes in because different individuals can become imprinted on different objects and because imprinting can sometimes be reversed.

Biologists suspect that imprinting can take place in baby birds because development "primes" their nervous systems to adapt to their particular circumstances shortly after hatching. Imprinting is thus a form of learned behavior, but it's limited to a narrow time period. People have narrow windows or **sensitive periods** (also called critical periods) for certain kinds of learning, including the grammar and pronunciation of whichever language(s) they acquire in childhood. This is why many people advocate foreign language instruction starting in elementary school rather than middle school or high school. Most kinds of learning, however, are not time-limited, and behavioral patterns can be modified during an individual's entire lifetime.

Learning: Behavior that Changes with Experience

Bonobos and other great apes—ourselves included—are masters at learning. For example, all the individual chimpanzees in one particular community use stones to crack open nuts. Observers have watched as youngsters in this group picked up this skill from their elders. In another community, however, chimpanzees do not crack open nuts with rocks, even though both nuts

Figure 26.6 **Imprinting: Hatchlings Bond to Biologist "Mom."** Imprinting has an inherent time limit; geese form parental attachments in the first two days after hatching. The attachments, however, are long-lived. Here, imprinted goslings trail after Konrad Lorenz.

and rocks are readily available. In fact, the animals seem unaware that nuts can even be used for food.

Biologists define **learning** as an adaptive and enduring change in an individual's behavior based on personal experiences in the environment. In this chimpanzee example, learning how to crack open nuts is adaptive because it gives animals access to a high-energy food source. The learning is based on a personal experience rather than an inherited trait. And the behavioral change endures for some period of time. As we'll see here, the many types of learning, including habituation, trial-and-error learning, classical conditioning, and insight learning, can all increase an animal's fitness—its ability to survive and reproduce.

Habituation

In a simple kind of learning called **habituation,** an animal experiences something again and again, receives neither reward nor punishment for it, and learns to ignore it. For example, bonobos living in the wild near Wamba and other study sites in the African country of Zaire become habituated to their human observers and eventually ignore them. As another example, young gull chicks crouch down when objects pass overhead (Fig. 26.7a), but this action wastes time and energy if the flying object is a harmless robin or fluttering leaf. It is adaptive, therefore, for chicks to habituate—to ignore silhouettes of common harmless species (Fig. 26.7b), but to continue to cower at predatory birds (Fig. 26.7c). Because these predators seldom pass overhead, the chicks never habituate to them and this is also clearly adaptive. Explorer 26.1 illustrates this and a number of other types of learning.

Trial-and-Error Learning

At some of the bonobo study areas in Zaire, biologists planted sugarcane for their study subjects. The animals associate the station with the chance to pick and eat sugarcane, and when they come to eat, biologists can observe them (Fig. 26.8). The kind of learning involved in this exchange is more complex than habituation. Behaviorists call it trial-and-error learning or **operant conditioning.** It is a form of learning in which an animal associates an action (an operant) with a reinforcer, a reward or punishment.

Renowned biologist B.F. Skinner studied operant conditioning in the laboratory and became convinced that virtually any action can be "conditioned," or trained, with the right kind of reinforcement. More recent experiments, however, show that the stimuli and responses must have meaning for an animal in nature. Rats rapidly learn to press a lever with their front paws to obtain a food pellet, but it is much harder for a rat to learn to push a lever to avoid an electric shock to their feet. In contrast, rats quickly learn to jump off the ground to escape a shock to the feet, but it's harder for them to learn to jump to earn a food pellet. These re-

GET THE PICTURE

Explorer 26.1

What Kind of Learning Is It?

Early:
Any object overhead

Later:
Common objects overhead

Rare objects overhead

Chick cowers

Chick habituates

Chick continues to cower

Figure 26.7 **Habituation in Gull Chicks.** (a) At first, a gull chick will cower no matter what object passes overhead. (b) Eventually, it habituates to the passage of common harmless objects like robins and leaves, but (c) continues to cower at rare ones like hawks, whether or not they are dangerous predators.

Figure 26.8 **Trial- and-Error Learning.** Bonobos have learned to associate the biology station at Wamba with the sugarcane that researchers planted for them. They come periodically to eat the sweet stalks, and this allows researchers to observe their behavior more closely.

sults may sound strange, but they make sense if you consider that rats obtain food by manipulating items with their paws, but avoid a harmful stimulus by jumping away from it. Animals are clearly prepared to learn some things more easily than others, and in ways that appear to increase survival and reproduction.

Classical Conditioning

One of history's most famous animal behavior experiments was the work of Russian biologist Ivan Pavlov, published in 1903. Pavlov repeatedly presented a dog with meat, a natural stimulus for salivation; but just before presenting the meat, Pavlov rang a bell—an arbitrary, unnatural stimulus. The dog learned to associate the natural stimulus (the meat) and the substitute (the bell), and began to salivate at the sound of the bell. In the operant conditioning we just talked about with rats and food pellets, the animal responds before receiving the reward or punishment (the reinforcer). But in the learning system Pavlov explored, **classical conditioning,** or associative learning, the animal responds *after* the reinforcement, and with more experience, the response shifts from one stimulus to another. In nature, a predatory hawk can learn to associate the presence of a broken eggshell near a gull's nest with the presence of a defenseless, tender gull chick and act to collect a reward: dinner. Advertising agencies manipulate our own species' susceptibility to classical conditioning by associating commercial products with positive images: fancy cars with financial success; quarter-pound hamburgers with happy family outings; or a toothpaste brand with successfully attracting a mate.

An animal's genes program it to learn different things in different ways that fit the animal to its environment. Many animals learn survival skills through both operant and classical conditioning. One final type of learning, however, is common only in bonobos and other primates and it has reached its peak in humans: reasoning.

Insight Learning

Picture a chimpanzee that enters a room where she sees several boxes and a long stick littering the floor and then notices bananas hanging from a hook on the ceiling. She stares at the distant bananas for some time, then, in a flash of insight—a so-called "Aha! Experience"—she stacks the boxes, climbs up, and then whacks down the bananas with the stick (Fig. 26.9 and *E* 26.1). The chimp has just displayed **insight learning,** or reasoning: formulating a course of action by understanding the re-

Figure 26.9 **Insight Learning and Problem Solving.** A (chimpanzee or bonobo) can plan a course of action to solve a problem, such as piling up boxes, climbing upon them, and reaching a suspended bunch of bananas. Here, a young female chimp has learned to choose a stick and use it as a tool to withdraw insects from a nest.

lationships between the parts of a problem. As you read, you are acquiring knowledge this way, as well. Insight learning allows an intelligent animal to encounter a new situation (that includes a new printed sentence) and, based on similar past experiences, figure out how to respond without actually trying out all possible solutions.

People often assume that their household pets are capable of reasoning, but in fact, dogs and cats are quite inept at solving even simple insight problems. For example, if a dog wraps its leash around a tree so that it can no longer reach its food and water, the dog can't conceptualize the cause of the problem and address it. You could train your dog to keep its leash untangled through operant conditioning. But a primate would conceive of the problem beforehand and simply avoid it.

Biologists suspect that animals have evolved particular forms of learning that provide the best chance of surviving and propagating in their own environments. Because there are so many animal species, so many environments, and so many selection pressures, nature displays an immense array of behaviors, as we'll see next.

26.3 Natural Selection and Behavior

At the start of the chapter, we saw some complex bonobo behaviors: problem solving, conveying information, deception, food sharing, and sexuality. But the bonobo's repertoire is far greater than this and includes several categories that are fundamental to animal survival: locating and defending a home territory, feeding and avoiding being eaten, reproducing, and communicating. In bonobos, as in thousands of other animal species, natural selection has shaped these behaviors, and they help determine the organism's chances of living day to day and passing along its genes.

Locating and Defending a Home Territory

Bonobo females cooperate to keep males away from limited food sources. Beavers locate the best site along a stream to build a lodge. Other animals travel huge distances to suitable territories. It's all in the interest of establishing and keeping a home territory and, with it, insuring access to desirable mates and food.

Migration and Homing   Many animals undergo migrations: They travel hundreds of kilometers each year in search of an appropriate habitat. Whales, sea turtles, eels, monarch butterflies, and caribou in the Arctic travel huge distances to spend winter and/or breed (Fig. 26.10). Salmon return to breed in the streams where they hatched. And many birds are champion voyagers. Black-headed gulls trek from Africa to northern Europe and back each year, and the small arctic tern flies 16,000 km (about 10,000 mi) round trip from Greenland and Alaska to Antarctica searching for endless summer.

Migrations like these are seasonal. On the other hand, **homing behavior** can take place any time certain kinds of animals are displaced from their home territories. You've probably heard of biologists trapping and relocating problem bears to get them away from campgrounds, picnic areas, and garbage dumps. This is usually a temporary solution, though, because a bear can travel 25 miles a day or more to get back to its home turf. This is homing.

When experimenters in the 1980s transferred a long-winged seabird called a shearwater from its home in Great Britain to a new location in Massachusetts, the remarkable bird was back on its nest in 12 days, having crossed about 4800 km (3000 mi) of open ocean! How do bears, birds, and other animals find their way, whether homing or migrating? Biologists are not entirely sure, but tests show that some animals can **orient,** or move in a set direction, and **navigate,** or move from one map point to another.

In daytime, birds, ants, and bees orient via the sun's position as it sweeps across the sky. For example, a pigeon released west of its birdhouse in the morning will fly toward the sun to get home, while a pigeon released west of its birdhouse in the evening will fly away from the sun to get home. When experimenters shifted the birds' sense of time by altering their exposure to light and dark in a closed room, the birds looked at the sun and flew in the wrong direction. This suggests that pigeons have an internal 24-hour clock (a **circadian clock**) that helps them compute orientation relative to the sun's path.

This is only part of the story, though, because homing pigeons can find their way home even in the dark or when researchers have fitted them with tiny, thickly frosted contact lenses that reduce their vision to a murky haze. Remarkably, birds wearing these lenses can still orient and navigate by detecting Earth's magnetic field. Biologists have discovered a few tiny crystals of magnetic iron oxide in tissues near the brains of birds, porpoises, and certain other animals. They suspect that these crystals tend to align with the planet's magnetic field, like the iron needle in a compass, and cause specific neurons to fire. The researchers concluded that the sun is a pigeon's primary cue for orientation,

Figure 26.10   **Caribou Migration in Arctic Alaska.** These caribou are swimming across the Kobuk river in Northern Alaska during their annual migration.

but that Earth's magnetic field acts as a substitute cue in the dark.

Migration is dangerous and exhausting and exacts a high toll, but it has benefits, too, including greener pastures for fattening up and protected regions with fewer predators for breeding or birthing. Once animals reach their destination, they often claim a territory and defend it, even against former traveling companions.

Territoriality Bonobos are peace-loving and use sex and vocal communication to avoid violence. They do not defend their feeding and living territories with the stylized posturings or brutal ferocity of chimpanzees. The males of their fiercer primate cousins, the chimps, will often display by screaming, uprooting trees, and throwing rocks. Sometimes they even fight to the death to defend living space—a behavior called **territoriality.**

A good example of animals competing for territories rich in food resources are the sleek male impalas of the African savanna (Fig. 26.11). The richer the territory an impala controls, the more females it entices, and the greater its chances of leaving progeny with its genes. Male impalas stake out the richest territories they can find, and when a competing male arrives, they clash, interlocking their graceful horns and thrusting furiously, trying to force each other out. Females choose to stay in each territory only so long as the food holds out, and during their stay, the resident male mates with all the receptive females. A male without rich territory to defend will have fewer contacts with females and thus leave far fewer offspring.

Figure 26.11 **Impala Males Defend Rich Territories.** Male impalas fight to defend their territories against other males. Here, male impalas practice sparring for harem leadership in Botswana's Okalango Delta. The best territories feed the most females and males are able to mate more frequently and leave more offspring.

Just as with migrating, defending a territory exacts a great cost: Constant battle readiness requires time and energy, and exhausting tournaments and injuries leave combatants vulnerable to predators. The rewards in terms of extra progeny produced, however, are tremendous. For its part, the loser will often give up fairly quickly; saving its skin may allow it to claim another piece of territory later.

Feeding Behavior

Is a bonobo better off traveling several kilometers to collect fruit from a particularly bountiful tree, or walking a shorter distance and expending less energy to gather food from a tree with less abundant or less nutritious fruit? The **optimality hypothesis** suggests that animals will automatically behave in ways that allow them to obtain the most food energy with the least effort and the least risk of falling prey to a predator.

Biologists studied crows, for example, as the birds preyed on whelks—large, snail-like mollusks. Crows search the intertidal zone and select a big whelk over several smaller ones, grasp it, fly it to a height of about 16 feet (5 m), then drop it onto the rocks below. If the shell shatters, the crow consumes the tender flesh; if it doesn't break, the crow lifts and drops it again. Based on the optimality hypothesis, researchers predicted (and tested for themselves) that large whelks would smash more easily than small ones; that a drop of 16 feet should be just enough to break most whelk shells; and that dropping an uncracked whelk a second time was as good a strategy as searching for a new whelk. Experiments verified each of these predictions. Crows don't get ahead by dropping small, light, unbreakable whelks from the wrong height. With optimal feeding behavior, less time and energy are spent on gathering food, leaving more time for producing offspring—and sending more genes into the next generation.

Reproductive Behavior

For nearly all animals, sexual behavior is exclusively associated with reproduction. We humans are exceptions, however, and so are bonobos, as we've seen. In many species, elaborate behaviors evolved around finding and choosing a mate, and producing and caring for offspring. Often, males and females go about these tasks in very different ways. Males tend to make large numbers of highly mobile, lightweight gametes (sperm) and behave in ways that increase the number of eggs they fertilize. In contrast, female animals tend to make fewer,

What might be the human equivalent of a bonobo's food optimization?

but larger gametes (eggs). A woman, for example, produces a few hundred eggs in her lifetime, while a man, in a similar period, produces enough sperm to inseminate all the women in the world. It's no surprise, then, that reproductive behavior should vary so much between the sexes.

Parental Investment

The small size of male gametes and the large size of female gametes influence how much time and energy each sex invests in parenting and in selecting a mate. Each large egg represents more resources than each small sperm; thus, in a physical sense, the female sacrifices opportunities to make more eggs in favor of increasing the chance that each egg will become a healthy offspring. Bonobos and other female mammals shelter and nourish their offspring in the uterus and feed them with milk long after birth. Bonobo females continue to support their sons by food sharing and helping to defend them for their entire lives (Fig. 26.12). These actions cost more in time and energy and exposure to predators. Yet on balance, the extra care usually pays off in terms of offspring with a greater chance of surviving and contributing genes to the next generation.

Figure 26.12 **Extensive Parental Investment.** This female bonobo is nursing her 2-year-old youngster and will continue to nurse him until he is 4. Males follow their mothers for life.

Sexual Selection in Members of the Same Sex

Where females make a greater parental investment than males, the males generally compete to inseminate as many females as possible. This can lead to a **dominance hierarchy**—a ranking of group members based on their past success in fights over food or mates. The results of those fights can lead to **sexual selection:** natural selection operating through mate choice or competition for mates. Bull elephant seals, for example, are massive beasts that threaten each other with tremendous roars, and bite and batter competitors. The loser of a skirmish lowers his head and retreats, with the victor in pursuit. A few years ago, a biologist ranked ten bulls living on a small island by their track records in winning such encounters, and then counted the successful copulations each achieved during the breeding season. The top-ranking male accomplished nearly 40 percent of all the copulations in the group, and all other males far fewer (Fig. 26.13). Natural selection certainly favored this dominant bull, as his genes most likely showed up in a high per-

Figure 26.13 **Mating and the Elephant Seal Dominance Hierarchy.** A casual advance by an amorous male—a flipper caress—sets the female Northern elephant seal (*Mirounga an gustirostris*) to screaming at high decibels. If the male suitor is subordinate, the king bull will probably come quickly and chase him off. Her call thus increases the likelihood that her sons will inherit alleles from the dominant male.

centage of the colony's offspring. A dominance hierarchy is an instance of intrasexual selection (selection among members of the same sex). Another example is the presence of the antlers and horns that male impalas and deer use for fighting. Individuals with larger antlers that can wield them successfully in contests are more likely to attract more mates and pass their alleles to future generations.

Male chimpanzees compete strongly for females and establish well-defined dominance hierarchies, as well. A dominant chimpanzee is a fearsome sight, screeching, baring his long canine teeth, rising on two legs, uprooting small trees, and smashing the ground with his hands, rocks, or tree limbs. Faced with a display like this, a subordinate male chimp will grovel on the ground and grunt (Fig. 26.14). And this response is well-justified: Two males fatally injured and castrated another male in a dominance dispute at the Arnhem Zoo in Holland. Dominant chimpanzees, like bull elephant seals, have far more opportunities to reproduce than do males low on the pecking order.

Sexual Selection and the Opposite Sex

Females generally invest more time and energy in each individual offspring than do males and tend to produce fewer offspring. As a result, the evolutionary penalty is higher for females who choose inappropriate mates. "Appropriate" generally means a male that can enhance the female's offspring through superior genes, protec-

Figure 26.14 **Dominance and Male Chimps.** The chimpanzee on the left clearly signals his dominant status by his upright stance and erect hair, creating an illusion of bulk. The subordinate chimpanzee on the right bows and pant-grunts in submission.

Figure 26.15 **The Widowbird's Courtship Flight.** The jet black male widowbird sports an improbably long tail that flutters as he performs a mesmerizing courtship flight. Experimentally shortened or lengthened tail feathers respectively decrease or increase his chances of attracting a mate.

tion, and a good territory with plenty of food. A female thus engages in sexual selection by choosing a desirable mate; she acts as an agent of natural selection, in other words, on the male's behavioral and physical traits.

By choosing dominant or propertied males, females increase their own biological fitness. Elephant seal cows, for example, create sexual selective pressure by simply screaming loudly whenever a bull attempts to copulate (see Fig. 26.13). Her cries alert the harem master, and if the courting male is low on the dominance hierarchy, the dominant bull sprints to the site of the tryst and chases away the intruder. The female's automatic cries greatly increase her chances of being inseminated by the dominant male rather than a subordinate and hence of having sons who will display the aggressive behavior that translates into high social rank among the elephant seals.

Males sometimes put on elaborate courtship displays, and the female's selectivity in choosing a mate probably led to the evolution of these rituals. For example, the female ptarmigan, a grouselike bird, gets information about the male's physiological condition when he does his strutting dance. Likewise, the female African widowbird gathers data as the male flaunts his remarkable 2-foot (60-cm) long tail in a showy courtship flight (Fig. 26.15). Experimenters have found that when they cut the tail feathers off a male widowbird, he is less likely to attract a mate, but when they glue on long feathers from another bird, making an extra-long tail, the bird can be quite irresistible to females. Experiments

like these suggest that even a potentially harmful trait, such as a long tail that gets in the way of flying ability—and escaping predators—may still be selected for if it gives the male extra reproductive success. How might this principle be applied to the fish-regurgitation ritual of the black-headed gull?

Communication

Birds call back and forth through the trees. Bees do buzzing dances inside the hive. Bonobos vocalize, grimace, and gesticulate. Social animals like these are continuously exchanging information. But even the most solitary animal has to find food and attract mates, and this often involves **communication,** a signal produced by one individual that alters a recipient's behavior and generally benefits both participants.

Communication by Sight and Sound

Animals live in all kinds of habitats—light, dark, open, densely wooded, and so on—and their ways of communicating must fit their ecological circumstances. The ptarmigan's courtship dances are visual displays that work well in the open fields where the birds live, but would fail in a forest full of underbrush. The widowbird's courtship flight is another example of visual communication. But what happens when an animal's line of sight is obstructed? Dense pond vegetation protects a bullfrog from the eyes of hungry predators but also hides the frog from potential mates. For frogs, booming calls and other auditory cues have evolved as the major form of communication. The trill of crickets on a summer night is the same—undercover auditory communication. Alarm calls are sound communication, too: Sea gulls, bonobos, and prairie dogs give off sharp warning signals to colony members who may not see a predator approach, and this benefits the entire colony, many members of which share genes with the caller.

Chemical Communication

For solitary animals unlikely to see or hear signals from other members of their species, odor molecules can convey information. A female moth gives off sex pheromones (see Chapter 18), chemicals that can travel for miles on the wind, announcing her location and sexual readiness. Chemical signals not only carry farther than sounds or sights, but they can persist longer in the environment, too. An experimenter can collect ant pheromones and paint an artificial trail with them. Ants are so programmed to pursue this chemical communication that they'll slavishly follow the pheromone trails in endless circles (Fig. 26.16). The persistence of chemical cues in the environment also helps explain why dogs urinate on fence posts and fire hydrants: the "message" remains long after the leaver exits the scene.

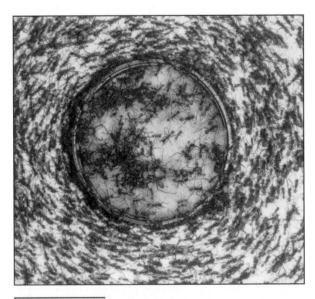

Figure 26.16   **Chemical Communication: Ants Following a Pheromone Trail.** When an experimenter placed a round dish amid a colony of army ants, a few explorers laid down a circular pheromone trail around it, and the rest followed endlessly like miniature automatons.

Communication by Touch

Groups of animals that live in dark hives, caves, or underground chambers often communicate by touch, as well as by sound and smell. In the 1940s, biologist Karl von Frisch discovered the classic example of touch communication: bee dances performed inside dark beehives that communicate the location of rich food sources (Fig. 26.17a. If a foraging bee finds flowers heavy with pollen and nectar near the hive, the bee flies home and immediately performs a *round dance* by walking in tight circles.

(a) Dancing bee

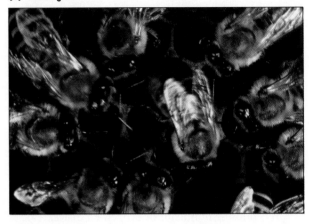

(b) Waggle dancing in the dark

(c) Dance oriented toward sunlight

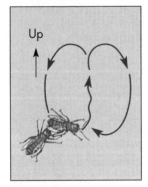
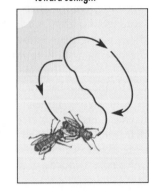

Figure 26.17   **Honeybee Dances.** Forager bees perform round dances (a) after discovering a food source close to the hive and waggle dances (b) after finding a more distant source. Hive mates gather like a front-row audience to get the information the dance conveys. The dancing bee's orientation on the hive wall signifies the direction to fly relative to the sun's position. If sunlight shines into the hive (c), the bee orients to the light. We know this because a partially blinded bee in a lighted hive dances as if it were dark, but observer bees interpret the dance according to the orientation of the light.

Observer bees follow the dancer in the dimly lit hive, sense its movements by touch, and learn that a nearby food source exists. Because worker bees can also smell bits of food clinging to the forager's body, they usually have no trouble locating the specific flowers.

If the forager discovers a food source farther from the nest, she returns with a rich load of pollen, climbs onto a vertical wall inside the hive, and begins a *waggle dance,* moving in a figure eight while wagging her abdomen from side to side and giving off an occasional burst of sound (Fig. 26.17b). An animated dance with many waggles and frequent sound bursts suggests a patch of flowers fairly close by. A lethargic dance with few waggles and little sound indicates a long or uphill trip to the food source. There's even information in the dance about map direction: The dancing bee's orientation on the hive wall reflects the angle of the necessary flight path with respect to the sun (Fig. 26.17c).

Von Frisch's translation of bee dances was brilliant, but do observer bees actually go out and find flowers on the basis of seeing and feeling a colony member's dance? An experimenter captured returning foragers and painted over their three simple eyes. This left them partially blind to light inside the hive, but left them still able to see sunlight through their two large compound eyes. The returning partially blinded forager danced as if the hive were dark, and that dance indicated a particular direction for the food. Her sisters, however, who could still see light in the hive, interpreted the dance as if the forager could see the light, too, and so they headed out in the wrong direction. These experiments proved that the sight and touch communication between bees, and not the scent of the pollen and flowers, guide the flight path of the colony bees to new flower patches.

Communication in Bonobos

The highly intelligent bonobo has a rich social life, and this depends in large part on facial expressions, gestures, vocal signals and other kinds of communication. Bonobos have wonderfully expressive faces. Through broad grins that show their teeth (Fig. 26.18), they convey fear or pleasure. During aggressive behaviors, their faces often look tense, with down-turned eyebrows and compressed lips. And bonobos even appear to pout, says Frans de Waal, who once observed Kalind, a young male, after a larger male drove him away from his favorite female. For a minute or two, Kalind stared into the distance with pursed lips—a behavior externally similar to a human child who has just dropped an ice cream cone.

Figure 26.18 **Facial Expressions and Visual Communication.** The female bonobo on the right is flashing a grin, and this probably communicates a clear message (about which we can only speculate) to the female on the left as she makes eye contract.

Behavioral biologists must be careful when interpreting animal behaviors. A bonobo's pursed lips and frown might not mean the same as a human child's. Primate grinning or smiling, for example, often indicates fear and submission rather than humor or pleasure. A question a behaviorist can confidently ask, however, is whether facial expressions actually convey information (whatever it might be) to other bonobos? A study by Sue Savage-Rumbaugh, a colleague of de Waal's at the Yerkes primate center, suggests that indeed, facial expressions do transmit information between bonobos. She was reviewing slow motion videotapes made of bonobos copulating and noticed that the speed and intensity of pelvic thrusts changed or stopped after one of the participants made particular facial expressions. For example, if partner A repeatedly broke off eye contact during face-to-face copulations or began to yawn, then partner B would often terminate thrusting.

Bonobos make characteristic hand and arm movements toward others, and these are apparently intended to alter the recipient's behavior. The simplest one is an outstretched open hand, an apparent request to another bonobo for food or contact (Fig. 26.19). De Waal notes that when they gesture, bonobos tend to use their right hands; this suggests a possible left-brain specialization for communication like our own left-brain speech localization (see Fig. 19.12).

Bonobos live in dense African forests, and vocal signals are often the only way an individual can send

Figure 26.19 **Communication by Gestures.** A bonobo youngster gestures with the right hand to invite an infant to play. Does the bonobo's right-hand dominance hint at left-brain specializations like those underlying human spoken language?

information about its needs or location to group members. Calls bring the group together at night and can drive rivals away. When playing, bonobos sometimes tickle each other in the armpit or on the belly, then both give off coarse laughs. A bonobo's vocal repertoire is much larger than a chimpanzee's, and the voice much higher. Bonobos often "comment" on happenings around them using high-pitched peeps and barks. In contrast, the more aggressive chimpanzees tend to vocalize when alarmed or fighting and make a frightful din with their hoo-hoo's and loud barks.

Language in Chimpanzees and Bonobos

For decades, researchers interested in the evolution of human language have trained apes to understand and use human words and symbols such as American sign language or pictograms. Sue Savage-Rumbaugh has trained a number of bonobos, including her biggest success, Kanzi, to read and type simple symbols for English words on a large keyboard. Savage-Rumbaugh was actually training Kanzi's adoptive mother when the youngster would voluntarily sit and watch. Amazingly, Kanzi picked up a large vocabulary without any rewards. This research and other ape language studies have shown that chimpanzees and bonobos can learn to understand speech and graphic symbols as long as they are exposed to the communication system by six months of age. After two years of age, chimpanzees can learn symbols, but can no longer learn to understand spoken speech.

Is Kanzi really understanding words and symbols or does he use subtle cues like his trainer's facial expressions or gestures? Kanzi will listen through earphones to words spoken by someone in another room then choose a corresponding picture from a pile. If he hears the word "melon," he picks up the picture of a melon; if he hears "cat," he chooses the cat photo. He can also understand word strings: If he hears "Give the dog a shot," he can pick out a toy syringe from among many objects and inject a stuffed dog. Experts are still debating the meaning of Kanzi's ability for the evolution of human language. His intelligence, however, is undisputed. What do bonobos use this brain power for in the forest? That's still a mystery. But Frans de Waal thinks that for the peace-loving bonobo, the "high points of…intellectual life are found not in cooperative hunting or strategies to achieve dominance but in conflict resolution and sensitivity to others."

Animal behaviorists wonder something else about bonobos and other social animals: Communication is most complicated in animals that live together in large social groupings. How could such complex societies have evolved?

26.4 Evolution of Social Behavior

Most animal species live solitary lives. Turtles, beetles, snails, and worms, for example, interact with others only to mate and have no long-term contact with parents or offspring. A fascinating minority of animal species, however, live in groups: giant colonies of ants, bees, and termites; large flocks of geese or communities of bonobos; small, stable wolf packs; and individuals that interact during mating season, like the graceful manatees of Florida. What biological mechanisms cause turtles to live as lone individuals, while bonobos live in societies? How do these mechanisms evolve and what advantages do they offer? And how does a colony member benefit from, say, emitting an alarm call when a predator comes near? Doesn't this put the caller himself in danger?

Social Living: Costs and Benefits

Biologists have long wondered what an individual gains by living in a group and what it sacrifices, and they've designed a number of studies to collect answers. One of the best case studies they've found is prairie dog societies. Observers have noticed that if a badger or other predator comes near a prairie dog colony, the first prairie dog to notice the intruder freezes for a moment, then gives a loud alarm call (Fig. 26.20). This dangerously betrays its own location to the predator, but it also warns family members and neighbors to dash off to the safety of their burrows. Because prairie dogs live in groups of different sizes, biologists had a way of investigating the effects of group size on behavior.

Experimenters pulled a stuffed badger at constant speed through different-sized prairie dog colonies and recorded the time it took before a prairie dog would sound the alarm. They found that in larger colonies, animals detected the predators sooner than in smaller colonies. The support of more watchful colony mates allowed individuals to spend more time foraging and less time scanning the surroundings for predators. Why, then, don't all prairie dogs live in large colonies?

The answer is that larger societies have disadvantages, too (as any urban freeway commuter can testify).

Figure 26.20 **Altruism in a Prairie Dog Warning.** A prairie dog throws back its head and gives a piercing warning sound as a predator nears. This act exposes the animal to danger but protects its kin, who carry many of its genes. This altruism improves their chances of reproducing in the future.

Prairie dogs in larger groups spend more time bickering over territory than those in smaller groups, and plague-carrying flea populations are much worse in large colonies than in smaller ones. You could predict that you'd find larger colonies in regions with more prairie dog predators. And you'd be right. Observations show that large colonies do tend to live exposed on the open plains, while smaller colonies inhabit scrubby land where individuals can hide in the bushes.

From other similar experiments and observations, biologists have concluded that animals will socialize only when the benefits of group living outweigh the costs. Occasionally, cooperating in a colony will cost an individual its feeding time or even its life. So how could selfless (altruistic) behaviors evolve if they sometimes harm the altruist?

Evolution of Altruistic Behavior

Social animals often cooperate and both helper and helped can benefit: When a lioness chases a wildebeest toward her hunting partners, for example, all share the kill. Sometimes, though, the helper suffers: When a raccoon comes to a beehive in search of honey, and a worker bee stings the masked intruder, this act protects the hive but the worker's stinger gets ripped out in the process and she dies. And when a bonobo female shares limited food with her adult son, she gets less to eat and could become more vulnerable to diseases and predators. Biologists have long wondered how animals could evolve such **altruism,** or seemingly unselfish acts for the welfare of other individuals when the selfless behavior can cost the giver its chance to reproduce. Why don't genetic alleles that promote altruistic behavior die out along with the selfless animals?

The answer involves a specific type of natural selection called **kin selection,** in which an individual helps its relatives—who share its alleles—to survive and reproduce, and in so doing, increases its own genetic contribution to the next generation.

Think for a minute about kin selection and the alleles you share with your family members. As Chapter 5 explained, you inherited half your genetic material from your mother and half from your father. Since you share the same parents, you'll also share half of your genes, on average, with each brother or sister. You share one-quarter with your aunts, uncles, and grandparents, and one-eighth with your cousins. It's not hard to see that you'd send equal numbers of your alleles into the next generation by producing two children yourself or by helping your siblings successfully raise four children.

Figure 26.21 **Males Ally in a Fight for Control.** Two adult male chimpanzees in the Arnheim Zoo, Nikkie and Yeroen, unite to confront a rival. Their erect hair makes them look bigger and fiercer, and they shout ferociously.

Geneticists discovered that in bees and wasps, females are diploid, with two copies of each chromosome, but males are haploid, with a single copy of each chromosome. As a result, the daughters of a single male share *all* their father's alleles instead of just half, as in most organisms. As a consequence, bee sisters are three-quarters alike genetically, instead of half alike as in most other animals. By helping the queen reproduce more sisters (including future queens), workers are really favoring the preservation of their own alleles.

In nature, the situation is a bit more complicated: some bees and wasps with haploid males don't live in societies, and some social insects—termites, for example—don't have haploid males. Biologists think that ecological factors also help explain why animal societies evolve, including our own species and our cousins, the bonobos and chimpanzees.

Social Apes: Bonobos and Chimpanzees

We saw in the chapter introduction that chimpanzees and bonobos share 98.5 percent of their DNA sequences with each other and with us. The two are extremely similar in many ways, but there are also striking contrasts in the makeup of their societies. As we saw, chimpanzee society is male-dominated and conflicts are resolved by fights among warring males, whereas bonobo society is female-centered and sex substitutes for aggression.

Both chimpanzees and bonobos forage for food during the day. Bonobos gather in groups of up to 20 individuals, but chimpanzees usually forage in smaller parties because their terrain is drier and food is harder to find. In each species, the larger community includes about a hundred animals and there is no casual mixing between members of different communities. Within bonobo parties, adult sons follow their mothers through the forest, and females form alliances.

In both species, males stay throughout life in their birth communities and males are often related to each other. But in chimpanzee society, males fight for dominance rank, and a would-be alpha male sometimes recruits other males to help him overthrow the current leader. If he wins this civil war, the new alpha male lets his allies, but not his rivals, mate with females. De Waal calls these complex, sometimes shifting alliances "chimpanzee politics." Bonobo males have a far milder way of competing for status: two rivals will chase each other around for awhile then reconcile by standing back to back and rubbing their scrotums together. Most of the time, the bonobo male's rank in society comes from his mother's status.

Chimpanzee brothers or cousins follow this strategy when one helps the other to become and stay the alpha male (Fig. 26.21). By defending his brother's or cousin's role as the dominant male, he helps him gain access to female chimpanzees and to reproduce. In this way, the helpful brother or cousin can be more successful than if an unrelated male was at the top of the hierarchy and got access to many more females. If certain genes shared by the brothers promote helping behavior, then kin selection will lead to a higher frequency of these altruistic alleles in the population.

A proponent of kin selection once quipped, "I'd lay down my life for two brothers or eight cousins." Through this theory we can see how something as drastic as the suicidal sting of a honeybee could have evolved.

Kin Selection and the Evolution of Social Insects

Can kin selection also account for the rigid caste system of social bees, wasps, and ants? Such societies often consist of a single reproducing female, the queen, plus a few drones, or males, and up to 80,000 sterile female helpers, or workers. But what advantage comes to the sterile worker who cannot reproduce herself and instead helps the queen reproduce?

Young bonobo females migrate to neighboring communities of unfamiliar and often hostile strangers. The young female newcomer will go up to an older female and invite mutual grooming and sexual contact (Fig. 26.22). This can then lead to a new close friendship and the eventual acceptance of the immigrant female.

Chimps and bonobos also differ dramatically in how the sexes interact. Researcher Amy Parish, working at Germany's Stuttgart Zoo in the 1970s, set up food competition experiments with separate sets of bonobos and chimps. She gave a tool (a short stick) to a male and two female chimps and a similar group of bonobos, and provided a hole through which they could collect honey. The male chimpanzee charged around the enclosure ferociously, then consumed all the honey he wanted before allowing the chimpanzee females to feed. In contrast, the bonobo females approached the honey together, rubbed genitals, then fed on the honey cooperatively. The bonobo male tried a few charging displays but the females ignored him and he fed after they walked away. From this and other observations, Frans De Waal concludes that "sex is the glue of bonobo society."

De Waal's work on bonobo and chimpanzee behavior has altered our view of primate evolution and allowed us to see several ways of acting—including warfare, peace making, erotic liaisons, and male and female dominance—within the lineage of humans and other apes. Neither the chimpanzee nor the bonobo is our ancestor, but their collective behaviors and those of the gorilla and orangutan can illuminate our own.

Humans as a Social Animal

We've seen many examples from the animal kingdom of social behavior that seems to promote an animal's current survival and its genetic endowment for future generations. But do similar principles apply to our own human behavior?

The tiny hand of a human baby grasps tightly to a parent's finger, its little lips suck rhythmically at a nipple, and it communicates forcefully with a cry that a parent cannot ignore. A baby born blind still flashes an alluring smile at the sound of its parent's voice. To some biologists, these very early behaviors suggest the natural selection for smiling and crying: adorable and demanding infants would be more likely to solicit parental care, to survive, and later to pass smiling and crying alleles to the next generation.

But in our own species, as in other animals, both heredity and environment surely help shape behavior. Genes help mold the native ability to learn language, for example, and this unfolds as neural pathways develop in the growing brain and make it capable of learning. Whether a child learns to speak English, Spanish, or Swahili, however, depends on its environment—the language its parents teach it.

This flexibility may help explain why people accept some cultural practices that help them survive and reject those customs that reduce their genetic success. For example, Brazilian tribes living in areas with poor soils and overhunted forests adopt Western farming techniques, while similar tribes living where the forests are less disturbed keep their traditional methods for growing and finding food. In both cases, the human nervous system responds flexibly to the environment in a way that promotes the survival of their families.

One familiar and effective way to study the role of genes in shaping human personality and behavior is twin studies. Psychologists compare identical and fraternal twins raised together in the same family or raised apart by adoptive families and look for similarities and differences in behavior (Fig. 26.23). Twin studies have shown that genes and the environment exert roughly equal influences over many human personality traits. These traits include social effectiveness, achievement, social closeness, alienation, aggression, respect for authority, and avoidance of danger. A single gene, of course, encodes a protein, not a trait like aggression or achievement. Researchers think that alleles of hundreds

Figure 26.22 **Sex as Appeasement in Female Bonobos.** After a dispute, female bonobos may embrace and rub their genital swellings against each other briefly, but intently. This appears to help soothe strained relationships. This pair was living in Wamba, Zaire.

Figure 26.23 **Reunited Twins.** Identical twins Jerry Levey and Jim Tedesco were separated as infants and raised by different families. Reunited in middle age, they discovered not only a close physical resemblance, but identical vocations—firefighting—and avocations—flirting, telling jokes, and drinking the same brand of beer.

We have been enormously—perhaps disastrously—successful as a species, due, in part, to the many behavioral patterns we've evolved in our short history. We are, for example, rapidly destroying the lush tropical forests that support our primate cousin, the bonobo. Today, probably fewer than 10,000 remain in the wild and there is grave concern about their ability to survive as a species. This destruction is being repeated in other regions, other biomes, and for other species, and Earth's biosphere threatens to collapse under the pressure of human population. Nevertheless, our behavioral plasticity remains. With it remains the hope that we will find answers to some of our greatest questions: What is life? Does it exist elsewhere in the universe? How did it originate? How can we cure and prevent AIDS, cancer, heart disease, diabetes, and other ailments? How can we harness genes wisely and safely? How can we identify all the other organisms that share our planet? What are the millions of precise interactions within ecosystems? How can we predict and lessen our impact on the biosphere? And how can we sustain and improve our own quality of life while allowing equal quality for bonobos, for marine mammals like those on our book cover, and for the millions of other living species? These questions and their future answers make the exploration of life a source of perpetual fascination and wonder.

of genes all interacting with a person's experiences combine to determine our tendencies toward anger and hostility, our capacity for sexual arousal, and our likelihood to learn one type of behavior versus another.

Chapter Summary and Selected Key Terms

Introduction   Bonobos and chimpanzees share 98.5 percent of their DNA sequences with each other and with us. Their behaviors, however, are very different from each other and from our own. Aggressive male chimpanzees usually dominate other males and females in chimpanzee societies, whereas in bonobo societies, females dominate males and use sexuality to lessen conflicts and resolve power issues. These observations are part of the study of **animal behavior** (p. 647), the responses animals make to stimuli from the environment.

Genetic Bases of Behavior   Some behaviors are genetically programmed in exact sequences such as the mating ritual of the black headed gull and its removal of broken eggshells from the nest. Other examples are the uncapping of diseased bee pupae and their ejection from the nest by "hygienic" bees governed by two separate genes.

Experience and Behavior   Some actions are little influenced by experience while others are strongly modified. Egg rolling in the greylag goose is a series of stereotyped physical movements called a **fixed action pattern** (p. 650). These behaviors are triggered by various environmental cues and proceed in an all or none way. Yawning and swallowing are additional examples. Young sea gull chicks peck at a red spot whether on a parent's bill or on a moving stick. But as the birds mature, they learn only to peck and beg from their parents. **Imprinting** (p. 650) is learning with a time limit: a young animal becomes "attached" to a particular adult or object but only within a specific developmental stage or **sensitive period** (p. 651). Goslings can imprint on a person and follow him or her like a "mom."

Many animals learn and this brings about behavioral change: **learning** (p. 652) is an adaptive, enduring change in behavior based on personal experiences in the environment. **Habituation** (p. 652) is simple learning based on repeated experience without consequences. Gull chicks learn to ignore harmless objects flying or moving overhead through habituation. Trial-and-error learning or **operant conditioning** (p. 652) is learning based on reward or punishment from experiences. Rats

learn to press levers to obtain food pellets this way. In **classical conditioning** (p. 653) or associative learning, the animal learns to associate an artificial stimulus such as a bell ringing with a natural stimulus like being fed. Pavlov trained dogs to salivate at the sound of a bell through this kind of training.

Insight learning (p. 653) involves reasoning, formulating a course of action by understanding the parts of a problem. A chimpanzee can pile up boxes and use a stick to reach a bunch of bananas dangling from a rope.

Natural Selection and Behavior   Several kinds of behavior are fundamental to animal survival: locating and defending home territory, feeding and avoiding being eaten, reproducing, and communicating. Locating home territory can involve seasonal **migrations** (p. 654) or **homing behavior** (p. 654), based on the abilities to **orient** (p. 654) or **navigate** (p. 654) and sometimes on an internal **circadian clock** (p. 654). The instinct to defend home territory often involves innate behavior and is called **territoriality** (p. 655), but can also involve learned defenses. When it comes to feeding behavior, many behaviorists accept the **opti-**

mality hypothesis (p. 655), the idea that natural selection has tuned animals to act automatically to obtain the most food energy with the least effort and risk. Reproductive behavior involves strong instincts but also elements of learning. The amount of parental investment made by females and males is a species specific trait. Where females make a greater parental investment than males, the males tend to compete to inseminate as many females as possible, often leading to a **dominance hierarchy** (p. 656), or ranking of group members based on success in fights over food and mates.

Females tend to select males that provide superior genes, protection, and good territory with available food and nesting sites. Through this **sexual selection** (p. 656), she acts as an agent of natural selection or male behavior and physical traits.

Animal interactions often involve **communication** (p. 657), or signals produced by one individual that alters a recipient's behavior and greatly benefits both. Communication can involve visual or sound displays, chemical signals, or touch. Bonobos have "squeal duels" and use other vocal sounds, and make gestures and specific facial expressions. Researchers have tried teaching pictograms and American sign language to apes with some interesting successes, especially in bonobos.

Evolution of Social Behavior
Most animals are solitary but some species are social—they live in groups and interact in ways that benefit others. Social behavior evolves only when the benefits of group living, such as protection from predators, outweigh the costs, such as competition for food and higher risk of disease. Generally, **altruism** (p. 661), or unselfish acts for the welfare of others, evolves through **kin selection** (p. 661): by helping its close relatives, even to its own detriment, an individual promotes the survival of its own genes. This is evident in primate behavior such as a male chimp defending his brother's role as dominant male. It's also evident in the caste system of social bees, wasps, and ants. Primate social behavior is particularly complex and fascinating. And the striking differences between bonobo and chimpanzee behavior can illuminate our own behavior and help researchers like Frans de Waal put human behavior in an evolutionary context. We display instincts, various forms of learning, reasoning, and great behavioral flexibility. Twin studies have helped show that genes and environment have roughly equal influences over many personality traits.

All of the following question sets also appear in the Explore Life *E* electronic component, where you will find a variety of additional questions as well.

Test Yourself on Vocabulary and Concepts

For each question set, complete or match the statements with the most appropriate term from the list. Each answer may be used once, more than once, or not at all.

SET I

(a) trial-and-error (b) habituation (c) classical conditioning (d) insight learning (e) sexual selection

1. Students ignore the fourth fire alarm in the same day in their classroom building.

2. Deciding not to read the directions, people learn how to turn on their VCR by pushing each of the buttons in turn until the machine is turned on.

3. Digestive juices begin to flow while people watch the on-screen advertising for refreshments before the movie begins.

4. Male stickleback fish with more red coloration attract more females to their nests.

5. Deciding how long it will take to reach a friend's house based on prior knowledge about traveling in the same area

SET II

(a) homing (b) migration (c) orientation
(d) navigation

6. The ability of an animal to travel along particular directional routes

7. Animal behavior that involves returning to the site of hatching or birth

8. The ability to travel from a particular map point to another; often tested by changing the starting point of an animal's voyage

9. Seasonal travel to feeding or breeding grounds

Integrate and Apply What You've Learned

10. When behavioral researchers keep blackcap warblers in a cage, the birds act restless during their normal migrating season. Birds from northern populations, which fly long distances, are restless for a longer time than birds from southern populations, which do not fly as far. Hybrid birds show intermediate periods of restlessness. What hypothesis do these data suggest? What additional experi-

ments would you want to perform as controls to help support or reject your hypothesis?

11. Discuss the proposition that in genetic terms altruism can be considered selfish.

12. When analyzing human behavior from a genetic point of view, what hypotheses might an animal behaviorist advance for why we have to work hard to teach our children to act with kindness and to "turn the other cheek"?

13. What observation led Nikolaas Tinbergen to believe that the removal of broken eggshells was a genetically determined pattern of behavior?
    (a) All gulls of that species do it.
    (b) He proved that gulls only remove jagged-edged objects.
    (c) He showed that gulls notice jagged-edged objects faster than other objects.
    (d) He proved that the nervous system of gulls directs the behavior.
    (e) He proved that the behavior was learned by imprinting.

Analyze and Evaluate the Concepts

14. What are some of the major behavioral differences between bonobos and chimpanzees? Why does Frans de Waal consider these differences to be important to understanding our own behavior?

15. During the hot, sunny summer, male lark buntings defend their nesting territories on the American prairie. Usually a single female builds a nest in a male's territory and he helps her raise a family. Sometimes a male's territory attracts a second female, and she mates with him but receives no help raising her offspring. That second reproduction therefore tends to be less successful. Devise an experiment to test the hypothesis that *polygyny* (having more than one female mate) arises when a male's territory contains enough shade that a female can rear more young by becoming his second female than by becoming the monogamous mate of a male defending a territory with no shade. What parameter in the male's territory will you manipulate and how? What data will you collect? What result would disprove the hypothesis?

16. In explaining the evolutionary cause of many behavior patterns, animal behaviorists rely on an analysis of the risks and rewards of the behavior to the individual's reproductive success. Give two ex-

amples of such behaviors. If animals do not consciously calculate the risks and rewards before undertaking a particular action, then what mechanism generates the decision?

17. Female langur monkeys have a gestation period of seven months and a nursing period of another eight months. While pregnant and during the nursing period, females do not ovulate. Langur monkeys travel in groups of 10 to 20 adult females, their young and 1 dominant male. If another male unseats the dominant male, he will attempt to kill all nursing offspring for six months after the takeover. He will not attempt to kill young who are weaned. How would an ethologist explain why he does not kill the weanlings?

(a) He is unsure of their parentage.

(b) The female weanlings represent an enlargement of his harem.

(c) After killing the newborns and nursing young, he has had enough of killing.

(d) The weanlings might not be genetically related to the previous dominant male.

(e) Killing weaned youngsters will not cause their mothers to become fertile sooner.

Appendix A:
The Properties of Water

Water, Temperature, and Life

Anyone who has had frostbite knows that water behaves differently at different temperatures. When a person is caught in a blizzard without gloves, the liquid water in his or her fingers may may turn into ice crystals, which can tear apart cells and kill them, perhaps leading to loss of the fingers. It is fortunate for living things that water usually remains liquid in the normal temperature range found on Earth, instead of assuming one of its other physical forms—gas (water vapor) or solid (ice).

Living things are also lucky that water is slow to heat; that is, the amount of heat needed to raise the temperature of a certain volume of water is greater than for most other liquids. This property is a direct result of water's hydrogen bonds. Much of the heat energy applied to water goes into stretching or breaking hydrogen bonds instead of raising the water temperature. This helps ensure the relatively constant external and internal environments living organisms need.

A tennis player perspiring at the end of a vigorous match can appreciate another important temperature-related property of water: An unusually high amount of heat is required to turn liquid water into water vapor. Before a water molecule can evaporate, it must jostle about rapidly enough to fly off the surface of a water droplet. Before it can really jostle, the hydrogen bonds linking the molecule to its neighbors must be broken by the absorption of heat energy.

In the process of absorbing heat, evaporating water cools its surroundings. For a wild horse escaping a predator in Nevada, this property provides a natural cooling system, and explains the evolution of sweat glands in horses, people, and some other mammals. Sweat pours from sweat glands in the skin and covers parts of the body surface. As the sweat evaporates, a large amount of body heat goes into breaking the hydrogen bonds in the water molecules. This cools the horse's body as sweat evaporates from the skin.

Physical Properties of Water

In addition to its properties related to temperature, water has several physical properties important to living things.

Water molecules exhibit **cohesion,** the tendency of like molecules to cling to each other, and **adhesion,** the tendency of unlike molecules to cling to each other—for example, water to paper, soil, or glass. Together, cohesion and adhesion account for **capillarity,** the tendency of a liquid substance to move upward through a narrow space against the pull of gravity, such as between the fibers of a paper towel being used to wipe up a spill, or up the inside of a narrow glass tube. Capillarity plays a part in the upward transport of water in many kinds of plants (see Chapter 23). Without the cohesion of water molecules to each other and adhesion to the walls of narrow tubes, there would be no tall trees such as redwoods or eucalyptus.

One physical property of water must be overcome the moment a newborn baby draws its first breath. This property, called **surface tension,** is the tendency of molecules at the surface of liquid water to cohere to each other rather than adhere to the air molecules above them. The water molecules in a newborn's lungs tend to bind to each other and to the tissue more strongly than to air; for this reason, they tend to pull the walls of the lungs together, which could cause those delicate organs to collapse. The body, however, produces special molecules called **surfactants,** which function like detergent molecules to lower the surface tension of the water in the lungs, thus preventing lung collapse. Premature babies born before their lungs are capable of generating surfactants sometimes die because their lungs collapse.

Water has yet another physical property with biological implications: the tendency of ice (solid water) to float in liquid water. The hydrogen bonds in ice are fairly rigid, creating an open latticework that holds water molecules farther apart than the more easily broken bonds of the liquid form. Thus, ice is less dense than liquid water, and floats in it. This is a unique property;

other substances become more dense when they freeze. If ice did not become less dense, many lakes would freeze solid in winter, and only the top few centimeters would thaw in summer. Instead, with a frozen top layer insulating the lower depths, plants and animals can survive winter in the chilly liquid water beneath the ice.

Have you ever suffered from acid indigestion after eating too much spicy food too fast? This uncomfortable condition is caused when cells in your stomach secrete large quantities of hydrochloric acid (HCl). While the acid speeds up the digestion of food, it can also irritate your stomach lining. When hydrochloric acid dissolves in water (in your stomach, or elsewhere), it releases H^+ into the solution:

$$HCl \longrightarrow H^+ + Cl^-$$

As an antidote to acid indigestion, you may have taken some bicarbonate (HCO_3^-), the active ingredient in many over-the-counter heartburn remedies. Bicarbonate can act as a **buffer,** a substance that regulates pH by "soaking up" or "doling out" hydrogen ions as needed: When hydrogen ion concentrations are high, buffers bind to H^+, and when hydrogen ion concentrations are low, buffers release H^+.

When basic, or **alkaline,** substances dissociate in water, instead of giving off hydrogen ions, they combine with them, resulting in an excess of hydroxide ions, OH^-. For example, the base NaOH dissociates into sodium and hydroxide ions:

$$NaOH \longrightarrow Na^+ + OH^-$$

The sodium ions dissolve in the water, and some of the freed OH^- ions combine with free H^+ to form $H-OH$ (or, more familiarly, H_2O), leaving fewer hydrogen ions than before NaOH was added to the water.

pH Why does the most acidic solution have the *lowest* number on the pH scale? And why does each pH unit represent a ten-fold difference in hydrogen ion concentration? Chemists measure the amount of substances in units called moles, and the concentration of substances in moles per liter (mol/L). The pH scale indicates the concentration of hydrogen ions in this concentration unit. A pH of 1 is 0.1 moles of H^+ per liter, a pH of 2 is 0.01 and a pH of l0 is 0.000,000,000,1 moles of H^+ per liter. The pH is equivalent to the number of decimal places to the right of the decimal point. This accounts for the factor-of-ten difference between pH units. Because low pH values have few decimal places to the right of the decimal point, they have great concentrations of hydrogen ions. In a solution, the concentration of hydrogen ion (H^+) times the concentration of hydroxide ion (OH^-) is always the same. Whenever the concentration of hydrogen ions goes up, the concentration of hydroxide ions goes down, and vice versa.

Appendix B: Microscopes

How do we know that eukaryotic cells have a membrane-bound nucleus, chloroplasts, or any of the other organelles discussed in this book? Much of what we know about cells comes from biologists peering into microscopes of three basic types and witnessing the beauty inside and on the surfaces of the cells.

Light microscopes are instruments containing optical lenses that refract, or bend, light rays so that an object appears larger than it really is (Figure a). Because it can illuminate and magnify, biologists use the light microscope extensively to locate cells in tissues, to observe the behavior of living cells (Figure b), and to detect cell organelles that can be stained bright colors, such as the nucleus, chloroplasts, and mitochon-

(a) Light microscope

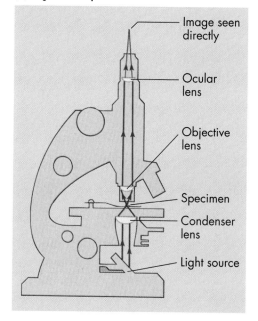

- Image seen directly
- Ocular lens
- Objective lens
- Specimen
- Condenser lens
- Light source

(e) Transmission electron microscope   **(g)** Scanning electron microscope

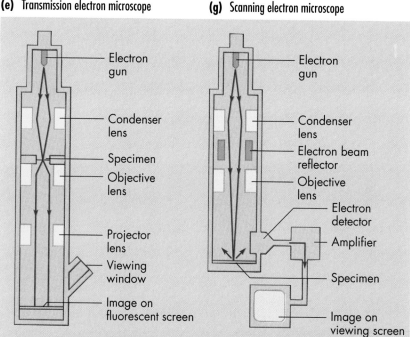

- Electron gun
- Condenser lens
- Specimen
- Objective lens
- Projector lens
- Viewing window
- Image on fluorescent screen

- Electron gun
- Condenser lens
- Electron beam reflector
- Objective lens
- Electron detector
- Amplifier
- Specimen
- Image on viewing screen

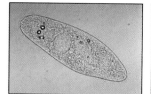

(b) Light microscopy of living *Paramecium* cell

(c) Light microscopy of stained *Paramecium* cell

(d) Differential interference microscopy of *Paramecium* cell

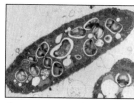

(f) Transmission electron microscopy of *Paramecium* cell

(h) Scanning electron microscopy of *Paramecium* cell

dria (Figure c). If a specimen is thin enough for light to pass through, a light microscope can magnify its physical details to more than 2000 times.

Colored stains can heighten the contrast between various structures in the specimen, making them easier to distinguish. Stains are usually used in combination with fixatives, agents that preserve the cells or tissues so that they remain unchanged. However, stains and fixatives kill cells, so to view living cells, biologists use differential interference contrast, or Nomarski light microscopes that augment the differences in light refraction between unstained structures so that the contrast between them is bright even without stains. (Figure d).

Electron microscopes (EMs) use electrons with wavelengths 100,000 times shorter than visible light; these short wavelengths provide greater resolution, which allows greater magnification. Electron microscopes also use magnets rather than ground glass lenses to focus the electron beams. Electron microscopes opened a new and marvelous realm to biologists and have added much of our current knowledge of cell structure. **Transmission electron microscopes (TEMs)** generate images of thin slices of a cell, tissue, or object. These have revealed a level of complexity and detail in the cell never suspected until the middle of the twentieth century (Figures e, f). And **scanning electron microscopes (SEMs)** generate three-dimensional surface images. These have shown us a beautiful and sometimes bizarre vision of the intricate outer shapes of living cells and organisms (Figures g, h).

Appendix C:
The Hardy-Weinberg Principle

The **Hardy-Weinberg principle** provides an idealized standard against which a geneticist can compare what happens in a real population and thus detect evolutionary change.

The principle has two main points:

1. **The allele principle:** If left undisturbed, the frequency of different alleles in a population remains unchanged over time.
2. **The genotype principle:** With no disturbing factors, the frequency of different genotypes will not change after the first generation.

We can understand the principle by considering a population of snails (see figure) that can fertilize themselves or each other at random (Step 1). These snails possess a gene for shell color that shows partial dominance, with genotype *AA* causing a blue shell, *aa* resulting in a yellow shell, and *Aa* giving a green shell. By analyzing changes in frequency of this shell color gene with the Hardy-Weinberg equations, we can determine whether the snail population is evolving.

Each of the five snails shown in the figure is a diploid with two copies of the color gene. In this population's gene pool, there are ten alleles: six *A* alleles and four *a* alleles. If the symbol p represents the fraction of *A* alleles, then $p = 6/10$, or 0.6. If the symbol q represents the fraction of *a* alleles, then $q = 4/10$, or 0.4.

Because the number of *A* alleles plus the number of *a* alleles represent all of the alleles of this gene in this snail population, $0.6 + 0.4 = 1$, or in symbols, $p + q = 1$. This is the **allele pool equation.**

To see if allele frequencies change and evolution occurs, we must examine what happens when the snails reproduce, keeping in mind Mendel's rule that alleles separate in the formation of egg and sperm. Notice that despite meiosis, the frequency of *A* and *a* alleles is the same in the gametes as it was in the original population, $6/10\ A$ and $4/10\ a$ (Step 2).

Now what happens to allele frequencies at the time of fertilization? If we assume random mating, then we can write the frequencies into a Punnett square (Step

(1) G_0 population

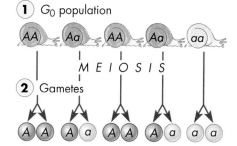

$M\ E\ I\ O\ S\ I\ S$

(2) Gametes

(3) Punnett square

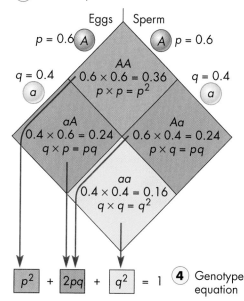

$$\boxed{p^2} + \boxed{2pq} + \boxed{q^2} = 1$$ **(4)** Genotype equation

(5) Expected G_1 genotype frequencies and expected number of snails given a G_1 population of 100 individuals

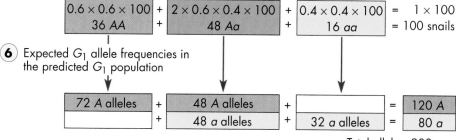

(6) Expected G_1 allele frequencies in the predicted G_1 population

Total alleles: 200

3). We can call $p^2 + 2pq + q^2 = 1$ the **genotype equation** (Step 4). This equation says that the sum of the individuals with *AA,* and *Aa,* and *aa* genotypes adds up to the entire population (the "1" in the equation).

To determine whether evolution has occurred in the snail population, we must look for a change in allele frequencies between generations. If five snails of the parental G_0 generation produce 100 snails in the G_1 generation, then we can expect the genotype frequencies and resulting genotype numbers shown in Step 5, barring outside influences. (Thus, the genotype equation predicts the number of each of the three different genotypes in the population.) What, then, are the frequencies of alleles in this new generation? Have they changed?

You can see from Step 6 that the frequency of *A* alleles is 120/200, or 0.6, and the frequency of *a* alleles is 80/200, or 0.4—the same as in the original generation.

From this generation on, the allele and genotype frequencies will remain the same, in the absence of outside influences, as you can prove for yourself by making another Punnett square and filling it in with the new data.

Applying the Hardy-Weinberg principle this way, we can predict that in populations without some outside influence, no evolution will occur over the generations. The Hardy-Weinberg equations provide a benchmark, a point of comparison for measuring any gene changes—or evolution—that might occur. If allele or genotype frequencies found in a population in nature are different from those predicted by the allele pool and the genotype equations, then there must be some outside influence, like mutation, natural selection, migration, nonrandom mating, or small populations. Biologists can then design experiments to see which outside influence may be important in any specific case.

Photo Credits

Figure 1.1, Earth Imaging / Stone; 1.2, NASA; 1.3(a), Juergen Berger, Max Planck Institute / SPL / Photo Researchers, Inc.; 1.3(b), Ray E. Ellis / Photo Researchers, Inc.; 1.4, NASA / SPL / Photo Researchers, Inc.; 1.5, Tim Davis / Photo Researchers, Inc.; 1.6, Jeff Lepore / Photo Researchers, Inc.; 1.7(a), CNRI / Science Photo Library / Photo Researchers Inc.; 1.7(b), J.A.L. Cooke, OSF / Animals Animals; 1.8(a-c), Sharon Amancher / UC Berkeley; 1.9, Amy Dunleavy; 1.10, Tom & Therisa Stack / Tom Stack & Associates; 1.13(a)(b), Maria Stenzel / National Geographic Image Sales; 1.14(a), Superstock; 1.14(b), Rick & Tora Bowers / Visuals Unlimited; 1.15, WHOL, D. Foster / Visuals Unlimited; 1.17, Andrew Errington / Stone.

Chapter Opener 2.3, Oliver Meckes / Gelderblom / Photo Researchers, Inc.; 2.1(a), Oliver Meckes / Gelderblom / Photo Researchers, Inc.; 2.16(a), Oliver Meckes / Gelderblom / Photo Researchers, Inc.; 2.16(b), CNRI / SPL / Photo Researchers, Inc.; 2.16(c), William Dentler / BPS / Stone; 2.18(a), Don Fawcett / Photo Researchers, Inc.; 2.20(a), Biology Media / Photo Researchers, Inc.; 2.22(a), David M. Phillips / Visual Unlimited; 2.22(c), Oliver Meckes / Photo Researchers, Inc.; 2.24(a), Don Fawcett / Photo Researchers, Inc.; 2.24(c), Professors P. Motta & T. Naguro / SPL / Photo Researchers, Inc.; 2.25(a), Don Fawcett / Photo Researchers, Inc.; 2.25(b), Superstock; 2.26(a), Dr. Stefano Fais; 2.26(c), JW Schuler / Photo Researchers, Inc.; 2.27(a), © Molecular Probes Inc./ Photo ID: G000441. FluoCells prepared microscope slide #1 (F.14780) showing bovine pulmonary artery endothelial cells (BPAEC) that have been labeled with MitoTracker Red CMXRox (M.7512) for mitochondria, and then fixed, permeabilized and stained with BODIPY FL phallacidin (B.607) for F.actin and DAPI (D.1306) for nuclei. This multiple exposure image was acquired using DAPI, fluorescein and Texas Red bandpass optical filter. Image contributed by Janell Bishop-Stewart, Molecular Probes, Inc.; 2.27(b), Don Fawcett / Photo Researchers, Inc.; 2.28(a), Biophoto Associates / Photo Researchers, Inc.; 2.28(c), David M. Phillips / Visual Unlimited; 2.29(b), Professors P. Motta / SPL / Photo Researchers, Inc.

Figure 3.3, G.C. Kelly / Photo Researchers, Inc.; 3.8, William McClure / Carnegie Mellon University; 3.10, Paraskevas Photography; 3.12(a), Photo Researchers, Inc.; 3.12(b), Photo courtesy of Odile Mathieu-Costello. Reprinted from Respiration Physiology, 89, Mathieu-Costello et al, 113, © 1992, with permission from Elsevier Science.; 3.15, T. Kanariki–D. Fawcett / Visuals Unlimited; 3.16, Paraskevas Photography; 3.21(c), Newcomb & Wergin / BPS / Stone.

Figure 4.1, John D. Cunningham / Visuals Unlimited; 4.2(a), Dr. P. Marazzi / SPL / Photo Researchers, Inc.; 4.2(b), Michael Abbey / Visuals Unlimited; 4.4(b), Gunther F. Bahr / AFIP / Stone; 4.6(a-h), Michael Abbey / Visuals Unlimited; 4.7(c), Dr. Gerald Schatten / SPL / Photo Researchers, Inc.; 4.15(a), Richard Hutchings / Photo Researchers, Inc.

Figure 5.1(a), Corbis-Bettmann; 5.1(b), Wally Eberhart / Visuals Unlimited; 5.2, John Willis, Duke University; 5.4(b), Cabisco / Visuals Unlimited; 5.9, Custom Medical Stock Photo; 5.10(a), Jackie Lewin / SPL / Photo Researchers, Inc.; 5.14, CNRI / Science Photo Library / Photo Researchers, Inc.; 5.16(a), Runk / Schoenberger from Grant Heilman; 5.17(a), Muscular Dystrophy Association; 5.19, Horst Schafer / Peter Arnold, Inc.; 5.20(b), Breast Screening Unit, Kings College Hospital, London / SPL / Photo Researchers, Inc.; 5.21, Yan DeSilva.

Figure 6.1, Ruth BreMiller, Institute of Neuroscience, University of Oregon; 6.2, Maclyn McCarty / Rockefeller University; 6.3(b), Biology Media / Science Source / Photo Researchers, Inc.; 6.4(a), A. Barrington Brown / Science Source / Photo Researchers, Inc.; 6.4(c), Science Photo Library / Photo Researchers, Inc.; 6.4(d), Cold Spring Harbor Laboratory; 6.5, R. Langridge, D. McCoy / Rainbow; 6.7(a), (b), Manfred Kage / Peter Arnold, Inc.; 6.11c, Dr. Menotti-Raymond.

Figure 7.2(b), Andrew Syred / SPL / Photo Researchers, Inc.; 7.9(b), E. Kiselva / D. Fawcett / Visuals Unlimited; 7.13(c), Brian Matthews / University of Oregon; 7.15, Dr. Gopal Murti / SPL / Photo Researchers, Inc.; 7.19, Ted Spiegel; 7.21, Mark Gibson / Visuals Unlimited.

Figure 8.1, Michael Fogden / DRK Photo; 8.4(c), Manfred Kage / Peter Arnold; 8.6, Dr. Gerald Schatten / SPL / Photo Researchers, Inc.; 8.7(b), © David M. Phillips / Visuals Unlimited; 8.7(c), © David M. Phillips / Visuals Unlimited; 8.12(a), Lennart Nilsson Boehringer Ingelheim International GmbH, from A Child is Born, p. 76, Dell Publishing Co., Inc.; 8.12(b), SIU / Visuals Unlimited; 8.14, Science, Vol. 280, 5 June 1988; 8.15, Dagmar K. Kalousek, M.D., University of British Columbia; 8.16(a), Cabisco / Visuals Unlimited; 8.16(b), CNRI / Phototake; 8.17, Reuters / Jeff Mitchell / Archive Photo; 8.18(a), (b), Oliver Meckes / Photo Researchers, Inc

Figure 9.2, The Natural History Museum, London; 9.3(a), (b), The Granger Collection, New York; 9.5, CNRI / Science Photo Library / Photo Researchers, Inc.; 9.6, Francois Gohier / Photo Researchers, Inc.; 9.7, S. Conway Morris; 9.12(a), Sharon Amancher / UC Berkeley; 9.12(b), Lennart Nilsson / A Child Is Born; 9.14(a), Ken Lucas / Visuals Unlimited; 9.14(b), Fritz Polking / Dembinsky Photo Associates; 9.14(c), Ed Kanze / Dembinsky Photo Associates; 9.17(a), ANT Photo Library / NHPA; 9.17(b), Nick Bergkessel / Photo Researchers, Inc.; 9.19, Eric Sander; 9.21(a), Superstock ; 9.21(b), Irven De Vore / Anthro-Photo; 9.21(c), Tim Davis / Photo Researchers, Inc.; 9.21(d), Keren Su / Stone; 9.21(e), Mike Yamashita / Woodfin Camp & Associates; 9.21(f), (g), Victor Engelbert / Photo Researchers, Inc; 9.21(h), Art Wolfe, © Tony Stone Images; 9.23, Gerard Lacz / Peter Arnold, Inc.; 9.24, Dr. Victor A. McKusick; 9.25, Frans Lanting / Minden Pictures; 9.26, OSF / Animals Animals; 9.28, William H. Mullins / Photo Researchers, Inc.

Figure 10.01(a), Mark Garlick / SPL / Photo Researchers, Inc.; 10.01(b), Earth Imaging / Stone; 10.01(c), Science VU / Visuals Unlimited; 10.01(d), Panterra; 10.01(e), Francois Gohier / Photo Researchers, Inc.; 10.01(f), Daniel J Cox; 10.2(a-c), Don Dixon / Cosmographica.com; 10.3(a), Fred Bavendam / Peter Arnold, Inc.; 10.3(b), Jeff Johansen / John Carroll University, Mark Schneegurt / Wichita State University and Cyanosite, www.cyanosite.bio.purdue.edu10.6(a), (b), Hawaii Undersea Research Laboratory; 10.7, Stanley Awramik / BPS / Stone; 10.11(a-c), Carr Botanical Consultation; 10.17, Greg Vaughn / Tom Stack & Associates.

Figure 11.2, Ken Graham / Accent Alaska; 11.3, Sylvia Coleman / Visuals Unlimited; 11.4, G.W. Willis / Visuals Unlimited; 11.5a, Oliver Meckes / Photo Researchers, Inc.; 11.5b, CNRI / SPL / Photo Researchers, Inc.; 11.5c, Oliver Meckes / Gelderblom / Photo Researchers, Inc.; 11.6, Barry Dowsett / SPL / Photo Researchers, Inc.; 11.7, Georg Gerster / Photo Researchers, Inc.; 11.8, Blga. Palmira Ventosilla, Instituto de Medicina Tropical Alexander von Humboldt, Universidad Peruana Cayetano Heredia; 11.9(a), IFA / Peter Arnold, Inc.; 11.9(b), Philip Sze / Visuals Unlimited; 11.10, Meckes / Gelderblom / Photo Researchers, Inc.; 11.11(a), (b), The Cohen Group, University of California, San Francisco; 11.11(c), Frank Hormann / Wide World Photos; 11.12, CNRI / SPL / Photo Researchers, Inc.; 11.14(a), Eric Grave / Photo Researchers, Inc.; 11.14(b), Oliver Meckes / Photo Researchers, Inc.; 11.14(c), P.M. Motta & F.M. Magliocca / SPL / Photo Researchers, Inc.; 11.15(b), Cabisco / Visuals Unlimited; 11.16, Daniel W. Gotshell / Visuals Unlimited; 11.18(a), (b), Eric Grave / Photo Researchers, Inc.; 11.18(c), Meckes / Ottawa / Eye of Science / Photo Researchers, Inc.; 11.19, Eric Grave / Photo Researchers, Inc.; 11.20(a), Stanley Flegler / Visuals Unlimited; 11.20(b), David Hall / Photo Researchers, Inc.; 11.20(c), Paul A. Souders / Corbis; 11.21 , D.P. Wilson / Eric & David Hosking / Photo Researchers, Inc.; 11.22(b) , Sinclair Stammers / SPL / Photo Researchers, Inc.

Figure 12.1, Runk/Schoenberger / Grant Heilman Photography; 12.3(b), James W. Richardson / Visuals Unlimited; 12.4, Ralph E. Williams / US Forest Service, Boise Field Office. Photo courtesy Terry Shaw & Susan Hagle.; 12.6, G. Shih.R. Kessel / Visuals Unlimited; 12.9(b), Mark Brundrett; 12.10(a), Breck P. Kent / Earth Scenes; 12.10(b), V. Ahmadjian / Visuals Unlimited; 12.12(a), Biophoto Associates / Photo Researchers, Inc.; 12.13(a), D. Cavaganaro / Visuals Unlimited; 12.13(b), Geoff Bryant / Photo Researchers, Inc.; 12.14(a), Biophoto Associates / Photo Researchers, Inc.; 12.16(b), Len Rue, Jr. / Earth Scenes; 12.16(c), Barry Runk / Grant Heilman Photography; 12.16(d), Ed Reschke / Peter Arnold, Inc.; 12.17(b), James W. Richardson / Visuals Unlimited; 12.17(c), Amy Dunleavy; 12.18(a), Walt Anderson / Visuals Unlimited; 12.18(b), Runk/Schoenberger / Grant Heilman Photography; 12.18(c), Renee Lynn / Photo Researchers, Inc.; 12.19, David L. Brown / Tom Stack & Associates; 12.21, Jeffrey Mitton / University of Colorado; 12.22(a), Merlin Tuttle, Bat Conservation International, Photo Researchers, Inc.; 12.22(b), Hal H. Harrison / Grant Heilman Photography; 12.22(c), Randall B. Henne / Dembinsky Photo Associates; 12.22(d), Fred Habegger / Grant Heilman Photography, Inc.

Figure 13.2, Ken Lucas / Visuals Unlimited; 13.4, Andrew J. Martinez / Photo Researchers, Inc.; 13.5(a), E.R. Degginger / Animals Animals; 13.5(b), Atkinson, OSF / Animals Animals; 13.5(c), Gregory Ochocki / Photo Researchers, Inc.; 13.5(d), Fred Bavendam / Peter Arnold, Inc.; 13.7, Jesse Cancelmo / Dembinsky Photo Associates; 13.8(a1), F.R. Turner; 13.8(b1), from Morrill & Santos, 1985; 13.8(a2), Courtesy of Dr. Michael M. Craig, Department of Biomedical Sciences, Southwest Missouri State University, Springfield, MO. Reprinted from Cell Vol. 68, cover, copyright 1992, with permission from Elsevier Science.; 13.8(b2), Shinya Inoue, MBL, WHOI; 13.9, Michael Abbey / Photo Researchers, Inc.; 13.10, Stanley Flegler / Visuals Unlimited; 13.11(a), Chris McLaughlin / Animals Animals; 13.11(b), E.R. Degginger / Dembinsky Photo Assoc.; 13.12(a), A. Kerstitch / Visuals Unlimited; 13.12(b), Dave B. Fleetham / Visuals Unlimited; 13.13(a), Juergen Berger, Max Planck Institute / SPL / Photo Researchers, Inc.; 13.13(b), Science VU / Visuals Unlimited; 13.15(a), Zig Leszczynski / Animals Animals; 13.15(b), Steinhart Aquarium / Photo Researchers, Inc.; 13.15(c), Stan Osolinski / Dembinsky Photo; 13.16(a-c), Armin P. Moczek / Duke University; 13.17(a), Randy Morse / Animals Animals; 13.17(b), Dave B. Fleetham / Visuals Unlimited; 13.21(a), David Hall / Photo Researchers, Inc.; 13.21(b), John Cunningham / Visuals Unlimited; 13.22(b), James L. Amos / Photo Researchers; 13.22(c), Tom McHugh / Photo Researchers, Inc.; 13.22(d), Tom McHugh / Photo Researchers, Inc.; 13.22(e), Norbert Wu / Peter Arnold, Inc.; 13.24(a), Zig Leszczynski / Animals Animals; 13.24(b), Tom McHugh / Photo Researchers, Inc.; 13.26(a), Ken Lucas / Visuals Unlimited; 13.26(b), Zig Leszczynski / Animals Animals; 13.28(a), Tom McHugh / Photo Researchers, Inc.; 13.28(b), Tom McHugh / Photo Researchers, Inc.; 13.30(a), David Haring / Animals Animals; 13.30(b), Ken Lucas / Visuals Unlimited; 13.30(c), Mickey Gibson / Animals Animals; 13.31, Anup Shah / Animals Animals; 13.32(a), Stan Osolinski / Dembinsky Photo Associates; 13.32(b), Erwin & Peggy Bauer / Animals Animals; 13.34, John Reader / SPL / Photo Researchers, Inc.; 13.35, Alan Walker / National Museums of Kenya; 13.36(a), The Granger Collection, New York; 13.36(b), Giraudon / Art Resource, NY.

Figure 14.3(c), Fred Hossler / Visuals Unlimited; 14.4(c), Cabisco / Visuals Unlimited; 14.5(a), Dennis Drenner / Visuals Unlimited; 14.5(b), John D. Cunningham / Visuals Unlimited; 14.6(c), Dee Colleni / Visuals Unlimited; 14.7, NASA; 14.8(a), AP / Wide World Photos; 14.8(b), Alan Carey / The Image Works; 14.9, Stan Osolinski / Dembinsky Photo Associates; 14.12(f), F. Spinelli. D.W. Fawcett / Visuals Unlimited; 14.15, NASA; 14.17, Werner H. Muller / Peter Arnold, Inc.

Glossary

A

9:3:3:1 ratio: The ratio of phenotypes found in the offspring of two individuals, both of whom are heterozygous for two traits whose alleles assort independently.

A, T, G, C (adenine, thymine, guanine, cytosine): The four types of nucleotides contained in DNA. They are identical except for the bases they contain.

abdomen: The posterior part of an arthropod's body; in vertebrates, the abdomen lies between the thorax and the pelvic girdle.

abdominals: Rectus abdominis; muscles of the stomach area.

abscisic acid (ABA): A plant hormone that promotes dormancy.

abscission: (L. *ab*, away, off + *scissio*, dividing) In vascular plants, the dropping of leaves or fruit at a particular time of year, usually the end of the growing season.

abscission zone: An area of relatively weak cells at the base of the stem, where the leaf or fruit will eventually separate from the branch.

absorption spectrum: The wavelengths of light absorbed by a pigment.

absorption: In this, the third step of the digestive process, small molecules pass across the gut wall into the animal's lymphatic system or bloodstream, which transports the small molecules to each cell in the body.

accessory organs: One of several organs that aid digestion, including the salivary glands, liver, gallbladder, and pancreas.

acetylcholine: Neurotransmitter that, along with transmitting nerve impulses within the brain, also relays nerve signals to the skeletal muscles. These signals, in turn, help maintain posture, breathing, and limb movements. Acetylcholine is also released by the vagus nerve, slowing the heart rate.

acid: A proton donor; any substance that gives off hydrogen ions when dissolved in water, causing an increase in the concentration of hydrogen ions. Acidity is measured on a pH scale, with acids having a pH less than 7; the opposite of a base.

acoelomate: The condition, in animals, of lacking a coelom.

ACTH (adrenocorticotropic hormone): A hormone produced by the pituitary in response to stress. ACTH acts on the cortex of the adrenal gland.

actin: (Gr. *actis*, a ray) A major component of microfilaments in contractile cells such as muscle.

actinomycetes: Single-celled organisms that produce antibiotic compounds.

action potential: A temporary all-or-nothing reversal of the electrical charge across a cell membrane; occurs when a stimulus of sufficient intensity strikes a neuron.

activation energy: The minimum amount of energy that molecules must have in order to undergo a chemical reaction.

active immunity: The production of antibodies or antigen-specific T cells by an individual's own immune system.

active site: A groove or a pocket on an enzyme's surface to which reactants bind. This binding lowers the activation energy required for a particular chemical reaction; thus, the enzyme speeds the reaction.

active transport: Movement of substances against a concentration gradient requiring the expenditure of energy by the cell.

acyclovir: An antiviral drug.

adaptation: (L. *adaptere*, to fit) A particular form of behavior, structure, or physiological process that makes an organism better able to survive and reproduce in a particular environment.

adaptive radiation: Evolutionary divergence of a single ancestral group into a variety of forms adapted to different resources or habitats.

addiction: Physical dependence upon a substance such as alcohol or other drugs, characterized by tolerance of increasingly larger doses, craving the drug, and withdrawal symptoms when use is discontinued.

ADH (Antidiuretic hormone): A water-balancing hormone produced in the posterior lobe of the pituitary that acts on the kidney to regulate the amount of water excreted in urine.

adhesion: The tendency of unlike molecules to cling to each other. A biological example is water adhering to the inner walls of xylem tubes in plants.

ADP (Adenosine diphosphate): An energy molecule related to ATP but having only two phosphate groups instead of three.

adrenal gland: (L. *ad*, to + *renes*, kidney) A hormone-producing endocrine gland located on top of the vertebrate kidney; secretes hormones mainly involved in the body's response to stress, such as epinephrine (adrenaline) and norepinephrine.

adrenaline: Also known as epinephrine, this hormone is derived from an amino acid and is released in response to stressful situations. Adrenaline causes the heart to beat faster, the breathing rate to increase, the blood sugar level to climb, and blood to be shunted to the brain and skeletal muscles, allowing the animal to defend itself or flee.

adventitious root: A new root that arises from an aboveground structure on a plant.

aerobic pathway: Energy harvest in the presence of oxygen. Also called cellular respiration.

age structure: The number of individuals in each age group of a given population.

aging: A progressive decline in the maximum functional level of individual cells and whole organs that occurs over time.

agnathan: A jawless fish.

agricultural revolution: The transition of a group of people from an often nomadic hunter-gatherer way of life to a usually more settled life dependent on raising crops, such as wheat or corn, and on livestock. It was under way in the Middle East by 8000 years ago.

AIDS (Acquired immune deficiency syndrome): A partial or total loss of immune function based on infection by the human immunodeficiency virus (HIV).

air sac: A thin-walled extension of the lungs; in birds, air flows through lungs and air sacs during respiration.

albumin: A group of large blood proteins that helps maintain osmotic pressure and binds to toxic substances in the blood.

aldosterone: A steroid hormone released by the cortex of the adrenal gland that regulates salt reabsorption in the kidney.

alga (pl. algae): Simple chlorophyll-containing organisms, often single-celled; were probably the ancestors to the land plants.

alignment: The positioning of chromosomes on the mitotic or meiotic spindle.

alimentary canal: The gastrointestinal tract; includes the mouth, esophagus, stomach, and small and large intestines.

alkaline: The characteristic of being basic or having pH greater than 7. Seawater (pH 8) has greater alkalinity than fresh water (pH 7); ammonia (pH 12) has greater alkalinity than seawater.

allele: One of the alternative forms of a gene.

allergy: An extreme reaction to foreign substances or allergens that mimics a body's nonspecific resistance reaction.

allopatric speciation: The divergence of new species as a result of geographical separation of populations of the same original species.

all-or-none response: A property of action potentials; they either don't occur at all, or they do occur and are always the same strength for a given neuron regardless of how powerful the stimulus may be.

alpha helix: One possible shape of amino acids in a protein that resembles a spiral staircase.

alternation of generations: The two different generations of multicellular individuals, one haploid and one diploid, that occur in the life cycle of a plant.

altruism: A behavior by one individual that benefits another even if the individual performing the act reduces its own fitness; e.g., a bird giving a warning cry that tells other birds of an approaching predator increases their chances of escape while drawing attention to itself and increasing its own chances of being eaten.

alveolates: Protists that include apicomplexans, dinoflagellates, foraminiferans, and ciliates.

alveoli: In protists, tiny membranous sacs under the plasma membrane.

alveoli (sing.; alveolus): In the lungs, tiny bubble-shaped sacs where gas exchange takes place.

Alzheimer's disease: A debilitating and irreversible disease that usually affects otherwise healthy middle-aged and older people and causes a progressive loss of mental function due to accumulations of twisted filaments within the brain cells and aggregations of proteins.

Ames test: A test for carcinogenicity (cancer-causing ability) of chemical and physical agents that exposes certain bacteria to suspected carcinogens and then measures the bacterial mutation rates.

amino acid: A molecule consisting of an amino group (NH_2) and an acid group (COOH). The 20 amino acids are the basic building blocks of proteins.

amino acid–derived hormone: A hormone produced when amino acids are chemically modified and act via second messengers; includes epinephrine.

amniocentesis: A procedure for obtaining fetal cells from amniotic fluid for the diagnosis of genetic and other abnormalities.

amnion: An important fetal membrane, resembling a hollow ball. The ball encloses a cavity (amniotic cavity) of fluid (amniotic fluid) where the embryo grows during pregnancy.

amniote egg: Eggs that encase an embryo in an amnion surrounding a pool of fluid and contain a yolk.

amoeboflagellates: Single-celled protists that live in water and soil, and usually display pseudopodia.

AMP (Adenosine monophosphate): an energy molecule related to ATP but having only one phosphate group instead of two (ADP) or three (ATP).

amphibian: A cold-blooded vertebrate that starts life as an aquatic larva, breathing through gills, and metamorphoses into an air-breathing adult; includes frogs, toads, newts, and salamanders.

amphioxus: See lancelet.

amylase: A digestive enzyme that begins the breakdown of carbohydrates.

amyloplasts: In plant cells, dense starch-storing granules.

anaerobe: Cells that don't require the presence of oxygen for every harvest.

anaerobic pathway: The series of metabolic reactions that results in energy harvest in the absence of oxygen.

analogous trait: A similar trait possessed by two different groups but not by ancestors of the groups.

anaphase: A period during nuclear division when the chromosomes move toward the poles of the cell.

anaphylaxis: A life-threatening drop in blood pressure brought on by an extreme allergic reaction.

anatomy: The study of biological structures such as bones and kidneys.

ancestral trait: A trait derived from an ancestral group.

androgens: Male steroid sex hormones.

anemia: Medical condition resulting from having too few red blood cells.

anemic: Lacking in sufficient red blood cells.

angina pectoris: A condition of squeezing chest pressure and pain caused by oxygen-deprived muscles of the heart.

angiosperm: A flowering plant.

angiotensin II: A hormone that causes blood vessels to constrict, counteracting falling blood pressure. It can also trigger the adrenal glands to secrete aldosterone, which results in sodium and water reabsorption and thus even higher blood volume and pressure.

animal behavior: Animals' responses to stimuli from the environment.

Animalia: The kingdom of animals.

Annelida: The phylum containing the segmented worms, worms with tiny ringlike external segments.

annual plant: (L. *annos,* year) A plant that completes its life cycle within one year and then dies.

anorexia nervosa: An eating disorder in which a person restricts food intake severely, becomes cadaverously thin, and yet still sees himself or herself as overweight.

antagonistic muscle pair: Two muscles that move the same object, such as a limb, in opposite directions.

anther: In flowers, the terminal portion of a stamen that contains grains of pollen in pollen sacs.

Anthropoid: A member of the suborder Anthropoides, order Primates; includes humans, apes, and New and Old World monkeys.

antibody: A protein produced by the immune system in response to the entry of specific antigens (foreign substances) into the body. A specific antibody binds to the antigen that stimulated the immune response.

antibody-mediated immune response: A type of immunity in which antibody proteins circulate freely in the body, bind to invaders, and help eliminate them.

anticodon: The three bases in a tRNA molecule that are complementary to three bases of a specific codon in mRNA.

antidiuretic hormone: ADH, a posterior pituitary gland hormone that regulates the amount of water passing from the kidney as urine.

antigen: (Gr. *anti,* against + *genos,* origin) Any substance, including toxins, foreign proteins, and bacteria, that when introduced into a vertebrate animal causes antibodies to form.

antivenin: An antibody that counteracts snake venom. Antivenins are prepared by technicians who inject horses or rabbits with inactivated venom and then collect the antibodies produced by the mammal to fight the antigen.

anus: The end of the digestive tract where waste products exit the body.

aorta: In a vertebrate animal, the main artery leading directly from the left ventricle of the heart; supplies blood to most of the animal's body.

apical dominance: The inhibition of lateral buds or meristems by the apical meristem of a plant.

apical meristem: The perpetual growth zone at the tip of a plant's roots and shoots.

Apicomplexa: A group of protists that includes the malarial parasite.

apodans: Legless members of class Amphibia.

appendicitis: A medical emergency characterized by an inflamed appendix, most often based on blockage by hardened waste matter.

appendix: Finger-shaped vestigial organ that plays no known role in digestion, but may help fight infections in the gut.

arachnid: Arthropods lacking antennae and usually having eight pairs of appendages; includes the spiders, ticks, mites, and scorpions.

Archaea: A domain including single-celled prokaryotes that have cell membranes and other biochemical and genetic traits different than those of the domain Bacteria.

Archean Era: The geological era spanning the time between Earth's origins and 2.5 billion years ago.

arteriole: A thinner-walled, smaller branch of an artery, which carries blood from arteries to the capillaries in the tissues.

arteriosclerosis: A condition of blood vessels characterized by thickened walls (*sclero* = hard), an accumulation of calcium deposits, a loss of elasticity, and a decrease in blood flow.

artery: A large blood vessel with thick, multilayered muscular walls, which carries blood from the heart to the body.

arthritis: An inflammation of joints such as the shoulder, hip, knee, or knuckles of the fingers and toes; can be based on wear-and-tear or on the body's own autoimmune attack.

arthropod: A joint-legged animal, including insects, spiders, ticks, centipedes, lobsters, and crabs.

Arthropoda: Earth's largest phylum. Member animals have legs with joints, an exoskeleton made of chitin, and specialized body segments including a head, thorax, and abdomen or a fused cephalothorax and abdomen.

artificial selection: A type of breeding in which farmers, breeders, or researchers choose individuals with desired phenotypes to become the parents of succeeding generations.

asexual reproduction: A type of reproduction in which new individuals arise directly from only one parent.

association cortex: Areas in the cerebral hemispheres that appear to integrate information from other areas and reconstruct it at different levels of consciousness.

asthma: Medical condition during which normal exhalation and inhalation can be reduced. This reduction can be due to inappropriate contractions of the smooth muscle cells that wrap the smaller bronchi and bronchioles; to increased mucus secretion; and to swelling of the tissues surrounding the airways. Allergies to pollen, molds, or house mites can also trigger asthma attacks.

atherosclerosis: The most common form of arteriosclerosis, and a formal name for cardiovascular disease; a condition during which waxy deposits or plaques build up inside the arteries and gradually obstruct blood flow.

atom: (Gr. *atomos,* indivisible) The smallest particle into which an element can be broken down and still retain the properties of that element.

atomic nucleus: The central core of an atom, containing protons and neutrons.

ATP (Adenosine triphosphate): A molecule consisting of adenine, ribose sugar, and three phosphate groups. ATP can transfer energy from one molecule to another. ATP hydrolyzes to form ADP, releasing energy in the process.

atrial natriuretic factor: A hormone secreted by the heart to keep blood pressure in check. Given off by the heart when blood volume and pressure increase, ANF causes sodium to appear in the urine.

atrioventricular (AV) node: A bundle of modified cardiac muscle cells between the heart's atria and ventricles that relays electrical activity to the ventricles.

atrium (pl. atria): (L., hallway) A chamber of the heart that receives blood from the veins. In mammals, birds, amphibians, and reptiles, the heart has two atria: the right, receiving deoxygenated blood from the body, and the left, receiving oxygenated blood from the lungs. The atria pass blood into the heart's ventricles, which in turn pump it to the lungs (right ventricle) or to the rest of the body (left ventricle).

Australopithecus afarensis: A hominid with an apelike skull and teeth, a brain the size of a chimp's, long arms and short legs, but with a fairly upright stance and bipedalism.

Australopithecus garhi: A hominid whose bones have been found with signs of butchered animal bones, suggesting a very ancient human ancestor that used tools.

autoimmune disease: A disease in which a person's immune system attacks the body's own cells or tissues.

autonomic nervous system: Motor neurons of the nervous system that regulate the heart, glands, and smooth muscles in the digestive and circulatory systems.

autosome: A chromosome other than a sex chromosome.

autotroph: (Gr. *auto,* self + *trophos,* feeder) An organism, such as a plant, that can manufacture its own food.

auxin: A plant hormone that stimulates cell elongation, among other activities.

axis: The main, usually vertical, region of a plant composed of the shoot and the root.

axon: The portion of a nerve cell that carries the impulse away from the cell body.

AZT: Azidothymidine, a drug often prescribed to AIDS patients that works by inhibiting the reverse transcription of the RNA genome of HIV into DNA.

B

B cell or B lymphocyte: A type of white blood cell that makes and secretes antibodies in response to foreign substances (antigens).

bacillus (pl. bacilli): (L. *bacilli,* a rod) A bacterial cell with a rod shape.

Bacteria: One domain including single-celled prokaryotes. The other such domain is Archaea.

bacteriophage: "Bacteria eater"; a virus that infects bacterial cells.

bark: Protective, corky tissue of dead cells present on the outside of older stems and roots of woody plants.

basal cell carcinomas: Generally nonpigmented, local overgrowth of cells in the lower skin layer; the most common form of skin cancer.

Basal Cell Nevus Syndrome (BCNS): A rare condition, inherited from one or both parents, that starts in adolescence and is characterized by numerous basal cell carcinomas (a type of skin cancer) throughout life.

basal cells: Cells that lie in a layer just below the epidermis.

basal layer: The deepest layer of the epidermis, the outer layer of the skin. The basal layer contains cells that divide and replace dead skin cells; also called germ layer.

basal metabolic rate: The rate at which the body uses energy while resting.

base: Any substance that accepts hydrogen ions when dissolved in water. Basic solutions have a pH greater than 7.

base substitution: A type of mutation that occurs when one base pair replaces another.

biconcave disk: The shape of normal red blood cells, which resembles a doughnut without a hole or a round throat lozenge with a central depression.

biennial: A plant that completes its life cycle in two years and then dies.

bilateral symmetry: A body plan in which the right and left sides of the body are mirror images of each other.

bile: A bitter, alkaline, yellow fluid produced by the liver, stored in the gallbladder, and released into the small intestine that aids in the digestion and absorption of fats.

bile salts: Molecules of modified cholesterol that break up fats in the small intestine.

binary fission: Asexual reproduction by division of a cell or body into two equivalent parts.

binomial nomenclature: A system of naming organisms that assigns each species a two-word name consisting of a genus name followed by a species name.

biofeedback: Physical training that uses electronic devices to indicate changes in blood pressure, blood flow to extremities, or other external states. Used to help lower blood pressure, warm hands and feet, reduce pain, induce deep relaxation, and so on. Biofeedback training can allow conscious control over parts of the autonomic nervous system.

biogeography: The study of the geographical location of living organisms.

biological magnification: The tendency for toxic substances to increase in concentration as they move up the food chain.

biological molecules: Molecules derived from living systems. The four major types are carbohydrates, lipids, proteins, and nucleic acids.

biomass: The total dry weight of organic matter present at a particular trophic level in a biological community.

biome: Large terrestrial geographic regions containing distinctive plant communities; major ecological community types, such as a desert, rain forest, or grassland.

bioremediation: Bacterial breakdown of environmental pollutants.

biosphere: (Gr. *bios,* life + *sphaira,* sphere) That part of the planet that supports life; includes the atmosphere, water, and the outer few meters of the Earth's crust.

biotic potential: An organism's capacity for reproduction under ideal conditions of growth and survival.

bipedal: Walking consistently on two legs.

bird: Winged vertebrates with feathers, air sacs, and a four-chambered heart.

birth control: also known as contraception: the prevention of birth by blocking fertilization, by interfering with implantation, or by terminating pregnancy (abortion).

bivalve: A mollusk whose body is enclosed within two valves or shells; includes mussels and clams.

blade: The broad, flat portion of a leaf.

blending model of heredity: The idea that maternal and paternal characteristics *blend* to produce the characteristics found in the offspring, just as cream mixes with dark-brown coffee to produce the beige-colored café au lait; disproved by Mendelian genetics.

blood pressure: A force of liquid blood within the circulatory system; created by blood moving within a closed system and propelled by contractions of the heart and muscular artery walls. Narrowing of the vessel walls due to arteriosclerosis can increase blood pressure; sudden loss of blood can reduce blood pressure.

blood-clotting cascade: A series of actions involved in blood coagulation that converts fibrinogen to fibrin, and leads to a blood clot.

blood-filled cavity: Area within the body of a mollusk where internal organs and tissues are bathed by blood.

blubber: A thick layer of fat acting as insulation in such mammals as whales, porpoises, and seals.

bolus: A soft, round mass of chewed food, shaped by the tongue and suitable for swallowing.

bone: The main supporting tissue of vertebrates, composed of a matrix of collagen hardened by calcium phosphate.

bony fish: Fish whose skeleton contains bone.

book lung: A chamber with leaflike plates that exchanges gases, found in spiders and other arthropods.

booster shot: A second dose of a vaccine; can induce memory cells to differentiate and form still more antibody-producing and memory cells.

bottleneck effect: The reduced genetic diversity that results from a drastic drop in a population's size.

bowel movement: The muscular expulsion of wastes through the rectum and out the anus; also called defecation.

Bowman's capsule: In the vertebrate kidney, the bulbous unit of the nephron that surrounds the blood capillaries of the glomerulus; the region that filters water and solutes from the blood.

Brachiopoda: The animal phylum that contains the lamp shells.

brain: The body organ that performs complex neural integration and, along with the spinal cord, forms the central nervous system.

brainstem: The most posterior portion of the vertebrate brain, relaying messages to and from the spinal cord and consisting of the medulla, pons, and midbrain.

bronchiole: One of the thousands of small branches from the bronchi that lead to the alveoli in the lungs.

bronchus (pl. bronchi): One of two hollow passageways branching off the trachea and entering the lungs.

brown algae: Chromista that inhabit cool, offshore waters and range from golden brown to dark brown to black.

bryophyte: The division of the plant kingdom comprising mosses, liverworts, and hornworts.

buffer: A substance that resists changes in pH when acids or bases are added to a chemical solution.

bulimia: An eating disorder in which a person secretly binges on high-calorie foods, then purges with self-induced vomiting, laxatives, diuretics, fasting, or vigorous exercise.

bundle sheath cells: Cells that protect the xylem and phloem from direct exposure to air.

bursae (sing. bursa): (L., *bursa* = purse) Small sacs filled with fluid that cushion the movement in places where skin, tendons, muscles, and bone rub against one another.

bursitis: Inflammation of bursa sacs, usually from overuse of joints.

C

caecum: A feature of the large intestine, this is a dead-end sac that is vestigial in the human but can be large in other animals, including the koala.

callus (pl. calli): An unorganized mass of cells that forms when scientists grow plant cells in a laboratory plate or tube.

Calvin-Benson cycle: A series of reactions in the light-independent phase of photosynthesis, during which chloroplasts fix carbon dioxide, form carbohydrates, and regenerate a starting material.

cambium: Plant cell layer between the bark and wood layers that produces the secondary xylem to the inside and the phloem to the outside.

Cambrian explosion: In just a few million years at the beginning of the Cambrian period, over 500 million years ago, all of the major animal phyla we see today began to leave preserved remains in the fossil record. In the hundreds of millions of years since, no new body plans have appeared.

camouflage: Body shapes, colors, or patterns that enable an organism to blend in with its environment and remain concealed from danger.

capillary: Tiny blood vessels with walls one cell thick that permeate the tissues and organs of the body. Substances such as oxygen diffuse out of capillaries to the tissues, and waste products diffuse into capillaries from the tissues.

capillary action: The tendency of water to move upward in a thin tube. Studies show that capillary action occurs in a plant's narrow xylem vessels.

capillary bed: A network of capillaries that links arterial and venous blood vessels.

capsid: The protein coat that encases a virus.

carapace: The exoskeleton covering the cephalothorax of many arthropods. Also refers to the tough outer coverings of the turtle and armadillo.

carbon cycle: The global flow of carbon atoms from plants through animals to the atmosphere, water, soil, and back to plants.

carbonaceous chondrite: A class of meteorites containing various kinds of organic molecules.

carbon-fixing: Reactions that convert atmospheric gaseous carbon dioxide into carbohydrates.

carbonic anhydrase: An enzyme contained by red blood cells that speeds the conversion of carbon dioxide (CO_2) and water (H_2O) to bicarbonate (HCO_3^-) and hydrogen ion (H^+).

carcinogen: A cancer-causing substance.

cardiac cycle: The contraction/relaxation (systolic/diastolic) sequence of atria and ventricles that makes up a single heartbeat.

cardiac muscle: The specialized muscle tissue of the heart.

cardiac sphincter: A ring of muscle located at the junction of the stomach and esophagus.

carotenoids: Plant pigments that absorb green, blue, and violet wavelengths and reflect red, yellow, and orange light.

carpel: A flower's female structure; the carpel is often shaped like a wine bottle and contains the ovary housing the ovules, a necklike style, and a sticky stigma on which pollen grains germinate.

carrier: [1] In genetics, a heterozygous individual not expressing a recessive trait but capable of passing it on to her or his offspring. [2] In biochemistry, a substance, often a protein, that transports another substance.

carrying capacity: The density at which growth of a population ceases due to the limitation imposed by resources.

cartilaginous fish: Fish whose skull, vertebrae, and other skeletal parts are made of cartilage instead of bone; includes skates and rays.

Casparian strip: A water-resistant belt encircling each endodermal cell of a plant, ensuring that any water or minerals entering the plant must flow through the cytoplasm of the endodermal cells.

castes: Subgroups of social insects that differ in appearance and behavior.

catalyst: A substance that speeds up the rate of a chemical reaction by lowering the activation energy without itself being permanently changed or used up during the reaction.

cell: The basic unit of life; cells are bounded by a lipid-containing membrane and are generally capable of independent reproduction.

cell body: The major portion of the neuron; houses the nucleus and other organelles.

cell cycle: The events that take place within the cell between one cell division and the next.

cell plate: In plant cells, a partition that arises during late cell division from vesicles at the center of the cell and that eventually separates the two daughter cells.

cell proliferation: The body's process of duplicating B cells for the purpose of fighting off invaders such as the rhinoviruses that cause the common cold.

cell theory: The biological doctrine stating that all living things are composed of cells; cells are the basic living units within organisms; the chemical reactions of life occur within cells and all cells arise from pre-existing cells.

cell wall: A fairly rigid structure that encloses prokaryotic, fungal, and plant cells.

cell-mediated immune response: A type of immunity in which T cells directly attack invaders.

cellular respiration: The harvest of energy from sugar molecules in the presence of oxygen.

cellular slime mold: A type of funguslike protist, usually existing as free-living amoebalike cells, but aggregating into a multicellular fruiting body before producing reproductive spores.

cellulose: (L. *cellula,* little cell) The chief constituent of the cell wall in green plants, some algae, and a few other organisms. Cellulose is an insoluble polysaccharide composed of straight chains of glucose molecules.

Cenozoic Era: The current geologic era, starting about 65 million years ago.

centipedes: Arthropods with a series of flattened body segments, each bearing a pair of jointed legs.

central nervous system (CNS): The part of the nervous system consisting of the brain and spinal cord, performing the most complex nervous system functions.

centriole: Pairs of short, rod-shaped organelles that organize the cytoskeletal fibers called microtubules into scaffolds. These intracellular frameworks help maintain cell shape and move chromosomes during cell division.

centromere: The point on the chromosome where the spindle attaches and also where the two chromatids are joined.

cephalization: (Gr. *kephale,* little head) A type of animal body plan or organization in which one end contains a nerve-rich region and functions as a head.

Cephalochordata: A subphylum containing the lancelets or amphioxus which, as adults, have the five characteristics of chordates.

cephalopod: A member of the class Cephalopoda, phylum Mollusca, including squids, octopuses, and the chambered nautilus.

cephalothorax: The fused head and thorax of many arthropods.

cerebellum: Anterior portion of the hindbrain region of the vertebrate brain lying posterior to the cerebrum; it integrates information about body position and motion, coordinates muscular activities, and maintains equilibrium.

cerebral cortex: The highly convoluted surface layers of the cerebrum, containing about 90 percent of a human brain's cell bodies. It is well-developed only in mammals and particularly prominent in humans, and is the region involved with our conscious sensations and voluntary muscular activity.

cerebrum: (L., brain) The forebrain of the vertebrate brain; coordinates and processes sensory input and controls motor responses.

chaparral: A biome that borders deserts and grasslands, characterized by hot, dry summers and cool, wet winters, and low woody shrubs that are often fragrant and have generally thick, waxy, evergreen leaves.

chelicerae: Poison fangs of an arachnid used in killing prey or during self-defense.

chemical bond: An attractive force that keeps atoms together in a molecule.

chemical reaction: The making or breaking of chemical bonds between atoms or molecules.

chemical senses: Taste and smell; considered to be "chemical senses" because they receive and detect flavor and odor molecules—actual chemical tidbits of the environment.

chemical warfare: A defense strategy of prey species in which these organisms produce distasteful oils or other toxic substances that kill or harm predators.

chemoautotroph: An organism that derives energy from a simple inorganic reaction.

chemosynthetic: A few kinds of prokaryotes are said to be chemosynthetic because they can synthesize the sugar molecules they need for glycolysis by extracting energy from inorganic chemicals such as hydrogen gas and hydrogen sulfide.

chemotherapy: A drug treatment for a disease such as cancer that involves the action of specific chemicals.

chitin: A complex nitrogen-containing polysaccharide that forms the cell walls of certain fungi, the major component of the exoskeleton of insects and some other arthropods, and the cuticle of some other invertebrates.

chlamydia: Bacterial species that live inside animal cells and lack an ability to make their own ATP; can cause a sexually transmitted disease.

chlorophyll: (Gr. *chloros,* green + *phyllon,* leaf) Light-trapping pigment molecules that act as electron donors during photosynthesis.

chloroplast: An organelle present in algae and plant cells that contains chlorophyll and is involved in photosynthesis.

choanocyte: Cells that line the pores perforating the body wall of sponges. Each cell has a flagellum that draws a current of water through the pore, and a sticky collar, which catches bacteria, protists, and other small organic particles suspended in the water; also called collar cells.

choanoflagellates: Single-celled or colonial protists living in fresh water and in the oceans. Each has a collar formed by a ring of microvilli.

cholecystokinin: A digestive hormone secreted by the small intestine that triggers the pancreas to release protein-digesting enzymes.

Chondrichthyes: A class of the phylum Chordata; includes cartilaginous fish, such as sharks and rays.

Chordata: A phylum that includes animals with a notocord and, often, a vertebral column.

chorion: A fluid-filled sac that surrounds the embryos of reptiles, birds, and mammals. In placental mammals it gives rise to part of the placenta and produces hCG, a hormone that maintains pregnancy by preventing menstruation.

chorionic villus sampling: A procedure for removing fetal cells from the chorion in the developing placenta for the diagnosis of genetic defects.

chromatid: A daughter strand of a duplicated chromosome. Duplication of a chromosome gives rise to two chromatids joined together at the centromere.

chromatin: The substance of a chromosome.

Chromista: Protists with golden, brownish, and greenish pigments.

chromosomal mutation: A genetic change that affects large regions of chromosomes or entire chromosomes.

chromosome: A self-duplicating body in the cell nucleus made up of DNA and proteins and containing genetic information. A human cell contains 23 pairs of chromosomes. In a prokaryote, the DNA circle that contains the cell's genetic information.

chyme: The semifluid contents of the stomach consisting of partially digested food and gastric secretions.

ciliate: A protozoan whose cells have rows of cilia that are used in locomotion and in sweeping food particles into the mouth.

cilium (pl. cilia): (L., eyelash) A short, centriole-based, hairlike organelle. Rows of cilia propel certain protists. Cilia also aid the movement of

substances across epithelial surfaces of animal cells.

circadian clock: An internal 24-hour clock.

circadian rhythms: (L. *circa,* about + *dies,* day) Regular rhythms of activity that occur on about a 24-hour cycle.

circulation: The path of blood or a blood equivalent such as hemolymph as it moves through an animal's body, carrying oxygen and nutrients to tissue cells and carrying wastes away; can be open or closed.

circulatory system: An organ system, generally consisting of a heart, blood vessels, and blood, that transports substances around the body of an animal.

clade: A taxonomic group made up of an ancestral organism and all the descendents derived from it.

cladistics: A method of reconstructing patterns of descent using traits or characteristics thought to have evolved only within the group under consideration; for example, hair evolved only in mammals.

class: A taxonomic group comprising members of similar orders.

classical conditioning: A type of learning in which an initially neutral stimulus (e.g., the sound of a bell) evokes a given response (e.g., salivation) by being paired repeatedly with a different stimulus that normally evokes that response (e.g., the presence of food).

cleavage: The early, rapid series of mitotic divisions of a fertilized egg resulting in a hollow sphere of cells known as the blastula.

climate: The accumulation of seasonal weather patterns in an area over a long period of time.

climax community: The most stable community in a habitat and one that tends to persist in the absence of a disturbance.

clonal selection mechanism: The combined process of the binding of a B cell to an antigen and proliferation into a clone of B cells.

clone: All cells derived from a single parent cell. A replica of a DNA sequence produced by recombinant DNA technologies.

closed circulatory system: A fluid transport system in segmented worms and vertebrates in which blood is completely contained inside vessels. This system is more efficient than an open circulatory system because the fluids can be shunted more easily to specific areas and the circulatory system can exert more pressure on the moving blood, forcibly distributing it to distant areas.

Cnidaria: The phylum containing animals with a radial body plan and nematocysts.

coagulation: The formation of a blood clot.

coccus (pl. cocci): A bacterial cell with a spherical shape.

coccyx: In vertebrates, the tailbone.

cochlea: A spiral tube in the inner ear that contains sensory cells involved in detecting sound and analyzing pitch.

codominance: The genetic situation in which both alleles in a heterozygous individual are fully expressed in the phenotype; this is a characteristic of human blood types.

codon: Three adjacent nucleotide bases in DNA or mRNA that code for a single amino acid.

coelacanth: A type of lobe-finned fish.

coelom: (Gr. *koiloma,* a hollow) The main body cavity of many animals, formed between layers of mesoderm, in which internal organs are suspended.

coevolution: The evolution of one species in concert with another as a result of their interrelationship within a biological community.

cohesion: The tendency of like molecules to cling to each other.

collagen: A fibrous protein found extensively in connective tissue and bone. Collagen also occurs in some invertebrates, as in the cuticle of nematode worms.

collecting duct: Any duct that drains an organ; in the kidney, the ducts that drain the renal pelvis of the kidney and carry urine to the ureters.

collenchyma: Plant tissue providing support to young, actively growing structures; it consists of closely packed cells with unevened, thickened corners.

colon (large intestine): The last 1.2 meters (4 feet) of the human alimentary canal, which leads to the rectum.

color blindness: Genetic condition that causes an inability to distinguish specific colors.

commensalism: A relationship between two species in which one species benefits and the other suffers no apparent harm.

communication: A signal produced by one individual that alters the behavior of a recipient individual and generally benefits both participants.

community: Two or more populations of different interacting species occupying the same area.

compact bone: Bone with regions of few, small spaces.

companion cell: In a flowering plant, a specialized cell associated with sieve tube members in phloem.

compartmentalization: The walling-off process of damaged areas of a plant that prevents invaders from gaining access to healthy tissues.

competitive exclusion: A situation in which one species eliminates another through competition.

complement system: A group of about 20 plasma proteins (a type of antimicrobial protein) that attack and destroy microbes and stimulate inflammation.

complementary base pairing: In nucleic acids, the hydrogen bonding of adenine with thymine or uracil, and guanine with cytosine. It holds two strands of DNA together and holds different parts of RNA molecules in specific shapes, and is fundamental to genetic replication, expression, and recombination.

compound leaves: Leaves that consist of many small leaflets.

condensation reaction: The forming of a molecule such as a disaccharide by means of a reaction during which an enzyme removes a hydrogen atom from one sugar and an —OH from the other sugar and joins them together to make water, while linking the two sugars together.

cone: One of the two types of photoreceptor (light-sensitive) cells in the retina of the vertebrate eye; cones are sensitive to high levels of light and to color. (*See also* rod.)

conifer: A cone-bearing tree.

coniferous forest: A biome that occurs across much of Canada, northern Europe, and Asia with vast forests of coniferous trees growing at latitudes with cold, snowy winters and short summers.

conjugation: In some prokaryotes, the temporary union of two unicellular organisms of different mating strains, during which time the genetic material is transferred from one to the other.

connective tissue: Animal tissue that connects or surrounds other tissues and whose cells are embedded in a collagen matrix; bones and tendons are mainly connective tissue.

consumer: In ecology, an organism that eats other organisms.

contractile ring: The ring of cytoskeletal elements (actin filaments) that separates one cell into two during the division of the cytoplasm.

control group: In a scientific experiment, a group of samples treated the same as the experimental group in all ways except not being subjected to the key manipulated variable. The control group provides a baseline for evaluating the effects of manipulation on the experimental group.

control: A check of a scientific experiment based on keeping all factors the same except for the one in question.

convergent evolution: Evolution of similar characteristics in two or more unrelated species; often found in organisms living in similar environments.

coral reefs: Underwater structures, sometimes enormous, produced by reef-building corals in the shallow ocean zones of the tropics.

coreceptor: In a T cell, a molecule associated with the T cell receptor that enhances the T cell's ability to bind specific antigens.

cork cambium: A type of lateral meristem that produces cork-forming cells.

cork: An external secondary plant tissue impermeable to water and gases.

cornea: The transparent outer portion of the vertebrate eye, through which light passes.

coronary arteries: The arteries of the heart.

corpus callosum: A tissue bridge between the left and right cerebral hemispheres.

corpus luteum: The group of follicle cells left behind in the ovary after an egg has been released. If the egg is fertilized, the corpus luteum remains active in the ovary, producing hormones that help to maintain the pregnancy. If the egg is not fertilized, the corpus luteum degenerates.

cortex: [1] The outer surface of an organ. In the kidney, the region where blood filtration takes place. [2] In plants, parenchyma tissue bounded externally by the epidermis and internally by phloem in the stem and pericycle in the root.

cortisol: A steroid hormone secreted by the adrenal cortex that helps an animal respond to stress by speeding the metabolism of sugars, proteins, and fats.

cotyledon: The first leaves of the embryonic plant, they often store food in dicotyledon plants and absorb food in monocotyledon plants.

countercurrent flow (or countercurrent exchange): A mechanism in which fluids with two different characteristics (such as temperature or solute concentrations) flow in opposite directions and can exchange energies or substances at points of contact along a concentration gradient.

covalent bond: (L. *co,* together + *volere,* sharing) A form of molecular bonding characterized by the sharing of a pair of electrons between atoms.

cranium: Part of the skull that encloses and protects the brain.

creatine phosphate: A high-energy molecule that can transfer energy to ATP, in the immediate energy system.

CRH (corticotropin-releasing hormone): A neurohormone secreted from the hypothalamus that diffuses into special blood vessels that carry it directly to the pituitary gland where it stimulates the release of ACTH or adrenocorticotropic hormone. This in turn induces the adrenal glands to secrete stress hormones such as cortisol.

crista (pl. cristae): A fold or folds formed by the inner membrane of a mitochondrion.

critical period: A specific developmental stage during which imprinting can occur in many kinds of animals and after which it cannot.

cross-fertilize: To deliberately cross two organisms; in plants, to transfer pollen from one self-fertilizing flower to another.

crossing over: The reciprocal exchange of DNA between homologous chromosomes during meiosis.

crustacean: Arthropods with a protective shell and two pairs of antennae, including crabs, shrimp, and lobsters.

ctenophores: The group of animals commonly known as comb jellies.

cultural evolution: The accumulation of useful skills and knowledge passed down through thousands of human generations.

cultural transmission: The learning of survival techniques from one's elders.

culture: Ways of living particular to a population.

cuticle: A waxy or fatty noncellular, waterproof outer layer on epidermal cells in plants and some invertebrates. In parasitic flatworms it prevents the worm from being digested in the gut of the host organism. In insects, the exoskeleton.

cyanobacteria (pl. cyanobacterium): One of the blue-green algae; a photosynthetic, oxygen-generating and nitrogen-fixing prokaryote.

cytokine: A short protein that acts as an inflammation mediator.

cytokinesis: The process of cytoplasmic division following nuclear division.

cytokinin: A growth-promoting plant hormone that regulates cell division, among other things.

cytoplasm: The part of a cell between the plasma membrane and the nuclear envelope.

cytoskeleton: Found in the cells of eukaryotes, an internal framework of microtubules, microfila-ments, and intermediate filaments that supports and moves the cell and its organelles.

cytotoxic T cell: A type of T cell that destroys foreign cells, including cells transplanted to a person from another individual or body cells transformed by virus infection or cancer.

D

Darwin's conclusion: Part of Darwin's theory of evolution and the result of his observations. Individuals whose hereditary traits allow them to cope more efficiently with the local environment are more likely to survive and produce offspring than individuals lacking those traits.

day-neutral plant: Plants that flower without regard to photoperiod.

deciduous: (L. *decidere,* to fall off) In plants, the property of shedding leaves at the end of the growing season.

decomposer: A type of consumer, also called a detritivore, that obtains energy and materials from organic wastes and dead organisms that accumulate from all trophic levels.

demographic transition: A changing pattern from a high birth rate and high death rate to a low birth rate and low death rate.

dendrite: (Gr. *dendron,* a tree) The branching projections of a neuron or nerve cell that transmit nerve impulses to the cell body.

depolarized: The state of a neuron after losing its electrical polarization.

derived trait: A newly originated inherited change.

dermal tissue system: The protective and water-proofing tissues of a plant, consisting of epidermis and bark.

dermis: The skin layer just below the epidermis or outer layer; the dermis contains tiny blood vessels, sweat glands, hair roots, and nerve endings.

descent with modification: The notion that all organisms are descended with changes from common ancestors.

desert: A very dry, often barren biome characterized by temperature extremes and by widely spaced plants with thick, waxy leaves and often protective spines.

detritivore: A consumer organism that obtains energy from dead organisms and/or organic waste matter.

detritus: A collective term for dead organisms and organic waste matter.

deuterostomes: An animal whose first opening in the embryo (blastopore) becomes the anus, while the second opening to develop becomes the mouth; includes echinoderms and chordates.

development: The process by which an offspring increases in size and complexity from a zygote to an adult.

diabetes mellitus: A hereditary condition in which insufficient insulin is produced and glucose accumulates in the blood.

diactomaceous earth: Crumbly white sediments made up of diatom shells.

dialysis: A filtration process that employs a medical device to mimic the function of a properly working kidney.

diaphragm: The muscle sheet that separates the thoracic cavity from the abdominal cavity and helps draw air into the lungs during breathing.

diastole: The relaxation phase of the heart muscle.

diatoms: Phytoplankton that are members of Chromista and usually contain golden pigments.

dicot: Short for dicotyledon, the larger of the two classes of angiosperms (flowering plants). Dicots are characterized by having two cotyledons (seed leaves); floral parts are usually in fours or fives; and leaves are typically net-veined. Maple trees and tomato plants are dicots.

diencephalon: (*di* = second + *cephalon* = brain) An extension of the brainstem anterior to the cerebellum that includes the thalamus, hypothalamus, and pineal gland.

dietary fiber (roughage): Cellulose plant fibers and cell walls that cannot be digested. Fiber is important for helping to propel wastes through the digestive system.

differentiation: The process by which a cell becomes specialized for a particular function.

diffusion: (L. *diffundere,* to pour out) The tendency of a substance to move from an area of high concentration to an area of low concentration.

digestion: The mechanical and chemical breakdown of food into small molecules that an organism can absorb and use; the second step of the digestive process.

dihybrid cross: A mating between two individuals in which the investigation follows the inheritance of only two traits.

dikaryon: In fungi, a single cell containing two haploid nuclei in one cytoplasm resulting from the union of a haploid cell of the plus mating type with a haploid cell of the minus mating type.

dinoflagellates: Protists with armorlike coverings and two flagella, one in a beltlike groove and the other trailing behind; often can cause red tides.

dinosaur: An extinct giant reptile; the dominant form of land vertebrate during the Jurassic and Cretaceous periods.

dioecious: (Gr., two houses) Species of plants, like willows and hemp, that have two types of individual plants, some with only male flowers and others with only female flowers.

diploid: A cell that contains two copies of each type of chromosome in its nucleus (except, perhaps, sex chromosomes).

Diplomonad: The group of protists that includes *Giardia*, a common human parasite.

directional selection: A type of natural selection in which an extreme form of a character is favored over all other forms.

disaccharide: A type of carbohydrate composed of two linked simple sugars.

disassembly: The end stage of protein synthesis during which the newly formed polypeptide falls away from the ribosome, and the ribosome's "workbench" components disassemble.

disordered loop: Regions of a less specific but constant shape that link other parts of a protein.

disruptive selection: A type of natural selection, in which two extreme and often very different phenotypes become *more* frequent in a population.

distal tubule: The convoluted tubule in the vertebrate kidney that receives the forming urine after it has passed through the loop of Henle.

diuresis: Abundant urine production.

divergent evolution: The splitting of a population into two reproductively isolated populations with different alleles accumulating in each one. Over geologic time, divergent evolution may lead to speciation.

division: A taxonomic group of similar classes belonging to the same phylum, which is often called a division in the kingdoms of plants or fungi.

division phase: During the cell cycle, the partitioning of the cells' internal organelles, and the dividing of the cell in two.

DNA: Deoxyribonucleic acid, the twisting, ladderlike molecule that stores genetic information in cells and transmits it during reproduction. DNA consists of two very long strands of alternating sugar and phosphate groups that form a double helix. Each "rung"

of the ladder is a base (either adenine, guanine, cytosine, or thymine) that extends from the sugar and bonds with a base from the other strand. Genes are made of DNA.

DNA ligase: A protein that joins DNA strands together end-to-end during DNA recombination and replication.

DNA polymerase: The enzyme that catalyzes the polymerization of DNA strands.

domain: A taxonomic group composed of members of similar kingdoms.

dominance hierarchy: A behavioral phenomenon in which members of a group are ranked from highest to lowest. Ranking is based on past success in aggressive encounters with the other members of the group; the individual winning all fights is the dominant member and has greater access to food and mates.

dominant: In genetics, an allele or corresponding phenotypic trait that is expressed in the heterozygote (in other words, that shows in the hybrid).

dopamine: A specific neurotransmitter in the brain area involved in integrating voluntary motor signals; its action can be prolonged by cocaine and certain other drugs and deficits of it are associated with Parkinson's disease.

dormancy: (L. *dormire,* to sleep) A state of reduced physiological activity that occurs in seeds and buds of plants, particularly over the winter. Normal physiological activities such as growth only resume if certain conditions, such as increased temperature, are met.

double fertilization: In flowering plants, the process in which one sperm from a pollen grain fuses with the egg cell of the gametophyte, while the second sperm penetrates the adjoining large endosperm cell and fuses with the two polar nuclei.

double helix: The term used to describe the physical structure of DNA, which resembles a ladder twisted along its long axis.

Down syndrome: A genetic condition resulting from having an extra chromosome 21; characterized by mental retardation, heart malformation, and certain external physical traits.

Duchenne muscular dystrophy (DMD): A degenerative muscle condition and the most common *X*-linked recessive genetic disease, striking one out of every 3500 boys.

duodenum: The upper section of the small intestine.

E

eardrum: A taut membrane that stretches like a drumskin across each ear canal.

early-loss survivorship curve: A plot of survivorship data indicating that most individuals in a population die young.

ecdysone: A steroid hormone that regulates the timing of molting of insects and other arthropods.

ecdysozoans: Molting animals, including the arthropods, nematodes, and several smaller groups.

Echinodermata: A phylum that includes "spiny-skinned" animals like the sea stars, brittle stars, sea urchins, and sea cucumbers.

ecology: The scientific study of how organisms interact with their environment and with each other and of the mechanisms that explain the distribution and abundance of organisms.

ecology of populations and communities: The study of how groups of organisms are distributed in a particular area at a particular time and how they interact with other species coexisting in the same locale.

ecosystem: A community of organisms interacting with a particular environment.

edema: Swelling.

EEG (electroencephalography): A tool used in studying the brain; involves a cap, electrodes, and a sophisticated monitoring and recording device to track brain wave activity.

effector: One of three primary elements characteristic of a homeostatic system; an effector executes the decisions. Also see integrator and receptor.

egg: The haploid female gamete.

egg sac: In flowering plants, a group of seven cells with eight nuclei that comprise the multicelled haploid generation (gametophyte).

electromagnetic spectrum: The full range of electromagnetic radiation in the universe, from highly energetic gamma rays to very low-energy radio waves.

electron: A negatively charged subatomic particle that orbits the nucleus of an atom. The negative charge of an electron is equal in magnitude to the proton's positive charge, but the electron has a much smaller mass.

electron carriers: Molecules that transfer electrons from one chemical reaction to another. Examples are NADH and $FADH_2$.

electron microscope: Instrument that magnifies objects greatly for viewing by focusing beams of electrons with wavelengths 100,000 times shorter than visible light. These microscopes use magnets instead of ground glass lenses to focus the electron beams, and these beams are generated in a vacuum.

electron transport chain: The second stage in aerobic respiration, in which electrons are passed down a series of molecules, gradually releasing energy that is harvested in the form of ATP.

element: A pure substance that cannot be broken down into simpler substances by chemical means.

elimination: In this last step of the digestive process, most of the water in intestinal wastes is absorbed into the circulation and undigested residues are eventually eliminated.

elongation: The second main stage of protein synthesis, during which the sequential enzymatic addition of amino acids builds the growing polypeptide chain.

embryo: An organism in the earliest stages of development. In humans this phase lasts from conception to about two months and ends when all the structures of a human have been formed; then the embryo becomes a fetus.

EMG (electromyograph): A tool used in studying muscle function; records electrical changes in the muscles during stimulation and plots the changes on paper.

emphysema: A degenerative disease in which the alveoli of the lung steadily deteriorate and breathing becomes difficult or impossible.

enamel: The substance covering teeth that is composed of calcium salts and protein; the hardest substance in the body.

endocrine gland: A gland that secretes hormones directly into the extracellular fluid or blood rather than into a duct, and includes the pituitary, adrenal, and thyroid glands; contrasts with exocrine glands.

endocrine system: The collection of endocrine glands of the body.

endocytosis: The process by which a cell membrane invaginates and forms a pocket around a cluster of molecules. This pocket pinches off and forms a vesicle that transports the molecules into the cell.

endodermis: A unicellular layer of tightly packed cells that surrounds the vascular tissue of plants. Water must pass through this layer to reach the rest of the plant.

endometriosis: A condition occurring when some of the endometrial tissue grows inappropriately outside the uterus on other pelvic or-

gans, such as the ovaries or oviducts. This condition can prevent pregnancy.

endoplasmic reticulum (ER): A system of membranous tubes, channels, and sacs that forms compartments within the cytoplasm of eukaryotic cells; functions in lipid synthesis and in the manufacture of proteins destined for secretion from the cell.

endorphins: Peptides found in the anterior lobe of the pituitary and in several other regions of the brain and spinal cord that can act as a natural pain killer and are the source of the "runner's high."

endoskeleton: (Gr. *endos,* within + *skeletos,* hard) An internal supporting structure, such as the bony skeleton of a vertebrate.

endosperm: In flowering plants, the triploid nutritive tissue of the developing embryo.

endospore: A heavily encapsulated resting cell formed within many types of bacterial cells during times of environmental stress.

endosymbiont hypothesis: The idea that mitochondria and chloroplasts originated from prokaryotes that fused with a nucleated cell.

end-Permian extinction: The mass extinction that took place at the end of the Permian period, about 213 million years ago.

energetics: The study of energy intake, processing, and expenditure.

energy: The power to perform chemical, mechanical, electrical, or heat-related work.

energy pyramid: The energy relationships between different trophic levels in an ecosystem.

energy shells: Energy levels occupied by electrons in orbit around an atomic nucleus. Each shell can contain a maximum number of electrons, for example, two electrons for the first shell, eight for the second.

energy-absorbing reactions: Reactions that will not proceed spontaneously and do not give off heat.

energy-releasing reactions: Reactions that proceed spontaneously or need a starting "push" but eventually release energy.

entropy: A measure of disorder or randomness in a system.

enzyme: A protein that facilitates chemical reactions by lowering the required activation energy but is not itself permanently altered in the process; also called a biological catalyst.

enzyme-substrate complex: In an enzymatic reaction, the unit formed by the binding of the substrate to the active site on the enzyme.

epicotyl: The portion of the plant embryonic axis above the cotyledons but below the next leaf.

epidermis: The outer layer of cells of an organism.

epiglottis: A flap of tissue just above the larynx that closes during swallowing and prevents food from entering the lungs.

epilepsy: Recurring nervous system attacks that often result in seizures and/or unconsciousness and can affect body movements and sensations.

epinephrine: A hormone secreted by the medulla of the adrenal glands, associated with the physiological responses to alarm such as increased concentration of sugar in the blood, raised blood pressure and heart rate, and increased muscle power and resistance to fatigue; also known as adrenaline.

epistasis: The interaction between two nonallelic genes in which one gene masks the expression of the other.

epithelial cell: A cell in epithelial tissue; usually lines ducts leading from the lungs, stomach, pancreas, sweat glands, and reproductive organs.

epithelial tissue: A major tissue type that covers the body surface and lines the body cavities, ducts, and vessels.

erythroblastosis fetalis: A serious anemia that a newborn can contract when mother and father have Rh blood type incompatibility.

erythrocyte: A red blood cell; its main function is transporting oxygen to the tissues.

erythropoietin: A hormone that stimulates the production of new red blood cells.

esophagus: The muscular tube leading from the pharynx to the stomach.

essential amino acid: Any one of the eight amino acids that the human body cannot manufacture and that must be obtained from food.

essential fatty acid: An unsaturated fatty acid that cannot be manufactured by the body and that therefore must be obtained from food. In humans, linoleic acid is an essential fatty acid.

estrogen: A female sex hormone produced by the ovary; prepares the uterus to receive an embryo, and causes secondary sex characteristics to develop.

estuary: The area where a river meets an ocean. Estuaries are some of Earth's most productive ecosystems.

ethylene: A gaseous plant hormone produced by ripening fruits that stimulates ripening in nearby fruits.

Eucarya: The domain including all organisms whose cells possess a true nucleus and other membrane-bound organelles.

eudicot: Formerly called "dicots," one group of angiosperms (flowering plants) that includes hemp, oak trees, peach trees, daisies, roses, and broccoli.

euglenoid: Green, spindle-shaped protists with eyespots.

eukaryote: An organism made up of one or more eukaryotic cells.

eukaryotic cell: A cell whose DNA is enclosed in a nucleus and associated with proteins; contains membrane-bound organelles.

Eumycota: The "true fungi" branch on the tree of life.

eutrophic: An aquatic environment with high phosphorous and other nutrient levels, characterized by dense blooms of algae and other aquatic plants and a decrease in dissolved oxygen.

eutrophication: In an aquatic environment, an oversupply of nutrients that support primary production.

evolution: Changes in gene frequencies in a population over time.

evolutionary fitness: An individual's ability to survive and reproduce in a particular environment relative to other members of its species.

excretion: The elimination of metabolic waste products from the body; in vertebrates, the main excretory organs are the kidneys.

excretory system: System that [1] cleanses the blood of organic waste molecules and [2] carries them out of the body as urine or its equivalent through a special set of excretory tubes.

exhalation: The passive release of air from the lungs.

exocrine gland: A gland with its own transporting duct that carries its secretions to a particular region of the body, and includes the digestive and sweat glands; contrasts with endocrine glands.

exocytosis: The process by which substances are moved out of a cell by cytoplasmic vesicles that merge with the plasma membrane.

exons: The portions of the gene that are transcribed and then spliced together and appear in the final RNA.

exoskeleton: The thick cuticle of arthropods, made of chitin.

experimental group: In a scientific experiment, the group of samples or individuals that is subjected to the key manipulated variable.

experimental: During the application of the scientific method, the phase involving the carefully planned and measured test of the hypothesis.

exponential growth: Growth of a population without any constraints; hence, the population will grow at an ever-increasing rate.

extracellular digestion: The enzymatic breakdown of nutrient molecules that occurs outside of cells.

extracellular fluid: The fluid within the body but outside of cells.

extracellular matrix: A meshwork of secreted molecules that act as a scaffold and a glue that anchors cells within multicellular organisms.

extremophiles: Prokaryotes that survive in Earth's most extreme environments.

F

facet: Six-sided segments that make up the compound eyes of arthropods.

facilitated diffusion: A type of transport in which a protein helps a substance pass across a cell membrane down its concentration gradient without energy expenditure by the cell.

family: A taxonomic group comprising members of similar genera.

fast oxidative-glycolytic muscle fibers: Muscle fibers that have characteristics midway between those of fast-twitch and slow-twitch muscle fibers; these fibers are moderately powerful and moderately resistant to fatigue.

fast-twitch muscle fibers: Muscle fibers that obtain most of their ATP from glycolysis; also called fast glycolytic muscle fibers. Fast-twitch fibers provide quick power of the type needed for the 50-yard dash.

fat: An energy storage molecule that contains a glycerol bonded to three fatty acids. Fats in the liquid state are known as oils.

fatty acid: A molecule composed of a long tail of carbon and hydrogen atoms, and a head consisting of an organic acid.

fatty acid hormone: A hormone derived from straight-chain fatty acids; for example, the structure of prostaglandin.

feather: A flat, light, waterproof epidermal structure on a bird; collectively functions as insulation and for flight.

feces: Semisolid undigested waste products that are stored in the rectum until excreted.

feedback inhibition: The buildup of a metabolic product that in turn inhibits the activity of an enzyme. Since this enzyme is involved in making the original product, the accumulation of the product turns off its own production.

feedback loop: A control system involving a series of steps that sense a change and, in turn, influence the functioning of the process; can be positive or negative.

femur: The large bone extending between the hip and knee joints.

fermentation: Extraction of energy from carbohydrates in the absence of oxygen, generally producing lactic acid or ethanol and CO_2 as byproducts.

fertility curve: Generally, graphs that plot reproduction rate versus the age of female population members.

fertilization: The fusion of two haploid gamete nuclei (egg and sperm), which forms a diploid zygote.

fetus: An unborn offspring after the embryonic period, possessing all organs in rudimentary form. In humans, an embryo that has reached its eighth week.

fibers: In plants, long, thin sclerenchyma cells that surround and support the internal transport vessels.

fibril: Long bundles found within a muscle cell, composed of sarcomeres.

fibrin: The activated form of the blood-clotting protein fibrinogen that combines into threads when forming a clot.

fibrinogen: A protein in blood plasma that is converted to fibrin during clotting.

fibrous root: One of a system of many branching narrow roots in which no one root is more prominent than another.

fight-or-flight response: In this rapid response to a sudden threat, nerve signals from the spinal cord cause the adrenal medulla to increase its secretion rate for epinephrine and norepinephrine.

filament: The "pin shaft," upon which sits the anther of a flower that aids pollen dispersal, especially in plants where wind, instead of insects, carries away the pollen.

filter feeders: Aquatic organisms with feeding structures that can strain out and collect tiny food particles suspended in water.

filtrate: In the vertebrate kidney, the fluid in the kidney tubule (nephron) that has been filtered through the Bowman's capsule.

filtration: The process in the vertebrate kidney in which blood fluid and small dissolved substances pass into the kidney tubule, but large proteins and blood cells are filtered out and remain in the blood.

first filial (F_1) generation: In Mendelian genetics, the first generation in the line of descent.

fixed-action pattern: In behavioral science, an action that continues to completion even if the stimulus causing the behavior is removed, e.g., a bird that rolls an empty eggshell out of its nest will continue the rolling process to the edge of the nest even if the eggshell is removed during the procedure.

flagellum (pl. flagella): Long whiplike organelle protruding from the surface of the cell that either propels the cell, acting as a locomotory device, or moves fluids past the cell, becoming a feeding apparatus.

flatworm: A member of the phylum Platyhelminthes, the simplest animal group to display bilateral symmetry and cephalization.

flower: A reproductive structure that contains the carpel and at least one stamen.

fMRI (functional magnetic resonance imaging): A sophisticated technique used in studying brain plasticity based on magnetic resonance imaging that creates high-resolution images of brain areas as they function.

follicle stimulating hormone (FSH): Hormone produced by the anterior pituitary that acts on the ovaries to simulate growth of the ovarian follicle and in testes, stimulating sperm production.

follicle: An oocyte (immature ovum) and its surrounding follicular cells.

food chains: The levels of feeding relationships among organisms in a community.

food web: Complex, interconnected feeding relationships between all the species in a community or ecosystem.

foot: The locomotive organ in mollusks.

foraminiferans: Delicately shaped protists that live in the oceans and secrete usually whitish, calcium-based shells.

fossil: Traces or remains of living things from a previous geologic time.

founder effect: In evolutionary biology, the principle that individuals founding a new colony carry only a fraction of the total gene pool present in the parent population.

four-chambered heart: A heart with four chambers that completely separates oxygenated and deoxygenated blood.

frond: The leaflike structure of an individual alga that collects sunlight and produces sugars. Also refers to the large divided leaf on a fern.

fruit: In flowering plants, a ripe, mature ovary containing seeds.

fruiting bodies: A spore-producing reproductive structure in many fungi.

functional group: A group of atoms that confers specific behavior to the (usually larger) molecules to which they are attached.

fundamental niche: The potential range of all environmental conditions under which an organism can thrive.

Fungi: The kingdom comprising multicellular heterotrophs such as mushrooms or molds that decompose other biological tissues.

fur: A body covering of hair.

G

G₁ phase: The portion of the cell cycle that follows mitosis but precedes DNA synthesis.

G₂ phase: The portion of the cell cycle that follows DNA synthesis but precedes mitosis.

gallbladder: The sac beneath the right lobe of the liver that stores bile.

gamete: (Gr., wife) A specialized sex cell, such as an ovum (egg) or sperm, that is haploid. A male gamete (sperm) and a female gamete (ovum) fuse and give rise to a diploid zygote, which develops into a new individual.

gametophyte: The haploid, gamete-producing phase in the life cycle of plants.

gamma globulin: A blood protein fraction that is rich in the class of antibody (IgG) effective at eliminating viruses and bacteria.

gastrin: A digestive hormone secreted in the stomach that causes the secretion of other digestive juices.

gastropod: A member of the class Gastropoda in the phylum Mollusca; includes snails, garden slugs, and sea slugs.

gastrovascular cavity: The digestive cavity in cnidarians.

gastrulation: The movement of cells in the embryo that generates three cell layers—the ectoderm, mesoderm, and endoderm—each layer in turn giving rise to specific body organs and tissues.

gel electrophoresis: A procedure that separates molecules on the basis of electric charge.

gene: (Gr. *genos,* birth or race) The biological unit of inheritance that transmits hereditary information from parent to offspring and controls the appearance of a physical, behavioral, or biochemical trait. A gene is a specific discrete portion of the DNA molecule in a chromosome that encodes an rRNA molecule.

gene duplication: One method that may lead to the evolution of new genes with new functions. This can occur when a chance error in DNA replication or recombination creates two identical copies of a gene.

gene flow: The incorporation into a population's gene pool of genes from one or more other populations through migration of individuals.

gene gun: Apparatus used in plant bioengineering that shoots gold particles covered with DNA at high speed into tissue culture. Inside the cells, the DNA detaches from the gold (which is inert inside the cell) and moves into the nucleus, occasionally integrating into the chromosomes.

gene mapping: The assignment of genes to specific locations along a chromosome.

gene pool: The sum of all alleles carried by the members of a population; the total genetic variability present in any population.

gene regulation: The process that controls how and when each gene is turned on and off in each living cell.

gene therapy: Altering a person's genes to combat disease.

genetic code: The specific sequence of three nucleotides in mRNA that encodes an individual amino acid in protein. For example, the code for methionine is AUG.

genetic drift: Unpredictable changes in allele frequency occurring in a population due to the small size of that population.

genetic marker: Any genetic difference between individuals that can be followed in a genetic cross. For example, areas of repeated base sequences along a DNA strand.

genetic recombination: The reshuffling of maternal and paternal chromosomes during meiosis, resulting in new genetic combinations.

genetic variation: Genetic differences among individuals of the same species.

genetics: The study of genes and inheritance.

genotype: The genetic makeup of an individual.

genus (pl. genera): A taxonomic group of very similar species of common descent.

germ cell: A sperm cell or ovum (egg cell) or their precursors; the haploid gametes produced by individuals that fuse to form a new individual.

germination: The resumption of growth in the seed after a period of dormancy.

germ-line gene therapy: The deliberate genetic alteration of sex cells to treat disease.

gibberellin: A specific hormone that regulates plant growth, for example, by increasing the length of plant stems.

gill: [1] A specialized structure that exchanges gases in water-living animals. [2] The plates under the cap of certain fungi.

gill filaments: Tiny subdivisions of the gill bar that support many thin, platelike structures, which help extract dissolved oxygen from water.

gill slits: In chordates, an opening through the pharynx to the exterior.

gizzard: A specialized region that grinds food in the digestive tracts of earthworms, chickens, and other animals.

gland: A group of cells organized into a discrete secretory organ.

glial cell: A non-neural cell of the nervous system that surrounds the neurons and provides them with protection and nutrients.

global warming: A slow but steady increase in Earth's average air and water temperatures.

globulin: One group of blood plasma proteins that includes antibodies—defensive molecules that attack invaders.

glomerulus: (L., a little hall) In the vertebrate kidney, a collection of tightly coiled capillaries enclosed by the Bowman's capsule.

glottis: The opening to the trachea, which is also known as the windpipe.

glucagon: A hormone made in the pancreas that causes glucose levels in the blood to rise, thus opposing the effect of insulin.

glycogen: A polysaccharide made up of branched chains of glucose; an energy-storing molecule in animals, found mainly in the liver and muscles.

glycolysis: The initial splitting of a glucose molecule into two molecules of pyruvate, resulting in the release of energy in the form of two ATP molecules. The series of reactions does not require the presence of oxygen to occur.

glycolytic energy system: Energy system based on the splitting of glucose by glycolysis in the muscles. The glycolytic system can sustain heavy exercise for a few minutes, as in a 200-m swim.

goiter: A large lump on the neck caused by an enlarged thyroid gland often due to an iodine deficiency.

Golgi apparatus: In eukaryotic cells, a collection of flat sacs that process proteins for export from the cell or for shunting to different parts of the cell.

gonad: An animal reproductive organ that generates gametes; testes and ovaries.

Gram's stain: A special stain that distinguishes gram-positive and gram-negative organisms.

gram-negative cells: Prokaryotes in which the peptidoglycan layer is covered by an outer sheet of proteins and lipopolysaccharides; don't pick up Gram's stain.

gram-positive cells: Prokaryotes containing peptidoglycans in a single broad layer; do pick up Gram's stain.

grassland: A treeless temperate region dominated by grass species; known as *prairies* in North America, *pampas* in South America, *steppes* in Asia, and *veldt* in Africa. This biome is wetter than deserts but drier than forests.

gravitropism: The orientation of a plant part toward the ground (positive gravitropism) or away from the ground (negative gravitropism).

great ape: A lineage of animals that includes the orangutans, the gorillas, the humans, and two species of chimpanzees (the common chimp and the bonobo).

green algae: Protists with green pigments that are closely related to plants; also called Chlorophyta.

greenhouse effect: The result of a buildup of carbon dioxide in the atmosphere (e.g., through the burning of fossil fuels) in which carbon dioxide traps solar heat beneath the atmospheric layers, leading to increased global temperatures and changes in climatic patterns.

gross primary productivity: The percent of sunlight that producers convert to chemical energy in the form of organic compounds.

ground tissue system: Plant tissues other than the epidermal tissue system and vascular tissue system; provides support and stores starch.

growth: An increase in size.

growth factors: Proteins that can enhance the growth and proliferation of specific cell types.

growth ring: Concentric rings of lighter and darker wood that occur because the generative cambium layer produces larger cells in the spring and summer (lighter areas), and smaller cells in the fall and winter (darker areas).

guard cells: Specialized kidney-shaped cells that enclose the pores (stomata) on the leaf epidermis; their opening and closing regulates water loss from the leaf.

gut (gastrointestinal tract): The food processing or alimentary canal made up of the mouth, esophagus, stomach, small intestine, and large intestine.

guttation: Upward pressure from plant roots producing tiny droplets of water that rim leaves early in the morning.

gymnosperm: (Gr. *gymnos*, naked + *sperma*, seed) Conifers and their allies; a primitive seed plant whose seeds are not enclosed in an ovary.

H

habitat: The physical place within a species' range where an organism actually lives.

habituation: A progressive decrease in the strength of a behavioral response to a constantly applied stimulus, when the stimulus has no negative effects.

hair cell: Box-shaped neurons containing cilia and attached to the cochlea. These cells are essential for hearing.

half-life: The length of time it takes for half of the total amount of radioactivity in an isotope to decay. As a result of such decay, for example, the concentration of carbon-14 relative to carbon-12 decreases. Half of the carbon-14 will decay in 5730 years; this is the half-life.

halophile: A type of archaebacterium that can tolerate extremely high salt concentrations.

haploid: Having only one copy of a chromosome set. A human haploid cell has 23 chromosomes.

Haplorhini: A suborder of primates including the tarsiers and the anthropoids—the New World monkeys, Old World monkeys, apes, and humans.

hard palate: The part of the oral cavity that forms the floor of the nasal cavity and most of the roof of the mouth.

Hardy-Weinberg equilibrium: A proposed state wherein a population, in the absence of external pressure, has both stable allele and stable genotype frequencies over many generations.

Hardy-Weinberg principle: In population genetics, the idea that in the absence of any outside forces, the frequency of each allele and the frequency of genotypes in a population will not change over generations.

head: The part of an animal that contains a major concentration of sense organs and generally encounters the environment first as the animal moves along.

heart: A muscular organ that pumps blood through vessels.

heart attack (myocardial infarction): Medical condition resulting from the blockage (from a blood clot) of the heart's arteries, causing complete cutoff of blood flow to a section of the heart muscle and subsequent damage or destruction of that muscle tissue.

heart murmur: In some people, a sound produced during a heartbeat when backwashing blood escapes through weak or defective valves.

heartwood: The nonconducting central wood of an older tree that becomes infiltrated with oils, gums, resins, and tannins.

heat: The random motion of atoms and molecules.

heavy chains: The two longest of the four polypeptides, or chains of amino acids, which compose immunoglobulins.

helix: A structure similar to a spiral staircase. DNA is a double helix.

helper T cells: A type of T cell that communicates via chemical signals with B cells and the other types of T cells to trigger or to augment an immune response.

heme: A series of chemical rings surrounding an iron atom.

Hemichordata: A phylum that includes the acorn worms.

hemocyanin: A large blue protein, containing many subunits, that transports oxygen within hemolymph.

hemoglobin: A blood protein with iron-containing heme groups that bind and transport oxygen.

hemolymph: In animals with open circulatory systems, the extracellular fluid in the body cavity that bathes the tissue cells.

hemophilia: Genetic condition in which sufferers lack a functional copy of a gene encoding a protein in the clotting cascade. As a result, they do not form normal blood clots.

heparin: A chemical released by mast cells that binds to clot-inhibiting proteins; part of the inflammatory response.

herbaceous: A plant with a mature stem that keeps its flexible green epidermis and does not produce wood.

herbicide: A chemical that kills plants.

heredity: The science of inheritance and variation.

heterotroph: (Gr. *heteros,* different + *trophos,* feeder) An organism, such as an animal, fungus, and most prokaryotes and protists, that takes in preformed nutrients from external sources.

heterozygote advantage: The greater fitness often conferred upon an organism by having heterozygous alleles of a gene rather than homozygous alleles.

heterozygote: An organism with two different alleles for a given trait.

heterozygous: (Gr. *heteros,* different + *zygotos,* pair) Having two different alleles for a specific trait.

hippocampus: A structure within the cerebrum of the brain that plays a crucial role in the formation of long-term memory.

histamine: A chemical released by mast cells that causes capillaries to dilate; part of the inflammatory response.

histone: Protein in the nucleus around which DNA molecules of the chromosomes wind, al-

lowing extremely long DNA molecules to be packed into a cell's nucleus.

HIV: Human immune deficiency virus, the causative agent in acquired immune deficiency syndrome (AIDS).

holdfast: A rootlike anchor that attaches an alga to its substrate, such as a rock on the ocean floor.

homeostasis: The maintenance of a constant internal balance despite fluctuations in the external environment.

homeotherm: A "warm-blooded animal" in which the internal body temperature is fairly constant, based on physiological mechanisms that also control the distribution of its large reservoir of warm, moving fluid when the environment is too hot or cold.

homeothermic: "Warm-blooded"; able to maintain a constant internal body temperature.

homing behavior: The tendency of displaced animals to travel (often great distances) to return to their home turf.

hominoids: The primate group including the apes and humans.

Homo erectus: "Erect human" species that evolved by 1.8 million years ago, used tools, and spread into northern Africa.

Homo habilis: Until the discovery of *Australopithecus garhi,* considered the earliest tool user in the genus *Homo.*

Homo neanderthalensis: An extinct species in the genus *Homo* that lived simultaneously with *Homo sapiens* for several thousand years and had a larger brain and heavier bones than our species.

Homo sapiens: The genus and species of modern humans.

homologous chromosomes: Chromosomes that pair up and separate during meiosis and generally have the same size, shape, and genetic information. One member of each pair of homologous chromosomes comes from the mother and the other comes from the father.

homologous element: Elements (such as a leg, flipper, or wing) in different species that derive from a single element in a common ancestor.

homologous trait: A similarity shared by two species due to inheritance from the last common ancestor.

homozygotes: An organism with two identical alleles for a given trait.

homozygous: Having two identical alleles for a specific trait.

hormone: A chemical messenger produced in one part of the body and transported, often in the

blood, to another region, where it exerts an effect.

human chorionic gonadotropin (hCG): a hormone produced by the chorion of a developing embryo that maintains pregnancy by maintaining the secretion of estrogen from the ovary and thus preventing menstruation.

Human Genome Project: The research effort to sequence the entire set of human genes and to understand their functions.

humus: The decomposing organic matter in soil.

hybrid: An offspring resulting from the mating between individuals of two different genetic constitutions.

hydrogen bond: A type of weak molecular bond in which a partially negatively charged atom (oxygen or nitrogen) bonds with the partial positive charge on a hydrogen atom when the hydrogen atom is already participating in a covalent bond.

hydrophilic: Compounds that dissolve readily in water, such as salt.

hydrophobic: Compounds that do not dissolve readily in water, such as oil.

hydroskeleton: A volume of fluid trapped within an animal's tissues that is noncompressible and serves as a firm mass against which opposing sets of muscles can act; for example, in earthworm segments.

hydrotropism: The growth of a plant root toward water.

hyoid: A bone contained within the skull that is the only human bone not jointed with other bones. It is suspended under the back of the mouth by ligaments and muscles.

hyperthermia: The state of elevated body temperature that sets in when environmental temperatures are too high and bodily exposure to heat is too prolonged.

hyperventilation: Very rapid, deep breathing.

hypha (pl. hyphae): One of many long, thin filaments of cells that make up a multicellular fungus.

hypocotyl: The portion of a plant embryo or seedling between the embryonic root (radicle) and the cotyledons.

hypothalamus: A collection of nerve cells at the base of the vertebrate brain just below the cerebral hemispheres; part of the diencephalon and responsible for regulating body temperature, many autonomic activities, and many endocrine functions.

hypothermia: The state that sets in when the

body's core temperature drops below 35 °C (90 °F).

hypothesis: A possible answer to a question about how the world works that can be tested by means of scientific experimentation.

I

ileum: The lower section of the small intestine.

immediate energy system: Energy in the body instantly available for a brief explosive action, such as one heave by a shot-putter.

immune system: The network of cells, tissues, and organs that defends the body against invaders.

immunity: Specific resistance, or immunity, is a slow series of reactions based on lymphocytes and antibodies (in vertebrates) that is more targeted to particular invaders or damaged cells within the organism than are nonspecific responses.

immunoglobulin: Globular proteins of the immune system that comprise antibodies. Classes of immunoglobulins include IgG, IgA, and IgE.

immunologic memory: The capability of memory B cells in vertebrates to produce a more rapid and vigorous immune response based on former encounters with antigens.

implantation: Attachment of a fertilized ovum to the uterine wall, where it is nourished and develops into an embryo and fetus.

imprinting: Attachment of young animals to a particular adult or object; in nature, the attachment of young to their parents. Animals may imprint on other objects or animals, such as young birds imprinting on humans, if they are exposed to that animal or object at the sensitive period.

***in vitro* fertilization:** The fusion of egg and sperm in a laboratory dish.

inbreeding depression: A situation of weakened genetic viability that occurs when a population has many more individuals that are less fit than in a normally breeding population.

inbreeding: Nonrandom mating that occurs when relatives mate with each other rather than with unrelated individuals.

incomplete dominance: The genetic situation in which the phenotype of the heterozygote is intermediate between the phenotypes of two homozygotes.

independent assortment: The random distribution of genes located on different chromosomes to the gametes; Mendel's second law, the principle of independent assortment.

indeterminate growth: The growth pattern of land plants that allows them to continue adding new branches, leaves, roots, and other organs throughout life.

industrial revolution: The replacement of hand tools with power-driven machines (like the steam engine) and the concentration of industry in factories beginning in England in the late 18th century.

inferior vena cava: One of the two largest veins in the human body; carries blood from the legs and most of the lower body to the right atrium of the heart.

inflammation mediators: Chemical signals released by damaged tissues that provoke inflammation reactions.

inflammation: The redness, heat, and swelling that is part of the body's second line of defense.

inflammatory response: The body's response to intruders, composed of a series of nonspecific internal resistance reactions including redness, pain, heat, and swelling.

ingestion: The taking of food into the mouth; the first step of the digestive process.

inhalation: The drawing in of fresh air.

inheritance of acquired characteristics: The long-disproved belief that changes in an organism's appearance or function during its lifetime could be inherited.

initiation: The first main stage of protein synthesis, during which tRNA associates with a ribosome and an mRNA, forming a complex of many molecules.

inorganic: Molecules that are not based on carbon.

insect: A member of class Insecta, arthropods having three main parts (head, thorax, and abdomen), three pairs of legs, and generally two pairs of wings; the largest class of animals.

insertion: The end of a skeletal muscle that attaches across a joint to a bone that moves.

insight learning: A type of learning in which an individual formulates a course of action by understanding the relationships between the parts of a problem; common only in higher primates.

insoluble dietary fiber: Dietary material, such as wheat bran, that is indigestible to certain animals and that does not dissolve in water, but binds to water molecules and helps substances pass through the digestive system more rapidly.

insulin: A hormone made in the pancreas that causes cells to remove the sugar glucose from the blood.

integrated pest management: An insect control system based on the principles of community ecology and aimed at keeping pest populations below economically harmful levels with a minimum of chemical pesticides.

integrator: One of three primary elements characteristic of a homeostatic system; an integrator evaluates the situation and makes decisions. Also see effector and receptor.

intercalated disk: An intercellular connection that links heart muscle cells electrically.

intercostal muscles: The muscles between ribs.

interferons: A class of antimicrobial proteins that are released by certain virus-infected cells that diffuse to other cells and, in turn, cause them to produce proteins that inhibit viral replication.

interleukin: A protein that speeds or slows the dividing and maturing of other immune cells.

interneuron: A nerve cell that relays messages between other nerve cells.

interphase: The period between cell divisions in a cell. During this period, the cell conducts its normal activities and DNA replication takes place in preparation for the next cell division. Interphase is divided into three periods: G_1, S, and G_2.

interspecific competition: Competition for resources — e.g., food or space — between individuals of different species.

intertidal zone: A region that lies at the ocean's edge that is underwater at high tide, exposed to air at low tide, and pounded by waves and wind.

intracellular digestion: The enzymatic breakdown of nutrient molecules that occurs within cells.

intracellular fluid: The fluid inside cells.

introns: The portions of the gene that are initially transcribed, but then are spliced out of the primary transcript.

ion: An atom that has gained or lost one or more electrons, thereby attaining a positive or negative electrical charge.

ionic bond: A type of molecular bond formed between ions of opposite charge.

iris: A pigmented ring of tissue in the vertebrate eye that regulates the amount of entering light.

islets of Langerhans: Endocrine cells in the pancreas that secrete the hormones glucagon and insulin.

isolation: Geographic separation of an ecological community (e.g., peninsulas, island chains, and other out-of-the-way locales) that influences species richness.

isotopes: An alternative form of an element having the same atomic number but a different atomic mass due to the different number of neutrons present in the nucleus. Some isotopes, such as carbon-14, are unstable and emit radiation.

J

jawed fish: Fish with hinged jaws.

jawless fish: Fish that lacked hinged jaws; agnathans.

jejunum: The central section of the small intestine.

joint: The hinge, or point of contact, between two bones.

J-shaped curve: A plot of population growth with an upsweeping curve that represents exponential growth.

K

kelp: One of the largest members of the algal world, a brown alga.

kidney: The main excretory organ of the vertebrate body; filters nitrogenous wastes from blood and regulates the balance of water and solutes in blood plasma.

kidney stones: Crystals usually formed of calcium and organic compounds that painfully block the ureter.

kin selection: A form of natural selection in which an individual increases its fitness by helping relatives, who share its genes.

kinetic energy: The energy possessed by a moving object.

kinetochore: A group of proteins located at the centromere that attaches to long fibers called spindle fibers.

Kinetoplastid: Primitive protists with long whiplike flagella; includes the protist causing African sleeping sickness.

kingdom: A taxonomic group composed of members of similar phyla, i.e., Animalia, Plantae, Fungi, and Protista.

Krebs cycle: The first stage of aerobic respiration, in which a two-carbon fragment is completely broken down into carbon dioxide and large amounts of energy are transferred to electron carriers; occurs in the mitochondrial matrix.

L

lamellae: Tiny platelike structures supported by gill filaments; help aquatic animals respire efficiently in water, even though most lake, river, or ocean water holds only one twentieth as much oxygen as air.

lancelet: A cephalochordate that lives as a small, streamlined, fishlike marine animal half-buried in sand.

larva: An immature form of an insect and many other animal types.

larynx: A cartilaginous structure containing the vocal cords; it is also known as the voice box.

late-loss survivorship curve: A plot of survivorship data indicating that an organism's life expectancy decreases with each passing year.

latency: The time delay it takes for a muscle to begin to respond to a nervous signal based on an internal state or a stimulus in the environment.

lateral bud: Cells produced by the apical meristem that form buds of embryonic tissue in the space between a leaf and the stem. These buds can give rise to branches or flowers, and are important for determining a plant's overall shape.

lateral meristem: An area of actively growing cells in the stems and roots of plants that causes these areas to thicken in secondary growth.

latitude: North-south position on Earth's surface that, because of sunshine levels, influences species richness.

leaf primordia: A portion of the cells left behind by the growing terminal bud that continues to divide and form leaves.

learning: An adaptive and enduring change in an individual's behavior based on its personal experiences in the environment.

leg: An appendage that supports or moves an animal.

legumes: Members of the pea family, including peas, beans, alfalfa, clover, and lupine.

lens: A circular, crystalline structure in the eye of certain animals that focuses light onto the retina.

leptin: A weight-regulating hormone produced by fat cells.

leukocyte: A white blood cell; functions in the body's defense against invading microorganisms or other foreign matters.

lichen: An association between a fungus and an alga, which live together symbiotically.

life expectancy: The maximum probable age an individual will reach.

life history strategy: The way an organism allocates energy to growth, survival, or reproduction.

ligament: A band of connective tissue that links bone to bone.

light chain: The two shortest of the four polypeptides, or chains of amino acids, which compose an immunoglobulin molecule.

light microscope: Instrument used often in the study of biology to magnify objects; it contains optical lenses that refract or bend light rays so that an object appears larger than it really is.

light-dependent reaction (or energy-trapping reaction): The first phase of photosynthesis, driven by light energy. Electrons that trap the sun's energy pass the energy to high-energy carriers such as ATP and NADPH, where it is stored in chemical bonds.

light-independent reaction: The second stage of photosynthesis (the Calvin cycle) in which the energy trapped and converted during the light-dependent reactions is used to combine carbon molecules into sugars.

lignin: A plant compound that gives cells and tissues strength and hardness.

limbic system: A series of small brain structures (including the hippocampus and hypothalamus) largely responsible for generating fear, rage, aggression, and pleasure and for regulating sex drives and reproductive cycles.

linkage: Alleles of two genes located so close to each other on the same chromosome that they fail to assort independently.

lipase: A pancreatic enzyme that digests fats.

liver: A large, lobed gland that destroys blood cells, stores glycogen, disperses glucose to the bloodstream, and produces bile.

loam: The best agricultural soil, consisting of deep layers of soil with a high mineral and humus content and a mixture of particle sizes.

lobe-finned fish: A member of the oldest of the groups of bony fishes; includes lungfishes and coelacanths.

locus (pl. loci): The location of a gene on a chromosome; sarcopterygians.

logistic growth: Growth of a population under environmental constraints that set a maximum population size.

long-day plant: Plants, like spinach, that require days longer than a certain length before they will bloom.

loop of Henle: U-shaped region of vertebrate kidney tubule chiefly responsible for reabsorption of water and salts from the filtrate by diffusion.

lophophore: Specialized feeding apparatus possessed by one group of protostomes.

lophotrochozoans: Animals with a specialized feeding apparatus called a lophophore or a specialized larval form called a trochophore. The group includes several kinds of worms and all of the mollusks.

lumen: The central cavity; in a blood vessel, the central space where the blood flows.

lung: The principal air-breathing organ of most land vertebrates.

lungfish: An air-breathing lobe-finned fish having a lunglike air bladder in addition to gills.

luteinizing hormone (LH): Hormone produced by the anterior pituitary that acts on the ovaries to stimulate ovulation and the synthesis of estrogen and progesterone and on the testes to stimulate testosterone production.

lymph: Fluid forced out of capillaries, occupying spaces between cells, and draining into the lymphatic system.

lymph node: A bean-shaped filtering organ and part of the immune system that stores lymphocytes and prevents debris in the lymph from mixing into the blood.

lymphatic system: In vertebrates, a collective term for a system of vessels carrying lymph (fluid that has been forced out of the capillaries) back to the bloodstream.

lymphocyte: A white blood cell formed in lymph tissue; active in the immune responses that protect the body against infectious diseases. There are two main classes of lymphocyte: B lymphocytes, involved in antibody formation, and T lymphocytes, involved in cell-mediated immunity.

lysosome: Spherical membrane-bound vesicles within the cell containing digestive (hydrolytic) enzymes that are released when the lysosome is ruptured; important in recycling worn-out mitochondria and other cell debris.

M

M phase: The portion of the cell cycle during which the nucleus divides by mitosis and the cytoplasm divides by cytokinesis.

macroevolution: Evolutionary changes above the level of species; evolutionary patterns viewed over geologic time spans.

macronucleus: In ciliates, a large nucleus containing many sets of chromosomes that control cell activities.

macronutrient: (Gr. *makros,* large + L. *nutrire,* to nourish) An inorganic chemical element, such as nitrogen, potassium, calcium, phosphorus, magnesium, and sulfur, required in large amounts for plant growth.

macrophage: An animal cell that ingests other organisms and substances; in the immune system, a cell that engulfs invaders and consumes debris.

major histocompatibility complex (MHC): A complex of proteins that are specific for each individual and are the factors that cause the body to reject transplanted tissues; their main function is to aid in communication among immune cells.

major histocompatibility proteins (MHC): (Gr. *histos,* web of a loom, or tissue) Cell surface proteins that mark an individual's cells for self-recognition and aid in communication among immune system cells.

major minerals: Elements we need in amounts greater than 0.1 g each day.

malting: The name for germination in grains like rice and barley.

mammal: A vertebrate animal of the class Mammalia, having the body generally covered with hair, nourishing young with milk from mammary glands, and generally giving birth to live young; includes lions, whales, rabbits, and kangaroos.

mammary gland: Gland that produces milk in mammals.

mantle: A thick fold of tissue found in mollusks that covers the visceral mass and that in some mollusks secretes the shell.

marijuana: One of the two forms of hemp (*Cannabis sativa*). Marijuana has high levels of a chemical called THC (tetrahydrocannabinol) that helps protect the plant from various environmental stresses and is also psychoactive in humans.

marine: Characteristic of oceans and seas.

marrow: The major site of red blood cell manufacture in adults, found in the spaces of some spongy bone tissue.

marrow cavity: The tissue down the center of the shaft in the femur and other bones that is a site of fat storage.

marsupial: A mammal having a pouch in which it carries its young, which are born in a small and undeveloped state. Found extensively in Australia, with a few representatives in America; includes kangaroos, opossums, and koala bears.

marsupium: An elastic pouch of skin that harbors newborn marsupials.

mass extinction: The disappearance of large numbers of species.

mast cell: A round cell that occupies connective tissue in the skin and mucous membranes, contains little packets of chemicals, and releases histamine and heparin when triggered by certain antigens.

mastication: The physical breakdown of food into small pieces in the mouth.

matrix: In HIV and many other viruses, a sphere of protein inside the envelope and outside the capsid. In a mitochondrion, the area enclosed by the inner membrane.

medulla oblongata: The lowest region of the brainstem, the most posterior part of the brain; involved in keeping body conditions constant.

medulla: The central portion of organs; in the kidney, the place that contains the loop of Henle; in the adrenal gland, the place where epinephrine (adrenaline) is made.

medusa: (Gr. mythology, a female monster with snake-entwined hair) A jellyfish, or the free-swimming stage in the life cycle of cnidarians; an inverted umbrella-shaped version of a polyp, with the mouth and tentacles pointing downward.

megaspore: In certain plants, a haploid spore that gives rise to the female gametophyte.

meiosis: The type of cell division that occurs during gamete formation; the diploid parent cell divides twice, giving rise to four cells, each of which is haploid.

meiosis I: The first division of meiosis, during which the number of chromosomes in a diploid cell is reduced from a diploid set of duplicated chromosomes to a haploid set of duplicated chromosomes.

meiosis II: The second division of meiosis, during which a haploid cell with duplicated chromosomes divides to form two haploid cells with unduplicated chromosomes.

melatonin: A hormone secreted by the pineal gland in the brain; in some animals, it affects the internal biological rhythms associated with night and day.

memory B cell: A type of B cell clone that can proliferate when stimulated by the type of antigen that triggered the original B cell, forming its own clone of both plasma and memory cells; also called memory cell. Memory B cells are responsible for the secondary immune response.

Mendel's principle of independent assortment: Mendel's second law; the random distribution of genes located on different chromosomes to the gametes.

menstrual cycle: In human females, generally a 28-day cycle, characterized by a gradual thickening of the lining of the uterus (womb) and the maturation of an ovarian follicle, from which a mature ovum is released. If the ovum is not fertilized, the inner lining of the uterus is

shed, a process known as menstruation, and the cycle starts again. In humans, the menstrual cycle normally starts between ages 12 and 14.

meristem: Plant tissue containing cells that can continually divide throughout the plant's life.

mesentery: The sheet of tissue that is joined to the serosa and attaches the gastrointestinal tract to the inner wall of the body cavity.

mesoderm: The middle cell layer of an embryo; gives rise to muscles, bones, connective tissue, and reproductive and excretory organs.

mesoglea: A jellylike substance lying between the epidermis and gastrodermis in cnidarians such as jellyfish.

mesophyll: The major photosynthetic tissue in a plant leaf, located between the upper and lower epidermal layers.

Mesozoic Era: The geologic era from about 245 million years ago to about 65 million years ago.

messenger RNA (mRNA): An RNA molecule that carries the information to make a specific protein. mRNA is transcribed from structural genes and is translated into protein by the ribosomes.

metabolic pathway: The chain of enzyme-catalyzed chemical reactions that converts energy and constructs needed biological molecules in cells and in which the product of one reaction serves as the starting substance for the next.

metabolism: (Gr. *metabole,* to change) The sum of all the chemical reactions that take place within the body; includes photosynthesis, respiration, digestion, and the synthesis of organic molecules.

metamorphosis: (Gr. *meta,* after + *morphe,* form + *osis,* state of) The process in which there is a marked change in morphology during postembryonic development; in insects the change in body form that takes place as the individual changes from a larva, such as a caterpillar, and emerges as an adult, such as a butterfly; in amphibians, the change from a tadpole to a frog.

metaphase: The period during nuclear division (mitosis) when the spindle microtubules cause the chromosomes to line up at the center of the cell.

metazoa: Multicelluar organisms that are not plants or fungi; animals.

methanogen: A type of archaebacterium that produce methane as a metabolic byproduct.

microevolution: Evolutionary changes in a species over geologically short time periods, such as the accumulation of black forms of certain moths in industrial England.

microgravity: State of near-weightlessness.

micronucleus: In ciliates, one of several nuclei that undergo meiosis and are exchanged during sexual reproduction.

micronutrient: (Gr. *mikros,* small + L. *nutrire,* to nourish) A nutrient, such as the minerals iron, copper, zinc, chlorine, manganese, molybdenum, and boron, required in small amounts for plant growth.

microspore: A pollen grain that gives rise to the male gametophyte in certain vascular plants.

microsporidia: Among the simplest of eukaryotic cells, they live only inside animal cells.

microvillus (pl. microvilli): One of hundreds of tiny fingerlike projections extending from the surface of cells lining the walls of the intestine that increases the surface area available for absorption.

midbrain: The part of the brainstem between the pons and the thalamus in humans; relays neural signals between different parts of the brain.

migration: The traveling of an animal each year in search of an appropriate habitat. Animals such whales, sea turtles, eels, monarch butterflies, and caribou in the Arctic travel huge distances each year to spend winter and/or breed.

milk: A fluid rich in fats and proteins produced in the mammary glands of mammals that nourishes newborns.

millipedes: Arthropods with a long series of body segments, each bearing two pairs of legs.

mimicry: The evolution of similar appearance in two or more species, which often gives one or all protection; for example, a nonpoisonous species may evolve protection from predators by its similarity to a poisonous model.

mineral: An inorganic element such as sodium or potassium that is essential for survival and is obtained in food.

minimal viable population: The smallest "bottleneck" a population can pass through and still have enough genetic variability to ensure the species' survival.

minor minerals: Elements we need in amounts less than 0.01 grams daily.

mitochondrion (pl. mitochondria): Organelle in eukaryotic cells that provides energy that fuels the cell's activities. Mitochondria are the sites of oxidative respiration; almost all of the ATP of nonphotosynthetic eukaryotic cells is produced in the mitochondria.

mitosis: The process of nuclear division in which replicated chromosomes separate and form two daughter nuclei genetically identical to each other and the parent nucleus. Mitosis is usually accompanied by cytokinesis (division of the cytoplasm).

mitotic spindle: A weblike structure of microtubules that suspends and moves the chromosomes; formed during prophase in mitosis.

molecular clock: The rate at which mutations accumulate in genes over evolutionary time.

molecular messenger: A chemical that regulates the activities of other cells.

molecule: A cluster of two or more atoms held together by specific chemical bonds.

Mollusca: The phylum containing the mollusks.

mollusk: A member of the phylum Mollusca; includes snails, slugs, clams, oysters, squid, and octopuses.

monoclonal antibodies: Antibodies secreted by and isolated from a single clone of B cells that biologists grow in a laboratory.

monocot: Short for monocotyledon, the smaller of two classes of angiosperms (flowering plants). Embryos have only one cotyledon, the floral parts are generally in threes, and leaves are typically parallel-veined. Corn and lilies are monocots.

monoecious: (Gr., one house) Species of plants, like corn and oaks, that have separate male and female flowers on the same plant.

monomer: A small molecule, many of which can be combined into a polymer.

monosaccharide: A simple sugar that cannot be decomposed into smaller sugar molecules. The most common forms are the hexoses, six-carbon sugars such as glucose, and the pentoses, five-carbon sugars such as ribose.

monotreme: An egg-laying mammal that also has many primitive or reptilian features. The only living forms are the spiny anteater and the duck-billed platypus.

monounsaturated fat: Fats with one double bond between carbons along the polymer chain of their fatty acid portion. They tend to lower the blood level of low-density lipoproteins.

morphogenesis: The growth, shaping, and arrangement of the organs and tissues in a developing embryo.

mosaic climax community: A community of few species that eventually establishes itself after the first plants to colonize a barren ecological space; a mosaic of early colonists and later-arriving species that outcompete some of the earlier species.

motility: The self-propelled movement of an individual or its parts.

motor cortex: The part of the cerebral cortex of the brain that controls the movement of body.

motor neuron: A neuron that sends messages from the brain or spinal cord to muscles or secretory glands.

motor unit: A group of skeletal muscle cells contracting at the same time because they are innervated by the same motor neuron coming from the spinal cord.

mountain sickness: Medical condition affecting people due to low oxygen levels at high altitudes; symptoms include headache, lethargy, dizziness, difficulty sleeping, and loss of appetite. In severe cases, the traveler can become uncoordinated, off-balance, confused, drowsy, nauseated, and irrational due to swelling around the brain, or can suffer fluid buildup in the lungs and fall into a coma.

mouth: The body opening where food enters the digestive tract.

mucosa: The innermost layer of the lining of many vertebrate canals and organs, such as the uterus, reproductive tracts, and alimentary canal; often consists of mucus-secreting cells. In the alimentary canal, the mucosa contains enzyme-secreting cells.

mucus: A slimy coating secreted by mucosa cells that lubricates food passing through the gastrointestinal tract and that coats the respiratory and reproductive tract linings.

multiple sclerosis (MS): An autoimmune disease in which a person's immune system reacts to and damages or destroys the myelin sheaths that surround and protect nerve cells.

muscle fiber: A muscle cell; a giant cell with many nuclei and numerous myofibrils, capable of contraction when stimulated.

muscle tissue: Tissue that enables an animal to move; the three types are smooth muscle, cardiac muscle, and skeletal muscle.

muscularis: A double layer outside the submucosa of the digestive tract that is made of muscle fibers, some encircling the gut (circular muscles) and other running along its length (longitudinal muscles).

mutagen: Physical and chemical agents, including ultraviolet rays from the sun, chemicals in cigarette smoke, and even many natural substances from plants or fungi, that can change DNA structure.

mutant: The allele that results from a mutation; also used to refer to the organism containing such a mutation.

mutation: Any heritable change in the base sequence of an organism's DNA.

mutualism: A symbiotic relationship between two species in which both species benefit.

myasthenia gravis: A medical condition occurring when muscle cells have few working receptors on their cell surfaces for one particular neurotransmitter; as a result, the muscle cells cannot respond normally to nerve signals, and affected people experience muscular weakness and fatigue.

mycelium: The mass of filaments (hyphae) that makes up the body of a fungus.

mycoplasma: A type of the smallest free-living cells, these simplified members of the domain Bacteria lack cell walls, live inside animals, plants, and sometimes other single-celled organisms, and can cause a dangerous form of pneumonia as well as infections of the urinary tract and other organs.

mycorrhiza (pl. mycorrhizae): An association between the root of a plant and a fungus that aids the plant in receiving water and nutrients and aids the fungus in receiving carbohydrates.

myelin sheath: A lipid membrane made up of Schwann cells that forms an insulating layer around neurons and speeds impulse travel.

myomere: Blocks of embryonic or larval animal tissue that generate the muscles and bones.

myosin: A muscle protein that interacts with actin and causes muscles to contract.

N

nastic response: The movement of a plant not specifically toward or away from a stimulus.

natural killer cell: A type of white blood cell that can kill many kinds of infectious microbes and some types of tumor cells; part of the nonspecific defense of the body.

natural selection: The increased survival and reproduction of individuals better adapted to the environment.

navigate: The ability to move from one map point to another.

nectar: A sugary fluid produced by flowers that attracts butterflies, aphids, ants, and other animals.

negative feedback loop: A series of steps used to resist change in a homeostatic system; first, by sensing a deviation from the baseline condition, then by turning on mechanisms that oppose that trend and thus bring things back toward the baseline.

nematocysts: A stinging capsule found in cnidari-

ans, which, when stimulated, shoots out a tiny barb containing a poisonous substance that immobilizes or kills the prey or predator.

Nematoda: The animal phylum containing roundworms, Earth's most abundant animals.

nephridia: Excretory units found in pairs on most segments of earthworms.

nephron: The functional unit of the vertebrate kidney; each of the million nephrons in a kidney consists of a glomerulus enclosed by a Bowman's capsule and a long attached tubule. The nephron removes waste from the blood.

nerve impulse: A change in ion permeabilities in a neuron's membrane that sweeps down the cell's axon to its terminal, where it can excite other cells.

nerve: A group of axons and/or dendrites from many different neurons operating together in a bundle and held together by connective tissue.

nervous system: An animal's network of nerves; includes the brain and spinal cord as well as the peripheral nerves, and integrates and coordinates the activities of all the body systems.

nervous tissue: Animal tissue containing neurons, cells whose main function is the transmission of electrochemical impulses.

net primary productivity: The amount of chemical energy producers actually store as organic molecules in the form of new leaves, roots, stems, flowers, fruits, and other structures and compounds.

net veins: Leaf veins that form a branching pattern; found in eudicots.

neurohormone: A hormone secreted by nerve cells.

neuron: A nerve cell that transmits messages throughout the body; made up of dendrites, cell body, and axon.

neurotransmitter: A chemical that transmits a nerve impulse across a synapse.

neutral mutation: Single-gene mutations that leave the gene's function basically intact and neither harm nor help the organism.

neutralists: Biologists who think that most genetic variation is unlinked to an organism's survival and reproduction.

neutron: (L. *neuter*, either) A subatomic particle without any electrical charge found in the nucleus of an atom.

New World monkeys: Primates with prehensile tails that live in the forests of Southern Mexico and Central and South America.

niche: The role, function, or position of an organism in a biological community.

nitric oxide: The simple chemical compound NO; it occurs as a gas under normal atmospheric conditions and has been found to act as a neurotransmitter during penile erection and certain other physiological processes.

nitrogen cycle: The cycling of nitrogen from the atmosphere to Earth's surface, through organisms, and back again.

nitrogen fixation: The assimilation of atmospheric nitrogen by certain prokaryotes into biologically usable nitrogenous compounds.

node: Attachment site between the leaf's petiole and the plant stem.

nodule: A swelling on the roots of legumes and some other plants that houses nitrogen-fixing bacteria.

nonpolar: Having a symmetrical distribution of electrical charge; i.e., a nonpolar molecule like most lipids will not dissolve readily in water.

nonspecific resistance: Cellular functions that fight disease in the same way regardless of the invader's characteristics.

norepinephrine: A hormone secreted by the adrenal medulla; raises blood sugar concentration, blood pressure, and heart rate and increases muscular power and resistance to fatigue; sometimes known as noradrenaline.

notochord: A rod of mesodermal cells in the chordate embryo that marks the location of the backbone in vertebrates.

nuclear envelope: Two bilipid membranes perforated by pores that enclose the nucleus of a eukaryotic cell.

nuclear pore: A hole that perforates the nuclear envelope.

nucleic acid: A polymer of nucleotides, e.g., DNA and RNA.

nucleolus (pl. nucleoli): A dark-staining region within the nucleus of a eukaryotic cell where ribosomal RNA is formed.

nucleosome: The basic packaging unit of eukaryotic chromosomes; a histone wrapped with two loops of DNA.

nucleotide joining: The last step of transcription from DNA to RNA, during this step, the enzyme RNA polymerase joins two adjacent RNA nucleotides together.

nucleotide: The basic chemical unit of DNA and RNA consisting of one of four nitrogenous bases (A, C, T, or G) (or in RNA, A, U, C, and G) linked to a sugar (a ribose or a deoxyribose), which is in turn linked to a phosphate.

nucleus: The membrane-enclosed region of a eukaryotic cell that contains the cell's DNA.

nutrition: The science concerned with the amounts and kinds of nutrients needed by the body.

O

obese: Having body weight that is more than 20 percent above ideal.

oceanic zone (open ocean): Water beyond the continental shelf and covering the deep abyssal plane, which is nearly deserted, even in the sunniest upper regions near the surface.

oil: A fluid lipid that is insoluble in water; often a prime form of energy storage in plants.

Old World monkeys: Monkeys lacking prehensile tails that live in Africa, India, and Southeast Asia.

olfactory epithelium: Button-sized patches of yellowish mucous membrane skin located high in the nasal passages. Embedded in the epithelium are olfactory neurons and their processes.

one gene–one enzyme hypothesis: The now modified idea that each gene regulates the production of a single enzyme.

one gene–one polypeptide hypothesis: The concept that for proteins made up of several polypeptide chains, a different gene encodes each polypeptide.

open circulatory system: A system of fluid transport in spiders, insects, and many other invertebrates in which a clear blood equivalent called hemolymph circulates partially in vessels and partially unconfined by tubes or vessels.

operant conditioning: A form of learning in which an animal associates an action (an operant; e.g., pressing a lever) with a reward or punishment (a reinforcer; e.g., food); also called trial-and-error learning.

operator: A series of nucleotides that bind to a repressor protein, thereby preventing the initiation of gene and transcription from the adjacent promoter.

operon: The collective term for a regulatory protein and a group of genes whose transcription it controls in bacterial cells.

opposable thumb: A first digit or thumb that can be held opposite the other digits; characteristic feature of some primates.

optic nerve: The nerve leading from the eye to the brain.

optimality hypothesis: The idea that animals act in ways that maximize energy intake and minimize the energy spent on gathering food.

order: A precise arrangement of structural units

and activities; also, in taxonomy, a taxonomic group comprising members of similar families.

organ: A body structure composed of two or more tissues that together perform a specific function.

organ system: A group of organs that carries out a particular function in an organism.

organelle: In eukaryotic cells, a complex cytoplasmic structure with a characteristic shape that performs one or more specialized functions.

organic: Molecules that are based on carbon and contain hydrogen.

organism: An individual that can independently carry out all life functions.

organismal respiration: The manner in which a whole animal exchanges carbon dioxide and oxygen with the atmosphere.

organogenesis: The formation of organs during embryonic development.

orient: The ability to move in a set direction.

origin: The end of a skeletal muscle that generally remains stationary during a contraction, like an anchor.

osmosis: The movement of water molecules across a selectively permeable membrane from a region of higher water concentration to one of lower water concentration.

osmoregulation: Maintenance of a constant internal salt and water concentration in an organism.

Osteichthyes: A class of the phylum Chordata that includes the bony fishes.

osteoporosis: A condition of bone thinning and weakness that occurs because cells specialized to remove stored calcium and phosphate from the existing bone matrix work faster than cells specialized to store the minerals and build bone.

outgroup: A closely related species whose lineage diverged before all of the members of the group in question; used to infer ancestral traits.

ovary: Egg-producing organ.

ovary wall: In a flowering plant, the structure that develops into a fruit.

overweight: Having body weight that is more than 10 percent above ideal.

oviducts (Fallopian tubes): The tubes along which ova travel from the ovary to the uterus; usually the site of fertilization.

ovulation: The release of a mature ovum from the ovary.

ovule: The structures that contain the egg and later mature into the plant seed.

ovum (pl. ova): An unfertilized egg cell.

oxidative energy system: The longest-sustaining energy system of the body that relies on the Krebs cycle and electron transport drain in mitochondria and its ability to use fats as fuel; typically fuels aerobic activity.

ozone layer: A zone encircling Earth several miles above its surface that absorbs about 99 percent of the ultraviolet light that would otherwise penetrate and destroy many biological molecules, including DNA.

P

pacemaker cell: Cells that set the rate of the heartbeat.

Paleozoic Era: The geologic era from about 580 million years ago to 245 million years ago.

palisade parenchyma: The main photosynthetic tissue of a leaf, which lies just beneath the upper epidermis.

pancreas: A gland located behind the stomach that secretes digestive enzymes into the small intestine and secretes the hormones insulin and glucagon into the blood.

Pangaea: A single supercontinent that existed on Earth about 245 million years ago.

Parabasalian: Ancient, primitive protists such as the organisms inside a termite's gut.

paracrine hormone: A type of hormone that acts on cells immediately adjacent to the ones that secreted it.

parallel vein: Leaf veins that run parallel to each other; found in monocots.

parasite: A type of predator that obtains benefits at the expense of another organism, its host. A parasite is usually smaller than its host, lives in close physical association with it, and generally saps its host's strength rather than killing it outright.

parasympathetic nerves: Neurons of the autonomic nervous system that emanate from the spinal cord and act on the respiratory, circulatory, digestive, and excretory systems as a "housekeeping" system, conserving and restoring body resources.

parathyroid gland: One of a set of four small endocrine glands located on the thyroid gland; it secretes parathyroid hormone, which controls blood calcium levels.

parenchyma: A plant tissue composed of living, loosely packed, thin-walled cells; the most abundant plant tissue.

parental (P₁) generation: In Mendelian genetics, the individuals that give rise to the first filial (F₁) generation.

parental type: An offspring having the characteristics of one of the parents.

Parkinson's disease: Medical condition occurring when neurons in the brain's substantia nigra region degenerate and dopamine levels in the brain fall below normal levels. This decrease in dopamine contributes to rigidity, weakness, and shaking.

particulate model of heredity: Mendel's idea that heredity could be governed by "particles" that retain their identity from generation to generation.

passive immunization: The injection of antibodies generated in one individual into another individual.

patella: Kneecap.

pathogen: A disease-causing agent.

pectoral (shoulder) girdle: The skeletal support to which front fins or limbs of vertebrates are attached.

pedigree: An orderly diagram of a family's relevant genetic history.

pelvic (hip) girdle: The skeletal support to which hind fins or limbs of vertebrates are attached.

penis: In mammals and reptiles, the male organ of copulation and urination.

pepsin: An enzyme secreted by the stomach that digests proteins.

pepsinogen: A precursor to the protein-digesting enzyme pepsin.

peptide bond: A bond joining two amino acids in a protein.

peptide hormone: A hormone made of strings of amino acids, either peptides (a few amino acids long) or polypeptides (many amino acids linked together). These hormones usually act by stimulating second messengers in target cells.

peptidoglycans: Sugar-protein complexes occurring in prokaryotic cell walls.

perennial: (L. *per,* through + *annus,* year) A plant that lives for many years, blooming and setting seeds several times before dying.

perforation plate: Holes in the end walls of vessel elements that allow water to flow through stacked vessel elements unimpeded.

perforins: A type of protein released by natural killer cells that perforate bacterial plasma membranes and kill bacteria.

pericycle: In roots of vascular plants, one or more layers of parenchyma tissue lying between the endodermis and the vascular layer.

periderm: In a plant, the outer protective layer, including cork and cork cambium, that is formed during secondary growth.

peripheral nervous system (PNS): The part of the nervous system consisting of the sensory and motor neurons, connecting the central nervous system with the sense organs, muscles, and glands of the body.

peristalsis: (Gr. *peristaltikos,* compressing around) In animals, successive waves of contraction and relaxation of muscles along the length of a tube, such as those in the digestive tract that help move food.

petal: A flower structure that can have a shape, fragrance, and showy color that attract animal pollinators. In dicots, petals occur just inside the ring of sepals.

petiole: (L. *petiolus,* a little foot) The stalk connecting the leaf to the stem of a plant.

pH scale: A logarithmic scale that measures hydrogen ion concentration; acids range from pH 1 to 7, water is neutral at a pH of 7, and bases range from pH 7 to 14.

phage: Another word for bacteriophage.

phagocyte: A specialized scavenger cell that devours debris.

phagocytosis: The type of endocytosis through which a cell takes in food particles.

pharynx: [1] In vertebrates, a tube leading from the nose and mouth to the larynx and esophagus; conducts air during breathing and food during swallowing; the *throat.* [2] In flatworms, a short tube connecting the mouth and intestine.

phenotype: The physical appearance of an organism controlled by its genes interacting with the environment.

pheromone: A compound produced by one individual that affects another individual at a distance; for example, female moths secrete pheromones that attract males.

phloem: Food-conducting tissue of plants, consisting of sieve tube members with companion cells, phloem parenchyma, and fibers.

phospholipid: A lipid composed of a phosphate functional group and two fatty acid chains attached to a glycerol molecule; a main component of cell membranes.

phosphorous cycle: The cycling of phosphorus between organisms and soil, rocks, and water.

photoautotroph: An organism that captures energy from light.

photons: A vibrating particle of light radiation that contains a specific quantity of energy.

photoperiod: The length of light and dark periods each day.

photoreceptor cells: Light-sensitive cells such as rods and cones in the eye.

photosynthesis: The metabolic process in which solar energy is trapped and converted to chemical energy (ATP and NADPH), which in turn is used in the manufacture of sugars from carbon dioxide and water.

phototropism: The orientation of a plant toward light.

phyletic gradualism: The concept that morphological changes occur gradually during evolution and are not always associated with speciation; distinct from punctuated equilibrium.

phylogeny: The study of the evolutionary history of different groups of organisms.

phylum (pl. phyla): A major taxonomic group just below the kingdom level, comprising members of similar classes, all with the same general body plan. Equivalent to the division in plants.

physiology: The study of how biological structures function.

phytochrome: Light-sensitive pigment found in plants that absorbs in the red or far-red wavelengths; associated with a number of timing processes, such as flowering, dormancy, leaf formation, and seed germination.

phytoplankton: Photosynthetic microorganisms that live near the surface of marine and fresh water.

pineal gland: An endocrine gland located in the vertebrate midbrain and part of the diencephalon; produces the hormone melatonin and probably involved in circadian rhythms.

pioneer community: The species that are first to colonize a habitat after a disturbance such as fire, plowing, or logging.

pit: A gap in the side walls of vessel elements that allows water to move through.

pit organ: Organs that allow snakes and certain other organisms to detect objects in the dark by sensing the heat patterns the objects give off.

pith: Parenchyma cells occupying the central portion of a stem or root and acting as storage cells for the plant.

pituitary gland: In vertebrates, an endocrine gland connected by a stalk to the hypothalamus at the base of the brain. The anterior lobe secretes growth hormone, prolactin, LH, FSH, ACTH, TSR, and MSH; the posterior lobe stores and releases oxytocin and ADH. Much of the functioning of the pituitary is under the control of the hypothalamus, and pituitary hormones control most of the other endocrine glands.

placenta: In mammals, the spongy organ rich in blood vessels by which the developing embryo receives nourishment from the mother.

placoderm: The ancient group of armor-plated fishes that gave rise to the modern fishes.

planaria: A group of flatworms that inhabit freshwater lakes, rivers, or bodies of salt water.

plant: A multicellular eukaryote that captures energy by photosynthesis and develops from an embryo.

plant anatomy: The study of the main parts and tissues of the plant body.

plant functions: How the parts of a plant work together in the living organism.

Plantae: The kingdom comprising multicellular photoautotrophs such as mosses, ferns, and flowering plants.

planula larva: In cnidarians, sexual reproduction produces so-called planula larva covered by cilia that allow the new organism to move about and disperse to new areas.

plaque: Waxy deposits that build up inside the arteries and gradually obstruct blood flow.

plasma: The liquid portion of blood or lymph.

plasma cell: A type of B cell clone that generates antibodies.

plasma membrane: The membrane that surrounds all cells, regulating entry and exit of substances; consists of a single lipid bilayer; also called the cell membrane and plasmalemma.

plasmid: A circular piece of DNA that can exist either inside or outside bacterial cells but can reproduce only inside a bacterial cell. Plasmids are used extensively in genetic engineering as carriers of foreign genes.

plasmodesmata: Specialized junctions between plant cells that facilitate cell-to-cell communication.

plasmodium: One form of a true slime mold that is a mass of continuous cytoplasm surrounded by one plasma membrane that moves slowly, like a giant amoeba. Also, the genus of malarial parasites.

plastid: An organelle found in plants and some protozoa that harvests solar energy and produces or stores carbohydrates or pigments.

plate tectonics: The geologic building and moving of crustal plates on Earth's surface.

platelet: A disk-shaped cell fragment in the blood important in blood clotting; also called a thrombocyte.

pleated sheet: Secondary protein structure in which hydrogen bonds link two parts of a protein into a shape like a corrugated sheet.

pleiotropy: The condition in which a single gene affects two or more distinct and seemingly unrelated traits.

pleural sac: One of two fluid-filled sacs that enclose the lungs.

plus/minus mating type: Fungi are neither male nor female, but each haploid individual is one of two mating types: plus or minus.

podocytes: The epithelial cells of each Bowman's capsule (literally, "foot cells").

poikilotherms: Animals with variable internal temperatures; colloquial synonym for poikilotherms is "cold-blooded animals."

polar: Having an asymmetrical distribution of electrical charge; i.e., a polar molecule like glucose will dissolve readily in water.

polar bodies: Haploid cells that form during meiosis in the female but do not become the egg.

polar ice caps: Icy treeless regions at our planet's highest latitudes.

polarized: The state of a cell when it has an imbalance of electrical charges across its plasma membrane.

pollen grain: The male gametophyte of seed plants.

pollination: The transfer of pollen grain to female flower parts.

polyploidy: An increase in the number of chromosome sets in a cell.

polymerase chain reaction (PCR): A laboratory procedure that employs DNA copying enzymes to greatly increase the amount of specific DNA fragments in a sample.

polymerization: The joining together of newly paired bases, creating a DNA strand identical to the original double helix strand of DNA.

polyp: (L. *palypus,* many-footed) The sedentary stage in the life cycle of cnidarians; a cylindrical organism with a whorl of tentacles surrounding a mouth at one end. Sea anemones and hydras are examples of polyps living alone; corals are examples of colonial polyps.

polypeptide: (Gr. *polys,* many + *peptin,* to digest) Amino acids joined together by peptide bonds into long chains. A protein consists of one or more polypeptides.

polysaccharide: A carbohydrate made up of many simple sugars linked together. Glycogen and cellulose are examples of polysaccharides.

polyunsaturated fat: Fat with several double bonds in the carbon chain of the molecule's fatty acid portion.

pons: The part of the brainstem involved in relaying sensory information to other areas of the brain.

population bottleneck: A situation arising when only a small number of individuals of a population survive and reproduce; therefore only a small percentage of the original gene pool remains.

population: A group of individuals of the same species living in a particular area.

population density: The number of individuals in a certain amount of space.

Porifera: The phylum containing the sponges.

positive feedback loop: A series of steps that brings about rapid change in a homeostatic system; amplifies an initial change of the external or internal environment in one direction further and further in the same direction.

postsynaptic cell: The nerve cell that receives the message that has crossed a synapse from a presynaptic cell.

potassium channel: A protein-lined pore in a cell's plasma membrane that permits potassium ions to flow in and out of the cell.

potato blight: A species of water mold that rots and kills growing potato vines.

potential energy: Energy that is stored and available to do work.

precapillary sphincters: Rings of smooth muscle around a capillary on the arterial (as opposed to venous) side of a capillary bed.

predation: The act of procurement and consumption of prey by predators.

predator: An organism, usually an animal, that obtains its food by eating other living organisms.

prediction: In the scientific method, an experimental result expected if a particular hypothesis is correct.

pregnancy: Begun by the implantation of the developing embryo in the uterine wall, a series of developmental events involving close cooperation between mother and embryo that transforms a fertilized egg (zygote) into a baby.

prehensile: Grasping (as in the tails of some species of monkeys).

pressure flow hypothesis: The best current hypothesis for phloem transport, which asserts that phloem transports substances from "sources," or areas of production, to "sinks," or areas of use, down a pressure gradient established via osmosis.

presynaptic cell: The neuron that sends a message down its axon and across a synapse to another cell, the postsynaptic cell.

prey: Living organisms that are food for other organisms.

primary consumer: In an ecosystem, an organism that eats producers; herbivores (plant eaters like cows and caterpillars) are primary consumers.

primary growth: Plant growth arising from the apical meristem.

primary phloem: Phloem that results from primary growth (growth of apical meristems).

primary producer: An organism that produces all the biological molecules required for its growth from nonliving substances taken directly from the environment; autotrophic organisms.

primary response: The initial immune response to an invader.

primary xylem: Xylem that results from primary growth (growth of apical meristems).

primate: The mammalian order including monkeys, apes, and humans.

primer: A molecule that defines the point at which DNA synthesis begins.

prion: An intracellular disease-causing entity apparently consisting only of protein and having no genetic material.

probe: A radioactive nucleic acid made in the laboratory that is complementary to a gene; a molecular genetic tool used during cloning.

process: A projecting part of an organism or organic structure, such as a bone.

product: A substance that results from a chemical reaction.

progesterone: A female sex hormone secreted by the corpus luteum in the ovary that stimulates uterine wall thickening and mammary duct growth.

prokaryote: An organism made up of a prokaryotic cell.

prokaryotic cell: A cell in which the DNA is loose in the cell. Eubacterial and archaebacterial cells are prokaryotic. Prokaryotic cells generally have no internal membranous organelles and evolved earlier than eukaryotic cells.

proliferation: An increase in cell number.

promoter: A series of nucleotides to which RNA polymerase binds and initiates transcription of the adjacent gene.

prophase: The first phase of nuclear division in mitosis or meiosis, when the chromosomes condense, the nucleolus disperses, and the spindle forms.

prostaglandins: A fatty acid–derived hormone secreted by many tissues; for example, prostaglandin from the uterus produces strong contractions in the uterine muscle, causing menstrual cramps or inducing labor.

protease: A pancreatic enzyme that breaks down proteins.

protein synthesis: Translation; the assembly of amino acids into a polypeptide chain, occurring in four stages: initiation, elongation, termination, and disassembly.

proteobacteria: The largest and most diverse group in the domain Bacteria.

Proterozoic Era: The geologic era from 2.5 billion years ago to about 580 million years ago.

prothrombin: The inactive precursor of the enzyme thrombin, which produces fibrin, during the process of blood clotting.

Protista: Generally single-celled organisms with nuclei, such as paramecia and amoebas.

proton: A positively charged subatomic particle found in the nucleus of an atom.

protostome: (Gr. *protos,* first + *stoma,* mouth) Any bilateral animal whose first opening in the embryo (blastopore) becomes the mouth, while the second opening becomes the anus; also characterized by spiral cleavage during development; includes annelids, mollusks, and arthropods.

proximal tubule: In the vertebrate kidney, the convoluted tube between the Bowman's capsule and the loop of Henle.

pseudocoelom: A "false" body cavity only partially covered with mesoderm.

pseudopodia: Limblike cellular extensions that help protists and certain blood cells move and feed.

puberty: The maturation of the sex organs and the development of secondary sex characteristics, such as breasts in females, and facial hair and a deep voice in males.

pulmonary artery: In amphibians, reptiles, birds, and mammals, the artery that carries blood from the heart's right ventricle to the lungs.

pulmonary circulation: The blood vessel system that carries oxygen-poor blood from the heart to the lungs, where gas exchange occurs, and carries oxygen-rich blood back to the heart; contrasts to systemic circulation.

pulmonary semilunar valve: One-way valve with three half-moon-shaped flaps through which the right ventricle of the heart pumps oxygen-poor blood toward the lungs.

pulmonary veins: Vessels that carry oxygenated blood from the lungs to the heart's left atrium; the only veins in an adult person that carry oxygen-rich blood.

punctuated equilibrium: The theory that morphological changes evolve rapidly in geologic time. In small populations, the resulting new species are distinct from the ancestral form. After speciation, species retain much the same form until extinction; distinct from phyletic gradualism.

Punnett square: In genetics, a diagrammatic way of presenting the results of random fertilization from a mating.

pupa: In insects with complete metamorphosis, the stage that intervenes between the larva and the adult; in some cases, as in moths, the pupa is encased inside a cocoon.

pupil: The central, shutterlike opening in the iris of the vertebrate eye, through which light passes to the lens and the retina.

pyloric sphincter: A muscular ring located at the junction of the stomach and small intestine where chyme passes from the stomach to the small intestine.

pyramid of biomass: The relationship between the total masses of various groups of organisms in a food chain, in which there is usually less mass, hence less stored energy, at each successive trophic level.

pyruvate: Ionized pyruvic acid; molecule formed during the process of glycolysis by the splitting of a six-carbon sugar (glucose) into two three-carbon compounds.

Q

quadriceps: Thigh muscles.

R

radial body plan: A body with a central axis with structures radiating outward like spokes of a wheel.

radiolarians: Single-celled protists that produce beautiful silicon-based shells.

radula: A rasping organ in mollusks that shreds plant material by rubbing it against the hardened surface of the mouth.

ray-finned fish: Fish whose fins are made of webs of skin over rays of bone.

reactant: The starting substance in a chemical reaction.

reading frame: Three ways a sequence of three nucleotides in a gene can be divided into contiguous, non-overlapping sets of three nucleotides in codons.

realized niche: The part of the fundamental niche that a species actually occupies in nature.

receptor: A protein of a specific shape that binds to a particular chemical. A second meaning, in physiology, is a structure that carries out one of three primary elements of a homeostatic system to sense environmental conditions. Also see effector and integrator.

recessive: An allele or corresponding pheno-

typic trait that is hidden by a dominant allele in a heterozygote.

recombinant DNA technology: The techniques involved in excising DNA from one genome and inserting it into a foreign genome.

recombinant type: An offspring in which characteristics of the parents are combined in new ways.

recombinant: Chromosomes of mixed ancestry contained within individual haploid cells after meiosis II.

recombination: The exchange of maternal and paternal DNA segments through genetic recombination.

rectum: The terminal portion of the colon that stores and helps remove solid waste by defecation.

red algae: Small delicate aquatic protists that occur as thin filaments or flat sheets and produce red pigments.

red tides: Dense blooms of certain dinoflagellates that tint water red and produce deadly toxins.

reflex arc: An automatic reaction, involving only a few neurons and requiring no input from the brain, in which a motor response quickly follows a sensory stimulus.

regulator cell: A cell that detects perturbations in the environment and secretes a molecular messenger in response.

renal cortex: One of the kidney's three distinct visible zones, where initial blood filtering takes place.

renal medulla: The kidney's central zone, divided into a number of pyramid-shaped regions; helps conserve water and valuable dissolved materials (solutes).

renal pelvis: A funnel-shaped, hollow inner compartment that serves as a repository for urine before it passes into the ureter and then collects in the bladder. Located inside the two zones consisting of the renal medulla and renal cortex.

renin: An enzyme released by the kidney that converts a blood protein into the hormone angiotensin II.

replication: The copying of one DNA molecule into two identical DNA molecules.

repressor: A protein that binds to an operator and blocks transcription of the adjacent gene.

reproduction: The method by which individuals give rise to other individuals of the same type.

reproductive isolating mechanisms: Any structural, behavioral, or biochemical feature that prevents individuals of a species from successfully breeding with individuals of another species.

reproductively isolated: Every true species in nature fails to generate fertile progeny with other species; the result is reproductive isolation.

reptile: A cold-blooded, scaly, lung-breathing vertebrate that lays large eggs that usually have a shell; the dominant group of animals in the Mesozoic Era; includes crocodiles, lizards, and tortoises.

resistance: The body's ability to fight off disease-causing agents.

respiration: The process by which oxygen is drawn into the body and transferred to blood, and carbon dioxide is expelled.

respiratory system: A group of organs specialized to extract oxygen from the environment.

responsiveness: The tendency of a living thing to sense and react to its surroundings.

resting potential: Potential energy that is the difference in electric charge between the cell's exterior and interior; measured in volts like a battery.

restriction enzyme: A naturally occurring enzyme that cuts DNA at precise points.

reticular formation: A network of neurons extending from the thalamus of the brain stem to the spinal cord; regulates respiration, cardiovascular centers, and awareness.

retina: (L., a small net) A multilayered region lining the back of the vertebrate eyeball containing light-sensitive cells.

retinol: A small ring-and-chain molecule cradled by rhodopsin; changes shape when light hits it, unleashing a cascade of reactions that we eventually perceive as light.

reverse transcriptase: The enzyme that facilitates "reverse" transcription of RNA to DNA.

Rh incompatibility: A situation in which a pregnant woman is Rh-negative but her fetus is Rh-positive, leading the mother's immune system to attack the fetus's red blood cells.

rheumatoid arthritis: An autoimmune disease in which joints swell painfully; in hands affected by this condition, the fingers can become gnarled and twisted, and everyday movements like buttoning a shirt can be painful or even impossible.

rhinovirus: A type of cold virus.

rhizoid: (Gr. *rhiza*, root) In bryophytes, some algae, and fungi, a hairlike structure that anchors the organism to the substrate.

rhizome: (Gr. *rhiza*, root) An elongated underground horizontal stem.

rhodopsin: The visual protein pigment in the rods and cones of vertebrate eyes that, when activated by light energy, begins a series of events resulting in vision.

ribosomal RNA (rRNA): An RNA molecule that is a structural component of ribosomes and is involved in protein synthesis.

ribosome: A structure in the cell that provides a site for protein synthesis. Ribosomes may lie freely in the cell or attach to the membranes of the endoplasmic reticulum.

ribozyme: An RNA molecule that can catalyze the cleavage or joining of itself and/or other RNAs.

rickettsias: Tiny, rod-shaped parasitic bacteria.

RNA: Ribonucleic acid; a nucleic acid similar to DNA except that it is generally single-stranded and contains the sugar ribose and the base uracil replaces thymine.

RNA world: Term used by some biologists to refer to a stage in the evolution of life during which RNA molecules would have catalyzed their own replication as well as joined amino acids into proteins.

rod: One of the two types of photoreceptor (light-sensitive) cells in the retina of the vertebrate eye; rods are sensitive to low levels of light and to movement but cannot distinguish color. (See also *cone*.)

root: The branching structure of a plant that grows downward into the soil. Roots anchor the plant and absorb and transport water and mineral nutrients.

root cap: A mass of cells covering and protecting the growing tissue of the root.

root hair: An extension from the root epidermis that increases the absorptive capacity of a root.

root pressure: Force that is generated when fluid moves up xylem cells. Root pressure can force water out of the leaves, where it accumulates in tiny droplets.

rotifer: A small common pond organism that has a cuticle formed within the epidermis.

rough ER: The part of the endoplasmic reticulum that is studded with ribosomes; synthesizes lysosomal and membrane proteins and secreted proteins.

roundworm: A member of the phylum Nematoda.

rumen: The sac in the gut of cows and other ruminants where bacteria help break down cellulose and provide energy.

S

S phase: The portion of the cell cycle during which the cell synthesizes DNA.

saliva: A watery liquid secreted by the salivary glands that moistens food particles for swallowing; contains the starch-digesting enzyme amylase.

salivary gland: A gland that secretes saliva.

saprobe: An organism that lives on decomposing organic matter.

sapwood: In the stem of a plant with secondary growth, younger wood around the outside of the heartwood that continues to transport water in the plant.

sarcomere: The contractile unit of a skeletal muscle, consisting of repeating bands of actin and myosin.

saturated fatty acid: Fatty acid portion of a lipid molecule containing only single covalent bonds between carbon atoms.

savanna: A tropical grassland biome, containing stunted, widely spaced trees, that is situated between tropical forests and deserts.

scanning electron microscope: A type of electron microscope that has a greater depth of field than a transmission electron microscope and thus allows scientists to see a specimen's outer surface in three dimensions. A specimen is coated with a thin layer of gold or other metal and electrons are then aimed to scan it rapidly.

schizophrenia: A condition characterized by psychosis, or false beliefs that can't be changed by evidence, and hallucinations, or phantom experiences perceived to have happened.

scientific method: A series of steps for understanding the natural world based on experimental testing of a hypothesis, a possible mechanism for how the world functions.

sclereid: "Stone cell;" hard, crystalline cells included in sclerenchyma tissue.

sclerenchyma: In plants, a supporting tissue composed of cells with lignified cell walls.

scolex: A tapeworm's head, which is little more than a knob with hooks or adhesive suckers around the mouth that attach to host tissues.

second filial (F$_2$) generation: In Mendelian genetics, the second generation in the line of descent.

second messenger: A molecule within a cell that causes cellular proteins to change behavior in response to a signal received at the cell surface; cyclic AMP and calcium ions are examples.

secondary compound: A molecule that helps ensure plant survival by repelling, killing, or interfering with the normal activities of plant-eating organisms.

secondary consumer: In an ecosystem, an organism that consumes herbivores; carnivores (meat eaters) are secondary consumers.

secondary growth: The enlarged diameter of a stem or root resulting from cell divisions in the lateral meristem.

secondary metabolite: Substances that are not directly needed for a plant's energy gathering and reproduction, but defend the plant against animals, fungi, and even other plants.

secondary phloem: Phloem that results from secondary growth, produced by the lateral meristems.

secondary response: A stronger, swifter reaction to an invader or other antigen when encountered a subsequent time.

secondary xylem: Xylem that results from secondary growth, produced by the lateral meristems.

secretin: A digestive hormone secreted by the small intestine that causes the pancreas to secrete bicarbonate, which neutralizes stomach acid.

sedimentary rock: Layer upon layer of sand and dirt accumulated over thousands or millions of years and eventually forming rock layers that sometimes entomb fossilized organisms from different eras.

seed: The product of a fertilized ovule of a seed plant, generally consisting of an embryo with its food reserves enclosed in a protective coat.

seed coat: The wall of the ovule.

seed-forming vascular plant: A plant with a vascular system that also produces seeds; the gymnosperms and angiosperms.

seedless vascular plant: A plant with an internal transport system but that requires standing water to reproduce because it does not produce seeds; includes horsetails and ferns.

seeds: The fertilized mature ovule of a gymnosperm or angiosperm, containing an embryonic plant.

segmented worm: A member of the phylum Annelida.

segregation principle: Mendel's first law, the principle of segregation, states that sexually reproducing diploid organisms have two alleles for each gene, and that during the gamete formation these two alleles separate from each other so that the resulting gametes have only one allele of each gene.

selectionists: Biologists who think that high rates of genetic variation exist because natural selection maintains genetic diversity in most populations.

selective advantage: An adaptation that increases the likelihood of an organism surviving the immediate environmental challenges.

selectively permeable: A property of the plasma membrane that allows some substances, but not others, to pass across it.

self-fertilization: The ability of a plant or animal to fertilize its own eggs.

self-pollination: The ability of the pollen from a single flower to fertilize an egg within the same flower; a type of self-fertilization.

self-tolerance: The lack of an immune system response to components of one's own body.

semiconservative replication: The universal mode of DNA replication, in which only one of the two strands of DNA of the parent molecule is passed to each daughter molecule. This inherited strand serves as the template for the synthesis of the other strand of the double helix.

senescence: A condition characterized by a profound decline in cell and organ function that occurs with aging.

sensitive period: The specific developmental stage during which a biological process such as imprinting can occur.

sensory cortex: The part of the cerebral cortex of the vertebrate brain that registers and integrates sensations from body parts.

sensory neuron: A nerve cell that receives information from the external or internal environment and transmits this information to the brain or spinal cord.

sepal: Flower structures that can have shapes, fragrances, and showy colors that attract animal pollinators. In dicots, they comprise the outmost of four rings of floral structures.

separation: The movement of chromatids or chromosomes away from each other toward opposite poles of the dividing cell during mitosis or meiosis.

septum (pl. septa): A dividing wall or partition between structures, such as the segments of an earthworm.

serosa: The outermost layer of the alimentary canal.

serotonin: Neurotransmitter associated with depression, insomnia, elation, and other mood swings.

sessile: In animals, the quality of being permanently attached to a fixed surface.

seta (pl. setae): Attached to segments of earthworms, a bristle that pushes against the ground and enables movement.

sex chromosomes: Pairs of chromosomes where the members of the pair are dissimilar in different sexes and are involved in sex determination, such as the *X* and *Y* chromosomes.

sex-limited: A trait that is confined to one sex.

sexual fusion (nuclear fusion): A process that produces a diploid cell or zygote through the fusing of two haploid nuclei; in fungi, this takes place within the fruiting body.

sexual reproduction: A type of reproduction in which new individuals arise from the mating of two parents.

sexual selection: A type of natural selection operating through mate choice or competition for mates.

sheath: An expanded portion at the base of the leaf encircling the stem that connects the blade to the plant stem at an attachment site called a node.

shock: Medical state involving the escape of fluid from the circulation system after an accident or allergic reaction; during this blood flow and blood pressure start to fall so low that blood flow to tissues is inadequate.

shoot: The main portion of a plant growing above ground.

short-day plant: Plants, such as rice, that require days shorter than a particular length to bloom.

sickle-cell anemia: A genetic condition inherited as a recessive mutation in a hemoglobin gene and characterized by pains in joints and abdomen, chronic fatigue, and shortness of breath. It currently affects about 60,000 people in the United States, including 1 out of 400 African-Americans.

sieve plate: One of the two perforated end walls of a sieve tube member.

sieve tube member: A conducting cell of phloem tissue in a vascular plant.

silversword alliance: Twenty-eight closely related species of plants that radiated from California tarweeds that were transported across the Pacific (probably by birds) and spread into the varied habitats of Hawaii as they formed and became available.

simple leaves: Leaves consisting of a single blade, such as an oak or maple leaf.

single-gene mutation: Changes in the base sequence of a single gene.

sinoatrial (SA) node: A lump of modified heart muscle cells near the upper right atrium that is spontaneously electrically excitable; the pacemaker that governs the basic rate of heart contractions in vertebrates.

siphon: In certain mollusks, the funnel through which water passes in the mantle cavity.

sister chromatids: The two rods of a replicated chromosome.

skeletal muscle: Muscle consisting of elongate, striated muscle cells; voluntary muscle.

skeleton: The rigid body support to which muscles attach and apply force, in vertebrates and invertebrates.

skull: The skeleton of the vertebrate head.

sliding filament mechanism: A mechanism of muscle contraction in which contraction is made possible by the unique arrangement and sliding of the thick filaments relative to the thin filaments, shortening the sarcomeres.

slow-twitch muscle fibers: Muscle fibers that obtain most of their ATP from the oxidative system; also called slow oxidative muscle fibers. Slow-twitch fibers provide power for most endurance sports.

small intestine: A coiled tube, about 6 meters (20 feet) long, that is the main site of carbohydrate and fat digestion and is also where protein digestion is completed, allowing nutrients to be absorbed into the blood.

smooth ER: The part of the endoplasmic reticulum folded into smooth sheets and tubules, containing no ribosomes; synthesizes lipids and detoxifies poisons.

smooth muscle: Muscle consisting of spindle-shaped, unstriated muscle cells; muscle type found in the digestive, reproductive, and circulatory systems.

social insect: A member of any of the insect groups, such as ants, bees, and termites, that lives in colonies of related individuals and exhibits complex social behaviors.

sodium channel: A protein-lined pore in the plasma membrane through which sodium ions can pass.

sodium-potassium pump: A protein that maintains the cell's osmotic balance by using ATP energy to transport sodium ions out of the cell and potassium ions into the cell.

soft palate: The posterior part of the roof of the mouth.

soil: The portion of Earth's surface consisting of disintegrated rock and decaying organic material.

soluble dietary fiber: Dietary material that dissolves in water and is found naturally in plant gums and in oats, peas, and beans.

solute: A substance that has been dissolved in a solvent.

solvent: A substance capable of dissolving other molecules.

somatic cell: (Gr. *soma,* body) A cell in an animal that is not a germ cell.

somatic gene therapy: The deliberate genetic alteration of body cells (not sex cells) to treat disease.

somatic nervous system: Motor neurons of the nervous system that are under voluntary control.

speciation: The emergence of a new species. Speciation is thought to occur mainly as a result of populations becoming geographically isolated from each other and evolving in different directions.

species: A taxonomic group of organisms whose members have very similar structural traits and who can interbreed with each other in nature.

species richness: The total number of species in a community.

species selection: (also known as **species replacement**) Natural selection acting at the level of species as members of different species compete for the same foods and habitats.

specific resistance: The body's ability to combat particular species or strains of pathogens and other invaders; works more slowly than nonspecific resistance.

sperm: The haploid male gamete.

spicule: A slender, spiky rod of silica or calcium carbonate found in sponges that supports the soft wall and provides some protection from predators.

spinal chord: A tube of nerve tissue that runs the length of a vertebrate animal, just above (dorsal to) the notochord.

spinal nerves: Pairs of nerves in which the sensory and motor neurons enter or leave the spinal cord.

spindle fibers: The cytoskeletal rods (microtubules) that move chromosomes towards the cell poles and that cause cell poles to separate during cell division.

spinneret: A tubular appendage in spiders and some insects that reels out silk threads.

spiracle: A hole in the body wall of insects that forms the opening of the air tubes (tracheae).

spirillum (pl. spirilla): A bacterial cell with a spiral shape.

spirochetes: Bacteria with a distinctive spiral shape.

spleen: Organ about the size and shape of a banana in an adult person that lies just behind the stomach. The spleen filters blood, produces phagocytic (debris-gobbling) white blood cells, and stores platelets and some blood cells.

spongy bone: Bone with regions with larger, more numerous spaces.

spongy parenchyma: A layer of loosely packed cells that lies between the palisade parenchyma and the lower epidermis in plants. Spongy parenchyma cells provide a huge surface area (analogous to an animal's lung) that absorbs

carbon dioxide from the air that enters the stomata.

spore: In eukaryotes, a reproductive cell that divides mitotically and produces a new individual. In prokaryotes, a resistant cell capable of surviving harsh conditions and germinating when conditions are once again favorable.

sporophyte: (Gr. *spora*, seed + *phyton*, plant) The diploid, spore-producing stage in the life cycle of many plants.

SRY: Specific gene on the *Y* chromosome that has been isolated as the "sex-determining region" and necessary for causing male development in mammals.

S-shaped curve: A plot of population growth with a flat section, a steeply rising section, and then a leveled off section that represents logistic growth.

stabilizing selection: A mode of natural selection that results in individuals with intermediate phenotypes; under these selection pressures, extreme forms are less successful at surviving and reproducing.

stamen: The male or pollen-producing structure within a dicot or monocot flower. Each stamen consists of the anther containing pollen sacs and a supporting filament.

starch: A polysaccharide composed of long chains of glucose subunits; the principal energy source of plants.

start codon: In DNA and mRNA, the codon that signals where the translation of a portion begins.

stem: The central, often elongated part of a plant, composed mainly of vascular tissue, that supports leaves and transports water and nutrients.

stem cells: A normal body (somatic) cell that can continue to divide, replacing cells that die during an animal's life.

stereocilia: Threadlike projections on the hair cells of the cochlea of the inner ear.

stereoscopic vision: Depth perception based on the orientation of the eyes and their brain connections.

sternum: The breastbone; in birds, a blade-shaped anchor for the pectoral muscles that enables flight.

steroid hormone: A hormone synthesized from cholesterol containing four joined rings of carbon atoms; e.g., estrogen, testosterone, and cortisol.

steroid: (Gr. *stereos*, solid + L. *oi*, having the form of + *oleum*, oil) A major class of lipids based on a 4-carbon-atom ring system and often a hydrocarbon tail. Cholesterol and sex hormones are steroids.

stigma: [1] The tiny, light-sensitive eyespot of a euglenoid. [2] The sticky top of a flower that serves as a pollen receptacle.

stipe: The stemlike structure that provides vertical support to an alga.

stoma: (pl. stomata): A tiny hole surrounded by two guard cells in the epidermis of plant leaves and stems, through which gases can pass.

stomach: An expandable, elastic-walled sac of the gut that receives food from the esophagus.

strand separation: The separation and unwinding of portions of the DNA double helix near one end of the gene; facilitated by specific enzymes.

Strepsirhini: A suborder of mostly tree-dwelling primates including the lemurs and lorises.

stress: The body's response to fear, pain, or other disturbances of our normal physiological equilibrium.

stressor: An activity or event that triggers a stress response in the body.

stroke: Paralysis or numbness caused by destruction of brain tissue due to a blood clot or blood vessel break.

stroma: In chloroplasts, the space between the inner membrane and the thylakoid membranes.

style: A structure leading from the ovary of flowering plants to the stigma (pollen receptacle).

subcutaneous: Literally below the skin; often refers to fat (adipose) cells that bind the skin to underlying organs found below the dermis layer.

submucosa: A connective tissue outside the mucosa of the digestive tract that is richly supplied with blood and lymph vessels and nerves.

substrate: A reactant in an enzyme-catalyzed reaction; fits into the active site of an enzyme.

succession: The process through which a regular progression of communities will regrow at a particular site.

superior vena cava: One of the two largest veins in the human body, carrying blood from the head, neck, and arms to the right atrium of the heart.

suppressor T cell: A type of T cell that can stop or slow ongoing immune responses.

surface tension: The tendency of molecules at the surface of liquid water to cohere to each other but not to the air molecules above them.

surfactants: Molecules that function like detergents to lower the surface tension of the water in lungs and other organs and in industrial applications.

survivorship curve: A plot of the data representing the proportion of a population that survives to a certain age.

sustainability: Balancing the human population and economic growth against the abilities of the environment to provide a constant level of resources.

swim bladder: A bag of gas that can be inflated or deflated slightly and helps a fish maintain its depth in the water without sinking or floating.

symbiont: An organism that lives in a close relationship with an organism of another species.

symbiosis: A close interrelationship of two different species.

symbolic thought: A cultural development attributed to the anatomically modern humans who lived 40,000 years ago and exemplified by sophisticated cave paintings, engravings, and sculptures, all of which suggest a major development of symbolic forms of communication, probably accompanied by music, increased language abilities, and a sense of mystery.

sympathetic nerves: Neurons of the autonomic nervous system that emanate from the central spinal cord and act on the respiratory, circulatory, digestive, and excretory systems in response to stress or emergency, preparing for flight or fight.

sympatric speciation: A situation in which a population diverges into two species after a genetic, behavioral, or ecological barrier to gene flow arises between subgroups of the population inhabiting the same region.

synapse: The region of communication between two neurons or between a neuron and a muscle cell.

synthetic theory of evolution: In the 1930s and 1940s, biologists began to combine evolutionary theory with genetics. The synthetic theory suggests that: (1) Gene mutations occur in reproductive cells at high enough frequencies to impact evolution. (2) Gene mutations occur in random directions unrelated to the organism's survival needs in its environment. (3) Natural selection acts on the genetic diversity brought about by such random mutations.

systematic circulation: The blood vessel system that carries oxygen-rich blood from the heart to the body and returns oxygen-poor blood to the heart; contrasts to pulmonary circulation.

systematics: The categorizing of organisms into taxonomic groups.

systole: The contraction phase of the heart muscle.

T

T cell or T lymphocyte: A type of white blood cell that kills foreign cells directly and also regulates the activities of other lymphocytes.

T cell receptor: The receptor protein of a T cell.

tail: In chordates, a structure that protrudes beyond the anus.

taproot: In plants, a root system with a prominent main root developing directly from the embryonic root, growing vertically downward and bearing lateral roots, e.g., the root of a carrot plant.

target cell: Any cell that responds to specific hormones.

taste buds: Sensory organs studding the tongue that consist of pores leading to a nerve cell. The nerve cell is surrounded by accessory cells arranged in an overlapping pattern that resembles an artichoke or onion bulb.

taxonomy: (Gr. *taxis*, arrangement + *nomos*, law) The science of classifying organisms into different categories.

Teleost: A modern bony fish and includes most living fish and characterized by the presence of a swim bladder and spiny fins.

telophase: The final phase of nuclear division when the chromosomes are at opposite poles of the cell, the nuclear membrane and nucleolus reappear, and the spindle disappears.

temperate forest: A biome that occurs north or south of subtropical latitudes, characterized by generally a mild climate and varied populations of evergreen and deciduous trees.

tendon: A band of connective tissue that connects bone to muscle.

tension: The force upon the unbroken chain of water molecules in the xylem tubes of plants as they are pulled upward from above.

terminal bud: The apical meristem at the tip of a branch.

termination factor: A protein that binds to the ribosome–mRNA complex during protein synthesis, bringing the growth of the polypeptide chain to a halt.

termination: The third main stage of translation (protein synthesis), during which the growth of the polypeptide chain comes to a halt when the ribosome reaches the stop codon.

territoriality: A form of behavior in which an individual defends its living or feeding space against any intruders.

tertiary consumers: In an ecosystem, a carnivore that eats other carnivores.

testis (pl. testes): The male reproductive organ that produces sperm and sex hormones.

testosterone: Hormone produced by the testes in vertebrate males; stimulates embryonic development of male sex organs, sperm production, male secondary sex characteristics, and male behaviors.

tetraploid: The condition of having four sets of chromosomes.

thalamus: Located beneath the cerebrum and above the hypothalamus of the vertebrate brain; makes up most of the diencephalon and relays and processes sensory impulses.

thallus: A flat, leaflike blade.

THC (tetrahydrocannibinol): A chemical produced in marijuana that helps protect the plant from various environmental stresses and is also psychoactive in humans.

thecodonts: Small extinct lizards that ran on two legs and gave rise to the dinosaurs.

theory: A general hypothesis that is repeatedly tested but never disproved.

therapsid: An extinct lineage of fierce, heavyset reptiles that gave rise to the mammals.

thermophile: Prokaryotes that thrive in very hot conditions.

thigmotropism: The orientation of a plant stem (as in vine tendrils) toward the touch pressure of a supporting object like a wall.

thoracic cavity: The region within the rib cage directly over the heart, in which the lungs and heart are suspended.

thorax: The central region of the body of an arthropod or vertebrate between the head and the abdomen.

thrombin: A type of enzyme created from the inactive protein prothrombin during the process of blood clotting.

thrombocyte: See platelet.

thylakoid: (Gr. *thylakos*, sac + *oides*, like) A stack of flattened membranous disks containing chlorophyll and found in the chloroplasts of eukaryotic cells.

thymus: A gland in the neck or thorax of many vertebrates; makes and stores lymphocytes in addition to secreting hormones.

thyroid: A gland in the neck, anterior to the trachea, that acts as the body's metabolic thermostat, regulating its use of energy as well as its growth.

thyroxine: The most abundant thyroid hormone, which governs metabolic and growth rates and stimulates nervous system function.

tibia: A large bone in the lower leg; shinbone.

tibialis: A shin muscle in the front of the lower leg.

tidal ventilation: The in-and-out flow of air that characterizes mammalian respiration.

tissue: A group of cells of the same type performing the same function within the body.

tissue culture: The growth of tissue cells in a laboratory plate or tube that often permits further differentiation and maturation of the cells. In plant breeding, the growing of new identical plants from somatic (body) cells.

tongue: A muscular organ on the floor of the mouth of most higher vertebrates that carries taste buds and manipulates food.

torpor: Metabolic state in which an animal's metabolism slows and its body temperature drops during the night, allowing the remaining energy stores in the animal's body to last until morning, when the animal is able to forage once again.

torsion: An internal twisting of the body mass during embryonic development.

trachea: The "windpipe," the major airway leading into the lungs of vertebrates.

tracheae: Branching networks of hollow air passages used for gas exchange in insects.

tracheid: An elongated hollow cell with thick, rigid, pitted walls; a basic unit of vascular tissue in most vascular plants.

trade winds: Predictable surface winds caused by a moving air mass which interacts with the rotating Earth.

transcription: (L. *trans,* across + *scribere,* to write). The transfer of information from a portion of a DNA molecule into an RNA molecule; the process is catalyzed by the enzyme RNA polymerase.

transduction: The transfer of genes from one bacterium to another via a virus.

transfer RNA (tRNA): A small RNA molecule that translates a codon in mRNA into an amino acid during protein synthesis.

transformation: The process of transferring an inherited trait by incorporating a piece of foreign DNA into a prokaryotic or eukaryotic cell.

transgenic animals: Animals bearing foreign genes inserted by researchers using recombinant DNA methodologies.

transition community: A community of organisms that establish themselves at a particular site based upon conditions produced by the activities of the pioneer community.

transition state: In a chemical reaction, the intermediate springlike state between reactant and product.

translation: The conversion of the information on a strand of RNA into a sequence of amino acids in a protein; occurs on ribosomes.

translocation: In plants, the transport of solutes in phloem cells.

transmission electron microscope: A type of electron microscope that greatly enlarges the view of an object by passing a beam of electrons through an ultrathin slice of the specimen. Some electrons are scattered or absorbed by structures in the specimen, while others pass through and strike a small fluorescent screen, creating an enlarged image of the object.

transpiration: The loss of water from plants by evaporation, mainly through the stomata on stems and leaves.

transpiration-cohesion-tension: The mechanism scientists propose for water transport in plants, which explains how a column of water molecules is "pulled" up the stem.

trichocyst: Tiny toxic darts in protists that spear and paralyze prey.

trichome: Little hairs formed by stem and leaf epidermal cells in certain plants.

triglyceride: A fat or oil composed of three fatty acids joined to three carbons of glycerol.

trochophore: A specialized larval form that occurs among many protosome animals, including several kinds of worms and all of the mollusks.

trophic level: A particular feeding level in a community or ecosystem.

tropical rain forest: A lush forest biome that occurs near the equator in Central and South America, Africa, and Southeast Asia, where rainfall is 200 to 400 cm (80 to 160 in) per year, and temperatures average about 25 °C (77 °F).

tropism: (Gr. *trope,* turning) The movement of a plant in response to environmental stimuli; e.g., phototropism, the response of plants to light.

true slime mold: A type of funguslike protist characterized by a plasmodium, a mass of continuous cytoplasm surrounded by one plasma membrane and containing many diploid nuclei.

true stem: An organ of usually vertical support housing vascular tissue.

tubular reabsorption: In the vertebrate kidney, the process whereby nutrients are returned from the kidney tubule to the blood.

tubular secretion: In the vertebrate kidney, the process whereby ions and drugs are secreted from the blood into the kidney tubule.

tumor: An uncontrolled or abnormal growth of cells.

tundra: A biome at the northern boundary of the coniferous forest characterized by low vegetation. This cold, treeless plain has annual temperatures of −5 °C or less.

tunic: The outer envelope of a sea squirt, enclosing the pharynx.

U

ulcers: Craterlike sores in the mucosa of the stomach or small intestine.

unsaturated fatty acid: Fatty acid portion of a lipid in which adjacent carbon atoms in the chain are linked by two covalent bonds, forming double bonds.

urea: A water-soluble nitrogenous waste product formed during the breakdown of proteins, nucleic acids, and other substances, excreted by mammals and some fishes.

ureter: The tube that carries urine from the kidney to the urinary bladder.

urethra: (Gr. *ouerin,* to urinate) The tube that carries urine and releases it to the outside; in males this tube also carries sperm.

uric acid: A water-insoluble nitrogenous waste product excreted by birds, land reptiles, and insects.

urinary bladder: A single storage sac into which the two ureters dump fluids for excretion. When full, it stretches local nerves, causing a feeling of urgency.

urine: A fluid that washes urea from the body.

Urochordata: A subphylum containing the sea squirts, which, as larvae, have the five features characteristic of chordates.

V

vaccination: A form of long-term protection produced by stimulating active immunity aimed at one microbe or toxin.

vacuole: Large fluid-filled sac inside cells surrounded by a single membrane; in plant cells, it is important for maintaining cell shape.

vagina: A hollow muscular tube that receives the penis during copulation and through which the fetus passes during birth.

valve: A tonguelike flap extending into the internal space of a vein that helps regulate the flow of blood.

varicose vein: A condition that results when venous valves become damaged and blood flows backward, causing the vein to distend and become visible as a large blue bulb.

vascular bundle: An organized grouping of xylem and phloem in the stem of an herbaceous annual such as hemp or a young perennial, such as an apple seedling.

vascular cambium: A type of lateral meristem that produces secondary phloem and secondary xylem, or wood.

vascular plant: A plant that possesses an internal transport system for water and food in the form of xylem and phloem cells.

vascular tissue system: Plant tissue that conducts fluid throughout the plant and helps strengthen roots, stems, and leaves; consists of xylem and phloem cells.

vasoconstriction: Contraction of blood vessel walls; regulates blood flow.

vasodilation: Relaxation of blood vessel walls; regulates blood flow.

vector: In molecular genetics, a virus or plasmid used to carry pieces of DNA into cells. In medicine, an organism such as a mosquito that transmits a disease-causing microbe such as the malarial protist.

vein: [I] A large thin-walled blood vessel that brings blood from the body to the heart. [2] In plant leaves, bundles of xylem and phloem cells that form a branching pattern in dicot leaves and run parallel to each other in monocot leaves.

ventilation: The inhalation and exhalation of the lungs of many vertebrates.

ventricle: A muscular chamber of the heart that pumps blood to the lungs or to the rest of the body.

venule: A small thin-walled blood vessel that arises from capillaries and carries blood from the tissues to the veins.

vertebra (pl. vertebrae): One of a series of interlocking bones that makes up the vertebral column, or backbone, of vertebrate animals.

vertebral column: The backbone.

Vertebrata: The subphylum containing the vertebrates or animals with vertebral columns.

vertebrate: An animal that possesses a vertebral column made of bony segments known as vertebrae.

vessel element: A special kind of cell in plant vascular tissue that transports water after the cells have died and hollowed out.

vestigial organs: A rudimentary structure with no apparent utility but bearing a strong resemblance to structures in probable ancestors.

vibrio (pl. vibrios): A bacterial cell with a curved rod shape.

villus: (L., a tuft of hair) A fingerlike projection of the intestinal wall that increases the surface area for absorption of nutrients.

viroid: An intracellular parasite that affects plants and consists only of small RNA molecules without any protein coat.

virus: An infectious agent consisting of RNA or DNA encased in a protein coat (capsid); incapable of metabolism or reproduction outside a host cell.

visceral mass: Most of the body of a mollusk, including the internal organs, and excluding the foot and mantle.

vitamin: (L. *vita,* life + *amine,* of chemical origin) An organic compound needed in small amounts for growth and metabolism and must be obtained in food.

W

water balance: A condition in which the amount of water entering the body equals the amount of water leaving.

water cycle: A sun-driven global exchange involving evaporation, precipitation, runoff, and transpiration that cycles water from the atmosphere to Earth's surface through organisms and back again.

water mold: A type of funguslike protist containing several nuclei within a common cytoplasm and forming relatively large immobile egg cells; members of the *Oomycota.*

water vascular system: In echinoderms, a system of fluid-filled canals that includes hundreds of short branches called tube feet that can attach to objects and thus aid in locomotion or feeding.

wax: A sticky, solid, waterproof lipid that forms the comb of bees and waterproofing of plant leaves.

weather: The condition of the atmosphere at any particular place and time, including its temperature, humidity, wind speed, and precipitation.

wood: The hard, cellulose-containing dead xylem cells of a perennial plant.

X

X-chromosome inactivation: A natural process of inactivating one X chromosome in each cell of a female embryo; takes place when the embryo (human or other mammal) consists of only about 1,000 cells. Once inactivated, the X chromosomes in those cells no longer read out any genetic information.

X chromosome: The sex chromosome found in two doses in female mammals, fruit flies, and many other species.

X-linked: Characteristic of a heritable trait that occurs on the X chromosome.

x-ray diffraction: A process in which a beam of x-rays is passed through a crystalline material to help determine its three-dimensional structure.

xylem: (Gr. *xylon,* wood) Water- and mineral-transporting tissue of plants composed of tracheids, vessel elements, parenchyma cells, and fibers; dead xylem cells form wood.

Y

Y chromosome: The sex chromosome found in a single dose in male mammals, fruit flies, and many other species.

yeasts: Generally single-celled fungi that reproduce by budding.

Z

zero population growth: In a population, the number of individuals gained is exactly equal to the number lost.

zoologists: Biologists who study animals.

zoology: The study of animals.

zygote: (Gr. *zygotos,* paired together) The diploid cell that results from the fusion of an egg and a sperm cell. A zygote may either form a line of diploid cells by a series of mitotic cell divisions or undergo meiosis and develop into haploid cells.

Index

Note: Italicized page numbers indicate a figure/illustration. The letter "t" after a page number indicates a table.

metaphase, 96
meteorites, *253,* 253–254
methacrine, 278
methane (CH_4), 619, 641
methanogens, 283
methionine, 168
microevolution, 244
microgravity, 374
micrometers, 39
micronutrients, 583–585
microspores, 317
microsporidia, 294, 295
microvilli, 461, *461*
midbrain, 497
migration, 237–238, 654–655
milk
    bovine growth hormone in, 180
    produced by mammals, 357
Miller, Stanley, 254, 255
millipedes, 345
mimicry, 610, *611*
minerals, 451–452
    in plants, 581–582
    from soil, 585
    transported in plants, 577
minimal medium diet, 159
minimum viale population, 239
minor minerals, 452
minus mating type, 304
Mir (Russian Space Station), 373
mitochondria, 51–52, *52,* 69–71, *70*
    absent in protists, 288–289
    endosymbiont hypothesis about, 260–261
    maternal inheritance and, 132
    obesity and, 78–79
mitochondrial DNA, 238–239, 366–367
"Mitochondrial Eve," 367
mitosis, 89, 558. *see also* interphase
    cancer therapy and, 97–98
    meiosis *vs.,* 108, *109*
    mitotic spindle, 96–97
    phases of, *94,* 94–97
mitotic spindle, 96–97
molds, 159–160
molecular clock, 242
molecular cloning, 176
molecular phylogenies, *269*
molecules, 25. *see also* biological molecules; macromolecules
    evolutionary evidence from, 226–227
    of immune system, 432–433
    movement in and out of cells, *44*
    self-replicating, 256–257
    size of, 37–38, *38*
Mollusca, 340
mollusks, *340,* 340–342
monarch butterflies, 610, *611*
Monera, 269
monoclonal antibodies, 436
monocots, 320–321, 532. *see also* plants
    leaves of, *547*
    roots of, *543*
    stems of, *546*
    structure of, *533*
Monod, Jacques, 174, 175
monoecious plants, 555
monomers, 32
monosaccharides, 32, 448
monotremes, 357, *358*
monounsaturated fatty acids, 35

Monsanto, 588
Morgan, Thomas Hunt, 128
"morning after pills," 206
morphogenesis, 201–202
mosaic climax community, 612
moths, *609,* 611–612, 658
motility, 3, 7–8
motor cortex, 498, 499, *499*
motor neurons, 494, 495–496, 521
motor unit, 523
mountain climbing, 419, 422
mountains, *635*
mountain sickness, 423
mountain streams, 639–640
mouth, 454, 457–458
M phase, 94
mRNA, *164, 165*
Mt. St. Helens, 612
mucosa, 456
mucus, 420, 456–457
mudsnails, *17,* 17–18
multicellular organisms, 327
multiple sclerosis (MS), 442
multiple traits and, 125–127
muscle fibers, 519
muscles, 375–376, 518–523, *519, 520*
    balance and, 511–512, *513*
    contraction of, 519–522, *522*
    fibers, 52–527, *525*
muscle tissue, 379
muscular dystrophy, Duchenne (DMD), 128–129, *130*
muscularis, 457
musculoskeletal system, 512,520. *see also* bones; muscles
    joints, 517–518
    skeleton, 512–518
mutagens, 173
mutants, 121
mutation, 128
    antibiotic resistance and, 246–247
    of DNA, 216, 219, *220*
    evolution and, 228–229, 237
    in genes, 159, 171–173, *172*
    single-gene, 171–173, 233, *233*
mutualism, 606, 611–612
myasthenia gravis, 492
mycelia, 302
mycoplasmas, 284
mycorrhizae, 306
myelin sheath, 491
myocardial infarction, 411
myoglobin, 524
myomeres, 349–350
myosin, 519, 520, *525*

NADH, 71
NADPH, *82,* 83
nasal lining cells, 433
NASA (National Aeronautics and Space Administration), 1
    Mars exploration by, 7
    Office of Planetary Protection, 2
nastic response, 563, *563*
National Biological Survey, 612
National Health and Nutrition Examination Surveys, 78
National Institutes of Health (NIH), 133–134, 136
natural killer cells, 430
natural selection, 13–14, 215, 218, *220,* 240–244. *see also* evolution
    behavior and, 648–649, 653–660
    eyes and, *228*

natural selection, *(Continued)*
    kin selection and, 661–662
    plant biotechnology and, 570
    sickle cell anemia and, 229
navigation, 654
neanderthals, *365,* 367
nectar, 321
negative feedback loops, 190, 193, 380–382, *381, 404*
nematocysts, 334
nematodes, 343
nemerterea, 338
nephridia, 339
nephrons, 387, *388*
    roles of, 389–391
    structure of, 387–389
nerve cells, *486,* 486–491
nerve impulse, 488
nerves, 462, *463*
nervous system, 379–380. *see also* brain
    of animals, 494–495
    central nervous system, 494, 496–500
    chemical synapses, 492–494
    hearing and, 484–486
    induction of, *201*
    of mollusks, 341
    nerve cells, *486,* 486–491
nervous tissue, 375–376, 379–380
net primary productivity, 624
neural crest, *202*
neural tube, 199, *199, 200*
neurons, 379, 486, *487*
    action potential of, *489,* 489–490
    resting potential of, 488, *488*
neuroscience, 484–485
Neurospora crassa, 159–160
neurotransmitters, 492, 493
neutralists, 242
neutrons, 24
neutrophil, 382
Neville, Helen, 484–486, 500–501
New World monkeys, *360,* 360–361
niche, 605–606
nicotine, 586
nitric oxide (NO), 190, 412–413, 424, 493
nitrogen cycle, 628–629, *629*
nitrogen fixation, 583–584, *584*
nitrous oxide (N_2O), 619, 642
node, plant, 548
nodules, root, 584–584
noncellular biological entities, 286–287
nonpolar molecules, 27
nonspecific resistance, to disease, 428–431, *429, 430*
norepinephrine, 493
nose, 433, *508*
notochord, 349
nuclear envelope, 46
nuclear fusion, 305
nuclear pores, 46
nuclei, *42,* 46–54, *47*
nucleic acids, *36,* 36–37. *see also* DNA (deoxyribonucleic acid); RNA (ribonucleic acid)
    as component of HIV, 24
    viruses and, 428
nucleoli, 47
nucleosomes, 147
nucleotides, *36,* 36–37. *see also* transcription
    codons, anticodons, 166–167
    in DNA structure, 144
    joining, 164